Isotope	NMR Frequency in MHz for a 10 Kilogauss Field	Natural Abundance %	Relative Sensitivity for Equal Numbers of Nuclei		Magnetic Moment, μ_N,[a] in multiples of nuclear magneton ($eh/4\pi mc$)	Spin, I, in multiples of $h/2\pi$	Electric Quadrupole Moment, Q, in multiples of 10^{-24} cm^2	Anisotropic Hyperfine Coupling, B, in MHz[b]	Isotropic Hyperfine Coupling, A_0, in MHz[c]
			at constant field	at constant frequency					
• ^{81}Br	11.498	49.43	9.84×10^{-2}	1.35	2.2626	3/2	0.28	696	23,432
• ^{85}Rb	4.111	72.8	1.05×10^{-2}	1.13	1.3483	5/2	0.31	—	1,012*
• ^{87}Rb	13.932	27.2	0.177	1.64	2.7415	3/2	0.15	—	3,417*
• ^{87}Sr	1.845	7.02	2.69×10^{-3}	1.43	−1.0893	9/2	—	—	—
• ^{89}Y	2.086	100.	1.17×10^{-4}	4.90×10^{-2}	−0.1368	1/2	—	—	—
^{91}Zr	4.0	11.23	9.4×10^{-3}	1.04	−1.3	5/2	—	—	—
• ^{93}Nb	10.407	100.	0.482	8.06	6.1435	9/2	−0.4±0.3	—	—
• ^{95}Mo	2.774	15.78	3.22×10^{-3}	0.761	−0.9099	5/2	—	—	−3,528
• ^{97}Mo	2.833	9.60	3.42×10^{-3}	0.776	−0.9290	5/2	—	—	−3,601
• ^{99}Tc†	9.583	—	0.376	7.43	5.6572	9/2	0.3	—	—
^{99}Ru	—	12.81	—	—	—	6/2	—	—	—
^{101}Ru	—	16.98	—	—	—	5/2	—	—	—
• ^{103}Rh	1.340	100.	3.12×10^{-5}	3.15×10^{-2}	−0.0879	1/2	—	—	—
^{105}Pd	1.74	22.23	7.79×10^{-4}	0.47	−0.57	5/2	—	—	—
• ^{107}Ag	1.722	51.35	6.69×10^{-5}	4.03×10^{-2}	−0.1130	1/2	—	—	−3,520
• ^{109}Ag	1.981	48.65	1.01×10^{-4}	4.66×10^{-2}	−0.1299	1/2	—	—	−4,044
• ^{111}Cd	9.028	12.86	9.54×10^{-3}	0.212	−0.5922	1/2	—	—	—
• ^{113}Cd	9.444	12.34	1.09×10^{-2}	0.222	−0.6195	1/2	—	—	—
• ^{115}In†	9.329	95.84	0.348	7.23	5.5072	9/2	1.161	—	—
• ^{117}Sn	15.77	7.67	4.53×10^{-2}	0.356	−0.9949	1/2	—	—	—
• ^{119}Sn	15.87	8.68	5.18×10^{-2}	0.373	−1.0409	1/2	—	—	—
• ^{121}Sb	10.19	57.25	0.160	2.79	3.3417	5/2	−0.8	—	—
• ^{123}Sb	5.518	42.75	4.57×10^{-2}	2.72	2.5334	7/2	−1.0	—	—
• ^{125}Te	13.45	7.03	3.16×10^{-2}	0.316	−0.8824	1/2	—	—	—
• ^{127}I	8.519	100.	9.35×10^{-2}	2.33	2.7939	5/2	−0.75	—	—
• ^{129}I†	5.669	—	4.96×10^{-2}	2.80	2.6030	7/2	−0.43	—	—
^{177}Hf	—	18.39	—	—	—	1/2 or 3/2	—	—	—
^{179}Hf	—	13.78	—	—	—	1/2 or 3/2	—	—	—
^{181}Ta	4.6	100.	2.60×10^{-2}	2.26	2.1	7/2	6.5	—	—
• ^{183}W	1.75	14.28	6.98×10^{-5}	4.12	0.115	1/2	—	—	—
• ^{185}Re	9.586	37.07	0.133	2.63	3.1437	5/2	2.8	—	—
• ^{187}Re	9.684	62.93	0.137	2.65	3.1760	5/2	2.6	—	—
• ^{189}Os	3.307	16.1	2.24×10^{-3}	0.385	0.6507	3/2	2.0	—	—
^{191}Ir	0.81	38.5	3.5×10^{-5}	9.5×10^{-2}	0.16	3/2	~1.2	—	—
^{193}Ir	0.86	61.5	4.2×10^{-5}	0.104	0.17	3/2	~1.0	—	—
• ^{195}Pt	9.153	33.7	9.94×10^{-3}	0.215	0.6004	1/2	—	—	—
^{197}Au	0.691	100.	2.14×10^{-5}	8.1×10^{-2}	0.136	3/2	0.56	—	—
• ^{199}Hg	7.612	16.86	5.72×10^{-3}	0.179	0.4993	1/2	—	—	—
^{201}Hg	3.08	13.24	1.90×10^{-3}	0.362	−0.607	3/2	0.5	—	—
• ^{203}Tl	24.33	29.52	0.187	0.571	1.5960	1/2	—	—	—
• ^{205}Tl	24.57	70.48	0.192	0.577	1.6114	1/2	—	—	—
• ^{207}Pb	8.899	21.11	9.13×10^{-3}	0.209	0.5837	1/2	—	—	—
• ^{209}Bi	6.842	100.	0.137	5.30	4.0389	9/2	−0.4	—	—
Free Electron $g = 2.00$	27.994	—	2.85×10^{8}	658	−1836	1/2	—	—	—

PHYSICAL METHODS IN CHEMISTRY

Saunders Golden Sunburst Series

RUSSELL S. DRAGO
University of Illinois, Urbana

1977

W. B. SAUNDERS COMPANY
Philadelphia · London · Toronto

W. B. Saunders Company: West Washington Square
Philadelphia, PA 19105

1 St. Anne's Road
Eastbourne, East Sussex BN21 3UN, England

1 Goldthorne Avenue
Toronto, Ontario M8Z 5T9, Canada

Library of Congress Cataloging in Publication Data

Drago, Russell S

Physical methods in chemistry.

(Saunders golden sunburst series)
Includes index.

1. Chemistry, physical and theoretical. 2. Spectrum analysis.
I. Title.

QD453.2.D7 543'.085 76-8572

ISBN 0-7216-3184-3

Physical Methods in Chemistry ISBN 0-7216-3184-3

Last digit is the print number: 9 8 7 6 5 4 3 2 1

Dedication

To Ruth
and our children

PREFACE

The rewrite of my earlier textbook, "Physical Methods in Inorganic Chemistry," was dictated by several reasons. The mathematical preparation and background of the users of this book (generally seniors and graduate students) have greatly increased since the original edition. Much of the material in the first three chapters of the first edition is now in general chemistry courses. Accordingly, these chapters have been replaced by a more advanced treatment of group theory and molecular orbital theory. This has permitted a more advanced treatment of the spectroscopic methods. Thus, the overall level of presentation has been increased. Topics have been subdivided into numbered paragraphs so the more theoretical ones can be dropped, if so desired, permitting use of the text in less theoretical, more applications-oriented courses.

Organometallic chemistry has done much to merge the old classical divisions of organic and inorganic chemistry. This has brought about increased appreciation of the fact that there is more to the application of physical methods than memorizing where functional groups appear. Accordingly, there is a core of material in the area of spectroscopy that is of importance to both organic and inorganic chemists. This text is designed to reflect this unity. A more fundamental subdivision of chemistry involves dividing it into systems that do and do not contain unpaired *d*-electrons. The material in this book has been so subdivided. Thus, in using this book, a chemist with interest in organic or organometallic synthesis can omit Chapters 9, 10, 11, and 12. To emphasize the broader scope of the text, the title has been changed to "Physical Methods in Chemistry."

This book contains more material than can be reasonably covered in a one-semester course. Accordingly, several of the methods which are in less common usage than IR, UV, NMR, and ESR have been presented at a very basic level, and students will hopefully find these self-explanatory.

The book continues in the philosophy of the earlier one, namely, that chemists without a highly formal mathematical background learn to use spectroscopic methods by reading about how problems have been solved with them. Thus, the treatment continues to rely heavily on describing examples which illustrate the kinds of information that can be obtained by application of the methods. Whenever a choice exists regarding the selection of an application, an example from my own research has been chosen, for I am most familiar with the details of this work. I have avoided those applications of spectroscopy in which I feel the literature report speculates well beyond what one can reasonably believe the results provide. Hopefully, more common knowledge of the material in this text will upgrade science in terms of this kind of speculation or at least make the reader more critical. I feel this was one of the important contributions of the original text.

This text has evolved from the author's presentation of a course in physical methods. He gratefully acknowledges the contributions to the development of this material by the teaching assistants and students involved in this course. David McMillin, Robert Richman, Ben Tovrog and Tom Kuechler are especially deserving of mention in this regard. Professors David Hendrickson and Jack Norton read the entire manuscript, and I am indebted to them for many constructive comments. Comments from Stuart Tobias on Chapters 1–6 and Linn Belford on the epr chapters are gratefully acknowledged. Professor Charles Root and his 1976 class at Bucknell offered many suggestions on Chapters 1–7. I thank Dante Gatteschi for his extensive contributions to the section on the angular overlap model. The superlative job done on the manuscript by Jay Freedman of the Saunders Editorial Department is deserving of special mention, as is the faith in the value of this effort by John Vondeling. Finally, I would like to apologize to my wife and children for the many hours of my time spent on chemistry instead of with them and thank them for their encouragement and unselfishness.

RUSSELL S. DRAGO

CONTENTS

4

5

1 SYMMETRY AND THE POINT GROUPS

1–1 DEFINITION OF SYMMETRY

Symmetry considerations are fundamental to many areas of chemical reactivity, electronic structure, and spectroscopy. It is customary to describe the structures of molecules in terms of the symmetry that the molecules possess. Spectroscopists have described molecular vibrations in terms of symmetry for many years. Modern applications of spectroscopic methods to the problem of structure determination require a knowledge of symmetry properties. We shall be concerned mainly with the description of the symmetry of an isolated molecule, the so-called point symmetry. Point symmetry refers to the set of operations transforming a system about a common point, which usually turns out to be the center of gravity of a molecule.

If a molecule has two or more orientations in space that are indistinguishable, the molecule possesses *symmetry*. Two possible orientations for the hydrogen molecule can be illustrated in Fig. 1–1 only by labeling the two equivalent hydrogen atoms in this figure with prime and double prime marks. Actually, the two hydrogens are indistinguishable, the two orientations are equivalent, and the molecule has symmetry. The two orientations in Fig. 1–1 can be obtained by rotation of the molecule through 180° about an axis through the center of and perpendicular to the hydrogen-hydrogen bond axis. This rotation is referred to as a *symmetry operation*, and the rotation axis is called a *symmetry element*. The terms symmetry element and symmetry operation should not be confused or used interchangeably. The symmetry element is the line, point, or plane about which the symmetry operation is carried out. The operation can be defined only with respect to the element, and the existence of the element can be shown only by carrying out the operation.

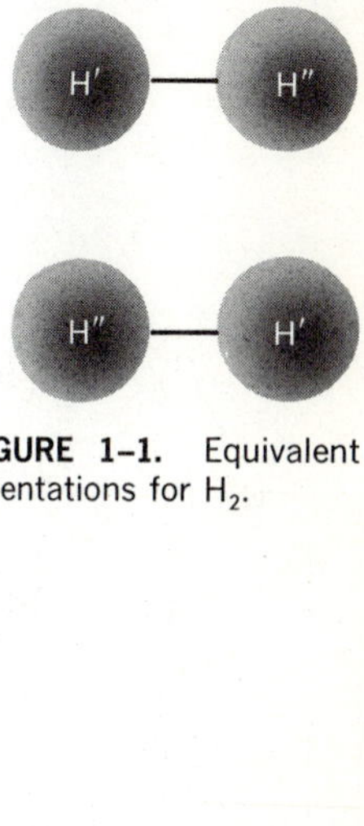

H′ — H″

H″ — H′

FIGURE 1–1. Equivalent orientations for H_2.

A simple test can be performed to verify the presence of symmetry. If you were to glance at a structure, turn your back and have someone perform a symmetry operation, on once again examining the molecule, you would not be able to determine that a change had been made. The symmetry operations of a molecule form a group and, as a result, are amenable to group theoretical procedures. Mastery of the material in this chapter is essential to the understanding of most of the material in this book.

1–2 SYMMETRY ELEMENTS

Five types of symmetry elements will be considered for point symmetry: (1) the center of symmetry (inversion center), (2) the identity, (3) the rotation axis, (4) the mirror plane, and (5) the rotation-reflection axis. The symmetry operations corresponding to these elements will be defined in the course of a further discussion of each element.

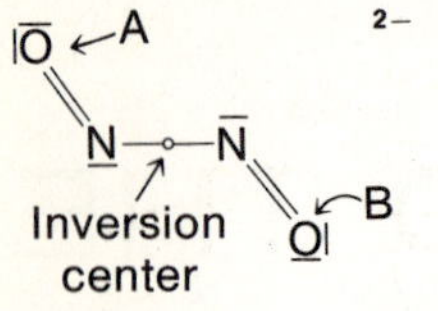

The Center of Symmetry, or Inversion Center

FIGURE 1–2. The center of symmetry in the hyponitrite ion.

A molecule is said to possess a center of symmetry or inversion center if every atom in the molecule, if moved in a straight line through this center and an equal distance on the other side of the center, encounters a like atom. If oxygen atom *A* of the hyponitrite ion (Fig. 1–2) is moved through the inversion center, it comes into coincidence with another oxygen atom, *B*. The same arguments must apply to atom *B* and also to both nitrogen atoms if the molecule is to possess a center of symmetry. This operation corresponds to placing this center at the center of a coordinate system, and taking every atom with coordinates (x, y, z) and changing its coordinates to $(-x, -y, -z)$. *At most* one atom can be at the center, and all other atoms in the molecule must exist in pairs. Neither of the two structures indicated in Fig. 1–3 nor any other tetrahedral molecule possesses an inversion center. Other molecules or ions that should be examined for the presence of an inversion center are 1,4-dioxane, tetracyanonickelate(II), *trans*-dichloroethylene, and *trans*-dichlorotetraammine cobalt(III). The molecule HCCl=CHBr (*cis* or *trans*) does not possess a center of symmetry because the symmetry operation on bromine or hydrogen does not result in coincidence. All points in the molecule must be inverted simultaneously in this operation, if the molecule as a whole is to possess this symmetry element. The symbol used to indicate an inversion center (center of symmetry) is *i*.

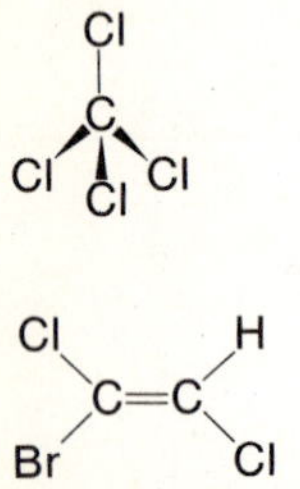

FIGURE 1–3. Absence of a center of symmetry in CCl_4 and HCCl=CBrCl.

The Identity

In the identity operation, no change is made in the molecule. Obviously, the result of this operation is not only to produce an equivalent orientation but an identical one; *i.e.*, even if similar atoms were labeled with prime or double prime marks, or any other notation, no change would be detected. All molecules possess this symmetry element, and it is indicated by the symbol *E*. This operation probably seems trivial at present, but, as will be seen in the section on group theory, this concept is required so that symmetry elements can be treated by this form of mathematics.

The Rotation Axis

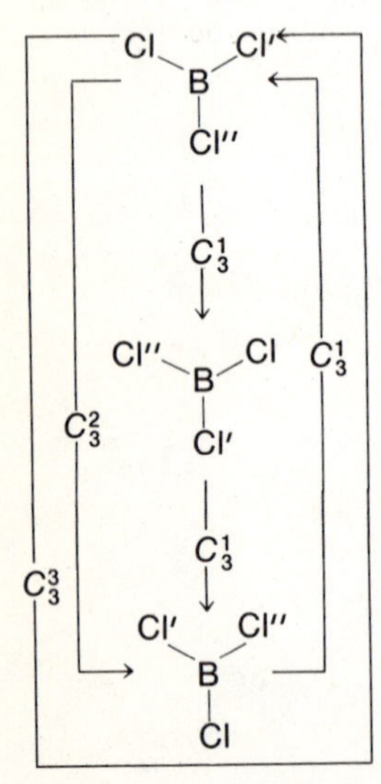

FIGURE 1–4. The results of rotations about the threefold rotation axis of BCl_3.

If an imaginary axis can be constructed in a molecule, around which the molecule can be rotated to produce an *equivalent* (*i.e.*, indistinguishable from the original) orientation, this molecule is said to possess a rotation axis. The symmetry element previously discussed for the hydrogen molecule is a rotation axis. This element is usually referred to as a *proper rotation axis*. It may be possible to carry out several symmetry operations around a single rotation axis. If the molecule can occupy *n* different equivalent positions about the axis, the axis is said to be of order *n*. For example, consider the axis through the center of the boron atom in BCl_3 perpendicular to the plane of the molecule. Rotation, which by convention is in a clockwise direction, about this axis two times through an angle of 120° each time produces two equivalent orientations. Taken with the initial orientation, we have the three different equivalent orientations, illustrated in Fig. 1–4. The order, *n*, of this axis is three, for three rotations are needed to return to the original position. The molecule is said to possess a threefold rotation axis, indicated by the symbol C_3. Rotation of the molecule through $2\pi/n$ (*i.e.*, 120°) produces equivalent orientations, and *n* operations produce the starting configuration referred to as the *identity*. The symbol C_3^2 is employed to indicate a rotation of 240° around a C_3 axis. The C_3^2 operation is identical to a counterclockwise rotation of 120°, which is indicated as C_3^-. It should be clear that an axis through the center and perpendicular to the plane of a benzene

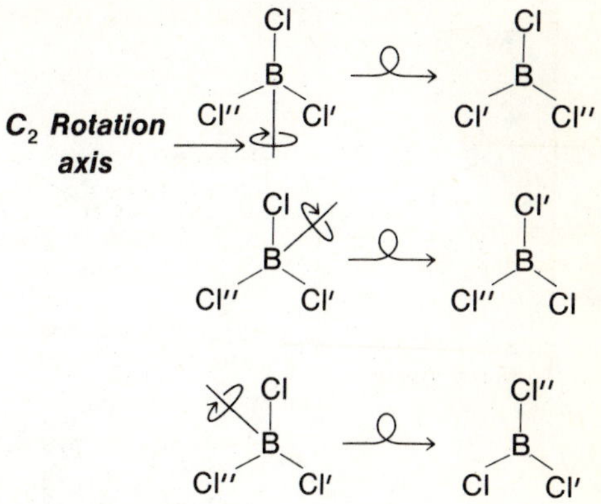

FIGURE 1–5. The results of rotations about the three twofold rotation axes in BCl_3. The 180° rotation on the left produces the result on the right.

ring is a sixfold axis, C_6. Since $n = 6$, rotation by 60° (= 360°/6) six times produces the six equivalent orientations. Further examination of the BCl_3 molecule indicates the lack of a center of symmetry and the presence of the three additional twofold rotation axes, C_2, illustrated in Fig. 1–5. One twofold axis, arbitrarily selected, is labeled C_2; the other two are labeled C_2' and C_2''. If $n = 1$, the molecule must be rotated 360° to produce an equivalent (and in this case identical) orientation. Consequently, the molecule is said to possess no symmetry if no elements other than the identity are present.

A rotation axis of order n generates n operations: $C_n, C_n^2, C_n^3, \ldots, C_n^{n-1}, C_n^n$. Furthermore, the operation C_4^2 is equivalent to C_2, the operation C_6^2 is equivalent to C_3, and C_n^n is the identity. *The highest-fold rotation axis is referred to as the principal axis in a molecule.* If all of the C_n axes are equivalent, any one may be chosen as the principal axis.

In Fig. 1–6, a rotation axis is illustrated for the H_2 molecule, for which $n = \infty$. The C_2 rotation axes are not illustrated but are perpendicular to the bond axis and centered between the two hydrogen atoms. There are an infinite number of these twofold axes. It should also be obvious that benzene possesses six twofold axes that lie in the plane of the molecule. Three pass through pairs of opposite carbon atoms, and the other three pass through the centers of C—C bonds.

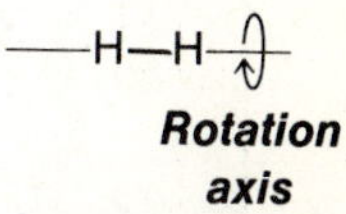

FIGURE 1–6. The C_∞ axis in H_2.

The molecule ClF_3 is illustrated in Fig. 1–7. The geometry is basically a trigonal bipyramid, with two lone pairs of electrons in the equatorial plane and with two fluorines bent down toward the equatorial fluorine. This molecule has only one rotation axis, the C_2 axis shown in Fig. 1–7. The reader should verify the absence of symmetry along the axes indicated by dashed lines.

For many purposes, it is convenient to locate a molecule and its symmetry elements in a Cartesian coordinate system. A right-handed coordinate system will be used. Movement of y to x is a positive rotation. The center of gravity of the molecule is located at the center of the coordinate system. If there is only one symmetry axis in the molecule, it is selected as the z-axis. If there is more than one symmetry axis, the principal rotation axis in the molecule is selected as the z-axis. If there is more than one highest-fold axis, the one connecting the most atoms is selected. If the molecule

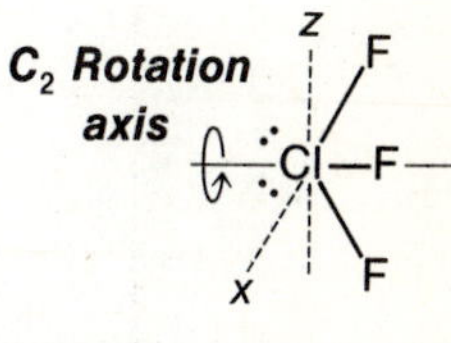

FIGURE 1–7. The symmetry axis in ClF_3.

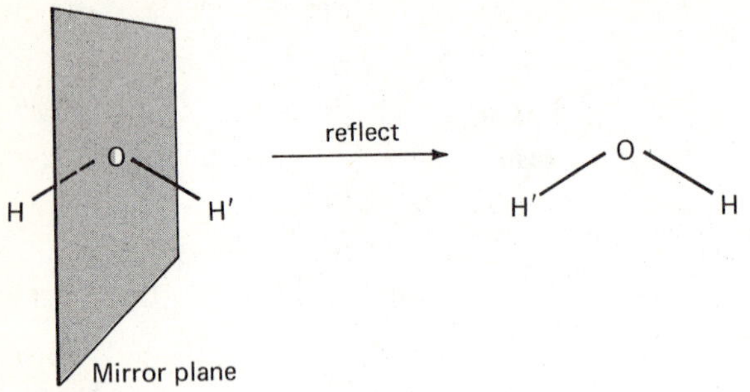

FIGURE 1-8. Operation of a mirror plane (σ) reflection on H_2O.

is planar and if the z-axis lies in this plane (as in the water molecule), the x-axis is chosen perpendicular to this plane. If the molecule is planar and the z-axis is perpendicular to this plane, *e.g.*, in (Cl)(H)C=C(H)(Cl), the x-axis is chosen to pass through the largest number of atoms. In *trans*-dichloroethylene the x-axis passes through the two carbons.

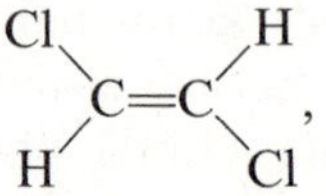

The Mirror Plane or Plane of Symmetry

If in a molecule there exists a plane that separates the molecule into two halves that are mirror images of each other, the molecule possesses the symmetry element of a *mirror plane*. This plane cannot lie outside the molecule but must pass through it. Another way of describing this operation involves selecting a plane, dropping a perpendicular from every atom in the molecule to the plane, and placing the atom at the end of the line an equal distance to the opposite side of the plane. If an equivalent configuration is obtained after this is done to all the atoms, the plane selected is a mirror plane. Reflection through the mirror plane is indicated by Fig. 1-8. A linear molecule possesses an infinite number of mirror planes.

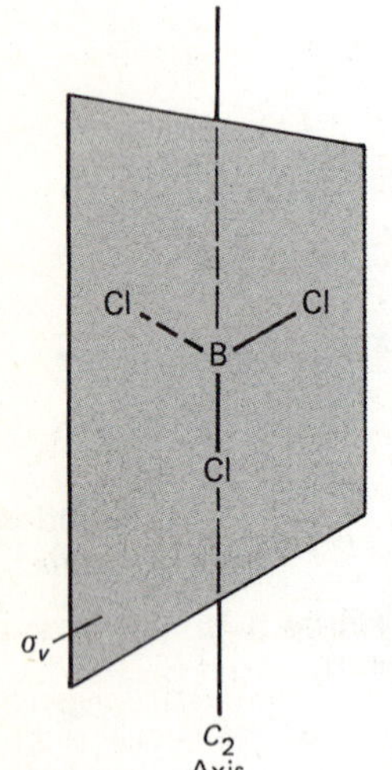

FIGURE 1-9. A mirror plane in BCl_3.

Often rotation axes lie in a mirror plane (see Fig. 1-9, for example) but there are examples in which this is not the case. The tetrahedral molecule $POBr_2Cl$, illustrated in Fig. 1-10, is an example of a molecule that contains no rotation axis but does contain a mirror plane. The atoms P, Cl, and O lie in the mirror plane. In general, the presence of a mirror plane is denoted by the symbol σ. In those molecules that contain more than one mirror plane (*e.g.*, BCl_3), the horizontal plane σ_h is taken as the one perpendicular to the principal (highest-fold rotation) axis. In Fig. 1-9, the plane of the paper is σ_h and there are then three vertical planes σ_v (two others, similar to the one illustrated, each containing the boron atom and one chlorine atom) perpendicular to σ_h.

When a molecule is located in a coordinate system, the z-axis always lies in the vertical plane(s).

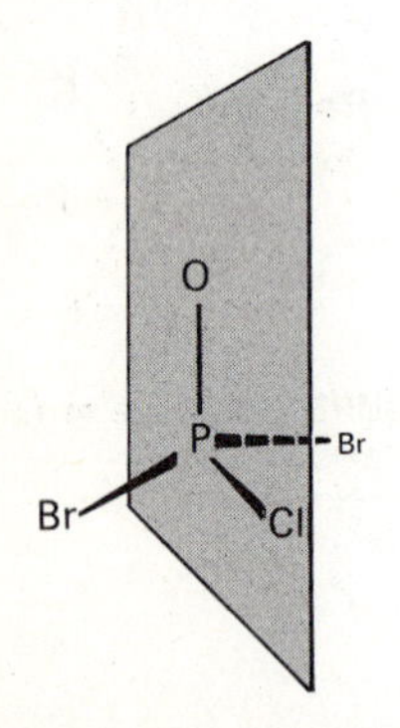

FIGURE 1-10. The mirror plane in $POBr_2Cl$.

Some molecules have mirror planes containing the principal axis but none of the perpendicular C_2 axes. These planes bisect the angle between two of the C_2 axes (in the xy plane); they are referred to as *dihedral planes* and are abbreviated as σ_d. Two σ_d planes are illustrated in Fig. 1-14, *vide infra.** In some molecules there is more than one set of mirror planes containing the highest-fold axes; *e.g.*, in $PtCl_4^{2-}$, one set consists of the xz and yz planes (z being the fourfold axis, and the Cl—Pt—Cl bond axes being x and y), and the other set bisects the angle between the x and y axes. The former set, including the four chlorine atoms, is called σ_v by convention, and the latter set is called σ_d. The diagonal plane is always taken as one that bisects x, y, z, or twofold axes in the molecule.

*We will use this term to indicate that a statement will be treated more fully later in the text.

The Rotation-Reflection Axis; Improper Rotations

This operation involves rotation about an axis followed by reflection through a mirror plane that is perpendicular to the rotation axis, or vice versa (*i.e.*, rotation-reflection is equivalent to reflection-rotation). When the result of the two operations produces an equivalent structure, the molecule is said to possess a rotation-reflection axis. This operation is referred to as an improper rotation, and the rotation-reflection axis is often called an alternating axis. The symbol S is used to indicate this symmetry element. The subscript n in S_n indicates rotation (clockwise by our convention) through $2\pi/n$.

Obviously, if an axis C_n exists and there is a σ perpendicular to it, C_n will also be an S_n. Now, we shall consider a case in which S_n exists when neither C_n or the mirror plane perpendicular to it exists separately. In the staggered form of ethane, Fig. 1–11,

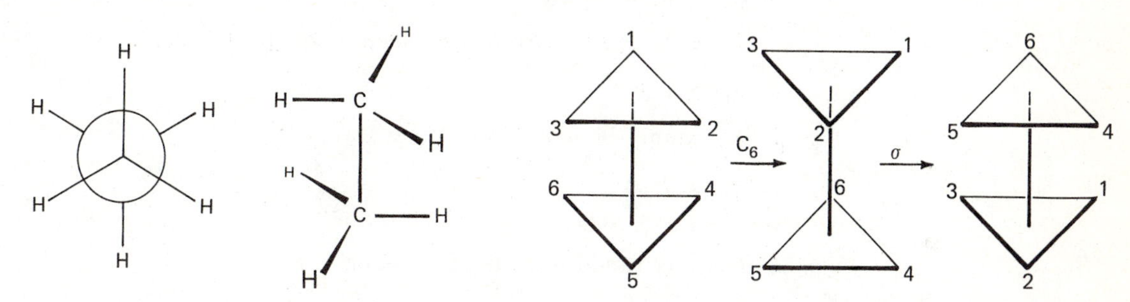

FIGURE 1–11. Improper rotation axis in the staggered form of ethane.

the C—C bond defines a C_3 axis, but there is no perpendicular mirror plane. However, if we rotate the molecule 60° and then reflect it through a plane perpendicular to the C—C bond, we have an equivalent configuration. Consequently, an S_6 axis exists and, clearly, there is no C_6.

The dotted line in Fig. 1–12 indicates the S_2 element in the molecule *trans*-dichloroethylene. The subscript two indicates clockwise rotation through 180°. S_2 is equivalent to i and, by convention, is usually called i.

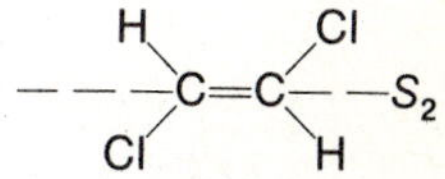

FIGURE 1–12. Rotation-reflection axis of symmetry.

A particular orientation of the molecule methane is illustrated in Fig. 1–13. Open

FIGURE 1–13. Effect of the operations C_4 and σ perpendicular to C_4 on the hydrogens of CH_4.

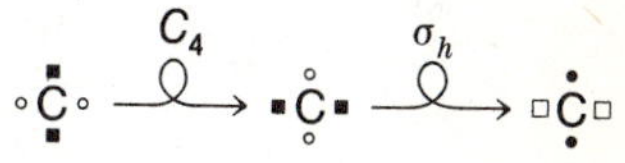

circles or squares represent hydrogen atoms in a plane parallel to but above the plane of the paper, and the solid squares or circles are those below the plane of the paper. The plane of the paper is the reflection plane, and it contains the carbon atom. The C_4 operation is straightforward. The operation of reflection, σ, moves the hydrogens below the plane to above the plane and vice versa. This is indicated by changing the solid squares to open squares and the open circles to solid circles. However, since all four hydrogens are identical, the initial and final orientations are equivalent. The molecule contains a fourfold rotation-reflection axis, abbreviated S_4. This operation can be repeated three more times, or four times in all. These four operations are indicated by the symbols S_4^1, S_4^2, S_4^3, and S_4^4. The reader should convince himself that S_4^2 is equivalent to a C_2 operation on this axis. It should also be mentioned that the molecule possesses two other rotation-reflection axes, and these symmetry elements are abbreviated by the symbols S' and S''. The operation S_4^3 is equivalent to a counterclockwise rotation of 90° followed by reflection. This is often indicated as S_4^-.

Next we shall consider some differences in improper rotation axes of even and odd order. With n even, $S_n{}^m$ generates the set $S_n{}^1, S_n{}^2, S_n{}^3, \ldots, S_n{}^n$. This is equivalent to $C_n{}^1\sigma^1, C_n{}^2\sigma^2, C_n{}^3\sigma^3, \ldots, C_n{}^n\sigma^n$. [Note: In carrying out $C_n{}^n\sigma^n$, one reflects first and then rotates, while in carrying out $\sigma^n C_n{}^n$, one rotates and then reflects.] We have the relation $\sigma^m = \sigma$ when m is odd, and $\sigma^m = E$ when m is even. The latter leads to the identities $S_n{}^n = C_n{}^n E = E$ and $S_n{}^m = C_n{}^m$.

Notice that the existence of an S_n axis of even order always requires a $C_{n/2}$, since $S_n{}^2 = C_n{}^2\sigma^2 = C_{n/2}$.

Consider the set S_6, $S_6{}^2$, $S_6{}^3$, $S_6{}^4$, $S_6{}^5$, $S_6{}^6$ by carrying out the operations described below on the staggered ethane example.

S_6 cannot be written any other way.

$S_6{}^2 = C_6{}^2\sigma^2 = C_3$

$S_6{}^3 = S_2 = i$ [An S_2 axis always equals i.]

$S_6{}^4 = C_3{}^2$

$S_6{}^5$ cannot be written any other way.

$S_6{}^6 = E$

The complete set would normally be written:

$$S_6 \quad C_3 \quad i \quad C_3{}^2 \quad S_6{}^5 \quad E$$

When n is odd, $S_n{}^m$ generates the set $S_n, S_n{}^2, S_n{}^3, S_n{}^4, \ldots, S_n{}^{2n}$. An odd-order S_n requires that C_n and a σ perpendicular to it exist. Note that the operation $S_n{}^n$ when n is odd is equivalent to $C_n{}^n\sigma^n = C_n{}^n\sigma = \sigma$. (Compare this with $S_n{}^n$ when n is even.)

For example, consider S_3. First one would reflect, producing a configuration that is equivalent to the starting configuration (because of the existence of σ in the molecule). Then one rotates by C_3 to give the configuration corresponding to $S_3{}^1$. Since $S_3{}^1$ is a symmetry operation for the molecule, the configuration after the C_3 rotation must be equivalent to the starting one, and therefore C_3 exists. In general, for any S_n (with n odd), C_n is also a symmetry operation. For practice, consider the $S_5{}^m$ operations:

$\boldsymbol{S_5 = C_5}$ **and then** $\boldsymbol{\sigma}$

$S_5{}^2 = C_5{}^2$

$\boldsymbol{S_5{}^3 = C_5{}^3}$ **and then** $\boldsymbol{\sigma}$

$S_5{}^4 = C_5{}^4$

$S_5{}^5 = \sigma$

$S_5{}^6 = C_5$

$\boldsymbol{S_5{}^7 = C_5{}^2}$ **and then** $\boldsymbol{\sigma}$

$S_5{}^8 = C_5{}^3$

$\boldsymbol{S_5{}^9 = C_5{}^4}$ **and then** $\boldsymbol{\sigma}$

$S_5{}^{10} = E$

$S_5{}^{11} = S_5$, etc.

TABLE 1–1. A SUMMARY OF SYMMETRY ELEMENTS AND SYMMETRY OPERATIONS

Symmetry Operation	Symmetry Element	Symbol	Examples
identity		E	all molecules
reflection	plane	σ	H_2O, BF_3 (planar)
inversion	point (center of symmetry)	i	$Cl_2B{-}BCl_2$ (planar, Cl atoms trans)
proper rotation	axis	C_n (n-order)	NH_3, H_2O
improper* rotation (rotation by $2\pi/n$ followed by reflection in plane perpendicular to axis)	axis *and* plane	S_n (n-order)	ethane, ferrocene (staggered structures)

*Ferrocene is staggered and possesses an S_{10} improper rotation axis.

The boldface indicates S_5 operations that cannot be written as a single operation in any other way. Thus, S_n with n odd generates $2n$ operations.

Table 1–1 summarizes the key aspects of our discussion of symmetry operations and elements.

1–3 POINT GROUPS

It is possible to classify any given molecule into one of the point groups. Each point group is a collection of all the symmetry operations that can be carried out on a molecule belonging to this group. We shall first present a general discussion in which molecules belonging to some of the very common, simple point groups are examined for the symmetry elements they possess. This will be followed by a general set of rules for assigning molecules to the appropriate point groups.

A molecule in the point group C_n has only one symmetry element, an n-fold rotation axis. (Of course, all molecules possess the element E, so this will be assumed in subsequent discussion.) A molecule such as *trans*-dichloroethylene, which has a horizontal mirror plane perpendicular to its C_n rotation axis, belongs to the point group C_{nh}. The point group C_{nv} includes molecules like water and sulfuryl chloride, which have n vertical mirror planes containing the rotation axis, but no horizontal mirror plane (the horizontal plane by definition must be perpendicular to the highest-fold rotation axis). The C_{2v} molecules, H_2O and SO_2Cl_2, contain only one rotation axis and two σ_v planes. The molecule is assigned to the point group that contains all of the symmetry elements in the molecule. For example, H_2O, which has a C_2 axis and two vertical planes, is assigned to the higher-order (more symmetry elements) point group C_{2v} rather than to C_2.

The symbol D_n is used for point groups that have, in addition to a C_n axis, n C_2 axes perpendicular to it. Therefore, the D_n point group has greater symmetry (*i.e.*, more symmetry operations) than the C_n group. A D_n molecule that also has a horizontal mirror plane perpendicular to the C_n axis belongs to the point group D_{nh}, and as a consequence will also have n vertical mirror planes. The addition of a horizontal mirror plane to the point group C_{nv} necessarily implies the presence of n C_2 axes in the horizontal plane, and the result is the point group D_{nh}. BCl_3 is an example of a molecule belonging to the D_{3h} point group. In the D_{nh} point groups, the

σ_h planes are perpendicular to the principal axis and contain all the C_2 axes. Each σ_v plane contains the principal axis and one of the C_2 axes.

D_n molecules may also have σ_d planes that contain the principal (C_n) axis but none of the perpendicular C_2 axes. As mentioned previously, these dihedral planes, σ_d, bisect the angle between two of the C_2 axes. The notation for a D_n molecule containing this symmetry element is D_{nd}. Such a molecule will contain an n-fold axis, n twofold axes perpendicular to C_n, and in addition n (vertical) planes of symmetry bisecting the angles between two twofold axes and containing the n-fold axis. Allene, $H_2C{=}C{=}CH_2$, is an example of a molecule belonging to the D_{2d} point group. Some of its symmetry elements are illustrated in Fig. 1-14. The two hydrogens on one

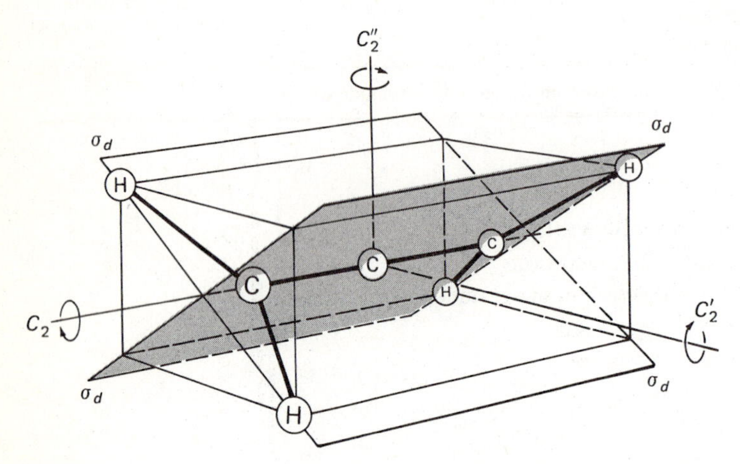

FIGURE 1-14. The C_2 axes and two σ_d planes in allene.

carbon are in a σ_d plane perpendicular to the σ_d plane containing the two hydrogens on the other carbon. The C_2 axes (the principal axis being labeled C_2, the others C_2' and C_2'') and the dihedral planes containing the principal axis are indicated. The two other C_2 axes, C_2' and C_2'', which are perpendicular to the principal C_2 axis, form 45° angles with the two dihedral planes. (Study the figure carefully to see this. Constructing a model may help.)

TABLE 1-2. SYMMETRY ELEMENTS IN SOME COMMON POINT GROUPS

Point Group	Symmetry Elements[a]	Examples
C_1	no symmetry	SiBrClFI
C_2	one C_2 axis	H_2O_2
C_{nh}	one n-fold axis and a horizontal plane σ_h which must be perpendicular to the n-fold axis	*trans*-$C_2H_2Cl_2$ (C_{2h})
C_{2v}	one C_2 axis and two σ_v planes	H_2O, SO_2Cl_2, $SiCl_2Br_2$
C_{3v}	one C_3 axis and three σ_v planes	NH_3, CH_3Cl, $POCl_3$
D_{2h}	three C_2 axes all $\perp$, two σ_v planes, one σ_h plane, and a center of symmetry	N_2O_4 (planar)
D_{3h}	one C_3, three C_2 axes $\perp$ to C_3, three σ_v planes, and one σ_h	BCl_3
D_{2d}	three C_2 axes, two σ_d planes, and one S_4 (coincident with one C_2)	$H_2C{=}C{=}CH_2$
T_d	three C_2 axes $\perp$ to each other, four C_3, six σ, and three S_4 containing C_2	CH_4, $SiCl_4$

[a] All point groups possess the identity element, E.

Table 1–2 contains the symmetry elements and examples of some of the more common point groups. The reader should examine the examples for the presence of the symmetry elements required for each point group and the absence of others. Remember that the symbol C indicates that the molecule has only one rotation axis. The symbol D_n indicates n C_2 axes in addition to the n-fold axis; *e.g.*, the point group D_4 contains a C_4 axis and four C_2 axes.

A more exhaustive compilation of the important point groups is contained in Herzberg[1]. In addition to the point groups listed in Table 1–2, the O_h point group [which includes molecules whose structures are perfect octahedra (SF_6, PCl_6^-)] is very common.

It is important to realize that although $CHCl_3$ has a tetrahedral geometry, it does not have tetrahedral symmetry, so it belongs to the point group C_{3v} and not T_d. A tetragonal complex, *trans*-dichlorotetraammine cobalt(III) ion, belongs to the point group D_{4h} (ignoring the hydrogens) and not O_h. Phosphorus pentachloride belongs to the point group D_{3h}, and not to C_{3h}, for it has three C_2 axes perpendicular to the C_3 axis. The structure of monomeric boric acid (assumed to be rigid), illustrated in Fig. 1–15, is an example of C_{3h} symmetry. It has a threefold axis and a σ_h plane but does not have the three C_2 axes or σ_v planes necessary for the D_{3h} point group.

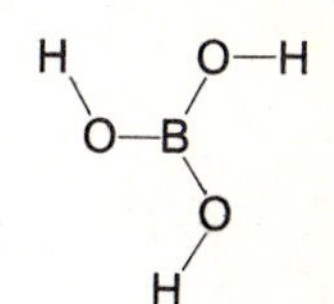

FIGURE 1–15. A structure with C_{3h} symmetry.

It is helpful but not always possible to classify additional structures by memorizing the above examples and using analogy as the basis for classification instead of searching for all the possible symmetry elements. The following sequence of steps has been proposed[2] for classifying molecules into point groups and is a more reliable procedure than analogy.

(1) Determine whether or not the molecule belongs to one of the special point groups, $C_{\infty v}$, $D_{\infty h}$, I_h, O_h, or T_d. The group I_h contains the regular dodecahedron and regular icosahedron. Only linear molecules belong to $C_{\infty v}$ or $D_{\infty h}$.

(2) If the molecule does not belong to any of the special groups, look for a proper rotation axis. If any are found, proceed to step (3); if not, look for a center of symmetry, i, or a mirror plane, σ. If the element i is present, the molecule belongs to the point group C_i; if a mirror is present, the molecule belongs to the point group C_s. If no symmetry other than E is present, the molecule belongs to C_1.

(3) Locate the principal axis, C_n. See if a rotation-reflection axis S_{2n} exists that is coincident with the principal axis. If this element exists and there are no other elements except possibly i, the molecule belongs to one of the S_n point groups (where n is even). If other elements are present or if the S_{2n} element is absent, proceed to step (4).

(4) Look for a set of n twofold axes lying in the plane perpendicular to C_n. If this set is found, the molecule belongs to one of the groups D_n, D_{nh}, or D_{nd}. Proceed to step (5). If not, the molecule must belong to either C_n, C_{nh}, or C_{nv}. Proceed to step (6) and skip (5).

(5) By virtue of having arrived at this step, the molecule must be assigned to D_n, D_{nh}, or D_{nd}. If the molecule contains the symmetry element σ_h, it belongs to D_{nh}. If this element is not present, look for a set of n σ_d's, the presence of which enables assignment of the molecule to D_{nd}. If σ_d and σ_h are both absent, the molecule belongs to D_n.

(6) By virtue of having arrived at this step, the molecule must be assigned to C_n, C_{nh}, or C_{nv}. If the molecule contains σ_h, the point group is C_{nh}. If σ_h is absent, look for a set of n σ_v's, which place the molecule in C_{nv}. If neither σ_v nor σ_h is present, the molecule belongs to the point group C_n.

---

[1] G. Herzberg, "Infrared and Raman Spectra," p. 11, Van Nostrand, New York (1945).

[2] F. A. Cotton, "Chemical Applications of Group Theory," p. 45, Wiley, New York (1971).

The following flow chart provides a systematic way to approach the classification of molecules by point groups:

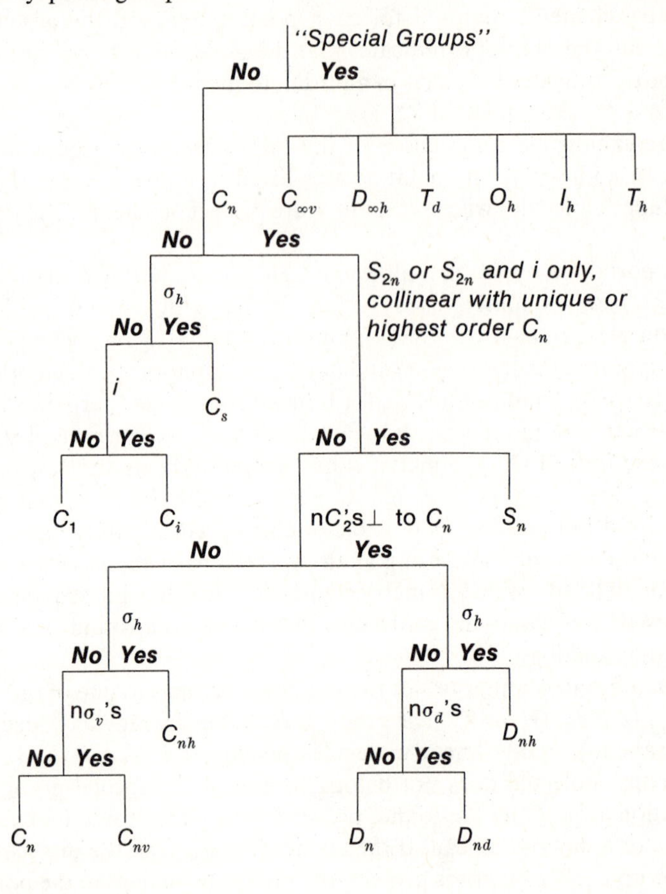

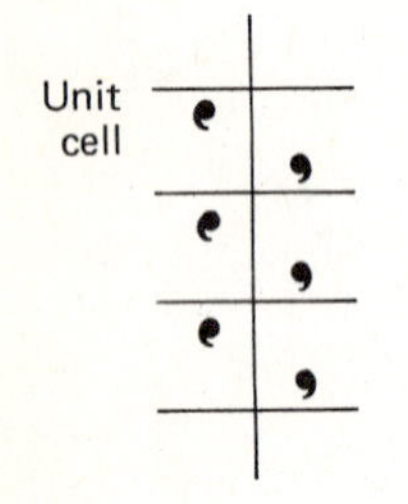

FIGURE 1-16. A glide plane. (The tails of these commas point into the page.)

1-4 SPACE SYMMETRY

In describing the symmetry of an arrangement of atoms in a crystal (essential to x-ray crystallography), two additional symmetry elements are required: (1) the glide plane and (2) the screw axis, corresponding to mixtures of point group operations and translation. A space group is a collection of symmetry operations referring to three-dimensional space. A glide plane is illustrated in Fig. 1-16. The unit (represented by a comma) is translated a fraction (in this case ½) of a unit cell dimension and is reflected. (A unit cell is the smallest repeating unit of the crystal.) The comma must lie in a plane perpendicular to the plane of the paper and all of the comma tails must point into the paper. A screw axis is illustrated in Fig. 1-17. Each position, as one proceeds from the top to bottom of the axis, represents rotation by 120° along with translation parallel to the axis. This is a threefold screw axis. Rotations by 90°, 60°, and other angles are found in other structures. There are 230 possible space groups. These are of importance in x-ray crystallography, and will be covered in more detail in that chapter.

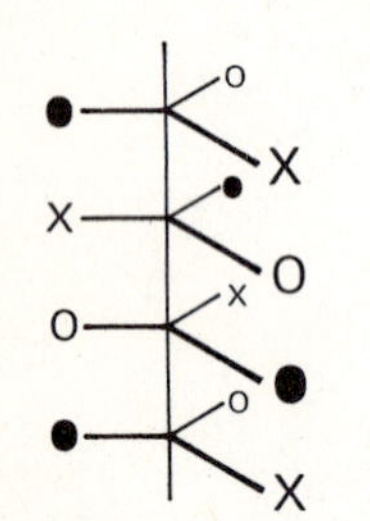

FIGURE 1-17. A screw axis.

1-5 SOME DEFINITIONS AND APPLICATIONS OF SYMMETRY CONSIDERATIONS

Products or Combinations of Symmetry Operations

The product of any two symmetry operations, defined as their consecutive application, must be a symmetry operation. Hence, the product of the C_2 and σ_v'

operations on a molecule with C_{2v} symmetry (*e.g.*, H_2O) is σ_v. (Recall that the C_2 axis is the z axis, the plane of the molecule is yz, and σ_v' is the yz plane.) This can be written as:

$$C_2 \times \sigma_v' = \sigma_v \text{ or}$$

$$C_2\sigma_v' = \sigma_v$$

Instead of the term *product,* the term *combination* is a better description of the above operations. The order in which the operations are written (*i.e.*, from left to right) is the reverse of the order in which they are applied. For the above example, the σ_v' operation is carried out first and is followed by C_2 to produce the same result as σ_v. In general, the final result depends on the order in which the operations are carried out, but, in some cases, it does not. When the result is independent of the order (*e.g.*, $C_2\sigma_v' = \sigma_v'C_2$), the two symmetry elements C_2 and σ_v' are said to *commute.* In a D_{3h} molecule (see, for example, Fig. 1–4), the two operations C_3 and σ_v do not commute; *i.e.*,

$$C_3\sigma_v \neq \sigma_v C_3$$

Equivalent Symmetry Elements and Equivalent Atoms

If a symmetry *element* A can be moved into the *element* B by an *operation* corresponding to the element X, then A and B are said to be *equivalent.* If we define X^{-1} as the reverse operation of X (*e.g.*, a counterclockwise rotation instead of a clockwise one), then X^{-1} will take B back into A. Furthermore, if A can be carried into C, then there must be a symmetry operation (or a sequence of symmetry operations) that carries B into C, since B can be carried into A. The elements A, B, and C are said to form an *equivalent set.*

Any set of symmetry elements chosen so that any member can be transformed into each and every other member of the set by application of some symmetry operation is said to be a set of *equivalent* symmetry elements. This collection of elements is said to constitute a *class.* To make this discussion clearer, consider the planar molecule $PtCl_4^{2-}$ illustrated in Fig. 1–18. The C_2' and C_2 axes form an equivalent set, as do the C_2'' and C_2'''. However, C_2' is not equivalent to C_2'', because the molecule possesses no symmetry operation that takes one into the other.

Equivalent atoms in a molecule are defined as those atoms that may be interchanged with one another by a symmetry operation that the molecule possesses. Accordingly, all of the chlorines in $PtCl_4^{2-}$ (Fig. 1–18) and all of the hydrogens in methane, benzene, cyclopropane, or ethane are equivalent. The fluorines in gaseous PF_5 (trigonal bipyramidal structure) are not all equivalent, but form two sets of equivalent atoms, one containing the three equatorial fluorines and one containing the two axial fluorine atoms. These considerations are very important when the topic of nuclear magnetic resonance is discussed, for, under favorable conditions, nonequivalent atoms give rise to separate peaks in the spectrum.

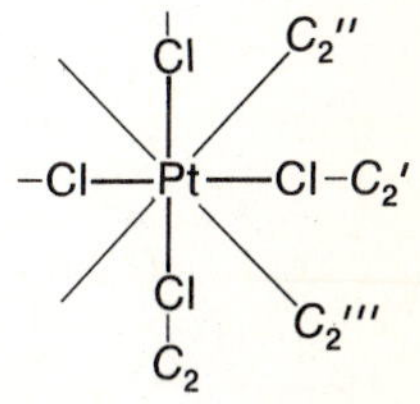

Cl C_2''
–Cl—Pt—Cl–C_2'
Cl C_2'''
C_2

FIGURE 1–18. Equivalent symmetry elements and atoms in the molecule $PtCl_4^{2-}$.

Optical Activity

If the mirror image of a molecule cannot be superimposed on the original, the molecule is optically active; if it can be superimposed, the molecule is optically inactive. In using this criterion, the mirror is understood to be external to the whole molecule, and reflection through the mirror gives an image of the whole molecule. With complicated molecules, the visualization of superimposability is difficult. Accordingly, it is to our advantage to have a symmetry basis for establishing the existence of optically active isomers. *Any molecule that has no improper rotation axis is said to be dissymmetric, and optically active molecules must be dissymmetric.* One often hears the incomplete statement that in order for optical isomerism to exist, the molecule must lack a plane or center of symmetry. Since $S_1 = \sigma$ (S_1 is a rotation by 360° followed by a reflection) and $S_2 = i$, if the molecule lacks an improper rotation axis, both i and σ must be lacking. To show the incompleteness of the earlier statement, we need to find a molecule that has neither σ or i, but does contain an S_n axis and is not optically active. Such a molecule is 1,3,5,7-tetramethylcycloocta-tetraene, shown in Fig. 1–19. This molecule does not have a plane or center of symmetry. However, since it has an S_4 axis, it is *not* optically active.

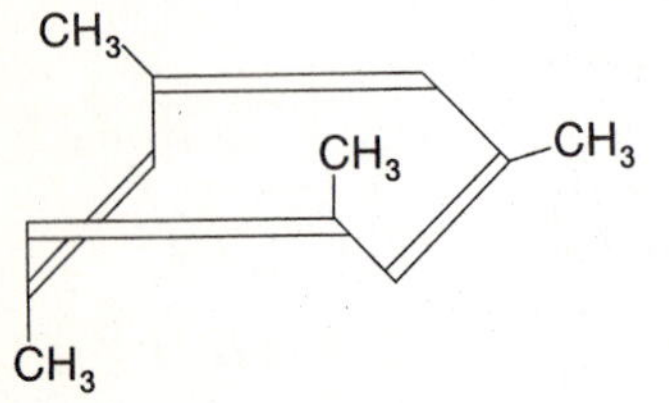

FIGURE 1–19. Structure of 1,3,5,7-tetramethylcyclooctatetraene.

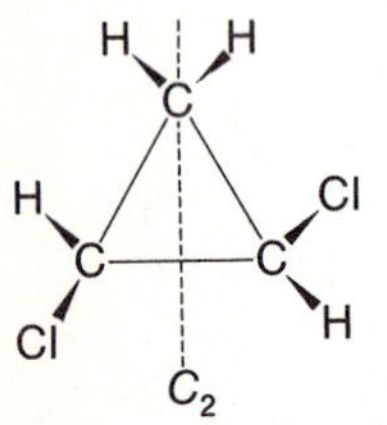

FIGURE 1–20. *Trans*-1,2-dichlorocyclopropane.

A dissymmetric molecule differs from an asymmetric one, for the latter type is completely lacking in symmetry. The molecule *trans*-1,2-dichlorocyclopropane, shown in Fig. 1–20, is dissymmetric (there is no S_n axis) and hence optically active, but it is not asymmetric, for it possesses a C_2 axis.

Nearly all molecules can exist in some conformation that is optically active. However, if rotation of the molecule about a bond produces a conformation with an improper axis, the molecule will not be optically active. If the conformation is frozen in a form that does not possess an improper axis, optical activity could result.

In summary, then, we can state that if a molecule possesses only C_n, it is dissymmetric and optically active. If $n = 1$, the molecule is asymmetric as well as dissymmetric; and if $n > 1$, the molecule is dissymmetric. If a molecule possesses S_n with any n, it cannot be optically active.

Dipole Moments

Molecules possess a center of gravity of positive charge, which is determined by the nuclear positions. When the center of gravity of the negative charge from the electrons is at some other point, the molecule has a dipole moment, which is related to the magnitude of the charge times the distance between the centers. The dipole moment is a vector property; that is, it has both a magnitude and a direction. For a fixed geometry (*i.e.*, a non-vibrating molecule) the dipole moment is a non-fluctuating property of the molecule; as a result it, like the total energy, must remain unchanged by the operation of every symmetry element of the molecule. In order for this to occur, the dipole moment vector must be coincident with *each* of the symmetry elements.

There are several obvious consequences of these principles. Molecules that have a center of symmetry cannot possess a dipole moment, because the vector cannot be a point, and any vector would be changed by inversion through the center. Molecules with more than one C_n axis cannot have a dipole moment, for a dipole moment vector could not be coincident with more than one axis. Thus, only the following types of molecules may have dipole moments: those with one C_n ($n > 1$), those with one σ and no C_n, those with a C_n and symmetry planes that include C_n, and those that have no symmetry. In all cases, where the molecule has symmetry, the direction of the dipole vector is determined; for it must lie in all of the symmetry elements that the molecule possesses.

ADDITIONAL READING REFERENCES

M. Tinkham, "Group Theory and Quantum Mechanics," McGraw-Hill, New York (1964).
F. A. Cotton, "Chemical Applications of Group Theory," 2nd Ed., Wiley-Interscience, New York (1971).
M. Orchin and H. H. Jaffé, J. Chem. Educ., *47,* 246, 372, 510 (1970).
M. Orchin and H. H. Jaffé, "Symmetry, Orbitals, and Spectra," Wiley-Interscience (1971).
C. D. H. Chisholm, "Group Theoretical Techniques in Quantum Chemistry," Academic Press, New York (1976).
For space groups, see: J. D. Donaldson and S. D. Ross, "Symmetry and Stereochemistry," Wiley (1972).

EXERCISES

1. Classify the following molecules in the appropriate point groups and, for c through k, indicate all symmetry elements except *E*.
   a. $CoCl_4^{2-}$
   b. $Ni(CN)_4^{2-}$
   c. *cis*-$CoCl_4(NH_3)_2^-$ (ignore the hydrogen atoms)
   d. C_6H_{12} (chair form)
   e. $Si(CH_3)_3 \cdot A \cdot B$ (with A and B *trans* in a trigonal bipyramid)
   f. PF_3
   g. $(CH_3)_2B(\mu\text{-}H)_2B(CH_3)_2$ (two bridging H atoms between the B atoms)
   h. $Cl\text{-}I\text{-}Cl^-$
   i. planar *cis*-$PdCl_2B_2$ (B=base)
   j. planar *trans*-$PdCl_2B_2$ (B=base)
   k. staggered configuration for C_2H_6
   l. ferrocene

2. a. How does a D_{2d} complex of formula MCl_4^{2-} differ from a T_d complex (*i.e.,* which symmetry elements differ)?
   b. What important symmetry element is absent in the PF_3 molecule that is present in the D_{3h} point group?
   c. If each Ni—C—N bond in planar $Ni(CN)_4^{2-}$ was not linear but was bent (Ni–C–N, bent at C), to what point group would this ion belong? What essential symmetry element present in D_{4h} is missing in C_{4h}?
   d. Indicate which operation is equivalent to the following products of the T_d point group
      (1) $S_4 \times S_4 = ?$
      (2) $C_3 \times C_2 = ?$
      (3) $\sigma_d \times C_2 = ?$

3. Does $POCl_2Br$ possess a rotation-reflection axis coincident with the P—O bond axis?

4. a. Rotate the molecule in Fig. 1–14, 180° around the S_4 axis, and draw the resulting structure.
   b. Locate the mirror plane of the S_4 operation and indicate on the structure resulting from part (a) where all the atoms are located after reflection through the plane. Use script letters to indicate the final location of each atom.
   c. Is S_2 a symmetry operation for the molecule?

5. To which point group does the molecule triethylenediamine (also called 1,4-diazabicyclo[2.2.2]octane, or dabco) belong?

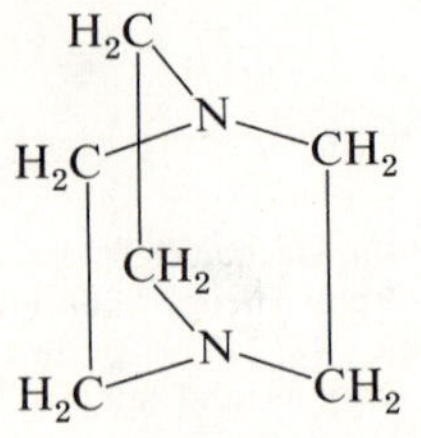

H_2C N CH_2 H_2C CH_2 N H_2C CH_2

6. Assign $PtCl_4^{2-}$ to a point group and locate one symmetry element in each class of this point group.

7. Give the point group for each of the following and indicate whether the entity could have a dipole moment:

   a. three-bladed propeller

   b. hexahelicene

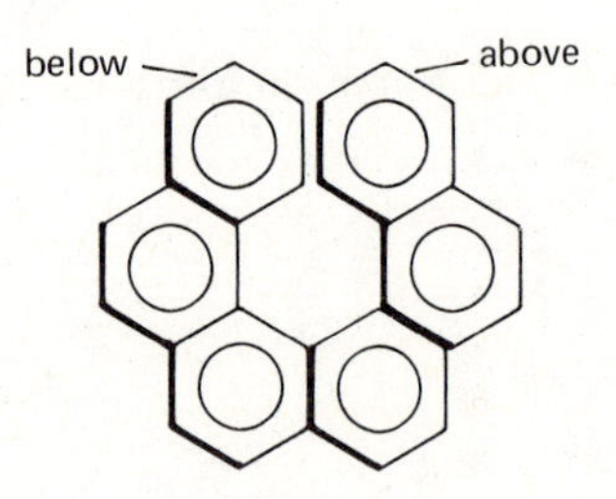

   c. 2,2-cyclophane

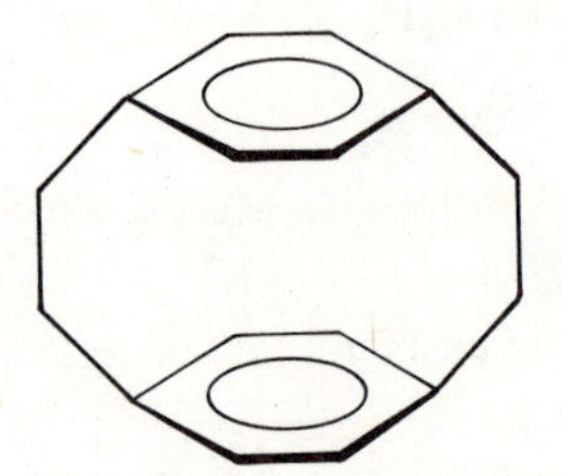

   d. hexadentate ligand on Fe^{2+}

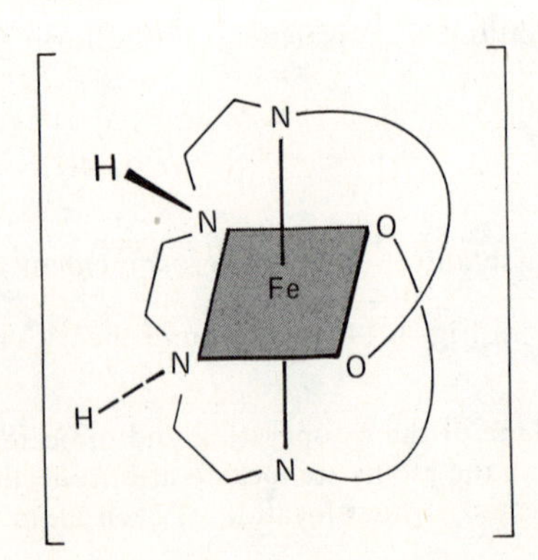

e. $Co(C_5H_5NO)_6^{2+}$ (C_5H_5NO is C—C, C, N—O, C—C ring)

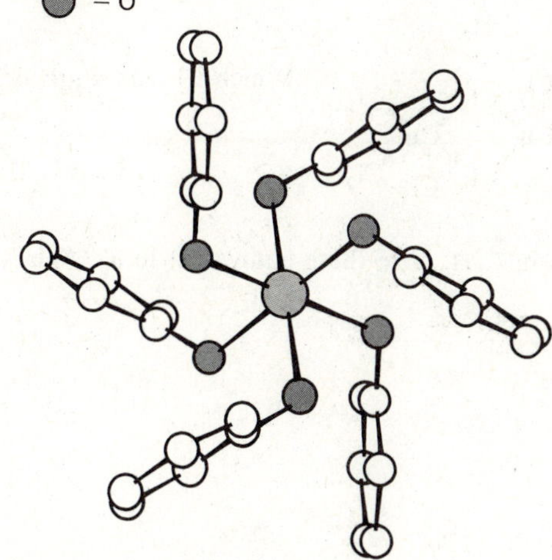

8. Is the following compound optically active?

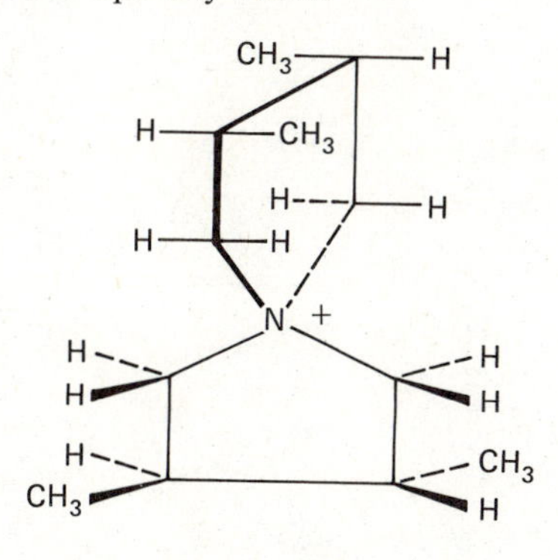

9. $Cr(en)_3^{3+}$ (where en = ethylenediamine) belongs to the D_3 point group. Is it optically active? Why?

10. Determine the point groups of the following molecules, and find the equivalent atoms within these molecules that are asked for.

a. Cl Cl H_3 H_1 H_2 Cl — Which protons, if any, are equivalent?

b. Cl H_8 H_5 H_7 H_6 Fe H_3 H_2 H_4 H_1 Cl — Which protons are equivalent?

c. 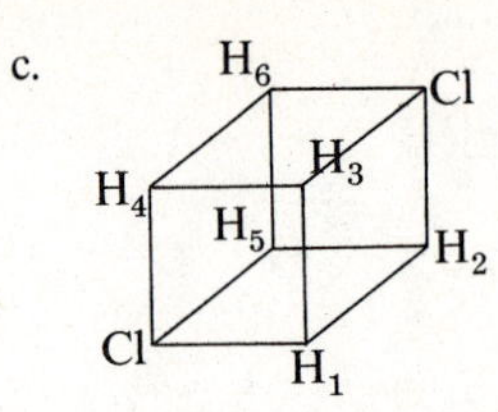

Which protons are equivalent?

d. Cl_8 Cl_6 Cl_5 Cl_1 Cu Cu Cl_7 Cl_4 Cl_3 Cl_2

Which Cl^- are equivalent?

11. Locate the σ_d planes in C_6H_6. Are these equivalent to σ_v? Why?

2 GROUP THEORY AND THE CHARACTER TABLES

2–1 INTRODUCTION

Group theory is a topic in abstract algebra that can be applied to certain systems if they meet specific requirements. There are many systems of interest to chemists that can be treated by these techniques, and by the end of this chapter you will only have begun to gain an appreciation for the power of this method. We shall use these concepts frequently throughout the rest of the book. We shall be involved in some abstract and seemingly irrelevant concepts in the course of our development of this topic. The reader is encouraged to persevere, for the rewards are many.

There are *two common major types* of applications of group theory.

(a) It can be used in the *generation of symmetry combinations*. If you are given some basis set, that is, a set of orbitals or mathematical functions pertaining to a molecule, you can use group theory to *construct linear combinations* of the things in the basis set that *reflect* the symmetry of the molecule. For example, if you are given a set of atomic orbitals for a molecule, you can use group theory to assist you in mapping out the shape of the molecular orbitals for the given molecule. There are many areas where such an application can be made, and a list of a few include our ability to:

(1) determine *hybrid orbitals* used in bonding the atoms in the molecule.
(2) determine which *atomic orbitals* can contribute to the various *molecular orbitals* in a molecule.
(3) determine the number and symmetries of *molecular vibrations.*
(4) predict how the *degeneracies of the d-orbitals* are removed by *crystal fields* of various symmetries.
(5) predict and generate the *spin functions* to be used in the Hamiltonian. (Used in EPR, NQR, Mössbauer.)
(6) construct symmetry adapted combinations of nuclear spin functions to work out nmr equations.

(b) It can be used to *ascertain which quantum mechanical integrals are zero; i.e.,* is $\int\psi^*$ op $\psi\, d\tau$ equal to zero? Such integrals are important to:

(1) determine the allowedness of electronic transitions.
(2) determine the activity of infrared and Raman vibrations.
(3) determine the allowedness of any given transition in nmr, esr, etc.

2–2 RULES FOR ELEMENTS THAT CONSTITUTE A GROUP

The symmetry operations can be treated with group theory, and we shall use them to illustrate the principles. In order for any set of elements to form a mathematical group, the following conditions must be satisfied.

1. The combination (Chapter 1; often referred to as the "product") of any two elements, and the square of each element, must produce an element of the group.

This combination is not the same thing as a product in arithmetic. We are referring to the consecutive operation of elements. It is a simple matter to square and take all possible combinations of the elements of the C_{2v} point group to illustrate the point that another element is obtained. The reader is encouraged to do so, using H_2O as an example. Unlike multiplication in ordinary arithmetic, we must worry about the order of the combination, *i.e.*, $C_2 \times \sigma_v$ or $\sigma_v \times C_2$.

2. One operation of the group must commute with all others and leave them unchanged. This is the *identity element.*

$$E\sigma_2 = \sigma_2 E = \sigma_2$$

3. The associative law of multiplication must hold, *i.e.*,

$$(XY)Z = X(YZ)$$

The result of combining three elements must be the same if the first element is combined with the product of the second two, [*i.e.*, $\sigma_v(C_2\sigma_v')$] or if the product of the first two is combined with the last element [*i.e.*, $(\sigma_v C_2)\sigma_v'$].

4. Every element must have a reciprocal that is also an element of the group. This requires that for every symmetry operation there be another operation that will undo what the first operation did. For any mirror plane, the inverse is the identical mirror plane, *e.g.*, $\sigma \times \sigma = E$. For a proper rotation $C_n{}^m$, the inverse is $C_n{}^{n-m}$; $C_n{}^m \times C_n{}^{n-m} = E$. If A has the reciprocal element B, then $AB = BA = E$. If B is the reciprocal of A, then A is the reciprocal of B. In general, the reciprocal of A can be written A^{-1}.

C_3 has the reciprocal element $C_3{}^2$:

$$C_3C_3{}^2 = C_3{}^2C_3 = E$$

We shall illustrate these rules in an abstract way in the next section.

2-3 GROUP MULTIPLICATION TABLES

Properties of the Multiplication Tables

If we have a complete and non-redundant list of all the elements in a finite group, and we know all the products, then the group is completely and uniquely defined. This information can be summarized with a *group multiplication table.* This table is simply an array in which each column and row is headed by an element. The matrix format is employed here because it is a convenient way to insure that we have taken all possible permutations of products of the elements. The number of elements in the group is referred to as the *order of the group, h.* The group multiplication table will consist of h rows and h columns. Certain groups have an infinite number of elements; *e.g.*, H_2 belongs to $D_{\infty h}$.

In order to illustrate simply the rules of group theory discussed earlier, we will consider a group of order three containing the abstract elements A and B as well as the identity E. We shall determine what requirements these elements must satisfy (*i.e.*, how their combinations must be defined) in order for these elements to constitute a group. Each matrix element is the result of the product of the column element times

	E	A	B
E			
A			
B			

the row element. Since multiplication is in general not commutative, we must *adhere to a consistent order for multiplication.* The convention is to carry out the operations in the order, *column element times row element; i.e.,* when we write a product RC, where C is column and R is row, we take the column element first followed by the row.

Each of the original h elements in the group must appear once and only once in each row and each column of the group multiplication table, and no two rows or columns may be alike. The following argument proves this. For a group of h different elements $E, A_1 \ldots A_n \ldots A_{h-1}$, the elements of the nth row (A_n) are $A_nE, A_nA_1 \ldots A_nA_{h-1}$. If any two products A_nA_i and A_nA_j were the same, left multiplication by the inverse of A_n would give $A_n^{-1}A_nA_i = A_i$ and $A_n^{-1}A_nA_j = A_j$ respectively, and thus require A_i and A_j to be the same. This conclusion is contrary to our original assumption that all elements of the group were different. Thus, all the products and all the h elements of the row must be different. Each of the original h elements in the group must thus appear once and only once in the row. A similar argument can be used for the columns. We can illustrate the requirements for the general, unspecified set of elements E, A, and B described above to constitute a group, and show how these rules define the multiplication table.

Since E is the identity, it is a simple matter to indicate the matrix elements for products involving E and write (REMEMBER column first, then row multiplication):

	E	A	B
E	E	A	B
A	A		
B	B		

There are only two ways to complete this table; these involve defining either $AA = E$ or $AA = B$. If $AA = E$, then according to rule 4, A is its own reciprocal, so $BB = E$. (A cannot be reciprocal to both B and A unless $A = B$, so B must be the reciprocal of B.) Then, if $BA = A$ or B, we would not be able to complete the table without repeating an element in a row or column; *i.e.,* $BA = A$ makes A repeat in a column, while $BA = B$ repeats B in the row. However, if we define $AA = B$, the following table results:

	E	A	B
E	E	A	B
A	A	B	E
B	B	E	A

If $AA = B$, we know that A must have a reciprocal, so $AB = BA = E$.

We can consider this entire group to be generated by taking an element and all its powers, *e.g.,* A, $A^2(= B)$, and $A^3(= E)$. Such a group is a *cyclic group*. For such a group, all multiplications must commute.

Thus, by adhering to the conditions specified for a collection of elements to constitute a group, we have been able to define our elements so as to construct a group multiplication table that specifies the result of all possible combinations of elements.

The very same procedures and rules can be used to construct group multiplication tables of higher order. When a table of order four is made, two possibilities result, which we shall label G_4^1 and G_4^2:

G_4^1	E	A	B	C
E	E	A	B	C
A	A	B	C	E
B	B	C	E	A
C	C	E	A	B

and

G_4^2	E	A	B	C
E	E	A	B	C
A	A	E	C	B
B	B	C	E	A
C	C	B	A	E

The superscripts arbitrarily number the two possibilities.

We can make these procedures more specific by considering a system in which the elements are symmetry operations. In so doing, our definition of the products will not seem so arbitrary, for we know physically what the results of the products of the symmetry operations are. We shall proceed by considering the C_{3v} ammonia molecule, whose point group contains the symmetry operations E, $2C_3$, and $3\sigma_v$. The three σ_v planes are labeled as in Fig. 2-1, and the group multiplication table is given as Table 2-1. The reader is encouraged to carry out and verify the combinations given in this table. All of the points made in conjunction with the discussion of the group A, B, E should also be verified.

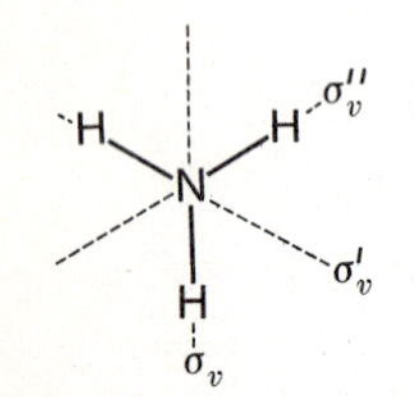

FIGURE 2-1. The σ_v planes in the C_{3v} ammonia molecule.

The group multiplication table (Table 2-1) contains sub-groups, which are smaller groups of elements in the point group that satisfy all of the requirements for constituting a group. E is such a sub-group, as is the collection of operations E, C_3, C_3^2. The order of any sub-group must be an integral divisor of the order, h, of the full group.

TABLE 2-1. GROUP MULTIPLICATION TABLE FOR THE C_{3v} POINT GROUP

	E	C_3	C_3^2	σ_v	σ_v'	σ_v''
E	E	C_3	C_3^2	σ_v	σ_v'	σ_v''
C_3	C_3	C_3^2	E	σ_v''	σ_v	σ_v'
C_3^2	C_3^2	E	C_3	σ_v'	σ_v''	σ_v
σ_v	σ_v	σ_v'	σ_v''	E	C_3	C_3^2
σ_v'	σ_v'	σ_v''	σ_v	C_3^2	E	C_3
σ_v''	σ_v''	σ_v	σ_v'	C_3	C_3^2	E

Similarity Transforms

A second way in which the symmetry operations of a group may be subdivided into smaller sets is by application of similarity transforms. If A and X are two operations of a group, then $X^{-1}AX$ will be equal to some operation of the group, say B:

$$B = X^{-1}AX$$

The operation B is said to be the similarity transform of A by X. In the general definition of a similarity transform, X does not necessarily represent an operation of the group. However, when taking the similarity transform of A in order for the result B to be some other operation of the group, *X must be an operation of the group.* We *then say that A and B are conjugate.* There are three important properties of conjugate operations:

1. Every operation is conjugate with itself. If we select one operation, say A, it must be possible to find at least one operation X such that $A = X^{-1}AX$.

Proof:

Left multiply by A^{-1}:

$$A^{-1}A = E = A^{-1}X^{-1}AX = (XA)^{-1}AX$$

(The reciprocal of the product of two or more operations is equal to the product of the reciprocals in reverse order; see Cotton, page 5, if you desire to see the proof). By definition, E must also equal $(AX)^{-1}(AX)$.

Both equations can be true only if A and X commute. Thus, the operation X may always be E, but it may also be any other operation that commutes with the selected operation A.

2. If A is conjugate with B, then B is conjugate with A; *i.e.*, if $A = X^{-1}BX$, then there must be some operation in the group, Y, such that $B = Y^{-1}AY$. Note that X equals Y^{-1} and vice versa.

3. If A is conjugate with both B and C, then B and C are conjugate with each other.

The use and meaning of the similarity transform will be made more concrete as we employ it in subsequent discussion in this chapter. For the moment, it may help to point out that a similarity transform can be used to change the coordinate system selected to describe a problem. A right-handed coordinate system is given by:

Z Y
X

or any rotation of these axes. A left-handed coordinate system is given by:

Z X
Y

The two coordinate systems are seen to be related to each other by a mirror plane σ_d that contains the Z-axis and bisects the X and Y-axes.

Now, consider a C_4^1 rotation about the Z-axis. A point with coordinates (a, b) (the X coordinate is always given first) in a right-handed system goes to $(b, -a)$ upon a 90° clockwise rotation, but to $(-b, a)$ with 270° rotation. In a left-handed system, C_4^1 takes (a, b) to $(-b, a)$, while $C_4{}^3$ yields $(b, -a)$. (Construct a figure if needed.) Thus, the roles of C_4 and $C_4{}^3$ are interchanged by changing the coordinate system. We can say that there is a similarity transform of $C_4{}^3$ to C_4 by σ_d; *i.e.*,

$$C_4 = \sigma_d{}^{-1}C_4{}^3\sigma_d$$

or, since $\sigma_d{}^{-1} = \sigma_d$ (note that $\sigma_d\sigma_d{}^{-1} = E$ but $\sigma_d\sigma_d = E$ so $\sigma_d = \sigma_d{}^{-1}$)

$$C_4 = \sigma_d C_4{}^3\sigma_d$$

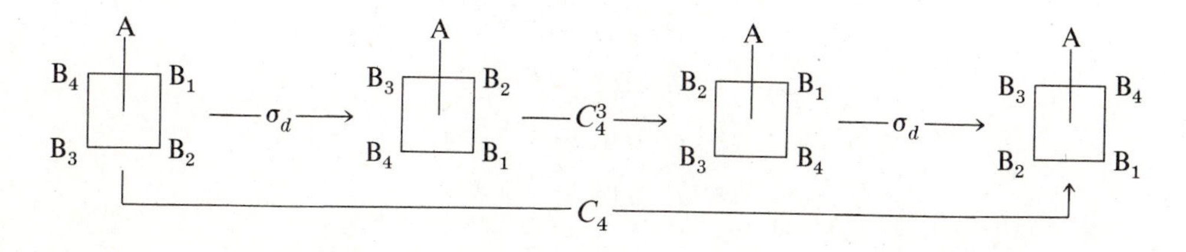

Classes of Elements

A complete set of elements that are conjugate to one another is called a *class* of the group. The order of each class must be an integral divisor of the order of the group. To determine the classes, begin with one element, say A, and work out all of its similarity transforms with all the elements of the group, including itself. Then find an element that is not one of those conjugate with A, and determine all of its transforms. Repeat this procedure until all of the elements in the group have been divided into classes. *E will always constitute a class by itself of order 1: $E^{-1}EE = E$.*

We shall demonstrate this procedure by working out all of the classes of elements in the group:

G_6^1	E	A	B	C	D	F
E	E	A	B	C	D	F
A	A	E	D	F	B	C
B	B	F	E	D	C	A
C	C	D	F	E	A	B
D	D	C	A	B	F	E
F	F	B	C	A	E	D

First of all, E is a class by itself. Next, work out all of the similarity transforms of A. These would be EAE, $A^{-1}AA$, $B^{-1}AB$, $C^{-1}AC$, $D^{-1}AD$, and $F^{-1}AF$. We already know that $EAE = A$. To carry out the combination $A^{-1}AA$, we first note from our table that $AA = E$. Thus, $A^{-1}AA = A^{-1}E$, so we have to find out what element is equal to A^{-1}. We know that $A^{-1}A = E$; accordingly, we go to the A column and determine which element, when combined with A, produces the result (matrix element) E. This element is A, and therefore $A^{-1} = A$. As a result, $A^{-1}AA = A^{-1}E = AE = A$. In a similar fashion, we can show that $B^{-1} = B$. To determine $B^{-1}AB$, we see that $AB = D$ and $B^{-1}D = BD = C$. Proceeding in a similar fashion, all of the similarity transforms of A can be determined:

$$E^{-1}AE = A$$

$$A^{-1}AA = A$$

$$B^{-1}AB = C$$

$$C^{-1}AC = B$$

$$D^{-1}AD = C$$

$$F^{-1}AF = B$$

Thus, we have determined that A, B, and C are all conjugate and members of the same class. We now know that all transforms of B and C are also A, B, or C (see rule 3, p. 21), so these do not have to be worked out.

Next, we find an element that is not conjugate with A and determine its transforms. Such an element is D, and its transforms are:

$$E^{-1}DE = D$$

$$A^{-1}DA = F$$

$$B^{-1}DB = F$$

$$C^{-1}DC = F$$

$$D^{-1}DD = D$$

$$F^{-1}DF = D$$

All transforms of D are either F or D, so these two elements constitute a class. Note that we have now placed every element in a class.

We can again make this procedure less abstract by working out the classes for the symmetry group of the ammonia molecule from Table 2–1. (You might try closing the book, working this out yourself, and then using the ensuing discussion to check your result.)

$$EEE = E$$
$$C_3^2EC_3 = E \text{ (note: } C_3^{-1} = C_3^2)$$
$$\vdots$$
$$\sigma_v''E\sigma_v'' = E$$

The first result is, of course, that E is always in a class by itself. Next, it can be shown that C_3 and C_3^2 form a class of the group.

$$\begin{aligned} E\,C_3E &= C_3 \\ C_3^2C_3C_3 &= C_3 \\ C_3C_3C_3^2 &= C_3 \\ \sigma_v C_3\sigma_v &= C_3^2 \\ \sigma_v'C_3\sigma_v' &= C_3^2 \\ \sigma_v''C_3\sigma_v'' &= C_3^2 \end{aligned}$$

Similarly, σ_v, σ_v', and σ_v'' form a class of the group:

$$\begin{aligned} E\sigma_vE &= \sigma_v \\ C_3^2\sigma_vC_3 &= \sigma_v'' \\ C_3\sigma_vC_3^2 &= \sigma_v' \\ \sigma_v\sigma_v\sigma_v &= \sigma_v \\ \sigma_v'\sigma_v\sigma_v' &= \sigma_v'' \\ \sigma_v''\sigma_v\sigma_v'' &= \sigma_v' \end{aligned}$$

In general, equivalent symmetry operations belong to the same class. (Recall that equivalent elements are ones that can be taken into one another by symmetry operations of the group.) For example, in the planar ion $PtCl_4^{2-}$ (D_{4h}), shown in Fig. 2–2, σ_v and σ_v' can never be taken into σ_d and σ_d'. Therefore, σ_v and σ_v' form one class of the group, while σ_d and σ_d' form another.

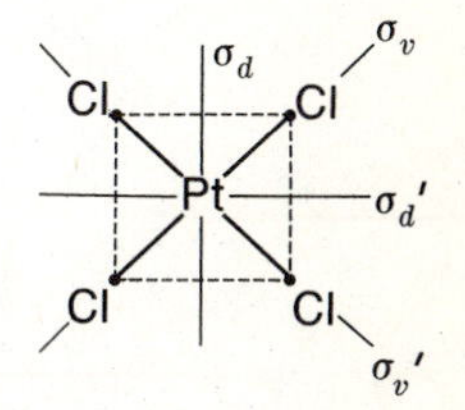

FIGURE 2–2. The σ_v and σ_d planes in the D_{4h} $PtCl_4^{2-}$ ion.

2–4 SUMMARY OF THE PROPERTIES OF VECTORS AND MATRICES

Vectors

A vector in three-dimensional space has a magnitude and a direction that can be specified by the lengths of its projections on the three orthogonal axes of a Cartesian

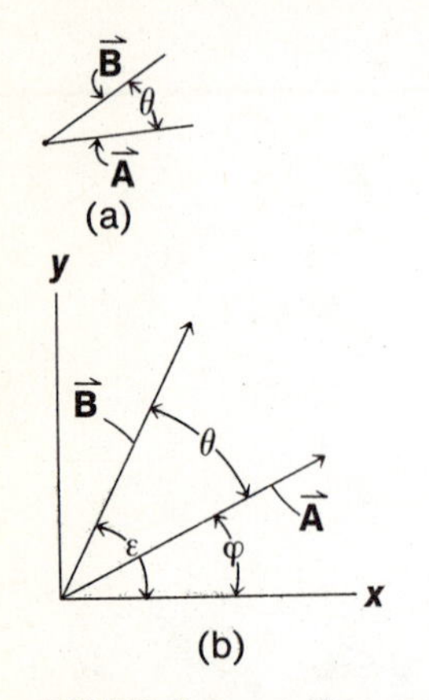

FIGURE 2–3. **a.** Two vectors, **A** and **B**, separated by an angle θ. **b.** The vectors placed in a two-dimensional *xy* coordinate system.

coordinate system. Vector properties may be more than three-dimensional, so the above statement can be extended to *p*-dimensional space and *p* orthogonal axes in *p*-space.

It is often necessary to take the product of vectors. One type of product that produces a number (*i.e.*, a scalar) is called the scalar or dot product. This is given by:

$$\vec{\mathbf{A}} \cdot \vec{\mathbf{B}} = AB \cos \theta \tag{2-1}$$

Here the boldface $\vec{\mathbf{A}}$ and $\vec{\mathbf{B}}$ refer to the two vectors, and the dot refers to their dot product. On the right-hand side of the equation, A and B refer to the lengths of the $\vec{\mathbf{A}}$ and $\vec{\mathbf{B}}$ vectors, and θ is the angle between them as shown in Fig. 2–3a. Accordingly, if the angle θ between two vectors is 90°, the dot product is zero ($\cos 90° = 0$) and the vectors are orthogonal.

It is necessary to reference our vectors to a coordinate system. This is done for a two-dimensional example (*i.e.*, the *xy* plane) in Fig. 2–3b. The angle θ is now seen to be $\varepsilon - \varphi$, so the dot product becomes

$$\vec{\mathbf{A}} \cdot \vec{\mathbf{B}} = AB \cos (\varepsilon - \varphi) \tag{2-2}$$

Simple trigonometry now gives us the projections of the $\vec{\mathbf{A}}$ and $\vec{\mathbf{B}}$ vectors on the *x* and *y* axes as:

$$A_x = A \cos \varphi \tag{2-3}$$

$$A_y = A \sin \varphi \tag{2-4}$$

$$B_x = B \cos \varepsilon \tag{2-5}$$

$$B_y = B \sin \varepsilon \tag{2-6}$$

Using a trigonometric identity, equation (2–2) can be rewritten as

$$\begin{aligned} \vec{\mathbf{A}} \cdot \vec{\mathbf{B}} &= AB(\cos \varphi \cos \varepsilon + \sin \varphi \sin \varepsilon) \\ &= A \cos \varphi \, B \cos \varepsilon + A \sin \varphi \, B \sin \varepsilon \end{aligned} \tag{2-7}$$

Substituting equations (2–3) through (2–6) into (2–7) produces

$$\vec{\mathbf{A}} \cdot \vec{\mathbf{B}} = A_x B_x + A_y B_y$$

Thus, the dot product of two vectors in two-dimensional space is the product of the components with all cross terms ($A_x B_y$ and so forth) absent. In *p*-space, the result obtained is

$$\vec{\mathbf{A}} \cdot \vec{\mathbf{B}} = \sum_{i=1}^{p} A_i B_i$$

where *i* ranges over the *p* orthogonal axes in *p*-space. Accordingly, the scalar square of a vector is given by

$$\vec{\mathbf{A}}^2 = \sum_{i=1}^{p} A_i^2$$

Matrices

A matrix is a rectangular array of numbers or symbols that has the following general form:

$$\begin{bmatrix} a_{11} & a_{12} & a_{13} \\ a_{21} & a_{22} & a_{23} \\ a_{31} & a_{32} & a_{33} \end{bmatrix}$$

The square brackets indicate that this is a matrix, as opposed to a determinant. The entire matrix can be abbreviated with a script letter or by the symbol $[a_{ij}]$. The symbol a_{ij} refers to the matrix element in the ith row and jth column. When the number of rows equals the number of columns, the matrix is called a *square matrix.* The elements a_{ij} of a square matrix for which $i = j$ (*i.e.*, a_{11}, a_{22}, a_{33}, etc.) are called the *diagonal elements,* and the other elements are called *off-diagonal.* When all of the off-diagonal elements of a matrix are zero, the matrix is said to be *diagonalized* or to be a diagonal matrix. When each of the diagonal elements of a square matrix equals 1 and all off-diagonal elements are zero, the matrix is called a *unit matrix.* The unit matrix is often abbreviated by the Kronecker delta,

$$\begin{bmatrix} 1 & 0 & 0 & 0 \\ 0 & 1 & 0 & 0 \\ 0 & 0 & 1 & 0 \\ 0 & 0 & 0 & 1 \end{bmatrix} = \delta_{ij}$$

Unless $i = j$, δ has a value of zero. The *trace* or *character* of a square matrix, an important property (*vide infra*), is simply the sum of the diagonal elements. A one-row matrix can be conveniently written on a single line. In order to write a one-column matrix on a single line, it is enclosed in braces, { }.

A vector is conveniently represented by a one-column matrix. In a three-dimensional orthogonal coordinate system, a vector initiating at the origin of the coordinate system is completely specified by the x, y, and z coordinates of the other end. Thus, the matrix $\{x, y, z\}$ is a one-column matrix that represents the vector. In p-space, a p by 1 column vector is needed. In both instances, the elements of the matrix give the projections of the vector on the orthogonal coordinates.

Matrices may be added, subtracted, multiplied, or divided by using the appropriate rules of matrix algebra. In order to add or subtract two matrices $\mathcal{A}$ and $\mathcal{B}$ to give a matrix $\mathcal{C}$, the matrices must all be of the same *dimension; i.e.,* they must contain the same number of rows and columns. The elements of the $\mathcal{C}$ matrix are given by:

$$c_{ij} = a_{ij} \pm b_{ij}$$

A matrix can be multiplied by a scalar (a single number). When multiplying by a scalar, each matrix element is multiplied by this scalar

$$k[a_{ij}] = [ka_{ij}]$$

The ijth matrix element of a product matrix is obtained by multiplying the ith row of the first matrix by the jth column of the second matrix, *i.e.,* row by column. That is, the matrix elements of $\mathcal{C}$ are given by

$$c_{ik} = \sum_{j=1}^{n} a_{ij} b_{jk}$$

where n is the number of elements in the ith row and in the jth column. This matrix multiplication is equivalent to taking the dot product of two vectors.

It should be clear that, in order to multiply a matrix by another matrix, the two matrices must be conformable; *i.e.*, if we wish to multiply $\mathcal{A}$ by $\mathcal{B}$ to give $\mathcal{C}$, the number of columns in $\mathcal{A}$ must equal the number of rows in $\mathcal{B}$. If the dimensionality of matrix $\mathcal{A}$ is i by j and that of $\mathcal{B}$ is j by k, then $\mathcal{C}$ will have a dimensionality of i by k. This can be seen by carrying out the following matrix multiplication:

$$\underset{3 \text{ by } 2}{\begin{bmatrix} a_{11} & a_{12} \\ a_{21} & a_{22} \\ a_{31} & a_{32} \end{bmatrix}} \underset{2 \text{ by } 3}{\begin{bmatrix} b_{11} & b_{12} & b_{13} \\ & & \\ b_{21} & b_{22} & b_{23} \end{bmatrix}} = \underset{3 \text{ by } 3}{\begin{bmatrix} c_{11} & c_{12} & c_{13} \\ c_{21} & c_{22} & c_{23} \\ c_{31} & c_{32} & c_{33} \end{bmatrix}}$$

The number of columns (two) in $\mathcal{A}$ equals the number of rows (two) in $\mathcal{B}$. The matrix elements of $\mathcal{C}$ are obtained by a row by column multiplication; *i.e.*,

$$c_{11} = a_{11}b_{11} + a_{12}b_{21}$$
$$c_{12} = a_{11}b_{12} + a_{12}b_{22}$$
$$c_{13} = a_{11}b_{13} + a_{12}b_{23}$$
$$c_{21} = a_{21}b_{11} + a_{22}b_{21}$$
$$c_{22} = a_{21}b_{12} + a_{22}b_{22}$$
$$c_{23} = a_{21}b_{13} + a_{22}b_{23}$$
$$c_{31} = a_{31}b_{11} + a_{32}b_{21}$$
$$c_{32} = a_{31}b_{12} + a_{32}b_{22}$$
$$c_{33} = a_{31}b_{13} + a_{32}b_{23}$$

Matrix multiplication always obeys the associative law, but is not necessarily commutative. Conformable matrices in the order $\mathcal{A}\mathcal{B}$ may not be conformable in the order $\mathcal{B}\mathcal{A}$.

Division of matrices is based on the fact that $\mathcal{A}$ divided by $\mathcal{B}$ equals $\mathcal{A}\mathcal{B}^{-1}$, where $\mathcal{B}^{-1}$ is defined as that matrix such that $\mathcal{B}\mathcal{B}^{-1} = \delta_{ij}$. Thus, the only new problem associated with division is the finding of an inverse. Only square matrices can have inverses. The procedure for obtaining the inverse is described in matrix algebra books, which should be consulted should the need arise.

Conjugate matrices deserve special mention. If two matrixes $\mathcal{A}$ and $\mathcal{B}$ are conjugate, they are related by a similarity transform just as conjugate elements of a group are; *i.e.*, there is a matrix $\mathcal{R}$ such that

$$\mathcal{A} = \mathcal{R}^{-1}\mathcal{B}\mathcal{R} \tag{2-8a}$$

One advantage of matrices that will be of significance to us is that they can be used in describing the transformations of points, vectors, functions, and other entities in space. The transformation of points will be discussed in the next section, where we deal with a very important concept in group theory: how symmetry operations (elements) can be represented by matrices and what advantages are thus obtained.

We can use matrices and vectors to gain a further appreciation of similarity transforms. Imagine three distinct fixed points O, P_1, and P_2 in three-dimensional space. Let $\vec{\mathbf{X}}$ be the vector from O to P_1, and let $\vec{\mathbf{Y}}$ be the vector from O to P_2. A set of three-dimensional Cartesian coordinates centered at the point O will be referred to as frame I. Suppose that an operator A, *also thought of as being associated with frame I,* transforms the vector $\vec{\mathbf{X}}$ in frame I into the vector $\vec{\mathbf{Y}}$ in frame I according to the equation

$$\vec{\mathbf{Y}} = A\vec{\mathbf{X}}$$

Next rotate frame I into a new position, keeping the origin of the coordinates fixed at O and leaving the points P_1 and P_2 fixed in space. Frame I in its new position is called frame II. Notice that the vectors $\vec{\mathbf{X}}$ and $\vec{\mathbf{Y}}$ have been unaffected by this procedure, since the three points O, P_1, and P_2 have remained fixed in space. However, the projections of $\vec{\mathbf{X}}$ and $\vec{\mathbf{Y}}$ on the coordinate axes of frame II will give numerical values different from those obtained when the vectors were projected on the coordinate axes of frame I. Thus, in frame II we write the vectors as $\vec{\mathbf{X}}'$ and $\vec{\mathbf{Y}}'$. We now want to find an operator A' *associated with frame II* that transforms the vector $\vec{\mathbf{X}}'$ in frame II into the vector $\vec{\mathbf{Y}}'$ in frame II according to

$$\vec{\mathbf{Y}}' = A'\vec{\mathbf{X}}'$$

The operator A' in frame II is said to be *similar* to the operator A in frame I. Thus, A' sends $\vec{\mathbf{X}}'$ (expressed in frame II) into $\vec{\mathbf{Y}}'$ (also expressed in frame II), while A does the same thing in frame I. In order to evaluate the operator A', we need to know the relationship between the components of $\vec{\mathbf{X}}$ and $\vec{\mathbf{Y}}$ in frame I and the components of $\vec{\mathbf{X}}'$ and $\vec{\mathbf{Y}}'$ in frame II. This information is given in the form of a transformation of coordinates expressed by a matrix S. Thus,

$$\vec{\mathbf{X}} = S\vec{\mathbf{X}}'$$

$$\vec{\mathbf{Y}} = S\vec{\mathbf{Y}}'$$

so that

$$\vec{\mathbf{Y}} = A\vec{\mathbf{X}}$$

becomes

$$S\vec{\mathbf{Y}}' = AS\vec{\mathbf{X}}'$$

Since $|S| \neq 0$, we know that S^{-1} exists, and

$$\vec{\mathbf{Y}}' = S^{-1}AS\vec{\mathbf{X}}'$$

Thus,

$$\vec{\mathbf{Y}}' = A'\vec{\mathbf{X}}'$$

if

$$A' = S^{-1}AS \qquad \text{(2-8b)}$$

The similar matrices A' and A are connected by equation (2-8b), which is known as a *similarity transformation. The trace of a matrix is invariant under a similarity transformation.*

2–5 REPRESENTATIONS; GEOMETRIC TRANSFORMATIONS

As discussed previously, E, σ, i, C_n and S_n describe the symmetry of objects. *Each of these operations can be described by a matrix*. Consider the point P in Fig. 2–4 with

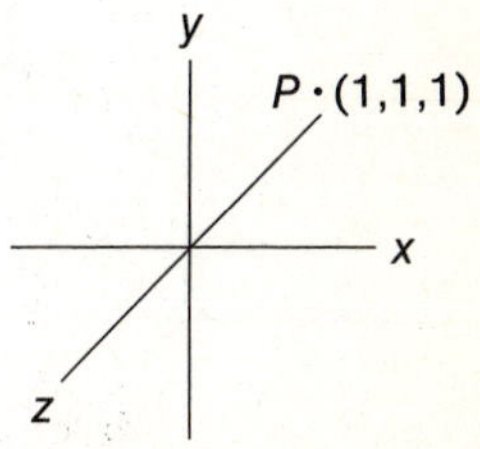

y
P·(1,1,1)
x
z

FIGURE 2–4. A point in the Cartesian coordinate system.

x, y, and z coordinates of 1, 1, and 1 corresponding to the projections of the point on these axes. The identity operation on this point corresponds to giving rise to a new set of coordinates that are the same as the old ones. The following matrix does this.

$$\begin{bmatrix} 1 & 0 & 0 \\ 0 & 1 & 0 \\ 0 & 0 & 1 \end{bmatrix} \begin{bmatrix} X \\ Y \\ Z \end{bmatrix} = \begin{bmatrix} X' \\ Y' \\ Z' \end{bmatrix}$$

Matrix multiplication yields:

$$X = X'$$
$$Y = Y'$$
$$Z = Z'$$

This unit matrix is said to be a representation of the identity operation.

A reflection in the xy plane, σ_{xy}, changes the sign of the z coordinate, but leaves x and y unchanged. The following matrix does this.

$$\begin{bmatrix} 1 & 0 & 0 \\ 0 & 1 & 0 \\ 0 & 0 & -1 \end{bmatrix} \begin{bmatrix} X \\ Y \\ Z \end{bmatrix} = \begin{bmatrix} X' \\ Y' \\ Z' \end{bmatrix}$$

Matrix multiplication gives $X = X'$, $Y = Y'$, and $Z = -Z'$, which is the result of reflection in the xy plane on the xyz coordinates of the point P. Similarly, the result of the operation σ_{yz} is given by

$$\begin{bmatrix} -1 & 0 & 0 \\ 0 & 1 & 0 \\ 0 & 0 & 1 \end{bmatrix} \begin{bmatrix} X \\ Y \\ Z \end{bmatrix} = \begin{bmatrix} X' \\ Y' \\ Z' \end{bmatrix}$$

and σ_{xz} is represented by

$$\begin{bmatrix} 1 & 0 & 1 \\ 0 & -1 & 0 \\ 0 & 0 & 1 \end{bmatrix} \begin{bmatrix} X \\ Y \\ Z \end{bmatrix} = \begin{bmatrix} X' \\ Y' \\ Z' \end{bmatrix}$$

The inversion operation is given by

$$\begin{bmatrix} -1 & 0 & 0 \\ 0 & -1 & 0 \\ 0 & 0 & -1 \end{bmatrix} \begin{bmatrix} X \\ Y \\ Z \end{bmatrix} = \begin{bmatrix} X' \\ Y' \\ Z' \end{bmatrix}$$

Call the z-axis the rotation axis; we shall derive the matrix for a clockwise (as viewed down the positive z-axis) rotation of the *point* by an angle φ. This rotation is illustrated in Fig. 2-5. The point (x_1, y_1) defines a vector $\mathbf{r}_1$ that connects it to the

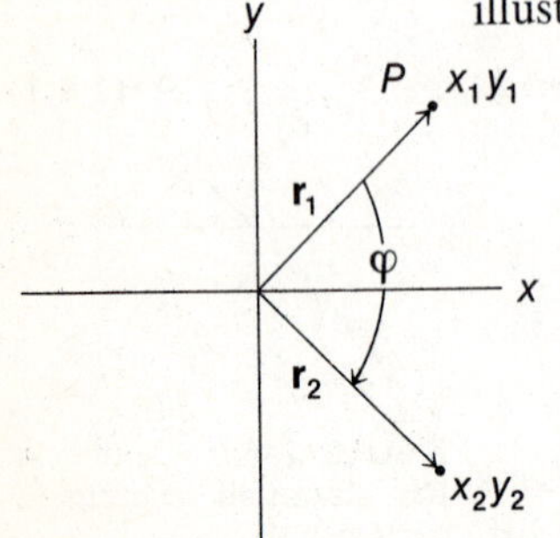

FIGURE 2-5. Clockwise rotation of the point (x_1, y_1) by the angle φ.

origin, and rotation leads to the vector $\mathbf{r}_2$. For any rotation of $\mathbf{r}_1$ about the z-axis by φ, the z component is unchanged, leading to:

$$\begin{bmatrix} & & 0 \\ & & 0 \\ 0 & 0 & 1 \end{bmatrix} \begin{bmatrix} x \\ y \\ z \end{bmatrix} = \begin{bmatrix} x' \\ y' \\ z' \end{bmatrix}$$

In Fig. 2–6, the rotation of $\mathbf{r}_1$ to $\mathbf{r}_2$ is shown. The problem is to determine the new coordinates for $\mathbf{r}_2$. The dots show rotation of the y_1 coordinate of $\mathbf{r}_1$ by φ, producing the vector $\mathbf{y}'$. The vector $\mathbf{y}'$ now has both x and y components, which are given by:

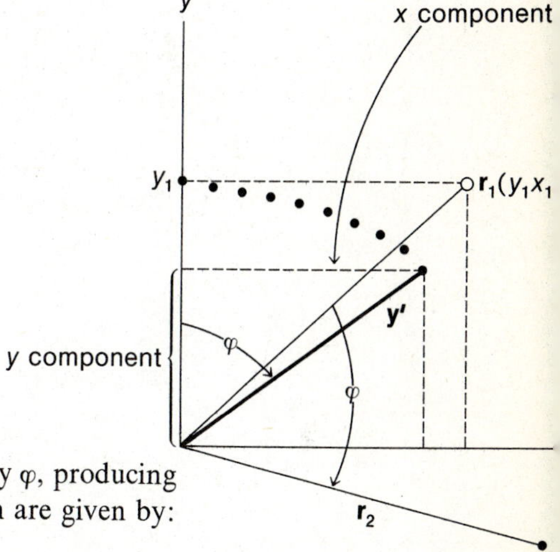

FIGURE 2–6. Rotation of the vector $\mathbf{r}_1$ and its y coordinate through the angle φ.

$$\mathbf{y}' = y_1 \sin \varphi + y_1 \cos \varphi$$

This result is obtained by the following trigonometric arguments: $\mathbf{y}'$ still has a length of y_1. Using trigonometry on the triangle defined by φ, $\mathbf{y}'$, and the y component of $\mathbf{y}'$, we see that

$$\cos \varphi = \frac{y \text{ component}}{y'} = \frac{y \text{ component}}{y_1}$$

or y component $= y_1 \cos \varphi$. Similarly, since $\sin \varphi = x$ component$/y_1$, we have x component $= y_1 \sin \varphi$. The vector $\mathbf{y}'$ then has components given by:

$$\mathbf{y}' = x \text{ component} + y \text{ component} = y_1 \sin \varphi + y_1 \cos \varphi$$

Figure 2–7 illustrates the rotation of the x coordinate of the vector $\mathbf{r}_1$, producing a new vector $\mathbf{x}'$ with both x and y components. As above, we can show that the y component is $-x_1 \sin \varphi$ and the x component is $x_1 \cos \varphi$. This leads to:

$$\mathbf{x}' = x_1 \cos \varphi - x_1 \sin \varphi$$

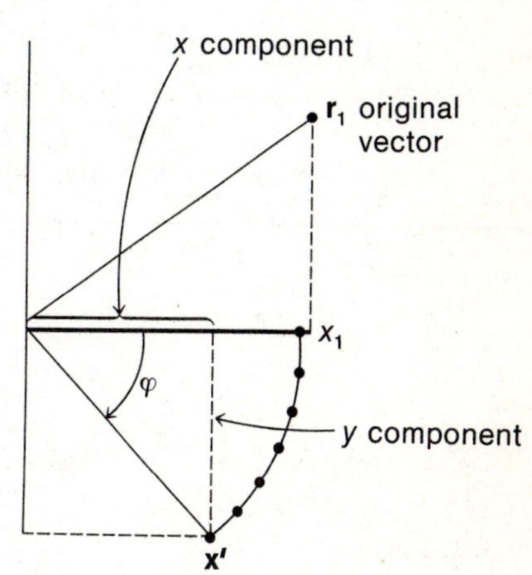

FIGURE 2–7. Rotation of the x coordinate of the vector $\mathbf{r}_1$ by the angle φ.

The x_2 and y_2 components of the rotated vector $\mathbf{r}_2$ must equal the sums of the x and y components of $\mathbf{x}'$ and $\mathbf{y}'$, so:

$$x_2 = x_1 \cos\varphi + y_1 \sin\varphi$$

$$y_2 = -x_1 \sin\varphi + y_1 \cos\varphi$$

Writing these equations in matrix form, we obtain for the clockwise rotation of a point in a fixed axis system:

$$\begin{bmatrix} \cos\varphi & \sin\varphi \\ -\sin\varphi & \cos\varphi \end{bmatrix} \begin{bmatrix} x_1 \\ y_1 \end{bmatrix} = \begin{bmatrix} x_2 \\ y_2 \end{bmatrix}$$

For the reverse transformation (*i.e.*, counterclockwise rotation of a point in a fixed axis system) the transformation matrix is given by:*

$$\begin{bmatrix} \cos\varphi & -\sin\varphi \\ \sin\varphi & \cos\varphi \end{bmatrix}$$

The total transformation matrix for a clockwise proper rotation is:

$$\begin{bmatrix} \cos\varphi & \sin\varphi & 0 \\ -\sin\varphi & \cos\varphi & 0 \\ 0 & 0 & 1 \end{bmatrix}$$

For a C_2 rotation we substitute $\varphi = -180°$ into the clockwise proper rotation matrix. A negative angle corresponds to a clockwise rotation according to trigonometric convention.

$$\begin{bmatrix} -1 & 0 & 0 \\ 0 & -1 & 0 \\ 0 & 0 & 1 \end{bmatrix}$$

A C_3 rotation matrix can be written by substituting $\varphi = -120°$ in the clockwise proper rotation matrix.

For an *improper* rotation around the z axis, we rotate in the xy plane and then reflect through it, leading to the matrix representation

$$\begin{bmatrix} \cos\varphi & \sin\varphi & 0 \\ -\sin\varphi & \cos\varphi & 0 \\ 0 & 0 & -1 \end{bmatrix}$$

The multiplication of these matrix representations of the symmetry operations produces the same result as the product of the symmetry operations, as they must if they are indeed the correct representations.

$$\sigma_{xz} \times \sigma_{xz} = E$$

$$\begin{bmatrix} 1 & 0 & 0 \\ 0 & -1 & 0 \\ 0 & 0 & 1 \end{bmatrix} \begin{bmatrix} 1 & 0 & 0 \\ 0 & -1 & 0 \\ 0 & 0 & 1 \end{bmatrix} = \begin{bmatrix} 1 & 0 & 0 \\ 0 & 1 & 0 \\ 0 & 0 & 1 \end{bmatrix}$$

*This is equivalent to a clockwise rotation of the axis system, which would be written as:

$$\begin{bmatrix} \cos\varphi & -\sin\varphi \\ \sin\varphi & \cos\varphi \end{bmatrix} \begin{bmatrix} \vec{\mathbf{X}} \\ \vec{\mathbf{Y}} \end{bmatrix} = \begin{bmatrix} \vec{\mathbf{X}}' \\ \vec{\mathbf{Y}}' \end{bmatrix}$$

All matrices that describe the transformations of a set of orthogonal coordinates* by the symmetry operations of a group are *orthogonal* matrices. Their inverses can be obtained by transposing rows and columns. For example, the inverse of the matrix that corresponds to a 30° rotation about the z-axis [cos (−30°) = $\sqrt{3}/2$]

$$\begin{bmatrix} \frac{\sqrt{3}}{2} & -\frac{1}{2} & 0 \\ \frac{1}{2} & \frac{\sqrt{3}}{2} & 0 \\ 0 & 0 & 1 \end{bmatrix} \quad \text{is} \quad \begin{bmatrix} \frac{\sqrt{3}}{2} & \frac{1}{2} & 0 \\ -\frac{1}{2} & \frac{\sqrt{3}}{2} & 0 \\ 0 & 0 & 1 \end{bmatrix}$$

2–6 IRREDUCIBLE REPRESENTATIONS

The total representation describing the effect of all of the symmetry operations in the C_{2v} point group on the point with coordinates x, y, and z is

$$\overset{E}{\begin{bmatrix} 1 & & \\ & 1 & \\ & & 1 \end{bmatrix}} \overset{C_2}{\begin{bmatrix} -1 & & \\ & -1 & \\ & & 1 \end{bmatrix}} \overset{\sigma_{xz}}{\begin{bmatrix} 1 & & \\ & -1 & \\ & & 1 \end{bmatrix}} \overset{\sigma_{yz}}{\begin{bmatrix} -1 & & \\ & 1 & \\ & & 1 \end{bmatrix}}$$

Notice each of these matrices is *block diagonalized; i.e.*, the total matrix can be broken up into blocks of smaller matrices with no off-diagonal elements between the blocks. The fact that it is block diagonalized indicates that the total representation described must consist of a set of so-called one-dimensional representations. We can see this as follows. If we were to be concerned only with operations on a point that had only an x coördinate (*i.e.*, the column matrix $\{x, 0, 0\}$), then only the first row of the total representation would be required (*i.e.*, 1, −1, 1, −1). This is an irreducible representation, which in this case is a set of one-dimensional matrices describing the symmetry properties of the one-dimensional x-vector in the specified point group. The symbol B_1 will be used to symbolize this irreducible representation. Do not concern yourself with the meaning of the B and 1 for now, but just think of this as a label. The irreducible representation for y is:

$$1 \quad -1 \quad -1 \quad 1$$

which is labeled B_2; and that for z is:

$$1 \quad 1 \quad 1 \quad 1$$

which is labeled A_1.

The total representation—the four 3 by 3 matrices—is a reducible representation. It is reducible, as discussed above, into three sets of 1 by 1 matrices, each set a representation by itself. The trace or character of each of the total representation matrices is the sum of the characters of each of the component irreducible representations; in order, these are 3, −1, 1, and 1.

We will show shortly that the block diagonalized 3 by 3 matrices for the effect of symmetry operations on our point resulted because the x, y, and z axes of our coordinate system were selected as the *basis set of vectors* for the symmetry operations, *i.e.*, z for rotation, xy for reflection. We shall subsequently consider a representation that is not block diagonalized.

* Orthogonal coordinates are those whose dot product is zero.

There is one additional irreducible representation in the C_{2v} point group. We can illustrate it by carrying out the operations of the group on a rotation R_z described by the sketch at left. The identity and C_2 rotation do not change the direction in which the arrow points, but reflection by σ_{xz} and σ_{yz} do. None of the irreducible representations we now have, nor any combination of these irreducible representations, describes the effect of the C_{2v} symmetry operations on this curved arrow representing rotation around the z-axis. The result is another irreducible representation:

$$1 \quad 1 \quad -1 \quad -1$$

which is labeled A_2.

2-7 CHARACTER TABLES

We now have empirically derived all of the irreducible representations of the C_{2v} point group, but we cannot be sure we have all of them with the procedures employed. There are rigorous procedures for deriving all of the irreducible representations of the various point groups, which are covered in many treatments of group theory. These procedures will not be covered here, for we will be more concerned with using the irreducible representations and will not have to generate them; they are readily available. The results of the preceding discussion for C_{2v} symmetry are summarized by *the character table* of the C_{2v} point group shown in Table 2-2.

Each entry in the character table is the character of the matrix for that operation in that representation. For 1×1 matrices, the character and the matrix are the same. There is a row for each irreducible representation.

The labels have the following general meaning:

(1) The symbol *A* indicates a singly degenerate state (*i.e.*, it consists of only one representation) that is symmetric about the principal axis; *i.e.*, the character table contains values of +1 under the column for the principal axis for all *A* species.

(2) The symbol *B* indicates a singly degenerate state that is antisymmetric about the principal axis; *i.e.*, the value −1 appears under the column for the principal axis for all *B* species.

(3) The subscripts 1 and 2 indicate symmetry or antisymmetry relative to a rotation axis other than the principal axis. If there is no second axis, the subscripts refer to the symmetry about a σ_v plane (*e.g.*, in the C_{2v} group, 1 is symmetric about the *xz* plane and 2 is antisymmetric).

All properties of a molecule with C_{2v} symmetry can be expressed in mathematical terms that will, for a basis set, form a general representation that consists of one (or a combination of more than one) of these irreducible representations. As we shall see, appropriate combinations of the irreducible representations summarize the results of our matrices operating on anything belonging to the C_{2v} point group.

TABLE 2-2. CHARACTER TABLE FOR THE C_{2v} POINT GROUP

	E	C_2	σ_{xz}	σ_{yz}	
A_1	+1	+1	+1	+1	z
A_2	+1	+1	−1	−1	R_z
B_1	+1	−1	+1	−1	x, R_y
B_2	+1	−1	−1	+1	y, R_x

Character tables are available for all of the point group symmetries, and are listed in Appendix A.

2–8 NON-DIAGONAL REPRESENTATIONS

If, instead of selecting the C_2 and σ elements to be coincident with the axes of our coordinate system, we selected as our basis the C_2 axis shown in Fig. 2–8 (and the

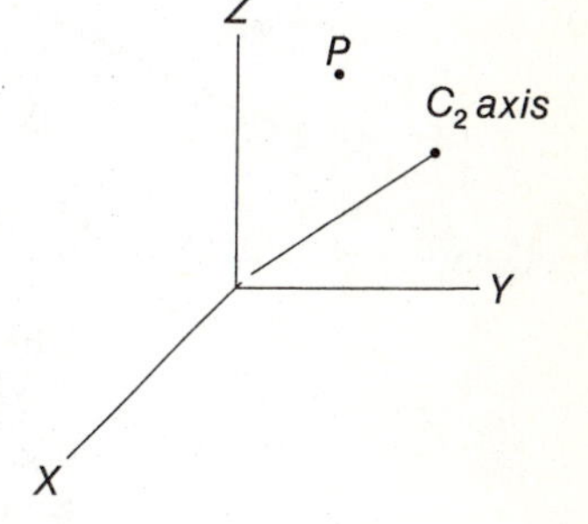

FIGURE 2–8. A basis set for geometric transformations on P that are *not* coincident with the Cartesian coordinates.

other elements appropriately), then our matrix representation would not be block diagonalized. For example, the C_2 rotation would not simply change the sign of the x coordinate as we saw before, but instead would have to keep the sign the same and change the magnitude of the coordinates. Instead of

$$\begin{bmatrix} -1 & 0 & 0 \\ 0 & -1 & 0 \\ 0 & 0 & 1 \end{bmatrix} \begin{bmatrix} X \\ Y \\ Z \end{bmatrix}$$

the new x coordinate would be dependent on what the y and z coordinates of the original point were; *i.e.*, there would be off-diagonal matrix elements. We shall not work this all out, for it is not worth the effort. However, the point should be made that *the trace of the matrices for all of the symmetry operations would be the same for any Cartesian basis set, i.e.*, 3, −1, 1, 1. The matrix representation for the axis selection we have made here can be converted into a block diagonal matrix by using an appropriate matrix, Q, for the similarity transform; *i.e.*,

$$Q^{-1}C_2Q,\ Q^{-1}\sigma Q, \text{ etc.}$$

In this case, the similarity transform is a rotation of the new basis set (for the symmetry operation with the rotation matrix) to the basis set of the original X, Y, Z axis system. The basis set is rotated, not the points or the molecule. In practice, when off-diagonal elements result, we realize that a poor selection of a coordinate system was made. We choose a new orthogonal coordinate system, each basis vector of which transforms as one of the irreducible representations, so that a block diagonal representation will result. This will be made clear by picking a simpler example that we can work through completely. Such a representation, which is not block diagonalized, can be obtained by considering the operations of the C_{2v} point group on two vectors **a** and **b**, using *these vectors as the basis set* to describe the representation. The problem is illustrated in Fig. 2–9. The coordinates were selected as $1/\sqrt{2}$ to produce normalized (*i.e.*, unit) vectors:

$$\left(\frac{1}{\sqrt{2}}\right)^2 + \left(\frac{1}{\sqrt{2}}\right)^2 = 1$$

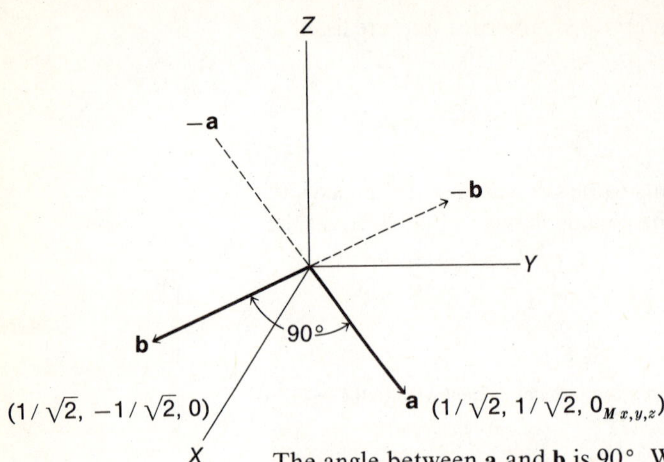

FIGURE 2–9. Geometric transformations in the **a**,**b** basis.

The angle between **a** and **b** is 90°. We are going to set up our matrices to operate on the **a**,**b** basis set, so these matrices will not work for the x,y,z basis set of the X,Y,Z coordinate system. They will tell us what **a** and **b** are changed to, in terms of **a** and **b**. Since we wish to illustrate with this example the dependence of our matrices on the location of our basis set in the coordinate system, we shall carry out all of the symmetry operations about axes and planes in the X,Y,Z coordinate system. In terms of the **a**,**b** basis, we have

$$E\mathbf{a} = \mathbf{a} \qquad \begin{bmatrix} 1 & 0 \\ 0 & 1 \end{bmatrix}$$
$$E\mathbf{b} = \mathbf{b}$$

Now, the C_2 operation changes **a** to $-\mathbf{a}$ and **b** to $-\mathbf{b}$

$$C_2\mathbf{a} = -\mathbf{a} \qquad \begin{bmatrix} -1 & 0 \\ 0 & -1 \end{bmatrix}$$
$$C_2\mathbf{b} = -\mathbf{b}$$

Reflection in the xz plane moves **a** into **b** and **b** into **a**:

$$\sigma_{xz}\mathbf{a} = \mathbf{b} \qquad \begin{bmatrix} 0 & 1 \\ 1 & 0 \end{bmatrix}$$
$$\sigma_{xz}\mathbf{b} = \mathbf{a}$$

Furthermore,

$$\sigma_{yz}\mathbf{a} = -\mathbf{b} \qquad \begin{bmatrix} 0 & -1 \\ -1 & 0 \end{bmatrix}$$
$$\sigma_{yz}\mathbf{b} = -\mathbf{a}$$

Note that these reflections have moved **a** into **b** and **b** into **a** with off-diagonal elements, and that the trace is zero. Further note that the trace of their representation matrices is the same as those for the x,y parts of the matrices generated earlier in the x,y basis for the C_{2v} point group. The off-diagonal elements immediately tell us that we made a bad choice of axes for this basis set. We know from earlier discussion that x and y form the basis of irreducible representations in C_{2v}; thus, if we locate our vectors along x and y, and use matrices that work on the x and y coördinates of the point defined by the vector (*i.e.*, an x,y basis set), then diagonal matrices will result. We can do this with a similarity transform.

For our first example of this procedure, we shall consider the operation σ_{yz}. The similarity transform is $Q^{-1}\sigma_{yz}Q$, where Q is a counterclockwise rotation matrix of 45° (this operation makes **a** and **b** coincide with x and y).

$$\begin{bmatrix} \cos\varphi & -\sin\varphi \\ \sin\varphi & \cos\varphi \end{bmatrix} = Q \text{ when } \varphi = 45°$$

so

$$Q = \begin{bmatrix} \frac{1}{\sqrt{2}} & -\frac{1}{\sqrt{2}} \\ \frac{1}{\sqrt{2}} & \frac{1}{\sqrt{2}} \end{bmatrix} \quad \text{and} \quad Q^{-1} = \begin{bmatrix} \frac{1}{\sqrt{2}} & \frac{1}{\sqrt{2}} \\ -\frac{1}{\sqrt{2}} & \frac{1}{\sqrt{2}} \end{bmatrix}$$

$$Q^{-1}\sigma_{yz}^{(ab)}Q = \begin{bmatrix} \frac{1}{\sqrt{2}} & \frac{1}{\sqrt{2}} \\ -\frac{1}{\sqrt{2}} & \frac{1}{\sqrt{2}} \end{bmatrix} \begin{bmatrix} 0 & -1 \\ -1 & 0 \end{bmatrix} \begin{bmatrix} \frac{1}{\sqrt{2}} & -\frac{1}{\sqrt{2}} \\ \frac{1}{\sqrt{2}} & \frac{1}{\sqrt{2}} \end{bmatrix} = \begin{bmatrix} -1 & 0 \\ 0 & 1 \end{bmatrix} = \sigma_{yz}$$

Note that the trace of the resulting matrix is the same as that of the starting one. The trace of a matrix is always invariant under a similarity transform. When we operate on all of the matrices for the various operations in C_{2v} with this similarity transform, we will have

$$\begin{array}{cccc} E & C_{2v} & \sigma_{v\,xz} & \sigma_{v\,yz} \\ \begin{bmatrix} 1 & 0 \\ 0 & 1 \end{bmatrix} & \begin{bmatrix} -1 & 0 \\ 0 & -1 \end{bmatrix} & \begin{bmatrix} 1 & 0 \\ 0 & -1 \end{bmatrix} & \begin{bmatrix} -1 & 0 \\ 0 & 1 \end{bmatrix} \end{array}$$

corresponding to the B_1 (upper left) and B_2 (lower right) representations for x and y, respectively; *i.e.*, our reducible representation in the **a,b** basis is decomposed into B_1 and B_2.

Degenerate Representations

For further practice, consider the ammonia molecule, represented in Fig. 2–10

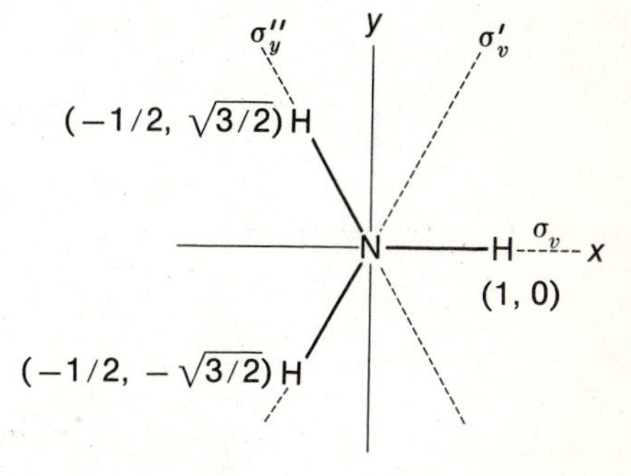

FIGURE 2–10. The ammonia molecule in a Cartesian coordinate system. (N is above the *xy* plane, and the projection of the N-H bond on the *x*-axis is taken as 1.)

with an x,y,z coordinate axis basis set. The symmetry operations are E, $2C_3$, and $3\sigma_v$. Consider the x, y, and z axes all together:

$$E = \begin{bmatrix} 1 & 0 & 0 \\ 0 & 1 & 0 \\ 0 & 0 & 1 \end{bmatrix} \qquad C_3 = \begin{bmatrix} -\frac{1}{2} & -\frac{\sqrt{3}}{2} & 0 \\ \frac{\sqrt{3}}{2} & -\frac{1}{2} & 0 \\ 0 & 0 & 1 \end{bmatrix}$$

The C_3 rotation matrix is obtained by substituting $\varphi = 2\pi/3$ ($= 120°$) into the general proper rotation matrix defined earlier (p. 30).

In order to determine the matrices for σ_v, we must work out the reflection matrix for each plane. The matrix for σ_v is quite simple:

$$\begin{bmatrix} 1 & 0 & 0 \\ 0 & -1 & 0 \\ 0 & 0 & 1 \end{bmatrix} \begin{bmatrix} X \\ Y \\ Z \end{bmatrix} = \begin{bmatrix} X' \\ Y' \\ Z' \end{bmatrix}$$

We see that reflection in σ_v just changes the sign of y. The following matrices give the new x and y coordinates after reflection by σ_v' and σ_v'':

$$\sigma_v' = \begin{bmatrix} -\frac{1}{2} & \frac{\sqrt{3}}{2} & 0 \\ \frac{\sqrt{3}}{2} & \frac{1}{2} & 0 \\ 0 & 0 & 1 \end{bmatrix} \qquad \sigma_v'' = \begin{bmatrix} -\frac{1}{2} & -\frac{\sqrt{3}}{2} & 0 \\ -\frac{\sqrt{3}}{2} & \frac{1}{2} & 0 \\ 0 & 0 & 1 \end{bmatrix}$$

To employ these matrices, substitute the x,y,z coordinates of one of the hydrogens for X, Y, and Z to get the new coordinates x', y', and z' after the operation.* Note that σ_v' does not change $(-1/2, -\sqrt{3}/2)$ but does interchange the other two hydrogens. For this symmetry we have off-diagonal elements that simultaneously change the old X vector into a new vector with x and y components.

Next, we shall work out the irreducible representations for the x, y, and z axes in the C_{3v} point group. The z-axis, which is the rotation axis, transforms unchanged and can be thought of as being described by the three 1×1 matrices [1].

E	$2C_3$	$3\sigma_v$
1	1	1

A 120° rotation of a vector on the x-axis produces a vector with both x and y components (Fig. 2-11). The same thing occurs for rotation of y. Substituting

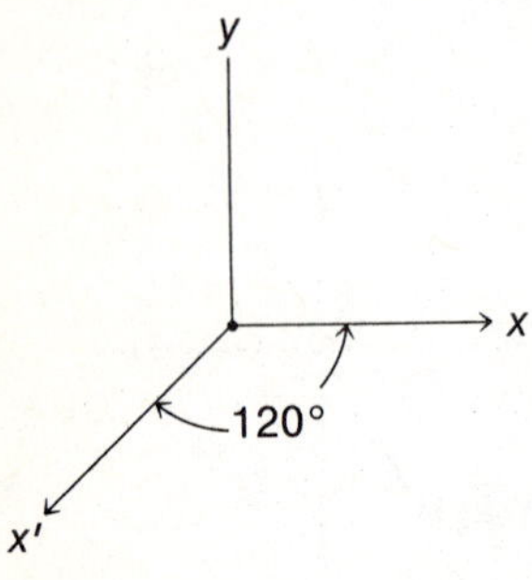

FIGURE 2–11. A rotation of the x-axis by 120° about a z-axis represented by the dot.

$\varphi = -120°$ into the rotation matrix and multiplying by $\{x,y\}$ gives x' and y' as:

$$x' = -\left(\frac{x}{2}\right) - \left(\frac{y\sqrt{3}}{2}\right)$$

$$y' = +\left(\frac{x\sqrt{3}}{2}\right) - \left(\frac{y}{2}\right)$$

---

*For example, to determine the effect of σ_v' on $(-1/2, \sqrt{3}/2)$, we see that the z-coordinate will not change, and x and y are given by

$$\begin{bmatrix} -\frac{1}{2} & \frac{\sqrt{3}}{2} \\ \frac{\sqrt{3}}{2} & \frac{1}{2} \end{bmatrix} \begin{bmatrix} -\frac{1}{2} \\ \frac{\sqrt{3}}{2} \end{bmatrix} = \begin{bmatrix} \frac{1}{4} + \frac{3}{4} \\ -\frac{\sqrt{3}}{4} + \frac{\sqrt{3}}{4} \end{bmatrix} = \begin{bmatrix} 1 \\ 0 \end{bmatrix}$$

As we have just shown, the two vectors, x and y, are related and cannot be transformed independently. The x and y vectors are represented in the character table by the symbol E, used to indicate double degeneracy. The trace of the matrix (obtained by summing the diagonal elements, and which has been shown to be independent of the basis selected) is the character reported in the character table. The character is invariant under any similarity transformation carried out on the matrix. The significance of this will be made clearer when we consider more complicated transformations.

In summary, the characters of all non-degenerate point groups (*i.e.*, those having a C_2 as the highest-fold rotation axis) are 1×1 matrices with characters of $+1$ or -1, which indicate symmetric or antisymmetric behavior of an appropriate basis function under the symmetry operation. A degenerate point group contains a character that is the trace of the transformation matrix and that summarizes how the degenerate basis set transform together.

Generally speaking, then, we shall end up with an $n \times n$ matrix for our representation (where n is the size of our basis set) for each operation in the point group; *e.g.*, in C_{2v}

$$\begin{matrix} E & C_{2v} & \sigma_{xz} & \sigma_{yz} \\ [\quad] & [\quad] & [\quad] & [\quad] \end{matrix}$$

If each of these is block diagonal, they correspond to irreducible representations. If not, we want a similarity transform to block diagonalize them.

2–9 MORE ON CHARACTER TABLES

We earlier defined a class as a complete set of elements that are conjugate to each other. We also mentioned that conjugate matrices have identical characters. (You can confirm this by checking the trace of the σ_v, σ_v', and σ_v'' matrices just given for the degenerate x and y vectors.) Thus, classes are grouped into one entry in the character tables; see, for example, $3\sigma_v$ of the C_{3v} character table.

The different irreducible representations may be thought of as a series of orthonormal vectors in h-dimensional space, where h is the order of the point group. Since vectors that form the bases for two different irreducible representations, χ_i and χ_j, are orthogonal, we have

$$\sum_R g\chi_i^*(R)\chi_j(R) = 0 \qquad \text{when } i \neq j \tag{2-9}$$

The sum is taken over all h symmetry operations (each denoted by R) in the point group, and g represents the number of elements in the class. The sum of the squares of the characters of any irreducible representation equals h,

$$\sum_R g[\chi_i(R)]^2 = h \tag{2-10}$$

where R is a sum over all operations, so $3\sigma_v$ *is counted as three times* σ_v in this summation for $g = 3$.

TABLE 2–3. CHARACTER TABLE FOR THE C_{3v} POINT GROUP

C_{3v}	E	$2C_3$	$3\sigma_v$		
A_1	1	1	1	z	$x^2 + y^2, z^2$
A_2	1	1	-1	R_z	
E	2	-1	0	(x, y) $R_x R_y$	$(x^2 - y^2, xy)(xz, yz)$
I		II		III	IV

A character table (Table 2–3) is divided into four main areas, as explained in the following paragraphs.

Area I: The meanings of A, B, E, and the 1 and 2 subscripts have been discussed. Other symbols are encountered in the character tables; see Appendix A for examples. The symbols E and T represent doubly and triply degenerate states, respectively. (F is often employed instead of T.) If the molecule has a center of symmetry (see the C_{2h}, D_{2h}, D_{4h}, and O_h character tables in the appendix), the subscript g is used to indicate symmetry $(+1)$ with respect to this center, while u indicates antisymmetry (-1). Prime and double prime marks are employed to indicate symmetry and antisymmetry, respectively, relative to a σ_h plane of symmetry.

Examples of the application of these rules can be obtained by referring to the character tables in Appendix A. All A species in the D_{3h} group have values of $+1$ in the column for the principal axis, $2C_3$. The species A_1 are symmetric (*i.e.*, have $+1$ values) to $3C_2$, and A_2 species are antisymmetric (-1) to these C_2 axes. In D_{4h}, all A species are symmetric with respect to the perpendicular C_2 axes. The prime species are symmetric $(+)$ to the horizontal plane σ_h, while the double prime species are antisymmetric $(-)$ to this plane. The u species are antisymmetric $(-)$ to the center of symmetry, while the g species are symmetric $(+)$.

Area II: This has already been discussed, but there are a few additional points to be made. Some character tables contain imaginary or complex characters. Whenever complex characters occur, they occur in pairs such that one is the complex conjugate of the other. While they are mathematically distinct irreducible representations, in all physical problems no difference is observed and the results are as if they had been a doubly degenerate irreducible representation. In the T_d point group, two triply degenerate species exist, T_1 and T_2. For T_2, x, y, and z form a basis; for T_1, all three of the rotations about these axes form a basis.

Area III: This area lists the transformation properties of vectors along the x, y, and z axes and rotations R_z, R_x, and R_y (represented as curved arrows) about the x, y, and z axes.

Area IV: This area indicates the transformation properties of the squares and binary products of the coordinates. Although there are six possible square and binary products (x^2, y^2, z^2, xy, xz, and yz), only five are ever indicated because $x^2 + y^2 + z^2 = r^2$ and, as a result, one of the linear combinations is redundant (r^2 is the square of the radius of the $x^2 + y^2 + z^2$ sphere). The direct product of two vectors is obtained by multiplying the species for each, *e.g.*, $\mathbf{XY} = B_1 \times B_2 = A_2$. To perform this multiplication, the following procedure is used:

$$E(B_1) \times E(B_2) \quad C_2(B_1) \times C_2(B_2) \quad \sigma_v(B_1) \times \sigma_v(B_2) \quad \sigma_v'(B_1) \times \sigma_v'(B_2)$$

or

$$(1)(1) \quad (-1)(-1) \quad (1)(-1) \quad (-1)(1)$$

giving the result

$$+1 \quad +1 \quad -1 \quad -1$$

The result $+1, +1, -1, -1$ is identical to the irreducible representation A_2; hence, $B_1 \times B_2 = A_2$. In subsequent chapters we shall have occasion to take the direct product of irreducible representations, and this procedure will be employed. Many of the combinations in this area have symmetry properties identical to those of the d-orbitals (*e.g.*, xy and d_{xy} are identical). Thus, these products can be used to indicate which orbitals will remain degenerate in transition metal ion complexes of various symmetries. The pair $x^2 - y^2$ and xy are degenerate in a C_{3v} molecule as are xz and yz, but xz and yz belong to a different class than do $x^2 - y^2$ and xy. The different classes can have different energies in complexes with different symmetries.

The wave functions of a molecule are an example of a basis for a representation; that is, they can serve as the basis set for the representation matrices of a group. These wave functions must possess the same transformation properties under the operations of the group as the irreducible representations. Thus, each molecular orbital in a molecule will have a symmetry given by one of the irreducible representations in the point group of the molecule.

2–10 MORE ON REPRESENTATIONS

If, instead of vectors, we choose to obtain a representation using a set of atom coordinates, the resulting matrix could have dimensions of $3N \times 3N$ associated with the three Cartesian coordinates for each of the N atoms in the molecule (Fig. 2–12).

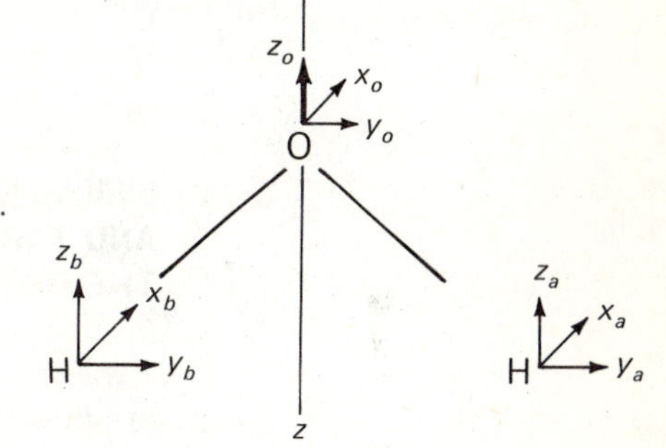

FIGURE 2–12. The 3*N* coordinates for H_2O in a Cartesian coordinate system.

For the E matrix, we have a 9×9 representation

$$\begin{bmatrix} 1 &&&&&&&& \\ & 1 &&&&&&& \\ && 1 &&&&&& \\ &&& 1 &&&&& \\ &&&& 1 &&&& \\ &&&&& 1 &&& \\ &&&&&& 1 && \\ &&&&&&& 1 & \\ &&&&&&&& 1 \end{bmatrix} \begin{bmatrix} x_o \\ y_o \\ z_o \\ x_a \\ y_a \\ z_a \\ x_b \\ y_b \\ z_b \end{bmatrix} = \begin{bmatrix} x_o' \\ y_o' \\ z_o' \\ x_a' \\ y_a' \\ z_a' \\ x_b' \\ y_b' \\ z_b' \end{bmatrix}$$

Any 1×1 diagonal matrix element in this large matrix corresponds to a non-degenerate irreducible representation in the final block diagonalized matrix. Accordingly, we shall need nine irreducible representations to describe this system.

For the C_2 symmetry operation on our nine dimensional basis set, we have

$$\begin{bmatrix} -1 & & & & & & & & \\ & -1 & & & & & & & \\ & & 1 & & & & & & \\ & & & 0 & 0 & 0 & -1 & 0 & 0 \\ & & & 0 & 0 & 0 & 0 & -1 & 0 \\ & & & 0 & 0 & 0 & 0 & 0 & 1 \\ & & & -1 & 0 & 0 & 0 & 0 & 0 \\ & & & 0 & -1 & 0 & 0 & 0 & 0 \\ & & & 0 & 0 & 1 & 0 & 0 & 0 \end{bmatrix} \begin{bmatrix} x_o \\ y_o \\ z_o \\ x_a \\ y_a \\ z_a \\ x_b \\ y_b \\ z_b \end{bmatrix} = \begin{bmatrix} x_o' \\ y_o' \\ z_o' \\ x_a' \\ y_a' \\ z_a' \\ x_b' \\ y_b' \\ z_b' \end{bmatrix}$$

Similar matrices can be constructed for $\sigma_{v_{xz}}(\sigma_v)$ and $\sigma_{v_{yz}}(\sigma_v')$.

These larger representations can be broken up, *i.e.*, are reducible into the sum of many of the smaller irreducible ones given earlier for the C_{2v} point group. We could proceed by looking for a similarity transform Q that block diagonalizes the 9×9 E, C_2, σ_v, and σ_v' matrices. Then the blocks in the diagonalized matrix would correspond to the irreducible representations for this molecule and in this basis set. Although this would not occur in the C_{2v} point group (or any point group with a lower than three-fold axis), matrices that cannot be reduced to a 1×1 matrix would correspond to degenerate representations. After diagonalization, the matrix would have the form shown in Fig. 2–13, and each block can be treated separately as a representation, provided that the reducible representations of the other symmetry operations are similarly block diagonalized. There is an easier way to solve this problem, and we shall describe it next.

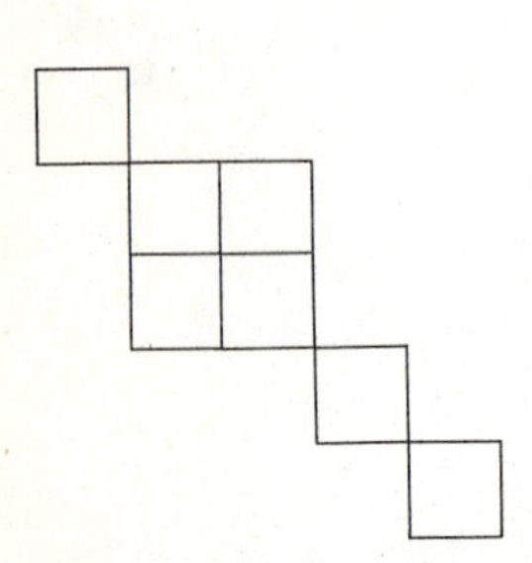

FIGURE 2–13. Form of a matrix after diagonalization.

2–11 SIMPLIFIED PROCEDURES FOR GENERATING AND FACTORING TOTAL REPRESENTATIONS; THE DECOMPOSITION FORMULA

The trace of the matrices for E, C_2, σ_v, and σ_v' operations for the problem described above is

	E	C_2	$\sigma_{v_{xz}}$	$\sigma_{v_{yz}}$
χ_T	9	-1	$+1$	3

In actual practice, similarity transforms are not used to find the irreducible representations that constitute this representation. Even the total representation is found by simpler procedures, and this is then factored into the individual irreducible representations by using the decomposition formula, *vide infra*.

With a little reflection, it can be seen that the following rules summarize what we have already said about generating the character of the total representation.

(1) Any vector (or part of the basis set) that is unchanged by a symmetry operation is assigned $+1$. (This diagonal matrix element left the coordinates unchanged in our earlier discussion.)

(2) Any vector or part of the basis set that is changed into the opposite direction is assigned -1. (The diagonal matrix element reflects this; in the C_{2v} example discussed earlier, all changes were either symmetric $(+1)$ or unsymmetric (-1)).

(3) A vector or part of the basis set that is moved onto another vector by a symmetry operation is counted as zero. Recall that in our discussion of the **a** and **b**

vectors, the vectors were interchanged in our earlier example of H_2O with off-diagonal elements, and the trace of the matrix dealing with H_a and H_b is zero. These rules give us directly the value of the trace of the matrices we generated in our earlier discussion of the nine coordinates ($3N$) of the water molecule.

Applying these rules to these nine coordinates, we see that E gives a trace for the total representation, χ_T, of 9. The C_2 operation moves all H_a coordinates into H_b and vice versa for a result of zero; while for oxygen, x goes to $-x$ (-1), y goes to $-y$ (-1), and z remains unchanged $(+1)$. Accordingly, χ_T for the C_2 operation is -1. With x perpendicular to the plane of the atoms, reflection in the xz plane moves all H_a and H_b vectors for a result of zero. The x and z vectors remain unchanged, while y goes to $-y$ for a net result of $+1$. Reflection in the yz plane does not change any of the six y or z vectors $(+6)$, while the three x vectors go to $-x$ (-3) for a total $(+6-3)$ of $+3$. Thus, $\chi_T = 9, -1, +1, 3$ as described before.

When an operation does not interchange atoms but moves them into a new position that is some combination of the old coordinates (*e.g.*, C_3 on $\vec{\mathbf{x}}$), the trace of the matrix must be determined by working out the matrix for the geometric transformation as was done earlier. For rotations, the rotation matrix can be employed.

Next, we shall describe how to determine all of the different irreducible representations that make up the total representation. The following formula shows how to do this; books on group theory should be consulted by those interested in the derivation. The formula summarizing how to factor a total representation, the so-called decomposition formula, is:

$$a_i = \frac{1}{h} \sum_R g\chi_i(R)\chi_T(R) \tag{2-11}$$

Here a_i is the number of contributions from the ith irreducible representation, R refers to a particular symmetry operation (or, if there is more than one element in the class, to the whole class); h, the order of the group, is given by the total number of symmetry operations in the point group (in determining this, be sure to count $3\sigma_v$'s as three); g is the number of elements in the class; $\chi_i(R)$ is the character of the irreducible representation; and $\chi_T(R)$ is the character for the analogous operation (R) in the total representation. The use of this formula is best demonstrated by decomposing the total character for the $3N$ coordinates of water,

$$\chi_T = 9 \quad -1 \quad +1 \quad 3$$

using the C_{2v} character table. The value of h is 4, and $g = 1$ for all symmetry operations. We sum over the R (in this case, 4) symmetry operations as follows:

$$a_{A_1} = \frac{1}{4}[g\chi_{A_1}(E)\chi_T(E) + g\chi_{A_1}(C_2)\chi_T(C_2) + g\chi_{A_1}(\sigma_v)\chi_T(\sigma_v) + g\chi_{A_1}(\sigma_v)\chi_T(\sigma_v)]$$

$$= \frac{1}{4}[1 \times 1 \times 9 + 1 \times 1 \times (-1) + 1 \times 1 \times 1 + 1 \times 1 \times 3] = 3$$

$$a_{A_2} = \frac{1}{4}[1 \times 1 \times 9 + 1 \times 1 \times (-1) + 1 \times (-1) \times 1 + 1 \times (-1) \times 3] = 1$$

$$a_{B_1} = \frac{1}{4}[1 \times 1 \times 9 + 1 \times (-1) \times (-1) + 1 \times 1 \times 1 + 1 \times (-1) \times 3] = 2$$

$$a_{B_2} = \frac{1}{4}[1 \times 1 \times 9 + 1 \times (-1) \times (-1) + 1 \times (-1) \times 1 + 1 \times 1 \times 3] = 3$$

Thus, we find that the total representation consists of the sum of the irreducible representations $3A_1$, $1A_2$, $2B_1$, and $3B_2$. The total representation above describes all of the degrees of freedom of the molecule. In infrared spectroscopy, we are concerned with all of the possible vibrations that a molecule can undergo. If one subtracts the irreducible representations for the three basic translational modes and three rotational modes from the total representation, the remaining irreducible representations describe the vibrations.

A physical view of the decomposition formula can be had by considering the irreducible representations as vectors in a given point group. The general representation is also taken as a vector. The dot product of the general representation vector with an irreducible representation vector that does not contribute will be zero. The group order is used as a normalization factor to insure that the dot product of a contributing irreducible representation vector will give the number of contributions from the vector.

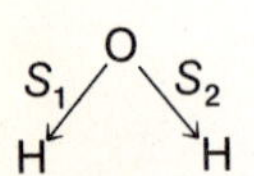

FIGURE 2-14. Vectors representing the two bonds in H_2O.

Another basis set, used to give a more transparent view of certain molecular vibrations, employs the bonds to be stretched during a vibration as vectors. Consider the two vectors for stretching the OH bonds of water, illustrated in Fig. 2-14. The matrices using these bond vectors (S_1 and S_2) as the basis set are:

$$E = \begin{bmatrix} 1 & 0 \\ 0 & 1 \end{bmatrix} \qquad \sigma_v = \begin{bmatrix} 1 & 0 \\ 0 & 1 \end{bmatrix}$$

$$C_2 = \begin{bmatrix} 0 & 1 \\ 1 & 0 \end{bmatrix} \qquad \sigma_v' = \begin{bmatrix} 0 & 1 \\ 1 & 0 \end{bmatrix}$$

The total representation χ_T is given by:

$$\begin{array}{rcccc} & E & C_2 & \sigma_v & \sigma_v' \\ \chi_T = & 2 & 0 & 2 & 0 \end{array}$$

Factoring the total representation for these bond stretches leads to the symmetry of the two stretching modes as A_1 and B_1, which correspond to the symmetric and asymmetric stretch, respectively.

2-12 DIRECT PRODUCTS

We have already mentioned that the characters of the representation of a direct product are obtained by taking the products of the characters of the individual sets of functions. In C_{2v},

$$B_1 \times B_1 = \underbrace{\chi_E(B_1) \times \chi_E(B_1)}_{1} \; \underbrace{\chi_{C_2}(B_1) \times \chi_{C_2}(B_1)}_{1} \; \underbrace{\chi_{\sigma_v}(B_1) \times \chi_{\sigma_v}(B_1)}_{1} \; \underbrace{\chi_{\sigma_v'}(B_1) \times \chi_{\sigma_v'}(B_1)}_{1}$$

$$= \quad 1 \quad 1 \quad 1 \quad 1$$

$$= A_1$$

That is, $B_1 \times B_1 = A_1$. When dealing with degenerate irreducible representations, remember that the degenerate bases are transformed in pairs, with the character corresponding to the trace of the transformation matrix for both. This can be illustrated by the product of $E \times E$ in the C_{4v} point group. In terms of the x,y basis for C_{4v}, we have

$$E = \begin{bmatrix} 1 & 0 \\ 0 & 1 \end{bmatrix} \quad C_4 = \begin{bmatrix} 0 & -1 \\ 1 & 0 \end{bmatrix} \quad C_2 = \begin{bmatrix} -1 & 0 \\ 0 & -1 \end{bmatrix}$$

$$E \times E = \begin{bmatrix} 1 & 0 \\ 0 & 1 \end{bmatrix} \times \begin{bmatrix} 1 & 0 \\ 0 & 1 \end{bmatrix} = \begin{bmatrix} 1 & 0 & 0 & 0 \\ 0 & 1 & 0 & 0 \\ 0 & 0 & 1 & 0 \\ 0 & 0 & 0 & 1 \end{bmatrix}$$

By analogy to taking the cross product, the identity gives a total representation of four because the cross product of two matrices increases the dimensionality of the matrix. The formula for taking the cross product of two matrices involves taking the a_{11} element (where a is the left matrix and b is the right) times the b matrix to give the upper left 2×2 part of the resulting 4×4 c matrix; that is, $a_{11}b_{11} = c_{11}$, $a_{11}b_{12} = c_{12}$, $a_{11}b_{21} = c_{21}$, and $a_{11}b_{22} = c_{22}$. The a_{12} element times the b matrix gives the upper right quarter of the c matrix, the a_{21} element times the b matrix gives the lower left quarter, and the a_{22} element times the b matrix gives the lower right quarter. The character of the direct product matrix is the product of the characters of the original matrices. In general, the direct products of irreducible representation matrices may be reducible.

ADDITIONAL READING REFERENCES

Same as for Chapter 1.

EXERCISES

1. Work out all the classes (of symmetry operations) in a two-dimensional equilateral triangle. (Only x and y dimensions apply, so there is only a three-fold rotation axis and $3\sigma_v$ planes.)
2. Work out the group multiplication table for NH_3. Determine all classes.
3. Identify the subgroups in G_4^1, G_4^2, and G_6^1. (The groups were given in this chapter.)
4. Demonstrate that the E_g and B_{2g} irreducible representations of the D_{4h} point group are orthogonal.
5. In the C_{4v} point group, indicate the operations that comprise
   a. the $2C_4$ given in the character table; that is, which of C_4^1, C_4^2, C_4^3, or C_4^4 comprise $2C_4$.
   b. the $2\sigma_v$.
   c. the $2\sigma_d$.
   d. Construct a group multiplication table for the symmetry operations in C_{4v}.
   e. Can a molecule with C_{4v} symmetry have a dipole moment? If it can, where is it located?
6. In addition to vectors and coordinates of a point, the orbitals of an atom can be assigned to irreducible representations. Using the character table for the D_{4h} point group, indicate

the representations for the p-orbitals and d-orbitals. Take the center of the orbitals as the point about which all operations are carried out. Why do the p-orbital representations have a u subscript?

7. For which irreducible representation do the following vibrational modes form a basis?

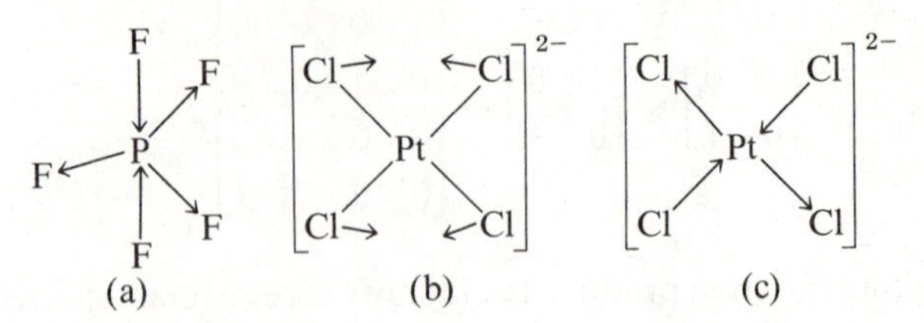

8. Recall the problem of a point in the Cartesian coordinate system with an x,y,z basis set.

   a. Write the matrix for an inversion operation.

   b. Perform the matrix multiplication that shows

$$i \times \sigma_h = C_2.$$

   c. Complete the multiplication table for the C_{2h} point group.

9. Consider the C_{2v} point group in the xy plane. Using unit vectors along the x and y axes as the basis, one finds the following representation:

$$\begin{matrix} E & C_2 & \sigma_{xz} & \sigma_{yz} \\ \begin{bmatrix} 1 & 0 \\ 0 & 1 \end{bmatrix} & \begin{bmatrix} -1 & 0 \\ 0 & -1 \end{bmatrix} & \begin{bmatrix} 1 & 0 \\ 0 & -1 \end{bmatrix} & \begin{bmatrix} -1 & 0 \\ 0 & 1 \end{bmatrix} \end{matrix}$$

where C_2 is along the z axis.

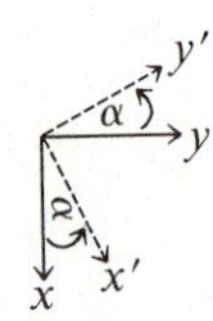

Consider the axes x' and y', which are at an angle of $\alpha = 30°$ counterclockwise with respect to the x and y axes and in the same plane. Keeping C_2 along the z axis and the σ planes in the xz and yz planes, give the representation for which unit vectors along x' and y' serve as a basis set. Show that the traces are the same in the new representation for each operation as those in the representation above.

10. Generate the matrices describing how the sets of functions (x, y, z), (xy, xz, yz), and $(2z^2 - x^2 - y^2, x^2 - y^2)$ are transformed by the various operations of the group O. Show how the trace transforms as irreducible matrix representations designated T_1, T_2, and E, respectively.

11. Write the matrices describing the effect on a point (x, y, z) of reflections in vertical planes that lie halfway between the xz and yz planes. By matrix methods, determine what operation results when each of these reflections is followed by reflection in the xy plane.

12. Consider $AuCl_4$ with D_{4h} symmetry. Use the x, y, and z Cartesian coordinate vectors at each atom as a basis set to form a total representation. Determine the irreducible representations.

13. In the groups specified, determine the direct product representations and reduce them when possible.

   a. $A_2 \times B_1$ in C_{2v}.

   b. $E \times E$ in C_{4v}.

   c. $E \times E$ in C_{3v}.

14. Consider the two complex ions *cis*-$[CoF_4Cl_2]^{3-}$ and *trans*-$[CoF_4Cl_2]^{3-}$. For each of these ions, find the symmetry designations of the *d*-orbitals.

15. In problem 10, we worked to find out that in the O point group the sets of functions (x, y, z), (xz, yz, xy), and $(x^2 - y^2, 2z^2 - x^2 - y^2)$ transformed as T_1, T_2, and E, respectively. The character table contains this information, but is truncated after giving information on product functions (xy, xz, etc.) and does not show how the triple product functions (*e.g.*, the xyz) transform. The triple products correspond to the *f*-orbital functions $(x^3, y^3, z^3, xyz, x(y^2 - z^2), y(z^2 - x^2), z(x^2 - y^2))$. Take all of these combinations to ascertain how the seven *f*-orbitals transform in the C_{2v} point group.

16. Use this square in the following problems. You are to assume that the square is like a table; *i.e.*, it cannot be turned upside down and still be the same. The two coordinate systems share a common z-axis perpendicular to the plane of the figure.

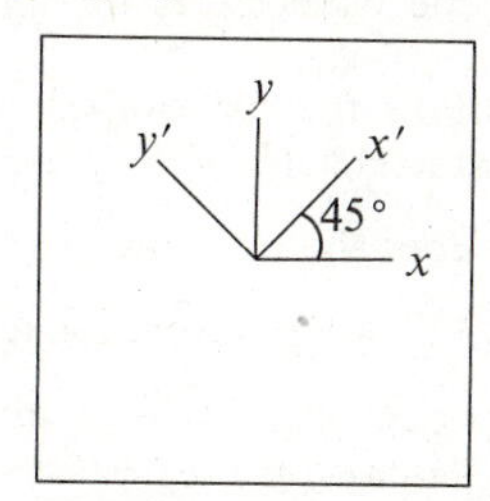

Consider a reflection, σ_{xz}. (Since the z-coordinate is always constant, work the following problems as two-dimensional cases.)

a. Derive the matrix corresponding to this reflection in the x, y axis system.

b. Derive the matrix corresponding to this reflection in the x', y' axis system.

c. Show that $\boldsymbol{\sigma}'_{xz}$ in the new axis system is related to $\boldsymbol{\sigma}_{xz}$ by a similarity transformation of the type

$$\boldsymbol{\sigma}'_{xz} = \mathbf{T}^{-1}\boldsymbol{\sigma}_{xz}\mathbf{T}$$

Derive the $\mathbf{T}$ matrix and verify this equation. The bold symbols represent matrices.

17. a. We discussed the problem of using the OH bonds in water to form a basis for 2×2 representation matrices in C_{2v} symmetry. Consider the H_2S molecule and assume a 90° bond angle. Write the four matrices for the symmetry operations in C_{2v} using the S—H bonds as a basis set.

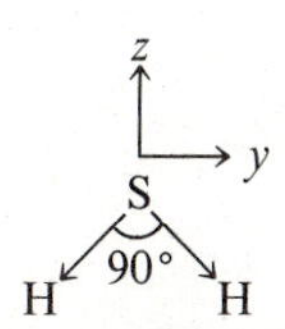

b. Actually, the same information is obtained from a set of axes $\vec{\mathbf{z}}$ and $\vec{\mathbf{y}}$ set up at the S nucleus as shown, with $\vec{\mathbf{z}}$ along the C_2 axis and $\vec{\mathbf{y}}$ in the molecular plane. Write out the matrices using this basis of $\vec{\mathbf{z}}$ and $\vec{\mathbf{y}}$.

c. Suppose that R is the matrix for some operation in the z, y frame and that R' is the corresponding matrix in the S—H bond frame. Then some transformation π of the points in the plane will convert R to R' by a similarity transformation, *i.e.*,

$$R' = \pi^{-1}R\pi$$

Write out the matrix π.

18. The following is excerpted from *J. Chem. Ed., 49,* 341 (1972):

In a recent issue it was pointed out by Schäfer and Cyvin that the well-known group theoretical equation

$$n(\gamma) = \frac{1}{g}\sum_R \chi_R \chi_R^{(\gamma)} \qquad (1)$$

fails to work when applied to linear groups. Equation (1) is used to determine the number of times, $n(\gamma)$, that a given symmetry species γ will occur in the reducible representation of a particular molecule. χ_R is the character of the reducible representation Γ; $\chi_R^{(\gamma)}$ is the character of the irreducible representation; and R is the index used to denote each of the symmetry operations of the group.

Outline of Method

(1) Assume a lower molecular symmetry which corresponds to a subgroup G of the molecular group G^0.

(2) Place a set of cartesian coordinate vectors on each atom. (The z axis must be placed along the maximum symmetry axis of the parent group.)

(3) Using standard methods obtain the characters of $\Gamma_{\text{reducible}}$.

(4) Calculate the values of each $n(\gamma)$ by application of eqn. (1).

(5) Finally compare the basis vectors of the irreducible representations under G^0 with those obtained for the molecule assuming G.

(By "basis vectors," rule 5 means the functions x, y, z, $x^2 - y^2$, etc., found in the right-hand columns of the character table.)

a. To what symmetry group does the molecule acetylene (C_2H_2) belong?

b. Why does equation (1) fail to work for groups to which linear molecules belong?

c. Use the method outlined (with $G = D_{2h}$) to obtain the irreducible representations (in the group to which C_2H_2 belongs) for a basis set consisting of x, y, and z vectors located at each atom.

3 MOLECULAR ORBITAL THEORY AND ITS SYMMETRY ASPECTS

3–1 OPERATORS

INTRODUCTION

As is true for many of the topics in this text, whole books have been written on the subject of this chapter. This chapter will be a very brief presentation of those fundamentals of molecular orbital theory that should be a part of the background of any modern chemist. We shall concentrate on those aspects that will be needed for our considerations of spectroscopy. One can well appreciate that a complete understanding of the spectroscopy, reactivity, and physical properties of substances will not be possible until we can interpret these phenomena in terms of the electron distribution in the molecule.

In principle, all the information about the properties of a system of N particles is contained in a wave function, ψ, which is a function of only the coordinates of the N particles and time. If time is included explicitly, ψ is called a time dependent wave function; if not, the system is said to be in a stationary state. The quantity $\psi^*\psi$ (where ψ^* is the complex conjugate of ψ) is proportional to the probability of finding the electron at a particular point.

For every observable property of a system, there exists a linear Hermitian operator, $\widehat{\alpha}$, and the observable can be inferred from the mathematical properties of its associated operator. A Hermitian operation is defined by

$$\int \psi_i{}^*\widehat{\alpha}\psi_j\, d\tau = \int \psi_j(\widehat{\alpha}\psi_i)^*\, d\tau$$

We shall abbreviate the above integrals by the so-called bra $\langle|$and ket$|\rangle$ notation as:

$$\langle\psi_i|\widehat{\alpha}|\psi_j\rangle$$

To construct an operator, $\widehat{\alpha}$, for a given observable, one first writes the classical expression for the observable of interest in terms of coordinates, q, momenta and time. Then the time and coordinates are left as they are and for Cartesian coordinates the momenta, p_q, are replaced by the differential operator, *i.e.*,

$$p_q = -i\hbar(\partial/\partial_q)$$

where q is the coordinate that is conjugate to p_q. For example, the quantum mechanical operator for kinetic energy, $\hat{T}$, can be constructed by first writing the classical expression for the kinetic energy:

$$T = \frac{1}{2}mv_x^2 + \frac{1}{2}mv_y^2 + \frac{1}{2}mv_z^2$$

Momenta $= mv = p$, so,

$$T = \frac{1}{2m}(p_x^2 + p_y^2 + p_z^2)$$

$$\hat{T} = \frac{1}{2m}\left[\left(-i\hbar\frac{\partial}{\partial x}\right)^2 + \left(-i\hbar\frac{\partial}{\partial y}\right)^2 + \left(-i\hbar\frac{\partial}{\partial z}\right)^2\right]$$

$$= -\frac{\hbar^2}{2m}\left(\frac{\partial^2}{\partial x^2} + \frac{\partial^2}{\partial y^2} + \frac{\partial^2}{\partial z^2}\right) = -\frac{\hbar^2}{2m}\nabla^2$$

where

$$\nabla^2 = \frac{\partial^2}{\partial x^2} + \frac{\partial^2}{\partial y^2} + \frac{\partial^2}{\partial z^2}.$$

If an operator $\hat{\alpha}$ corresponds to an observable, and *if* ψ_s, *the wave function for the state, is an eigenfunction of the operator* $\hat{\alpha}$, *then* $\hat{\alpha}\psi_s = a_s\psi_s$ *where* a_s *is a number. An experimentalist making a series of measurements of the quantity corresponding to* $\hat{\alpha}$ *will always get* a_s. The wave functions are eigenfunctions of the Hamiltonian operator which yields energies as *eigenvalues, i.e.,* numbers;

$$\hat{H}\psi_n = \mathbf{E}\psi_n \tag{3-1}$$

The operator for angular momentum about the Z-axis is also such an operator:

$$\hat{L}_z\psi_n = \ell_z\psi_n \text{ where } \ell_z = 0, 1, 2 \cdots (n-1) \tag{3-2}$$

Some properties of a system are not characterized by an eigenfunction for the appropriate operator which describes that property. Then a series of measurements of this property will not give the same result, but instead a distribution of results. The *average value* or expectation value, $\langle a_s \rangle$, is given by:

$$\langle a_s \rangle = \frac{\langle \psi_s | \hat{\alpha} | \psi_s \rangle}{\langle \psi_s | \psi_s \rangle} \tag{3-3}$$

ψ is generally not an eigenfunction of the operator $\hat{\alpha}$, but using equation (3–3), one can obtain the average value.

In the course of this book, we shall have occasion to consider many operators. We shall discuss them as the need arises. Our immediate concern will be with the Hamiltonian operator and the Schrodinger equation:

$$\hat{H}\psi = E\psi$$

This equation cannot be solved directly for the energy, E, on molecular systems because one cannot find a function ψ such that $\hat{H}\psi/\psi$ is a constant and not a variable function of the position of the electron. Consequently, we average $\hat{H}\psi$ over all space by multiplying by ψ^* and integrating over all space:

$$\int \psi^* \hat{H}\psi \, d\tau = \int \psi^* E\psi \, d\tau = E \int \psi^*\psi \, d\tau$$

so

$$E = \int \frac{\psi^* \hat{H}\psi \, d\tau}{\psi^*\psi \, d\tau} = \frac{\langle\psi|\hat{H}|\psi\rangle}{\langle\psi|\psi\rangle} \tag{3-4}$$

The Hamiltonian operator for most systems is easily written as

$$\hat{H} = -(h^2/8\pi^2 m)\nabla^2 + V,$$

where V is the potential energy of the system. Other operators are just as easily written by writing the classical expression for the property of interest and replacing the momentum by $-i\hbar\partial/\partial_x$. Thus, if we had accurate wave functions and could solve the resulting integrals, we could calculate all of the properties of a molecule. Unfortunately, such calculations are rarely practical.

In the LCAO–MO method (linear combination of atomic orbitals–molecular orbital) of solving this problem, the wave functions, ψ_j, of the resulting j molecular orbitals are approximated as linear combinations of atomic orbitals:

$$\psi_j = \sum_r C_{jr}\varphi_r \tag{3-5}$$

where C_{jr} is the coefficient describing the contribution of φ_r to the jth molecular orbital. The total electronic ground state wave function Ψ is a product of the individual molecular orbital wave functions. As we shall show shortly, the ψ_j's form the basis for irreducible representations of the molecular point group. Therefore, the symmetry of the total electronic ground state wave function is obtained by taking the direct product of the irreducible representations of the occupied ψ_j's.

The various ways of carrying out molecular orbital calculations correspond to different approximations that are used in the solution of the problem. One such approximation ignores the correlation of the motion of the electrons in the molecule. If there are two electrons in a molecule, we know that at any instant one of the electrons will tend to avoid the region occupied by the other. Over a longer time period, the motion of one electron will be correlated with the motion of the second electron. When this correlation is ignored, the resulting ψ_j's describe the system in terms of the coordinates of a single electron. Such wave functions are labeled *one electron molecular orbitals.* In describing the molecule, two electrons are added to each of the one electron m.o.'s.

For clarity, we can proceed with the problem of determining our wave functions by using a two-term basis set φ_1 and φ_2 (*i.e.,* two atomic orbitals on two different atoms) without any loss of generality, *i.e.,* equation (3–5) for our wave function is

$$\psi = C_1\varphi_1 + C_2\varphi_2$$

Substituting this equation for ψ into (3–4) produces

$$E = \frac{\int (C_1\varphi_1 + C_2\varphi_2)\hat{H}(C_1\varphi_1 + C_2\varphi_2)\, d\tau}{\int (C_1\varphi_1 + C_2\varphi_2)(C_1\varphi_1 + C_2\varphi_2)\, d\tau}$$

$$= \frac{\int [C_1\varphi_1\hat{H}C_1\varphi_1 + C_1\varphi_1\hat{H}C_2\varphi_2 + C_2\varphi_2\hat{H}C_1\varphi_1 + C_2\varphi_2\hat{H}C_2\varphi_2]\, d\tau}{\int [C_1\varphi_1C_1\varphi_1 + C_2\varphi_2C_1\varphi_1 + C_1\varphi_1C_2\varphi_2 + C_2\varphi_2C_2\varphi_2]\, d\tau}$$

Since C_1 and C_2 are constants, they can be taken out of the integrals. We can abbreviate $\int \varphi_i \hat{H} \varphi_j \, d\tau$ as H_{ij} and $\int \varphi_i \varphi_j \, d\tau$ as S_{ij}. Furthermore, $H_{21} = H_{12}$ and $S_{12} = S_{21}$, leading to equation (3-6).

$$E = \frac{C_1^2 H_{11} + 2C_1C_2H_{12} + C_2^2 H_{22}}{C_1^2 S_{11} + 2C_1C_2S_{12} + C_2^2 S_{22}} \quad (3\text{-}6)$$

The various types of integrals are given different names:

H_{ii} integrals are referred to as *Coulomb integrals;*

H_{ij} as *resonance integrals;* and

S_{ij} as *overlap integrals.* The value of $S_{ii} = 1$ for normalized a.o.'s.

The *variational principle* is employed to find the set of coefficients C_1 and C_2 that minimizes the energy, *i.e.*, gives the most stable system. The energy we get, called the *variation energy,* will depend upon how good our wave function is.

$$E_{\text{var}} > E_{0(\text{ground state energy})} \quad (3\text{-}7)$$

This minimization (*i.e.*, the variation procedure) is carried out by setting the partial derivative of E with respect to each of the coefficients (C's) equal to zero. Taking the partial derivative of equation (3-6) with respect to each of the coefficients, setting $\partial E/\partial C_1$ and $\partial E/\partial C_2$ both equal to zero, and collecting terms, we get

$$\begin{aligned} C_1(H_{11} - ES_{11}) + C_2(H_{12} - ES_{12}) &= 0 \\ C_1(H_{21} - ES_{21}) + C_2(H_{22} - ES_{22}) &= 0 \end{aligned} \quad (3\text{-}8)$$

This gives us two homogeneous linear equations in two unknowns. The trivial solution is $C_1 = C_2 = 0$, but this would require $\psi = 0$ (where $\psi = C_1\varphi_1 + C_2\varphi_2$). Other solutions exist only for certain values of E, and these values of E can be found by solving the secular determinant:

$$\begin{vmatrix} H_{11} - ES_{11} & H_{12} - ES_{12} \\ H_{21} - ES_{21} & H_{22} - ES_{22} \end{vmatrix} = 0 \quad (3\text{-}9)$$

The determinant is solved by standard mathematical procedures. Essentially, we are solving two linear, homogeneous equations in two unknowns. An $n \times n$ determinant gives an nth order polynomial equation in energy with n roots. Since the size of the determinant, n, equals the number of atomic orbitals, we get n energies or n molecular orbitals from n atomic orbitals. In the case where the two atoms are the same, we get

$$(H_{11} - E)^2 - (H_{12} - S_{12}E)^2 = 0$$

(with S_{11} and $S_{22} = 1$). The two roots for this equation are:

$$E_1 = \frac{H_{11} + H_{12}}{1 + S_{12}} \quad \text{and} \quad E_2 = \frac{H_{11} - H_{12}}{1 - S_{12}} \quad (3\text{-}10)$$

Solving or approximating the integrals enables one to solve for the energies. The energies so obtained can be substituted one at a time back into (3-8), and the coefficients are then gotten by using a third equation: $C_1^2 + C_2^2 = 1$. The two equations that result are not independent (the determinant equals 0), so we need the third equation.

The evaluation of the S_{12} integrals is straightforward when the structure of the molecule is known. The integrals H_{11} and H_{12} cannot be determined accurately for complex molecules.

For the hydrogen molecule, the Hamiltonian is

$$(-(h^2/8\pi^2 m)\nabla^2 - e^2/r_{A,1} - e^2/r_{B,1} - e^2/r_{A,2} - e^2/r_{B,2} + e^2/r_{A,B} + e^2/r_{12})$$

The two hydrogen nuclei are labeled A and B and the two electrons 1 and 2. The charge on the electron is e, the distance from nucleus A to electron 1 is $r_{A,1}$, and so forth. We solve the integrals by using the various individual terms in the Hamiltonian separately. Accordingly, for one $\int \varphi_i \hat{H} \varphi_j \, d\tau$, there are many integrals to be evaluated. The electron-electron repulsion integrals $\langle \varphi_i | e^2/r_{12} | \varphi_j \rangle$ are particularly difficult. You can well imagine how complicated the Hamiltonian is and how many integrals would be needed even for PCl_5. *Ab initio* calculations do all the integrals and thus are practical only for small molecules or wealthy chemists.

The various ways to "guesstimate" values for these integrals correspond to various ways to do molecular orbital calculations. We shall return to a more detailed description of some of these procedures subsequently.

3–2 A MATRIX FORMULATION OF MOLECULAR ORBITAL CALCULATIONS

Modern approaches to the solution of eigenvalue problems employ a matrix formulation. Since the quantum mechanical description of many of the spectroscopic methods treated in this text is discussed in these terms, the previous section will be reformulated in these terms to introduce the general approach. The solutions of the secular equations are eigenvalues of the Hamiltonian operator $\hat{H}$, so $\hat{H}\psi = E\psi$ can be written in the LCAO approximation in matrix form as

$$[\hat{H}_{ij}][C] = [S][C][E] \tag{3-11}$$

where $[C]$ is the coefficient matrix, $[\hat{H}_{ij}]$ is the energy matrix of elements $\langle \varphi_i | \hat{H} | \varphi_j \rangle$, $[S]$ is the overlap matrix, and $[E]$ is the diagonal matrix of orbital energies, one for each m.o. The $[E]$ and $[C]$ matrices are to be determined.

Dividing both sides by $[C]$, we obtain

$$[C]^{-1}[\hat{H}_{ij}][C] = [S][E] \tag{3-12}$$

The coefficient matrix is, in effect, a similarity transform that diagonalizes the $[\hat{H}_{ij}]$ matrix to produce a unit $[S]$ matrix. The energies result directly from the diagonalized $[\hat{H}]$ matrix, as do the LCAO coefficients of the corresponding molecular orbital.

3–3 PERTURBATION THEORY

Perturbation theory is a useful technique in quantum mechanics when one is trying to ascertain the effect of a small perturbation on a system that has been solved in the absence of this perturbation. Basically, we have the solution to

$$\hat{H}^\circ \psi_i^\circ = E_i^\circ \psi_i^\circ$$

We wish to determine the effect of some perturbation on this ith molecular orbital, *i.e.*,

$$\hat{H}\psi_i = E_i\psi_i \qquad (3\text{-}13)$$

where

$$\hat{H} = \hat{H}^\circ + \hat{H}' \qquad (3\text{-}14)$$

and the term $\hat{H}'$ is a small perturbation on the system. It can be shown that the first order correction to the energy is given by:

$$E_i^1 = \langle\psi_i^\circ|\hat{H}'|\psi_i^\circ\rangle \qquad (3\text{-}15)$$

The correction to the old wave function, ψ_i^1, can be written as a linear combination of the k unperturbed wave functions of the original system:

$$\psi_i^1 = \sum_{k\neq i} a_{ik}\psi_k^\circ \qquad (3\text{-}16)$$

By mixing the appropriate empty excited-state wave function into the ground state, it is possible to polarize or distort the ground state molecule in any way we desire. The contributions that the various ψ_k° functions make in our perturbed system are given by the coefficients a_{ik}, where

$$a_{ik} = -\frac{\langle\psi_k^\circ|\hat{H}'|\psi_i^\circ\rangle}{E_k^\circ - E_i^\circ} \qquad (k \neq i) \qquad (3\text{-}17)$$

Thus, our final expressions for the energy and wave function of the perturbed orbital, E_i and ψ_i, correct to first order, are

$$E_i = E_i^\circ + \langle\psi_i^\circ|\hat{H}'|\psi_i^\circ\rangle \qquad (3\text{-}18)$$

and

$$\psi_i = \psi_i^\circ + \sum_{k\neq i}\frac{\langle\psi_k^\circ|\hat{H}'|\psi_i^\circ\rangle}{E_i^\circ - E_k^\circ}\psi_k^\circ \qquad (3\text{-}19)$$

Note that the zero order wave function gives us the first order energy.

The formula for the second order correction to the energy is given by

$$E_i^{(2)} = \sum_{k\neq i}\frac{\langle\psi_i^\circ|\hat{H}'|\psi_k^\circ\rangle\langle\psi_k^\circ|\hat{H}'|\psi_i^\circ\rangle}{E_i^\circ - E_k^\circ} \qquad (3\text{-}20)$$

It is informative to examine what we have done in the context of a rigorous solution of a 2×2 secular determinant. A 2×2 problem is selected because its secular determinant can be solved easily. We shall again use our original wave functions and the new Hamiltonian $\hat{H}$. We write $\langle\psi_1^\circ|\hat{H}|\psi_1^\circ\rangle$ as H_{11} and $\langle\psi_2^\circ|\hat{H}|\psi_2^\circ\rangle$ as H_{22}, leading to

$$\begin{vmatrix} H_{11} - E & H_{12} \\ H_{21} & H_{22} - E \end{vmatrix} = 0$$

which in turn leads to

$$E^2 - (H_{11} + H_{22})E + (H_{11}H_{22} - H_{12}^2) = 0$$

with roots

$$E = \frac{H_{11} + H_{22} \pm \sqrt{H_{11}^2 + H_{22}^2 - 2H_{11}H_{22} + 4H_{12}^2}}{2}$$

Ignoring H_{12}, we obtain the solutions

$$E_1 = H_{11} = E_1^\circ + \langle\psi_1^\circ|\hat{H}'|\psi_1^\circ\rangle \quad \text{and} \quad E_2 = H_{22} = E_2^\circ + \langle\psi_2^\circ|\hat{H}'|\psi_2^\circ\rangle$$

Using the perturbation equations given above, we write directly

$$E_1 = E_1^\circ + \langle\psi_1^\circ|\hat{H}'|\psi_1^\circ\rangle \text{ and } E_2 = E_2^\circ + \langle\psi_2^\circ|\hat{H}'|\psi_2^\circ\rangle$$

We now see that first order perturbation theory is permissible only when the off-diagonal elements can be ignored, compared with the diagonal ones.

The second order correction to the energy gives us:

$$E_1 = H_{11} - \frac{H_{12}^2}{H_{22} - H_{11}}$$

$$E_2 = H_{22} - \frac{H_{12}^2}{H_{11} - H_{22}}$$

which is an even closer approximation to the exact solution.

3–4 WAVE FUNCTIONS AS A BASIS FOR IRREDUCIBLE REPRESENTATIONS

SYMMETRY IN QUANTUM MECHANICS

Since the energy of a molecule is time invariant and independent of the position of the molecule, the Hamiltonian must be unchanged by carrying out a symmetry operation on the system. (Recall that the symmetry operation produces an equivalent configuration that is physically indistinguishable from the original.) Thus, any symmetry operation must commute with the Hamiltonian; *i.e.*, $\hat{H}\hat{R} = \hat{R}\hat{H}$. Like the irreducible representations, the eigenfunctions are constructed to be orthonormal:

$$\int \psi_i^*\psi_j \, d\tau = \delta_{ij}$$

i.e., $\delta = 0$ when $i \neq j$ and $\delta = 1$ when $i = j$. It is a simple matter to show that *the wave functions form a basis for irreducible representations.*

If we take the wave equation:

$$\hat{H}\psi_i = E_i\psi_i$$

and multiply each side by the symmetry operation $\hat{R}$, we get (for non-degenerate eigenvalues):

$$\hat{H}\hat{R}\psi_i = E_i\hat{R}\psi_i$$

where $\hat{R}\hat{H}\psi = \hat{H}\hat{R}\psi$, and E commutes with $\hat{R}$ because E_i is a constant. Consequently, $\hat{R}\psi_i$ is an eigenfunction. Since ψ_i is normalized, $\hat{R}\psi_i$ must also be, for we do not change the length of a vector by a symmetry operation; therefore,

$$\hat{R}\psi_i = \pm 1\psi_i$$

We have just shown that by applying each operation of the point group to ψ_i (non-degenerate), we generate a representation of the group with all characters equal to ± 1, that is, a one-dimensional and therefore irreducible representation. Thus, the eigenfunctions (non-degenerate) for a molecule are a basis for irreducible representations. It can also be shown that this is true for degenerate eigenfunctions.

Any integrals involving products of molecular orbitals that belong to two different irreducible representations must be zero. To show this, let ψ_a and ψ_b be such molecular orbitals that are eigenfunctions of $\hat{R}$ with eigenvalues a and b. Thus, $\langle\psi_a|\hat{H}\hat{R}|\psi_b\rangle = b\langle\psi_a|\hat{H}|\psi_b\rangle$. Since $\hat{H}\hat{R} = \hat{R}\hat{H}$, we have

$$\langle\psi_a|\hat{H}\hat{R}|\psi_b\rangle = \langle\psi_a|\hat{R}\hat{H}|\psi_b\rangle = \langle\hat{R}\psi_a|\hat{H}|\psi_b\rangle$$

$\hat{R}$ (like other operators with physical significance) is Hermitian, so

$$\langle\hat{R}\psi_a|\hat{H}|\psi_b\rangle = a\langle\psi_a|\hat{H}|\psi_b\rangle$$

We have just shown that

$a\langle\psi_a|\hat{H}|\psi_b\rangle = b\langle\psi_a|\hat{H}|\psi_b\rangle$; since $a \neq b$, the integral $\langle\psi_a|\hat{H}|\psi_b\rangle$ must be zero.

3-5 PROJECTING MOLECULAR ORBITALS

In quantum mechanical calculations, we begin with LCAO's that themselves form a basis for an irreducible representation. Although this is not necessary, it is clearly often convenient to do so. We shall show next how symmetry considerations alone provide much information about the atomic orbitals that contribute to the various molecular orbitals in a molecule. In the last chapter, it was shown that any mathematical function (*e.g.*, points, $3N$ Cartesian vectors, or bond vectors) could be used as a basis set for a reducible representation. It is a simple matter to extrapolate to the case in which the valence atomic orbitals for a given molecule are used as a basis set for a reducible representation. As a simple example, we shall work out the traces of the reducible representation for the three a.o.'s comprising the π-orbitals of the nitrite ion, shown in a coordinate system in Fig. 3-1.

Recall that earlier, when we carried out symmetry operations on the two vectors **a** and **b**, many symmetry operations transformed one vector into the other and *the trace of the representation matrix was thus zero, the only non-zero elements being off-diagonal.* Accordingly, referring to Fig. 3-1, O_1 is moved into O_3 with C_2 and σ_v, so contributions to the total character from the oxygen orbitals are zero. This is true whenever atoms are interchanged. The C_2 and σ_{xz} operations on the nitrogen p_z orbital lead to -1 and $+1$, respectively. Reflection in the yz plane changes the sign of the p-orbitals, making a contribution of -1 for each of the three p-orbitals. The total representation is (C_{2v} point group; recall that x is perpendicular to plane)

C_{2v}	E	C_2	$\sigma_v(xz)$	$\sigma_v'(yz)$
	3	-1	$+1$	-3

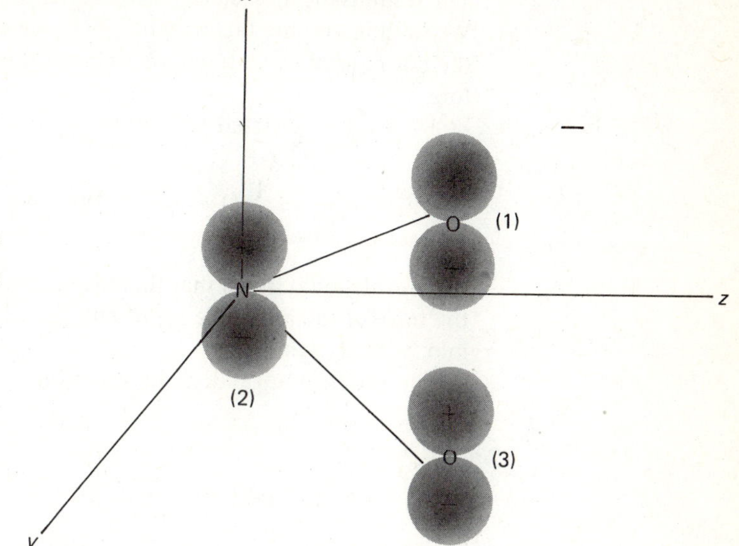

FIGURE 3–1. A.O. basis set for the π-orbitals of the nitrite ion. (x is perpendicular to the NO_2 plane.)

Factoring the total representation gives A_2 and $2B_1$; *i.e.*, the symmetries of the three m.o.'s (from three a.o.'s, we must get three m.o.'s) that result from these a.o.'s must be A_2, B_1, and B_1.

We can now use a technique involving *projection operators*, $\hat{P}$, to give us more information about the wave functions. The projection operators work on the basis set (in this example, the three *p*-orbitals) and annihilate (convert to zero) any element in the basis set that does not contribute to a given irreducible representation. In using the projection operators, we generally work on one atomic orbital (or a linear combination of them) (*e.g.*, φ_1, the *p*-orbital of atom 1 of Fig. 3–1) at a time with the operator

$$\hat{P} = \frac{\ell}{h} \sum_R \chi_R \hat{R} \tag{3-21}$$

where $\hat{R}$ is the symmetry operation, χ_R is the character of that operation in the appropriate point group, ℓ is the dimension of the irreducible representation, and h is the order of the group. The sum is taken over all symmetry operations in the point group. Accordingly, if we operate on φ_1 of the nitrite ion with the projection operator of the A_2 irreducible representation, we get:

$$\psi \text{ (for } A_2) = \hat{P}(A_2)\varphi_1 = \tfrac{1}{4}[(1)E\varphi_1 + (1)C_2\varphi_1 + (-1)\sigma_v\varphi_1 + (-1)\sigma_v{}'\varphi_1].$$

The numbers in parentheses are characters of the A_2 irreducible representation. Next, we carry out the specified operations on φ_1, and indicate the result in terms of what the orbital φ_1 is changed into. The result is

$$\hat{P}(A_2)\varphi_1 = \frac{1}{4}[(1)(1)\varphi_1 + (1)(-1)\varphi_3 + (-1)(1)\varphi_3 + (-1)(-1)\varphi_1]$$

$$= N[\varphi_1 - \varphi_3]$$

The result is the molecular orbital corresponding to the A_2 irreducible representation. We shall ignore the factor $\frac{1}{4}$ for, later, we will want to obtain the absolute values for the normalized wave function, and not the relative values resulting from the operators.

If we had selected the orbital φ_2 to operate on, we would have gotten

$$\widehat{P}(A_2)\varphi_2 = \frac{1}{4}[(1)(1)\varphi_2 + (1)(-1)\varphi_2 + (-1)(1)\varphi_2 + (-1)(-1)\varphi_2] = 0$$

This result could mean that there is no A_2, but we know better, from having factored the trace of the reducible representation. In this case, a node exists at atom 2 in the m.o. with A_2 symmetry.

In order to normalize the wave function that we have just obtained, we realize that

$$\int (C_1\varphi_1 - C_1\varphi_3)^2\,d\tau = C_1^2\int \varphi_1^2\,d\tau - 2C_1^2\int \varphi_1\varphi_3\,d\tau + C_1^2\int \varphi_3^2\,d\tau = 1$$

Recall that φ_1 and φ_3 have the same coefficient. Using an orthonormal basis set,

$$\int \varphi_1^2\,d\tau = 1;\ \int \varphi_1\varphi_3\,d\tau = 0;\ \text{and}\ \int \varphi_3^2\,d\tau = 1.$$

These substitutions convert the normalization equation to

$$2C_1^2 = 1 \qquad \text{or} \qquad C_1 = \frac{1}{\sqrt{2}}$$

The normalized function is

$$\frac{1}{\sqrt{2}}(\varphi_1 - \varphi_3)$$

From factoring the total representation, we know that there is no A_1 symmetry m.o. We shall use $\widehat{P}(A_1)$ to show that a consistent result is obtained. First, consider:

$$\widehat{P}(A_1)\varphi_1 = \frac{1}{4}[(1)(1)\varphi_1 + (1)(-1)\varphi_3 + (1)(1)\varphi_3 + (1)(-1)\varphi_1] = 0$$

Try the other p-orbitals, and you will see that there is no A_1 symmetry m.o. Next, try $\widehat{P}(B_2)$ on φ_1:

$$\widehat{P}(B_2)\varphi_1 = \frac{1}{4}[(1)\varphi_1 + (+1)\varphi_3 + (-1)\varphi_3 + (-1)\varphi_1] = 0$$

We know that there are two m.o.'s with B_1 symmetry. Accordingly,

$$\widehat{P}(B_1)\varphi_1 = \frac{1}{4}[(1)(1)\varphi_1 + (-1)(-1)\varphi_3 + (1)(1)\varphi_3 + (-1)(-1)\varphi_1]$$

$$= N[\varphi_1 + \varphi_3]$$

When we use $\widehat{P}(B_1)$ on φ_2, the following result is obtained:

$$\widehat{P}(B_1)\varphi_2 = \frac{1}{4}[(1)(1)\varphi_2 + (-1)(-1)\varphi_2 + (1)(1)\varphi_2 + (-1)(-1)\varphi_2] = \varphi_2$$

Thus, the projection operators produce two m.o.'s of B_1 symmetry, $\psi = N[\varphi_1 + \varphi_3]$ and $\psi = \varphi_2$. The A_2 and two B_1 molecular orbital wave functions that we have generated are referred to as *symmetry adapted linear combinations* of atomic orbitals. Since the B_1 m.o.'s have similar symmetries, they can mix so that any linear combination of these two m.o.'s also has the appropriate symmetry to be one of the two B_1 m.o.'s; *i.e.*, the two m.o.'s can be:

$$\psi_1 = a\varphi_1 + b\varphi_2 + a\varphi_3$$
$$\psi_2 = a'\varphi_1 - b'\varphi_2 + a'\varphi_3$$

The magnitudes of a and b will depend on the energies of the φ_1 and φ_2 atomic orbitals and their overlap integrals. Thus, we cannot go any further in determining the wave function without a molecular orbital calculation. We shall indicate approximate ways of doing this in the next sections. We can at this point make one further generalization. Since there is no symmetry operation that interchanges the nitrogen orbital with oxygen, the nitrogen must form a separate irreducible representation.

We are now in a position to draw pictures of the m.o.'s we have constructed. They are shown in Fig. 3–2.

If one carries out the operations of the C_{2v} point group on the m.o.'s shown in Fig. 3–2, it will be found that these shapes transform according to the symmetry labels.

We shall work through the C_{3v} ammonia molecule as our next example, for it has more than one element in some classes and also has a doubly degenerate irreducible

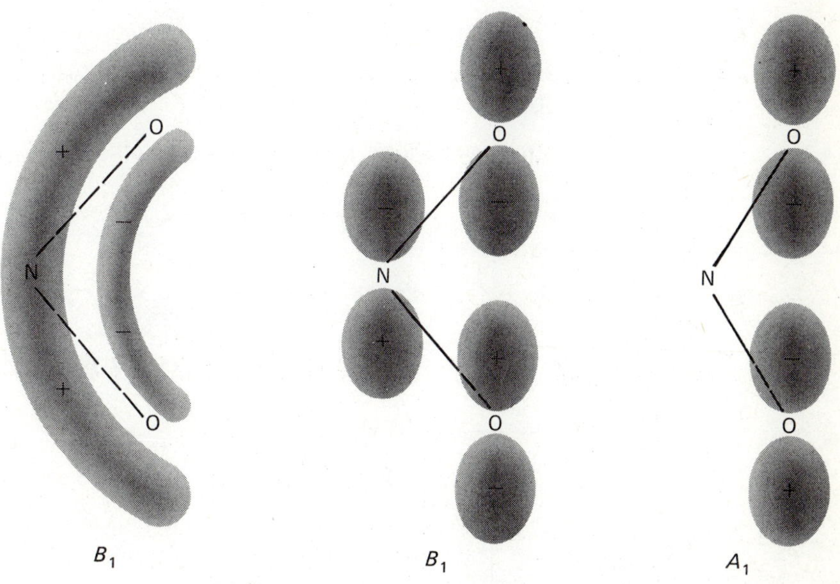

FIGURE 3–2. Shapes of the π molecular orbitals of the nitrite ion.

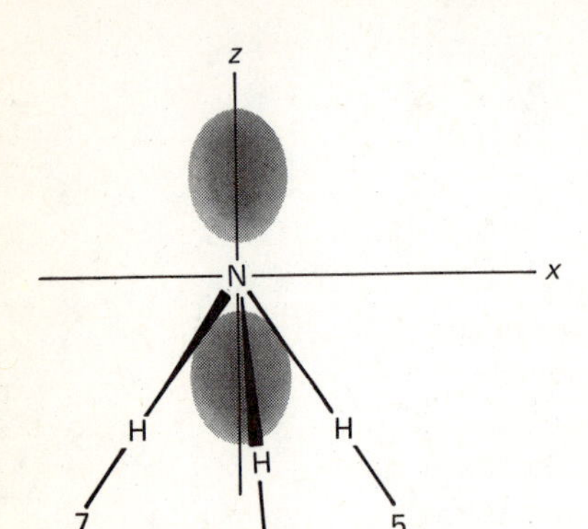

FIGURE 3–3. The location of NH_3 in a coordinate system. The xz plane passes through H_5 and N.

representation. The coordinate system is given in Fig. 3–3. Our procedure is as follows:

(1) First work out the total representation and factor it:

Orbital Shapes	E	$2C_3$	$3\sigma_v$
N 2s	1	1	1
z, y, x	3	0*	1
N, H, H, H	3	0	1
Total	7	1	3

Factoring the total representation, we obtain:

$$a_{A_1} = \frac{1}{6}[1 \times 1 \times 7 + 2 \times 1 \times 1 + 3 \times 1 \times 3] = 3$$

$$a_{A_2} = \frac{1}{6}[1 \times 1 \times 7 + 2 \times 1 \times 1 + 3 \times (-1) \times 3] = 0$$

$$a_E = \frac{1}{6}[1 \times 2 \times 7 + 2 \times (-1) \times 1 + 0] = 2$$

$$^*C_3\begin{bmatrix} x \\ y \\ z \end{bmatrix} = \begin{bmatrix} -\frac{1}{2} & -\frac{\sqrt{3}}{2} & 0 \\ \frac{\sqrt{3}}{2} & -\frac{1}{2} & 0 \\ 0 & 0 & 1 \end{bmatrix}\begin{bmatrix} x \\ y \\ z \end{bmatrix}$$

(trace is 0)

No symmetry operation can take a hydrogen atom into a nitrogen atom, so we can factor the total reducible representation formed from the hydrogen orbitals (3, 0, 1) and find an A_1 and E. In this way we can work on the nitrogen atom and hydrogen atoms separately.

(2) Next, work with projection operators $\hat{P}_{A_1}$ and $\hat{P}_E$. There is no need to work with $\hat{P}_{A_2}$, for factoring the total representation told us that there are no molecular orbitals with this symmetry. For this purpose, we label our atomic orbitals as follows:

$$\varphi_1 = \text{N}2s$$

$$\varphi_2 = \text{N}2p_x$$

$$\varphi_3 = \text{N}2p_y$$

$$\varphi_4 = \text{N}2p_z$$

$$\varphi_5 = \text{H}_5 1s \quad (\sigma_v \text{ contains } \text{H}_5)$$

$$\varphi_6 = \text{H}_6 1s \quad (\sigma_v' \text{ contains } \text{H}_6)$$

$$\varphi_7 = \text{H}_7 1s \quad (\sigma_v'' \text{ contains } \text{H}_7)$$

$\hat{P}_{A_1}$ on φ_5.

$$\psi_1 = \frac{1}{6}[1E\varphi_5 + 1C_3\varphi_5 + 1C_3^2\varphi_5 + 1\sigma_v\varphi_5 + 1\sigma_v'\varphi_5 + 1\sigma_v''\varphi_5]$$

$$= \frac{1}{6}[\varphi_5 + \varphi_7 + \varphi_6 + \varphi_5 + \varphi_6 + \varphi_7] = \frac{1}{6}[2\varphi_5 + 2\varphi_6 + 2\varphi_7]$$

$\hat{P}_{A_1}$ operating on φ_6 and φ_7 will, of course, give the same wave function, so there is no need to carry these out. We need linear combinations that are not identical; *i.e.*, they must not be related to one another by multiplication by -1 or by interchange of the labels of equivalent atoms.

$\hat{P}_{A_1}$ on φ_4. None of the operations take φ_4 into anything else, so it is not necessary to use the projection operator on φ_4. It is a basis set by itself and must belong to the irreducible representation A_1, for it is not changed by any of the symmetry operations. If we use the projection operator, we obtain:

$$\psi_2 = \frac{1}{6}[\varphi_4 + \varphi_4 + \varphi_4 + \varphi_4]$$

$$\psi_2 = \frac{4}{6}\varphi_4$$

$\hat{P}_{A_1}$ on φ_1. We recognize by inspection that φ_1 is also a basis set by itself and must belong to the irreducible representation A_1.

$$\psi_3 = \frac{4}{6}\varphi_1$$

$\hat{P}_{A_1}$ on φ_2. We know by inspection that φ_2 and φ_3 form a doubly degenerate basis set that must belong to an E irreducible representation. Thus, it is not necessary to carry out this operation. We shall do so here to show that we get the right answer with the operator and to demonstrate how the symmetry operations work on a doubly degenerate basis set.

$$\frac{1}{6}[1E\varphi_2 + 1C_3\varphi_2 + 1C_3^2\varphi_2 + 1\sigma_v\varphi_2 + 1\sigma_v'\varphi_2 + 1\sigma_v''\varphi_2]$$

$$= \frac{1}{6}\left[\varphi_2 \overbrace{- \frac{1}{2}\varphi_2 - \frac{\sqrt{3}}{2}\varphi_3} \overbrace{- \frac{1}{2}\varphi_2 + \frac{\sqrt{3}}{2}\varphi_3} + \varphi_2 \overbrace{- \frac{1}{2}\varphi_2 + \frac{\sqrt{3}}{2}\varphi_3} \overbrace{- \frac{1}{2}\varphi_2 - \frac{\sqrt{3}}{2}\varphi_3}\right]$$

$$= 0$$

For C_3, we use

$$\begin{bmatrix} \cos\varphi & \sin\varphi \\ -\sin\varphi & \cos\varphi \end{bmatrix} \begin{bmatrix} \vec{\mathbf{x}} \\ \vec{\mathbf{y}} \end{bmatrix} = \begin{bmatrix} \vec{\mathbf{x}}' \\ \vec{\mathbf{y}}' \end{bmatrix}$$

For a clockwise C_3 rotation, $\varphi = -120°$; for C_3^2, we have $\varphi = -240°$.

$$\vec{\mathbf{x}}' = \vec{\mathbf{x}}\cos\varphi + \vec{\mathbf{y}}\sin\varphi$$
$$\vec{\mathbf{y}}' = -\vec{\mathbf{x}}\sin\varphi + \vec{\mathbf{y}}\cos\varphi$$

For the mirror planes, we use the matrices developed in Chapter 2 and the coordinate system shown in Fig. 3–4.

Likewise, the result of $\widehat{P}_{A_1}$ on φ_3 is 0.

Now we apply the projection operator for the degenerate irreducible representation E.

$$\widehat{P}_E \text{ on } \varphi_2 = \psi_4 = \frac{1}{3}[2E\varphi_2 - 1C_3\varphi_2 - 1C_3^2\varphi_2]$$
$$= \frac{1}{3}\left\{2\varphi_2 - \left[-\frac{1}{2}\varphi_2 - \frac{\sqrt{3}}{2}\varphi_3\right] - \left[-\frac{1}{2}\varphi_2 + \frac{\sqrt{3}}{2}\varphi_3\right]\right\} = \varphi_2$$

$$\widehat{P}_E \text{ on } \varphi_3 = \psi_5 = \varphi_3$$
$$\widehat{P}_E \text{ on } \varphi_5 = \frac{1}{3}[2E\varphi_5 - 1C_3\varphi_5 - 1C_3^2\varphi_5]$$
$$= \frac{1}{3}[2\varphi_5 - \varphi_6 - \varphi_7]$$
$$\widehat{P}_E \text{ on } \varphi_6 = \frac{1}{3}[2\varphi_6 - \varphi_7 - \varphi_5]$$
$$\widehat{P}_E \text{ on } \varphi_7 = \frac{1}{3}[2\varphi_7 - \varphi_5 - \varphi_6]$$

Our resulting orbitals from φ_5, φ_6 and φ_7 are not orthogonal, and we know that there are only two wave functions in E. Appropriate linear combinations must be taken to get an orthonormal basis. There are various ways to do this. One that works for this

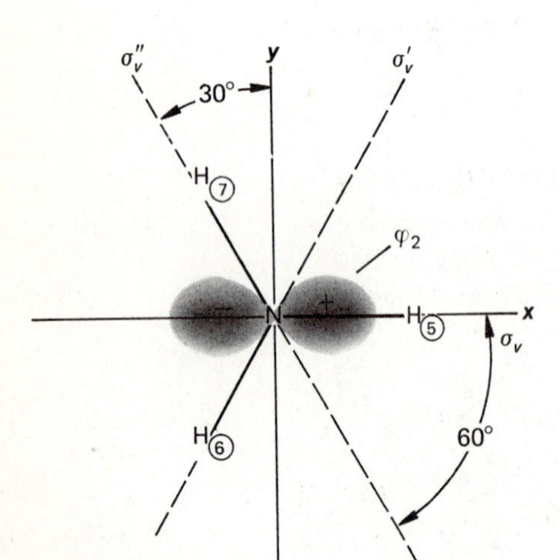

FIGURE 3–4. Effect of reflections through σ_v on φ_2 and the hydrogens of ammonia.

system involves working with the matrix elements, instead of with the trace as we did above. The diagonal elements are:

$$\begin{matrix} E & C_3 & C_3^2 & \sigma_v & \sigma_v' & \sigma_v'' \\ \begin{bmatrix} 1 & 0 \\ 0 & 1 \end{bmatrix} & \begin{bmatrix} -\frac{1}{2} & \\ & -\frac{1}{2} \end{bmatrix} & \begin{bmatrix} -\frac{1}{2} & \\ & -\frac{1}{2} \end{bmatrix} & \begin{bmatrix} 1 & \\ & -1 \end{bmatrix} & \begin{bmatrix} -\frac{1}{2} & \\ & \frac{1}{2} \end{bmatrix} & \begin{bmatrix} -\frac{1}{2} & \\ & \frac{1}{2} \end{bmatrix} \end{matrix}$$

where a_{11} corresponds to $\widehat{P}_{E_a}$ and a_{22} to $\widehat{P}_{E_b}$.

$\widehat{P}_{E_a}$ on φ_5 gives:

$$\psi_a = N\left[E\varphi_5 - \frac{1}{2}C_3\varphi_5 - \frac{1}{2}C_3^2\varphi_5 + 1\sigma_v\varphi_5 - \frac{1}{2}\sigma_v'\varphi_5 - \frac{1}{2}\sigma_v''\varphi_5\right] = N[2\varphi_5 - \varphi_6 - \varphi_7]$$

$\widehat{P}_{E_b}$ on φ_5 gives:

$$\psi_b = N\left[E\varphi_5 - \frac{1}{2}C_3\varphi_5 - \frac{1}{2}C_3^2\varphi_5 - 1\sigma_v\varphi_5 + \frac{1}{2}\sigma_v'\varphi_5 + \frac{1}{2}\sigma_v''\varphi_5\right] = 0$$

There is a node in ψ_b at φ_5. The result of similar operations with $\widehat{P}_{E_b}$ on φ_6 leads to the other wave function:

$$\psi_b = N(\varphi_6 - \varphi_7)$$

One additional way to get wave functions for degenerate irreducible representations involves working with the character table for the rotation group involving only the principal axis, *i.e.*, C_n.

For ammonia, we would use C_3. In C_3, the E representations are given terms of the imaginaries; *e.g.*, for C_3,

$$\begin{matrix} & E & C_3 & C_3^2 \\ E & \begin{Bmatrix} 1 \\ 1 \end{Bmatrix} & \begin{matrix} \varepsilon \\ \varepsilon^* \end{matrix} & \begin{Bmatrix} \varepsilon^* \\ \varepsilon \end{Bmatrix} \end{matrix}$$

where $\varepsilon = \exp(2\pi i/3)$. Thus,

$$\widehat{P}_a\varphi_6 = 1E\varphi_6 + \varepsilon C_3\varphi_6 + \varepsilon^* C_3^2\varphi_6 = \varphi_6 + \varepsilon\varphi_7 + \varepsilon^*\varphi_5 = \psi_a'$$
$$\widehat{P}_b\varphi_6 = 1E\varphi_6 + \varepsilon^* C_3\varphi_6 + \varepsilon C_3^2\varphi_6 = \varphi_6 + \varepsilon^*\varphi_7 + \varepsilon\varphi_5 = \psi_b'$$

If we add the two together, we get one wave function; when we subtract the two and divide by i, we get the second wave function. Upon addition, we get

$$2\varphi_6 + (\varepsilon^* + \varepsilon)\varphi_7 + (\varepsilon + \varepsilon^*)\varphi_5$$

where $\varepsilon = \exp(2\pi i/h)$ and h is the order of the rotation group. Abbreviating these as $\varepsilon = \exp(i\Phi)$ and $\varepsilon^* = \exp(-i\Phi)$ and using the trigonometric identities

$$\exp(i\Phi) = \cos\Phi + i\sin\Phi$$
$$\exp(-i\Phi) = \cos\Phi - i\sin\Phi$$

we see that $\varepsilon + \varepsilon^* = 2\cos\Phi$.

In the C_3 group, the rotation axis is 120°, so $2\cos(-120°) = 2(-½) = -1$. Accordingly, we have

$$\psi_a = 2\varphi_6 - \varphi_7 - \varphi_5$$

(This is equivalent to our previous result, $\psi_a = 2\varphi_5 - \varphi_6 - \varphi_7$, obtained by using the C_{3v} point group.)

To get the other degenerate wave function, we subtract ψ_a' from ψ_b' and divide by i:

$$\frac{\psi_b' - \psi_a'}{i} = 0 + \frac{(\varepsilon^* - \varepsilon)}{i}\varphi_7 + \frac{(\varepsilon - \varepsilon^*)}{i}\varphi_5$$

Substituting the trigonometric functions for ε and ε^* produces

$$\frac{(\varepsilon - \varepsilon^*)}{i} = \frac{2i}{i}\sin\Phi \quad \text{and} \quad \frac{(\varepsilon^* - \varepsilon)}{i} = -\frac{2i}{i}\sin\Phi$$

Since $\sin(-120°) = -\sqrt{3}/2$, we obtain

$$\psi_b = \sqrt{3}\varphi_7 - \sqrt{3}\varphi_5$$

The final wave functions should be normalized to complete the problem. Upon normalization of ψ_b, we obtain:

$$\psi_b = \frac{1}{\sqrt{2}}(\varphi_7 - \varphi_5)$$

The two E functions from φ_5, φ_6, and φ_7 can be obtained in yet another way. First, one projects $\hat{P}_E$ on φ_5 to obtain $2\varphi_5 - \varphi_6 - \varphi_7$. Now, if one can judiciously select for projection another function involving φ_6 and φ_7 (or sometimes φ_5, φ_6, and φ_7) that is orthogonal to the one just obtained, the desired function will result. First, we try φ_6 and φ_7 and realize that $\varphi_6 - \varphi_7$ will be orthogonal to $2\varphi_5 - \varphi_6 - \varphi_7$.

$$\hat{P}_E(\varphi_6 - \varphi_7) = \hat{P}_E\varphi_6 - \hat{P}_E\varphi_7 = \varphi_6 - \varphi_7$$

With these two examples, we see that the symmetry of the molecule determines many aspects of our resulting wave functions. Next, we shall consider some approximate molecular orbital calculations that complete the problem (but not always necessarily correctly).

MOLECULAR ORBITAL CALCULATIONS

3–6 HÜCKEL PROCEDURE

The various ways to guesstimate values for the integrals presented in the introduction correspond to various ways to do m.o. calculations. First, we shall consider some very crude approximations, the so-called Hückel m.o. calculations, which give a fair account of the properties of many hydrocarbons and which will make the solution of the quantum mechanical problem in the introduction more specific while the math remains simple. *Generally, only the π orbitals of hydrocarbon* *systems are treated.* The secular equations are written directly by generalizing our earlier result for a two-atomic-orbital system as:

$$\begin{matrix} C_{11}(H_{11}-ES_{11}) + \cdots C_{1n}(H_{1n}-ES_{1n}) & = 0 \\ \vdots \qquad\qquad \vdots & \\ C_{n1}(H_{n1}-ES_{n1}) + \cdots C_{nn}(H_{nn}-ES_{nn}) & = 0 \end{matrix}$$

The following Hückel Approximations are then introduced:

(1) When two atoms i and j are not directly bonded, we take $H_{ij} = 0$.

(2) When all atoms are alike and if all bond distances are equal (*e.g.*, the carbon $2p_z$ orbitals in benzene), all neighboring atom H_{ij} values are taken as equal and symbolized by β (*i.e.*, $H_{12} = H_{23} = \beta$).

(3) If all atoms are alike, the H_{ii} integrals are symbolized by α ($\int \varphi_1 \hat{H} \varphi_1 \, d\tau = \int \varphi_2 \hat{H} \varphi_2 \, d\tau = \alpha$).

(4) Since we use normalized atomic orbitals in the basis set, $S_{ii} = 1$.

(5) All S_{ij}'s are set equal to zero.

The secular equations become

$$\begin{matrix} C_{11}(\alpha - E) + C_{12}\beta_{12} \cdots\cdots & C_{1n}\beta_{1n} = 0 \\ \vdots & \vdots \\ C_{n1}\beta_{n1} + \cdots\cdots\cdots & C_{nn}(\alpha - E) = 0 \end{matrix}$$

The secular determinant now becomes

$$\begin{vmatrix} \alpha - E & \beta_{12} \cdots\cdots \beta_{1n} \\ \vdots & \vdots \\ \beta_{n1} \cdots\cdots\cdots & \alpha - E \end{vmatrix} = 0 \qquad (3\text{-}22)$$

where all but directly bonded β's are zero. As a specific example, we shall consider the π-system of the allyl radical in Fig. 3-5. (All sigma bonds are ignored.)

The 3×3 secular determinant can be directly written from equation (3-22) as:

$$\begin{vmatrix} \alpha - E & \beta & 0 \\ \beta & \alpha - E & \beta \\ 0 & \beta & \alpha - E \end{vmatrix} = 0$$

Zeroes arise for the β_{13} matrix element, for the carbon atoms 1 and 3 are not bonded. Dividing by β and letting $(\alpha - E)/\beta = X$, we obtain the determinant:

$$\begin{vmatrix} X & 1 & 0 \\ 1 & X & 1 \\ 0 & 1 & X \end{vmatrix} = 0$$

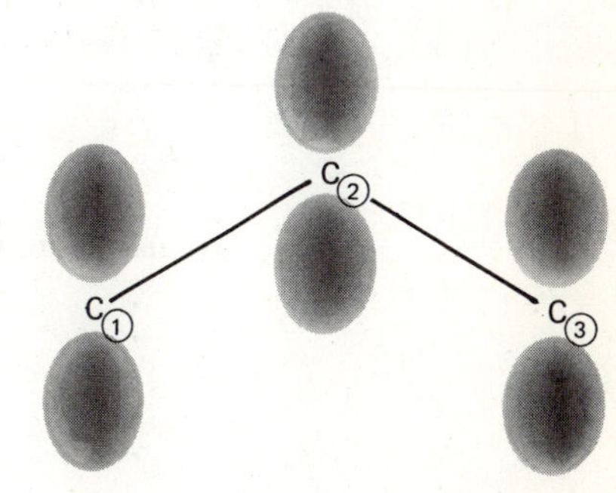

FIGURE 3–5. The π-system of the allyl radical.

This determinant can be solved by expanding by the method of cofactors. The procedure involves an expansion as shown for the general case:

$$\begin{vmatrix} a_1 & b_1 & c_1 \\ a_2 & b_2 & c_2 \\ a_3 & b_3 & c_3 \end{vmatrix} = a_1 \begin{vmatrix} b_2 & c_2 \\ b_3 & c_3 \end{vmatrix} - b_1 \begin{vmatrix} a_2 & c_2 \\ a_3 & c_3 \end{vmatrix} + c_1 \begin{vmatrix} a_2 & b_2 \\ a_3 & b_3 \end{vmatrix} = 0$$

The expansion can be done by row or column. In expanding by row, we take each element separately and multiply it by the determinant that results when the row and column containing the element are eliminated. Labeling the row i and the column j, the sign is positive whenever $i \times j$ is odd and negative when $i \times j$ is even.

For the allyl radical determinant, we expand to get:

$$X \begin{vmatrix} X & 1 \\ 1 & X \end{vmatrix} - 1 \begin{vmatrix} 1 & 1 \\ 0 & X \end{vmatrix} = 0$$

Solving these determinants gives $X^3 - 2X = 0$ with solutions $X = 0, \pm\sqrt{2}$, *i.e.*,

$$\frac{\alpha - E}{\beta} = 0,\ \sqrt{2},\ -\sqrt{2}.$$

These three roots produce the three energies:

$$E_1 = \alpha + \sqrt{2}\beta$$
$$E_2 = \alpha$$
$$E_3 = \alpha - \sqrt{2}\beta$$

For the allyl radical, the secular equations are easily written by taking the general set of secular equations and writing explicit ones for the three-atom allyl π-system or by multiplying the column matrix $\{C_1, C_2, C_3\}$ by the 3×3 secular determinant:

$$C_1X + C_2 = 0 \tag{3-23}$$
$$C_1 + C_2X + C_3 = 0 \tag{3-24}$$
$$C_2 + C_3X = 0 \tag{3-25}$$

where $X = (\alpha - E)/\beta$.

For ψ_1, the bonding molecular orbital, $X = -\sqrt{2}$, so substituting this into (3-23) and (3-25) gives

$$-\sqrt{2}C_1 + C_2 = 0 \quad \text{and} \quad C_2 - \sqrt{2}C_3 = 0$$

Rearranging, we get:

$$C_1 = \frac{C_2}{\sqrt{2}} \quad \text{and} \quad C_3 = \frac{C_2}{\sqrt{2}}$$

Using the normalization criterion:

$$C_1^2 + C_2^2 + C_3^2 = \left(\frac{C_2}{\sqrt{2}}\right)^2 + C_2^2 + \left(\frac{C_2}{\sqrt{2}}\right)^2 = 1$$

or

$$2C_2^2 = 1 \qquad \text{and} \qquad C_2 = \frac{1}{\sqrt{2}}$$

Now

$$C_1 = C_3 = \frac{C_2}{\sqrt{2}}$$

and since

$$C_2 = \frac{1}{\sqrt{2}}$$

we obtain

$$C_1 = C_3 = \frac{1}{2}.$$

With the coefficients defined, we can write

$$\psi_1 = \frac{1}{2}\varphi_1 + \frac{1}{\sqrt{2}}\varphi_2 + \frac{1}{2}\varphi_3 \quad \text{(Bonding)}$$

For ψ_2

$$X = 0 \qquad C_2 = 0 \qquad \text{and} \qquad C_1 = -C_3$$

$$C_1^2 + C_3^2 = 1 \qquad \text{or} \qquad C_1 = \frac{1}{\sqrt{2}}$$

Accordingly, we can write

$$\psi_2 = \frac{1}{\sqrt{2}}\varphi_1 - \frac{1}{\sqrt{2}}\varphi_3 \quad \text{(Non-bonding)}$$

Similarly

$$\psi_3 = \frac{1}{2}\varphi_1 - \frac{1}{\sqrt{2}}\varphi_2 + \frac{1}{2}\varphi_3 \quad \text{(Antibonding)}$$

The radical, cation and anion of allyl are all described by adding the appropriate number of electrons to these molecular orbitals.

Pictures of these orbitals can be drawn by realizing that a node exists every time the wave function changes sign. In general, as the number of nodes in the wave function increases, its energy increases. Furthermore, we see that all of the symmetry information about the molecule is built into the molecular orbital calculation. Using projection operators on the allyl radical, one can show that:

$$\psi_2(A_2) = \frac{1}{\sqrt{2}}(\varphi_1 - \varphi_3) \qquad \text{and}$$

$$\psi_1(B_1) = \frac{1}{\sqrt{2}}(\varphi_1 + \varphi_3) \qquad \text{and} \qquad \psi_3(B_1) = \varphi_2$$

The two B_1 m.o.'s have the same symmetry and can mix. Projection operators cannot tell us the extent of mixing. Using a and b to describe the mixing, we can write from symmetry the two wave functions:

$$a\varphi_1 + b\varphi_2 + a\varphi_3 \quad \text{and} \quad a'\varphi_1 - b'\varphi_2 + a'\varphi_3$$

The Hückel calculation gives the a's and b's, and also gives the same form for the wave functions as the projection operator because all of the symmetry is automatically included in the Hückel Hamiltonian.

The symmetry considerations for the nitrite ion are the same as for allyl, but the molecular orbital coefficients are different. Hückel calculations have been carried out on systems containing heteroatoms, X. These are treated by using

$$\alpha_x = \alpha_c + k\beta_{c-c}$$

and

$$\beta_{c-x} = k'\beta_{c-c}$$

Tables of values to use for the quantities k and k' are listed in books on Hückel m.o. calculations but will not be presented here because the general agreement between fact and prediction deteriorates rapidly when Hückel calculations are employed on molecules containing carbon bound to other kinds of atoms.

3-7 PROPERTIES DERIVED FROM WAVE FUNCTIONS

There are many quantities of chemical interest that can be calculated from the wave functions. A few are listed and defined:

(1) Electron Density at an Atom r, q_r

The electron density q_r at an atom r in a molecule is given by:

$$q_r = -\sum_j n_j C_{jr}^2 \tag{3-26}$$

where n_j is the number of electrons in jth m.o. The summation is carried out over all of the j occupied m.o.'s and C_{jr} is the coefficient of the atomic orbital r in the jth m.o.

Using the Hückel wave functions for the filled molecular orbitals of the allyl anion (two electrons in each):

$$\psi_1 = \frac{1}{2}\varphi_1 + \frac{1}{\sqrt{2}}\varphi_2 + \frac{1}{2}\varphi_3$$

$$\psi_2 = \frac{1}{\sqrt{2}}\varphi_1 - \frac{1}{\sqrt{2}}\varphi_3$$

we obtain for the electron density on carbon 1 (with $n = 2$) for the bonding and nonbonding m.o.'s over which we sum:

$$q_1 = -2\left(\frac{1}{2}\right)^2 - 2\left(\frac{1}{\sqrt{2}}\right)^2 = -1\frac{1}{2}$$

At carbon 2 we find:

$$q_2 = -2\left(\frac{1}{\sqrt{2}}\right)^2 - 2(0) = -1$$

The sum of all the q's equals the total number of electrons in the system.

(2) FORMAL CHARGE, δ

The electron density at an atom in the molecule minus that on a free atom is referred to as the formal charge on the atom in the molecule. Considering carbon, we place one electron in each of the orbitals of the atom to form four bonds. In Hückel theory, where we worry only about the π-system, we consider only the p_z orbital on a carbon. Since it contains one electron, the charge on the "free atom" is -1. If the charge on this atom in a molecule (*i.e.*, the electron density at an atom r) is -1.4, δ, the formal charge, is -0.4, that is, the difference between the charge on this atom in the molecule and that on the neutral atom. The sum of all the formal charges equals the total charge on the molecule.

For the allyl anion we obtain, using the charge density calculated above, a charge of -0.5 at carbon 1 and a charge of zero at carbon 2.

(3) BOND ORDER, P_{AB}

The bond order between two adjacent atoms A and B is given by:

$$P_{AB} = \sum_j n_j C_{Aj} C_{Bj} \tag{3-27}$$

i.e., just take the product of the coefficients of atoms A and B in each m.o. times the number of electrons in the m.o. and sum over all of the m.o.'s.

Using this formula, the pi-bond order between carbon atoms 1 and 2 of the allyl anion is given by

$$P_{12} = 2\left(\frac{1}{2}\right)\left(\frac{1}{\sqrt{2}}\right) + 2\left(\frac{1}{\sqrt{2}}\right)(0) = \frac{\sqrt{2}}{2} = 0.7$$

Hückel theory contains many crude approximations which can be justified* well for hydrocarbon molecules. With more polar molecules, they are very crude. There have been many attempts to improve on these calculations. Extended Hückel, CNDO, INDO, and others constitute various semi-empirical procedures that have been used to approximate various integrals. In all of these procedures, the overlap integral is evaluated and not set equal to zero. To go through the various procedures in detail would take more time than we can allot to this topic in this text. We will provide only a very rough overview of the calculations, and will concern ourselves more with what can be done with the output.

---

*For further details see, M. J. S. Dewar, "The Molecular Orbital Theory of Organic Chemistry," McGraw-Hill Publishing Co., New York, Chap. 5.

3-8 EXTENDED HÜCKEL PROCEDURE

In the extended Hückel method, the H_{ii}'s are approximated by valence state ionization potentials. The H_{ij}'s, given by the Wolfsberg-Helmholz formula, are:

$$H_{ij} = \frac{1}{2} k S_{ij}(H_{ii} + H_{jj}) \tag{3-28}$$

where k is a constant usually taken to be 1.8. Cusach's formula for estimating H_{ij}'s is:

$$H_{ij} = \frac{1}{2}(2 - S_{ij}) S_{ij}(H_{ii} + H_{jj})$$

All overlaps are explicitly evaluated. Slater or Clementi atomic orbitals are generally used.

We shall briefly describe the criteria used to describe the electron density in a molecule when $S \neq 0$ for $i \neq j$. With a LCAO wave function

$$\psi_1 = C_1\varphi_1 + C_2\varphi_2$$

we see that the total electron density, $\psi_1{}^2$, is given by

$$\int \psi_1{}^2 \, d\tau = C_1{}^2 \int \varphi_1{}^2 \, d\tau + 2C_1C_2 \int \varphi_1\varphi_2 \, d\tau + C_2{}^2 \int \varphi_2{}^2 \, d\tau = C_1{}^2 + C_2{}^2 + 2C_1C_2S_{12}$$

Using normalized a.o.'s, $\varphi_1{}^2 = 1$, we also see that $\int \varphi_1\varphi_2 \, d\tau$ is the overlap integral S_{12}. We shall now define some quantities that are important results of calculations in which $S \neq 0$.

(1) NET ATOMIC POPULATION

The quantity $n_i C_i{}^2$ is referred to as the net atomic population (n_i equals the number of electrons in the m.o.). It is that part of the total electron density in the m.o. that can be assigned to atom 1.

(2) OVERLAP POPULATION

The quantity $2nC_1C_2S_{12}$ is referred to as the overlap population between *atomic orbitals 1 and 2* for one molecular orbital. In a polyatomic molecule, the overlap population between atomic orbitals i and j is obtained by summing over all occupied molecular orbitals k:

$$\sum_k 2nC_iC_jS_{ij} \tag{3-29}$$

It is often convenient to have the *atom-atom* overlap population. This corresponds to the total overlap of all the atomic orbitals on atom x with all the atomic orbitals on atom y. Many m.o. programs present this as a reduced overlap population matrix. This is

$$\sum_{k(xy)} 2nC_iC_jS_{ij} \tag{3-30}$$

where $k(xy)$ refers to the xy overlap population of every atomic orbital on x and y summed over all filled molecular orbitals. When one compares bonds with similar ionic character, the overlap population is related to the bond strength.

(3) GROSS ATOMIC POPULATION

This is a procedure for assigning all the electron density in a molecule to each of the individual atoms. Half of the bonding electron density (*e.g.*, half of $2nC_1C_2S_{12}$) is given to each of the atoms involved in bonding.

$$\text{Gross atomic population} = \sum_{\substack{\text{all} \\ \text{m.o.s}}} (nC_i^2 + nC_iC_jS_{ij}) \qquad (3\text{-}31)$$

(4) FORMAL CHARGE

The formal charge on an atom is equal to the core charge (nucleus plus all non-valence electrons) minus the gross atomic population.

To give you some appreciation for the kind of information available from a semi-empirical molecular orbital calculation, the output from an extended Hückel m.o. calculation is presented in Table 3–1. The matrix summarizes the wave function; *e.g.*, from m.o. 12, we see that

$$\psi_{12} = -0.40\ C(2)2p_z - 0.56\ C(1)2p_z - 0.40\ C(3)2p_z$$

(Numbers in parentheses refer to the numbering in the figure of allyl on the second page of computer output.) This is the lowest-energy filled pi-orbital. The odd electron in m.o. 9 is delocalized over both of the terminal carbons, and these are the only two orbitals contributing to this molecular orbital.

One important concept to be gained from this calculation is the extent of delocalization of the bonding in the sigma system, as can be seen by looking at a sigma m.o. For example, none of the filled m.o.'s even vaguely resemble a localized carbon-hydrogen sigma bond. All of the σ m.o.'s are very extensively delocalized. The π-type atomic orbitals are orthogonal to the σ m.o.'s, and they contain only contributions from the carbon p_z orbitals. Net charges, orbital occupations, and an atom-by-atom overlap population are evaluated in this program from the wave functions calculated.

Extended Hückel calculations give reasonable wave functions for systems in which the formal charges on the atoms of the molecule are small, $\sim \pm 0.5$. As an example of the utility of such calculations, the results from a study of several metallocenes* and bis-arene complexes† will be discussed. In all systems studied, the correct ground state results; *i.e.*, the unpaired electrons end up being placed in molecular orbitals that are essentially the correct *d*-orbitals. Certain ground state spectroscopic properties (nmr and epr) are also predicted. This agreement with experiment suggests that the wave functions are reasonably good. They are then analyzed in detail to provide information about the bonding interactions responsible for bonding the hydrocarbon ring to the metal. An interesting result involved the extensive mixing of the ring sigma orbitals with the metal orbitals in these systems.

---

*M. F. Rettig and R. S. Drago, J. Amer. Chem. Soc., *91,* 3432 (1969).
†S. E. Anderson, Jr., and R. S. Drago, Inorg. Chem., *11,* 1564 (1972).

TABLE 3–1. WAVE FUNCTION FROM EXTENDED HÜCKEL CALCULATION OF THE ALLYL RADICAL

(The vertical numbers correspond to A.O.'s and the horizontal ones to M.O.'s)

A.O.'s ↓ \ M.O.'s →	1	2	3	4	5	6	7	8_π	9_π
1	−1.80	−0.00	0.24	−0.00	0.27	−0.00	0.25	−0.00	0.00
2	0.00	0.00	0.00	−0.00	0.00	−0.00	0.00	−0.97	−0.00
3	−0.51	−0.00	−0.35	0.00	−1.31	0.00	−0.24	0.00	−0.00
4	0.00	−1.28	−0.00	−0.68	−0.00	−0.29	−0.00	−0.00	−0.00
5	0.76	1.12	0.87	−0.38	−0.01	0.22	0.01	−0.00	0.00
6	−0.00	−0.00	−0.00	0.00	0.00	−0.00	0.00	0.65	−0.74
7	−0.43	−0.53	0.31	−0.58	0.30	0.62	−0.65	0.00	−0.00
8	−0.68	−0.13	0.58	−0.89	−0.53	−0.54	0.15	−0.00	0.00
9	0.76	−1.12	0.87	0.33	−0.01	−0.22	0.01	−0.00	−0.00
10	−0.00	−0.00	−0.00	0.00	−0.00	0.00	−0.00	0.65	0.74
11	−0.43	0.53	0.31	0.58	0.30	−0.62	−0.65	−0.00	−0.00
12	0.68	−0.13	−0.58	−0.89	0.53	−0.54	−0.15	0.00	0.00
13	0.28	0.00	−0.30	0.00	−1.17	0.00	−0.61	−0.00	0.00
14	−0.07	−0.48	−0.61	0.49	0.46	0.68	−0.59	0.00	0.00
15	0.15	−0.09	−0.75	0.82	−0.03	−0.55	0.61	−0.00	−0.00
16	0.15	0.09	−0.75	−0.82	−0.03	0.55	0.61	0.00	0.00
17	−0.07	0.48	−0.61	−0.49	0.46	−0.68	−0.59	0.00	−0.00

Atom	A.O.	10	11	12_π	13	14	15	16	17
C	1	−0.00	0.01	−0.00	0.00	−0.09	0.33	−0.00	0.48
C	2	−0.00	−0.00	−0.56	0.00	−0.00	0.00	−0.00	−0.00
C	3	−0.00	0.33	−0.00	−0.00	−0.36	−0.17	−0.00	0.01
C	4	−0.49	−0.00	−0.00	−0.30	0.00	−0.00	0.16	−0.00
C	5	−0.14	0.05	0.00	−0.03	−0.03	−0.20	0.44	0.36
C	6	−0.00	0.00	−0.40	−0.00	0.00	−0.00	0.00	0.00
C	7	−0.12	−0.35	−0.00	0.37	−0.12	−0.12	0.01	−0.00
C	8	0.41	0.00	−0.00	0.07	0.29	−0.14	0.02	−0.01
C	9	0.14	0.05	−0.00	0.03	−0.03	−0.20	−0.44	0.36
C	10	0.00	0.00	−0.40	−0.00	0.00	−0.00	0.00	0.00
C	11	0.12	−0.35	−0.00	−0.37	−0.12	−0.12	−0.01	−0.00
C	12	0.41	−0.00	−0.00	0.07	−0.29	0.14	0.02	0.01
H	13	0.00	−0.40	−0.00	0.00	0.27	0.32	0.00	0.08
H	14	0.37	0.26	0.00	−0.15	0.27	−0.15	0.17	0.04
H	15	−0.07	−0.33	0.00	0.36	0.02	−0.23	0.19	0.05
H	16	0.07	−0.33	0.00	−0.36	0.02	−0.23	−0.19	0.05
H	17	−0.37	0.26	−0.00	0.15	0.27	−0.15	−0.17	0.04

M.O.'s 10 to 17 have $2e^-$
" 9 has $1e^-$
" 1 to 8 empty
Atomic Orbitals 1–4, 5–8 and 9–12 are carbon orbitals in the order $2s$, $2p_z$, $2p_x$, $2p_y$ for each atom.
A.O.'s 13 to 17 are hydrogen $1s$.

TABLE 3–1. *(Cont'd.)*

M.O.	Energy (eV)	Occ.
1	53.22	0
2	36.55	0
3	28.31	0
4	22.97	0
5	18.76	0
6	7.30	0
7	5.19	0
8	−4.63	0
9	−10.16	1
10	−12.67	2
11	−13.83	2
12	−14.03	2
13	−15.01	2
14	−15.96	2
15	−19.40	2
16	−22.83	2
17	−27.54	2

Gross Atomic Population

Atomic Orbital	Occupations
1	1.13
2	0.98
3	0.91
4	0.96
5	1.17
6	1.01
7	0.96
8	0.89
9	1.17
10	1.01
11	0.96
12	0.89
13	0.98
14	1.00
15	0.98
16	0.98
17	1.00

Coordinate System

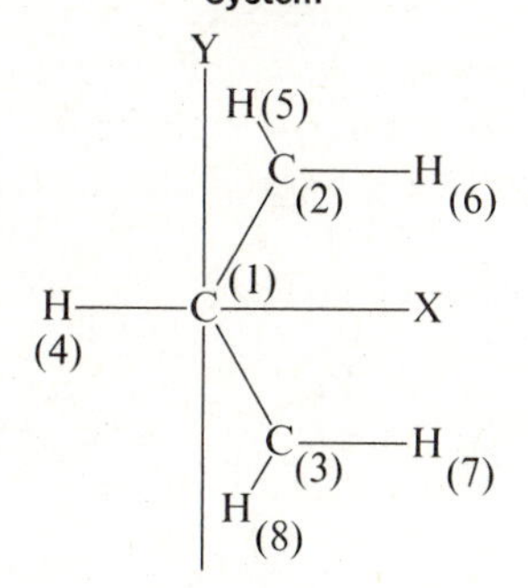

Atom Number	Net Charges
1	0.013
2	−0.020
3	−0.030
4	0.015
5	0.000
6	0.016
7	0.017
8	0.000

Reduced Overlap Population Matrix Atom by Atom

	1	2	3	4	5	6	7	8
1	5.236	1.141	1.141	0.836	−0.117	−0.075	−0.075	−0.117
2	1.141	5.604	−0.216	−0.097	0.821	0.812	−0.013	0.009
3	1.141	−0.216	5.604	−0.097	0.009	−0.013	0.812	0.821
4	0.836	−0.097	−0.097	1.360	−0.020	0.005	0.005	−0.020
5	−0.117	0.821	0.009	−0.020	1.402	−0.094	0.000	−0.000
6	−0.075	0.812	−0.013	0.005	−0.094	1.334	−0.000	0.000
7	−0.075	−0.013	0.812	0.005	0.000	−0.000	1.334	−0.094
8	−0.117	0.009	0.821	−0.020	−0.000	0.000	−0.094	1.402

This run cost $3.81

3–9 SCF-INDO (INTERMEDIATE NEGLECT OF DIFFERENTIAL OVERLAP)

Before discussing the INDO calculation, it is appropriate to discuss in more detail some of the approximations that have been employed without mention in the methods discussed for many-electron systems. The wave function will be considered first. The ground state of our molecule is described by a wave function Ψ, which, when substituted into equation (3–4), would give us the energy, if the equation could be rigorously solved. In the LCAO approach, we are attempting to represent Ψ by a combination of one-electron functions (*i.e.*, molecular orbitals), each of which depends only on the coordinates of one electron. For an n-electron system, the ground state wave function is given by simply taking the product of the occupied one-electron orbitals, *i.e.*,

$$\Psi(1, 2 \cdots n) = \psi_1(1)\psi_2(2) \cdots \psi_n(n) \tag{3-32}$$

The subscript on ψ labels the orbital used and the number in parentheses describes the electron. Physically, this amounts to making the untenable assumption that electron 1 is independent of the $n - 1$ other electrons in the system. We shall have to correct or pay the consequences for this later.

If the $\hat{H}$'s were written as the sum of one-electron operators, the Schrödinger equation could be solved by a simple separation of variables. This would amount to ignoring completely the existence of the other electrons; *e.g.*, the e^2/r_{ij} electron repulsion terms would not be included. What one does is to use one-electron operators, and correct for the potential experienced at an electron p from the field of the atomic nuclei and the average field of the other $n - 1$ electrons. We label this field as $V_{(p)}$, and it can be approximated as will be discussed shortly. Essentially, then, our Hamiltonian is $\mathfrak{F}_{(p)}$, given by

$$\mathfrak{F}_{(p)} = \sum_p F_{(p)} = \sum_p \left[-\frac{1}{2}\frac{\hbar^2}{m}\nabla_{(p)}{}^2 + V_{(p)} \right] \tag{3-33}$$

When $\mathfrak{F}_{(p)}$ is used in a Schrödinger-type equation, we have:

$$\mathfrak{F}_{(p)}(1, 2 \cdots n)\Psi(1, 2 \cdots n) = \mathcal{E}\Psi(1, 2 \cdots)$$

Note that our operator $\mathfrak{F}_{(p)}$ can be separated (separation of variables) into a sum of one-electron operators $F_{(p)}$, which give rise to a set of equations of the form

$$F_{(1)}\psi_i(1) = E_i\psi_i(1) \tag{3-34}$$

The ψ_i functions are the one-electron molecular orbitals obtained by the variation procedure. These are the wave functions we have been talking about when we do Hückel or extended Hückel calculations. In these cases, all of the electron-electron repulsions are very indirectly lumped into the parameters employed, the charge correction and the Wolfsberg-Helmholz type of approximation. The Hamiltonian is not explicitly defined.

In self-consistent field (SCF) methods, the Hamiltonian is explicitly written and used. The $V_{(p)}$ term of equation (3–33) is an e^2/r_{ij} operator. In an INDO calculation, the formalism is basically SCF (*vide infra*), but many of the resulting integrals are ignored. In SCF calculations, the Hamiltonian for an n-electron problem is written as:

$$\hat{H} = \sum_{i=1}^{n} \hat{H}_i + \sum_{i<j} \frac{e^2}{r_{ij}} \tag{3-35}$$

Here, r_{ij} is the interelectronic distance and the $i < j$ index ensures that electron-electron interactions are counted only once. $\hat{H}_i$, the operator for electron i in the field of the nuclei, consists of a kinetic energy term and an electron-nuclear attraction term:

$$\hat{H}_i = -\frac{\hbar^2}{2m}\nabla_i^2 - \sum_k \frac{Z_k e^2}{r_{ik}} \tag{3-36}$$

where ∇_i^2 is the familiar kinetic energy operator for electron i, r_{ik} is the electron-nucleus distance, and Z_k is the charge on the kth nucleus. The energy is found by a variational procedure, and successive sets of trial wave functions are used until this energy is minimized. These calculations are made more difficult because the operators depend on the wave functions. That is, in order to evaluate e^2/r_{ij} we must know the electron distribution, so we must know the wave function before we calculate the wave function. A starting wave function is guessed, the calculations are carried out, and a new wave function is obtained. The new wave function is then used to repeat the calculation. The procedure is iterated until there are no further changes in the wave functions, *i.e.*, we have a self-consistent field. This still is not a rigorous procedure because we are treating the electrons in a molecule as if they experience a charge cloud from all the other electrons. In actual fact, the electron motion is correlated; *i.e.*, the motion of one electron is governed by the instantaneous positions of the other electrons and not their average. This effect, called *electron correlation,* is often ignored, leading to erroneous wave functions.

In any molecule of more than four or five atoms, even a limited basis set (*i.e.*, one in which only the valence atomic orbitals on each atom are employed) gives rise to a huge number of integrals when all interactions (kinetic energy, nuclear-electron, nuclear-nuclear, and electron-electron) are taken into account. The various calculation procedures CNDO (complete neglect of differential overlap) CNDO-2, INDO (intermediate neglect of differential overlap), and so forth, differ in the types of integrals that are ignored. At present, INDO is the most popular of these approximate methods.

We can illustrate the types of integrals encountered by considering the following four orbitals taken from a large molecule:

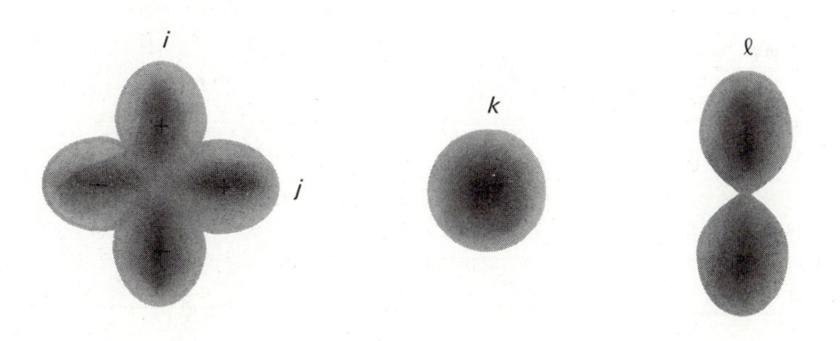

Electron repulsion integrals of the form

$$\iint \varphi_i^*(1)\varphi_j^*(1)\frac{1}{r_{12}}\varphi_k(2)\varphi_l(2)\, d\tau_1\, d\tau_2$$

are the most numerous and require the most computer time to integrate numerically. This integral represents the $1/r$ repulsion between the electron density in the region where φ_i and φ_j have differential overlap and the electron density in the region where φ_k and φ_l have differential overlap. The *differential overlap* for electron 1, for example, is simply defined as

$$\psi_i(1)\psi_j(1)\, d\tau_1$$

TABLE 3–2. INDO CALCULATION FOR ALLYL

(One atomic unit of energy is defined as twice the energy of the first Bohr orbit in the hydrogen atom; i.e., twice the ionization potential of a hydrogen atom.)

Electronic Energy	−60.0089
Electronic Energy	−60.0340
Electronic Energy	−60.0352
Electronic Energy	−60.0353
Electronic Energy	−60.0353
Electronic Energy	−60.0353
Electronic Energy	−60.0353
Energy Satisfied	

Eigenvalues and Eigenvectors

			eV →	**−52.19**	**−42.62**	**−36.32**	**−33.74**	**−30.21**	**−26.56**	**−23.36**	**−23.17**
Eigenvalues A.U.			→	**−1.915**	**−1.566**	**−1.335**	**−1.240**	**−1.110**	**−0.976**	**−0.859**	**−0.851**
(orbital energies)				**1**	**2**	**3**	**4**	**5**	**6**	**7**	**8**
1	1	C	S	−0.62	−0.00	−0.34	0.02	0.00	0.00	0.00	−0.10
2	1	C	Px	−0.13	−0.00	0.29	0.48	−0.00	−0.00	−0.00	0.40
3	1	C	Py	−0.00	0.42	−0.00	0.00	−0.45	−0.00	0.33	0.00
4	1	C	Pz	−0.00	0.00	0.00	−0.00	0.00	−0.72	0.00	−0.00
5	2	C	S	−0.45	0.51	0.27	−0.06	0.03	−0.00	0.16	0.06
6	2	C	Px	0.08	−0.08	0.30	0.20	0.44	0.00	0.24	−0.36
7	2	C	Py	0.19	0.11	0.21	−0.43	0.25	0.00	−0.42	0.02
8	2	C	Pz	0.00	0.00	−0.00	−0.00	−0.00	−0.49	−0.00	−0.00
9	3	C	S	−0.45	−0.51	0.27	−0.06	−0.03	−0.00	−0.16	0.06
10	3	C	Px	0.08	0.08	0.30	0.20	−0.44	0.00	−0.24	−0.36
11	3	C	Py	−0.19	0.11	−0.22	0.43	0.25	−0.00	−0.42	−0.02
12	3	C	Pz	0.00	−0.00	−0.00	−0.00	0.00	−0.49	0.00	−0.00
13	4	H	S	−0.20	−0.00	−0.33	−0.28	0.00	0.00	0.00	−0.47
14	5	H	S	−0.13	0.30	0.15	0.33	0.00	−0.00	−0.40	0.27
15	6	H	S	−0.13	0.23	0.34	−0.05	0.38	0.00	0.17	−0.31
16	7	H	S	−0.13	0.23	0.34	−0.05	−0.38	0.00	−0.17	−0.31
17	8	H	S	−0.13	0.30	0.15	−0.33	−0.00	0.00	0.40	0.27

It is important to emphasize that although an overlap integral may be zero,

$$\int \psi_k(1)\psi_l(1)\,d\tau_1 = 0$$

an integral such as

$$\int \hat{O}_p \psi_k(1)\psi_l(1)\,d\tau_1$$

is not necessarily zero. For example, when the operator $\hat{O}_p$ is $1/r_{1N}$, we have the electron-nuclear attraction integral, which cannot be assumed to be zero.

Note that φ_k and φ_l overlap very little, so the differential overlap $\int \varphi_k(2)\varphi_l(2)\,d\tau_2$ is almost zero everywhere. The value of the integral is thus very small. If a systematic method were developed to set certain of these differential overlaps to zero, the number of such three- and four-center integrals would be drastically reduced. This is just what the various NDO schemes do.

For purposes of comparison with the results from the Hückel calculation and the extended Hückel output, the computer output from the INDO calculation on allyl is presented in Table 3–2. The eigenvalues of the energy are given above each molecular orbital in units of electron volts and atomic units.

It is of interest to examine ψ_9 in the INDO output:

$$\psi_9 = 0.71(\varphi_8 - \varphi_{12})$$

Eigenvalues →				**−8.53 −0.314**	**−7.05 −0.068**	**−7.07 −0.051**	**−0.08 −0.033**	**−0.20 −0.029**	**0.81 0.030**	**1.79 0.066**	**4.48 0.165**	**5.76 0.212**
				9	**10**	**11**	**12**	**13**	**14**	**15**	**16**	**17**
1	1	C	S	0.00	0.43	0.00	−0.19	0.00	0.00	0.43	0.30	−0.00
2	1	C	Px	0.00	−0.09	−0.00	−0.20	0.00	0.00	−0.31	0.60	−0.00
3	1	C	Py	−0.00	0.00	−0.29	−0.00	0.00	−0.12	0.00	0.00	0.65
4	1	C	Pz	0.00	−0.00	0.00	0.00	0.69	0.00	−0.00	0.00	0.00
5	2	C	S	0.00	−0.26	0.35	0.37	−0.00	0.25	0.07	−0.10	−0.18
6	2	C	Px	0.00	−0.22	0.29	−0.24	0.00	−0.23	0.37	0.02	0.32
7	2	C	Py	−0.00	0.08	−0.08	0.14	−0.00	0.37	0.16	0.42	0.34
8	2	C	Pz	0.71	0.00	−0.00	−0.00	−0.51	0.00	0.00	−0.00	0.00
9	3	C	S	−0.00	−0.26	−0.35	0.37	−0.00	−0.25	0.07	−0.10	0.18
10	3	C	Px	−0.00	−0.22	−0.29	−0.24	0.00	0.23	0.36	0.02	−0.32
11	3	C	Py	−0.00	−0.08	−0.08	−0.14	0.00	0.37	−0.16	−0.42	0.34
12	3	C	Pz	−0.71	0.00	0.00	−0.00	−0.51	−0.00	0.00	−0.00	−0.00
13	4	H	S	−0.00	−0.50	−0.00	−0.16	0.00	0.00	−0.43	0.33	−0.00
14	5	H	S	0.00	−0.08	0.08	−0.48	0.00	−0.49	0.01	−0.16	−0.04
15	6	H	S	−0.00	0.39	−0.49	−0.04	0.00	−0.06	−0.33	−0.08	−0.19
16	7	H	S	0.00	0.39	0.49	−0.04	−0.00	0.06	−0.33	−0.08	0.19
17	8	H	S	−0.00	−0.08	−0.08	−0.48	0.00	0.49	0.01	−0.19	0.04

Binding Energy = −2.6884 A.U.
= −73.15 eV

This run cost $5.75.
Dipole Moment = 3.09 Debyes

It is normalized, since

$$\int \psi_9{}^2\, d\tau = (0.71)^2\left[\int \varphi_8\varphi_8\, d\tau - 2\int \varphi_8\varphi_{12}\, d\tau + \int \varphi_{12}\varphi_{12}\, d\tau\right]$$
$$= (0.71)^2(1 - 0 + 1)$$
$$= 1$$

Now look at ψ_9 in the extended Hückel calculation:

$$\psi_9 = 0.74(\varphi_{10} - \varphi_6)$$

It is the same orbital with the same symmetry, but it has a different coefficient. Is it still normalized? Yes, because extended Hückel does not neglect differential overlaps, so

$$\int \varphi_{10}\varphi_6\, d\tau \neq 0$$

and the normalization procedure as used above will yield a different coefficient.

3–10 SOME PREDICTIONS FROM M.O. THEORY ON ALTERNATELY DOUBLY BONDED HYDROCARBONS

You probably have seen examples of the successes of m.o. theory for homonuclear diatomics, *e.g.*, the prediction of paramagnetic O_2. One further success involves the prediction of the structures of even numbered, alternately double bonded hydrocarbons. The systems C_4H_4, C_8H_8, and $C_{16}H_{16}$ are more stable as a set of alternating double bonds than they are as a delocalized pi-system. On the other hand, C_6H_6, $C_{10}H_{10}$, $C_{14}H_{14}$, and $C_{18}H_{18}$ (*i.e.*, $4n + 2$, when $n = 1, 2, 3, \ldots$) are aromatic.

The m.o. results for these even numbered, alternately double bonded hydrocarbons show a singly degenerate low energy orbital and a singly degenerate high energy level with doubly degenerate levels in between:

C_4	C_6	C_8	C_{10}	C_{12}	C_{14}	etc.
—	—	—	—	—	—	
↑ ↑	— —	— —	— —	— —	— —	
↑↓	↑↓ ↑↓	↑ ↑	— —	— —	— —	
	↑↓	↑↓ ↑↓	↑↓ ↑↓	↑ ↑	— —	
		↑↓	↑↓ ↑↓	↑↓ ↑↓	↑↓ ↑↓	
			↑↓	↑↓ ↑↓	↑↓ ↑↓	
				↑↓	↑↓ ↑↓	
					↑↓	

Note that the molecules with two unpaired electrons are stabilized much less than the other molecules because the highest energy electrons are in non-bonding m.o.'s. These molecules will in fact be more stable if they undergo a distortion in which the delocalized molecular orbital is converted into two localized double bonds, as shown below:

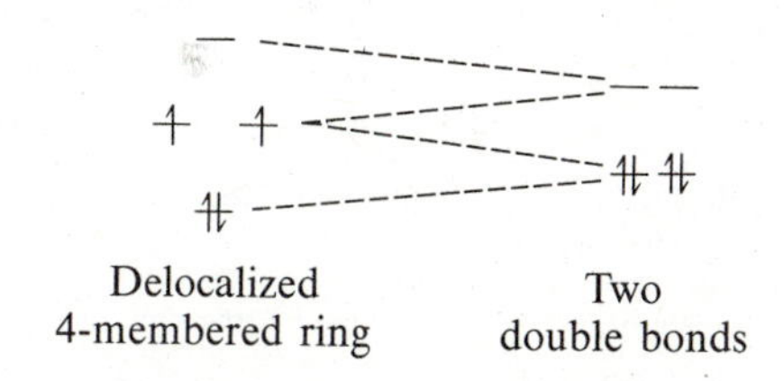

3–11 MORE ON PRODUCT GROUND STATE WAVE FUNCTIONS

The product wave function written in equation (3–32) is not complete. If we write a wave function, interchange of the labels on the electrons 1, 2, . . . n cannot change any physical property associated with the electron density, Ψ^2, of the system. In order for Ψ^2 to be unaffected, the effect of the interchange on Ψ must produce $\pm\Psi$ so that Ψ^2 is unchanged. To demonstrate the problem, we shall define a permutation operator which switches the electron labels. For a two-electron system α and β in orbital ψ_1, we can write

$$\Psi = \psi_1(1)\alpha(1)\psi_1(2)\beta(2) \qquad (3\text{-}37)$$

The permutation operator switches the electron labels in parentheses to:

$$\hat{P}_{12}\Psi(1, 2) = \psi_1(2)\alpha(2)\psi_1(1)\beta(1) \qquad (3\text{-}38)$$

The result is not related to our initial wave function by ± 1.

However, consider the wave function

$$\Psi = \psi_1(1)\alpha(1)\psi_1(2)\beta(2) - \psi_1(2)\alpha(2)\psi_1(1)\beta(1) \qquad (3\text{-}39)$$

Now

$$\hat{P}_{12}\Psi = \psi_1(2)\alpha(2)\psi_1(1)\beta(1) - \psi_1(1)\alpha(1)\psi_1(2)\beta(2) \qquad (3\text{-}40)$$

This result (3–40) is related to our starting wave function by -1. Since we have the result $\hat{P}\Psi = -1\Psi$, we have an *antisymmetrized* wave function. Had we chosen

$$\Psi = \psi_1(1)\alpha(1)\psi_1(2)\beta(2) + \psi_1(2)\alpha(2)\psi_1(1)\beta(1)$$

the result would have been $+1$. This is a *symmetrized* wave function. Later we shall show that the antisymmetrized function must be chosen in order that the Pauli exclusion principle apply.

One can write the correct antisymmetrized wave function for a $2n$-electron system by constructing a Slater determinant and expanding it. The determinant is constructed as follows:

$$\Psi = N\begin{vmatrix} \psi_1(1)\alpha(1) & \psi_1(1)\beta(1) & \psi_2(1)\alpha(1) & \cdots\cdots & \psi_n(1)\beta(1) \\ \psi_1(2)\alpha(2) & \psi_1(2)\beta(2) & ------ & & \\ ----- & ----- & ----- & ----- & ----- \\ \psi_1(2n)\alpha(2n) & \cdots\cdots & & & \psi_n(2n)\beta(2n) \end{vmatrix}$$

Expanding this determinant by the method of cofactors produces the antisymmetrized wave function. Try it for the two-electron system discussed above.

The properties of determinants are consistent with several properties of electrons in molecules. For example, if rows of a determinant are changed, so is the sign. Exchanging rows corresponds to the use of the permutation operator, *e.g.*, equation (3-40). Thus, the antisymmetry property is built into the determinant. Further, there is a theorem stating that if two columns of a determinant are identical, the determinant vanishes. Thus, a non-zero wave function cannot be constructed if two electrons with the same spin are placed in the same orbital.

ADDITIONAL READING REFERENCES

M. J. S. Dewar, "The Molecular Orbital Theory of Organic Chemistry," McGraw-Hill Book Co., New York (1969).

J. M. Anderson, "Introduction to Quantum Chemistry," W. A. Benjamin, Inc., New York (1969).

A. Streitwieser, "Molecular Orbital Theory for Organic Chemists," John Wiley and Sons, Inc., New York (1961).

A. Liberles, "Introduction to Molecular Orbital Theory," Holt, Rinehart and Winston, Inc., New York (1966). (A simple approach to Hückel theory with many worked examples.)

W. Kauzman, "Quantum Chemistry," Academic Press, New York (1957).

J. A. Pople and D. L. Beveridge, "Approximate Molecular Orbital Theory," McGraw-Hill Book Co., New York (1970).

W. G. Richards and J. A. Horsley, "Ab Initio Molecular Orbital Calculations for Chemists," Oxford University Press, New York (1970).

COMPILATIONS

W. G. Richards, T. E. H. Walker and R. K. Hinkley, "A Bibliography of Ab Initio Molecular Wave Functions," Oxford University Press, New York (1971).

W. G. Richards, et al., "Bibliography of Ab Initio Molecular Wave Functions—Supplement for 1970–73," Oxford University Press, New York (1974).

A. Hammett, Specialist Periodical Reports, The Chemical Society, London, England.

H. Hulbronner and C. Straub, "Hückel Molecular Orbitals, Springer-Verlag, New York (1966).

A. Streitwiser and J. Brauman, "Supplemental Tables of Molecular Orbital Calculations," Volumes I and II, Pergamon Press, New York (1965).

EXERCISES

1. Using the Hückel wave functions reported in this chapter, calculate the formal charges on the carbons and the C—C bond orders for
   a. the allyl cation
   b. the allyl radical

2. a. Using projection operators, work out the symmetry-adapted linear combinations of pi orbitals for the allyl radical.
   b. Use these linear combinations to obtain a set of orthonormal molecular orbitals.
   c. Sketch the shapes of the π and π^* molecular orbitals.
   d. Compare your results with those obtained from the Hückel calculation described in this chapter.

3. a. Using projection operators, work out the symmetry-adapted linear combinations of pi orbitals for cyclobutadiene.
   b. Use these linear combinations to obtain a set of orthonormal molecular orbitals.
   c. Sketch the shapes of the π and π^* molecular orbitals.
   d. Would any more information be obtained from a Hückel calculation? If so, what?

4. For the π *system* of borazole, $B_3N_3H_6$, obtain:

(1) B; (2) N; (3) B; (4) N; (5) B; (6) N

a. the total representation

b. all of the wave functions

c. the normalization factor for each wave function; and indicate which wave functions can mix.

5. Propargylene is a linear diradical with the following valence bond structure:

$$\text{H—C}\equiv\text{C—}\ddot{\text{C}}\text{—H} \longrightarrow Z\text{-axis}$$

Using the three $2p_x$ and three $2p_y$ orbitals of the carbons as a basis set,

a. how many m.o.'s will be produced?

b. do a simple Hückel calculation to determine the energies of the m.o.'s from part a.

c. what are the a.o. coefficients of the two lowest energy m.o.'s?

d. calculate the bond order between two adjacent carbons.

6. The ion $Re_2Cl_8^{=}$ is made up of two square planar $ReCl_4$ units joined by a Re—Re bond. The two planes are parallel and the chlorines are eclipsed (as viewed down the Re—Re bond, which we define as the z-axis). The symmetry is therefore D_{4h}.

a. Determine the symmetries of the m.o.'s obtained using the two Re d_{z^2} orbitals as the basis set. *Use projection operators* to determine these m.o.'s and sketch them.

b. Repeat part a using the two $d_{x^2-y^2}$ orbitals as the basis set. These orbitals point directly at the chlorines.

c. Applying this same procedure to the d_{xy} orbitals, one obtains

$$\psi(b_{2g}) = \frac{1}{\sqrt{2}}(d_{xy}{}^{(1)} + d_{xy}{}^{(2)})$$

$$\psi(b_{1u}) = \frac{1}{\sqrt{2}}(d_{xy}{}^{(1)} - d_{xy}{}^{(2)})$$

For d_{xz} and d_{yz}:

$$\psi_1(e_u) = \frac{1}{\sqrt{2}}(d_{xz}{}^{(1)} + d_{xz}{}^{(2)})$$

$$\psi_2(e_u) = \frac{1}{\sqrt{2}}(d_{yz}{}^{(1)} + d_{yz}{}^{(2)})$$

$$\psi_1(e_g) = \frac{1}{\sqrt{2}}(d_{xz}{}^{(1)} - d_{xz}{}^{(2)})$$

$$\psi_2(e_g) = \frac{1}{\sqrt{2}}(d_{yz}{}^{(1)} - d_{yz}{}^{(2)})$$

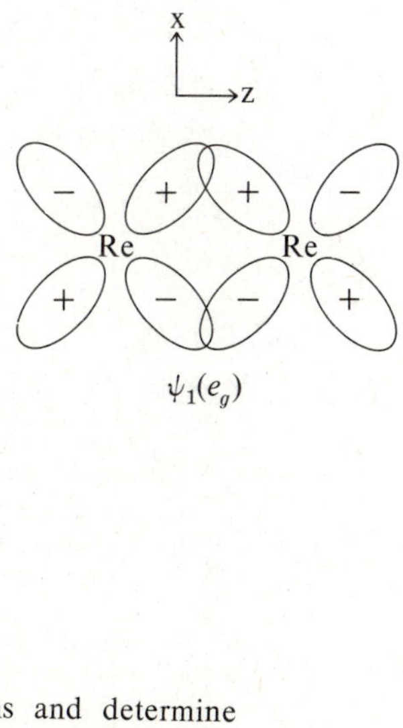

X
Z
− + + −
Re Re
+ − − +
$\psi_1(e_g)$

Sketch a plausible energy level diagram. Fill in the electrons and determine the Re—Re bond order. (Hint: the bonding orbital obtained in part b participates in dsp^2 hybrids that are filled by chlorine electron pairs. The other orbitals must be filled by Re electrons.)

d. Ignore the chlorines and treat the ion as if it belonged to $D_{\infty h}$. What are the symmetries of the ten m.o.'s obtained above? (Note: this illustrates the origin of the terms σ-bond, π-bond, and so forth.)

7. Consider the H_3^+ molecule in both the linear $(\text{H—H—H})^+$ and triangular $(\text{H}_3\ \text{triangle})^+$ states.

a. Utilizing simple Hückel m.o. theory with the hydrogen $1s$ orbitals as a basis set, set up the secular determinant and compute the energies of the m.o.'s for each geometry.

b. From your calculation, predict which geometry would be most stable for H_3^+ and for H_3^-.

c. Consider only the linear geometry again, using the hydrogen 1*s* orbitals as a basis set. How many molecular orbitals will be formed? Substitute the energies found for the m.o.'s in part a into the secular determinant and find the *orthonormal* set of LCAO m.o.'s.

8. a. Two possible structures for the ion I_3^- are

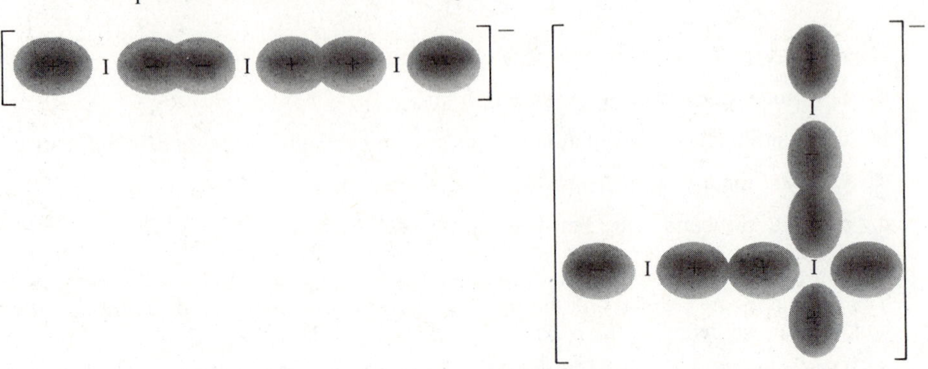

Using a basis set of p_z and p_y orbitals, simple Hückel molecular orbitals can be calculated for the two structures. Ignoring any π bonding, calculate the energies of the three sigma m.o.'s for the linear case, using the three atomic p_z orbitals shown above. (Do not calculate coefficients.)

b. Making the assumptions that $\beta = \int \varphi_i \hat{H}_{\text{bent}} \varphi_j = \int \varphi_i \hat{H}_{\text{linear}} \varphi_j$ and $\alpha = \int \varphi_j \hat{H}_{\text{bent}} \varphi_j = \int \varphi_j \hat{H}_{\text{linear}} \varphi_j$, which structure is more stable if the total energy of the *p*-orbital lone pairs and the m.o.'s in the bent configuration is $16\alpha + 2\beta$? (Hint: There will be six nonbonding *p*-orbital lone pairs, which will contribute 12α to the total energy of the linear structure.)

9. Consider the trigonal bipyramidal $CuCl_5^{3-}$ anion:

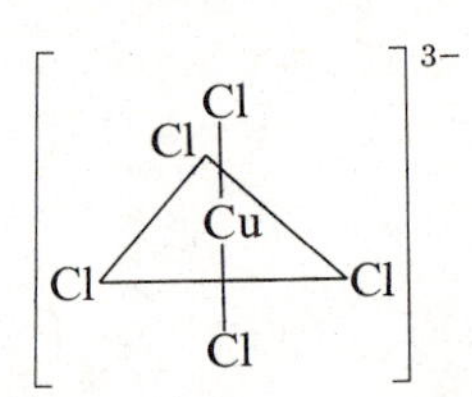

Choose axes at the Cl^- ligands such that z points upward, parallel to C_3. For the in-plane Cl^- nuclei, let x point toward the Cu^{2+}. Thus, from the bottom the xy system for the ligands looks like the following:

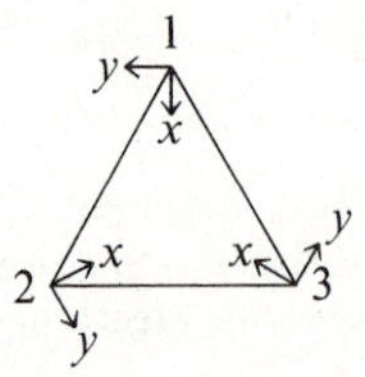

Choose the x and y axes as you wish for the axial Cl^- ligands.

a. Find the total representation of the *p*-orbitals of the five chlorines and decompose it into irreducible representations.

b. Use *a priori* reasoning to determine the effect on an equatorial p_x or p_y orbital of the projection operator $\hat{P}_{\chi''}$, where χ'' denotes any doubly primed irreducible representation. (Hint: What do these irreducible representations have in common?)

c. Operate on an equatorial p_x with $\hat{P}_{E'}$ to find a symmetry-adapted linear combination with E' symmetry.

4 GENERAL INTRODUCTION TO SPECTROSCOPY

4–1 NATURE OF RADIATION

There are many apparently different forms of radiation, *e.g.*, visible light, radio waves, infrared, x-rays, and gamma rays. According to the wave model, all of these kinds of radiation may be described as oscillating electric and magnetic fields. Radiation, traveling in the z direction for example, consists of electric and magnetic fields perpendicular to each other and to the z direction. These fields are represented in Fig. 4–1 for plane-polarized radiation. Polarized radiation was selected for simplicity of representation, since all other components of the electric field except those in the x-z plane have been filtered out. The wave travels in the z direction with the velocity of light, c (3×10^{10} cm sec^{-1}). The intensity of the radiation is proportional to the amplitude of the wave given by the projection on the x- and y-axes. At any given time, the wave has different electric and magnetic field strengths at different points along the z-axis. The wavelength, λ, of the radiation is indicated in Fig. 4–1, and the variation in the magnitude of this quantity accounts for the apparently different forms of radiation listed above. If the radiation consists of only one wavelength, it is said to be monochromatic. Polychromatic radiation can be separated into essentially monochromatic beams. For visible, UV, or IR radiation, prisms or gratings are employed for this purpose.

Radiation consists of energy packets called photons, which travel with the velocity of light. The different forms of radiation have different energies.

In our discussion of rotational, vibrational, and electronic spectroscopy, our

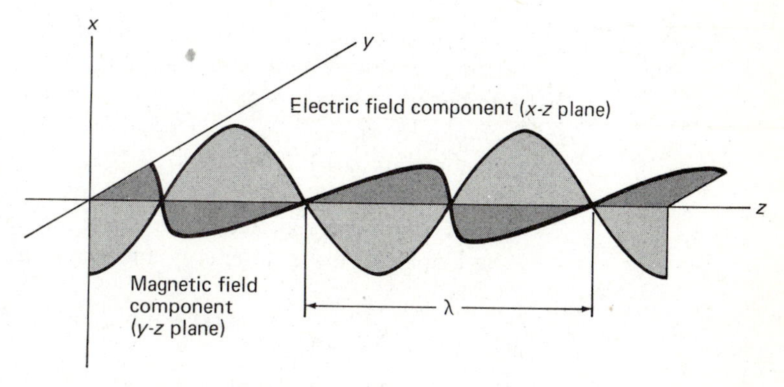

FIGURE 4–1. Electric and magnetic field components of plane polarized electromagnetic radiation.

concern will be with the interaction of the electric field component of radiation with the molecular system. This interaction results in the absorption of radiation by the molecule. In epr and nmr the concern is with the interaction with the magnetic component of radiation.

In order for absorption to occur, the energy of the radiation must match the energy difference between the quantized energy levels that correspond to different states of the molecule. If the energy difference between two of these states is represented by ΔE, the wavelength of the radiation, λ, necessary for matching is given by the equation:

$$\Delta E = hc/\lambda \qquad \text{or} \qquad \lambda = hc/\Delta E \tag{4-1}$$

where h is Planck's constant, 6.623×10^{-27} erg sec molecule^{-1}, and c is the speed of light in cm sec^{-1}, giving ΔE in units of erg molecule^{-1}. Equation (4–1) relates the wave and corpuscular models for radiation. Absorption of one quantum of energy, hc/λ, will raise one molecule to the higher energy state.

As indicated by equation (4–1), the different forms of electromagnetic radiation (*i.e.*, different λ) differ in energy. By considering the energies corresponding to various kinds of radiation and comparing these with the energies corresponding to the different changes in state that a molecule can undergo, an appreciation can be obtained for the different kinds of spectroscopic methods.

4–2 ENERGIES CORRESPONDING TO VARIOUS KINDS OF RADIATION

Radiation can be characterized by its wavelength, λ, its wave number, $\bar{\nu}$, or its frequency, ν. The relationship between these quantities is given by equations (4–2a) and (4–2b):

$$\nu(\text{sec}^{-1}) = \frac{c(\text{cm sec}^{-1})}{\lambda(\text{cm})} \tag{4-2a}$$

$$\bar{\nu}(\text{cm}^{-1}) = \frac{1}{\lambda(\text{cm})} \tag{4-2b}$$

The quantity $\bar{\nu}$ has units of reciprocal centimeters, for which the official IUPAC nomenclature is a Kayser; 1000 cm^{-1} are equal to a kiloKayser (kK). From equations (4–1) and (4–2), the relationship of energy to frequency, wave number, and wavelength is:

$$\Delta E(\text{ergs molecule}^{-1}) = h\nu = hc/\lambda = hc\bar{\nu} \tag{4-3}$$

In describing an absorption band, one commonly finds several different units being employed by different authors. Wave numbers, $\bar{\nu}$, which are most commonly employed, have units of cm^{-1} and are defined by equation (4–2b). Various units are employed for λ. These are related as follows: 1 cm = 10^8 Å (Ångstroms) = 10^7 nanometers = 10^4 μ (microns) = 10^7 mμ (millimicrons). The relationship to various common energy units is given by: 1 cm^{-1} = 2.858 cal/mole of particles = 1.986×10^{-16} erg/molecule = 1.24×10^{-4} eV/mole. These conversion units can be employed to relate energy and wavelength; or the equation

$$\Delta E(\text{kcal mole}^{-1}) \times \lambda(\text{Å}) = 2.858 \times 10^5 \tag{4-4}$$

can be derived to simplify the calculation of energy from wavelength.

Wave numbers corresponding to various types of radiation are indicated in Fig. 4–2. The small region of the total spectrum occupied by the visible portion is demonstrated by this figure. The higher energy radiation has the smaller wavelength and the larger frequency and wave number [equation (4–3)]. The following sequence represents decreasing energy:

$$\text{ultraviolet} > \text{visible} > \text{infrared} > \text{microwave} > \text{radio-frequency}$$

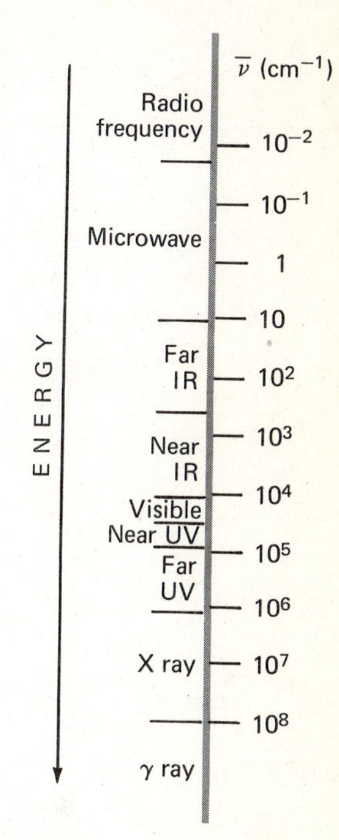

FIGURE 4–2. Wave numbers of various types of radiation.

4–3 ATOMIC AND MOLECULAR TRANSITIONS

In an atom, the change in state induced by the quantized absorption of radiation can be regarded as the excitation of an electron from one energy state to another. The change in state is from the ground state to an excited state. In most cases, the energy required for such excitation is in the range from 60 to 150 kcal mole^{-1}. Calculation employing equation (4–3) readily shows that radiation in the ultraviolet and visible regions will be involved. Atomic spectra are often examined as emission spectra. Electrons are excited to higher states by thermal or electrical energy, and the energy emitted as the atoms return to the ground state is measured.

In molecular spectroscopy, absorption of energy is usually measured. Our concern will be with three types of molecular transitions induced by electromagnetic radiation: electronic, vibrational, and rotational. A change in *electronic state* of a molecule occurs when a bonding or nonbonding electron of the molecule in the ground state is excited into a higher-energy empty molecular orbital. For example, an electron in a π-bonding orbital of a carbonyl group can be excited into a π^* orbital, producing an excited state with configuration $\sigma^2\pi^1\pi^{*1}$. The electron distributions in the two states (ground and excited) involved in an electronic transition are different.

The *vibrational energy states* are characterized by the directions, frequencies, and amplitudes of the motions that the atoms in a molecule undergo. As an example, two different kinds of vibrations for the SO_2 molecule are illustrated in Fig. 4–3. The

FIGURE 4–3. Two different vibrations for the SO_2 molecule. (The amplitudes and angle deformations are exaggerated to illustrate the mode.)

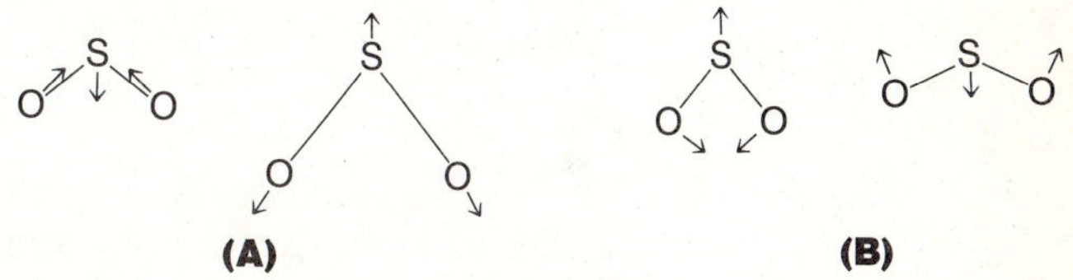

(A) **(B)**

atoms in the molecule vibrate (relative to their center of mass) in the directions indicated by the arrows, and the two extremes in each vibrational mode are indicated. In the vibration indicated in (a), the sulfur-oxygen bond length is varying, and this is referred to as the *stretching vibration.* In (b), the motion is perpendicular to the bond axis and the bond length is essentially constant. This is referred to as a *bending vibration.* In these vibrations, the net effect of all atomic motion is to preserve the center of mass of the molecule so that there will be no net translational motion. The vibrations indicated in Fig. 4–3 are drawn to satisfy this requirement. Certain vibrations in a molecule are referred to as *normal vibrations* or *normal modes.* These are independent, self-repeating displacements of the atoms that preserve the center of mass. In a normal vibration, all the atoms vibrate in phase and with the same frequency. It is possible to resolve the most complex molecular vibration into a relatively small number of such normal modes. For a non-linear molecule, there are $3N - 6$ such modes, where N is the number of atoms in the molecule. In Chapter 6, the procedure for calculating the total number of normal modes and their symmetries will be given. The normal modes can be considered as the $3N - 6$ internal degrees of freedom that (in the absence of anharmonicity) could take up energy independently of each other. The motion of the atoms of a molecule in the different normal modes can be described by a set of *normal coordinates.* These are a set of coordinates defined so as to describe the normal vibration most simply. They often are complicated functions of angles and distances.

The products of the normal modes of vibration are related to the total vibrational state, ψ_v, of a non-linear molecule as shown in equation (4–5).

$$\psi_v = \prod_{n=1}^{3N-6} \psi_n \tag{4-5}$$

where Π indicates that the product of the n vibrational modes is to be taken and ψ_n is the wave function for a given normal mode. There exists for each normal vibration (ψ_n) a whole series of excited vibrational states, i, whose harmonic oscillator wave functions are given by:

$$\psi_i = N_i \exp\left[\left(-\frac{1}{2}\right)a_i q^2\right] H_i(\sqrt{a_i}q) \tag{4-6}$$

where $i = 0, 1, 2 \ldots$; H_i is the Hermite polynomial of degree i; $a_i = 2\pi\nu_i/h$; $N_i = [\sqrt{a_i}/(2^i i! \sqrt{\pi})]^{1/2}$; and q is the normal coordinate.

From this, the following wave functions are obtained for the ground, first, and second excited states:

$$\psi_0 = \left(\frac{a_0}{\pi}\right)^{1/4} \exp\left[\left(-\frac{1}{2}\right)a_0 q^2\right] \tag{4-7}$$

$$\psi_1 = 2\left(\frac{a_1}{\pi}\right)^{1/4} q \exp\left[\left(-\frac{1}{2}\right)a_1 q^2\right] \tag{4-8}$$

$$\psi_2 = 2\left(\frac{a_2}{\pi}\right)^{1/4} (2a_2 q^2 - 1) \exp\left[\left(-\frac{1}{2}\right)a_2 q^2\right] \tag{4-9}$$

When ψ_0 is plotted as a function of displacements about the normal coordinate, q, with zero taken as the equilibrium internuclear distance, Fig. 4–4 is obtained. This

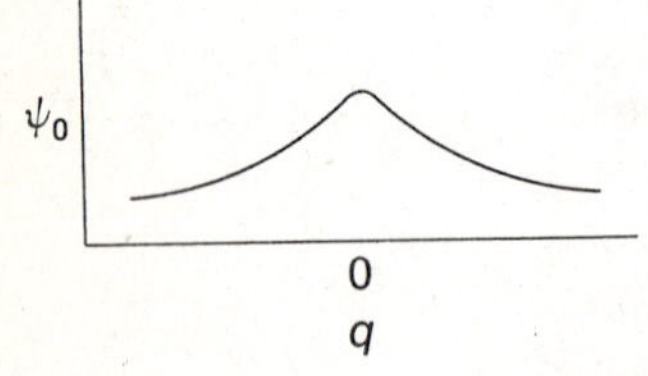

FIGURE 4–4. A plot of the ground vibrational wave function versus the normal coordinate.

symmetric function results because q appears only as q^2. The same type of plot is obtained for the ground state of all normal modes. It is totally symmetric; accordingly, the total vibrational ground state of a molecule must belong to the totally symmetric irreducible representation, for it is the product of only totally symmetric vibrational wave functions.

The vibrational excited state function ψ_1 has a functional dependence on q that is not of an even power, and accordingly it does not necessarily have a_1 symmetry. When one normal mode is excited in a vibrational transition, the resulting total state is a product of all the other totally symmetric wave functions and the wave function for this first vibrational excited state for the excited normal mode. Thus, the total vibrational state has symmetry corresponding to the normal mode excited.

The *rotational states* correspond to quantized molecular rotation around an axis without any appreciable change in bond lengths or angles. Different rotational states correspond to different angular momenta of rotation or to rotations about different axes. Rotation about the C_2 axis in SO_2 is an example of rotational motion.

In the treatment of molecular spectra, the *Born-Oppenheimer approximation* is invoked. This approximation proposes that the total energy of a system may be regarded as the sum of three independent energies: electronic, vibrational, and rotational. For example, the electronic energy of the system does not change as vibration of the nuclei occurs. The wave function for a given molecular state can then be described by the product of three independent wave functions: ψ_{el}, ψ_{vib}, and ψ_{rot}. As we shall see later, this approximation is not absolutely valid.

The relative energies of these different molecular energy states in typical molecules are represented in Fig. 4-5. Rotational energy states are more closely spaced

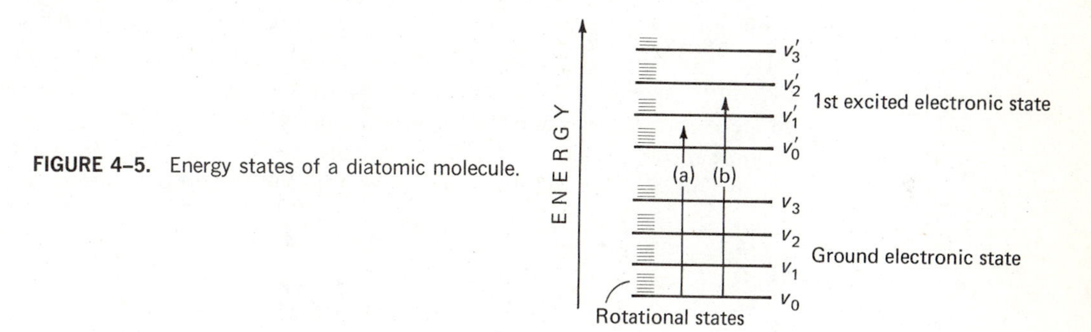

FIGURE 4-5. Energy states of a diatomic molecule.

than are vibrational states, which, in turn, have smaller energy differences than electronic states. The letters ν_0, ν_1, etc., and ν_0', ν_1', etc., represent the vibrational levels of one vibrational mode in the ground and first electronic excited states, respectively. *Ultraviolet or visible radiation is commonly required to excite the molecule into the excited electronic states. Lower-energy infrared suffices for vibrational transitions, while pure rotational transitions are observed in the still lower-energy microwave and radio-frequency regions.*

Electronic transitions are usually accompanied by changes in vibration and rotation. Two such transitions are indicated by arrows (a) and (b) of Fig. 4-5. In the vibrational spectrum, transitions to different rotational levels also occur. As a result, vibrational fine structure is often detected in electronic transitions. Rotational fine structure in electronic transitions can be detected in high resolution work in the gas phase. Rotational fine structure in vibrational transitions is sometimes observed in the liquid state and generally in the gaseous state.

The energy level diagram in Fig. 4-5 is that for a diatomic molecule. For a polyatomic molecule, the individual observed transitions can often be described by diagrams of this type, each transition in effect being described by a different diagram.

4-4 SELECTION RULES

In order for matter to absorb the electric field component of radiation, another requirement in addition to energy matching must be met. The energy transition in the molecule must be accompanied by a change in the electrical center of the molecule in order that electrical work can be done on the molecule by the electromagnetic radiation field. Only if this condition is satisfied can absorption occur. Requirements for the absorption of light by matter are summarized in the *selection rules*. Transitions that are possible according to these rules are referred to as *allowed* transitions, and those not possible as *forbidden* transitions. It should be noted, however, that the term "forbidden" refers to rules set up for a simple model and, while the model is a good one, "forbidden" transitions may occur by mechanisms not included in the simple model. The intensity of absorption or emission accompanying a transition is related to the probability of the transition, the more probable transitions giving rise to more intense absorption. Forbidden transitions have low probability and give absorptions of very low intensity. This topic and the symmetry aspects of this topic will be treated in detail when we discuss the various spectroscopic methods.

4-5 CHEMICAL PROCESSES AFFECTING THE NATURAL LINE WIDTH OF A SPECTRAL LINE

When two or more chemically distinct species coexist in rapid equilibrium (two conformations, or rapid chemical exchange, *e.g.*, proton exchange between NH_3 and NH_4^+, or other equilibria), one often sees absorption peaks corresponding to the individual species in some forms of spectroscopy; but with other methods, only a single average peak is detected. A process may cause a broadening of the spectral line in some spectral regions, while the same process in some other spectral region has no effect. This behavior can be understood from the Uncertainty Principle, which states

$$\Delta E \Delta t \sim \hbar$$

or

$$\Delta \nu \Delta t \sim \frac{1}{2\pi} \tag{4-10}$$

Consider the case in which two different sites give rise to two distinct peaks. As the rate of the chemical exchange comes close to the frequency $\Delta\nu$ of the spectroscopic method, the two peaks begin to broaden. As the rate becomes faster, they move together, merge, and then sharpen into a single peak. In subsequent chapters, we shall discuss procedures for extracting rate data over this entire range. Here we shall attempt to gain an appreciation for the time scales corresponding to the various methods by examining the rates that result in line broadening of a sharp single resonance and merging of the two distinct resonances.

First, we shall be concerned with the broadening of a resonance line of an individual species. If Δt is the lifetime of the excited state, we have:

$$\Delta\nu(\text{sec}^{-1}) \sim \frac{1}{2\pi\Delta t(\text{sec})}$$

If there is some chemical or physical process that is fast compared to the excited state lifetime, Δt can be shortened by this process and the line will be broadened. In infrared, for example, it is possible to resolve* two bands corresponding to different sites that are separated by 0.1 cm^{-1}, so we can substitute this value into equation (4-10) and find that lifetimes, Δt, corresponding to this are given by:

$$\Delta\nu = 0.1\ \text{cm}^{-1} \times 3 \times 10^{10}\ \text{cm/sec} = 3 \times 10^{9}\ \text{sec}^{-1}$$

$$\Delta t = \frac{1}{(2)(\pi)(3 \times 10^{9}\ \text{sec}^{-1})} = 5 \times 10^{-11}\ \text{sec}$$

---

*By resolve, we mean that one can detect the existence of two distinct maxima. We are using this quantity to roughly estimate $\Delta\nu$, the time scale of the spectroscopic method (that is, the natural line width).

Therefore, we need a process that will give rise to lifetimes of $\sim 10^{-11}$ sec or less to cause broadening. Since lifetimes are the reciprocals of first order rate constants, this means we need a process whose rate constant is at least 10^{11} sec^{-1} in order to detect a broadening in the infrared band. Diffusion-controlled chemical reactions have rate constants of only 10^{10}, so for systems undergoing chemical exchange, the chemical process will have no influence on the infrared line shape. Rotational motion occurs on a time scale that causes broadening of an infrared or Raman line, and it is possible to obtain information about the motion from the line shape.[1] The inversion doubling of ammonia is also fast enough to affect the Raman and microwave spectra.

In nmr, typical resolution is ~ 0.1 Hz (cycle per second), which, when substituted into the above equation, shows that a process with a lifetime of about 2 sec (or less) or a rate constant of about 0.5 sec^{-1} (or more) is needed to broaden the spectral line. This is in the range of many chemical exchange reactions.

Next, we shall consider the rate at which processes must occur to result in a spectrum in which only an average line is detected for two species. In order for this to occur, the species must be interconverting so fast that the lifetime in either one of the states is less than Δt, so that only an average line results. This is calculated by substituting the difference between the frequencies of the two states for $\Delta\nu$ in equation (4–10). In the infrared, a rate constant of 5×10^{13} sec^{-1} or greater would be required to merge two peaks that are separated by 300 cm^{-1}.

$$\Delta t = \frac{1}{(2)(\pi)(9.0 \times 10^{12}\ \text{sec}^{-1})} = 2 \times 10^{-14}\ \text{sec}$$

In nmr, for a system in which, for example, the two proton peaks are separated by 100 Hz, one obtains

$$\Delta t = \frac{1}{(2)(\pi)(100\ \text{sec}^{-1})} = 2 \times 10^{-3}\ \text{sec}$$

Thus, an exchange process involving this proton that had a rate constant of 5×10^2 sec^{-1} would cause these two resonances to appear as one broad line. As the rate constant becomes much larger than 5×10^2 (*e.g.*, by raising the temperature for a positive activation enthalpy process), the broadened single resonance begins to sharpen. Eventually, as the process becomes very fast, the line width is no longer influenced by the chemical process. The radiation is too slow to detect any chemical changes that are occurring and a sharp average line results. This is analogous to the eye being too slow to detect the electronic sweep that produces a television picture.

In x-ray diffraction, the frequency of the radiation is 10^{18} sec^{-1}. Since this is so fast, compared with molecular rearrangements, all we would detect for a dynamic system would be disorder, *i.e.*, contributions from each of the dominant conformations.

In Mössbauer spectroscopy, we observe a very high energy process in which a nucleus in the sample absorbs a γ-ray from the source. There are no chemical processes that affect the lifetime of the nuclear excited state, which for iron is 10^{-7} sec. Thus, if there is a chemical process occurring in an iron sample that equilibrates two iron atoms with a first order rate constant greater than 10^7 sec^{-1}, the Mössbauer spectrum will reveal only an average peak. In dichlorobisphenanthroline iron (III), the equilibrium mixture of high spin (five unpaired electrons) and low spin (one unpaired electron) iron complexes interconvert so rapidly (they have a lifetime $< 10^{-7}$ sec) that only a single iron species is observed in the spectrum. For a ruthenium nucleus, the lifetime of the nuclear excited state is 10^{-9} sec; this determines the time scale for experiments with this nucleus. A similar situation pertains in x-ray photoelectron

TABLE 4–1. KINETIC TECHNIQUES AND THE APPROPRIATE LIFETIMES*

conventional kinetics	10 seconds or longer (determined by ability to mix reagents)
stop flow	10^{-3} sec or longer (faster mixing is possible)
nmr	10^{-1} to 10^{-5} sec
esr	10^{-4} to 10^{-8} sec
Temperature jump	0.1 to 10^{-6} sec
Mössbauer (iron)	10^{-7} sec
ultrasonics	10^{-4} to 10^{-8} sec
fluorescent polarization (the depolarization is measured)	10^{-8} to 10^{-9} sec
IR and Raman line shapes	10^{-11} sec
photoelectron spectroscopy	10^{-18} sec

* If some of these techniques are unfamiliar at present, do not be concerned. This will provide a ready reference after you encounter them in your study.

spectroscopy, in that the time scale for ejection of an electron cannot be influenced by chemical processes. The lifetime for the resulting excited state is about 10^{-18} sec.

Table 4–1 summarizes the methods used in kinetic studies and the ranges of lifetimes that can be investigated.

GENERAL APPLICATIONS

The following general applications of spectroscopy are elementary and pertain to both vibrational and electronic spectroscopic methods: (1) determination of concentration, (2) "finger-printing," and (3) determination of the number of species in solution by the use of isosbestic points.

4–6 DETERMINATION OF CONCENTRATION

Measurement of the concentrations of species has several important applications. If the system is measured at equilibrium, equilibrium constants can be determined. By evaluating the equilibrium constant, K, at several temperatures, the enthalpy $\Delta H°$ for the equilibrium reaction can be calculated from the van't Hoff equation:

$$\log K = \frac{-\Delta H^0}{2.3RT} + C \tag{4-11}$$

Determination of the change in concentration of materials with time is the basis of kinetic studies that give information about reaction mechanisms. In view of the contribution of results from equilibrium and kinetic studies to our understanding of chemical reactivity, the determination of concentrations by spectroscopic methods will be discussed.

The relationship between the amount of light absorbed by certain systems and the concentration of the absorbing species is expressed by the Beer-Lambert law:

$$A = \log_{10}\frac{I_0}{I} = \varepsilon cb \tag{4-12}$$

where A is the absorbance, I_0 is the intensity of the incident light, I is the intensity of the transmitted light, ε is the molar absorptivity (sometimes called extinction coefficient) at a given wavelength and temperature, c is the concentration (the molarity if ε is the molar absorptivity), and b is the length of the absorbing system. The molar absorptivity varies with both wavelength and temperature, so these must be held constant when using equation (4–12). When matched cells are employed to eliminate scattering of the incident beam, there are no exceptions to the relationship between absorbance and b (the Lambert law). For a given concentration of a certain substance, the absorbance is always directly proportional to the length of the cell. The part of equation (4–12) relating absorbance and concentration ($\log_{10} I_0/I = \varepsilon c$, for a constant cell length) is referred to as Beer's law. Many systems have been found that do not obey Beer's law. The anomalies can be attributed to changes in the composition of the system with concentration (*e.g.*, different degrees of ionization or dissociation of a solute at different concentrations). For all systems, the Beer's law relationship must be demonstrated, rather than assumed, over the entire concentration range to be considered. If Beer's law is obeyed, it becomes a simple matter to use equation (4–12) for the determination of the concentration of a known substance if it is the only material present that is absorbing in a particular region of the spectrum. The Beer's law relationship is tested and ε determined by measuring the absorbances of several solutions of different known concentrations covering the range to be considered. For each solution, ε can be calculated from:

$$\varepsilon = \frac{A(1\text{ cm cell})}{c,\ \text{molarity}} \tag{4-13}$$

where the units of ε are liters mole^{-1} cm^{-1}. For a solution of this material of unknown concentration, the absorbance is measured, ε is known (4–13), and c is calculated from equation (4–12).

There are some interesting variations on the application of Beer's law. If the absorption of two species should overlap, this overlap can be resolved mathematically and the concentrations determined. This is possible as long as the two ε values are not identical at all wavelengths. Consider the case in which the ε values for two compounds whose spectra overlap can be measured for the pure compounds. The concentration of each component in a mixture of the two compounds can be obtained by measuring the absorbance at two different wavelengths, one at which both compounds absorb strongly and a second at which there is a large difference in the absorptions. Both wavelengths should be selected at reasonably flat regions of the absorption curves of the pure compounds, if possible. Consider two such species B and C, and wavelengths λ_1 and λ_2. There are molar absorptivities of $\varepsilon_{B_{\lambda_1}}$ for B at λ_1, $\varepsilon_{B_{\lambda_2}}$ for B at λ_2, and similar quantities $\varepsilon_{C_{\lambda_1}}$ and $\varepsilon_{C_{\lambda_2}}$ for C. The total absorbance of the mixture at λ_1 is A_1, and that at λ_2 is A_2. It follows that:

$$A_1 = x\varepsilon_{B_{\lambda_1}} + y\varepsilon_{C_{\lambda_1}} \tag{4-14}$$

and

$$A_2 = x\varepsilon_{B_{\lambda_2}} + y\varepsilon_{C_{\lambda_2}} \tag{4-15}$$

where x = molarity of B and y = molarity of C. The two simultaneous equations (4–14) and (4–15) have only two unknowns, x and y, and can be solved.

A situation often encountered in practice involves an equilibrium of the type

$$D + E \rightleftharpoons DE$$

Here the spectra of, say, D and DE overlap but E does not absorb; if the equilibrium constant is small, pure DE cannot be obtained, so its molar absorptivity cannot be determined directly. It is possible to solve this problem for the equilibrium concentrations of all species, *i,e.*, to determine the equilibrium constant:

$$K = \frac{[DE]}{[D][E]} \tag{4-16}$$

Let $[D]_0$ be the initial concentration of D. This can be measured, as can $[E]_0$, the initial concentration of E. The molar absorptivity of DE, ε_{DE}, cannot be determined directly, but it is assumed that Beer's law is obeyed. It can be seen from a material balance that:

$$[D] = [D]_0 - [DE] \tag{4-17}$$

$$[E] = [E]_0 - [DE] \tag{4-18}$$

so

$$K = \frac{[DE]}{([D]_0 - [DE])([E]_0 - [DE])} \tag{4-19}$$

The total absorbance consists of contributions from [D] and [DE] according to

$$A = \varepsilon_D[D] + \varepsilon_{DE}[DE] \tag{4-20}$$

Substituting equation (4–17) into (4–20) and solving for [DE], one obtains

$$[DE] = \frac{A - \varepsilon_D[D]_0}{\varepsilon_{DE} - \varepsilon_D}$$

but $\varepsilon_D[D]_0$ is the initial absorbance, A^0, of a solution of D with concentration $[D]_0$ without any E in it, so

$$[DE] = \frac{A - A^0}{\varepsilon_{DE} - \varepsilon_D} \tag{4-21}$$

When equation (4–21) is substituted into the equilibrium constant expression, equation (4–19), and this is rearranged, we have[2]

$$K^{-1} = \frac{A - A^0}{\varepsilon_{DE} - \varepsilon_D} - [D]_0 - [E]_0 + \frac{[D]_0[E]_0(\varepsilon_{DE} - \varepsilon_D)}{A - A^0} \tag{4-22}$$

The advantage of this equation is that it contains only two unknown quantities, ε_{DE} and K^{-1}. Furthermore, these unknowns are constant for any solution of different concentrations $[D]_0$ and $[E]_0$ that we care to make up. For two different sets of experimental conditions (different values of $[D]_0$ and $[E]_0$) two simultaneous equations can be solved; ε_{DE} is eliminated, and K is obtained. If several sets of experimental conditions are employed, all possible combinations of simultaneous equations can be considered by employing a least-squares computer analysis[3-7] that finds the best values of K^{-1} and ε_{DE} to reproduce the experimental absorbances. These least-squares procedures usually provide an error analysis that is most important to have (*vide infra*). There are several advantages to a graphical display of the simultaneous equations that are being analyzed. The graph is constructed[2] by taking a solution of known $[D]_0$, $[E]_0$, and A, and calculating the values of K^{-1} that would result by selecting several different values of ε_{DE} near the expected value. These results are plotted, as in Fig. 4–6, by the line $[D]_0[E]_0{}^*$. The procedure is repeated for other initial concentrations, *e.g.*, $[D]_0'[E]_0'$ and $[D]_0''[E]_0''$. The intersection of any two of

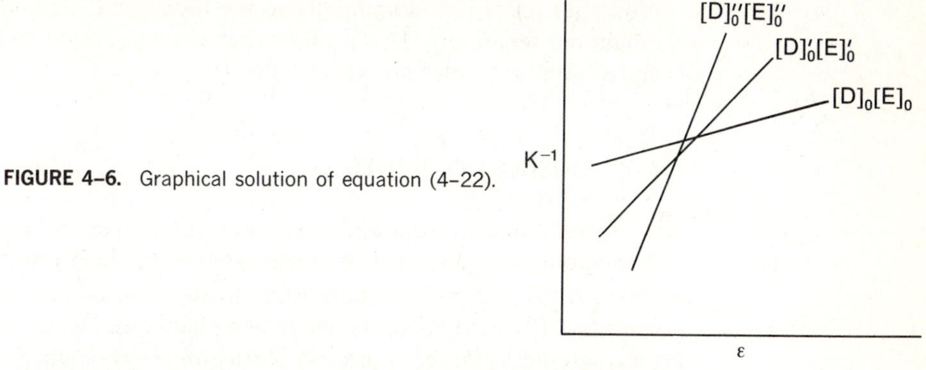

FIGURE 4–6. Graphical solution of equation (4–22).

these lines is a graphical representation of the solution of two simultaneous equations. The intersection of all the calculated curves should occur at a point whose values of K^{-1} and ε satisfy all the experimental data. This common intersection justifies the Beer's law assumption used in the derivation because it indicates a unique value for ε for all concentrations. As a result of experimental error, a triangle usually results instead of a point. The best K^{-1} and ε to fit the data are then found by the least-squares procedure. References 6 and 7 describe the least-squares procedure and error analysis. When the experiment described above is not designed properly, a set of concentrations of D and E are employed that result in a set of parallel lines for the K^{-1} *vs.* ε plots. The slope of one of these lines is given by taking the partial derivative of the K^{-1} expression in (4–22) with respect to $\varepsilon_{DE} - \varepsilon_D$:

$$\frac{\partial K^{-1}}{\partial \varepsilon_{DE} - \varepsilon_D} = -\frac{A - A^0}{(\varepsilon_{DE} - \varepsilon_D)^2} + \frac{[D]_0[E]_0}{A - A^0} \qquad (4\text{-}23)$$

Since the first term is generally small, we see from equation (4–23) that if experimental conditions are selected for a series of experiments in which [DE] or the value of A nearly doubles every time $[D]_0$ or $[E]_0$ is doubled, then the various resulting K^{-1} *vs.* ε plots will be nearly parallel. The values of K^{-1} and ε obtained from the computer analysis of this kind of data[6] are thus highly correlated and should be considered undefined, for the simultaneous equations are essentially dependent ones. The K^{-1} *vs.* ε plots have been described in terms of the results from the error analysis of the least-squares calculation, and the reader is referred to reference 7 for details.

The approach described above for the evaluation of equilibrium constants is a general one that applies to any form of spectroscopy or any kind of measurement in which the measured quantity is linearly related to concentration (*i.e.*, in which a counterpart to equation (4–20) exists). Similar analyses have been described for calorimetric data[8] and for nmr spectral data.[7]

There have been many reports in the literature of attempts to solve simultaneous equations for equilibrium constants for two or more consecutive equilibria:

$$A + B \rightleftharpoons AB \qquad \text{step 1}$$

$$AB + B \rightleftharpoons AB_2 \qquad \text{step 2}$$

Usually ε_{AB}, ε_{AB_2}, and the stepwise equilibrium constants K_1 and K_2 are unknown. In most instances, even though the reported parameters fit the experimental data well, the system is undefined. Careful examination often shows that many other, very different combinations of parameters also fit the data. To solve this problem, one must find a region of the spectrum in which AB makes the main contribution to the absorbance and another in which AB_2 makes the predominant contribution. Only by working at both wavelengths can one solve for all the unknowns in a rigorous fashion. This is impossible in many nmr, uv-visible, and calorimetric experiments because properties related predominantly to the individual AB and AB_2 species cannot be monitored separately. This problem has been discussed in detail in the literature[4] and several examples are given there.

4-7 ISOSBESTIC POINTS

If two substances at equal concentrations have absorption bands that overlap, there will be some wavelength at which the molar absorptivities of the two species are equal. If the sum of the concentrations of these two materials in solution is held constant, there will be no change in absorbance at this wavelength as the ratio of the two materials is varied. For example, assume a reaction $A + B \longrightarrow AB$, in which only A and AB absorb; the sum of [A] plus [AB] will be constant as long as the initial concentration of A is held constant as B is varied. Since the absorbance of the solution is given by

$$\text{abs} = \varepsilon_A[A] + \varepsilon_{AB}[AB] \quad (4\text{-}24)$$

the absorbance will be constant when $\varepsilon_A = \varepsilon_{AB}$ and [A] plus [AB] is held constant. The invariant point obtained for this system is referred to as the *isosbestic point*. The existence of one or more isosbestic points in a system provides information regarding the number of species present. For example, the spectra in Fig. 4-7 were obtained[9] by keeping the total iron concentration constant in a system consisting of:

Curve 1 2.1 moles of LiCl per $Fe(DMA)_6(ClO_4)_3$
Curve 2 2.6 moles of LiCl per $Fe(DMA)_6(ClO_4)_3$
Curve 3 3.1 moles of LiCl per $Fe(DMA)_6(ClO_4)_3$
Curve 4 4.1 moles of LiCl per $Fe(DMA)_6(ClO_4)_3$

where DMA is the abbreviation for *N,N*-dimethylacetamide. Points *A* and *B* are isosbestic points; their existence suggests that the absorption in this region is essentially accounted for by two species. Curve 4 is characteristic of $FeCl_4^-$. A study of solutions more dilute in chloride establishes the existence of a species $Fe(DMA)_4Cl_2^+$, which exists at a 2:1 ratio of Cl^- to Fe^{III} and absorbs in this region. The isosbestic points indicate that the system can be described by the species $Fe(DMA)_4Cl_2^+$ and $FeCl_4^-$ over the region from 2:1 to 4:1 ratios of Cl^- to Fe^{III}. The addition compound $FeCl_3 \cdot DMA$ probably does not exist in appreciable concentration in this system, *i.e.*, at total Fe^{III} concentrations of $2 \times 10^{-4}M$. Curve 4 in this spectrum does not pass through the isosbestic points. Small deviations (*e.g.*, curve 4 at point *A*) may be due to experimental inaccuracy, changes in solvent properties in different solutions, or a small concentration of a third species, probably $FeCl_3 \cdot DMA$, present in all systems except that represented by curve 4.

The conclusion that only two species are present in appreciable concentrations could be in error if $FeCl_3 \cdot DMA$ or other species were present that had molar absorptivities identical to that of the above two ions at the isosbestic points. However, if equilibrium constants for the equilibrium

$$Fe(DMA)_4Cl_2^+ + 2Cl^- \rightleftharpoons FeCl_4^- + 4DMA$$

are calculated from these different curves at different wavelengths, the possibility of a third species is eliminated if the constants agree.

An interesting situation results in which an isosbestic point can be obtained in solution when more than two species with differing extinction coefficients exist if the base (or acid) employed has two donor (or acceptor) sites. For example, if DMA were

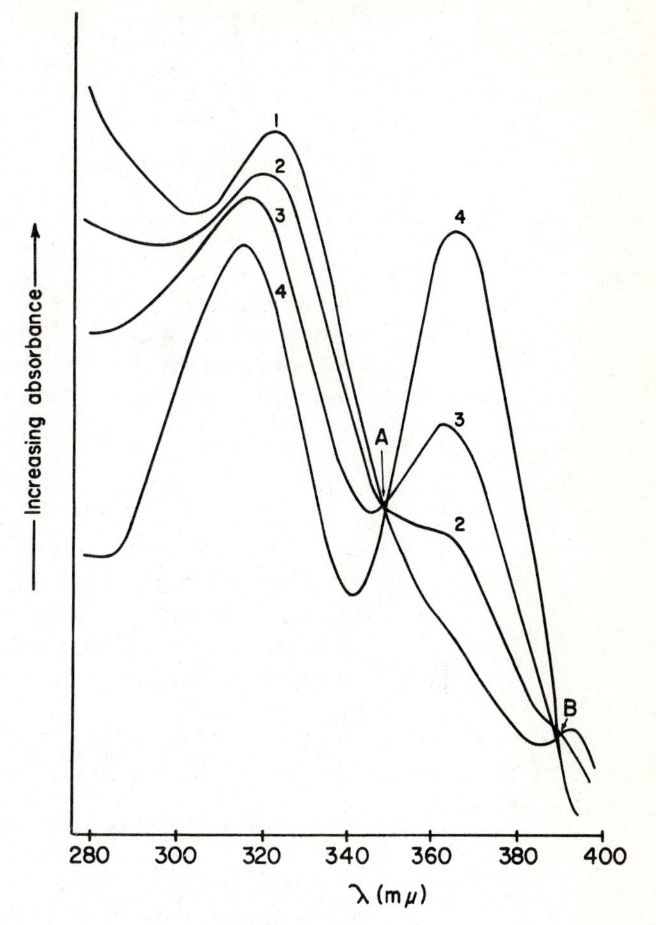

FIGURE 4-7. Spectra of the $Fe(DMA)_4Cl_2^+$–LiCl system in *N,N*-dimethylacetamide as solvent.

to coordinate to an acid A to produce an oxygen-bound complex and a nitrogen-bound one, the mixture of complexes AN (nitrogen bound), AO (oxygen bound), and free A (an absorbing Lewis acid) will give rise to an isosbestic point.[10] The absorbance for such a system is given by equation (4–25):

$$\text{abs} = \varepsilon_A[\text{A}] + \varepsilon_{AO}[\text{AO}] + \varepsilon_{AN}[\text{AN}] \tag{4-25}$$

The equilibrium constant expressions are given by

$$K_O = \frac{[\text{AO}]}{[\text{A}][\text{B}]} \quad \text{and} \quad K_N = \frac{[\text{AN}]}{[\text{A}][\text{B}]}$$

$$K_O + K_N = \frac{[\text{AO}] + [\text{AN}]}{[\text{A}][\text{B}]} = \frac{[\text{AB}]}{[\text{A}][\text{B}]} \tag{4-26}$$

where we define [AO] + [AN] = [AB]. The fraction of complex that is oxygen-coordinated, X_{AO}, is given by

$$x_{AO} = \frac{K_O}{K_O + K_N} = \frac{\dfrac{[\text{AO}]}{[\text{A}][\text{B}]}}{\dfrac{[\text{AB}]}{[\text{A}][\text{B}]}} = \frac{[\text{AO}]}{[\text{AB}]} \tag{4-27}$$

The fraction that is nitrogen-coordinated is similarly derived as:

$$X_{AN} = \frac{[AN]}{[AB]} \tag{4-28}$$

Now the total absorbance, abs, becomes

$$abs = \varepsilon_A[A] + \varepsilon_{AO}X_{AO}[AB] + \varepsilon_{AN}X_{AN}[AB]$$

or

$$abs = \varepsilon_A[A] + (\varepsilon_{AO}X_{AO} + \varepsilon_{AN}X_{AN})[AB] = \varepsilon_A[A] + \varepsilon'[AB] \tag{4-29}$$

Since the sum of [A] and [AB] is constant, and there must be a point in the overlapping spectra at which $\varepsilon = \varepsilon'$, an isosbestic point will be obtained even though three absorbing species exist. This is true because

$$\frac{X_{AN}}{X_{AO}} = \frac{K_N}{K_O} = \text{a constant}$$

We have taken two absorbing species whose ratio is a constant, independent of the parameter being varied, and we have translated them into what is effectively a single absorbing species *via* equation (4–29). The general criterion is thus that $2 + N$ absorbing species will give rise to an isosbestic point if there are N *independent* equations of the form

$$\frac{[Y]}{[Z]} = k \tag{4-30}$$

where Y and Z are two of the absorbing species in the system and the value of k is independent of the parameter being varied.

As an example of the application of this criterion, consider an acid that can form two isomers with a base, AB and A*B (*e.g.*, $Cu(hfac)_2$ forming basal and apical adducts); let this acid form an adduct with a base that can form two isomers with an acid, AB and AB* (*e.g.*, *N*-methyl imidazole bound through the amine and imine nitrogens). Five absorbing species (A, AB, A*B, AB*, and A*B*) can exist,

$$A + B \longrightarrow AB$$

$$A + B \longrightarrow A^*B$$

$$A + B \longrightarrow AB^*$$

$$A + B \longrightarrow A^*B^*$$

so three constants are needed. The equilibrium constants are

$$K_1 = \frac{[AB]}{[A][B]} \qquad K_2 = \frac{[A^*B]}{[A][B]} \qquad K_3 = \frac{[AB^*]}{[A][B]} \qquad K_4 = \frac{[A^*B^*]}{[A][B]}$$

We now note that three independent ratios exist, which are independent of [B]:

$$\frac{K_1}{K_2} = \frac{[AB]}{[A^*B]} \qquad \frac{K_2}{K_3} = \frac{[A^*B]}{[AB^*]} \qquad \frac{K_3}{K_4} = \frac{[AB^*]}{[A^*B^*]}$$

Therefore, an isosbestic point is expected if $[A]_0$ is held constant and $[B]_0$ is varied. Any other ratio of equilibrium constants, *e.g.*, K_1/K_3, is not independent of the three ratios written. The general rules presented here apply to a large number of systems. However, rote application of rules is no substitute for an understanding of the systems under consideration.

4-8 JOB'S METHOD OF ISOMOLAR SOLUTIONS

By examining the spectra of a series of solutions of widely varying mole ratios of A to B, but with the same total number of moles of A + B, the stoichiometry of complexes formed between A and B in solution can often be determined. Absorbance at a wavelength of maximum change is plotted against the mole ratio of A to B; the latter is usually used as the abscissa. The plots have at least one extremum, often a maximum. In simple cases, the extrema occur at mole fractions corresponding to the stoichiometry of the complexes that form in solution.[11-13]

4-9 "FINGERPRINTING"

This technique is useful for the identification of an unknown compound that is suspected to be the same as a known compound. The spectra are compared with respect to ε values, wavelengths of maximum absorption, and band shapes.

In addition to this direct comparison, certain functional groups have characteristic absorptions in various regions of the spectrum. For example, the carbonyl group will generally absorb at certain wavelengths in the ultraviolet and infrared spectra. Its presence in an unknown compound can be determined from these absorptions. Often one can even determine whether or not the carbonyl group is in a conjugated system. These details will be considered later when the spectroscopic methods are discussed individually.

Spectroscopic methods provide a convenient way of detecting certain impurities in a sample. For example, the presence of water in a system can easily be detected by its characteristic infrared absorption. Similarly, a product can be tested for absence of starting material if the starting material has a functional group with a characteristic absorption that disappears during the reaction. Spectral procedures are speedier and less costly than elemental analyses. The presence of contaminants in small amounts can be detected if their molar absorptivities are large enough.

EXERCISES

1. A series of different molecular transitions require the following energies in order to occur. Indicate the spectral region in which you would expect absorption of radiation to occur, and the wavelength of the radiation.
   a. 0.001 kcal.
   b. 100 kcal.
   c. 30 kcal.
2. Convert the wavelength units in part a of exercise 1 to wavenumbers (cm^{-1}).
3. Convert the following wavenumbers to μ and Å:
   a. 3600 cm^{-1}.
   b. 1200 cm^{-1}.
4. Convert 800 mμ to cm^{-1}.
5. The ε value for compound X is 9000 liters $mole^{-1}$ cm^{-1}. A 0.1 molar solution of X in water, measured in a 1 cm cell, has an absorbance of 0.542. It is known that X reacts according to the equation:

$$X \rightleftharpoons Y + Z$$

X obeys Beer's law, and Y and Z do not absorb in this region. What is the equilibrium constant for this reaction?

6. For the equilibrium

$$A + B \rightleftharpoons C$$

assume that only C absorbs and that ε_C cannot be measured. Derive an equation similar to (4–22) for this system.

7. a. Use equations (4–14) and (4–15) and values of $A_1 = 0.3, A_2 = 0.7, \varepsilon_{B_{\lambda_1}} = 5, \varepsilon_{B_{\lambda_2}} = 12, \varepsilon_{C_{\lambda_1}} = 13$, and $\varepsilon_{C_{\lambda_2}} = 2$ to construct a plot similar to that in Fig. 4–6 to illustrate the graphical solution of the two simultaneous equations.

   b. Construct the same plot as in part a, using $A_1 = 0.3$, $A_2 = 0.7$, $\varepsilon_{B_{\lambda_1}} = 5$, $\varepsilon_{B_{\lambda_2}} = 12$, $\varepsilon_{C_{\lambda_1}} = 3$, and $\varepsilon_{C_{\lambda_2}} = 5$. Compare your confidence in the two answers, realizing that there is 2 to 3 per cent error in the A and ε values.

   c. Equation (4–22) is somewhat analogous to that used in parts a and b, for the first term is generally small under the conditions used in a spectroscopic study. Often investigators work in a concentration range in which, holding $[D]_0$ constant and doubling $[E]_0$, they double the amount of complex formed or they double $A - A°$. What would the two lines in the K^{-1} *vs.* ε plot look like?

8. A proton is being rapidly exchanged between B and B′, *i.e.*,

$$BH^+ + B' \rightleftharpoons B'H^+ + B$$

The frequency difference between the H^+ peaks of BH^+ and $B'H^+$ in the nmr spectrum is 100 Hz (cycles per second).

   a. What rate of exchange would be required for the two peaks to appear as a single peak?

   b. Would a slower or faster rate be required to observe two separate resonances that are broadened by the exchange process?

   c. What is the approximate maximum rate of chemical exchange that could occur and still have no effect on the nmr spectrum?

9. A system $A + B \rightleftharpoons C + D$ is investigated spectroscopically. Species A, C, and D all have absorbances that overlap. Since there are more than two absorbing species, would one expect to find an isosbestic point if this were the only equilibrium involved? Prove your answer.

REFERENCES CITED

1. E. F. Johnson and R. S. Drago, J. Amer. Chem. Soc., *95,* 1391 (1973), and references therein.
2. N. J. Rose and R. S. Drago, J. Amer. Chem. Soc., *81,* 6138 (1959).
3. T. D. Epley and R. S. Drago, J. Amer. Chem. Soc., *91,* 2883 (1969), and references therein.
4. T. O. Maier and R. S. Drago, Inorg. Chem., *11,* 1861 (1972).
5. P. J. Lingane and Z. Z. Hugus, Jr., Inorg. Chem., *9,* 757 (1970).
6. R. M. Guidry and R. S. Drago, J. Amer. Chem. Soc., *95,* 6645 (1973).
7. F. L. Slejko, R. S. Drago, and D. G. Brown, J. Amer. Chem. Soc., *94,* 9210 (1972).
8. T. F. Bolles and R. S. Drago, J. Amer. Chem. Soc., *87,* 5015 (1965).
9. R. L. Carlson, K. F. Purcell, and R. S. Drago, Inorg. Chem., *4,* 15 (1965).
10. R. G. Mayer and R. S. Drago, Inorg., Chem., *15,* 2010 (1976).
11. W. Likussar and D. F. Boltz, Anal. Chem., *43,* 1265 (1971).
12. M. M. Jones and K. K. Innes, J. Phys. Chem., *62,* 1005 (1958).
13. E. Asmus, Z. Anal. Chem., *183,* 321, 401 (1961).

5 ELECTRONIC ABSORPTION SPECTROSCOPY

INTRODUCTION

5–1 VIBRATIONAL AND ELECTRONIC ENERGY LEVELS IN A DIATOMIC MOLECULE

Prior to a discussion of electronic absorption spectroscopy, the information summarized by a potential energy curve for a diatomic molecule (indicated in Fig. 5–1) will be reviewed. Fig. 5–1 is a plot of E, the total energy of the system, *versus* r, the internuclear distance, and is one of many types of potential functions referred to as a Morse potential. The curve is expressed mathematically by

$$V = D\{1 - \exp[-\nu_0(2\pi^2\mu D)^{1/2}(r - r_e)]\}^2$$

All terms are defined in Fig. 5–1 except μ, which is the reduced mass $(m_1m_2)/(m_1 + m_2)$.

As the bond distance is varied in a given vibrational state, *e.g.*, along A–B in the v_2 level, the molecule is in a constant vibrational energy level. At A and B we have, respectively, the minimum and maximum values for the bond distance in this vibrational level. At these points the atoms are changing direction, so the vibrational kinetic energy is zero and the total vibrational energy of the system is potential. At r_e, the equilibrium internuclear distance, the vibrational kinetic energy is a maximum and the vibrational potential energy is zero. Each horizontal line represents a different vibrational energy state. The ground state is v_0 and excited states are represented as v_1, v_2, etc. Eventually, if enough energy can be absorbed in vibrational

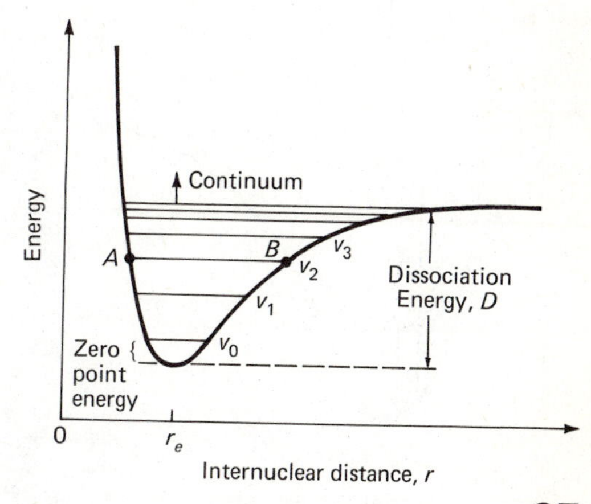

FIGURE 5–1. Morse energy curve for a diatomic molecule.

modes, the molecule is excited into the continuum and it dissociates. For most compounds, nearly all the molecules are in the v_0 level at room temperature because the energy difference $v_1 - v_0$ is usually much larger than kT (thermal energy), which has a value of 200 cm^{-1} at 300°K. (Recall the Boltzmann expression.)

Each excited electronic state also contains a series of different vibrational energy levels and may be represented by a potential energy curve. The ground electronic state and one of the many excited electronic states for a typical diatomic molecule are illustrated in Fig. 5–2. Each vibrational level, v_n, is described by a vibrational wave function, ψ_{vib}. For simplicity only four levels are indicated. The square of a wave function gives the probability distribution, and in this case ψ_{vib}^2 indicates probable internuclear distances for a particular vibrational state. This function, ψ_{vib}^2, is indicated for the various levels by the dotted lines in Fig. 5–2. The dotted line is not related to the energy axis. The higher this line, the more probable the corresponding internuclear distance. The most probable distance for a molecule in the ground state is r_e, while there are two most probable distances corresponding to the two maxima in the next vibrational energy level of the ground electronic state, three in the third, etc. In the excited vibrational levels of the ground and excited electronic states, there is high probability of the molecule having an internuclear distance at the ends of the potential function.

5–2 RELATIONSHIP OF POTENTIAL ENERGY CURVES TO ELECTRONIC SPECTRA

An understanding of electronic absorption spectroscopy requires consideration of three additional principles:

(1) In the very short time required for an electronic transition to take place

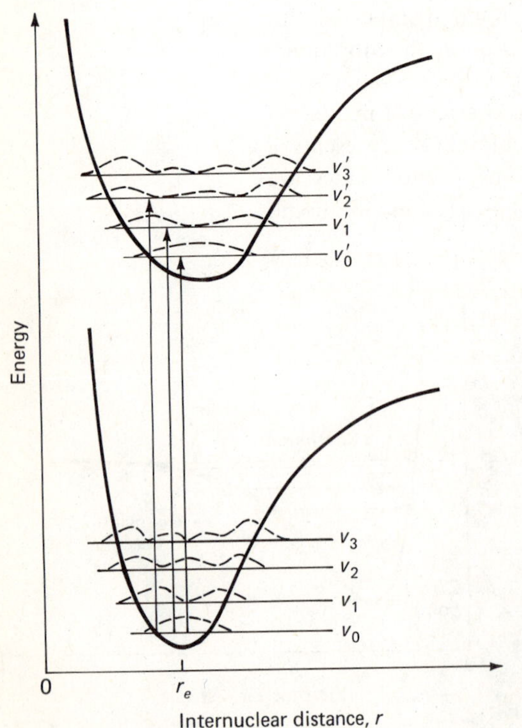

FIGURE 5–2. Morse curves for a ground and excited state of a diatomic molecule, showing vibrational probability functions, ψ_{vib}^2, as dotted lines.

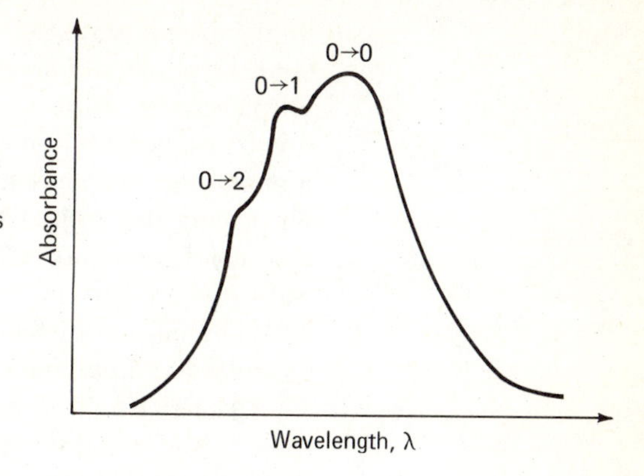

FIGURE 5–3. Spectrum corresponding to the potential energy curves indicated in Fig. 5–2.

(about 10^{-15} sec), the atoms in a molecule do not have time to change position appreciably. This statement is referred to as the *Franck-Condon principle.** Since the electronic transition is rapid, the molecule will find itself with the same molecular configuration and vibrational kinetic energy in the excited state that it had in the ground state at the moment of absorption of the photon. As a result, all electronic transitions are indicated by a vertical line on the Morse potential energy diagram of the ground and excited states (see the arrows in Fig. 5–2); *i.e.*, there is no change in internuclear distance during the transition.

(2) There is no general selection rule that restricts the change in vibrational state accompanying an electronic transition. Frequently transitions occur from the ground vibrational level of the ground electronic state to many different vibrational levels of a particular excited electronic state. Such transitions may give rise to vibrational fine structure in the main peak of the electronic transition.

The three transitions indicated by arrows in Fig. 5–2 could give rise to three peaks. Since nearly all of the molecules are present in the ground vibrational level, nearly all transitions that give rise to a peak in the absorption spectrum will arise from v_0.† Transitions from this ground level (v_0) to v_0', v_1', or v_2' are referred to as $0 \rightarrow 0$, $0 \rightarrow 1$, or $0 \rightarrow 2$ transitions, respectively. It can be shown[1] that the relative intensity of the various vibrational sub-bands depends upon the vibrational wave function for the various levels. A transition is favored if the probabilities of the ground and excited states of the molecule are both large for the same internuclear distance. Three such transitions are indicated by arrows in Fig. 5–2. The spectrum in Fig. 5–3 could result from a substance in solution undergoing the three transitions indicated in Fig. 5–2. The $0 \rightarrow 0$ transition is the lowest energy–longest wave-length transition. The differences in wavelength at which the peaks occur represent the energy differences of the vibrational levels in the excited state of the molecule. Much information about the structure and configuration of the excited state can be obtained from the fine structure.

Electronic transitions from bonding to antibonding molecular orbitals are often encountered. In this case the potential energy curve for the ground state will be quite different from that of the excited state because there is less bonding electron density

---

*This principle was originally proposed by Franck. It was given a quantum mechanical interpretation and extended by Condon.

†In the gas phase, various rotational levels in the ground vibrational state will be populated and transitions to various rotational levels in the excited state will occur, giving rise to fine structure in the spectrum. This fine structure is absent in solution because collision of the solute with a solvent molecule occurs before a rotation is completed. Rotational fine structure will not be discussed.

in the excited state. As a result, the equilibrium internuclear distance will be greater and the potential energy curve will be broader for the excited state. Because of this displacement of the excited state potential energy curve, the $0 \rightarrow 0$ and transitions to other low vibrational levels may not be observed. A transition to a higher vibrational level becomes more probable. This can be visualized by broadening and displacing the excited state represented in Fig. 5–2.

(3) There is an additional *symmetry requirement,* which was neglected for the sake of simplicity in the above discussion. It has been assumed that this symmetry requirement is satisfied for the transitions involved in this discussion. This will be discussed subsequently in the section on selection rules.

The above discussion pertains to a diatomic molecule, but the general principles also apply to a polyatomic molecule. Often the functional group in a polyatomic molecule can be treated as a diatomic molecule (for example, $\gt$C=O in a ketone or aldehyde). The electronic transition may occur in the functional group between orbitals that are approximated by a combination of atomic orbitals of the two atoms as in a diatomic molecule. The actual energies of the resulting molecular orbitals of the functional group will, of course, be affected by electronic, conjugative, and steric effects arising from the other atoms. This situation can be understood qualitatively in terms of potential energy curves similar to those discussed for the diatomic molecules. For more complex cases in which several atoms in the molecule are involved (*i.e.*, a delocalized system), a polydimensional surface is required to represent the potential energy curves.

The energies required for electronic transitions generally occur in the far uv, uv, visible, and near infrared regions of the spectrum, depending upon the energies of the molecular orbitals in a molecule. For molecules that contain only strong sigma bonds (*e.g.*, CH_4 and H_2O), the energy required for electronic transitions occurs in the far uv region of the spectrum and requires specialized instrumentation for detection. In fact, in the selection of suitable transparent solvents for study of the u.v. spectrum of the dissolved solute, such transitions come into play in determining the cutoff point of the solvent. On the other hand, the colored dyes used in clothing have extensively conjugated pi-systems and undergo electronic transitions in the visible region of the spectrum. Standard instruments cover the region from 50,000 cm^{-1} to 5,000 cm^{-1}. This spectral region is subdivided as follows:

Ultraviolet	50,000 cm^{-1} to 26,300 cm^{-1} (2000 to 3800 Å)
Visible	26,300 cm^{-1} to 12,800 cm^{-1} (3800 to 7800 Å)
Near infrared	12,800 cm^{-1} to 5,000 cm^{-1} (7800 to 20,000 Å)

5–3 NOMENCLATURE

In our previous discussion we were concerned only with transitions of an electron from a given ground state to a given excited state. In an actual molecule there are electrons in different kinds of orbitals (σ bonding, nonbonding, π bonding) with different energies in the ground state. Electrons from these different orbitals can be excited to higher-energy molecular orbitals, giving rise to many possible excited states. Thus, many transitions from the ground state to different excited states (each of which can be described by a different potential energy curve) are possible in one molecule.

There are several conventions[2,3] used to designate these different electronic transitions. A simple representation introduced by Kasha[2] will be illustrated for the

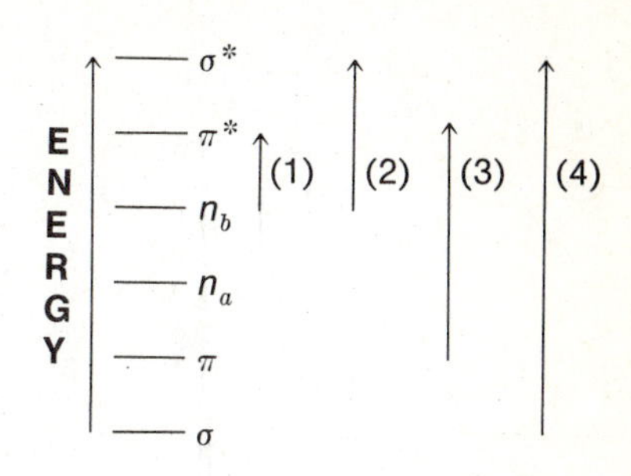

FIGURE 5-4. Relative energies of the carbonyl molecular orbitals in H_2CO.

carbonyl group in formaldehyde. A molecular orbital description of the valence electrons in this molecule is:

$$\sigma^2 \quad \pi^2 \quad n_a^{\ 2} \quad n_b^{\ 2} \quad (\pi^*)^0 \quad (\sigma^*)^0$$

The n_a and n_b orbitals are the two non-bonding molecular orbitals containing the lone pairs on oxygen. Symmetry considerations do not require the lone pairs to be degenerate, for there are no doubly degenerate irreducible representations in C_{2v}. They are not accidentally degenerate either, but differ in energy. The relative energies of these orbitals are indicated in Fig. 5-4. The ordering of these orbitals can often be arrived at by intelligent guesses and by looking at the spectra of analogous compounds. Also indicated with arrows are some transitions chosen to illustrate the nomenclature. The transitions (1), (2), (3), and (4) are referred to as $n \rightarrow \pi^*$, $n \rightarrow \sigma^*$, $\pi \rightarrow \pi^*$, and $\sigma \rightarrow \sigma^*$, respectively. The $n \rightarrow \pi^*$ transition is the lowest energy-highest wavelength transition that occurs in formaldehyde and most carbonyl compounds.

Electron excitations can occur with or without a change in the spin of the electron. If the spin is not changed in a molecule containing no unpaired electrons, both the excited state and ground state have a multiplicity of one; these states are referred to as singlets. The *multiplicity* is given by two times the sum of the individual spins, m_s, plus one: $2S + 1 = 2\Sigma m_s + 1$. If the spin of the electron is changed in the transition, the excited state contains two unpaired electrons with identical magnetic spin quantum numbers, has a multiplicity of 3, and is referred to as a triplet state.

There are some shortcomings of this simple nomenclature for electronic transitions. It has been assumed that these transitions involve a simple transfer of an electron from the ground state level to an empty excited state level of our ground state wave function. For many applications, this description is precise enough. In actual fact, such transitions occur between states, and the excited state is not actually described by moving an electron into an empty molecular orbital of the ground state. The excited state has, among other things, different electron-electron repulsions than those in the system represented by simple excitation of an electron into an empty ground state orbital. In most molecules, various kinds of electron-electron interactions in the excited state complicate the problem; in addition to affecting the energy, they give rise to many more transitions than predicted by the simple picture of electron promotion because levels that would otherwise be degenerate are split by electronic interactions. This problem is particularly important in transition metal ion complexes. We shall return to discuss this problem more fully in the section on configuration interaction.

In a more accurate[4] system of nomenclature the symmetry, configuration, and multiplicity of the states involved in the transition are utilized in describing the transition. This system of nomenclature can be briefly demonstrated by again considering the molecular orbitals of formaldehyde. The diagrams in Fig. 5-5

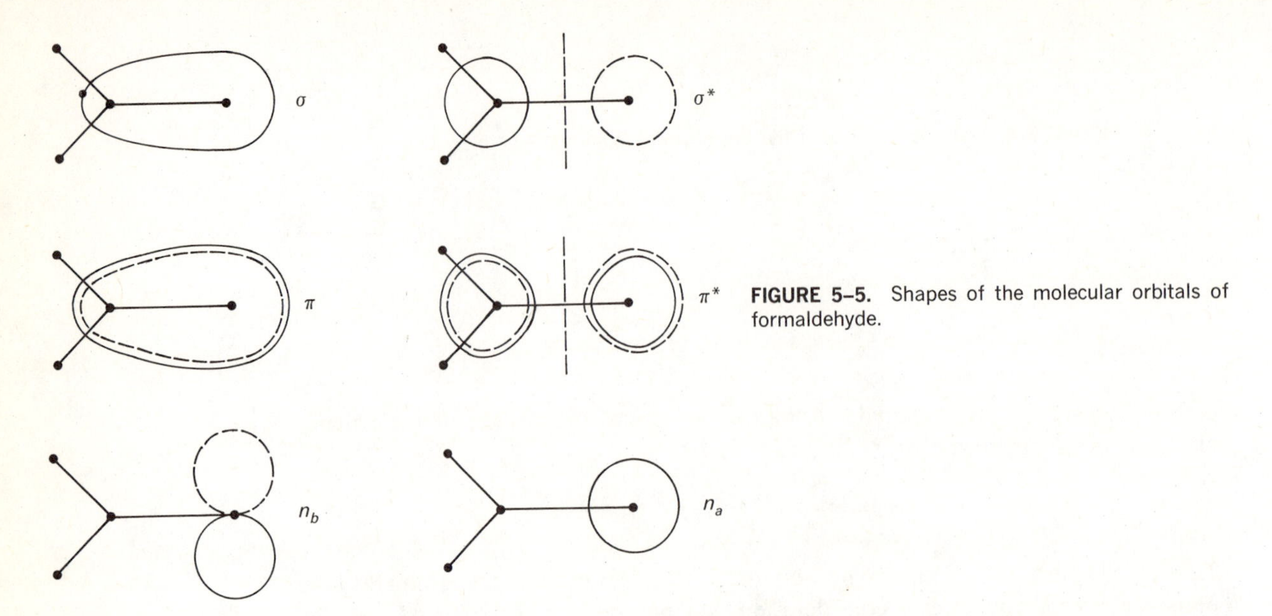

FIGURE 5-5. Shapes of the molecular orbitals of formaldehyde.

qualitatively represent the boundary contours of the molecular orbitals. The solid line encloses the positive lobe and the dashed line the negative lobe. The larger π and π^* lobes indicate lobes above the plane of the paper, and the smaller ones represent those below the plane; the two lobes actually have identical sizes. To classify these orbitals it is necessary first to determine the overall symmetry of the molecule, which is C_{2v}. The next step is to consult the C_{2v} character table. Character tables for several common point groups are listed in Appendix A. The C_{2v} character table is duplicated here in Table 5-1. By convention, the yz plane is selected to contain the four atoms of formaldehyde. The symmetry operations E, C_2, $\sigma_{v(xz)}$, and $\sigma_v{}'_{(yz)}$ performed on the π orbital produce the result $+1$, -1, $+1$, -1. This result is identical to that listed for the irreducible representation B_1 in the table. The orbital is said to belong to (or to transform as) the symmetry species b_1, the lower case letter being employed for an orbital and the upper case letters being reserved to describe the symmetry of the entire ground or excited state. Similarly, if the n_a, n_b, π^*, σ, and σ^* orbitals of formaldehyde are subjected to the above symmetry operations, it can be shown that these orbitals belong to the irreducible representations a_1, b_2, b_1, a_1, and a_1, respectively. The two n orbitals can be viewed as s and p_y orbitals (p_z is used in the σ bond and p_x in the π). As a result, they lie in the yz plane and possess a_1 and b_2 symmetry. (The a_1 s orbital can mix with p_z in the sigma bonding.) The difference in s orbital character causes the energies of the a_1 and b_2 lone pairs to differ. Molecular orbital calculations are consistent with these ideas. As mentioned in Chapter 2, a and b indicate single degeneracy. The a representation does not change sign with rotation about the n-fold axis, but b does.

TABLE 5-1. CHARACTER TABLE FOR THE C_{2v} POINT GROUP*

	E	C_2	$\sigma_{v(xz)}$	$\sigma_v{}'_{(yz)}$	
A_1	1	1	1	1	z
A_2	1	1	-1	-1	R_z
B_1	1	-1	1	-1	x, R_y
B_2	1	-1	-1	1	y, R_x

* x-axis is perpendicular to the plane of the molecule.

The symmetry species of a state is the product of the symmetry species of each of the odd electron orbitals. In the state that results from the $n \rightarrow \pi^*$ transition in formaldehyde, there is one unpaired electron in the n orbital with b_2 symmetry and one in the π^* orbital with b_1 symmetry. The direct product is given by:

	E	C_2	$\sigma_{v(xz)}$	$\sigma_v{}'_{(yz)}$	
$b_1 \times b_2 =$	$(1)(1)$	$(-1)(-1)$	$(1)(-1)$	$(-1)(+1)$	
result $=$	$+1$	$+1$	-1	-1	$= A_2$

The resulting irreducible representation is A_2.* The excited state from this transition is thus described as A_2 and the transition as $A_1 \rightarrow A_2$. A common convention involves writing the high energy state first and labeling the transition $A_2 \leftarrow A_1$. The spin multiplicity is usually included, so the complete designation becomes ${}^1A_2 \leftarrow {}^1A_1$. The ground state is A_1 because there is a pair of electrons in each orbital. Commonly, the orbitals involved in the transitions are indicated, and the symbol becomes ${}^1A_2(n, \pi^*) \leftarrow {}^1A_1$. If a general symbol is needed for a state symmetry species, Γ is employed.

Instead of representing the orbitals of formaldehyde symbolically, as in Fig. 5–5, we could simply have been given the wave functions. We can deduce the symmetry from ψ by converting the wave functions into a physical picture. The following equations describe the formaldehyde π and π^* orbitals:

$$\psi_\pi = a\varphi_{p^o} + b\varphi_{p^c}$$

$$\psi_{\pi^*} = b'\varphi_{p^o} - a'\varphi_{p^c}$$

where φ_{p^o} and φ_{p^c} are the wave functions for the atomic oxygen and carbon p orbitals, respectively. The atomic orbitals are mathematically combined to produce the π and π^* orbitals. Since oxygen is more electronegative (*i.e.*, $a > b$), it becomes clear why it is often stated that an electron is transferred from oxygen to carbon in the $\pi \rightarrow \pi^*$ transition. We shall return to a discussion of the experimental spectrum of formaldehyde shortly.

Some of the molecular orbitals for benzene are represented mathematically and pictorially in Fig. 5–6. Notice that a difference in sign between adjacent atomic orbitals of the wave function represents a node (point of zero probability) in the molecular orbital. Using the D_{6h} character table, the symmetries of the orbitals can be shown to be a_{2u}, e_{1g}, e_{2u}, and b_{2g}, for ψ_1, $\psi_2 + \psi_3$, $\psi_4 + \psi_5$, and ψ_6, respectively.

In addition to the above conventions used to label ground and excited states, a convention used for diatomic molecules will be described. With this terminology the electronic arrangement is indicated by summing up the contributions of the separate atoms to obtain the net orbital angular momentum. If all electrons are paired, the sum is zero. Contributions are counted as follows: one unpaired electron in a σ orbital is zero, one in a π orbital one, and one in a δ orbital two. For more than one electron, the total is $|\Sigma m_l|$. If the total is zero, the state is described as Σ, one as Π, and two as Δ. The multiplicity is indicated by a superscript, e.g., the ground state of NO is ${}^2\Pi$. A plus or minus sign often follows the symbol to illustrate, respectively, whether the molecular orbitals are symmetric or antisymmetric to a plane through the molecular axis.

---

*The actual procedure for determining the symmetry of this state involves multiplication of the orbital symmetries for all the electrons in the carbonyl group. However, all filled orbitals contain two electrons whose product must be A_1. Thus, paired electrons have no effect on the final result for the excited state symmetry.

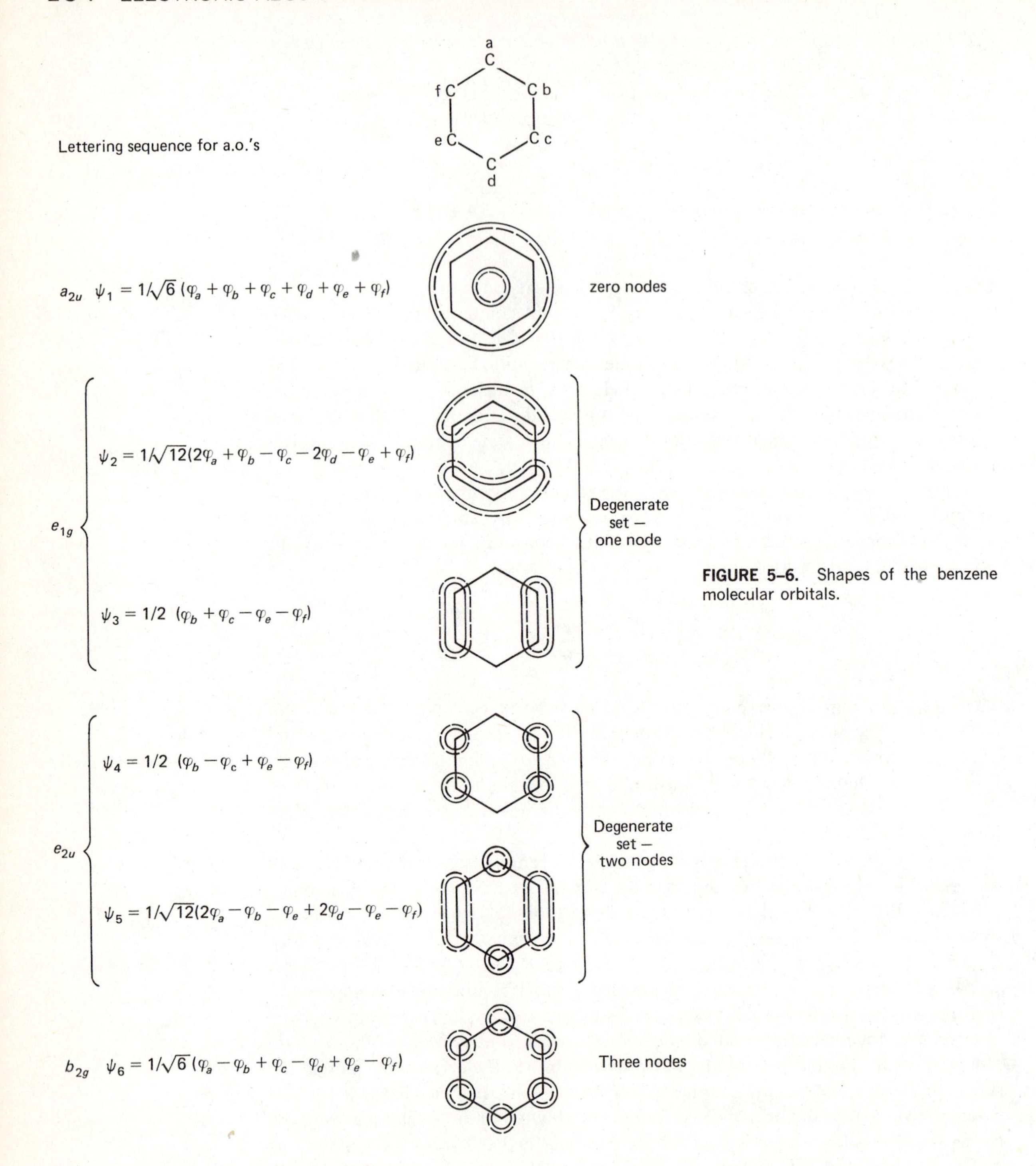

FIGURE 5–6. Shapes of the benzene molecular orbitals.

ASSIGNMENT OF TRANSITIONS

If the energies of electronic transitions were related to the ground state molecular orbital energies, the assignment of transitions to the observed bands would be simple. In formaldehyde (Fig. 5–4) the $n \rightarrow \pi^*$ would be lower in energy than the $\pi \rightarrow \pi^*$, which in turn would be lower in energy than $\sigma \rightarrow \pi^*$. In addition to different electron-electron repulsions in different states, two other effects complicate the picture by affecting the energies and the degeneracy of the various excited states. These effects are *spin-orbit coupling* and *higher state mixing*.

5–4 SPIN-ORBIT COUPLING

There is a magnetic interaction between the electron spin magnetic moment (signified by quantum number $m_s = \pm\frac{1}{2}$) and the magnetic moment due to the orbital motion of an electron. To understand the nature of this effect, consider the nucleus as though it were moving about the electron (this is equivalent to being on earth and thinking of the sun moving across the sky). We consider the motion from this reference point because we are interested in effects at the electron. The charged nucleus circles the electron, and this is equivalent in effect to placing the electron in the middle of a coil of wire carrying current. As moving charge in a solenoid creates a magnetic field in the center, the orbital motion described above causes a magnetic field at the electron position. This magnetic field can interact with the spin magnetic moment of the electron, giving rise to spin-orbit interaction. The orbital moment may either complement or oppose the spin moment, giving rise to two different energy states. The doubly degenerate energy state of the electron (previously designated by the spin quantum numbers $\pm\frac{1}{2}$) is split, lowering the energy of one and raising the energy of the other. Whenever an electron can occupy a set of degenerate orbitals that permit circulation about the nucleus, this interaction is possible. For example, if an electron can occupy the d_{yz} and d_{xz} orbitals of a metal ion, it can circle the nucleus around the z axis. (See Chapter 11 for a more complete discussion.)

5–5 CONFIGURATION INTERACTION

As mentioned before, electronic transitions do not occur between empty molecular orbitals of the ground state configuration, but between states. The energies of these states are different from those of configurations derived by placing electrons in the empty orbitals of the ground state, because electron-electron repulsions in the excited state differ from those in the simplified "ground state orbital description" of the excited state. A further complication arises from configuration interaction. One could attempt to account for the different electron-electron repulsions in the excited state by doing a molecular orbital calculation on a molecule having the ground state geometry but with the electron arrangement of the excited configuration. This would not give the correct energy of the state because this configuration can mix with all of the other configurations in the molecule of the same symmetry by configuration interaction.

This mixing is similar in a mathematical sense (though much smaller in magnitude) to the interaction of two hydrogen atoms in forming the H_2 molecule. Thus, we could write a secular determinant to account for the mixing of two B_1 states, B'_{1a} and B'_{1b}, as:

$$\begin{vmatrix} E_1{}^0 - E & H_{12} \\ H_{12} & E_2{}^0 - E \end{vmatrix} = 0$$

where H_{12} is the difficult-to-solve integral $\int \psi(B_{1a})\hat{H}\psi(B_{1b})\, d\tau$, whose value is dependent upon the interelectron repulsions in the various states. The closer in energy the initial states $E_1{}^0$ and $E_2{}^0$, the more mixing occurs. Solution of the secular determinant gives us the two new energies after mixing:

$$E_1 = \frac{1}{2}[E_1{}^0 + E_2{}^0 + ((E_1{}^0)^2 + (E_2{}^0)^2 - 2E_1{}^0E_2{}^0 + 4H_{12}{}^2)^{1/2}]$$

$$E_2 = \frac{1}{2}[E_1{}^0 + E_2{}^0 - ((E_1{}^0)^2 + (E_2{}^0)^2 - 2E_1{}^0E_2{}^0 + 4H_{12}{}^2)^{1/2}]$$

The energies of the initial (B'_{1a}, B'_{1b}) and final (B_{1a}, B_{1b}) states are illustrated in Fig. 5–7 along with the new wave functions to describe the final state. The wave function is seen to be a linear combination of the two initial molecular orbitals. Interactions of this sort can occur with all the molecular orbitals of B_1 symmetry in the molecule, complicating the problem even further than is illustrated in Fig. 5–7.

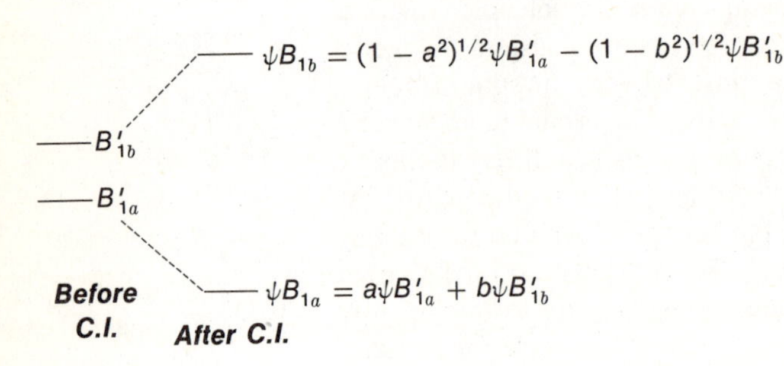

FIGURE 5–7. Energy levels before and after configuration interaction.

5–6 CRITERIA TO AID IN BAND ASSIGNMENT

An appreciation for some of the difficulties encountered in assigning transitions can be obtained from reading the literature[5] and noting the changes in the assignments that have been made over the years. Accordingly, many independent criteria are used in making the assignment. These include the intensity of the transition and the behavior of the absorption band when polarized radiation is employed. Both of these topics will be considered in detail shortly. Next we shall describe some simple observations that aid in the assignment of the $n \rightarrow \pi^*$ and $\pi \rightarrow \pi^*$ transitions.

For $n \rightarrow \pi^*$ transitions, one observes the following characteristics:

(1) The *molar absorptivity* of the transition is generally less than 2000. An explanation for this is offered in the section on intensity.

(2) A *blue shift* (hypsochromic shift, or shift toward shorter wavelengths) is observed for this transition in high dielectric or hydrogen-bonding solvents. This indicates that the energy difference between the ground and excited state is increased in a high dielectric or hydrogen-bonding solvent. In general, for solvent shifts it is often difficult to ascertain whether the excited state is raised in energy or the ground state lowered. A blue shift may result from a greater lowering of the ground state

relative to the excited state or a greater elevation of the excited state relative to the ground state. It is thought that the solvent shift in the $n \rightarrow \pi^*$ transition results from a lowering of the energy of the ground state and an elevation of the energy of the excited state. In a high dielectric solvent the molecules arrange themselves about the absorbing solute so that the dipoles are properly oriented for maximum interaction (*i.e.*, solvation that lowers the energy of the ground state). When the excited state is produced, its dipole will have an orientation different from that of the ground state. Since solvent molecules cannot rearrange to solvate the excited state during the time of a transition, the excited state energy is raised in a high dielectric solvent.[6]

Hydrogen bonding solvents cause pronounced blue shifts. This is reported to be due to hydrogen bonding of the solvent hydrogen with the lone pair of electrons in the n orbital undergoing the transition. In the excited state there is only one electron in the n orbital, the hydrogen bond is weaker and, as a result, the solvent does not lower the energy of this state nearly as much as that of the ground state. In these hydrogen bonding systems an adduct is formed, and this specific solute-solvent interaction is the main cause of the blue shift.[7] If there is more than one lone pair of electrons on the donor, the shift can be accounted for by the inductive effect of hydrogen bonding to one electron pair on the energy of the other pair.

(3) The $n \rightarrow \pi^*$ band often disappears in acidic media owing to protonation or upon formation of an adduct that ties up the lone pair, *e.g.*, $BCH_3^+I^-$ (as in $C_5H_5NCH_3^+I^-$), where B is the base molecule containing the n electrons. This behavior is very characteristic if there is only one pair of n electrons on B.

(4) Blue shifts occur upon the attachment of an electron-donating group to the chromophore (*i.e.*, the series $CH_3\overset{O}{\overset{/\!/}{C}}{-}H$, $CH_3\overset{O}{\overset{/\!/}{C}}{-}CH_3$, $CH_3\overset{O}{\overset{/\!/}{C}}{-}OCH_3$, and $CH_3\overset{O}{\overset{/\!/}{C}}{-}N(CH_3)_2$ represents an increasing blue shift in the carbonyl absorption band). A molecular orbital treatment[8] indicates that this shift results from raising the excited π^* level relative to the n level.

(5) The absorption band corresponding to the $n \rightarrow \pi^*$ transition is absent in the hydrocarbon analogue of the compound. This would involve, for example, comparison of the spectra of benzene and pyridine or of $H_2C{=}O$ and $H_2C{=}CH_2$.

(6) Usually, but not always, the $n \rightarrow \pi^*$ transition gives rise to the lowest energy singlet-singlet transition.

In contrast to the above behavior, $\pi \rightarrow \pi^*$ transitions have a high intensity. A slight red (bathochromic) shift is observed in high dielectric solvents and upon introduction of an electron-donating group. It should be emphasized that in the above systems only the difference in energy between the ground and excited states can be measured from the frequency of the transition, so only the relative energies of the two levels can be measured. Other considerations must be invoked to determine the actual change in energy of an individual state.

5-7 OSCILLATOR STRENGTHS

THE INTENSITY OF ELECTRONIC TRANSITIONS

As mentioned in the previous chapter, the intensity of an absorption band can be indicated by the molar absorptivity, commonly called the extinction coefficient. A parameter of greater theoretical significance is f, the *oscillator strength* of integrated intensity, often simply called the *integrated intensity:*

$$f = 4.315 \times 10^{-9} \int \varepsilon d\bar{\nu} \qquad (5\text{-}1)$$

In equation (5–1), ε is the molar absorptivity and $\bar{\nu}$ is the frequency expressed in wave numbers. The concept of oscillator strength is based on a simple classical model for an electronic transition. The derivation* indicates that $f = 1$ for a fully allowed transition. The quantity f is evaluated graphically from equation (5–1) by plotting ε, on a linear scale, *versus* the wavenumber $\bar{\nu}$ in cm^{-1}, and calculating the area of the band. Values of f from 0.1 to 1 correspond to molar absorptivities in the range from 10,000 to 100,000, depending on the width of the peak.

For a single, symmetrical peak, f can be approximated by the expression:

$$f \approx (4.6 \times 10^{-9})\varepsilon_{max}\,\Delta\nu_{1/2} \qquad (5\text{-}2)$$

where ε_{max} is the molar absorptivity of the peak maximum and $\Delta\nu_{1/2}$ is the half intensity band width, *i.e.*, the width at $\varepsilon_{max}/2$.

5–8 TRANSITION MOMENT INTEGRAL

The integrated intensity, f, of an absorption band is related to the transition moment integral as follows:

$$f \propto \left| \int_{-\infty}^{+\infty} \psi_{el}\hat{M}\psi_{el}{}^{ex}\, dv \right|^2 = D \qquad (5\text{-}3)$$

where D is called the dipole strength, ψ_{el} and $\psi_{el}{}^{ex}$ are electronic wave functions for the ground and excited states respectively, $\hat{M}$ is the electric dipole moment operator (*vide infra*), and the entire integral is referred to as the *transition moment integral.* To describe $\hat{M}$, one should recall that the electric dipole moment is defined as the distance between the centers of gravity of the positive and negative charges times the magnitude of these charges. The center of gravity of the positive charges in a molecule is fixed by the nuclei, but the center of gravity of the electrons is an average over the probability function. The vector for the average distance from the nuclei to the electron is represented as $\vec{\mathbf{r}}$. The electric dipole moment vector, $\vec{\mathbf{M}}$, is given by $\vec{\mathbf{M}} = \Sigma e\vec{\mathbf{r}}$, with the summation carried out over all the electrons in the molecule. For the ground state, the electric dipole moment is given by the average over the probability function, or

$$\int \psi_g \sum e\vec{\mathbf{r}}\psi_g\, d\tau$$

By comparison of this equation for the ground state dipole moment with the transition moment integral:

$$\int \psi_g \hat{M}\psi^{ex}\, d\tau$$

this integral can be seen roughly to represent charge migration or displacement during the transition.

*For this derivation, see General References, Barrow, pages 80 and 81.

When the integral in equation (5–3) is zero, the intensity will be zero and, to a first approximation, the transition will be forbidden. In general, we do not have good wave functions for the states to substitute into this equation to calculate the intensity, for reasons discussed in the previous section. However, symmetry can often tell us if integrals are zero, so it is important to examine the symmetry properties of the integrand of equation (5–3), for they enable us to make some important predictions. These symmetry considerations will also enable us to derive some selection rules for electronic transitions. The quantity $\widehat{M}$ is a vector quantity and can be resolved into x, y, and z components. The integral in equation (5–3) then has the components:

$$\int \psi_{el} \widehat{M}_x \psi_{el}^{ex} \, dv \tag{5-4}$$

$$\int \psi_{el} \widehat{M}_y \psi_{el}^{ex} \, dv \tag{5-5}$$

$$\int \psi_{el} \widehat{M}_z \psi_{el}^{ex} \, dv \tag{5-6}$$

In order to have an allowed transition, at least one of the integrals in equations (5–4) to (5–6) must be non-zero. If all three of these integrals are zero, the transition is called forbidden and, according to approximate theory, should not occur at all. Forbidden transitions do occur; more refined theories (discussed in the section on Spin-Orbit and Vibronic Coupling) give small values to the intensity integrals.

Symmetry considerations can tell us if the integral is zero in the following way. An integral can be non-zero only if the direct product of the integrand belongs to symmetry species A_1. Another way of saying this is that such integrals are different from zero only when the integrand remains unchanged for any of the symmetry operations permitted by the symmetry of the molecule. The reason for the above statements can be seen by looking at some simple mathematical functions. First, consider a plot of the curve $y = x$, shown in Fig. 5–8A. Symmetry considerations tell us that this function does not have A_1 symmetry. The integral $\int y(x)\,dx$ represents the area under the curve. Since $y(x)$ is positive in one quadrant and negative in the other, the area from $+a$ to $-a$ or from $+\infty$ to $-\infty$ is zero. Now, plot a function $y(x)$ such that y is symmetric in x. One of many such functions is $y = x^2$, shown in Fig. 5–8B. This function is called "even" and does have A_1 symmetry, because it is

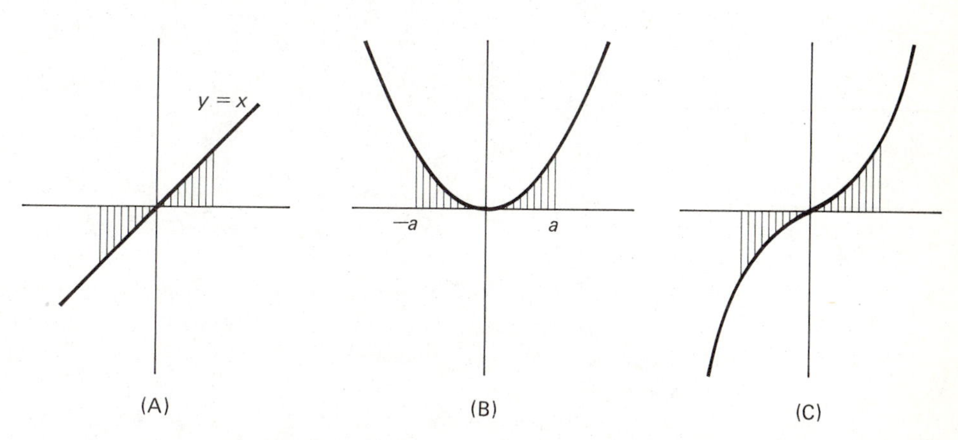

FIGURE 5–8. Plots of some simple functions, $y = f(x)$, and their symmetries.

unchanged by any of the operations of the symmetry group to which the function belongs. This can be shown by carrying out the symmetry operations of the C_{2v} point group on Fig. 5–8B. The integral does not vanish. The shaded area gives the value for the integral $\int_{-a}^{a} y\,dx$ for $y = x^2$, and it is seen not to cancel to zero. Since y is a function of x^2, we can take the direct product of the irreducible representation of x with itself, $\Gamma_x \times \Gamma_x$, and reduce this to demonstrate that this is a totally symmetric irreducible representation. Next, the function $y = x^3$, in Fig. 5–8C, will be examined. Symmetry considerations tell us that this is an odd function, and we see that $\int y\,dx$ (for $y = x^3$) $= 0$. Does $\Gamma_x \times \Gamma_x \times \Gamma_x = A_1$? The integral over all space of an odd function always vanishes, as shown for the examples in Figs. 5–8A and 5–8C. The integral over all space of an even function *generally* does not vanish.

Thus, to determine whether or not the integral is zero, we take the direct product of the irreducible representations of everything in the integrand. If the direct product is or contains A_1, the integral is non-zero and the transition is allowed.

The application of these ideas is best illustrated by treating an example. Consider the $\pi \rightarrow \pi^*$ transition in formaldehyde. The ground state, like all ground states containing no unpaired electrons, is A_1. The excited state is also A_1 ($b_1 \times b_1 = A_1$). The components $\hat{M}_x$, $\hat{M}_y$, and $\hat{M}_z$ transform as the x, y, and z vectors of the point group. The table for the C_{2v} point group indicates that $\hat{M}_z$, the dipole moment vector lying along the z-axis, is A_1. Since $A_1 \times A_1 \times A_1 = A_1$, the integrand $\psi_{el}\hat{M}_z\psi_{el}^{ex}$ for the $\pi \rightarrow \pi^*$ is A_1, and the $\pi \rightarrow \pi^*$ transition is allowed.

For an $n \rightarrow \pi^*$ transition, the ground state is A_1 and the excited state is A_2. The character table indicates that no dipole moment component has symmetry A_2. Therefore, none of the three integrals [equations (5–4) to (5–6)] can be A_1, and the transition is forbidden ($A_2 \times A_2$ is the only product of A_2 that is A_1).

As was mentioned when we introduced this topic, the transition moment integral can be used to derive some important selection rules for electronic transitions.

5–9 DERIVATION OF SOME SELECTION RULES

(1) For molecules with a center of symmetry, allowed transitions are $g \rightarrow u$ or $u \rightarrow g$. (The abbreviations g and u refer to *gerade* and *ungerade,* which are German for even and odd, respectively.) The d and s orbitals are g, and p orbitals are u. All wave functions in a molecule with a center of symmetry are g or u. All components of the vector $\hat{M}$ in a point group containing an inversion center are necessarily *ungerade*.

$$\Gamma\psi_g \times \Gamma_{op} \times \Gamma\psi_{ex} = \Gamma$$

$u \times u \times u = u$	forbidden
$u \times u \times g = g$	allowed
$g \times u \times g = u$	forbidden
$g \times u \times u = g$	allowed

This leads to the selection rule that $g \rightarrow u$ and $u \rightarrow g$ are allowed, but $g \rightarrow g$ and $u \rightarrow u$ are forbidden. Therefore, $d \rightarrow d$ transitions in transition metal complexes with a center of symmetry are forbidden. Values of ε for the d-d transitions in $Ni(H_2O)_6^{2+}$ are ~20.

(2) Transitions between states of different multiplicity are forbidden. Consider a singlet → triplet transition. Focusing on the electron being excited, we have in

the singlet ground state $\psi\alpha\psi\beta$ and, in the excited state, $\psi\alpha\psi\alpha$ or $\psi\beta\psi\beta$, where α and β are the spin coordinates. The dipole strength is given by

$$D = \left| \int \psi_i \alpha \hat{M} \psi_f \beta \, d\tau \, d\sigma \right|^2$$

(where $d\sigma$ is the volume element in the spin coordinates and the i and f subscripts refer to initial and final states). We can rewrite the integral corresponding to D as

$$\left| \int \psi_i \hat{M} \psi_f \, d\tau \int \alpha\beta \, d\sigma \right|^2$$

Since the second term is the product of $+1/2$ and $-1/2$ spins, it is always odd and zero, *i.e.*, the spins are orthogonal. Since $\int \alpha\alpha \, d\sigma = 1$ and $\int \beta\beta \, d\sigma = 1$, in working out the intensity integral we only have to worry about the electron that is undergoing the transition, and we can ignore all the electrons in the molecule that do not change spin. The ε for absorption bands involving transitions between states of different multiplicity is generally less than one.

(3) Transitions in molecules without a center of symmetry depend upon the symmetries of the initial and final states. If the direct product of these and any one of $\hat{M}_x$, $\hat{M}_y$, or $\hat{M}_z$ is A_1, the transition is allowed. If all integrals are odd, the transition is forbidden.

5–10 SPECTRUM OF FORMALDEHYDE

We can summarize the above ideas and illustrate their utility by returning again to the ultraviolet spectrum of formaldehyde. The various possible excited states arising from electron excitations from the highest-energy filled orbitals (n_a, n_b, and π) are given by:

$$a_1{}^2b_1{}^2b_2{}^1b_1{}^{1*} = {}^1A_2 \qquad (n_b \rightarrow \pi^*)$$
$$a_1{}^2b_1{}^2b_2{}^1a_1{}^{1*} = {}^1B_2 \qquad (n_b \rightarrow \sigma^*)$$
$$a_1{}^2b_1{}^1b_2{}^2b_1{}^{1*} = {}^1A_1 \qquad (\pi \rightarrow \pi^*)$$
$$a_1{}^2b_1{}^1b_2{}^2a_1{}^{1*} = {}^1B_1 \qquad (\pi \rightarrow \sigma^*)$$
$$a_1{}^1b_1{}^2b_2{}^2b_1{}^{1*} = {}^1B_1 \qquad (n_a \rightarrow \pi^*)$$
$$a_1{}^1b_1{}^2b_1{}^2a_1{}^{1*} = {}^1A_1 \qquad (n_a \rightarrow \sigma^*)$$

Two bands are observed, one with $\varepsilon = 100$ at 2700 Å and an extremely intense one at 1850 Å. We see from Fig. 5–4 that the lowest-energy transitions are $n_b \rightarrow \pi^*$ and $\pi \rightarrow \pi^*$, better expressed as ${}^1A_1 \rightarrow {}^1A_2$ and ${}^1A_1 \rightarrow {}^1A_1$. The ${}^1A_1 \rightarrow {}^1A_2(n_b \rightarrow \pi^*)$ is forbidden, and accordingly is assigned to the band at 2700 Å. Both ${}^1A_1 \rightarrow {}^1B_1(n_a \rightarrow \pi^*)$ and ${}^1A_1 \rightarrow {}^1A_1(\pi \rightarrow \pi^*)$ are allowed. The former may contribute to the observed band at 1850 Å or may be in the far uv.

The integrands for ${}^1A_1 \rightarrow {}^1B_1(\pi \rightarrow \sigma^*)$, $({}^1A_1 \rightarrow {}^1B_2(n_b \rightarrow \sigma^*)$, and ${}^1A_1 \rightarrow {}^1A_1(n_a \rightarrow \sigma^*)$ are all A_1, leading to allowed transitions. These are expected to occur at very short wavelengths in the far ultraviolet region. This is the presently held view of the assignment of this spectrum, and it can be seen that the arguments are not rigorous. We shall subsequently show how polarization studies aid in making assignments more rigorous.

Next, it is informative to discuss the uv spectrum of acetaldehyde, which is quite similar to that of formaldehyde. The $n_b \rightarrow \pi^*$ transition has very low intensity. However, acetaldehyde has C_s symmetry; this point group has only two irreducible representations, A and B, with the x and y vectors transforming as A and the z vector as B. Accordingly, *all* transitions will have an integrand with A_1 symmetry and will be allowed. Though the $n_b \rightarrow \pi^*$ transition is allowed by symmetry, the value of the transition moment integral is very small and the intensity is low. The intensity of this band in acetaldehyde is greater than that in formaldehyde. We can well appreciate the fact that although monodeuteroformaldehyde [DC(O)H] does not have C_{2v} symmetry, it will have an electronic spectrum practically identical to that of formaldehyde. These are examples of a rather general type of result, which leads to the idea of *local symmetry*. According to this concept, even though a molecule does not have the symmetry of a particular point group, if the groups attached to the chromophore have similar bonding interactions, the molecule for many purposes can be treated as though it had this higher symmetry.

5-11 SPIN-ORBIT AND VIBRONIC COUPLING CONTRIBUTIONS TO INTENSITY

The discrepancy between the theoretical prediction that a transition is forbidden and the experimental detection of a weak band assignable to this transition is attributable to the approximations of the theory. More refined calculations that include effects from *spin-orbit coupling* (review p. 105) often predict low intensities for otherwise forbidden transitions. For example, a transition between a pure singlet state and a pure triplet state is forbidden. However, if spin-orbit coupling is present, the singlet could have the same total angular momentum as the triplet and the two states could interact. The interaction is indicated by equation (5-7):

$$\psi = a\,{}^1\psi + b\,{}^3\psi \tag{5-7}$$

where ${}^1\psi$ and ${}^3\psi$ correspond to pure singlet and triplet states, respectively, ψ represents the actual ground state, and a and b are coefficients indicating the relative contributions of the pure states. If $a \gg b$, the ground state is essentially singlet with a slight amount of triplet character and the excited state will be essentially triplet. This slight amount of singlet character in the predominantly triplet excited state leads to an intensity integral for the singlet-triplet transition that is not zero; this explains why a weak peak corresponding to a multiplicity-forbidden transition can occur.

Another phenomenon that gives intensity to some forbidden transitions is *vibronic coupling*. We have assumed until now that the wave function for a molecule can be factored into an electronic part and a vibrational part, and we have ignored the vibrational part. When we applied symmetry considerations to our molecule, we assumed some symmetrical, equilibrium internuclear configuration. This is not correct, for the molecules in our system are undergoing vibrations and during certain vibrations the molecular symmetry changes considerably. For example, in an octahedral complex, the T_{1u} and T_{2u} vibrations shown in Fig. 5-9 remove the center of symmetry of the molecule. Since electronic transitions occur much more rapidly than molecular vibrations, we detect transitions occurring in our sample from many geometries that do not have high symmetry, *e.g.*, the vibrationally distorted molecules of the octahedral complex shown in Fig. 5-9. The local symmetry is still very close to octahedral, so the intensity gained this way is not very great; but it is large enough to allow a forbidden transition to occur with weak intensity.

The electronic transition can become allowed by certain vibrational modes but

FIGURE 5-9. T_{1u} and T_{2u} vibrations of an octahedral complex.

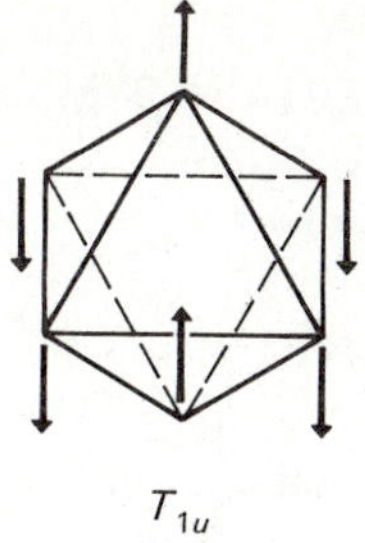

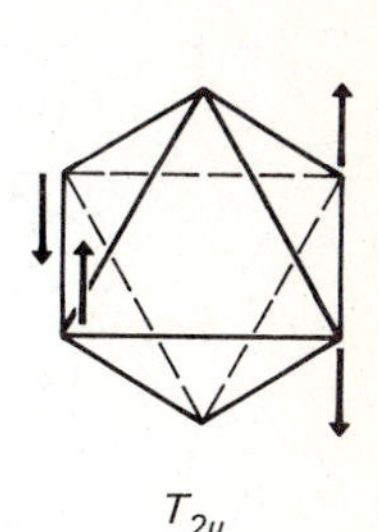

not by all. We can understand this by rewriting the transition moment integral to include both the electronic and the vibrational components of the wave function as in equation (5-8):

$$f \propto D = \left| \int \psi_{\text{el}}{}^* \psi_{\text{vib}}{}^* \hat{M} \psi_{\text{el}}{}^{\text{ex}} \psi_{\text{vib}}{}^{\text{ex}} \, d\tau \right|^2 \tag{5-8}$$

As we mentioned in Chapter 4, all ground vibrational wave functions are A_1, so the symmetry of $\psi_{\text{el}}\psi_{\text{vib}}$ becomes that of ψ_{el}, which is also A_1 for molecules with no unpaired electrons. (In general discussion, we shall use the symbol A_1 to represent the totally symmetric irreducible representation, even though this is not the appropriate label in some point groups.) To use this equation to see whether a forbidden transition can gain intensity by vibronic coupling, we must take a product $\hat{M}_{(x, y, \text{ or } z)}\psi_{\text{el}}{}^{\text{ex}}$ that is not A_1 and see whether there is a vibrational mode with symmetry that makes the product $\hat{M}_{(x, y \text{ or } z)}\psi_{\text{el}}{}^{\text{ex}}\psi_{\text{vib}}{}^{\text{ex}}$ equal to A_1. When $\psi_{\text{vib}}{}^{\text{ex}}$ has the same symmetry as the product $\hat{M}_{(x, y, \text{ or } z)}\psi_{\text{el}}{}^{\text{ex}}$, the product will be A_1.

This discussion can be made clearer by considering some examples. We shall consider vibrational spectroscopy in more detail in the next chapter. A non-linear molecule has $3N - 6$ internal vibrations; for formaldehyde these are $3a_1$, b_1, and $2b_2$. For the forbidden transition ${}^1A_1 \rightarrow {}^1A_2(n_b \rightarrow \pi^*)$, the vibrational wave function of a_1 symmetry does not change the direct product $\hat{M}_{(x, y, \text{ or } z)}{}^1A_2$ so no intensity can be gained by this mode. Excitation of the b_1 vibrational mode leads to a direct product $\psi_{\text{el}}{}^{\text{ex}}\psi_{\text{vib}}{}^{\text{ex}}$ of $b_1 \times A_2 = B_2$. Since $\hat{M}_y$ has B_2 symmetry, the total integral $(\int \psi_{\text{el}}\psi_{\text{vib}}\hat{M}_y\psi_{\text{el}}{}^{\text{ex}}\psi_{\text{vib}}{}^{\text{ex}}\, d\tau)$ has A_1 symmetry, and the electronic transition becomes allowed by vibronic coupling to the b_1 mode.

It is informative to consider $Co(NH_3)_6{}^{3+}$ as an example, for it contains triply degenerate irreducible representations. The ground state is ${}^1A_{1g}$ (a strong field O_h d^6 complex). The excited states from d-d transitions are ${}^1T_{1g}$ and ${}^1T_{2g}$. $\hat{M}_x$, $\hat{M}_y$, and $\hat{M}_z$ transform as T_{1u}. For the ${}^1A_{1g} \rightarrow {}^1T_{1g}$ transition one obtains:

$$A_{1g} \times T_{1u} \times T_{1g}$$

The resulting direct product representation has a dimensionality of nine (the identity is $1 \times 3 \times 3 = 9$) and the total representation is reduced into a linear combination of $A_{1u} + E_u + T_{1u} + T_{2u}$ irreducible representations. With no A_{1g} component, the ${}^1A_{1g} \rightarrow {}^1T_{1g}$ transition is forbidden. However, the vibrations for an octahedral complex have the symmetries of a_{1g}, e_g, $2t_{1u}$, t_{2g}, t_{2u}. Since the direct products $t_{1u} \times T_{1u}$ and $t_{2u} \times T_{2u}$ have A_{1g} components, this transition becomes allowed by vibronic coupling. For practice, the reader should take the direct products and factor the reducible representations discussed above.

5-12 MIXING OF *d* AND *p* ORBITALS IN CERTAIN SYMMETRIES

There is one further aspect of the intensity of electronic transitions that can be understood *via* the symmetry aspects of electronic transitions. The electronic spectra of tetrahedral complexes of cobalt(II) contain two bands assigned to *d-d* transitions at $\sim$20,000 cm^{-1} and $\sim$6000 cm^{-1}, assigned as $A_2 \rightarrow T_1$ and $A_2 \rightarrow T_2$ transitions respectively, with molar absorptivities of 600 and 50. Since the $\widehat{M}$ components transform as T_2, we obtain for the $A_2 \rightarrow T_1$ transition

$$A_2 \times T_2 \times T_1 = A_1 + E + T_1 + T_2$$

so the transition is allowed. However, if only the *d*-orbitals were involved in this transition, the intensity would be zero for the integrals

$$\int \psi_{d_{xy}} \widehat{M} \psi_{d_{xz}} \, d\tau = 0$$

However, in the T_d point group, the d_{xy}, d_{xz}, and d_{yz} orbitals and the *p*-orbitals transform as T_2 and therefore can mix. If the two states involved in the transition, A_2 and T_1, have differing amounts of *p*-character, intensity is gained by having some of the highly allowed $p \rightarrow d$ or $d \rightarrow p$ character associated with the transition.

Consider the consequences of this mixing on the $A_2 \rightarrow T_2$ transition. The transition moment integrand for this transition is

$$A_2 \times T_2 \times T_2$$

which, as the reader should verify, can be reduced to $A_2 + E + T_1 + T_2$. Since there is no A_1 component, the transition is forbidden. Mixing *p*-character into the wave functions will not help, for this type of transition is still forbidden. Accordingly, the ε for the $A_2 \rightarrow T_1$ transition is ten times greater than that of $A_2 \rightarrow T_2$. The latter transition gains most of its intensity by vibronic coupling.

5-13 MAGNETIC DIPOLE AND ELECTRIC QUADRUPOLE CONTRIBUTIONS TO THE INTENSITY

So far, our discussion of the intensity of electronic transitions has centered on the electric dipole component of the radiation, for we concerned ourselves with the transition moment integral with an electric dipole operator, $e\vec{r}$. There is also a magnetic dipole component. The magnetic dipole operator transforms as a rotation $R_xR_yR_z$, and the intensity from this effect may be regarded as arising from the rotation of electron density. Transition moment integrals similar to those for electric dipole transitions can be written for the contribution from both magnetic dipole and electric quadrupole effects. In a molecule with a center of symmetry, both of these operators are symmetric with respect to inversion, so $g \rightarrow g$ and $u \rightarrow u$ transitions are allowed. Approximate values of the transition moment integral for allowed transitions for these different operators are: 6×10^{-36} cgs units for an electric dipole transition, 9×10^{-41} cgs units for a magnetic dipole transition, and 7×10^{-43} cgs units for a quadrupole transition. Thus, we can see that these latter two effects will be important only when electric dipole transitions are forbidden. They do complicate the assignment of very weak bands in the spectrum.

5-14 CHARGE TRANSFER TRANSITIONS

A transition in which an electron is transferred from one atom or group in the molecule to another is called a *charge-transfer* transition. More accurately stated, the transition occurs between molecular orbitals that are essentially centered on different atoms. Very intense bands result, with molar absorptivities of 10^4 or greater. The frequency at maximum absorbancy, ν_{max}, often, but not always, occurs in the ultraviolet region. The anions ClO_4^- and SO_4^{2-} show very intense bands. Since MnO_4^- and CrO_4^{2-} have no *d* electrons, the intense colors of these ions cannot be explained on the basis of *d-d* transitions; they are attributed to charge transfer transitions.[9] The transitions in MnO_4^- and CrO_4^{2-} are most simply visualized as an electron transfer from a nonbonding orbital of an oxygen atom to the manganese or chromium ($n \rightarrow \pi^*$), in effect reducing these metals in the excited state.[10] An alternate description for this transition involves excitation of an electron from a π bonding molecular orbital, consisting essentially of oxygen atomic orbitals, to a molecular orbital that is essentially the metal atomic orbital.

In the case of a pyridine complex of iridium(III), a charge transfer transition that involves oxidation of the metal has been reported.[11] A metal electron is transferred from an orbital that is essentially an iridium atomic orbital to an empty π^* antibonding orbital in pyridine.

In gaseous sodium chloride, a charge transfer absorption occurs from the ion pair Na^+Cl^- to an excited state described as sodium and chlorine atoms having the same internuclear distance as the ion pair. A charge transfer absorption also occurs in the ion pair, *N*-methylpyridinium iodide[36] (see Fig. 5-13) in which an electron is transferred from I^- to a ring antibonding orbital. The excited state is represented in Fig. 5-13. A very intense charge transfer absorption is observed in addition compounds formed between iodine and several Lewis bases. This phenomenon will be discussed in more detail in a later section.

5-15 POLARIZED ABSORPTION SPECTRA

If the incident radiation employed in an absorption experiment is polarized, only those transitions with similarly oriented dipole moment vectors will occur. In a powder, the molecules or complex ions are randomly oriented. All allowed transitions will be observed, for there will be a statistical distribution of crystals with dipole moment vectors aligned with the polarized radiation. However, suppose, for example, that a formaldehyde crystal, with all molecules arranged so that their *z*-axes are parallel, is examined. As indicated in the previous section, the integrand $\psi^*\widehat{M}_z\psi$ has appropriate symmetry for the ${}^1A_{(\pi,\pi^*)} \leftarrow {}^1A$ transition, but $\psi^*\widehat{M}_x\psi$ and $\psi^*\widehat{M}_y\psi$ do not. When the *z*-axes of the molecules in the crystal are aligned parallel to light that has its electric vector polarized in the *z*-direction, light will be absorbed for the ${}^1A_{(\pi,\pi^*)} \leftarrow {}^1A$ transition. Light of this wavelength polarized in other planes will not be absorbed. If this crystal is rotated so that the *z*-axis is perpendicular to the plane of polarization of the light, no light is absorbed. This behavior supports the assignment of this band to the transition ${}^1A_1 \leftarrow {}^1A_1$. To determine the expected polarization of any band, the symmetry species of the product $\psi_a\psi_b$ is compared with the components of $\widehat{M}$, as was done above for formaldehyde. The polarization experiment is schematically illustrated in Fig. 5-10.

In Fig. 5-10A, absorption of radiation will occur if $\widehat{M}_z$ results in an A_1 transition moment integrand for equation (5-6). No absorption will occur if it is not A_1 regardless of the symmetries of the integrand for the $\widehat{M}_x$ or $\widehat{M}_y$ components [*i.e.*,

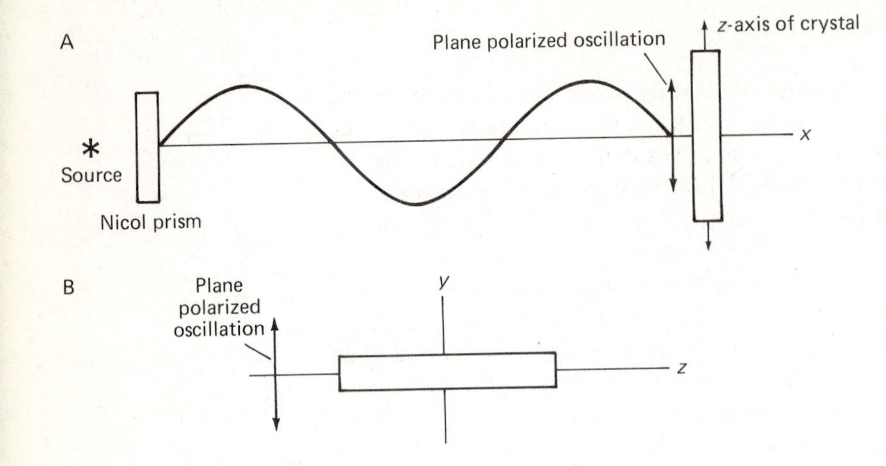

FIGURE 5–10. Schematic illustration of a polarized single crystal study. (A) The z-axis of the crystal is parallel to the oscillating electromagnetic plane polarized component. (B) The z-axis of the crystal is perpendicular to the oscillating electromagnetic plane polarized component, and the y-axis is parallel to it.

equations (5–4) and (5–5)]. In Fig. 5–10B, absorption will occur if the $\hat{M}_y$ component gives an A_1 transition moment integral. Even if $\hat{M}_z$ has an integrand with A_1 symmetry, no absorption of the z-component will occur for this orientation and no absorption at all will occur if the $\hat{M}_y$ integrand is not A_1.

We can further illustrate these ideas by considering the electronic absorption spectrum of $PtCl_4{}^{2-}$. The transitions are charge transfers involving electron excitation from a mainly chlorine m.o. to the empty $d_{x^2-y^2}$ orbital on Pt(II). The symmetry is D_{4h}; using the same approach as that employed in Chapter 3 on the $NO_2{}^-$ ion, on a basis set of four p_z orbitals on chlorine, we obtain symmetry orbitals for chlorine of b_{2u}, e_u, and a_{2u} symmetry. This leads to the following possible charge transfer transitions:

$b_{2u}(\pi) \rightarrow b_{1g}(d_{x^2-y^2})$ with state labels ${}^1A_{1g} \rightarrow {}^1A_{2u}$ (here A_{2u} is the direct product of $b_{2u} \times b_{1g}$)
$e_u(\pi) \rightarrow b_{1g}(d_{x^2-y^2})$ with state labels ${}^1A_{1g} \rightarrow {}^1E_u$
$a_{2u}(\pi) \rightarrow b_{1g}(d_{x^2-y^2})$ with state labels ${}^1A_{1g} \rightarrow {}^1B_{2u}$

In the D_{4h} point group, $\hat{M}_x$ and $\hat{M}_y$ transform as E_u, and $\hat{M}_z$ as A_{2u}. If we first consider the $A_{1g} \rightarrow A_{2u}$ transition, we get for $\hat{M}_z$:

$$A_{1g}A_{2u}A_{2u} = A_{1g}$$

and for $\hat{M}_x$ and $\hat{M}_y$ we get:

$$A_{1g}E_uA_{2u} \neq A_{1g}$$

Accordingly, this transition is allowed and is polarized in the z-direction. If we use polarized light and a single crystal, light will be absorbed when the z-axis of the crystal is parallel to the z-direction of the light; but there will be no absorption when the z-axis is perpendicular to the light because the $\hat{M}_x$ and $\hat{M}_y$ integrands are not A_1.

For the band assigned to ${}^1A_{1g} \rightarrow {}^1E_u$, the $A_{1g}E_uE_u$ product has an A_{1g} component, so this transition is also allowed. Since $\hat{M}_z$ yields $A_{1g}A_{2u}E_u$, which does not have an A_{1g} component, there will be no absorption when the z-component is parallel to the plane of the polarized light but absorption will occur when the x- and y-axes of the crystal are parallel to the light.

The $A_{1g} \rightarrow B_{2u}$ transition turns out to be forbidden. Thus, we see that by employing polarized single crystal spectroscopy, we can rigorously assign the two intense charge transfer bands observed in the electronic spectrum of $PtCl_4^{2-}$. *If the single crystal employed in these experiments did not have all of the molecular z-axes aligned, the polarization experiments would not work.*

APPLICATIONS

Most applications of electronic spectroscopy have been made in the wavelength range from 2100 to 7500 Å, for this is the range accessible with most recording spectrophotometers. Relatively inexpensive commercial instruments can now be obtained to cover the range from 1900 to 8000 Å. The near infrared region, from 8000 to 25,000 Å, has also provided much useful information. Spectra can be examined through the 1900 to 25,000 Å region on samples of vapors, pure liquids, or solutions. Solids can be examined as single crystals or as discs formed by mixing the material with KCl or NaCl and pressing with a hydraulic press until a clear disk is formed.[12] Spectra of powdered solids can also be examined over a more limited region (4000 to 25,000 Å) as reflectance spectra or on mulls of the solid compounds.[12]

5–16 FINGERPRINTING

Since many different substances have very similar ultraviolet and visible spectra, this is a poor region for product identification by the "fingerprinting" technique. Information obtained from this region should be used in conjunction with other evidence to confirm the identity of the compound. Evidence for the presence of functional groups can be obtained by comparison of the spectra with reported data. For this purpose, ν_{max}, ε_{max}, and band shapes can be employed. It is also important that the spectra be examined in a variety of solvents to be sure that the band shifts are in accord with expectations (see discussion of blue shifts).

Spectral data have been compiled by Sadtler (see General References), Lang,[13] and Hershenson,[14] and in "Organic Electronic Spectral Data"[15] and the ASTM Coded IBM Cards.[16] A review article by Mason[17] and the text by Jaffe and Orchin[1] are excellent for this type of application. If a functional group (chromophore) is involved in conjugation or steric interactions, or is attached to electron-releasing groups, its spectral properties are often different from those of an isolated functional group. These differences can often be predicted semiquantitatively for molecules in which such effects are expected to exist.[17]

The spectra of some representative compounds and examples of the effect of substituents on the wavelength of a transition will be described briefly.

Saturated Molecules. Saturated molecules without lone pair electrons undergo high-energy $\sigma \rightarrow \sigma^*$ transitions in the far ultraviolet. For example, methane has a maximum at 1219 Å and ethane at 1350 Å corresponding to this transition. When lone pair electrons are available, a lower-energy $n \rightarrow \sigma^*$ transition is often detected in addition to the $\sigma \rightarrow \sigma^*$. For example, in triethylamine two transitions are observed, at 2273 and 1990 Å.

Table 5–2 contains a listing of absorption maxima for some saturated compounds and gives some indication of the variation in the range and intensity of transitions in saturated molecules.

Carbonyl Compounds. The carbonyl chromophore has been very extensively studied. Upon conjugation of the carbonyl group with a vinyl group, four π energy levels are formed. The highest occupied π level has a higher energy, and one of the lowest empty π^* levels has a lower energy, than the corresponding levels in a nonconjugated carbonyl group. The lone pair and σ electrons are relatively unaf-

TABLE 5–2. FREQUENCIES OF ELECTRONIC TRANSITIONS IN SOME SATURATED MOLECULES

Compound	λ_{max} Å	ϵ_{max}	Medium
H_2O	1667	1480	vapor
MeOH	1835	150	vapor
Me_2O	1838	2520	vapor
Me_2S	2290,2100	140,1020	ethanol
S_8	2750	8000	ethanol
F_2	2845	6	vapor
Cl_2	3300	66	vapor
Br_2	4200	200	vapor
I_2	5200	950	vapor
ICl	~4600	153	CCl_4
SCl_2	3040	1150	CCl_4
PI_3	3600	8800	Et_2O
AsI_3	3780	1600	pet. ether

fected by conjugation. As a result, the $\pi \rightarrow \pi^*$ and $n \rightarrow \pi^*$ transition energies are lowered and the absorption maxima are shifted to longer wavelengths when the carbonyl is conjugated. The difference is greater for the $\pi \rightarrow \pi^*$ than for the $n \rightarrow \pi^*$ transition. The $n \rightarrow \sigma^*$ band is not affected appreciably and often lies beneath the shifted $\pi \rightarrow \pi^*$ absorption band. As stated earlier, electron-donating groups attached to the carbonyl cause a blue shift in the $n \rightarrow \pi^*$ transition and a red shift in $\pi \rightarrow \pi^*$.

It is of interest to compare the spectra of thiocarbonyl compounds with those of carbonyl compounds. In the sulfur compounds, the carbon-sulfur π interaction is weaker and, as a result, the energy difference between the π and π^* orbitals is smaller than in the oxygen compounds. In addition, the ionization potential of the sulfur electrons in the thiocarbonyl group is less than the ionization potential of oxygen electrons in a carbonyl. The n electrons are of higher energy in the thiocarbonyl and the $n \rightarrow \pi^*$ transition requires less energy in these compounds than in carbonyls. The absorption maximum in thiocarbonyls occurs at longer wavelengths and in some compounds is shifted into the visible region.

Inorganic Systems. The SO_2 molecule has two absorption bands in the near ultraviolet at 3600 Å ($\varepsilon = 0.05$) and 2900 Å ($\varepsilon = 340$) corresponding to a triplet and singlet $n \rightarrow \pi^*$ transition. The gaseous spectrum shows considerable vibrational fine structure, and analysis has produced information concerning the structure of the excited state.[18]

In nitroso compounds, an $n \rightarrow \pi^*$ transition involving the lone pair electrons on the nitrogen occurs in the visible region. An $n \rightarrow \pi^*$ transition involving an oxygen lone pair occurs in the ultraviolet.

The nitrite ion in water has two main absorption bands at 3546 Å ($\varepsilon = 23$) and 2100 Å ($\varepsilon = 5380$) and a weak band at 2870 Å ($\varepsilon = 9$). The assignment of these bands has been reported,[19] and this article is an excellent reference for gaining an appreciation of how the concepts discussed in this chapter are used in band assignments. The band at 3546 Å is an $n \rightarrow \pi^*$ transition ($^1B_1 \leftarrow {}^1A_1$) involving the oxygen lone pair. The band at 2100 Å is assigned as $\pi \rightarrow \pi^*(^1B_2 \leftarrow {}^1A_1)$, and the band at 2870 Å is assigned to an $n \rightarrow \pi^*$ transition ($^1A_2 \leftarrow {}^1A_1$) involving the oxygen lone pair.

The absorption peaks obtained for various inorganic anions in water or alcohol solution are listed in Table 5–3. For the simple ions (Br^-, Cl^-, OH^-) the absorption is attributed to charge transfer in which the electron is transferred to the solvent.

TABLE 5-3.
CHARACTERISTIC ABSORPTION MAXIMA FOR SOME INORGANIC ANIONS

	λ (Å)	ε
Cl^-	1810	10^4
Br^-	1995	11,000
	1900	12,000
I^-	2260	12,600
	1940	12,600
OH^-	1870	5000
SH^-	2300	8000
$S_2O_3^{-2}$	2200	4000
$S_2O_8^{-2}$	2540	22
NO_2^-	3546	23
	2100	5380
	2870	9
NO_3^-	3025	7
	1936	8800
$N_2O_2^{-2}$	2480	4000

There are many more examples of applications of electronic spectroscopy to inorganic and organometallic systems. These are reviewed on a regular basis in the Specialist Periodical Reports of the Chemical Society (London). Well armed with the fundamentals, the interested reader is referred to this source[20] for more examples.

5-17 MOLECULAR ADDITION COMPOUNDS OF IODINE

The absorption band maximum for iodine (core plus $\sigma^2\pi^4n^4\pi^{*4}$) occurs at about 5200 Å in the solvent CCl_4 and is assigned to a $\pi^* \rightarrow \sigma^*$ transition. When a donor molecule is added to the above solution, two pronounced changes in the spectrum occur (see Fig. 5-11).

A blue shift is detected in the iodine peak, and a new peak arises in the ultraviolet region that is due to a charge transfer transition.[21] The existence of an isosbestic point at 490 mμ indicates that there are only two absorbing species in the system; namely, free iodine and the complex B:I—I. As indicated in Chapter 4, a 1:1 equilibrium constant can be calculated from absorbance measurements for this system. The constant value for K obtained over a wide range of donor concentrations is evidence for the existence of a 1:1 addition compound.

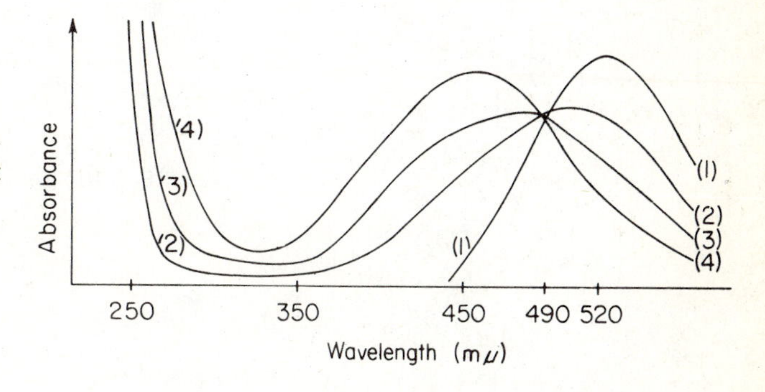

FIGURE 5-11. Spectra of iodine and base-iodine solutions. (1) I_2 in CCl_4; (2, 3, 4) same I_2 concentration but increasing base concentration in the order $2 < 3 < 4$.

The bonding in iodine addition compounds can be described by the following equation:

$$\psi^0 = a\psi_{\text{cov}} + b\psi_{\text{el}}$$

where ψ_{el} includes contributions from purely electrostatic forces while ψ_{cov} includes contributions from covalent interactions (these are described as charge transfer interactions). In many of these complexes, $b > a$ in the ground state. In these cases, the band around 2500 Å arises from a charge transfer transition in which an electron from this ground state is promoted to an excited state in which $a > b$. In view of these coefficients, the charge transfer band assignment can be approximated by a transfer of a base electron, n_b, to the iodine σ^* orbital. These facts and the blue shift that occurs in the normal $\pi^* \rightarrow \sigma^*$ iodine transition upon complexation can be explained by consideration of the relative energies of the molecular orbitals of iodine and the complex (Fig. 5–12). In Fig. 5–12, n_b refers to the donor orbital on the base, and $\sigma_{I_2}{}^*$ and $\pi_{I_2}{}^*$ refer to the free iodine antibonding orbitals involved in the transition leading to iodine absorption. The σ_c, $\pi_c{}^*$, and $\sigma_c{}^*$ orbitals are molecular orbitals in the complex that are very much like the original base and iodine orbitals because of the weak Lewis acid-base interaction (2 to 10 kcal). The orbitals n_b and $\sigma_{I_2}{}^*$ combine to form molecular orbitals in the complex, σ_c and $\sigma_c{}^*$, in which σ_c, the bonding orbital, is essentially n_b and $\sigma_c{}^*$ is essentially $\sigma_{I_2}{}^*$. Since $\sigma_c{}^*$ is slightly higher in energy than the corresponding $\sigma_{I_2}{}^*$, the transition in complexed iodine [arrow (2) in Fig. 5–12] requires slightly more energy than the corresponding transition in free I_2 [arrow (1)] and a blue shift is observed. The charge-transfer transition occurs at higher energy in the ultraviolet region and is designated in Fig. 5–12 by arrow (3). Some interesting correlations have been reported, which claim that the blue shift is related to the magnitude of the base-iodine interaction, *i.e.*, the enthalpy of adduct formation.[21] This would be expected qualitatively from the treatment in Fig. 5–12 as long as the energy of $\pi_c{}^*$ differs very little from that of $\pi_{I_2}{}^*$ or else its energy changes in a linear manner with the enthalpy, ΔH. A rigorous evaluation of this correlation with accurate data on a wide range of different types of Lewis bases indicates that a rough general trend exists, but that a quantitative relation (as good as the accuracy of the data) does not exist. A relationship involving the charge transfer band, the ionization potential of the base, I_b, and the electron affinity of the acid, E_a, is also reported:[22,23]

$$\nu = I_b - E_a - \Delta \tag{5-9}$$

where Δ is an empirically determined constant for a related series of bases.

The enthalpies for the formation of these charge transfer complexes are of interest and significance to both inorganic and organic chemists. For many inorganic systems, especially in the areas of coordination chemistry and nonaqueous solvents,

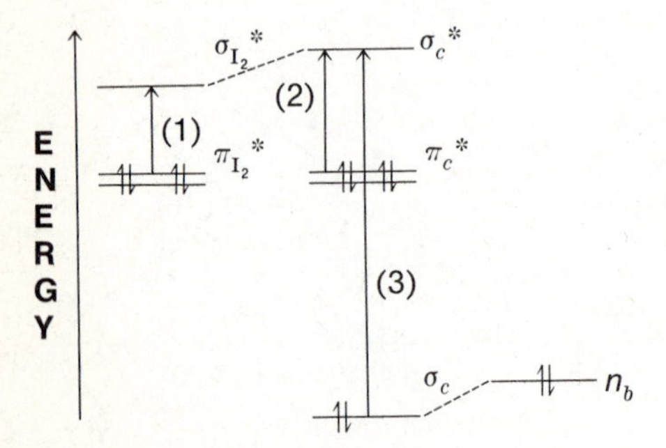

FIGURE 5–12. Some of the molecular orbitals in a base-iodine addition compound.

TABLE 5-4. EQUILIBRIUM CONSTANTS AND ENTHALPIES OF FORMATION FOR SOME DONOR-I_2 ADDUCTS

Donor	K(liter mole^{-1})	$-\Delta H$(kcal mole^{-1})
C_6H_6	0.15 (25°)	1.4
Toluene	0.16 (25°)	1.8
CH_3OH	0.47 (20°)	1.9
Dioxane	1.14 (17°)	3.5
$(C_2H_5)_2O$	0.97 (20°)	4.3
$(C_2H_5)_2S$	180 (25°)	8.3
$CH_3C(O)N(CH_3)_2$	6.1 (25°)	4.7
Pyridine	270 (20°)	7.8
$(C_2H_5)_3N$	5130 (25°)	12.0

information about donor and acceptor interactions is essential to an understanding of many phenomena (for example, catalysis and metalloproteins). Since the above adducts are soluble in CCl_4 or hexane, the thermodynamic data can be interpreted more readily than results obtained in polar solvents, where large solvation enthalpies and entropies are encountered. As a result of these solvation effects, structural interpretations of the effect of substituents on pK_b values of bases and on stability constant data for various ligands and metal ions are often highly questionable. Some typical results from donor-I_2 systems in which such solvation effects are minimal illustrate the wide range of systems that can be studied, and are contained in Table 5-4.

The following few examples illustrate the information that can be obtained by studying enthalpies of association in non-polar, weakly basic solvents.

(1) The donor properties of the π electron systems of alkyl substituted benzenes have been reported.[24]

(2) A correlation of the heat of formation of iodine adducts of a series of *para*-substituted benzamides with the Hammett substituent constants[25] of the benzamides is reported.

(3) The donor properties of a series of carbonyl compounds [$(CH_3)_2CO$, $CH_3C(O)N(CH_3)_2$, $(CH_3)_2NC(O)N(CH_3)_2$, $CH_3C(O)OCH_3$, $CH_3C(O)SCH_3$] have been evaluated and interpreted [26] in terms of conjugative and inductive effects of the group attached to the carbonyl functional group.

(4) The donor properties of sulfoxides, sulfones, and sulfites have been investigated.[27] The results are interpreted to indicate that sulfur-oxygen π bonding is less effective in these systems than carbon-oxygen π bonding is in ketones and acetates.

(5) The effect of ring size on the donor properties of cyclic ethers and sulfides has been investigated.[28] It was found that for saturated cyclic sulfides, of general formula $(CH_2)_nS$, the donor properties of sulfur are in the order $n = 5 > 6 > 4 > 3$. The order for the analogous ether compounds is $4 > 5 > 6 > 3$. Explanations of these effects are offered.

(6) The donor properties of a series of primary, secondary, and tertiary amines have been evaluated.[29,30] The order of donor strength of amines varies with the acid studied. Explanations have been offered, which are based upon the relative importance of covalent and electrostatic contributions to the bonding in various adducts.

In addition to iodine, several other Lewis acids form charge transfer complexes that absorb in the ultraviolet or visible regions. For example, the relative acidities of I_2, ICl, Br_2, SO_2, and phenol toward the donor *N,N*-dimethylacetamide have been evaluated. Factors affecting the magnitude of the interaction[31] and information regarding the bonding in the adducts are reported. Good general reviews of charge-transfer complexes are available.[9,32,33,34,35]

5-18 EFFECT OF SOLVENT POLARITY ON CHARGE TRANSFER SPECTRA

The ion pair *N*-methylpyridinium iodide undergoes a charge-transfer transition that can be represented[36] as in Fig. 5-13. It has been found that the position of the

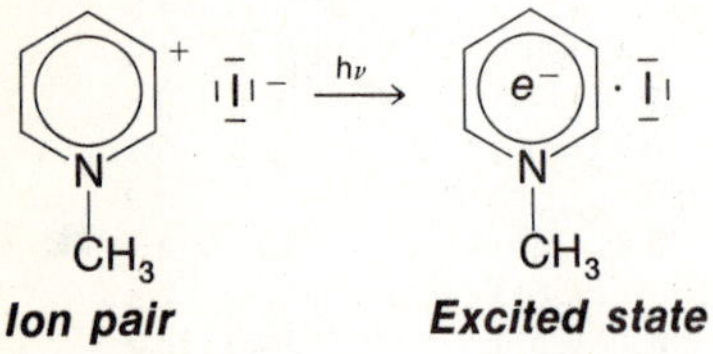

Ion pair **Excited state**

FIGURE 5-13. Ion pair and charge transfer excited state of *N*-methylpyridinium iodide.

charge-transfer band is a function of the solvating ability of the solvent. A shift to lower wavelengths is detected in the better solvating solvents. The positions of the bands are reported as transition energies, E_T. Transition energies (kcal mole^{-1}) are calculated from the frequency as described in Chapter 4. The transition energy is referred to as the Z value. Some typical data are reported in Table 5-5. An explanation for the observed shift has been proposed.[36] The dipole moment of the ion pair, $C_5H_5NCH_3{}^+I^-$, is reported to be perpendicular to the dipole moment of the excited state (Fig. 5-13). Polar solvent molecules will align their dipole moments for maximum interaction with the ground state, lowering the energy of the ground state by solvation. The dipole moment of the solvent molecules will be perpendicular to the dipole moment of the excited state, producing a higher energy for the excited state than would be found in the gas phase. Since solvent molecules cannot rearrange in the time required for a transition, the relative lowering of the ground state and raising of the excited state increases the energy of the transition, E_T, over that in the gas phase (Fig. 5-14), shifting the wavelength of absorption to higher frequencies. Hydrogen-bonding solvents are often found to increase E_T more than would be expected by comparing their dielectric constants with those of other solvents. This is due to the formation of hydrogen bonds with the solute. The use of the dielectric constant to infer solvating ability can lead to difficulty because the local dielectric constant in the vicinity of the ion may be very different from the bulk dielectric constant.

The data obtained from these spectral shifts are employed as an empirical measure of the ionizing power of the solvent. The results can be correlated with a

TABLE 5-5. Z VALUES FOR SOME COMMON SOLVENTS

Solvent	E_T or Z value[a]
H_2O	94.6
CH_3OH	83.6
C_2H_5OH	79.6
CH_3COCH_3	65.7
$(CH_3)_2NCHO$	68.5
CH_3CN	71.3
Pyridine	64.0
CH_3SOCH_3	71.1
H_2NCHO	83.3
CH_2Cl_2	64.2
Isooctane	60.1

[a] The E_T or Z value is the transition energy in kcal mole^{-1} at 25°C, 1 atm pressure, for the compound 1-ethyl-4-carbomethoxypyridinium iodide.

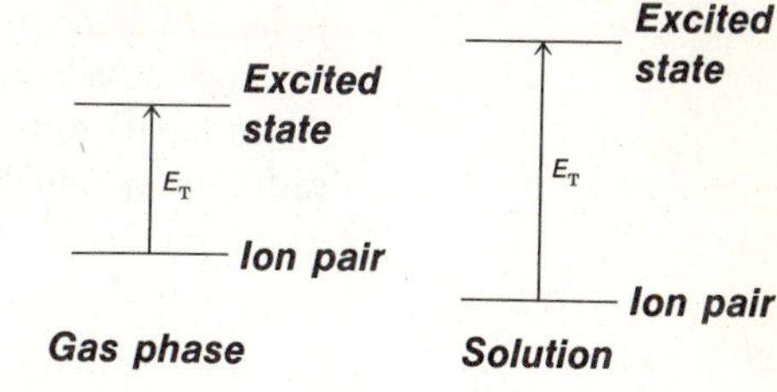

FIGURE 5-14. Effect of solvent on the transition energy, E_T.

scale of "solvent polarities" determined from the effect of solvent on the rate of solvolysis of *t*-butyl chloride.[36] Other applications of these data to kinetic and spectral studies are reported.[36] Solvent effects are quite complicated, and these correlations at best provide a semiquantitative indication of the trends expected.

Significant differences exist between "solvating power" inferred from the dielectric constant and the results from spectral and kinetic parameters. Although methanol and formamide are found to have similar Z values, the dielectric constants are 32.6 and 109.5, respectively. Solvent effects cannot be understood solely on the basis of the dielectric constant. The extent of ion-pair association of the pyridinium salts in various solvents can be approximated from the apparent molar absorptivities of the charge-transfer absorption, because the dissociated ion pair is not expected to contribute to the charge transfer absorption. The tendencies of ion pairs to dissociate in solvents estimated in this way do not correlate with transition energies. The dissociating tendency is expected to be more closely related to the dielectric constant ($F = q_1q_2/Dr^2$, where F is the force between two ions with charges q_1 and q_2 separated by a distance r, and D is the dielectric constant). Specific Lewis acid-base interactions make the problem more complex than the simple dielectric model.

The band positions for the $n \rightarrow \pi^*$ transitions in certain ketones in various solvents are found to be linearly related to the Z values for the solvents. A constant slope is obtained for a plot of E_T versus Z for many ketones. Deviations from linearity by certain ketones in this plot can be employed to provide interesting structural information about the molecular conformation. Cycloheptanone, for example, does not give a linear plot of E_T versus Z. The deviation is attributed to solvent effects on the relative proportion of the conformers present in solution.

5-19 STRUCTURES OF EXCITED STATES

Considerable information is available about the structure of excited states of molecules from analysis of the rotational band contours in the electronic spectra. Both geometrical information and vibrational information about the excited states of large molecules can be obtained.[37] By studying and analyzing the perturbation made on the vibrational fine structure of an electronic transition by an electric field, the dipole moment of the excited state can be obtained.[38]

OPTICAL ROTATORY DISPERSION, CIRCULAR DICHROISM AND MAGNETOCIRCULAR DICHROISM

5-20 INTRODUCTION

Plane polarized light consists of two circularly polarized components of equal intensity. The two types of circularly polarized light correspond to right-handed and left-handed springs. Circularly polarized light is defined as right-handed when its electric or magnetic vector rotates clockwise as viewed by an observer facing the direction of the light propagation (*i.e.*, the source). The frequency of the rotation is related to the frequency of the light. Plane polarized light can be resolved into its two circular components, and the two components when added together produce plane polarized light in an optically isotropic medium. If plane polarized light is passed through a sample for which the refractive indices of the left and right polarized components differ, the components will, upon recombination, give plane polarized radiation in which the plane of the polarization has been rotated through an angle α, given by

$$\alpha = \frac{n_\ell - n_r}{\lambda} \tag{5-10}$$

where the subscripts refer to left and right, n is the appropriate refractive index, and λ is the wavelength of light employed. The units are radians per unit length, with the length units given by those used for λ.

If the concentration of an optically active substance, c', is expressed in units of g cm^{-3} (corresponding to the density for a pure substance), the specific rotation $[\alpha]$ is defined as:

$$[\alpha] = \frac{\alpha}{c'd'} \tag{5-11}$$

where d' is the thickness of the sample in decimeters. The molar rotation $[M]$ is defined as:

$$[M] = M[\alpha] \times 10^{-2} = M\alpha \times 10^{-2}/c'd' \tag{5-12}$$

where M is the molecular weight of the optically active component. (The quantity 10^{-2} is subject to convention and not always included in $[M]$.)

The *optical rotatory dispersion* curve, ORD, is a plot of the molar rotation, $[\alpha]$ or $[M]$, against λ. When the plane of polarization rotates clockwise as viewed by an observer facing the direction of propagation of the radiation, $[\alpha]$ or $[M]$ is defined as positive; a counterclockwise rotation is defined as negative.

The technique whereby one determines that an optically active substance *absorbs* right and left circularly polarized light differently is called *circular dichroism, CD.* All optically active substances exhibit CD in the region of appropriate electronic absorption bands. The molar circular dichroism $\varepsilon_\ell - \varepsilon_r$ is defined as

$$\varepsilon_\ell - \varepsilon_r = \frac{k_\ell - k_r}{c} \tag{5-13}$$

where k, the absorption coefficient, is defined by $I = I_0\, 10^{-kd}$ with I_0 and I being the intensity of the incident and resultant light and d being the cell thickness. By plotting $\varepsilon_\ell - \varepsilon_r$ versus λ, the CD curve results.

Wherever circular dichroism is observed in a sample, the resulting radiation is not plane polarized, but is elliptically polarized. The quantity α in the above equations is then the angle between the initial plane of polarization and the major axis of the ellipse of the resultant light. One can define a quantity φ' (in radians), the tangent of which is the ratio of the major to minor axes of the ellipse. The quantity φ' is used to approximate the ellipticity; when it is expressed in degrees, it can be converted to a specific ellipticity $[\varphi]$ or molar ellipticity $[\theta]$ by

$$[\varphi] = \frac{\varphi'}{c'd'} \tag{5-14}$$

and

$$[\theta] = M[\varphi]10^{-2} \qquad (5\text{-}15)$$

where the symbols are as defined in equations (5–11) and (5–12). The quantity $[\theta]$ is related to $\varepsilon_\ell - \varepsilon_r$ by the following equation:

$$\varepsilon_\ell - \varepsilon_r = 0.3032 \times 10^{-3}[\theta] \qquad (5\text{-}16)$$

Thus, one often sees the CD curve plotted as $[\theta]$ versus λ.

With CD one can measure only the optical activity if there is an accompanying electronic absorption band. On the other hand, ORD is measurable both inside and outside the absorption band.* The ORD and CD curves of D-(−)-$[Rh(en)_3]^{3+}$ are illustrated [39] in Fig. 5–15. Throughout most of the visible region, the ORD curve is negative. However, the CD curve associated with the visible *d-d* transitions at ~300 mμ is clearly positive. All the chromophores in a molecule contribute to the rotatory power at a given wavelength, but only the chromophore that absorbs at the given wavelength contributes to the CD. Thus, a transition in the far uv can make a significant contribution to the rotation in the region of *d-d* transitions in the ORD. The negative effect in the uv dominates the *d-d* contribution through most of the visible region, and the negative ORD curve results. For most of the applications to be discussed here, CD is the method of choice.

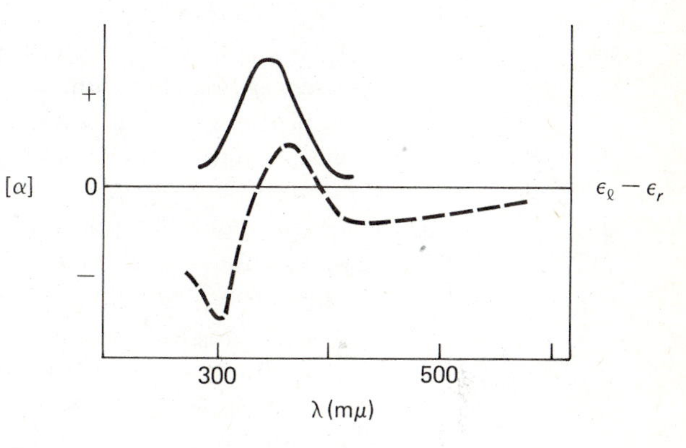

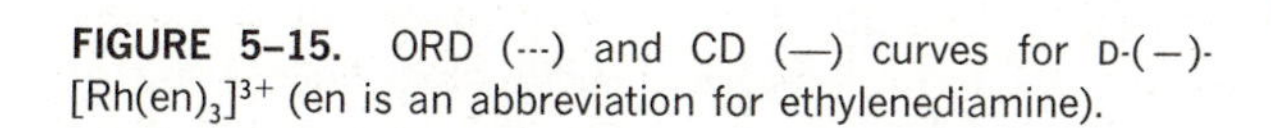
FIGURE 5–15. ORD (---) and CD (—) curves for D-(−)-$[Rh(en)_3]^{3+}$ (en is an abbreviation for ethylenediamine).

5–21 SELECTION RULES

We have previously given [equation (5–3)] the transition moment integral for an electric dipole transition, and we mentioned that a magnetic dipole transition integral has a similar form. In order for an electronic transition to give rise to optical activity, the transition must be both electric and magnetic dipole allowed, *i.e.*,

$$R \propto \left[\int \psi_{el}\widehat{M}\psi_{el}{}^{ex}\,dv\right]\left[\int \psi_{el}\widehat{MD}\psi_{el}{}^{ex}\,dv\right]$$

where R is the rotational strength, $\widehat{MD}$ is the magnetic dipole operator, and $\widehat{M}$ is the electric dipole operator. If an electronic absorption band is observed, there must be some mechanism for making this allowed; it then becomes important to be concerned with the magnetic dipole selection rules. The sign and magnitude of the activity can be calculated[40] by evaluating both the electric and magnetic dipole integrals.

*CD is the absorption difference between left and right circularly polarized light for an electronic transition. ORD is related to the difference between the indices of refraction for left and right circularly polarized light. The two effects are interrelated via the Kronig-Kramers relation.[46]

5-22 APPLICATIONS

The use of CD in band assignments is an obvious application of our previous discussion of the selection rules. For example, octahedral nickel(II) complexes have three bands assigned as $^3A_{2g}$ to $^3T_{2g}$, $^3T_{1g}$(F), and $^3T_{1g}$(P) in order of increasing energy. Since the magnetic dipole operator (the magnetic dipole transforms like the rotations $R_xR_yR_z$) in O_h is T_{1g}, only the $^3A_{2g} \rightarrow {}^3T_{2g}$ transition is magnetic dipole allowed. Accordingly, it is found in the CD spectrum of $Ni(pn)_3^{2+}$ (pn = propylene diamine) that the low-energy band has a maximum $(\varepsilon_l - \varepsilon_r)$ value of 0.8, while those for the other bands are less than 0.04. This confirms the original band assignments.

A second type of application involves using CD to show that certain absorption bands have contributions from more than one electronic transition. The CD bands are usually narrower and can be positive or negative. This idea is illustrated in Fig. 5-16, where the absorption curve and CD curve for $\Delta(+)$-$Co(en)_3^{3+}$ in aqueous solution are shown.[41]

Co(III) complexes with O_h symmetry commonly have two absorptions assigned to $^1T_{1g}$ and $^1T_{2g}$. The symmetry of $Co(en)_3^{3+}$ is D_3, so the transitions to the T states are expected to show some splittings. This is not detected in the absorption spectrum, as seen in Fig. 5-16. However, the effects of lower symmetry are observed in the CD spectrum. The magnetic dipole selection rules for D_3 predict that the low-energy band will have two magnetic dipole allowed components, $^1A_1 \rightarrow {}^1A_2$ and $^1A_1 \rightarrow {}^1E_a$. The $^1A_1 \rightarrow {}^1E_b$ transition of the high-energy band is magnetic dipole allowed, but the $^1A_1 \rightarrow {}^1A_1$ transition is not. In the CD, two components (+ and −) are seen in the low-energy band and one in the high-energy band, as predicted for a D_3 distortion. Applications of these ideas can be used to indicate the symmetry of molecules in solution and in uniaxial single crystals.

Another application involves using the sign of the CD to obtain the absolute configuration of a molecule.[42] This application has been particularly successful for organic compounds.[43] In inorganic systems, absolute configurations are often assigned by analogy to known systems. Particular care must be employed in determining what constitutes an analogous complex.[44,45] However, by using complete operator-matrices for the electric and magnetic dipole components, the signs of the trigonal components in some Co(III) and Cr(III) complexes have been related to the absolute configurations of the complexes.[40]

Finally, optically active transitions are polarized, and the polarization information can be used to support the assignment of the electronic spectrum.

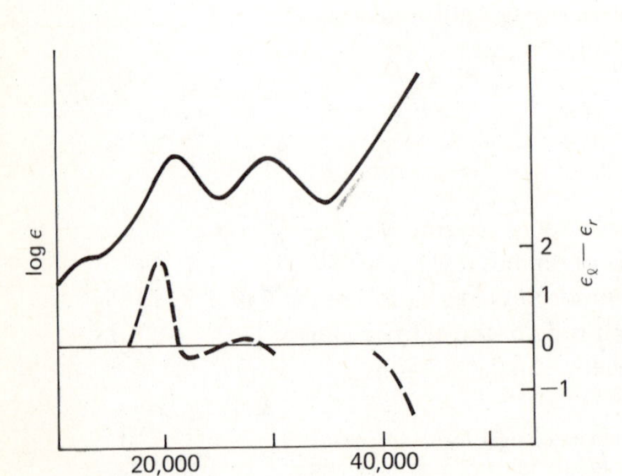

FIGURE 5-16. The absorption spectrum (solid line) and CD curve (dashed line) of $(+)$-$Co(en)_3^{3+}$ in aqueous solution.

5-23 MAGNETOCIRCULAR DICHROISM

When plane-polarized light is passed through any substance in a magnetic field, H_0, whose component in the direction of the light propagation is non-zero, the substance appears to be optically active. Left and right circularly polarized light do not interact in equivalent ways. For atoms, for example, left circularly polarized (lcp) light induces a transition in which Δm_J is -1, while for right circularly polarized (rcp) light Δm_J is $+1$. Note that m_J has the same relationship to J as m_ℓ has to ℓ. If one observed a transition from an S state where $J = 0$ to a ^1P state where $J = 1$, the two transitions in Fig. 5-17 would occur as a consequence of this selection rule.[46]

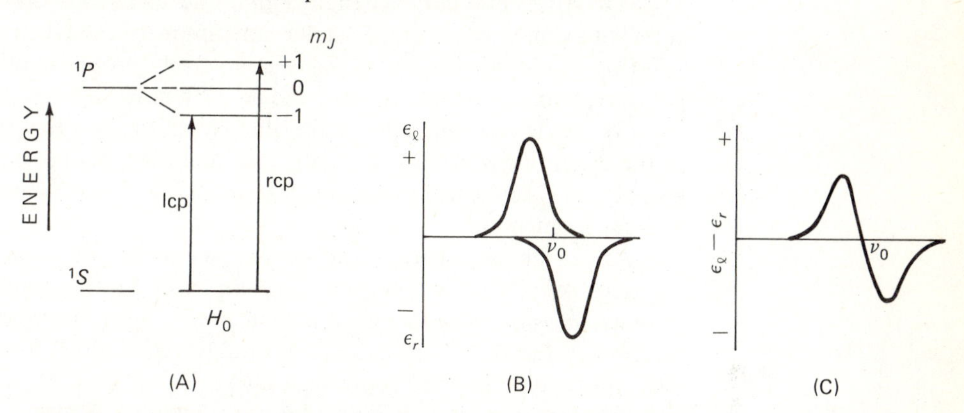

FIGURE 5-17. The transitions and expected spectrum for $^1S \rightarrow {}^1P$ in the mcd experiment. (A) the transitions; (B) spectra for left (ε_ℓ) and right (ε_r) circularly polarized radiation; (C) the mcd spectrum ($\varepsilon_\ell - \varepsilon_r$); A-term behavior.

If the absorptions of left and right circularly polarized light corresponding to these transitions are measured separately, the curves in Fig. 5-17(B) are obtained. Here ν_0 is the band maximum for the absorption band. When the mcd curve is plotted, the result in Fig. 5-17(C) is obtained, provided that the band width is much greater than the Zeeman splitting of the excited state. A curve of this sort is referred to as an A-term and can arise only for a transition in which $J > 0$ for one of the states involved. The sign of the A-term in molecules depends upon the sign of the Zeeman splitting and the molecular selection rules for circularly polarized light.

Next consider a transition from a ^1P to a ^1S state. Fig. 5-18 summarizes this situation. The $\Delta m_J = +1$ transition for rcp light is now that of m_J from -1 to 0. Because the m_J states are not equally populated but Boltzmann populated, the two transitions will not have equal intensity, as shown in Fig. 5-18(B). The relative intensities will be very much temperature dependent. The resultant mcd curve, shown in Fig. 5-18(C), is referred to as a C-term. The band shape and intensity are very temperature dependent. An A-term curve usually occurs superimposed upon a C-term curve.

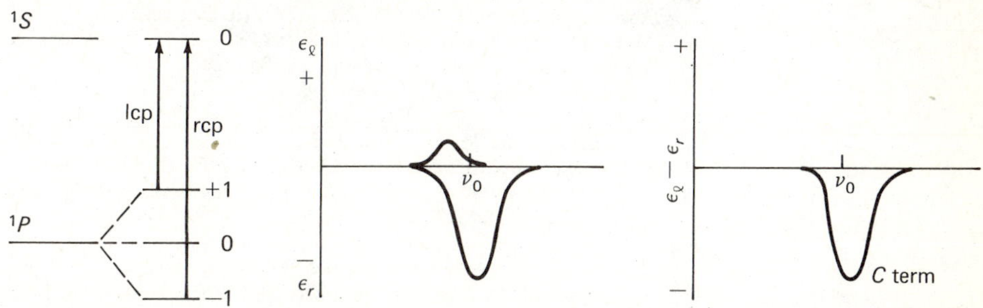

FIGURE 5-18. The transitions and expected spectrum for $^1P \rightarrow {}^1S$ in the mcd experiment. (A) the transitions; (B) spectra for left and right circularly polarized radiation; (C) the mcd spectrum; C-term behavior.

A third type of curve (*B*-term) results when there is a field-induced mixing of the states involved (this phenomenon also creates temperature independent paramagnetism, TIP, and will be discussed in more detail in the chapter on magnetism). This is manifested in a curve that looks like a *C*-curve but that is temperature independent. Since this mixing is present to some extent in all molecules, *all substances have mcd activity*. The magnitude of the external magnetic field intensity will determine whether or not the signal is observed.

The following characteristics summarize the basis for detecting and qualitatively interpreting mcd curves:

(1) An *A*-term curve changes sign at the absorption maximum, while *B* and *C* curves maximize or minimize at the maximum of the electronic absorption band.

(2) A *C*-term curve's intensity is inversely proportional to the absolute temperature, while a *B*-term is independent of temperature.

(3) An *A*-term spectrum is possible only if the ground or excited state involved in the electronic transition is degenerate and has angular momentum.

(4) A *C*-term spectrum is possible only if the ground state is degenerate and has angular momentum.

As can be anticipated, mcd measurements are of considerable utility[46,47] in assigning the electronic spectrum of a compound. Furthermore, the magnitude of the parameters is of considerable utility in providing information about many subtle electronic effects.[47] The molecular orbital origin of an electron involved in a transition can be determined. The lowest energy band in RuO_4 is clearly an oxygen to ruthenium charge-transfer band. One cannot determine from the electronic spectrum whether the oxygen electron involved in the transition came from a $t_1\pi$ or $t_2\pi$ type of oxygen molecular orbital. The sign of the mcd *A*-term established[48] the transition as $t_1\pi$ (oxygen) $\rightarrow e_{x^2-y^2,z^2}$ (ruthenium). Another significant advantage of mcd is in the assignment of spin-forbidden electronic transitions that have very low intensity in the electronic absorption spectrum. The assignments of the components in a six-coordinate chromium(III) complex have been made with this technique.[49] Other applications have been summarized in review articles.[46,47] Mcd has been extensively applied to provide information about the properties of excited electronic states.[46b] Deductions regarding their symmetries, angular momenta, electronic splittings, and vibrational-electronic interactions are possible.

REFERENCES CITED

1. H. H. Jaffe and M. Orchin, "Theory and Applications of Ultraviolet Spectroscopy," Wiley, New York, 1962.
2. M. Kasha, Discussions Faraday Soc., *9,* 14 (1950).
3. M. Kasha, Chem. Revs., *41,* 401 (1947); J. R. Platt, J. Opt. Soc. Amer., *43,* 252 (1953): A. Terenin, Acta Physics Chim. USSR, *18,* 210 (1943) (in English).
4. M. Kasha, "The Nature and Significance of $n \longrightarrow \pi^*$ Transitions," in Light and Life, ed. by W. D. McElroy and B. Glass, Johns Hopkins, Baltimore, 1961.
5. J. Sidman, Chem. Revs., *58,* 689 (1958).
6. H. McConnell, J. Chem. Phys., *20,* 700 (1952).
7. G. J. Brealey and M. Kasha, J. Amer. Chem. Soc., *77,* 4462 (1955).
8. S. Nagakura, Bull. Chem. Soc. Japan, *25,* 164 (1952) (in English).
9. L. E. Orgel, Quart. Revs., *8,* 422 (1954). For a good discussion of charge transfer transitions, see A. B. Lever, J. Chem. Ed., *51,* 612 (1974).
10. S. P. McGlyn and M. Kasha, J. Chem. Phys., *24,* 481 (1956).
11. C. K. Jorgenson, Acta Chem. Scand., *11,* 166 (1957); R. J. P. Williams, J. Chem. Soc., *1955,* 137.
12. R. P. Bauman, "Absorption Spectroscopy," Wiley, New York, 1962.
13. L. Lang, "Absorption Spectra in the Ultraviolet and Visible Region," Vols. 1–20, Academic, New York, 1961–76.
14. H. M. Hershenson, "Ultraviolet and Visible Absorption Spectra—Index for 1930–54," Academic, New York, 1956.

15. "Organic Electronic Spectral Data," Vols. 1–10, Interscience, New York. Compilation of spectral data from 1946 to 1968.
16. "ASTM (American Society for Testing Materials) Coded IBM Cards for Ultraviolet and Visible Spectra" ASTM, Philadelphia, 1961.
17. S. F. Mason, Quart. Revs., *15,* 287 (1961).
18. N. Metropolis, Phys. Rev., *60,* 283, 295 (1941).
19. S. J. Strickler and M. Kasha, J. Amer. Chem. Soc., *85,* 2899 (1963).
20. "Spectroscopic Properties of Inorganic and Organometallic Compounds." Volumes 1– (1967–), Specialist Periodical Report, Chemical Society, London.
21. R. Foster, "Organic Charge Transfer Complexes," Academic Press, London, 1969.
22. H. McConnell, J. S. Ham and J. R. Platt, J. Chem. Phys., *21,* 66 (1953).
23. G. Briegleb and J. Czekalla, Z. Phys. Chem. (Frankfurt), *24,* 37 (1960).
24. R. M. Keefer and L. J. Andrews, J. Amer. Chem. Soc., *77,* 2164 (1955).
25. R. L. Carlson and R. S. Drago, J. Amer. Chem. Soc., *85,* 505 (1963).
26. R. L. Middaugh, R. S. Drago, and R. J. Niedzielski, J. Amer. Chem. Soc., *86,* 388 (1964).
27. R. S. Drago, B. Wayland, and R. L. Carlson, J. Amer. Chem. Soc., *85,* 3125 (1963).
28. Sister M. Brandon, O. P., M. Tamres, and S. Searles, Jr., J. Amer. Chem. Soc., *82,* 2129 (1960); M. Tamres and S. Searles, Jr., J. Phys. Chem., *66,* 1099 (1962).
29. H. Yada, J. Tanaka, and S. Nagakura, Bull. Chem. Soc. Japan, *33,* 1660 (1960).
30. R. S. Drago, D. W. Meek, R. Longhi, and M. Joesten, Inorg. Chem., *2,* 1056 (1963).
31. R. S. Drago and D. A. Wenz, J. Amer. Chem. Soc., *84,* 526 (1962). Other donors studied toward these acids are summarized here.
32. L. J. Andrews and R. M. Keefer, "Advances in Inorganic Chemistry and Radiochemistry," Vol. 3, eds. H. J. Emeleus and A. G. Sharpe, pp. 91–128, Academic, New York, 1961.
33. O. Hassel and Chr. Rømming, Quart. Revs., *16,* 1 (1962).
34. R. Foster, "Molecular Complexes," Crane, Russak and Co., New York, 1974.
35. G. Briegleb, "Elektronen Donator-Acceptor-Komplexes," Springer Verlag, Berlin, 1963.
36. E. M. Kosower, et al., J. Amer. Chem. Soc., *83,* 3142, 3147 (1961); E. M. Kosower, J. Amer. Chem. Soc., *80,* 3253, 3261, 3267 (1958); E. M. Kosower, "Charge Transfer Complexes," in "The Enzymes," Vol. 3, eds. Boyer, Lardy, and Myrbach, p. 171, Academic Press, New York, 1960. The Spectrum of *N*-methylpyridinium iodide in many solvents is treated exhaustively in these references.
37. J. M. Hollas, "Molecular Spectroscopy," Vol. 1, pp. 62–112, Specialist Periodical Reports, Chemical Society, London (1972).
38. D. E. Freeman and W. Klemperer, J. Chem. Phys., *45,* 52 (1966).
39. J. P. Mathieu, J. Chim. Phys., *33,* 78 (1936).
40. R. S. Evans, A. F. Schreiner, and J. Hauser, Inorg. Chem., *13,* 2185 (1974) and references therein.
41. A. J. McCaffery and S. F. Mason, Mol. Phys., *6,* 359 (1963).
42. E. J. Corey and J. C. Bailar, Jr., J. Amer. Chem. Soc., *81,* 2620 (1959).
43. L. Velluz, M. Legrand, and M. Grosjean, "Optical Circular Dichroism," Academic Press, New York, 1965.
44. F. Woldbye, "Technique of Inorganic Chemistry," Vol. 4, eds. H. B. Jonassen and A. Weissberger, p. 249, Interscience, New York, 1965.
45. A. M. Sargeson, Transition Metal Chemistry, *3,* 303 (1966).
46. (a) A. D. Buckingham and P. J. Stephens, Ann. Rev. Phys. Chem., *17,* 399 (1966); (b) P. J. Stephens, Ann. Rev. Phys. Chem., *25,* 201 (1974).
47. P. N. Schatz and A. J. McCaffery, Quart. Revs., *23,* 552 (1969), and references therein.
48. A. H. Bowman, R. S. Evans, and A. F. Schreiner, Chem. Phys. Letters, *29,* 140 (1974), and references therein.
49. P. J. Hauser, A. F. Schreiner, and R. S. Evans, Inorg. Chem., *13,* 1925 (1974), and references therein.

GENERAL REFERENCES

A. The following references may be consulted for further study:

H. H. Jaffe and M. Orchin, "Theory and Applications of Ultraviolet Spectroscopy," Wiley, New York, 1962.

M. Orchin and H. H. Jaffe, "Symmetry, Orbitals and Spectra," Wiley Interscience, New York, 1971.

C. Sandorfy, "Electronic Spectra and Quantum Chemistry," Prentice-Hall, Englewood Cliffs, N. J., 1964.

R. P. Bauman, "Absorption Spectroscopy," Wiley, New York, 1962.

E. A. Braude and F. C. Nachod, "Determination of Organic Structures by Physical Methods," pp. 131–195, Academic, New York, 1955.

G. L. Clark, "The Encyclopedia of Spectroscopy," Reinhold, New York, 1960.

G. Herzberg, "Molecular Spectra and Molecular Structure," Vols. 1 and 3, D. Van Nostrand Co., Princeton, N.J., 1966.

J. N. Murrell, "The Theory of Electronic Spectra of Organic Compounds," Wiley, New York, 1963.

A. E. Gillam and E. S. Stern, "Electronic Absorption Spectroscopy," 2nd ed., Arnold, London, 1957.

G. W. King, "Spectroscopy and Molecular Structure," Holt, Rinehart, and Winston, New York, 1964.

A. Walsh, J. Chem. Soc., *1953,* 2260.

G. M. Barrow, "Introduction to Molecular Spectroscopy," McGraw-Hill, New York, 1962.

B. Spectral curves, wavelengths of the absorption peaks, and molar absorptivities for many compounds are tabulated in the following references:

"Sadtler Standard Spectra," Sadtler Research Laboratories, Philadelphia, Pa.

ASTM (American Society for Testing Materials), "Coded IBM Cards for Ultraviolet and Visible Spectra," ASTM, 1916 Race Street, Philadelphia.

H. M. Hershenson, "Ultraviolet and Visible Absorption Spectra—Index for 1930–54," Academic, New York, 1956.

L. Lang, "Absorption Spectra in the Ultraviolet and Visible Region," Vols. 1–20, Academic, New York, 1961–76.

L. Lang, "Absorption Spectra in the Ultraviolet and Visible Region, Cumulative Index (XVI–XX)," Academic, New York, 1976.

Organic Electronic Spectral Data," Vols. 1–10, Interscience, New York. Compilation of spectral data from 1946 to 1968.

EXERCISES

1. The compound $(C_6H_5)_3As$ is reported [J. Mol. Spectr., *5*, 118–132 (1960)] to have two absorption bands, one near 2700 Å and a second around 2300 Å. One band is due to the $\pi \rightarrow \pi^*$ transition of the phenyl ring and the other is a charge-transfer transition from the lone pair electrons on arsenic to the ring. The 2300 band is solvent dependent, and the 2700 band is not.

   a. Which band do you suspect is $\pi \rightarrow \pi^*$?

   b. What effect would substitution of one of the phenyl groups by CF_3 have on the frequency of the charge-transfer transition?

   c. In which of the following compounds would the charge-transfer band occur at highest frequency, $(C_6H_5)_3P$, $(C_6H_5)_3Sb$, or $(C_6H_5)_3Bi$?

2. What effect would changing the solvent from a nonpolar to a polar one have on the frequency for the following:

   a. Both ground and excited states are neutral (*i.e.*, there is no charge separation)?

   b. The ground state is neutral and the excited state is polar?

   c. The ground state is polar, the excited state has greater charge separation, and the dipole moment vector in the excited state is perpendicular to the ground state moment?

   d. The ground state is polar and the excited state is neutral?

3. Would you have a better chance of detecting vibrational fine structure in an electronic transition of a solute in liquid CCl_4 or CH_3CN solution? Why?

4. Do the excited states that result from $n \rightarrow \pi^*$ and $\pi \rightarrow \pi^*$ transitions in pyridine belong to the same irreducible representation? To which species do they belong?

5. Under what conditions can electronic transitions occur in the infrared spectrum? Which compounds would you examine to find an example of this?

6. Explain why the transition that occurs in the ion pair *N*-methylpyridinium iodide does not occur in the solvent-separated ion pair.

7. Recall the center of gravity rule and explain why the blue shift in the iodine transition should be related to the heat of interaction of iodine with a donor (see Fig. 5–12).

8. What is the polarization expected in the $^1A_1 \rightarrow {}^1A_2(n \rightarrow \pi^*)$ transition of formaldehyde from vibronic coupling? The vibrational modes have a_1, b_1, and b_2 symmetry.

9. Two observed transitions in octahedral cobalt(II) complexes occur at 20,000 cm^{-1} and 8000 cm^{-1} with extinction coefficients of $\varepsilon \sim 50$ and 8, respectively. These have been assigned to $^4T_{1g} \rightarrow {}^4T_{1g}$ and $^4T_{1g} \rightarrow {}^4T_{2g}$. Indicate whether or not vibronic coupling can account for the intensity difference. (The vibrations are a_{1g}, e_g, t_{1u}, t_{2g}, and t_{2u}).

10. The vibrations of $CoCl_4^{2-}$ have symmetries corresponding to a_1, e, and $2t_2$. Explain whether or not the $A_2 \rightarrow T_2$ and $A_2 \rightarrow T_1$ *d-d* transitions can gain intensity by vibronic coupling.

11. Consider the cyclopropene cation shown below:

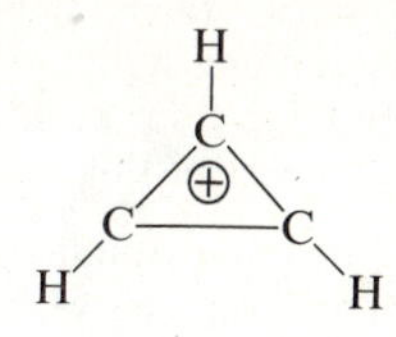

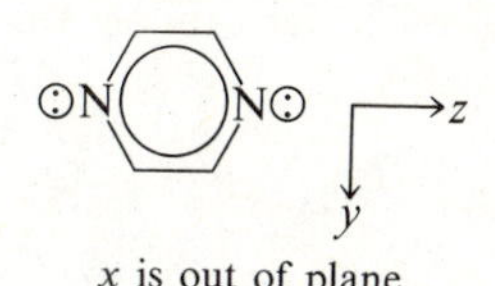

x is out of plane

a. Using the three out-of-plane p-orbitals (labeled φ_1, φ_2, and φ_3) as a basis set, do a simple Hückel calculation to obtain the energies of the resultant m.o.'s in terms of α and β.

b. The symmetries of these m.o.'s in the D_{3h} point group are A_2'' and E''. Briefly describe two methods you could use to find $\psi(A_2'')$ in terms of φ_1, φ_2, and φ_3.

c. Show whether the electronic transition to the first excited state is allowed. If so, what is its polarization?

d. What is the energy of this transition (in terms of α and β)?

e. What complications are likely in the approach used in part d?

12. Pyrazine has D_{2h} symmetry. The six pi levels are shown below with a rough order of energies. The two unlabeled levels are the non-bonding m.o.'s from the nitrogen lone pairs.

b_{2g} ———————

a_u ———————

b_{3u} ———————

———————

- - - - -

———————

b_{2g} ———————

b_{1g} ———————

b_{3u} ———————

a. What are the symmetries of the two non-bonding m.o.'s?

b. What is the symmetry of the state of lowest energy that arises from a $\pi \rightarrow \pi^*$ transition? Is it allowed? If so, what is its polarization?

c. What is the symmetry of the states that arise from $n \rightarrow \pi^*$ transitions from the two non-bonding levels to the lowest π^* level? Determine whether either is allowed, and, if so, give the polarization. If either is forbidden, give the symmetry that a vibrational mode would need in order to lend intensity via the vibronic mechanism.

13. In the $Re_2Cl_8^{2-}$ ion, the transition of an electron from the b_{2g} orbital to the b_{1u} orbital is a d-d transition in a molecule with a center of inversion. Is it allowed? Explain.

$$\psi(b_{2g}) = \frac{1}{\sqrt{2}}(d_{xy}^{(1)} + d_{xy}^{(2)})$$

$$\psi(b_{1u}) = \frac{1}{\sqrt{2}}(d_{xy}^{(1)} - d_{xy}^{(2)})$$

14. In several coordination complexes of *cis*-butadiene, the 2,3 carbon-carbon bond distance was observed to be appreciably shorter than the normal single bond distance. The possibility of back donation of electron density from the metal into the π^* orbital of the butadiene was proposed to account for this shortening. This exercise is designed to provide some insights into the chemistry of this ligand.

    a. Do a simple Hückel calculation on the π orbitals of *cis*-butadiene. Obtain the energies of the π-symmetry orbitals. (Hint: set $y = x^2$.)

    b. Work out the symmetries of the π and π^* *orbitals.*

    c. The coefficients of the π molecular orbitals are given below.

	C_1	C_2	C_3	C_4
ψ_a	.60	−.37	−.37	.60
ψ_b	.37	.60	.60	.37
ψ_c	.37	−.60	.60	−.37
ψ_d	.60	.37	−.37	−.60

    (1) Assign the symmetry species of each molecular orbital.
    (2) Arrange them in order of increasing energy.

    d. Determine the symmetries of all four singly excited *states* of *cis*-butadiene. Which transitions are allowed? Give their polarizations.

    e. The lowest-energy $\pi \rightarrow \pi^*$ transition is observed at $\lambda = 217$ mμ in *cis*-butadiene. Calculate β.

    f. Calculate the 2,3 bond order for *cis*-butadiene and then calculate the 2,3 bond order with two additional electrons in the lowest π^* orbital. Do your results agree with the conclusions regarding π back bonding?

15. For *trans*-butadiene,

    a. determine the symmetries of the π molecular orbitals.

    b. determine the symmetries of all four singly excited states. Which transitions are allowed? Give their polarizations.

    c. compare the results in parts a and b with those obtained for *cis*-butadiene. Would an HMO calculation differentiate between these two rotamers?

16. Consider the π molecular orbitals obtained from the p_z atomic orbitals in NO_3^-. The ground state orbital energies and symmetries are:

Energy ↑

———— a_2''

———— ———— e''

———— a_2''

    a. Obtain the state symmetries of the singly excited states.

    b. Which electronic transitions are allowed? Explain.

    c. NO_3^- has four vibrational modes, A_1', $2E'$, and A_2''. Does vibronic coupling provide a mechanism by which any forbidden transitions become allowed? Explain.

17. Show that the following two statements are equivalent for the point groups D_4, C_{3h}, and T_d:
1) In order to observe circular dichroism, the transition from ground state to excited state must be simultaneously electric dipole allowed and magnetic dipole allowed. (The magnetic dipole moment along the α axis transforms as a rotation about the α axis, R_α.)
2) In order to observe circular dichroism, the molecule must be optically active.

6 VIBRATION AND ROTATION SPECTROSCOPY: INFRARED, RAMAN AND MICROWAVE

6–1 HARMONIC AND ANHARMONIC VIBRATIONS

INTRODUCTION

As discussed earlier (Chapter 4), quanta of radiation in the infrared region have energies comparable to those required for vibrational transitions in molecules. Let us begin this discussion by considering the classical description of the vibrational motion of a diatomic molecule. For this purpose it is convenient to consider the diatomic molecule as two masses, A and B, connected by a spring. In Fig. 6–1(A) the equilibrium position is indicated. If a displacement of A and B is carried out, moving them to A' and B', [as in Fig. 6–1(B)], there will be a force acting to return the system to the equilibrium position. If the restoring force exerted by the spring, f, is proportional to the displacement Δr, *i.e.*,

$$f = -k\,\Delta r \tag{6-1}$$

the resultant motion that develops when A' and B' are released and allowed to oscillate is described as *simple harmonic motion*. In equation (6–1), the Hooke's law constant for the spring, k, is called the *force constant* for a molecular system held together by a chemical bond.

For harmonic oscillation of two atoms connected by a bond, the potential energy, V, is given by

$$V = \frac{1}{2}kX^2$$

where X is the displacement of the two masses from their equilibrium position. A plot of the potential energy of the system as a function of the distance x between the masses is thus a parabola that is symmetrical about the equilibrium internuclear distance, r_e, as the minimum (see Fig. 6–2). The force constant, k, is a measure of the curvature of the potential well near r_e.

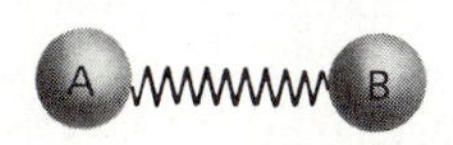

(A)

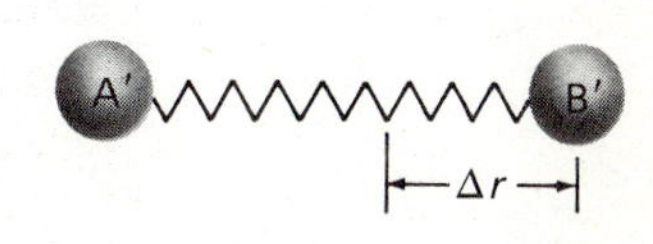

(B)

FIGURE 6–1. Displacement of the equilibrium position of two masses connected by a spring.

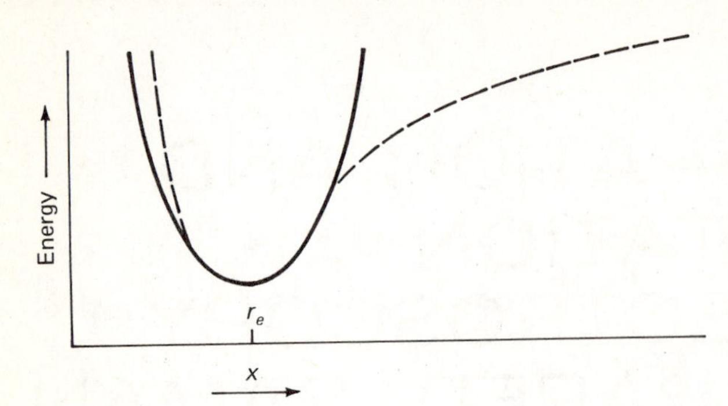

FIGURE 6–2. Potential energy versus distance, x, (A) for a harmonic oscillator (solid line) and (B) for an anharmonic oscillator (dotted line).

This classical spring-like model does not hold for a molecule because a molecular system cannot occupy a continuum of energy states, but can occupy only discrete, quantized energy levels. A quantum mechanical treatment of the molecular system yields the following equation for the permitted energy states of a molecule that is a simple harmonic oscillator:

$$E_v = h\nu(v + \tfrac{1}{2}) \tag{6-2}$$

where v is an integer 0, 1, 2, . . . , representing the vibrational quantum number of the various states, E_v is the energy of the vth state, h is Planck's constant, and ν is the *fundamental vibration frequency* (sec^{-1}) (*i.e.*, the frequency for the transition from state $v = 0$ to $v = 1$). These states are indicated for a harmonic oscillator in Fig. 6–3.

The potential energy curve of a real molecule (see Fig. 5–1), reproduced as a dotted line in Fig. 6–2, is not a perfect parabola. The vibrational energy levels are indicated in Fig. 5–1; they are not equally spaced, as equation (6–2) required, but converge. The levels converge because the molecule undergoes anharmonic rather than harmonic oscillation, *i.e.*, at large displacements the restoring force is less than predicted by equation (6–1). Note that as the molecule approaches dissociation, the bond becomes easier to stretch than the harmonic oscillator function would predict. This deviation from harmonic oscillation occurs in all molecules and becomes greater as the vibrational quantum number increases. As will be seen later, the assumption of harmonic oscillation will be sufficiently accurate for certain purposes (*e.g.*, the description of fundamental vibrations) and is introduced here for this reason.

6–2 ABSORPTION OF RADIATION BY MOLECULAR VIBRATIONS—SELECTION RULES

The interaction of electromagnetic radiation in the infrared region with a molecule involves interaction of the oscillating electric field component of the radiation with an oscillating electric dipole moment in the molecule. Thus, *in order for molecules to absorb infrared radiation as vibrational excitation energy, there must be a change in the dipole moment of the molecule as it vibrates.* Consequently, the stretching of homonuclear diatomic molecules will not give rise to infrared absorptions. According to this selection rule, any change in direction or magnitude of the dipole during a vibration gives rise to an oscillating dipole that can interact with the oscillating electric field component of infrared radiation, giving rise to absorption of radiation. A vibration that results in a change in direction of the dipole is illustrated by the N—C—H bending mode of HCN. There is little change in the magnitude of the dipole, but an appreciable change in direction occurs when the molecule bends.

The second selection rule can be derived from the harmonic oscillator approximation. This selection rule, which is rigorous for a harmonic oscillator, states that in the absorption of radiation only transitions for which $\Delta v = +1$ can occur. Since most molecules are in the v_0 vibrational level at room temperature, most transitions will occur from the state v_0 to v_1. This transition is indicated by arrow (1) of Fig. 6–3. The frequency corresponding to this energy is called the *fundamental frequency*. According to this selection rule, radiation with energy corresponding to transitions indicated by arrows (2) and (3) in Fig. 6–3 will not induce transitions in the molecule. Since most molecules are not perfect harmonic oscillators, this selection rule breaks down and transitions corresponding to (2) and (3) do occur. The transition designated as (2) occurs at a frequency about twice that of the fundamental (1), while (3) occurs at a frequency about three times that of the fundamental. Transitions (2) and (3) are referred to as the *first* and *second overtones,* respectively. The intensity of the first overtone is often an order of magnitude less than that of the fundamental, and that of the second overtone is an order of magnitude less than the first overtone.

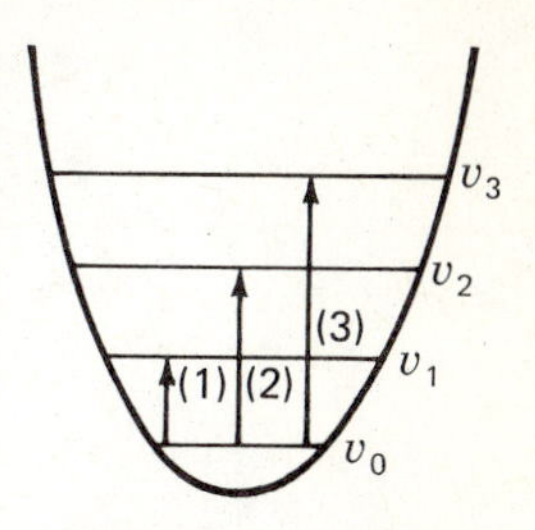

FIGURE 6–3. Vibrational states corresponding to a normal vibrational mode in a harmonic oscillator.

6–3 FORCE CONSTANT

The difference in energy, ΔE, between two adjacent levels, E_v and E_{v+1}, is given by equation (6–3) for a harmonic oscillator:

$$\Delta E = \left(\frac{h}{2\pi}\right)\left(\frac{k}{\mu}\right)^{1/2} \tag{6-3}$$

where k is the stretching force constant and μ is the reduced mass [$\mu = m_A m_B/(m_A + m_B)$ for the diatomic molecule A—B]. The relationship between energy and frequency, $\Delta E = h\nu = hc\bar{\nu}$, was presented in Chapter 4. The symbol ν will be used interchangeably for frequency (sec^{-1}) or wavenumber (cm^{-1}), but the units will be indicated when necessary. In the HCl molecule, the absorption of infrared radiation with $\nu = 2890$ cm^{-1} corresponds to a transition from the ground state to the first excited vibrational state. This excited state corresponds to a greater amplitude and frequency for the stretching of the H—Cl bond. Converting ν to energy produces ΔE of equation (6–3). Since all other quantities are known, this equation can be solved to produce: $k = 4.84 \times 10^5$ dynes cm^{-1} or, in other commonly used units, 4.84 md Å^{-1}. Stretching force constants for various diatomic molecules are summarized in Table 6–1.

The force constants in Table 6–1 are calculated by using equation (6–3), which was derived from the harmonic oscillator approximation. When an anharmonic oscillator model is employed, somewhat different values are obtained. For example, a force constant of 5.157×10^5 dynes cm^{-1} results for HCl. The latter value is obtained by measuring the first, second, and third overtones and evaluating the anharmonicity

TABLE 6–1. STRETCHING FORCE CONSTANTS FOR VARIOUS DIATOMIC MOLECULES (CALCULATED BY THE HARMONIC OSCILLATOR APPROXIMATION)

Molecule	ν (cm^{-1})	k (dynes cm^{-1})
HF	3958	8.8×10^5
HCl	2885	4.8×10^5
HBr	2559	3.8×10^5
HI	2230	2.9×10^5
F_2[a]	892	4.5×10^5
Cl_2[a]	557	3.2×10^5
Br_2[a]	321	2.4×10^5
I_2[a]	213	1.7×10^5
CO	2143	18.7×10^5
NO	1876	15.5×10^5

[a] Observed by Raman spectroscopy.

from the deviation of these frequencies from 2, 3, and 4 times the fundamental, respectively. Since these overtones are often not detected in the larger molecules more commonly encountered, we shall not be concerned with the details of the anharmonicity calculation.

The force constants for some other stretching vibrations of interest are listed in Table 6–2. For larger molecules, the nature of the vibration that gives rise to a particular peak in the spectrum is quite complex. Accordingly, one cannot calculate a force constant for a bond by substituting the "carbonyl frequency," for example, of a complex molecule into equation (6–3). This will become clearer as we proceed and is mentioned here as a note of caution. The force constants in Table 6–2 result from a normal coordinate analysis, which will also be discussed in more detail shortly. A larger force constant is often interpreted as being indicative of a stronger bond, but *there is no simple relation between bond dissociation energy and force constant.* We defined the force constant earlier as a measure of the curvature of the potential well near the equilibrium internuclear configuration. The curvature is the rate of change of the slope, so the force constant is the second derivative of the potential energy as a function of distance:

$$k = \left(\frac{\partial^2 V}{\partial r^2}\right)_{r \to 0} \tag{6-4}$$

Here V is the potential energy and r is the deviation of the internuclear distance from the equilibrium internuclear distance, at which $r = 0$. In a more complicated molecule, r is replaced by q, which is a composite coordinate that describes the vibration.

Triple bonds have stretching force constants of 13 to 18 $\times$ 10^5, double bonds about 8 to 12 $\times$ 10^5, and single bonds below 8 $\times$ 10^5 dynes cm^{-1}. In general, force constants for bending modes are often about a tenth as large as those for stretching modes.

The bands in the 4000 cm^{-1} to 600 cm^{-1} region of the spectrum mostly involve stretching and bending vibrations. Most of the intense bands above 2900 cm^{-1} involve hydrogen stretching vibrations for hydrogen bound to a low-mass atom. This frequency range decreases as the X—H bond becomes weaker and the atomic weight of X increases. Triple bond stretches occur in the 2000 to 2700 cm^{-1} region. Absorption bands assigned to double bond stretches occur in the 1500 to 1700 cm^{-1} region. Bands in this region of the spectrum of an unknown molecule are easily recognized and are a considerable aid in determining what the material is.

Metal-ligand vibrations usually occur below 400 cm^{-1} and into the far infrared

TABLE 6–2.
STRETCHING FORCE CONSTANTS FOR VARIOUS STRETCHING VIBRATIONS (HARMONIC OSCILLATOR APPROXIMATION)

Bond	k (dynes cm^{-1})
⋛C—C≡	4.5 × 10^5
⋛C—C≡	5.2 × 10^5
>C=C<	9.6 × 10^5
—C≡C—	15.6 × 10^5
>C=O	12.1 × 10^5
—C≡N	17.7 × 10^5
≡C—H	5.9 × 10^5
⋛C—H	4.8 × 10^5

region. They are very hard to assign, since many ligand ring deformation and rocking vibrations as well as lattice modes (vibrations involving the whole crystal) occur in this region.

As indicated by equation (6–3), the reduced mass is important in determining the frequency of a vibration. If, for example, a hydrogen bonded to carbon is replaced by deuterium, there will be a *negligible change in the force constant but an appreciable change in the reduced mass.* As indicated by equation (6–3), the frequency should be lower by a factor of about $1/\sqrt{2}$. The frequencies for C-H and C-D vibrations are proportional to $[12 \times 1/(12 + 1)]^{-1/2}$ and $[12 \times 2/(12 + 2)]^{-1/2}$, respectively. Normally, a vibration involving hydrogen will occur at 1.3 to 1.4 times the frequency of the corresponding vibration in the deuterated molecule. This is of considerable utility in confirming assignments that involve a hydrogen atom. The presence of the natural abundance of ^{13}C in a metal carbonyl also is found to give separate bands due to $^{13}C{\equiv}O$ stretching vibrations in metal carbonyls (*vide infra*) because of the difference in reduced mass. Use of metal isotopes also has utility in confirming the assignment of vibrations involving the metal-ligand bond.(1)

6–4 THE 3N – 6(5) RULE

VIBRATIONS IN A POLYATOMIC MOLECULE

The positions of the N atoms in a molecule can be described by a set of Cartesian coordinates, and the general motion of each atom can be described by utilizing three displacement coordinates. The molecule is said, therefore, to have $3N$ degrees of freedom. Certain combinations of these individual degrees of freedom correspond to translational motion of the molecule as a whole without any change in interatomic dimensions. There are three such combinations which represent the x, y, and z components of the translational motion, respectively. For a nonlinear molecule there are three combinations that correspond to rotation about the three principal axes of the molecule. Therefore, *for a nonlinear molecule there are* $3N - 6$ *normal modes of vibration* that result in a change in bond lengths or angles in the molecule. *Normal modes represent independent self-repeating motions in a molecule.* They correspond to $3N - 6$ degrees of freedom that, in the absence of anharmonicity, could take up energy independently of each other. *These modes form the bases for irreducible representations.* Since a molecule is fundamentally not changed by applying a symmetry operation R, the normal mode $R\hat{Q}$ must have the same frequency as the normal mode $\hat{Q}$. Thus, if $\hat{Q}$ is non-degenerate, $R\hat{Q} = \pm 1\hat{Q}$ for all R's. Consequently, $\hat{Q}$ forms the basis for a one-dimensional representation in the molecular symmetry group. It can be shown that degenerate modes transform according to irreducible representations of dimensionality greater than one. The center of mass of the molecule does not change in the vibrations associated with the normal mode, nor is angular momentum involved in these vibrations. *All general vibrational motion that a molecule may undergo can be resolved into either one or a combination of these normal modes.*

For a linear molecule all the vibrations can be resolved into 3N – 5 normal modes. The additional mode obtained for a linear molecule is indicated in Fig. 6–4(B), where

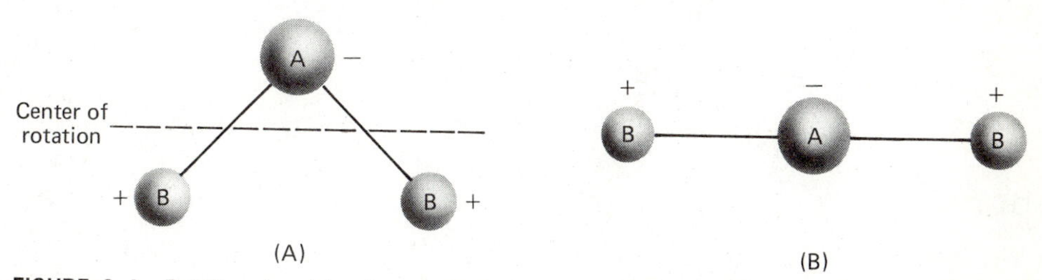

FIGURE 6–4. Rotational and bending modes for (A) non-linear and (B) linear molecules.

TABLE 6–3.
INFRARED SPECTRUM OF SO_2

ν (cm^{-1})	Assignment
519	ν_2
606	$\nu_1 - \nu_2$
1151	ν_1
1361	ν_3
1871	$\nu_2 + \nu_3$
2305	$2\nu_1$
2499	$\nu_1 + \nu_3$

plus signs indicate motion of the atoms into the paper and minus signs represent motion out of the paper. For the nonlinear molecule [Fig. 6–4(A)] the motion indicated corresponds to a rotation. For a linear molecule a similar motion corresponds to a bending of the bonds, and hence this molecule has an additional normal vibrational mode ($3N - 5$ for a linear molecule vs. $3N - 6$ for a nonlinear one).

As will be seen later, there are many applications for which we need to know which bands correspond to the fundamental vibrations.

6–5 EFFECTS GIVING RISE TO ABSORPTION BANDS

Sulfur dioxide is predicted to have three normal modes from the $3N - 6$ rule. The spectral data (Table 6–3) show the presence of more than three bands. The three bands at 1361, 1151, and 519 cm^{-1} are the fundamentals and are referred to as the ν_3, ν_1, and ν_2 bands, respectively (see Fig. 6–5). The ν_n symbolism is used to label the various frequencies of fundamental vibrations and should not be confused with the symbols v_0, v_1, v_2, etc., used to designate various vibrational levels of one mode in a molecule. By convention the highest-frequency totally symmetric vibration is called ν_1, the second highest totally symmetric vibration ν_2, etc. When the symmetric vibrations have all been assigned, the highest-frequency asymmetric vibration is counted next, followed by the remaining asymmetric vibrations in order of decreasing frequency. An exception is made to this rule for the bending vibration of a linear molecule, which is labeled ν_2. Another common convention involves labeling stretching vibrations ν, bending vibrations δ, and out-of-plane bending vibrations π. Subscripts, *as*, for asymmetric; *s*, for symmetric; and *d*, for degenerate, are employed with these symbols.

The ν_1 mode in SO_2 is described as the *symmetric stretch,* ν_3 as the *asymmetric stretch,* and ν_2 as *the* O—S—O *bending mode.* In general, the asymmetric stretch will occur at higher frequency than the symmetric stretch, and stretching modes occur at much higher frequencies than bending modes. There is a slight angle change in the stretching vibrations in order for the molecule to retain its center of mass. The other

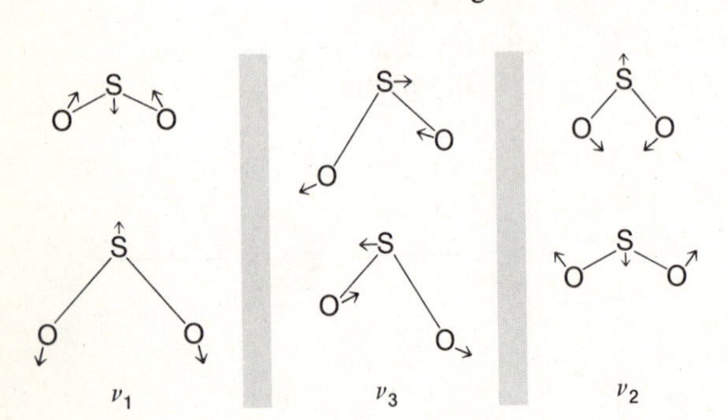

FIGURE 6–5. The three fundamental vibrations for sulfur dioxide. (The amplitudes are exaggerated to illustrate the motion.)

FIGURE 6–6. Carbon dioxide fundamental vibration modes.

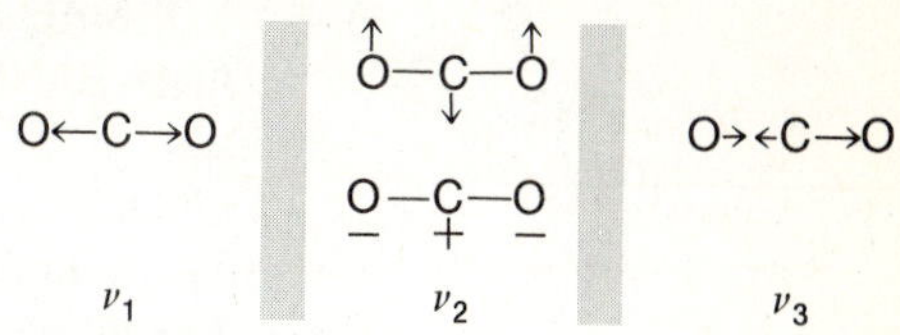

O←C→O	O—C—O (↑ ↓ ↑) / O—C—O (− + −)	O→←C→O
ν_1	ν_2	ν_3

absorption frequencies in Table 6–3 are assigned as indicated. The overtone of ν_1 occurs at about $2\nu_1$ or 2305 cm^{-1}. The bands at 1871 and 2499 cm^{-1} are referred to as *combination bands.* Absorption of radiation of these energies occurs with the simultaneous excitation of both vibrational modes of the combination. The 606 cm^{-1} band is a *difference band,* which involves a transition originating from the state in which the ν_2 mode is excited and changing to that in which the ν_1 mode is excited. Note that all the bands in the spectrum are accounted for by these assignments. Making the assignments is seldom this simple; as will be shown later, much other information, including a normal coordinate analysis, is required to substantiate these assignments.

A more complicated case is the CO_2 molecule, for which four fundamentals are predicted by the $3N - 5$ rule. A single band results from the two degenerate vibrations ν_2 of Fig. 6–6, which correspond to bending modes at right angles to each other. Later we shall see how symmetry considerations aid in predicting the number of degenerate bands to be expected. In more complex molecules some of the fundamentals may be accidentally degenerate because two vibration frequencies just happen to be equal. This is not easily predicted, and the occurrence of this phenomenon introduces a serious complication. The assignment of the fundamentals for CO_2 is more difficult than for SO_2 because many more bands appear in the infrared and Raman spectra. Bands at 2349, 1340 and 667 cm^{-1} have been assigned to ν_3, ν_1, and ν_2, respectively. The tests of these assignments have been described in detail by Herzberg (see General References) and will not be repeated here. In this example the fundamentals are the three most intense bands in the spectrum. In some cases, there is only a small dipole moment change in a fundamental vibration, and the corresponding absorption band is weak (see the first selection rule).

The above discussion of the band at 1340 cm^{-1} has been simplified. Actually, it is an intense doublet with band maxima at 1286 and 1388 cm^{-1}. This splitting is due to a phenomenon known as *Fermi resonance.* The overtone $2\nu_2$ ($2 \times 667 = 1334$ cm^{-1}) and the fundamental ν_1 should occur at almost the same frequency. The two vibrations interact by a typical quantum mechanical resonance, and the frequency of one is raised while the frequency of the other is lowered. The wave function describing these states corresponds to a mixing of the wave functions of the two vibrational excited states (ν_1 and $2\nu_2$) that arise from the *harmonic oscillator approximation.* We cannot say that one line corresponds to ν_1 and the other to $2\nu_2$, for both are mixtures of ν_1 and $2\nu_2$. This interaction also accounts for the high intensity of what, in the absence of interaction, would have been a weak overtone ($2\nu_2$). The intensity of the fundamental is distributed between the two bands, for both bands consist partly of the fundamental vibration.

The presence of Fermi resonance can sometimes be detected in more complex molecules by examining deuterated molecules or by determining the spectrum in various solvents. Since the Fermi resonance interaction requires that the vibrations involved have nearly the same frequency, the interaction will be affected if one mode undergoes a frequency shift from deuteration or a solvent effect while the other mode does not. The two frequencies will no longer be equivalent, and the weak overtone will revert to a weak band or not be observed in the spectrum. Other requirements for the Fermi resonance interaction will be discussed in the section on symmetry considerations.

6-6 NORMAL COORDINATE ANALYSES AND BAND ASSIGNMENTS

Degeneracy, vibration frequencies outside the range of the instruments, low intensity fundamentals, overtones, combination bands, difference bands, and Fermi resonance all complicate the assignment of fundamentals. The problem can sometimes be resolved for simple molecules by a technique known as a *normal coordinate analysis.* A normal coordinate analysis involves solving the classical mechanical problem of the vibrating molecule, assuming a particular form of the potential energy (usually the valence force field). The details of this calculation are beyond the scope of this text,[2,3] but it is informative to outline the problem briefly so the reader can assess the value and the limitations of the approach. Furthermore, several important qualitative ideas will be developed that we shall use in subsequent discussion. When we have finished, you will not know how to do a normal coordinate analysis, but hopefully you will have a rough idea of what is involved.

Just as the electronic energy and electronic wave functions of a molecule are related by a secular determinant and secular equations (Chapter 3), the vibrational energies, vibrational wave functions, and force constants are related by a secular determinant and a series of secular equations. The vibrational secular determinant will be given here as (for the derivation see references 2 and 3):

$$\begin{vmatrix} F_{11} - (G^{-1})_{11}\lambda & F_{12} - (G^{-1})_{12}\lambda & \cdots\cdots & F_{1n} - (G^{-1})_{1n}\lambda \\ \vdots & & & \vdots \\ F_{n1} - (G^{-1})_{n1}\lambda & F_{n2} - (G^{-1})_{n2}\lambda & \cdots\cdots & F_{nn} - (G^{-1})_{nn}\lambda \end{vmatrix} = 0$$

Here $\lambda = 4\pi^2\nu^2$, where ν is the vibrational frequency (note the resemblance of this $G\lambda$ term to the energy term in a molecular orbital calculation); this quantity is known, for one begins by making a tentative assignment of all the normal modes to the bands in the spectrum. The basis set for this calculation (*i.e.,* the counterpart of atomic orbitals in an m.o. calculation) is the set of internal coordinates of the molecule expressed in terms of $3N - 6$ atom displacements, L. Three such coordinates, needed to describe the normal modes for the water molecule, are illustrated in Fig. 6-7. The selection of these internal coordinates is complicated in larger highly symmetric molecules because redundant coordinates can result. One proceeds by selecting an internal displacement vector for a change in bond length for each bond in the molecule and then selecting *independent* bond angles to give the $3N - 6$ basis set. The normal mode is going to be some combination of this basis set of internal displacement coordinates, and the vibrational wave function will tell us what this combination will be (again note the resemblance of this to an electronic wave function for a molecule consisting of the atomic orbital basis set). In the secular determinant given above, F_{11} is the force constant for stretching the O—H bond along L_{11}; F_{22} is the force constant for stretching along L_{22}, and F_θ is related to the bend along L_θ. The off-diagonal element F_{21} in the vibrational secular determinant is called an *interaction force constant,* and it indicates how the two isolated stretches interact with one another. When, for example, L_{22} is subjected to a unit displacement, the bond along L_{11} will distort to minimize the potential energy of the strained molecule. F_{21} is roughly proportional to the displacement of the bond along L_{22} resulting from minimization of the energy of the molecule after displacement along L_{11}. Part of the interaction relates to how the bond strength along L_{22} changes as the oxygen rehybridizes when the bond along L_{11} is stretched. F_{ij} is not necessarily identical to F_{ji}.

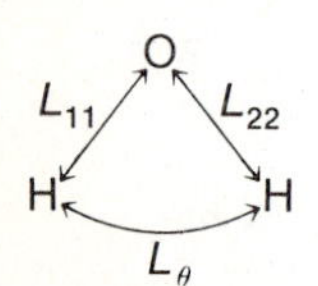

FIGURE 6-7. Internal coordinates for the H_2O molecule.

The F-matrix elements account for the potential energies of the vibration. The G matrix elements contain information about the kinetic energy of the molecule. The latter can be written exactly for a molecule from formulas given by Wilson, Decius, and Cross[2] if we know the atom masses, the molecular bond distances, and the bond angles.* This is a symmetric matrix. In practice, the force constants are usually the only unknowns in the secular determinant, and they can be determined. (Now the problem differs from the format of the molecular orbital calculation.)

The secular determinant given above can be written in matrix notation as:

$$|\mathbf{F} - \mathbf{G}^{-1}\lambda| = 0$$

If we multiply by **G** we get

$$|\mathbf{GF} - \mathbf{E}\lambda| = 0 \tag{6-5}$$

where **E** is the unit matrix. Furthermore, the G matrix has the property that the λ's in the off-diagonal elements of the previous secular determinant have been eliminated. The problem now is to solve

$$\begin{vmatrix} G_{11} & \cdots & G_{1n} \\ \vdots & & \vdots \\ G_{n1} & & G_{nn} \end{vmatrix} \begin{vmatrix} F_{11} & \cdots & F_{1n} \\ \vdots & & \vdots \\ F_{n1} & & F_{nn} \end{vmatrix} - \begin{vmatrix} 1 & \cdots & 0 & \cdots & 0 \\ \vdots & & 1 & & \\ 0 & & & & 1 \end{vmatrix} \lambda = 0$$

where the G elements are all known, as are the λ's. There are in general n values of λ that satisfy this equation, and they are known. The difficulty with normal coordinate analyses on most molecules is that there are more unknown force constants than there are frequencies. For the water molecule, the force constant matrix is 3×3, and there are four unknowns† for the three frequencies. If one deuterates the water molecule, new frequencies are obtained, but no new force constants are introduced. The ^{16}O, ^{17}O and ^{18}O molecules could also be studied. The problem here is whether or not the new equations we obtain in trying to get more simultaneous equations than force constants are different enough to allow for a unique, meaningful solution. (This is similar to the K^{-1} vs. ε problem discussed in Chapter 4. Two parallel or nearly parallel equations do not solve the problem, even though in the latter case a computer can pick a minimum.) In the simpler molecules, this is often not a problem; but in more complex molecules, the analysis is not to be believed unless errors are reported and some statistical criterion of the significance of the fit is presented. This is readily accomplished with a correlation matrix that indicates how extensively the individual force constants are correlated. This problem is potentially so severe that one should not accept the results from a force constant analysis unless the correlation matrix is reported.

Suppose we were to do a normal coordinate analysis for $Mn(CO)_5X$. We would have 30 internal displacement coordinates ($3N - 6$) and a 30×30 force constant matrix, or many force constants. Symmetry can reduce this matrix to smaller blocks by requiring certain interaction constants to be zero (*vide infra*), but there remain several approximations that must be introduced to solve the problem. In one analysis of the problem, one could assume that the carbonyl stretches are so far removed in frequency from any of the other vibrations that they can be treated separately.[4] This is equivalent to claiming that the metal-carbon stretch does not influence the C—O vibration and setting all interaction constants of the C≡O stretches with anything else equal to zero. This leads to a 5×5 block for the total secular determinant of force constants. Isotopic substitution is then employed to solve the problem. In a

*The evaluation of the G-matrix for a bent XY_2 molecule has been worked out in detail in reference 18, Section I-11. The reader is referred to this treatment for further details.

†The symmetry of the water molecule leads to this.

more thorough analysis,[5] the interaction constants between M—C and C—O stretch coordinates that are *trans* to each other have been determined and found to be significant.

Cotton and Kraihanzel[4] have proposed a crude approximation for these systems. They arbitrarily set any interaction force constant F_{trans} for carbonyl moieties that are trans to one another equal to $2F_{cis}$, where F_{cis} corresponds to interaction constants involving groups that are *cis*. The numbers of unknowns and frequencies become comparable in many carbonyl compounds when this is done. Jones[5] has carried out a more rigorous evaluation of this problem and suggests that severe limitations be placed on systems for which the Cotton-Kraihanzel approximations are invoked. It has been shown in the more complete study that F_{cis} is considerably larger than F_{trans}. Another approach that is being investigated to facilitate normal coordinate analyses involves transferring interaction force constants from a simple molecule in which they are well known to similar systems.

Let us next consider a simple system, which we assume to have been solved for meaningful force constants. The force constant values can be substituted into the **G** and **F** matrices given above, and the secular equations are now written by multiplying by a matrix of the basis set **L**, the internal displacements; *i.e.*,

$$|\mathbf{GF} - \mathbf{E}\lambda_n||\mathbf{L}| = 0 \tag{6-6}$$

In the case of water, **F** is 3 × 3, **G** is 3 × 3, **E** is a 3 × 3 unit matrix, **L** is 3 × 1 (L_{11}, L_{22}, and L_{33}), and λ_n is the eigenvalue for which we wish to determine the normal mode. The energies are now substituted one at a time. Matrix multiplications then yield the contribution of the basis set (*i.e.*, the contribution of the individual internal displacements) to the vibrational wave function for the normal mode corresponding to that frequency. This is done three times for the water molecule with the three frequencies, leading to the wave functions for the three normal modes.

For the water molecule, the wave functions for the $3N - 6$ vibrations are:

$$\psi_1 = NL_\theta + N'(L_{11} + L_{22}) \qquad N \gg N'$$

$$\psi_2 = N(L_{11} + L_{22}) + N'L_\theta$$

$$\psi_3 = \frac{1}{\sqrt{2}}(L_{11} - L_{22})$$

The first two have A_1 symmetry, and the last has B_1 symmetry.

If we had a diatomic molecule (*e.g.*, HCl), all of the above matrices would be 1 × 1 and we would have from equation (6–6)

$$F_{11}G_{11} - \lambda = 0$$

where the G matrix element is the reduced mass and F_{11} is the only unknown. Rearranging, we see that

$$\frac{F_{11}}{\mu_{HCl}} = 4\pi^2\nu^2$$

or

$$\nu = \sqrt{\frac{F_{11}}{4\pi^2\mu}} \tag{6-7}$$

The resemblance of this result to equation (6–3) is clear for $\Delta E = h\nu$. Now we see why equation (6–3) can be used to give the force constant for a bond in a diatomic molecule but not in a more complex molecule. In larger, *more complex* molecules, the observed frequency corresponds to a more complex vibration that depends upon several bond force constants. The results of this involved calculation produce the force constants and indicate exactly the form of each normal mode in terms of the internal coordinates. For water, for example, the internal coordinates in Fig. 6–7 are combined to produce normal modes that are similar in form to those shown for SO_2 in Fig. 6–5. This is expected because the internal displacement coordinates (which are of the same form for SO_2 and H_2O) can form the basis for producing a total representation by employing the symmetry operations of the point group on them. Factoring the total representation produces the symmetry of the irreducible representations, which in this case are the normal modes. For the water molecule, the total representation obtained by operating with the C_{2v} point group on the internal displacements shown in Fig. 6–7 is 3 1 3 1 (the angle θ is not changed by σ_v or C_2). Factoring the total representation produces $2A_1 + B_1$. Next, projection operators can be used on this basis set just as on an a.o. basis set to produce

$$\begin{aligned} \hat{P}(A_1)L_{11} &\cong L_{11} + L_{22} \\ \hat{P}(A_1)L_\theta &\cong L_\theta \\ \hat{P}(B_1)L_{11} &\cong L_{11} - L_{22} \end{aligned} \tag{6-8}$$

This problem nicely illustrates the analogy between the symmetry aspects of the molecular orbital problem and those of the vibrational problem.

The inverse of the force constant matrix produces a matrix of *compliance constants*.[6] A diagonal compliance constant is a measure of the displacement that will take place in a coordinate as a result of a force imposed upon this coordinate, if the other coordinates are allowed to adjust to minimize the energy. There are reported advantages to employing compliance constants instead of force constants, and the reader is referred to the literature for details.[6]

6–7 GROUP VIBRATIONS AND THE LIMITATIONS OF THIS IDEA

The idea that we can look at the spectrum of a complex molecule and assign bands in the spectrum to various functional groups in the molecule is called the *group vibration* concept. This approach arose from the experimental observation that many functional groups absorb in a narrow region of the spectrum regardless of the molecule that contains the group. In the acetone molecule,* for example, one of the normal modes of vibration consists of a C—O stretching motion with negligible motion of the other atoms in the molecule. Similarly, the methyl groups can be considered to undergo vibrations that are independent of the motions of the carbonyl group. In various molecules it is found that carbonyl absorptions due to the stretching vibration occur in roughly the same spectral region ($\sim$1700 cm^{-1}). As will be seen in the section on applications, the position of the band does vary slightly ($\pm$150 cm^{-1}) because of the mass, inductive, and conjugative effects of the groups attached. The methyl group has five characteristic absorption bands: two bands in the region from 3000 to 2860 (from the asymmetric and symmetric stretch), one around 1470 to 1400 (the asymmetric bend), one around 1380 to 1200 (the symmetric bend), and one in the region from 1200 to 800 cm^{-1} (the rocking mode, in which the whole methyl group twists about the C—C bond, undergoing a rocking-chair type of motion). The concept

*This molecule has been analyzed by a normal coordinate treatment.

of group vibrations involves dissecting the molecule into groups and assigning one or more bands in the spectrum to the vibrations of each group. Many functional groups in unknown compounds have been identified by using this assumption. Unfortunately, in many complicated molecules there are many group vibrations that overlap, and assignment of the bands in a spectrum becomes difficult. It is possible to perform experiments that help resolve this problem: deuteration will cause a shift in vibrations involving hydrogen (*e.g.*, C—H, O—H, or N—H stretches and bends) by a factor of 1.3 to 1.4; characteristic shifts, which aid in assignments, occur with certain donor functional groups (*e.g.*, C=O) when the spectrum is examined in hydrogen-bonding solvents or in the presence of acidic solutes.

The limitations of the group vibration concept should be emphasized so that incorrect interpretations of infrared spectral data can be recognized. The group vibration concept implies that the vibrations of a particular group are relatively independent of the rest of the molecule. If the center of mass is not to move, this is impossible. All of the nuclei in a molecule must undergo their harmonic oscillations in a synchronous manner in normal vibrations. In view of the discussion in the previous section, we can see that the group vibration concept will be good when the normal mode wave function consists largely (80 to 90 per cent) of the internal displacement in which only the group is involved, *i.e.*, the group vibration is the main internal displacement coordinate. This will occur when the symmetry properties of the molecule are such as to restrict the combination of the internal displacements of the group with other internal displacement coordinates; *e.g.*, L_θ in the water molecule cannot contribute to the B_1 asymmetric stretch. When the vibrational motions involving the two internal displacement coordinates are very different in energy, then the off-diagonal interaction constants are small, the two internal displacement coordinates are not mixed extensively in the vibrational wave functions, and the vibrations can be treated separately as group vibrations. This is equivalent to the assumption made when we treated the carbonyl vibrations in $M(CO)_5X$ separately to obtain the 5×5 matrix.

When the atoms in a molecule are of similar mass and are connected by bonds of comparable strength (*e.g.*, all single bonds as in $BF_3 \cdot NH_3$), all the normal modes will be mixtures of several internal displacement vectors. For example, in $BF_3 \cdot NH_3$, it would be impossible to assign a band to the B—N stretching vibration because none of the normal modes correspond predominantly to this sort of motion. When this occurs, the various group vibrations are said to be coupled. This term simply describes a situation in which the very crude group vibration concept is not applicable to the description of the normal mode corresponding to a particular absorption band. This discussion should enable the reader to distinguish coupling from the phenomena of Fermi resonance or combination bands.

The HCN molecule is interesting to consider in the context of the above discussion, for the frequencies of a pure C—H and pure C—N stretch are similar in energy. The C—H and C—N group stretching vibrations in the H—C—N molecule are coupled. The absorption band attributed to the C—H stretch actually involves C—N motion to some extent and vice versa; *i.e.*, the observed frequencies cannot be described as being a pure C—H and a pure C—N stretch. Evidence to support this comes from deuteration studies, in which it is found that deuteration affects the frequency of the band that might otherwise be assigned to the C—N stretch. This absorption occurs at 2089 and 1906 cm^{-1} in the molecules H—C—N and D—C—N, respectively. Equation (6–3) would predict that deuteration should have very little effect on the C—N frequency, for it has little effect on the C—N reduced mass. The only way the frequency can be affected is to have a coupling of the C—D stretch with the C—N stretch. Upon deuteration, the C—H stretch at 3312 cm^{-1} is replaced by a C—D stretch at 2629 cm^{-1}. Since the C—D frequency is closer to the C—N frequency than is that of C—H, there is more extensive coupling in the deuterated compound. Because the symmetries of the normal modes differ, the C—D bending vibration does not couple with the "C—N stretch."

It would be improper to draw conclusions concerning the strength of the C—H bond from comparison of the frequency for the C—H stretch in H—CN with data for other C—H vibrations. The force constants for the C—H bonds can be calculated for the various compounds by a normal coordinate analysis. These values should be employed for such comparisons.

As another example of the difficulties encountered in the interpretation of frequencies, consider the stretching vibration for the carbonyl group in a series of compounds. The absorption bands occur at 1928, 1827, and 1828 cm^{-1} in F_2CO, Cl_2CO, and Br_2CO, respectively.[7] One might be tempted to conclude from the frequencies that the high electronegativity of fluorine causes a pronounced increase in the C—O force constant. However, a normal coordinate analysis[7] shows that the C—O and C—F vibrations are coupled and that the C—O stretch also involves considerable C—F motion. The frequency of this normal mode is higher than would be expected for an isolated C—O stretching vibration with an equivalent force constant. There is a corresponding lowering of the C—F stretching frequency. Since chlorine and bromine are heavier, they make little contribution to the carbonyl stretch. The normal coordinate analysis indicates that the C—O stretching force constants are 12.85, 12.61, and 12.83 millidynes/Å in F_2CO, Cl_2CO, and Br_2CO, respectively.

A normal coordinate analysis and infrared studies of ^{12}C-, ^{13}C-, ^{16}O-, ^{18}O-, H-, and D-substituted ketones containing an ethyl group attached to the carbonyl group indicate[8] that the band near 1750 cm^{-1} assigned to C—O stretch in these ketones corresponds to a normal mode that consists of about 75 per cent C—O and 25 per cent C—C stretching motions. These examples should be kept in mind whenever one is tempted to interpret the frequencies of infrared bands in terms of electronic effects. Force constants obtained from normal coordinate analyses should be compared; frequencies read directly from spectra should not be. Unfortunately, reliable force constants are in limited supply.

RAMAN SPECTROSCOPY

6–8 INTRODUCTION

Raman spectroscopy is concerned with vibrational and rotational transitions, and in this respect it is similar to infrared spectroscopy. Since the selection rules are different, the information obtained from the Raman spectrum often complements that obtained from an infrared study and provides valuable structural information.

In a Raman experiment, a monochromatic beam of light illuminates the sample, and observations are made on the scattered light at right angles to the incident beam (Fig. 6–8). Monochromatic light sources of different frequencies can be employed; gas lasers are commonly used for this purpose. Various sample holder configurations are employed; the only requirement is that the detector must be at right angles to the source. This configuration is illustrated in Fig. 6–8 for a capillary cell. Absorption of the monochromatic light beam (leading to decomposition) can be a problem, as can fluorescence. These problems are minimized by choice of an appropriate gas laser line: the He—Ne laser gives a line at 6328 Å (red); the Ar laser gives lines at 4579, 4658, 4765, 4880, 4915 and 5145 Å (blue-green) (the Ar^+ 4880 Å and 5145 Å lines are usually used); and the Kr laser gives 5682 and 6471 Å lines. The introduction of tunable dye lasers has extended the range of usable wavelengths even further. Use of a narrow bandpass filter for each laser line reduces effects from the large number of

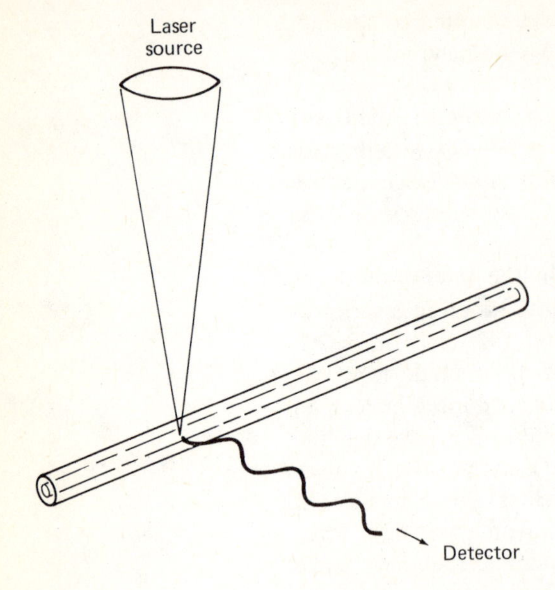

FIGURE 6–8. The Raman experiment (schematic).

laser ghosts. A valuable discussion of various aspects of this experiment is contained in reference 11.

We should wonder how we obtain vibrational transitions when we employ a source with visible or uv emission. A quantum of the incident light of frequency $\nu°$ and energy $h\nu°$ can collide with a molecule and be scattered with unchanged frequency. This is referred to as Rayleigh scattering. The mechanism involves inducing a dipole moment, D, in the molecule when it is in the field of the electric vector of the radiation. The electrons in the molecule are forced into oscillations of the same frequency as the radiation. This oscillating dipole radiates energy in all directions and accounts for the Rayleigh scattering. If the photon is actually absorbed in the process and re-emitted, this is fluorescence. The difference between scattering and fluorescence is thus a subtle one, having to do with the lifetime of the species formed in the photon-molecule collision.

The scattering process just described corresponds to an elastic collision of the molecule with the photon. In an inelastic collision, Raman scattering, the molecule in its ground vibrational state accepts energy from the photon being scattered, exciting the molecule into a higher vibrational state, while the incident radiation now becomes scattered with energy $h(\nu° - \nu_v)$. In the scattered light measured at right angles, we now detect a frequency $\nu° - \nu_v$, called the Stokes line, in the spectrum indicated in Fig. 6-9. The measured value of ν_v is identical to the infrared frequency that would excite this vibrational mode if it were infrared-active.

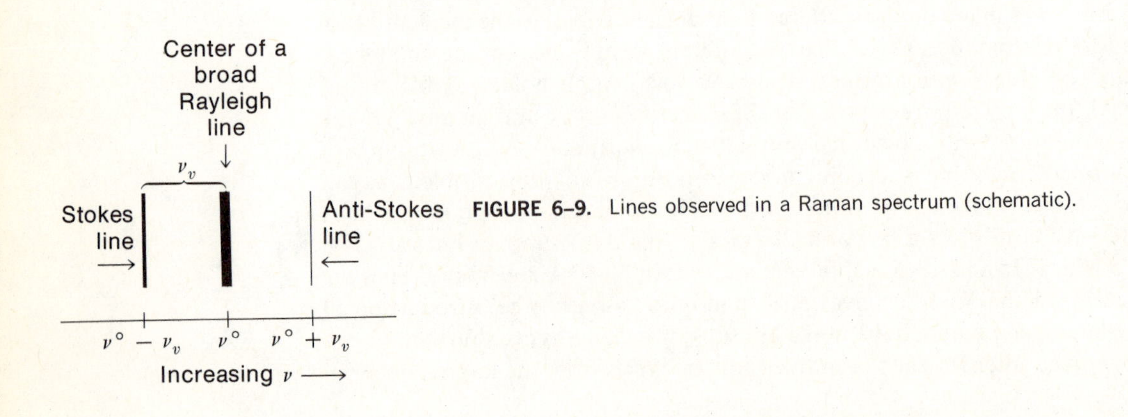

FIGURE 6–9. Lines observed in a Raman spectrum (schematic).

A molecule in the vibrationally excited state $v = 1$ can collide with an incident light quantum of frequency $\nu°$. The molecule can return to the ground state by giving its additional energy $h\nu_v$ to the photon. This photon, when scattered, will have a frequency $\nu° + \nu_v$. The spectral line with this frequency is referred to as an anti-Stokes line (see Fig. 6–9). Because of the Boltzmann distribution, there are fewer molecules in the $v = 1$ state than in $v = 0$, and the intensity of the anti-Stokes line is much lower than that of the Stokes line.

Both Rayleigh and Raman scattering are relatively inefficient processes. Only about 10^{-3} of the intensity of the incident exciting frequency will appear as Rayleigh scattering, and only about 10^{-6} as Raman scattering. As a result, very intense sources are required in this experiment. Laser beams provide the required intensity ($\sim$100 milliwatts to $\sim$1 watt of power) and produce good spectra even with very small samples.

6–9 RAMAN SELECTION RULES

The molecule interacts with electromagnetic radiation in the Raman experiment via the oscillating induced dipole moment or, more accurately, the oscillating molecular polarizability. The intensity, I, of the scattered radiation, ν_s, is a function of the polarizability, α_{ij}, of the molecule which, for a randomly oriented solid, is given by:

$$I = \frac{2^7\pi^5}{3^2c^4} I_0 \nu_s{}^4 \sum_{ij} \alpha_{ij} \tag{6-9}$$

where I_0 is the intensity of the incident radiation and α_{ij} is an element of the molecular polarizability tensor. Molecular polarizability can be represented by an ellipsoid; the change in polarizability (as the CO_2 molecule, for example, undergoes the symmetric stretch) can be schematically illustrated as in Fig. 6–10.

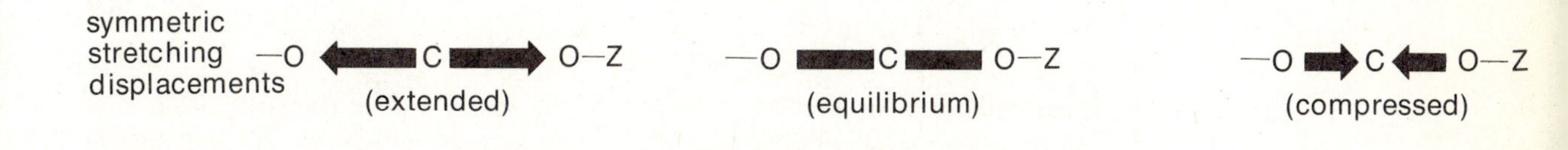

corresponding
polarizability
ellipsoid

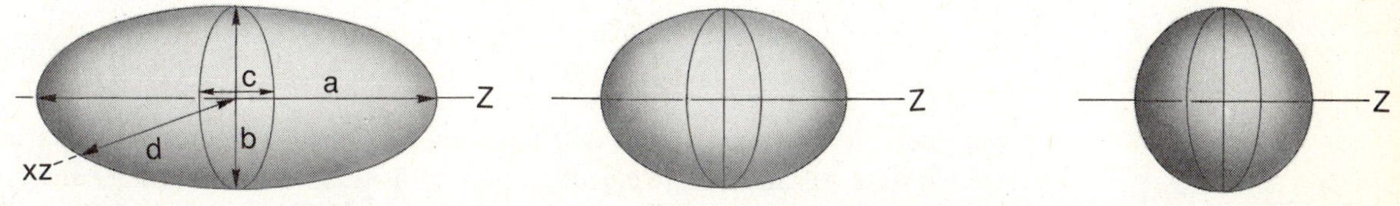

FIGURE 6–10. The change in polarizability for the CO_2 molecule for various displacements of the atoms during the symmetric stretching vibration. The double-headed arrows indicate (*a*) the *z*-axes of the ellipsoid, (*b*) the *x*, (*c*) the *y* and (*d*) the *xz*.

When a molecule is placed in a static electric field, the nuclei are attracted toward the negative pole and the electrons toward the positive pole, inducing a dipole moment in the molecule. If D represents the induced dipole moment and E the electric field, the polarizability α is defined by the following equation:

$$D = \alpha E \tag{6-10}$$

The magnitude of this polarizability depends upon the orientation of the bonds in the molecule with respect to the electric field direction; *i.e.*, it is anisotropic. In most molecules it is a tensor quantity. The induced dipole moment in the molecule will depend upon the orientation of the molecule with respect to the field and, since the dipole moment is a vector, it will have three components D_x, D_y, and D_z. The polarizability tensor can be represented physically as an ellipsoid (see Fig. 6–10), and it has nine components that can be described by a 3 × 3 matrix. The induced dipole moments are related to the polarizability tensor by the following equations:

$$\begin{aligned} D_x &= \alpha_{xx}E_x + \alpha_{xy}E_y + \alpha_{xz}E_z \\ D_y &= \alpha_{yx}E_x + \alpha_{yy}E_y + \alpha_{yz}E_z \\ D_z &= \alpha_{zx}E_x + \alpha_{zy}E_y + \alpha_{zz}E_z \end{aligned} \tag{6-11}$$

These equations state that if we locate a molecule in an electric field so that the ellipsoid axes point between the Cartesian axes, then the x-component of the induced dipole moment will depend upon the magnitude of the electric field along the y-axis because there is a polarizability component along xy that can interact with E_y and make a contribution to the change observed along x. The same is true for the interaction of E_z with α_{xz}. Equations (6–11) can be written in matrix form as

$$\begin{bmatrix} D_x \\ D_y \\ D_z \end{bmatrix} = \begin{bmatrix} \alpha_{xx} & \alpha_{xy} & \alpha_{xz} \\ \alpha_{yx} & \alpha_{yy} & \alpha_{yz} \\ \alpha_{zx} & \alpha_{zy} & \alpha_{zz} \end{bmatrix} \begin{bmatrix} E_x \\ E_y \\ E_z \end{bmatrix}$$

For an optically inactive molecule, the α tensor is symmetric and

$$\alpha_{xy} = \alpha_{yx}{}^*$$

If we choose a proper set of axes, we can carry out a transformation that diagonalizes the tensor previously given. The new axes are called the principal axes of the polarizability tensor with components $\alpha_{x'x'}$, $\alpha_{y'y'}$, and $\alpha_{z'z'}$ whose trace is equal to that in the previous coordinate system.

Quantum mechanically, the polarizability of a molecule by electromagnetic radiation along the direction ij is given by

$$(\alpha_{ij})_{mn} = \frac{1}{h}\sum_e \left[\frac{(M_j)_{me}(M_i)_{en}}{\nu_e - \nu_0} + \frac{(M_i)_{me}(M_j)_{en}}{\nu_e + \nu_s}\right] \tag{6-12}$$

where m and n are the initial and final states of the molecule and e represents an excited state. M_i and M_j are electric dipole transition moments along i and j, while ν_e is the energy of the transition to e and ν_0 and ν_s are the frequencies of the incident and scattered radiation.

As previously mentioned, the selection rules for Raman spectroscopy are different from those for infrared. *In order for a vibration to be Raman active, the change in polarizability of the molecule with respect to vibrational motion must not be zero at the equilibrium position of the normal vibration; i.e.,*

$$\left(\frac{\partial \alpha}{\partial r}\right)_{r_e} \neq 0 \qquad (6\text{-}13)$$

where α is the polarizability and r represents the distance along the normal coordinate. If the plot of polarizability, α, versus distance from the equilibrium distance, r_e, along the normal coordinate is that represented by Fig. 6–11(A), the vibration will be Raman active. If the plot is represented by the curves *1* or *2* of Fig. 6–11(B), $\partial\alpha/\partial r$ will be zero at or near the equilibrium distance, r_e, and the vibration will be Raman inactive.

Small amplitudes of vibration (as are normally encountered in a vibration mode) are indicated by the region on the distance axis on each side of zero and between the dotted lines. As can be seen from Fig. 6–11, the vibration in (A) corresponds to an appreciable change in polarizability occurring in this region, while that in (B) corresponds to practically no change. Hence, the Raman selection rule (6–13) is often stated: *In order for a vibration to be Raman active, there must be a change in polarizability during the vibration.*

The completely symmetric stretch for CO_2 is represented in Fig. 6–10. The points labeled *A* and *C* in Fig. 6–11(A) correspond to extreme stretching of the bonds for the symmetric stretch. *A* and *C* represent, respectively, more and less polarizable structures than the equilibrium structure. A plot of α versus r for the symmetric stretch is represented by Fig. 6–11(A). The change in polarizability for the asymmetric stretch is represented by one of the curves in Fig. 6–11(B). As a result, this vibration is Raman inactive. Infrared activity is just the opposite of Raman activity in this case. *In general, for any molecule that possesses a center of symmetry, there will be no fundamental lines in common in the infrared and Raman spectra.* This is a very valuable generalization for structure determination. If the same absorption band is found in both the infrared and Raman spectra, it is reasonably certain that the

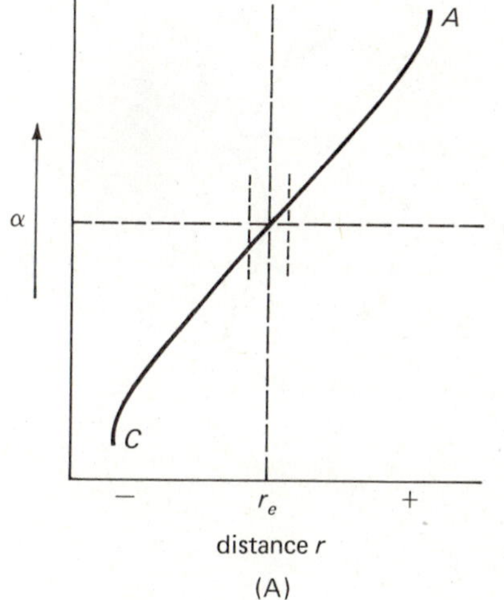

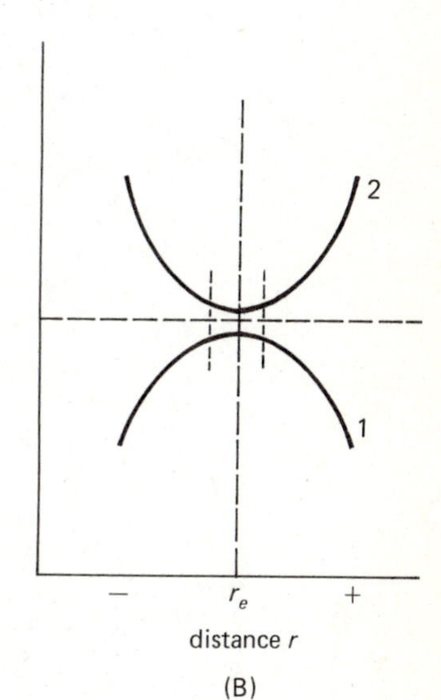

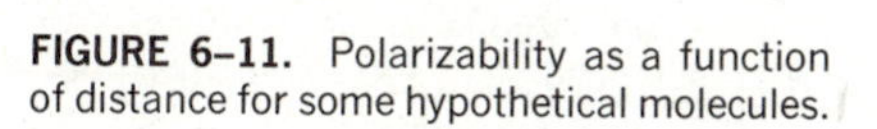
FIGURE 6–11. Polarizability as a function of distance for some hypothetical molecules.

molecule lacks a center of symmetry. It is possible that in a molecule lacking a center of symmetry no identical lines appear because of the low intensity of one of the corresponding lines in one of the spectra.

This discussion serves to illustrate qualitatively the selection rules for Raman spectroscopy. It is often difficult to tell by inspection the form of the polarizability curve for a given vibration. As a result, this concept is more difficult to utilize qualitatively than is the dipole moment selection rule in infrared. In a later section it will be shown how character tables and symmetry arguments can provide information concerning the infrared and Raman activity of vibrations. For now, we summarize the above discussion by stating that in order for a vibration to be Raman active, the shape, size, or orientation of the ellipsoid must change during the vibration as shown in Fig. 6–11(A).

6–10 POLARIZED AND DEPOLARIZED RAMAN LINES

Valuable information can be obtained by studying the polarized components of a Raman line. In order to understand how the Stokes lines can have a polarization, we will first consider some scattering experiments with polarized light. In Fig. 6–12 the interaction of yz polarized radiation with a molecule that has an isotropic polarizability tensor is indicated. The oscillation induced in the molecule will be in the same plane as the electric field and will have a z-component as shown by the double-headed arrow. If we observe emitted radiation along the x-axis, it will have to be xz polarized. Since α is isotropic, the direction of the induced dipole is unchanged as the molecule tumbles, and only xz scattered radiation results. If we had used xy polarized radiation as our source, the oscillating induced dipoles in a molecule with an isotropic polarizability would be oscillating parallel to the x-axis. In Fig. 6–13 the x-axis is perpendicular to the page, so the xy plane polarized radiation is also perpendicular to the page. No component is detected along x, for the molecular induced dipole moment oscillates parallel to x for all orientations of the molecule.

As a result of the above discussion, it should now be clear that—even if our source is non-polarized—the scattered radiation will be *polarized* if our molecule is isotropic in α, for there will be no xy component.

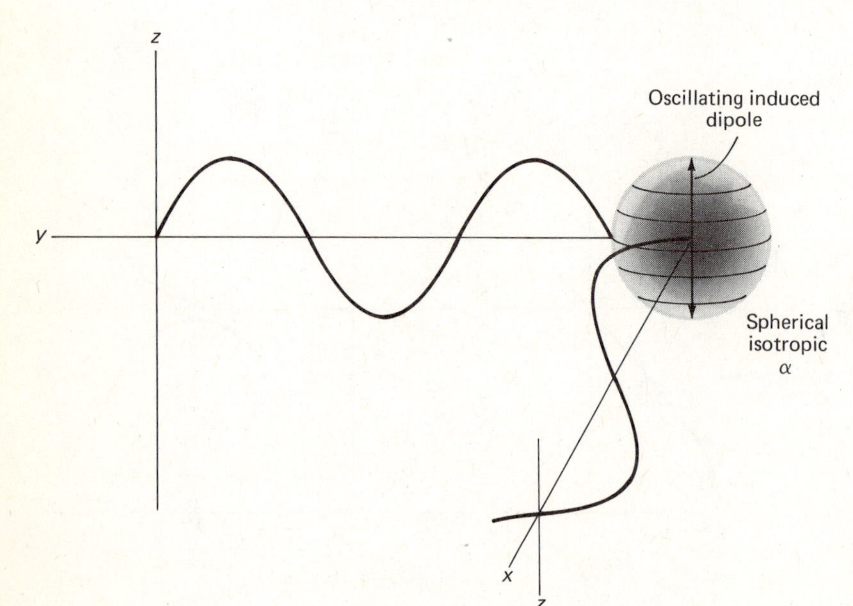

FIGURE 6–12. Raman scattering of yz plane polarized radiation producing an xz polarized scattered wave when the polarizability is isotropic.

FIGURE 6–13. Raman scattering of *xy* plane polarized radiation produces no *xz* or *yz* components for the scattered wave when the polarizability is isotropic.

The polarizability tensor is usually anisotropic, as shown in Fig. 6–14. Accordingly, the induced moment will not be coincident with the plane of the electric field but will tend to be oriented in the direction of greatest polarizability. The scattered light vibrates in the same plane as the induced dipole. As the molecule tumbles, the orientation of the induced dipole moment relative to the x-axis changes. Therefore, there are both xz and xy components in the scattered beam giving rise to radiation that is *depolarized.* Even if the incident radiation is polarized, an anisotropic polarizability tensor will give rise to scattered radiation that is depolarized. In the actual experiment, polarized radiation from a laser source is employed and the polarization of the radiation scattered along the x-axis is determined.

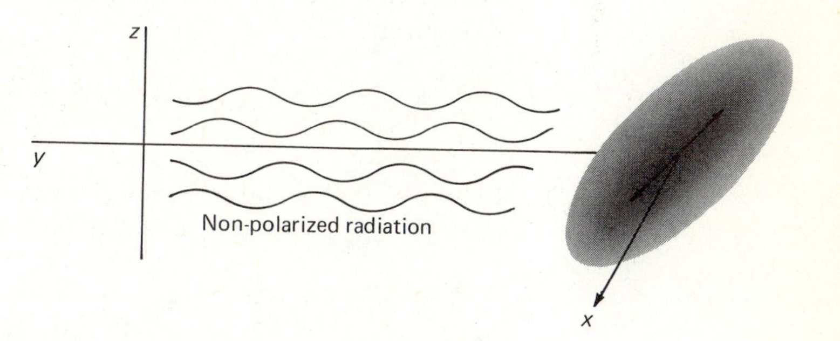

FIGURE 6–14. Raman scattering of nonpolarized radiation from an anisotropic polarizability tensor.

The above considerations apply to the Stokes lines, for which it is found that a *totally symmetric vibration mode gives rise to a polarized scattered line and that a vibration with lower symmetry is depolarized. This property can be used to confirm whether or not a vibration has A_1 symmetry.* Thus, by using an analyzer, the scattered Stokes radiation traveling along the x-axis is resolved into two polarized components: radiation polarized in the y direction and that in the z direction. Radiation traveling along x and polarized in the y direction is polarized parallel to the direction in which the incident light is traveling, and that component polarized in the z direction is polarized perpendicular to the incident radiation. The *depolarization ratio,* ρ, is defined as the ratio of the intensity, I, of the parallel, $y(\|)$, to perpendicular, $z(\perp)$, components of the Stokes line:

$$\rho = \frac{I_{y(\|)}}{I_{z(\perp)}} \tag{6-14}$$

Those Raman lines for which $\rho = \frac{3}{4}$ are referred to as depolarized lines and correspond to vibrations of the molecule that are not totally symmetric. Those Raman lines for which $0 < \rho < \frac{3}{4}$ are referred to as polarized lines. The vibrations of the molecule must transform as A_1 in order to be polarized. The use of this information in making band assignments will be made clearer after a discussion of symmetry considerations.

6–11 RESONANCE RAMAN SPECTROSCOPY

As the laser exciting frequency in the Raman experiment approaches an allowed electronic transition in the molecule being investigated, those normal modes that are vibronically active in the electronic transition exhibit a pronounced enhancement in their Raman intensities.[12] There are two applications of this fact. One involves utilizing the intensity enhancement to study materials that can be obtained only at low concentrations. The second application involves the fact that only those few vibrational modes near the site of an electronic transition in a complex molecule will be enhanced. Most examples of resonance Raman spectroscopy involve the enhancement of totally symmetric modes and arise from intense electronic transitions. There are recent examples in which vibronically allowed electronic transitions enhance the Raman bands of modes that are not totally symmetric, making them vibronically allowed.

It is possible for more than one electronic transition to be responsible for the resonance Raman effect, complicating the symmetry-based application just discussed. By studying the intensity of the resonance Raman bands as a function of the frequency of the exciting laser line, one can determine whether one electronic transition (the so-called *A*-mechanism) or two (*B*-mechanism) are involved. The intensity of a particular band is proportional to the square of the frequency factor, *F*. For the two mechanisms above, equations (6–15) and (6–16) have been proposed, respectively:

$$F_A = \frac{\nu^2(\nu_e^2 + \nu_0^2)}{(\nu_e^2 - \nu_0^2)^2} \tag{6-15}$$

where ν_e is the frequency for the electronic transition, ν_0 is the source frequency and $\nu = \nu_0 - \Delta\nu_{m,n}$ with $\Delta\nu_{m,n}$ equal to the frequency for the vibrational transition; and

$$F_B = \frac{2\nu^2(\nu_e\nu_s + \nu_0^2)}{(\nu_e^2 - \nu_0^2)(\nu_s^2 - \nu_0^2)} \tag{6-16}$$

where ν_e and ν_s are the frequencies for the two electronic transitions involved and the other terms are as defined for equation (6–15). A large number of other mechanisms and equations have recently been proposed.

An example of the enhanced intensity[13] from the resonance Raman effect is shown in Fig. 6–15. In (A), a typical electronic spectrum for a heme chromophore is shown. In (B), parallel and perpendicular components of the resonance Raman spectrum of a 5×10^{-4} M solution of oxyhemoglobin are illustrated. Without the resonance enhancement associated with the 568 nm exciting wavelength, this concentration of material would not have yielded a Raman spectrum. The label dp refers to a depolarized band (depolarization ratio of ¾), p to a polarized band, and ip to a band in which the depolarization ratio is greater than one (*i.e.*, the scattered light is polarized perpendicular to the polarization of the incident light). An ip band requires that the scattering tensor be antisymmetric (*i.e.*, $\alpha_{ij} = -\alpha_{ji}$), which is possible when ν_0 approaches ν_e. When the exciting line is far from an electronic transition, the ratio cannot be larger than ¾. The scattering tensor elements discussed here and symbolized by α are not to be confused with the polarizability tensor elements discussed earlier.

An assignment of the heme Raman spectrum has been made that is consistent with the assigned electronic transitions. The spectral features common to many heme proteins are discussed, as is the use of the depolarization ratios to infer local symmetry of the heme unit.[13]

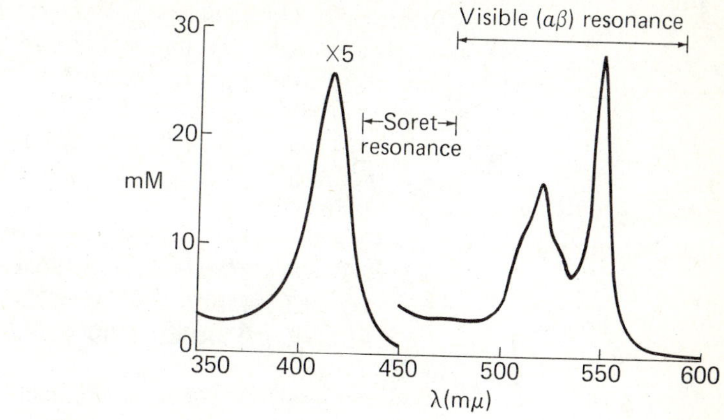

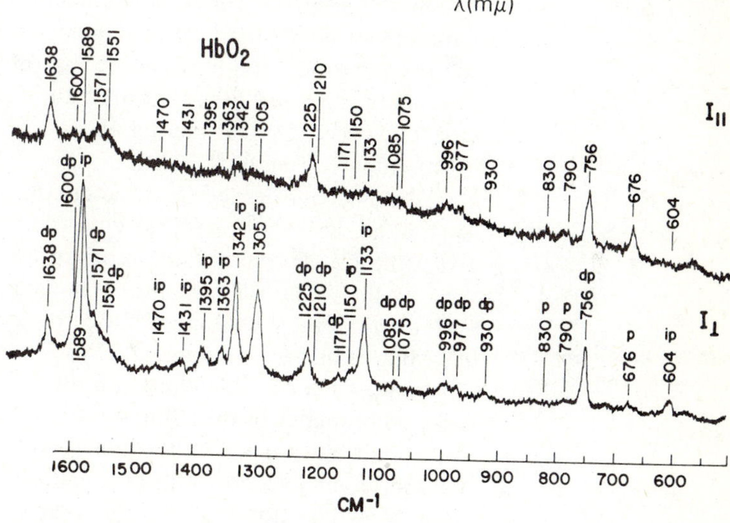

FIGURE 6–15. (A) Ultraviolet (Soret) and visible (α–β) chromophores in hemes. Spectrum of ferrocytochrome c. (B) Resonance Raman spectra of oxyhemoglobin. Both the direction and the polarization vector of the incident laser radiation are perpendicular to the scattering direction. The scattered radiation is analyzed into components perpendicular ($I_\perp$) and parallel ($I_\parallel$) to the incident polarization vector. The exciting wavelength was 568.2 nm (5682 Å) for HbO_2. Concentrations about 0.5 mM for HbO_2. [From T. G. Spiro and T. C. Strekas, *Proc. Natl. Acad. Sci.*, **69**, 2622 (1972).]

In another application of this technique,[14] a resonance Raman analysis of oxyhemerythrin was carried out using the oxygen $\longrightarrow$ iron charge-transfer band at 5000 Å to intensify the O—O stretching vibrations. Two Raman frequencies were observed at 844 cm^{-1} ($\rho = 0.33$) and at 500 cm^{-1} ($\rho = 0.4$). When $^{18}O_2$ is employed, the bands shift to 798 cm^{-1} and 478 cm^{-1}, respectively. These bands do not appear in the deoxygenated protein. The band at 844 cm^{-1} has been assigned to the O—O stretch and is not coupled to the other vibrational modes in the molecule, since the predicted frequency for isotopic substitution (796 cm^{-1}) is observed. The 500 cm^{-1} band is assigned to an Fe—O—O stretching mode. The spectra indicate only one type of O_2 complex.

6–12 SIGNIFICANCE OF THE NOMENCLATURE USED TO DESCRIBE VARIOUS VIBRATIONS

SYMMETRY ASPECTS OF MOLECULAR VIBRATIONS

In this section the conventions used to label the vibrations A_1, A_2, etc., will be reviewed to refresh the reader's memory. If the vibration is symmetric with respect to the highest-fold rotation axis, the vibration is designated by the letter A. If it is antisymmetric, the letter B is employed. E stands for a doubly degenerate vibration and T for a triply degenerate mode (F is commonly used instead of T in vibrational

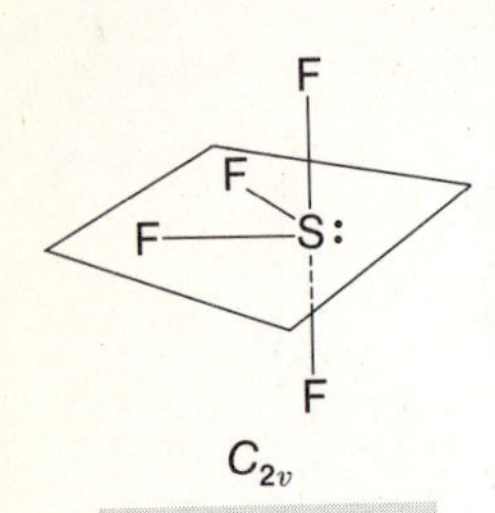

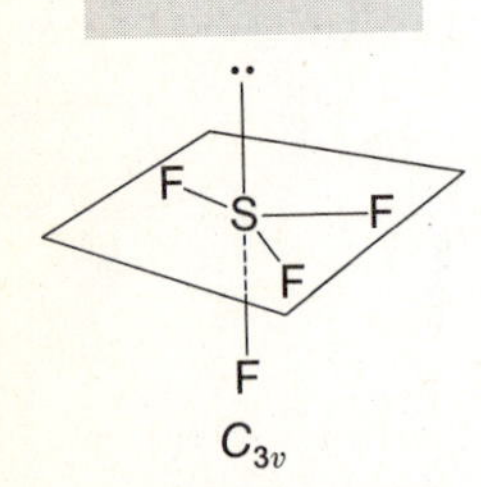

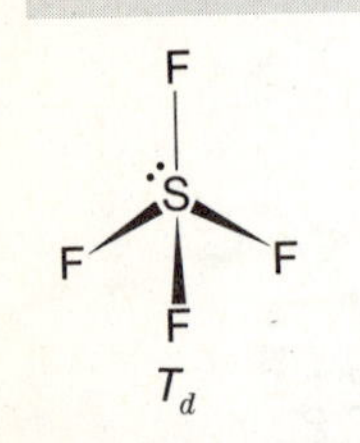

FIGURE 6–16. Some possible structures for SF_4.

spectroscopy). The subscripts g and u refer to symmetry with respect to an inversion through a center of symmetry and are used only for molecules with a center of symmetry. If a vibration is symmetric with respect to the horizontal plane of symmetry, this is designated by ′, while ″ indicates antisymmetry with respect to this plane. Subscripts 1 and 2 (as in A_1 and A_2) indicate symmetry and antisymmetry, respectively, toward a twofold axis that is perpendicular to the principal axis.

6–13 USE OF SYMMETRY CONSIDERATIONS TO DETERMINE THE NUMBER OF ACTIVE INFRARED AND RAMAN LINES

In this section we shall be concerned with classifying the $3N - 6$ (or $3N - 5$ for a linear molecule) vibrations in a molecule to the various irreducible representations in the point group of the molecule. This information will then be used to indicate the degeneracy and number of infrared and Raman active vibrations. The procedure is best illustrated by considering the possible structures for SF_4(C_{2v}, C_{3v}, and T_d) represented in Fig. 6–16.

We first consider the C_{2v} structure. An x, y, z coordinate system is constructed for each atom, and we proceed to determine the total representation for this basis set using the C_{2v} character table reproduced in Table 6–4.

We know that when an atom is moved by a symmetry operation, the contribution to the total representation will be zero; so we need only worry about the x, y, and z coordinates on atoms not moved. For E, none of the atoms or coordinates are moved, giving $\chi_T(E) = 15$. Only the sulfur atom is not moved by the C_2 operation, but the x and y coordinates of the sulfur each contribute -1 to the total representation while the z contributes $+1$. This leads to $\chi_T(C_2) = -1$. Sulfur and two of the fluorines are not moved by reflection in the σ_{xz} plane, while two fluorines are moved. On the sulfur atom this reflection changes the sign of y but leaves x and z unchanged for a contribution of $+1$. The same is true for each of the unshifted fluorines, leading to $\chi_T(\sigma_{v(xz)}) = 3$. Reflection through yz changes the x coordinate to -1 for all three atoms not moved but does not change y or z coordinates, leading to $\chi_T(\sigma'_{(yz)}) = 3$. Summing up these results, we obtain:

$$\chi_T = 15 \quad -1 \quad 3 \quad 3$$

We can factor the total representation as follows:

$$\begin{array}{lllll} & \text{for } E & \text{for } C_2 & \text{for } \sigma_{v(xz)} & \text{for } \sigma'_{v(yz)} \\ n^{A_1} = \frac{1}{4}[& 1\cdot 1\cdot 15 + & 1\cdot 1\cdot(-1) + & 1\cdot 1\cdot 3 + & 1\cdot 1\cdot 3] = 5 \\ n^{A_2} = \frac{1}{4}[& 1\cdot 1\cdot 15 + & 1\cdot 1\cdot(-1) + & 1\cdot(-1)\cdot 3 + & 1\cdot(-1)\cdot 3] = 2 \end{array}$$

TABLE 6–4. C_{2v} CHARACTER TABLE*

	E	C_2	$\sigma_{v(xz)}$	$\sigma'_{v(yz)}$	
A_1	1	1	1	1	z, α_{x^2}, α_{y^2}, α_{z^2}
A_2	1	1	−1	−1	R_z, α_{xy}
B_1	1	−1	1	−1	x, R_y, α_{zx}
B_2	1	−1	−1	1	y, R_x, α_{yz}

*For planar molecules the x-axis is perpendicular to the plane.

$$n^{B_1} = \frac{1}{4}[1 \cdot 1 \cdot 15 + 1 \cdot (-1) \cdot (-1) + 1 \cdot (-1) \cdot 3 + 1 \cdot 1 \cdot 3] \quad = 4$$

$$n^{B_2} = \frac{1}{4}[1 \cdot 1 \cdot 15 + 1 \cdot (-1) \cdot (-1) + 1 \cdot 1 \cdot 3 \quad + 1 \cdot (-1) \cdot 3] = 4$$

The results $n^{A_1} = 5$, $n^{A_2} = 2$, $n^{B_1} = 4$, and $n^{B_2} = 4$ represent the $3N$ degrees of freedom for a pentatomic C_{2v} structure.

To get the total number of vibrations, the three degrees of translational freedom and three of rotational freedom must be subtracted (giving $3N - 6$ for a nonlinear molecule). Translation along the x-axis can be represented by an arrow lying along this axis, which as seen from the C_{2v} character table belongs to the species B_1. Similarly, translation along the z- and y-axes transforms according to species A_1 and B_2. Rotations along the x-, y-, and z-axes are represented by the symbols R_x, R_y, and R_z and belong to species A_2, B_1, and B_2, respectively. These six degrees of freedom ($1A_1$, $1A_2$, $2B_1$, and $2B_2$) are subtracted from the representation of the total degrees of freedom, producing $n^{A_1} = 4$, $n^{A_2} = 1$, $n^{B_1} = 2$, and $n^{B_2} = 2$ corresponding to the nine vibrations predicted from the $3N - 6$ rule.

We have now calculated the species to which the $3N - 6$ vibrations of SF_4 belong if it has a C_{2v} structure. Four are of species A_1, one of A_2, two of B_1, and two of B_2. The next job is to determine which vibrations are infrared active and which are Raman active. For a fundamental transition to occur by absorption of infrared electromagnetic radiation, one of the three integrals

$$\int \psi_v{}^{\circ} \hat{x} \psi_v{}^{\text{ex}}\, d\tau \qquad \int \psi_v{}^{\circ} \hat{y} \psi_v{}^{\text{ex}}\, d\tau \qquad \int \psi_v{}^{\circ} \hat{z} \psi_v{}^{\text{ex}}\, d\tau \tag{6-17}$$

must be non-zero or A_1. The $\hat{x}$, $\hat{y}$, and $\hat{z}$ operators correspond to the orientation of the electric field vector relative to a molecular Cartesian coordinate system. The operator is identical to that discussed in electronic absorption spectroscopy. Since all ground vibrational wave functions are A_1 (if necessary, re-read Chapter 4, the section on atomic and molecular transitions), this amounts to having a component of the transition dipole operator $\hat{x}$, $\hat{y}$, or $\hat{z}$ that has the same symmetry as $\psi_{\text{vib}}{}^*$, where the total vibrational excited state for a fundamental transition has the same symmetry as the normal mode excited. Accordingly, *a fundamental will be infrared active if the excited normal mode belongs to one of the irreducible representations corresponding to the x, y and z vectors, and will be inactive if it does not.*

In order for a vibration to be Raman active, it is necessary that one of the integrals of the type

$$\int \psi_v{}^{\circ} \hat{P} \psi_v{}^{\text{ex}}\, d\tau \tag{6-18}$$

be non-zero, *i.e.*, the integrand A_1. Here the operator $\hat{P}$ is one of the quadratic functions of the x, y, and z vectors (*i.e.*, x^2, y^2, z^2, xy, yz, xz), simply or in combination (*i.e.*, $x^2 - y^2$). The symmetries of these functions are listed in the character tables opposite their irreducible representations. Since these quantities are components of the polarizability tensor, we find the following rule stated: *A fundamental transition will be Raman active if the corresponding normal mode belongs to the same irreducible representations as one or more of the components of the polarizability tensor.*

For the C_{2v} structure of SF_4, the A_1, B_2, and B_1 modes are infrared active, while A_2 is infrared inactive (neither x, y, nor z has A_2 symmetry). Eight infrared bands are expected: $4A_1$, $2B_1$, and $2B_2$. All nine of the fundamental vibrations are Raman active, and the four A_1 modes will be polarized.

When we carry out the above procedure on the T_d structure for SF_4, we obtain a total representation $\chi_T = 15(E)\ 0(8C_3)\ 3(6\sigma_d) - 1(6S_4) - 1(3S_4^2 = 3C_2)$. Factoring, we obtain:

$$n^{A_1} = \frac{1}{24}[1 \cdot 1 \cdot 15 + 8 \cdot 1 \cdot 0 \quad + 6 \cdot 1 \cdot 3 \quad + 6 \cdot 1 \cdot (-1) \quad + 3 \cdot 1 \cdot (-1)] \quad = 1$$

$$n^{A_2} = \frac{1}{24}[1 \cdot 1 \cdot 15 + 8 \cdot 1 \cdot 0 \quad + 6 \cdot (-1) \cdot 3 + 6 \cdot (-1) \cdot (-1) + 3 \cdot 1 \cdot (-1)] \quad = 0$$

$$n^{E} = \frac{1}{24}[1 \cdot 2 \cdot 15 + 8 \cdot (-1) \cdot 0 + 6 \cdot 0 \cdot 3 \quad + 6 \cdot 0 \cdot (-1) \quad + 3 \cdot 2 \cdot (-1)] \quad = 1$$

$$n^{T_1} = \frac{1}{24}[1 \cdot 3 \cdot 15 + 8 \cdot 0 \cdot 0 \quad + 6 \cdot (-1) \cdot 3 + 6 \cdot 1 \cdot (-1) \quad + 3 \cdot (-1) \cdot (-1)] = 1$$

$$n^{T_2} = \frac{1}{24}[1 \cdot 3 \cdot 15 + 8 \cdot 0 \cdot 0 \quad + 6 \cdot 1 \cdot 3 \quad + 6 \cdot (-1) \cdot (-1) + 3 \cdot (-1) \cdot (-1)] = 3$$

The T_d character table indicates that the three translations are of species T_2 and the three rotations of species T_1. Since T_1 and T_2 are triply degenerate, we need subtract only one of each from the total degrees of freedom to remove three degrees of translation and three of rotation. The total number of vibrations belong to species A_1, E, and $2T_2$ giving rise to the nine modes predicted from the $3N - 6$ rule. The six T_2 vibrations (two triply degenerate sets) are infrared active and give rise to two fundamental bands in the infrared spectrum. All modes are Raman active, giving rise to four spectral bands corresponding to fundamentals. Of these, A_1 is polarized. It can be shown similarly that the C_{3v} structure of SF_4 leads to six infrared lines (three A_1 and three E) and six Raman lines (three A_1 and three E), three of which are polarized (A_1).

The actual spectrum is found to contain five infrared fundamentals, and five Raman lines, one of which is polarized. These results, summarized in Table 6–5, eliminate the T_d structure, leaving either C_{2v} or C_{3v} of the structures considered as possible ones. This example demonstrates the point that although there cannot be more fundamental vibrations than allowed by a symmetry type, often not all vibrations are detected. The problems in application of these concepts involve separating overtones and combinations from the fundamentals and, as is often the case, not finding some "active" fundamental vibrations because they have low intensities.

Table 6–5 should be verified by the reader by using the character tables contained in Appendix A and the procedure described above. By a detailed analysis of the band contours and an assignment of the frequencies, it was concluded that the C_{2v} structure best fits the observed spectrum. This structure was subsequently confirmed by other physical methods.

TABLE 6–5. SUMMARY OF ACTIVE MODES EXPECTED FOR VARIOUS CONFIGURATIONS OF SF_4

	C_{2v}	T_d	C_{3v}	**Found**[a]
Infrared active modes	$8(4A_1, 2B_1, 2B_2)$	$2(2T_2)$	$6(3A_1, 3E)$	5(or 7)
Raman modes	$9(4A_1, A_2, 2B_1, 2B_2)$	$4(A_1, E, 2T_2)$	$6(3A_1, 3E)$	5(or 8)
Polarized modes	$4(4A_1)$	$1(A_1)$	$3(3A_1)$	1

[a] R. E. Dodd, L. A. Woodward, and H. L. Roberts, *Trans. Faraday Soc.*, **52**, 1052 (1956).

TABLE 6–6. INFRARED AND RAMAN FUNDAMENTALS FOR $CHCl_3$

IR Active Vibrations for $CHCl_3$ (cm^{-1})	Raman Active Vibrations for $CHCl_3$ (cm^{-1})	Raman Spectra of $CDCl_3$ (cm^{-1})	Designation
260	262	262	ν_6
364	366 (polarized)	367	ν_3
667	668 (polarized)	651	ν_2
760	761	738	ν_5
1205	1216	908	ν_4
3033	3019 (polarized)	2256	ν_1

The data available on $CHCl_3$ and $CDCl_3$ are summarized in Table 6–6 and will be discussed briefly because this example serves to review many of the principles discussed. The $3N - 6$ rule predicts nine normal vibrations. Just as with the C_{3v} structure of SF_4, the total representation shows that these nine vibrations consist of three A_1 and three E species or a total of six fundamental frequencies. All six fundamentals will be IR and Raman active, and the three A_1 fundamentals will give polarized Raman lines. The observed bands that have been assigned to fundamental frequencies are summarized in Table 6–6. These data indicate that the 3033, 667, and 364 cm^{-1} bands belong to the totally symmetric vibrations (species A_1), for these bands are Raman polarized. They are labeled ν_1, ν_2, and ν_3, respectively. The other three bands are of species E, and the 1205, 760, and 260 cm^{-1} bands are ν_4, ν_5, and ν_6, respectively. Note that since the molecule does not have a center of symmetry, the IR and Raman spectra have lines in common. The exact form of these fundamental vibrations is illustrated in Appendix C. The bands ν_1 and ν_4 are the C—H stretching and bending vibrations, respectively. Note how deuteration has a pronounced effect on these frequencies but almost no effect on the others. The vibrations at 3033 and 1205 cm^{-1} are almost pure C—H modes. Since the very light hydrogen or deuterium atom is moving, very little C—Cl motion is necessary to retain the center of mass (so that there will be no net translation of the molecule).

The above manipulations are quite simple and yet yield valuable information about the structures of simple molecules. There are many applications of these concepts. It is strongly recommended that the reader carry out the exercises at the end of this chapter.

Spectroscopists often use the ν_n symbolism in assigning vibrations. This can be translated into the language of stretches, bends, and twists by referring to the diagrams in Appendix C or in Herzberg (see General References) for an example of a molecule with similar symmetry and the same number of atoms. The irreducible representation of a given normal vibration can be determined from the diagrams by the procedure outlined in Chapter 2.

6–14 SYMMETRY REQUIREMENTS FOR COUPLING, COMBINATION BANDS, AND FERMI RESONANCE

There are some important symmetry requirements regarding selection rules for overtone and combination bands. These can be demonstrated by considering BF_3, a D_{3h} molecule, as an example. The D_{3h} character table indicates that the symmetric stretch, ν_1, of species A_1' is infrared inactive (there is no dipole moment change). The species of the combination band $\nu_1 + \nu_3$ (where ν_3 is of species E') is given by the product of $A_1' \times E' = E'$. The combination band is infrared active. The ν_2 vibration is of symmetry A_2'' and is infrared active. The overtone $2\nu_2$ is of species $A_2'' \times$

$A_2'' = A_1'$, which is infrared inactive; $3\nu_2$ is of species A_2'' and is observed in the infrared spectrum. This behavior is strong support for the initial structural assignment, that of a planar molecule, and serves here as a nice example to demonstrate the symmetry requirements for overtones and combination bands (recall the discussion in Section 6–5).

In the case of Fermi resonance it is necessary that the states which are participating have the same symmetry. For example, if $2\nu_2$ is to undergo Fermi resonance with ν_1, one of the irreducible representations for the direct product $2\nu_2$ must be the same as for ν_1.

Coupling of group vibrations was mentioned earlier in this chapter. In order for coupling to occur, the vibrations must be of the same symmetry type. For example, in acetylene the symmetric C—H stretching vibration and the C—C stretching vibrations are of the same symmetry type and are highly coupled. The observed decrease in the frequency of the symmetric C—H stretch upon deuteration is smaller than expected, because of this coupling. Since the asymmetric C—H stretch and the "C—C" stretch are of different symmetry, they do not couple, and deuteration has the normal effect on the asymmetric C—D stretch (a decrease from 3287 to 2427 cm^{-1} is observed).

6–15 MICROWAVE SPECTROSCOPY

Pure rotational transitions in a molecule can be induced by radiation in the far infrared and microwave regions of the spectrum. Extremely good precision for frequency determination is possible in the microwave region. Compared to the infrared region of the spectrum, where measurements are routinely made to about 1 cm^{-1}, one can get resolution of about 10^{-8} cm^{-1} in the microwave region. A wide spectral range plus resolution and accuracy to 10^{-8} cm^{-1} make this a very valuable region for fingerprint applications. A list of frequencies has been tabulated[15,16] consisting of 1800 lines for about 90 different substances covering a span of 200,000 MHz.* Only in 10 of the cases reported were two of the 1800 lines closer than 0.25 MHz.

There are two requirements that impose limitations on microwave studies. (1) The spectrum must be obtained on the material in the gaseous state. For conventional instrumentation, a vapor pressure of at least 10^{-3} torr is required. (2) The molecule must have a permanent dipole moment in the ground state in order to absorb microwave radiation, since rotation alone cannot create a dipole moment in a molecule.

In addition to the fingerprint application, other useful data can be obtained from the microwave spectrum of a compound. Some of the most accurate bond distance and bond angle data available have been obtained from these studies. Let us first consider a diatomic molecule. The rotational energy, E, for a diatomic molecule is given by the equation:

$$E = hBJ(J + 1) \tag{6-19}$$

where J is an integer, the rotational quantum number; h is Planck's constant (6.62×10^{-27} erg sec); and $B = h/8\pi^2 I$ where I is the moment of inertia. Since $E = h\nu$, we obtain the relationship for the frequency, ν, corresponding to this energy:

$$\nu = BJ(J + 1) \tag{6-20}$$

* A wave number of 1 mm^{-1} is equivalent to a frequency of 299,800 MHz.

The energy of the transition from state J to $J+1$ can be determined by substitution in equation (6-19), leading to

$$\Delta E = 2Bh(J+1)$$

or

$$\Delta\nu = 2B(J+1)$$

There is another selection rule (in addition to the dipole moment requirement) for microwave absorption, which states that $\Delta J = \pm 1$. Therefore, the longest wavelength (lowest energy) absorption band in the spectrum will correspond to the transition from $J = 0$ to $J = 1$, for which the frequency of the absorbed energy is:

$$\Delta\nu = 2B \doteq \frac{2h}{8\pi^2 I} \tag{6-21}$$

$\Delta\nu$ is measured and all other quantities in equation (6-21) are known except I, which can then be calculated.

All other lines in the spectrum will occur at shorter wavelengths and will be separated from each other by $2B$ [*i.e.*, $\Delta\Delta\nu = 2B(J+2) - 2B(J+1) = 2B$, where $(J+1)$ is the quantum number for the higher rotational state and J that for the lower; see Fig. 6-17]. Once the moment of inertia is determined, the equilibrium internuclear separation, r_0, in the diatomic molecule can be calculated with equation (6-22):

$$I = \mu r_0^2 \tag{6-22}$$

where μ is the reduced mass.

In the above discussions we have assumed a rigid rotor as a model. This means that the atom positions in the molecule are not influenced by the rotation. Because of the influence of centrifugal force on the molecule during the rotation, there will be a small amount of distortion. As a result, the bond distance (and I) will be larger for high values of J, and the spacings between the peaks will decrease slightly as J increases.

In a more complex molecule the moment of inertia is related to the bond lengths and angles by a more complex relationship than (6-22). A whole series of simultaneous equations has to be solved to determine all the structural parameters. In order to get enough experimental observations to solve all these equations, isotopic substitution is employed. For pyridine, the microwave spectra of six isotopically substituted compounds (pyridine, 2-deuteropyridine, 3-deuteropyridine, 4-deuteropyridine, pyridine-2-^{13}C, and pyridine-3-^{13}C) were employed. For a more complete discussion of this problem, see Gordy, Smith, and Trambarulo.[15]

By a detailed analysis of the microwave spectrum, one can obtain information about internal motions in molecules. The torsional barrier frequency and energy can be obtained.[17a]

Double resonance experiments have also proved valuable for assigning peaks and for determining relaxation times of states. A given transition is saturated with microwave power, producing a non-equilibrium distribution in the populations of the two levels involved. One then looks for an increase or decrease in other transitions that are thought to involve one of these levels.[17b]

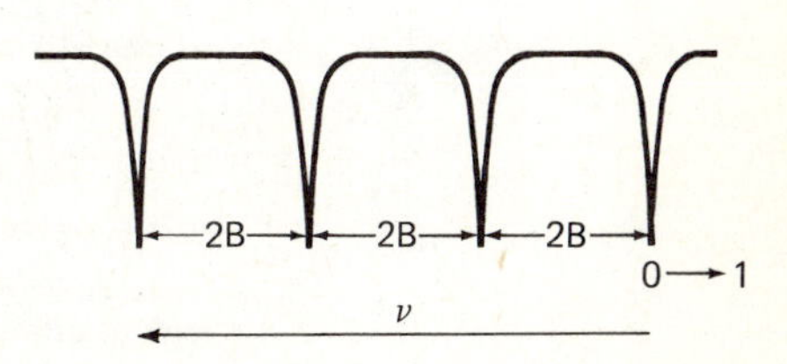

FIGURE 6-17. Illustration of the 0 ⟶ 1 and other rotational transitions (idealized).

Accurate dipole moment measurements can be made from microwave experiments. When an electric field is applied to the sample being studied, the rotational line is split (the Stark effect). The magnitude of the splitting depends upon the product of the dipole moment (to be evaluated) and the electric field strength (which is known).

Much information is available from the Zeeman splitting of rotational lines.[17c,d] The components of the magnetic susceptibility anisotropy can be evaluated, as can the expectation value $\langle r^2 \rangle$. As we shall see in the chapter on nuclear quadrupole resonance, nqr, much the same information obtained on solids by nqr can be obtained on gases by microwave spectroscopy.

6–16 ROTATIONAL RAMAN SPECTRA

Information equivalent to that obtained in the microwave region can be obtained from the rotational Raman spectrum, for which the permanent dipole selection rule does not hold. As a result, very accurate data on homonuclear diatomic molecules can be obtained from the rotational Raman spectrum. Experimentally, the bands are detected as Stokes lines with frequencies corresponding to rotational transitions.

In order for a molecule to exhibit a rotational Raman spectrum, the polarizability perpendicular to the axis of rotation must be anisotropic; *i.e.*, the polarizability must be different in different directions in the plane perpendicular to the axis. If the molecule has a threefold or higher axis of symmetry, the polarizability will be the same in all directions and rotational modes about this axis will be Raman inactive. Other rotations in the molecule may be Raman active.

For diatomic molecules the selection rule $\Delta J = \pm 2$ now applies:

$$\Delta E = Bh(4J' - 2) \tag{6-23}$$

and the frequency separation between the lines is $4B$. If this separation is not at least $\sim 0.1 \text{ cm}^{-1}$ the lines cannot be resolved in the Raman spectrum. The rest of the calculation is identical to that described previously. There has been only limited application of this technique to more complex molecules.

APPLICATIONS OF INFRARED AND RAMAN SPECTROSCOPY*

6–17 PROCEDURES

a. In Infrared

The most common infrared equipment covers the wavelength region from 5000 to 250 cm^{-1}. The limiting factor is the grating in the instrument. The grating resolves polychromatic radiation into monochromatic radiation so that variations in absorption of a sample with change in wavelength can be studied. The resulting spectrum is a plot of sample absorbance or per cent transmission versus wavelength (see Fig. 6–18).

The cells that hold the sample commonly are constructed of NaCl, AgCl, CaF_2, BaF_2, or TlCl. A special optical material called "Irtran-2" can be employed to construct cells for solvents (such as water) that dissolve the common sodium chloride

---

*The reader is referred to the text by Nakamoto[18] and to reference 11 for extensive compilations of examples of infrared and Raman spectroscopy, respectively.

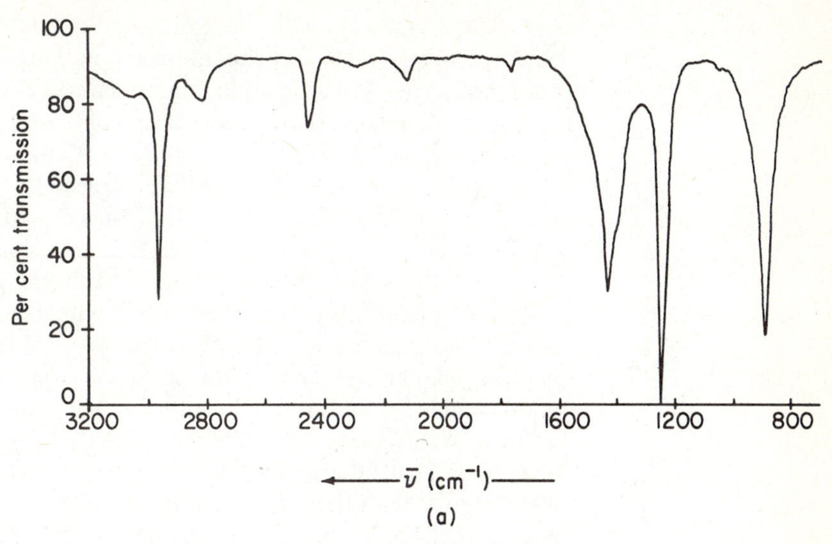

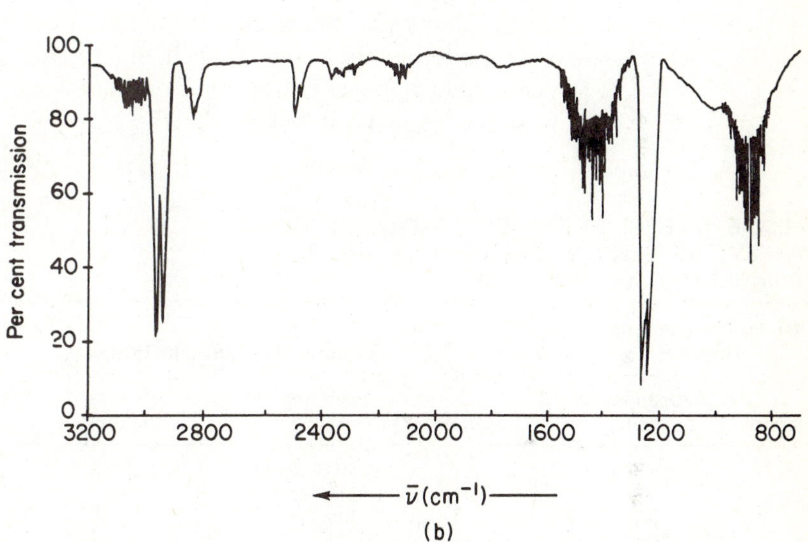

FIGURE 6–18. The infrared spectra of (A) liquid CH_3I and (B) gaseous CH_3I. [From G. M. Barrow: *Introduction to Molecular Spectroscopy.* Copyright © 1962 by McGraw-Hill, Inc. Used by permission of McGraw-Hill Book Company.]

cells. The silver chloride cells have the disadvantage of darkening on exposure to light.

The wavelength scale on the instrument usually changes with use, so frequent calibration is necessary. In the 5000 to 650 cm^{-1} region, polystyrene, ammonia, and water vapor are standards commonly used to calibrate the instrument.

In Chapter 7, we shall discuss Fourier transform techniques as applied to nmr spectroscopy. Here it should be mentioned that such an approach offers considerable advantage in infrared spectroscopy as well.[24] The method greatly decreases the amount of time needed to accumulate a spectrum. Thus, multiple scanning of the spectrum and computer storage of the scans become feasible, greatly enhancing the signal-to-noise ratio. Furthermore, filtering is not required, eliminating the need for energy-wasting slits. Enhanced resolution in a shorter time period is possible. Experimental details have been summarized in reference 24.

Samples can be examined as gases, liquids, or solids, or in solution. Special cells with long path length are needed for most gaseous samples. Solid samples are often examined as mulls in either Nujol (paraffin oil) or hexachlorobutadiene. The mull is prepared by first grinding the sample to a fine particle size and then adding enough oil or mulling agent to make a paste. The paste is examined as a thin layer between sodium chloride (or other optical material) plates. The quality of the spectrum obtained is very much dependent on the mulling technique. When the spectrum is examined, peaks from the mulling material will appear in the spectrum and possibly mask sample peaks. If two spectra are obtained, one in Nujol and one in hexachlorobutadiene, all wavelengths in the 5000 to 650 cm^{-1} region can be examined. Solid samples are sometimes examined as KBr discs. The sample and KBr are intimately mixed, ground, and pressed into a clear disc that is mounted and examined directly. Care should be exercised in this procedure, for anion exchange and other reactions may occur with the bromide ion during grinding or pressing. Reflectance spectra of solids can be obtained with commercially available attachments.

Solutions are most easily examined by use of the double beam technique, which in effect subtracts out solvent absorption. A solution of the sample is placed in a cell in the sample beam. A cell of the same path length, containing pure solvent, is placed in the reference beam. The difference in absorption is measured, in effect canceling absorption bands due to the solvent. In regions where the solvent absorption is very large, almost all of the radiation is absorbed by the two solutions, insufficient energy passes through the sample to activate the instrument, and the pen is "dead." No sample absorptions can be detected in these blank regions. As a matter of fact, if one were to interrupt the sample beam in these regions of the spectrum with his hand, the pen would not move. The open regions for several common liquids used for solution work are tabulated in Table 6–7.

TABLE 6–7. REGIONS IN WHICH VARIOUS SOLVENTS TRANSMIT AT LEAST 25 PER CENT OF THE INCIDENT RADIATION (*i.e.*, OPEN REGIONS)[a]

(A) For Cells of 1 mm Thickness		**(B) For Cells of 0.1 mm Thickness**			
Solvent	*Open Region* (cm^{-1})	*Solvent*	*Open Region* (cm^{-1})	*Solvent*	*Open Region* (cm^{-1})
CCl_4	4000–1610	CS_2	4000–2200	CH_3CN	4000–3700
	1500–1270		2140–1595		3500–2350
	1200–1020		1460– 650		2250–1500
	960– 860				1350–1060
		C_6H_6	4000–3100		1030– 930
$CHCl_3$	4000–3100		3000–1820		910– 650
	2980–2450		1800–1490		
	2380–1520		1450–1050	CH_3NO_2	4000–3100
	1410–1290		1020– 680		2800–1770
	1155– 940				1070– 925
	910– 860	CCl_4	4000– 820		910– 690
			720– 650		
CH_2Cl_2	4000–3180			C_5H_5N	4000–3500
	2900–2340	$CHCl_3$	4000–3020		3000–1620
	2290–1500		3000–1240		1400–1230
	1130– 935		1200– 805		980– 780
CS_2	4000–2350	CH_2Cl_2	4000–1285	$HC(O)N(CH_3)_2$	4000–3000
	2100–1640		1245– 900		2700–1780
	1385– 875		890– 780		1020– 870
	845– 650		750– 650		860– 680
$Cl_2C{=}CCl_2$	4000–1375	$Cl_2C{=}CCl_2$	4000– 935	CH_3OH[b]	2800–1500
	1340–1180		875– 820		1370–1150
	1090–1015		745– 650		970– 700

[a]The infrared spectra of some solvents in the 600 to 450 cm^{-1} region are contained in references 18 and 19, and F. F. Bentley et al., *Spectrochim. Acta,* **13,** 1 (1958).

[b]Not to be used in NaCl cells.

TABLE 6-8. FREQUENCIES (IN cm^{-1}) FOR SO_2 FUNDAMENTALS IN DIFFERENT PHYSICAL STATES

	Gas	Liquid[a]	Solid
$\nu_2(a_1)$	518	525	528
$\nu_1(a_1)$	1151	1144	1144
$\nu_3(b_1)$	1362	1336	1322
			1310

[a] Raman spectrum.

Most pure liquids can be examined in standard sodium chloride cells. If the liquid attacks the sample cells slightly or else has a very high absorption, the sample can be examined as a smear between two sodium chloride plates. Plates of NaCl can be purchased for a very reasonable price. Reference 19 provides a good description of the techniques for this method.

The spectrum obtained for a given sample depends upon the physical state of the sample. Gaseous samples usually exhibit rotational fine structure. This fine structure is damped in solution spectra because collisions of molecules in the condensed phase occur before a rotation is completed. This difference between gaseous and liquid spectra is illustrated by the spectra in Fig. 6–18.

In addition to the difference in resolved fine structure, the number of absorption bands and the frequencies of the vibrations vary in the different states. As an illustration, the fundamentals of SO_2 are reported in Table 6–8 for the different physical states.

Other effects not illustrated by the data in Table 6–8 are often encountered. Often there are more bands in the liquid state than in the gaseous state of a substance. Frequently, new bands below 300 cm^{-1} appear in the solid state spectrum. The causes of these frequency shifts, band splittings, and new bands in the condensed states are well understood[(20)] and will be discussed briefly here.

The stronger intermolecular forces that exist in the solid and liquid states compared with those in the gaseous state are the cause of slight shifts (assuming that there are no pronounced structural changes with change of phase) in the frequencies. The bands below 300 cm^{-1} are often caused by *lattice vibrations, i.e.,* translational and torsional motions of the molecules in the lattice. These are called *phonon modes* and are responsible for the infrared cutoff of NaCl cells at low wave numbers. Heavier atom alkali halides have phonon modes at lower energies. These vibrations can form combination bands with the intramolecular vibrations and cause pronounced frequency shifts in higher frequency regions of the spectrum. An additional complication arises if the unit cell of the crystal contains more than one chemically equivalent molecule. When this is the case, the vibrations in the individual molecules can couple with each other. This intermolecular coupling can give rise to frequency shifts and band splitting.

As mentioned earlier, molecular symmetry is very important in determining the infrared activity and degeneracy of a molecular vibration. When a molecule is present in a crystal, the symmetry of the surroundings of the molecule in the unit cell, the so-called *site symmetry,* determines the selection rules. Often bands forbidden in the gaseous state or in solution appear in the solid, and degenerate vibrations in the gaseous state are split in the solid. The general problem of the effect of site symmetry on the selection rules has been treated theoretically.[(21)] As a simple illustration of this effect, the infrared spectra of materials containing carbonate ion will be considered.

The infrared and Raman spectra of $CaCO_3$ in calcite, where the carbonate ion is in a site of D_3 symmetry, contain the following bands (in cm^{-1}): ν_1, 1087 (R); ν_2, 879 (IR); ν_3, 1432 (IR, R); ν_4, 710 (IR, R). The (IR) indicates infrared activity and the (R) Raman activity. The infrared spectra of $CaCO_3$ in aragonite, where the site symmetry for carbonate is C_s, differs in that ν_1 becomes infrared active and ν_3 and ν_4 each split into two bands. By using the symmetry considerations previously discussed on the CO_3^{2-} ion, the following results can be obtained:

	ν_1	ν_2	ν_3	ν_4
D_{3h}	$A_1'(R)$	$A_2''(IR)$	$E'(IR, R)$	$E'(IR, R)$
D_3	$A_1(R)$	$A_2(IR)$	$E(IR, R)$	$E(IR, R)$
C_s	$A'(IR, R)$	$A''(IR, R)$	$A'(IR, R) + A'(IR, R)$	$A'(IR, R) + A'(IR, R)$

As a result of all these possible complications, the interpretation of spectra obtained on solids is difficult.

The spectrum of a given solute often varies in different solvents. In hydrogen-bonding solvents the shifts are, in part, due to specific solute-solvent interactions that cause changes in the electron distribution in the solute. The frequency of the band is also dependent upon the refractive index of the solvent[22] and other effects.[23]

Matrix isolation experiments combined with infrared and Raman studies have led to interesting developments. Unstable compounds, radicals, and intermediates are trapped in an inert or reactive solid matrix by co-condensing the matrix (*e.g.*, argon) and the species to be studied at low temperatures (often 4.2 to 20°K). Uranium, platinum, and palladium carbonyls have been prepared[25a,b] by allowing controlled diffusion of CO into an argon matrix of the metal. The compounds LiO_2, NaO_2, KO_2, RbO_2, and LiN_2 have also been made[25c] and investigated by infrared. The findings are consistent with C_{2v} symmetry. Several interesting species containing bound O_2, CO, and N_2 have been made.[25d]

b. In Raman

Raman spectra are routinely run on gases, liquids, solutions, or pressed pellets. Liquids can be purified for spectral studies by distillation into a sample cell. Removal of dust is essential, since such contamination increases the background near the exciting line. Turbid solutions are to be avoided. Water and D_2O are excellent Raman solvents because, in contrast to their behavior in the infrared, they have good transparency in the Raman vibrational region. Fig. 6–19 illustrates the cutoff region

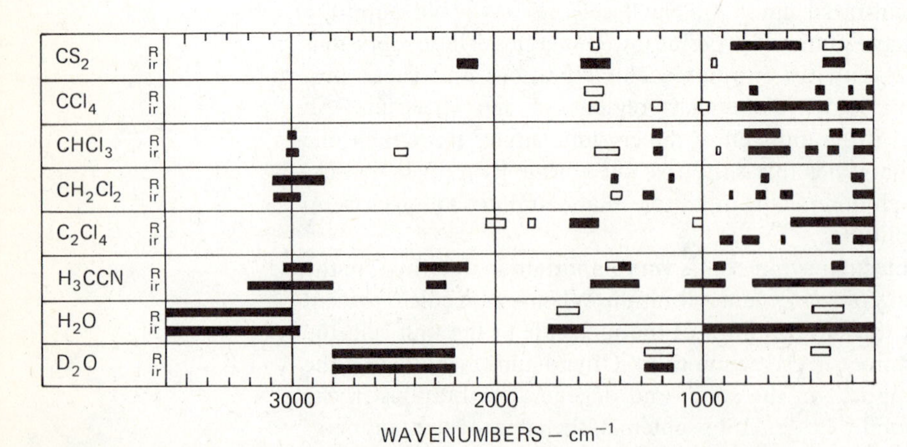

FIGURE 6–19. Solvent interference regions in the Raman region. Solid regions are completely obscured, open rectangles partially obscured.

for various solvents in the Raman region. A complete Raman spectrum can be obtained by using CS_2, $CHCl_3$, and C_2Cl_4. For colored solutions, an exciting line should be used that is not absorbed. Fluorescence and photolysis are problems when the sample absorbs the exciting line.

Solids are best investigated as pellets or mulls. The techniques for preparing these are the same as those for infrared. If powders are examined, the larger the crystals the better. For colored materials, rotating the sample during the experiment prevents sample decomposition and permits one to obtain a good spectrum. Surface scanning techniques are also helpful.[9] Experimental aspects are described in more detail in references 10 and 11.

6-18 FINGERPRINTING

If an unknown is suspected to be a known compound, its spectrum can be compared directly with that of the known. The more bands the sample contains, the more reliable the comparison.

The presence of water in a sample can be detected by its two characteristic absorption bands in the 3600 to 3200 cm^{-1} region and in the 1650 cm^{-1} region. If the water is present as lattice water, these two bands and one in the 600 to 300 cm^{-1} region are observed. If the water is coordinated to a metal ion, an additional band in the 880 to 650 cm^{-1} region is often observed.[18] In the clathrate compound $Ni(CN)_2NH_3 \cdot C_6H_6 \cdot xH_2O$, the infrared spectrum[23b] clearly showed the presence of water, although it was not detected in a single crystal x-ray study. In connection with the earlier discussion on the effect of physical state on the spectra of various substances, it is of interest to mention in passing that the benzene is so located in the crystal lattice of this clathrate that the frequencies of the out-of-plane vibrations are increased over those in the free molecule, while the in-plane vibrations are not affected.

If the product of a reaction is suspected of being contaminated by starting material, this can be confirmed by the presence of a band in the product due to the starting material that is known to be absent in the pure product. Even if spectra are not known, procedures aimed at purification may be attempted. Changes in relative band intensities may then indicate that partial purification has been achieved and may be used as a check on completeness of purification.

A more common analysis employs the group vibration concept to ascertain the presence or absence of various functional groups in the molecule. The following generalizations aid in this application:

(1) Above 2500 cm^{-1} nearly all fundamental vibrations involve a hydrogen stretching mode. The O—H stretching vibration occurs around 3600 cm^{-1}. Hydrogen bonding lowers the frequency and broadens the band. The N—H stretch occurs in the 3300 to 3400 cm^{-1} region. These bands often overlap the hydrogen bonded O—H bands, but the N—H peaks are usually sharper. The N—H stretch in ammonium and alkylammonium ions occurs at lower frequencies (2900 to 3200 cm^{-1}). The C—H stretch occurs in the 2850 to 3000 cm^{-1} region for an aliphatic compound and in the 3000 to 3100 cm^{-1} region for an aromatic compound. Absorptions corresponding to S—H, P—H, and Si—H occur around 2500, 2400, and 2300 cm^{-1}, respectively.

(2) The 2500 to 2000 cm^{-1} region involves the stretching vibration for triply-bonded molecules. The C≡N group gives rise to a strong, sharp absorption that occurs in the 2200 to 2300 cm^{-1} region.

(3) The 2000 to 1600 cm^{-1} region contains stretching vibrations for doubly-bonded molecules and bending vibrations for the O—H, C—H, and N—H groups. The carbonyl group in a ketone absorbs around 1700 cm^{-1}. Conjugation in an amide

$\left(\mathrm{R\overset{\overset{\displaystyle O}{\|}}{C}{-}N(CH_3)_2} \longleftrightarrow \mathrm{R\overset{\overset{\displaystyle O}{|}}{C}{=}N(CH_3)_2}\right)$ decreases the C—O force constant and lowers the highly coupled (with C—N) carbonyl absorption to the 1650 cm^{-1} region. Hydrogen bonding lowers the carbonyl vibration frequency. Stretching vibrations from C=C and C=N occur in this region.

(4) The region below 1600 cm^{-1} is referred to as the fingerprint region for many organic compounds. In this region significant differences occur in the spectra of substances that are very much alike. This is the single bond region and, as mentioned in the section on coupling, it is very common to get coupling of individual single bonds that have similar force constants and connect similar masses (*e.g.*, C—O, C—C, and C—N stretches often couple). The absorption bands in this region for a given functional group occur at different frequencies depending upon the skeleton of the molecule, because each vibration often involves oscillation of a considerable number of atoms of the molecular skeleton.

Many characteristic infrared group frequencies have been compiled to aid in this application. Fig. 6–20 shows several typical plots.

In the far infrared region, it becomes very difficult to assign the observed spectrum to group frequencies. Fig. 6–21 is a compilation that illustrates typical regions in which absorption bands are found for molecules containing particular functional groups.

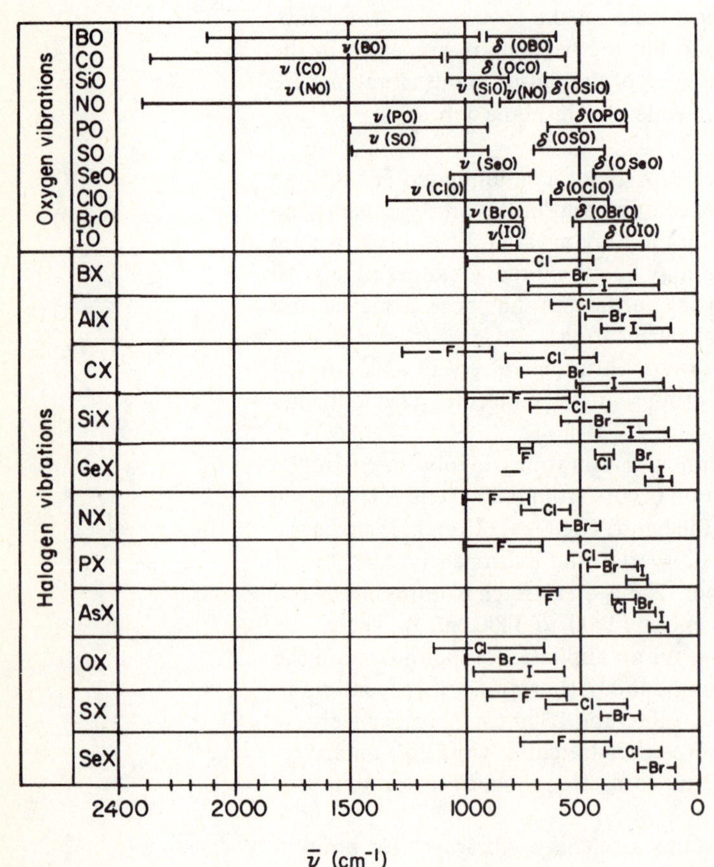

FIGURE 6–20(A), (B), and (C). Range of infrared group frequencies for some inorganic and organic materials. The symbol $\nu_{\text{M-X}}$, where M and X are general symbols for the atoms involved, corresponds to a stretching vibration. The symbols ν_1, ν_2, etc. were previously defined. The symbol δ corresponds to an in-plane bending vibration (δ_s is symmetric bend, δ_d is asymmetric bend), π to an out-of-plane bend, ρ to rocking and wagging vibrations.

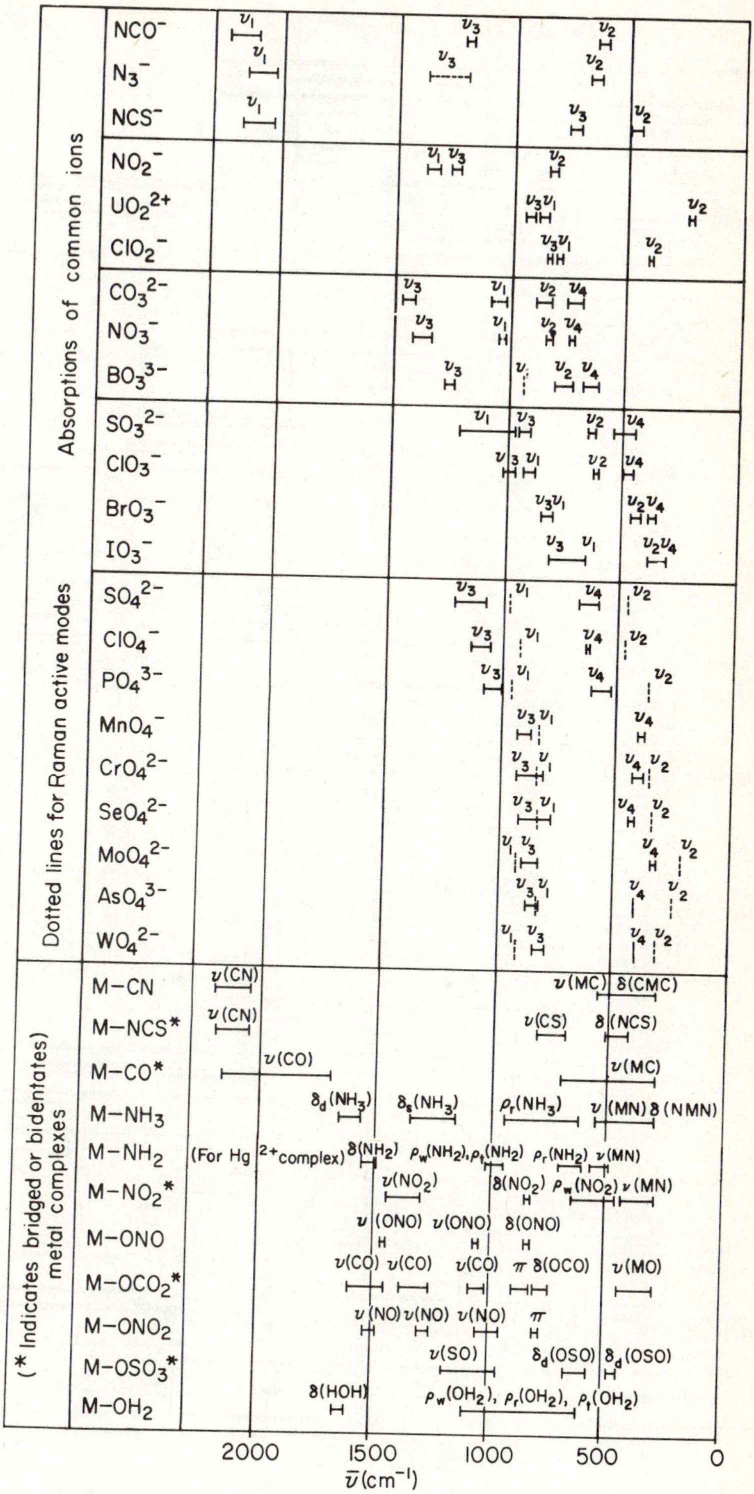

FIGURE 6-20. *Continued.*

(B)

Absorptions of common ions

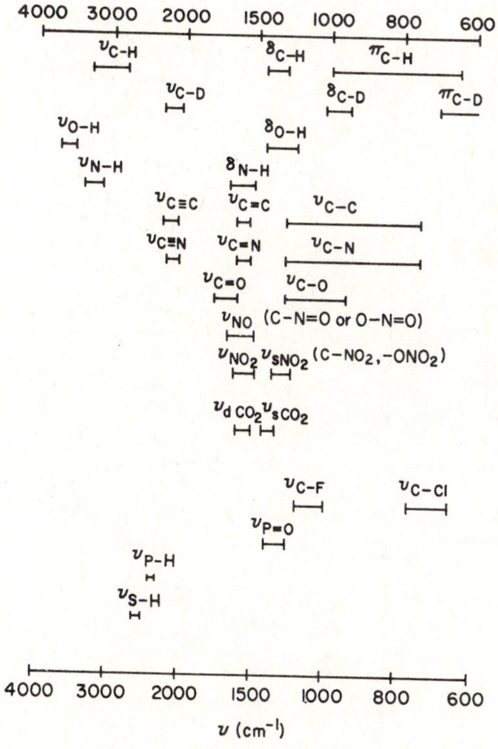

Dotted lines for Raman active modes

NCO⁻: ν_1, ν_3, ν_2
N_3^-: ν_1, ν_3, ν_2
NCS^-: ν_1, ν_3, ν_2
NO_2^-: ν_1, ν_3, ν_2
UO_2^{2+}: $\nu_3\nu_1$, ν_2
ClO_2^-: $\nu_3\nu_1$, ν_2
CO_3^{2-}: ν_3, ν_1, ν_2, ν_4
NO_3^-: ν_3, ν_1, ν_2, ν_4
BO_3^{3-}: ν_3, ν_1, ν_2, ν_4
SO_3^{2-}: ν_1, ν_3, ν_2, ν_4
ClO_3^-: ν_3, ν_1, ν_2, ν_4
BrO_3^-: $\nu_3\nu_1$, $\nu_2\nu_4$
IO_3^-: ν_3, ν_1, $\nu_2\nu_4$
SO_4^{2-}: ν_3, ν_1, ν_4, ν_2
ClO_4^-: ν_3, ν_1, ν_4, ν_2
PO_4^{3-}: ν_3, ν_1, ν_4, ν_2
MnO_4^-: ν_3, ν_1, ν_4
CrO_4^{2-}: ν_3, ν_1, ν_4, ν_2
SeO_4^{2-}: ν_3, ν_1, ν_4, ν_2
MoO_4^{2-}: ν_1, ν_3, ν_4, ν_2
AsO_4^{3-}: ν_3, ν_1, ν_4, ν_2
WO_4^{2-}: ν_1, ν_3, ν_4, ν_2

(* Indicates bridged or bidentates) metal complexes

M–CN: ν(CN), ν(MC), δ(CMC)
M–NCS*: ν(CN), ν(CS), δ(NCS)
M–CO*: ν(CO), ν(MC)
M–NH_3: $\delta_d(NH_3)$, $\delta_s(NH_3)$, $\rho_r(NH_3)$, ν(MN), δ(NMN)
M–NH_2: (For Hg^{2+} complex) $\delta(NH_2)$, $\rho_w(NH_2)$, $\rho_t(NH_2)$, $\rho_r(NH_2)$, ν(MN)
M–NO_2*: $\nu(NO_2)$, $\delta(NO_2)$, $\rho_w(NO_2)$, ν(MN)
M–ONO: ν(ONO), ν(ONO), δ(ONO)
M–OCO_2*: ν(CO), ν(CO), ν(CO), π, δ(OCO), ν(MO)
M–ONO_2: ν(NO), ν(NO), ν(NO), π
M–OSO_3*: ν(SO), δ_d(OSO), δ_d(OSO)
M–OH_2: δ(HOH), $\rho_w(OH_2)$, $\rho_r(OH_2)$, $\rho_t(OH_2)$

2000 1500 1000 500 0
$\bar{\nu}$ (cm^{-1})

(C)

4000 3000 2000 1500 1000 800 600

ν_{C-H}, δ_{C-H}, π_{C-H}
ν_{C-D}, δ_{C-D}, π_{C-D}
ν_{O-H}, δ_{O-H}
ν_{N-H}, δ_{N-H}
$\nu_{C\equiv C}$, $\nu_{C=C}$, ν_{C-C}
$\nu_{C\equiv N}$, $\nu_{C=N}$, ν_{C-N}
$\nu_{C=O}$, ν_{C-O}
ν_{NO} (C–N=O or O–N=O)
ν_{NO_2} ν_{sNO_2} (C–NO_2, –ONO_2)
ν_d CO_2 ν_s CO_2
ν_{C-F}, ν_{C-Cl}
$\nu_{P=O}$
ν_{P-H}
ν_{S-H}

4000 3000 2000 1500 1000 800 600
ν (cm^{-1})

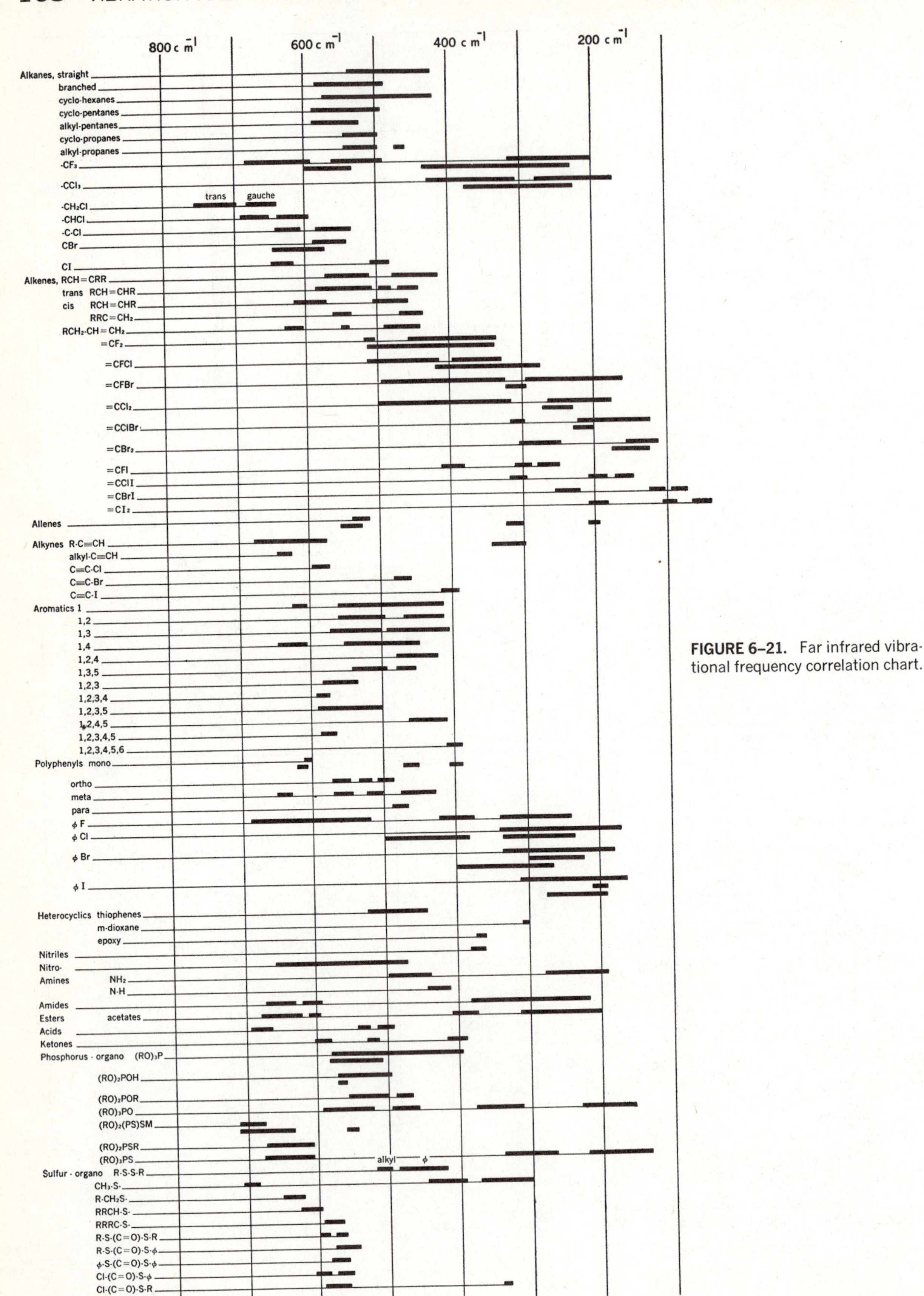

FIGURE 6–21. Far infrared vibrational frequency correlation chart.

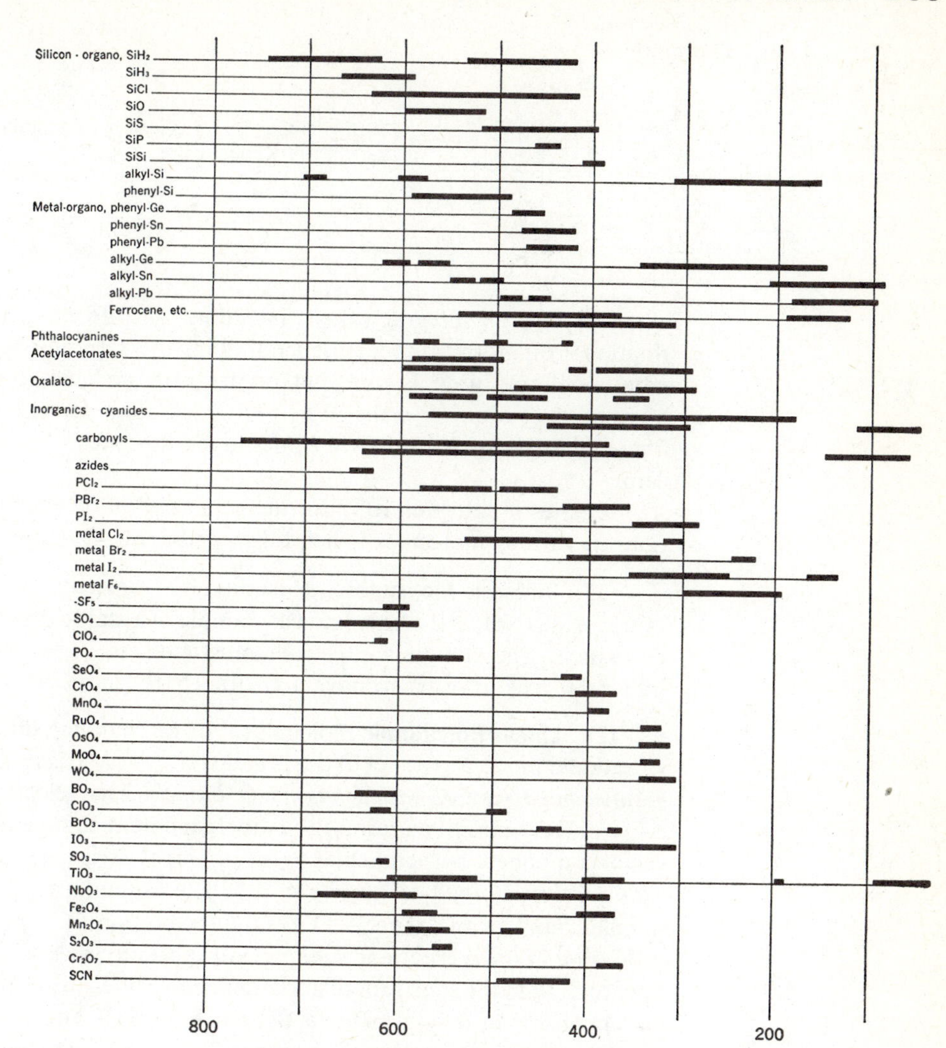

FIGURE 6–21. *Continued.*

6–19 SPECTRA OF GASES

The spectra of gases are often very much different from the spectra of materials in the condensed phase or in solution. As can be seen in Fig. 6–18, a significant difference is the presence of considerable fine structure in the gaseous spectrum. The fine structure is due to a combination of vibrational and rotational transitions. For example, in a diatomic molecule, there are not only transitions corresponding to the pure vibrational mode, ν_0, but also absorption corresponding to $\nu_0 \pm \nu_r$ where ν_r represents the rotational frequency detected. Since any finite sample contains many molecules, many different rotational states will be populated and there will be a whole series of lines corresponding to different ν_r values (*i.e.*, transitions between many different rotational states). This phenomenon is illustrated in Fig. 6–22. The Q branch corresponds to the transition in which ν_r is zero (*i.e.*, a transition with no change in rotational quantum number), the R branch to $\nu_0 + \nu_r$, and the P branch to $\nu_0 - \nu_r$.

The frequencies of all the bands in Fig. 6–22 can be expressed by the following equation:

$$\nu = \nu_0 + 2hm_J/8\pi^2 I \qquad (6\text{-}24)$$

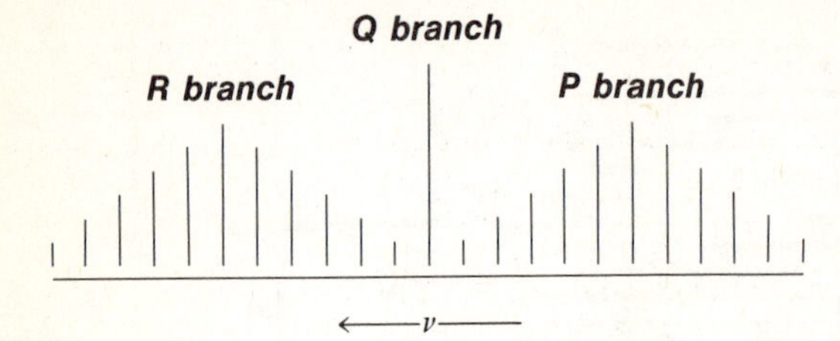

FIGURE 6–22. Schematic of the transitions giving rise to P, Q, and R branches in a spectrum of a gas.

where m_J has all integral values, including zero, from $+J$ to $-J$ (where J is the rotational quantum number) depending on the selection rules. When $m_J = 0$, the vibrational transition is one that occurs with no change in rotational quantum number. A Q branch results. When m_J is less than zero ($J_{n+1} \longrightarrow J_n$ transitions), lines in the P branch result, while lines in the R branch correspond to m_J greater than zero.

A series of selection rules for the combination of vibration and rotation transitions in various molecules is helpful in deducing structure:

(1) Diatomic Molecules. Most diatomic molecules do not possess a Q branch. Nitric oxide (NO) is the only known example of a stable diatomic molecule that has a Q branch. The diatomic molecule must possess angular momentum about the molecular axis in order to have a Q branch. (Σ states, $L = 0$, have no Q branch.)

(2) Linear Polyatomic Molecules. If the changing dipole moment for a given vibrational mode is parallel to the principal rotation axis in the molecule, a so-called *parallel band* results, which *has no Q branch.* The selection rule for this case is $\Delta J = \pm 1$; ΔJ cannot be zero. If the dipole moment change for the vibration has any vector component perpendicular to the principal axis, a *perpendicular band* will result *with a Q branch,* and ΔJ can be 0, ± 1. The asymmetric C—O stretch in CO_2 is a parallel band, while the C—O bending vibration is a perpendicular band. The utilization of these criteria in making and substantiating assignments of vibrations is apparent. If, in the spectrum of a triatomic molecule, any of the infrared bands (ν_1, ν_2, or ν_3) consists of a single P and R branch without any Q branch (*i.e.,* a zero gap separation), the molecule must be linear. This type of evidence can be used to support linear structures for N_2O, HCN, and CO_2. The rotational spacings can be employed, as discussed under the section on microwave spectroscopy, to evaluate the moment of inertia. Note that even though CO_2 does not have a permanent dipole (and hence is microwave inactive), the rotational spacings can be obtained from the fine structure in the infrared spectrum.

(3) Non-linear Polyatomic Molecules. For the discussion of vibration-rotation coupling in non-linear polyatomic molecules it is necessary to define the terms *spherical, symmetric,* and *asymmetric top.* Every non-linear molecule has three finite moments of inertia. In a spherical top, *e.g.,* CCl_4, all three moments are equal. In a symmetric top, two of the three moments are equal. For example, if one selects the C_3 axis of CH_3Br as the z-axis, then the moment of inertia along the x-axis is equal to that along the y-axis. Any molecule with a threefold or higher rotation axis is a symmetric top molecule (unless it is a spherical top). In an asymmetric top, no two of the three moments are equal. Molecules with no symmetry or those with only a twofold rotation axis (H_2O, for example) are asymmetric top molecules.

For symmetric top molecules, those vibrations having oscillating dipole moments that are parallel to the principal axis produce a *parallel absorption band with P, Q, and R branches.* The symmetric C—H stretching and bending frequencies in CH_3Br are examples of parallel bands. The type of spectrum obtained for a parallel band is illustrated in Fig. 6–23. In this example the rotational fine structure in the R branch is

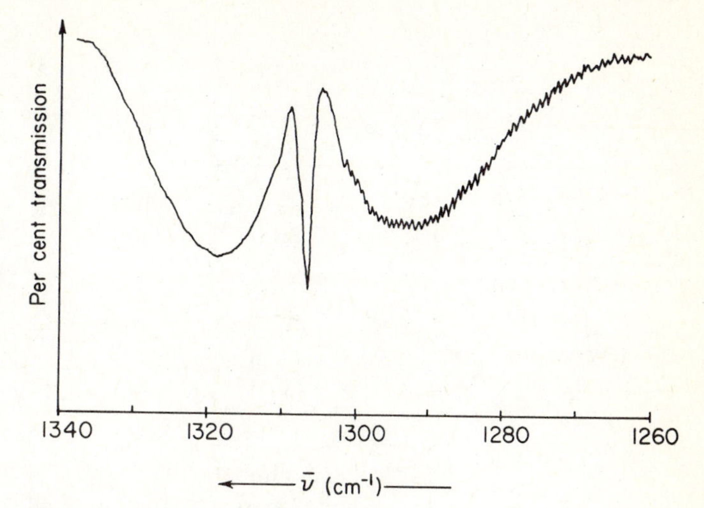

FIGURE 6-23. A parallel band for CH_3Br (symmetric top). [From G. M. Barrow: *Introduction to Molecular Spectroscopy*. Copyright © 1962 by McGraw-Hill, Inc. Used by permission of McGraw-Hill Book Company.]

not resolved. The parallel band for a symmetric top molecule is similar to the perpendicular band for a linear molecule. For a perpendicular absorption band in a symmetric top molecule, several Q bands are detected, often with overlapping unresolved P and R branches. The C—Cl bending vibration in CH_3Cl is an example of a perpendicular band in a symmetric top molecule. A typical spectrum for this case is illustrated in Fig. 6–24. In a spherical top, the selection rule for a *perpendicular band is* $\Delta J = 0, \pm 1$. This information can be employed to support band assignments.

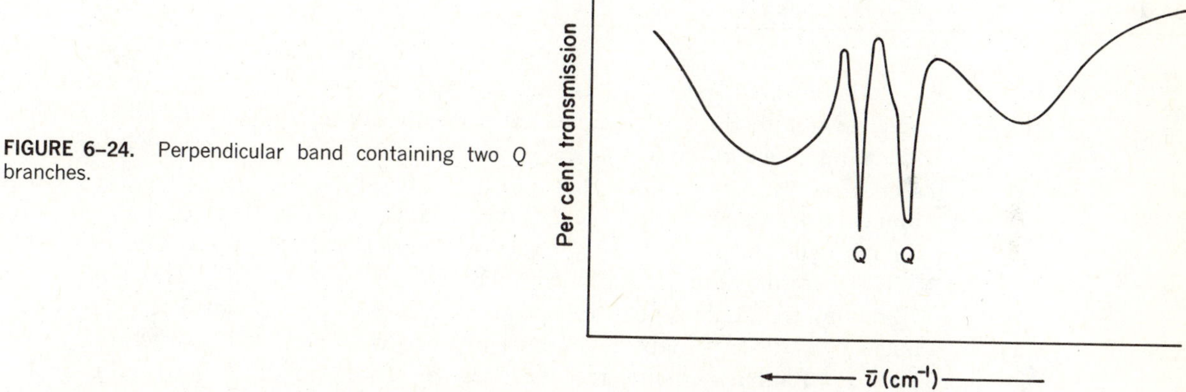

FIGURE 6-24. Perpendicular band containing two Q branches.

It becomes very difficult in some cases to distinguish between symmetric and asymmetric top molecules. As the symmetry of the asymmetric top becomes lower, the task becomes easier. Considerable information can be obtained from the band shape to support vibrational assignments.*

6-20 APPLICATION OF RAMAN AND INFRARED SELECTION RULES TO THE DETERMINATION OF INORGANIC STRUCTURES

(1) The structures in Fig. 6–25 represent some of the possibilities that should be considered for a material with an empirical formula NSF_3. Table 6–9 summarizes the

*See Herzberg (in General References), pp. 380–390, for an explanation of the phenomena discussed above. Many applications of these principles are discussed in this reference.

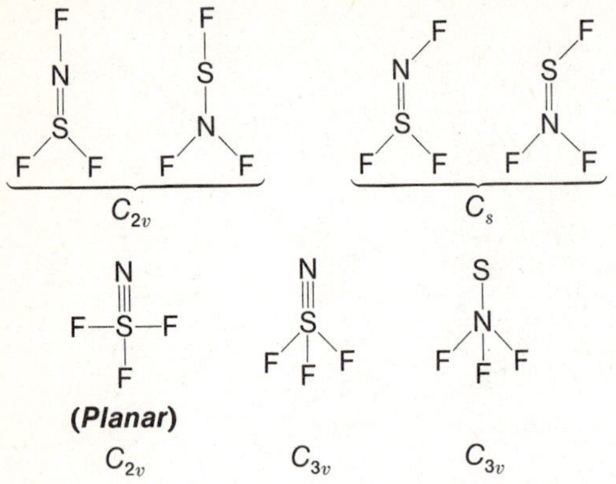

C_{2v} C_s

(Planar) C_{2v} C_{3v} C_{3v}

FIGURE 6–25. Some possible structures for NSF_3.

calculated number and symmetry of bands for these possible structures. These results and the infrared activity of the bands should be determined for practice, employing the procedure in the section on symmetry considerations and the character tables in Appendix A.

TABLE 6–9. CALCULATED NUMBER OF BANDS FOR VARIOUS STRUCTURES OF NSF_3

C_{3v}	C_{2v} (C_2-axis is the z-axis)	C_s
$3A_1$ (IR active)	$4A_1$ (IR active)	$7A'$ (IR active)
	$3B_1$ (IR active)	$2A''$ (IR active)
$3E$ (IR active)	$2B_2$ (IR active)	

The infrared spectrum has been reported.[26a] Six intense bands are found, in agreement with the C_{3v} structure but certainly not in conflict with other possible structures. Some of the fundamentals may be of such low intensity that they are not seen. However, it was found that four of the bands ($\nu_1, \nu_2, \nu_3, \nu_5$) have P, Q, and R branches. This is good evidence that the molecule being investigated is a symmetric top molecule and supports a C_{3v} structure. The spectral data are contained in Table 6–10, where they are compared with data reported for POF_3.[26b] The forms of the vibrations are diagrammed in Appendix C (see ZXY_3 molecule). These data, combined with an NMR study, have been employed to support the C_{3v} structure $F_3S{-}N$. Polarized Raman studies were not carried out. These would not add a great deal of structural information but would help confirm the assignments. The A_1 bands should be polarized and the E bands depolarized.

TABLE 6–10. FUNDAMENTAL VIBRATION FREQUENCIES FOR NSF_3 AND OPF_3

	NSF_3	OPF_3
ν_1	1515	1415
ν_2	775	873
ν_3	521	473
ν_4	811	990
ν_5	429	485
ν_6	342	345

The F_3N—S structure has not been eliminated by these data. The frequencies of the observed bands provide information to favor the F_3SN structure. The S—N stretch is the only fundamental vibration that could be expected at a frequency of 1515 cm^{-1}. [A very rough approximation for this frequency can be obtained by employing the C≡C force constant and the S and N masses in equation (6–3).] The NF stretching vibrations in NF_3 occur at 1031 cm^{-1} and would not be expected to be as high as 1515 cm^{-1} in F_3N—S. The F_3SN structure receives further support from microwave and mass spectroscopic studies. The low moment of inertia calculated from the microwave studies cannot be explained unless the sulfur atom is at the center of mass.

(2) The same approach as that outlined above was employed to prove that perchlorylfluoride, ClO_3F, has a C_{3v} structure instead of O_2ClOF.[27]

(3) An appreciation for some of the difficulties encountered in making spectral assignments can be obtained from an article by Wilson and Hunt[28a] on the reassignment of the spectrum of SO_2F_2. Often a set of band assignments may appear self-consistent, but these assignments may not be a unique set. The effect of isotopic substitution on spectral shifts can be employed to provide additional support for the assignments made. A Raman study has also been reported.[28b]

(4) The infrared and Raman spectra have been interpreted to indicate a planar structure for N_2O_4.[29] Structures corresponding to the planar D_{2h} model and the staggered D_{2d} configuration were considered. Since the infrared and Raman spectra have no lines in common, it is proposed that the molecule has a center of symmetry. A rigorous assignment of the bands could not be made with the available data.

(5) The spectral data for B_2H_6 support the bridged hydrogen structure.[30] The entire spectrum is analyzed, assignments are made, and conclusions regarding the structure are drawn from the rotational fine structure of some of the perpendicular bands.

(6) An infrared and Raman study[31] supports a C_{2v} planar T-shaped structure for ClF_3. Too many fundamentals were observed for either a D_{3h} or C_{3v} structure. Some of the bands had *P*, *Q*, and *R* branches, while others had only *P* and *R* branches. Since all fundamentals for a tetraatomic C_{3v} molecule (*e.g.*, PF_3) have *P*, *Q*, and *R* branches, this structure is eliminated.

(7) The infrared and Raman spectra of $B_2(OCH_3)_4$ have been interpreted to indicate[32] a planar arrangement for the boron and oxygen atoms. Many bands were found in the spectrum. Assignments were made that are consistent with a planar C_{2h} molecule. More fundamentals were found than would be predicted from a D_{2d} model. The infrared and Raman spectra do not have a coincidence of any of the vibrations assigned to B—B or B—O modes; but, as would be expected for the proposed structure, the CH_3 and O—C vibrations are found in both the infrared and Raman. Good agreement is found for the bands assigned to the O—C and —CH_3 modes in this molecule and in the $B(OCH_3)_3$ molecule, illustrating the applicability of the group frequency concept. The above assignments indicate a planar structure, but since the interpretation is very much dependent on the correctness of the assignments, they should be confirmed by isotopic substitution. The band assignments made are consistent with those calculated from a normal coordinate analysis assuming a C_{2h} model.

One final advantage of combining Raman studies with infrared will be mentioned. Often, in the infrared spectra, it is difficult to distinguish combination bands or overtones from fundamentals. This is usually not a problem in the Raman, for the fundamentals are much more intense. An extensive review of the inorganic applications of Raman spectroscopy[11] should be consulted for other applications of this technique.

BOND STRENGTH-FREQUENCY SHIFT RELATIONS

There are many examples in the literature of attempts to infer the bond strength for coordination of a Lewis acid to Lewis bases from the magnitude of the infrared shift of some acid (or base) functional group upon coordination. Usually a relationship between these two quantities is tacitly assumed, and explanations for its magnitude are offered in terms of the electronic structures of the acid and base. This approach is to be discouraged, for the problem is too complex to permit such an assumption.

The systems that have been most thoroughly studied are the hydrogen bonding ones. In general, hydrogen bonding[33] to an X—H molecule results in a decrease in the frequency and a broadening of the absorption band that is assigned to the X—H stretching vibration. The spectra of free phenol, A (where X = $C_6H_5O^-$), and a hydrogen-bonded phenol, B, are indicated in Fig. 6–26. The magnitude of $\Delta\nu_{OH}$, the frequency shift upon formation of a 1:1 complex with a base, is related to $-\Delta H$ for the

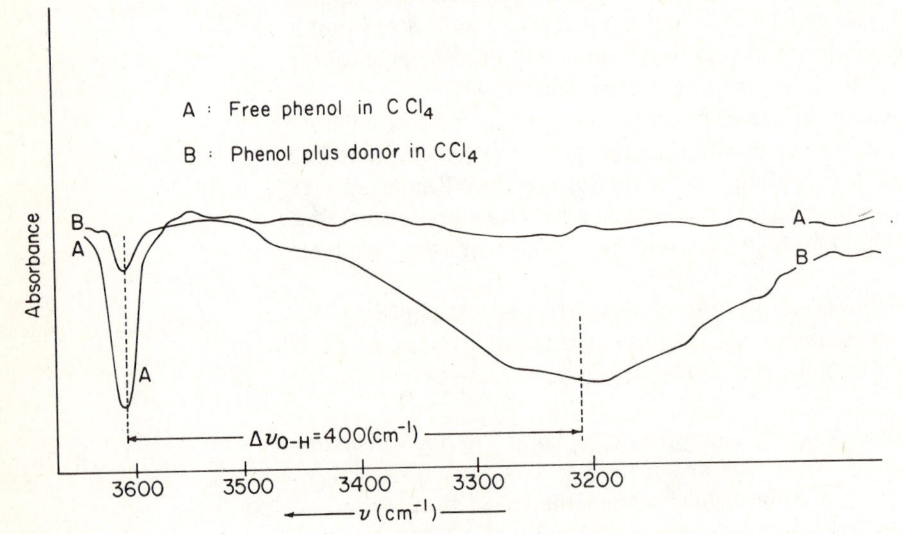

FIGURE 6–26. Infrared spectra of phenol and a hydrogen bonded phenol.

formation of the 1:1 adduct measured in a poorly solvating solvent, for a large series of different bases as shown[34] in Fig. 6–27. Different alcohols require different lines, as illustrated by the separate solid lines in this figure. The solid lines are referred to as constant-acid lines. The dotted lines represent data in which the base is held constant and the acid is varied. Over this range of enthalpies, a linear constant-acid equation exists for various alcohols:

$$-\Delta H(\text{kcal mole}^{-1}) = (0.0105 \pm 0.0007)\,\Delta\nu_{OH} + 3.0(\pm 0.2)(\text{phenols}) \quad (6\text{-}25a)$$

$$-\Delta H(\text{kcal mole}^{-1}) = (0.0115 \pm 0.0008)\,\Delta\nu_{OH} + 3.6(\pm 0.03)[(CF_3)_2CHOH] \quad (6\text{-}25b)$$

$$-\Delta H(\text{kcal mole}^{-1}) = (0.0106 \pm 0.0005)\,\Delta\nu_{OH} + 1.65(\pm 0.09)(t\text{-butanol}) \quad (6\text{-}25c)$$

$$-\Delta H(\text{kcal mole}^{-1}) = (0.0123 \pm 0.0006)\,\Delta\nu_{OH} + 1.8(\pm 0.1)(\text{pyrrole}^{34}) \quad (6\text{-}25d)$$

The authors[34-38] caution against the use of this relationship for a new class of donors (*i.e.*, a functional group not yet studied) without first evaluating $-\Delta H$ and $\Delta\nu_{OH}$ for at least one system involving this functional group. The enthalpy for a given acid has been shown to be related to two different properties[35] of the base via the following relation:

$$-\Delta H = E_A E_B + C_A C_B \quad (6\text{-}26)$$

where E_A and E_B are roughly related to the tendency of the acid and base to undergo

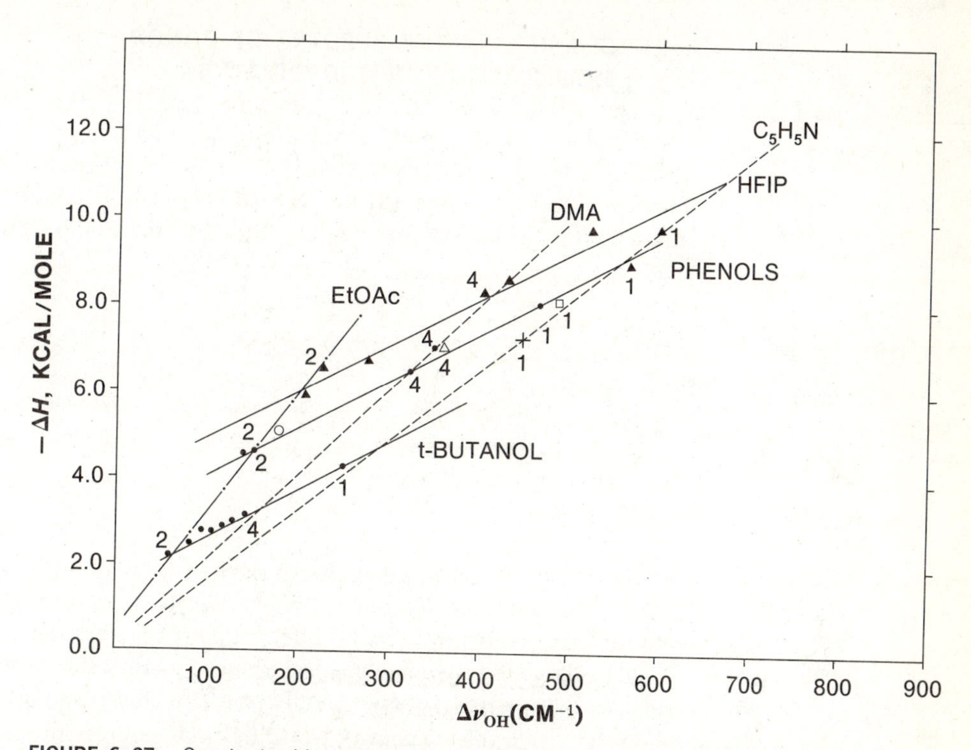

FIGURE 6–27. Constant-acid–constant-base frequency shift–enthalpy relations. Solid lines are constant-acid lines. The phenol line also contains values for *p-t*-butylphenol (+), phenol (○), *p*-chlorophenol (□), and *m*-trifluoromethylphenol (△). The acid butanol is represented by ● and $(CF_3)_2CHOH$ by ▲. The number 1 represents all points on the pyridine constant-base line; 2, ethyl acetate; and 4, *N,N*-dimethylacetamide. (The pyridine–$(CF_3)_2CHOH$ $\Delta\nu$ is estimated from the $\Delta\nu$ vs. $\Delta\nu$ plot of this acid and phenol.)

electrostatic interaction and C_A and C_B to their tendency to undergo covalent interaction. Parameters for many different acids and bases have been reported. When the range of C_B/E_B is between 1.5 and 6.0, the enthalpy-frequency shift correlations hold to within 0.3 kcal mole^{-1}. For sulfur donors, where this ratio is ~20, the plots of Fig. 6–27 do not hold.[36] Unless $\Delta\nu_{OH}$ should have the same functional dependence on E_B and C_B as $-\Delta H$ does, there will have to be a limitation on the C/E ratio of the base that can be used. Furthermore, there is an experimental point at which ΔH and $\Delta\nu_{OH}$ are both equal to zero. Accordingly, the line will have to curve; this curvature is predicted by theory.[37] The theoretical treatment also predicts curvature on the high frequency end of the curve. As a result, extrapolation of the curve in either direction is dangerous. The analysis[37] indicates that frequency shifts can be employed instead of force constants because the high-energy O—H stretching vibration is far enough removed in energy from other vibrations that the band assigned to it is a nearly pure O—H stretching vibration.

In view of the considerable effort involved in an enthalpy measurement (often one or two days of work), this relationship is a welcome one for ascertaining the strength of the hydrogen bond. However, much confusion has been generated by the inappropriate use of this relation. Inappropriate solvent selection, and enthalpies that are a composite of things other than the simple 1:1 adduct formation reaction (*e.g.*, two different donor sites in the same molecule[38]), should be avoided. The reader is referred to the original literature for description of the appropriate use.

Additional correlations of this sort involving different types of acids, and others involving some spectroscopic change in the base upon coordination, would be most useful. However, they cannot be naively assumed but must be tested against measured enthalpies and the limitations sought. In a study of the Lewis acid chloroform,[39] it was found that the enthalpy of adduct formation did not correlate with the observed frequency shift.

6–21 CHANGES IN THE SPECTRA OF DONOR MOLECULES UPON COORDINATION

The infrared spectrum of *N,N*-dimethylacetamide in the solvent CCl_4 has an absorption band at 1662 cm^{-1} that is due to a highly coupled carbonyl absorption. The low frequency compared to acetone (1715 cm^{-1}) is attributed to a resonance interaction with the lone pair on the nitrogen (see Fig. 6–28). Upon complexation

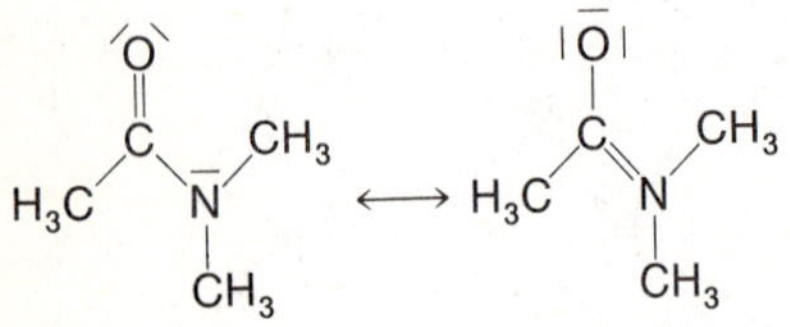

FIGURE 6–28. Resonance structures for *N,N*-dimethylacetamide.

with several Lewis acids, a decrease in the frequency of this band is observed.[40] This decrease has been attributed to the effect of oxygen coordination to the acid. Oxygen coordination could have several effects upon the vibration:

(1) Since the oxygen atom has to move against the atom to which it is coordinated, an increase in frequency should result. In effect this is to say that for the system X—O=C<, the C—O and X—O vibrations couple, producing a higher-energy carbonyl absorption.

(2) A change in oxygen hybridization could increase (or decrease) the C—O σ bond strength and increase (or decrease) the force constant of the C—O bond.

(3) The most important effect in this case involves decreasing the carbonyl force constant by draining π electron density out of the carbonyl group. This causes the observed decrease in the carbonyl frequency and indicates oxygen coordination. The absence of any absorption in the carbonyl region on the high-frequency side of the uncomplexed carbonyl band is further support for oxygen coordination. If there were nitrogen coordination in the complexes, the nitrogen lone pair would be involved, resulting in a decreased C—N vibration frequency and a higher-energy carbonyl absorption.

The decrease in the carbonyl stretching frequency of urea upon complexation to Fe^{3+}, Cr^{3+}, Zn^{2+} or Cu^{2+} is interpreted as indicating oxygen coordination in these complexes.[41] The explanation is similar to that described for the amides. This conclusion is supported by x-ray studies on the structure of the iron and chromium complexes.[42] Nitrogen coordination is observed in the compounds $Pd(NH_2CONH_2)_2Cl_2$ and $Pt(NH_2CONH_2)_2Cl_2$, and the spectra show the expected increase in the C—O stretching frequency as well as a decrease in the C—N frequency.

Decreases in the P—O stretching frequencies indicative of oxygen coordination are observed when triphenylphosphine oxide[43] and hexamethylphosphoramide, $OP[N(CH_3)_2]_3$,[44] are coordinated to metal ions, phenol, or iodine. A decrease in the S—O stretching frequency, indicative of oxygen coordination, is observed when dimethyl sulfoxide or tetramethylene sulfoxide is complexed to many metal ions, iodine, and phenol.[45] The S—O stretching frequency increases in the palladium complex of dimethyl sulfoxide, compared to free sulfoxide. This is an indication of sulfur coordination in this complex. The N—O stretching frequency of pyridine *N*-oxide is decreased upon complexation.[46]

The infrared spectra of some ethylenediaminetetraacetic acid complexes indicate that this ligand behaves as a tetradentate, pentadentate, or hexadentate ligand in various complexes. The interpretation is based upon absorption bands in the carbonyl region corresponding to free and complexed carbonyl groups. The procedure is outlined in detail in the literature.[47]

The change in C≡N stretching frequency of nitriles[48] and metal cyanides[49] resulting from their interaction with Lewis acids has attracted considerable interest. When such compounds are coordinated to Lewis acids that are not generally involved in π back-bonding, the C≡N stretching frequency increases. This was originally thought to be due to the coupling effect described above in the discussion of the coordination of amides. A combined molecular orbital and normal coordinate analysis of acetonitrile and some of its adducts[48] indicates that a slight increase is expected from this effect but that the principal contribution to the observed shift arises from an increase in the C≡N force constant. This increase is mainly due to an increase in the C≡N sigma bond strength from nitrogen rehybridization.[48] In those systems where there is extensive π back-bonding from the acid into the π^* orbitals of the nitrile group, the decreased pi bond energy accounts for the decreased frequency.

There have been many papers[52] on the assignment of the absorption bands in simple ammine complexes. The spectra have been analyzed in terms of a single $M{-}NH_3$ group, *i.e.*, a local C_{3v} structure. In the sodium chloride range, absorption is usually found in four regions near 3300, 1600, 1300, and 825 cm^{-1}. These have been assigned to N—H stretching, NH_3 degenerate deformation, NH_3 symmetric deformation, and $-NH_3$ rocking, respectively. Deuteration studies[51] support these assignments. It was concluded that no metal-nitrogen vibrations in these transition metal ion complexes occur in the sodium chloride region. The frequencies of the bands are dependent upon the metal ion, the type of crystal lattice, and the anion.[52] The anion effect is attributed to hydrogen bonding. References by Chatt et al.[52] are highly recommended readings for more details on this subject.

Infrared spectroscopy has been very valuable[53,54,55] in determining whether the thiocyanate ion, NCS^-, is bonded in a metal complex via the nitrogen atom or the sulfur atom. Since both modes of bonding have been detected, the anion is said to be *ambidentate*. The CN stretch generally occurs at a lower frequency in the N-bonded isomer than in the S-bonded one. The vibration assigned to "C—S" is a better guide, occurring in the 780 to 760 cm^{-1} region for nitrogen-coordinated complexes and in the 690 to 720 cm^{-1} region for sulfur-bonded ones. The "NCS bend" is also diagnostic, occurring in the 450 to 440 cm^{-1} region for the N-bonded isomer and in the 400 to 440 cm^{-1} region for the S-bonded isomer. The observed intensity differences for the different isomers are also diagnostic. The reader is referred to literature reviews[56,57] for many examples of ambidentate coordination by this anion.

It should be pointed out that solid materials can be obtained that contain noncoordinated donor molecules trapped in the crystal lattice. By virtue of the lattice environment, changes in the spectrum of the donor can occur in these instances without coordination. It is thus not possible to prove coordination by a frequency shift in the infrared spectrum of the donor taken on a mull of a solid. This problem can be resolved if the solid can be put in solution and the spectrum compared with that of the donor in this solvent.

Caution should also be exercised in drawing conclusions regarding the strength of the interaction between a donor and a Lewis acid (including metal ions) from the magnitude of the frequency shift of the donor. Usually, this relationship has not been adequately demonstrated with reliable, easily interpreted enthalpy data. There will very probably be cases in which good correlations are obtained, but no general statements can be made at present.

Excellent discussions of the important role infrared spectroscopy has played in elucidating the structure of many transition metal complexes are available.[18] The spectra of metal carbonyls, metal acetylacetonates, cyanide, thiocyanate, and several other types of complexes are discussed.

6–22 CHANGE IN SPECTRA ACCOMPANYING CHANGE IN SYMMETRY UPON COORDINATION

Some of the most useful applications of infrared spectroscopy in the area of coordination and organometallic chemistry occur when the symmetry of a ligand is changed upon complexation. For example, the symmetry change accompanying the binding of small molecules (*e.g.*, N_2, O_2, and H_2) to transition metal ions has a dramatic influence on the infrared spectra of these materials. When azide ion is added to $Ru(NH_3)_5H_2O^{3+}$, a product containing coordinated dinitrogen, N_2, is obtained.[58] The stretching vibration of free N_2 is infrared inactive but Raman active, with a band at 2331 cm^{-1}. In $Ru(NH_3)_5N_2^{2+}$, a sharp intense line appears[59] at 2130 cm^{-1} in the infrared spectrum that is attributed to the N—N stretching vibration. In $IrN_2Cl[P(C_6H_5)_3]_2$ and $HCoN_2[P(C_6H_5)_3]_3$,[60] the same vibration gives rise to bands at 2095 cm^{-1} and 2088 cm^{-1}, respectively.

The complex $IrCOCl[P(C_6H_5)_3]_2$ picks up O_2 and binds it in such a way that the O—O bond axis is perpendicular to a line drawn from iridium to the center of the O—O bond. An infrared band assigned[61] to the O—O stretching vibration in the complex occurs at 857 cm^{-1}. This corresponds to an infrared inactive stretch in free O_2, which does have a Raman band at 1555 cm^{-1}.

When H_2 is added to $IrCOCl[P(C_6H_5)_3]_2$, two new sharp absorption bands arise in the infrared at 2190 and 2100 cm^{-1}. The infrared bands at 2160 and 2101 cm^{-1} in $IrH_2CO[P(C_6H_5)_3]_2$ are shifted to 1620 and 1548 cm^{-1} in the deuterated analogue.[62] The agreement of these shifts with those predicted from the change in reduced mass indicates that these vibrations are mainly metal-hydrogen modes.

Evidence to support coordination of the nitrate ion can also be obtained from infrared studies[63,64] because of the symmetry change accompanying coordination. Free nitrate ion has D_{3h} symmetry, but upon coordination of one of the oxygens, the symmetry is lowered to C_{2v} or to C_s as shown in Fig. 6–29.

M—O—N(=O)(O) or M(O)(O)N—O

C_s C_{2v}

FIGURE 6–29. C_s and C_{2v} structures of coordinated nitrate.

By determining the total representation and the symmetries of the irreducible representations comprising the $3N - 6$ vibrations in D_{3h} $NaNO_3$, one finds A_1'(R), A_2''(IR), $2E'$(R)(IR). The Raman and infrared activity is indicated in parentheses. The bands in D_{3h} $NaNO_3$ are assigned as ν_1 1068 cm^{-1} (R)A_1', ν_2 831 cm^{-1} (IR)$A^{2''}$, ν_3 1400 cm^{-1} (R)(IR)E', and ν_4 710 cm^{-1} (R)(IR)E'. The forms of these vibrations are shown in Appendix B. In C_{2v} and C_s symmetry, A_1 becomes infrared active and each of the E' modes splits into two bands, as shown in Table 6–11.

TABLE 6–11. CORRELATION CHART CONNECTING THE VIBRATIONAL MODE SYMMETRIES FOR VARIOUS KINDS OF NO_3^- COORDINATION

	ν_1	ν_2	ν_3	ν_4
D_{3h}	A_1(R)	A_2''(IR)	E'(IR, R)	E'(IR, R)
C_{2v}	A_1(IR, R)	B_1(IR, R)	A_1(IR, R) + B_2(IR, R)	A_1(IR, R) + B_2(IR, R)
C_s	A_1'(IR, R)	A''(IR, R)	A'(IR, R) + A'(IR, R)	A'(IR, R) + A'(IR, R)

The E' asymmetric stretch in the nitrate ion is split into a high-frequency N—O asymmetric stretch and a lower-frequency symmetric stretch (see Appendix B for planar XY_3 and ZXY_2 molecules) upon coordination. These occur at 1460 and 1280 cm^{-1} (± 20 cm^{-1}), respectively. An additional NO stretching frequency, corresponding roughly to the inactive symmetric stretch in nitrate ion, also appears. Thus, in the absence of complications from site symmetry problems, one can readily tell from the infrared spectrum whether or not the nitrate ion is coordinated. The magnitude of the splitting of the E' mode is generally greater in C_{2v} symmetry than in C_s, and this criterion has been used to support this mode of coordination.

The sulfate ion is another good example to demonstrate the effect of change in symmetry on spectra. The sulfate ion (T_d symmetry) has two infrared bands in the sodium chloride region: one assigned to ν_3 at 1104 cm^{-1} and one to ν_4 at 613 cm^{-1}. (See Appendix C for the forms of these vibrations.) In the complex $[Co(NH_3)_5OSO_3]Br$ the coordinated sulfate group has lower symmetry, C_{3v}. Six bands now appear at 970 (ν_1), 438 (ν_2), 1032 to 1044 and 1117 to 1143 (from ν_3), and 645 and 604 cm^{-1} (from ν_4). In a bridged sulfate group the symmetry is lowered to C_{2v} and even more bands appear. For a bridged group the ν_3 band of SO_4^{2-} is split into three peaks and the ν_4 band into three peaks.(65) Infrared spectroscopy is thus a very effective tool for determining the nature of the bonding of sulfate ion in complexes.

The infrared spectra of various materials containing perchlorate ion have been interpreted(66) to indicate the existence of coordinated perchlorate. As above, the change in symmetry brought about by coordination increases the number of bands in the spectrum. References to similar studies on other anions are contained in the text by Nakamoto.(18)

In another application of this general idea, it was proposed that the five-coordinate addition compound, $(CH_3)_3SnCl \cdot (CH_3)_2SO$, had a structure in which the three methyl groups were in the equatorial positions because the symmetric Sn—C stretch present in $(CH_3)_3SnCl$ disappeared in the addition compound.(67) In $(CH_3)_3SnCl$ the asymmetric and symmetric stretches occur at 545 cm^{-1} and 514 cm^{-1}, respectively.(68) In the adduct a single Sn—C vibration due to the asymmetric Sn—C stretch is detected at 551 cm^{-1}. This is expected for the isomer with three methyl groups in the equatorial position because a small dipole moment change is associated with the symmetric stretch in this isomer. For all other possible structures that can be written, at least two Sn—C vibrations should be observed. There is also a pronounced decrease in the frequency of the Sn—Cl stretching mode in the addition compound.

Infrared spectroscopy has been used to excellent advantage in a study of the influence of pressure on molecular structural transformations. Several systems were found to change reversibly with pressure. Changes in the infrared spectrum accompanying high spin-low spin interconversions of transition metal ion systems with pressure are also reported.(69)

A very elegant illustration of the consequences of symmetry change on the infrared spectrum involves the metal carbonyl compounds of general formula $M(CO)_5X$ and the ^{13}C substituted derivatives.(70) Our main interest is the high wavenumber region associated mainly with the carbonyl stretching vibrations. This, coupled with the fact that the C≡O stretch and the M—C stretch are widely separated in energy, enables us to use as a crude basis set for the vibrational problem the five C—O bond displacement vectors shown in Fig. 6-30(A). The operations of the C_{4v} point group for the all-^{12}C—O compound produce the result $2A_1$ (one radial and one axial), B_1 (radial), and E (radial). Three are infrared active and four are Raman active. The forms of these vibrations are illustrated in Fig. 6-30(B).

The infrared spectrum of $Mo(CO)_5Br$ is shown in Fig. 6-31. The band labeled a is assigned to A_1 (radial), e to E_1, and g to A_1 (axial). The assignment of e to E_1 is confirmed by studying(71) derivatives in which Br is replaced by an asymmetric group that lowers the symmetry from C_{4v}. This results in splitting of this band into two

(A)

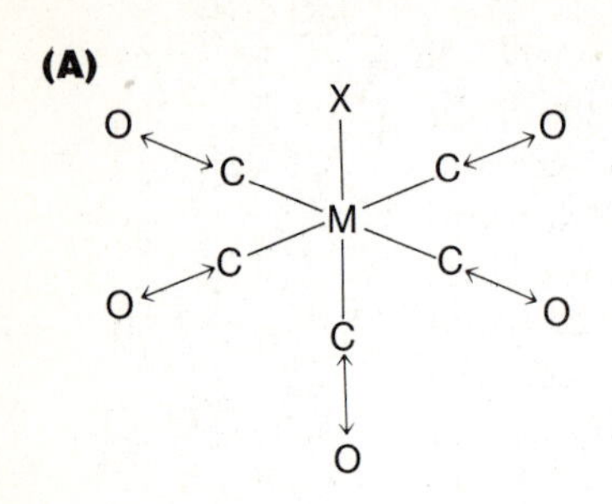

(B)

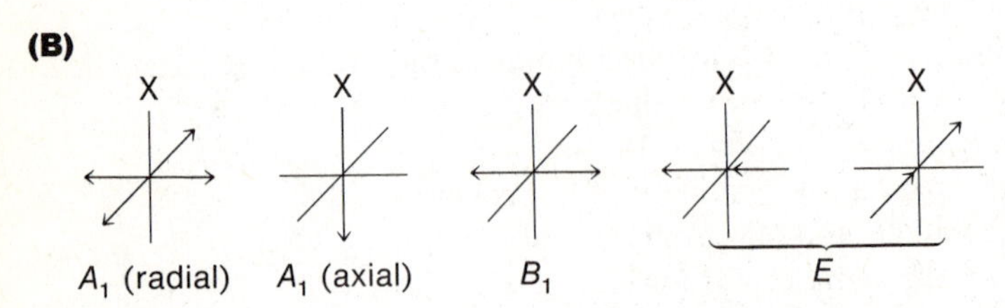

FIGURE 6–30. (A) CO basis set for carbonyl stretches. (B) The resulting normal modes.

peaks separated by 3 to 12 cm^{-1} in various derivatives. The other bands and shoulders in the spectrum have been assigned to ^{13}C derivatives, for there is 1.1% ^{13}C in naturally occurring carbon compounds. Consequently, 1.1% of the molecules have an axial ^{13}CO and 4.4% have a radial ^{13}CO. There will be very few molecules with two ^{13}CO groups in them.

First, we shall consider the consequences of having an axial ^{13}CO in the molecule. The symmetry is still C_{4v}, and the B_1 and E modes (which have no contribution from the axial group) will be unaffected by the mass change. The two A_1 modes having the same symmetry will mix, so isotopic substitution could affect both; but it will have a major effect on A_1 (axial). The band h (at 1958 cm^{-1}) in the spectrum is assigned to A_1 (axial), for it is roughly what can be expected for the change in g from the mass change.

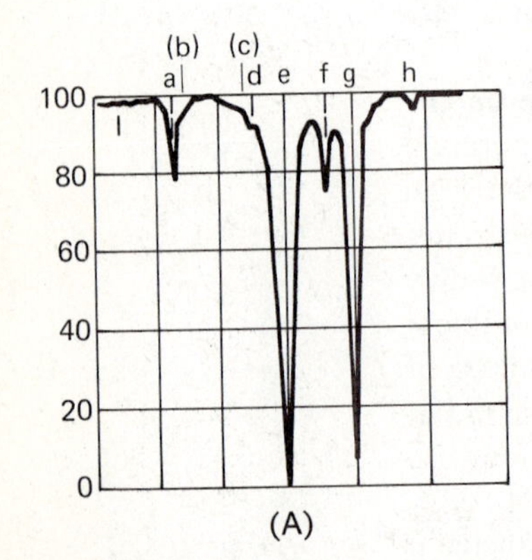

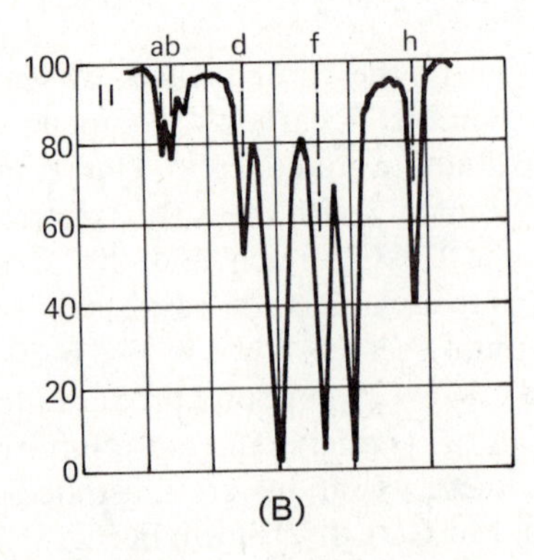

FIGURE 6–31. (A) High resolution infrared spectrum of $Mo(CO)_5Br$ in the 2200 to 1900 cm^{-1} region. (B) Spectrum after 3 hours of exchange with ^{13}CO.

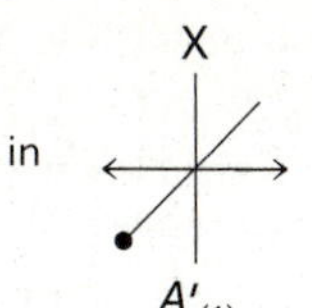

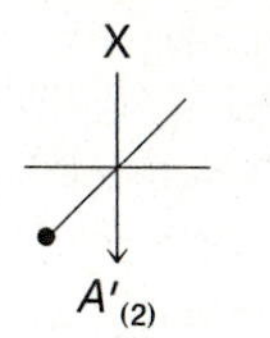

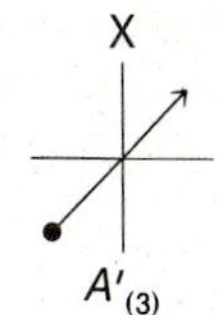

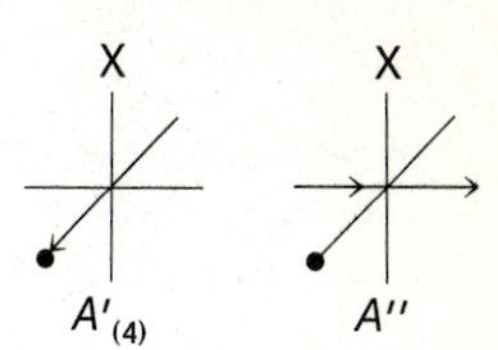

FIGURE 6–32. Infrared allowed vibrations in $M(CO)_4(^{13}CO)X$ with ^{13}CO in the radial position.

For molecules in which a ^{13}CO is located in the radial position, the symmetry is lowered to C_s, leading to the possible vibrations (the solid circle indicates ^{13}CO) illustrated in Fig. 6–32. Only the A'' will rigorously be unaffected by the mass change, and it resembles the E mode in the all-^{12}CO molecule. All of the other four modes can mix and could be shifted by distributing the mass effect over all four modes. This would shift the vibrations to lower frequencies than those in the all-^{12}CO molecule. Band f is assigned as an A' mode in the radial ^{13}CO monosubstituted molecules that is largely A'. Band d is either a radial A_1 mode of the axially substituted molecule or the highest A' of the C_s molecule. The former assignment would involve an intensity that is about 1% of the A_1 in the all-^{12}CO molecule; since the intensity is found to be about 10%, it is assigned to an A' mode in the C_s molecule. A normal coordinate analysis[4] has been carried out to substantiate these assignments. An important part of this work is the assignment of a particular band to radial ligands and another to axial ligands. When ^{13}CO is added, the change in intensity of these bands can be followed and the relative rates of axial *vs.* radial substitution determined.

Jones et al. have carried out a normal coordinate analysis on systems of this type, employing a large number of isotopes. They conclude that the π-bond orders on these systems obtained by using the Cotton-Kraihanzel[4] approximations must be viewed with suspicion. The more complete analysis indicates that as the formal positive charge on the central atom increases in forming $M(CO)_5Br$ from $M(CO)_6$, the σ-bonding increases and the extent of π back-bonding decreases.[5] This study[5] provides a good reference to illustrate the information that can be reliably obtained from a normal coordinate analysis.

EXERCISES

1. Consider the C_{3v} structure of $N{-}SF_3$. (a) Show that the total representation is $E = 15$; $2C_3 = 0$; $3\sigma_v = 3$. (b) Indicate the procedure that shows that the irreducible representations $4A_1$, $1A_2$ and $5E$ result from this total representation. (c) Show that $4A_1 + 1A_2 + 5E$ equals $E = 15$; $2C_3 = 0$; $3\sigma_v = 3$. (d) Indicate the species for the allowed: (e) infrared, (f) Raman, and (g) Raman polarized bands. (h) How many bands would be seen for (e), (f), and (g), assuming all allowed fundamentals are observed?

2. Consider XeF_4 as having D_{4h} symmetry. Using the procedure in Exercise 1, show that the total number of vibrations is indicated by $A_{1g} + B_{1g} + B_{2g} + A_{2u} + B_{2u} + 2E_u$. Also indicate the species for the allowed (a) infrared, (b) Raman, and (c) Raman polarized bands. (d) How many bands would be seen for (a), (b), and (c), assuming all allowed fundamentals are observed?

3. Consider the molecule *trans*-N_2F_2.
   a. To which point group does it belong?
   b. How many fundamental vibrations are expected?
   c. To what irreducible representations do these belong?
   d. What is the difference between the A_u and B_u vibrations?
   e. Which vibrations are infrared active and which are Raman active?
   f. How many polarized lines are expected?

g. How many lines are coincident in the infrared and Raman? Does this agree with the center of symmetry rule?

h. To which species do the following vibrations belong? N—N stretch, symmetric N—F stretch, and asymmetric N—F stretch.

i. Can the N—N stretch and N—F stretch couple?

4. Indicate to which irreducible representations the vibrations belong and indicate how many active IR, active Raman, and polarized Raman lines are expected for:

   a. *cis*-N_2F_2[(C_{2v}) 5 IR active, ($3A_1$, $2B_2$); 6 Raman active ($3A_1$, A_2, $2B_1$)]

   b. A linear N_2F_2 molecule ($D_{\infty h}$)

5. a. Why is *cis*-N_2F_2 in the point group C_{2v} instead of D_{2h}?

   b. What symmetry element is missing from the pyramidal structure of SF_4 that is required for D_{4h} (*i.e.*, which one element would give rise to all missing operations)?

6. The infrared spectrum of gaseous HCl consists of a series of lines spaced 20.68 cm^{-1} apart. (Recall that wavenumbers must be converted to frequency to employ the equations given in this text.)

   a. Calculate the moment of inertia of HCl.

   b. Calculate the equilibrium internuclear separation.

7. Suppose that valence considerations enabled one to conclude that the following structures were possible for the hypothetical molecule X_2Y_2:

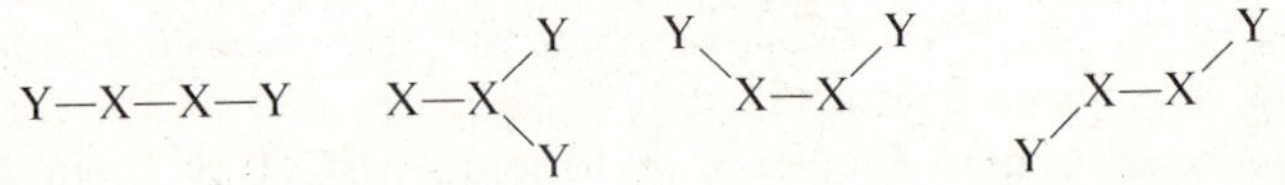

   Y—X—X—Y X—X(Y)(Y) (Y)X—X(Y) [both Y on same side] (Y)X—X(Y) [Y on opposite sides]

   The infrared spectrum of the gas has several bands with *P*, *Q*, and *R* branches. Which structures are eliminated?

8. An X—H fundamental vibration in the linear molecule A—X—H is found to occur at 3025 cm^{-1}. At what frequency is the X—D vibration expected? It is found that the X—D vibration is lowered by only about half the expected amount, and the A—X stretching frequency is affected by deuteration. Explain. Would you expect the A—X—H bending frequency to be affected by deuteration? Why?

9. How many normal modes of vibration does a planar BCl_3 molecule have? Refer to Appendix C and illustrate these modes. Indicate which are infrared inactive. Confirm these conclusions by employing the procedures in the section, Use of Symmetry Considerations to Determine the Number of Active Infrared and Raman Lines. Indicate the Raman polarized and depolarized lines and the parallel and perpendicular vibrations.

10. The following assignments are reported for the spectrum of GeH_4: 2114, T_2; 2106, A_1; 931, E; 819, T_2. Label these using the ν_n symbolism. (E modes are numbered after singly degenerate symmetric and asymmetric vibrations, and T modes are numbered after E modes.) Refer to Appendix C to illustrate the form for each of these vibrations and label them as bends or stretches.

11. Refer to the normal modes for a planar ZXY_2 molecule (Appendix C). Verify the assignment of ν_5 as B_2 and ν_6 as B_1 by use of the character tables. Explain the procedure.

12. a. Indicate the infrared and Raman activities for the modes of the trigonal bipyramidal XY_5 molecule (Appendix C).

    b. Does this molecule possess a center of symmetry?

13. The spectrum of $Co(NH_3)_6(ClO_4)_3$ has absorption bands at 3320, 3240, 1630, 1352, and 803 cm^{-1}. For purposes of assignment of the ammonia vibrations, the molecule can be treated as a C_{3v} molecule, $[Co—NH_3]^{3+}$ (Co—N bonded to three H). Refer to Appendix C for a diagram of these

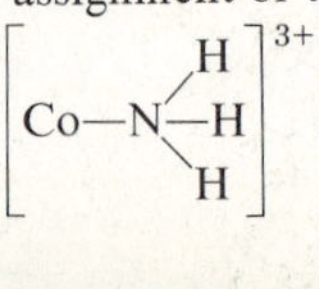

modes; utilizing the material in this chapter and Chapter 4, assign these modes. Use the ν_n symbolism to label the bands and also describe them as bends, stretches, etc.

14. What changes would you expect to see in the infrared spectrum of CH_3COSCH_3 if
    a. Coordination occurred on oxygen?
    b. Coordination occurred on sulfur?

15. In which case would the spectrum of coordinated $[(CH_3)_2N]_3PO$ be most likely to resemble the spectrum of the free ligand:
    a. Oxygen coordination?
    b. Nitrogen coordination?

    Why?

16. Using a value of $k = 4.5 \times 10^5$ dynes cm^{-1} for the C—C bond stretching force constant, calculate the wavenumber (cm^{-1}) for the C—C stretching vibration.

17. The compound $HgCl_2 \cdot O\langle{}^{CH_2-CH_2}_{CH_2-CH_2}\rangle O$ could be a linear polymer with dioxane in the chair form or a monomer with bidentate dioxane in the boat form. How could infrared and Raman be used to distinguish between these possibilities?

18. Which complex would have more N—O vibrations: $(NH_3)_5CoNO_2^{2+}$ or $(NH_3)_5CoONO^{2+}$?

19. What differences (number of bands and frequencies) are expected between the C—O absorptions of the following:

    CH_3COO^-, $Zn\left(\begin{matrix}O\\ \\O\end{matrix}\!\!>\!C-CH_3\right)_2\cdot 2H_2O$, CH_3COOH, and $CH_3C\langle{}^{O}_{O}\rangle Ag$?

20. Given the NH_3 molecule below, determine the total representation for the vibrations using r's and θ's (θ_2 not shown). Use projection operators to determine the vibrational wave functions.

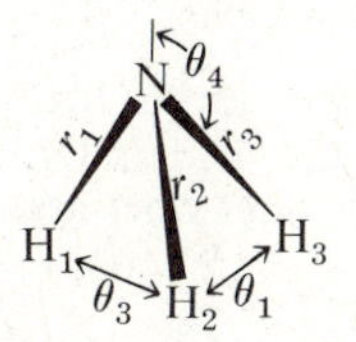

21. Consider the tetrahedral anion BCl_4^-.
    a. How many bands do you expect in the infrared spectrum?
    b. How many bands do you expect in the Raman spectrum?
    c. How many Raman polarized bands do you expect?
    d. How many B—Cl stretches do you expect in the IR?

22. The hydrogen bonding Lewis acid $CHCl_3$ has a C—H stretching frequency that is lowered upon deuteration.
    a. What is responsible for the theoretical factor of 1.4?
    b. What could be responsible if the observed factor were less than 1.4?
    c. When the compound is dissolved in the Lewis base NEt_3, the C—H stretching frequency is lowered. Rationalize the lowering.
    d. If one examines the C—D shift upon adduct formation with Et_3N, the percentage lowering is smaller than that in (b). Rationalize this finding.

23. The compound $(CH_3)_3SnCl$ exhibits two infrared active Sn—C stretches at 545 cm^{-1} and 514 cm^{-1}. The IR spectrum on the five-coordinate adduct $B \cdot (CH_3)_3SnCl$ (where B is a base) shows a single Sn—C stretching band around 550 cm^{-1}.

    a. Consider a tetrahedral structure for $(CH_3)_3SnCl$ and predict the number of Sn—C stretches expected.

    b. Assume a trigonal bipyramidal structure for the adduct with the methyl groups lying in a plane. How many Sn—C stretching bands are expected?

    c. How do you explain the actual spectrum?

24. a. How many infrared active normal modes containing Sn—Cl stretching would be expected for the six-coordinate species *cis*-$SnCl_4X_2$?

    b. Describe briefly the selection rules for Raman spectroscopy and for infrared spectroscopy.

25. Iron pentacarbonyl, $Fe(CO)_5$, possesses D_{3h} symmetry.

    a. Determine the number and irreducible representations of the IR active and Raman active fundamentals to be expected for this compound.

    b. Determine the number of IR active carbonyl stretching bands to be expected.

26. Consider the following IR data, taken from Nakamoto's book:[18]

$K_3[Mn(CN)_6]$	2125 cm^{-1}
$K_4[Mn(CN)_6]$	2060 cm^{-1}
$K_5[Mn(CN)_6)]$	2048 cm^{-1}

Discuss this trend in CN stretching frequencies in terms of bonding in the complexes, and the relationship this has on force constants and frequencies. What do you predict for the Mn—C force constants in this series?

27. Use of IR and Raman spectroscopy is valuable in determining the stereochemistry of metal carbonyl complexes and various substituted metal carbonyls. C=O stretches are often analyzed separately from the rest of the molecule (explain why this assumption is or is not valid).

    a. For $Cr(CO)_6$ and $Cr(CO)_5L$ (treat L as a point ligand), work out the symmetry of the IR and Raman allowed CO stretches.

    b. Do the same as in (a) for $Mo(CO)_4DTH$ (DTH is the bidentate ligand $CH_3SCH_2CH_2SCH_3$), and for *trans*-$Mo(CO)_4[P(OC_6H_5)_3]_2$. Do your findings explain the IR spectra below? (Hint: remember assumptions implicit in the analysis.)

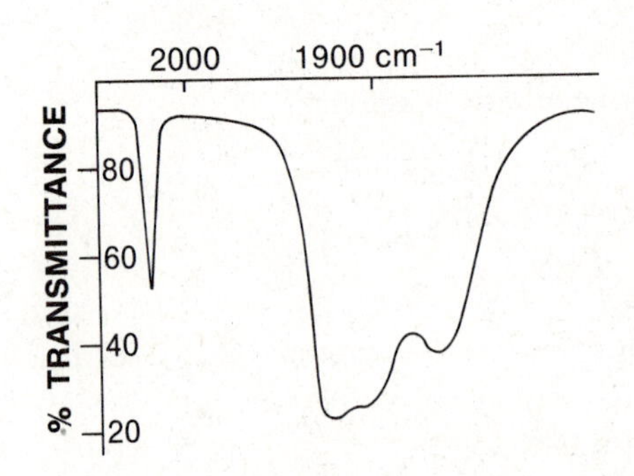

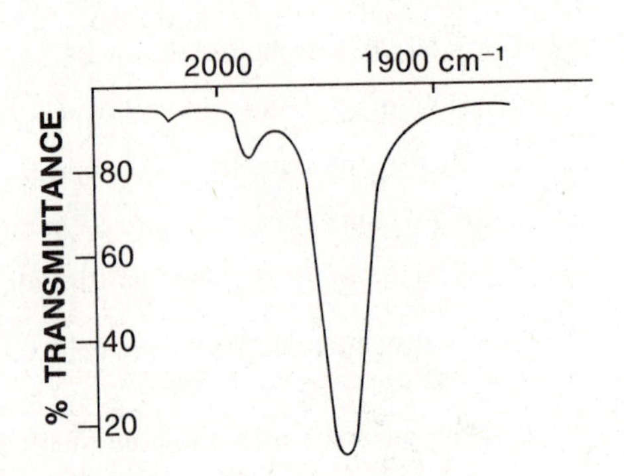

REFERENCES CITED

1. K. Nakamoto, et. al., Chem. Comm., 1451 (1969).
2. E. B. Wilson, J. C. Decius, and P. C. Cross, "Molecular Vibrations," McGraw-Hill, New York, 1955.
3. G. Barrow, "Introduction to Molecular Spectroscopy," pp. 146–130, McGraw-Hill, New York, 1962.
4. F. A. Cotton and C. Kraihanzel, J. Amer. Chem. Soc., *84,* 4432 (1962); Inorg. Chem., *2,* 533 (1963).
5. L. H. Jones, et al., Inorg. Chem., *6,* 1269 (1967); ibid., *7,* 1681 (1968); ibid., *8,* 2349 (1969); ibid., *12,* 1051 (1973).
6. a. L. H. Jones and R. R. Ryan, J. Chem. Phys., *52,* 2003 (1970).
   b. L. H. Jones, "Inorganic Vibrational Spectroscopy," p. 40, M. Dekker, New York, 1971.
   c. L. H. Jones and B. I. Swanson, Accts. Chem. Res., *9,* 128 (1976).
7. J. Overend and J. R. Scherer, J. Chem. Phys., *32,* 1296 (1960).
8. J. O. Halford, J. Chem. Phys., *24,* 830 (1956); S. A. Francis, J. Chem. Phys., *19,* 942 (1951); G. Karabatsos, J. Org. Chem., *25,* 315 (1960).
9. J. A. Gachter and B. F. Koningstein, J. Opt. Soc. Amer., *63,* 892 (1973).
10. J. A. Loader, "Basic Laser Raman Spectroscopy," Heyden-Sadtler, London, 1970.
11. a. S. Tobias, "The Raman Effect," Volume 2, Chapter 7, ed. A. Anderson, M. Dekker, New York, 1973.
    b. T. G. Spiro, "Chemical and Biochemical Applications of Lasers," Chapter 2, C. B. Moore, Ed., Academic Press, New York, 1974.
12. J. Tang and A. C. Albrecht, "Raman Spectroscopy," Chapter 2, ed. H. A. Syzmanski, Plenum Press, New York, 1970.
13. T. G. Spiro and T. C. Strekas, Proc. Nat. Acad. Sci., *69,* 2622 (1972); Accts. Chem. Res., *7,* 339 (1974).
14. J. B. Dunn, D. F. Shriver, and I. M. Klotz, Proc. Nat. Acad. Sci., *70,* 2582 (1973).
15. National Bureau of Standards, Monograph 70, I to IV.
16. C. H. Townes and A. L. Schawlow, "Microwave Spectroscopy," pp. 56–59, 110–114, McGraw-Hill, New York, 1955; Ann. Rev. Phys. Chem.—see the indices for all volumes.
17. a. W. Gordy and R. L. Cook, "Technique of Organic Chemistry," Volume IX, Part II, Microwave Molecular Spectra, ed. A. Weissberger, Interscience, New York, 1970.
    b. J. E. Parkin, Ann. Reports Chem. Soc., *64,* 181 (1967); ibid., *65,* 111 (1968).
    c. W. H. Flygare, Ann. Rev. Phys. Chem., *18,* 325 (1967).
    d. H. D. Rudolph, Ann. Rev. Phys. Chem., *21,* 73 (1970).
18. K. Nakamoto, "Infrared Spectra of Inorganic and Coordination Compounds," 2nd Ed., Wiley, New York, 1970. An excellent reference for inorganic compounds.
19. a. W. J. Potts, Jr., "Chemical Infrared Spectroscopy," Volume I, Wiley, New York, 1963.
    b. R. G. Miller, "Laboratory Methods in Infrared Spectroscopy," Heyden, London, 1965.
20. S. Krimm, "Infrared Spectra of Solids," in "Infrared Spectroscopy and Molecular Structure," Chapter 8, ed. M. Davies, Elsevier, New York, 1963.
21. a. H. Winston and R. S. Halford, J. Chem. Phys., *17,* 607 (1949).
    b. D. F. Hornig, J. Chem. Phys., *16,* 1063 (1948).
22. M.-L. Josien and N. Fuson, J. Chem. Phys., *22,* 1264 (1954).
23. a. A. Allerhand and P. R. Schleyer, J. Amer. Chem. Soc., *85,* 371 (1963).
    b. R. S. Drago, J. T. Kwon, and R. D. Archer, J. Amer. Chem. Soc., *80,* 2667 (1958).
24. M. D. Low, J. Chem. Educ., *47,* **A** 349, **A** 415 (1970).
25. a. J. L. Slater, et al., J. Chem. Phys., *55,* 5129 (1971).
    b. H. Huber, et al., Nature Phys. Sci., *235,* 98 (1972).
    c. R. C. Spiker, et al., J. Amer. Chem. Soc., *94,* 2401 (1972).
    d. See G. A. Ozin and A. V. Voet, Accts. Chem. Res., *6,* 313 (1973) for a review.
26. a. H. Richert and O. Glemser, Z. anorg. allgem. Chem., *307,* 328–344 (1961).
    b. N. J. Hawkins, V. W. Coher, and W. S. Koski, J. Chem. Phys., *20,* 258 (1952).
27. F. X. Powell and E. R. Lippincott, J. Chem. Phys., *32,* 1883 (1960).
28. a. M. K. Wilson and G. R. Hunt, Spectrochim. Acta, *16,* 570 (1960).
    b. E. L. Pace and H. V. Samuelson, J. Chem. Phys., *44,* 3682 (1966).
29. R. G. Snyder and J. C. Hisatsune, J. Mol. Spectr., *1,* 139 (1957).
30. W. C. Price, J. Chem. Phys., *16,* 894 (1948).
31. H. H. Claassen, B. Weinstock, and J. G. Malm, J. Chem. Phys., *28,* 285 (1958).
32. H. Becher, W. Sawodny, H. Nöth, and W. Meister, Z. anorg. allgem. Chem., *314,* 226 (1962).
33. An excellent treatment of the work on hydrogen bonding is presented in: M. D. Joesten and L. J. Schaad, "Hydrogen Bonding," M. Dekker, New York, 1974.
34. M. S. Nozari and R. S. Drago, J. Amer. Chem. Soc., *92,* 7086 (1970) and references therein.
35. R. S. Drago, G. C. Vogel, and T. E. Needham, J. Amer. Chem. Soc., *93,* 6014 (1971); A. P. Marks and R. S. Drago, J. Amer. Chem. Soc., *97,* 3324 (1975).
36. G. C. Vogel and R. S. Drago, J. Amer. Chem. Soc., *92,* 5347 (1970).
37. K. F. Purcell and R. S. Drago, J. Amer. Chem. Soc., *89,* 2874 (1967).
38. R. S. Drago, B. Wayland, and R. L. Carlson, J. Amer. Chem. Soc., *85,* 3125 (1963).
39. F. L. Slejko, R. S. Drago, and D. G. Brown, J. Amer. Chem. Soc., *94,* 9210 (1972).
40. C. D. Schmulbach and R. S. Drago, J. Amer. Chem. Soc., *82,* 4484 (1960); R. S. Drago and D. A. Wenz, J. Amer. Chem. Soc., *84,* 526 (1962).
41. R. B. Penland, S. Mizushima, C. Curran, and J. V. Quagliano, J. Amer. Chem. Soc., *79,* 1575 (1957).
42. Y. Okaya, et al., "Abstracts of Papers of 4th International Congress of the International Union of Crystallography," p. 69, Montreal, 1957.
43. F. A. Cotton, et al., J. Chem. Soc., *1961,* 2298, 3735, and references contained therein.

44. J. T. Donoghue and R. S. Drago, Inorg. Chem., *1,* 866 (1962).
45. R. S. Drago and D. W. Meek, J. Phys. Chem., *65,* 1446 (1961), and papers cited therein.
46. J. V. Quagliano, et al., J. Amer. Chem. Soc., *83,* 3770 (1961).
47. M. L. Morris and D. H. Busch, J. Amer. Chem. Soc., *78,* 5178 (1956).
48. K. F. Purcell and R. S. Drago, J. Amer. Chem. Soc., *88,* 919 (1966).
49. D. F. Shriver and J. Posner, J. Amer. Chem. Soc., *88,* 1672 (1966).
50. M. G. Miles, et al., Inorg. Chem., *7,* 1721 (1968), and references therein.
51. S. Mizushima, I. Nakagawa, and J. V. Quagliano, J. Chem. Phys., *23,* 1367 (1955); G. Barrow, R. Krueger, and F. Basolo, J. Inorg. Nucl. Chem., *2,* 340 (1956).
52. J. Chatt, L. A. Duncanson, and L. M. Venanzi, J. Chem. Soc., *1955,* 4461; *1956;* 2712.
53. J. Lewis, R. S. Nyholm, and P. W. Smith, J. Chem. Soc., *1961,* 4590.
54. A. Sabatini and I. Bertini, Inorg. Chem., *4,* 959 (1965).
55. N. J. DeStefano and J. L. Burmeister, Inorg. Chem., *10,* 998 (1971).
56. J. L. Burmeister, Coord. Chem. Rev., *3,* 225 (1968).
57. A. H. Norbury and A. I. P. Sinha, Quart. Rev. Chem. Soc., *24,* 69 (1970).
58. A. D. Allen and C. V. Senoff, Chem. Comm., 621 (1965); A. D. Allen, et al., J. Amer. Chem. Soc., *89,* 5595 (1967); A. D. Allen and F. Bottomley, Accts. Chem. Res., *1,* 360 (1968).
59. J. P. Collman and Y. Kang, J. Amer. Chem. Soc., *88,* 3459 (1966).
60. I. Yamamoto, et al., Chem. Comm., *79* (1967); Bull. Chem. Soc. Japan, *40,* 700 (1967).
61. P. B. Chock and J. Halpern, J. Amer. Chem. Soc., *88,* 3511 (1966).
62. L. Vaska, Chem. Comm., 614 (1960).
63. B. M. Gatehouse, S. E. Livingston, and R. S. Nyholm, J. Chem. Soc., *1957,* 4222; C. C. Addison and B. M. Gatehouse, J. Chem. Soc., *1960,* 613.
64. J. R. Ferraro, J. Inorg. Nucl. Chem., *10,* 319 (1959).
65. K. Nakamoto, et al., J. Amer. Chem. Soc., *79,* 4904 (1957); C. G. Barraclough and M. L. Tobe, J. Chem. Soc., *1961,* 1993.
66. B. J. Hathaway and A. E. Underhill, J. Chem. Soc., *1961,* 3091.
67. N. A. Matwiyoff and R. S. Drago, Inorg. Chem., *3,* 337 (1964).
68. H. Kriegsman and S. Pischtschan, Z. anorg. allgem. Chem., *308,* 212 (1961); W. F. Edgell and C. H. Ward, J. Mol. Spectr., *8,* 343 (1962).
69. J. R. Ferraro and G. J. Long, Accts. Chem. Res., *8,* 171 (1975).
70. H. D. Kaesz, et al., J. Amer. Chem. Soc., *89,* 2844 (1967).
71. J. B. Wilford and F. G. A. Stone, Inorg. Chem., *4,* 389 (1965).

GENERAL REFERENCES

A. In the area of infrared, highly recommended references dealing with theory are:

INTRODUCTORY

L. H. Jones, "Inorganic Vibrational Spectroscopy," Volume 1, M. Dekker, New York, 1971.
W. G. Fately, et al., "Infrared and Raman Selection Rules for Molecular and Lattice Vibrations: The Correlation Method," Wiley-Interscience, New York, 1972.
L. A. Woodward, "Introduction to the Theory of Molecular Vibrations and Vibrational Spectroscopy," Oxford University Press, New York, 1972.
G. Barrow, "Introduction to Molecular Spectroscopy," McGraw-Hill, New York, 1962.
N. B. Colthup, L. H. Daly, and S. E. Wiberley, "Introduction to Infrared and Raman Spectroscopy," 2nd ed., Academic Press, New York, 1975.

ADVANCED

M. Davies, ed., "Infrared and Molecular Structure," Elsevier, New York, 1963.
A. B. F. Duncan, "Theory of Infrared and Raman Spectra" in "Chemical Applications of Spectroscopy," ed. W. West (vol. IX, "Technique of Organic Chemistry," ed. A. Weissberger) Interscience, New York, 1956.
G. Herzberg, "Spectra of Diatomic Molecules," 2nd ed., Van Nostrand, Princeton, 1950.
G. Herzberg, "Infrared and Raman Spectra of Polyatomic Molecules," Van Nostrand, Princeton, 1945.
E. B. Wilson, J. C. Decius, and P. C. Cross, "Molecular Vibrations," McGraw-Hill, New York, 1955.
D. Steele, "Theory of Vibrational Spectroscopy," W. B. Saunders Company, Philadelphia, 1971.

B. References dealing with applications are:

L. H. Jones, "Inorganic Vibrational Spectroscopy," Vol. 1, M. Dekker, New York, 1971.
M. Davies, ed., "Infrared Spectroscopy and Molecular Structure," Elsevier, Amsterdam, 1963.
K. Nakamoto, "Infrared Spectra of Inorganic and Coordination Compounds," 2nd ed., Wiley, New York, 1970.
K. Nakanishi, "Infrared Absorption Spectroscopy–Practical," Holden-Day, San Francisco, 1962. This reference contains many problems (with answers) involving structure determination of organic compounds from infrared spectra.
A. Finch, et al., "Chemical Applications of Far Infrared Spectroscopy," Academic Press, New York, 1970.
L. J. Bellamy, "The Infrared Spectra of Complex Molecules," 2nd Ed., Methuen, London, 1958.
L. J. Bellamy, "Advances in Infrared Group Frequencies," Methuen, London, 1968.
J. R. Ferraro, "Low Frequency Vibrations of Inorganic Coordination Compounds," Plenum Press, New York, 1971.

"Spectroscopic Properties of Inorganic and Organometallic Compounds," Volumes 1- (1968-), Specialist Periodical Reports of the Chemical Society, London.

COMPILATIONS

"Sadtler Standard Spectra," Sadtler Research Laboratories, Philadelphia, Pa. (A collection of spectra for use in identification, determination of the purity of compounds, etc.)

RAMAN SPECTROSCOPY

H. A. Szymanski, ed., "Raman Spectroscopy," Plenum Press, New York, 1967; Volume 2, 1970.
J. Brandmüller and H. Moser, "Binführung in die Ramanspektroskopie," Darmstadt, 1962.
A. Anderson, ed., "The Raman Effect," Volumes 1 and 2, M. Dekker, New York, 1971.
M. C. Tobin, "Laser Raman Spectroscopy," Wiley, New York, 1971; Volume 35 in "Chemical Analysis," P. J. Elving and I. M. Kolthoff, ed.
R. S. Tobias, J. Chem. Educ., *44,* 2, 70 (1967).
J. A. Koningstein, "Introduction to the Theory of the Raman Effect," D. Reidel, Dordrecht, 1972.
C. B. Moore, ed., "Chemical and Biochemical Applications of Lasers," Academic Press, New York, 1974.

7 NUCLEAR MAGNETIC RESONANCE SPECTROSCOPY— ELEMENTARY ASPECTS

INTRODUCTION[1,2,3] In this chapter, the principles necessary for an elementary appreciation of nmr will be presented. The reader mastering this chapter will have a minimum knowledge of those principles required for spectral interpretation of the results from slow passage as well as Fourier transform experiments. The next chapter expands upon several of the important concepts introduced here and illustrates several other types of applications of nmr. Varying appreciations of this subject can be obtained by complete reading of Chapter 7 and selected readings from Chapter 8.

Protons and neutrons both have a spin quantum number of $\frac{1}{2}$ and, depending on how these particles pair up in the nucleus, the resultant nucleus may or may not have a net non-zero nuclear spin quantum number, I. If the spins of all the particles are paired, there will be no net spin and the nuclear spin quantum number I will be zero. This type of nucleus is said to have zero spin and is represented in Fig. 7–1(A). When I is $\frac{1}{2}$, there is one net unpaired spin and *this unpaired spin imparts a nuclear magnetic moment, μ, to the nucleus*. The distribution of positive charge in a nucleus of this type is spherical. The properties for $I = \frac{1}{2}$ are represented symbolically in Fig. 7–1(B) as a spinning sphere (*vide infra*). When $I \geqslant 1$, the nucleus has spin associated with it and the nuclear charge distribution is non-spherical; see Fig. 7–1(C). The nucleus is said to possess a quadrupole moment eQ, where e is the unit of electrostatic charge and Q is a measure of the deviation of the nuclear charge distribution from spherical symmetry. For a spherical nucleus, eQ is zero. A positive value of Q

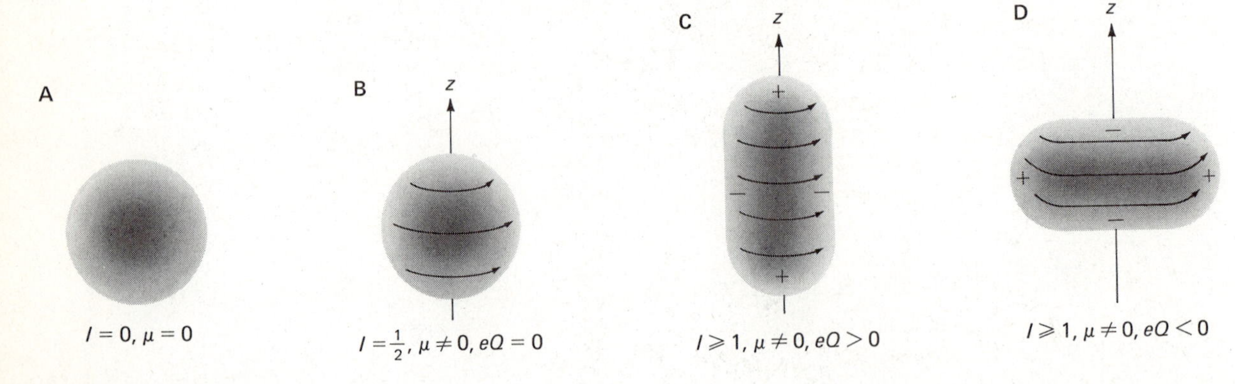

FIGURE 7–1. Various representations of nuclei.

indicates that charge is oriented along the direction of the principal axis (Fig. 7–1(C)), while a negative value for Q indicates charge accumulation perpendicular to the principal axis (Fig. 7–1(D)).

Nuclei with even numbers of both protons and neutrons all belong to the type represented in Fig. 7–1(A). Typical examples include ^{16}O, ^{12}C, and ^{32}S. The values for I, μ, and eQ for many other nuclei have been tabulated,[1,2,3] and are more easily looked up than figured out. A table summarizing these properties is included inside the back cover.

NMR spectroscopy is most often concerned with nuclei with $I = \frac{1}{2}$, examples of which include ^1H, ^{13}C, ^{31}P, and ^{19}F. Spectra often result from nuclei for which $I \geqslant 1$, but cannot be obtained on nuclei with $I = 0$.

Unpaired nuclear spin leads to a nuclear magnetic moment. The allowed orientations of the nuclear magnetic moment vector in a magnetic field are indicated by the *nuclear spin angular momentum quantum number,* m_I. This quantum number takes on values $I, I - 1, \ldots, (-I + 1), -I$. When $I = \frac{1}{2}$, $m_I = \pm\frac{1}{2}$ corresponding to alignments of the magnetic moment with and opposed to the field. When $I = 1$, m_I has values of 1, 0, and -1, corresponding, respectively, to alignments with, perpendicular to, and opposed to the field. In the absence of a magnetic field, all orientations of the nuclear moment are degenerate. In the presence of an external field, however, this degeneracy will be removed. For a nucleus with $I = \frac{1}{2}$, the $m_I = +\frac{1}{2}$ state will be lower in energy and the $-\frac{1}{2}$ state higher, as indicated in Fig. 7–2. The energy

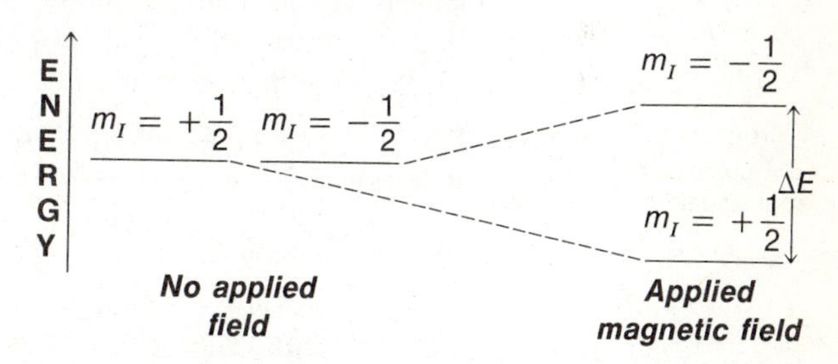

FIGURE 7–2. Splitting of the $m = \pm\frac{1}{2}$ states in a magnetic field.

difference between the two states, at magnetic field strengths commonly employed, corresponds to radio frequency radiation; it is this transition that occurs in the nmr experiment.

CLASSICAL DESCRIPTION OF THE NMR EXPERIMENT—THE BLOCH EQUATIONS

7–1 SOME DEFINITIONS

We shall begin our classical description of nmr by reviewing some of the background physics of magnetism and defining a few terms necessary to understand nmr. *It is very important to appreciate what is meant by angular momentum.* Circulating charges have angular momentum associated with them. Planar angular momentum, $\vec{\boldsymbol{\rho}}(\varphi)$ is given by $\vec{\boldsymbol{\rho}}(\varphi) = \vec{\mathbf{r}} \times m\vec{\mathbf{v}}$ where $\vec{\mathbf{r}}$, shown in Fig. 7–3, is the position vector of the particle e; $\vec{\mathbf{v}}$ its linear momentum vector; m is the mass; and φ is the angular change that signifies an angular momentum. The angular momentum $\vec{\boldsymbol{\rho}}$ is perpendicular to the plane of the circulating charge, and the direction of the angular momentum vector is given by the right hand rule. For the situation shown in Fig. 7–3, the vector points into the page.

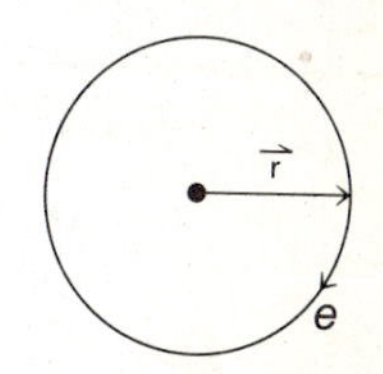

FIGURE 7–3. Circulating charge giving rise to planar angular momentum.

The nucleus is a more complicated three-dimensional problem. The total angular momentum of a nucleus is given by $\vec{\mathbf{J}}$, but it is convenient to define a dimensionless angular momentum operator $\hat{I}$ by

$$\hat{J} = \hbar\hat{I} \tag{7-1}$$

Associated with the angular momentum is a classical magnetic moment, $\vec{\mu}_N$, which can be taken as parallel to $\vec{\mathbf{J}}$, so:

$$\vec{\mu}_N = \gamma\vec{\mathbf{J}} = \gamma\hbar\vec{\mathbf{I}} \tag{7-2}$$

where γ, the magnetogyric (sometimes called gyromagnetic) ratio is a constant characteristic of the nucleus. From (7–2), we see that the magnetogyric ratio represents the ratio of the nuclear magnetic moment to the nuclear angular momentum.

7–2 BEHAVIOR OF A BAR MAGNET IN A MAGNETIC FIELD

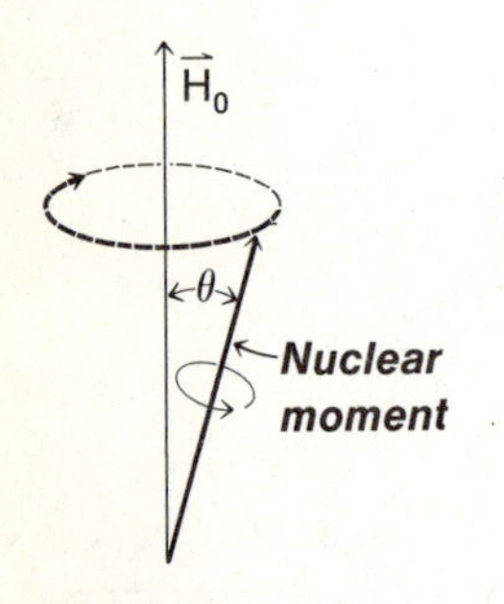

FIGURE 7–4. Precession of a nuclear moment in an applied field of strength H_0.

A nucleus with a magnetic moment can be treated as though it were a bar magnet. If a bar magnet were placed in a magnetic field, the magnet would precess about the applied field, $\vec{\mathbf{H}}_0$, as shown for a spinning nuclear moment in Fig. 7–4. Here θ is the angle that the magnetic moment vector makes with the applied field; and ω, the Larmor frequency, is the frequency of the nuclear moment precession. The instantaneous motion of the nuclear moment (indicated by the arrowhead on the dashed circle) is tangential to the circle and perpendicular to $\vec{\mu}$ and $\vec{\mathbf{H}}_0$. The magnetic field is exerting a force or torque on the nuclear moment, causing it to precess about the applied field. For use later, we would like to write an equation to describe the precession of a magnet in a magnetic field. The applied field $\vec{\mathbf{H}}_0$ exerts a torque $\vec{\tau}$ on $\vec{\mu}$ which is given by the *cross product:*

$$\vec{\tau} = \vec{\mu} \times \vec{\mathbf{H}}_0 \tag{7-3}$$

The right hand rule tells you that *the moment is precessing clockwise in Fig. 7–4, with the torque and instantaneous motion perpendicular to* $\vec{\mu}$ and $\vec{\mathbf{H}}_0$. According to Newton's Law, force is equal to the time derivative of the momentum, so the torque is the time derivative of the angular momentum.

$$\vec{\tau} = \frac{d}{dt}(\hbar\vec{\mathbf{I}})$$

With the equation $\vec{\mu} = \gamma\hbar\vec{\mathbf{I}}$, *the equation for precession of the moment is given by:*

$$\frac{d\vec{\mu}}{dt} = \frac{\gamma d(\hbar\vec{\mathbf{I}})}{dt} = \gamma\vec{\tau}$$

or

$$\frac{d\vec{\mu}}{dt} = \dot{\vec{\mu}} = -\gamma\vec{\mathbf{H}}_0 \times \vec{\mu} \tag{7-4}$$

(The dot is an abbreviation for the time derivative of some property, in this case $\vec{\mu}$; and the minus sign arises because we have changed the order for taking the cross product.) Thus, ω, the precession frequency or the *Larmor frequency,* is given by:

$$\omega = |\gamma| H_0 \tag{7-5}$$

where the sign of γ determines the sense of the precession. According to equation (7–5), *the frequency of the precession depends upon the applied field strength and the* *magnetogyric ratio of the nucleus. The energy of this system is given by the dot product of $\vec{\mu}$ and $\vec{\mathbf{H}}_0$.*

$$E = -\vec{\mu} \cdot \vec{\mathbf{H}}_0 = -|\mu|\,|H_0| \cos\theta \tag{7-6}$$

7–3 ROTATING AXIS SYSTEMS

There is one more mathematical construct that greatly aids the analysis of the nmr experiment, and this is the idea of a rotating coordinate system or *rotating frame.* In Fig. 7–5, a set of x, y, and z coordinates is illustrated. The rotating frame is

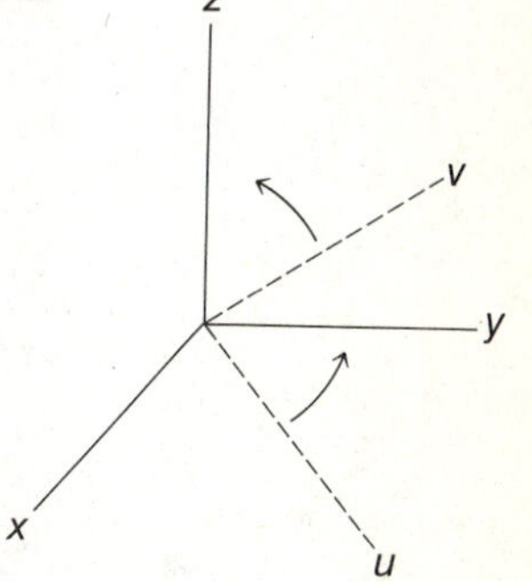

FIGURE 7–5. The rotating frame u,v,z in a Cartesian coordinate system x,y,z.

described by the rotating axes u and v, which rotate at some frequency ω_1 in the xy-plane. The z-axis is common to both coordinate systems. If the rotating frame rotates at some frequency less than the Larmor frequency, it would appear to an observer on the frame that the precessional frequency has slowed, which would correspond to a weakening of the applied field. Labeling this apparently weaker field as $\vec{\mathbf{H}}_{\text{eff}}$, our precessing moment is described in the rotating frame by

$$\dot{\vec{\mu}} = -\gamma \vec{\mathbf{H}}_{\text{eff}} \times \vec{\mu} \tag{7-7}$$

where $|H_{\text{eff}}| < |H_0|$. $\vec{\mathbf{H}}_{\text{eff}}$ is seen to be a function of ω_1, and if ω_1 is faster than the precessional frequency of the moment, ω, it will appear as though the z-field has changed direction. The effective field, $\vec{\mathbf{H}}_{\text{eff}}$, can be written as:

$$\vec{\mathbf{H}}_{\text{eff}} = \left[H_0 - \frac{\omega_1}{\gamma}\right]\vec{\mathbf{e}}_z \tag{7-8}$$

where $\vec{\mathbf{e}}_z$ is a unit vector along the z-axis. The time dependence of the moment, $\vec{\mu}$, can be rewritten as

$$\dot{\vec{\mu}} = -\gamma\left\{\left(H_0 - \frac{\omega_1}{\gamma}\right)\vec{\mathbf{e}}_z\right\} \times \vec{\mu} \tag{7-9}$$

When $\omega_1 = \omega = \gamma H_0$, then $|H_{\text{eff}}| = 0$ and $\vec{\mu}$ appears to be stationary. Thus, in a frame rotating with the Larmor frequency, we have eliminated time dependency and, in so doing, greatly simplified the problem.

7–4 MAGNETIZATION VECTORS AND RELAXATION

The concepts in this section are vital to an understanding of nmr. We shall develop the concept of relaxation in stages as the necessary background information is covered. For practical purposes, it is necessary to consider an ensemble or large number of moments because the nmr experiment is done with bulk samples. The individual moments in the sample add vectorially to give a *net magnetization,* $\vec{\mathbf{M}}$.

$$\vec{\mathbf{M}} = \sum_i \vec{\boldsymbol{\mu}}_i \tag{7-10}$$

In an ensemble of spins in a field, those orientations aligned with the field will be lower in energy and preferred. However, thermal energies oppose total alignment; experimentally, only a small net magnetization is observed. The equation for the motion of precession of $\vec{\mathbf{M}}$ is similar to that for $\vec{\boldsymbol{\mu}}$, *i.e.,*

$$\dot{\vec{\mathbf{M}}} = -\gamma\vec{\mathbf{H}}_0 \times \vec{\mathbf{M}} \tag{7-11}$$

If we place a sample in a magnetic field at constant temperature and allow the system to come to equilibrium, the resulting system is said to be at thermal equilibrium. At thermal equilibrium, the magnetization, M_0, is given by

$$|M_0| = \frac{N_0\mu^2}{3kT} I(I + 1)H_0 \tag{7-12}$$

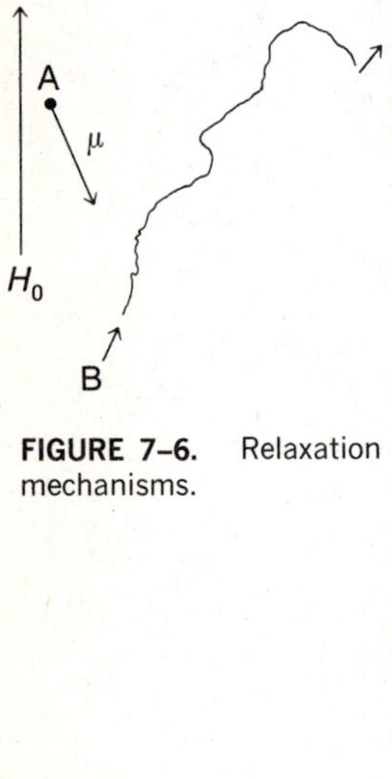

where k is the Boltzmann constant and N_0 is the number of magnetic moments per unit volume (*i.e.,* the number of spins per unit volume). This equation results from equation (7–6) and the assumption of a Boltzmann distribution. For H_2O at 27° and $H_0 = 10{,}000$ gauss, $M_0 \approx 3 \times 10^{-6}$ gauss.

FIGURE 7–6. Relaxation mechanisms.

The description of the magnetization is not yet complete for, at equilibrium, we have a dynamic situation in which any one given nuclear moment is rapidly changing its orientation with respect to the field. The mechanism for reorientation involves time dependent fields that arise from the molecular motion of magnetic nuclei in the sample. Suppose, as in Fig. 7–6, that nucleus B, a nucleus that had a magnetic moment, passed by nucleus A, whose nuclear moment is opposed to the field. The spin of A could be oriented with the field via the fluctuating field from moving B. In the process, the translational or rotational energy of the molecule containing B would be increased, since A gives up energy on going to a more stable orientation aligned with the field. This process is referred to as *spin-lattice relaxation.* Nucleus B is the lattice; it can be a magnetic nucleus in the same or another sample molecule or in the solvent. These nuclei do not have to be hydrogens and may even be unpaired electrons in the same or other molecules. The fluctuating field from motion of the magnetic nucleus must have the same frequency as that of the nmr transition in order for this relaxation process to occur. However, moving magnetic nuclei have a wide distribution of frequency components, and the required one is usually included.

Another process can also occur when the two nuclei interact, whereby nucleus B, with $m_I = +\frac{1}{2}$ goes to the higher-energy $m_I = -\frac{1}{2}$ state while nucleus A changes from $-\frac{1}{2}$ to $+\frac{1}{2}$. There is no net change in spin from this process, and it is referred to as a *spin-spin relaxation* mechanism.

To gain more insight into the nature of relaxation processes, let us examine the decay in magnetization as a function of time. Consider an experiment in which we have our magnetization aligned along the u axis as in Fig. 7–7(A). We then switch on a field aligned with the z-axis; M_u diminishes and M_z grows, generally at different rates. In Fig. 7–7(E), for example, the u-component has completely decayed, but the z-component has not yet reached equilibrium, for it gets larger in Fig. 7–7(F). The

FIGURE 7–7. The decay of the M_u component and growth of the M_z component when a field in the u-direction is turned off and one in the z-direction is turned on.

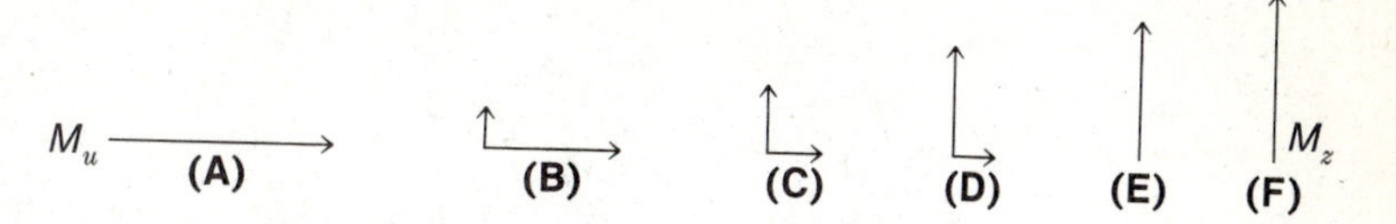

growth and decay are first order processes, and Bloch proposed the following equation for the three-dimensional case (*i.e.*, u, v, and z magnetizations are involved and the field is located along the z-axis):

$$\dot{M}_u = -\frac{M_u}{T_2} \qquad \dot{M}_v = -\frac{M_v}{T_2} \tag{7-13}$$

and

$$\dot{M}_z = -\frac{1}{T_1}(M_z - M_0) \tag{7-14}$$

where $1/T_2$ and $1/T_1$ are first order rate constants, M_0 is the equilibrium value of the z magnetization, and u- and v-components vanish at equilibrium. In general, it is found that $1/T_2 \geqslant 1/T_1$. Instead of rate constants, $1/T$, it is more usual to refer to their reciprocals (*i.e.*, the T's), which are lifetimes or relaxation times. Since T_1 refers to the z-component, it is called the *longitudinal relaxation time*, while T_2 is called the *transverse relaxation time*. The spin-lattice mechanism contributes to T_1, and the spin-spin mechanism is one of several contributions to T_2.

7–5 THE NMR TRANSITION

Before proceeding with our classical description of the nmr experiment, it is advantageous to introduce a few concepts from the quantum mechanical description of the experiment. When the bare nucleus (no electrons around it) is placed in a magnetic field, H_0, the field and the nuclear moment interact [see equation (7–6)] as described by the Hamiltonian for the system

$$\hat{H} = -\vec{\mu}_N \cdot \vec{\mathbf{H}}_0 \tag{7-15}$$

where, as shown in equation (7–2), $\vec{\mu}_N = \gamma\hbar\vec{\mathbf{I}}$ with the N subscript denoting a nuclear moment. When it is obvious that we are referring to a nuclear moment, the N subscript will be dropped. Combining (7–2)* and (7–15) yields

$$\hat{H} = -\gamma_N\hbar H_0\hat{I}_z = -g_N\beta_N H_0\hat{I}_z \tag{7-16}$$

where γ_N is constant for a given nucleus, g_N is the nuclear g factor and:

$$\beta_N = e\hbar/2Mc \tag{7-17}$$

In equation (7–17), M is the mass of the proton, e is the charge of the proton, and c is the speed of light.

The expectation values of the operator $\hat{I}_z$, where z is selected as the applied field direction, are given by m_I where $m_I = I, I-1, \ldots, -I$. The degeneracy of the m_I states that existed in the absence of the field is removed by the interaction between

*We locate the field H_0 along the z-axis and, since H_y and H_x are zero, only $H_0\hat{I}_z$ is non-zero; *i.e.*, $H_x\hat{I}_x$ and $H_y\hat{I}_y$ are zero because H_x and H_y are zero.

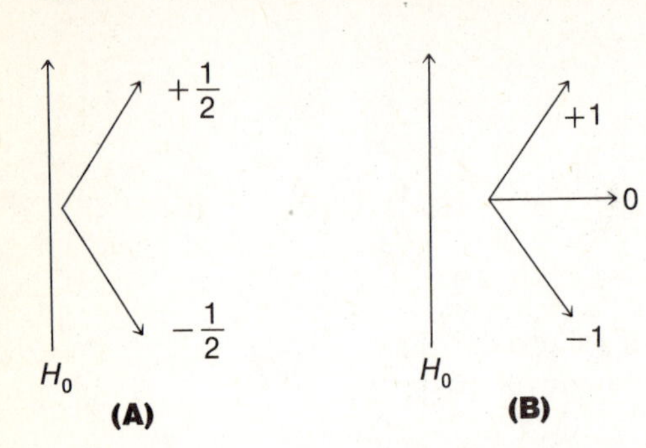

FIGURE 7–8. Quantized orientation of m_I for (A) $I = \frac{1}{2}$ and (B) $I = 1$.

the field H_0 and the nuclear magnetic moment μ_N. The quantized orientations of these nuclear moments relative to an applied field $\vec{\mathbf{H}}_0$ are shown in Fig. 7–8 for $I = \frac{1}{2}$ and $I = 1$. Since the eigenvalues of the operator $\hat{I}_z$ are m_I, the eigenvalues of $\hat{H}$ (*i.e.*, the energy levels) are given by equation (7–18).

$$E = -\gamma\hbar m_I H_0 \tag{7-18}$$

The energy as a function of the field is illustrated for $I = \frac{1}{2}$ in Fig. 7–9. The quantity

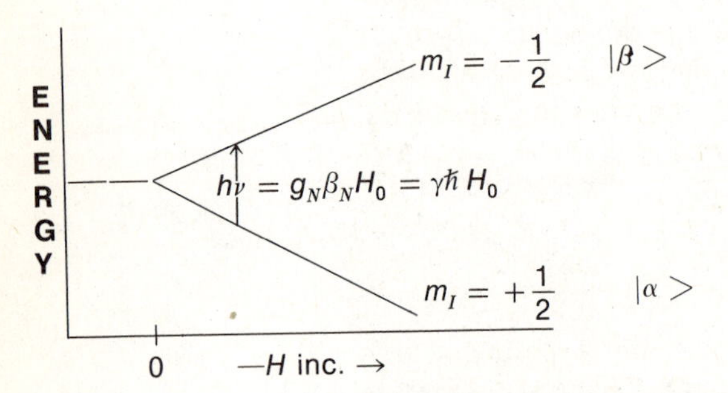

FIGURE 7–9. The field dependence of the energies of $m_I = \pm\frac{1}{2}$.

g_N is positive for a proton, as is β_N, so the positive m_I value is lowest in energy. The nuclear wave functions are abbreviated as $|\alpha\rangle$ and $|\beta\rangle$ for the $+\frac{1}{2}$ and $-\frac{1}{2}$ states, respectively. By inserting values of $m_I = +\frac{1}{2}$ and $m_I = -\frac{1}{2}$ into equation (7–18), we find that the energy difference, ΔE, between these states or the energy of the transition $h\nu$ for a nucleus of spin $\frac{1}{2}$ is given by $g_N\beta_N H_0$ or $\gamma\hbar H_0$.

Since $m_I = +\frac{1}{2}$ is lower in energy than $m_I = -\frac{1}{2}$, there will be a slight excess population of the low energy state at room temperature, as described by the Boltzmann distribution expression in equation (7–19):

$$\frac{N(-\frac{1}{2})}{N(+\frac{1}{2})} = \exp(-\Delta E/kt) \cong 1 - \frac{\Delta E}{kT} \text{ when } \Delta E \ll kT \tag{7-19}$$

ΔE is $\sim 10^{-3}$ cm^{-1} in a 10,000 gauss field, and $kT \sim 200$ cm^{-1}. At room temperature, there is a ratio of 1.0000066 $(+\frac{1}{2})$ spins to one $(-\frac{1}{2})$. The probability of a nuclear moment being in the $+\frac{1}{2}$ state is $(\frac{1}{2})\left(1 + \frac{\mu H_0}{kT}\right)$ and that of it being in the $-\frac{1}{2}$ state is $(\frac{1}{2})\left(1 - \frac{\mu H_0}{kT}\right)$ [recall our discussion of equation (7–12)].

The energy separation corresponding to $h\nu$ occurs in the radio-frequency region of the spectrum at the magnetic field strengths usually employed in the experiment. One applies a circularly polarized radio frequency (r.f.) field at right angles to $\vec{\mathbf{H}}_0$ (*vide infra*), and the magnetic component of this electromagnetic field, $\vec{\mathbf{H}}_1$, provides a torque to flip the moments from $m_I = +\frac{1}{2}$ to $-\frac{1}{2}$, causing transitions to occur.

7-6 THE BLOCH EQUATIONS

In order to understand many of the applications of nmr, it is necessary to appreciate the change in magnetization of the system with time as the H_1 field is applied. This result is provided with the Bloch equation. Incorporating equations (7-13) and (7-14), describing relaxation processes, into (7-11), which describes the precession of the magnetization, and converting to the rotating frame gives the Bloch equation:

$$\dot{\vec{\mathbf{M}}} = \underbrace{-\gamma\vec{\mathbf{H}}_{\text{eff}} \times \vec{\mathbf{M}}}_{\text{torque from the magnetic field}} \underbrace{- \frac{1}{T_2}(M_u\vec{\mathbf{e}}_u + M_v\vec{\mathbf{e}}_v) - \frac{1}{T_1}(M_z - M_0)\vec{\mathbf{e}}_z}_{\text{relaxation effects}} \qquad (7\text{-}20)$$

In the presence of H_1 and in the rotating frame, equation (7-8)—which described $\vec{\mathbf{H}}_{\text{eff}}$ in the rotating frame—becomes:

$$\vec{\mathbf{H}}_{\text{eff}} = \left(H_0 - \frac{\omega_1}{\gamma}\right)\vec{\mathbf{e}}_z + H_1\vec{\mathbf{e}}_u \qquad (7\text{-}21)$$

where the frame is rotating with the frequency ω_1 corresponding to the frequency of H_1, the oscillating field at right angles to $\vec{\mathbf{H}}_0$. Equation (7-20) is a vector equation in the rotating frame that can best be written in terms of the components of $\vec{\mathbf{M}}$, which are M_u, M_v, and M_z.

The three components of the Bloch equation are:

$$\dot{M}_u = \frac{dM_u}{dt} = -(\omega_1 - \omega_0)M_v - \frac{M_u}{T_2} \qquad (7\text{-}22)$$

$$\dot{M}_v = \frac{dM_v}{dt} = (\omega_1 - \omega_0)M_u - \frac{M_v}{T_2} + \gamma H_1 M_z \qquad (7\text{-}23)$$

$$\dot{M}_z = \frac{dM_z}{dt} = -\gamma H_1 M_v + (M_0 - M_z)/T_1 \qquad (7\text{-}24)$$

In these equations, ω_0 is the Larmor frequency, which equals γH_0; and the u, v reference frame is rotating at angular velocity ω_1.

Experimentally, we monitor the magnetization in the xy plane, referred to as the transverse component. Using a phase-sensitive detector, we monitor the component of magnetization induced along the u-axis. In the normal slow passage or steady state experiment, only a u-component of magnetization exists; but because of a 90° phase lag associated with the electronic detection system, a component 90° out of phase with u is measured. Slow passage or steady state conditions require H_1 to be weak (of the order of milligauss) compared to H_0 (which is of the order of kilogauss). Then, according to equation (7-21), the z-component dominates unless ω_1 is close to ω_0 so that the first term ($H_0 = \omega_0/\gamma$) becomes small. (Note that ω_1 is the frequency of the r.f. field and *not* the Larmor frequency of precession about H_1; *i.e.*, $\omega_1 \neq \gamma H_1$.) When ω_1 is close to the Larmor frequency, ω_0, then $\vec{\mathbf{H}}_{\text{eff}}$ is tipped toward the u-axis. Since H_0 is being changed slowly, the net effect is to change H_{eff} slowly. The individual moments continue to precess about $\vec{H}_{eff}$ as a consequence of the torque, which is perpendicular to $\vec{H}_{eff}$. As a result, $\vec{M}$ remains parallel to $\vec{H}_{eff}$ and a u-component results. Fig. 7–10 represents the tipping of $\vec{M}$ to remain aligned with $\vec{H}_{eff}$ as we pass

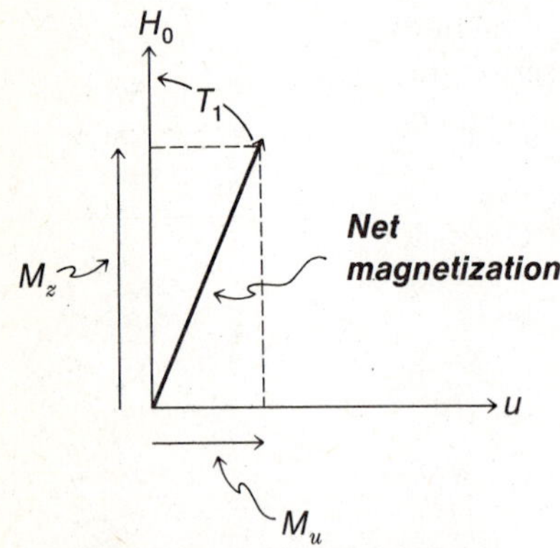

FIGURE 7–10. Effect of H_1 on the net magnetization of an H_0 field.

through resonance. The $\vec{H}_1$ field is along the u-axis. It is also an alternating field with the frequency of the rotating frame. A static field will not tip the magnetization vector significantly because H_1 is so small.

Relaxation processes, T_1, are trying to preserve the orientation along the strong z-field, as shown by the arrow labeled T_1 in Fig. 7–10. Under steady state conditions, all time derivatives are zero, so equations (7–22), (7–23), and (7–24) are all equal to zero and can be solved to produce:

$$M_u = M_0 \frac{\gamma H_1 T_2^2(\omega_0 - \omega_1)}{1 + T_2^2(\omega_0 - \omega_1)^2 + \gamma^2 H_1^2 T_1 T_2} \tag{7-25}$$

$$M_v = M_0 \frac{\gamma H_1 T_2}{1 + T_2^2(\omega_0 - \omega_1)^2 + \gamma^2 H_1^2 T_1 T_2} \tag{7-26}$$

$$M_z = M_0 \frac{1 + T_2^2(\omega_0 - \omega_1)^2}{1 + T_2^2(\omega_0 - \omega_1)^2 + \gamma^2 H_1^2 T_1 T_2} \tag{7-27}$$

7–7 THE NMR EXPERIMENT

Equations (7–25) to (7–27) describe the magnetization of our sample in the so-called slow passage experiment, which is schematically illustrated in Fig. 7–11. In this method, one applies a strong homogeneous magnetic field, causing the nuclei to precess. Radiation of energy comparable to ΔE is then imposed with a radio frequency transmitter, producing H_1. When the applied frequency from the radio transmitter is equal to the Larmor frequency, the two are said to be in resonance, and a u,v-component is induced which can be detected. This is the condition in (7–21) when $H_0 \approx \omega_1/\gamma$. Quantum mechanically, this is equivalent to some nuclei being excited from the low energy state ($m_I = +½$) to the high energy state ($m_I = -½$; see Fig. 7–9) by absorption of energy from the r.f. source at a frequency equal to the Larmor frequency. Since $\Delta E = h\nu$ and $\omega = 2\pi\nu$, ΔE is proportional to the Larmor frequency, ω. Energy will be extracted from the r.f. source only when this resonance

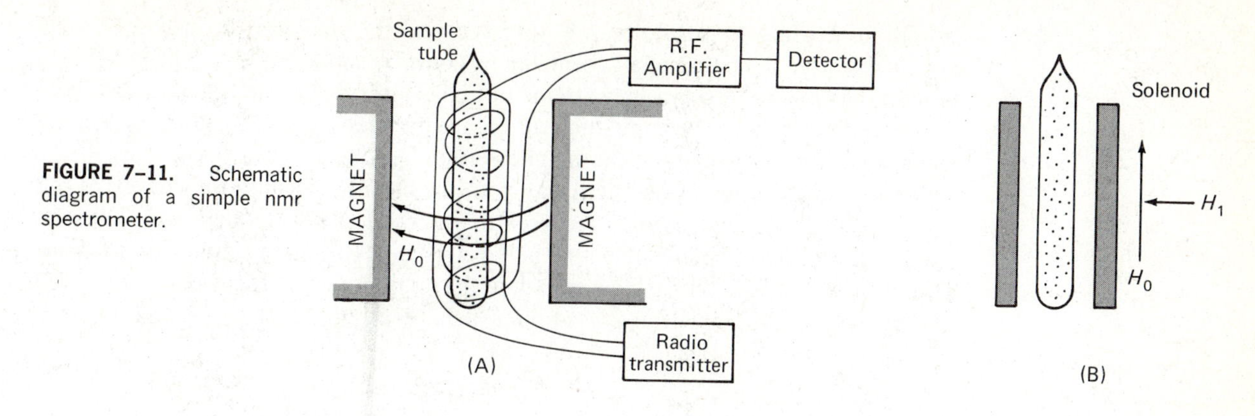

FIGURE 7–11. Schematic diagram of a simple nmr spectrometer.

condition ($\omega = 2\pi\nu$) is fulfilled. With an electronic detector (see Fig. 7–11), one can observe the frequency at which a *u,v*-component is induced or at which the loss of energy from the transmitter occurs, allowing the resonance frequency to be measured. For a more complete discussion of the experimental procedure, see Pople, *et al.*[2]

It is possible to match the Larmor frequency and the applied radio frequency either by holding the field strength constant (and hence ω constant) and scanning a variable applied radio frequency until matching occurs or, as is done in the more common nmr apparatus, by varying the field strength until ω becomes equal to a fixed applied frequency. In the latter method, fixed frequency probes (source and detector coils) are employed and the field strength at which resonance occurs is measured. Two experimental configurations are used in this experiment. In Fig. 7–11(A) the H_0 field direction (*z*-direction) is perpendicular to the sample tube, while in Fig. 7–11(B) the H_0 field and the sample are coaxial. There are some applications in which this difference is important (*vide infra*).

We can now see why the relaxation processes discussed in equations (7–13) and (7–14) had to be added to our complete equation [(7–20), which led to (7–25) to (7–27) for the slow passage experiment] for the behavior of the magnetization. If the populations of nuclei in the ground and excited states were equal, then the probability that the nucleus would emit energy under the resonance condition would equal the probability that the nucleus would absorb energy [*i.e.*, transitions $m_I(+\frac{1}{2}) \longrightarrow m_I(-\frac{1}{2})$ would be as probable as $m_I(-\frac{1}{2}) \longrightarrow m_I(+\frac{1}{2})$]. No net change would then be detected by the radio frequency probe. As mentioned earlier, in a strong magnetic field there will be a slight excess of nuclei aligned with the field (lower energy state) and consequently a net absorption of energy results. As energy is absorbed from the r.f. signal, enough nuclei could be excited after a finite period of time so that the population in the lower state would be equal to that in the higher state. Initially, absorption might be detected but this absorption would gradually disappear as the populations of ground and excited states became equal. When this occurs, the sample is said to be saturated. If the nmr instrument is operated properly, saturation usually does not occur, because the relaxation mechanisms allow nuclei to return to the lower energy state without emitting radiation. As a result there is always an excess of nuclei in the lower energy state, and a continuous absorption of energy from the r.f. source by the sample occurs.

Remembering that the *u*-component of the magnetization is induced, equation (7–25) predicts that an absorption band in the nmr spectrum will have the general form of a Lorentzian function, $A\left(\frac{1}{a + bx^2}\right)$, for when H_1 is small the last term in the denominator is negligible. Thus, as we sweep through resonance, the magnitude of M_u gives a Lorentzian plot which is the nmr spectrum. When a large H_1 is employed, conditions leading to saturation are observed and the shape of the spectral line is grossly distorted from that of a Lorentzian by contributions to M_u from the $\gamma^2 H_1^2 T_1 T_2$ term of the denominator. In the extreme of high source power, no signal is observed, for we in effect destroy the population difference between $m_I = +\frac{1}{2}$ and $-\frac{1}{2}$ and the resulting net z-component of the magnetization.

The nmr experiment has significance to the chemist because the energy of the resonance (*i.e.*, the field strength required to attain a Larmor frequency equal to the fixed frequency) is dependent upon the electronic environment about the nucleus. The electrons shield the nucleus, so that the magnitude of the field seen at the nucleus, H_N, is different from the applied field, H_0:

$$H_N = H_0(1 - \sigma) \tag{7-28}$$

where σ, the shielding constant, is a dimensionless quantity that represents the shielding of the nucleus by the electrons. The value of the shielding constant depends on several factors, which will be discussed in detail later. Equation (7–28) states that the value of H_0 in equation (7–21) is different from the H applied for different nuclei in the molecule, and H_0 should then be replaced by H_N. Consequently, in a slow sweep of H_0, the various magnetizations of the different nuclei are sampled individually, because when H_1 dominates for one kind of nucleus the H_N term will still be dominant for others. For a molecule with two different kinds of hydrogen atoms this will lead to a spectrum like that shown in Fig. 7–12.

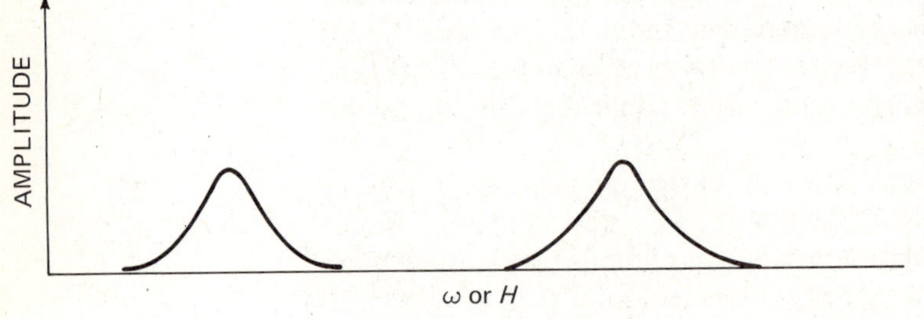

FIGURE 7–12. A low resolution nmr spectrum of a sample containing two different kinds of protons.

One other point is worth making here. The differences in the magnetogyric ratios of different kinds of nuclei are much larger than the effects from σ, so there is no trouble distinguishing signals from the different kinds of nuclei in a sample; *e.g.*, ^{19}F and ^1H are never confused. The ranges are so vastly separated that different instrumentation is required to study different kinds of nuclei. In Table 7–1, the resonance conditions ($h\nu = g_N \beta_N H_0$) for various nuclei are given for an applied field of 10,000 gauss.

The frequencies in Table 7–1 are in megaHertz (10^6 Hertz or 10^6 c/s), and the variation in the proton resonances of typical organic compounds from different shielding constants is only 600 Hz. In some paramagnetic complexes, shifts of the order of magnitude of 840,000 Hz have been observed; but even these could not be confused with a fluorine resonance. The relative sensitivities of various nuclei in the nmr experiment are also listed in Table 7–1.

Another limiting experiment is the pulse experiment. We shall discuss this experiment in more detail later, but now wish to briefly consider it in the context of equation (7–21). Suppose that a single strong pulse is imposed (*e.g.*, $H_1 \approx 100$ gauss $\gg H_0 - \omega_1/\gamma$ for all of the protons). Referring to the equation for $\vec{\mathbf{H}}_{\text{eff}}$,

$$\vec{\mathbf{H}}_{\text{eff}} = H_1 \vec{\mathbf{e}}_u + \left(H_0 - \frac{\omega_1}{\gamma}\right)\vec{\mathbf{e}}_z$$

TABLE 7–1. IMPORTANT NUCLEI IN NMR

Isotope	Abundance (per cent)	NMR Frequency in 10 Kilogauss Field[d]	Relative[a] Sensitivity (constant H_0)	Magnetic[b] Moment (μ)	Spin[c] (I)
^{1}H	99.9844%	42.577	1.0000	2.7927	$\frac{1}{2}$
^{2}H (D)	0.0156	6.536	0.0096	0.8574	1
^{10}B	18.83	4.575	0.0199	1.8006	3
^{11}B	81.17	13.660	0.165	2.6880	$\frac{3}{2}$
^{13}C	1.108	10.705	0.0159	0.7022	$\frac{1}{2}$
^{14}N	99.635	3.076	0.0010	0.4036	1
^{15}N	0.365	4.315	0.0010	−0.2830	$\frac{1}{2}$
^{19}F	100.	40.055	0.834	2.6273	$\frac{1}{2}$
^{29}Si	4.70	8.460	0.0785	−0.5548	$\frac{1}{2}$
^{31}P	100.	17.235	0.0664	1.1305	$\frac{1}{2}$
^{117}Sn	7.67	15.77	0.0453	−0.9949	$\frac{1}{2}$
^{119}Sn	8.68	15.87	0.0518	−1.0409	$\frac{1}{2}$

[a] For equal numbers of nuclei, where ^{1}H equals one.
[b] In multiples of the nuclear magneton, $eh/4\pi Mc$.
[c] In multiples of $h/2\pi$.
[d] MHz.

the local H_0 will vary slightly from nucleus to nucleus, but $\vec{\mathbf{H}}_1$ is so large that the term $H_1\vec{\mathbf{e}}_u$ is completely dominant, and $\vec{\mathbf{H}}_{\text{eff}}$ is almost the same for all nuclei present. All nuclei are sampled simultaneously. Since $\vec{\mathbf{H}}_1$ is directed along $\vec{\mathbf{u}}$ and the net magnetization along $\vec{\mathbf{z}}$, the cross product requires that $\vec{\mathbf{M}}$ begin to precess about $\vec{\mathbf{u}}$. The pulse time is so short, however, that the magnetization does not have time to precess around $\vec{\mathbf{u}}$, but merely tips toward $\vec{\mathbf{v}}$. The pulse duration is shorter than the time corresponding to one full precession about $\vec{\mathbf{u}}$. Accordingly, the v-component is measured in this experiment as $\vec{\mathbf{M}}$ is tipped toward $\vec{\mathbf{v}}$. This is how ^{13}C experiments, for example, are now being done by Fourier transform procedures. If a strong pulse were employed for a very long time, the nuclei would precess around $\vec{\mathbf{H}}_{\text{eff}}$ as in the slow passage experiment; a u-component would arise and v would disappear. Pulse experiments cannot be done this way. When a strong pulse is employed for a short enough duration* that there is no time for any relaxation to occur during the pulse ($t_p \approx 10$ microseconds), the Bloch equations predict the following expression for the v-component, as a function of time, t:

$$M_v = \sin(\gamma H_1 t_p) M_0 \exp(-t/T_2) \qquad (7\text{-}29)$$

This expression results from the Fourier transform of equation (7–23), which allows conversion from the frequency domain (7–23) to the time domain (7–29). We shall discuss this topic in more detail later in this chapter. The quantity dM_v/dt cannot be set equal to zero in this experiment. The v-component of the magnetization decays with time according to equation (7–29), as shown in Fig. 7–13. This is called a free induction decay curve and it can be analyzed by computer (via Fourier transformation, *vide infra*) to give a frequency spectrum identical to the Lorentzian slow passage result. A comparison of the slow passage and Fourier transform experiments is provided in Table 7–2.

---

* Later, we shall provide an *equivalent* description of this experiment in terms of the distribution of frequencies that make up the square wave produced from this pulsing.

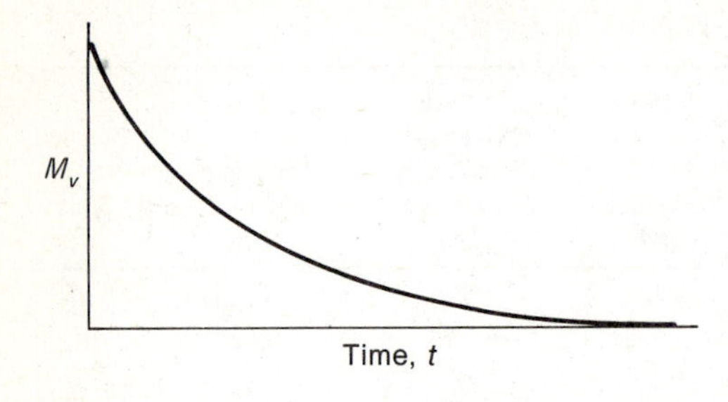

FIGURE 7–13. Free induction decay curve, a plot of the change in magnetization of the sample with time.

TABLE 7–2. A COMPARISON OF A TYPICAL 100 MHz EXPERIMENT OF THE FOURIER TRANSFORM AND THE SLOW PASSAGE EXPERIMENT IN TERMS OF THE EQUATION $\gamma\vec{H}_{eff}{}^{a} = (\gamma H_0 - \omega_1)\vec{e}_z + \gamma H_1\vec{e}_u$

Experiment	Order of magnitude of frequencies in sec^{-1} (Hz) γH_0 and ω_1	$\gamma H_0 - \omega_1{}^{b}$	γH_1	What happens?	What is measured?
Slow Passage	10^8	10^0 (at resonance)	10^0	$\vec{\mu}_i$'s precess about $\vec{H}_{eff}$ many times as we sweep through resonance, so $\vec{M}$ tips toward $\vec{u}$.	M_u
Fourier Transform (FT)	10^8	10^4	10^6	$\vec{\mu}_i$'s (and hence $\vec{M}$) precess about $\vec{H}_{eff}$ $\frac{1}{4}$ of a revolution or less during the pulse time, so $\vec{M}$ tips toward $\vec{v}$.	M_v

[a] Multiplying by γ puts H_{eff} in units of frequency (Hz).
[b] If you are accustomed to thinking in terms of parts per million, note that $10^4/10^8 = 100 \times 10^{-6} = 100$ ppm.

THE QUANTUM MECHANICAL DESCRIPTION OF THE NMR EXPERIMENT

7–8 PROPERTIES OF $\hat{I}$

Now, to gain valuable insight, let us reexamine this whole problem using a quantum mechanical instead of a classical mechanical approach. Quantum mechanics shows us that for $I = \frac{1}{2}$, there are two allowed orientations of the spin angular momentum vector in a magnetic field (Fig. 7–8) and will also indicate the necessary requirements to induce transitions between these energy states by an appropriate perturbation, *i.e.*, the application of an oscillating magnetic field with energies corresponding to r.f. radiation. The necessary direction for this field can be determined from a consideration of the spin angular momentum operators.

The $\hat{I}^2$ operator has eigenvalues $I(I+1)$ (as in an atom, where the orbital angular momentum operator $\hat{L}^2\psi = \ell(\ell+1)\hbar^2\psi$). Any one of the components of $\hat{I}$ (*e.g.*, $\hat{I}_z$) commutes with $\hat{I}^2$, so we can specify eigenvalues of both $\hat{I}^2$ and $\hat{I}_z$. The eigenvalues of $\hat{I}_z$ are given by $I, I-1, \ldots -I$ (like m_ℓ values in an atom, where the z-component of angular momentum, $\hat{L}_z$, is given by $\hat{L}_z\psi = m_\ell\hbar\psi$ with $m_\ell = \pm\ell, \ldots, 0$). *In general, if two operators commute, then there exist simultaneous eigenfunctions of both operators for which eigenvalues can be specified.*

It is a simple matter to determine whether two operators commute. If they do, then by our earlier definition of commutation (Chapter 2)

$$\hat{I}^2\hat{I}_z - \hat{I}_z\hat{I}^2 = 0$$

The above difference is abbreviated by the symbol $[\hat{I}^2, \hat{I}_z]$. Similar equations can be written for $\hat{I}_x$ and $\hat{I}_y$, *i.e.*,

$$[\hat{I}^2, \hat{I}_x] = [\hat{I}^2, \hat{I}_y] = 0$$

However, $\hat{I}_z$ does not commute with $\hat{I}_x$ or $\hat{I}_y$, *e.g.*,

$$\hat{I}_z\hat{I}_y - \hat{I}_y\hat{I}_z \neq 0$$

Eigenvalues for $\hat{I}^2$ exist, and if we decide to specify eigenvalues for $\hat{I}_z$, then eigenvalues for $\hat{I}_x$ and $\hat{I}_y$ do not exist.

Average values (Chapter 3) could be calculated for the $\hat{I}_x$ and $\hat{I}_y$ operators. We shall make these concepts more specific by applying these operators to the spin wave functions α and β for $I = ½$ nuclei and showing the results:

$$\hat{I}_z\alpha = (+½)\alpha \qquad \hat{I}_z\beta = (-½)\beta \tag{7-30}$$

$$\hat{I}_x\alpha = (½)\beta \qquad \hat{I}_x\beta = (-½)\alpha \qquad \hat{I}_y\alpha = (½)i\beta \qquad \hat{I}_y\beta = (-½)i\alpha \tag{7-31}$$

Thus, the $\hat{I}_z$ operator yields eigenvalues, since operation on α gives back α and operation on β gives β. The $\hat{I}_x$ and $\hat{I}_y$ operators do not yield eigenvalues, since operation on α produces β and operation on β yields α. The average value for the property $\hat{I}_x$ or $\hat{I}_y$ is given by an equation of the sort $\int\psi^*\text{op}\psi d\tau/\int\psi^2 d\tau$. The following relations hold for α and β (as they do for orthonormal electronic wave functions):

$$\int \alpha^2 d\tau = \int \beta^2 d\tau = 1 \quad \text{and} \quad \int \alpha\beta d\tau = 0$$

As mentioned in Chapter 3, the integrals encountered in quantum mechanical descriptions of systems are written employing the *bra* and *ket notation* for convenience. Recall that the symbol $|\ \rangle$ is referred to as a ket and $\langle\ |$ as a bra. An integral of the form $\int(\psi^*\ \text{Operator}\ \psi)d\tau$ is written as $\langle\psi|\text{Op}|\psi\rangle$, while an integral of the form $\int\psi_1^*\psi_2\, d\tau$ is written as $\langle\psi_1|\psi_2\rangle$.

7-9 TRANSITION PROBABILITIES

Consider the effect of the H_1 field in the quantum mechanical description. If the alternating field is written in terms of an amplitude $H_x°$, we get a perturbing term in the Hamiltonian of the form of equation (7-32):

$$\hat{H}_{\text{pert}} = -\gamma\hbar H_x°\hat{I}_x \cos \omega_1 t \tag{7-32}$$

Recall from equation (7-16) that $\hat{H}$ for a nucleus in a z-field was $\hat{H} = -\gamma\hbar H_0\hat{I}_z = -g_N\beta_N H_0\hat{I}_z$, so now the perturbation is of a similar form but for an x-field that is alternating.

The equation describing the probability of a transition in the nmr, P, is similar to that in u.v. and i.r., and is given by

$$P = 2\pi\gamma_N{}^2 H_1{}^2 \left|\langle\varphi^{ex}|\hat{I}_x|\varphi\rangle\right|^2 g(\omega) \tag{7-33}$$

where $g(\omega)$ is a general line shape function, which is an empirical function that describes how the absorption varies near resonance. To apply equation (7–33), we need to evaluate matrix elements of the form $\langle\varphi^{ex}|\hat{I}_x|\varphi\rangle$ and determine whether they are zero or non-zero.

The solution is best accomplished by constructing a matrix that summarizes all of the integrals that must be evaluated to describe the system in a magnetic field with and without H_1. Rows and columns are constructed, which are headed by the basis set. In this case, we have one nucleus with α and β nuclear spin wave functions leading to:

	α	β
α	$\langle\alpha\|\hat{H} + \hat{H}_{pert}\|\alpha\rangle$	$\langle\alpha\|\hat{H} + \hat{H}_{pert}\|\beta\rangle$
β	$\langle\beta\|\hat{H} + \hat{H}_{pert}\|\alpha\rangle$	$\langle\beta\|\hat{H} + \hat{H}_{pert}\|\beta\rangle$

By this procedure, we have considered all possible matrix elements and also have them in such a form that if E were subtracted from the diagonal elements we would have the secular determinant, which can be solved to give the energies of these states in an applied field $(H_0 + H_1)$.* The resulting energies could then be used in the secular equations (produced by matrix multiplication of the secular determinant with a matrix of the basis set) to give the wave functions in the field. (Notice the analogy of this to our handling of the secular determinant and equations in the section on Hückel calculations in Chapter 3.)

We begin by evaluating the elements in the secular determinant when the applied field is H_0 (with a z-component only), *i.e.*, the Zeeman experiment. Since there is no x or y field component (only z), there are no $\hat{I}_x$ or $\hat{I}_y$ operators and all matrix elements of the form $\langle|\hat{I}_x|\rangle$ or $\langle|\hat{I}_y|\rangle$ are zero. The off-diagonal elements, $\langle\alpha|\hat{I}_z|\beta\rangle$ and $\langle\beta|\hat{I}_z|\alpha\rangle$, are also zero because $\hat{I}_z|\beta\rangle = -\frac{1}{2}\beta$, and $\langle\alpha|\beta\rangle$ and $\langle\beta|\alpha\rangle$ are zero. The only nonzero matrix elements are $\langle\alpha|\hat{I}_z|\alpha\rangle$ and $\langle\beta|\hat{I}_z|\beta\rangle$, with the former corresponding to stabilization, $+\frac{1}{2}$ (*i.e.*, $\frac{1}{2}\langle\alpha|\alpha\rangle$), and the latter to destabilization, $-\frac{1}{2}$. With no off-diagonal elements, the eigenvalues are directly obtained and the basis set is not mixed, so the two wave functions are α and β. When these are substituted into equation (7–33) for φ and φ^{ex}, the matrix element is zero and the transition is not allowed, $\langle\alpha|\beta\rangle = 0$.

Next, we shall consider what happens when an H_1 field along the x-axis is added to the Zeeman experiment described above. We must now worry about matrix elements involving $\hat{I}_x$. The diagonal elements $\langle\alpha|\hat{I}_x|\alpha\rangle$ are zero $[\langle\alpha|\hat{I}_x|\alpha\rangle = \frac{1}{2}\langle\alpha|\beta\rangle = 0]$, but the off-diagonal elements $\langle\alpha|\hat{I}_x|\beta\rangle$ are non-zero. Since the H_1 field is small compared to H_0 (z-component), these off-diagonal matrix elements are so small as to have a negligible effect on the energies (the effect of $\hat{I}_z$ on the diagonal is the same as in the Zeeman experiment). However, the small off-diagonal matrix elements are important because they provide a mechanism for inducing transitions

*Recall that $\langle\psi|\hat{H}|\psi\rangle = E\langle\psi|\psi\rangle$. For an orthonormal basis set, $\langle\psi_n|\psi_m\rangle$ equals 1 when $n = m$ but equals zero when $n \neq m$. Thus, E appears only on the diagonal of the energy determinant.

from α to β *because the new wave functions for the system with H_1 present* (*i.e.*, obtained after diagonalizing our matrix) *mix a little β character into the α-Zeeman state and a little α into the β-Zeeman state.* The new wave function for the α-Zeeman state now is $\varphi = \sqrt{1 - a^2}|\alpha\rangle + a|\beta\rangle$, where $a \lll 1$, and a similar change occurs in the β state. When these new wave functions φ are substituted into equation (7–33), P is non-zero, making the transition allowed. This corresponds to the classical picture of H_1 exerting a torque to give a transverse component to the magnetization. Thus, we see that the probability, P, of a transition occurring depends upon the off-diagonal matrix elements in the α,β basis being non-zero, so that equation (7–33) is non-zero.

Next, consider what would happen if the r.f. perturbing field H_1 was placed along the z-axis. The off-diagonal matrix elements would again be zero, so equation (7–33) becomes

$$P = \langle\alpha|\hat{I}_z|\beta\rangle = 0$$

This would correspond to a slight reinforcement of H_0, but would not allow transitions. This exercise shows that H_1 cannot be collinear with H_0 to get an nmr transition. (In the classical model, there has to be a perpendicular component for H_1 to exert a torque.)

Finally, consider the case in which $I = 1$. Matrix elements $\langle m'|\hat{I}_x|m\rangle$ vanish unless $m' = m \pm 1$. Consequently, for $I = 1$, the allowed transitions are between adjacent levels with $\Delta m = \pm 1$, giving $\hbar\omega = \Delta E = \gamma\hbar H_0$.

In summary, simply placing a sample in a magnetic field, H_0, removes the degeneracy of the m_I states. Now, a radio frequency source is needed to provide $h\nu$ to induce the transition. Absorption of energy occurs provided that the magnetic vector of the oscillating electromagnetic field, H_1, has a component perpendicular to the steady field, H_0, of the magnet. Otherwise (*i.e.*, if H_1 is parallel to H_0), the oscillating field simply modulates the applied field, slightly changing the energy levels of the spin system, but no energy absorption occurs.

RELAXATION EFFECTS AND MECHANISMS

The influence of relaxation effects on nmr line shapes leads to some very important applications of nmr spectroscopy. Accordingly, it is worthwhile to summarize and extend our understanding of these phenomena. We begin with a discussion of relaxation processes and their effect on the shapes of our resonance line. The lifetime of a given spin state influences the spectral line width via the Uncertainty Principle, which is given in equation (7–34):

$$\Delta E\,\Delta t \approx \hbar \tag{7–34}$$

Since $\Delta E = h\Delta\nu$ and $\Delta t = T_2$, the lifetime of the excited state, the range of frequencies is given by $\Delta\nu \approx 1/T_2$. The quantity $1/T_2$ as employed here lumps together all of the factors influencing the line width (*i.e.*, all the relaxation processes) and is simply one-half the width of the spectral line at half-height. When the only contribution to T_2 is from spin-lattice effects, then $T_2 = T_1$. Most molecules contain magnetic nuclei; in the spin-lattice mechanism a local fluctuating field arising from the motion of magnetic nuclei in the lattice (where the lattice refers to other atoms in the molecule or other molecules including the solvent) couples the energy of the nuclear spin to other degrees of freedom in the sample, *e.g.*, translational or rotational energy. For liquids, T_1 values are usually between 10^{-2} and 10^2 seconds but approach values of 10^{-4} seconds if certain paramagnetic ions are present. The extent of spin-lattice relaxation depends upon (1) the magnitude of the local field and (2) the rate of fluctuation. Paramagnetic ions have much more intense magnetic fields associated with them and are very efficient at causing relaxation. The water proton signal in a 0.1 M solution of $Mn(H_2O)_6^{2+}$ is so extensively broadened by efficient relaxation from Mn^{2+} that a proton signal cannot be detected in the nmr spectrum. As mentioned before, the spin-lattice process can be described by a first order rate constant $(1/T_1)$ for the decay of the z-component of the magnetization, say, after the field is turned off, and is referred to as *longitudinal relaxation*.

Next, we shall discuss some processes that affect the xy-components of the magnetization and are referred to as *transverse relaxation processes*. The spin-lattice effect discussed above always contributes to randomization of the xy-component; therefore,

$$\frac{1}{T_2} = \frac{1}{T_1} + \frac{1}{T_2'}$$

where $1/T_2'$ includes all effects other than the spin-lattice mechanism. When the dominant relaxation mechanism is spin-lattice for both longitudinal and transverse processes, *i.e.*, $1/T_1 = 1/T_2$, then $1/T_2'$ is ignored. The quantity we used in the Bloch equations is $1/T_2$ and not $1/T_2'$. In solution, these other effects, $1/T_2'$, are small for a proton compared to $1/T_1$, so $(1/T_2) = (1/T_1)$. The other effects include field inhomogeneity, spin-spin exchange, and the interaction between nuclear moments. We shall discuss these in detail, beginning with field inhomogeneity. When the field is not homogeneous, protons of the same type in different parts of the sample experience different fields and give rise to a distribution of frequencies, as shown in Fig. 7–14. This causes a broad band. The effect of inhomogeneity can be minimized by spinning the sample.

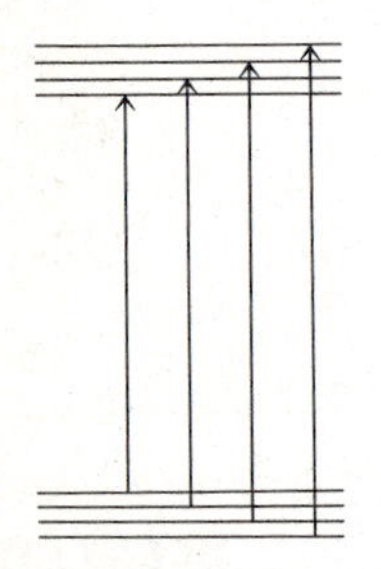

FIGURE 7–14. The same nmr transition occurring in a bulk sample in an inhomogeneous field.

Interaction between nuclear moments is also included in T_2'. When a neighboring magnetic nucleus stays in a given relative position for a long time, as in solids or viscous liquids, the local field felt by the proton has a *zero frequency* contribution; *i.e.*, it is not a fluctuating field as in the T_1 process from the field of the neighboring magnetic dipoles. A given type of proton could have neighbors with, for instance, a $+ + - + -$ combination of nuclear moments in one molecule, $+ - + - -$ in another, and so forth. Variability in the static field experienced by different protons of the same type causes broadening just like field inhomogeneity did. To give you some appreciation for the magnitude of this effect, a proton, for example, creates a field of about 10 gauss when it is 1 Å away. This could cause a broadening of 10^5 Hz. As a result of this effect, extremely broad lines are observed when the nmr spectra of solids are taken, but this effect is averaged to zero in non-viscous solutions.

The process of spin-spin exchange contributes to T_2, but not to T_1, for it does not influence the z-component of the magnetization. In this process, a nucleus in an excited state transfers its energy to a nucleus in the ground state. The excited nucleus returns to the ground state in the process and simultaneously converts the ground state nucleus to the excited state. No net change in the z-component results, but the u and v-components are randomized.

In common practice, $1/T_2'$ is never really used. Either $1/T_2 = 1/T_1$ (with $1/T_2'$ negligible) or we just discuss $1/T_2$. In a typical nmr spectrum, $1/T_2$ is obtained from the line shape (as will be shown) and it either equals $1/T_1$ or it does not.

As mentioned earlier, the true form of a broadened line is described empirically by a shape function $g(\omega)$, which describes how the absorption of energy varies near resonance according to equation (7–33). Since magnetic resonance lines in solution have a Lorentzian line shape:

$$g(\omega) = \frac{T_2}{\pi} \frac{1}{1 + T_2^2(\omega - \omega_0)^2} \tag{7-35}$$

The width of the band between the points where absorption is half its maximum height is $2/T_2$ in units of radians sec^{-1}. In units of Hz, the full band width at half height is given by $1/\pi T_2$.

7-10 MEASURING THE CHEMICAL SHIFT

The discussion of equation (7-28) indicated how the shielding constant results in the spectrum shown in Fig. 7-12 and contributes to making nmr of interest to a chemist. We now must consider ways of making the abscissa quantitative. In the typical nmr spectrum, the magnetic field is varied until all of the protons in the sample have undergone resonance. This is illustrated by the low resolution spectrum of ethanol shown in Fig. 7-15. The least shielded proton (smallest σ) on the electro-

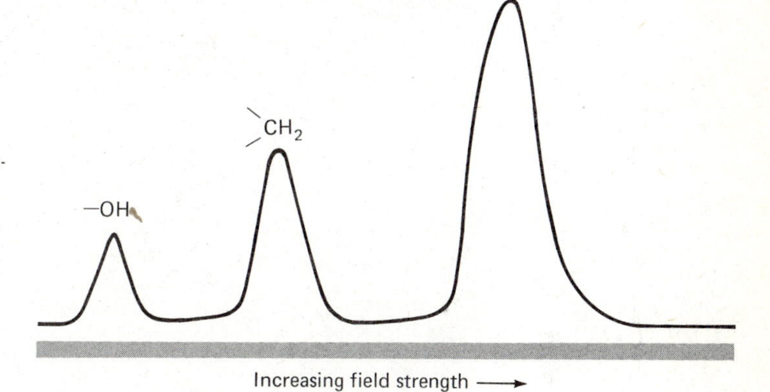

FIGURE 7-15. Low resolution proton nmr spectrum of C_2H_5OH.

negative oxygen atom interacts with the field at lowest applied field strength. The areas under the peaks are in direct proportion to the numbers of equivalent hydrogens, 1:2:3, on the hydroxyl, methylene, and methyl groups. Note that the separation of absorption peaks from $—CH_3$ and $—CH_2—$ hydrogens in this spectrum is much greater than that in the infrared spectrum.

We wish to calibrate the horizontal axis of Fig. 7-15 so that the field strength (or some function of it) at which the protons absorb energy from the radio frequency probe can be recorded. Equation (7-28) could be employed, but accurate measurement of H_N and H_0 is difficult. Instead, a reference material is employed, and the difference between the field strength at which the sample nucleus absorbs and that at which the nucleus in the reference compound absorbs is measured.

The reference compound is added to the sample (*vide infra*), so it must be unreactive. Furthermore, it is convenient if its resonance is in a region that does not overlap other resonances in the molecules typically studied. Tetramethylsilane, TMS, has both properties and is a very common reference material for non-aqueous solvents. In view of the limited solubility of TMS in water, the salt $(CH_3)_3SiCD_2CD_2CO_2^-$ is commonly used in this solvent. The position of its resonance is set at zero on the chart paper. The field strength is swept linearly, and the sweep is geared to a recorder producing a spectrum that at low resolution would resemble that in Fig. 7-15.

The magnetic field differences indicated by the peaks in Fig. 7–15 are very small, and it is difficult to construct a magnet that does not drift on this scale. Accordingly, most instruments pick a resonance and electronically adjust the field circuit to maintain or lock this peak at a constant position. This can be accomplished by having some material in a sealed capillary in the sample tube or in the instrument to lock on, or else by picking a resonance in the spectrum of the solution being studied for this purpose. The former procedure is described as an *external lock* and the latter as an *internal lock*. The internal lock produces more accurate results, for the field is being locked on a resonance that has all of the advantages of an internal standard (*vide infra*). In a typical experiment employing an internal lock, TMS is added to the sample for this purpose.

With TMS at zero, it is possible to measure the differences in peak maxima, Δ. Although the field is being varied in this experiment, the abscissa and hence the differences in peak positions are calibrated in a frequency unit of cycles per second, referred to as Hertz. This should cause no confusion, for frequency and field strength are related by equation (7–5):

$$\omega = \gamma H$$

According to equation (7–28), the shielding of the various nuclei, σH_0, depends upon the field strength, H_0. If a fixed frequency probe of 60 MHz is employed, the field utilized will be different than if a 100 MHz probe is employed. As shown in Fig. 7–16, the peak separations are 10/6 as large at 100 MHz as at 60 MHz. To overcome

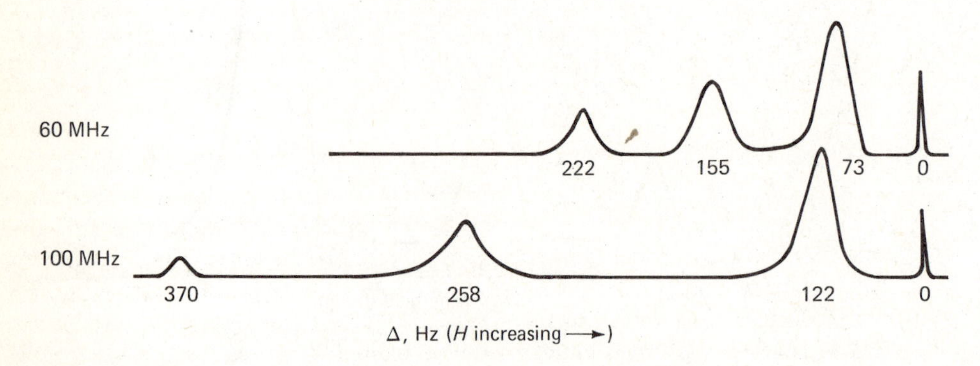

FIGURE 7–16. The nmr spectrum of CH_3CH_2OH at 60 MHz and at 100 MHz.

this problem and to obtain values for the peak positions, which are independent of field strength, the chemical shift, δ, is defined as

$$\delta = \frac{\Delta \times 10^6}{\text{fixed frequency of the probe, } \nu_0} \quad (7\text{–}36)$$

Since the probe frequency of ν_0 is in units of MHz, and Δ has units of Hertz, the fraction is multiplied by the factor 10^6 to give convenient numbers for δ in units of parts per million, ppm. The chemical shift, δ, is independent of the probe frequency employed.

If the sample resonance peak occurs at a lower field strength than the reference peak, Δ is positive* by convention. When $Si(CH_3)_4$ is employed as a standard, almost

---

* In the early literature and even as late as 1970 in the area of ^{19}F and ^{31}P nmr, the convention used is the opposite of that described here. The reader must exercise considerable care in ascertaining the convention used.

all δ values for organic compounds are positive and the larger positive numbers refer to lesser shielding. The τ scale has been commonly used in the past in organic chemistry for chemical shifts relative to TMS measured in CCl_4 as solvent. It is defined by the equation

$$\tau = 10 - \delta \tag{7-37}$$

In recent work, its use has been de-emphasized.

For very accurate chemical shift determination, the side band technique can be employed to insure that the field is being swept linearly. In this experiment, the field axis is calibrated by displacing part of this spectrum electronically, by imposing on the r.f. field a fixed audio frequency field. Part of the intensity of all the lines in the spectrum will be displaced a certain number of cycles per second, equal to the audio frequency. The spectrum indicated by the dashed line in Fig. 7-17 will result. If the

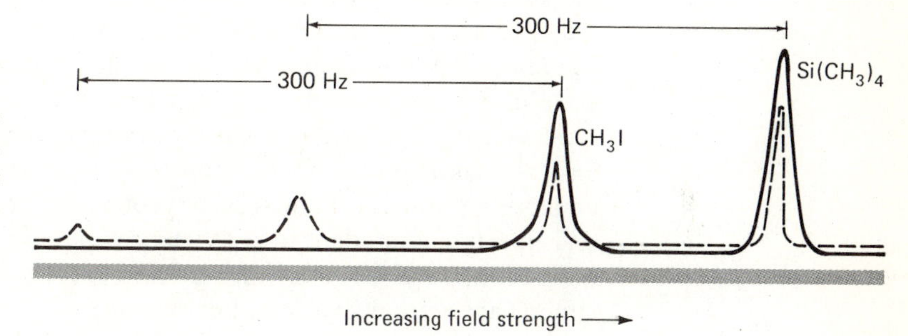

FIGURE 7-17. Proton nmr spectrum of CH_3I with $Si(CH_3)_4$ as an internal standard (solid line). The dashed line is the same spectrum with a 300 Hz side band.

audio frequency were 300 Hz, the distance on the chart paper between the original and the displaced peaks would be 300 Hz. The displaced peaks are referred to as side bands. In Fig. 7-17, for example, the distance between the $Si(CH_3)_4$ and CH_3I peaks is divided by the distance between one of the main peaks and its side band. Multiplication of this ratio by 300 Hz (or by whatever imposed frequency is selected) gives the value, Δ, in Hertz for the difference between the sample signal (CH_3I) and the reference signal ($Si(CH_3)_4$). The ordinate is calibrated very precisely by this procedure. By working with several peaks, the linearity of the sweep can be verified. The spectrum in Fig. 7-17 has been drawn to scale (a 60 MHz probe was employed), so that with a ruler and the information given, the reader should be able to calculate the value: $\delta = 2.2$. Appendix F contains data to permit conversion of δ relative to various standards to δ relative to $Si(CH_3)_4$ as the reference.

For several reasons, it is advantageous to add the TMS directly to the solution being studied as opposed to having it in a separate sealed capillary tube, *i.e.*, an *internal* vs. an *external standard.* Magnetic field-induced circulations of the paired electron density in a molecule give rise to a magnetic moment that is opposed to the applied field. In diamagnetic substances, this effect accounts for the repulsion, or diamagnetism, experienced by these materials when placed in a magnetic field. The magnitude of this effect varies in different substances, giving rise to varying diamagnetic susceptibilities. This diamagnetic susceptibility in turn gives rise to magnetic shielding of a molecule in a solvent, and is variable for different solvents. The effect is referred to as the volume diamagnetic susceptibility of the solvent. Thus, the chemical shift of a solute molecule in a solvent will be influenced not only by shielding of electrons but also by the volume diamagnetic susceptibility of the solvent. If the solute were liquid, the δ value obtained for a *solution* of the liquid relative to another liquid external standard (Fig. 7-18) would be different from that obtained for the

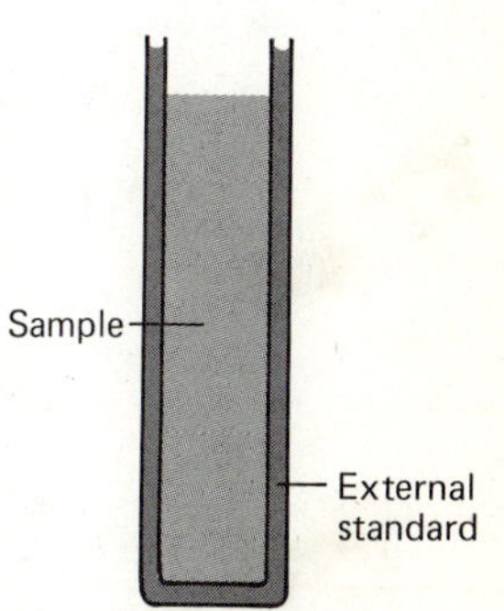

FIGURE 7-18. Concentric nmr sample tube containing an external standard. For illustrative purposes the external standard compartment size is exaggerated. This is really just large enough for 2 to 3 drops of standard.

pure liquid solute (this is often referred to as the neat liquid) to the extent that the volume diamagnetic susceptibilities of the solvent and neat liquid were different.

Because of the variation in the contribution from the volume diamagnetic susceptibility, the chemical shifts of neat liquids relative to an external standard are difficult to interpret. Problems are also encountered when substances are examined in solution. The diamagnetic contributions to the shielding of the solute depend upon the average number of solute and solvent molecules, *i.e.*, the number of solvent and solute neighbors. Consequently, the chemical shift will be concentration dependent. To get a meaningful value for δ, it is therefore necessary to eliminate or keep constant the contribution to δ from the diamagnetic susceptibility of the solvent. This can be accomplished by measuring δ at different concentrations and extrapolating to infinite dilution, in effect producing a value for δ under volume susceptibility conditions of the pure solvent. If all values for different solutes are compared in this solvent, the effect is constant.

Contributions to the measured δ from the volume diamagnetic susceptibility of the solvent can be more easily minimized by using an *internal standard.* The internal standard must, of course, be unreactive with the solvent and sample. Under these conditions the standard is subjected to the same volume susceptibility (from solvent molecules) as is the solute, and the effects will tend to cancel when the difference, Δ, is calculated. (Due to variation in the arrangement of solvent around different solutes, an exact cancellation is often not obtained.) Cyclohexane and $Si(CH_3)_4$ are commonly employed as internal standards for protons. In order for the results from an internal and an external standard to be strictly comparable, δ for the standard as a pure liquid must be identical with δ for the standard in the solvent. It is now common practice to employ an internal standard. For accurate work, spectra at two different concentrations should be run as checks and the results relative to two internal standards compared to insure that the internal standard approximations are working.

The solvent employed for an nmr experiment should be chemically inert, and should have a symmetrical electron distribution. Carbon tetrachloride is ideal for proton resonance. Chemical shift values close to those obtained in carbon tetrachloride are obtained in chloroform, deuterochloroform, and carbon disulfide. Acetonitrile, dimethylformamide, and acetone are frequently used but can be employed satisfactorily only if there are no specific interactions (H bonding or other Lewis acid-base interactions) and if an internal standard is employed. It is also possible to measure δ in such a solvent for compounds whose chemical shifts in CCl_4 are known, to determine a correction factor for the volume susceptibility of the solvent. This is then applied to the chemical shifts of other solutes in that solvent.

As seen in the above discussion, the solvent in which the chemical shift is examined has a pronounced effect on the value obtained. It is found that even the relative values of δ for the different protons in a given molecule may vary in different solvents. A quantity σ_{solv} has been proposed[3] to encompass all types of shielding from solvent effects. This quantity consists of the following:

$$\sigma_{solv} = \sigma_B + \sigma_A + \sigma_W + \sigma_E$$

where σ_B is the shielding contribution from the bulk magnetic susceptibility of the solvent, σ_A arises from anisotropy of the solvent, σ_W arises from van der Waals interactions between the solvent and solute, and σ_E is the shielding contribution from a polar effect caused by charge distribution induced in neighboring solvent molecules by polar solutes (*i.e.*, an induced solvent dipole–solute interaction). As mentioned above, σ_B is eliminated by the use of internal standards. Some indication of the importance of the other effects for a given solute can be obtained by determining δ in different solvents.

7-11 SIMPLE APPLICATIONS OF THE CHEMICAL SHIFT

The main concern in this section will be with demonstrating the wide range of systems that can be studied by nmr. Only the simplest kinds of applications will be discussed. The nuclei that have been most frequently studied are ^{1}H, ^{19}F, ^{13}C, and ^{31}P. There are also extensive reports of ^{11}B, ^{17}O, ^{15}N, and ^{59}Co nmr spectra.

The range of proton chemical shifts measured on pure liquids[(4)] for a series of organic compounds is illustrated in Fig. 7-19. Proton shifts outside this total range have been reported, and sometimes shifts outside the range indicated for a given functional group occur. In general, the data in Fig. 7-19 serve to give a fairly reliable means of distinguishing protons on quite similar functional groups. Note the difference in CH_3—C, CH_3C=, CH_3—O, HC=, HCO, etc. Very extensive compilations of proton chemical shifts have been reported,[(6)] which can be employed for the fingerprint type of application. Care must be exercised with —OH groups, for the shift is very concentration- and temperature-dependent. For example, when the spectrum of ethanol is examined as a function of concentration in an inert solvent (*e.g.*, CCl_4), the total change in chemical shift in going from concentrated to dilute solution amounts to about 5 ppm. When the data are extrapolated to infinite dilution, the hydroxyl proton appears at a higher field than the methyl protons, in contrast to the spectrum of pure ethanol represented in Fig. 7-16. These changes are due to differing degrees of hydrogen bonding. There is more hydrogen bonding in concentrated than in dilute solutions. The effect of this interaction is to reduce the screening of the proton, causing a shift to lower field. This behavior on dilution can be employed to verify the assignment of a peak to a hydroxyl group. This dilution technique has been used to investigate the existence of steric effects in hydrogen bonding[(5)] and should aid in distinguishing between intermolecular and intramolecular hydrogen bonding. Solvent effects are quite large whenever hydrogen bonding or other specific interactions occur and, as will be shown later, nmr has been very valuable in establishing the existence or absence of these interactions.

For a limited number of compounds, a correlation has been reported between the electronegativity of X and the proton chemical shift of the CH_3 group in CH_3X compounds,[(7)] and also between the electronegativity of X and δCH_3—δCH_2 for a series of C_2H_5X compounds.[(8)] The more electronegative X is, the less shielded are the protons. The severe limitations of such a correlation are discussed in the section on chemical shift interpretation.

Protons attached to metal ions are in general very highly shielded, the resonance often occurring 5 to 15 ppm to the high field side of TMS and, in some cases, occurring over 60 ppm upfield from TMS. The shifts are attributed to paramagnetic shielding (*vide infra,* Chapter 8) of the proton by the filled *d* orbital electrons of the metal.[(9)] The proton nmr in $HRh(CN)_5^{3-}$ is doublet, the center of which occurs at 10.6 ppm on the high field side of TMS. This splitting (*vide infra*) and shift establish the existence of a bond between rhodium and hydrogen.

The range of fluorine chemical shifts is an order of magnitude larger than that normally encountered for protons. The difference between the fluorine resonances in F_2 and in F^-(aq) is 542 ppm, compared to the range of about 12 ppm for proton shifts (see Fig. 7-19). A brief compilation of fluorine shifts is contained in Table 7-3 to illustrate this point. A wide range (~500 ppm) of phosphorus chemical shifts has also been reported. Fluck[(10)], Advances in Magnetic Resonance, Topics in Phosphorus Chemistry (Volume 5), and several of the recent ^{13}C texts referenced in Chapter 8 can be consulted for compilations. The fingerprint application is immediately obvious for compounds containing these elements. Before additional applications are discussed, it is necessary to consider in more detail some other effects influencing the nmr spectrum.

SPIN-SPIN SPLITTING

7-11 EFFECT OF SPIN-SPIN SPLITTING ON THE SPECTRUM

When an nmr spectrum is examined under high resolution, considerable fine structure is often observed. The difference in the high and low resolution spectra of ethanol can be seen by comparing Fig. 7-20 with Fig. 7-15.

The chemical shift of the CH_2 group relative to the CH_3 group in Fig. 7-20 is indicated by Δ measured from the band centers. The fine structure in the $—CH_3$ and $—CH_2—$ peaks arises from the phenomenon known as *spin-spin splitting,* and the separation, J, between the peaks comprising the fine structure is referred to as the

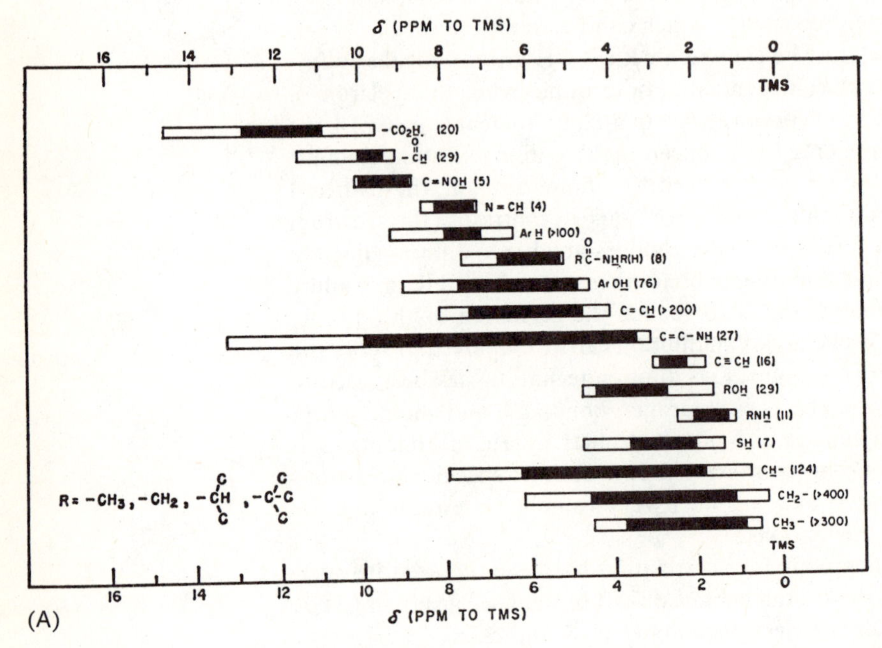

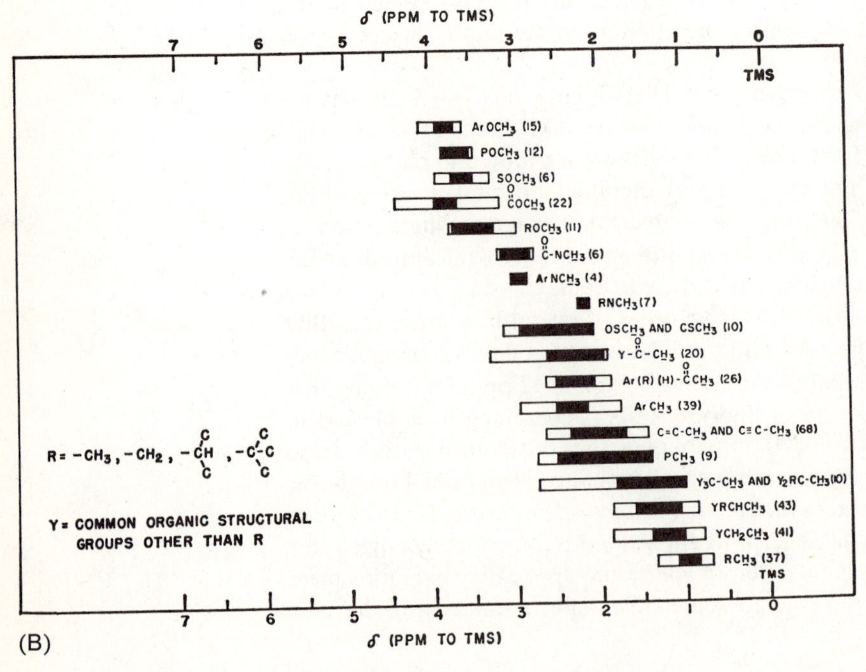

FIGURE 7-19. Hydrogen nuclear magnetic resonance chemical shift correlation charts, (A) for main proton groups, (B) for CH, OH, and NH subgroups, and (C) for CH_3 subgroups. Open bar denotes extreme 10% of data. [From M. W. Dietrich and R. E. Keller, Anal. Chem., *36,* 258 (1964).] *Illustration continued on opposite page.*

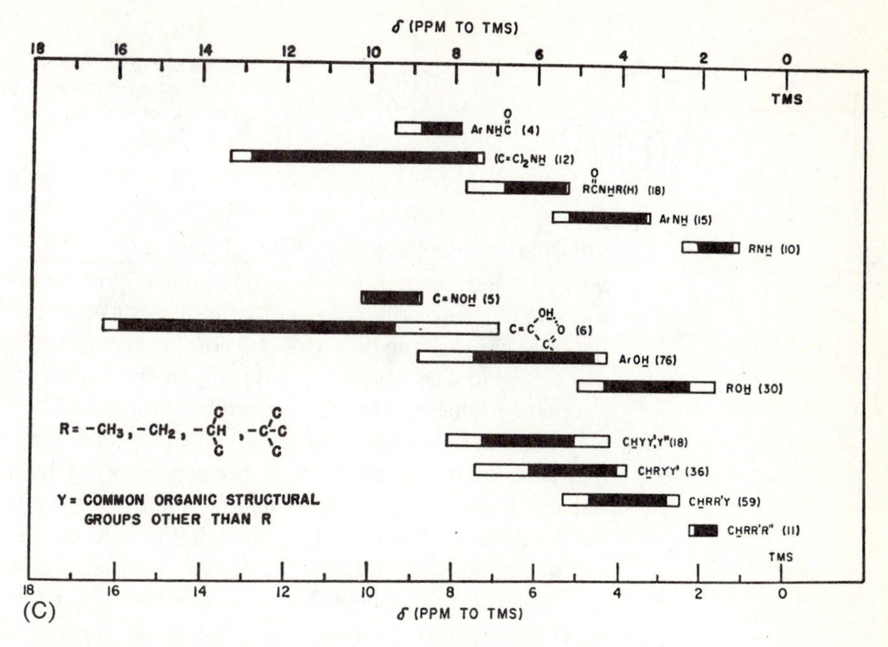

FIGURE 7–19. *Continued.*

TABLE 7–3. FLUORINE CHEMICAL SHIFTS OF SELECTED COMPOUNDS (IN ppm RELATIVE TO $CFCl_3$—LARGER NEGATIVE NUMBERS INDICATE HIGHER FIELD)

CH_3F	−278	CF_4	−70	SOF_2	+70
HF	−203	TeF_6	−64	TiF_6^{2-}	+75
$(CF_3)_3CF$	−188	$(CF_3)_4C$	−61.5	XeF_6	+78
XeF_4	−182.3	AsF_3	−48	ClF_3	+80
SiF_4	−177	BrF_3	−38	$XeOF_4$	+89
BeF_2	−155	CF_3Cl	−32	BrF_5	+132, +269
BF_3OEt_2	−153	CF_2Cl_2	−9	NF_3	+140
BF_4^-	−149	CF_3I	−4	CF_3OF	+142
BF_3	−133	$CFCl_3$	0	SF_5OF	+178
SiF_6^{2-}	−129	IF_5	+4	$FOClO_3$	+225
$(CF_3CF_2)_2$	−127	SO_2F_2	+36	IF_7	+238, +274
F^-(aq)	−120	SF_6	+42	OF_2	+250
SbF_3	−117	SF_5OF	+48	$FClO_2$	+288
CF_3H	−88	SeF_6	+50	FNO_2	+394
SbF_3	−86	IF_5	+53	F_2	+422
$(CF_3)_2CO$	−82	N_2F_4	+60	FNO	+478
CF_3COOH	−77	NSF_3	+66	UF_6	+746

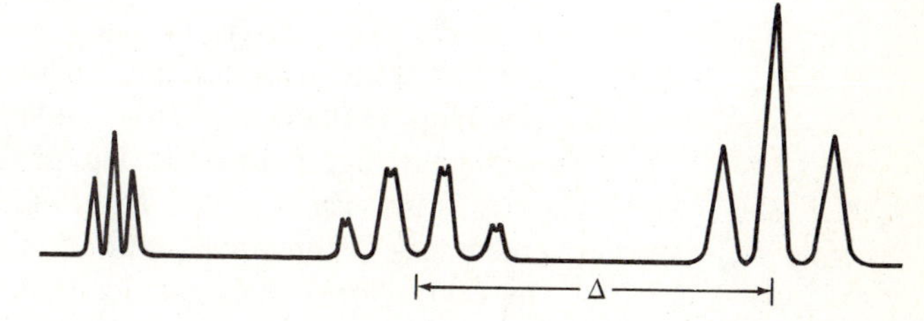

FIGURE 7–20. High resolution nmr spectrum of ethanol (facsimile). Compare with Fig. 7–15, p. 205.

FIGURE 7-21. The interpretation in terms of the appropriate coupling constants of the spectrum in Fig. 7-20. The meaning of this stick-type spectrum will be made clear shortly in the text.

spin-spin coupling constant. This parameter is usually expressed in Hertz. As mentioned earlier, the magnitude of Δ depends upon the applied field strength. However, the magnitude of the spin-spin coupling constant in Hz is field independent.

The cause of the fine structure and the reason for the field-independent character of J can be understood by considering an H—D molecule. If by some mechanism the magnetic moment of the proton can be transmitted to the deuteron, the field strength at which the deuteron precesses at the probe frequency will depend upon the magnetic quantum number of the neighboring hydrogen nucleus. If the proton nucleus has a spin of $+\frac{1}{2}$, its magnetic moment is aligned with the field so that the field experienced by the deuterium is the sum of the proton and applied fields. A lower applied field strength, H_0, will be required to attain the precession frequency of the deuterium nucleus in this molecule than in the one in which the hydrogen has a magnetic quantum number of $-\frac{1}{2}$. In the latter case the field from the proton opposes the applied field and must be overcome by the applied field to attain a precessional frequency equal to the probe frequency. The simulated spectrum that would result for the deuterium resonance is indicated in Fig. 7-22(A). The m_I values

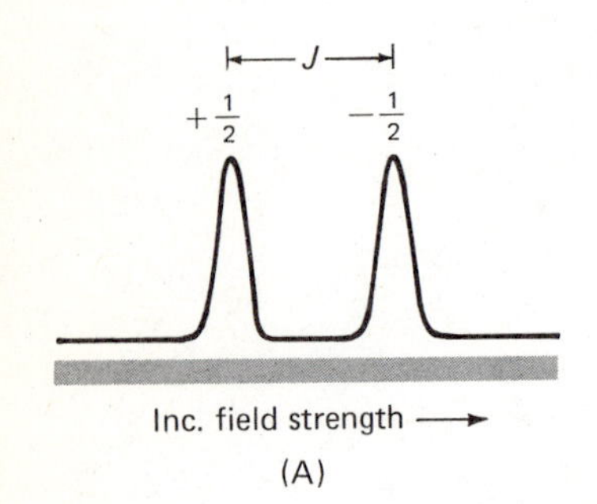

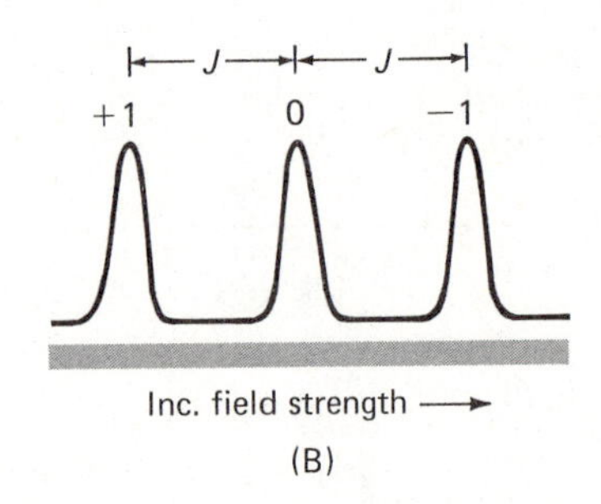

FIGURE 7-22. NMR spectra for the hypothetical HD experiment. (A) Deuterium resonance for HD; (B) proton resonance for HD.

for the hydrogen nuclei in the different molecules that give rise to different peaks are indicated above the respective peaks. The two peaks are of equal intensity because there is practically equal probability that the hydrogen will have $+\frac{1}{2}$ or $-\frac{1}{2}$ magnetic quantum numbers. The proton resonance spectrum is indicated in Fig. 7-22(B). These proton resonance peaks correspond to magnetic quantum numbers of $+1$, 0, and -1 for the attached deuterium nuclei in different molecules. The spin-spin coupling constants J_{DH} and J_{HD} in Fig. 7-22(A) and 7-22(B), respectively, have the same value. Subsequently, we shall discuss mechanisms for transmitting the magnetic moment of a neighboring atom to the nucleus undergoing resonance.

Returning to ethyl alcohol, we shall examine the splitting of the methyl protons by the methylene protons. The two equivalent protons on the $—CH_2—$ group can have the various possible combinations of nuclear orientations indicated by the arrows in Fig. 7-23(A). In case 1, both nuclei have m_I values of $+\frac{1}{2}$, giving a sum of $+1$ and accounting for the low field peak of the $—CH_3$ resonance. Case 2 is the combination of $—CH_2—$ nuclear spins that gives rise to the middle peak, and case 3 causes the high field peak. The probability that the spins of both nuclei will cancel (case 2) is twice as great as that of either of the combinations represented by case 1 and case 3. (There are equal numbers of $+\frac{1}{2}$ and $-\frac{1}{2}$ spins.) As a result, the area of the central peak will be twice that of the others (see Fig. 7-20).

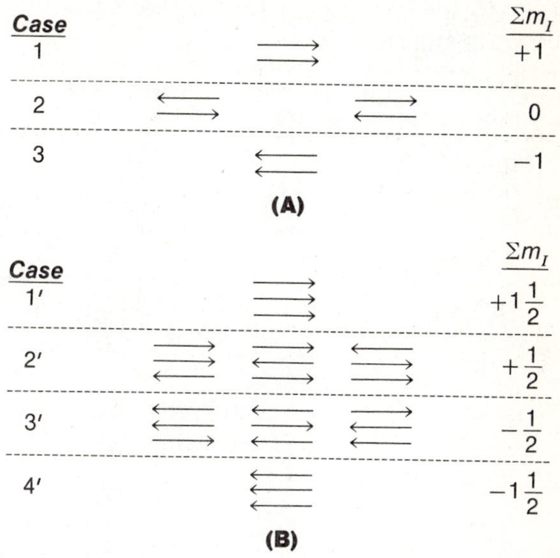

Case			Σm_I
1		→ →	+1
2	← →		→ ←	0
3		← ←	−1

(A)

Case				Σm_I
1′		→ → →		$+1\frac{1}{2}$
2′	→ → ←	→ ← →	← → →	$+\frac{1}{2}$
3′	← ← →	← → ←	→ ← ←	$-\frac{1}{2}$
4′		← ← ←		$-1\frac{1}{2}$

(B)

FIGURE 7–23. Possible orientations of proton nuclear magnetic moments for (A) —CH_2— and (B) —CH_3 groups.

The nuclear configurations of the —CH_3 group that cause splitting of the —CH_2— group are indicated in Fig. 7–23(B). The four different total net spins give rise to the four peaks with the largest separation in this multiplet, shown in Fig. 7–20. The relative areas are in the ratio 1:3:3:1. The separation between these peaks in units of Hz is referred to as $J_{H—C—C—H}$ or as ${}^3J_{H—H}$. The latter symbol indicates H—H coupling between three (the superscript) bonds. The peak separation in the methylene resonance from this coupling is equal to the peak separation in the methyl group. The spectrum of the CH_2 resonance is further complicated by the fact that each of the peaks in the quartet from the methyl splitting is further split into a doublet by the hydroxyl proton. In the actual spectrum, some of the eight peaks expected from this splitting overlap, so they are not all clearly seen.

The OH peak is split into a triplet by the CH_2 protons with the same separation as $J_{H—C—OH}$ in the methylene resonance. Usually, though not always, the effects of spin-spin coupling are not seen over more than three bonds. Accordingly, the interaction of the —OH proton with the methyl group is not seen. The spectral interpretation is illustrated in Fig. 7–21 by showing which lines arise from which couplings. This "stick-type spectrum" is constructed by drawing a line for each chemically shifted different nucleus. On the next row, the effect of the largest J is shown. Additional lines are added for each J until the final spectrum is obtained.

Because of the selection rules for this process (*vide infra*), equivalent nuclei do not split each other; *e.g.*, one of the protons in the —CH_3 group cannot be split by the other two protons. A general rule for splitting can be formulated that eliminates the necessity for going through a procedure such as that in Fig. 7–23. For the general case of the peak from nucleus A being split by a non-equivalent nucleus B, the number of peaks, n, in the spectrum due to A is given by the formula

$$n_A = 2\Sigma S_B + 1 \qquad (7\text{-}38)$$

where ΣS_B equals the sum of the spins of equivalent B nuclei. The application of this formula is illustrated by the examples in Table 7–4. The relative intensities of the peaks can be obtained from the coefficients of the terms that result from the binomial expansion of $(r + 1)^m$, where $m = n - 1$ and r is an undefined variable; *e.g.*, when there are four peaks, $n = 4$ and $m = 3$, leading to $r^3 + 3r^2 + 3r + 1$. The coefficients

TABLE 7-4. SPIN-SPIN SPLITTING IN VARIOUS MOLECULES[a]

Molecule	Groups Being Split (A)	Groups Doing the Splitting (B)	ΣS_B	n
CH_3CH_2OH	CH_3	CH_2	1	3
CH_3CH_2OH	CH_2	CH_3	$3/2$	4
PF_3	P	F	$3/2$	4
PF_3	F	P	$1/2$	2
$(CH_3)_4N^+$	CH_3	N	1	3

[a] Spin quantum numbers for the most abundant isotopes of nuclei contained in the above compounds are H = ½, P = ½, F = ½, N = 1.

1:3:3:1 produce the relative intensities. The Pascal triangle is a convenient device for remembering the coefficients of the binomial expansion for nuclei with $I = \frac{1}{2}$.

```
            1
          1   1
        1   2   1
      1   3   3   1
    1   4   6   4   1
  1   5  10  10   5   1
1   6  15  20  15   6   1
```

The triangle is readily constructed, for the sum of any two elements in a row equals the element between them in the row below.

When two groups of non-equivalent nuclei B and C split a third nucleus A, the number of peaks in the A resonance is given by

$$n_A = (2\Sigma S_B + 1)(2\Sigma S_C + 1) \tag{7-39}$$

This is equivalent to what was done in the discussion of the methylene resonance of ethanol, when each of the four peaks from spin coupling by $—CH_3$ was further split into a doublet from coupling to —OH, leading to a total of eight peaks. When a nucleus with $I \geqslant 1$ is coupled to the observed nucleus, the number of lines is still given by equation (7-38); however, the intensities are no longer given by the binomial expansion. For instance, in Fig. 7-22(B), the hydrogen signal is split into three lines by the deuterium, for which $I = 1$. The intensities of the three lines are all equal and not in the ratio 1:2:1, since there is near equal probability that the deuterium will have +1, 0, and −1 magnetic quantum numbers. The same situation occurs with splitting from ^{14}N.

For both chemical shift and spin-spin coupling applications, one must recognize when nuclei in a molecule are non-equivalent. This will be the subject of the next section.

7-13 DISCOVERING NON-EQUIVALENT PROTONS

In some molecules, a certain proton appears to be non-equivalent to others in a particular rotamer, but becomes equivalent when rapid rotation occurs; *e.g.*, consider the non-equivalence of the three methyl protons in a staggered configuration of CH_3CHCl_2. Gutowsky[11] has treated an interesting problem in which the non-equivalence is not removed by rapid rotation. This is observed in the chemical shift of substituted ethanes of the type:

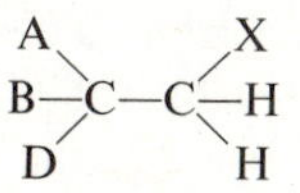

```
A         X
 \       /
B—C—C—H
 /       \
D         H
```

The two protons are not equivalent from symmetry arguments, and *no conformation can be found in which the two protons can be interchanged by a symmetry operation.* This intrinsic asymmetry will result in different chemical shifts which, if large enough, will be observable. This asymmetry can be seen if you label the two hydrogens and make a table indicating which atoms are staggered on each side of the two hydrogens for all possible staggered rotamers of (ABD)—C—CH_2X.

Early arguments were based on the different shifts resulting from different rotamer populations, but it should be pointed out that even in the absence of population differences the protons are intrinsically non-equivalent in all rotamers.

The criterion to be applied in recognizing this phenomenon is the absence, in any conformer that can be drawn, of a symmetry operation that interchanges the two protons. In CH_3CH_2Cl, all the methyl protons are interconvertible in the various rotamers that can be drawn, as are the methylene protons. Another interesting example[12] of non-equivalent protons involves the methylene protons of $(CH_3CH_2O)_2SO$. The two protons of a given methylene group are not stereochemically equivalent because of the lack of symmetry of the sulfur atom with respect to rotation about the S—O—C bonds. One of the possible rotamers is depicted in Fig. 7-24. The small dot in the center represents the sulfur with lines connecting the

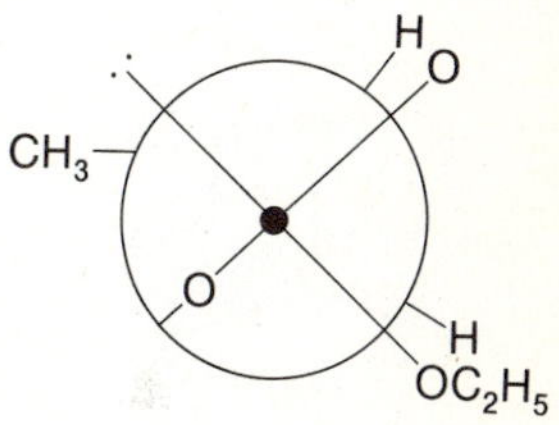

FIGURE 7-24. Non-equivalence of the two protons of a given methylene in one of the rotamers of $(CH_3CH_2O)_2SO$.

oxygen, the lone pair, and the ethoxyl groups. The large circle represents the methylene carbon with two hydrogens, an oxygen, and a methyl group attached. The molecule is so oriented that we are looking along a line from sulfur to carbon. The non-equivalence of the two methylene hydrogens of a given —CH_2— group is seen in this rotamer, and they cannot be interchanged by a symmetry operation in any other rotamer that can be drawn. Two non-equivalent nuclei, which cannot be interchanged by any symmetry element the molecule possesses, are called *diastereotopic* nuclei. Nuclei that are equivalent and have identical chemical shifts are called isochronous nuclei.

7-14 EFFECT OF THE NUMBER AND NATURE OF THE BONDS ON SPIN-SPIN COUPLING

The nature of the spectra of complex molecules will depend on the number of bonds through which spin-spin coupling can be transmitted. For proton-proton coupling in saturated molecules of the light elements, the magnitude of J falls off rapidly as the number of bonds between the two nuclei increases and, as mentioned earlier, usually is negligible for coupling of nuclei separated by more than three bonds. Long-range coupling (coupling over more than three bonds) is often observed

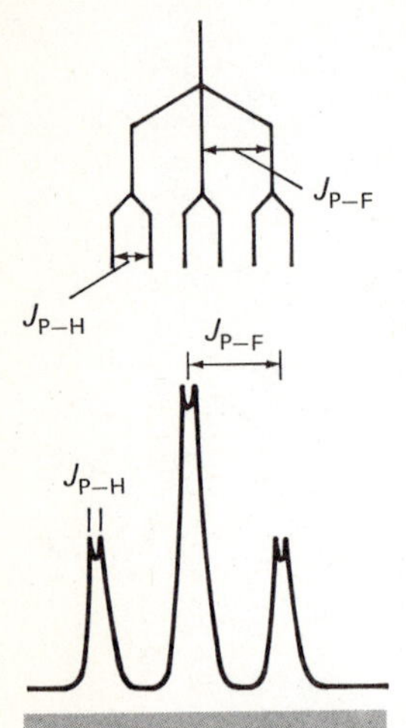

FIGURE 7-25. ^{31}P nmr spectrum expected for HPF_2 if $^1J_{P-F} > {}^1J_{P-H}$.

in unsaturated molecules. A number of examples are contained and discussed in articles by Hoffman and Gronowitz.[13] In unsaturated molecules, the effects of nuclear spin are transmitted from the C—H σ bond by coupling of the resulting electron spin on carbon with the π electrons. This π electron spin polarization is easily spread over the whole molecule because of the extensive delocalization of the π electrons. Spin polarization at an atom in the π-system can couple back into the σ-system, in the same way that it coupled into the π-system initially. In this way, the effect of a nuclear spin is transmitted several atoms away from the splitting nucleus. This problem has been treated quantitatively.[14]

When spin-spin coupling involves an atom other than hydrogen, long-range coupling can occur by through-space coupling. This interaction is again one of spin polarization, but it occurs through non-bonding pairs of electrons, and the coupling is through space instead of through σ or π bonds. In the compound $CF_3CF_2SF_5$ the coupling constant for the fluorines of the CF_2 group and the *trans* fluorine of the SF_5 group is about 5 Hz, much smaller than that for the CF_2 group coupling to the *cis* fluorines ($\sim$16 Hz), because of the reported contribution from through-space coupling in the latter case.[15a] This effect has been invoked rather loosely, and it is suggested that an abnormal temperature dependence be demonstrated before such an explanation is accepted.[15b]

In order to demonstrate one further point, it is interesting to consider the phosphorus nuclear magnetic resonance spectrum of the compound HPF_2. The "stick spectrum" in Fig. 7-25 represents the splitting of the phosphorus signal by the two fluorine atoms followed by the smaller $^1J_{P-H}$ splitting. The resulting spectrum is indicated.

If instead of $^1J_{P-F} > {}^1J_{P-H}$ we have $^1J_{P-H} > {}^1J_{P-F}$, the spectrum in Fig. 7-26 arises. The fact that $^1J_{P-H}$ is larger than $^1J_{P-F}$ is illustrated in the "stick spectrum."

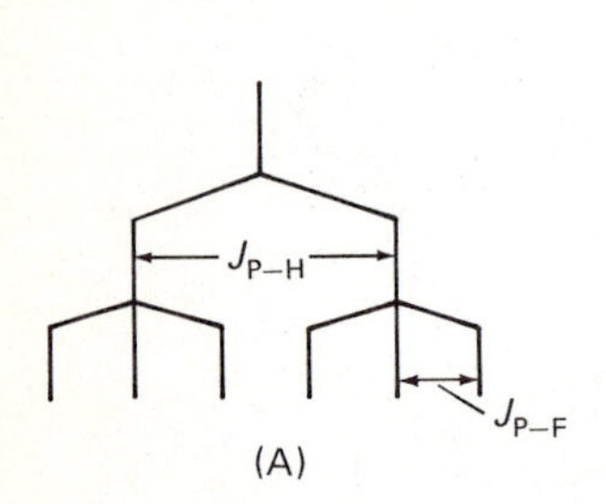

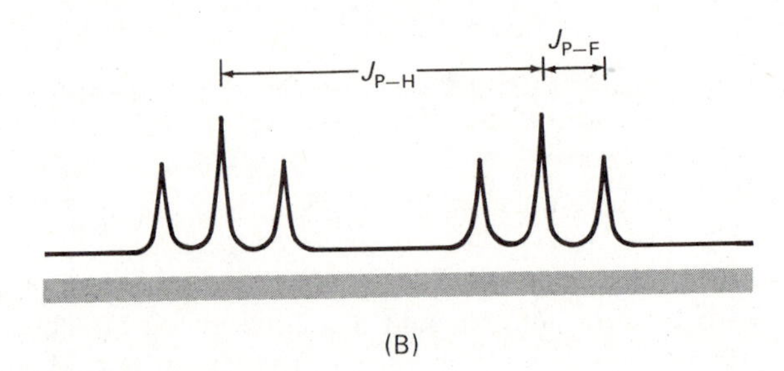

FIGURE 7-26. ^{31}P nmr spectrum expected for HPF_2 if $^1J_{P-H} > {}^1J_{P-F}$.

The spectrum is expected to be one of those shown in Figs. 7-25 or 7-26, depending upon the magnitude of $^1J_{P-F}$ *vs.* $^1J_{P-H}$. If the P—F coupling constant is greater than the P—H coupling constant, Fig. 7-25 will result; while Fig. 7-26 results if $^1J_{P-H}$ is much greater than $^1J_{P-F}$. If the two coupling constants were similar, the spectrum would be a complex pattern intermediate between those shown. In most compounds studied, $^1J_{P-F} > {}^1J_{P-H}$ ($\sim$1500 Hz *vs.* $\sim$200 Hz), and as expected, the spectrum in Fig. 7-25 is found experimentally.

If two nuclei splitting another group in a molecule are magnetically non-equivalent, the spectrum will be very much different from that in which there is splitting by two like nuclei. Four lines of equal intensity will result from splitting by two non-equivalent nuclei with $I = \frac{1}{2}$, and three lines with an intensity ratio 1:2:1 will be observed for splitting by two equivalent nuclei. Consequently, we must determine what constitutes magnetically non-equivalent nuclei. Non-equivalence may arise because of differences in the chemical shifts of the two splitting nuclei or because of differences in the J values of the two splitting nuclei with the nucleus being split. An example of the latter is found in the molecule $H_2C{=}CF_2$, where neither the two hydrogens nor the two fluorines can be considered equivalent. Each proton sees two non-equivalent fluorines (one *cis* and one *trans*) which have identical chemical shifts but different J_{H-F} coupling constants. The simple 1:2:1 triplet expected when both protons and both fluorines are equivalent is not observed in either the fluorine or the proton nmr because of the non-equivalence that exists in the J values. We shall consider this effect in more detail in the section on second order effects in the next chapter.

Another effect that gives rise to spectra other than those predicted by equation (7–38) is nuclear quadrupole relaxation. Often splittings do not occur because the quadrupolar nucleus to which the element being investigated is attached undergoes rapid relaxation, which causes a rapid change in the spin state of the quadrupolar nucleus. Only the average spin state is detected; in some instances this relaxing quadrupolar nucleus gives rise to very broad lines in the nmr spectrum of the $I = \frac{1}{2}$ nucleus bonded to it, and sometimes a proton resonance absorption is broadened by this effect to such an extent that the signal cannot be distinguished from the background. We shall discuss this in more detail shortly in Section 7–19.

7–15 QUALITATIVE DESCRIPTION OF A MECHANISM FOR SPIN-SPIN COUPLING

In the preceding section the splitting in the spectrum of the HD molecule by the magnetic moment of the attached nucleus was discussed. We shall be concerned here briefly, and in more detail later, with the mechanism whereby information regarding the spin of the nucleus causing the splitting is transmitted to the nucleus whose resonance is split. Consider a proton split by the nucleus X (which has a spin of $\frac{1}{2}$) in the hypothetical molecule HX. The nuclear spin of X is transmitted to H by polarization of the bonding electrons. Various processes cause polarization, and they constitute the mechanisms for spin-spin splitting of hydrogen in solution.

We shall discuss one of these processes here and consider it in more detail with other mechanisms in Chapter 8. If X has a nuclear magnetic quantum number $+\frac{1}{2}$, the electron with spin of $-\frac{1}{2}$ will have the highest probability of being found near X. Since all the electron spins are paired, the electron with spin $+\frac{1}{2}$ must have a slightly higher probability of being near the proton. Consequently, the spin of X is transmitted to H. The two different spin states of X give rise to two different energy transitions. The nuclear moment of X is transmitted through the bonding electrons and affects the proton regardless of the orientation of the molecule in the field. This process is referred to as the *Fermi contact* mechanism. It can also be appreciated that the value of J will simply depend upon the energy difference of the two different kinds of molecules containing X in different spin states. This energy difference will be independent of the field strength. The field strength independence of J provides a criterion for determining whether two peaks in a spectrum are the result of two non-equivalent protons or spin-spin coupling. The peak separation in the spectrum will be different when 60 MHz and 100 MHz probes are employed if the two peaks are due to non-equivalent protons, but the separation will be the same at the two different frequencies if the two peaks arise from spin-spin splitting.

7–16 APPLICATIONS OF SPIN-SPIN COUPLING TO STRUCTURE DETERMINATION

There have been many applications of spin-spin coupling to the determination of structures. For example, if the spectrum of a sample contains the very characteristic fine structure of the $—CH_2—$ and $—CH_3$ resonances of an ethyl group, this is a good indication of the presence of this group in the molecule. Other applications involve variations in the magnitude of J in different types of compounds. For two non-equivalent protons on the same carbon, *e.g.*, $ClBrC{=}CH_2$, proton-proton spin coupling constants, $J_{H—H}$, of 1 to 3 Hz are observed. Coupling constants for non-equivalent *trans* ethylenic protons have values in the range of 17 to 18 Hz, while *cis* protons give rise to coupling constants of 8 to 11 Hz. These differences aid in determining the structures of isomers.

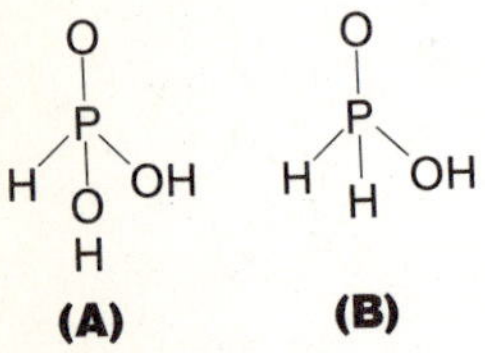

(A) (B)

FIGURE 7-27. Structures of (A) phosphorous and (B) hypophosphorous acids.

The characteristic chemical shift of hydrogens attached to phosphorus occurs in a limited range, and the peaks have fine structure corresponding to ${}^1J_{P—H}$ (phosphorus $I = \frac{1}{2}$). The phosphorus resonance[17] of $HPO(OH)_2$ and $H_2PO(OH)$ is reported to be a doublet in the former and a triplet in the latter compound, supporting the structures in Fig. 7-27. The coupling of the hydroxyl protons with phosphorus is either too small to be resolved, or it is not observed because of a fast proton exchange reaction, a phenomenon to be discussed shortly. Similar results obtained from phosphorus nmr establish the structures of $FPO(OH)_2$ and $F_2PO(OH)$ as containing, respectively, one and two fluorines attached to phosphorus. The ${}^{31}P$ resonance in P_4S_3 consists of two peaks with intensity ratios of three to one.[18] The more intense peak is a doublet and the less intense is a quadruplet. Since $I = 0$ for ${}^{32}S$, both the spin-spin splitting and the relative intensities of the peaks indicate three equivalent phosphorus atoms and a unique one. It is concluded that P_4S_3 has the structure in Fig. 7-28.

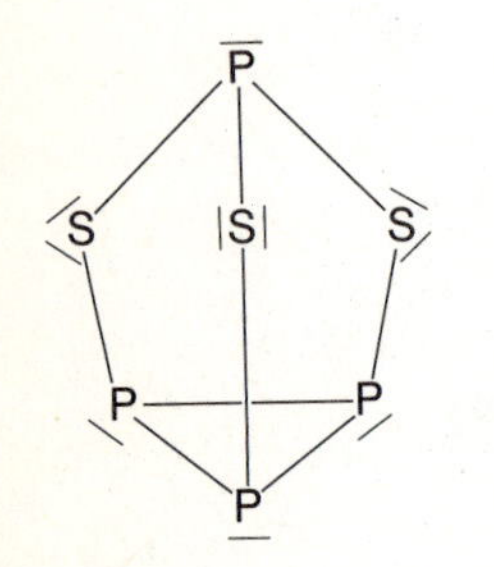

FIGURE 7-28. The structure of P_4S_3.

The fluorine resonance in BrF_5 consists of two peaks with an intensity ratio of four to one. The intense line is a doublet and the weak line is a quintet (1:4:6:4:1). This indicates that the molecule is a symmetrical tetragonal pyramid.

Solutions of equimolar quantities of TiF_6^{2-} and TiF_4 in ethanol give fluorine nmr spectra[19] consisting of two peaks with the intensity ratio 4:1 ($I = 0$ for ${}^{48}Ti$). The low intensity peak is a quintuplet and the more intense peak a doublet. The structure $[TiF_5(HOC_2H_5)]^-$ containing octahedrally coordinated titanium was proposed.

The factors that determine the magnitude of the coupling constant are not well understood for most systems. It has been shown that in the Hamiltonian describing the interaction between the ${}^{13}C$ nucleus and a directly bonded proton, the *Fermi contact term* is the dominant one. Qualitatively, this term is a measure of the probability of the bonding pair of electrons existing at both nuclei. The need for this can be appreciated from the earlier discussion of the Fermi contact coupling mechanism. The greater the electron density at both nuclei, the greater will be the interaction of the nuclear moments with the bonding electrons and hence with each other through electron spin polarization. Since an *s* orbital has a finite probability at the nucleus and *p*, *d*, and higher orbitals have nodes (zero probability) at the nucleus, the Fermi contact term will be a measure of the *s* character of the bond at the two nuclei.

Since the *s* orbital of hydrogen accommodates all of the proton electron density, the magnitude of ${}^1J_{{}^{13}C—H}$ for directly bonded carbon and hydrogen will depend upon the fraction of *s* character, ρ, in the carbon hybrid orbital bonding the hydrogen. The following equation permits calculation of ρ from ${}^1J_{{}^{13}C—H}$ data.

$$ {}^1J_{{}^{13}C—H}(\text{Hz}) = 500\rho_{C—H} \tag{7-40} $$

Since $\rho = 0.33$ for an sp^2 hybrid and 0.25 for an sp^3 hybrid, it can be seen that ${}^1J_{{}^{13}C—H}$ should be a sensitive measure of carbon hybridization. The natural abundance of ${}^{13}C$ ($I = \frac{1}{2}$) is 1.1 per cent, and the two peaks resulting from this splitting can be seen in spectra of concentrated solutions of unenriched compounds. The use of ${}^1J_{{}^{13}C—H}$ to measure the hybridization of carbon in a C—H bond was studied in detail by Muller[20] and is supported by valence bond calculations.[21] As expected from equation (7-40), the ${}^{13}C$—H coupling constant for a hydrogen in a saturated hydrocarbon is about 125 Hz, that for the ethylenic hydrogen of a hydrocarbon is around 160 Hz, and that for an acetylenic hydrogen is around 250 Hz, corresponding to ρ's of 0.25, 0.33, and 0.50, respectively.

It has been shown that a linear relationship exists between $^1J_{^{13}C-H}$ and the proton shift, τ, for a series of methyl derivatives in which the contribution to τ from magnetic anisotropy is approximately constant,[22,23] or varies with electronegativity of the group attached to methyl. The plot[23] is contained in Fig. 7-29, where the plus

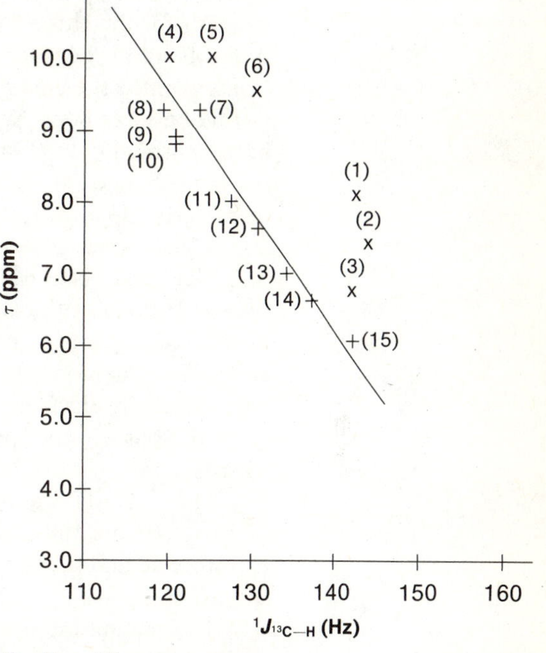

FIGURE 7-29. Relationship between $^1J_{^{13}C-H}$ and τ. X indicates points for compounds known to contain anisotropic contributions to the shift: (1) CH_3I, (2) CH_3Br, (3) CH_3Cl, (4) $Ge(CH_3)_4$, (5) $Sn(CH_3)_4$, and (6) $Pb(CH_3)_4$. + indicates points used to construct the line: (7) C_2H_6, (8) $(CH_3)_4C$, (9) $(CH_3CH_2)_2O$, (10) CH_3CH_2OH, (11) $(CH_3)_3N$, (12) $(CH_3)_2NH$, (13) $(CH_3)_2O$, (14) CH_3OH, and (15) CH_3F.

sign is used to indicate those points employed to construct the line. There are many compounds that would fall close to this line and consequently do not have abnormal anisotropic contributions to the chemical shift, but this could not be determined *a priori* with confidence so they were not employed to construct the line. Those compounds represented by points that deviate from this line have appreciable contributions to τ from anisotropy. The following explanation is offered for the existence of this relationship. Isovalent hybridization arguments[24] indicate that as X becomes more electronegative in the compounds CH_3X, more p character is employed in the C—X orbital, and there is a corresponding increase in the s character in the C—H orbitals. It has been shown that carbon electronegativity increases as the hybridization changes from sp^3 to sp^2 to sp.[25] If there are no anisotropic contributions (or if the contribution varies with electronegativity), a correlation would be expected between *carbon* electronegativity and τ. Since carbon electronegativity is related to carbon hybridization, which is in turn related to $^1J_{^{13}C-H}$, a correlation is expected between $^1J_{^{13}C-H}$ and τ for compounds in which the contributions to τ from magnetic anisotropy are constant (or vary linearly with the electronegativity of group X). In view of the difficulty of assessing the existence of even large anisotropic contributions to the chemical shift (see section on chemical shifts which are greatly influenced by anisotropy), this relationship is of considerable utility. Certainly the chemical shifts of compounds that do not fall on the line cannot be interpreted in terms of inductive or electronegativity arguments, *i.e.*, a local diamagnetic effect. It is not possible to check calculated anisotropies rigorously by the deviation of a compound from the line because the possibility does exist that the line in Fig. 7-29 contains contributions from neighbor anisotropy that vary linearly as the electronegativity of X varies.

The constancy of the $^1J_{^{13}C-H}$ coupling constants of the alkyl halides (149, 150, 152, and 152 Hz respectively for F, Cl, Br, and I) is interesting to note. Even though there is a pronounced change in the electronegativity of F, Cl, Br, and I, $^1J_{^{13}C-H}$ does not change[20] as would be predicted from the isovalent hybridization argument. In this series, there is an effect on the carbon hybridization that is opposite in direction to the electronegativity change of the halogen. These two effects cause the hybridization and electronegativity of carbon to remain constant. This other effect is based on orbital overlap, and one should think in terms of the *s* orbital dependence of the bond strength for the three hydrogens and X. As X becomes larger, the orbital used in sigma bonding becomes more diffuse and the overlap integral becomes smaller. Consequently, the energy of the C—X bond becomes less sensitive to the amount of *s* orbital used to bond X, and the total energy of the system is lowered by using more of the *s* orbital in the bonds to hydrogen. Thus, as the principal quantum number of the atom bonded to carbon increases, we expect to see, from this latter effect, an increase in $^1J_{^{13}C-H}$.

It has been shown that the carbon hybridization in $Si(CH_3)_4$, $Ge(CH_3)_4$, $Sn(CH_3)_4$, and $Pb(CH_3)_4$ changes in such a way as to introduce more *s* character into the C—H bond as the atomic number of the central element increases. An explanation for this change in terms of changing bond energies is proposed, and it is conclusively demonstrated that chemical shift data for these compounds cannot be correlated with electronegativities of the central element,[23] because large anisotropic contributions to the proton chemical shift exist in $Ge(CH_3)_4$, $Sn(CH_3)_4$, and $Pb(CH_3)_4$.

The ^{13}C—H coupling constant in the carbonium ions $(CH_3)_2CH^+$ and $(C_6H_5)_2CH^+$ are 168 and 164 Hz, respectively, compared to values of 123 and 126 Hz for propane and $(C_6H_5)_2CH_2$. This is consistent with a planar sp^2 hybridized carbonium ion.

The magnitude of the coupling constant between two hydrogens on adjacent carbons, $^3J_{H-H}$, is a function of the dihedral angle between them[26,27,28] as shown in Fig. 7-30. The dihedral angle θ is shown at the top of Fig. 7-30, where the C—C bond is being viewed end-on with a dot representing the front carbon and the circle representing the back one. The function is of the form:

$$^3J_{H-H} = A + B\cos\theta + C\cos 2\theta \tag{7-41}$$

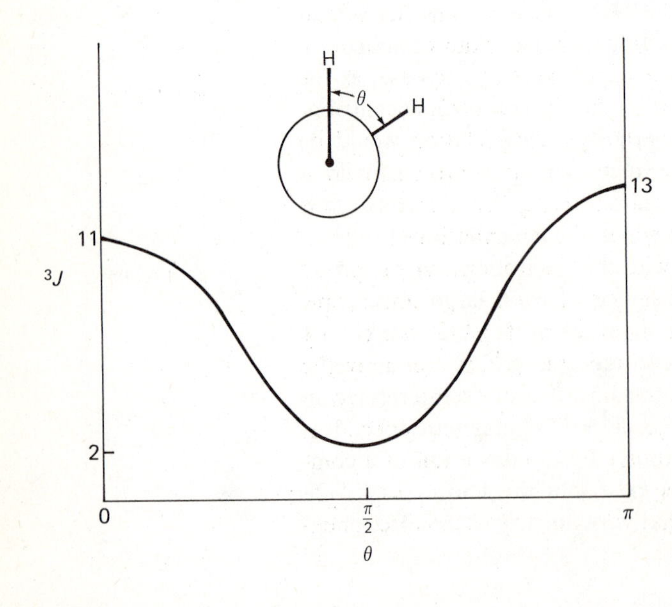

FIGURE 7-30. Variation in $^3J_{H-H}$ with the dihedral angle θ shown at the top.

For hydrocarbons, A is 7 Hz, B is 1 Hz, and C is 5 Hz. The equation has several limitations. Difficulties are encountered with electronegative substituents and with systems in which there are changes in the C—C bond order.

The $^1J_{^{13}C-H}$ values can be of use in assigning peaks in a spectrum.(29) For example, the proton chemical shifts of the two methyl resonances in $CH_3C(O)SCH_3$ are very similar, as shown in Fig. 7–31, and impossible to assign directly to methyl

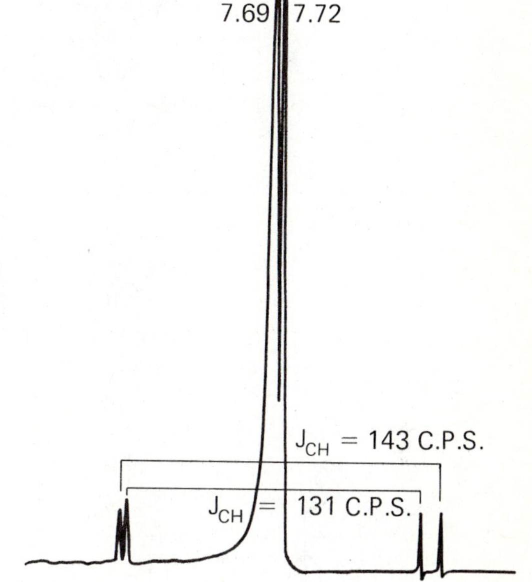

FIGURE 7–31. Proton nmr spectrum of S-methyl thioacetate at high spectrum amplitude showing satellites arising from ^{13}C. [From R. L. Middaugh and R. S. Drago, J. Amer. Chem. Soc., *85*, 2575 (1963).]

groups in the compound. However, by virtue of the fact that sulfur is a larger atom than carbon and has a comparable electronegativity, the $^1J_{^{13}C-H}$ values for methyl groups attached to sulfur [J = 138 to 140 Hz in CH_3SH, $(CH_3)_2SO$, $(CH_3)_2SO_2$] are larger than those for methyl groups attached to carbonyls (J = 125 to 130 Hz). Since the proton resonance on a ^{12}C will be in the center of the ^{13}C satellites, the downfield peak is unequivocally assigned(29) to the acetyl methyl.

It has been proposed(30,31) that a relationship exists between J_{Sn-H} and the hybridization of tin in the tin-carbon bonds of compounds of the type $(CH_3)_{4-n}SnX_n$. This relationship and results from other physical methods were employed to establish the existence and structure of five-coordinate tin addition compounds $(CH_3)_3SnCl \cdot B$,(31) where B is a Lewis base [*e.g.*, $(CH_3)_2SO$ or $(CH_3)C(O)N(CH_3)_2$]. The $^2J_{Sn-H}$ values suggest a trigonal bipyramidal geometry in which the tin employs essentially sp^2 orbitals in bonding to carbon, and, consequently, a three-center molecular orbital using a tin p or p—d orbital in bonding to the Lewis base and chlorine. Similar results were obtained for $^2J_{Pb-H}$ in the analogous lead compounds.(32)

Investigation(33) of a whole series of addition compounds between $(CH_3)_3SnCl$ and various bases indicated that the $^2J_{Sn-H}$ coupling constant changed in direct proportion to $-\Delta H$ of adduct formation with the base. This was interpreted to indicate that as the tin-base bond became stronger, the hybrids used to bond the methyl groups approached sp^2 from the $\sim sp^3$ hybrids used in $(CH_3)_3SnCl$. The relation found was:

$$^2J_{^{119}Sn-H} = 216\,\rho_{Sn} \tag{7-42}$$

The $^2J_{Sn-H}$ values for $Sn(CH_3)_4$, $Sn(CH_3)_3Cl$, and $Sn(CH_3)_3Cl_2^-$ were 54, 57.6 (in keeping with the isovalent hybridization prediction), and 72 [which when substituted into equation (7–42) gives $\rho_{Sn} = 0.33$]. The coupling constants for the adducts ranged from 64.2 to 71.6; the weakest base studied was CH_3CN, and the strongest was $[(CH_3)_2N]_3PO$. The plot of $-\Delta H$ *vs.* $^2J_{Sn-H}$ was extrapolated to the value for free trimethyltin chloride at $-\Delta H = 0$.

Coupling constants of equivalent hydrogens cannot be determined directly. Deuteration of a molecule has only a slight effect on the molecular wave function, so this technique can be used to gain information about the coupling of equivalent protons. For example, the H—H coupling in H_2 cannot be directly measured, but that in HD is shown to be 45.3 Hz. Since the effect of isotopic substitution on the magnitude of J is proportional to the magnetogyric ratio, we have

$$\frac{J_{HH}}{J_{HD}} = \frac{\gamma_H}{\gamma_D} = 6.515 \qquad (7\text{-}43)$$

From this, we calculate $^1J_{H-H} = 295.1$ Hz.

There have been several very interesting applications of phosphorus nmr in the determination of structures of complexes of phosphorus ligands.[34] The nmr spectrum and its interpretation are illustrated in Fig. 7–32 for the complex $Rh(\varphi_3P)_3Cl_3$

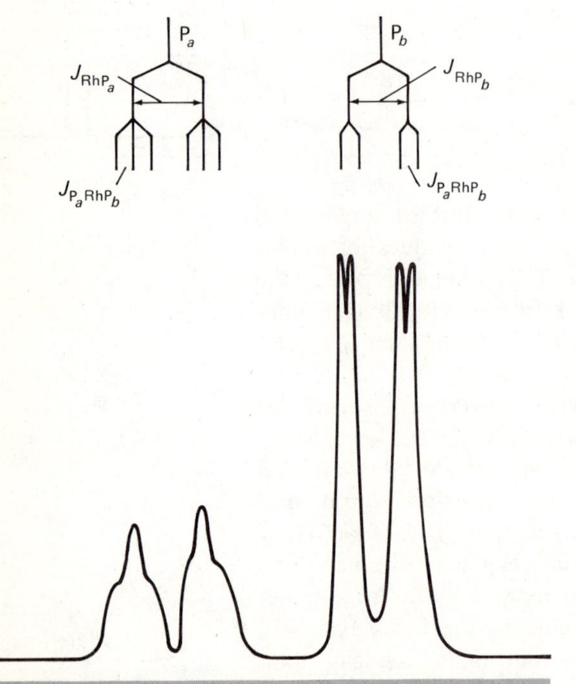

FIGURE 7–32. The ^{31}P nmr spectrum of $Rh(\varphi_3P)_3Cl_3$.

$[I(Rh) = \frac{1}{2}]$. Two isomers are possible, facial and meridional. All of the phosphorus ligands are equivalent in the facial isomer, so the spectrum in Fig. 7–32 substantiates that the complex studied was meridional. In this isomer two phosphorus atoms are *trans* to one another (labeled P_b), and one is *trans* to a chlorine (labeled P_a). The splittings are interpreted in the stick diagram at the top of Figure 7–32.

The magnitude of $^2J_{P-P}$ often provides interesting information about the stereochemistry of complexes. The magnitude of this coupling is usually much larger when two phosphorus atoms are *trans* to one another than when they are *cis*. In *cis*-$PtCl_2(bu_3P)(C_6H_5O)_3P$, the value of $^2J_{P-P}$ is 20 Hz, while a value of 758 Hz is found in *trans*-$PdI_2(bu_3P)(C_6H_5O)_3P$ and 565 Hz is found in *cis*-$PdI_2(CH_3)_3P(C_2H_5)_3P$. Some exceptions to this rule are found in Cr^0, Mo^0, W^0 and Mn(I) compounds. For example, in *trans*-$W(CO)_4(C_6H_5)_3P(bu_3P)$, $^2J_{P-P}$ is found to be 65 Hz.

A very interesting result is obtained in the proton or carbon-13 nmr spectra of phosphorus complexes. In the proton nmr of *trans*-$PdI_2[P(CH_3)_3]_2$, we might expect to find a doublet methyl resonance with perhaps a small splitting of each peak from the second phosphorus. Although the two phosphorus atoms are *chemically equivalent,* they are not *magnetically equivalent*. Any given methyl group would experience two different phosphorus couplings, $^2J_{P-H}$ and $^4J_{P-H}$. When the $^2J_{P-P}$ is much larger than $^2J_{P-H}$, as it is in the above complex, the proton nmr spectrum observed is a 1:2:1 triplet; *i.e.,* the two phosphorus nuclei behave as though they were two equivalent nuclei splitting the proton resonance. The two phosphorus nuclei are said to be *virtually coupled.*[35] In a similar *cis* complex where $^2J_{P-P} \ll {}^2J_{P-H}$, a doublet is obtained. In view of the very common, large $^2J_{P-P}$ for *trans* phosphorus ligands and the small $^2J_{P-P}$ values for *cis* phosphines, this phenomenon can be used to distinguish these two kinds of isomers. The use of ^{13}C nmr in this type of application[36a] has been criticized.[36b] Triplets occur in the ^{13}C nmr with much smaller values of J_{P-P}, so it is more probable that *cis* complexes will show virtual coupling. The reasons for virtual coupling have been discussed in the literature and will be considered in more detail in the section on second order effects in the next chapter. We will also be in a position to critically discuss other applications of coupling constants that have been reported in the literature after considering the contributions to the coupling constant in the next chapter.

7-17 EFFECT OF FAST CHEMICAL REACTIONS ON THE SPECTRUM

FACTORS INFLUENCING THE APPEARANCE OF THE NMR SPECTRUM

If one examines the high resolution spectrum of ethanol in an acidified solution, the result illustrated in Fig. 7-33 is obtained, in contrast to the spectrum shown in Fig. 7-20. The difference is that the spin-spin splitting from the hydroxyl proton has disappeared. Acid catalyzes a very rapid exchange of the hydroxyl proton. In the time

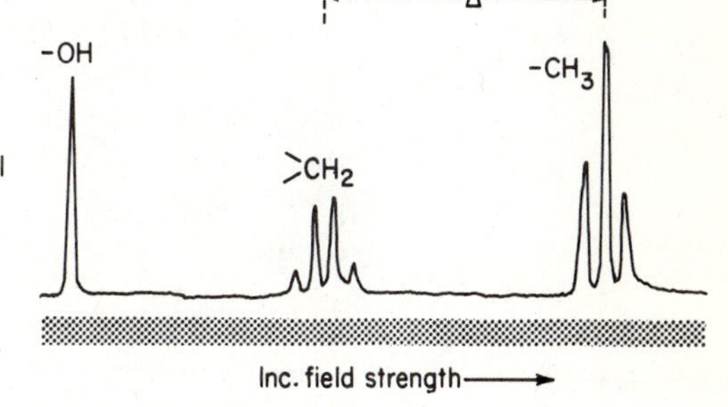

FIGURE 7-33. High resolution nmr spectrum of a sample of ethanol (acidified).

it takes for a methylene proton to undergo resonance, many different hydrogen nuclei have been attached to the oxygen. As a result, the methylene proton experiences a field averaged to zero from the O—H nuclear moment, and the $^3J_{HCOH}$ coupling disappears. In a similar fashion, the hydroxyl proton is attached to many different ethanol molecules, averaging to zero the field it experiences from —CH_2— protons, and only a single resonance is observed.

A very dramatic illustration of this effect is the spectrum of a solution of aqueous ammonia in which one does not see separate N—H and O—H protons, but only a single exchange-averaged line. When exchange is rapid, the chemical shift of this exchange-averaged line is found to be a mole-fraction-weighted average of the shifts of the different types of protons being exchanged:

$$\delta_{AVG} = N_{NH_3}\delta_{NH_3} + N_{H_2O}\delta_{H_2O} \tag{7-44}$$

It is important to emphasize that N_{NH_3} is not the mole fraction of ammonia, but the mole fraction of N—H protons, *i.e.*, $N_{NH_3} = 3[NH_3]/(3[NH_3] + 2[H_2O])$. The ^{19}F spectra of solutions of TiF_4 in donor solvents taken at −30° C consist of two triplets of equal intensity.[37] Six-coordinate complexes form by coordinating two solvent molecules, and the spectrum obtained is that expected for the *cis* structure. This structure contains two sets of non-equivalent fluorine atoms, with two equivalent fluorines in each set. At 0° C only a single fluorine peak is obtained. It is proposed that a rapid dissociation reaction occurs at 0° C, making all fluorines equivalent;

$$TiF_4 \cdot 2B \rightleftharpoons TiF_4B + B$$

at −30° C this reaction is slowed down so that the non-equivalence can be detected by nmr. Internal rearrangements and ionic exchange mechanisms are also possible.

This example illustrates one of the possible pitfalls in structure determination using nmr spectroscopy. If only the high temperature (0° C) spectrum had been investigated or if rapid exchange occurred at −30° C, it could have been incorrectly assumed that the adduct had the *trans* structure on the basis of the single nmr peak. If the actual structure were *trans,* only a singlet fluorine resonance would be detected at all temperatures and it would have been difficult to draw any structural conclusion because of the possibility that rapid exchange might be occurring at both temperatures. Even in the present case, the possibility exists, on the basis of these data alone, that the *cis* isomer is the structure at −30° C and that the *trans* isomer predominates at 0° C.

The fluorine nmr spectra of a large number of compounds of general formula $R_{5-n}PF_n$ (where R is a hydrocarbon, fluorocarbon, or halide other than fluorine) have been reported.[38] The number of peaks in the spectrum and the magnitude of the coupling constants are employed to deduce structures. For a series of compounds of the type R_2PF_3, a trigonal bipyramidal structure is proposed, and it is found that the J_{P-F} values for apical fluorines are ~170 Hz less than those for equatorial ones. The most electronegative groups are found in the apical positions. The spectra and coupling constants, obtained on these compounds at low temperature, indicate that the two methyl groups in $(CH_3)_2PF_3$ are equatorial and the two trifluoromethyl groups in $(CF_3)_2PF_3$ are axial. At room temperature, rapid intramolecular exchange occurs and the effect of this exchange is to average the coupling constants.

7-18 SPECTRA OF SYSTEMS WHERE THE MAGNITUDE OF J IS ABOUT THAT OF Δ

When the magnitude of the separation between two peaks in the nmr is of the same order of magnitude as the coupling constant, so-called *second order spectra* result. We shall discuss the reason for this in the next chapter. It is mentioned here simply to point out that when this occurs, the peaks in the resulting spectrum cannot be assigned by inspection as we have done before. This is illustrated in Fig. 7–34, where the spectra of ClF_3, obtained by using two different probes, demonstrate these complications.

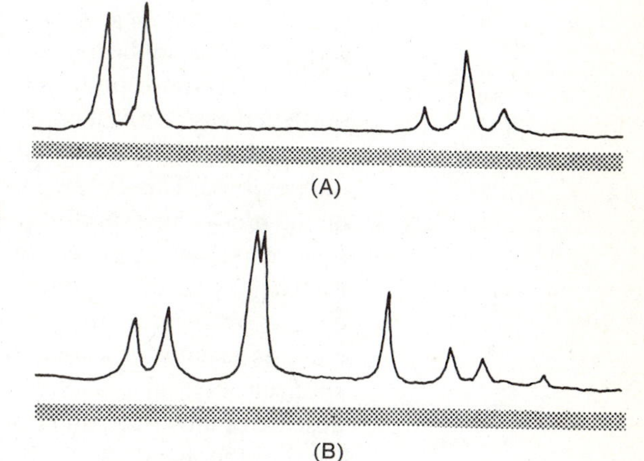

FIGURE 7–34. Fluorine nmr spectra of ClF_3. (A) at 40 MHz and (B) at 10 MHz.

With a higher frequency probe (40 MHz), $J \ll \Delta$ and the spectrum in Fig. 7–34(A) is obtained. The molecule ClF_3 has two long Cl—F bonds and one short one, giving rise to non-equivalent fluorines. The spectrum obtained at 40 MHz is that expected for non-equivalent fluorines splitting each other. As expected, the triplet is half the intensity of the doublet. Using a lower frequency probe (10 MHz), the difference in Δ for the non-equivalent fluorines is of the order of magnitude of J_{F-F} (recall that Δ is field dependent) and the complex spectrum in Fig. 7–34(B) is obtained. Complex patterns of this sort result whenever the coupling constants between non-equivalent nuclei are of the order of magnitude of the chemical shift. It is possible to calculate the J and Δ values of the molecule from these complex patterns by procedures that will be treated in more detail in the next chapter.

7–19 EFFECTS ON THE SPECTRUM OF NUCLEI WITH QUADRUPOLE MOMENTS

Quadrupolar nuclei are often very efficiently relaxed by the fluctuating electric fields that arise from the dipolar solvent and solute molecules. The mechanism of quadrupole relaxation depends upon the interaction of the quadrupolar nucleus with the electric field gradient at the nucleus. This gradient arises when the quadrupolar nucleus is in a molecule in which it is surrounded by a non-spherical distribution of electron density.

The *field gradient* q is used to describe the deviation of the electronic charge cloud about the nucleus from spherical symmetry. If the groups about the nucleus in question have cubic symmetry (*e.g.*, T_d or O_h point groups), the charge cloud is spherical and the value of q is zero. If the molecule has cylindrical symmetry (a threefold or higher symmetry axis), the deviation from spherical symmetry is expressed by the magnitude of q. If the molecule has less than cylindrical symmetry, two parameters are usually needed, q and η. The quantity η is referred to as the *asymmetry parameter*. The word "usually" is inserted because certain combinations of angles and charges can cause fortuitous cancellations of effects leading to $\eta = 0$. The axis of largest q is labeled z and is described by q_{zz}. The other axes, described by field gradients q_{xx} and q_{yy}, are described by the asymmetry parameter, which is defined as:

$$\eta = (q_{xx} - q_{yy})/q_{zz}$$

The effectiveness of the relaxation depends upon the magnitude of the field gradient. *Rapid nuclear quadrupole relaxation* has a pronounced effect on the line-width obtained in the nmr spectrum of the quadrupolar nucleus, and it also influences the nmr spectra of protons or other nuclei attached to this quadrupolar nucleus. In the latter case, splittings of a proton from the quadrupolar nucleus may not be observed or the proton signal may be so extensively broadened that the signal itself is not observed. This can be understood by analogy to the effect of chemical exchange on the proton nmr spectra. Either rapid chemical exchange or rapid nuclear quadrupole relaxation in effect places the proton on a nucleus (or nuclei, for chemical exchange) whose spin state is rapidly changing. Nuclear quadrupole relaxation rates often correspond to an intermediate rate of chemical exchange, so extensive broadening is usually observed. As a result of quadrupole relaxation, the proton nmr spectrum of $^{14}NH_3$ (^{14}N, $I = 1$) consists of three very broad signals; while in the absence of this effect, the spectrum of $^{15}NH_3$ (^{15}N, $I = \frac{1}{2}$) consists of a sharp doublet. On the other hand, in $^{14}NH_4^+$, where a spherical distribution of electron density gives rise to a zero field gradient, a sharp three-line spectrum results. In a molecule with a very large field gradient, a broad signal with no fine structure is commonly obtained.

When one attempts to obtain an nmr spectrum of a nucleus with a quadrupole moment (*e.g.*, ^{35}Cl and ^{14}N) that undergoes relaxation readily, the signals are sometimes broadened so extensively that no spectrum is obtained. This is the case for most halogen (except fluorine) compounds. Sharp signals have been obtained for the halide ions and symmetrical compounds of the halogens (*e.g.*, ClO_4^-), where the spherical charge distribution gives rise to only small field gradients at the nucleus, leading to larger values for T_1.

Solutions of I^- (^{127}I, $I = \frac{5}{2}$) give rise to an nmr signal. When iodine is added, the triiodide ion, I_3^-, is formed, destroying the cubic symmetry of the iodide ion so that quadrupole broadening becomes effective and the signal disappears. Small amounts of iodine result in a broadening of the iodide resonance, and the rate constant for the reaction $I^- + I_2 \longrightarrow I_3^-$ can be calculated from the broadening.[39] It is interesting to note that chlorine chemical shifts have been observed[40] for the compounds: $SiCl_4$, CrO_2Cl_2, $VOCl_3$, and $TiCl_4$, where the chlorine is in an environment of lower than cubic symmetry.

An interesting effect has been reported for the fluorine nmr spectrum of NF_3. The changes in a series of spectra obtained as a function of temperature are opposite to those normally obtained for exchange processes. At −205° C a sharp single peak is obtained for NF_3; as the temperature is raised the line broadens and a spectrum consisting of a sharp triplet ($I = 1$ for ^{14}N) results at 20° C. It is proposed that at low temperature the slow molecular motions are most effective for quadrupole relaxation of ^{14}N; as a result, a single line is obtained. At higher temperatures, relaxation is not as effective and the lifetime of a given state for the ^{14}N nucleus is sufficient to cause spin-spin splitting. A similar effect is observed for pyrrole.[41] The ^{14}N spectrum of azoxybenzene exhibits only a singlet. The nitrogens are not equivalent, and it is

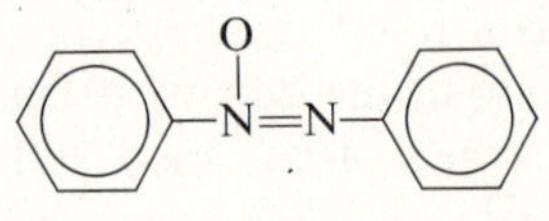

O
|
—N=N—

proposed that the field gradient at the N—O nitrogen is so large as to make this resonance unobservable.

7-20 PRINCIPLES

FOURIER TRANSFORM NMR

Many magnetic nuclei are present in nature in low abundance and also have low sensitivity (*e.g.*, ^{13}C). In examining the spectra of these materials, the tendency is to increase the rf power, but this often saturates the signal. Alternatively, one can sweep the spectrum many times at low power and store the spectra in a computer. This is called a CAT (computer averaged transients) experiment. The noise is random and cancels, but the signal is reinforced by each sweep and, eventually, on adding many sweeps, the spectrum emerges from the background. This process is time consuming. With limited computer storage facilities, the experiment becomes even more difficult if it is not known at which frequency the signals will occur. Pulse techniques, to be described next, are very advantageous for this situation, as we shall show.

The rf pulse is applied for a time t_p (typical times are 10^{-5} to 10^{-6} sec) with a fixed frequency ω_0 and with very high power (200 to 1000 watts). After waiting a couple of seconds, the pulse is applied again, as shown in Fig. 7-35.

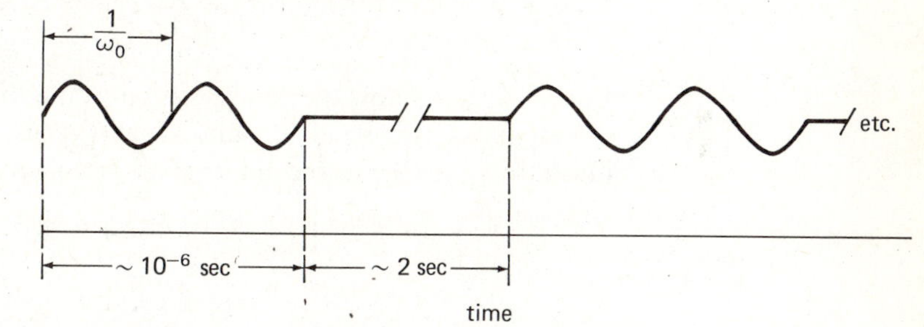

FIGURE 7-35. A sequence of rf pulses.

The pulse, if properly selected, causes all of the magnetic nuclei of a certain element in the sample to absorb and eventually (after a transformation) leads to a typical nmr spectrum. The problem now is to show how all these nuclei with different Larmor frequencies can be made to undergo transitions by a pulse of a single frequency ω_0. The mathematical function of time, $f(t)$ in Fig. 7-35, corresponding to a single pulse can be reproduced by summing together a series of sine and cosine functions with different ω's. This can be accomplished with a Fourier transform. A Fourier transform corresponds to a generalized transformation in function space. In this case, we wish to convert the time plot in Fig. 7-35 to the corresponding frequency plot, using equation (7-45)

$$F(\omega) = 2\pi \int_{-\infty}^{\infty} e^{i\omega t} f(t)\, dt \qquad (7\text{-}45)$$

where $f(t)$ is in terms of ω_0, t_p and t. The result in Fig. 7-36 is obtained, where Δ is

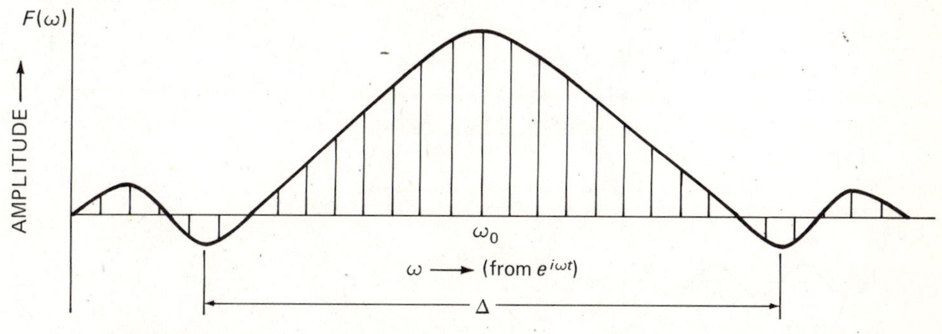

FIGURE 7-36. The frequency plot corresponding to the time plot of Fig. 7-35.

the range of the principal frequency components. Consequently, pulsing our frequency ω_0 is comparable to having a whole distribution of frequencies available to us from the rf source.

Probably the easiest way to see this is by considering how a square wave, $f(t)$, can be produced with a Fourier series. A square wave is reproduced in Fig. 7-37.

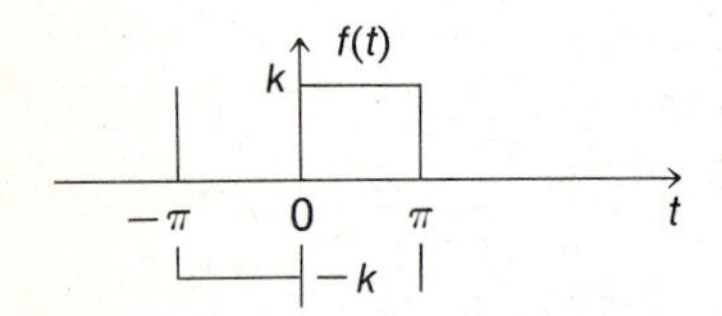

FIGURE 7-37. Graphical representation of a square wave.

In the limit as $n \rightarrow \infty$, the following Fourier series will reproduce the square wave function:

$$f(t) = a_0 + \sum_{1}^{n} (a_n \cos 2\pi nt + b_n \sin 2\pi nt) \tag{7-46}$$

Fig. 7-38 shows how the superposition of the first three partial sums S_1, S_2, and S_3 corresponding to $n = 1$, 2, and 3 approaches this function. In Fig. 7-38(A), we illustrate S_1; in (B) the second term has been added to give S_2. Adding the third term

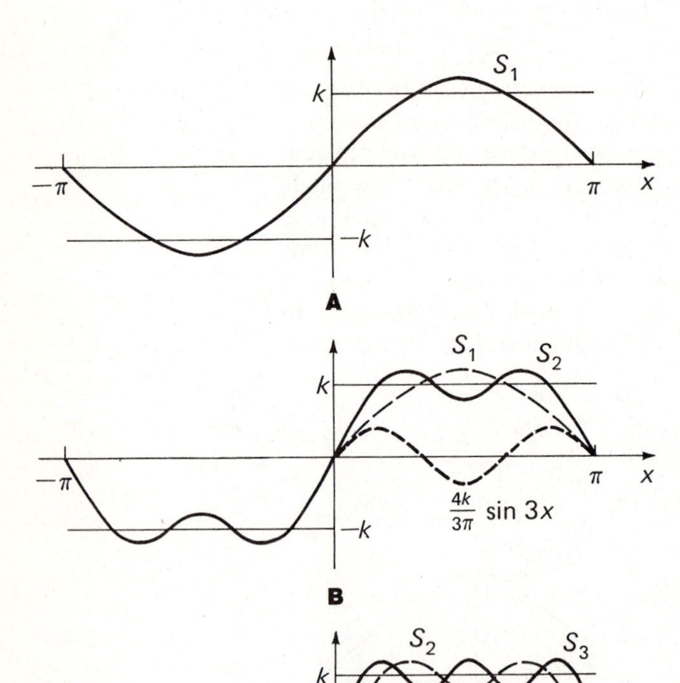

FIGURE 7-38. Addition of the first three waves in the Fourier series leading to a square wave.

in (C) produces S_3, which is beginning to resemble a square wave. As n becomes very large, the resemblance becomes better.

In a similar way to that described above for a square wave, the distribution of frequencies in Fig. 7–36 can be converted to A *vs.* t plots and added to give the curve in Fig. 7–35. If there were no pulse, but just one continuous wave approaching infinite time (the slow passage limit), only one frequency would be required to describe this continuous wave, ω_0. As the time, t_p, of a single pulse decreases, the span of frequencies needed to describe this pulse increases. The range of frequencies, Δ, in Fig. 7–36 is obtained from the Fourier transform of the wave in Fig. 7–35 and is given by:

$$\Delta = 4\pi/t_p \tag{7-47}$$

where t_p is the duration of the pulse. In the description of the single pulse, the wave stops after time t_p. A given frequency continues for infinite time. When the pulse is passed through the sample, the appropriate frequencies in Fig. 7–36 are absorbed by the sample, causing transitions. Therefore, the pulse must be short enough to cover the distribution of expected spectral frequencies with similar intensity frequency components. Accordingly, $t_p \ll 4\pi/\Delta$. For a typical compound, we pulse for 10^{-6} sec, measure the free induction decay curve, store it in the computer, pulse again, measure the FID curve, add it to the other one in the computer, and so forth, for many pulses. The free induction decay curve is simply a plot of the decay of the x-y component of the magnetization with time. The Fourier transform of the resulting FID curve gives the frequency spectrum. If the nucleus were sensitive enough, just one pulse would give an FID curve that could be Fourier transformed to give the frequency spectrum. When it is not sensitive enough, the entire spectrum is run many times and stored in a computer. Since one FID curve is run in a few seconds, several spectra can be obtained in the time required for a slow passage experiment. Basically, then, we measure the decay of the magnetization in the u,v plane in which the net magnetizations of the different kinds of nuclei are precessing at their Larmor frequencies. For a single line in the frequency spectrum, the FID curve looks like Fig. 7–39. When ω_0 equals the chemical shift of the proton in the sample, the dotted line is obtained. When a value of ω_0 is selected that is off resonance, the solid line is obtained. The latter selection is usually made. The dotted line is obtained when the rotating reference frame is stationary with respect to the rotation of the x-y component of the

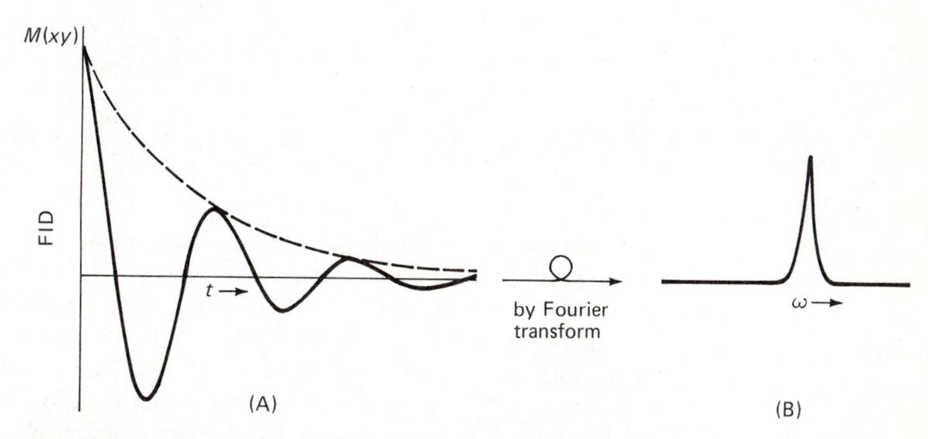

FIGURE 7–39. A free induction decay curve (A) and its Fourier transform for a single absorbing nucleus (B). (The dashed line is obtained when ω_0 equals the chemical shift, and the solid line when ω_0 is off resonance.)

magnetization vector of the bulk sample. The solid line is obtained when these two are not moving in phase. The damped oscillation that is observed corresponds to the Larmor frequency and the rotating frame moving in and out of phase.

The FID spectrum of $^{13}CH_3I$ (^{13}C spectra) is shown in Fig. 7-40 (the spin-spin

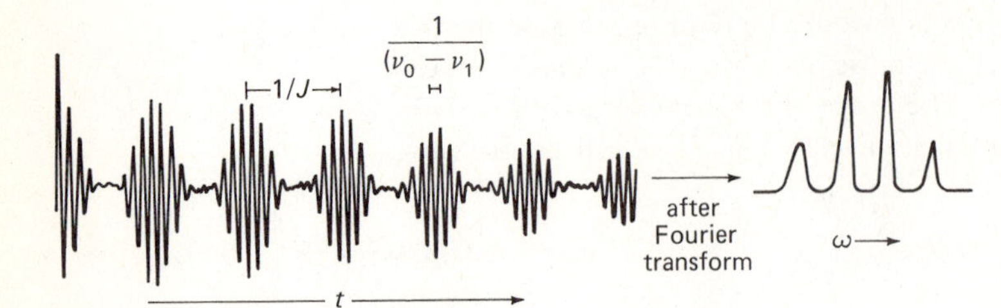

FIGURE 7-40. FID spectrum and its Fourier transform for $^{13}CH_3I$. [From T. C. Farrar and E. D. Becker, "Introduction to Pulse and Fourier Transform NMR Methods," Academic Press, New York (1971).]

splitting from the protons gives the quartet).[42] The time spectrum is due to the addition of the various carbon resonances whose magnetizations precess at different frequencies, giving rise to an interferogram; *i.e.*, four decay curves like the off-resonance curves in Fig. 7-39 are added together to produce the interferogram in Fig. 7-40.

The FID ^{13}C spectra of molecules containing more nuclei are even more complicated.[43,44] That of progesterone, with the protons decoupled, is given in Fig. 7-41.

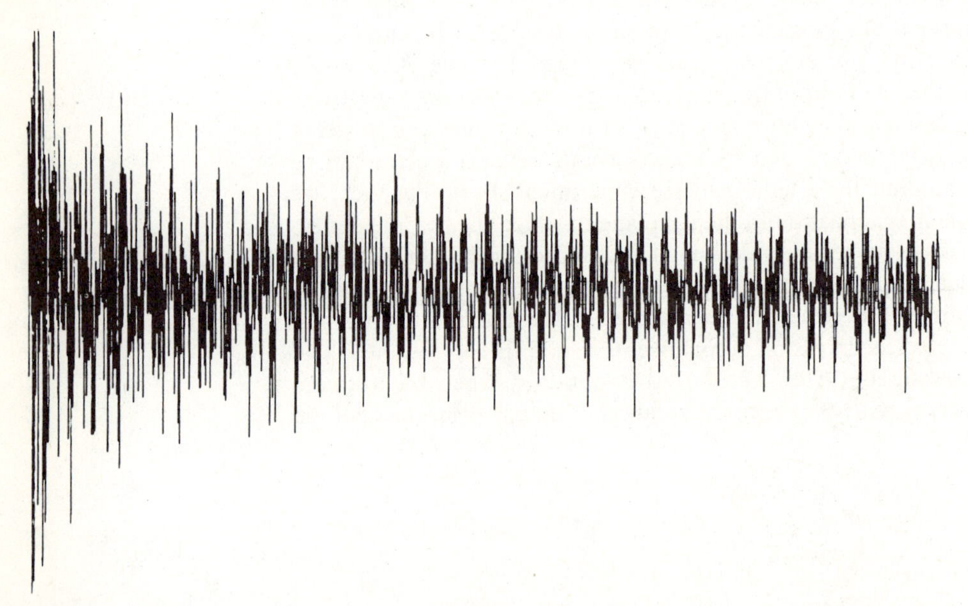

FIGURE 7-41. ^{13}C resonance of progesterone (proton noise decoupled). [From T. C. Farrar and E. D. Becker, "Introduction to Pulse and Fourier Transform NMR Methods," Academic Press, New York (1971).]

In view of the complexity of this pattern, the FID interferogram is seldom reported. Because of the computer time required to do a Fourier transform, the FID curve is often stored in the computer.

Spin-spin coupling from ^{13}C—^{13}C is not seen in the spectrum of a molecule containing several C—C bonds if it is not enriched in ^{13}C, because with a 1.1% natural abundance the probability that two ^{13}C nuclei would be next to each other in the molecule is very low. Because of a phenomenon referred to as the nuclear Overhauser effect, to be discussed in the next chapter, the integrated intensity of a resonance is not necessarily proportional to the number of carbons under the peak. In some cases, this problem can be overcome by adding a free radical or a paramagnetic ion to the solution.[45,46]

7-21 THE MEASUREMENT OF T_1 BY FTS (FOURIER TRANSFORM SPECTROSCOPY)

In a static field, the nuclear moments precess about the field direction as shown in Fig. 7-42. As described earlier, when a secondary field H_1 is applied, which is in phase with the Larmor frequency, a torque is exerted that tends to make the moment precess about H_1. If we define a rotating coordinate system that rotates at the Larmor frequency, we only have to worry about the torque from H_1. If we consider the H_1 direction to be perpendicular to the page, the cone is so tipped that projection of the magnetic moment vectors in the *x-y* plane gives a net *x-y* component (see Fig. 7-43). Relaxation tends to restore the system to the situation in Fig. 7-42. The torque is the cross product of the magnetic moment vector and $\vec{\mathbf{H}}_1$, so at resonance in the slow passage experiment, H_1 tends to tip the net magnetic moment vector (which has no *x-y* component in the absence of H_1), inducing an *x-y* component. (When H_1 is applied, α and β are not eigenfunctions, but some linear combination of them is.) This net *x-y* magnetization is detected when one passes through resonance in the slow passage experiment.

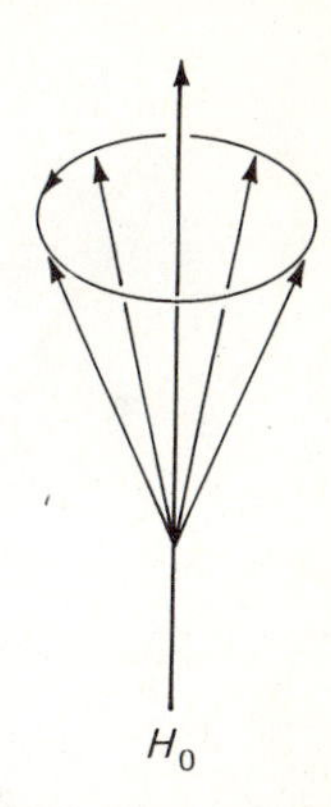

FIGURE 7-42. Precession of nuclear moments in a Zeeman experiment. [From T. C. Farrar and E. D. Becker, "Introduction to Pulse and Fourier Transform NMR Methods," Academic Press, New York (1971).]

In a pulse experiment, it is possible to tip the magnetization vector 90°, 180°, or n°, depending on the duration of the pulse. The angle θ (in radians) through which the magnetization is tipped is given by:

$$\theta = \gamma H_1 t_p \tag{7-48}$$

In all but the 180° pulse experiments, an *x-y* component is induced. In a 180° pulse experiment, we invert the magnetic moment vector (180° inversion) from the position where $H_1 = 0$ (*i.e.,* from positive m_z to negative m_z) and do not generate any *x-y* component. Any individual moment that is turned 180°, as well as a 180° turn in the net moment, still has the same projection on *xy*, so there is no change in the net *x-y* component as can be seen in Fig. 7-44.

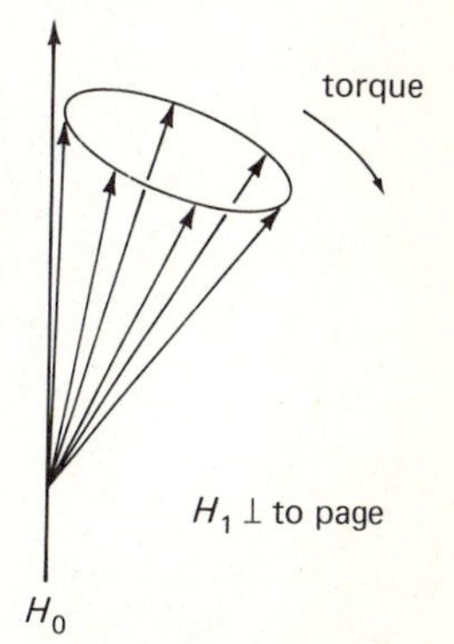

FIGURE 7-43. Effect of a secondary field H_1.

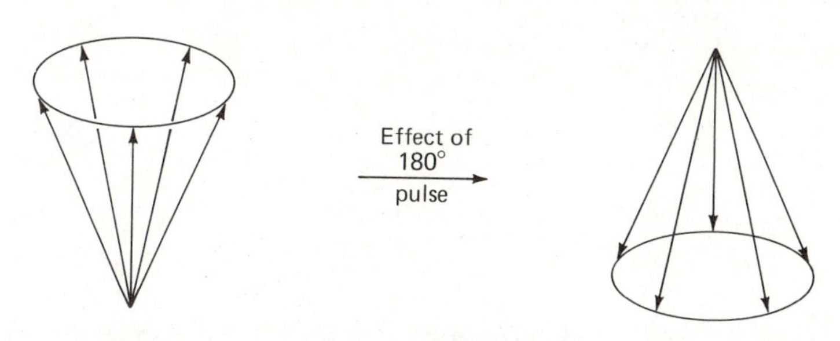

FIGURE 7-44. The effect on the magnetization of a 180° pulse.

After the pulse is turned off and decay occurs, the magnitude of M_z just decreases at a rate governed by the longitudinal relaxation time T_1. The series of arrows in Fig. 7-45 represent the decay of the $\vec{\mathbf{M}}_z$ vector with time. *Since the individual moments relax in a purely random manner, no x-y component results.*

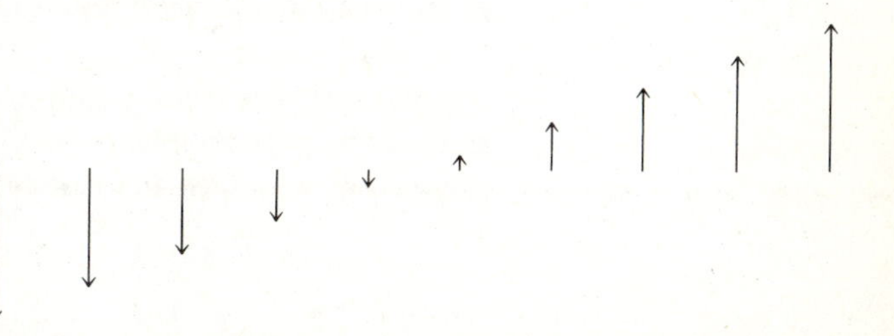

FIGURE 7-45. Decay of the M_z component with time after a 180° pulse.

It is impossible to detect this decay of M_z for it has no x-y component. Now consider an experiment in which we hit the sample with a 180° pulse, followed by a 90° monitoring pulse. We can then detect the magnetization. After waiting for thermal equilibrium to be reestablished (usually a time corresponding to $5T_1$ is employed), we can again subject the sample to the same 180° pulse, wait a while, and then follow with a 90° monitoring pulse. Such an experiment is referred to by the symbolism 180-τ-90. The process can be repeated by waiting longer before applying the 90° monitoring pulse. The series of spectra that result from a sample with a single resonance is shown in Fig. 7–46, where the delay time increases from left to right.

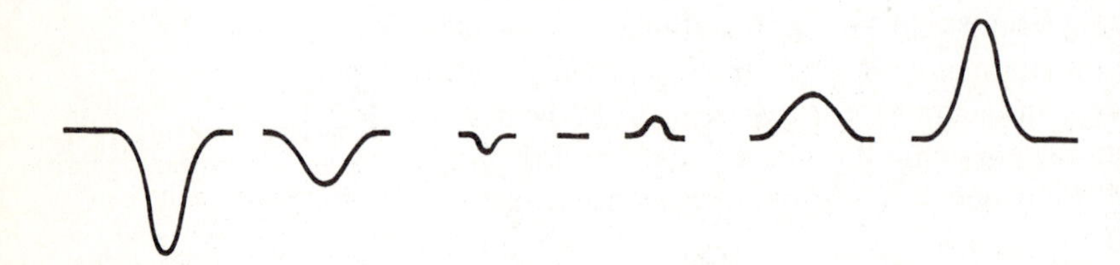

FIGURE 7–46. Spectra resulting from a series of 180° and 90° pulses. (The time between 180° and 90° pulses increases as we proceed from left to right. The negative peak corresponds to a large 270° component; *i.e.*, a negative z-component gives a 270° component after a 90° pulse.)

If one plots the area of the band in Fig. 7–46, A, versus time between pulses, one obtains the first order decay curve shown in Fig. 7–47, from which T_1 can be calculated. At zero signal, $t_{1/2} = T_1 \ln 2$.

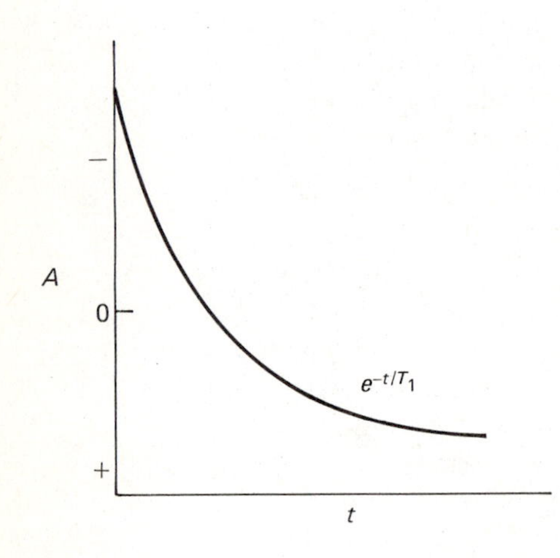

FIGURE 7–47. First order decay curve of the z-component corresponding to the 180° pulse.

An inverted resonance is obtained whenever the net magnetization of the sample is opposed to the field, for the detection system senses this as an emission (transition from a state opposed to the field to one aligned), and vice versa for net magnetization with the field. Another way of looking at this is that a 90° clockwise rotation about $\vec{\mathbf{H}}_1$ of a magnetic moment vector opposed to the field gives rise to a different phase (180° different) than rotation by 90° of a vector aligned with the field. This is illustrated in Fig. 7–48, where $\vec{\mathbf{H}}_1$ is to be considered perpendicular to the page.

7–22 USE OF T_1 FOR PEAK ASSIGNMENTS

We will digress for a moment from our discussion of nmr to make clear what is meant by a *correlation function*. Correlation functions are employed in the description of processes in which the value of x does not depend on t in a completely definite

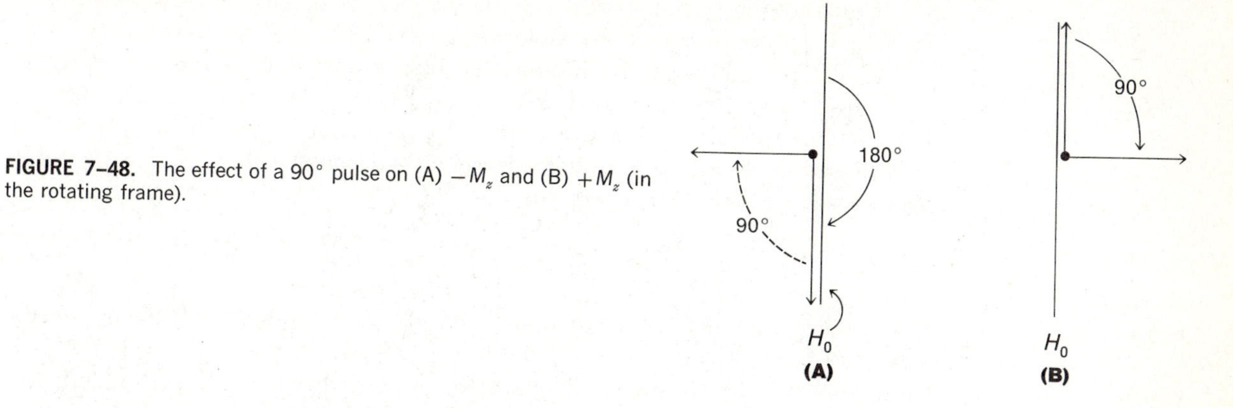

FIGURE 7–48. The effect of a 90° pulse on (A) $-M_z$ and (B) $+M_z$ (in the rotating frame).

way, *i.e.*, a random time process. However, the average dependence of $x(t)$ on t can be written in terms of probability distributions. When discussing the self-correlation of a variable, the function is referred to as an *autocorrelation function.*

The autocorrelation function for the displacement of a particle undergoing Brownian motion is given by equation (7–49)

$$R(\tau) = \lim_{T\to\infty} \frac{1}{2T} = \int_{-T}^{T} x'(t + \tau)\, x(t)\, dt \tag{7-49}$$

where $x(t)$ is the value of x at time t and $x'(t + \tau)$ is the value of x at time t plus a time interval τ between the two values of the displacement to be compared, and where the average value of the displacement is zero. The symbol T refers to the time limits of integration.

A correlation function displays the decay in correlation with increasing time interval τ. At short times, the value of x' at $(t + \tau)$ is highly correlated with the initial value of x; but at large values of τ, x and x' are independent random variables and are not correlated.

A plot of a correlation function for some molecular process that gives a Lorentzian line in the frequency domain is given in Fig. 7–49. The correlation time τ_c is defined as the time required to get through $1/e$ of the curve (curve is $R(\tau) = e^{-\tau/\tau_c}$) for a process in which $R(\tau)$ is an exponential in the time domain. As we shall see, these definitions of correlation functions will be needed in order to interpret the values obtained from T_1 measurements.

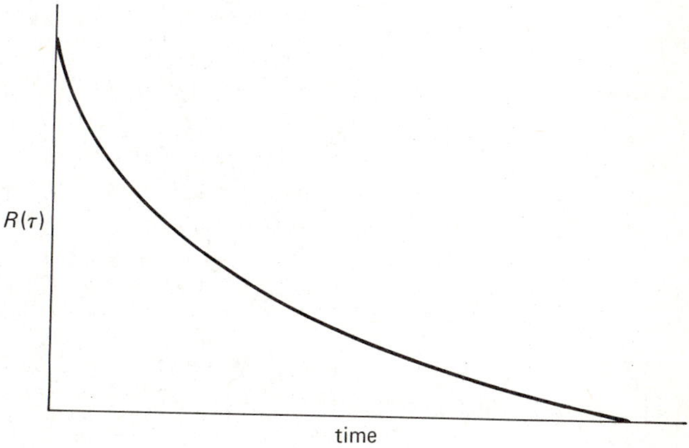

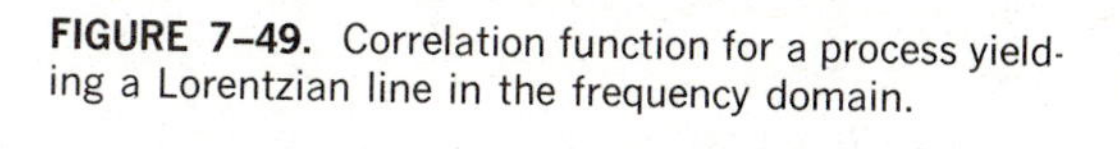

FIGURE 7–49. Correlation function for a process yielding a Lorentzian line in the frequency domain.

The following principles underlie the use of T_1 in making peak assignments in the nmr spectrum of a complex molecule.

1. ^{13}C relaxation times of protonated carbons in large or asymmetric molecules are dominated by dipolar interactions with the directly bonded protons. The value of $1/T_1$ for ^{13}C relaxation by a hydrogen comes from use of equation (7–50) when molecular rotation is isotropic, the hydrogens are decoupled, and $(\omega_C + \omega_H)\tau_{eff} \ll 1$ (the so-called extreme narrowing limit).

$$1/T_1 = N\hbar^2\gamma_C{}^2\gamma_H{}^2 r_{CH}{}^{-6}\tau_{eff} \qquad (7\text{-}50)$$

Here γ_C and γ_H are the gyromagnetic ratios of ^{13}C and ^1H, N is the number of directly attached hydrogens, r_{CH} is the CH distance, and τ_{eff} is the effective correlation time for rotational reorientation. Assume that only directly bound hydrogens contribute. Otherwise, we have to sum the $r_{CH}{}^{-6}\tau_{eff}$ terms for any other proton that could contribute. In order for the limit $(\omega_C + \omega_H)\tau_{eff} \ll 1$ to hold, the molecule must be rapidly rotating. The equation applies only under conditions of proton decoupling where the scalar coupling is removed. However, under these conditions, it should be emphasized that the magnetic moments of the protons are not undergoing transitions *fast enough to average the moment to zero* in the time required for a rotation, so dipolar relaxation still occurs. Under these decoupled conditions, the Fermi contact coupling makes no contribution to the relaxation mechanism and the dipolar effect dominates. It should be emphasized that because of the inverse sixth power in r_{CH}, dipolar contributions to $1/T_1$ from atoms a long distance away will be negligible.

2. If two carbon atoms in a molecule have the same τ_{eff}, but one of them is not protonated and the other is, then the non-protonated one will have a much longer T_1 than the protonated one.

3. Differences in τ_{eff} for different carbons in the same molecule may arise from anisotropic rotation of the molecule in solution and from the effects of internal reorientation. For a more complete discussion of the effects of internal rotation and of proton decoupling, the reader is referred to references 47 and 48.

Applications of these principles are illustrated in the T_1 measurements of some of the ^{13}C resonances of cholesteryl chloride.[43] The results are summarized in Fig. 7–50. The protonated carbons on the ring backbone (not shown) and other protonated carbons (also not shown) all have the same T_1 and hence the same τ_{eff}. The overall reorientation of the molecule is isotropic, and T_1 values can be used to distinguish CH and CH_2 protons. The carbon with no protons attached, C(13), is seen to have a much larger T_1 (*i.e.*, a sharper line when $T_1 = T_2$) than others in the molecule. The methyl carbons have long T_1's, considering that there are three protons on such a carbon. This is due to the rapid rotation about the C—C bond that decreases τ_{eff}; as seen in

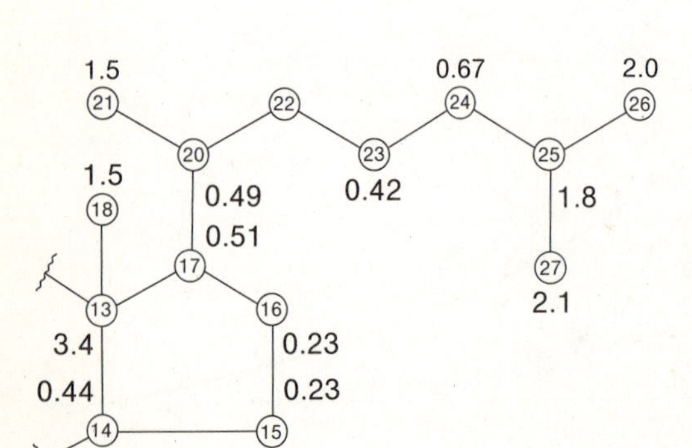

FIGURE 7–50. T_1 values (sec) for various ^{13}C atoms in cholesteryl chloride in CCl_4 at 42°. [Reprinted with permission from A. Allerhand and D. Doddrell, J. Amer. Chem. Soc., 93, 2777 (1971). Copyright by the American Chemical Society.]

equation (7–49), this decreases $1/T_1$ or increases T_1. By comparing T_1's down the long side chain at the top of the figure, one sees that the effect of internal motion increases toward the free end of the chain as expected.

The proton T_1 values of vinyl acetate are shown in Fig. 7–51. The r^{-6} dipolar effects of the protons on each other are seen in all three values.

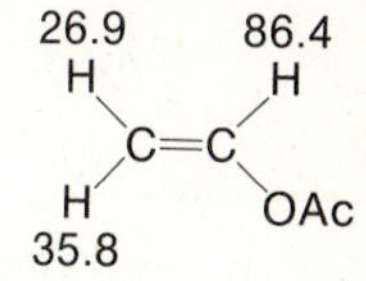

Advantage can be taken of the large differences in the T_1 values of protons to simplify the nmr spectrum. For a particular delay time following the initial 180° pulse, the magnetization vector can decay to zero intensity. No intensity will then be detected when this is followed by a monitoring 90° pulse (see Fig. 7–46 and the discussion of it). Thus, if there are two overlapping peaks with different T_1's, a 180° pulse followed by a properly timed wait before the 90° pulse will remove the peak with the shorter T_1. A three-fold or greater differential in the relaxation times of the overlapping protons is ideal for this application.

26.9 H, 86.4 H, C=C, H 35.8, OAc

FIGURE 7–51. T_1 values for the protons of vinyl acetate.

7–23 MEASUREMENT OF T_2

Theoretically, T_2 is obtainable directly from the nmr linewidth, but in practice, magnetic field inhomogeneities often dominate these linewidths. In other words, the slight differences in the H_0 field at different places in the sample tube lead to different Larmor frequencies and make up the major mechanism of fanning out the magnetic moments in the xy plane. This is a problem, since the chemist is interested in the kinetic and molecular effects on T_2. To combat this, Hahn designed the ingenious spin-echo experiment, described in Fig. 7–52.

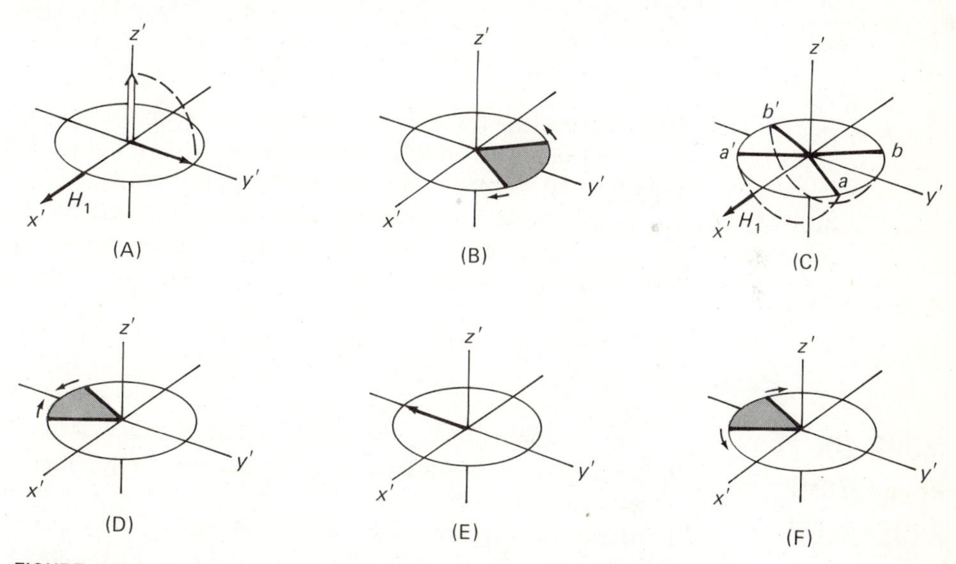

FIGURE 7–52. The Hahn spin-echo experiment. (a) A 90° pulse applied along x' at time 0 causes M to tip to the positive y' axis. (b) During the time τ, the microscopic moments in the xy plane dephase because of field, H_0, inhomogeneity. (c) A 180° pulse is applied. (d) The moments regroup. (e) After a time τ following the 180° pulse, the moments rephase. (f) At subsequent times, the moments dephase again. [From T C. Farrar and E. D. Becker, "Introduction to Pulse and Fourier Transform NMR Methods," Academic Press, New York, 1971.]

A 90° pulse is applied along x', causing $\vec{\mathbf{M}}$ to tip to y'. We wait for a time τ during which the nuclei in different parts of the sample see a different field and precess at different frequencies. Those nuclei precessing faster than average (the rotation rate of the rotating frame) appear in the rotating frame to move toward the observer (*i.e.*, looking down the z' axis they appear to move clockwise), while those slower than average move away (counterclockwise). A 180° pulse is then applied along x', moving all the moments 180° about the x' axis (*i.e.,* they are now fanned out about $-y'$). It should be emphasized that a 180° rotation about x' is different from a 180° rotation about z. The former causes a to become a' and b to become b'. While we are waiting another time period corresponding to τ, the faster nuclei move clockwise and the slower nuclei move counterclockwise in the rotating frame, as shown by the arrows in Fig. 7–52(D). However, now the slower nuclei move toward the observer and the faster ones away from the observer, causing them to get back in phase. After the same time τ that was used above, they have regrouped along y'; if there had been no T_2 relaxation by other mechanisms, we would have the same signal intensity, in effect eliminating the field inhomogeneity contribution to T_2. At longer times, the moments again dephase. In reality, there are two effects that reduce this amplitude: (1) T_2 of the nucleus, which fans out the magnetization randomly as opposed to the systematic nature of the applied field, and (2) diffusion of molecules to different parts of the sample tube in the time 2τ. The first effect goes as $e^{-\tau}$ and the second goes as $e^{-\tau^3}$, so it is possible to obtain molecular diffusion times as well as T_2 by measuring the amplitude as a function of τ.

7–24 NMR OF QUADRUPOLAR NUCLEI

The nmr lines of quadrupolar nuclei are very broad. For spherical rotation, the width is a function of the nuclear quadrupole moment and the correlation time τ_c as given in equation (7–51):

$$\frac{1}{T_1} = \frac{3}{40} \cdot \frac{2I+3}{I^2(2I-1)} \left(1 + \frac{1}{3}\eta^2\right) \left(\frac{e^2qQ}{\hbar}\right)^2 \tau_c \qquad (7\text{-}51)$$

Since the nuclear quadrupole effect on the relaxation process dominates the line width and since this effect is intramolecular, only the rotational contribution to T_1 is important. However, it is possible to have both isotropic and anisotropic rotation, and this complicates the line width interpretation. For example, in $CHCl_3$, the rotational diffusion constant at room temperature perpendicular to the C_3 axis, $D_\perp$, is 0.96×10^{11} sec^{-1} while that parallel to this axis, $D_\parallel$, is 1.8×10^{11} sec^{-1}.

MORE ON RELAXATION PROCESSES

In equation (7–49), the autocorrelation function for a particle undergoing Brownian motion was discussed. As mentioned earlier, the Fourier transform of the correlation function converts it from the time domain to the frequency domain. The result is a quantity called the *spectral density,* $J(\omega)$, which is defined by:

$$J(\omega) = \int_{-\infty}^{\infty} R(\tau_c)\, e^{i\omega\tau_c}\, d\tau \qquad (7\text{-}52)$$

The spectral density for any given frequency, $J(\omega)$, gives the intensity of the fluctuation at that particular frequency. The fluctuations of importance in nmr include the relative diffusion of one molecule past another, described by a correlation time, τ_D; the rotation of a molecule about its rotation axes, τ_R; a rapidly relaxing nucleus, τ_S; and chemical exchange, τ_E. All of these phenomena cause a given nucleus to experience a fluctuating field from another nucleus in solution. The efficiency of nuclear relaxation will depend upon the intensity of the frequency arising from the fluctuation, which in turn depends upon the correlation times τ_c for the various processes.

Accordingly, it is appropriate to examine in more detail the relationship between spectral density and τ_c.

If we plot the spectral density versus ω, curves comparable to those shown in Fig. 7–53 are obtained. These different curves correspond to different correlation times, τ_c,

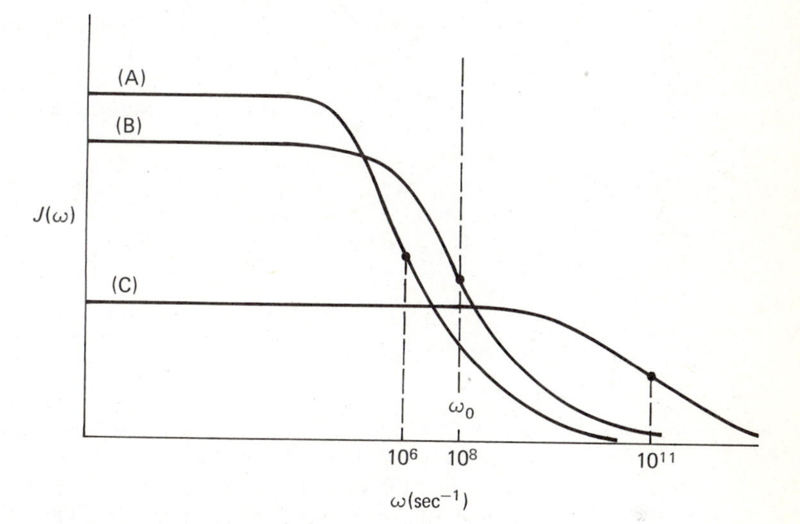

FIGURE 7–53. A plot of the spectral density of various frequencies for systems (A), (B), and (C) corresponding to different correlation times.

for our systems. (Recall that R depends upon the value of τ_c.) The half-intensity height indicated by the solid dots corresponds to the frequency $\omega = \tau_c^{-1}$. The flat portion of the curve corresponds to $\omega\tau_c \ll 1$, while the region to the right of the dot corresponds to $\omega\tau_c \gg 1$. The area under each of these curves is the same. This is equivalent to saying that the kinetic energy in any two samples at the same temperature is the same. This plot gives an intensity distribution for any particular frequency; *e.g.*, in the case of a magnetic nucleus undergoing molecular motion, we have the intensity distribution for each frequency of the oscillating fields arising from the complex motion that an assemblage of molecules undergoes.

For the middle curve (B), a correlation time has been selected to maximize $J(\omega)$ at ω_0 (*i.e.*, τ_c^{-1} is selected to be ω_0). Any curve with a larger or smaller τ_c will have a smaller value of $J(\omega)$ at ω_0. If we choose ω_0 at the Larmor frequency, the magnitude of $J(\omega)$ at ω_0 will be proportional to the relaxation efficiency; *i.e.*, the greater the intensity of the fluctuating moment, the more effective the relaxation. Correlation times corresponding to curve (B) are most efficient, (C) next most of those shown, and (A) the least efficient.

The T_1 relaxation is caused by frequencies that correspond to ω_0, while T_2 is caused by $\omega = 0$ and ω_0 frequency components. This is consistent with our earlier description of the fluctuating field causing T_1 relaxation and the static ($\omega = 0$) field in solids causing T_2 relaxation. If we were to plot the behavior of T_1 as a function of τ_c, we would obtain the curve labeled T_1 in Fig. 7–54. The minimum corresponds to the solid dot of curve (B) in Fig. 7–53. Larger or smaller correlation times give a smaller spectral density at this frequency. Curve (C) of Fig. 7–53, corresponding to the dashed line labeled (C) in Fig. 7–54, gives the behavior at shorter correlation times; a longer relaxation time results. It is interesting to point out that T_1 is a double-valued function; *i.e.*, a given T_1 can correspond to two possible values of τ_c. To obtain τ_c, one must know on what part of the curve the system is located.

The behavior of T_2 is described by the curve so labeled in Fig. 7–54. Note that for short correlation times, T_2 parallels T_1; *i.e.*, the spin-lattice relaxation mechanism is randomizing the z-component and the xy-component equally. In curve (A) of Fig. 7–53, we have a situation that would correspond to correlation times in a viscous

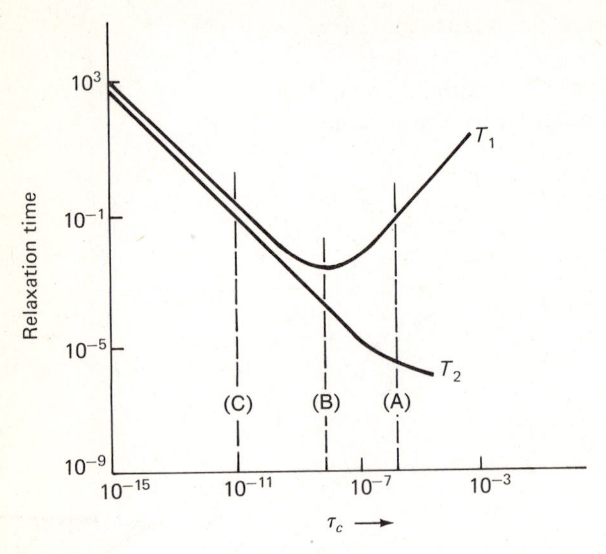

FIGURE 7–54. Dependences of the T_1 and T_2 relaxations.

liquid or solid. There is a greatly reduced intensity of the frequency component ω_0 needed for longitudinal relaxation. However, the $\omega = 0$ component is very large and the transverse relation is enhanced. The zero frequency dipolar broadening thus decreases T_2, causing line broadening in viscous liquids and solids. The increased T_1 associated with the increased correlation time in this region of the curve explains why spectra of solids and viscous liquid are easily saturated.

Often in the following pages we will note that processes that increase or decrease the correlation time can sharpen nmr peaks, *i.e.*, increase T_1 for the protons. One should be able to deduce whether T_1 will increase or decrease by knowing whether the spectral density change caused by the perturbation moves the curve away from (B) toward (A) or (C). If the curve is broadened, the perturbation moves the system from the (A) or (C) direction toward (B).

REFERENCES CITED

1. H. S. Gutowsky, "Physical Methods of Organic Chemistry," 3rd Ed., part 4, ed. A. Weissberger (Vol. 1 of "Technique of Organic Chemistry"), Interscience, New York, 1960.
2. J. A. Pople, W. G. Schneider, and A. J. Bernstein, "High Resolution Nuclear Magnetic Resonance," McGraw-Hill, New York, 1959.
3. (a) J. W. Emsley, J. Feeney, and L. H. Sutcliffe, "High Resolution Nuclear Resonance Spectroscopy," Vol. 1, Pergamon Press, New York, 1966.
   (b) Same authors and title; Vol. 2.
4. M. W. Dietrich and R. E. Keller, Anal. Chem., *36,* 258 (1964) and references therein.
5. L. Eberson and S. Farsen, J. Phys. Chem., *64,* 767 (1960).
6. N. S. Bhacca, L. F. Johnson, and J. N. Shoolery, "NMR Spectra Catalog," Varian Associates, Palo Alto, Calif., 1962; "The Sadtler Standard Spectra; NMR," Sadtler Research Laboratories, Philadelphia, 1972; L. F. Johnson and W. C. Jankowski, "C-13 NMR Spectra" (a collection), Interscience, New York, 1972.
7. A. L. Allred and E. G. Rochow, J. Amer. Chem. Soc., *79,* 5361 (1957).
8. B. P. Dailey and J. N. Shoolery, J. Amer. Chem. Soc., *77,* 3977 (1955).
9. A. D. Buckingham and P. J. Stephens, J. Chem. Soc., *1964,* 4583.
10. E. Fluck, "Anorganische und Allgemeine Chemie in Einzeldarstellungen. Band V. Die Kernmagnetische Resonanz und Ihre Anwendung in der Anorganische Chemie," Springer-Verlag, Berlin, 1963.
11. H. S. Gutowsky, J. Chem. Phys., *37,* 2136 (1962).
12. J. S. Waugh and F. A. Cotton, J. Phys. Chem., *65,* 562 (1961).
13. R. A. Hoffman and S. Gronowitz, Arkiv. Kemi., *16,* 471, 563, (1961); *17,* 1 (1961); Acta Chem. Scand., *13,* 1477 (1959).
14. M. Karplus, J. Chem. Phys., *33,* 1842 (1960).

15. (a) M. T. Rogers and J. D. Graham, J. Amer. Chem. Soc., *84,* 3666 (1962); (b) J. D. Kennedy *et al.,* Inorg. Chem., *12,* 2742 (1973).
16. W. P. Griffith and G. Wilkinson, J. Chem. Soc., *1959,* 2757.
17. H. S. Gutowsky, D. W. McCall, and C. P. Slichter, J. Chem. Phys., *21,* 279 (1953).
18. C. F. Callis, J. R. Van Wazer, J. N. Shoolery, and W. A. Anderson, J. Amer. Chem. Soc., *79,* 2719 (1957).
19. R. O. Ragsdale, and B. B. Stewart, Inorg. Chem., *2,* 1002 (1963).
20. N. Muller, J. Chem. Phys., *36,* 359 (1962).
21. H. S. Gutowsky, and C. Juan, J. Amer. Chem. Soc., *84,* 307 (1962); J. Chem. Phys., *37,* 2198 (1962).
22. J. H. Goldstein and G. S. Reddy, J. Chem. Phys., *36,* 2644 (1962).
23. R. S. Drago and N. A. Matwiyoff, J. Chem. Phys., *38,* 2583 (1963); J. Organomet. Chem., *3,* 62 (1965).
24. H. A. Bent, Chem. Revs., *61,* 275 (1961).
25. J. Hinze and H. H. Jaffe, J. Amer. Chem. Soc., *84,* 540 (1962).
26. M. Karplus, J. Amer. Chem. Soc., *85,* 2870 (1963).
27. S. Sternhall, Quart. Revs. (London), *23,* 236 (1969) and references therein.
28. M. A. Cooper and S. L. Manatt, J. Amer. Chem. Soc., *92,* 1605 (1970).
29. R. L. Middaugh and R. S. Drago, J. Amer. Chem. Soc., *85,* 2575 (1963).
30. J. R. Holmes and H. D. Kaesz, J. Amer. Chem. Soc., *83,* 3903 (1961).
31. N. A. Matwiyoff and R. S. Drago, Inorg. Chem., *3,* 337 (1964).
32. G. D. Shier and R. S. Drago, J. Organomet. Chem., *6,* 359 (1966).
33. T. F. Bolles and R. S. Drago, J. Amer. Chem. Soc., *88,* 5730 (1966).
34. J. F. Nixon and A. Pidcock, Ann. Rev. NMR Spectroscopy, *2,* 345 (1969).
35. R. K. Harris, Canad. J. Chem., *42,* 2275, 2575 (1964).
36. (a) B. E. Mann, B. L. Shaw, and K. E. Stainbank, Chem. Comm., 151 (1972);
(b) D. E. Axelson and C. E. Holloway, Chem. Comm., 455 (1973).
37. E. L. Muetterties, J. Amer. Chem. Soc., *82,* 1082 (1960).
38. E. L. Muetterties, W. Mahler, and R. S. Schmutzler, Inorg. Chem., *2,* 613 (1963).
39. O. E. Myers, J. Chem. Phys., *28,* 1027 (1958).
40. Y. Masuda, J. Phys. Soc. Japan, *11,* 670 (1956).
41. J. D. Roberts, J. Amer. Chem. Soc., *78,* 4495 (1956).
42. T. C. Farrar and E. D. Becker, "Introduction to Pulse and Fourier Transform NMR Methods," Academic Press, New York, 1971.
43. A. Allerhand and D. Doddrell, J. Amer. Chem. Soc., *93,* 2777 (1971).
44. A. Allerhand, *et al.,* J. Chem. Phys., *55,* 189 (1971).
45. G. N. LaMar, J. Amer. Chem. Soc., *93,* 1040 (1971).
46. D. F. S. Natusch, J. Amer. Chem. Soc., *93,* 2566 (1971).
47. K. F. Kuhlmann, D. M. Grant, and R. K. Harris, J. Chem. Phys., *52,* 3439 (1970).
48. T. D. Alger, S. W. Collins, and D. M. Grant, J. Chem. Phys., *54,* 2820 (1971).

EXERCISES

1. The δ value of a substance relative to the external standard, methylene chloride, is -2.5. Calculate δ relative to external standards (a) benzene and (b) water using the data in Appendix F.

2. Assuming the relationship discussed between J_{Sn-H} and hybridization, what would be the ratio of the coupling constants in a five-coordinate and six-coordinate complex of $(CH_3)_3SnCl$ [*i.e.,* $(CH_3)_3SnCl \cdot B$ and $(CH_3)_3SnCl \cdot 2B$]?

3. The compound $B[N(CH_3)_2]_3$ is prepared and dissolved in a wide number of different solvents. Propose a method of determining in which ones the solvent is coordinated to the compound.

4. Consider the diamagnetic complex $(Me_3P)_4Pt^{2+}$. Sketch the phosphorus resonance signal if

   a. $J_{P-H} > J_{P-Pt}$.

   b. $J_{P-Pt} > J_{P-H}$.

5. Indicate the number of isomers for cyclic compounds of formulas $P_3N_3(CH_3)_2Cl_4$, and sketch the phosphorus resonance spectrum of each (assume $\Delta > J$, J_{P-H} is small, and J_{P-H} can be ignored for phosphorus atoms which do not contain methyl groups).

6. Would you expect the ^{14}N NMR spectrum to be sharper in NH_3 or NH_4^+? (For ^{14}N, $I = 1$.) Explain.

7. Consider all possible isomers that could be obtained for the eight-membered ring compound $P_4N_4Cl_6(NHR)_2$ and indicate the ideal phosphorus resonance spectrum expected for each. Which of the above are definitely eliminated if the phosphorus resonance consists of two triplets of equal intensity?

8. The proton nmr spectrum of $\begin{matrix} CH_2\text{—}O \\ | \\ CH_2\text{—}O \end{matrix}\!\!>\!S\text{—}O$ is not a singlet. Is the SO_3 group planar?

   What would the spectrum look like if the sulfur underwent rapid inversion?

9. It is found that the methylene groups in $(CH_3CH_2)_2SBF_3$ give rise to a single methylene resonance. Explain.

10. a. Are the methyl groups in $(CH_3)_2\overset{H}{C}\text{—}P(O)(Cl)\text{—}C_6H_5$ equivalent?

    b. Ignore the splitting of the methyl groups by the phenyl protons in the above compound and assume $J < \Delta$. What would the spectrum of the methyl protons look like?

11. The proton nmr spectra of a series of compounds are given below. Assign their geometries and interpret the spectra.

    a. The compound is $ReCl_3[P(CH_3)_2C_6H_5]_3$. Curve (1) is the full spectrum, (2) is a more intense sweep of the 6.0 to 13.0 ppm region, and (3) is the 100 MHz spectrum of the 7.0 to 10 ppm region.

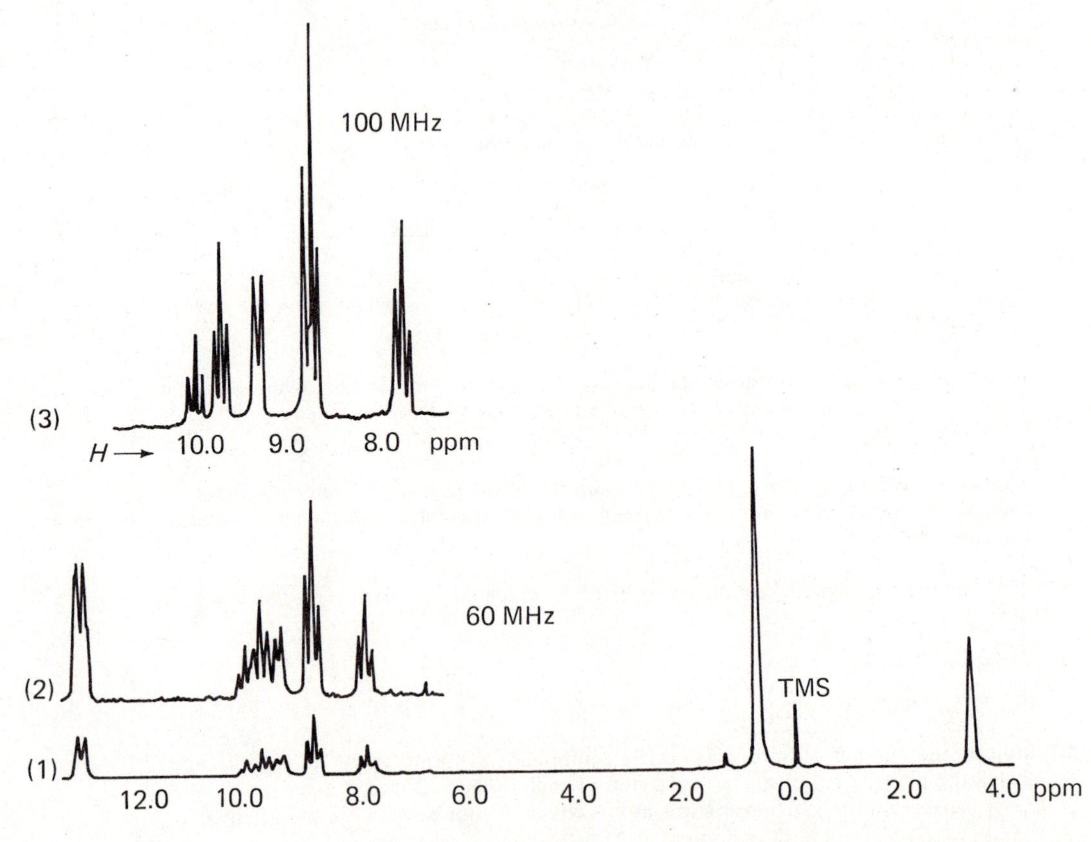

    b. Why is the spectrum in part (a) at 100 MHz different from that at 60 MHz? Why are the chemical shifts in ppm relative to TMS the same?

c. A compound with empirical formula $C_4H_{11}N$.

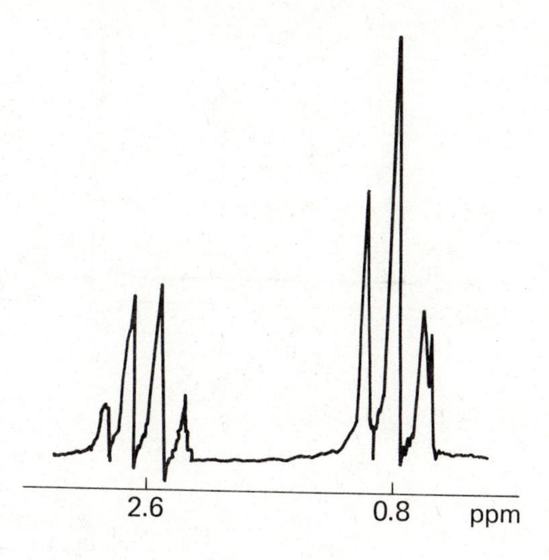

d. A compound with empirical formula $C_2H_3F_3O$.

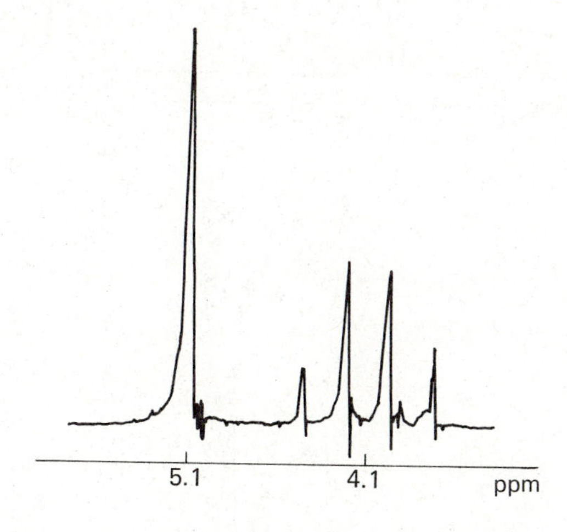

e. A compound with empirical formula C_2H_4O.

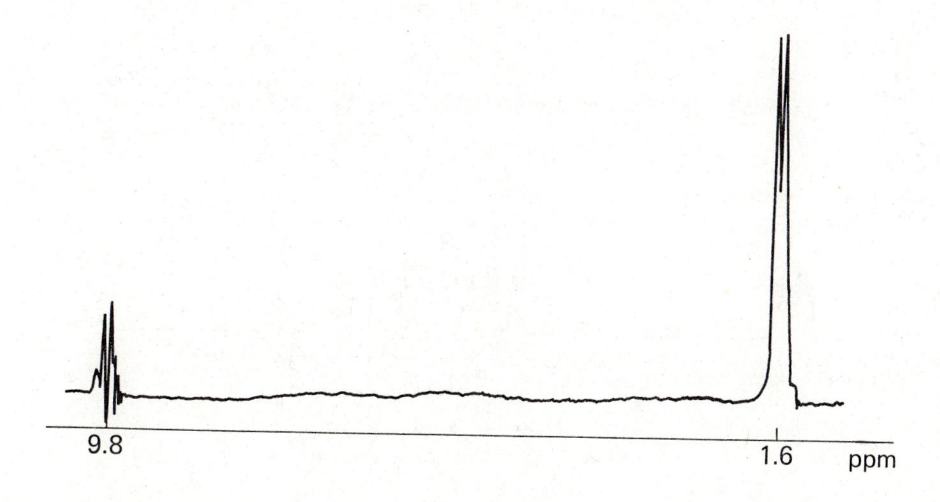

f. A compound with empirical formula C_4H_6.

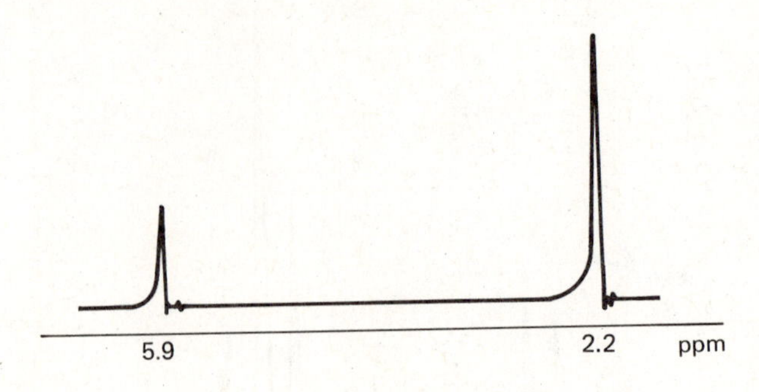

g. A compound with empirical formula $C_4H_8O_2$.

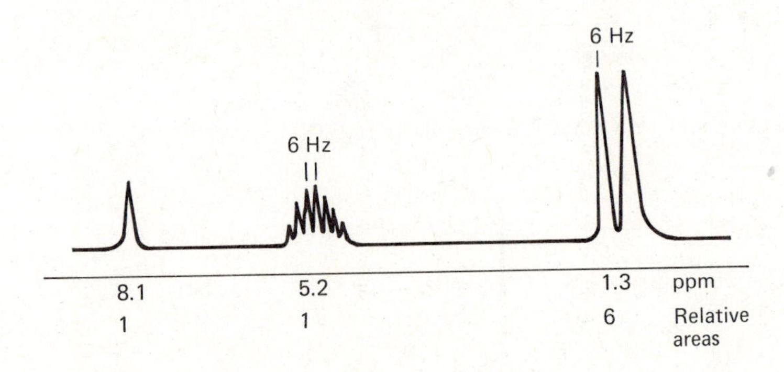

h. A compound with empirical formula $C_3H_5ClF_2$.

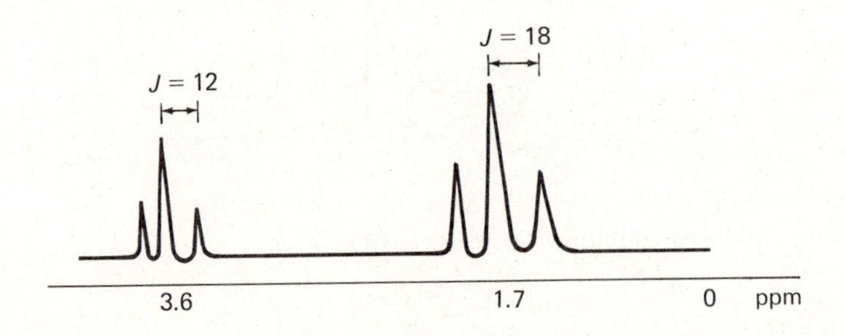

i. A compound with empirical formula $C_8H_{10}O$. The peak at 2.4 ppm vanishes in D_2O.

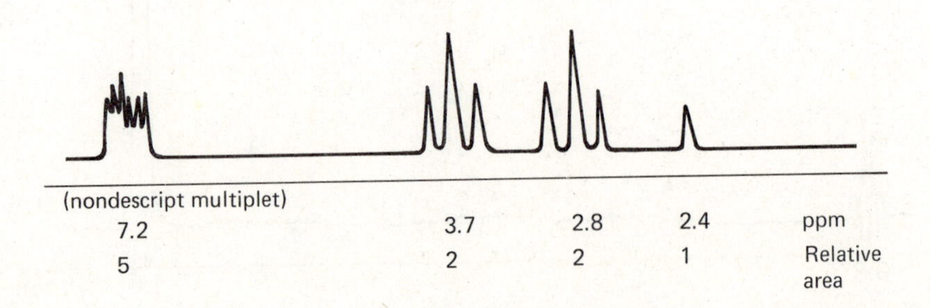

j. A compound with empirical formula $C_6H_{15}O_4P$.

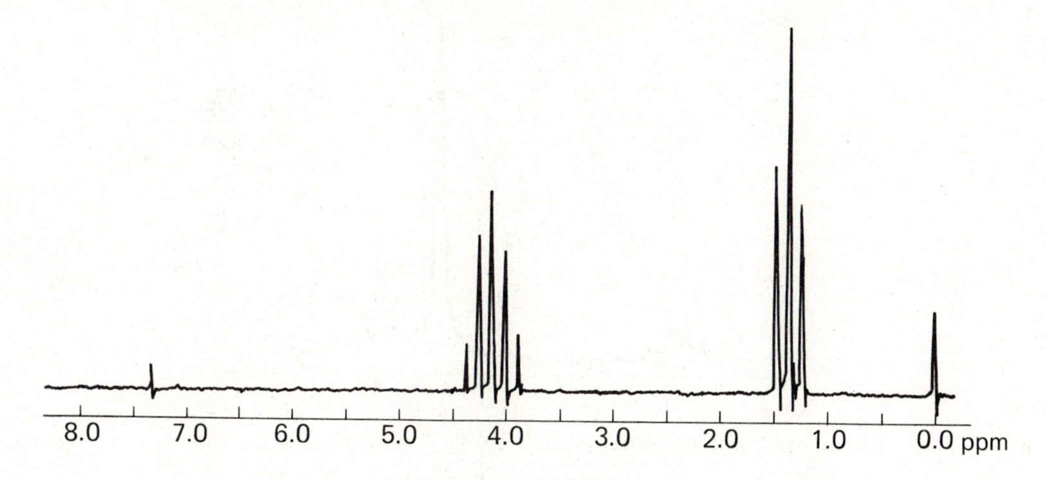

k. A compound with empirical formula $C_3H_5O_2Cl$.

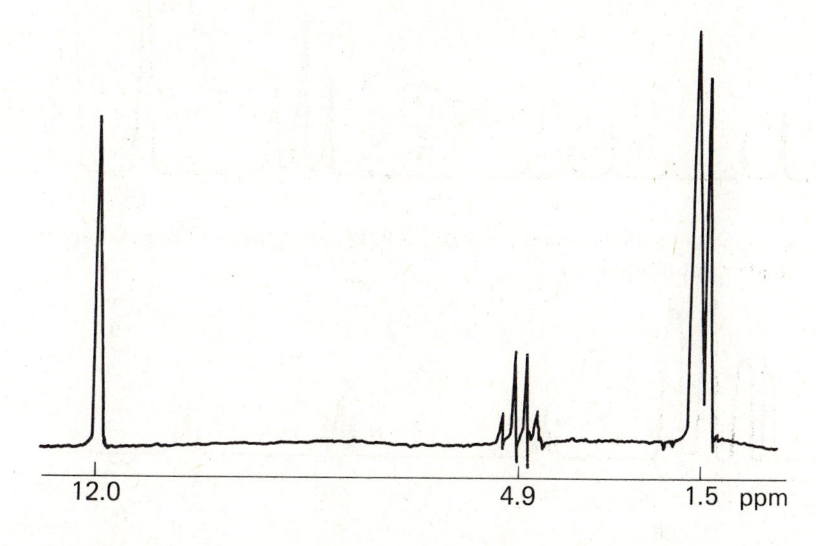

l. A compound with empirical formula C_2H_8NCl dissolved in water.

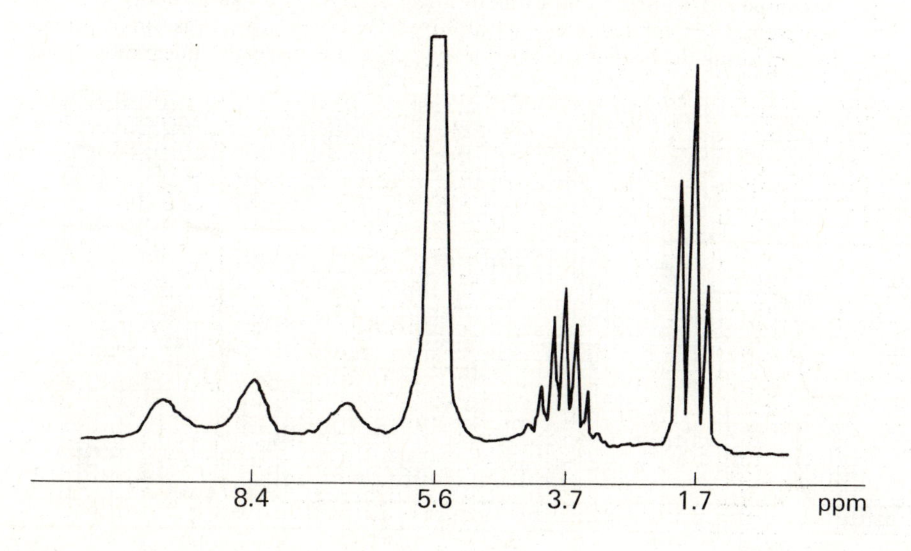

m. A compound with empirical formula $C_{14}H_{22}O_4$.

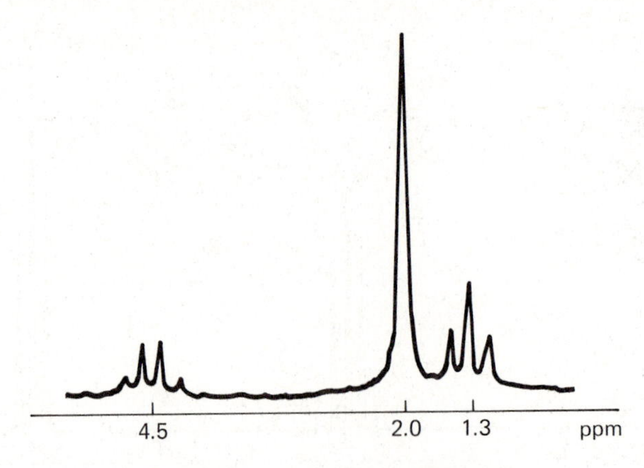

n. A compound with empirical formula $(SiH_3)_2PSiH_2CH_3$.

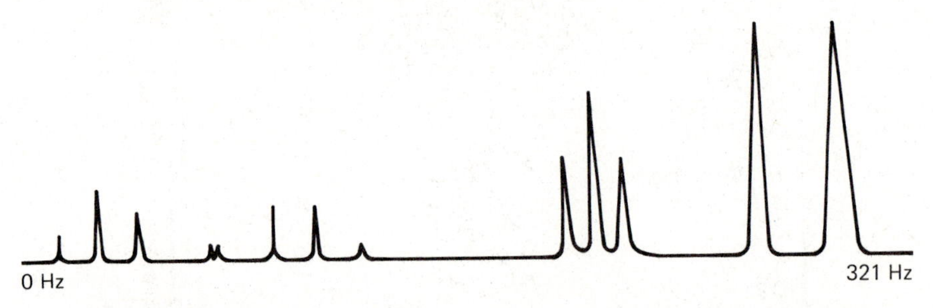

o. A compound with the empirical formula $Pt[P(C_2H_5)_3]_2HCl$. (The peaks at 8 ppm are nondescript multiplets.)

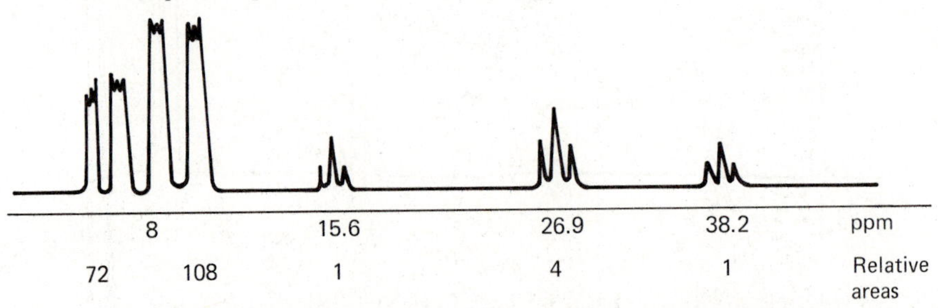

p. A compound with the empirical formula $C_{11}H_{16}$. The curves above the peaks represent the integrated intensities of the peaks. The relative areas can be obtained by comparing the heights (number of squares) of the respective integration curves.

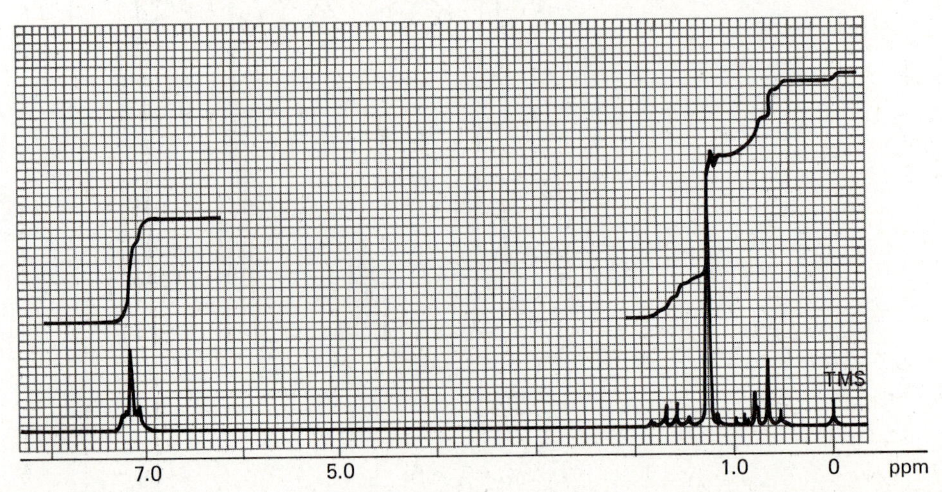

q. A compound with empirical formula $C_8H_{10}O$. The curves above the peaks represent the integrated intensities of the peaks. The relative areas can be obtained by comparing the heights, in squares, of the respective integration curves.

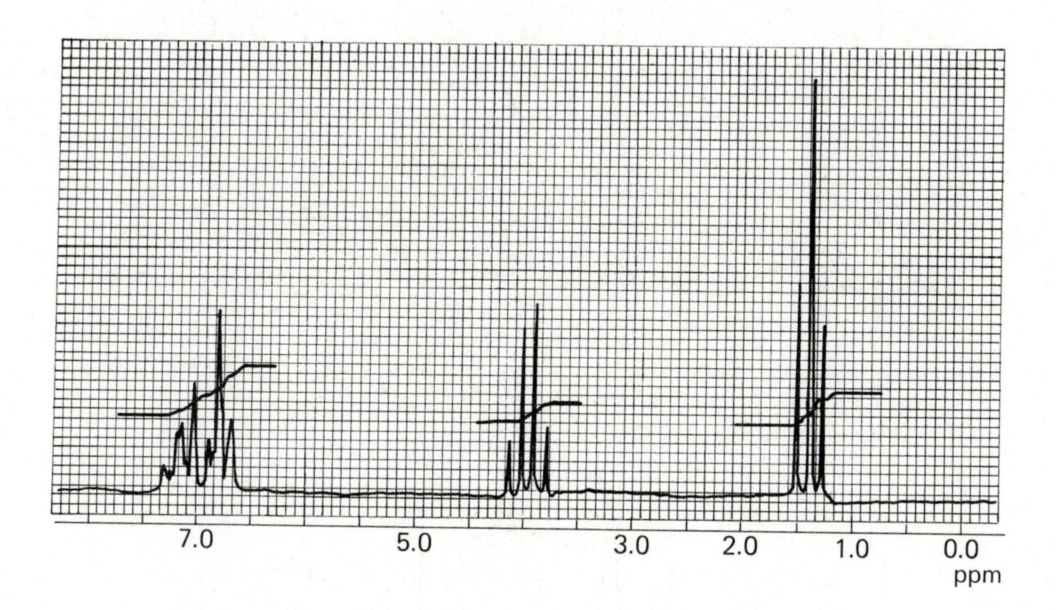

r. A compound with empirical formula $C_5H_{10}O$. The curves above the peaks represent the integrated intensities of the peaks. The relative areas can be obtained by comparing the heights, in squares, of the respective integration curves.

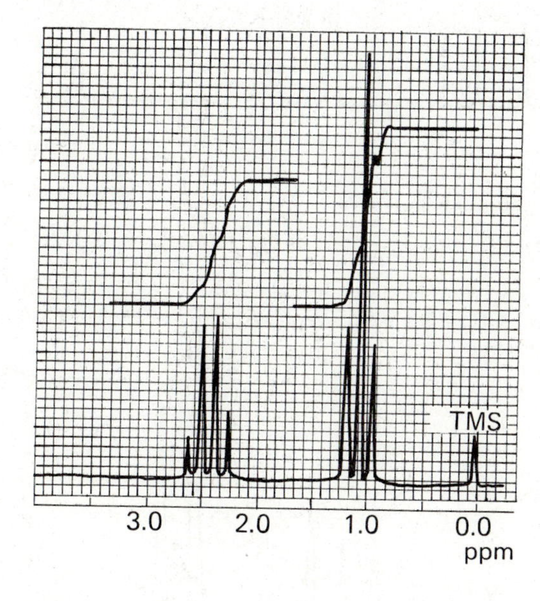

s. A compound with empirical formula C_8H_7OF.

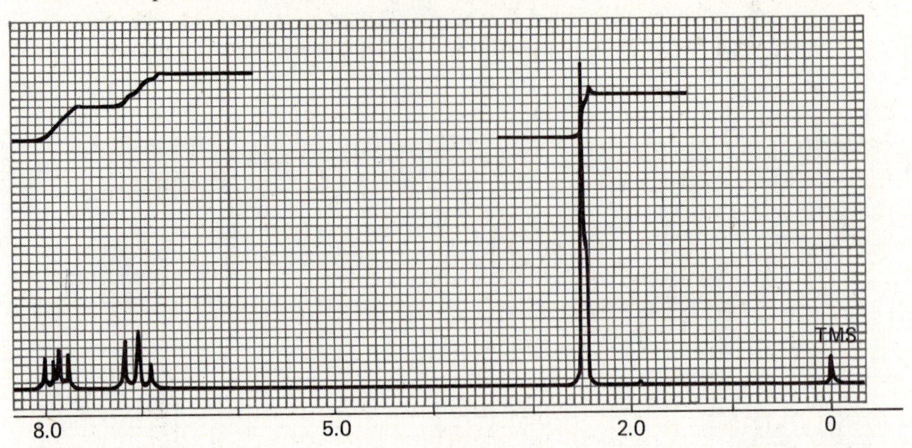

12. Spectrum (a) results for the methyl region of one isomer of $Pd[P(CH_3)_2C_6H_5]_2I_2$. Spectrum (b) is obtained in the methyl region for the other isomer of $Pt[P(CH_3)_2C_6H_5]_2I_2$.

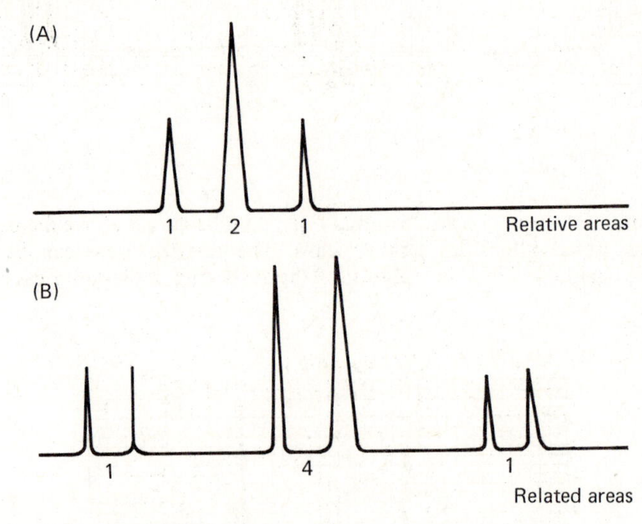

a. Which is *cis* and which is *trans?*

b. Explain why the areas in (b) are in the ratio of 1 to 4 to 1.

13. The ^{11}B nmr spectrum below is obtained for 5—Cl—2,4—$C_2B_5H_6$. Interpret the spectrum. (Hint: $B_7H_7^{2-}$ is a pentagonal bipyramid of B—H subunits in which the axial borons are numbered 1 and 7.)

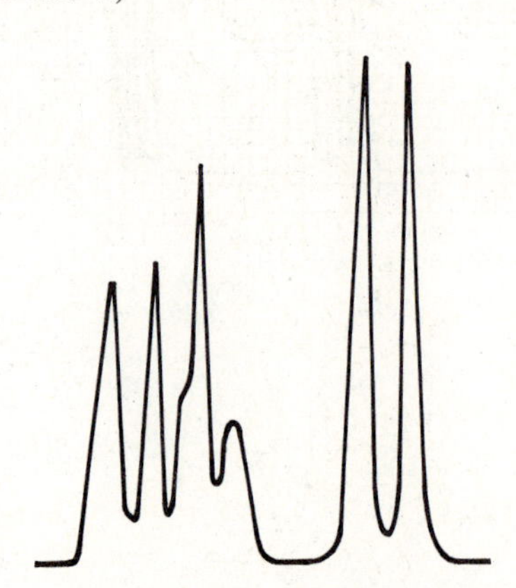

14. The ^{19}F nmr spectrum of $PF_3(NH_2)_2$ is reported in (a), and expanded spectra of the individual peaks are presented in (b) through (e). Propose a structure and interpret the spectrum.

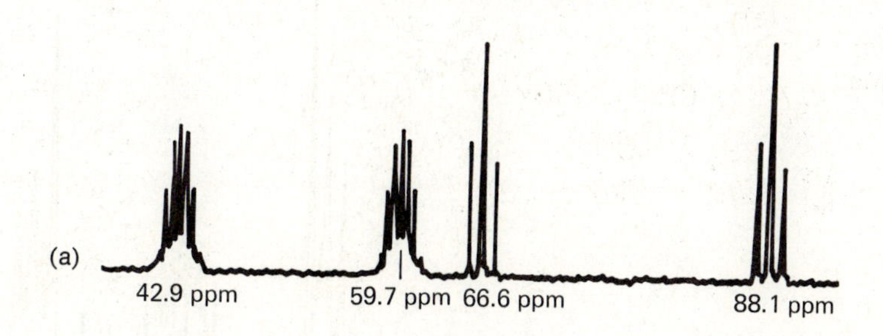

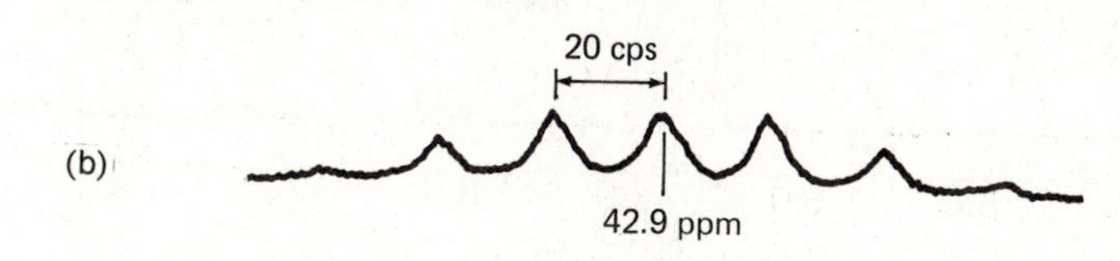

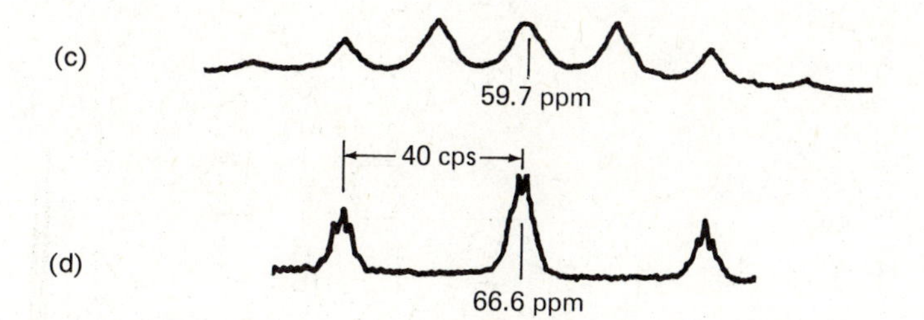

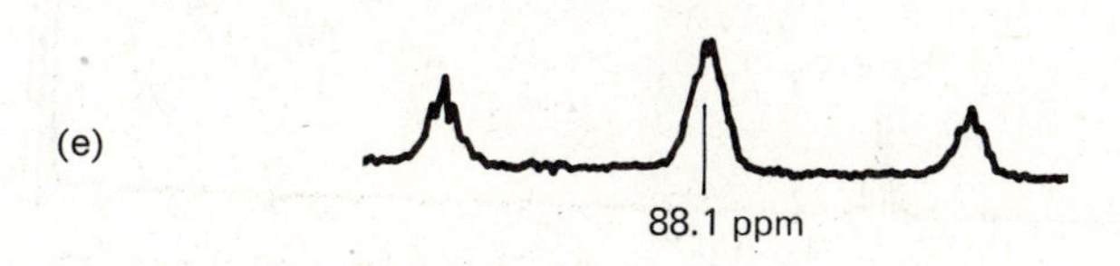

15. The proton resonance of the hydridic hydrogen of $HNi[OP(C_2H_5)_3]_4^+$ is given below. Propose a structure for this complex.

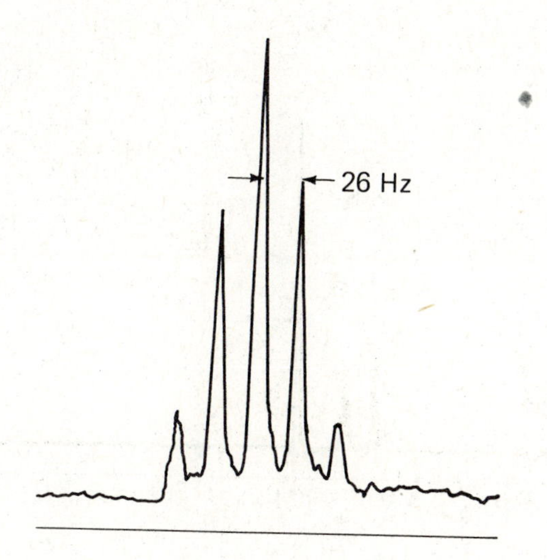

16. Given the following proton nmr spectra and molecular structures, assign all peaks:

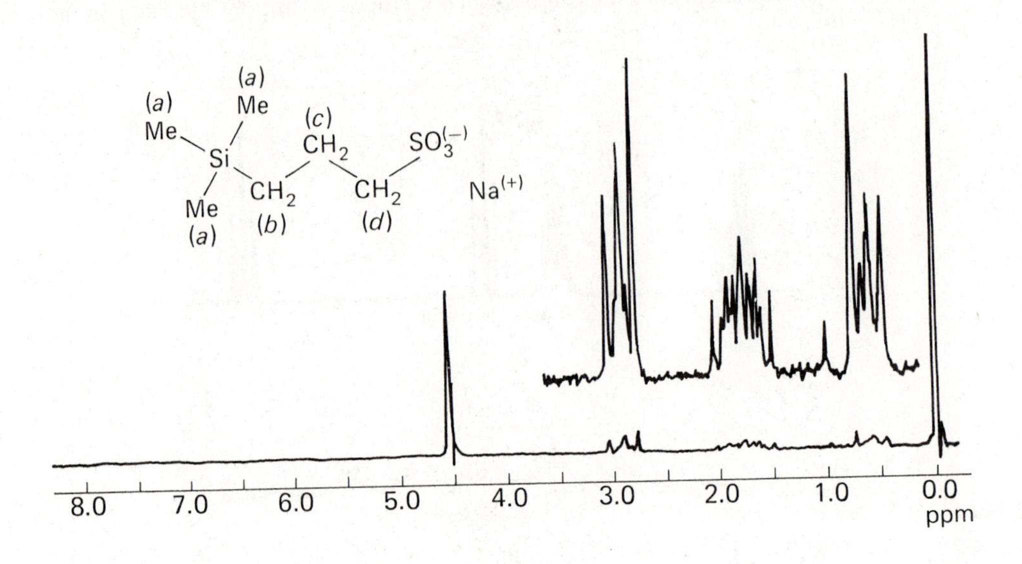

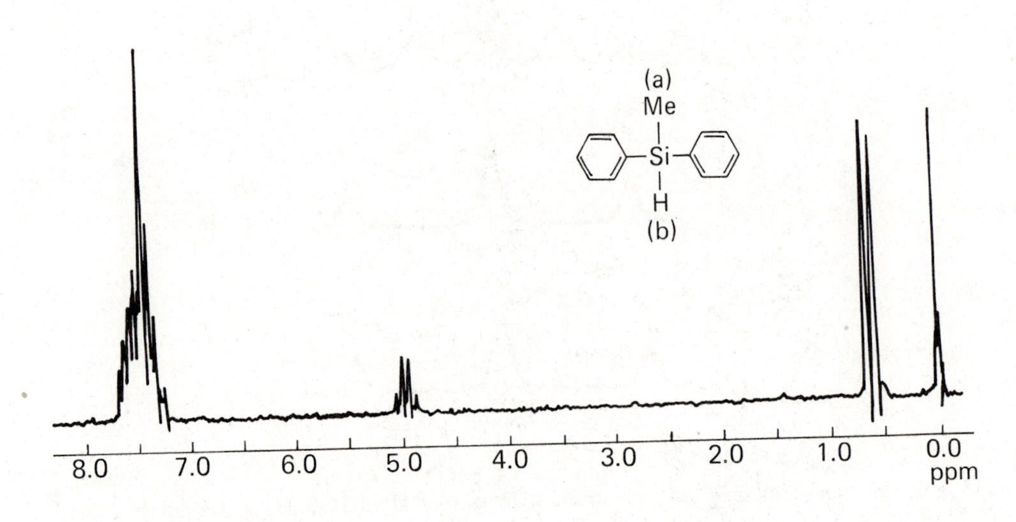

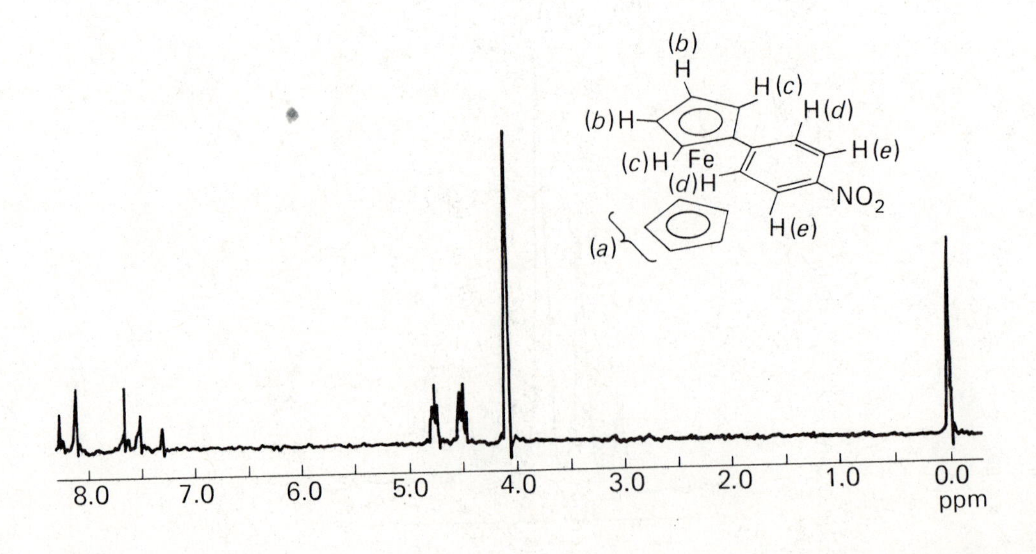

17. What properties in a functional group X (which can consist of more than one atom) attached to an isopropyl group can cause non-equivalent methyls to occur in the isopropyl group?

18. The following proton nmr spectrum is that of a compound with the molecular formula $C_8H_{11}NO$. Propose a structure consistent with the formula and spectrum. Insofar as possible, assign the peaks. Numbers in parentheses refer to relative peak areas.

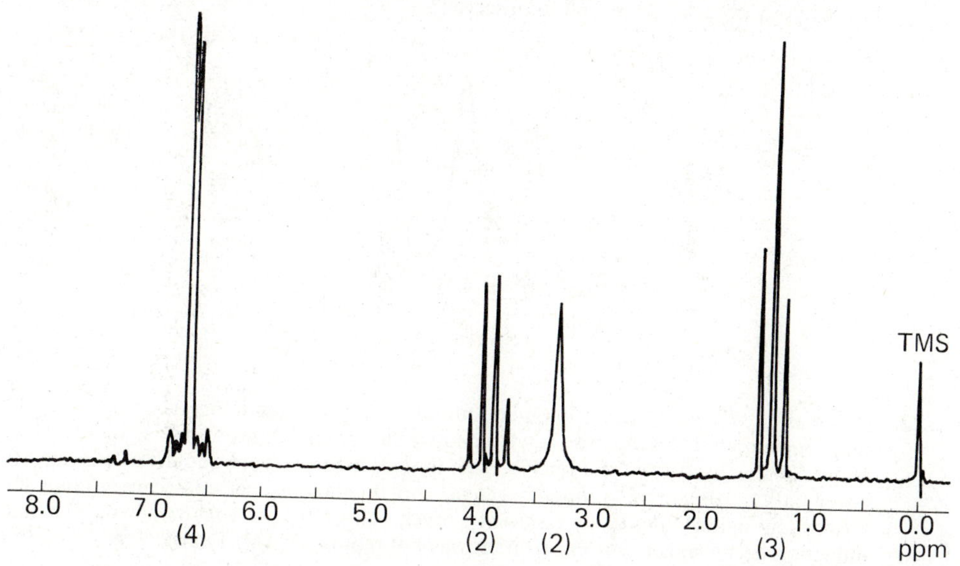

19. Interpret the spectrum below. Justify all splittings. Is this the *cis* or *trans* isomer? (Hint: For ^{31}P, $I = \frac{1}{2}$, abundance = 100%. For ^{195}Pt, $I = \frac{1}{2}$, abundance = 34%.) [See Inorganic Chemistry, *12,* 994 (1973).]

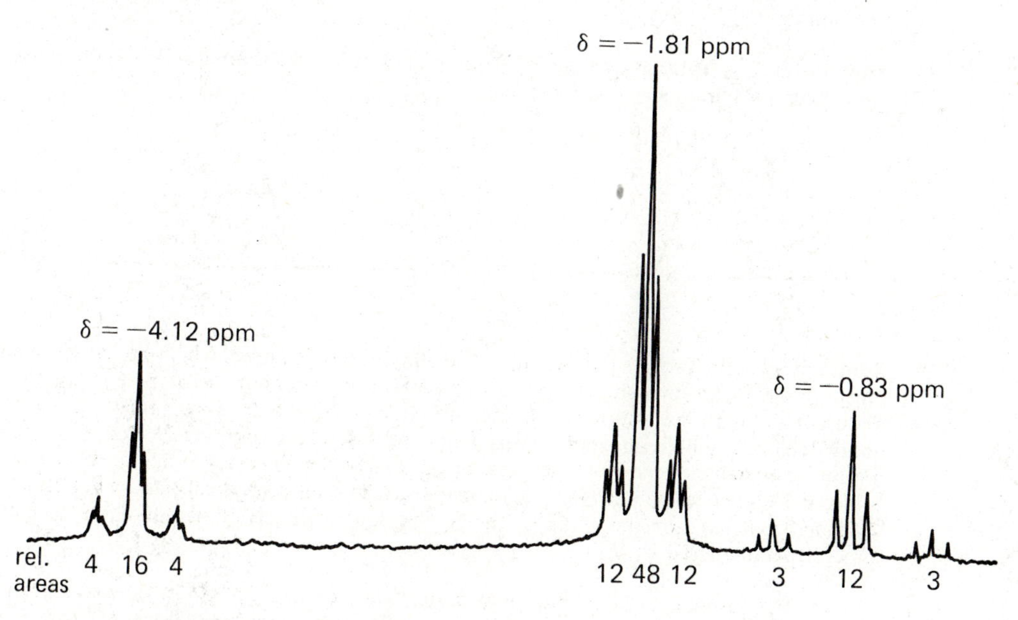

20. Sketch the spectra for the following molecules and conditions (assume 100% abundances):

a. H_2PF_3, structure F—P(H)(H)(F)(F) $J_{P-F} > J_{P-H} > J_{H-F}$ $I_H = \frac{1}{2}$, $I_F = \frac{1}{2}$, $I_P = \frac{1}{2}$

^1H, ^{19}F, and ^{31}P resonances

b. $N(CH_2F)_4^+$ $\quad J_{FCH} > J_{FCN};\ J_{FCNCH} = 0$
^{19}F *only*

21. The following proton nmr spectrum is reported [Inorganic Chemistry, *2*, 939 (1963)] for $((NH_2)_3P)_2Fe(CO)_3$ (the two phosphorus ligands are *trans*).

    a. Explain the splittings, assuming that there is no proton-nitrogen coupling observed.

    b. Why isn't the N—H coupling observed?

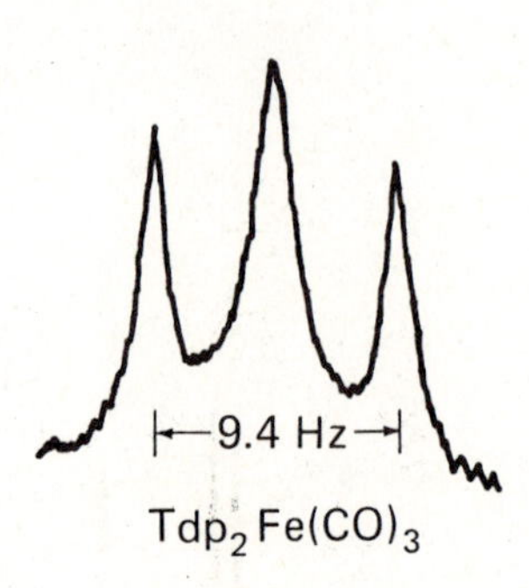

$Tdp_2Fe(CO)_3$

22. ^{57}Fe has an extremely small magnetic moment, and its abundance is only 2.2%. The magnetic moment is not known to very good accuracy, and almost no information exists about chemical shifts of ^{57}Fe nuclei. Even though the resonance signal was expected to cover only a very small frequency range, and thus a slow passage scan through the absorption signal would not take very much longer to observe than would a free induction decay signal, this experiment was not reported until Fourier transform nmr became available.

    a. Why was Fourier transform nmr advantageous for this experiment?

    b. Why wasn't any ^{57}Fe-^{13}C spin-spin splitting observed in the natural-abundance sample?

23. Suppose that a sample has two sites with different chemical shifts. In the CW experiment, we expect a separate signal for each site.

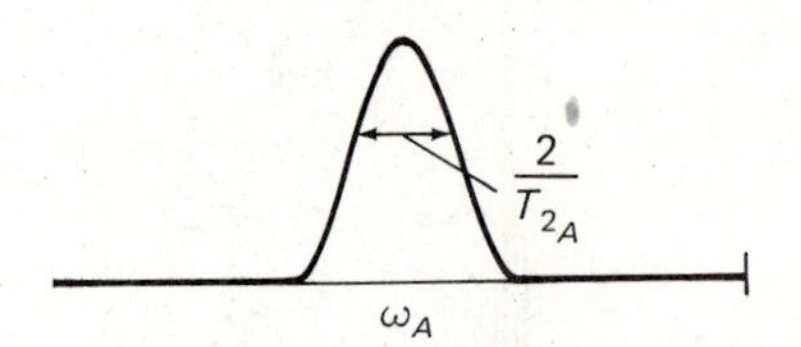

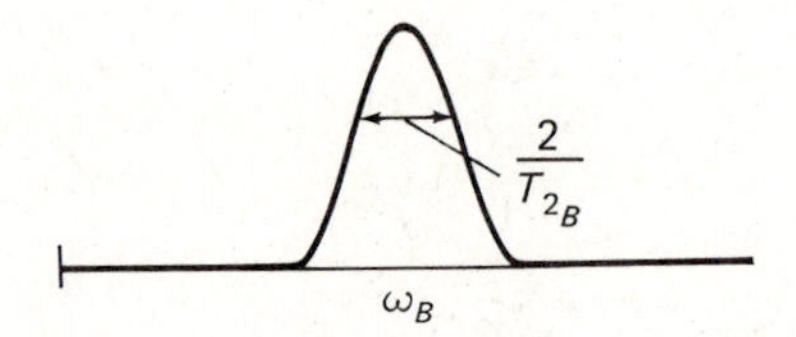

Consider a kinetic process that sets in as the temperature is raised, which exchanges the nuclei between the two sites A and B. (In general, the problem is complex, since the differential equations, one for each site, are coupled by the dynamic process and the individual component magnetizations begin to lose their identities.) Consider the limiting case in which the kinetic process begins to have an observable effect, yet one can still consider the individual M_A magnetization. Argue on a physical basis what effect exchange will have on the width of the A-type resonance. (Hint: Recall the physical process parametrized by T_{2_A}).

24. a. Let the resonance frequency of a water sample be ω_{H_2O}. Show how one can determine the pulse duration τ and a field strength H_1 for an oscillator tuned to ω_{H_2O} that will invert the H_2O magnetization.

    b. Suppose that at a later time a $3\pi/2$ pulse was applied and that no free induction decay curve was observed. What explains this? Of what use is this experiment?

    c. A solute HX is added to the sample, and proton exchange occurs at a rate such that separate but broadened resonances are observed for H_2O and HX. Discuss what effect is expected on solute signal intensity observed in a CW experiment if, while scanning through it, one strongly irradiates the sample of ω_{H_2O}, saturating the H_2O spin system.

25. In the experiment described in the text for measuring T_1, why must the 180° pulse be followed by a 90° pulse?

26. In the slow passage nmr experiment, one frequently observes the following type of pattern when sweeping the field from left to right:

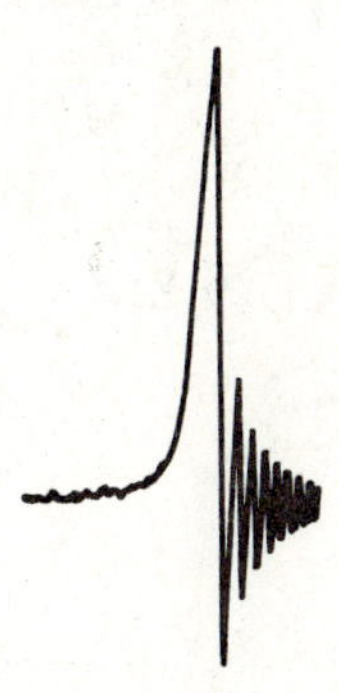

The wiggles to the right of the absorption peak are referred to as "ringing" and are found only when the nucleus observed has a long T_2. Explain ringing in terms of the behavior of the bulk magnetization of the sample in this experiment. Be sure to specify your frame of reference.

27. In the section on the effects of nuclei with quadrupole moments, there was a discussion of the width of the ^{19}F resonance in the spectrum of NF_3 as a function of temperature.

    a. Explain the quadrupole moment relaxation in terms of the concept of spectral density.

    b. Explain the resulting nmr spectrum in terms of spectral density.

28. Consider a frame of reference rotating at a frequency $\omega_1 = 1.0000 \times 10^8$ Hz and a sample of nuclei with a Larmor frequency of $\omega_0 = 1.0001 \times 10^8$ Hz, with $T_1 = 100$ sec and $T_2 = 0.01$ sec. At $t = 0$, a 90° pulse is applied along the u-axis using a strong rf field of frequency ω_1, in a negligible length of time. Using a u-z or v-z axis system in each case, draw the net magnetization vector

    a. just before the pulse.

    b. just after the pulse.

    c. at $t = 2.5 \times 10^{-5}$ sec.

    d. at $t = 1$ sec.

    e. at $t = 10^4$ sec.

8 NUCLEAR MAGNETIC RESONANCE SPECTROSCOPY—ADDITIONAL PRINCIPLES AND APPLICATIONS

INTRODUCTION

In this chapter we shall consider some of the more subtle principles and applications of nmr. A third chapter (Chapter 12), devoted to the nmr spectra of paramagnetic ions, will complete the topic of nmr. The nmr and esr of paramagnetic transition metal complexes will not be discussed in this text until after some fundamental background in magnetism is developed in Chapter 11.

EVALUATION OF THERMODYNAMIC DATA WITH NMR

As mentioned in Chapter 7, when two species undergo rapid exchange on the nmr time scale, the chemical shift observed is a mole-fraction weighted average of the two resonances. For example, with rapid exchange occurring in the system:

$$A + B \rightleftharpoons AB \tag{8-1}$$

the chemical shift of the A (or B) resonance will be a mole-fraction weighted average of the resonance of the free A (or B) and that of the analogous atom in the AB adduct, as shown in equation (8-2).

$$\delta_{obs} = N_A\delta_A + N_{AB}\delta_{AB} \tag{8-2}$$

where N refers to mole fraction. An analogous equation could be written for a resonance in B.

This exchange averaging can be used[1] to evaluate the equilibrium constant for a reaction. The approach will be illustrated by deriving an expression for the equilibrium constant of the reaction illustrated in equation (8-1). At the outset, we shall assume that we are observing the chemical shift of a proton of molecule A which is shifted considerably upon forming the complex AB and that there is rapid exchange between A and AB giving a mole-fraction weighted shift as in equation (8-2).

Expressing equation (8-2) in molarity units, we obtain for the reaction in equation (8-1):

$$\delta_{obs} = \frac{[A]}{[A] + [AB]}\delta_A + \frac{[AB]}{[A] + [AB]}\delta_{AB}$$

Rearranging, collecting terms, and subtracting $[AB]\delta_A$ from both sides of the equation produces:

$$[A](\delta_{obs} - \delta_A) + [AB](\delta_{obs} - \delta_A) = [AB](\delta_{AB} - \delta_A)$$

Defining $\Delta\delta_{obs}$ as $(\delta_{obs} - \delta_A)$ and $\Delta\delta_{CA}$ as $(\delta_{AB} - \delta_A)$, we can write:

$$[AB] = \frac{[A^\circ]\,\Delta\delta_{obs}}{\Delta\delta_{CA}} \tag{8-3}$$

where $[A^\circ]$, the initial concentration of A, equals $[A] + [AB]$.

Substituting equation (8–3) into the equilibrium constant expression

$$K = \frac{[AB]}{([A^\circ] - [AB])([B^\circ] - [AB])}$$

one obtains equation (8–4)

$$K = \frac{\Delta\delta_{obs}}{(\Delta\delta_{CA} - \Delta\delta_{obs})\left([B^\circ] - \dfrac{\Delta\delta_{obs}[A^\circ]}{\Delta\delta_{CA}}\right)} \tag{8-4}$$

In equation (8–4), all quantities are known except K and $\Delta\delta_{CA}$. The two unknowns are constant at a given temperature and can be obtained[1] from a series of simultaneous equations that result from measuring $\Delta\delta_{obs}$ in a series of experiments in which $[B^\circ]$ and $[A^\circ]$ are varied. This aspect of the problem is similar to that described in Chapter 4 for systems that obey Beer's law.

8–1 RATE CONSTANTS AND ACTIVATION ENTHALPIES FROM NMR

NMR KINETICS

The nmr spectrum of the deuterated cyclohexane molecule (shown in Fig. 8–1(A)) as a function of temperature is shown in Fig. 8–1(B) to (G). Two isomers are possible for this molecule, in which the hydrogen is in either the axial or equatorial position of the cyclohexane ring. At room temperature, the two forms are rapidly interconverting and there is no contribution to the observed line width from exchange effects. As the temperature is lowered, the band begins to broaden from chemical exchange contributions to the nuclear excited state lifetime. This range, down to the temperature at which the two separate peaks are just beginning to be resolved, is referred to as the *near fast exchange region.* As the temperature is lowered further, two peaks become detectable and their chemical shifts change as a function of temperature until curve (F) is obtained. This temperature range is referred to as the *intermediate exchange region.* As the temperature is lowered still further, the chemical shift of the peaks no longer changes, but the resonances sharpen throughout the so-called *slow exchange region.* Finally, at -79° and lower, there are no kinetic contributions to the shape of the spectrum, and this is referred to as the *stopped exchange region.*

By a full analysis of the influence of chemical exchange on the magnetization via the Bloch equations, it is possible to derive equations for the evaluation of rate constants from the nmr spectra. The derivations are beyond the scope of this treatment and the reader is referred to Emsley, Feeney, and Sutcliffe, Volume 1, Chapter

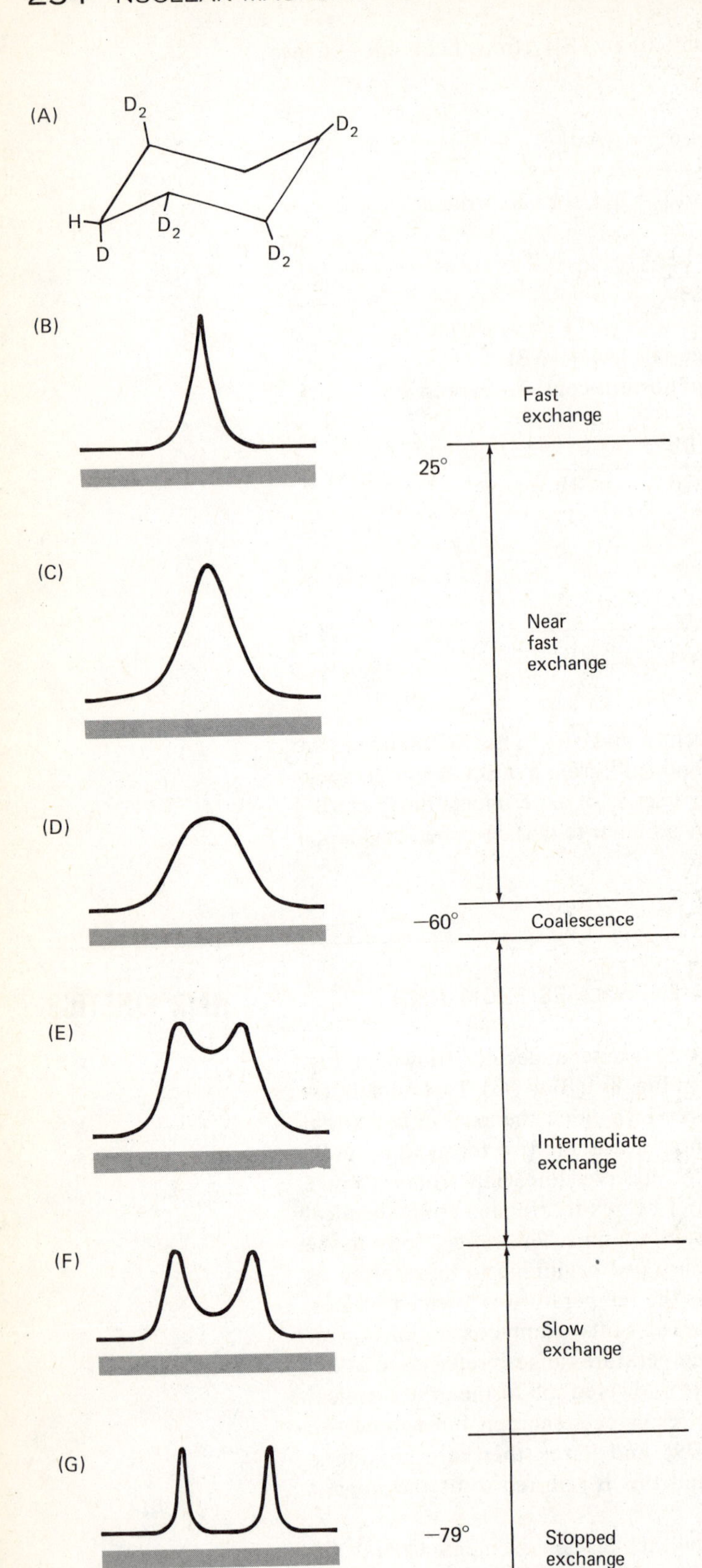

FIGURE 8–1. Deuterated cyclohexane (A) and the temperature dependence of its nmr spectrum (B to G).

9, or Pople, Schneider, and Bernstein, p. 218, for more details. We shall present the results of these derivations here in a form useful for a kinetic analysis and comment on the shortcomings of the various approaches that have been employed.

One of the simplest systems to treat is one in which a given proton can be at either one of two molecular sites; the probability that it will be at one site is equal to the probability that it will be at the other, and it has the same lifetime at each. The cyclohexane interconversion and many other isomer problems satisfy these criteria. If one works with the chemical shift changes, rate data can be extracted from any two-site (A and B) equal-lifetime process both in the intermediate exchange region and at the temperature at which the peaks have just merged by using equations (8–5) and (8–6), respectively. Equation (8–5) gives very crude results.

$$\frac{(\nu_A - \nu_B)_{obs}}{\nu_A{}^0 - \nu_B{}^0} = \left[1 - \frac{1}{2\pi^2\tau'^2(\nu_A{}^0 - \nu_B{}^0)^2}\right]^{1/2} \quad (8\text{–}5)$$

In this equation $(\nu_A{}^0 - \nu_B{}^0)$ is the separation of peaks in the stopped exchange region in Hz; $(\nu_A - \nu_B)_{obs}$ is the separation of peaks in the intermediate exchange region. Recall that we stipulated that the two lifetimes, τ_A' and τ_B', are equal. The lifetime τ' then is the sum of the two, or the lifetime at site A is simply $\tau'/2$.

In the spectrum in which the two peaks have just merged, so that $(\nu_A - \nu_B)_{obs}$ equals zero, the approximations used to derive equation (8–5) no longer apply, and equation (8–6) can be used to obtain the lifetime:

$$\tau' = \frac{\sqrt{2}}{2\pi(\nu_A{}^0 - \nu_B{}^0)} \quad (8\text{–}6)$$

Equation (8–6) shows that the necessary condition for detecting two exchanging nuclei as separate resonances is given by:

$$\tau' > \frac{\sqrt{2}}{2\pi(\nu_A{}^0 - \nu_B{}^0)} \quad (8\text{–}7)$$

Thus, the farther apart the chemical shifts at sites A and B, the shorter the lifetime or the faster the kinetic process will have to be to average them. Recall the discussion of this effect in Chapter 4.

It should be pointed out that utilization of the above equations [especially (8–5)] wherein chemical shifts are employed leads to rate constants with large error limits. Activation enthalpies are obtained from the temperature dependence of the rate constant via the Arrhenius equation. Since the temperature range corresponding to the intermediate exchange region usually is very narrow, huge errors in the activation enthalpies can result. There is considerably more information available about the kinetic process from line width changes that are observed over the entire temperature region in which the spectrum is influenced by the kinetic process. In Fig. 8–2, a typical plot is shown for the change in the full width of the resonance line at half height as a function of $1/T$.

In the slow exchange region, the observed line width, $1/T_{2A}'$, of a given peak for site A has contributions to it from the natural line width, $1/T_{2A}$, and from chemical exchange, τ_A, according to:

$$\pi\,\Delta\nu_{1/2}(A) = \frac{1}{T_{2A}'} = \frac{1}{T_{2A}} + \frac{1}{\tau_A} \quad (8\text{–}8a)$$

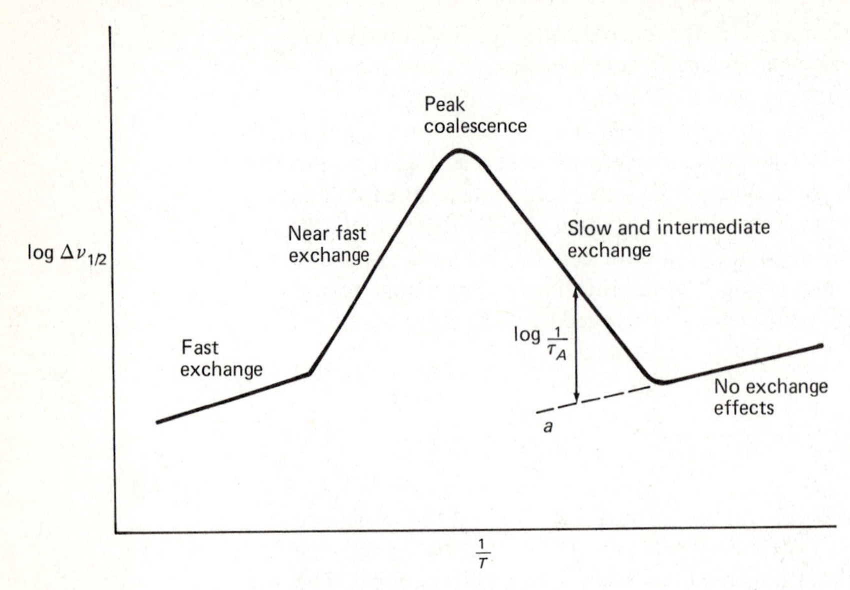

FIGURE 8–2. The influence of chemical exchange on the line width, $\Delta\nu_{1/2}$, of a spectral band as a function of absolute temperature, T. (The log is plotted to give linear activation energy plots for various regions.)

Since this equation employs units for $\Delta\nu$ of Hz, $\Delta\nu_{1/2}$ represents the full line width at half height.

This equation and all those in the (8–8) series are for a two-site problem with equal lifetimes at both sites, but not necessarily equal populations. This equation does not apply if there is spin-spin coupling of the protons, for changes in the contribution of this process to $1/T_2$ are not included. If one extrapolates the line for no exchange effects (dotted line a), the $1/T_{2A}$ contribution to the width in the slow and intermediate exchange regions can be determined. By difference, $(1/T_{2A}' - 1/T_{2A})$, one obtains $1/\tau_A$. The same calculation could be carried out with the B resonance.

In the near fast exchange region, the contributions to the line width are given by

$$\frac{1}{T_{2A}'} = \frac{N_A}{T_{2A}} + \frac{N_B}{T_{2B}} + N_A{}^2 N_B{}^2(\omega_A{}^0 - \omega_B{}^0)^2(\tau_A + \tau_B) \tag{8-8b}$$

where $\omega_A = 2\pi\nu_A$, and $\omega_A{}^0$ and $\omega_B{}^0$ are the chemical shifts in the absence of exchange. The first two terms on the right-hand side of the equality are mole-fraction weighted averages of $1/T_{2A}$ and $1/T_{2B}$ for the A and B sites in the absence of exchange. The last term is the exchange-broadened contribution. The quantities N_A/T_{2A} and N_B/T_{2B} can also be obtained by virtue of the fact that in the fast exchange region

$$\pi\,\Delta\nu_{1/2} = \frac{1}{T_{2A}'} = \frac{N_A}{T_{2A}} + \frac{N_B}{T_{2B}} \tag{8-8c}$$

Because one can work over a larger temperature range with the line width equations, one can obtain more accurate values for the activation enthalpies.

At present, most researchers would not use either of the above approaches (chemical shifts or line widths) to evaluate rate constants. Instead, the entire spectral line shape would be calculated using the following equation for a two-site (but not necessarily equal-population) system with no spin-spin coupling:

$$g(\omega) = \frac{\left\{P\left[1 + \tau\left(\frac{N_B}{T_{2A}} + \frac{N_A}{T_{2B}}\right)\right] + Q\left[R + \tau\left(\frac{1}{T_{2B}} - \frac{1}{T_{2A}}\right)\frac{\delta\omega}{2}\right]\right\}\omega_1 M_0}{P^2 + R^2 + \tau^2\left(\frac{1}{T_{2B}} - \frac{1}{T_{2A}}\right)^2\left(\frac{\delta\omega}{2}\right)^2 + 2R\tau\left(\frac{1}{T_{2B}} - \frac{1}{T_{2A}}\right)\frac{\delta\omega}{2}} \tag{8-9}$$

where

$$P = \tau\left[\frac{1}{T_{2A}T_{2B}} - \Delta\omega^2 + \left(\frac{\delta\omega}{2}\right)^2\right] + \frac{N_A}{T_{2A}} + \frac{N_B}{T_{2B}}$$

$$Q = \tau\left[\Delta\omega - \frac{\delta\omega}{2}(N_A - N_B)\right]$$

$$R = \Delta\omega\left[1 + \frac{\tau}{T_{2A}} + \frac{\tau}{T_{2B}}\right] + \frac{\delta\omega}{2}(N_A - N_B)$$

$$\tau = \frac{\tau_A\tau_B}{\tau_A + \tau_B}$$

Also, ω_1 is the frequency of the applied rf field, $\Delta\omega$ is the difference between frequency ω and the frequency at the center of the two resonance components, $\delta\omega$ is the chemical shift between magnetic sites A and B, N_A and N_B are the mole fractions of nuclei in sites A and B, respectively, and M_0 is the equilibrium value of the magnetization. Thus, one has an expression for the complete line shape of a simple AB exchanging system, which is a function of many parameters including the rate of exchange. If the values for all of the parameters in equation (8-9) except the exchange rate ($1/\tau$) can be determined, one can obtain the exchange rate by varying τ until the calculated line shape fits the experimental curve. This analysis is generally carried out by a computer that calculates the line shape using an estimated value of τ, *i.e.*, one from a rough $\Delta\nu_{1/2}$ calculation. The difference between the calculated and experimental intensities is determined. The computer then varies τ until the difference is minimized. Thus, from the line shape at a given temperature, the rate constant can be determined and the activation parameters obtained in the usual manner from the temperature dependence of the rate constant. A typical comparison between a calculated and an experimental spectrum is illustrated in Fig. 8-3. The spectrum[2] is that of the methyl proton of 2-picoline. The picoline is undergoing exchange with $Co(2\text{-pic})_2Cl_2$ in $(CD_3)_2CO$ solvent and the region is the near fast exchange.

When one wishes to study a multisite problem or when there is spin-spin coupling to one of the protons involved in the exchange, equation (8-9) can be modified to treat these systems, but the result is very cumbersome. A better method, even for the two-site no-coupling problem, involves use of the *density matrix approach.*[3-5] This approach is beyond the scope of our treatment and the reader is referred to references 3 to 5 for details.

8-2 DETERMINATION OF REACTION ORDERS BY NMR

We return now to a more complete discussion of the τ' values in order to understand how to obtain information about reaction orders. In the nmr experiment, we are studying a reaction that is at chemical equilibrium, which causes a decay in the net magnetization of the sample as the reaction occurs. Suppose, for example, that the

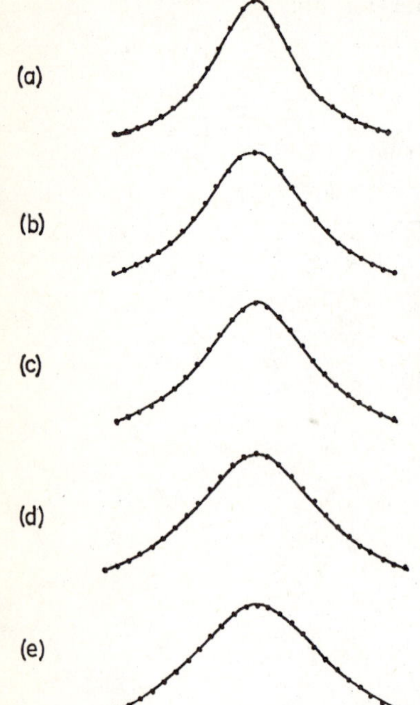

FIGURE 8–3. Comparison of experimental (solid line) and theoretical (dots) nmr methyl resonance spectra at $T = -44°$. All solutions are 0.09 *M* in Co(2-pic)$_2$Cl$_2$ and contain the following concentrations of excess 2-picoline: (a) 3.54 *M*, (b) 2.64 *M*, (c) 2.40 *M*, (d) 2.11 *M*, (e) 1.85 *M*. [From S. S. Zumdahl and R. S. Drago, J. Amer. Chem. Soc., *89*, 4319 (1967).]

decay at site A is observed. The magnetization decays by a first order decay process just like the decay of radioactive material. The rate constant $1/\tau_A$ is a first order rate constant for the decay of the initial number of protons in our sample at site A, *i.e.*,

$$-\frac{d[A]}{dt} = \frac{1}{\tau_A}[A]$$

or

$$-\frac{1}{[A]}\frac{d[A]}{dt} = \frac{1}{\tau_A} \tag{8-10}$$

In these expressions [A] is the initial concentration of protons at this site and $d[A]/dt$ is the rate at which this initial concentration "disappears." Thus, we are concerned with the lifetime of the initial A and not with the bulk concentration of A. What we observe is $1/\tau_A$ and not the rate, so in a particular experiment a first order decay constant is always observed.

Now we have to consider the mechanism for the chemical reaction that is occurring and causing A to leave site A. It could be a first order process for which

$$-\frac{d[A]}{dt} = k[A] \quad \text{or} \quad k = -\frac{1}{[A]}\frac{d[A]}{dt} = \frac{1}{\tau_A}$$

Therefore, for a first order reaction, the rate constant k equals the $1/\tau_A$ that is measured. If several experiments are carried out at different concentrations of A for a process that is first order in A, there will be a change in the rate; but there will be no change in the observable from the nmr experiment, which is $1/\tau_A$. Thus, an observation of no change in lifetime with a change in concentration corresponds to a first order process.

Next consider a process that is second order in A:

$$-\frac{d[\mathrm{A}]}{dt} = k[\mathrm{A}]^2$$

or

$$-\frac{1}{[\mathrm{A}]}\frac{d[\mathrm{A}]}{dt} = k[\mathrm{A}] = 1/\tau_{\mathrm{A}} \tag{8-11}$$

Now in a series of experiments in which [A] is changed, $1/\tau_A$ will also change linearly with the concentration. Observation of this behavior suggests a second order reaction.

Next consider a reaction that is first order in A and first order in B:

$$-\frac{d[\mathrm{A}]}{dt} = k[\mathrm{A}][\mathrm{B}]$$

Now we must specify whether the A resonance or the B resonance is being studied. If we are studying the A resonance, then we must use

$$-\frac{1}{[\mathrm{A}]}\frac{d[\mathrm{A}]}{dt} = k[\mathrm{B}] = 1/\tau_{\mathrm{A}} \tag{8-12}$$

Now as the concentration of A is changed in a series of experiments, the value of $1/\tau_A$ remains constant; but as the concentration of B is changed in a set of experiments, the value of $1/\tau_A$ changes in direct proportion to the concentration of B. The opposite result is obtained if the B resonance is examined.

In the system illustrated in Fig. 8–3, the exchange of free 2-picoline with the four-coordinate complex, $Co(2\text{-pic})_2Cl_2$, was studied[2] in deuterated acetone as a function of the concentration of 2-picoline. The lifetime of the picoline on the metal, τ_M, was independent of the concentration of $Co(2\text{-pic})_2Cl_2$, but varied linearly with the free ligand concentration. A plot of $1/\tau_M$ *vs.* [2-picoline] passed through the origin. This indicates a reaction mechanism that is second order overall: first order in $Co(2\text{-pic})_2Cl_2$ and first order in 2-picoline. The activation parameters calculated in this article from a full line shape analysis were compared[2] with those from the line width equations. Errors in τ ranging from 5 to 15 per cent were introduced by the line width approach.

8–3 SOME APPLICATIONS OF NMR KINETIC STUDIES

Ligand exchange reactions have been studied for many transition metal ion complexes. In many cases, the same compound serves as both the ligand and the solvent. One cannot determine the order of the reaction in ligand when the ligand and the solvent are the same compound since its concentration cannot be varied. Furthermore, if one wishes to study a series of ligands in a series of experiments, one has varying contributions to the thermodynamic parameters obtained from changing the solvent when the ligand is changed, since the same compound serves both purposes. As a result, very little fundamental information can be obtained from these systems. These problems are avoided, as in the system reported in Figure 8–3, when the complex has solubility in a non-coordinating solvent.

In a study[6] of the exchange of free $[(CH_3)_2N]_3PO$ (free L) with the ligand, L, of CoL_2Cl_2 in $CDCl_3$, the resulting kinetic data indicated that the reaction proceeded by both first (*i.e.*, dependent only on CoL_2Cl_2) and second (*i.e.*, first order in CoL_2Cl_2

and first order in L) order reaction paths. Since $CDCl_3$ is a non-coordinating solvent, the first order path provides evidence for the existence, as an intermediate, of a three-coordinate cobalt(II) complex.

The mechanisms for ligand substitution reactions in octahedral transition metal ion complexes are difficult to ascertain. A recent study[7a] of the exchange of CH_3OH with $[Co(CH_3OH)_6^{2+}](BF_4^-)_2$ illustrates the problem. The value of $1/\tau_M$ is independent of the metal complex or methanol concentration, supporting a rate law:

$$\text{Rate} = nk\ [\text{complex}]$$

where n represents the coordination number of the exchanging ligand. The following mechanisms[7a] are consistent with this rate law and should be considered:

Mechanism 1: S_N1(lim) or D-type

$$Co(MeOH)_6^{2+} \underset{k_{-1}}{\overset{k_1}{\rightleftarrows}} Co(MeOH)_5^{2+} + MeOH \qquad \text{slow}$$

$$Co(MeOH)_5^{2+} + MeOH^* \xrightarrow{k_2} Co(MeOH)_5(MeOH^*)^{2+} \qquad \text{fast}$$

for which the rate law is

$$\text{Rate} = k_1\,[Co(MeOH)_6^{2+}]$$

Mechanism 2: S_N(IP) or I_d

$$Co(MeOH)_6^{2+} + MeOH^* \overset{K_{OS}}{\rightleftarrows} \{Co(MeOH)_6^{2+}\}\,\{MeOH^*\} \qquad \text{fast}$$

$$\{Co(MeOH)_6^{2+}\}\,\{MeOH^*\} \xrightarrow{k_1} Co(MeOH)_5(MeOH^*) + MeOH \qquad \text{slow}$$

which has the rate law

$$\text{Rate} = \frac{k_1 K_{OS}\,[Co]_T[MeOH]}{K_{OS}[MeOH] + 1}$$

where

$$[Co]_T = [Co(MeOH)_6^{2+}] + [\{Co(MeOH)_6^{2+}\}\,\{MeOH^*\}]$$

In the second mechanism, K_{OS} is the outer sphere complex formation constant. When $K_{OS}[CH_3OH] \gg 1$, the rate law for mechanism 2 reduces to a form identical to that for mechanism 1, and the two cannot be distinguished. If the $[CH_3OH]$ could be reduced to a small value, then $1 \gg K_{OS}[CH_3OH]$ and second order kinetics would be observed, differentiating the two mechanisms. For a K_{OS} of about 1, changing the $[CH_3OH]$ from 3 M to 8 M would increase the rate constant by only 10% and experimental error would make this change hard to detect. A larger value of K_{OS} would make the difference even smaller. At lower $[CH_3OH]$, dissociation of some methanol from the complex accompanied by anion coordination becomes a problem. These problems are described here because they are common to many nmr studies on systems of this type. In order to truly distinguish these types of mechanisms, it will be necessary to study systems in which the free ligand concentration, [L], can be made sufficiently low so that $1 \gg K_{OS}[L]$.

In a study[7b] of the kinetic parameters obtained by nmr using a density matrix analysis and the thermodynamics of adduct formation, it was shown that the reaction

$$Ni(SDPT) \cdot B + B' \longrightarrow Ni(SDPT) \cdot B' + B$$

(where SDPT is a pentadentate ligand with two negative charges, B and B′ is 4-methyl pyridine) proceeds by a pure dissociative mechanism involving a five-coordinate, NiSDPT, intermediate. By comparing data obtained in toluene and in CH_2Cl_2 as solvent, the role played by the latter solvent in the reaction was clearly established.

The spectrum of *N,N*-dimethylacetamide, $CH_3C(O)N(CH_3)_2$, has three peaks at room temperature, two of which correspond to the different environments of the two methyl groups on the nitrogen (one *cis* and one *trans* to oxygen). The C—N bond has multiple bond character and gives rise to an appreciable barrier to rotation about this bond. As a result of this barrier, the two non-equivalent N—CH_3 groups are detected. As the temperature is increased, the rate of rotation about the C—N bond increases and the *N*-methyl resonances merge, giving rise to a series of spectra similar to those in Fig. 8–1. The lifetime of a particular configuration can be determined as a function of temperature, and the activation energy for the barrier to rotation can be evaluated.[8] Similar studies have been carried out on other amides[9] and on nitrosamines.[10] The quadrupolar ^{14}N nucleus makes a temperature dependent contribution to the line widths in some of these systems, introducing error into the resulting parameters. The rates of inversion of substituted 1,2-dithianes and 1,2-dioxanes[11] are among the many other rates that have been studied.

The mechanism of proton exchange for solutions of methyl ammonium chloride[12] in water as a function of pH was evaluated by an nmr procedure. In acidic solution (pH = 1.0) the nmr spectra of these solutions consist of a quadruplet methyl peak (split by the three ammonium protons), a sharp water peak, and three broad peaks from the ammonium protons. The triplet for the ammonium protons results from nitrogen splitting. No fine structure is observed in the ammonium proton peak from the expected coupling to the methyl protons. Partial quadrupole relaxation by the nitrogen causes these peaks to be broader than $^3J_{H(CH_3)H(NH_3)}$. As the pH is increased, rapid proton exchange reactions begin to occur and the —CH_3, H_2O, and —NH_3^+ bands begin to broaden. Eventually, at about pH 5, two peaks with no fine structure remain, one from the protons on the —CH_3 group and the other a broad peak from an average of all other proton shifts. As the pH is raised to 8, the broad proton peak sharpens again. The —CH_3 broadening yields the exchange rate of protons on the amine nitrogen; the broadening of the water line measures the lifetime of the proton on water, and the broadening of the —NH_3^+ triplet measures the lifetime of the proton on the ammonium nitrogen. An analysis of the kinetic data yields: (1) the rate law and a consistent mechanism for the exchange

$$CH_3NH_3^+ + B \xrightarrow{k_1} CH_3NH_2 + BH^+$$

where the base B = H_2O, CH_3NH_2, or OH^-; (2) the fraction of the above protolysis that involves water; and (3) the contribution to the exchange reaction from:

$$CH_3NH_2 + BH^+ \longrightarrow CH_3NH_3^+ + B$$

and

$$CH_3NH_2 + B \longrightarrow CH_3NH_2 + B'$$

(B′ is B with one hydrogen replaced by a different hydrogen). The details of this analysis can be obtained from the reference by Grunwald *et al.*,[12] which is highly recommended reading. In subsequent studies the rates of proton exchange in aqueous solutions containing NH_4^+, $(CH_3)_2NH_2^+$, and $(CH_3)_3NH^+$ were measured and compared.[13]

Another example of information obtained from nmr rate studies is reported by Muetterties and Phillips.[14] They found that SiF_6^{2-} in aqueous solution gives rise to

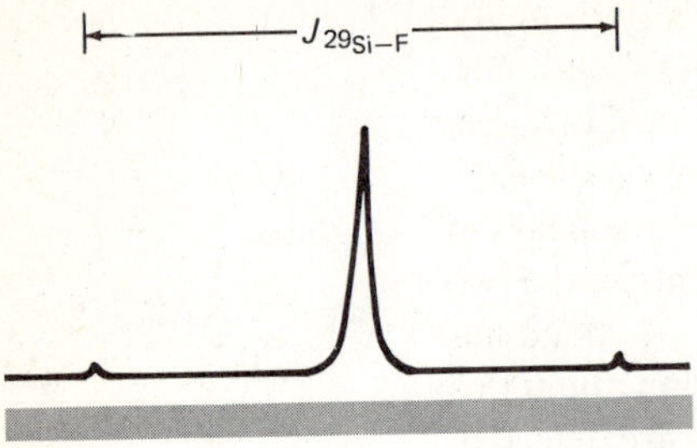

$J_{^{29}Si-F}$

FIGURE 8-4. The fluorine nmr spectrum of SiF_6^{2-}. See table inside back cover (Properties of Selected Nuclei) for ^{29}Si natural abundance.

the spectrum in Fig. 8-4. The main peak arises from fluorines on ^{28}Si ($I = 0$), and the two small peaks result from spin-spin splitting of the fluorine by ^{29}Si ($I = \frac{1}{2}$). The appearance of the satellites corresponding to J_{Si-F} indicates that the rate of exchange of fluorine atoms must be less than 10^3 sec^{-1} [$\tau' > 1/(\nu_A - \nu_B)$]. It is also found that the spectrum of solutions of SiF_6^{2-} containing added F^- contains two separate fluorine resonances. These are assigned to F^- and SiF_6^{2-}. There are satellites (J_{Si-F}) on the SiF_6^{2-} peak. When a solution of SiF_6^{2-} is acidified, rapid exchange occurs, the satellites disappear, and the central peak broadens. The following reactions are proposed:

$$SiF_6^{2-} + H_3O^+ \longrightarrow HFSiF_5^- \xrightarrow{H_2O} H_2OSiF_5^- \xrightarrow{HF} SiF_6^{2-} + H_3O^+$$

The ^{13}C spectrum of CO_2 in water gives rise to two peaks,[15] one from dissolved CO_2 and a second from H_2CO_3, HCO_3^-, and CO_3^{2-}. Rapid proton exchange gives rise to a single ^{13}C peak for these latter three species. The reaction $CO_2 + H_2O \longrightarrow H_2CO_3$ has a half-life of about 20 sec, so a separate peak for dissolved CO_2 is detected.

8-4 INTRAMOLECULAR REARRANGEMENTS STUDIED BY NMR—FLUXIONAL BEHAVIOR

The nmr spectrum of $B_3H_8^-$ is of interest because it demonstrates the effect of intramolecular exchange on the nmr spectrum. The structure of $B_3H_8^-$ is illustrated in Fig. 8-5, along with a possible mechanism for the intramolecular hydrogen exchange.[16-18] The ^{11}B spectrum is a nonet that results from a splitting of three equivalent borons by eight equivalent protons. The process represented in Fig. 8-5 occurs very rapidly, making the three borons equivalent and the eight protons equivalent in the nmr spectrum. In this example, the eight hydrogen atoms remain attached to the boron atoms of the same molecule during the kinetic process and the splitting does not disappear. Contrast this to rapid intermolecular exchange, in which a single boron resonance signal would result if exchange made all protons equivalent.

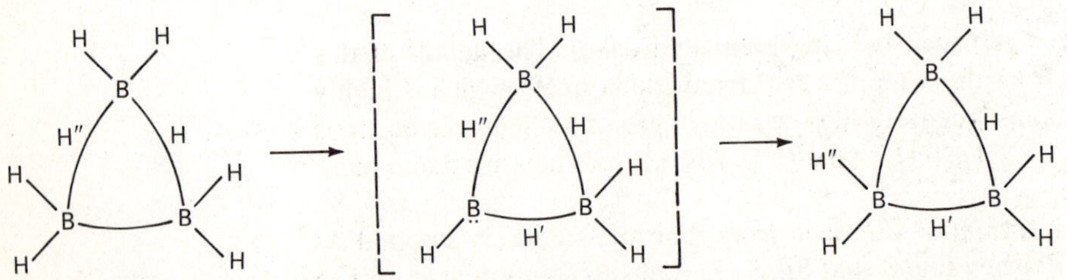

FIGURE 8-5. The structure of $B_3H_8^-$ and a proposed intermediate for the intramolecular exchange.

Another possible explanation leading to the observed spectrum that does not involve exchange is based on virtual coupling. Virtual coupling of the eight protons would give rise to the observed nonet in the ^{11}B resonance. Strong boron coupling has not been observed in the spectra of other boron hydrides.

Considerable information is available regarding the mechanism of exchange processes from the non-symmetrical collapse of an nmr spectrum. An illustration of this basic idea is provided[19] by the temperature dependence of the nmr spectrum of $(\pi\text{-}C_5H_5)Fe(CO)_2C_5H_5$, whose structure is shown in Fig. 8–6(A). The spectrum

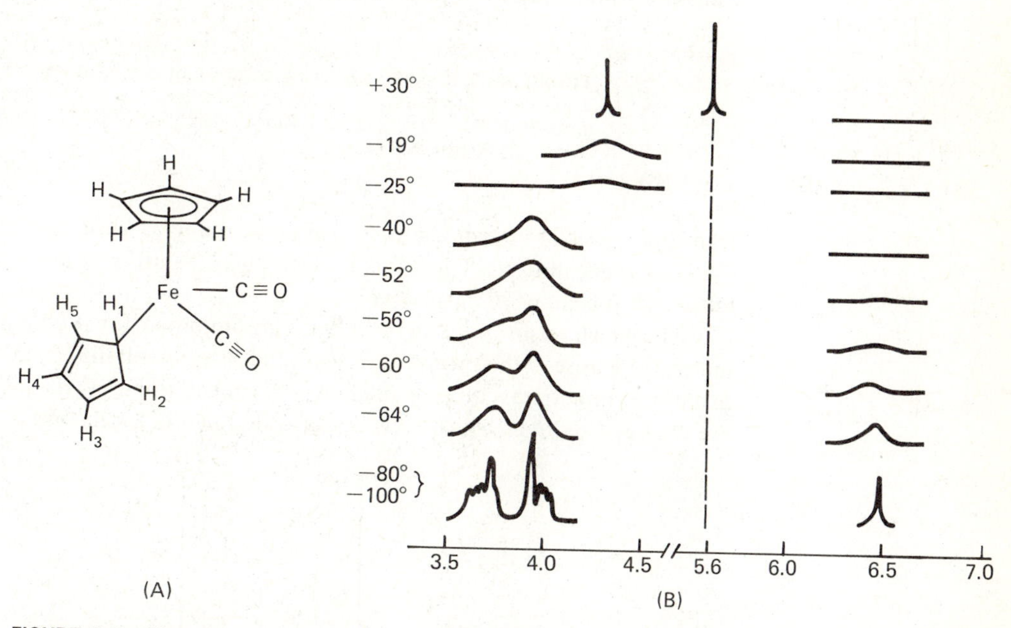

FIGURE 8–6. The proton magnetic resonance spectra of $(\pi\text{-}C_5H_5)Fe(CO)_2C_5H_5$ in CS_2 at various temperatures. The dotted line represents the resonance position of the $\pi\text{-}C_5H_5$ protons at each temperature. The amplitude of the +30° spectrum is shown ×0.1 relative to the others. [Part (B) reprinted with permission from M. J. Bennett, Jr., *et al.*, J. Amer. Chem. Soc., *88*, 4371 (1966). Copyright by the American Chemical Society.]

obtained as a function of temperature is shown in Fig. 8–6(B). The pi-bonded cyclopentadiene resonance gives rise to a single sharp peak at $\tau = 5.6$ at all temperatures because rapid rotation of the ring results in the equivalence of the five protons. The remaining peaks in the spectrum arise from the sigma-bonded cyclopentadiene ring. At −100°C, three distinct groups of resonances are observed at ~3.7, ~4.0, and 6.5 τ with relative intensities 2:2:1. The upfield resonance ($\tau = 6.5$) is assigned to H_1. The H_2 and H_5 protons are equivalent, as are H_3 and H_4. The two sets differ only slightly in chemical shift relative to the magnitude of their coupling constant and, as shown in a subsequent section, this situation gives rise to complicated *second order spectra* that often contain more peaks than a simple analysis of the chemical shift differences and spin-spin coupling would produce. The peaks at ~3.7 and ~4.0 are due to the 2, 3, 4, and 5 protons. The splitting from the H_1 proton is small; in the limit where $^3J_{H_1-H_{2,5}}$ equals $^4J_{H_1-H_{3,4}}$ the two resonances at 3.7 and 4.0 would be mirror images. However, since $^3J_{H_1-H_{2,5}}$ would be expected to be larger than $^4J_{H_1-H_{3,4}}$, additional fine structure would be expected in the peak assigned to H_2 and H_5. With $^3J_{H_1-H_{2,5}}$ small, the additional fine structure would not be resolved but could be manifested in a broadening of the resonances. Using this criterion, the multiplet with the broader resonance at low field ($\tau \approx 3.7$) is assigned to the 2 and 5 protons. Since these protons are equivalent, we shall label them as A and the 3,4 pair as B. As the temperature is raised, the A resonance collapses faster than the B resonance. At

−25 °C, they are broadened and have completely collapsed. At +30 °C, a sharp peak corresponding to a mole-fraction weighted average resonance of the *three* types of protons is obtained.

Several mechanisms for this *fluxional behavior* are possible. The non-symmetrical collapse of the spectrum rules out any bimolecular process (dissociative or exchange in nature). A first order dissociative process is ruled out because the experimentally found activation enthalpy is too low. This low activation enthalpy suggests some interaction between the pi orbitals of the ring and the metal in the transition state. The three structures shown in Fig. 8–7 are considered as possible

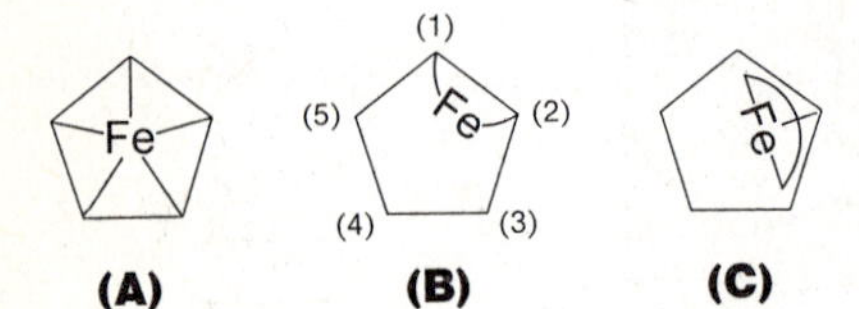

FIGURE 8–7. Diagrams showing possible intramolecular paths leading to the nmr equivalence of the σ-C_5H_5 protons at room temperature. [Reprinted with permission from M. J. Bennett, Jr., *et al.*, J. Amer. Chem. Soc., *88*, 4371 (1966). Copyright by the American Chemical Society.]

transition states or intermediates and are referred to as the π-cp (A), π-olefin (B) and π-allyl (C) mechanisms. The terms 1,2 shift and 1,3 shift have also been applied to the latter two mechanisms respectively.

The mechanism involving (A) can be eliminated for it would result in a symmetrical collapse of the spectrum. The mechanisms involving (B) and (C) would result in a non-symmetrical collapse, as shown by considering how the labels in Fig. 8–7(B) are changed by a 1,2 shift. This is illustrated symbolically by

$$\begin{bmatrix} H_a \\ A \\ B \\ B \\ A \end{bmatrix} \longrightarrow \begin{bmatrix} A \\ H_a \\ A \\ B \\ B \end{bmatrix}$$

This analysis shows that all A's are changed into different types of protons, but only half the B's are, so the A resonance should collapse more rapidly.

On the other hand, the 1,3 shift is described by

$$\begin{bmatrix} H_a \\ A \\ B \\ B \\ A \end{bmatrix} \longrightarrow \begin{bmatrix} B \\ A \\ H_a \\ A \\ B \end{bmatrix}$$

For this mechanism, the B resonance should collapse faster. Thus, to the extent that the original assignment of the A and B resonances is correct, the 1,2 shift is established. The intermediate or transition state in the 1,2 shift could be thought of as one in which the π-electron density of the ring is arranged in a two-center π bond coordinated to the iron, while the other three carbon atoms have an allyl anion distribution of electron density.

In spite of *extensive* further investigations of these processes by F. A. Cotton and his students, practically no underlying principles have resulted from this work and they have not been able to demonstrate that *their* findings have any consequences on chemical reactivity. There have been those who have wondered to themselves, and occasionally out loud, just when these extensions of the original discovery of Piper and Wilkinson become too quixotic to be valuable. It is not impossible that the venerable sport of jousting at windmills is now being practiced, but only time will tell for certain.

The preceding paragraph is not intended to imply the non-existence of notable contributions in this area by various groups. Many fluxional systems have been discovered and are summarized in reference 20. A procedure for a complete line shape analysis on systems of this sort has been reported.[21] The line shape equations for the various protons are written in a matrix formulation; *i.e.*, the τ's, T_2's, etc. are expressed as matrices. These matrices are multiplied by an exchange matrix, which is constructed to indicate how the protons are permuted by the exchange reaction. This can be illustrated for the 1,2 shift using the numbering system in Fig. 8–7(B). A matrix of the numbered protons is written, in which the row indicates how a given proton is changed by a 1,2 shift. For example, in row 1, the number -1 indicates that a given proton is lost from site 1; the $+\frac{1}{2}$ entries in columns 2 and 5 indicate that there is a 50 per cent probability that this proton will end up at either of those sites. The other rows in the matrix can be confirmed by similar reasoning.

$$\begin{array}{c} \\ H_1 \\ H_2 \\ H_3 \\ H_4 \\ H_5 \end{array} \begin{array}{c} \begin{array}{ccccc} H_1 & H_2 & H_3 & H_4 & H_5 \end{array} \\ \begin{bmatrix} -1 & \frac{1}{2} & 0 & 0 & \frac{1}{2} \\ \frac{1}{2} & -1 & \frac{1}{2} & 0 & 0 \\ 0 & \frac{1}{2} & -1 & \frac{1}{2} & 0 \\ 0 & 0 & \frac{1}{2} & -1 & \frac{1}{2} \\ \frac{1}{2} & 0 & 0 & \frac{1}{2} & -1 \end{bmatrix} \end{array}$$

There has been a very considerable effort devoted to nmr studies of the intramolecular rearrangements of trigonal bipyramidal and octahedral complexes. In 1951, Gutowsky and Hoffman[22] reported that the ^{19}F nmr spectrum of PF_5 was a doublet, even though electron diffraction had established the trigonal bipyramidal structure of this molecule. Since P-F coupling is maintained, an intramolecular process is required to equilibrate the fluorine atoms. Since this time, many other systems that show similar behavior have been observed, *e.g.*, $Fe(CO)_5$ (^{13}C nmr), $CF_3Co(CO)_3PF_3$, several $HM(PF_3)_4$ species, and $HIr(CO)_2[P(C_6H_5)_3]_2$. There have been several attempts[23,24] to systematically enumerate all of the physically distinguishable intramolecular modes for interconverting groups on a trigonal bipyramid. The problem is a complex one, for it is necessary to insure that apparently different pathways are physically distinguishable. The reader is referred to references 23 and 24 for details.

We shall briefly consider the fluxional behavior of $CH_3Ir(COD)[C_6H_5P(CH_3)_2]_2$ to provide an illustration of the detailed mechanistic information obtainable from work in this area of fluxionality.[25,26] The static structure of the molecule at low temperature is shown in Fig. 8–8(A). At $-3°$, the resonances are assigned as in Fig. 8–8(C). The two methyls on each dimethylphenylphosphine group are not equivalent (*i.e.*, they are diastereotopic). Should the phosphorus ligands change sites, they would become equivalent. The vinyl hydrogens, H_1 and H_2, are also non-equivalent. The phenyl resonances are not shown. As the temperature is raised, the vinyl proton signal collapses into a singlet, but the P-CH_3 resonance is not affected up to $67°$.

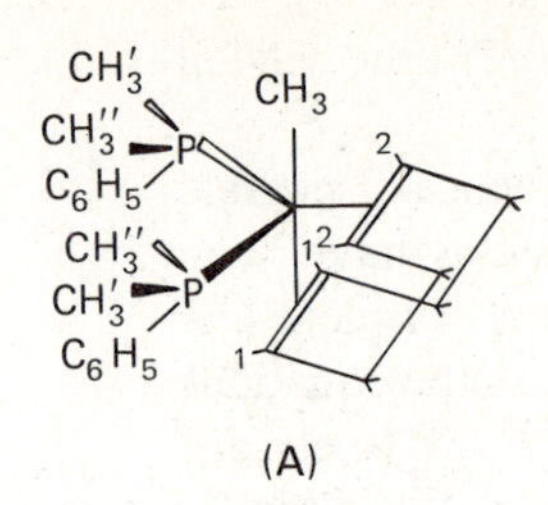

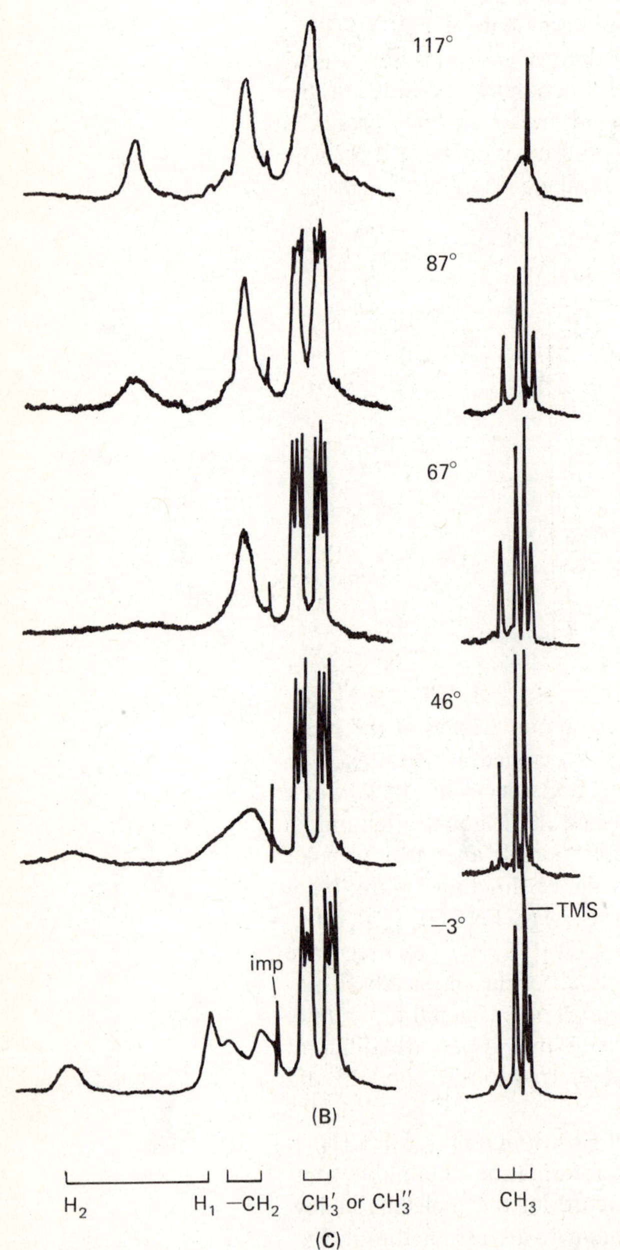

FIGURE 8–8. (A) Structure, (B) temperature dependent proton nmr spectra, and (C) assignment of low temperature spectrum of $CH_3Ir(COD)[(P(C_6H_5)(CH_3)_2]_2$. Solvent is chlorobenzene, and IMP refers to acetone impurity present from recrystallization. [Reprinted with permission from J. R. Shapley and J. A. Osborn, Accts. Chem. Res., 6, 305 (1973). Copyright by the American Chemical Society.]

The reasonable mechanistic paths that could exchange axial and equatorial positions in the trigonal bipyramid are illustrated in Fig. 8–9. Scheme A involves a twist of the diene about a pseudo-twofold axis through the metal and through a point midway between the two double bonds. The intervening intermediate or transition state is the distorted tetragonal pyramid labeled I.

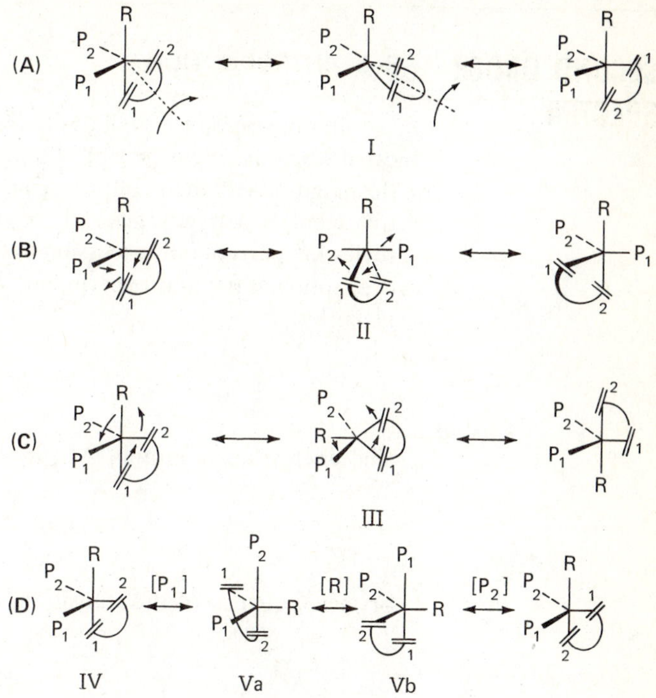

FIGURE 8–9. Mechanistic schemes to account for axial-equatorial equilibration of COD vinyl protons in the complexes $RIr(COD)P_2$, where R is CH_3. [Reprinted with permission from J. R. Shapley and J. A. Osborn, Accts. Chem. Res., 6, 305 (1973). Copyright by the American Chemical Society.]

Schemes B and C involve permutations of three sites. In B, one axial and two equatorial ligands are interchanged by a rotation about a pseudo-threefold axis constructed by drawing a line from the metal to the center of the face of the trigonal bipyramid defined by double bond 1, double bond 2, and P_1. This rotation leads to the transition state labeled II. The process in C is best described by considering the group R as being located in the center of the face of a tetrahedron formed by P_1, P_2, double bond 1, and double bond 2. The R group then moves through an edge and into the center of the face formed by P_1, P_2, and double bond 1. The resulting structure has interchanged the two double bonds. Scheme D proceeds through two trigonal bipyramid intermediate structures, Va and Vb, each of which involves a change of four ligands. Each change occurs by a so-called Berry mechanism: two equatorial groups open up their angle and the two axial groups move together in the direction in which the equatorial angle is increasing, to form a distorted tetragonal pyramid. The motion continues in this direction until the two axial bonds are equatorial and the two equatorial bonds become axial. Structures Va and Vb are enantiomers, and ready interconversion is expected. Paths in which the diene spans equatorial sites have been eliminated as energetically unfavorable because of the geometrical preference of this ligand for a 90° chelate angle.

Examination of these schemes indicate that B and C interchange the two phosphorus ligands, but A and D do not. Thus, only the latter modes are consistent with the observed spectral behavior. One cannot distinguish between the A and D modes with nmr. There is a very extensive literature on this subject and, with the example discussed here, we have indicated the kind of information that can be obtained. For more details, the reader should consult references 25 and 26. Studies involving fluxional behavior in six-coordinate complexes have also been reported.[27,28,29]

SECOND ORDER SPECTRA

8–5 INTRODUCTION

In this section we shall develop in detail the reason why spectra become complicated when the chemical shift difference of non-equivalent protons becomes equal to the magnitude of their coupling constant. Our discussion will involve protons, but the treatment is perfectly general for any spin ½ nucleus. We shall begin by briefly reviewing parts of our discussion in the previous chapter on the quantum mechanical description of the nmr experiment. The energy of a bare proton in a magnetic field is given by:

$$E = -g_N \beta_N H_0 m_I$$

The energy for a proton, A, surrounded by paired electron density is given by:

$$E = -g_N \beta_N H_0(1 - \sigma_A) m_I$$

or the frequency of the transition, ν, is given by:

$$\nu = \nu_0(1 - \sigma_A)|\Delta m_I| \tag{8–13}$$

where ν_0 is the resonance frequency for a bare proton. The shielding constant, σ_A, accounts for the chemical shift differences of hydrogen atoms in different environments in the molecule.

Next consider a molecule containing two different hydrogens that are involved in spin-spin coupling. We shall focus attention only on the two hydrogens and ignore the rest of the molecule. When the chemical shift difference of the two hydrogens is very large compared to their J, we shall label such a system AX. The energy of a system of X_j protons whose shift differences are larger than J is given by equation (8–14).

$$E = -h \sum_j \nu_0(1 - \sigma_j) m_j + h \sum_{j<k} J_{jk} m_j m_k \tag{8–14}$$

The first summation on the right gives a different chemical shift term for each of the j different types of nuclei in the molecule. The second summation is taken only for pairs of nuclei in which $j < k$, to insure that each pair is considered only once. It is assumed in writing this equation that J is isotropic. The significance of equation (8-14) is illustrated in Fig. 8–10, where the energies of the various nuclear configurations of an AX system in a magnetic field are illustrated for the case in which J_{AX} is zero and for the case in which it is finite.

In Fig. 8–10(A), where $J_{AX} = 0$, the energy levels for the $\alpha\alpha$, $\alpha\beta$, $\beta\alpha$, and $\beta\beta$ nuclear spin configurations are shown. The A nuclear spin state is listed first and the X nucleus second. The $\alpha\alpha$ energy is the sum of $-h[\nu_0(1 - \sigma_A)(½) + \nu_0(1 - \sigma_X)(½)] = [-1 + (½)(\sigma_A + \sigma_X)]h\nu_0$. The energies of the other levels are calculated similarly by substituting the appropriate m_I values. The solid arrows indicate the transitions for the A type of nucleus. One arrow corresponds to A nuclei bonded to X nuclei with α spin and the other to those on X nuclei with β spin. The selection rule is $\Delta m_I = 1$, so the A and X nuclear spins cannot change simultaneously. The dashed arrows indicate the transitions of the X nucleus. By referring to equation (8–14) and Fig. 8–10, we see

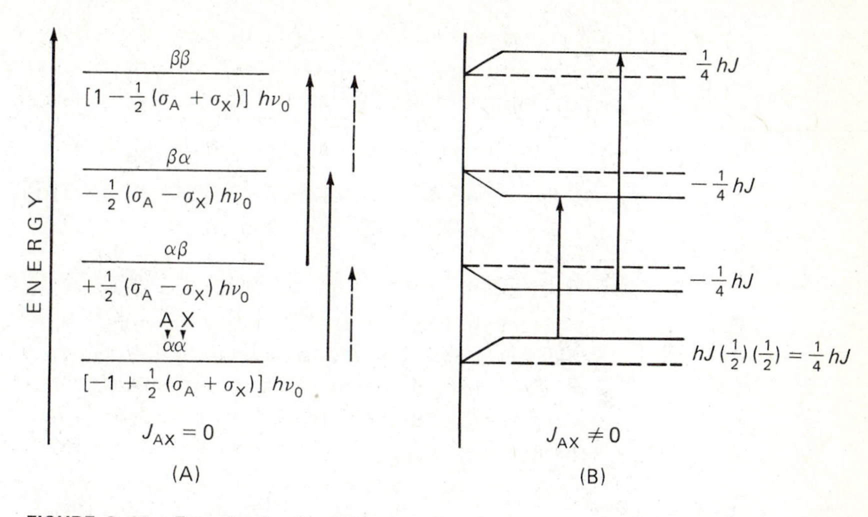

FIGURE 8–10. Energies for the AX system in a magnetic field [I(A) = ½, I(X) = ½]. (A) $J_{AX} = 0$. The wave function, written in the order AX, is written above the energy level; the energy is written below the level. (B) $J_{AX} \neq 0$. The J labeling indicates the change in energy that occurs from the dotted line, which represents the energy when $J_{AX} = 0$.

that the frequency difference between $\alpha\alpha$ and $\beta\alpha$ ($\nu_0 - \sigma_A\nu_0$) is equal to the difference between $\alpha\beta$ and $\beta\beta$; *i.e.*, the two transitions of nucleus A are degenerate. Thus, without any coupling, a single peak would be observed for A. By similar reasoning, we expect a single peak for X.

Where J_{AX} is finite, and considerably smaller than the chemical shift difference of A and X, the $J_{jk}m_jm_k$ terms modify the energies shown in Fig. 8–10(A), producing those shown in (B). Since the $\alpha\alpha$ energy is raised by (¼) Jh and $\beta\alpha$ is lowered by (¼) Jh, the $\alpha\alpha \rightarrow \beta\alpha$ transition occurs at a frequency $J/2$ lower than the corresponding transition in the $J_{AX} = 0$ case. The $\alpha\beta \rightarrow \beta\beta$ transition is at a frequency $J/2$ higher than that of the $J_{AX} = 0$ transition. Thus, the frequency separation between the two peaks is equal to J in the $J_{AX} \neq 0$ case.

Spin-spin coupling constants can be positive or negative. In the above problem, a positive J was assumed. If J was taken as negative, the $\alpha\alpha$ energy would be lowered and the $\beta\alpha$ energy would be raised. This would shift the peak for the $\alpha\alpha \rightarrow \beta\alpha$ transition to a frequency $J/2$ higher than in the $J_{AX} = 0$ case, while the other transition would occur at a frequency $J/2$ lower. (A greater frequency corresponds to a lower field.) There would be no change in the appearance of the spectrum, so there would be no way to determine the sign of J as illustrated by referring to Fig. 8–11. In Fig. 8–11(A), line I represents the chemical shift difference between A and X and line

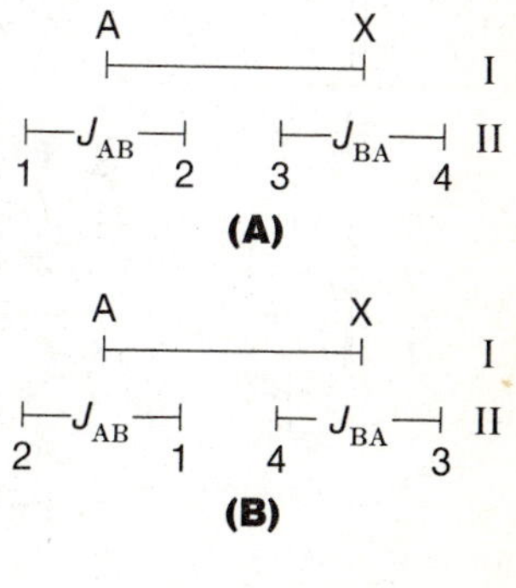

A X I
J_{AB} J_{BA} II
1 2 3 4
(A)

A X I
J_{AB} J_{BA} II
2 1 4 3
(B)

FIGURE 8–11. Effect of the sign of J_{AX} on the splitting of an A—X system. Here 1 refers to the $\alpha\alpha$ to $\beta\alpha$ transition, and 2 refers to the $\alpha\beta$ to $\beta\beta$ transition.

II shows the spin-spin splitting of A by X when J_{AX} is positive. If J_{AX} were negative the result in Fig. 8–11(B) would be obtained. One cannot distinguish the two possibilities by examining the spectrum obtained in the normal nmr experiment.

8–6 COMPLETE QUANTUM MECHANICAL TREATMENT OF COUPLING

Equation (8–14) is the so-called first order solution of the chemical shift–coupling constant problem. It applies only when the chemical shift difference, Δ, of the two nuclei is large compared to their coupling constant, J. We now treat the more general problem using a system where Δ and J have comparable magnitudes. The label AB will be used to describe this case, where $I(\text{A}) = \frac{1}{2}$ and $I(\text{B}) = \frac{1}{2}$. We shall solve for the energies using:

$$\int \psi^* \hat{H} \psi \, d\tau = \frac{E}{h} \int \psi^* \psi \, d\tau$$

With E/h being a frequency, the Hamiltonian is written in frequency units as:

$$\hat{H} = -\nu_0(1 - \sigma_A)\hat{I}_{ZA} - \nu_0(1 - \sigma_B)\hat{I}_{ZB} + J_{AB}\hat{I}_A \cdot \hat{I}_B \tag{8-15}$$

Our basis set will be the nuclear spin functions $\varphi_1 = |\alpha\alpha\rangle$, $\varphi_2 = |\alpha\beta\rangle$, $\varphi_3 = |\beta\alpha\rangle$, and $\varphi_4 = |\beta\beta\rangle$. To obtain the four energies, we need to solve for all the matrix elements in the secular determinant:

$$\begin{array}{c} \\ |\alpha\alpha\rangle \\ |\alpha\beta\rangle \\ |\beta\alpha\rangle \\ |\beta\beta\rangle \end{array}
\begin{array}{cccc} |\alpha\alpha\rangle & |\alpha\beta\rangle & |\beta\alpha\rangle & |\beta\beta\rangle \end{array}$$

$$\begin{vmatrix}
\langle\alpha\alpha|\hat{H}|\alpha\alpha\rangle - \frac{E}{h} & \langle\alpha\alpha|\hat{H}|\alpha\beta\rangle & \langle\alpha\alpha|\hat{H}|\beta\alpha\rangle & \langle\alpha\alpha|\hat{H}|\beta\beta\rangle \\
\langle\alpha\beta|\hat{H}|\alpha\alpha\rangle & \langle\alpha\beta|\hat{H}|\alpha\beta\rangle - \frac{E}{h} & \langle\alpha\beta|\hat{H}|\beta\alpha\rangle & \langle\alpha\beta|\hat{H}|\beta\beta\rangle \\
\langle\beta\alpha|\hat{H}|\alpha\alpha\rangle & \langle\beta\alpha|\hat{H}|\alpha\beta\rangle & \langle\beta\alpha|\hat{H}|\beta\alpha\rangle - \frac{E}{h} & \langle\beta\alpha|\hat{H}|\beta\beta\rangle \\
\langle\beta\beta|\hat{H}|\alpha\alpha\rangle & \langle\beta\beta|\hat{H}|\alpha\beta\rangle & \langle\beta\beta|\hat{H}|\beta\alpha\rangle & \langle\beta\beta|\hat{H}|\beta\beta\rangle - \frac{E}{h}
\end{vmatrix} = 0$$

The term $-E/h$ appears on the diagonal because only the diagonal $[-(E/h)\langle\varphi_n|\varphi_n\rangle]$ terms are non-zero; for example,

$$-(E/h)\langle\alpha\alpha|\alpha\alpha\rangle = -E/h$$

while

$$-(E/h)\langle\alpha\beta|\alpha\alpha\rangle = 0$$

To evaluate these matrix elements, we need to know how the $\hat{I}_A \cdot \hat{I}_B$ operator works. This problem is simplified by defining the so-called *raising and lowering operators*. We shall have occasion to use these operators in other problems. Remember that

$$\hat{I}^2 = \hat{I}_X^2 + \hat{I}_Y^2 + \hat{I}_Z^2$$

but that we can only find simultaneously $\hat{I}^2$ and the component in one direction. The raising and lowering operators, $\hat{I}_+$ and $\hat{I}_-$, are defined by taking linear combinations of $\hat{I}_X$ and $\hat{I}_Y$ such that:

$$\hat{I}_+ = \hat{I}_X + i\hat{I}_Y \tag{8-16}$$

$$\hat{I}_- = \hat{I}_X - i\hat{I}_Y \tag{8-17}$$

These operators have the property that when they operate on a wave function $|I,m_I\rangle$ (*i.e.*, one defined by quantum numbers I and m_I) we get:

$$\hat{I}_+|I,m_I\rangle = [I(I+1) - m_I(m_I+1)]^{1/2}|I,m_I+1\rangle$$

$$\hat{I}_-|I,m_I\rangle = [(I(I+1) - m_I(m_I-1)]^{1/2}|I,m_I-1\rangle$$

Then we find that these operators work on $|\alpha\rangle$ and $|\beta\rangle$ as follows:

$\hat{I}_+|\alpha\rangle = 0$ (we cannot raise a $+\frac{1}{2}$ spin by 1 when $I = \frac{1}{2}$)

$\hat{I}_-|\alpha\rangle = |\beta\rangle$

$\hat{I}_+|\beta\rangle = |\alpha\rangle$

$\hat{I}_-|\beta\rangle = 0$ (a $-\frac{1}{2}$ spin cannot be lowered by 1 when $I = \frac{1}{2}$)

The $\hat{I}_A \cdot \hat{I}_B$ operator for a two-spin AB system is given by:

$$\begin{aligned}\hat{I}_A \cdot \hat{I}_B &= \hat{I}_{ZA} \cdot \hat{I}_{ZB} + \hat{I}_{XA} \cdot \hat{I}_{XB} + \hat{I}_{YA} \cdot \hat{I}_{YB} \\ &= \hat{I}_{ZA} \cdot \hat{I}_{ZB} + (\tfrac{1}{2})(\hat{I}_{+A}\hat{I}_{-B} + \hat{I}_{-A}\hat{I}_{+B})\end{aligned} \tag{8-18}$$

Equation (8–18) in terms of the raising and lowering operators can be derived by solving equations (8–16) and (8–17) for $\hat{I}_X$ and $\hat{I}_Y$ and substituting this result into the equation for $\hat{I}_A \cdot \hat{I}_B$ in terms of X, Y, and Z components. In operating on the basis set ($\alpha\alpha$, etc.) with the $\hat{I}_A \cdot \hat{I}_B$ operator, the $\hat{I}_A$ spin operator acts only on the A nucleus (the first spin function listed) and the $\hat{I}_B$ operator acts only on the B nucleus (the second one listed). Accordingly,

$$\hat{I}_{ZA}|\alpha\beta\rangle = (\tfrac{1}{2})|\alpha\beta\rangle$$

while

$$\hat{I}_{ZB}|\alpha\beta\rangle = -(\tfrac{1}{2})|\alpha\beta\rangle$$

The general Hamiltonian for the coupled AB system, equation (8–15), becomes equation (8–19) when expressed in terms of the raising and lowering operators.

$$\begin{aligned}\hat{H}_{AB} = {} & -\nu_0(1-\sigma_A)\hat{I}_{ZA} - \nu_0(1-\sigma_B)\hat{I}_{ZB} + J_{AB}\hat{I}_{ZA} \cdot \hat{I}_{ZB} \\ & + (\tfrac{1}{2})J_{AB}(\hat{I}_{+A}\hat{I}_{-B} + \hat{I}_{-A}\hat{I}_{+B})\end{aligned} \tag{8-19}$$

Now we return to the evaluation of the matrix elements in the secular determinant given earlier. To evaluate $\langle\alpha\alpha|\hat{H}|\alpha\alpha\rangle$, we need to evaluate the effect of $\hat{H}$ on $|\alpha\alpha\rangle$ and then simply multiply by $\langle\alpha\alpha|$. Thus, we shall proceed by first evaluating $\hat{H}|\alpha\alpha\rangle$, $\hat{H}|\alpha\beta\rangle$, $\hat{H}|\beta\alpha\rangle$, and $\hat{H}|\beta\beta\rangle$.

$$\hat{H}|\alpha\alpha\rangle = [-\nu_0(1-\sigma_A)(\tfrac{1}{2}) - \nu_0(1-\sigma_B)(\tfrac{1}{2}) + (\tfrac{1}{4})J]|\alpha\alpha\rangle$$

The $(\frac{1}{4})J$ arises from $\hat{I}_{ZA} \cdot \hat{I}_{ZB}|\alpha\alpha\rangle$ for both $\hat{I}_{+A}\hat{I}_{-B}|\alpha\alpha\rangle$ and $\hat{I}_{-A}\hat{I}_{+B}|\alpha\alpha\rangle$ equal zero, since $\hat{I}_{+A}|\alpha\alpha\rangle = 0$ and $\hat{I}_{+B}|\alpha\alpha\rangle = 0$. Rearranging the above expression for $\hat{H}|\alpha\alpha\rangle$, we have

$$\hat{H}|\alpha\alpha\rangle = \left[\nu_0\left(-1 + \frac{1}{2}\sigma_A + \frac{1}{2}\sigma_B\right) + \frac{1}{4}J\right]|\alpha\alpha\rangle$$

$$\hat{H}|\alpha\beta\rangle = \left[-\nu_0(1-\sigma_A)\frac{1}{2} - \nu_0(1-\sigma_B)\left(-\frac{1}{2}\right) - \frac{1}{4}J\right]|\alpha\beta\rangle + \frac{1}{2}J[0 + |\beta\alpha\rangle]$$

$$= \left[+\frac{1}{2}\nu_0(\sigma_A - \sigma_B) - \frac{1}{4}J\right]|\alpha\beta\rangle + \frac{1}{2}J|\beta\alpha\rangle$$

$$\hat{H}|\beta\alpha\rangle = \left[-\nu_0(1-\sigma_A)\left(-\frac{1}{2}\right) - \nu_0(1-\sigma_B)\frac{1}{2} - \frac{1}{4}J\right]|\beta\alpha\rangle + \frac{1}{2}J[|\alpha\beta\rangle + 0]$$

$$= \left[-\frac{1}{2}\nu_0(\sigma_A - \sigma_B) - \frac{1}{4}J\right]|\beta\alpha\rangle + \frac{1}{2}J|\alpha\beta\rangle$$

$$\hat{H}|\beta\beta\rangle = \left[-\nu_0(1-\sigma_A)\left(-\frac{1}{2}\right) - \nu_0(1-\sigma_B)\left(-\frac{1}{2}\right) + \frac{1}{4}J\right]|\beta\beta\rangle$$

$$= \left[\nu_0\left(1 - \frac{1}{2}\sigma_A - \frac{1}{2}\sigma_B\right) + \frac{1}{4}J\right]|\beta\beta\rangle$$

With these quantities evaluated, it is a simple matter to multiply them by $\alpha\alpha$, $\alpha\beta$, etc., and thus to evaluate all the matrix elements in the secular determinant. For example, $\langle\beta\alpha|\hat{H}|\alpha\beta\rangle = 0 + \frac{1}{2}J$. The results for all the matrix elements are contained in Table 8–1. The result of $\hat{H}|\alpha\alpha\rangle$ is seen to yield $|\alpha\alpha\rangle$ back again, so this wave function is an eigenfunction of the Hamiltonian, as is $|\beta\beta\rangle$. However, $|\beta\alpha\rangle$ and $|\alpha\beta\rangle$ are not eigenfunctions. Accordingly, we find off-diagonal values of $(\frac{1}{2})J$ connecting $\alpha\beta$ and $\beta\alpha$. The determinant is block diagonalized, with two frequencies given as:

$$\frac{E_1}{h} = \nu_0\left[-1 + \frac{1}{2}\sigma_A + \frac{1}{2}\sigma_B\right] + \frac{J}{4} \tag{8-20}$$

$$\frac{E_4}{h} = \nu_0\left[1 - \frac{1}{2}\sigma_A - \frac{1}{2}\sigma_B\right] + \frac{J}{4} \tag{8-21}$$

The energies of E_2 and E_3 are obtained by solving the 2×2 block. This solution can be simplified by making a few definitions. The definitions to be made are not obvious

TABLE 8–1. EVALUATION OF THE MATRIX ELEMENTS FOR THE COUPLED AB SYSTEM

	$\alpha\alpha$	$\alpha\beta$	$\beta\alpha$	$\beta\beta$
$\alpha\alpha$	$\nu_0\left[-1 + \frac{\sigma_A}{2} + \frac{\sigma_B}{2}\right] + \frac{J}{4} - \frac{E}{h}$	0	0	0
$\alpha\beta$	0	$+\frac{1}{2}\nu_0(\sigma_A - \sigma_B) - \frac{J}{4} - \frac{E}{h}$	$\frac{J}{2}$	0
$\beta\alpha$	0	$\frac{J}{2}$	$-\frac{1}{2}\nu_0(\sigma_A - \sigma_B) - \frac{J}{4} - \frac{E}{h}$	0
$\beta\beta$	0	0	0	$\nu_0\left[1 - \frac{\sigma_A}{2} - \frac{\sigma_B}{2}\right] + \frac{J}{4} - \frac{E}{h}$

a priori choices. They come with experience in solving determinantal equations and trying to relate the results to observed spectra. We will substitute:

$$\left(\frac{1}{2}\right)\nu_0(\sigma_A - \sigma_B) = \frac{\Delta}{2} = C\cos 2\theta$$

and

$$\frac{J}{2} = C\sin 2\theta$$

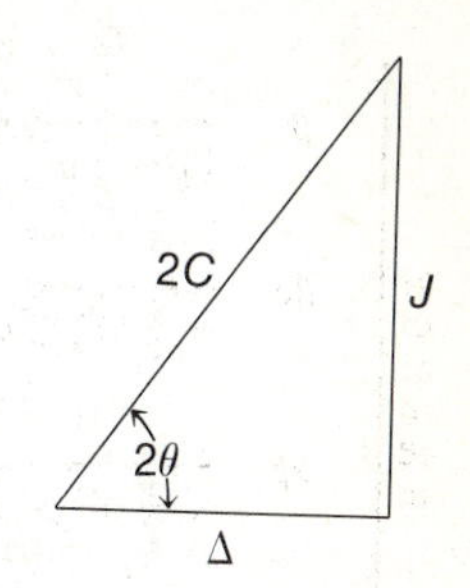

FIGURE 8–12. Geometric relation of Δ, θ, J, and C.

where $C = (\frac{1}{2})(J^2 + \Delta^2)^{1/2}$. The geometrical relation of these defined quantities is shown in Fig. 8–12. With these definitions the 2 × 2 block of the secular determinant becomes:

$$\begin{vmatrix} -C\cos 2\theta - \frac{1}{4}J - \frac{E}{h} & C\sin 2\theta \\ C\sin 2\theta & C\cos 2\theta - \frac{1}{4}J - \frac{E}{h} \end{vmatrix} = 0$$

The solutions to this determinant are

$$\frac{E_2}{h} = -\frac{1}{4}J - C \qquad (8\text{-}22)$$

$$\frac{E_3}{h} = -\frac{1}{4}J + C \qquad (8\text{-}23)$$

To get the wave functions for these two states, we substitute the energies one at a time into the determinant given above. The matrix corresponding to this determinant (this matrix has the same form as the determinant) is multiplied by a column matrix (vector) of the coefficients. The product is zero. Expansion produces the two resulting, linearly dependent equations, which can be solved for C_1 and C_2 when the normalization requirement is imposed. In general, this is done for each energy. The procedure is similar to that used in Chapter 3 when Hückel theory was discussed. The wave functions ψ_2 and ψ_3 so obtained are listed below along with ψ_1 and ψ_4.

$$\psi_1 = |\alpha\alpha\rangle \qquad (8\text{-}24)$$

$$\psi_2 = \cos\theta|\alpha\beta\rangle - \sin\theta|\beta\alpha\rangle \qquad (8\text{-}25)$$

$$\psi_3 = \sin\theta|\alpha\beta\rangle + \cos\theta|\beta\alpha\rangle \qquad (8\text{-}26)$$

$$\psi_4 = |\beta\beta\rangle \qquad (8\text{-}27)$$

We can summarize the results by constructing an energy diagram for this AB system in Fig. 8–13, as was previously done in Fig. 8–10 for the AX system. In Fig. 8–13(A), the situation with $J_{AB} = 0$ [which is identical to Fig. 8–10(A)] is presented as a starting place. The contribution to the total energy of the various levels from J_{AB} is shown in Fig. 8–13(B).

The arrows for the four transitions are indicated in Fig. 8–13. When A and B have similar chemical shifts, all four transitions will appear in a narrow region of the spectrum. The center of the spectrum is given by the average of the $E_1 \rightarrow E_2$ and $E_2 \rightarrow E_4$ transition energies or:

$$\frac{1}{2}(E_2 - E_1 + E_4 - E_2) = \frac{E_4 - E_1}{2} = \nu_0\left[1 - \frac{1}{2}\sigma_A - \frac{1}{2}\sigma_B\right]$$

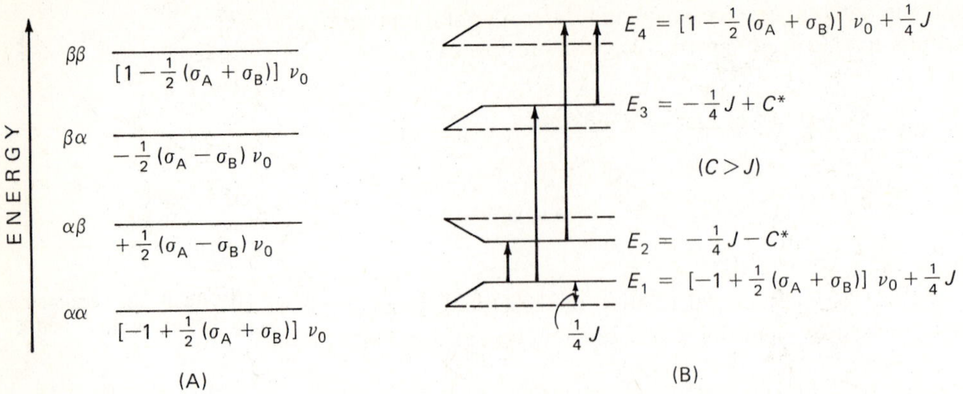

ENERGY

$\beta\beta$ $[1 - \frac{1}{2}(\sigma_A + \sigma_B)]\,\nu_0$

$\beta\alpha$ $-\frac{1}{2}(\sigma_A - \sigma_B)\,\nu_0$

$\alpha\beta$ $+\frac{1}{2}(\sigma_A - \sigma_B)\,\nu_0$

$\alpha\alpha$ $[-1 + \frac{1}{2}(\sigma_A + \sigma_B)]\,\nu_0$

(A)

$E_4 = [1 - \frac{1}{2}(\sigma_A + \sigma_B)]\,\nu_0 + \frac{1}{4}J$

$E_3 = -\frac{1}{4}J + C^*$

$(C > J)$

$E_2 = -\frac{1}{4}J - C^*$

$E_1 = [-1 + \frac{1}{2}(\sigma_A + \sigma_B)]\,\nu_0 + \frac{1}{4}J$

$\frac{1}{4}J$

(B)

FIGURE 8–13. The second order energies (written in frequency units) of an AB spin system. (A) $J_{AB} = 0$. (B) The second-order result. (*These are the total energies, where the quantity C contains the shielding constant terms.)

The transition energies relative to this center can then be found using the energies given above and illustrated in Fig. 8–13(B). The results are presented in Table 8–2

TABLE 8–2. ENERGIES AND INTENSITIES FOR THE TRANSITIONS IN AN AB MOLECULE

Transition	Separation from Center	Relative Intensity
$1 \longrightarrow 2$	$-(J/2) - C$	$1 - \sin 2\theta$
$1 \longrightarrow 3$	$-(J/2) + C$	$1 + \sin 2\theta$
$2 \longrightarrow 4$	$(J/2) + C$	$1 - \sin 2\theta$
$3 \longrightarrow 4$	$(J/2) - C$	$1 + \sin 2\theta$

where $\sin 2\theta = J/2C$ and $C = (1/2)(J^2 + \Delta^2)^{1/2}$

along with the intensities of the various transitions. The intensities are dependent on $\hat{I}_X$ (recall Chapter 7, where we discussed the fact that H_1 made the $\hat{I}_X$ matrix elements non-zero) and are determined from the evaluation of integrals of the form

$$|\langle\psi_j|\hat{I}_{XA} + \hat{I}_{XB}|\psi_i\rangle|^2 \tag{8-28}$$

For example, the $1 \rightarrow 3$ (*i.e.*, E_1 to E_3) transition yields the result:

$$\begin{aligned}|\langle\alpha\alpha|\hat{I}_{XA} + \hat{I}_{XB}|\beta\alpha\cos\theta + \alpha\beta\sin\theta\rangle|^2 &= (1/4)(\cos\theta + \sin\theta)^2 \\ &= (1/4)(1 + \sin 2\theta)\end{aligned}$$

8–7 EFFECTS OF THE RELATIVE MAGNITUDES OF J AND Δ ON THE SPECTRUM OF AN AB MOLECULE

We are now in a position to use the results summarized in Table 8–2 to see what influence the relative magnitudes of Δ and J have on the appearance of the spectrum.

A. $J = 0$ and $\Delta \neq 0$

The $1 \rightarrow 2$ and $3 \rightarrow 4$ transitions are degenerate, as are $1 \rightarrow 3$ and $2 \rightarrow 4$. The spectrum consists of two peaks, one from A and one from B, as in the earlier discussion of Fig. 8–10(A).

B. $J \neq 0$, $\Delta \neq 0$, but $J \ll \Delta$—An AX System

When Δ dominates C, we have $\sin 2\theta = J/\Delta \approx 0$. With the $\sin 2\theta$ equal to zero, all four transitions have equal intensity. The energies become those shown in Fig. 8-10(B). Relative to the center, we have $1 \rightarrow 2$ at $(-\frac{1}{2})J - (\frac{1}{2})\,\Delta$, $1 \rightarrow 3$ at $(-\frac{1}{2})J + (\frac{1}{2})\,\Delta$, $2 \rightarrow 4$ at $(\frac{1}{2})J + (\frac{1}{2})\,\Delta$, and $3 \rightarrow 4$ at $(\frac{1}{2})J - (\frac{1}{2}\,\Delta)$ for a spectrum that looks like that shown in Fig. 8-14.

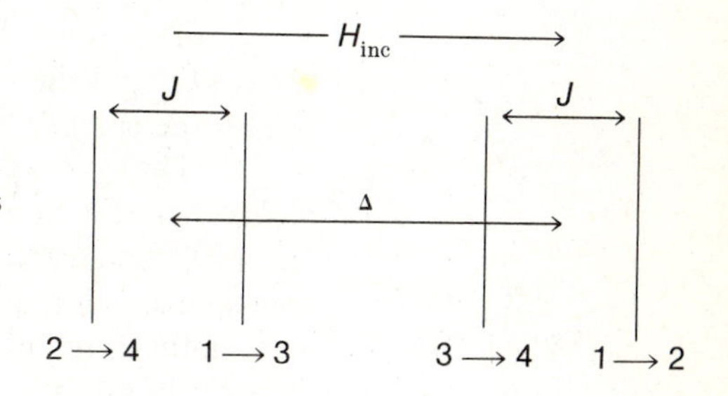

FIGURE 8-14. Spectrum of an AX system. (Greater frequency corresponds to lower field.)

C. $\Delta \neq 0$ and $J \neq 0$, but $J \approx \Delta$

When $J \approx \Delta$, $\sin 2\theta$ is appreciable and positive. Therefore, the $1 \rightarrow 3$ and $3 \rightarrow 4$ transitions will have equal intensities which are greater than those of $1 \rightarrow 2$ and $2 \rightarrow 4$. The $1 \rightarrow 2$ and $3 \rightarrow 4$ transitions will differ by J, as will the $2 \rightarrow 4$ and $1 \rightarrow 3$; also, the $3 \rightarrow 4$ and $2 \rightarrow 4$ will differ by $2C$, as will the $1 \rightarrow 3$ and $1 \rightarrow 2$. The spectrum will look like that in Fig. 8-15.

FIGURE 8-15. Spectrum of an AB system.

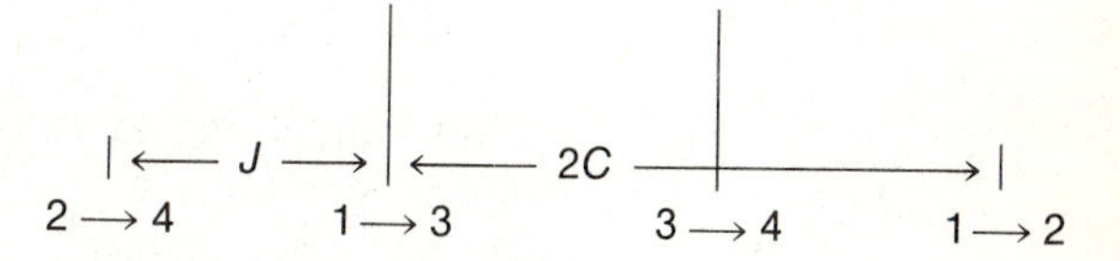

D. $J \neq 0$, Δ very small ($\Delta < J$)

When σ_A and σ_B are almost equal, the value of $\sin 2\theta$ approaches unity and the $1 \rightarrow 2$ and $2 \rightarrow 4$ transitions almost disappear. The spectrum resembles that shown in Fig. 8-16.

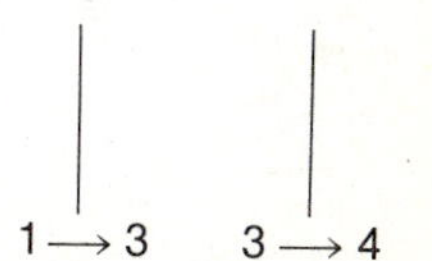

FIGURE 8-16. Spectrum of an AB molecule when σ_A and σ_B are close.

E. $J \neq 0$, but $\sigma_A = \sigma_B$, so $\Delta = 0$

This is the case of equivalent protons, and we are now in a position to see why equivalent protons do not split each other. When $\Delta = 0$, $\sin 2\theta$ equals unity and, as can be seen from Table 8-2, the $1 \rightarrow 2$ and $2 \rightarrow 4$ transitions have zero intensity. The $1 \rightarrow 3$ and $3 \rightarrow 4$ transitions have intensity, but they have the same energy and occur at the center of the spectrum; *i.e.*, $-(\frac{1}{2})J + C = (\frac{1}{2})J - C = 0$. This is a manifestation of a general rule: magnetically equivalent nuclei do not split each other.

8-8 MORE COMPLICATED SECOND ORDER SYSTEMS

It is convenient to classify molecules according to the general type of Δ and J systems to which they belong, for this suffices to define the Hamiltonian for the coupling. The conventions that will be given are normally applied to spin $\frac{1}{2}$ systems.

A. Each type of magnetic nucleus in the molecule is assigned a capital letter of the Roman alphabet. All of one equivalent set of nuclei are given a single letter, and a subscript is used to indicate the number of such nuclei in the set; *e.g.*, benzene is A_6.

B. Roman letters are assigned to different nuclei in order of decreased shielding; *i.e.*, three different nuclei with similar shifts listed in order of decreased shielding would be labeled A, B, and C. If the second set of nuclei has a very different shift than the first (*i.e.*, $\Delta \gg J$), a letter far removed in the alphabet is used (X, Y, Z) to label the low field peaks. For example, acetaldehyde, CH_3CHO, is an A_3X system. This implies a first order system. HF would be an AX system. The molecule $HPFCl_2$ would be an AMX system. When one is not sure of the chemical shift difference, nearby letters are used to be safe, for this implies that the system could be second order.

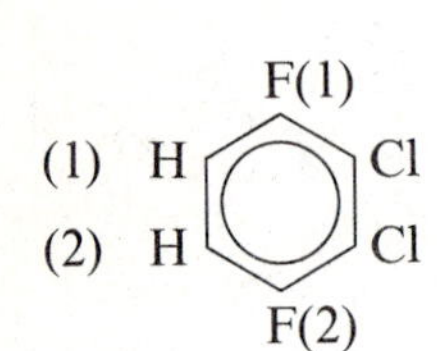

C. The system of labeling must provide information about the magnetic, as well as the chemical, non-equivalence of nuclei in the molecule. If two atoms are equivalent by symmetry, they have the same chemical shift, but they could be involved in spin-spin coupling to other nuclei such that the J to one nucleus is different from the J to another; *e.g.*, in $C_6H_2F_2Cl_2$ the two hydrogens are equivalent, but two different HF couplings are needed for the full analysis. The H(1)-F(1) coupling constant is different from the H(1)-F(2) coupling constant. The protons and the fluorines are said to be chemically equivalent, but *magnetically non-equivalent. All that is meant by this* is that, in the Hamiltonian for this system, two different HF couplings must be considered even though the two hydrogens and two fluorines are equivalent by symmetry. The dichloro-difluoro-benzene isomer discussed above is thus referred to as an AA′XX′ system. The reader should now realize that in the non-planar (D_{2d}) molecule

$H_{(a)}$ $\quad$ $H_{(c)}$
$^{13}C{=}C{=}^{13}C$
$H_{(b)}$ $\quad$ $H_{(d)}$

the ab pair is magnetically non-equivalent to the cd pair because two different coupling constants to the ^{13}C atoms are involved. The system will be labeled $A_2A_2'XX'$. The molecule ClF_3 discussed in the previous chapter (consider only fluorines) would be classified as an AB_2 molecule (at 10 MHz). The phosphorus nmr of the tetrapolyphosphate anion, $P_4O_{13}^{6-}$, would yield an A_2B_2 spectrum, and P_4S_3 is an A_3X system.

If the only magnetic nuclei in the molecule are all symmetry equivalent (*i.e.*, *isochronous*), there cannot be any magnetic non-equivalence, for, as shown in our discussion of the AB system, equivalent nuclei cannot split one another. Benzene, methyl iodide, and tetramethylsilane are all examples of molecules in which the protons are isochronous.

The energy levels of various kinds of coupled systems have been worked out in detail [(30–32)] in terms of the shift differences and coupling constants, and are reported in the literature. To use these results, one classifies the molecule of interest according to the scheme described above and looks up the analysis for this type of system; *e.g.*, one could find the energy levels for any AA′XX′ system.[(32)] Computer programs are also available that find the best values of the J's and Δ's which reproduce the chemical shifts and intensities of all the peaks in the experimental spectrum. The interpretation of a relatively simple second order spectrum is illustrated for the ^{31}P nmr spectrum of the anion

O O $^{(3-)}$
H—P—P—O
O O

indicated in Fig. 8-17. The actual spectrum, IV, is interpreted by generating it in three stages. Consideration of the first stage, I, yields two lines from the two non-equivalent phosphorus atoms, $P_{(a)}$ and $P_{(b)}$, and their separation is the chemical shift difference. The second consideration, II,

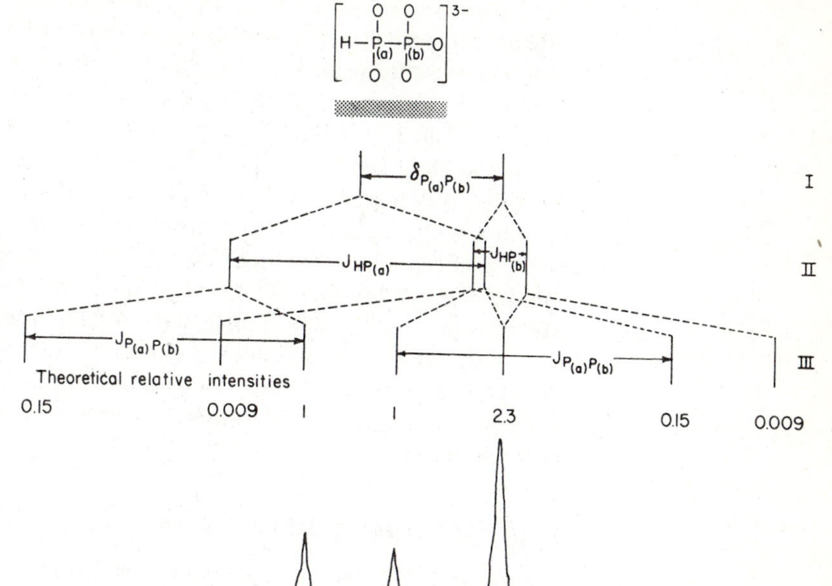

FIGURE 8-17. The phosphorus nmr spectrum of the diphosphate anion $HP_2O_5^{3-}$ and its interpretation. I, Chemical shift differences in $P_{(a)}$ and $P_{(b)}$. II, H—P splitting. III, P—P splitting. IV, Observed spectrum. [Reprinted with permission from C. F. Callis, *et al.*, J. Amer. Chem. Soc., *79*, 2722 (1957). Copyright by the American Chemical Society.]

includes splitting by hydrogen. Since the hydrogen is on $P_{(a)}$, $J_{P_{(a)}-H} > J_{P_{(b)}-H}$. The four lines which result are included in II. The third consideration, III, included $P_{(a)}$-$P_{(b)}$ splitting and accounts for the final spectrum for this ABX case. Two of the expected lines are not detected in the final spectrum because they are too weak to be detected and another pair fall so close together that they appear as a single peak (the most intense peak). The analysis of the $HP_2O_5^{3-}$ spectrum to yield the interpretation contained in Fig. 8-17 was carried out [33] with a computer analysis that fitted the intensities and chemical shifts of the experimental spectrum.

Problem 24 in the exercise section of this chapter has been designed to give further practice on the analysis of complex spectra.

The phenomenon of virtual coupling, which we discussed earlier, is a magnetically non-equivalent system of, for example, the type XAA′X′ where $J_{AA'}$ is large.

8-9 INTRODUCTION

DOUBLE RESONANCE AND SPIN TICKLING EXPERIMENTS

In a double resonance experiment, the sample is subjected to a second rf source whose frequency corresponds to the Larmor frequency of one of the nuclei in the sample. This field causes the contribution to the spectrum from this nucleus to disappear, and this nucleus is said to be *decoupled*. The second rf field is applied with a large amplitude, at right angles to H_1 and orthogonal to the pickup coils. The net effect of this field is to cause nuclear transitions in, say, B of the AB system. Decoupling of the B nucleus is achieved, and B makes no contribution to the A spectrum. In practice, decoupling can be accomplished only if $\nu_A - \nu_B > 5|J_{AB}|$. When the spectrum of A is examined with B being irradiated, the symbol $A \rightarrow \{B\}$ is employed to indicate this fact.

For systems where $J \approx \Delta$, one can perform a *spin tickling experiment*. In this experiment, a weak rf field is employed and all transitions having an evergy level in common with the peak being irradiated will undergo a change. Referring to Figs. 8–13 and 8–15, the $1 \rightarrow 2$ transition is seen to have an energy level in common with the $1 \rightarrow 3$ and $2 \rightarrow 4$ transitions, but not the $3 \rightarrow 4$. Thus, the latter peak is not split by spin tickling the $1 \rightarrow 2$ peak, but the other spectral lines will be split. Experimentally, it is found that the tickling splits the lines that have an energy level in common with the line being saturated. Furthermore, if the transitions for the two peaks connected by a common energy level correspond to a consecutive change in spin of both nuclei, each by 1, as in $1 \rightarrow 2$ and $2 \rightarrow 4$ (that is, $\alpha\alpha$ to $\alpha\beta$ and $\alpha\beta$ to $\beta\beta$) or as in $1 \rightarrow 3$ and $3 \rightarrow 4$, a sharp doublet results. When this is not the case, as in $1 \rightarrow 2$ and $1 \rightarrow 3$ (that is, $\alpha\alpha$ to $\alpha\beta$ and $\alpha\alpha$ to $\beta\alpha$) or $3 \rightarrow 4$ and $2 \rightarrow 4$, a broad doublet results. We shall not go into the reasons for this, but simply point out that this is a valuable technique for spectral assignment and energy level ordering in second order systems.

8–10 SPECTRAL SIMPLIFICATION

In an $A \rightarrow \{B\}$ spectrum of an AB molecule, only A is detected, so its chemical shift is accurately determined. Proton spectra are greatly simplified by irradiating a set of proton nuclei that differ in chemical shift from another set and that are undergoing second order effects with this set (*i.e.*, A or B of AB_2X, A_2B_3, etc.). The chemical shifts of A and B can be determined by a series of decoupling experiments. This information is of considerable utility in the analysis of a second order spectrum and as a check on the analysis. In the AB example described above, if the chemical shifts of A and B were known, the only unknowns in equations (8–20) to (8–23) would be J_{A-B}. This information is not all that helpful in the analysis of a simple AB spectrum, but is very valuable in more complex systems where many unknowns, corresponding to the many coupling constants and chemical shifts influencing the spectrum, have to be determined.

The double resonance technique can be employed to evaluate chemical shifts for nuclei other than protons by using a proton probe.[66] If nucleus Y is splitting a proton, the frequency of the rf field that is most effective for decoupling Y from the protons is measured, and thus the chemical shift of Y is determined using a proton probe.

The proton nmr spectrum of diborane is illustrated in Fig. 8–18(A). This spectrum results from two sets of non-equivalent protons (bridge and terminal protons) being split by the ^{11}B nuclei. The asterisks indicate fine structure arising from the smaller abundance of protons on ^{10}B nuclei. (^{10}B has a natural abundance of 18.83 per cent and $I = 3$ compared to 81.17 per cent for ^{11}B with $I = 3/2$.) In Fig. 8–18(B), the splitting caused by ^{11}B has been removed by saturation of the boron nuclei by the double resonance technique. Two peaks of intensity ratio 2:1 are obtained, corresponding to the four terminal and two bridge protons.[34]

The double resonance technique has been successfully used on the proton nmr spectrum of $Al(BH_4)_3$. This molecule contains six Al—H—B bridge bonds. Both B and ^{27}Al ($I = 5/2$) have quadrupole moments. The proton nmr at 30 MHz consists[35] of a single broad line [Fig. 8–19(A)]. When the ^{11}B nucleus is saturated ($^1H \rightarrow \{^{11}B\}$), the proton resonance spectrum in Fig. 8–19(B) results. Fig. 8–19(C) represents the proton nmr spectrum when the sample is irradiated with frequency corresponding to that of ^{27}Al ($^1H \rightarrow \{^{27}Al\}$). The four large peaks in (C) arise from ^{11}B splitting of the proton and the smaller peaks from ^{10}B splitting. The bridging and terminal hydrogens are not distinguished because of a rapid proton exchange reaction which makes all hydrogens magnetically equivalent.

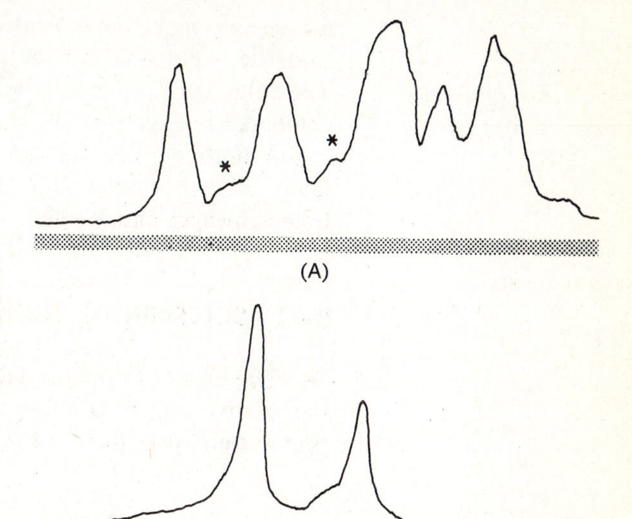

FIGURE 8-18. Proton nmr spectrum of B_2H_6. (A) Proton nmr with ^{11}B and ^{10}B splitting. (B) Proton nmr with ^{11}B nucleus saturated. [From J. N. Schoolery, Disc. Faraday Soc., *19*, 215 (1955).]

(B)

Two isomers have been obtained in the preparation of N_2F_2. One definitely has a *trans* structure with one fluorine on each nitrogen. In conflicting reports, the structure of the second isomer has been reported to be the *cis* isomer and also $F_2N{=}N$. An excellent discussion of the results obtained by employing several different physical methods in an attempt to resolve this problem has been reported along with the fluorine nmr spectrum and results from a double resonance experiment.[36] Saturation of the ^{14}N nucleus in this second isomer with a strong rf field causes collapse of all nitrogen splitting. It is concluded that the chemical shift of the two

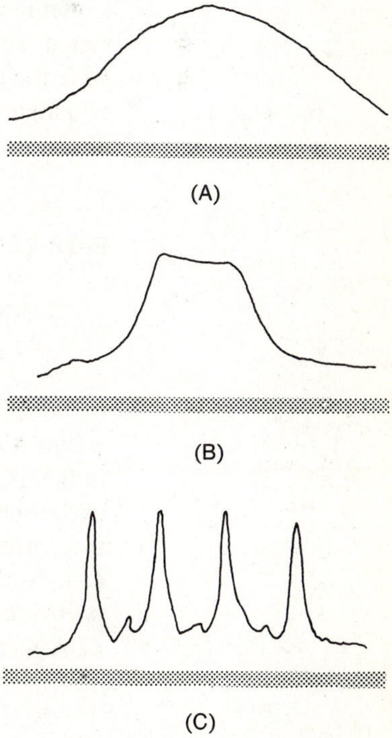

(A)

(B)

(C)

FIGURE 8-19. Proton nmr of $Al(BH_4)_3$. (A) Proton resonance. (B) Proton resonance, ^{11}B saturated. (C) Proton resonance, ^{27}Al saturated. [From R. A. Ogg, Jr., and J. D. Ray, Disc. Faraday Soc., *19*, 239 (1955).]

nitrogens must be equivalent, and this eliminates the $F_2N{=}N$ structure. Additional evidence is obtained for the *cis* structure from a complete spectral interpretation. The value for J_{N-F} calculated in this study for a *cis* structure is reasonable when compared to J_{N-F} for NF_3.

A more complete discussion of the theory of the double resonance technique and many more examples of its application are contained in a review article[37] by Baldeschwieler and Randall.

8-11 DETERMINING SIGNS OF COUPLING CONSTANTS

The double resonance technique has been successfully employed to determine the relative sign of coupling constants. This can be illustrated by considering the proton nmr spectrum[38] of $(C_2H_5)_2Tl^+$ in Fig. 8-20 ($I = \frac{1}{2}$ for Tl). If J_{Tl-CH_3} and

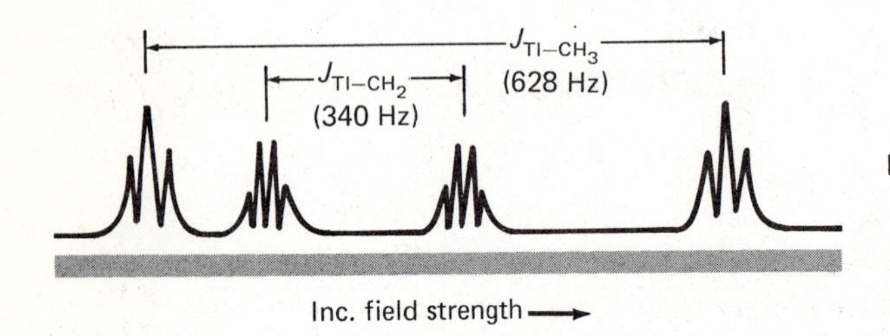

FIGURE 8-20. NMR spectrum of $(C_2H_5)_2\,Tl^+$ (facsimile).

J_{Tl-CH_2} are both positive, both low field peaks correspond to interaction with positive nuclear magnetic quantum numbers of Tl. If the signs of J are different, one low field peak corresponds to interaction with the moment from thallium nuclei where $m_I = +\frac{1}{2}$ and the other to $-\frac{1}{2}$. By irradiation at the center of each of the multiplets, it was shown that each CH_3 triplet was coupled to the distant methylene quartet and vice versa. For example, irradiation with a frequency corresponding to the low field triplet resulted in the disappearance of the fine structure of the high field methylene signal. This result indicates that J_{Tl-CH_3} and J_{Tl-CH_2} have opposite signs, for if the sign were the same, the two low field multiplets would be coupled together as would the two high field multiplets, and saturation of the low field triplet would cause collapse of the fine structure in the low field methylene signal.

8-12 SPIN SATURATION LABELING[39]

Consider a system

$$AB + A' \longrightarrow A'B + A$$

where A and A′ are the same and the exchange rate is slow enough to give two peaks in the nmr. If we saturate a proton resonance of A′, this peak will disappear, but the exchange process will also cause a partial saturation, *i.e.*, a decrease in the intensity of the corresponding proton resonance in AB, if the exchange rate of A′ with AB is comparable to the relaxation rate at the two sites. The lifetime for A, leaving a *particular spin state in AB*, τ_{1AB}, then has contributions from T_{1AB} and τ_{AB} (the lifetime of A at AB)

$$\tau_{1AB} = (T_{1AB}{}^{-1} + \tau_{AB}{}^{-1})^{-1} \tag{8-29}$$

If a saturating rf field is turned on at resonance A′, saturation of this resonance occurs immediately and one can observe (by sitting on the resonance of AB) an asymptotic approach to a new equilibrium value of the magnetization at this point. The plot of intensity versus time and the equilibrium value for the magnetization can be analyzed for τ_{1AB}. When T_{1AB} is known, τ_{AB} can be calculated simply from the equilibrium value of the magnetization of AB.

When the field is turned off, the resonance at AB returns to its initial intensity asymptotically, and this curve can also be analyzed to yield τ_{1AB}. This procedure is suited for the determination of reaction rates in the range between 10^{-3} and 1 sec^{-1}. It thus can provide data on the slow side of the nmr line shape experiment and is complimentary to the line shape technique.

Spin saturation can also be thought of as equivalent to a deuterium labeling experiment, and the same mechanistic information is available from this technique as from the labeling experiment. For example, the heptamethylbenzenonium ion is fluxional:

(1) (1) H_3C CH_3; (2) H_3C, (2) CH_3; (3) H_3C, (3) CH_3; (4) CH_3 ⇄ [rearranged ion] ⇄ etc.

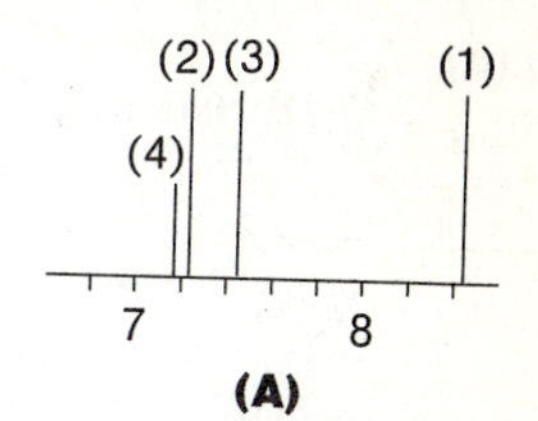

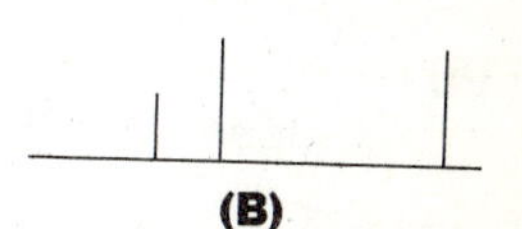

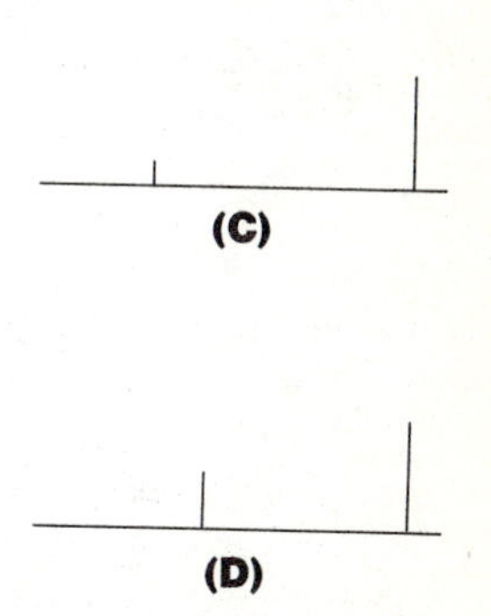

FIGURE 8–21. The spectrum of the heptamethylbenzenonium ion; (B) spectrum (A) with the (2) proton decoupled; (C) spectrum (A) with the (2) and (3) protons decoupled; (D) spectrum (A) with the (2) and (4) protons decoupled.

Does the interchange proceed by a 1,2 shift or by a random migration process? At 28°, four fairly sharp resonances result, as shown in Fig. 8–21(A). [There are two type (1) CH_3 groups, two orthos, two metas, and one para.] Upon raising the temperature, the resonances broaden, etc., and a complete line shape analysis suggests a 1,2 shift; but the differences between the expected spectrum for a 1,2 shift and that for the random process are subtle, and a simultaneous operation of both processes was not ruled out. Saturation of the methyl resonance at site (2) produced spectrum (B) in Fig. 8–21, in which the intensity at site (1) is decreased. No additional decrease in intensity is observed at site (1) when (2) and (3) are saturated or when (2) and (4) are saturated, as shown in Fig. 8–21(C) and (D). If the random mechanism were operative, a further reduction in intensity at site (1) of 22% would be observed upon saturation at sites (3) and (4). The intensity doesn't change, within the 1% accuracy of its determination, so the random process is minor or absent.

An interesting phenomenon[40] associated with the double resonance experiment is the Overhauser effect, discovered in the course of studying free radicals. When there is a coupling of the nuclear and electron spins, an enhancement in the intensity of the nmr transitions is observed when the esr transitions are saturated. Most commonly, this coupling is Fermi contact in nature (see section 8–18) and is represented by the Hamiltonian

$$\hat{H} = a\vec{\mathbf{I}}\cdot\vec{\mathbf{S}} = a\left(\hat{I}_z\hat{S}_z + \frac{1}{2}\hat{I}_+\hat{S}_- + \frac{1}{2}\hat{I}_-\hat{S}_+\right)$$

The energy level diagram for this system is shown in Fig. 8–22(A). The population of each state *at equilibrium* is given in parentheses, given that the population of the lowest level is N. As usual, this is given by Boltzmann statistics, with population decreasing as energy increases. The nmr transitions are shown by the dotted arrows II → I and IV → III, and the intensity of each is proportional to the population difference between ground and excited states involved in the transition. Next, let's

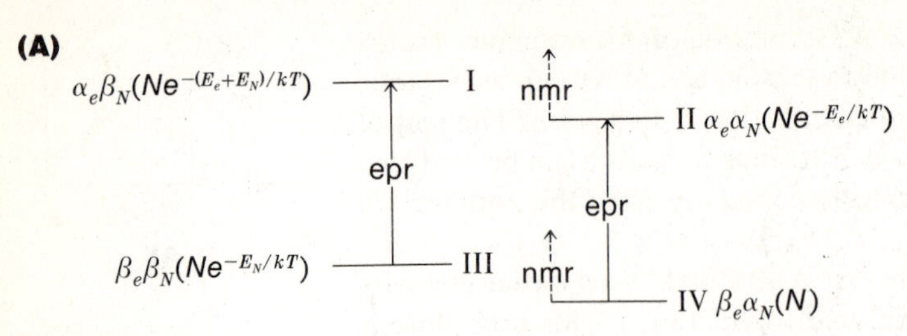

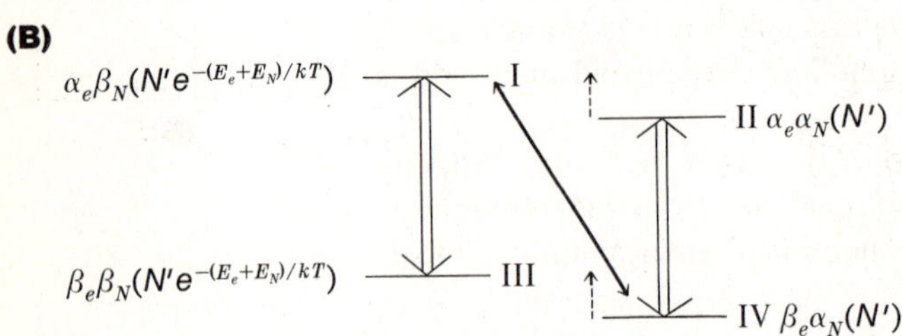

FIGURE 8–22. The spin levels, populations, and allowed transitions for a two-spin system (A) without and (B) with electron spin decoupling. E_N and E_e are the nuclear and electron Zeeman energies, $g_N\beta_N H$ and $g\beta H$ respectively. In view of the different signs of the magnetogyric ratios for the electrons and proton, α_N (corresponding to $m_I = +\frac{1}{2}$) is lower for the nuclear moment but β_e (corresponding to $m_S = -\frac{1}{2}$) is lower for the electron.

consider how this situation changes when we saturate the esr resonances [solid arrows III → I and IV → II in Fig. 8–22(A)]. A non-equilibrium situation results, as shown in Fig. 8–22(B). We force the population of III to equal that of I and we force IV to equal II when we saturate. However, for purposes of the nmr experiment, we must focus on the relationship of the population of IV to that of III (and similarly II to I), for this controls the intensity of the nmr transition. The relationship is determined by the fact that the relaxation process denoted by the solid arrow connecting states I and IV is the dominant relaxation process. (This is due to the I_+S_- and I_-S_+ terms in the coupling Hamiltonian; *i.e.*, the terms consisting of I_+S_- operating on $\alpha_e\alpha_N$ and $\beta_e\beta_N$ are zero.) Since I and IV are connected by the fastest relaxation process, the populations of states I and IV must therefore be related by a Boltzmann factor of $e^{-(E_N+E_e)/kT}$.* Hence, IV and III must also be related by that factor since I and III have the same population when the electron transition is saturated. The populations of II and I are similarly related by the same factor. Hence, the nmr intensity (which depends upon the IV → III and II → I populations) is greatly enhanced, by a factor of $(E_N + E_e)/E_N$ to be exact. This factor actually represents an upper limit attainable only when other relaxation processes are negligible.

This same effect occurs in a nuclear-nuclear double resonance experiment and is called the nuclear Overhauser effect. However, there are two important differences. First, most nuclei including ^1H and ^{13}C have positive g_N values, so that the energy level diagram resembles that of Fig. 8–23. Second, the major mechanism for coupling

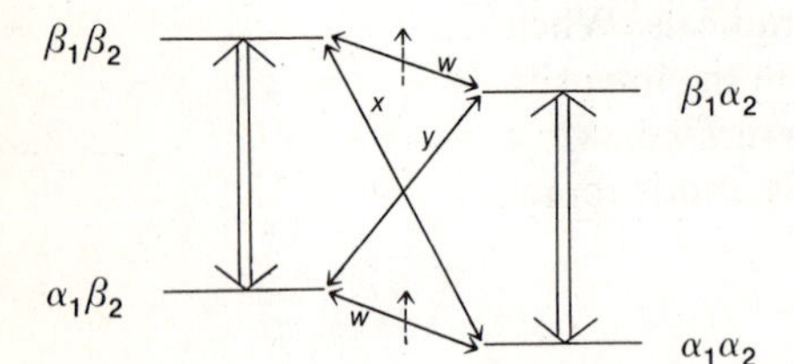

FIGURE 8–23. Energy level diagram showing transitions and possible relaxation processes in the nuclear Overhauser effect. The double arrow is the decoupling frequency and *w*, *x*, and *y* are relaxation processes.

the two nuclei is generally dipolar, given by the Hamiltonian

$$\hat{H} = g_N{}^2\beta_N{}^2 \left\{ \frac{\vec{I}_1 \cdot \vec{I}_2}{r^3} - \frac{3(\vec{I}_1 \cdot \vec{r})(\vec{I}_2 \cdot \vec{r})}{r^5} \right\}$$

* An equivalent way to look at this is that the population of state I times the I → IV relaxation probability must equal the population of IV times the IV → I probability. Since the probabilities are related by a Boltzmann factor, the populations must also be so related.

This gives rise to terms of the form $I_{1+}I_{2-}$, $I_{1+}I_{2+}$, $I_{1-}I_{2-}$, etc. In other words, all the processes labeled x, y, and w in Fig. 8–23 are induced. In the limit of fast tumbling, the ratio of $x:y:w$ is 12:2:3. Because the x process is fastest, we again obtain signal enhancement.

These are the most common cases, but there are other possibilities. Depending on the relative sizes and magnitudes of the g_N values, there may be signal reduction instead of enhancement, and even negative Overhauser enhancements (emission instead of absorption). The enhancement has been treated quantitatively,[41] and it can be shown that for the direct coupling mechanism the theoretical enhancement, f, is given by:

$$f = \left(1 + \frac{g_1\beta_1}{g_2\beta_2}\right) = \left(1 + \frac{\gamma_1}{\gamma_2}\right) \tag{8-30}$$

where the 1 subscript refers to the spin being saturated and the 2 subscript refers to the spin being observed. For electron saturation during a proton nmr observation, the theoretical enhancement is 659. The ideal enhancement is seldom observed because of incomplete saturation and relaxation by other processes, *e.g.*, the T_1 nuclear relaxation I to II and III to IV.

This effect can be used to indicate whether the coupling mechanism is direct or indirect. Furthermore, by systematically observing the intensity changes in a proton nmr spectrum as various protons are saturated, one can determine which nuclei are in close proximity. The magnitude of the indirect coupling decreases with the sixth power of the distance. Thus, one can distinguish *cis-trans* isomers this way.

When pairs of free radicals are produced in solution, one may observe an effect called chemically induced dynamic nuclear polarization (CIDNP) without the need of saturating the electron spin transition to attain equal electron spin state populations. The mechanism for the CIDNP process is involved, and the reader is referred to reference 42 for more details.

8–13 LOCAL CONTRIBUTIONS TO THE CHEMICAL SHIFT

CHEMICAL SHIFT CONTRIBUTIONS

Local effects are those that arise on the atom that is undergoing the nmr transition. These are best understood in terms of the Ramsey shielding equation, which evaluates the field created by the electron density in a molecule when the molecule is placed in a magnetic field. This equation will not be derived here (see reference 32 if interested), but will be presented and used. Equation (8–31) arises from perturbation theory by treating the magnetic field as a perturbation on the ground state, field-free molecular wave function. The shielding is a tensor quantity. The zz component gives the contribution parallel to the field when the z-axis of the molecule is aligned with the field, and it is calculated from:

$$\sigma_{zz} = \frac{e^2}{2mc^2}\left\langle 0\left|\frac{x^2+y^2}{r^3}\right|0\right\rangle - \left(\frac{e\hbar}{2mc}\right)^2 \sum_n \left\{ \frac{\langle 0|\hat{L}_z|n\rangle \left\langle n\left|\frac{2\hat{L}_z}{r^3}\right|0\right\rangle}{E_n - E_0} + \frac{\left\langle 0\left|\frac{2\hat{L}_z}{r^3}\right|n\right\rangle \langle n|\hat{L}_z|0\rangle}{E_n - E_0} \right\} \tag{8-31}$$

This equation is similar to the Lamb equation for an atom, and it tells us what the applied field does to our molecule. A positive sign indicates shielding, *i.e.*, an upfield or a diamagnetic contribution. The negative sign indicates deshielding, *i.e.*, a downfield or a paramagnetic contribution.

The first term on the right-hand side of the equality is the so-called diamagnetic term. The symbol $\langle 0|$ corresponds to the ground state wave function, and r is the distance from the electron to the nucleus undergoing the transition. Since only $\langle 0|$, the ground state wave function, is involved in this first term, no excited states are mixed in by this term. The field does not distort the electron distribution in the molecule, but just induces a spherical electron circulation. If this were the only effect, the molecular wave function would be independent of the magnetic field.

For simplicity, we have treated only the σ_{zz} component to illustrate the factors that contribute to the shielding. There are similar expressions for σ_{xx} and σ_{yy} which are not necessarily equal to σ_{zz} (for σ_{xx}, the matrix element is $\langle 0|(y^2 + z^2)/r^3|0\rangle$) and the three quantities must be averaged to produce the shielding observed when the molecule is rapidly tumbling.

$$\sigma = \frac{1}{3}(\sigma_{xx} + \sigma_{yy} + \sigma_{zz})$$

We shall now attempt to gain some physical insight into the shielding mechanisms, using equation (8-31). The effect from the first term of the Ramsey equation gives rise to the normal diamagnetism observed for $S = 0$ molecules. It is a very important contribution to the total shielding observed in pmr (proton magnetic resonance). The field-induced, *spherical* circulation of electron density around the nucleus from this term shields the nucleus, as illustrated in Fig. 8-24. The magnitude

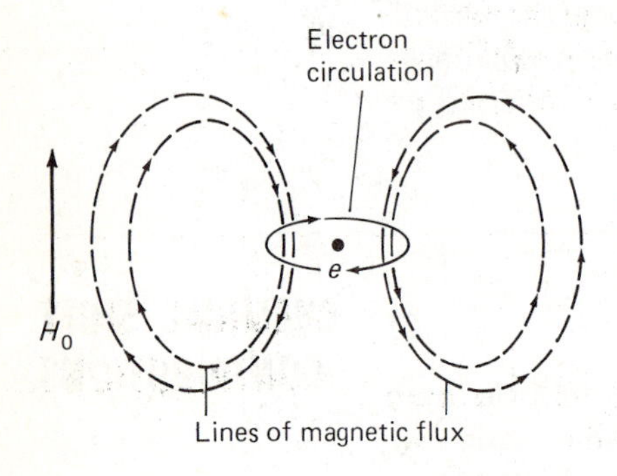

FIGURE 8-24. Local diamagnetic contribution to the shielding. Electron circulations occur in a plane perpendicular to the plane of the page. The induced field given by the arrow in the middle is opposed to H_0.

of the field generated at the nucleus by electron circulations is directly proportional to the strength of the applied field, and also will depend on the nature of the electron density surrounding the nucleus. As a consequence of the former effect, the chemical shift is field dependent. Evaluating the matrix elements for the local diamagnetic part, we have

$$\frac{e^2}{2mc^2}\left\langle 0\left|\frac{x^2 + y^2}{r^3}\right|0\right\rangle$$

Since $x^2 + y^2 + z^2 = r^2$ and since for a spherical H atom we can write

$$x^2 = y^2 = z^2$$

we find

$$\frac{x^2 + y^2}{r^3} = \frac{\frac{2}{3}r^2}{r^3} = \frac{2}{3r}$$

This matrix element reduces to $(\frac{2}{3})(e^2/2mc^2)\lambda\int(\psi_{1s}{}^2/r)\,d\tau$. We have in effect averaged ψ^2 over $\frac{2}{3}$ of the coordinate system, *i.e.*, x and y. The term λ refers to the effective number of electrons on hydrogen (in the $1s$ orbital). If this were the only effect of the magnetic field on a molecule, the chemical shift of a hydrogen nucleus would parallel the electronegativity of the groups attached. This correlation is not observed because of the other effects to be discussed in this section. The integral, $\int(\psi_{1s}{}^2/r)\,d\tau$, can be evaluated for a hydrogen atom. Using a Slater $1s$ hydrogen orbital, $\psi_{1s} = \pi^{-1}\exp(-1.2r)$, we obtain (in atomic units) $\sigma_{LD} = 21.4\lambda \times 10^{-6}$, where σ_{LD} represents the local diamagnetic contribution of the observed shielding.

In a molecule, spherical symmetry does not usually exist, and the total induced electron circulations do not follow simple paths. The electron cloud surrounding the nucleus being investigated is arbitrarily divided into an isotropic and an anisotropic part. The isotropic part is described by the diamagnetic term of equation (8–31). The anisotropic contribution is included in the local paramagnetic shielding, which is described by the paramagnetic term of equation (8–31).

The second local effect is the so-called *paramagnetic* term.* We are *not* using the word paramagnetic here to connote the same thing as when it is used in connection with molecules containing unpaired electrons. It is used here to describe the contributions from the field-induced nonspherical circulation of the electron density, *i.e.*, the second term on the right-hand side of the Ramsey equation. The term involves field-induced mixing in of excited states with the ground state, and this gives rise to a mechanism for non-spherical electron circulation and the accompanying paramagnetic contribution to the shielding. The equation for the paramagnetic effect corresponds very closely to the paramagnetic term in the Van Vleck equation for magnetic susceptibility (*vide infra*). The following quantities appear: $\langle 0|$ ground state and $\langle n|$ excited state wave functions; $\hat{L}_z$, the orbital angular momentum operator; and $E_n - E_0$, the energy difference between the ground state and the excited state being mixed in. The summation is carried out over all excited states; thus, to use the equation one needs wave functions and energies of all excited states, including those in the continuum. The energies of virtual orbitals (empty ones) are difficult to calculate, and thus it becomes impossible to employ the Ramsey equation in the rigorous calculation of the chemical shift of most molecules. However, there are many qualitative features of the diamagnetic and paramagnetic contributions to observed chemical shifts about which we can gain a good appreciation by considering this equation in more detail. This can be accomplished by using the Ramsey equation to evaluate the paramagnetic contribution to the ^{19}F nmr shift of F_2.

We begin by summarizing the behavior of the $\hat{L}_z$ operator.

$$\hat{L}_z|p_z\rangle = m_l|p_z\rangle = 0|p_z\rangle$$
$$\hat{L}_z|p_x\rangle = i|p_y\rangle$$
$$\hat{L}_z|p_y\rangle = -i|p_x\rangle$$
$$\hat{L}_z|p_{+1}\rangle = 1|p_{+1}\rangle$$
$$\hat{L}_z|p_{-1}\rangle = -1|p_{-1}\rangle$$

With $p_{+1} = p_x + ip_y$ and $p_{-1} = p_x - ip_y$, we obtain

$$\hat{L}_z(p_x \pm ip_y) = \pm 1(p_x \pm ip_y)$$

*The distinction between diamagnetic and paramagnetic contributions in equation (8–31) is somewhat artificial, for it depends upon the gauge selected for the vector potential (see Slichter text, p. 76, under General References). The traditional gauge corresponds to the measurement of angular momentum about the nucleus whose chemical shift is being calculated. The total shielding is gauge invariant.

The above equations can be verified by substituting the wave functions for the p-orbitals and realizing that

$$\hat{L}_z = -i\hbar\left(x\frac{\partial}{\partial y} - y\frac{\partial}{\partial x}\right)$$

The results of the $\hat{L}_x$ and $\hat{L}_y$ operators will be presented when needed.

The molecular orbital wave functions for the F_2 molecule are needed to substitute into equation (8–31). The ground state is:

$$\sigma_{1s}{}^2\sigma_{1s}{}^{*2}\sigma_{2s}{}^2\sigma_{2s}{}^{*2}\sigma_{2p_z}{}^2(\pi_x = \pi_y)^4(\pi_x{}^* = \pi_y{}^*)^4$$

We see from equation (8–31) that we need to evaluate matrix elements of the type

$$\left\langle 0 \left| \frac{\hat{L}_\alpha}{r^3} \right| n \right\rangle \quad \langle n|\hat{L}_\alpha|0\rangle$$

where $\alpha = z, x$, and y for the σ_{zz}, σ_{xx}, and σ_{yy} components of the shielding tensor, for all possible excitations of an electron from the ground state m.o.'s to $\sigma_{2p_z}{}^*$. (To simplify the problem, higher excited states will be omitted because the $E_n - E_0$ term becomes very large for molecular orbitals derived from $3s$, $3p$, and higher atomic orbitals.)

It is convenient to define the origin of the molecular coordinate system at one atom and to calculate the paramagnetic contribution to the shift at this atom. In this coordinate system, Fig. 8–25(A), as opposed to the local coordinate system on each

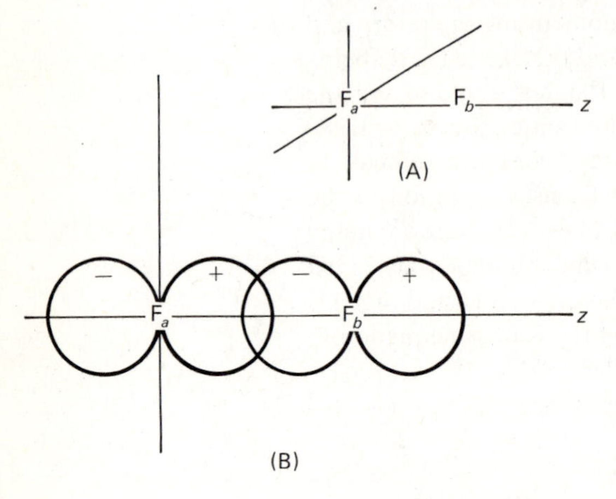

FIGURE 8–25. (A) The coordinate system selected for the chemical shift discussion of F_2; (B) p_z orbitals in a coordinate system centered on F_a.

atom commonly used in molecular orbital calculations, the p_z orbitals on two atoms are as shown in Fig. 8–25(B). The bonding molecular orbital now is described as:

$$\sigma_{2p_z} = \frac{1}{\sqrt{2}}(p_{za} - p_{zb})$$

and the anti-bonding molecular orbital as:

$$\sigma_{2p_z}{}^* = \frac{1}{\sqrt{2}}(p_{za} + p_{zb})$$

We should consider matrix elements for all possible states arising from one electron excitation, *i.e.*, $\sigma^2 \rightarrow \sigma\sigma^*$, $\pi^{*4} \rightarrow \pi^{*3}\sigma^*$, etc. The matrix elements of general form $\langle\sigma^2|\widehat{\text{op}}|\sigma\sigma^*\rangle$ are the same as $\langle\sigma|\widehat{\text{op}}|\sigma^*\rangle$, because we have a one-electron operator. To be systematic, we shall consider the three different orientations in which the molecular axes are aligned with the z-axis parallel to the field, with the x-axis parallel to the field, and with the y-axis parallel to the field to evaluate σ_{zz}, σ_{xx}, and σ_{yy}, respectively. This requires the use of the $\hat{L}_z$, $\hat{L}_x$, and $\hat{L}_y$ operators, respectively, in the Ramsey equation.

Consider nucleus a and the cases where:

(1) **The z-axis of the molecule is aligned with the field.** When the z-axis of the molecule is parallel to the applied field, the molecule senses no H_x or H_y field, so there are no field-induced $\hat{L}_x$ or $\hat{L}_y$ components. All one-electron excitations place an electron into the anti-bonding molecular orbital, leading to:

$$\langle\sigma|\hat{L}_z|\sigma^*\rangle,\quad \langle 2s|\hat{L}_z|\sigma^*\rangle,\quad \langle\pi_x|\hat{L}_z|\sigma^*\rangle \quad\text{or}\quad \langle\pi_y|\hat{L}_z|\sigma^*\rangle$$

and the corresponding π^* transitions where $\hat{L}_z = \hat{L}_z^{\,a}$ for nucleus a, $\langle\sigma|\hat{L}_z|\sigma^*\rangle = (\tfrac{1}{2})\langle(p_{za} - p_{zb})|\hat{L}_z|(p_{za} + p_{zb})\rangle$. However, we know that $\hat{L}_z|p_z\rangle = 0$. Thus, all the matrix elements for this orientation of the molecule are zero, and σ_{zz} is zero. This leads to a generalization that will become important in subsequent discussion of the chemical shift. Namely, *the contribution to the chemical shift from paramagnetic terms, σ_p, is zero when the highest-fold symmetry axis (z-axis) is parallel to the field.*

(2) **The x-axis of the molecule is aligned with the field;** *i.e.*, the z-axis is perpendicular to the field. This corresponds to evaluating σ_{xx}, and the $\hat{L}_x$ operator in the molecular coordinate system must be employed. Considering one of the possible transitions, we have

$$\langle\sigma|\hat{L}_x|\sigma^*\rangle = (\tfrac{1}{2})[\langle p_{za}|\hat{L}_x|p_{za}\rangle + \langle p_{za}|\hat{L}_x|p_{zb}\rangle - \langle p_{zb}|\hat{L}_x|p_{za}\rangle - \langle p_{zb}|\hat{L}_x|p_{zb}\rangle]$$

Since the last term involves the $\hat{L}_x$ operator centered on a operating on b, integrals of this sort will generally be small; they are dropped here and in subsequent discussion. The $\langle p_{zb}|\hat{L}_x|p_{za}\rangle$ and $\langle p_{za}|\hat{L}_x|p_{zb}\rangle$ terms are two-center integrals; they, too, generally are small and will not be considered in future discussion. In the case under consideration, all of these matrix elements are zero because $\hat{L}_x|p_z\rangle$ equals $i|p_x\rangle$, so:

$$\langle p_z|\hat{L}_x|p_z\rangle = \langle p_z|ip_x\rangle = 0$$

Only those matrix elements corresponding to $\langle p_x|\hat{L}_x|p_z\rangle$ can be non-zero. As a result, we have only to evaluate the matrix elements

$$\langle\pi|\hat{L}_x|\sigma^*\rangle \qquad\text{and}\qquad \langle\pi^*|\hat{L}_x|\sigma^*\rangle$$

There is only one such matrix element to evaluate for F_2, namely, $(\tfrac{1}{2})\langle p_{xa}|\hat{L}_x|p_{za}\rangle$ (recall that $\langle p_{xa}|\hat{L}_x|p_{zb}\rangle$ is a two-center integral and is small in comparison), which is equal to $i/2$. This value corresponds to the matrix element $\langle\pi|\hat{L}_x|\sigma^*\rangle$. The element $\langle\pi^*|\hat{L}_x|\sigma^*\rangle$ gives the same result. Thus, *there is a contribution to the paramagnetic term for the field along the x-axis. We see that the field has coupled the ground state with the excited state; i.e., we have field-induced mixing.*

(3) **The y-axis of the molecule is aligned with the field;** *i.e.*, the z-axis is perpendicular to the field again, but we must now concern ourselves with $\hat{L}_y$ in the molecular coordinate system. The result is

$$\hat{L}_y|p_z\rangle = i|p_y\rangle$$

By analogy with our discussion of $\hat{L}_x$, only the matrix element $\langle p_{ya}|\hat{L}_y|p_{za}\rangle$ is non-zero, and we also expect a paramagnetic contribution to the shielding at fluorine when the field is along the y-axis. As before, the other matrix elements are zero.

Evaluation of the entire second term of equation (8–31) and replacement of the $E_n - E_0$ term in the denominator by an average energy for the excited states in the molecule (*i.e.*, use of some electronic transition in the molecule that is felt to be an average of all possible transitions to excited states) yields the paramagnetic contribution, σ_p, by averaging σ_{xx}, σ_{yy}, and σ_{zz}:

$$\sigma_p = -\frac{2}{3}\left(\frac{e\hbar}{mc}\right)^2 \left\langle\frac{1}{r^3}\right\rangle \frac{1}{\Delta E} \tag{8-32}$$

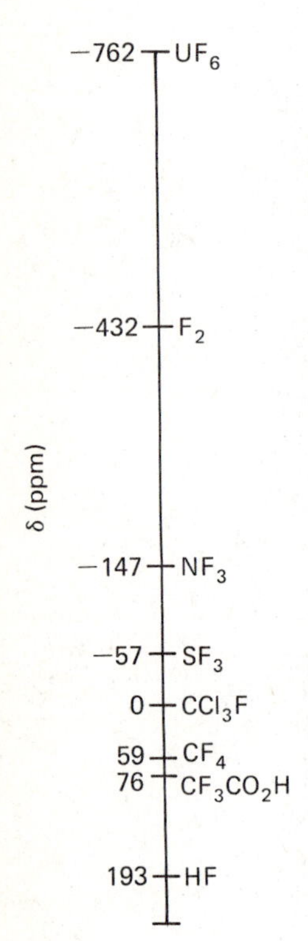

FIGURE 8–26. Some representative ^{19}F shifts relative to CCl_3F.

The average energy approximation is often referred to as the *closure approximation.* Substituting appropriate values for F_2 into equation (8–32) (*i.e.*, the energy of 4.3 eV for the $\pi \rightarrow \sigma^*$ transition and $\langle 1/r^3 \rangle$ for a fluorine atom $2p$ orbital) one obtains $\sigma_p \approx 2000$ ppm. Since F^- has spherical symmetry, there can be no angular momentum associated with the electron density in this species, and all the paramagnetic terms in the Ramsey equation must be zero. A chemical shift difference between F^- and F_2 of about 2000 ppm is expected. It is impossible to obtain a fluoride ion that is not solvated or ion-paired. The F_2 chemical shift is found to be 625 ppm downfield from HF and is estimated to be ~800 ppm downfield from the fluoride ion, in relatively poor agreement with the rough calculation carried out above. This is due in part to the poor nature of the average energy approximation and in part the use of an atomic $\langle 1/r^3 \rangle$ value. The calculated shift difference between F_2 and HF is ~1400 ppm, which is also too large.

In spite of the poor quantitative calculation of σ_p, there are several important generalizations that can be drawn from this discussion. Since $\hat{L}_z$ is zero for an s orbital in hydrogen compounds, one must invoke transitions to the $2p$ orbital in order to obtain a paramagnetic contribution. Now the $E_n - E_0$ value is so large that paramagnetic contributions are not very important in the proton nmr spectrum. These paramagnetic contributions become very large when there are both an asymmetric distribution of p and d electrons in the molecule and low-lying excited states. The paramagnetic term gives rise to the principal contribution to the shift in ^{19}F and ^{13}C nmr, and a large chemical shift range is observed as shown for ^{19}F compounds in Fig. 8–26. The chemical shifts in ^{11}B nmr have significant contributions from both local paramagnetic and local diamagnetic effects.

8–14 NEIGHBOR ANISOTROPIC CONTRIBUTIONS TO THE CHEMICAL SHIFT

In the previous section we discussed local diamagnetic and local paramagnetic screening effects. The existence of these field-induced moments, at atoms in the molecule other than the one undergoing the nmr transition, can be felt at the nucleus being investigated. The effects on other atoms are referred to as *remote effects* or as *neighbor anisotropic contributions.* The field at this remote atom may be dominated by either a paramagnetic or a diamagnetic effect but may have a different direction at the atom being studied. We shall work through the case of the HX molecule first, assuming that the diamagnetism of the remote atom X is dominant. As illustrated in Fig. 8–27, the field at the proton in HX from the diamagnetic effect at X will be strongly dependent upon the orientation of the HX molecule with respect to the direction of the applied field, H_0. When the applied field is parallel to the internuclear axis, the magnetic field generated from diamagnetic electron circulations on X (indi-

FIGURE 8–27. Neighbor anisotropy in HX for different orientations.

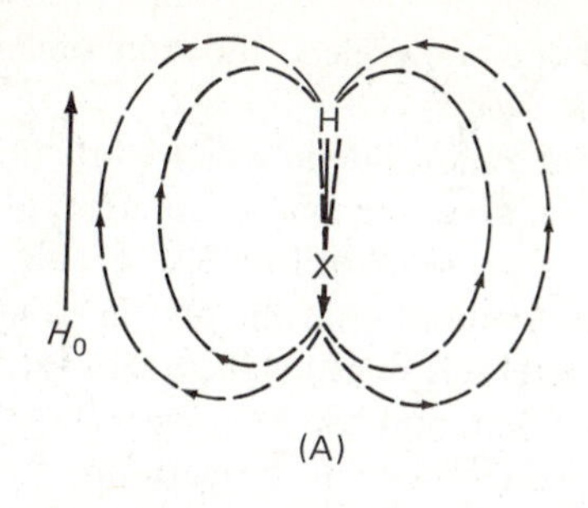

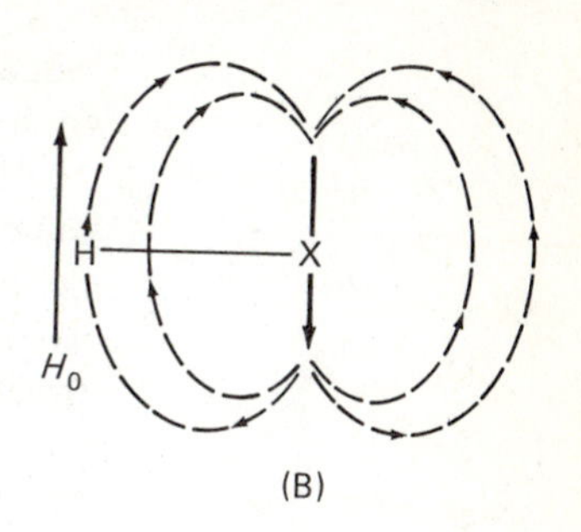

cated by the dotted lines) will shield the proton [Fig. 8–27(A)], while in a perpendicular orientation [Fig. 8–27(B)] this same effect at X will result in deshielding at the proton. The magnitude of the induced moment on atom X (and hence the field at the proton from this neighbor effect) for the parallel and perpendicular orientations will depend upon the susceptibility of X for parallel and perpendicular orientations, $\chi_{\parallel}$ and $\chi_{\perp}$ respectively. The susceptibility, χ, is related to the intensity of the magnetization, M, by

$$\chi = \frac{M}{H_0} \tag{8-33}$$

When HX is parallel to the field, the induced moment on X is given by $\chi_{\parallel(X)}H_0$. The contribution this remote effect makes to the shielding *at the proton,* $\sigma_{\parallel}$, is given by $-\sigma_{\parallel}H_0$; expressed in terms of the susceptibility of X, $\sigma_{\parallel}$ is:

$$\sigma_{\parallel} = -2R^{-3}\chi_{\parallel(X)} \tag{8-34}$$

where R is the distance from X to the proton, χ is negative (and consequently the proton is shielded), and $\chi_{\parallel(X)}$ is the parallel component of the susceptibility at X.

For the perpendicular ($\perp$) orientation of HX, the contribution from X at the proton is given by

$$\sigma_{\perp} = R^{-3}\chi_{\perp(X)} \tag{8-35}$$

Note that equations (8–34) and (8–35) give the correct signs for the shielding at the proton, as illustrated in Fig. 8–27 ($\chi_{\parallel}$ and $\chi_{\perp}$ are both negative). We have two perpendicular components to consider, χ_x and χ_y, which are equal here and in any molecule with a threefold or higher symmetry axis. In solution, the molecule is rapidly tumbling, so the concern is with the average value of σ given by equation (8–36):

$$\sigma = -\frac{1}{3}R^{-3}(2\chi_{\parallel} - \chi_{\perp} - \chi_{\perp}) \tag{8-36}$$

According to equation (8–36), if the susceptibility is isotropic (*i.e.*, $\chi_{\parallel} = \chi_{\perp}$), there will be no contribution to the shielding at the nucleus of interest from diamagnetic effects on the neighbor. In CH_4, for example, $\chi_{\parallel} = \chi_{\perp}$ and neither the carbon nor the C—H bonds can make a neighbor contribution to the proton shift. In HX molecules, the remote contribution from X arises because $\chi_{\parallel} \neq \chi_{\perp}$. The value to use for R in equations (8–35) and (8–36) is a problem, for one must decide whether the susceptibility arises on the remote atom or in the bond. This question is not a real one, but results from our arbitrary factoring of the molecular susceptibility into parts due to different atoms in the molecule. In spite of this simplification, the approach is valuable and one often selects the atom or center of the bond as the point from which to measure R.

If one had a molecule in which paramagnetic contributions were dominant, the same equations would be employed, except that the signs of χ would be positive; *i.e.*, the boldface arrow on X of Fig. 8–27 would point in the opposite direction. Now, as we saw in our evaluation of the matrix elements for F_2, paramagnetic contributions from X (or the H—X bond) will be zero for the orientation in which the bond axis is parallel to the field and will be a maximum for the perpendicular orientation. Thus, the paramagnetic term will be very anisotropic and often dominates the neighbor anisotropic contributions to the proton shifts. Refer to Fig. 8–27(B) and change the direction of the arrows for this paramagnetic effect at X. The paramagnetic effect at X is shielding (diamagnetic) at the proton. When discussing a remote effect, we label it according to the direction of the effect on the remote atom. If $\chi_\parallel$ and $\chi_\perp$ were known, we would know its sign and which effect is dominant. This information is seldom available and both paramagnetic and diamagnetic neighbor effects must be qualitatively considered when interpreting proton shifts.

Equations (8–34) and (8–35) apply only for linear molecules. For the general case

$$\sigma = \frac{1}{3}R^{-3}[(1 - 3\cos^2\theta_x)\chi_{xx} + (1 - 3\cos^2\theta_y)\chi_{yy} + (1 - 3\cos^2\theta_z)\chi_{zz}] \quad (8\text{–}37)$$

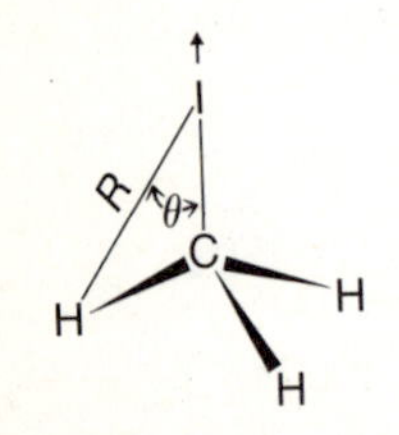

FIGURE 8–28. Illustration of the parameters θ and R of equation (8–37).

where the χ_{xx}, χ_{yy}, and χ_{zz} are the values along the three principal axes of the susceptibility tensor and θ_x, θ_y, and θ_z correspond to the angles between these axes and a line drawn from the center of the anisotropic contributor (the neighbor atom or bond) to the atom being investigated. The angle is illustrated for a molecule of C_{3v} symmetry in Fig. 8–28, where the source of the neighbor effect has been taken at the iodine atom, χ_{zz} is taken along the threefold axis, and $\chi_{xx} = \chi_{yy}$. When $\chi_{xx} \neq \chi_{yy} \neq \chi_{zz}$, we can select our coordinate system such that the radius vector R lies in the yz (or xz) plane. Equation (8–37) then becomes

$$\sigma = \frac{1}{3}R^{-3}(2\Delta\chi_1 - \Delta\chi_2 - \Delta\chi_1 3\cos^2\theta_z) \quad (8\text{–}38)$$

where $\Delta\chi_1 = \chi_{zz} - \chi_{yy}$ and $\Delta\chi_2 = \chi_{zz} - \chi_{xx}$. When axial symmetry pertains, $\chi_{zz} \neq \chi_{xx} = \chi_{yy}$ and equation (8–38) becomes (8–39):

$$\sigma = \frac{1}{3}R^{-3}(\chi_\parallel - \chi_\perp)(1 - 3\cos^2\theta_z) = \frac{1}{3}R^{-3}\Delta\chi(1 - 3\cos^2\theta_z) \quad (8\text{–}39)$$

Values for $\Delta\chi$ for various bonds can be found in the literature.

8–15 INTERATOMIC RING CURRENTS

Interatomic ring currents develop in cyclic, conjugated systems. Field-induced electron circulations occur in a loop around the ring and extend over a number of atoms. Analogous to the circulation of electrons in a wire, a magnetic moment is induced by this effect. The moment induced at the center of the ring is opposed to the field, but in benzene, for example, the magnetic lines of flux at the protons are parallel to the applied field, and these protons are deshielded as shown in Fig. 8–29.

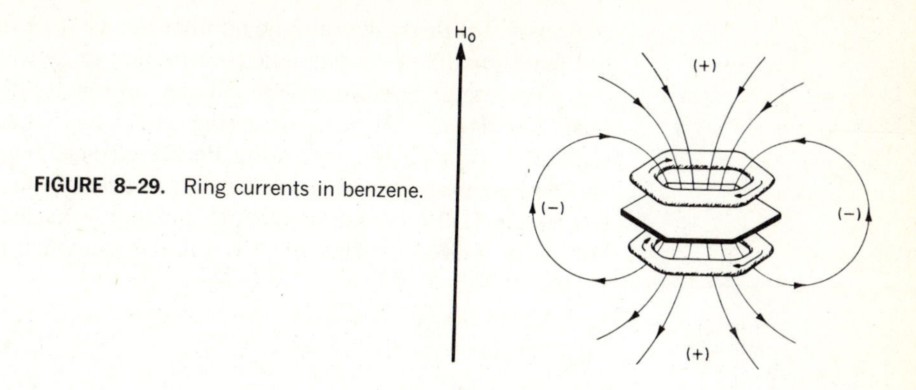

FIGURE 8–29. Ring currents in benzene.

This is the explanation for the large downfield value of the chemical shift observed for the protons in benzene.[43] Johnson and Bovey[43] have calculated the ring current contribution to the chemical shift for a proton located at any position relative to the benzene ring.

8–16 CHEMICAL SHIFT INTERPRETATION

One important generalization can be drawn from the preceding and subsequent discussion, *viz.*, chemical shift data are not reliable indications of the electron density around the nucleus being measured. Many effects contribute to δ. The next step, the quantitative quantum mechanical evaluation of shielding in various molecules, appears formidable at present. Recall, from the previous chapter, that a chemical shift *vs.* $J_{^{13}C-H}$ relation was presented to give a rough indication of the magnitude and sign of the neighbor anisotropic effect. For systems that cannot be evaluated with this empirical correlation, contributions from neighbor anisotropy are qualitatively invoked when needed to account for differences between measured δ values and those expected on the basis of chemical behavior. When suspected deviations are encountered, the molecule is examined to see what property it possesses that could account for the observed discrepancies.

We can clarify this discussion by considering a few examples. The δ value for HCl (in the vapor phase) indicates that this proton is more shielded than those in methane. This is in contrast to the greater formal positive charge on the proton of HCl than on that of CH_4. Thus, the local diamagnetic effect does not explain this behavior. The *local* paramagnetic effect is not expected to be significant for a proton. Since HCl is linear and cylindrically symmetric about the hydrogen-chlorine bond, a neighbor paramagnetic contribution is expected when the molecule is perpendicular to the applied field but not when it is parallel to the field. The net effect will be a shielding of the proton in HCl from a paramagnetic, neighbor anisotropic effect. The effect is paramagnetic (deshielding) at X but shielding at the proton. Considering HCl, we see that the shielding from the neighbor effect is greater than and in a direction opposite to the local (proton) diamagnetic effect, which is related to electronegativity of the attached atom and the removal of electrons from the hydrogen orbitals. The neighbor anisotropic effect dominates the difference in CH_4 and HCl shifts.

In the series HF, HCl, HBr, and HI, the magnitude of the neighbor paramagnetic contribution, $\Delta\sigma_p$, giving rise to anisotropy increases with increasing atomic number of the halogen. This is due to the decreased difference in the energy of the ground and excited states, ΔE, making a field-induced mixing of ground and excited states in the compounds of the higher atomic number atoms more favorable. The Ramsey equation states that the paramagnetic contribution to $\Delta\sigma$ will be proportional to $-1/\Delta E$.

The formal positive charge on the proton for the compounds in the series $C_2H_6 < C_2H_4 < C_2H_2$ increases in the order listed. The δ values increase in the order $C_2H_6 < C_2H_2 < C_2H_4$, indicating decreased shielding. Acetylene is more highly shielded than is expected on the basis of its acidity, and the shielding is attributed to two effects: (1) Remote diamagnetic shielding from electron circulations in the triple bond [Fig. 8-30(A)]. (This effect is a maximum when the molecular axis is aligned

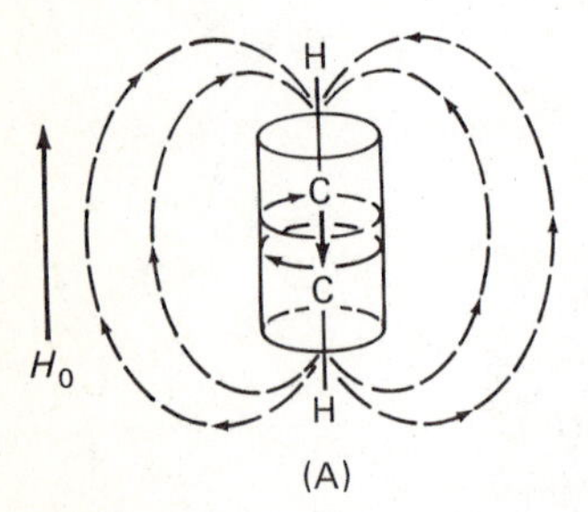

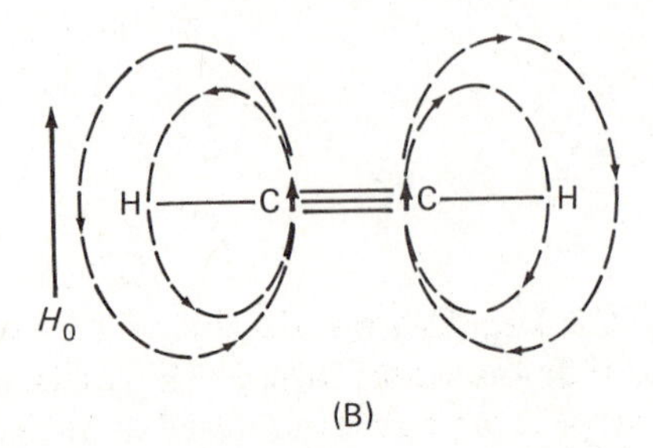

FIGURE 8–30. Contributions to the shielding in acetylene. (A) The effect here is a diamagnetic effect at carbon, which is a maximum for this orientation. (B) The effect here is a paramagnetic effect at carbon, which is a maximum for this orientation.

with the field, and it is classified here as a diamagnetic effect because the moment arising from the electron circulations opposes the applied field. Recall that the paramagnetic contribution is zero for this orientation and is a maximum when the axis is perpendicular to field.) (2) Remote paramagnetic shielding from higher state mixing [Fig. 8-30(B)]. Both the remote diamagnetic and paramagnetic effects influence $\Delta\chi$ so as to provide shielding at the proton. The measured proton chemical shift of acetylene relative to ethane is thus explained by the remote diamagnetic and paramagnetic contributions arising from the triple bond.

Even in those systems where the local diamagnetic term dominates the observed proton shift, there are alternative interpretations of the trends. In addition to the trends in the population of the hydrogen 1*s* orbital as a consequence of the electronegativity of the attached group, Buckingham has proposed[44a] an *electric field model* to account for changes in the electron density of a bound hydrogen atom. In Fig. 8-31, an H—F bond has been located in an electric field whose lines of flux have

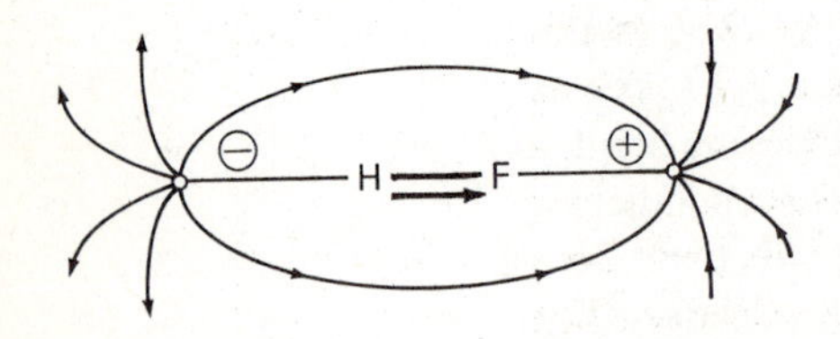

FIGURE 8–31. The HF molecule in an electric field. (The ⊕ and ⊖ indicate the electric field direction, and the heavy arrow indicates the polarization of the bonding electron density by the field.)

been indicated. The boldface arrow indicates the direction in which the bonding electron density will be distorted in an attempt to avoid the negative region of the field. If this field is a fixed one, tumbling of the molecule will average this effect to zero. However, if the electric field arises from an electric dipole in the molecule (*i.e.*, a polar functional group), rotation will not average out this field and a distortion of the bonding electron density can result. This effect is illustrated in Fig. 8-32 for the ethanol molecule, where the electric field from the polar C—O bond is seen to deshield the methylene proton. The heavy arrow indicates the drift in electron density in the C—H bond induced by the polarity of the electric field from the polar C—O

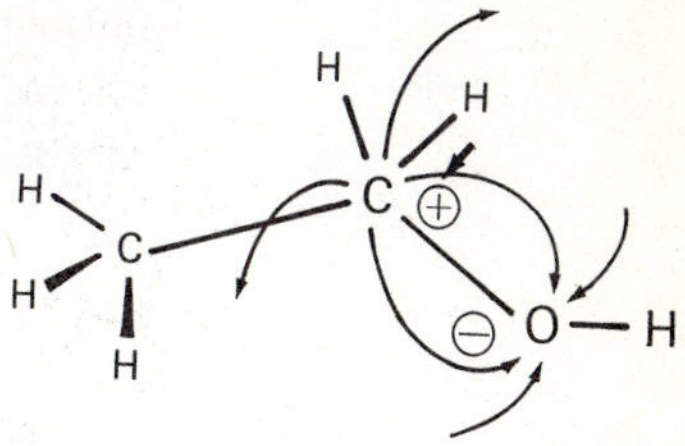

FIGURE 8–32. The electric field effect in ethanol.

bond. The contribution[44] to the proton shielding from this effect, σ_E, is given by equation (8–40):

$$\sigma_E = 2 \times 10^{-12} E_z - 10^{-18} E^2 \quad (8\text{-}40)$$

where E_z is the z-component of the total electric field, E, directed along the C—H bond. This model has been successfully employed to interpret the chemical shift changes of a series of hydrogen bonding acids interacting with a series of Lewis bases.[45]

An additional complication associated with the measurement of the chemical shift of a compound is the *reaction field effect*[44a] associated with the solvent in which the compound is dissolved. This is a shielding effect felt at the proton of interest by a field from the solvent, which has been induced by a dipole in the solute. Consider a polar solute molecule in a solvent cavity. This solute induces a dipole moment in the solvent, which creates a reaction field at the center of the cavity. As the solute molecule tumbles, the reaction field changes with it, so the effect is not averaged to zero. This effect is difficult to evaluate quantitatively. A few assumptions can be made that enable one to estimate its order of magnitude and elucidate factors to emphasize in solvent selection to minimize it. The simplifying assumptions, which are very crude, are that the solvent cavity is a sphere surrounded by a homogeneous, polarizable, continuous medium of dielectric constant ε. The reaction field $\vec{\mathbf{R}}$ is then parallel to the dipole moment vector of the solute and is given[44a] by equation (8–41):

$$\vec{\mathbf{R}} = \frac{\varepsilon - 1}{2\varepsilon + 2.5} \frac{\vec{\mu}}{\alpha} \quad (8\text{-}41)$$

where α is the polarizability of the sphere and $\vec{\mu}$ is the dipole moment of the solvent. The shielding contribution at a given proton is given by:

$$\sigma_{RF} = -3.0 \times 10^{-12} |\vec{\mathbf{R}}| \cos \varphi \quad (8\text{-}42)$$

where φ is the angle between the bond vector (*e.g.*, C—H) and the molecular dipole moment. The effect is minimized by working in low dielectric constant solvents.

In view of the many contributions described above, even the qualitative interpretation of proton chemical shifts is filled with complication. When ^{13}C and ^{19}F chemical shifts are considered, the dominance of the local paramagnetic term gives rise to such large shift differences in series of molecules that reaction field and neighbor anisotropy effects become small in comparison.[46] The qualitative interpretation of the shifts in these compounds has met with more success.[47] The problem is still not simple or quantitative by any means. Changes in the average excitation energy, one-center terms, and two-center terms in a molecular orbital description are all important,[48-50] leading to the situation where there are many more parameters influencing the measurement than there are experimental observables. The reader is referred to references 46, 48 and 49 in the event one is ever tempted to propose a rationalization of such data.

8–17 CHEMICAL SHIFTS OF OPTICAL ISOMERS

The chemical shifts of analogous atoms in two different enantiomers are identical. However, as with many physical properties, differences are detected in diastereoisomers. Advantage has been taken of this fact[50] to utilize nmr for the determination of optical purity. For example, when a *d,l* mixture of the phosphine oxide, $C_6H_5CH_2-P(=O)(C_6H_5)(CH_3)$, is added to the resolved solvent *d*-$C_6H_5-C(H)(CF_3)(OH)$ (2,2,2-trifluoro-1-phenyl ethanol), the hydrogen bonded adducts are diastereoisomers in which the chemical shift of the methyl group differs by 1.4 Hz. These peaks can be integrated and the relative amounts of the *d,l* isomers in the phosphine oxide determined. When resolution is desired, the optical purity of the product can be ascertained by the absence of one of the peaks. The experiment can also be carried out in an inert solvent employing a large enough excess of optically pure reagent to insure nearly complete complexation of the substance being tested. Since this shift difference is observed under conditions where both the *d* and *l* forms are nearly completely complexed, the difference in shift must arise from differences in the neighbor anisotropic contributions in the diastereoisomers.

SCALAR SPIN-SPIN COUPLING MECHANISMS

8–18 NATURE OF THE COUPLING[51]

We have already mentioned in Chapter 7 and in the discussion of the Overhauser effect that there are various contributions to the magnitude of the spin-spin coupling constant, J, in various molecules. Contributions to the scalar coupling, J, are transmitted via the electron density in the molecule and consequently are not averaged to zero as the molecule tumbles. Three contributions will be considered:

(1) Spin-orbital effects
(2) Dipolar coupling, or indirect or through-space coupling
(3) Fermi contact coupling

It is worth emphasizing that all three effects are transmitted via the electron density in the molecule.

Spin-orbital effects involve the perturbation that the nuclear spin moment makes on the orbital magnetic moments of the electrons around the nucleus. For an $I = ½$ nucleus which we label B, for example, the magnetic field of the nuclear dipole interacts differently for $m_I = +½$ than for $m_I = -½$, with the orbital magnetic moment of an appropriate electron around B causing a change in the magnetic field from the orbital contribution of these electrons. The new field from the electrons at B produces a field at the nucleus being split, A, that depends upon whether the nuclear moment at B is $+½$ or $-½$. Consequently, a splitting of the nmr resonance of A results. The Hamiltonian for the interaction on B that is felt at A is

$$g_N\beta_N\beta_e \frac{2\hat{L}\cdot\hat{I}}{r^3} \tag{8-43}$$

where $\hat{L}$ is the electron orbital angular momentum operator, $\hat{I}$ is the nuclear spin moment operator, and r is the distance from A to B.

Dipolar coupling, often referred to as indirect coupling, is analogous to the classical dipolar interaction of two bar magnets. Since the classical situation is simpler, we shall treat it first. It is essential that you obtain a complete understanding of this interaction, for we shall encounter in the remainder of this text many phenomena in which this type of interaction is important. The classical interaction energy, E, between two magnetic moments (which we shall label $\vec{\mu}_e$ and $\vec{\mu}_N$) is given by:

$$E = \frac{\vec{\mu}_N \cdot \vec{\mu}_e}{r^3} - \frac{3(\vec{\mu}_N \cdot \vec{\mathbf{r}})(\vec{\mu}_e \cdot \vec{\mathbf{r}})}{r^5} \tag{8-44}$$

where $\vec{\mathbf{r}}$ is a radius vector from $\vec{\mu}_e$ to $\vec{\mu}_N$ and r is the distance between the two moments.

The indirect dipolar coupling mechanism corresponds to a polarization of the paired electron density in a molecule by the nuclear moment. The polarization of this electron density depends on whether $m_I = +\frac{1}{2}$ or $-\frac{1}{2}$, and the modified electron moment is felt through space by the second nucleus. Replacing $\vec{\mu}_e$ with $-g\beta\hat{S}$ for the electron magnetic moment and $\vec{\mu}_N$ with $g_N\beta_N\hat{I}$ for the nuclear magnetic moment gives the dipolar interaction Hamiltonian:

$$\hat{H} = -g\beta g_N\beta_N \left\{ \frac{\hat{I} \cdot \hat{S}}{r^3} - \frac{3(\hat{I} \cdot \vec{\mathbf{r}})(\hat{S} \cdot \vec{\mathbf{r}})}{r^5} \right\} \tag{8-45}$$

The interaction between the electron spin moment and the nuclear moment polarizes the spin in the parts of the molecule near the splitting nucleus B. When this effect is averaged over the entire wave function, the field at B is modified and this modified field of the electron density acts directly through space on the nucleus A, which is being split. The direction of the effect depends upon the m_I value of the B nucleus, and thus this effect makes a contribution to J at the A nucleus.

The important point is that this indirect dipolar coupling of the nucleus (B) and the paired electron density, which modifies the electron moment felt at A, is not averaged to zero by rotation of the molecule. This is illustrated in Fig. 8–33(I), where

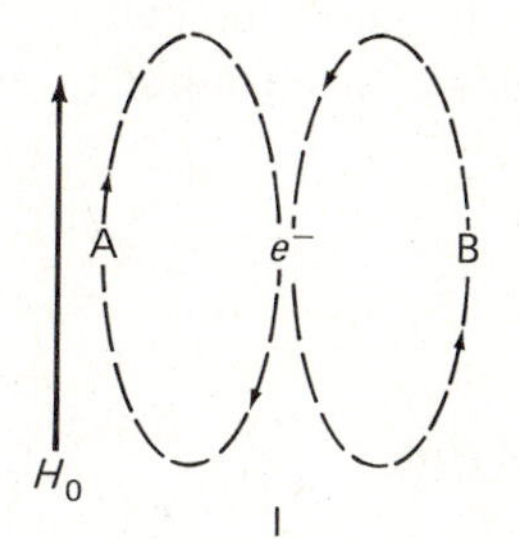

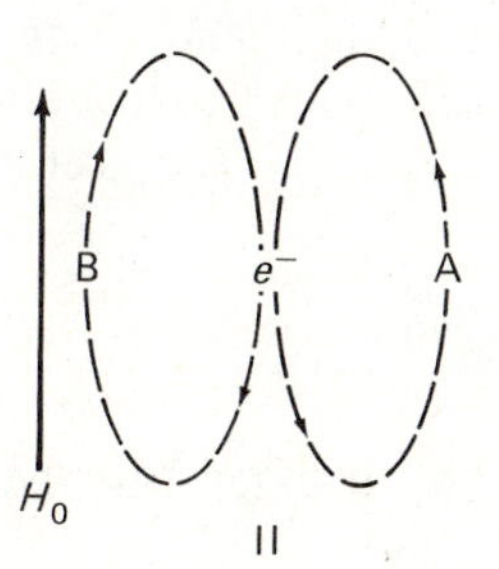

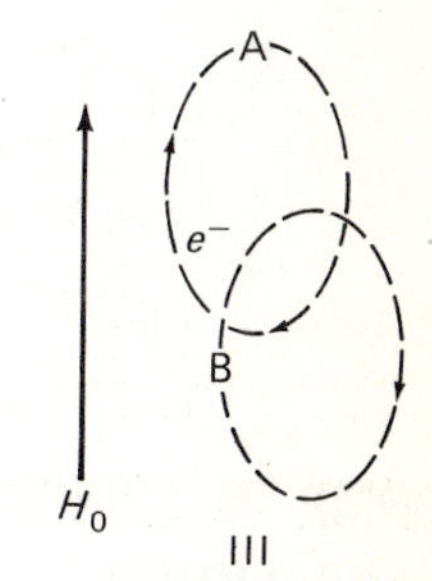

FIGURE 8–33. The indirect dipolar coupling of the nuclear moment B to A for different orientations of the molecule.

an $m_I = +\frac{1}{2}$ value is illustrated at B. Three different orientations of the molecule are shown. The lines of flux from the moment on B are shown affecting the moment at the electron. The lines of flux associated with the change in the moment at the electron from B are shown at nucleus A. For the three orientations shown, the direction at A is the same.

The *Fermi contact term* is the final coupling mechanism we shall consider for molecules rapidly rotating in solution. This mechanism involves a direct interaction of the nuclear spin moment with the electron spin moment such that there is increased

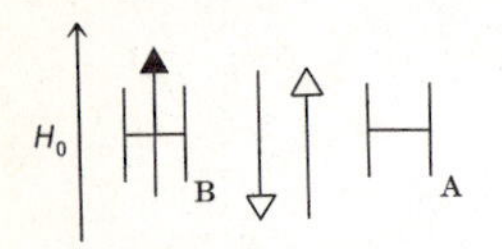

FIGURE 8-34. The antiparallel alignment of the nuclear moment (solid arrow) and the electron spin moment (hollow arrow).

probability that the electron near B will have spin that is antiparallel to the nuclear spin. Since the electron pair in the bond have their spins paired, a slight increase in the probability of finding an electron of one spin near B will result in there being an increased probability of finding an electron of opposite spin near A, as shown in Fig. 8-34. Thus, A receives information about the spin of nucleus B; since the effect at A is in opposite directions when m_I equals $+\frac{1}{2}$ and $-\frac{1}{2}$, it contributes to the magnitude of J. The effect is a direct interaction of the nuclear spin and the electron spin moment, $a\hat{I} \cdot \hat{S}$, where a is the coupling constant.

If we consider the spin of nucleus B to be quantized along the z-axis for simplicity of presentation, we can describe this effect in more detail. The contact Hamiltonian becomes:

$$\hat{H}_{A} = \frac{8\pi}{3} g\beta g_N \beta_N \hat{I}_{zB} \hat{S}_{zB} \tag{8-46}$$

where $\hat{S}_{zB}$ has contributions from both electrons 1 and 2 at B given by

$$\hat{S}_{zB} = \hat{S}_{z1}\delta(r_1 - r_B) + \hat{S}_{z2}\delta(r_2 - r_B) \tag{8-47}$$

Here r_B is the radius of the nucleus and r_1 is the distance from nucleus B to the electron. The symbol $\delta(r_1 - r_B)$ is the Dirac delta function and has a value of zero unless electron 1 is at nucleus B. Thus, if p-orbitals were used to bond two atoms together, there would be nodes at the nuclei and this term would be zero. Accordingly, in molecules where the Fermi contact term dominates, correlations of J with the amount of s character in the bond have been reported.

The mechanism for uncoupling the spins in the σ-bond involves field-induced mixing of ground and excited states. Perturbation theory produces:

$$J = -\left(\frac{8\pi}{3} g\beta g_N \beta_N\right)^2 \sum_n \frac{\langle 0|\hat{S}_{zA}|n\rangle\langle n|\hat{S}_{zB}|0\rangle + \langle 0|\hat{S}_{zB}|n\rangle\langle n|\hat{S}_{zA}|0\rangle}{E_n - E_0} \tag{8-48}$$

In the H_2 molecule, J_{H-H} (estimated from J_{H-D}) has an experimental value of 280 Hz. About 200 Hz comes from the Fermi contact term, 20 from the dipolar contribution, and 3 from the nuclear spin-orbital effect. In general, it is felt that the Fermi contact contribution dominates most coupling constants involving hydrogen, *e.g.*, ${}^1J_{^{13}C-H}$, ${}^2J_{Sn-H}$, etc.

NMR IN SOLIDS AND LIQUID CRYSTALS

8-19 DIRECT DIPOLAR COUPLING

We shall begin this discussion by considering the direct dipole-dipole interaction between two nuclei. This is done by considering *fixed orientations* of two hydrogen atoms in a molecule relative to an external field, as shown in Fig. 8-35. The dashed line is the internuclear axis connecting the two hydrogen atoms, the boldface arrow indicates the orientation of the nuclear moment on b relative to the field, and the curved arrow represents the lines of flux arising from this nuclear moment. We see that for this fixed orientation of the molecule, we would obtain two different peaks in the nmr of the H_a resonance as a consequence of the two different fields from b. This is a through-space effect which, in contrast to the mechanisms discussed in the preceding section, does not involve the electron density in the molecule. The peak

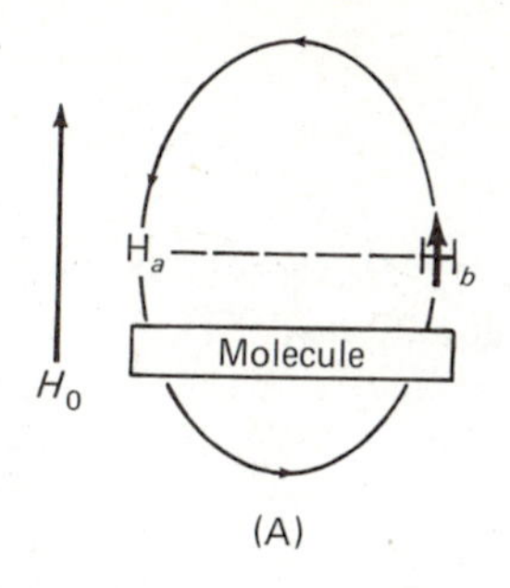

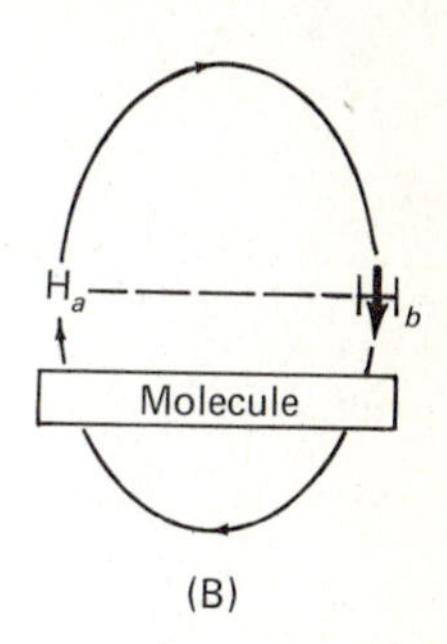

FIGURE 8–35. The direct dipolar interaction of two hydrogens in a molecule.

separation is indicated by a coupling constant B^{dir} (where dir stands for direct). B can be very large and, for example, is about 120,000 Hz for two protons separated by 1 Å when the H—H internuclear axis is aligned with the field.

The magnitude of B will vary with the position of the internuclear axis in relation to the field. The mathematical expression for the magnitude of the direct dipolar interaction between two nuclei p and q, $B_{pq}{}^{dir}$, is derived from the expression for the magnetic field due to a point magnetic dipole (*i.e.*, the other nucleus) and the expression for the potential energy of a point dipole (the proton of interest) in a magnetic field. The result is:

$$B_{pq}{}^{dir}(\text{Hz}) = -\frac{h}{4\pi^2}\gamma_p\gamma_q \frac{1}{2}\left[\frac{3\cos^2\theta_{pq} - 1}{r_{pq}{}^3}\right] \tag{8-49}$$

where θ_{pq} is the angle that the pq internuclear axis makes with the external field direction. For a complicated molecule, there is one B^{dir} for every pair of magnetic nuclei in the sample.

In a normal solvent, any given internuclear axis is randomly oriented with respect to the external field, and the direct dipole-dipole interaction is averaged to zero. Since the indirect coupling constants proceed through the electron density of the molecule, they are not averaged to zero in a solvent. However, the direct dipole-dipole interaction is through space; it is given by equation (8–49) and is averaged out for a random orientation of the internuclear axis.

8–20 NMR STUDIES OF SOLIDS

In single crystals, the H—H internuclear axes in the molecules have a fixed orientation relative to the applied field and B^{dir} is not averaged out. According to equation (8–49), if one were to study the angular variation of B^{dir} by investigating different orientations of a single crystal, it would be possible to solve the resulting simultaneous equations for $1/r^3$ and to find the magnitude and direction of the H—H internuclear axes. Unfortunately, only rarely are the protons in the crystal few enough in number and far enough apart to permit resolution of the spectral lines and determination of B^{dir}. Generally, the spectra of solids consist of very broad, poorly resolved bands because of the direct dipolar interaction between protons in the molecule and between those from the nearby molecules. However, structural information can be obtained from the broadened resonance line of single crystals or powders by the so-called method of second moments.[52] The second moment is the mean square width $\overline{(\Delta H)^2}$ measured from the center of the resonance line. The center of the resonance line is the average magnetic field, as seen in equation (8–50).

$$H_{av} = \int_0^\infty Hf(H)\,dH \tag{8-50}$$

where $f(H)$ represents the normalized line shape. The second moment is then given by equation (8–51),

$$(\overline{\Delta H})^2 = \int_0^\infty (H - H_{av})^2 f(H)\,dH \tag{8-51}$$

which can be evaluated graphically from the observed line shape. Most observed peaks have a Gaussian shape and are described by

$$f(H) = \frac{1}{\Delta H\sqrt{2\pi}} e^{-\{(H-H_{av})^2/2(\Delta H)^2\}} \tag{8-52}$$

The second moment is then determined from the integration of the analytical expression obtained from equations (8–51) and (8–52). Normally, the $(\Delta H)^2$ is determined directly from the experimental curve. The center of the band is selected as H_{av}. The spectrum is divided into equal increments $H_1, H_2, \ldots H_n$. The area of the H_{av}-H_1 increment is obtained using Simpson's rule or some other method. This is multiplied by the square of the difference between H_1 and H_{av}. The procedure is repeated for the H_1-H_2 increment, and so on, until the entire band has been covered. The second moment is the sum of these quantities divided by the band area (to normalize it), *i.e.*,

$$(\Delta H)^2 = \frac{\int_0^\infty (H - H_{av})^2 f(H)\,dH}{\int_0^\infty f(H)\,dH}$$

The second moment for a single crystal is related to r and θ, defined in the previous section, by

$$(\Delta H)^2 = \frac{3}{4} g_N{}^2 \beta_N{}^2 I(I+1)\left(\frac{1}{n}\right)\sum_{jk}\frac{(1 - 3\cos^2\theta_{jk})^2}{r_{jk}{}^6} + \frac{1}{3} g_N{}'^2 \beta_N{}'^2 I'(I'+1)\left(\frac{1}{n}\right)\sum_{jf}\frac{(1 - 3\cos^2\theta_{jf})^2}{r_{jf}{}^6} \tag{8-53}$$

where j and k refer to n identical nuclei in the unit cell that are at resonance. The sum runs over each nucleus j in the unit cell and all its neighbors in the crystal, including those in the unit cell. The index f refers to other magnetic nuclei in the crystal that are not at resonance.

In a powdered sample the equation becomes:

$$(\Delta H)^2 = \frac{3}{5} g_N{}^2 \beta_N{}^2 \left(\frac{1}{n}\right) I(I+1)\sum_{jk}\frac{1}{r_{jk}{}^6} + \frac{4}{15} g_N{}'^2 \beta_N{}'^2 \left(\frac{1}{n}\right) I'(I'+1)\sum_{jf}\frac{1}{r_{jf}{}^6} \tag{8-54}$$

The derivation of these equations is involved, and the reader is referred to references 52 through 54 for more details.

In solving problems by employing this method, the various possible structures are listed. For any given structural possibility, the appropriate distances are substituted into equation (8–54) and second moments are calculated. This is done for all possible structures, and the results are compared with experiment. Those that do not agree can be eliminated. This technique has been used to show that the infusible white precipitate from the reaction of NH_3 with $HgCl_2$ is NH_2HgCl and not NHg_2ClNH_4Cl or $XHgO(1 - X)\ HgCl_2 \cdot 2NH_3$.[55] These studies can be carried out on more complex molecules in conjunction with deuteration studies. The magnetic moment of the deuteron is very small, and dipolar interactions that involve deuterium can usually be neglected.

The temperature dependence of the second moment has also been employed to provide information on molecular motion in solids. For continuous motion, the $1 - 3\cos^2\theta_{jk}$ term must be averaged over the entire motion. It has been shown that benzene is fixed in the solid below 90°K, but rotates rapidly about the sixfold axis between 120° and 280°K. The second moment changes gradually from 9.7 gauss^2 below 90°K to 1.6 gauss^2 at 120°K, and remains at this value to 280°K. The magnitude of the change can be reproduced by calculation of the second moment for these two structures. Rotation of cyclohexane about the S_6 axis has also been demonstrated. This is an extremely sensitive technique, for the rate of rotation required to narrow the resonance is not much higher than the proton resonance frequency in a field of a few gauss, *i.e.*, 10^4 Hz. Second moment studies demonstrated[56] the rotation of the benzene and cyclopentadiene rings in dibenzene-chromium and ferrocene as well as[57] rotation of benzene in the benzene-silver perchlorate complex. A study[58] of $[Me_3SiNSiMe_2]_2$ demonstrated that at 77° the methyl groups rotate about the Si—C bond, and this is the only motion in the molecule. At room temperature, methyl groups rotate about the C—Si bond, $(CH_3)_3Si$ groups rotate about the Si—N bond, and the whole molecule rotates about a molecular axis. Thermodynamic data can be obtained for the various motional processes from the temperature dependence of the spectrum.

In a clever experiment, Waugh *et al.*[59,60] have employed double resonance and decoupling techniques to sharpen the resonance lines of solids. By studying the single crystal nmr spectrum as a function of angle, the six independent elements of the shielding tensor can be evaluated. In solution, an isotropic spectrum results, and only the average value of the trace of the shielding tensor can be obtained. The additional information available from these studies on the solid has considerable potential for providing structural information and testing quantum mechanical wave functions.

8–21 NMR STUDIES IN LIQUID CRYSTAL SOLVENTS

Certain materials, *e.g.*,

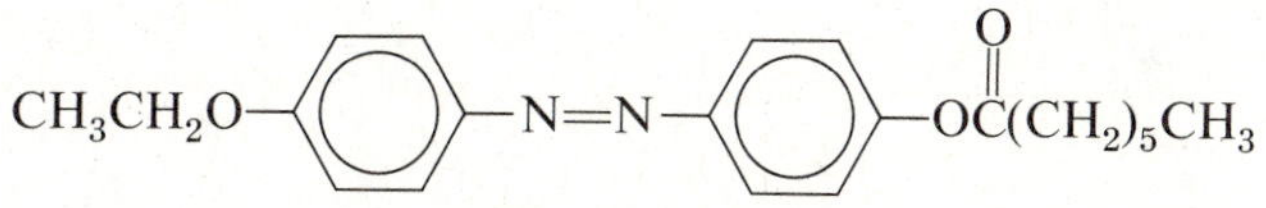

$CH_3CH_2O-C_6H_4-N{=}N-C_6H_4-OC(=O)(CH_2)_5CH_3$

and

$CH_3(CH_2)_6-C_6H_4-N(\rightarrow O){=}N-C_6H_4-O(CH_2)_6CH_3$

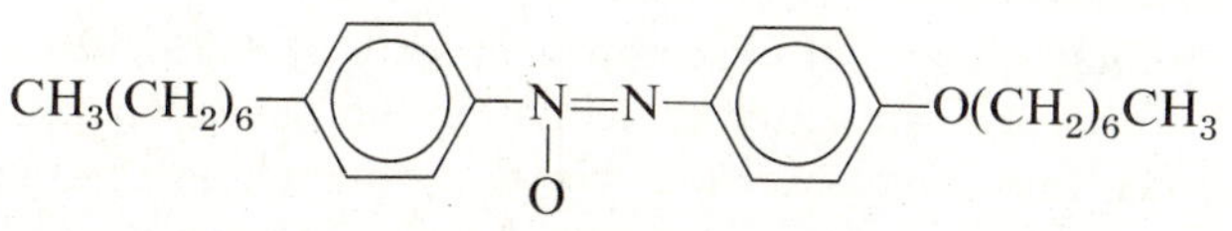

melt to produce a turbid fluid in which certain domains exist where there is considerable ordering of the molecules. The resulting fluid has some properties of both liquids and crystals. The liquid is strongly anisotropic in many of its properties,[61] but further heating produces an isotropic liquid. This general class of materials is referred to as *thermotropic liquid crystals* (*i.e.*, produced by melting solid). Three subclassifications are illustrated in Fig. 8–36 by schematically indicating the type of ordering in the domains. The representations in Fig. 8–36 are of small domains in the whole liquid sample. These domains tend to become aligned in the presence of a magnetic field.

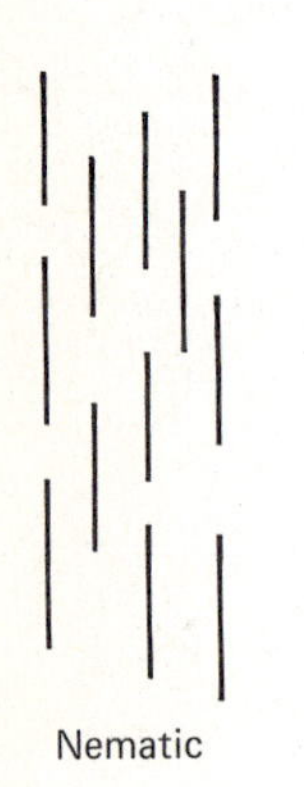

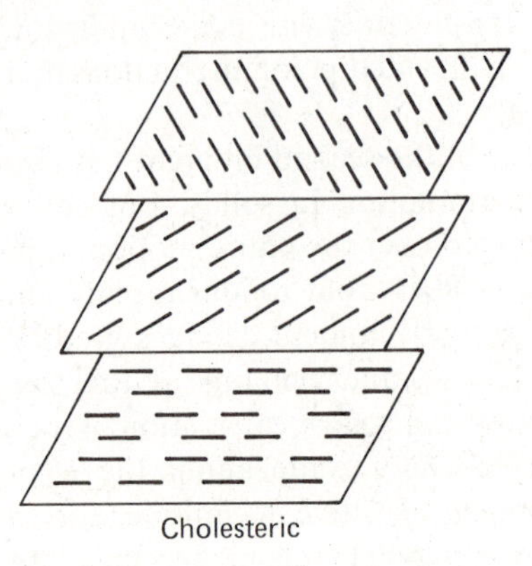

FIGURE 8–36. Types of ordered domains in a liquid crystal.

The nmr spectra of the molecules that constitute the liquid crystal are very broad, nondescript resonances that sometimes are so broad as not to be observed. The broadening results because the viscous solvent molecules contain many hydrogens and consequently there are a large number of direct H-H dipole-dipole interactions that are not averaged out in the partially ordered solvent.

If some benzene is dissolved in the liquid crystal as solvent, the spectrum illustrated in Fig. 8–37(A) results.[62] This spectrum of benzene consists of a large

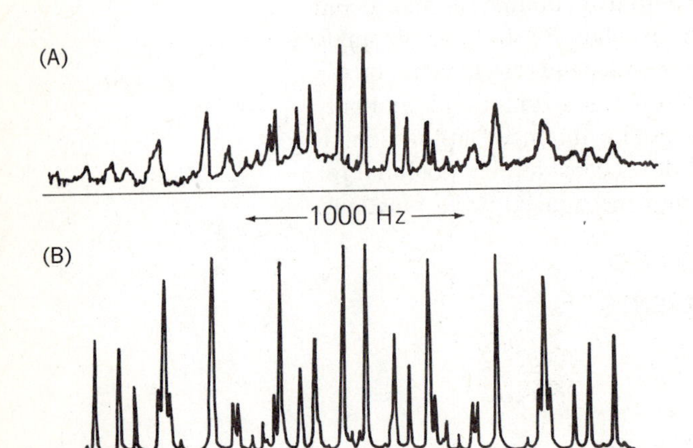

FIGURE 8–37. (A) Liquid crystal nmr spectrum of benzene; (B) simulated spectrum.

number of sharp lines spread out over approximately 2400 Hz. Some of the broad resonances of the solvent are discernible under the sharp benzene peaks. The benzene spectrum is thus somewhere between that obtained in a nonviscous solvent (a single sharp line) and that of a solid (a broad line spread over a wide field). In an isotropic solvent, the various axes in a solute molecule assume all possible orientations with respect to the magnetic field with equal probability, and B is averaged to zero. As mentioned before, the indirect couplings proceed via the electron density in the molecule and are not averaged to zero. When all orientations of an axis with respect to the field are not equally probable, the contribution to the field at a proton from each surrounding nucleus depends on the orientation of the nuclear moment in the neighbor, and the position of that neighbor nucleus relative to the one being observed. In a solid, where the nuclei are fixed and where so many protons affect the field of any one kind of proton, a continuous very broad single peak results. We now must explain why we see such sharp peaks for the solute in the liquid crystal spectrum and why there are so many peaks.

In a liquid crystal, certain orientations of the solute are more favored than others (long axes line up with the solvent long axis) because the magnetic field tends to align the solvent and solute molecules in the field direction. Solute molecules diffuse freely and tumble freely enough so that there is no contribution to B^{dir} from the solvent or other solute molecules. All the couplings are intramolecular. Since the molecule is tumbling rapidly, sharp lines are observed in the nmr. We see many lines in the spectrum because $B_{pq}{}^{dir}$ [equation (8–49)] makes an observed contribution to the resonance line positions, since the anisotropic motion does not average this quantity to zero. Each θ_{pq} of equation (8–49) now becomes some average value for the net orientation of each of the respective H-H axes in this rapidly and anisotropically tumbling molecule.

In benzene, for example, we have six protons and need a matrix of all possible combinations of six spins (*i.e.*, a basis set $\alpha\alpha\alpha\alpha\alpha\alpha$, $\alpha\alpha\alpha\alpha\alpha\beta$, etc.) to describe this system. Thus, we have peaks corresponding to molecules with all these different permutations of spins with B and J values for all pairs of hydrogens. Often, second order spectra result, further complicating the appearance of the spectrum. The energies are described by the spin Hamiltonian matrix for this system, which is similar to that discussed in treating second order spectra except that, for every J on the diagonal in the solution problem, we now have a $J + 2B$; and for every J previously on the off-diagonal we now have a $J - B$. This problem is solved by finding J and B values [$J_{\text{ortho}(o)}$, $J_{\text{meta}(m)}$, $J_{\text{para}(p)}$, B_o, B_m, and B_p] that will reproduce the experimental spectrum. The calculated spectrum is shown in Fig. 8-37(B). For benzene, the resulting values are $B_o = -639.5$ Hz, $B_m = -123.1$ Hz, $B_p = -79.93$ Hz, $J_o = 6.0$ Hz, $J_m = 2.0$ Hz, and $J_p = 1.0$ Hz.

Our next concern is to show how this information can be used to obtain structural information. If the structure of the molecule is rigid, the values of r_{pq} will be fixed and we define a quantity S_{pq} for any pq axis as the average value of $(\frac{1}{2})(3\cos^2\theta_{pq} - 1)$ for that axis, leading to:

$$B_{pq}{}^{dir} = \frac{-h}{4\pi^2}\gamma_p\gamma_q\frac{S_{pq}}{r_{pq}{}^3} \tag{8-55}$$

If the axis is completely aligned with the field, S_{pq} is 1. If the axis is aligned perpendicular to the field, then $S_{pq} = -\frac{1}{2}$. For a random orientation, $S_{pq} = 0$.

If we knew S_{pq}, we could calculate r_{pq} from the experimental $B_{pq}{}^{dir}$ and get distances and geometries of molecules in solution. In a rigid molecule, the values of S_{pq} for the various axes must be interrelated. This interrelationship between the axes and the orientation of the molecule in three dimensions can be described with a tensor $\vec{\mathbf{S}}$.

The $\vec{\mathbf{S}}$ tensor is 3×3, symmetric, and traceless, so only five of the nine elements are independent. As discussed in Chapter 2, we can define a molecular coordinate system that diagonalizes the tensor. This usually corresponds to symmetry axes if the molecule has symmetry; then there will be only two independent elements. (All off-diagonal elements are zero and the diagonal ones are traceless, *i.e.*, their sum equals zero.) For a molecule with a three-fold or higher axis, $S_{xx} = S_{yy} = -(\frac{1}{2})S_{zz}$ and there is only one independent tensor element. Note that the tensor is independent of r. We shall consider the p-dichlorobenzene molecule shown in Fig.

8–38 as an example. There are three dipole-dipole coupling constants that can be measured. We have as unknowns two S tensor elements (there is no three-fold axis) and two distances corresponding to the two sides of the triangle made by B_p, B_m, and H-H in Fig. 8–38. We are thus confronted with three equations and four unknowns.

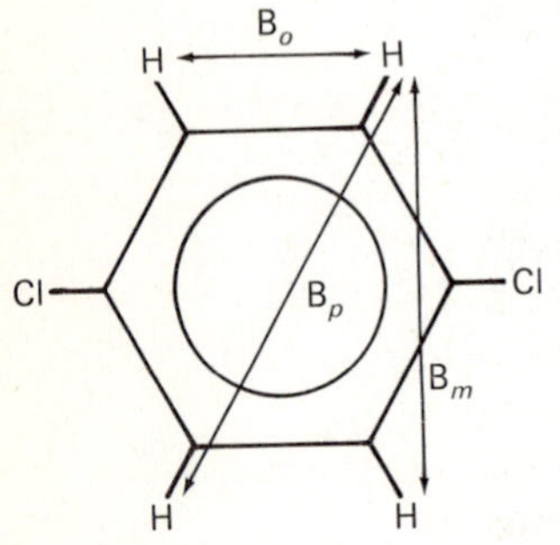

FIGURE 8–38. The B_{pq} interactions in p-$Cl_2C_6H_4$.

Therefore, it is only possible to get the ratios of all the distances; or, if we can assume one, the others can be calculated. It is often the case in these systems that there is one more unknown than there are knowns. However, even if a particular system has more coupling constants than unknowns, we cannot obtain a unique solution. This can be seen by referring to equation (8–55). For a given S and r that satisfy equation (8–55), we could multiply r by a factor and S by the cube of that factor and get the same B. This corresponds to a uniform expansion of the structure without any change in the ratios of the distances. In spite of this severe limitation, a very considerable amount of information can still be obtained from nmr studies in liquid crystals. The fitting obtained for benzene indicates that only three distances are needed to interpret the spectrum, so it must be a regular hexagon in solution—a not too surprising result. In a study[63] of the liquid crystal nmr spectrum of *trans*-$[Cl_2Pt(C_2H_4)C_5H_5N]$, the influences that various dynamic processes in solution have on the resulting spectra are discussed. The ordering factor S is very low and severe overlapping of the resonances results. A method was developed in which the spectrum intensities are fitted in the spectral analysis. The ratio of the r_{gem}/r_{cis} protons in the coordinated ethylene is consistent with a structure in which the C—H bonds are bent back away from the metal. Except for the dynamic process occurring in solution (the ethylene rotates rapidly about the bond to platinum), the solution structure of the ethylene fragment is similar to that in the solid.

When the chemical shift differences of the protons involved are large and first order spectra are obtained, the spectrum is readily interpreted by inspection and information about the symmetry of the molecule can be obtained from the spectrum in a straightforward manner. Consider the results reported[64] for $Ru_3(CO)_9C_2H_6$, which led to the proposed structure shown in Fig. 8–39 along with the spectrum observed for this molecule in a liquid crystal. The methyl group is rapidly rotating about the C—C bond axis. The quartets arise from the three equivalent methyl protons splitting the three bridging protons and vice versa. The three methyl protons are equivalent with respect to the bridging hydrides because of the rapid rotation. In a fixed staggered configuration (*i.e.*, CH_3 versus metal hydrides), the hydride protons would be split by a set of two equivalent protons and one non-equivalent methyl proton, producing six peaks in each triplet component, *i.e.*, 36 peaks total. One of the two groups of triplets arises from direct dipolar coupling of the methyl protons with each other, and the second (more closely spaced) triplet arises from the dipolar coupling of the three bridged hydrides with each other. If we consider one of the protons of the methyl group, it can be split by the other two because $++$, $+-$, $-+$ and $--$ combinations of nuclear moments in different molecules cause this proton to experi-

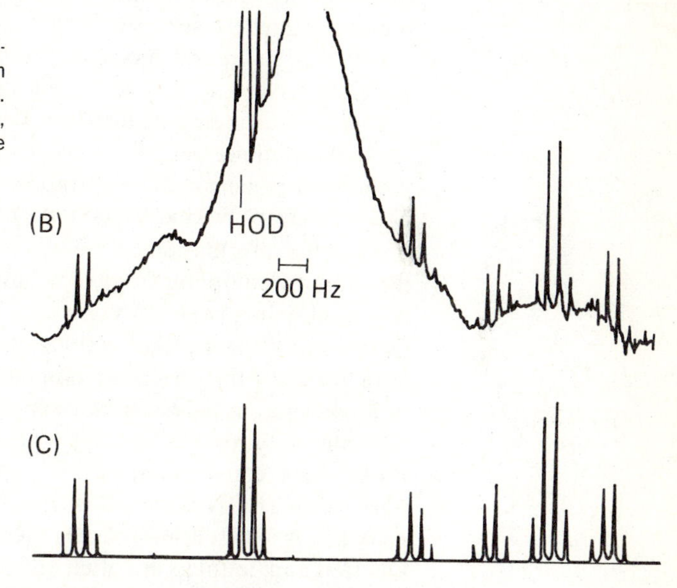

FIGURE 8–39. (A) Structure of $Ru_3(CO)_9C_2H_6$. (B) Experimental liquid crystal spectrum (broad absorption is from the liquid crystal). (C) Simulated liquid crystal spectrum. [Reprinted with permission from A. D. Buckingham, *et al.*, J. Amer. Chem. Soc., *95*, 2732 (1973). Copyright by the American Chemical Society.]

ence different fields, since the molecule is not undergoing isotropic rotation and the direct dipolar coupling is not averaged to zero. The larger coupling constants arise when the protons causing the splitting are closer to each other.

Liquid crystal work in which nuclei heavier than hydrogen are employed (*e.g.*, ^{13}C or ^{19}F) is complicated by the fact that the indirect coupling constants (J's) are orientation dependent and cannot be factored out of the observed couplings easily. When hydrogen-hydrogen couplings exist, the isotropic value from the normal solution nmr is used in the computer analysis of the data.

APPLICATIONS AND STRATEGIES IN ¹³C MAGNETIC RESONANCE

A brief discussion of ^{13}C magnetic resonance, abbreviated cmr, will give us an opportunity to review and apply several of the principles and phenomena we have developed. The reader is referred back to the previous chapter for the discussion of the Fourier transform technique used in these measurements, the use of T_1 data to assign the resonances, and the discussion of the nuclear Overhauser enhancement that results in the ^{13}C spectrum by decoupling the proton nmr.

The total signal-to-noise gain when all of the techniques we have been discussing are employed is interesting to note. We will make our comparison to a single sweep slow passage ^{13}C spectrum of a CH_2 peak with the sample in a 5 mm tube. In Fourier transform spectroscopy, the multiplet collapse from double resonance gives us a gain of a factor of two, the nuclear Overhauser effect gives us a factor of three, the larger size of sample tube used yields a factor of four, the accumulation that can be done in the same time results in a gain of forty, and the gain of ten from the Fourier technique leads to a total gain of $\sim 10^4$. Routine high resolution spectra of organic molecules can be obtained on 0.5 M solutions in 20 minutes and on as low a concentration as 0.05 M in 20 hours.

The cmr spectra of various molecules are dominated by the local paramagnetic term and occur over a wide range of field strength in different kinds of compounds. The shifts observed in a particular region of the spectrum provide a very good indication of the functional groups present in a molecule, as can be seen from the correlation chart in Fig. 8–40. The fact that a given carbon-containing functional group (*e.g.*, a substituted benzene or heterocycle containing ring carbons) shows a very characteristic shift has made the fingerprint type of application even more successful than in proton magnetic resonance.[65,66]

In a large organic molecule, there are many ^{13}C resonances to be assigned. The correlation chart in Fig. 8–40 will obviously not permit an assignment of all the resonances. As mentioned earlier, T_1 measurements are a considerable aid. There is another technique, referred to as *off-center double resonance,* that is also valuable. As mentioned earlier, if all the protons are decoupled in an nmr experiment, a considerable nuclear Overhauser enhancement (*i.e.*, a factor of about three) results. However, all the information potentially available regarding the proton splitting is lost. If the proton decoupling frequency employed is off-center with respect to the proton frequencies involved, only a partial collapse of the multiplet results and some Overhauser enhancement is obtained. The resulting coupling constants are smaller than those in the spectrum obtained without any decoupling, so the overlap of multiplets in the off-resonance experiment is not as extensive and does not confuse the interpretation. In instances of strongly coupled nuclei, however, misleading information can be obtained.[67] The result of this experiment on the molecule $(CH_3)_2CHCH_2OH$ is shown in Fig. 8–41. The triplet structure of the low field ^{13}C peak in Fig. 8–41(B) establishes the resonance as coming from the —CH_2— group. Quartets and doublets are often difficult to distinguish, but the upfield resonance here is assigned to the —CH_3 group. The relative intensities of the peaks are badly distorted from relaxation effects and cannot be used to assign the —CH_3 resonances; *i.e.*, a 1:2:1 ratio is not obtained. Since a major relaxation path involves the dipolar coupling of bound protons, it is often found (particularly for non-protonated carbons, *e.g.*, a metal carbonyl) that a carbon signal can be saturated and is totally lacking from the spectrum. In these instances, an intermolecular relaxation agent can be added to speed up the relaxation rate and produce a normal spectrum. Trisacetylacetonato chromium(III) is a paramagnetic complex that is quite unreactive and has been successfully employed this way.[68]

The absence of protons in metal carbonyls results in these compounds having very long T_1's. Accordingly, small flip angles [θ of equation (7–48)] are used in running the spectra and long pauses between pulses must be employed to obtain the free induction decay curve. Poor spectra are often obtained. It has been shown that T_1 can be drastically reduced by adding paramagnetic complexes, as mentioned earlier, for large fluctuating magnetic fields arise from the paramagnetic complex moving through the solvent and these lead to the very effective T_1 relaxation. A series of spectra in which the concentration of the relaxation agent is varied should be studied to determine whether the chemical shift of the ^{13}C resonance is being influenced by the paramagnetic species. Doddrell and Allerhand[69] have utilized a combination of these techniques in the assignment of the resonances of vitamin B_{12}, coenzyme B_{12}, and other corrinoids. The reader is referred to the discussion in this reference for a practical illustration of ^{13}C peak assignments.

Another interesting spectral enhancement technique is referred to as *gated decoupling.*[70] When proton decoupling is terminated immediately before a field

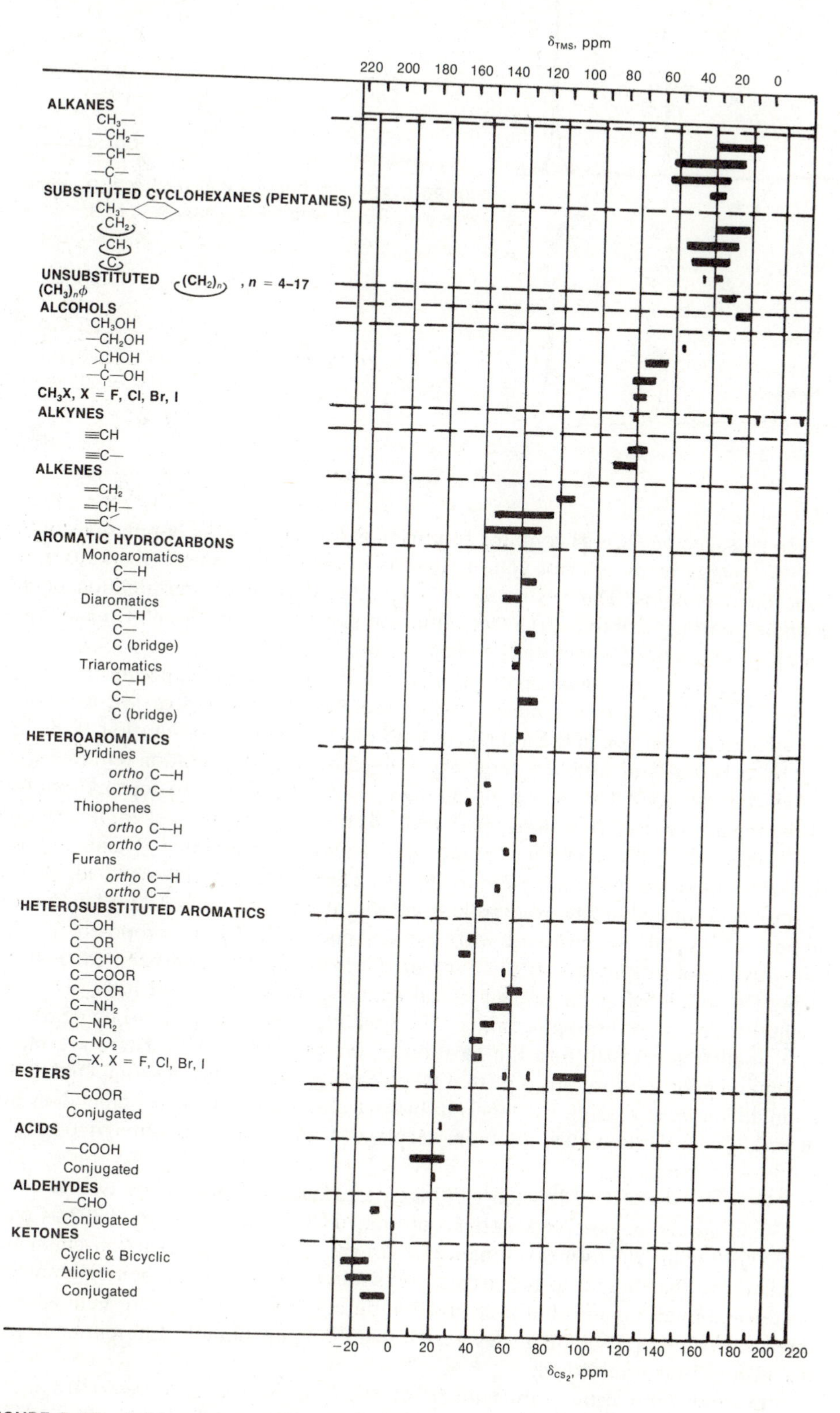

FIGURE 8–40. A ¹³C correlation chart. [From R. K. Jensen and L. Petrakis, J. Mag. Res., *7*, 106 (1972).]

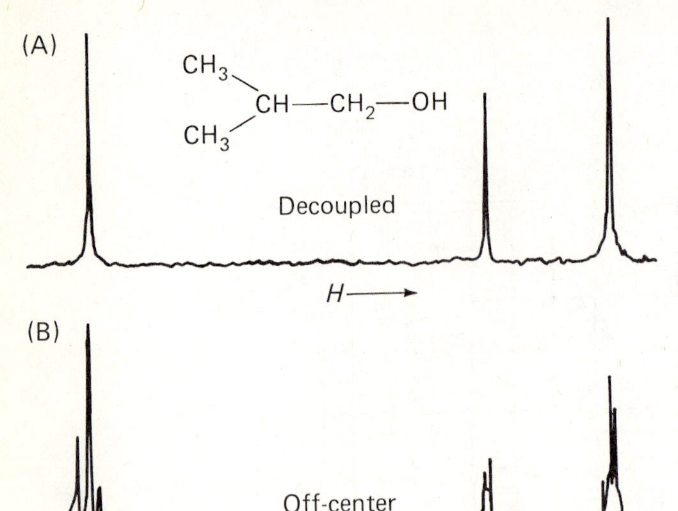

FIGURE 8–41. ^{13}C FTS of isobutanol: (A) all protons decoupled; (B) off-center decoupling.

sweep passage, the ^{13}C—H coupling returns immediately, but the populations of the nuclear energy levels are not equilibrated as rapidly. Thus, some Overhauser enhancement remains. The technique thus involves decoupling, termination of decoupling, pulsing, storage of the free induction decay, and repetition of this process for all pulse cycles. Correct values of $J_{^{13}C-H}$ are obtained.

A very exciting application of cmr involves its use as a probe, permitting one to employ stable isotopes as tracers.[(71)] Because the ^{13}C resonances are extremely sensitive to the location of the atom in a molecule, the site of incorporation of the ^{13}C can often be easily identified without the time-consuming degradation required with radioactive isotopes. The incorporation of ^{13}C-labeled glycine into coproporphyrin-III by the purple bacterium *Rhodopseudomonas spheroides* is an excellent illustration of the method.[(71)] The bacteria were grown on a medium containing glycine that was 93 per cent labeled in the α-position with ^{13}C. The resulting porphyrins had the ^{13}C atoms incorporated solely as pyrrole α-carbon atoms and methine bridge carbon atoms. This result is consistent with the sequence of reactions proposed for the biosynthesis of porphyrins. If this type of tracer application had been carried out using ^{14}C and subsequent radiochemical analysis, the study would have been less definitive and more time-consuming. The product would have had to be degraded into small fragments to find the location of the radioactive ^{14}C. There are many potential applications of this type that involve using ^{13}C as a tracer atom. One of the main advantages to using ^{13}C labels in biological systems is that this label does not disturb the conformation of the biomolecule as a spin label or paramagnetic probe might.

In a ^{13}C study[(72)] of labeled CO binding to human hemoglobin, two separate resonances could be observed for the coordinated CO. These occurred at 207.04 ppm and 206.60 ppm. The former resonance was assigned to CO bound to the α-chain and the latter to that bound to the β-chain by studying an abnormal hemoglobin that contains normal β-chains but α-chains that do not bind CO. In rabbit hemoglobin, three distinct iron(II) binding sites were found.[(73)] A functionally different hemoglobin subunit was established.

In another interesting application[(74)] of ^{13}C, T_1 measurements were carried out on selectively ^{13}C-labeled histidine bound to intracellular and extracellular mouse hemoglobin. The intracellular and extracellular T_1 values differed by only 25 per cent, suggesting that the viscosity of the intracellular fluid is not unusually large (at least in this system). There had been considerable controversy regarding this problem.

When molecules containing two directly bonded ^{13}C atoms are investigated, spin-spin splitting results. The deviation of the intensity of the multiplet pattern from statistical considerations in a biosynthetic experiment can provide a measure of the correlation in the enrichment.[75] Such information has important mechanistic implications. For example, mixtures of doubly labeled and unlabeled material can be studied. Dilution in the $^{13}C—^{13}C$ interaction would indicate cleavage of this bond in the biosynthesis.

Cmr has been of considerable utility[76] in identifying the existence and structure of carbonium ions and carbocations in solution. Typically, a positively charged carbon will be deshielded compared to the analogous carbon in a neutral reference compound, and it will cause an inductive deshielding of 5 to 15 ppm in a neighbor atom.

The cmr spectra of many organometallic compounds have been reported.[77-81] Complexation of olefins to $AgNO_3$ results [77] in 1 to 4 ppm shifts, while shifts of 30 to 110 ppm have been observed upon coordination to rhodium.[78] The benzene resonances in $C_6H_6Cr(CO)_3$ are shielded[79] some 30 ppm relative to free benzene, while those of bound CO are usually downfield from free CO. The ^{13}CO resonances in diamagnetic metal carbonyls are relatively insensitive to substituent and metal. Some typical results[77] are listed in Table 8–3. The interpretation of the ^{13}C chemical

TABLE 8–3. SOME TYPICAL ^{13}C RESONANCES IN METAL CARBONYLS

Complex	δ ^{13}C of Carbonyl
$Ni(CO)_4$	−191.6
$Fe(CO)_5$	−212
$Cr(CO)_6$	−211.3
$Mo(CO)_6$	−200.8
$W(CO)_6$	−191.4
$V(CO)_6$	−225.7
$CpCr(CO)_3^-$	−246.8
$CpMn(CO)_3$	−225.1
$(C_6H_5)_3PW(CO)_5$	−221.3 *cis*; −216.5 *trans*
$(C_6H_5O)_3PW(CO)_5$	−217.6 *cis*; −213.9 *trans*
(Cp = cyclopentadienide)	

shifts in organometallic compounds has recently been criticized by Norton.[82] A thorough analysis of the shift interpretation and some results on the calculation of the shielding constant have recently been summarized.[83] The interpretation of small shift differences is very difficult.[46-49,80] Consistent interpretations based on increased deshielding with increased metal-to-ligand π back-bonding have been offered.

REFERENCES

1. F. L. Slejko, R. S. Drago, and D. G. Brown, J. Amer. Chem. Soc., *94* 9210 (1972).
2. S. S. Zumdahl and R. S. Drago, J. Amer. Chem. Soc., *89,* 4319 (1967).
3. S. Alexander, J. Chem. Phys., *37,* 967 (1962); Rev. Mol. Physics, *24,* 74 (1957).
4. C. S. Johnson, Adv. Mag. Res., *1,* 33 (1965).
5. G. Fano, Rev. Mod. Physics, *24,* 74 (1957).
6. S. S. Zumdahl and R. S. Drago, Inorg. Chem., *7,* 2162 (1968).
7. (a) W. D. Perry, R. S. Drago, and N. K. Kildahl, J. Coord. Chem., *3,* 203 (1974); (b) N. K. Kildahl and R. S. Drago, J. Amer. Chem. Soc., *95,* 6245 (1973).
8. H. S. Gutowsky and C. H. Holm, J. Chem. Phys., *25,* 1228 (1956).
9. M. Rogers and J. C. Woodbrey, J. Phys. Chem., *66,* 540 (1962).

10. C. Looney, W. D. Phillips, and E. L. Reilly, J. Amer. Chem. Soc., *79,* 6136 (1957).
11. G. Claeson, G. Androes, and M. Calvin, J. Amer. Chem. Soc., *83,* 4357 (1961).
12. E. Grunwald, A. Loewenstein, and S. Meiboom, J. Chem. Phys., *27,* 630 (1957).
13. S. Meiboom, *et al.*, J. Chem. Phys., *29,* 969 (1958) and references therein.
14. E. L. Muetterties and D. W. Phillips, J. Amer. Chem. Soc., *81,* 1084 (1959).
15. A. Patterson, Jr., and R. Ettinger, Z. Elektrochem., *64,* 98 (1960).
16. W. D. Phillips, *et al.*, J. Amer. Chem. Soc., *81,* 4496 (1959).
17. W. N. Lipscomb, Adv. Inorg. Chem. Radiochem., *1,* 132 (1959).
18. B. M. Graybill, J. K. Ruff, and M. F. Hawthorne, J. Amer. Chem. Soc., *83,* 2669 (1961).
19. M. J. Bennett, Jr., *et al.*, J. Amer. Chem. Soc., *88,* 4371 (1966); F. A. Cotton, Accts. Chem. Res., *1,* 257 (1968).
20. (a) K. Vrieze and P. W. N. M. van Leeueven, Prog. Inorg. Chem., *14,* 1 (1971); (b) "Dynamic Nuclear Magnetic Resonance Spectroscopy," ed. by L. M. Jackman and F. A. Cotton, Academic Press, New York (1975).
21. L. W. Reeves and K. N. Shaw, Can. J. Chem., *48,* 3641 (1970).
22. H. S. Gutowsky and C. J. Hoffman, J. Chem. Phys., *19,* 1259 (1951).
23. J. I. Musher, J. Amer. Chem. Soc., *94,* 5662 (1972) and references therein.
24. (a) W. G. Klemperer, J. Amer. Chem. Soc., *95,* 380 (1973) and references therein; (b) J. P. Jesson and P. Meakin, J. Amer. Chem. Soc., *96,* 5760 (1974).
25. G. M. Whitesides and H. L. Mitchell, J. Amer. Chem. Soc., *91,* 5384 (1969).
26. J. R. Shapley and J. A. Osborn, Accts. Chem. Res., *6,* 305 (1973).
27. J. P. Jesson, in "Transition Metal Hydrides," ed. by E. L. Muetterties, Marcel Dekker, New York (1971).
28. S. S. Eaton, J. R. Hutchison, R. H. Holm, and E. L. Muetterties, J. Amer. Chem. Soc., *94,* 6411 (1972).
29. D. J. Duffy and L. H. Pignolet, Inorg. Chem., *11,* 2843 (1972).
30. J. D. Roberts, "An Introduction to the Analysis of Spin-Spin Splitting in High Resolution NMR Spectra," W. A. Benjamin, New York (1961).
31. P. L. Corio, Chem. Revs., *60,* 363 (1960).
32. (a) J. W. Emsley, *et al.*, "High Resolution Nuclear Magnetic Resonance Spectroscopy," Vol. 1, Pergamon Press, New York (1966); (b) P. Diehl, E. Fluck, and R. Kosfeld, eds., "NMR Basic Principles and Progress," Vol. 5, Springer-Verlag, New York (1971).
33. C. F. Callis, *et al.*, J. Amer. Chem. Soc., *79,* 2719 (1957).
34. J. N. Shoolery, Disc. Faraday Soc., *19,* 215 (1955).
35. R. A. Ogg, Jr., and J. D. Ray, Disc. Faraday Soc., *19,* 239 (1955).
36. J. H. Noggle, J. D. Baldeschwieler, and C. B. Colburn, J. Chem. Phys., *37,* 182 (1962).
37. J. D. Baldeschwieler and E. W. Randall, Chem. Rev., *63,* 82 (1963).
38. J. P. Maher and D. F. Evans, Proc. Chem. Soc., *1961,* 208; D. W. Turner, J. Chem. Soc., *1962,* 847.
39. J. W. Faller, in "Determination of Organic Structures by Physical Methods," *Vol. 5,* ed. by F. Nachod and J. Zuckerman, Academic Press, New York (1973) and references therein.
40. P. D. Kennewell, J. Chem. Educ., *47,* 278 (1970).
41. See, for example, A. Carrington and A. D. McLachlan, "Introduction to Magnetic Resonance," Chapt. 13, Harper and Row, New York (1967).
42. (a) G. L. Closs, *et al.*, J. Amer. Chem. Soc., *92,* 2183, 2185, 2186 (1970); (b) S. H. Pine, J. Chem. Educ., *49,* 664 (1972).
43. C. E. Johnson and F. A. Bovey, J. Chem. Phys., *29,* 1012 (1958) and references therein.
44. (a) A. D. Buckingham, Can. J. Chem., *38,* 300 (1960); (b) J. I. Musher, J. Chem. Phys., *37,* 34 (1962).
45. F. L. Slejko and R. S. Drago, J. Amer. Chem. Soc., *95,* 6935 (1973).
46. W. N. Lipscomb, *et al.*, J. Amer. Chem. Soc., *88,* 5340 (1966); Adv. Mag. Res., *2,* 137 (1966).
47. D. Gurka and R. W. Taft, J. Amer. Chem. Soc., *91,* 4794 (1969).
48. F. Prosser and L. Goodman, J. Chem. Phys., *38,* 374 (1963).
49. (a) M. Karplus and T. P. Das, J. Chem. Phys., *34,* 1683 (1961); (b) R. J. Pugmire and D. M. Grant, J. Amer. Chem. Soc., *90,* 697 (1968).
50. W. H. Pirkle, S. D. Beare, and R. L. Muntz, J. Amer. Chem. Soc., *91,* 4575 (1969); *ibid., 93,* 2817 (1971) and references therein.
51. (a) N. F. Ramsey, Phys. Rev., *91,* 303 (1953); (b) J. A. Pople and D. P. Santry, Mol. Phys., *8,* 1 (1964); (c) C. J. Jameson and H. S. Gutowsky, J. Chem. Phys., *51,* 2790 (1969).
52. C. P. Slichter, "Principles of Magnetic Resonance," Chapt. 3, Harper and Row, New York (1963).
53. A. Abragam, "The Principles of Nuclear Magnetism," Chapt. 4, Clarendon Press, Oxford (1961).
54. E. R. Andrew, "Nuclear Magnetic Resonance," Chapt. 6, Cambridge University Press, New York (1955); E. R. Andrew and P. S. Allen, J. Chem. Phys., *63*(1), 85–91 (1966).
55. C. M. Deely and R. E. Richards, J. Chem. Soc., 3697 (1954).
56. L. N. Mulay, E. G. Rochow, and E. O. Fischer, J. Inorg. Nucl. Chem., *4,* 231 (1957).
57. D. F. R. Gilson and C. A. McDowell, J. Chem. Phys., *40,* 2413 (1964) and references therein.
58. H. Levy and W. E. Grizzle, J. Chem. Phys., *45,* 1954 (1966).
59. S. Pausak, A. Pines and J. S. Waugh, J. Chem. Phys., *59,* 591 (1973).
60. A. Pines, M. G. Gibby, and J. S. Waugh, J. Chem. Phys., *59,* 569 (1973).
61. P. Diehl, *et al.*, "NMR Basic Principles and Progress," Vol. 1, Springer-Verlag, New York (1969).
62. L. C. Snyder and E. W. Anderson, J. Amer. Chem. Soc., *86,* 5023 (1964).
63. D. R. McMillin and R. S. Drago, Inorg. Chem., *13,* 546 (1974).
64. A. D. Buckingham, *et al.*, J. Amer. Chem. Soc., *95,* 2732 (1973).
65. J. B. Stothers, "Carbon-13 NMR Spectroscopy," Academic Press, New York (1972).

66. G. C. Levy and G. L. Nelson, "Carbon-13 Nuclear Magnetic Resonance for Organic Chemists," Wiley-Interscience, New York (1972).
67. R. A. Newmark and J. R. Hill, J. Amer. Chem. Soc., *95,* 4435 (1973).
68. (a) O. A. Gansow, A. R. Burke, and G. N. LaMar, Chem. Comm., 456 (1972); (b) O. A. Gansow, *et al.,* J. Amer. Chem. Soc., *94,* 2550 (1972); (c) D. F. S. Natusch, J. Amer. Chem. Soc., *93,* 2566 (1971).
69. D. Doddrell and A. Allerhand, Proc. Nat. Acad. Sci. (USA), *68,* 1083 (1971).
70. O. A. Gansow and W. Schittenhelm, J. Amer. Chem. Soc., *93,* 4294 (1971).
71. N. A. Matwiyoff and D. G. Ott, Science, *181,* 1125 (1973).
72. P. J. Vergamini, N. A. Matwiyoff, R. C. Wohl, and T. Bradley, Biochem. Biophys. Res. Comm., *55,* 453 (1973).
73. N. A. Matwiyoff, *et al.,* J. Amer. Chem. Soc., *95,* 4429 (1973).
74. R. E. London, C. T. Gregg, and N. A. Matwiyoff, Science, *188,* 266 (1975).
75. R. E. London, V. H. Kollman, and N. A. Matwiyoff, J. Amer. Chem. Soc., *97,* 3565 (1975).
76. G. A. Olah, *et al.,* J. Amer. Chem. Soc., *93,* 4219 (1971).
77. R. G. Parker and J. D. Roberts, J. Amer. Chem. Soc., *92,* 743 (1970).
78. G. M. Bodner, *et al.,* Chem. Comm., *1970,* 1530.
79. G. M. Bodner and L. J. Todd, Inorg. Chem., *13,* 360 (1974); *ibid., 13,* 1335 (1974).
80. L. J. Todd and J. R. Wilkinson, J. Org. Met. Chem., *77,* 1 (1974).
81. B. E. Mann, Adv. Org. Met. Chem., *12,* 135 (1974).
82. J. Evans and J. R. Norton, Inorg. Chem., *13,* 3043 (1974).
83. R. Ditchfield and P. D. Ellis, in "Topics in Carbon-13 NMR Spectroscopy," Vol. 1, ed. by G. C. Levy, Wiley-Interscience, New York (1974).

GENERAL REFERENCES FOR NMR

J. W. Emsley, *et al.,* "High Resolution Nuclear Magnetic Resonance Spectroscopy," Vols. 1 and 2, Pergamon Press, New York (1966).
T. C. Farrar and E. D. Becker, "Introduction to Pulse and Fourier Transform NMR Methods," Academic Press, New York (1971).
C. P. Slichter, "Principles of Magnetic Resonance," Harper and Row, New York (1963).
A. Abragam, "The Principles of Nuclear Magnetism," Clarendon Press, Oxford (1961).
E. D. Becker, "High Resolution NMR," Academic Press, New York, N.Y.
G. C. Levy and G. L. Nelson, "Carbon-13 Nuclear Magnetic Resonance for Organic Chemists," Wiley-Interscience, New York (1972).
J. B. Stothers, "Carbon-13 NMR Spectroscopy," Academic Press, New York (1972).
R. A. Dwek, "Nuclear Magnetic Resonance in Biochemistry," Oxford University Press, New York (1974).
R. J. Myers, "Molecular Magnetism and Magnetic Resonance Spectroscopy," Prentice-Hall, Englewood Cliffs, N.J. (1973).
"Nuclear Magnetic Resonance Shift Reagents," ed. by R. E. Sievers, Academic Press, New York (1973).
C. P. Poole, Jr., and H. A. Farach, "The Theory of Magnetic Resonance," Wiley-Interscience, New York (1972).
"Dynamic Nuclear Magnetic Resonance Spectroscopy," ed. by L. M. Jackman and F. A. Cotton, Academic Press, New York (1975).

Series

"Annual Reports on NMR Spectroscopy," Academic Press, New York, 1968–.
"Advances in Magnetic Resonance," ed. by J. S. Waugh, Academic Press, New York, 1965–.
"Advances in Magnetic Resonance, Supplement 1: High Resolution NMR in Solids," U. Haeberlen, Academic Press, New York, 1976.
"NMR Abstracts and Index," Preston Technical Abstracts Co., Evanston, Ill., 1968.
"Nuclear Magnetic Resonance," The Chemical Society (Specialist Periodical Report), London, 1972–.
"Topics in Carbon-13 NMR Spectroscopy," ed. by G. C. Levy, Wiley-Interscience, New York, 1974–.

Compilations of Spectra

"The Sadtler Standard Spectra; NMR," Sadtler Research Laboratory, Philadelphia (1972).
L. F. Johnson and W. C. Jankowski, "C-13 NMR Spectra," Wiley-Interscience, New York (1972).
F. A. Bovey, "NMR Data Tables for Organic Compounds," Wiley-Interscience, New York (1967).

EXERCISES

1. a. Would you expect the difference $\delta_{CH_3} - \delta_{CH_2}$ in C_2H_5I to be independent of the remote anisotropy in the C—I bond? Why?

   b. Will this difference have a greater or smaller contribution from anisotropy than δ_{CH_2}?

   c. If in a series of compounds the δ_{CH_2} values showed a different trend than the $\delta_{CH_3} - \delta_{CH_2}$ values, what could you conclude about anisotropic contributions?

2. Combine in your thinking (reread if necessary) the discussion on interatomic ring currents with the relationship between $J_{^{13}C-H}$ and τ (discussed in section 7–15 on Applications of Spin-Spin Coupling to Structure Determination, p. 217). Using these concepts, propose a method for determining if there is anisotropy in the methyl chemical shift of *B*-trimethyl borazine.

3. Consider the molecule

```
X     X     X
  \  / \  /
   M     M
  /  \ /  \
X     X     X
```

where M and X have $I = \frac{1}{2}$. Sketch the nmr spectra for the following conditions, assuming $\Delta > J$ in all cases (when necessary to assume orders of magnitude for various coupling constants, state your assumption). Ignore coupling if the atoms are not directly bonded.

   a. nmr spectrum of X, no exchange.

   b. nmr spectrum of M, no exchange.

   c. nmr spectrum of X with rapid intermolecular exchange of all X groups.

   d. nmr spectrum of M with rapid $[\tau' < 1/(\nu_A - \nu_B)]$ intermolecular exchange of all X groups.

   e. nmr spectrum of X with rapid intramolecular exchange.

   f. nmr spectrum of M with rapid intramolecular exchange.

4. In the absence of any exchange, two peaks A—H and B—H are separated by 250 Hz. At room temperature, exchange occurs and the peaks are separated by 25 Hz. The spin-lattice relaxation of A—H and B—H is long, and there are equal concentrations (0.2 M) of each. Calculate the lifetime of a proton on A, and from this the rate constant for the exchange (specify units).

5. In a given compound MF_4 (for M, $I = \frac{1}{2}$) the J_{M-F} value is 150 Hz. In the absence of chemical exchange, the F^- and M—F signals are separated by 400 Hz. At room temperature the F^- and MF_4 exchange at a rate such that the fine structure just disappears. Assuming equal concentrations of M—F and F^- species and no stable intermediates, calculate τ' for F^-. What will be the separation of the M—F and F^- peaks under these conditions?

6. The spectrum of the following compound is a complex type. Explain how the double resonance technique could be employed to aid in interpreting the spectrum.

```
   H       H
    \     /
     C—C
    //   \\
H—C       C—H
    \    /
      O
```

7. The spectrum in the following figure is that of an AB type of molecule.

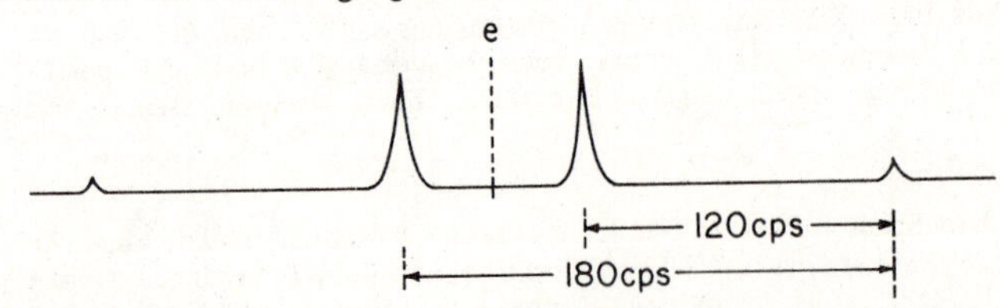

   a. Calculate J and Δ.

   b. A 40 MHz probe was employed. Calculate the difference in δ for the two peaks.

   c. If e occurs at a δ value of 2.8, what are the δ values for A and B?

8. Consider the series *i*-propyl X (where X = Cl, Br, I).

   a. In which compound would the remote paramagnetic effect from the halogen be largest? Why?

   b. In which would the remote diamagnetic effect be largest?

9. What would the nmr spectrum of PF_5 look like under the following conditions $(\Delta_{F(a)-F(b)} > J_{F(a)-F(b)})$?

a. Very slow fluorine exchange.

b. Rapid intermolecular fluorine exchange.

c. Rapid intramolecular fluorine exchange.

10. Consider the molecule shown in problem 8 of Chapter 7. Using the A, B, X, . . . terminology, classify this molecule and indicate the non-equivalent protons.

11. a. List and briefly describe in your own words the factors that influence the magnitudes of proton chemical shifts.

b. Why do ^{19}F and ^{13}C chemical shifts cover a much larger range than do those of protons?

12. Ramsey's formula is used to calculate the local contributions to the chemical shielding of a nucleus; it is given as equation (8–31) on page 283. The nuclear magnetic resonance spectrum of ^{59}Co has been observed in a variety of environments. A correlation has been proposed relating the chemical shift to the wavelength of an electronic absorption observed in these complexes.

Complex	λ (Ångstroms)
$K_3[Co(CN)_6]$	3110
$[Co(en)_3]Cl_3$	4700
$[Co(NH_3)_6]Cl_3$	4750
(en = ethylenediamine)	

a. Do you expect the diamagnetic term in Ramsey's formula to account for an 8150 ppm variation in the chemical shift observed in this series of complexes? Explain briefly.

b. Do you expect the paramagnetic contribution to the chemical shielding to vary in this series of complexes? Why?

c. List these cobalt complexes in order of increasing magnetic field for resonance. Give your reasoning.

13. Ramsey's equation [equation (8–31), p. 283] is used to calculate local contributions to chemical shifts. Consider the following boron compounds and their ^{11}B chemical shifts (in ppm relative to BF_3 etherate).

Three-coordinate	Four-coordinate
$BI_{3(melt)} = +5.5$	$NaBF_4 = +2.3$
$BF_{3(gas)} = -9.4$	$B(OH)_4^- = -1.8$
$BCl_3 = -47.7$	$BF_3 \cdot$ piperidine $= +2.3$
$BBr_3 = -40.1$	$NaB(C_6H_5)_4 = +8.2$
$B(OCH_3)_3 = -18.1$	$LiB(OCH_3)_4 = -2.9$
$B(C_2H_5)_3 = -85.0$	$BF_3 \cdot P(C_6H_5)_3 = -0.4$

The generalized m.o. description for three-coordinate boron compounds consists of bonding, non-bonding (empty boron a.o.) and antibonding molecular orbitals (I). Four-coordinate boron compounds are described by bonding and antibonding m.o.'s (no non-bonding m.o.) (II).

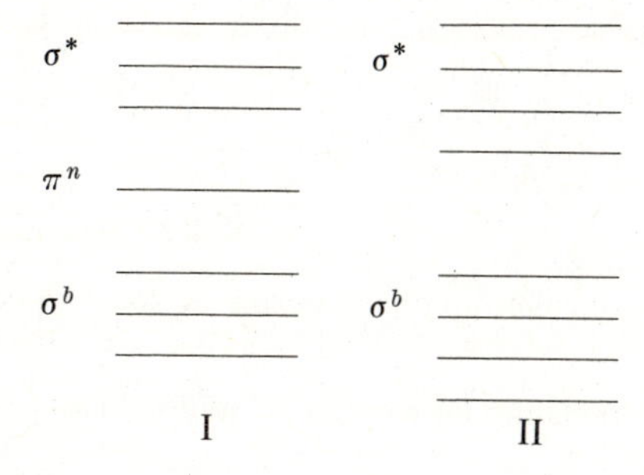

a. Can the first term of the Ramsey equation explain the ^{11}B chemical shifts of the three-coordinate compounds? Why or why not?

b. Rationalize from the m.o. description the fact that the ^{11}B chemical shift of the three-coordinate complexes varies 125 ppm, while that of the four-coordinate species varies 11.1 ppm.

14. It was said that there was no contribution to the local paramagnetic shielding in a linear molecule aligned parallel to the applied magnetic field. Construct molecular orbitals for the molecule HF and show that matrix elements resulting from the Ramsey equation lead to the above result.

15. Consider benzene and cyclohexane.

a. In which do you think the proton resonances would be further upfield? Why?

b. In which do you think the ^{13}C resonances would be further upfield? Why?

c. In which would $J_{^{13}C-H}$ be larger? Why?

16. In compounds of the type CH_3HgX, the ^{199}Hg-^{1}H coupling constant is observed to vary by more than a factor of two, depending on the substituent X. Some examples are given below:

X	J(Hz)
CH_3	104
I	200
Br	212
Cl	215
ClO_4	233

Propose an explanation for the observed splittings.

17. A number of ^{15}N-labeled aminophosphines have been synthesized and the ^{15}N-^{1}H nmr coupling constants have been measured. A sample of the results is as follows:

Compound	J(Hz)
$F_4P^{15}NH_2$	90.3
$(CF_3)_2P^{15}NH_2$	85.6
$F_2P^{15}NH_2$	82.7

There are two possible explanations in terms of the bonding in this series.

a. Give the explanation that is based on the extent of N $\longrightarrow$ P π-bonding.

b. Offer another explanation that was presented, which also accounts for these results.

18. The following ^{13}C T_1's have been determined by Freeman [J. Chem. Phys., *54,* 3367 (1971)] on 3,5-dimethylcyclohex-2-ene-1-one:

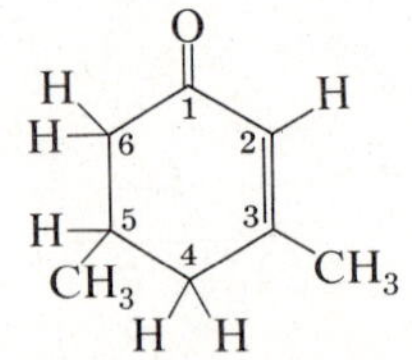

	T_1 (sec)
C_1	37
C_3	33
3-Me	5.9
C_2	5.4
C_5	5.3
C_6	3.1
C_4	3.1
5-Me	2.7

a. What appears to be the dominant means of spin-lattice relaxation for these ^{13}C atoms?

b. In light of your explanation, why do C_1 and C_3 have such long T_1's?

19. The following are ^{31}P nmr spectra, $I_P = I_F = \frac{1}{2}$. No J_{P-H} coupling is ever resolved.

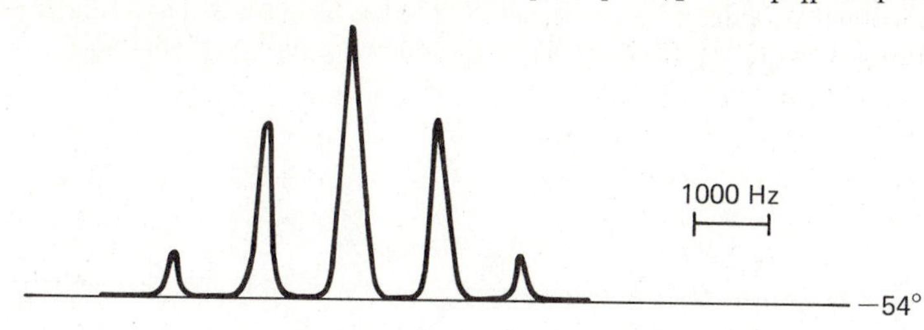

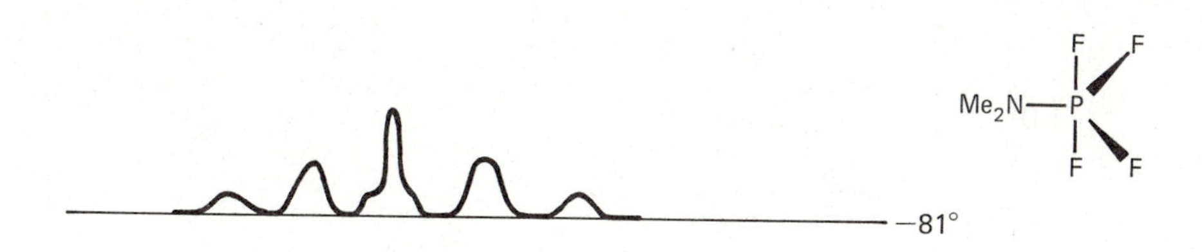

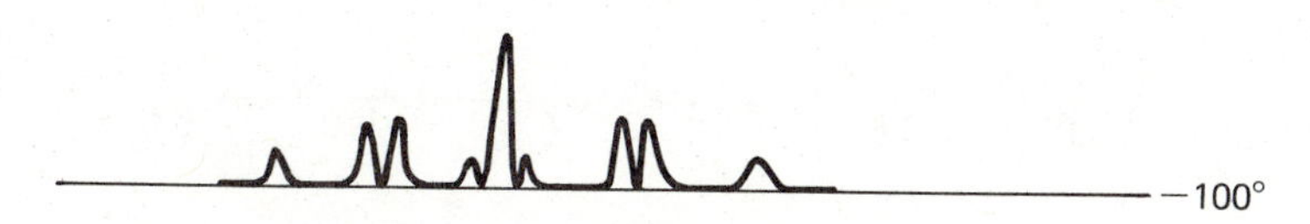

a. Explain the low temperature spectrum.

b. Explain the high temperature spectrum.

c. On the basis of these spectra, what can you say about the mechanism of exchange?

d. What is the relationship between coupling constants in the slow and fast exchange regions?

20. The complex $[(CF_3)Co(CO)_3(PF_3)]$ [J. Amer. Chem. Soc., *91*, 526 (1969)] is assumed to be trigonal bipyramidal. The ^{19}F nmr spectra at +30°C and −70°C are given below. Splittings from ^{59}Co are not observed.

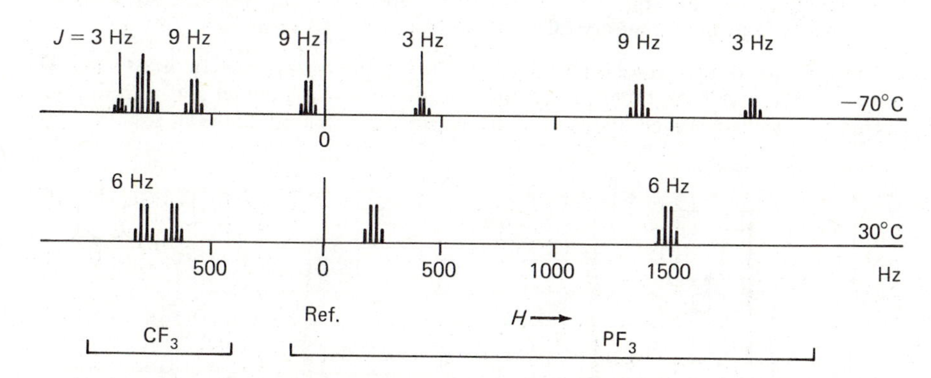

First consider the −70°C spectrum.

a. Explain the reason why four quartets are observed for the fluorines bonded to phosphorus. (No ^{59}Co spin-spin coupling is detected.)

b. Explain how the CF_3 resonances are consistent with your explanation in part a.

c. Describe the reason for the smaller number of PF_3 resonances at 30°C.

d. What is the significance of the PF_3 resonances being observed as quartets in the 30°C spectrum?

21. The line width of the methyl proton resonance of 3-picoline-*N*-oxide in solutions containing (3-picoline-*N*-oxide)$_6$Ni^{2+} and excess ligand has been studied as a function of temperature [Inorg. Chem., *10,* 1212 (1971)]. The following plot was obtained.

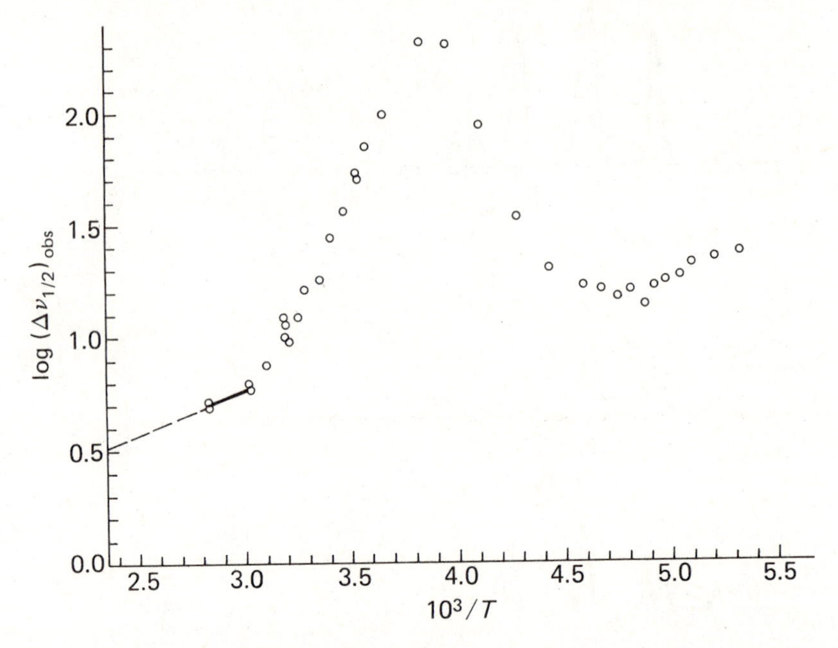

a. Label the fast, near fast, intermediate, and slow exchange regions. Do this by giving approximate boundaries to the region in units of $10^3/T$; *e.g.*, fast: x_1 to x_2 units.

b. These data were analyzed using full line shape analysis from the classical treatment. Why wasn't pyridine-*N*-oxide used as the ligand?

22. a. When the nmr spectrum of benzonitrile in a liquid crystal is observed, why do solvent resonances not appear?

b. Why are the benzonitrile lines in a liquid crystal much sharper than those in solid benzonitrile?

c. Compared to an nmr spectrum of benzonitrile in CCl_4, why are so many lines observed in the liquid crystal nmr spectrum of benzonitrile?

23. The spectrum of pyrogallol is given below. The OH protons are not coupled to any other nuclei. Why don't the three phenyl protons give rise to a doublet and a triplet? (The group of peaks centered about 6.7 δ are expanded in the offset sweep.)

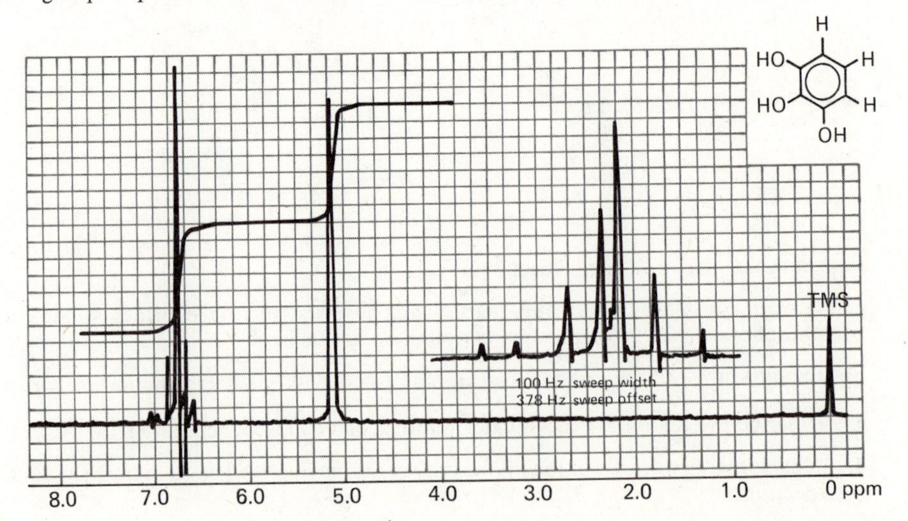

24.

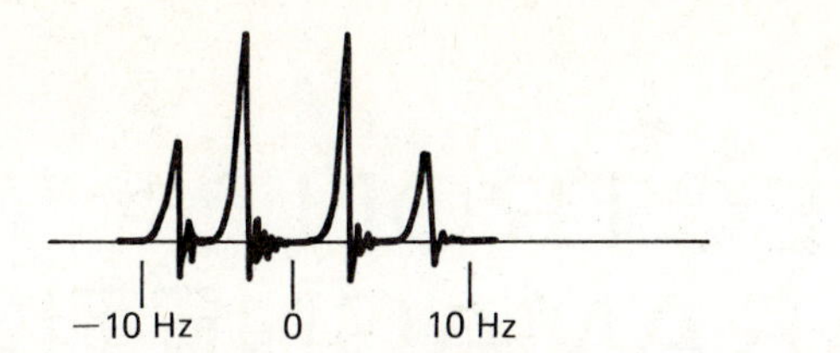

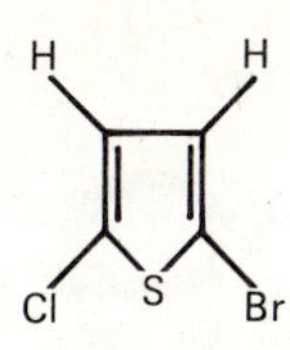

Above is the proton nmr spectrum of 2-bromo-5-chlorothiophene at 60 MHz. For a field impressed along the z direction:

$$\hat{H} = (1 - \sigma_1)g\beta H_z \hat{I}_{1z} + (1 - \sigma_2)g\beta H_z \hat{I}_{2z} + J\hat{I}_{1z}\hat{I}_{2z} + \frac{1}{2}J(\hat{I}_{1+}\hat{I}_{2-} + \hat{I}_{1-}\hat{I}_{2+})$$

$$\varphi_1 = |\alpha\alpha\rangle \qquad \varphi_2 = |\alpha\beta\rangle \qquad \varphi_3 = |\beta\alpha\rangle \qquad \varphi_4 = |\beta\beta\rangle$$

a. Compute the diagonal matrix element for the function φ_1. Compute the off-diagonal element between φ_2 and φ_3.

b. Briefly, why is the spectrum not two separated doublets of equal intensity? (*I.e.*, under what conditions would it be two doublets of equal intensity and what is evidently different here?)

c. Suggest a physical method that would simplify the spectrum into two doublets.

25. a. Write the complete spin Hamiltonian for HD, neglecting quadrupole effects of D ($I = 1$).

b. Set up the complete secular determinant and indicate non-zero matrix elements by (+)'s for chemical shift terms and (−)'s for coupling terms. (Label basis functions *clearly*.)

c. Given that $g_D\beta H_z = -9.21 \times 10^6$ Hz and $g_H\beta H_z = -60.00 \times 10^6$ Hz, would you expect a first or second order spectrum for the proton resonance if $J_{HD} = 15$ Hz? If $J_{HD} = 100$ Hz? Why?

26. Write the spin Hamiltonian matrix for the molecule

H F
C=C
Cl H

ignoring the quadrupolar chlorine nucleus. Consider the spectrum to be first order, find the allowed transitions, and show what the proton and fluorine spectra will look like.

9 ELECTRON PARAMAGNETIC RESONANCE SPECTROSCOPY*

INTRODUCTION

9-1 PRINCIPLES

Electron paramagnetic resonance is a branch of spectroscopy in which radiation of microwave frequency is absorbed by molecules, ions, or atoms possessing electrons with unpaired spins. This phenomenon has been designated by different names: "electron paramagnetic resonance" (epr), "electron spin resonance" (esr), and "electron magnetic resonance." These are equivalent and merely emphasize different aspects of the same phenomenon. There are some similarities between nmr and epr spectroscopy that are of help in understanding epr. In nmr spectroscopy, the two different energy states (when $I = 1/2$) arise from the alignment of the nuclear magnetic moments relative to the applied field, and a transition between them occurs upon the application of radio-frequency radiation. In epr, different energy states arise from the interaction of the unpaired electron spin moment (given by $m_s = \pm 1/2$ for a free electron) with the magnetic field, the so-called electronic Zeeman effect. The Zeeman Hamiltonian for the interaction of an electron with the magnetic field is given by equation (9–1):

$$\hat{H} = g\beta H\hat{S}_z \qquad (9\text{-}1)$$

where g for a free electron has the value 2.0023193; β is the electron Bohr magneton, $e\hbar/2m_ec$, which has the value $9.274096 \pm (0.000050) \times 10^{-21}$ erg gauss^{-1}; $\hat{S}_z$ is the spin operator; and H is the applied field strength. This Hamiltonian operating on the *electron* spin functions α and β corresponding to $m_s = +1/2$ and $-1/2$, respectively,

*The General References contain reviews of epr. The reference by J. E. Wertz and J. R. Bolton is especially recommended.

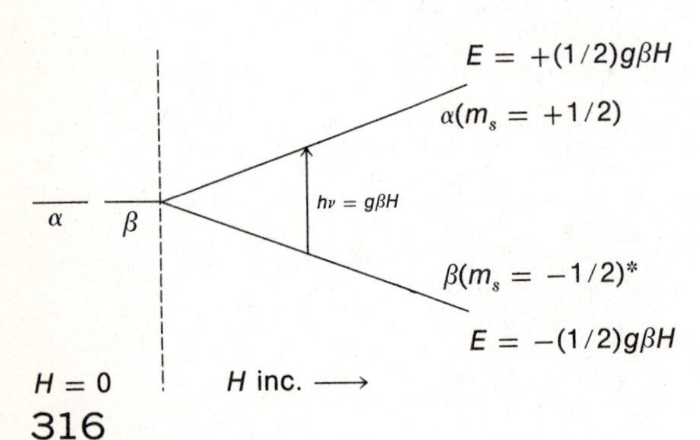

FIGURE 9–1. The removal of the degeneracy of the α and β electron spin states by a magnetic field. *Note the difference in the ground state from nmr.

produces the result illustrated in Fig. 9–1. The β spin state has its moment aligned with the field, in contrast to nmr, where the lowest energy state corresponds to $m_I = +1/2$ (α_N). The lowest energy state in epr corresponds to $m_s = -1/2$ because the sign of the charge on the electron is opposite that on the proton. The transition energy is given by equation (9–2):

$$\Delta E = g\beta H \tag{9-2}$$

The energy difference between the α and β spin states in magnetic fields of strengths commonly used in the epr experiment (several thousand gauss) corresponds to frequencies in the microwave region. For a field strength of 10,000 gauss, one can calculate from equation (9–2) that ΔE is 28,026 MHz. This is to be contrasted with the energy for the transition of the proton nuclear moment of 42.58 MHz (*i.e.*, radiation in the radio-frequency region) in a magnetic field of identical strength. The magnetic moment for the electron is -9.2849×10^{-21} erg gauss^{-1} compared to 1.4106×10^{-23} erg gauss^{-1} for a proton nuclear moment.

The epr experiment is generally carried out at a fixed frequency. Two common frequencies are in the X-band frequency range (about 9500 MHz or 9.5 gigahertz, GHz*, where a field strength of about 3400 gauss is employed) and the so-called Q-band frequency (35 GHz, where a field strength of about 12,500 gauss is used). Since the sensitivity of the instrument increases roughly as ν^2 and better spectral resolution also results, the higher frequency is to be preferred. There are several limitations on the use of the Q-band frequency. Smaller samples are required for Q-band, so the sensitivity is not as much greater as one would predict from ν^2. It is more difficult to attain the higher field homogeneity ($\delta H/H$) that is required at higher frequencies. Finally, for aqueous samples, dielectric absorption by the solvent becomes more serious as the frequency increases and this results in decreased sensitivity.

Water, alcohols, and other high dielectric constant solvents are not the solvents of choice for epr because they strongly absorb microwave power. They can be used when the sample has a strong resonance and is contained in a specially designed cell (a very narrow sample tube). EPR measurements on gases, solutions, powders, single crystals, and frozen solutions can be carried out. The best frozen solution results are obtained when the solvent freezes to form a glass. Symmetrical molecules or those that hydrogen bond extensively often do not form good glasses. For example, cyclohexane does not form a good glass, but methylcyclohexane does. Some solvents and mixtures that form good glasses are listed in Table 9–1.

The sample tube employed is also important. If the signal-to-noise ratio is low, a quartz sample tube is preferred, because Pyrex absorbs more of the microwave power and also exhibits an epr signal.

There are many effects that modify the electron energy states in a magnetic field. We shall consider these factors one at a time by discussing the epr spectra of increasingly complex systems.

In the way of introduction, it will simply be mentioned that differences in the energy of the epr transition for different molecules are described by changing the value of g in equation (9–2). This is to be contrasted with nmr, where one customarily holds g_N fixed and introduces the shielding constant to describe the different resonance energies, *i.e.*,

$$\Delta E = -g_N\beta_N(1 - \sigma)H\Delta m_I \tag{9-3}$$

As we proceed to more complex systems, we shall discuss the factors that influence the magnitude of g.

---

*The prefix mega indicates 10^6, and giga indicates 10^9.

TABLE 9–1. COMMONLY USED GLASSES*

Pure Substances

3-Methylpentane	Sulfuric acid	Sugar
Methylcyclopentane	Phosphoric acid	Triethanolamine
Nujol (paraffin oil)	Ethanol	2-Methyltetrahydrofuran
Isopentane	Isopropanol	Di-*n*-propyl ether
Methylcyclohexane	1-Propanol	*cis-trans* Decalin
Isooctane	1-Butanol	Triacetin
Boric acid	Glycerol	Toluene

Mixtures

Components	*Ratio A/B*
3-Methylpentane/isopentane	1/1
Isopentane/methylcyclohexane	1/6
Methylcyclopentane/methylcyclohexane	1/1
Pentene-2(*cis*)/pentene-2(*trans*)	
Propane/propene	1/1
Isopropyl benzene/propane/propene	2/9/9
Ethanol/methanol	4/1, 5/2, 1/9
Isopropyl alcohol/isopentane	3/7
Ethanol/isopentane/diethyl ether	2/5/5
Alphanol 79′/mixture of primary alcohols	
Isopentane/*n*-butyl alcohol	7/3
Isopentane/isopropyl alcohol	8/2
Isopentane/*n*-propyl alcohol	8/2
Diethyl ether/isooctane/isopropyl alcohol	3/3/1
Diethyl ether/isooctane/ethyl alcohol	3/3/1
Diethyl ether/isopropyl alcohol	3/1
Diethyl ether/ethanol	3/1
Isooctane/methylcyclohexane/isopropyl alcohol	3/3/1
Diethyl ether/toluene/ethanol	2/1/1
Isopropyl alcohol/isopentane	2/5/5
Propanol/diethyl ether	2/5
Butanol/diethyl ether	2/5
Diethyl ether/isopentane/dimethyl formamide/ethanol	12/10/6/1
Water/propylene glycol	1/1
Ethylene glycol/water	2/1
Trimethylamine/isopentane/diethyl ether	2/5/5
Triethylamine/isopentane/diethyl ether	3/1/3
Methylhydrazine/methylamine/trimethylamine	2/4/4
Diethyl ether/isopentane/ethanol/pyridine	12/10/6/1
Di-*n*-butyl ether/diisopropyl ether/dimethyl ether	3/5/12
Diphenyl ether/1,1-diphenylethane/triphenylmethane	3/3/1
Diethyl ether/isopentane	1/1 to 1/2
Dipropyl ether/isopentane	3/1
Dipropyl ether/methylcyclohexane	3/1
Diethyl ether/pentene-2(*cis*)-pentene-2(*trans*)	2/1
Ethyl iodide/isopentane/diethyl ether	1/2/2
Ethyl bromide/methylcyclohexane/isopentane/methylcyclopentane	1/4/7/7
Ethanol/methanol/ethyl iodide	16/4/1
Ethanol/methanol/propyl iodide	16/4/1
Ethanol/methanol/propyl chloride	16/4/1
Ethanol/methanol/propyl bromide	16/4/1
Diethyl ether/isopentane/ethanol/1-chloronaphthalene	8/6/2/2
3-Methylpentane/isopentane	1/2
Propyl alcohol/propane/propene	2/9/9
Diisopropylamine/propane/propene	2/9/9
Dipropyl ether/propane/propene	2/4/4
Toluene/methylene chloride	1/1 or excess toluene
Toluene/acetone	1/1 or excess toluene
Toluene/methanol or ethanol	1/1 or excess toluene
Toluene/acetonitrile	1/1 or excess toluene
Toluene/chloroform	1/1 or excess toluene
2-Methyltetrahydrofuran/methanol	2/1
2-Methyltetrahydrofuran/proprionitrile	2/1
2-Methyltetrahydrofuran/methylene chloride	1/1

*Abstracted in part, with permission, from B. Meyer, "Low Temperature Spectroscopy," American Elsevier Publishing Co., New York (1971).

9–2 THE HYDROGEN ATOM

NUCLEAR HYPERFINE SPLITTING

The first contribution to epr transition energies that will be introduced is the electron-nuclear hyperfine interaction. The hydrogen atom (in free space) is a simple system to discuss because, by virtue of its spherical symmetry, anisotropic effects are absent. In the development of epr, we shall employ the Hamiltonian to quantitatively describe the effects being considered. The full interpretation of the esr spectrum of a system is given in terms of an *effective spin Hamiltonian*. This is a Hamiltonian that contains those effects, of the many to be described, which are used to interpret the particular spectrum of the compound studied.

The complete spin Hamiltonian for the hydrogen atom (in free space) is

$$\begin{aligned}\hat{H} &= g\beta H\cdot\hat{S} - g_N\beta_N H\cdot\hat{I} + a\hat{I}\cdot\hat{S}\\ &= g\beta(H_x\hat{S}_x + H_y\hat{S}_y + H_z\hat{S}_z) - g_N\beta_N(H_x\hat{I}_x + H_y\hat{I}_y + H_z\hat{I}_z) + a\hat{I}_x\hat{S}_x\\ &\qquad + a\hat{I}_y\hat{S}_y + a\hat{I}_z\hat{S}_z\end{aligned}$$

For a spherical system in a magnetic field that is defined as the z-axis, this simplifies* to

$$\hat{H} = g\beta H\hat{S}_z - g_N\beta_N H\hat{I}_z + a\hat{I}\cdot\hat{S} \tag{9-4}$$

The first term of this Hamiltonian has been discussed [equation (9–1)] and led to the energy-field relation shown in Fig. 9–1. The second term of the Hamiltonian is familiar from our discussion of nmr; it describes the interaction of the nuclear moment of the hydrogen atom with a magnetic field. It is of opposite sign (the state with $m_I = +1/2$ is lowest) and smaller in magnitude than the first term. The combined effect of the first two terms in equation (9–4) upon the energies of the spin states of the hydrogen atom in a magnetic field is shown in Fig. 9–2(C). The field strength is fixed in this figure, and the dashed lines simply show the energy changes incurred by adding a new term in the Hamiltonian. In order to determine the energy of the hydrogen atom in a magnetic field, we employ a basis set for this Hamiltonian [equation (9–4)] that consists of the four possible electron and nuclear spin functions. Such a basis set is $\varphi_1 = |\alpha_e\alpha_N\rangle$, $\varphi_2 = |\alpha_e\beta_N\rangle$, $\varphi_3 = |\beta_e\alpha_N\rangle$ and $\varphi_4 = |\beta_e\beta_N\rangle$. Let us begin by calculating the energies arising from the first two terms in the Hamiltonian, $\hat{H}_0$. We must solve the simultaneous equations $\langle\varphi_n|\hat{H}_0|\varphi_m\rangle - E\langle\varphi_n|\varphi_m\rangle = 0$, where n and m may or may not be equal. Thus, the 4×4 secular determinant in this basis set contains diagonal terms of the type:

$$\begin{aligned}&\langle\alpha_e\beta_N|g\beta H\hat{S}_z - g_N\beta_N H\hat{I}_z|\alpha_e\beta_N\rangle - E\langle\alpha_e\beta_N|\alpha_e\beta_N\rangle\\ &\qquad = \langle\alpha_e\beta_N|g\beta H\hat{S}_z|\alpha_e\rangle\cdot|\beta_N\rangle - \langle\alpha_e\beta_N|g_N\beta_N H\hat{I}_z|\beta_N\rangle\cdot|\alpha_e\rangle - E = 0.\end{aligned}$$

Since for this problem the operators are $\hat{I}_z$ and $\hat{S}_z$, α and β are eigenfunctions; *i.e.*, $\hat{S}_z|\alpha_e\rangle = \frac{1}{2}\alpha_e$, $\hat{S}_z|\beta_e\rangle = -\frac{1}{2}\beta_e$, $\hat{I}_z|\alpha_N\rangle = \frac{1}{2}\alpha_N$, and $\hat{I}_z|\beta_N\rangle = -\frac{1}{2}\beta_N$. Furthermore, $\hat{S}_z$ does not operate on the nuclear spin function and $\hat{I}_z$ does not operate on the electron spin function, leading to:

$$\langle\alpha_e\beta_N|\hat{H}_0|\alpha_e\beta_N\rangle - E\langle\alpha_e\beta_N|\alpha_e\beta_N\rangle = \frac{1}{2}g\beta H + \frac{1}{2}g_N\beta_N H - E = 0$$

It should now be clear that all of the off-diagonal elements will be zero with this Hamiltonian, for all of the off-diagonal elements are of the form $\langle\varphi_n|H_0|\varphi_m\rangle - \langle\varphi_n|\varphi_m\rangle$, which equals zero when $n \neq m$. Since the Hamiltonian matrix is diagonal,

* H_x and H_y are zero and $H = H_z$. The effects of S_x, S_y, I_x, and I_y are not necessarily zero.

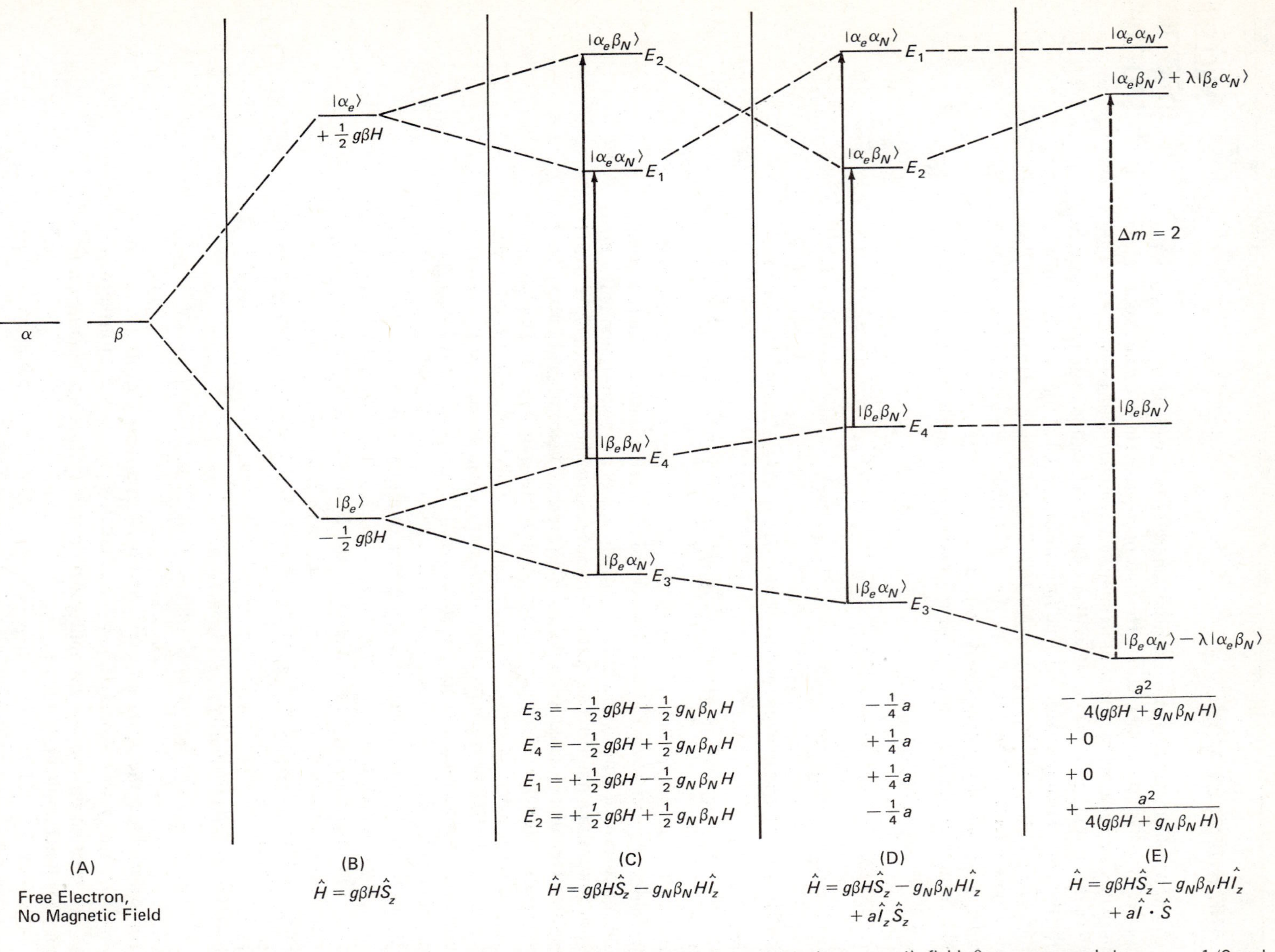

FIGURE 9–2. The influence of various terms in the Hamiltonian on the energy of a hydrogen atom in a magnetic field. $\beta_e\alpha_N$ corresponds to $m_S = -1/2$ and $m_I = +1/2$; $\beta_e\beta_N$ to $m_S = -1/2$ and $m_I = -1/2$; $\alpha_e\beta_N$ to $m_S = +1/2$ and $m_I = -1/2$; and $\alpha_e\alpha_N$ to $m_S = +1/2$ and $m_I = +1/2$.

the determinant is already factored and we get the four energies directly, as shown above for $\alpha_e\beta_N$. They are indicated for E_1, E_2, E_3, and E_4 in Fig. 9–2(C). These results can be verified for practice. The usual selection rules in epr are $\Delta m_I = 0$ and $\Delta m_S = \pm 1$. It will be noticed that the two esr transitions ($\Delta m_I = 0$) illustrated in Fig. 9–2(C) have the same energy. Considering only the first two terms of the Hamiltonian, the epr spectrum of the hydrogen atom would be the same as that of a free electron, *i.e.*, one line at a field $h\nu/g\beta$ or $g = 2.0023$.

Next, we must concern ourselves with the $a\hat{I}\cdot\hat{S}$ term in the Hamiltonian. This term describes the coupling of the electron and nuclear spin moments, which classically corresponds to the dot product of these two vectors. The quantity a indicates the magnitude of the interaction and has the dimensions of energy. This is referred to as the Fermi contact contribution to the coupling, and its magnitude depends upon the amount of electron density at the nucleus, $\psi_{(0)}{}^2$, according to:

$$a = \frac{8\pi}{3} g\beta g_N\beta_N|\psi_{(0)}|^2 \tag{9-5}$$

For a hydrogen atom, the Slater orbital function is $\psi_{1s} = (1/\pi a_0{}^3)\exp(-r/a_0)$ where a_0, the Bohr radius, equals $h^2/me^2 = 0.52918$ Å. Substituting into equation (9–5), the value of ψ for a hydrogen s-orbital at $r = 0$ yields $a/h = 1422.74$ MHz. Since the nuclear hyperfine interaction we are talking about involves the dot product of the nuclear and spin moments, it has x, y, and z components, so

$$a\hat{I}\cdot\hat{S} = a(\hat{I}_x\hat{S}_x + \hat{I}_y\hat{S}_y + \hat{I}_z\hat{S}_z) \tag{9-6}$$

The elements generated by the $\hat{I}_z\hat{S}_z$ term operating on the basis $\alpha_e\alpha_N$, etc., are again only diagonal elements because $\langle\varphi_n|a\hat{I}_z\hat{S}_z|\varphi_m\rangle = 0$ when $m \neq n$; for example, $\langle\alpha_e\beta_N|a\hat{I}_z|\alpha_N\rangle = 0$. The following results are obtained for the diagonal elements:

$$\langle\alpha_e\alpha_N|a\hat{S}_z\hat{I}_z|\alpha_e\alpha_N\rangle = \frac{1}{4}a$$

$$\langle\alpha_e\beta_N|a\hat{S}_z\hat{I}_z|\alpha_e\beta_N\rangle = -\frac{1}{4}a$$

$$\langle\beta_e\alpha_N|a\hat{S}_z\hat{I}_z|\beta_e\alpha_N\rangle = -\frac{1}{4}a$$

$$\langle\beta_e\beta_N|a\hat{S}_z\hat{I}_z|\beta_e\beta_N\rangle = \frac{1}{4}a$$

These have to be added to the energies in Fig. 9–2(C), modifying the energies as shown in Fig. 9–2(D). The contributions of $\pm\frac{1}{4}a$ to E_1, E_2, etc., from $a\hat{I}_z\hat{S}_z$ are indicated at the bottom of (D). Now we see by looking at the arrows for the two electron spin changes that *the transition energies are no longer equal.* One transition gives rise to a spectral peak at lower energy than that corresponding to $g = 2.0023$ [see Fig. 9–2(D)] by ($\frac{1}{2}a$), and the other occurs at an energy that is higher by ($\frac{1}{2}a$). The energy separation of the two peaks is a.

To complete the problem, we now have to add the effects of $\hat{I}_x\hat{S}_x$ and $\hat{I}_y\hat{S}_y$. This is best done in terms of the raising and lowering operators, which work in a similar fashion to $\hat{I}_+$ and $\hat{I}_-$ that were discussed earlier (p. 271). For the electron spin operators, we define:

$$\hat{S}_+ = \hat{S}_x + i\hat{S}_y$$
$$\hat{S}_- = \hat{S}_x - i\hat{S}_y$$

Thus $$\hat{S}_+\hat{I}_- = (\hat{S}_x\hat{I}_x + \hat{S}_y\hat{I}_y) + i(\hat{S}_y\hat{I}_x - \hat{S}_x\hat{I}_y)$$

and $$\hat{S}_-\hat{I}_+ = \hat{S}_x\hat{I}_x + \hat{S}_y\hat{I}_y - i(\hat{S}_y\hat{I}_x - \hat{S}_x\hat{I}_y).$$

Combining these equations, we see that:

$$\hat{S}_x\hat{I}_x + \hat{S}_y\hat{I}_y = \frac{1}{2}(\hat{S}_+\hat{I}_- + \hat{S}_-\hat{I}_+)$$

The following results are obtained by analogy to our earlier discussion of $\hat{I}_+$ and $\hat{I}_-$:

$$\hat{S}_+\hat{I}_-|\beta_e\alpha_N\rangle = |\alpha_e\beta_N\rangle$$
$$\hat{S}_-\hat{I}_+|\alpha_e\beta_N\rangle = |\beta_e\alpha_N\rangle$$
$$\hat{S}_-\hat{I}_+|\alpha_e\alpha_N\rangle = 0$$

All other operations of $\hat{S}_-\hat{I}_+$ or $\hat{S}_+\hat{I}_-$ upon the basis set produce zero. Thus, if we consider the 4×4 matrix shown in Fig. 9-3, the only nonvanishing matrix elements from $\hat{S}_+\hat{I}_-$ and $\hat{S}_-\hat{I}_+$ are

$$\langle\alpha_e\beta_N|a\hat{S}_+\hat{I}_-|\beta_e\alpha_N\rangle = a$$
$$\langle\beta_e\alpha_N|a\hat{S}_-\hat{I}_+|\alpha_e\beta_N\rangle = a$$

	$\lvert\alpha_e\alpha_N\rangle$	$\lvert\alpha_e\beta_N\rangle$	$\lvert\beta_e\alpha_N\rangle$	$\lvert\beta_e\beta_N\rangle$
$\lvert\alpha_e\alpha_N\rangle$	$\frac{1}{2}(g\beta H - g_N\beta_N H) + \frac{1}{4}a - E$	0	0	0
$\lvert\alpha_e\beta_N\rangle$	0	$\frac{1}{2}(g\beta H + g_N\beta_N H) - \frac{1}{4}a - E$	$\frac{1}{2}a$	0
$\lvert\beta_e\alpha_N\rangle$	0	$\frac{1}{2}a$	$-\frac{1}{2}(g\beta H + g_N\beta_N H) - \frac{1}{4}a - E$	0
$\lvert\beta_e\beta_N\rangle$	0	0	0	$-\frac{1}{2}(g\beta H - g_N\beta_N H) + \frac{1}{4}a - E$

FIGURE 9-3. Secular determinant for the field-free hydrogen atom.

The matrix element

$$\langle\alpha_e\beta_N|a\hat{S}_x\hat{I}_x + a\hat{S}_y\hat{I}_y|\beta_e\alpha_N\rangle = \frac{1}{2}a$$

and $$\langle\beta_e\alpha_N|a\hat{S}_x\hat{I}_x + a\hat{S}_y\hat{I}_y|\alpha_e\beta_N\rangle = \frac{1}{2}a$$

We can summarize this entire section by completing the full determinant for the original spin Hamiltonian, equation (9-4), operating on the φ basis set to give energies $\langle\varphi_m|\hat{H}|\varphi_n\rangle = E\langle\varphi_m|\varphi_n\rangle$. The determinant shown in Fig. 9-3 equals zero. Note that it is block-diagonal so that two of the energies, E_1 and E_4, are obtained directly. We also see that $\hat{I}_x\hat{S}_x$ and $\hat{I}_y\hat{S}_y$ lead to off-diagonal elements that mix φ_2 and φ_3. A perturbation theory solution* of the resulting 2×2 determinant gives (to second order):

$$E_2 = \frac{1}{2}g\beta H + \frac{1}{2}g_N\beta_N H - \frac{1}{4}a + \frac{a^2}{4(g\beta H + g_N\beta_N H)}$$

$$E_3 = -\left(\frac{1}{2}g\beta H + \frac{1}{2}g_N\beta_N H\right) - \frac{1}{4}a - \frac{a^2}{4(g\beta H + g_N\beta_N H)}$$

*The exact solution is $E = -\frac{1}{4}a \pm \frac{1}{2}[(g\beta + g_N\beta_N)^2H^2 + a^2]^{1-2}$.

We can see in Fig. 9–2(E), where the effects of these off-diagonal elements on the energy levels are illustrated, that the energies of both transitions are increased by the same amount. Since the off-diagonal elements are small compared to the diagonal, effects arising from this part of the Hamiltonian are referred to as second order effects. Second order effects thus have no influence on the value of a read off the spectrum, but will change the value you read off the spectrum for g. A more interesting contribution to the spectral appearance is that the previously forbidden transition, $E_3 \rightarrow E_2$ (the simultaneous electron and nuclear spin flip), now becomes allowed because of the mixing of the basis set.*

9–3 PRESENTATION OF THE SPECTRUM

As in nmr, the epr spectrum can be represented by plotting intensity, I, against the strength of the applied field; but epr spectra are commonly presented as derivative curves, *i.e.*, the first derivative (the slope) of the absorption curve is plotted against the strength of the magnetic field. It is easier to discern features in a derivative presentation if the absorption lines are broad. The two modes of presentation are easily interconverted, and the relationship between the two kinds of spectra is illustrated in Fig. 9–4. In (A), a single absorption peak with no fine structure is represented; (B) is the derivative curve that corresponds to (A). The derivative curve crosses the abscissa at a maximum in the absorption curve, for the slope changes sign at a maximum. Curve (C) is the absorption counterpart of curve (D). Note that the shoulders in (C) never pass through a maximum and as a result the derivative peaks in (D) corresponding to these shoulders do not cross the abscissa. The number of peaks and shoulders in the absorption curve can be determined from the number of minima (marked with an asterisk in (D)) or maxima in the derivative curve.

The epr spectrum of the hydrogen atom is illustrated in Fig. 9–5. To a good approximation, the g value is measured at x, which is midway between the two solid circles corresponding to absorption peak maxima. The hyperfine splitting, $a/g\beta$, is the separation between the solid circles in gauss. The sign of a generally cannot be obtained directly from the spectrum. The splitting indicated in Fig. 9–2 implies a

*This allowedness is derived in Carrington and McLachlan (see General References) on page 22.

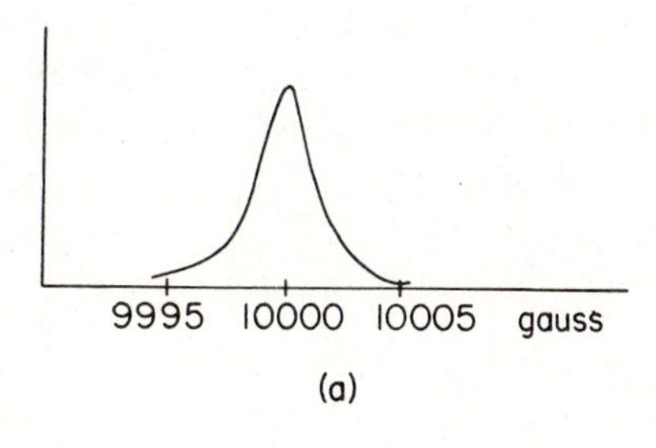

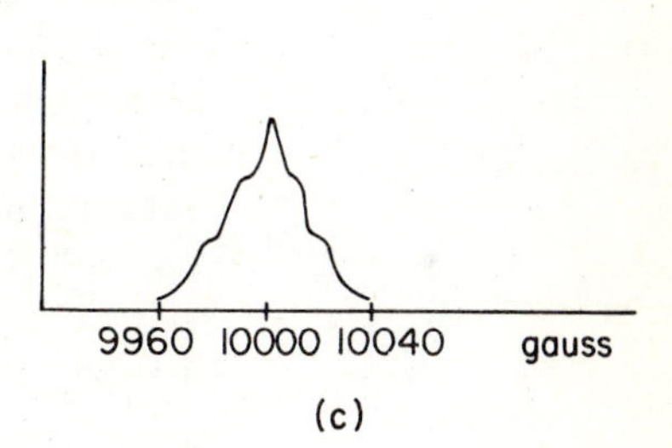

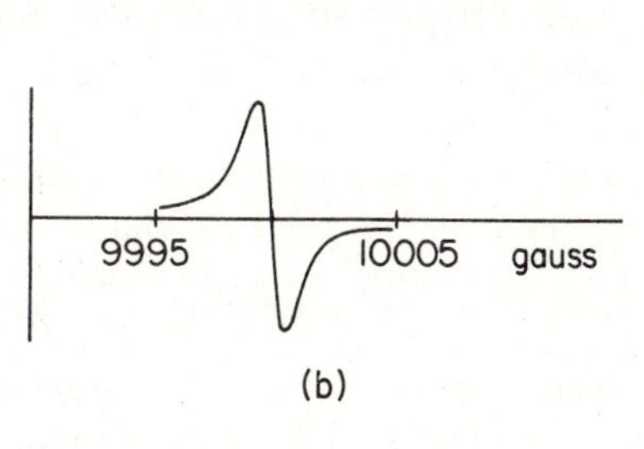

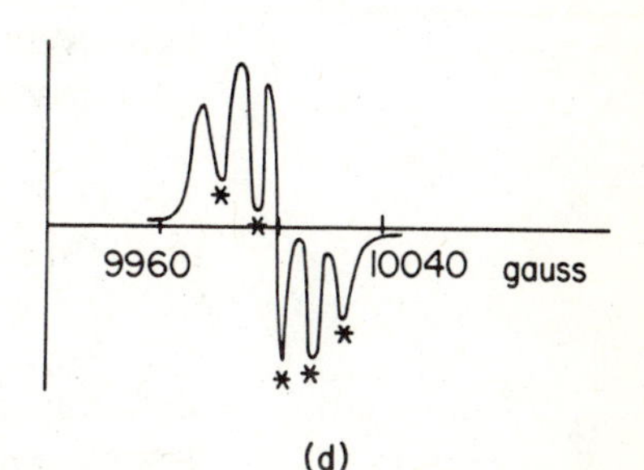

FIGURE 9–4. Comparison of spectral presentation as absorption (A and C) and derivative (B and D) curves.

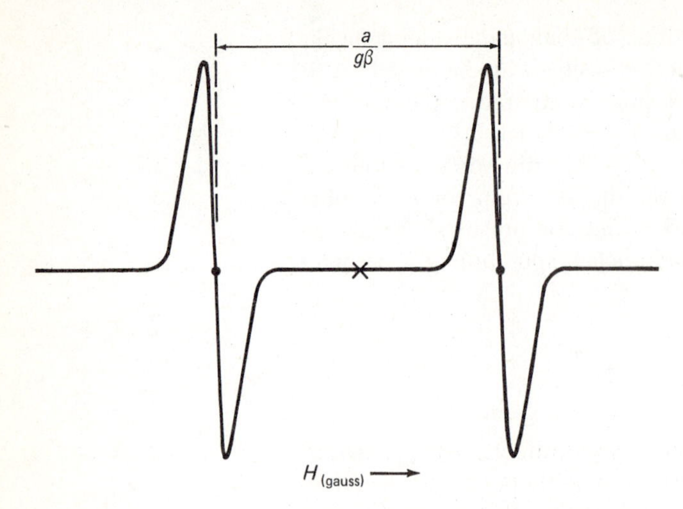

FIGURE 9-5. The epr spectrum of a hydrogen atom (β has units of ergs/gauss).

positive value for a. If a were negative, then $m_s = -\frac{1}{2}$ and $m_I = -\frac{1}{2}$; that is, $(\beta_e\beta_N)$ would be the low energy state.

The epr spectrometer is designed to operate at a fixed microwave source frequency. The magnetic field is swept, and the horizontal axis in Fig. 9-5 is in units of gauss. One can set the field at any position using the field dial and sweep from that place. For a fingerprint type of identification, greater accuracy is needed than can be obtained using the instrument dials. For this purpose, an external standard, diphenylpicrylhydrazide, DPPH, is used together with a microwave frequency counter. *DPPH has a g-value of 2.0037 ± 0.0002.* The field sweep is assumed to be linear, and the g values of other peaks are calculated relative to this standard. The field axis is in units of gauss, and the g value is reported as a dimensionless quantity using

$$g = \frac{h\nu}{\beta H}$$

where ν is the fixed frequency of the probe and H (which is being swept) is obtained from the spectrum. A frequency counter should be used to measure the probe frequency ν.

The value of a is sometimes reported in units of gauss, MHz, or cm^{-1}. It is to be emphasized that the line separation in a spectrum in units of gauss is given by $a/g\beta$ (where a has units of ergs and β has units of ergs G^{-1}). When $g \neq 2$, it is incorrect to report this separation as a in units of gauss. One would have to multiply the line separation by $g\beta$ and divide by $g_e\beta$ (where g_e is the free electron value of 2.0023193) to report a correct value for a in gauss. Since a is an energy, it is best to report its value as an energy. This is simply done by multiplying the line separation in gauss by $g\beta$, with β in units of cm^{-1} G^{-1}. There is no g-value dependence for this unit. The value of a in MHz is obtained by multiplying $a(cm^{-1})$ by $c(3 \times 10^{10}$ cm $sec^{-1})$ and dividing by 10^6.

9-4 HYPERFINE SPLITTINGS IN ISOTROPIC SYSTEMS INVOLVING MORE THAN ONE NUCLEUS

The first order energies of the levels in the hydrogen atom are given by equation (9-7), which ignores the small nuclear Zeeman interaction.

$$E = g\beta H m_s + a m_s m_I \tag{9-7}$$

Substituting the values of m_s and m_I into this equation enables one to reproduce the energies given in Fig. 9-2(D). For a nucleus with any nuclear spin, the projection of the nuclear magnetic moment along the effective field direction at the nucleus can take any of the $2I + 1$ values corresponding to the quantum numbers $-I$, $-I + 1$, $\ldots, I - 1, I$. These orientations give rise to $2I + 1$ different nuclear energy states (one for every value of m_I); and when each of these couples with the electron moment, $2I + 1$ lines result in the epr experiment. Since these energy differences are small, all levels with the same m_s value are equally populated for practical purposes and the epr absorption lines will usually be of equal intensity and equal spacing. For example, three lines are expected for an unpaired electron of ^{14}N, where $I = 1$.

We shall next consider the effect on the spectrum when the electron interacts with (*i.e.*, is delocalized onto) several nuclei. For simplicity, assume that the species is rotating very rapidly in all directions, and that the g value is close to the free electron value. As an example, the methyl radical will be discussed. As illustrated in Fig. 9–6, addition of the nuclear spin angular momentum quantum numbers of the individual protons results in four different values for the total nuclear spin moment, M_I. As indicated in Fig. 9–7, this gives rise to four transitions ($\Delta M_I = 0$, $\Delta m_s = \pm 1$). Since there are three different ways to obtain a total of $M_I = +\frac{1}{2}$ or $-\frac{1}{2}$ (see Fig. 9–6), but only one way to obtain $M_I = +\frac{3}{2}$ or $-\frac{3}{2}$, the former system is three times more probable than the latter and the observed relative intensities for the corresponding transitions (Fig. 9–7) are in the ratio 1:3:3:1.

			M_I
	↑↑↑		$+\frac{3}{2}$
↑↑↓	↑↓↑	↓↑↑	$+\frac{1}{2}$
↓↓↑	↓↑↓	↑↓↓	$-\frac{1}{2}$
	↓↓↓		$-\frac{3}{2}$

FIGURE 9–6. Possible nuclear spin arrangements of the protons in a methyl radical.

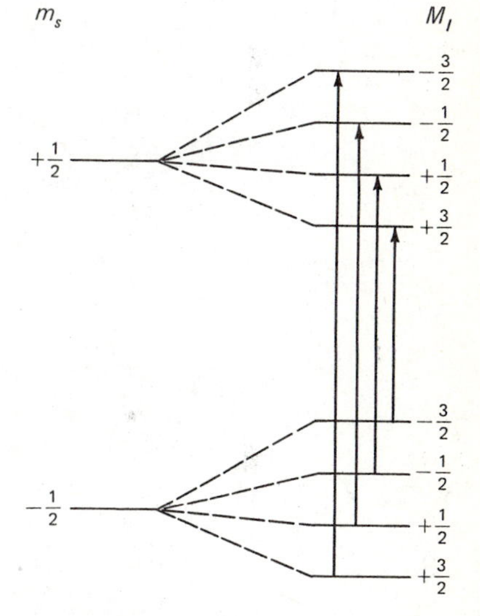

FIGURE 9–7. The four transitions that occur in the epr spectrum of the methyl radical (see Fig. 9–8 for the spectrum). (As with the H atom, the $+m_I$ state is lowest for $m_s = -\frac{1}{2}$ and the $-m_I$ state is lowest for $m_s = +\frac{1}{2}$, from the $\hat{I}\cdot\hat{S}$ term.)

In general, when the absorption spectrum is split by n *equivalent* nuclei of equal spin I_i, the number of lines is given by $2nI_i + 1$. When the splitting is caused by both a set of n equivalent nuclei of spin I_i and a set of m equivalent nuclei of spin I_j, the number of lines is given by $(2nI_i + 1)(2mI_j + 1)$. The following specific cases illustrate the use of these general rules.

(1) If a radical contains n non-equivalent protons onto which the electron is delocalized, a spectrum consisting of 2^n lines will arise.

(2) If the odd electron is delocalized over a number, n, of equivalent protons, a total of $n + 1$ lines, $(2nI + 1)$, will appear in the spectrum. This number is less than the number of lines expected for non-equivalent protons (*i.e.*, 2^n) because several of the possible arrangements of the nuclear spins are degenerate (see Fig. 9–6). The spectrum of the methyl radical illustrated in Fig. 9–8 contains the four peaks expected from these considerations.

The spectra expected for differing numbers of equivalent protons can easily be predicted by considering the splitting due to each proton in turn, as illustrated in Fig. 9–9. When the signal is split by two equivalent protons, the total M_I for the three levels can have the values $\Sigma m_I = +1, 0$, and -1. Since there are two ways in which

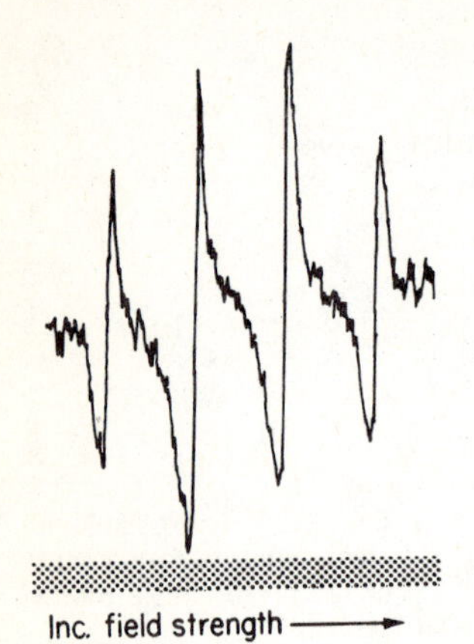

FIGURE 9–8. The derivative spectrum of the methyl radical in a CH_4 matrix at 4.2° K.

we can arrange the separate m_I's to give a total $M_I = 0$ (namely $+\frac{1}{2}$, $-\frac{1}{2}$, and $-\frac{1}{2}$, $+\frac{1}{2}$), the center level is doubly degenerate. Three peaks are observed in the spectrum ($\Delta M_I = 0$, $\Delta m_S = \pm 1$), and the intensity ratio is 1:2:1. The case of three protons (*e.g.*, the methyl radical) was discussed above, and similar considerations are employed for the systems represented in Fig. 9–9 with more than three protons.

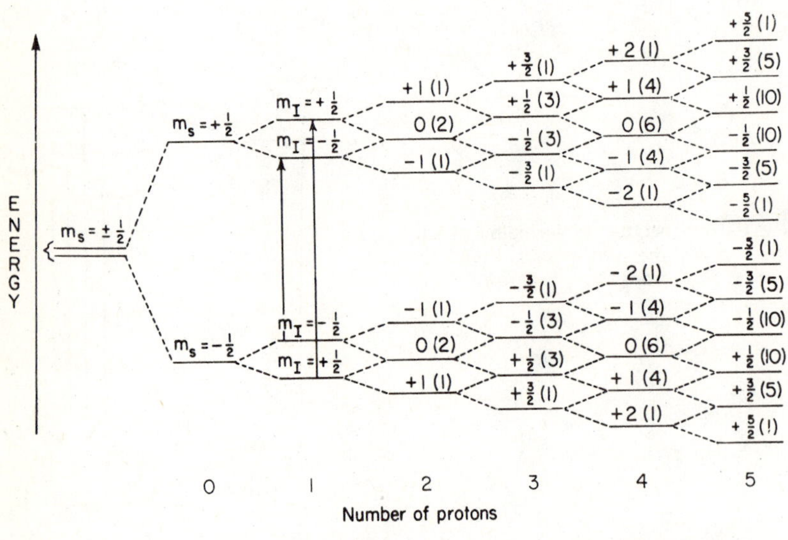

FIGURE 9–9. Hyperfine energy levels resulting from interaction of an unpaired electron with varying numbers of equivalent protons.[2] Each number in parentheses gives the degeneracy of the level to which it refers, and hence the relative peak intensities for the corresponding transitions.

The relative intensities of the peaks are given by the coefficients of the binomial expansion. It should be remembered in applying this formula that it is restricted to equivalent protons or other nuclei having $I = \frac{1}{2}$.

(3) If the odd electron is delocalized over two sets of non-equivalent protons, the number of lines expected is the product of the number expected for each set $[(2nI_i + 1)(2mI_j + 1)]$. The naphthalene negative ion, which can be prepared by adding sodium to naphthalene, contains an odd electron that is delocalized over the entire naphthalene ring. Naphthalene contains two different sets of four equivalent protons. A total of $n + 1$, or five peaks, is expected for an electron delocalized on either set of four equivalent protons. In the naphthalene negative ion, the two sets of four equivalent protons should give a total of twenty-five lines in the epr spectrum. This is found experimentally.

(4) If the electron is delocalized on nuclei with spin greater than $\frac{1}{2}$, a procedure similar to that for protons can be applied to calculate the number of peaks expected. If the electron is delocalized over several equivalent nuclei that have spins greater than $\frac{1}{2}$, the number of peaks expected in the spectrum is predicted from the formula

$2nI + 1$. For example, five peaks are expected for an electron delocalized on two equivalent nitrogen atoms. A procedure similar to that in Fig. 9–6 shows that the intensities of the five peaks will be in the ratio 1:2:3:2:1.

(5) If the electron is delocalized over several non-equivalent atoms, the total number of peaks expected is obtained by taking the product of the number expected for each atom. The scheme illustrated in Fig. 9–10 for an electron delocalized onto

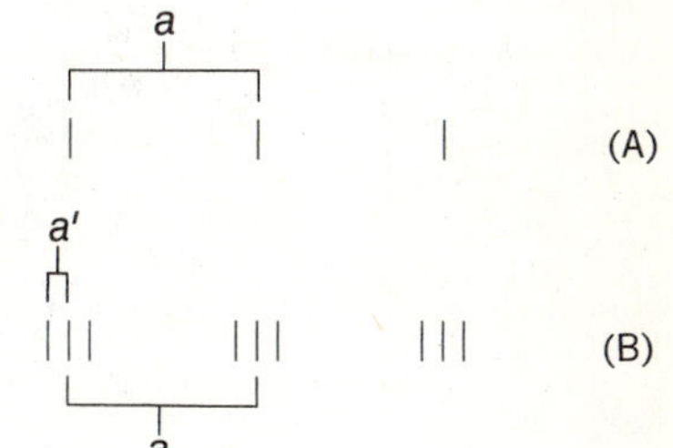

FIGURE 9–10. (A) Three lines expected from an electron on a nucleus with $I = 1$, and (B) nine lines resulting from the splitting by a second non-equivalent nucleus with $I = 1$.

two non-equivalent nuclei with $I = 1$ is often employed to indicate the splitting expected. The three lines in (A) represent splitting of an epr peak by a nucleus with $I = 1$ and a hyperfine coupling constant a. Each of these lines is split into three components as a result of delocalization of the electron on a second, non-equivalent nucleus with $I = 1$, and a hyperfine coupling constant a', producing in (B) a total of nine lines. In subsequent discussions a scheme similar to that in Fig. 9–10 will be employed for the interpretation of spectra. The shape of the spectrum and the separations of the peaks will depend upon the resonant field, g, and the coupling constants a and a'. Frequently the measured spectrum will not reveal all the lines expected, because the line widths are large compared to $a/g\beta$ and two close lines are not resolved. For example, the spectrum in Fig. 9–11 could result for the hypothetical radical $H—\dot{X}^+ \leftrightarrow \dot{H}—X^+$ where $I = 1$ for X. The two lines in (A) result from the proton splitting. In (B) each line in turn is split into three components owing to interaction with nucleus X; thus, we would expect six lines, all of equal intensity. However, it is possible to detect only five lines, if the two innermost components are not resolved. They would give rise to a single peak with twice the area of the other peaks (see Fig. 9–11).

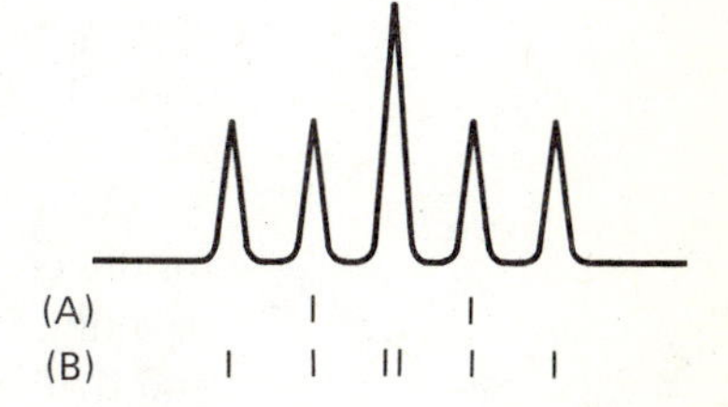

FIGURE 9–11. Hypothetical absorption spectrum for the radical $H \cdot X^+$ ($I = 1$ for X).

The epr spectrum[3] of bis-salicylaldimine copper(II) in Fig. 9–12 is an interesting example summarizing this discussion of nuclear coupling. This spectrum was obtained on a solid and is not isotropic; this aspect will be discussed shortly. Four main groups of lines result from coupling of the ^{63}Cu nucleus ($I = \frac{3}{2}$) with the electron. The hyperfine structure in each of the four groups consists of eleven peaks of intensity ratio 1:2:3:4:5:6:5:4:3:2:1. These peaks result from splitting by the two equivalent nitrogens and two hydrogens, H′ in Fig. 9–12. The total number of peaks expected is fifteen; $(2n_N I_N + 1)(2n_H I_H + 1) = 5 \times 3 = 15$. The eleven peaks found for each subgroup in the actual spectrum result from overlap of some of the fifteen peaks as indicated in Fig. 9–13. The line for an electron not split by a nucleus is

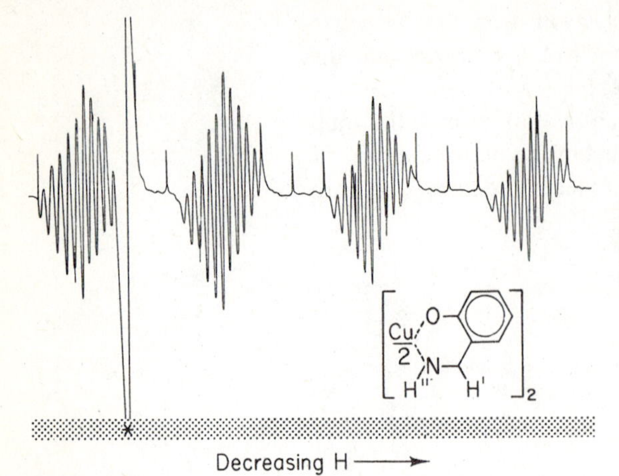

FIGURE 9-12. EPR derivative spectrum of bis-salicylaldimine copper(II) with isotopically pure ^{63}Cu. Asterisk indicates calibration peak from DPPH. From A. H. Maki and B. R. McGarvey, J. Chem. Phys., *29*, 35 (1958).

shown in (A). The splittings by the two equivalent nitrogens are indicated in (B) and the subsequent splitting by two equivalent protons is indicated in (C). The two nitrogens split the resonance into five peaks of relative intensity 1:2:3:2:1. These

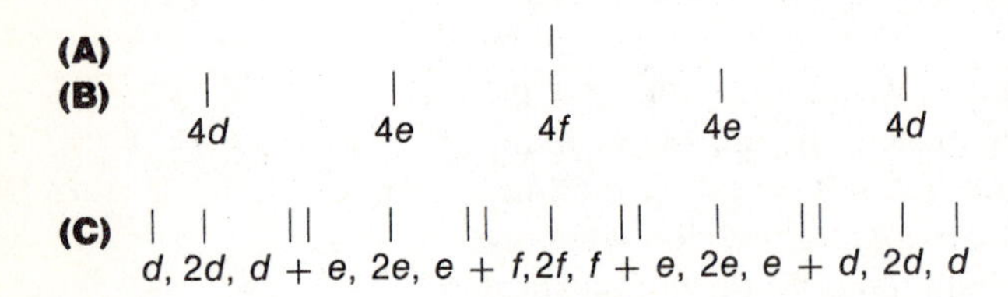

FIGURE 9-13. Interpretation of the epr spectrum of bis-salicylaldimine copper(II). (A) An unsplit transition. (B) Splitting by two equivalent nitrogens, with d, e, and f indicating relative intensities of 1, 2, and 3. (C) Further splitting by two equivalent protons.

values are denoted in (B) by $4d$, $4e$, and $4f$ where the intensities correspond to $d = 1$, $e = 2$, and $f = 3$. The splitting by two equivalent protons will give rise to three lines for each line in (B), with an intensity ratio of 1:2:1. The intensities indicated by letters underneath the lines in (C) result from the summation of the expected intensities. Since the relative intensities are $d = 1$, $e = 2$, and $f = 3$, the ratio of the intensity of the bands in (C) is 1:2:3:4:5:6:5:4:3:2:1. The experimental spectra agree with this interpretation, which is further substantiated by the following results:

(1) Deuteration of the N—H″ groups (see Fig. 9-12) produced a compound which gave an identical spectrum.

(2) When the H′ hydrogens were replaced by methyl groups, the epr spectrum for this compound consisted of four main groups, each of which consisted of five lines resulting from nitrogen splitting only. The hyperfine splitting by the N—H″ proton and that by the protons on the methyl group are either too small to be detected or nonexistent.

This spectrum furnishes conclusive proof of the delocalization of the odd electron in this complex onto the ligand. This can be interpreted only as covalence in the metal-ligand interaction, for only by mixing the metal ion and ligand wave functions can we get ligand contributions to the molecular orbital in the complex that contains the unpaired electron.

Another interesting application of epr[4] involves the spectrum of $Co_3(CO)_9Se$, whose structure and epr spectrum are shown in Fig. 9-14. The 22-line spectrum indicates that the one unpaired electron in this system is completely delocalized over the three cobalt atoms ($I_{Co} = 7/2$). This in effect gives rise to an oxidation state of $+2/3$ for each cobalt atom.

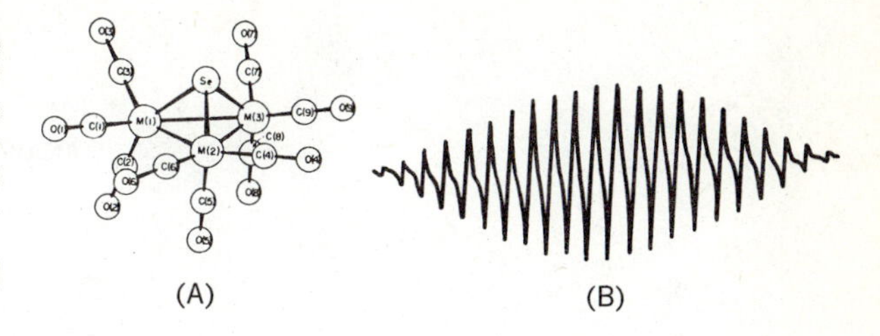

FIGURE 9–14. (A) Basic molecular geometry of the $Co_3(CO)_9Se$ complex. (B) The epr spectrum of a single crystal of $FeCo_2(CO)_9Se$ doped with about 0.5% of paramagnetic $Co_3(CO)_9Se$. This spectrum containing 22 hyperfine components was recorded at 77°K with the molecular threefold axis parallel to the magnetic field direction. [Reprinted with permission from C. E. Strouse and L. F. Dahl, J. Amer. Chem. Soc., 93, 6032 (1971). Copyright by the American Chemical Society.]

9–5 CONTRIBUTIONS TO THE HYPERFINE COUPLING CONSTANT IN ISOTROPIC SYSTEMS

In Equation (9–5), we saw that:

$$a = \frac{8\pi}{3} g\beta g_N \beta_N |\psi_0|^2$$

In a molecule, the hyperfine splitting from delocalization of unpaired spin density $\rho(r_N)$ onto a hydrogen atom of the molecule is given by equation (9–8):

$$a = \frac{8\pi}{3} g\beta g_N \beta_N \rho_H(r_N) \qquad (9\text{–}8)$$

where $\rho(r_N)$ can be crudely thought of as the difference in the average numbers of electrons at the nucleus that have spin moments characterized by $m_s = +\frac{1}{2}$ and $-\frac{1}{2}$. When there is an excess of $+\frac{1}{2}$ spin density (*i.e.*, the electron spin moment opposed to the field) at the nucleus, this nucleus is said to experience *negative spin density*. An excess of electron density with $m_s = -\frac{1}{2}$ is called *positive spin density*. Positive spin density is represented by an arrow that is aligned with the external field, and negative spin density is represented by one opposed to the field. To further complicate matters, positive spin density is often referred to as α spin in the literature, even though the wave function for evaluation of the matrix elements for this electron is represented by β_e. Thus, the common convention for labeling spin density is exactly the opposite of that used to label the electron spin wave functions. We will avoid the α and β labels of spin density and use these as symbols for the spin wave functions in this book.

We should also be careful to point out that the amount of unpaired spin density on an atom in the molecule does not correspond directly to the atom contributions in the molecular orbital containing the unpaired electron. We shall refer to the latter effect as *unpaired electron density*. An unpaired electron in an orbital of one atom in a molecule can polarize the paired spins in an orthogonal sigma bond so that one of the electrons is more often in the vicinity of one atom than in the vicinity of the other. This puts unpaired spin density at the nucleus of the atom even though there is no unpaired electron density delocalized onto it. We can make this more specific with the following example.

Some of the first attempts at interpreting hyperfine couplings involved aromatic radicals with the unpaired spin in the π-system, *e.g.*, $C_6H_5NO_2^-$. Hückel calculations were carried out and the squares of the various carbon p_z coefficients in the m.o. containing the unpaired electron were employed to give the amount of unpaired electron density on the various carbon atoms. The hyperfine splittings observed experimentally were from the ring hydrogens, which are orthogonal to the π-system. No unpaired electron density can be delocalized *directly* onto them, but unpaired spin density is felt at the hydrogen nucleus by the so-called *spin-polarization* or *indirect mechanism*. We shall attempt to give a grossly simplified view of this effect using a

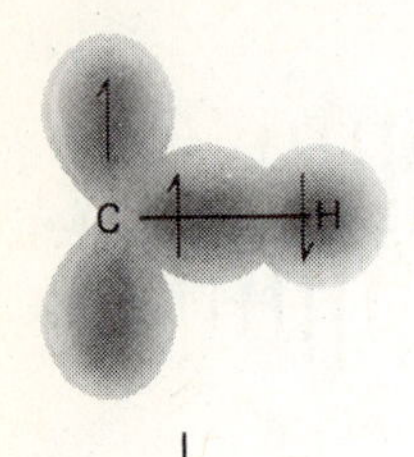

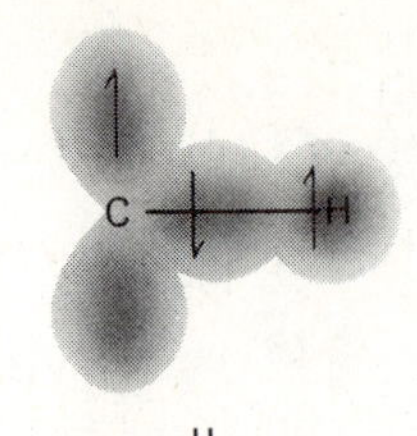

FIGURE 9–15. Resonance forms for a C—H sigma bond fragment with an unpaired electron on the carbon.

valence bond formalism. Consider the two resonance forms in Fig. 9–15 for a C—H bond in a system with an unpaired electron in a carbon p_π orbital. In the absence of any interaction between the π and σ systems, the so-called perfect pairing approximation, we can write a valence bond description of the bonding and antibonding sigma orbital wave functions:

$$\psi^\circ = \frac{1}{\sqrt{2}}(\psi_{\text{I}} + \psi_{\text{II}}) \qquad \text{and} \qquad \psi^* = \frac{1}{\sqrt{2}}(\psi_{\text{I}} - \psi_{\text{II}})$$

Here ψ_{I} and ψ_{II} represent wave functions for structures I and II in Fig. 9–15, which we shall not attempt to specify in terms of an a.o. and spin basis set. When interaction of the π and σ systems is considered, we find that I is a more stable structure than II and, accordingly, it contributes to the ground state more than does II, resulting in valence bond functions:

$$\psi^\circ = a\psi_{\text{I}} + b\psi_{\text{II}} \qquad (a > b)$$

and

$$\psi^* = a'\psi_{\text{I}} - b'\psi_{\text{II}} \qquad (a' < b')$$

This in effect polarizes the electrons in the C—H sigma bond, leaving spin density on the hydrogen that is opposite to the unpaired spin density in the carbon p_z orbital. The stabilizing feature of structure I over structure II is the *electron exchange interaction.* It is the same effect that causes the lowest-energy excited state for helium, $1s^12s^1$, to be a triplet instead of a singlet. If we label the two electrons that are mainly on carbon in each of the structures in Fig. 9–15, we see that they can be interchanged in structure I,

$$\overset{a\uparrow}{\underset{b}{\text{C}\uparrow\downarrow\text{H}}} \qquad \overset{b\uparrow}{\underset{a}{\text{C}\uparrow\downarrow\text{H}}}$$

but not in structure II. Thus, I is stabilized by a quantum mechanical interaction analogous to resonance.

A more complete molecular orbital description of this effect is presented in Dewar's text.[5] In treating the e^2/r_{ij} interactions in a molecule, one type of integral that is obtained describes a repulsive interaction and has the form

$$J_{mn} = \int\int \psi_m{}^*(i)\,\psi_m(i)\frac{e^2}{r_{ij}}\psi_n{}^*(j)\psi_n(j)\,d\tau_i\,d\tau_j \tag{9-9}$$

This represents the Coulomb repulsion of the electron density [$\int \psi_m{}^*(i)\psi_m(i)\,d\tau_i$ is the density in i] from electrons i and j in orbitals m and n, where m may or may not equal

n. There are other integrals that we shall label K_{mn}, which are zero when the electron spins are paired and are non-zero when the spins are parallel. The K_{mn} integrals are:

$$K_{mn} = \int\int \psi_m{}^*(i)\psi_n(i)\frac{e^2}{r_{ij}}\psi_m(j)\psi_n{}^*(j)\, d\tau_i\, d\tau_j \tag{9-10}$$

This is called an exchange integral because it corresponds to an exchange of the orbitals containing electrons i and j. When the spins of the two electrons are parallel, the squares of these wave functions show that the two electrons have a drastically reduced probability of being near each other when compared to two electrons with opposite spins. Thus, the Coulomb repulsion is decreased when the spins are parallel because the electrons stay away from each other. The magnitude of the K_{mn} integral depends upon the overlap of the two orbitals m and n. The quantity $\psi_m(j)\psi_n(j)\, d\tau_j$ was referred to earlier as the differential overlap. The overlap integral of the differential overlap of two orthogonal orbitals is zero; *i.e.*,

$$\int \psi_m(j)\psi_n{}^*(j)\, d\tau_j = 0$$

However, when the differential overlap in the volume element $d\tau_j$ is operated upon by e^2/r_{ij}, multiplied by $\psi_m{}^*(i)\psi_n(i)$, and integrated over all volume elements $d\tau_i$ and $d\tau_j$, the result is not zero, but is the exchange integral. These exchange effects not only give rise to the spin polarization described above and cause the first excited triplet state of helium to be lower than the first excited singlet, but they also lead to Hund's rule.

The spin density experienced at the hydrogen atom of the C—H bond when there is unpaired electron density in the π-orbital ($2p_z$) is expressed by equation (9-11):

$$a_{\mathrm{H}} = Q\rho_{\mathrm{C}} \tag{9-11}$$

where ρ_{C} is the *unpaired electron density* in the carbon $2p_z$ orbital and Q is the value of a_{H} when there is a full electron on the carbon. The spin density at the proton is negative, so a_{H} is negative and Q must be negative. In systems where ρ_{C} is known, Q can be calculated and is found experimentally to vary from -22 to -27 gauss. A rough value of -23 gauss for aromatic radicals treated by Hückel theory suffices for most purposes that will concern us. When an extended Hückel calculation is used, the overlap is not set equal to zero and the molecular orbital coefficients are normalized to include overlap. Accordingly, the value of Q used depends on the m.o. calculation used; that is, on the definition of ρ_{C}.

When there is a node in the m.o. containing the unpaired electron at one of the carbon atoms in a π-system, similar exchange interactions with lower-energy filled pi-molecular orbitals operate to place negative spin density on this carbon. We shall make this more specific in the discussion of the allyl radical below. Resulting exchange interaction of this unpaired spin with the C—H sigma bond places positive spin density on the hydrogen. The Hückel, extended Hückel, or any restricted m.o. calculation (*i.e.*, one where two electrons are fed into each molecular orbital) do not include these exchange interactions. They simply indicate a node at the carbon atom or the hydrogen atom. One attempts to correct for this shortcoming, for example, at a hydrogen directly bonded to a carbon containing unpaired electron density in an orthogonal C_{2p} orbital by employing equation (9-11). Often other polarization effects in a molecule are qualitatively discussed, and the protons or carbons where these effects dominate are ignored when one attempts a quantitative fit of the calculated and experimental coupling constants in a molecule. This discussion can be made

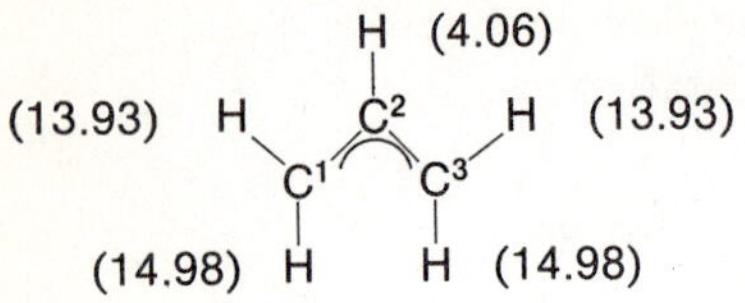

H (4.06)
(13.93) H C2 H (13.93)
C1 C3
(14.98) H H (14.98)

FIGURE 9-16. Proton hyperfine splittings in the allyl radical.

more specific by considering the observed[6] proton hyperfine splittings in the allyl radical shown in Fig. 9-16. The radical contains three electrons in the π-system whose wave functions are given by:

$$\psi_1 = \frac{1}{2}(\varphi_1 + \sqrt{2}\varphi_2 + \varphi_3) \quad \text{(bonding)} \tag{9-12}$$

$$\psi_2 = \frac{1}{\sqrt{2}}(\varphi_1 - \varphi_3) \quad \text{(non-bonding)} \tag{9-13}$$

$$\psi_3 = \frac{1}{2}(\varphi_1 - \sqrt{2}\varphi_2 + \varphi_3) \quad \text{(anti-bonding)} \tag{9-14}$$

The odd electron is placed in ψ_2, so one would predict the unpaired density at C_1 to be $\rho_{C_1} = (1/\sqrt{2})^2 = 0.5$, where $1/\sqrt{2}$ is the C_1 coefficient in the m.o. containing the unpaired electron. Using equation (9-11) with $Q = -23$, one would predict a_H to be equal to -11.5. Furthermore, there would be no unpaired electron density at C_2 and, without some kind of spin polarization involving the carbon π electron density, one would predict a zero coupling constant for this hydrogen or for a ^{13}C at this position. This is not observed. Two polarizations are needed to account for a hydrogen coupling constant from this middle hydrogen. The qualitative explanation involves taking the filled orbital ψ_1 and writing two separate spin orbitals for it, ψ_{1a} and ψ_{1b}. Only one electron is placed in each spin orbital. The wave functions in terms of the above wave functions for allyl become

$$\psi_{1a} = \psi_1 + \lambda\psi_3$$
$$\psi_{1b} = \psi_1 - \lambda\psi_3 \quad (\text{where } \lambda \ll 1)$$

For spin aligned with the field in the lower energy ψ_{1a} and opposed to the field in ψ_{1b}, there will be increased spin density aligned with the field on atoms 1 and 3 relative to that on atom 2. In ψ_{1b}, with spin density opposed to the field, there will be more negative spin density on atom 2 than on carbon atoms 1 and 3. By this mechanism, we are not introducing any unpaired electrons into the old ψ_1 orbital, but we simply are influencing the distribution of the paired spins over the three atoms giving rise to negative (opposed to the applied field) spin density on C_2. This negative spin density then undergoes spin polarization with the electron pair in the C—H bond [see the discussion of equation (9-11)] to place spin density on the hydrogen. The exchange interaction of the unpaired electron in ψ_2 (mainly on C_1 and C_3) with the pair in ψ_1 is the effect that lowers the energy of ψ_{1a} relative to ψ_{1b}. The two hydrogens on one of the terminal carbons are not equivalent by symmetry, but our discussion so far has not introduced any effects that would make them non-equivalent from the standpoint of spin distribution. Exchange polarization involving filled sigma molecular orbitals is required.

The phenomena discussed above are all *indirect* mechanisms for placing unpaired spin density on the hydrogen. When the free radical is a sigma radical, *e.g.*, the vinyl radical $H_2C{=}\dot{C}{-}H$, the protons in the molecule make a contribution to the sigma molecular orbital containing the unpaired electron. Thus, the unpaired electron is delocalized directly onto the proton and a_H is proportional to ψ^2. Since Hückel calculations are inappropriate for sigma systems of this sort, the initial work in this area utilized extended Hückel molecular orbital calculations. Procedures have been reported[7] for evaluating ψ^2 at the hydrogen nucleus from the wave function, and a_H is calculated by:

$$a_H \text{ (gauss)} = 1887 \, \psi_{(H)}^2 \tag{9-15}$$

Again, spin polarization is foreign to this calculation. Often the majority of the unpaired electron density resides at a given atom in the radical. When this is the case, spin polarization makes a large contribution at protons directly bound to this atom. Poor agreement between calculated and experimental results can be expected for this atom. Generally, for other protons in the molecule (with a few exceptions), spin polarization effects make a relatively insignificant contribution when direct delocalization is appreciable.

Two semi-empirical quantitative approaches have been reported to incorporate the effects of spin-polarization. In one described by McLachlan,[8] wave functions from a restricted calculation are employed and exchange effects are added on. This has met with limited success. In a second approach, an unrestricted molecular orbital calculation, *i.e.*, one utilizing spin orbitals, is employed. The most common one at present is the so-called INDO calculation,[9] which has been parameterized by Pople *et al.* to calculate spin densities. It has not been extensively tested on atoms for which spin polarization dominates, but is certainly the method of choice for this calculation at present. The output consists of one-electron orbitals, and all the positive and negative spin densities at an atom in all the filled molecular orbitals are summed to produce the net spin density at the atom. Poor agreement of calculated (INDO) and experimental results is reported[15] for the pyridinium radical.

The application of the results from molecular orbital calculations to the hyperfine splittings from atoms other than hydrogen is considerably more complex. In contrast to protons, which have only the direct and indirect mechanisms described above, ^{13}C hyperfine splittings have contributions from other sources. (1) Unpaired electrons in a $p(\pi)$ orbital can polarize the filled $2s$ and the filled $1s$ orbitals on the same atom. (2) There can be direct delocalization of electron density into the $2s$ orbital in a sigma radical. (3) Spin density on a neighbor carbon, by polarizing the C—C sigma bond, can place spin density into the $2s$ and $2p$ orbitals of the carbon whose resonance is being interpreted. The calculations[10-13] of ^{14}N, ^{33}S, and ^{17}O hyperfine couplings have been more successful than those for ^{13}C. Silicon-containing radicals have also been successfully treated.[13] The effects from spin densities on neighboring atoms are found to be less important for these nuclei than for ^{13}C.

The application of the results from molecular orbital calculations to the assignment of the esr spectrum of an organic radical is an obvious application of the above discussion. Another interesting application involves determining the geometry of free radicals. For instance, is $CH_3\cdot$ planar, or does the C—H bond of the vinyl radical lie along the C—C bond axis? When the calculated hyperfine coupling constants are found to vary considerably with geometry (*i.e.*, a whole series of molecular orbital calculations are performed for different geometries), the fit of calculated and experimental results can be used to suggest the actual geometry.[14] In several examples, molecular orbital calculations have provided evidence about the nature of a radical produced in an experiment.[11a, 14] For example, the γ irradiation of pyridine produced a radical believed to be the pyridine cation; *i.e.*, one of the lone pair electrons was removed. The results shown in Table 9–2 indicate that the 2-pyridyl radical was actually formed.[11a]

TABLE 9-2. EXTENDED HÜCKEL AND EXPERIMENTAL ISOTROPIC HYPERFINE COUPLING CONSTANTS (GAUSS)

	Calculated for $C_5H_5N^+$	Calculated for 2-pyridyl C_5H_4N	Experimental
a_N	52.5	33.8	29.7
a_H	27.0 (2,6 H)	5.3	4.3
	9.7 (3,5 H)	9.3	
	38.8 (4 H)	7.6	
		0.0	

One other interesting result that comes out of the calculations and esr experiments is the very extensive amount of electron delocalization that occurs in the sigma system. For example, significant amounts of unpaired electron density are found on the methyl group protons of $CH_3\dot{N}H_2^+$ and $C_2H_5\dot{N}H_2^+$. The lone pair molecular orbital is not a localized nitrogen lone pair orbital, but is a delocalized molecular orbital. The extent of delocalization onto the protons varies in the different rotamers.[16]

ANISOTROPIC EFFECTS

9-6 ANISOTROPY IN THE *g* VALUE

The next feature of esr spectroscopy can be introduced by describing the *g* values for the NO_2 radical trapped[17] in a single crystal of KNO_3. When the crystal is mounted with the field parallel to the *z*-axis of NO_2 (the twofold rotation axis), a *g* value of 2.006 is obtained. When the crystal is mounted with the *x*- or *y*-axis (the plane of the molecule containing *y*) parallel to the field, a *g* value of 1.996 is obtained. The molecule is rapidly rotating about the *z*-axis in the solid, so the same result is obtained for *x* or *y* parallel to the field. The differences in the *g* values with orientation are even more pronounced in transition metal ion complexes (*vide infra*) and in complexes of the lanthanides and actinides.

The treatment so far has involved so-called isotropic spectra. These are obtained when the radical under consideration has spherical or cubic symmetry. For radicals with lower symmetry, anisotropic effects are manifested in the solid spectra for both the *g* values and the *a* values. Usually, for these lower symmetry systems, the solution spectra qualitatively appear as isotropic spectra because the anisotropic effects are averaged to zero by the rapid rotation of the molecules. Our concern here is how these anisotropic effects arise and how they can be determined. Later (Chapter 13), we shall see how the anisotropy in *g* and *a* can be used to provide information about the electronic ground state of transition metal ion complexes.

Anisotropy in g arises from coupling of the spin angular momentum with the orbital angular momentum. The spin angular momentum is oriented with the field, but the orbital angular momentum, which is associated with electrons moving in molecular orbitals, is locked to the molecular wave function. Consider a case where there is an orbital contribution to the moment from an electron in a circular molecular orbit that can precess about the *z*-axis of the molecule. In Fig. 9-17, two different orientations of this orbital in the molecule relative to the field are schematically indicated by ellipses. In Fig. 9-17(A), $\vec{\mu}_L$ (the orbital magnetic moment vector*) and $\vec{\mu}_s$ (the spin

*The direction of the orbital angular momentum given by the right-hand rule is opposite to that of the magnetic moment vector of the electron.

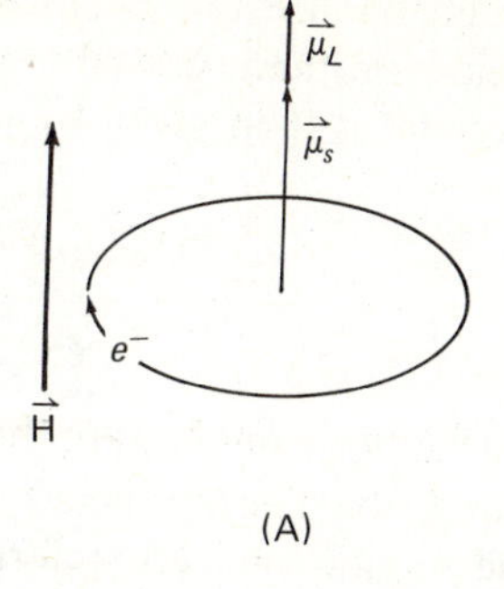

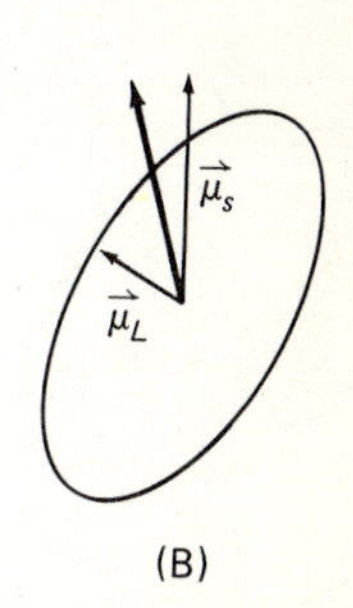

FIGURE 9–17. Coupling of the projections of the spin and orbital angular momentum for two different molecular orientations relative to the applied field *H*.

magnetic moment vector) are in the same direction. In Fig. 9–17(B), a different orientation of the molecule is shown. The electron moment $\vec{\mu}_s$ has the same magnitude as before but now the net moment, indicated by the boldface arrow, results because $\vec{\mu}_L$ and $\vec{\mu}_s$ do not point in the same direction. If it were not for the orbital contribution, the moment from the electron would be isotropic. *When the effects of the orbital moment are small, they are incorporated into the g value, and now we see that this g value will be anisotropic.* A tensor, equation (9–16), is needed to describe it. The *g*-tensor then gives us an effective spin, *i.e.*, $\mathbf{g} \cdot S = S_{\text{eff}}$. For different orientations, the *g*-tensor lengthens and shortens S_{eff} to incorporate orbital effects. It should be emphasized that even when the ground state of a molecule has no orbital angular momentum associated with it, field-induced mixing in of an excited state that does have orbital angular momentum can lead to anisotropy in *g*. The information that *g*-tensor anisotropy provides about the electronic structure of the molecule will be discussed in Chapter 13 on the epr of transition metal ion complexes.

For an isotropic system, we wrote our Hamiltonian to describe the interaction of an electron spin moment with the magnetic field and with a magnetic nucleus in equation (9–4) as:

$$\hat{H} = g\beta H\hat{S}_z - g_N\beta_N H\hat{I}_z + a\hat{I}\cdot\hat{S}$$

Both *g* and *a* were scalar quantities. When an anisotropic free radical is investigated in the solid state, both *g* and *a* have to be replaced by tensors or matrices. The $g\beta H\hat{S}_z$ term in the Hamiltonian becomes $\beta\hat{S}\cdot\mathbf{g}\cdot H$, which can be represented with matrices as:

$$\beta[S_x\ S_y\ S_z]\begin{bmatrix} g_{xx} & g_{xy} & g_{xz} \\ g_{yx} & g_{yy} & g_{yz} \\ g_{zx} & g_{zy} & g_{zz} \end{bmatrix}\begin{bmatrix} H_x \\ H_y \\ H_z \end{bmatrix} \tag{9-16}$$

and *a* in the $a\hat{I}\cdot\hat{S}$ term is also replaced by a tensor. Here *x*, *y*, and *z* are defined in the laboratory frame; *i.e.*, they are crystal axes. The off-diagonal element g_{zx} gives the contribution to *g* along the *z*-axis of the crystal when the field is applied along the *x*-axis. *This matrix is diagonal when the crystal axes are coincident with the molecular coordinate system that diagonalizes* **g**. When they are not coincident and the crystal is studied along the *x*, *y*, and *z* crystal axes, we get off-diagonal contributions, as we shall show subsequently. The *g* matrix can be made diagonal by a suitable choice of coordinates.

If one studies the esr of a single crystal with anisotropy in *g*, the measured *g* value is a function of the orientation of the crystal with the field because we measure an effective *g* value oriented along the field. If we define molecular axes *X*, *Y*, and *Z* that diagonalize the g-tensor and pick as an example a case in which they are coincident with the crystal axes, the effective value of g for an arbitrary orientation of the crystal is then given by

$$\begin{aligned} g_{\text{eff}}^2 &= (g^2)_{xx}\cos^2\theta_{Hx} + (g^2)_{yy}\cos^2\theta_{Hy} + (g^2)_{zz}\cos^2\theta_{Hz} \\ &= (g^2)_{xx}l_x^2 + (g^2)_{yy}l_y^2 + (g^2)_{zz}l_z^2 \end{aligned} \tag{9-17}$$

Here θ_{Hx}, θ_{Hy}, and θ_{Hz} are the angles between the field H and the X, Y, and Z axes, respectively. The symbols l_x, l_y, and l_z are often used to represent the cosines of these angles and are referred to as the *direction cosines*. From trigonometry, $l_x^2 + l_y^2 + l_z^2 = 1$, so two parameters suffice to specify the direction. The above equation can be indicated in matrix notation as

$$g_{\text{eff}}^2 = [l_x\ l_y\ l_z]\begin{bmatrix} (g^2)_{xx} & 0 & 0 \\ 0 & (g^2)_{yy} & 0 \\ 0 & 0 & (g^2)_{zz} \end{bmatrix}\begin{bmatrix} l_x \\ l_y \\ l_z \end{bmatrix} \tag{9-18}$$

Since S_x, S_y, and S_z as well as H_x, H_y, and H_z are defined in terms of the molecular coordinate x,y,z system, they can be replaced by the same direction cosines. The molecular coordinate system that diagonalizes the g-tensor may not be coincident with the arbitrary axes associated with the crystal morphology. Since this experiment is carried out using the easily observed axes of the bulk crystal, the above equation has to be rewritten in non-diagonal form as

$$g_{\text{eff}}^2 = [l_x\ l_y\ l_z]\begin{bmatrix} (g^2)_{xx} & (g^2)_{xy} & (g^2)_{xz} \\ (g^2)_{yx} & (g^2)_{yy} & (g^2)_{yz} \\ (g^2)_{zx} & (g^2)_{zy} & (g^2)_{zz} \end{bmatrix}\begin{bmatrix} l_x \\ l_y \\ l_z \end{bmatrix} \tag{9-19}$$

Equation (9–19) can be used to evaluate all the tensor components. The matrix is symmetric [*i.e.*, $(g^2)_{yx} = (g^2)_{xy}$], so only six independent components need to be evaluated. Since it is most convenient to orient the crystal in the magnetic field relative to the observed crystal axes, the x, y, and z axes are defined in terms of these observed axes of the bulk crystal. S_x, S_y, and S_z, as well as H_x, H_y, and H_z, are defined in terms of these axes. Consider the first case where the crystal is mounted, as shown in Fig. 9–18, with the y-axis perpendicular to the field so that the crystal can be

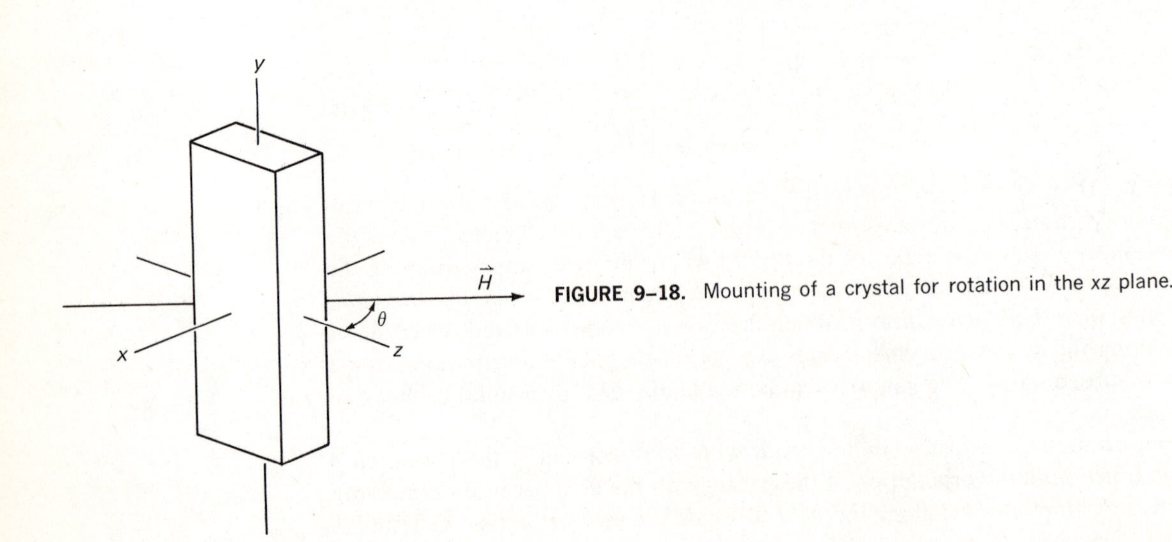

FIGURE 9–18. Mounting of a crystal for rotation in the xz plane.

rotated around y with $\vec{\mathbf{H}}$ making different angles, θ, to z in the xz plane. Now, l_z equals $\cos\theta$ and l_x equals $\sin\theta$, where θ is the angle between $\vec{\mathbf{H}}$ and the z-axis. Substituting these quantities into equation (9–19) and carrying out the matrix multiplication yields

$$g_{\text{eff}}^2 = (g^2)_{xx}\sin^2\theta + 2(g^2)_{xz}\sin\theta\cos\theta + (g^2)_{zz}\cos^2\theta \tag{9-20}$$

For rotations in the yz plane, we have $l_x = 0$, $l_y = \sin\theta$, and $l_z = \cos\theta$. Substitution into (9–19) and matrix multiplication then yields

$$g_{\text{eff}}^2 = (g^2)_{yy}\sin^2\theta + 2(g^2)_{yz}\sin\theta\cos\theta + (g^2)_{zz}\cos^2\theta \tag{9-21}$$

In a similar fashion, rotation in the xy plane yields

$$g_{\text{eff}}^2 = (g^2)_{xx}\cos^2\theta + 2(g^2)_{xy}\sin\theta\cos\theta + (g^2)_{yy}\sin^2\theta \tag{9-22}$$

These equations thus tie our matrix in equation (9–19) to experimental observables, g_{eff}^2. In our experiment, a g value is obtained, but we do not know its sign. The measured g value is squared and used in this analysis. For rotations in any one plane, only three measurements of g_{eff}^2 need be made (corresponding to three different θ values) to solve for the three components of the g^2-tensor in the respective equations. For the xz plane, one measures $(g^2)_{zz}$ at $\theta = 0$ and $(g^2)_{xx}$ at $\theta = 90°$. With these values and g_{eff}^2 at $\theta = 45°$, one can solve for $(g^2)_{xz}$. In this way, the six independent components of the g^2-tensor can be measured. In practice, many measurements are made and the data are analyzed by the least squares method. One then solves for a transformation matrix that rotates the coordinate system and diagonalizes the g^2-tensor. This produces the molecular coordinate system for diagonalizing the g^2-tensor, and the square roots of the individual diagonal g^2 matrix elements produce g_{xx}, g_{yy}, and g_{zz} in this special coordinate system. In order for this procedure to work as described, it is necessary that all the molecules in the unit cell have the same orientation of their molecular axes relative to the crystal axes. Thus, these measurements are often carried out in conjunction with a single crystal x-ray determination.

9–7 ANISOTROPY IN THE HYPERFINE COUPLING

We introduced the previous section by describing the anisotropy in g when a single crystal of NO_2 in KNO_3 was examined at different orientations relative to the field. The a values of this system are also very anisotropic. When the molecular twofold axis is parallel to the applied field, the observed nitrogen hyperfine coupling constant is 176 MHz, while a value of 139 MHz is observed for the orientation in which this axis is perpendicular to the field. In rigid systems, interactions between the electron and nuclear dipoles give rise to anisotropic components in the electron nuclear hyperfine interaction. The classical expression for the interaction of two dipoles was treated in Chapter 8, and the same basic considerations apply here. For the interaction of an electron moment and a nuclear moment, the Hamiltonian is:

$$\hat{H}_{\text{dipolar}} = -g\beta g_N\beta_N\left[\frac{\hat{S}\cdot\hat{I}}{r^3} - \frac{3(\hat{S}\cdot\vec{\mathbf{r}})(\hat{I}\cdot\vec{\mathbf{r}})}{r^5}\right] \tag{9-23}$$

The sign is opposite that employed for the interaction of two nuclear dipoles, which was the problem treated in the section on liquid crystal and solid nmr. Substituting

$\hat{S} = \hat{S}_x + \hat{S}_y + \hat{S}_z$, $\hat{I} = \hat{I}_x + \hat{I}_y + \hat{I}_z$, and $r = x + y + z$, and expanding these vectors, leads to

$$\hat{H}_{\text{dipolar}} = -g\beta g_N\beta_N\left\{\left[\frac{r^2 - 3x^2}{r^5}\right]\hat{S}_x\hat{I}_x + \left[\frac{r^2 - 3y^2}{r^5}\right]\hat{S}_y\hat{I}_y + \left[\frac{r^2 - 3z^2}{r^5}\right]\hat{S}_z\hat{I}_z - \left[\frac{3xy}{r^5}\right](\hat{S}_x\hat{I}_y + \hat{S}_y\hat{I}_x) - \left[\frac{3xz}{r^5}\right](\hat{S}_x\hat{I}_z + \hat{S}_z\hat{I}_x) - \left[\frac{3yz}{r^5}\right](\hat{S}_y\hat{I}_z + \hat{S}_z\hat{I}_y)\right. \tag{9-24}$$

When this Hamiltonian is applied to an electron in an orbital, the quantities in brackets must be replaced by average values; we employ angular brackets to refer to the average value over the electronic wave function. In matrix notation, we then have

$$\hat{H}_{\text{dipolar}} = -(g\beta g_N\beta_N)[\hat{S}_x\ \hat{S}_y\ \hat{S}_z]\begin{bmatrix} \left\langle\frac{r^2 - 3x^2}{r^5}\right\rangle & \left\langle\frac{-3xy}{r^5}\right\rangle & \left\langle\frac{-3xz}{r^5}\right\rangle \\ \left\langle\frac{-3xy}{r^5}\right\rangle & \left\langle\frac{r^2 - 3y^2}{r^5}\right\rangle & \left\langle\frac{-3yz}{r^5}\right\rangle \\ \left\langle\frac{-3xz}{r^5}\right\rangle & \left\langle\frac{-3yz}{r^5}\right\rangle & \left\langle\frac{r^2 - 3z^2}{r^5}\right\rangle \end{bmatrix}\begin{bmatrix}\hat{I}_x \\ \hat{I}_y \\ \hat{I}_z\end{bmatrix} \tag{9-25}$$

This equation is abbreviated as

$$\hat{H}_{\text{dipolar}} = h\hat{S}\cdot\mathbf{T}\cdot\hat{I} \tag{9-26}$$

where **T** is the dipolar interaction tensor (in units of Hz) that gauges the anisotropic nuclear hyperfine interaction. The Hamiltonian now becomes

$$\hat{H} = \beta\hat{S}\cdot\mathbf{g}\cdot H - g_N\beta_N H\cdot\hat{I} + h\hat{S}\cdot\mathbf{A}\cdot\hat{I} \tag{9-27}$$

where the first term on the right is the electron Zeeman term, the second is the nuclear Zeeman term, and the third is the hyperfine interaction term. The quantity **A** in the third term includes both the isotropic and the anisotropic components of the hyperfine interaction; *i.e.*,

$$\mathbf{A} = \mathbf{T} + a\mathbf{1} \tag{9-28}$$

In the application of the Hamiltonian given in equation (9–27) to organic free radicals, several simplifying assumptions can be introduced. First, $g_N\beta_N H\cdot\hat{I}$, the nuclear Zeeman effect, usually gives rise to a small energy term compared to the others. (Recall our earlier discussion about the energies of the esr and nmr transitions.) Second, g-anisotropy is small, and we shall assume that g is isotropic in treating the hyperfine interaction.* (This would be a particularly bad assumption for certain transition metal complexes, *vide infra.*) The electron Zeeman term is assumed to be the dominant energy term, so $\hat{S}$ is quantized along $\vec{\mathbf{H}}$, which we label as the z-axis. We see in this example, as we shall see over and over again, that *it is often convenient to define the coordinate system to be consistent with the largest energy effect.* Next we have to worry about the orientation of the nuclear moment relative to the z-field. Our discussion is general, but it may help to consider an ethyl radical

*If g-tensor anisotropy is comparable to hyperfine anisotropy, this assumption cannot be made. The reader is referred to Chapter 13 and to A. Abragam and B. Bleany, "EPR of Transition Ions," p. 167, Clarendon Press, Oxford, England (1970) for a discussion of this situation.

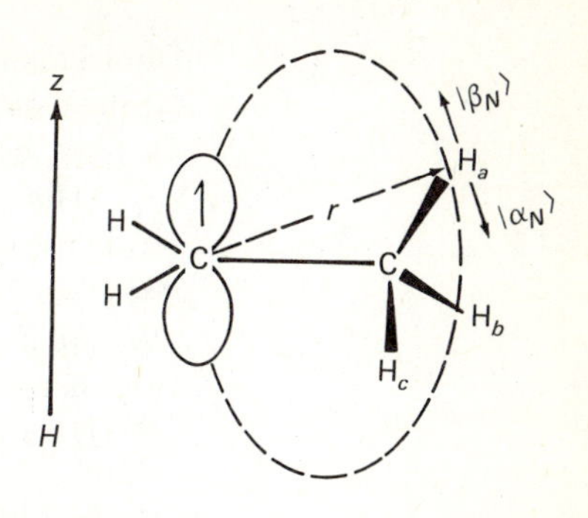

FIGURE 9–19. The orientation of the spin and nuclear moments in an applied field.

oriented as shown in Fig. 9–19, with the —CH_3 group not undergoing rotation. To make this point, focus attention on nucleus H_a involved in dipolar coupling to the electron. The nuclear moment will not be quantized along z, but along an effective field, $\vec{H}_{eff}$, which is the vector sum of the direct external field $\vec{H}$ (nuclear Zeeman) and the hyperfine field produced by the nearby electron. If the hyperfine interaction is large (~100 gauss), the hyperfine field at this nucleus (*i.e.*, the field from the electron magnetic moment felt at the hydrogen nucleus) is about 11,700 G. (This is to be contrasted to the field of ~3000 G from the magnet and the field of ~18 G felt at the electron from the considerably smaller nuclear moment.) Thus, we may be somewhat justified in ignoring the nuclear Zeeman term, $g_N\beta_N H \cdot \hat{I}$, in equation (9–27). The Hamiltonian for most organic free radicals (where g is isotropic) is then considerably simplified from the form in equation (9–27), and becomes

$$\hat{H} = g\beta H\hat{S}_z + h\hat{S}_z(A_{zx}\hat{I}_x + A_{zy}\hat{I}_y + A_{zz}\hat{I}_z) \tag{9-29}$$

The terms on the far right give the z-component of the electron-nuclear hyperfine interaction with contributions from $\hat{I}_x$ and $\hat{I}_y$ as well as from $\hat{I}_z$, for the z-field does not quantize $\hat{I}$, but does quantize $\hat{S}$. When this Hamiltonian operates on the $|\alpha_e\alpha_N\rangle$ and other wave functions, off-diagonal matrix elements in the secular determinant result. When it is diagonalized and solved for energy, the following results are obtained:

$$\left.\begin{matrix}\alpha_e\alpha_N\\ \alpha_e\beta_N\end{matrix}\right\} E = \frac{1}{2}g\beta H \pm \frac{1}{4}\sqrt{A_{zx}^2 + A_{zy}^2 + A_{zz}^2}$$

$$\left.\begin{matrix}\beta_e\alpha_N\\ \beta_e\beta_N\end{matrix}\right\} E = -\frac{1}{2}g\beta H \pm \frac{1}{4}\sqrt{A_{zx}^2 + A_{zy}^2 + A_{zz}^2}$$

The term containing the square root replaces the $\frac{1}{4}a$ obtained from the evaluation of the $a\hat{I} \cdot \hat{S}$ term for the hydrogen atom. The energy for the hyperfine coupling is thus given by

$$\Delta E_{hf} = \frac{1}{2}\sqrt{A_{zx}^2 + A_{zy}^2 + A_{zz}^2} \tag{9-30}$$

The quantity **A** contains both the isotropic, a, and anisotropic, **T**, components of the hyperfine interaction. Since in solution the anisotropic components are averaged to zero, it becomes a simple matter to take one-third of the trace of **A** to decompose **A** into **T** and a. (This assumes that the solvent or solid lattice has no effect on the electronic structure.)

These expressions apply for any orientation of the molecule relative to the applied field. In a single-crystal experiment, in which the crystal and molecular axes are not aligned, we proceed as in the case of the evaluation of the g-tensor to determine all of the components of the hyperfine tensor. *The coordinate system that diagonalizes the g-tensor need not be the same one that diagonalizes the A-tensor, and neither one of these need be the apparent molecular coordinate system.*[17b] If the molecule has overall symmetry (*i.e.*, the full ligand environment included) such that it possesses an n-fold rotation axis, the same axis will be diagonal for g and A, and it must be coincident with the molecular z-axis.

The angular dependence of the hyperfine interaction for the case where the field from the hyperfine interaction is large, $I = \frac{1}{2}$, and the system has axial symmetry can be expressed by substituting $r \cos\theta$ for z and $r \sin\theta$ for x and y into equation (9–24). We are in effect resolving the nuclear moment in Fig. 9–19 into components parallel and perpendicular to the field. The Hamiltonian including the electron Zeeman term ($g\beta H\hat{S}_z$) becomes

$$\hat{H} = g\beta H\hat{S}_z + h\hat{S}_z\{[a + B(3\cos^2\theta - 1)]\hat{I}_z + 3B\cos\theta\sin\theta\hat{I}_x\} \qquad (9\text{-}31)$$

The result of this Hamiltonian operating on the basis set produces the energies given by

$$E = g\beta HM_s \pm \frac{hM_s}{2}[(a - B)^2 + 3B(2a + B)\cos^2\theta]^{1/2} \qquad (9\text{-}32)$$

where a is the isotropic hyperfine coupling constant, B is the anisotropic hyperfine coupling constant, and θ is the angle that the z-axis of the molecule makes with the field. The hyperfine coupling constant A observed experimentally is the difference between the energies of the appropriate levels and is given (in cm^{-1}) by

$$A = h[(a - B)^2 + 3B(2a + B)\cos^2\theta]^{1/2}$$

One often sees the following equation presented in the literature to describe the anisotropy of g and A for an axial system:

$$\Delta E = h\nu = \left(\frac{1}{3}g_{\parallel} + \frac{2}{3}g_{\perp}\right)\beta H_0 + am_I + \left[\frac{1}{3}(g_{\parallel} - g_{\perp})\beta H_0 + Bm_I\right](3\cos^2\theta - 1) \qquad (9\text{-}33)$$

where θ is the angle between the z-axis and the magnetic field, a is the isotropic coupling constant, and B is the anisotropic coupling constant. The equation results from equation (9–32) by adding the anisotropy in g and by assuming that both the g- and A-tensors are diagonal in the same axis system.

Since in the analysis of the anisotropy in the hyperfine coupling we deal with $\mathbf{A}^2$ [see equation (9–30)], usually one cannot obtain the sign of the coupling constant from the esr experiment.* However, it can be readily predicted for an organic radical that contains an electron in a p-orbital (there is no anisotropic contribution from unpaired electron density in a spherical $2s$ orbital). We shall begin by returning to equation (9–25) to further explore its meaning. Assume that the electron is in a hypothetical orbital that can be represented by a unit vector. When this hypothetical orbital lies along z, we have $z = r$, $x = 0$, and $y = 0$, and all off-diagonal terms are zero. We observe, on substitution into equation (9–25), that $T_{xx} = k\langle 1/r^3\rangle$, $T_{yy} = k\langle 1/r^3\rangle$, and $T_{zz} = -k\langle 2/r^3\rangle$.† [The matrix elements as written in equation (9–25) have the opposite sign, but the whole term is negative to describe the interaction of the positive nuclear moment and the negative electron moment.] Note that the trace is zero.

---

* For certain systems, one can determine the sign of the coupling constant by using nmr, because spin aligned with the field causes a downfield shift and spin opposed to the field causes an upfield shift.

† The proportionality constant is $g\beta_N g_N \beta/h$.

To make the problem more realistic, we shall next consider the electron to be in a p_z orbital. It is convenient to convert to spherical polar coordinates to solve this problem by substituting $z = r\cos\theta$, $x = r\sin\theta\cos\varphi$, and $y = r\sin\theta\sin\varphi$. The result for an electron at a specific (r,θ) (after consideration of the negative sign) is

$$T_{zz} = g\beta g_N\beta_N\langle(3\cos^2\theta - 1)/r^3\rangle$$

$$T_{yy} = -\frac{1}{2}g\beta g_N\beta_N\langle(3\cos^2\theta - 1)/r^3\rangle$$

$$T_{xx} = -\frac{1}{2}g\beta g_N\beta_N\langle(3\cos^2\theta - 1)/r^3\rangle$$

The latter two matrix elements are readily obtained by substituting for x and y in equation (9–25) and substituting $\langle\cos^2\varphi\rangle = \frac{1}{2}$ for an axial system. Note that the trace is zero. Now, consider that the electron can be located at any place in the p-orbital. Thus, we have to integrate over all possible angles for the radius vector to the electron in this orbital and then over all radii r. In doing so, we get

$$T_{zz} = \frac{4}{5}g\beta g_N\beta_N\langle 1/r^3\rangle \tag{9-34}$$

where $\langle 1/r^3\rangle$ is the average value of the quantity $1/r^3$. Abbreviating equation (9–34) as $T_{zz} = \frac{4}{5}P_p$, we have

$$T_{xx} = -\frac{2}{5}P_p \quad \text{and} \quad T_{yy} = -\frac{2}{5}P_p \tag{9-35}$$

These considerations are valuable in predicting the signs of the anisotropic components of the hyperfine coupling constant.

It is informative to apply these equations to the anisotropic hyperfine tensor of a ^{13}C nucleus, which depends mainly on the unpaired electron density in the p-orbital of this atom. We wish to consider the signs of T_{xx}, T_{yy}, and T_{zz} for this system. The three orientations of the p-orbital in the molecule relative to the applied field are indicated in Fig. 9–20. The dotted lines indicate the regions where the $(3\cos^2\theta - 1)$ function is zero. This corresponds to plotting the signs for the various regions of the lines of flux emanating from the nuclear moment. Accordingly, by visual inspection, we can tell whether T_{zz} expressed in equation (9–34) will be positive

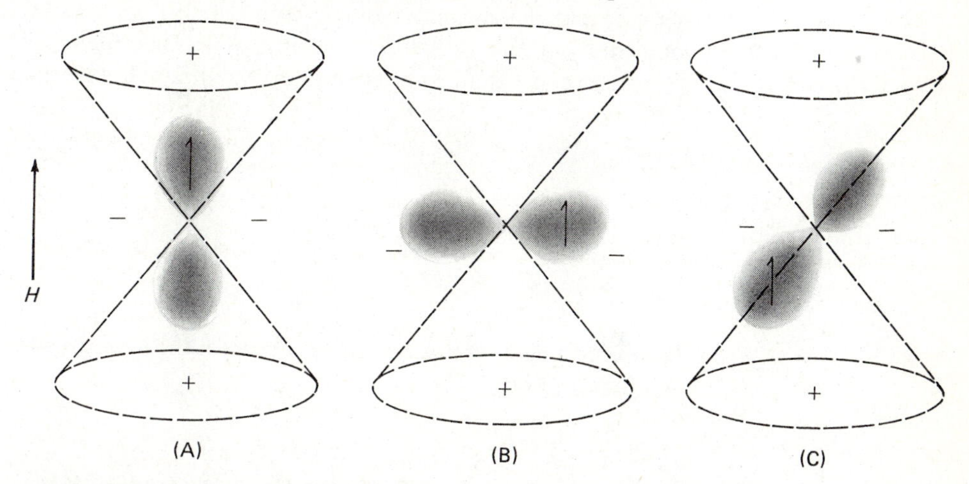

FIGURE 9–20. Visual representation of dipolar averaging of the electron and nuclear moments: (A) p-orbital oriented along the field; (B) and (C) p-orbital perpendicular to the field.

or negative. For example, as we see in Fig. 9–20(A), when the p_z orbital is aligned with the field almost the entire averaging of the dipolar interaction of the nuclear moment over the p_z orbital will occur in the positive part of the cone. A large, positive T_{zz} value is thus expected. For the orientation along the x-axis shown in Fig. 9–20(B), the dipolar interaction, T_{xx}, will be large and negative; the same is true of T_{yy} for the orientation shown in Fig. 9–20(C). Analysis[18,19] of the ^{13}C hyperfine structure of the isotopically enriched malonic acid radical, $H^{13}\dot{C}(COOH)_2$, produces $a_c = 92.6$ MHz, $T_{xx} = -50.4$ MHz, $T_{yy} = -59.8$ MHz, and $T_{zz} = +120.1$ MHz. After the hyperfine tensor is diagonalized, the relative signs of T_{xx}, T_{yy}, and T_{zz} are known (the trace must be zero), but the absolute signs are not; *i.e.*, all those given above could be reversed. However, the arguments based on Fig. 9–20 provide us with good reason to think that the signs presented above are correct.

It is informative to predict the signs of the anisotropic hydrogen hyperfine components of a C—H radical. By analogy to our discussion above, the three orientations of the p_π orbital of this radical shown in Fig. 9–21 predict that T_{zz} is small

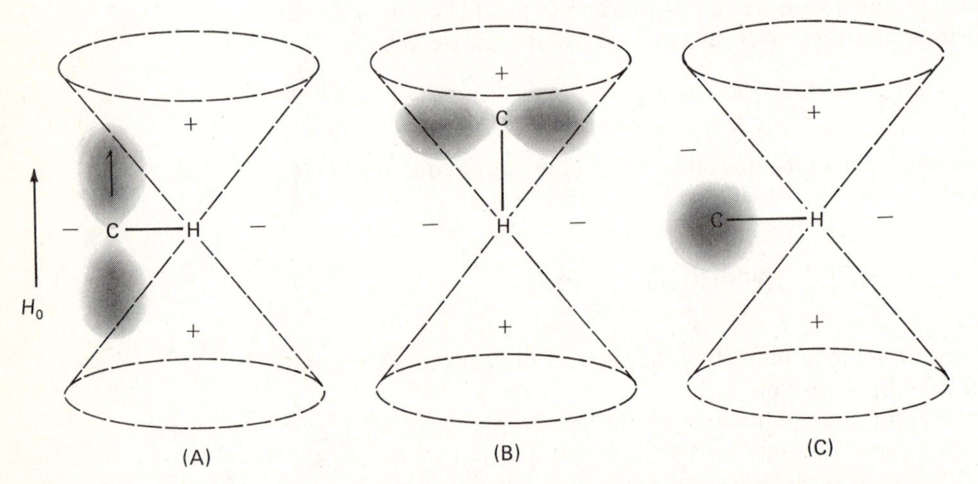

FIGURE 9–21. Visual representation of the dipolar averaging of the nuclear moment on the hydrogen with the electron moment in a *p*-orbital of carbon. (A) H_0 is parallel to the *z* crystal axis; (B) H_0 is parallel to the *y* crystal axis; (C) H_0 is parallel to the *x* crystal axis, and we are looking down the axis of the *p*-orbital.

while T_{yy} is positive and T_{xx} is negative. Visual averaging of the p-orbital with our cone of magnetic nuclear flux also suggests that T_{zz} will be small. Note that the cones representing the nuclear moment lines of flux are drawn at the nucleus whose moment is causing the splitting via the dipolar interaction with the electron. If the x, y, and z axes are defined relative to the fixed crystal axes (which are coincident with the molecular axes) as in Fig. 9–21, calculation[20] shows that a full unpaired electron in the carbon p-orbital would lead to an anisotropic hyperfine tensor of

$$\mathbf{T}_H = \begin{bmatrix} -38 & 0 & 0 \\ 0 & +43 & 0 \\ 0 & 0 & -5 \end{bmatrix} \text{ MHz}$$

The experimental proton hyperfine tensor for the α proton of the malonic acid radical was found to be

$$\mathbf{A}_H = \begin{bmatrix} \pm 91 & 0 & 0 \\ 0 & \pm 29 & 0 \\ 0 & 0 & \pm 58 \end{bmatrix} \text{ MHz}$$

Since the isotropic hyperfine coupling constant, a, is one-third the trace of $\mathbf{A}_H$, it equals ± 59 MHz. Accordingly, the anisotropic hyperfine tensor $\mathbf{T}$ must be:

$$\mathbf{T}_H = \begin{bmatrix} -32 & 0 & 0 \\ 0 & +30 & 0 \\ 0 & 0 & +1 \end{bmatrix} \text{MHz or} \begin{bmatrix} +32 & 0 & 0 \\ 0 & -30 & 0 \\ 0 & 0 & -1 \end{bmatrix} \text{MHz}$$

The arguments presented in the discussion of Fig. 9-21 and comparison to the theoretical tensor lead us to predict the tensor on the left to be correct. This matrix arose from $a = -59$ MHz. Since a positive a would have given tensor components that correspond to values greater than that for a full electron, the isotropic hyperfine coupling constant must be negative.

Anisotropic and isotropic hyperfine coupling constants have been measured in several organic and inorganic radicals and have provided considerable information about the molecular orbital containing the unpaired electron. The value of B for one electron in a p-orbital of various atoms can be evaluated using an SCF wave function from

$$B = \frac{2}{5} h^{-1} g\beta g_N \beta_N \langle r^{-3} \rangle \tag{9-36}$$

For ^{13}C, the anisotropic hyperfine coupling constants are given by

$$\begin{bmatrix} -B & & \\ & -B & \\ & & +2B \end{bmatrix}$$

where B is calculated from SCF wave functions to be 91 MHz. For $H\dot{C}(COOH)_2$, the experimental value of T_{zz} for ^{13}C is found to be $+120.1$ MHz compared to the value of 182 MHz for an electron localized in a C_{2p} orbital. Accordingly, it is concluded that ρ_c is 0.66. From ^{13}C enrichment, it is found that $a_c = +92.6$ MHz. A full electron in an s-orbital has an isotropic hyperfine coupling constant of 3110 MHz. The measured a_c corresponds to a value of 0.03 for C_{2s} spin density. The radical is expected to be nearly planar.

The magnitude of the isotropic ^{13}C hyperfine coupling constant supports a planar[21] CH_3 radical, $a_c = 38.5$ G, but the value of $a_c = 271.6$ G in the CF_3 radical indicates[21] that it is pyramidal with s-character in the orbital containing the unpaired electron.

The isotropic ^{14}N hyperfine coupling constant in NO_2 is 151 MHz, and the maximum value for the anisotropic hyperfine coupling constant is 12 MHz. With 1540 MHz expected for one electron in a nitrogen $2s$ orbital and 48 for an electron in a $2p$ orbital, ρ_s is calculated to be 0.10 and ρ_p is found to be 0.25, for a $2p/2s$ ratio of 2.5. An sp^2 orbital would have a ratio of 2.0, so this suggests that more p character is being used in the orbitals to bond oxygen and an angle greater than 120° is predicted. Microwave results gave a value of 134° for NO_2 in the gas phase.

9-8 THE EPR OF TRIPLET STATES

The next complication we shall discuss arises when there is more than one unpaired electron in the molecule. An example is provided by the triplet state that is formed upon u.v. irradiation of naphthalene. The single crystal epr spectrum was studied for a sample doped into durene. The similar shapes of these two molecules allowed the naphthalene to be trapped in the durene lattice; dilution of the naphthalene greatly increases the lifetime of the triplet state. The spectrum consists of three peaks, which change resonance fields drastically with orientation of the crystal. The changes could not be fitted with the anisotropic g- and a-tensors we have developed. The anisotropy in this system arises from electron-electron spin interaction and is described by the spin Hamiltonian given in equation (9–37). This Hamiltonian is seen to be very similar to that for the dipolar interaction of an electron and nuclear spin [equation (9–23)].

$$\hat{H} = g\beta H \cdot (\hat{S}_1 + \hat{S}_2) + g^2\beta^2 \left\{ \frac{\hat{S}_1 \cdot \hat{S}_2}{r^3} - \frac{3(\hat{S}_1 \cdot \vec{r})(\hat{S}_2 \cdot \vec{r})}{r^5} \right\} \tag{9-37}$$

where $\vec{r}$ is the vector joining the two electrons labeled 1 and 2.

The magnitude of the contribution to the epr spectrum from these effects depends upon the extent of the interaction of the two spins. We discussed the Coulomb (J) and exchange (K) integrals earlier in this chapter. When two molecular orbitals are closer in energy than the difference between exchange and repulsive energies, a stable triplet state arises. In the case of naphthalene, the ground state is a singlet, and the excited triplet state has an appreciable lifetime because of the forbidden nature of the triplet-singlet transition. For the ground state $S = 0$ and $m_s = 0$, and this configuration is illustrated in Fig. 9–22(A). For the triplet state we

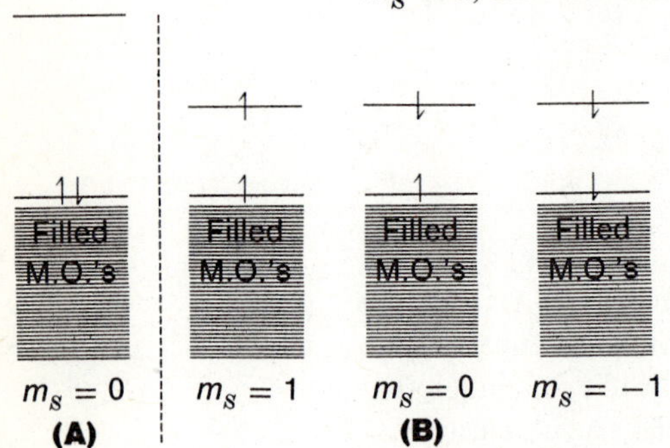

FIGURE 9–22. (A) Singlet ground term; (B) $m_s = 1, 0, -1$ components of the $S = 1$ state.

have $S = 1$ and $m_s = 1, 0, -1$. These electronic configurations are illustrated in Fig. 9–22(B). If only exchange and electrostatic interactions existed in the molecule, the three configurations $m_s = 1$, 0, and -1 would be degenerate in the absence of a magnetic field. The magnetic field would remove this degeneracy as shown in Fig. 9–23(A), and only a single transition would be observed as was the case for $S = 1/2$. However, magnetic dipole-dipole interaction between the two unpaired electrons removes the degeneracy of the m_s components of $S = 1$ *even in the absence of an external field,* as shown in Fig. 9–23(B). This removal of the degeneracy in the absence of the field is called *zero-field splitting*. When a magnetic field is applied, the levels are split so that two $\Delta m_s = \pm 1$ transitions can be detected, as illustrated in Fig. 9–23(B). Earlier we mentioned that three peaks were observed for the epr spectrum of triplet naphthalene. Two of these are the $\Delta m_s = \pm 1$ transitions shown in Fig. 9–23(B); the third is the $\Delta m_s = \pm 2$ (between $m_s = -1$ and $m_s = +1$) transition, which becomes allowed when the zero-field splitting is small compared to the microwave frequency. When this is the case, one cannot assign precise $m_s = +1, 0$, or -1 values to the states that exist. When the zero-field splitting is very large, as in Fig. 9–23(C), the m_s values become valid quantum numbers and the energies for the $\Delta m_s = \pm 1$ allowed transitions become too large to be observed in the microwave region; accordingly, no spectrum is seen.

Since the electron-electron interaction is dipolar [equation (9–37)], it is expected to be described by a symmetric tensor, the so-called zero-field splitting tensor, **D**.

$$\hat{H}_D = \hat{S} \cdot \mathbf{D} \cdot \hat{S} \tag{9-38}$$

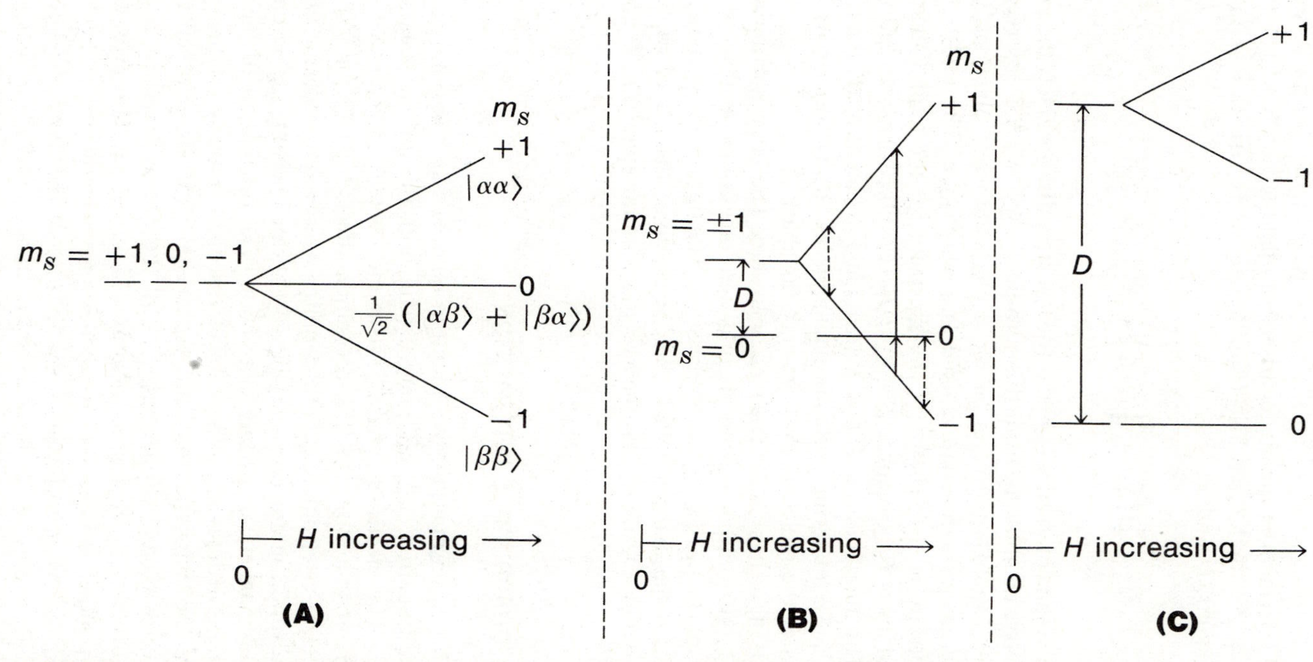

FIGURE 9–23. The effects of zero-field splitting on the expected epr transitions. (A) No zero-field effects. (B) Moderate zero-field splittings. The dashed arrows show the fixed frequency result, and the solid arrows show the fixed field result. (C) Large zero-field effects. The magnetic field is assumed to be parallel to the dipolar axis in the molecule.

The **D** tensor elements have the same form as those for **T** given in equation (9–25). This dipolar **D** tensor accounts for the large anisotropy observed in the spectrum of a bi-radical. In terms of the principal axes that diagonalize the zero-field splitting tensor, we can write

$$\hat{H}_D = -X\hat{S}_x^{\,2} - Y\hat{S}_y^{\,2} - Z\hat{S}_z^{\,2}$$

where X is the D_{xx} element in the diagonalized zero-field tensor. Since the tensor is traceless, $X + Y + Z = 0$, the zero-field Hamiltonian can be written in terms of two independent constants, D and E:

$$\hat{H}_D = D\left(\hat{S}_z^{\,2} - \frac{1}{3}\hat{S}\cdot\hat{S}\right) + E(\hat{S}_x^{\,2} - \hat{S}_y^{\,2})$$

Since $X = Y$ in a system of axial symmetry, the last term disappears. If the molecule has cubic symmetry, no splitting from this zero-field effect will be observed.

Operating with this Hamiltonian on the triplet state wave functions, one can calculate the energies as a function of field and orientation. The results indicate substantial anisotropy in the spectrum. The spectrum for the naphthalene triplet state is described by g (isotropic) $= 2.0030$, $D/hc = +0.1012\ \text{cm}^{-1}$, and $E/hc = -0.0141\ \text{cm}^{-1}$. The magnitudes of D and E are related to how strongly the two spins are interacting. These quantities become negligible as the two electrons become localized in parts of the molecule that are very far apart.

In liquids, the traceless tensor **D** averages to zero. The large fluctuating fields arising from the large, rotating anisotropic spin-spin forces in molecules with appreciable zero-field splitting cause effective relaxation. The lines in the spectra are thus usually too broad to be detected. With some exceptions, the epr of triplet states cannot be observed in solution unless the two spins are far apart (*i.e.*, D and E are small).

9–9 NUCLEAR QUADRUPOLE INTERACTION

A nucleus that has a nuclear spin quantum number $I \geqq 1$ also has an electric moment, and the unpaired electron interacts with both the nuclear magnetic and electric moments. The electric field gradient at the nucleus can interact with the quadrupole moment as in nqr, and this interaction affects the electron spin energy states via the nuclear-electronic magnetic coupling as a second order perturbation. The effect of quadrupole interaction is usually complicated because it is accompanied by a much larger magnetic hyperfine interaction. The orientation of the nuclear moment is quantized with respect to both the electric field gradient and the magnetic field axis. When the magnetic field and the crystal axes are parallel, the only quadrupole effect is a small displacement of all the energy levels by a constant amount, which produces no change in the observed transitions. However, when the two axes are not parallel, the effect is a competition between the electric field and the magnetic field. This has two effects on the spacing of the hyperfine lines: (1) a displacement of all energy levels by a constant amount and (2) a change in the separation of the energy levels that causes the spacing between adjacent epr lines to be greater at the ends of the spectrum than in the middle.

This quadrupole effect can easily be distinguished from another second order effect that produces a gradual increase or decrease in the spacing from one end of the spectrum to the other. The variation in the spacing from this other second order effect occurs when the magnetic field produced by the nucleus becomes comparable in magnitude to the external field. In this case, the unequal spacing can be eliminated by increasing the applied magnetic field.

A further effect of this competition between the quadrupolar electric field and the magnetic field is the appearance of additional lines that are normally forbidden by the selection rule $\Delta m_I = 0$. Both $\Delta m_I = \pm 1$ and $\Delta m_I = \pm 2$ transitions are sometimes observed.[22] An analysis of the forbidden lines gives the nuclear quadrupole coupling constant. The approach involves a single-crystal epr study of a compound doped into a diamagnetic host. A spectrum containing these transitions is illustrated in Fig. 9–24 for bis(2,4-pentanedionato)copper(II) [^{63}Cu(acac)$_2$], doped into Pd(acac)$_2$. The forbidden transitions are marked in Fig. 9–24(A) with arrows and

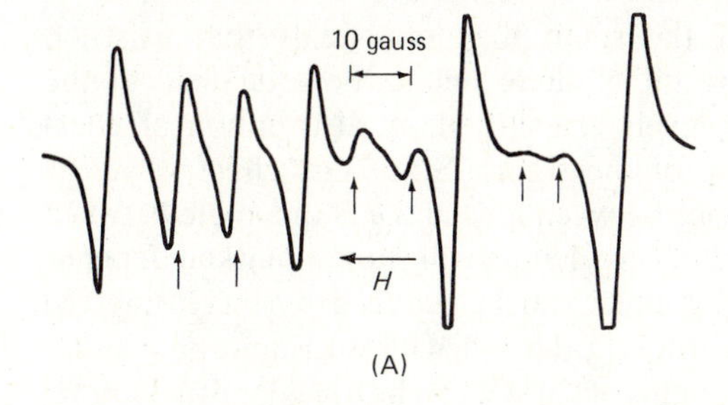

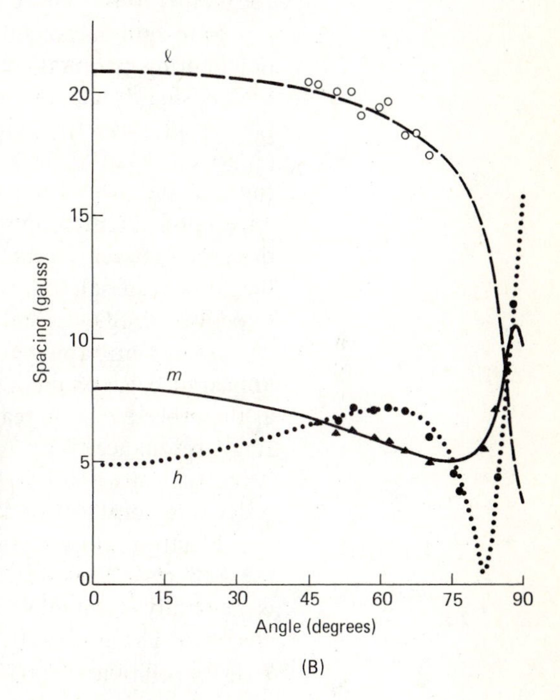

FIGURE 9–24. (A) The Q-band spectrum of Cu(acac)$_2$ at $\theta = 87°$. Forbidden lines are indicated by arrows. (B) Angular dependence of the spacing between each pair of forbidden lines ($\Delta m_I = \pm 1$) in the spectrum in (A). The curves are calculated values using $Q' = 3.4 \times 10^{-4}$ cm^{-1}, and the symbols are experimental values. The letters ℓ, m, and h represent low-field, medium-field, and high-field pairs, respectively. [Reprinted with permission from H. So and R. L. Belford, J. Amer. Chem. Soc., *91*, 2392 (1969). Copyright by the American Chemical Society.]

the other bands are the four allowed transitions [$I(^{63}\text{Cu}) = 3/2$]. The spacing and intensity of the forbidden lines vary considerably with the angle θ. The variation in the spacing is shown in Fig. 9–24(B). By matrix diagonalization, the spectra could be computer-fitted to a spin Hamiltonian:

$$\hat{H} = \beta[g_{\parallel}H_z\hat{S}_z + g_{\perp}(H_x\hat{S}_x + H_y\hat{S}_y)] + A\hat{S}_z\hat{I}_z + B(\hat{S}_x\hat{I}_x + \hat{S}_y\hat{I}_y)$$
$$+ Q'\left[\hat{I}_z^2 - \frac{1}{3}I(I+1)\right] - g_N\beta_N H\cdot\hat{I}$$

The term $Q'[I_z^2 - \frac{1}{3}I(I+1)]$ accounts for the quadrupole effects, and all other symbols have been defined previously. A value of $Q'/hc = (3.4 \pm 0.2) \times 10^{-4}$ cm^{-1} fits all the data; and for $I = 3/2$, $4Q'$ is the quantity related to the field gradient e^2Qq (see Chapter 14) at the copper nucleus, *i.e.*, $Q' = 3e^2qQ/4I(2I+1)$. In similar studies of Cu(bzac)$_2$ doped into Pd(bzac)$_2$ and of bis-dithiocarbamato copper(II) [Cu(dtc)$_2$] in diamagnetic Ni(dtc)$_2$, Q' values of $(3.3 \pm 0.21) \times 10^{-4}$ cm^{-1} and $(0.7 \pm 0.1) \times 10^{-4}$ cm^{-1}, respectively, were obtained. A Q' value of about 16×10^{-4} cm^{-1} is expected for a hole in a $d_{x^2-y^2}$ orbital of a free copper ion, and covalency should lower the value found (Q' measures the electric field gradient in the direction of the symmetry axis). The charge distribution in the sulfur complex is very close to symmetrical and was interpreted to indicate a very considerable amount of covalency in the copper-ligand bond.

9–10 LINE WIDTHS IN EPR

In this section, we shall discuss briefly a number of factors that influence the epr line width. Many of these are similar to effects we have discussed in nmr.

Broadening due to spin-lattice relaxation results from the interaction of the paramagnetic ions with the thermal vibrations of the lattice. The variation in spin-lattice relaxation times in different systems is quite large. For some compounds it is sufficiently long to allow the observation of spectra at room temperature, while for others this is not possible. Since relaxation times usually increase as the temperature decreases, many salts of the transition metals need to be cooled to liquid N_2, H_2, or He temperatures before well-resolved spectra are observed.

Spin-spin interaction results from the small magnetic fields that exist on neighboring paramagnetic ions. As a result of these fields, the total field at the ions is slightly altered and the energy levels are shifted. A distribution of energies results, which produces broadening of the signal. Since this effect varies as $(1/r^3)(1 - 3\cos^2\theta)$, where r is the distance between ions and θ is the angle between the field and the symmetry axis, this kind of broadening will show a marked dependence upon the direction of the field. The effect can be reduced by increasing the distance between paramagnetic ions by diluting the salt with an isomorphous diamagnetic material; for example, small amounts of $CuSO_4$ can be doped into a diamagnetic host $ZnSO_4$ crystal.

As in nmr, rapid chemical processes also influence the spectral line widths and appearance. Resonances that are separate in the stopped exchange limit will broaden as the process rate increases, and they will then coalesce to give a weighted-average single resonance. With one-half the width at half height of an organic free radical being typically ~0.1 G, significant line broadening occurs for processes with first order rate constants of $5 \times 10^7\ sec^{-1}$.

Electron spin exchange processes are very common in free radical systems, and these effects drastically influence line width and spectral appearance. In solution, this is generally a bimolecular process in which two radicals collide and exchange electrons. The effects are similar to those observed in nmr and are illustrated in Fig. 9–25 for solutions of varying concentrations of di-*t*-butyl nitroxide, $[(CH_3)_3C]_2NO$. As

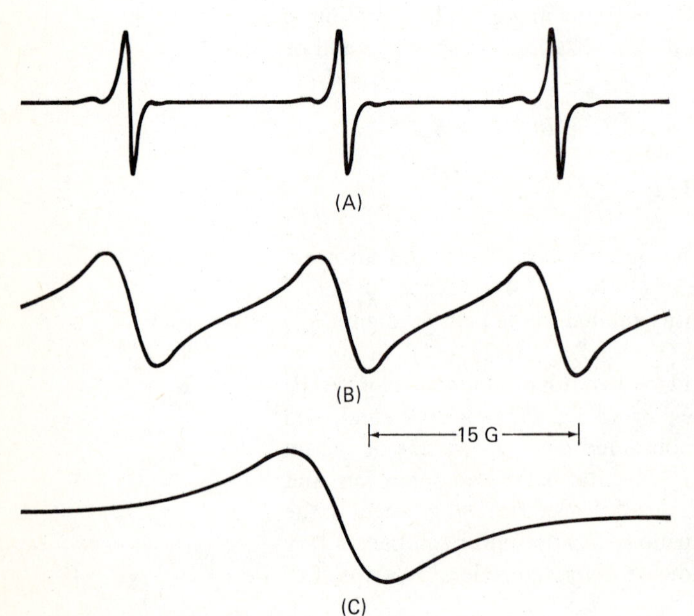

FIGURE 9–25. EPR spectra of di-*t*-butyl nitroxide radical in C_2H_5OH at 25°C. (A) 10^{-4} M; (B) 10^{-2} M; (C) 10^{-1} M.

the concentration increases on going from (A) to (C), the rate of bimolecular exchange increases and the resonances broaden. The rate constant for this process has been evaluated in the solvent *N,N*-dimethylformamide and was found to be 7×10^9 l mole^{-1} sec^{-1}, a value corresponding to the diffusion controlled limit. As the solution whose spectrum is shown in Fig. 9–25(C) becomes considerably more concentrated, the single line resonance sharpens as in the very fast exchange limit in nmr. This sharpening is referred to as *exchange narrowing*.

Electron transfer between a radical and a diamagnetic species can also occur at a rate that causes line broadening of the epr spectrum. One of the first systems investigated involved the electron exchange between naphthalene and the naphthalene negative ion. A second order electron transfer rate constant of 6×10^7 l mole^{-1} sec^{-1} was found[25a] in the solvent THF. This is a factor of one hundred slower than the diffusion controlled rate constant; it is thought to be slower because the positive counter-ion of the negative ion radical in the ion pair must also be transferred with the electron.

Many effects cause the line widths of one band to differ from those of another in a given spectrum. We mentioned earlier that the spin density on the CH_3 proton of ethyl amine is conformation-dependent. The time dependency of this type of process can influence the line widths of different protons in a molecule differently. Rapid interchange between various configurations of an ion pair with an anion or cation radical can also lead to greater line broadening of certain resonances than of others.[25b,26]

9–11 THE SPIN HAMILTONIAN

The spin Hamiltonian operates only on the spin variables and describes the different interactions that exist in systems containing unpaired electrons. It can be thought of as a shorthand way of representing the interactions described above. The epr spin Hamiltonian for an ion in an axially symmetric field (*e.g.*, tetragonal or trigonal) is:

$$\hat{H} = D\left[\hat{S}_z^{\,2} - \frac{1}{3}S(S+1)\right] + g_{\parallel}\beta H_z\hat{S}_z + g_{\perp}\beta(H_x\hat{S}_x + H_y\hat{S}_y) \tag{9-39}$$
$$+ A_{\parallel}\hat{S}_z\hat{I}_z + A_{\perp}(\hat{S}_x\hat{I}_x + \hat{S}_y\hat{I}_y) + Q'\left[\hat{I}_z^{\,2} - \frac{1}{3}I(I+1)\right] - g_N\beta_N H_0\cdot\hat{I}$$

The first term describes the zero-field splitting, the next two terms describe the effect of the magnetic field on the spin degeneracy remaining after zero-field splitting, the terms $A_{\parallel}$ and $A_{\perp}$ measure the hyperfine splitting parallel and perpendicular to the unique axis, and Q' measures the changes in the spectrum produced by the quadrupole interaction. All of these effects have been discussed previously. The final term takes into account the fact that the nuclear magnetic moment μ_N can interact directly with the external field $\mu_N H_0 = g_N\beta_N H_0 I$. This interaction can affect the paramagnetic resonance only when the unpaired electrons are coupled to the nucleus by nuclear hyperfine or quadrupole interactions. Even when such coupling occurs, the effect is often negligible in comparison with the other terms.

In the case of a distortion of lower symmetry, the principal g values become g_x, g_y, and g_z; the hyperfine coupling constants become A_x, A_y, and A_z; and two additional terms need to be included, *i.e.*, $E(\hat{S}_x^{\,2} - \hat{S}_y^{\,2})$ as an additional zero-field splitting and $Q''(\hat{I}_x^{\,2} - \hat{I}_y^{\,2})$ as a further quadrupole interaction. The symbols P and P' are often employed for Q' and Q'', respectively.

The importance of the spin Hamiltonian is that it provides a standard phenomenological way in which the epr spectrum can be described in terms of a small number of constants. Once the values for the constants have been determined from experiment, calculations relating these parameters back to the electronic configurations and the energy states of the ion are possible. It should be pointed out that not all terms in equation (9–39) are of importance for any given system. For a nucleus with no spin, all terms containing I are zero. In the absence of zero-field splitting, the first term is equal to zero.

9–12 MISCELLANEOUS APPLICATIONS

When the epr spectrum for $CuSiF_6 \cdot 6H_2O$, diluted with the corresponding diamagnetic Zn salt, was obtained at 90° K, the spectrum was found to consist of one band with partially resolved hyperfine structure and a nearly isotropic g value.[27] In a cubic field, the ground state of Cu^{2+} is orbitally doubly degenerate. Although $Cu(H_2O)_6SiF_6$ has trigonal rather than cubic symmetry, this orbital degeneracy is not destroyed. Thus, Jahn-Teller distortion will occur. However, there are three distortions with the same energy that will resolve the orbital degeneracy. These are three tetragonal distortions with mutually perpendicular axes (elongation along the three axes connecting *trans* ligands). As a result, three epr transitions are expected, one for each species. Since only one transition was found, it was proposed that the crystal field resonates among the three distortions.[28] When the temperature is lowered, the spectrum becomes anisotropic and consists of three sets of lines corresponding to three different copper ions distorted by three different tetragonal distortions.[29] The transition takes place between 50° and 12° K; the three perpendicular tetragonal axes form the edges of a unit rectangular solid and the trigonal axis is the body diagonal. Other mixed copper salts have been found to undergo similar transitions: $(Cu,Mg)_3La_2(NO_3)_{12} \cdot 24D_2O$ between 33° and 45° K, and $(Zn,Cu)(BrO_3)_2 \cdot 6H_2O$, incomplete below 7° K. The following parameters were reported for $CuSiF_6 \cdot 6H_2O$:[30]

90° K	20° K
$g_{\parallel} = 2.221 \pm 0.005$	$g_z = 2.46 \pm 0.01$
$g_{\perp} = 2.230 \pm 0.005$	$g_x = 2.10 \pm 0.01$
$A = 0.0021 \pm 0.0005\ cm^{-1}$	$g_y = 2.10 \pm 0.01$
$B = 0.0028 \pm 0.0005\ cm^{-1}$	$A_z = 0.0110 \pm 0.0003\ cm^{-1}$
	$\left.\begin{matrix} A_x \\ A_y \end{matrix}\right\} < 0.0030\ cm^{-1}$

No quadrupole interaction was resolved. A similar behavior (*i.e.*, a resonating crystal field at elevated temperatures) was detected in the spectra of some tris complexes of copper(II) with 2,2′-dipyridine and 1,10-phenanthroline.[31]

The epr spectrum of the complex $[(NH_3)_5Co{-}O{-}O{-}Co(NH_3)_5]^{5+}$ is an interesting example to demonstrate how structural information can be derived from spin density information and from hyperfine splitting. This complex can be formulated as: (1) two cobalt(III) atoms connected by an O_2^- bridge; (2) cobalt(III) and cobalt(IV) atoms connected by a peroxy, O_2^{2-} bridge; (3) two equivalent cobalt atoms, owing to equal interaction of one unpaired electron with both cobalt atoms; (4) interaction of the electron with both cobalt atoms, but more with one than with the other.

If (1) were the structure, a single line would result, while (2) would give rise to eight lines ($I = \frac{7}{2}$ for Co). Structure (3) would result in fifteen lines and (4) in sixty-four. It was found[32] that the spectrum consists of fifteen lines, eliminating the unlikely structures (2) and (4) and supporting structures (1) or (3) or a mixture of both. An ^{17}O hyperfine result would be required to determine the importance of structure (1).

A study[33] of the 1:1 adducts of cobalt(II) complexes with dioxygen has led to an internally consistent interpretation of the ^{17}O and ^{59}Co isotropic and anisotropic hyperfine coupling constants. Depending on the ligands attached to cobalt, the adducts are described as consisting of bound O_2 or O_2^-.

One of the advantages of epr is its extreme sensitivity to very small amounts of paramagnetic materials. For example, under favorable conditions a signal for diphenylpicrylhydrazyl (DPPH) radical can be detected if there is 10^{-12} gram of material in the spectrometer. This great sensitivity has been exploited in a study of the radicals formed by heating sulfur. When sulfur is heated, the diamagnetic S_8 ring is cleaved to produce high molecular weight S_x chains that have one unpaired electron at each end. The chains are so long that the concentration of radicals is low, and paramagnetism cannot be detected with a Gouy balance. An epr signal was detected,[34] and the number of unpaired electrons (which is proportional to the area under the absorption curve) was determined by comparing the area of this peak with the area of a peak resulting from a known concentration of added radicals from DPPH. The total number of radicals in the system is thus determined, and since the total amount of sulfur used is also known, the average molecular weight of the species $\cdot SS_xS\cdot$ can be calculated. The radical concentration at 300° C was $1.1 \times 10^{-3} M$, and the average chain length at 171° C was 1.5×10^6 atoms. By studying the radical concentration as a function of temperature, a heat of dissociation of the S—S bond of 33.4 kcal mole^{-1} per bond was obtained.

Copper(II) forms complexes of varying geometries, which have similar electronic spectra and magnetic susceptibilities. Thus, it is often difficult to infer the geometries of these materials in solution or in media other than the solid, where single crystal x-ray studies can be used. A recent ^{17}O study of the five-coordinate adducts formed by various Lewis bases and hexafluoroacetylacetonate copper(II) describes[35] a procedure for determining whether apical or basal isomers of a square pyramid are formed.

This high sensitivity of epr measurements has been of great practical utility in biological systems.[36] Many metalloproteins have been studied in order to determine the metal's oxidation state, the coordination number of the metal, and the kinds of ligands attached. The measurements are generally made on frozen solutions. The interpretation of the results is difficult, and conclusions are based upon analogies between these spectra and those of model compounds. These applications are more appropriately considered in the chapter on the epr spectra of transition metal ions.

There have been several studies reported in which epr has been employed to identify and provide structural information about radicals generated with high energy radiation.[37] The materials O_2^-, ClO, ClO_2, PO_3^{2-}, SO_3^-, and ClO_3 are among the many interesting radicals produced by this technique.

Two types of double resonance techniques are available, electron-nuclear double resonance (ENDOR) and electron-electron double resonance (ELDOR). The main applications of ENDOR to date involve enhanced accuracy and resolution in determining small hyperfine couplings, in determining which nuclei are responsible for splitting, and in determining nuclear *g*-values. ELDOR can be used to aid in assigning overlapping resonance peaks in much the same way that double resonance is used in nmr. These are specialized uses, and the reader is referred to Wertz and Bolton[38] for details.

REFERENCES

1. F. J. Adrian, *et al.,* Adv. Chem. Ser., *36,* 50 (1960).
2. A. Carrington, Quart. Revs., *17,* 67 (1963).
3. A. H. Maki and B. R. McGarvey, J. Chem. Phys., *29,* 35 (1958).
4. C. E. Strouse and L. F. Dahl, J. Amer. Chem. Soc., *93,* 6032 (1971).
5. M. J. S. Dewar, "Molecular Orbital Theory of Organic Chemistry," McGraw-Hill Book Co., New York (1969).
6. R. W. Fessenden and R. H. Schuler, J. Chem. Phys., *39,* 2147 (1963).
7. R. S. Drago and H. Petersen, Jr., J. Amer. Chem. Soc., *89,* 3978 (1967).
8. A. D. McLachlan, Mol. Phys., *3,* 233 (1960).
9. J. A. Pople, *et al.*, J. Amer. Chem. Soc., *90,* 4201 (1968) and references therein; J. A. Pople and D. L. Beveridge, "Approximate Molecular Orbital Theory," McGraw-Hill BookCo., New York (1970).
10. M. Karplus and G. K. Fraenkel, J. Chem. Phys., *35,* 1312 (1961).
11. a. R. E. Cramer and R. S. Drago, J. Amer. Chem. Soc., *90,* 4790 (1968).
    b. P. D. Sullivan, J. Amer. Chem. Soc., *90,* 3618 (1968).
12. M. Broze, Z. Luz and B. L. Silver, J. Chem. Phys., *46,* 4891 (1967).
13. I. Biddles and A. Hudson, Mol. Phys., *25,* 707 (1973).
14. a. R. S. Drago and H. Petersen, Jr., J. Amer. Chem. Soc., *89,* 5774 (1967).
    b. R. E. Cramer and R. S. Drago, J. Chem. Phys., *51,* 464 (1969).
15. R. W. Fessenden and P. Neta, Chem. Phys. Letters, *18,* 14 (1973).
16. R. Fitzgerald and R. S. Drago, J. Amer. Chem. Soc., *90,* 2523 (1968).
17. a. R. Livingston and H. Zeldes, J. Chem. Phys., *41,* 4011 (1964).
    b. J. C. W. Chien and L. C. Dickerson, Proc. Natl. Acad. Sci., USA, *69,* 2783 (1972).
18. H. M. McConnell and R. W. Fessenden, J. Chem. Phys., *31,* 1688 (1959).
19. T. Cole and C. Heller, J. Chem. Phys., *34,* 1085 (1961).
20. H. M. McConnell, *et al.,* J. Amer. Chem. Soc., *82,* 766 (1960).
21. R. W. Fessenden and R. H. Schuler, J. Chem. Phys., *43,* 2704 (1965).
22. P. W. Atkins and M. C. R. Symons, J. Chem. Soc., 4794 (1962).
23. N. Hirota, C. A. Hutchison, Jr., and P. Palmer, J. Chem. Phys., *40,* 3717 (1964) and references therein.
24. H. So and R. L. Belford, J. Amer. Chem. Soc., *91,* 2392 (1969); R. L. Belford, D. T. Huang and H. So, Chem. Phys. Letters, *14,* 592 (1972) and references therein.
25. a. R. L. Ward and S. I. Weissman, J. Amer. Chem. Soc., *79,* 2086 (1957); T. A. Miller and R. N. Adams, J. Amer. Chem. Soc., *88,* 5713 (1966).
    b. J. R. Bolton and A. Carrington, Mol. Phys., *5,* 161 (1962).
26. J. H. Freed and G. K. Fraenkel, J. Chem. Phys., *37,* 1156 (1962).
27. B. Bleaney and D. J. E. Ingram, Proc. Phys. Soc., *63,* 408 (1950).
28. A. Abragam and M. H. L. Pryce, Proc. Phys. Soc., *63,* 409 (1950).
29. B. Bleaney and K. D. Bowers, Proc. Phys. Soc., *65,* 667 (1952).
30. B. Bleaney, K. D. Bowers, and R. S. Trenam, Proc. Roy. Soc. (London) *A228,* 157 (1955).
31. H. C. Allen, Jr., G. E. Kokoszka, and R. G. Inskeep, J. Amer. Chem. Soc., *86,* 1023 (1964).
32. E. A. V. Ebeworth and J. A. Weil, J. Phys. Chem., *63,* 1890 (1959).
33. B. Tovrog, D. Kitko, and R. S. Drago, J. Amer. Chem. Soc., *98,* 5144 (1976).
34. D. M. Gardner and G. K. Fraenkel, J. Amer. Chem. Soc., *78,* 3279 (1956).
35. D. McMillin, R. S. Drago, and J. A. Nusz, J. Amer. Chem. Soc., *98,* 3120 (1976).
36. See, for example, C. S. Yang and F. M. Heuennekens, Biochemistry, *9,* 2127 (1970).
37. P. W. Atkins and M. C. R. Symons, "The Structure of Inorganic Radicals," Elsevier Publishing Co., Amsterdam (1967).
38. J. E. Wertz and J. R. Bolton, "Electron Spin Resonance," Mc-Graw Hill Book Co., New York (1972), Chapter 13.

GENERAL REFERENCES*

J. E. Wertz and J. R. Bolton, "Electron Spin Resonance," McGraw-Hill Book Co., New York (1972).

C. P. Poole and H. A. Farach, "The Theory of Magnetic Resonance," Wiley-Interscience, New York (1972).

A. Carrington and A. D. McLachlan, "Introduction to Magnetic Resonance," Harper and Row, New York (1967).

C. P. Poole, Jr., "Electron Spin Resonance—A Comprehensive Treatise on Experimental Techniques," Interscience Publishers, Inc., New York (1967).

B. A. Goodman and J. B. Raynor, Adv. Inorg. Chem. Radiochem., *13,* 135 (1970).

R. Bersohn, "Determination of Organic Structures by Physical Methods," Vol. 2, eds. F. C. Nachod and W. D. Phillips, Academic Press, New York (1962).

A. Carrington, Quart. Revs., *17,* 67 (1963).

N. M. Atherton, "Electron Spin Resonance," Halsted Press, London (1973).

*For references heavily oriented toward the epr of transition metal ions, see Chapter 13.

P. W. Atkins and M. C. R. Symons, "The Structure of Inorganic Radicals," Elsevier Publishing Co., Amsterdam (1967).
C. P. Slichter, "Principles of Magnetic Resonance," Harper and Row, Publishers, New York (1963).
R. M. Golding, "Applied Wave Mechanics," Van Nostrand-Reinhold, New York (1964).
A. Abragam, "The Principles of Nuclear Magnetism," Clarendon Press, Oxford (1961).
G. E. Pake and T. L. Estle, "Paramagnetic Resonance," 2nd Ed., W. A. Benjamin, Inc., New York (1973).
H. Fischer, "Magnetic Properties of Free Radicals," Landolt-Bernstein Tables, New Series Group II, Vol. I, Springer-Verlag, Berlin (1965).

EXERCISES

1. Convert the derivative curves below to absorption curves:

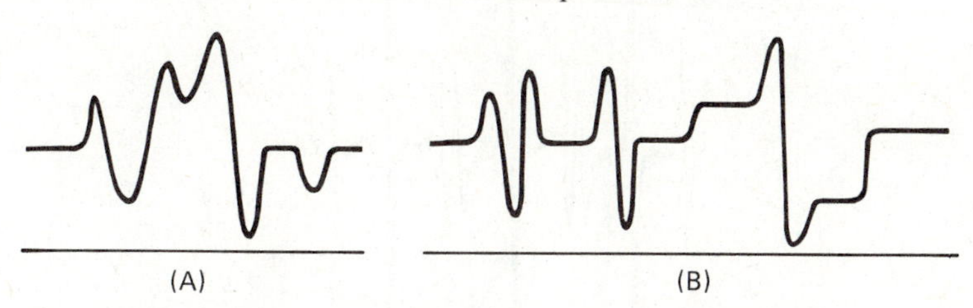

2. a. How many hyperfine peaks would be expected from delocalization of the odd electron in dibenzene chromium cation onto the rings?

   b. Using a procedure similar to that in Fig. 9–6, explain how the number of peaks arises and what the relative intensities would be.

3. a. Copper(II) acetate is a dimer, and the two copper atoms are strongly interacting. The epr spectrum consists of seven lines with intensity ratios 1:2:3:4:3:2:1. Copper nuclei have an I value of $^3/_2$, and copper acetate consists of a ground state that is a singlet and an excited state that is a triplet. Explain the number and relative intensity of the lines in the spectrum. [For answer, see B. Bleaney and K. D. Bowers, Proc. Roy. Soc. (London), *A214,* 451 (1952).]

   b. What would you expect to happen to the signal intensity as a sample of copper acetate is cooled? Why?

4. Predict the epr spectrum for $(SO_3)_2^{\bullet}NO^{2-}$.

5. The mono negative ion $[O{-}C_6H_4{-}O\cdot]^-$ (the ring bearing H, H above and H, H below) can be prepared.

   a. How many lines are expected in the spectrum, and what would be the relative intensities of these?

   b. What evidence would you employ and what experiments could be carried out to indicate electron delocalization onto the oxygen?

   c. The magnitude of a_H in this material is 2.37 gauss. Compare the spin density on hydrogen in this molecule with that on a hydrogen atom.

   d. How would the sign of the proton hyperfine coupling constant indicate whether the odd electron was in a sigma or pi molecular orbital?

   e. Using the value of a_H given above and the fact that the unpaired electron is in the π-system, calculate the spin density on the nearest neighbor carbon.

6. The ^{13}C hyperfine coupling in the methyl radical is 41 gauss, and the proton hyperfine coupling is 23 gauss. Sketch the spectrum expected for $^{13}\cdot CH_3$ radical. [For answer, see T. Cole, *et al.,* Mol. Phys., *1,* 406 (1958).]

7. Assume that all hyperfine lines can be resolved and sketch the spectrum for the chlorobenzene anion radical.

8. Assuming all other factors constant, would line broadening be greater for a bimolecular process with a rate constant of 10^7 or with a rate constant of 10^{10}?

9. How many lines would you expect in the epr spectrum of $(CN)_5CoO_2Co(NH_3)_5$? Explain.

10. The spectrum below is obtained for the NH_2 radical:

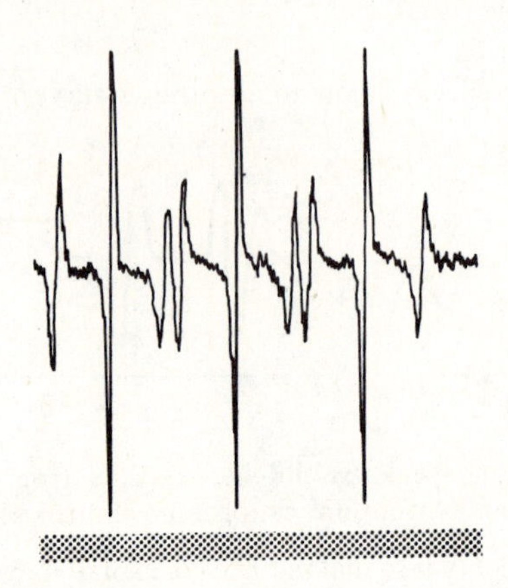

   a. Convert it to an absorption spectrum.

   b. How could you determine whether the larger or smaller splitting is due to hydrogen?

   c. Assume that the larger splitting is due to nitrogen. Construct a diagram similar to Fig. 9–10 to explain the spectrum.

11. a. How many lines would you expect in the spectrum of the hypothetical molecule SCl_3 (I for S = 0 and Cl = $\frac{3}{2}$)?

   b. Using a procedure similar to that in Fig. 9–6 and Fig. 9–7, explain how this number arises and indicate the transitions with arrows. State what the expected relative intensities would be.

12. The epr spectrum of the cyclopentadiene radical ($C_5H_5\cdot$) rapidly rotating in a single crystal of cyctopentadiene is given below.

   a. Write the appropriate spin Hamiltonian.

   b. Interpret the spectrum.

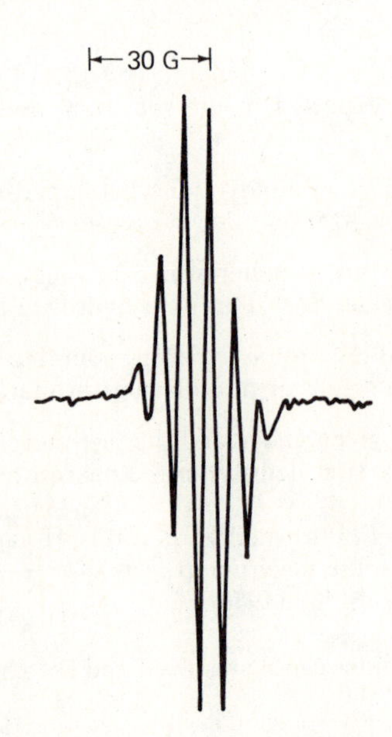

13. Interpret the epr spectrum of $\cdot CH_2OH$ given below.

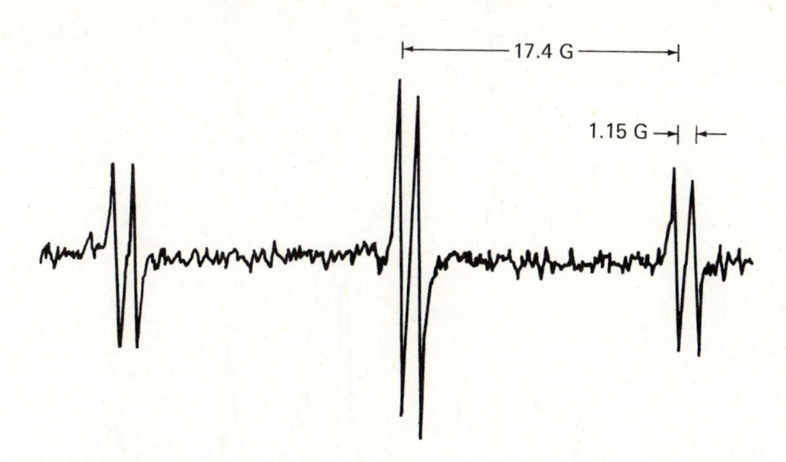

14. The epr spectrum of $C_6H_5Ge(CH_3)_3^-$ is given below. Interpret this spectrum, given the fact that all of the splittings arise from the phenyl ring protons. Calculate the *a*-values.

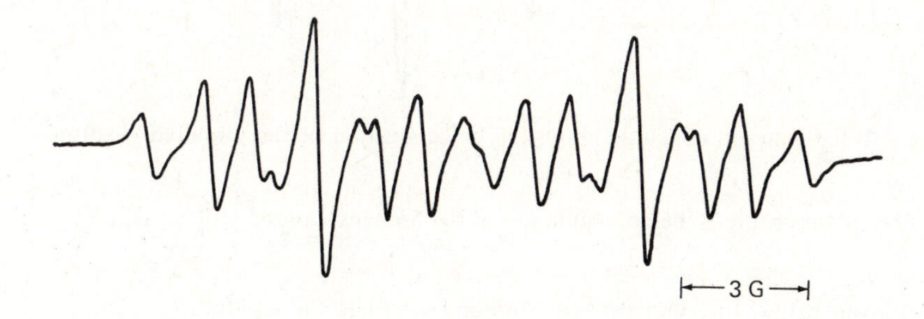

15. Given below is the epr spectrum of

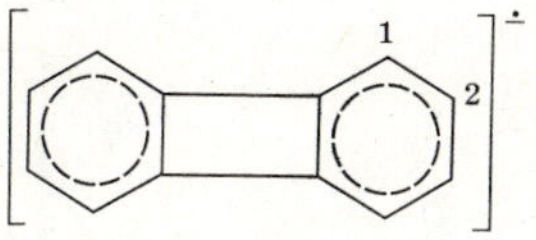

Write the spin Hamiltonian, interpret the spectrum, and report the *a* values.

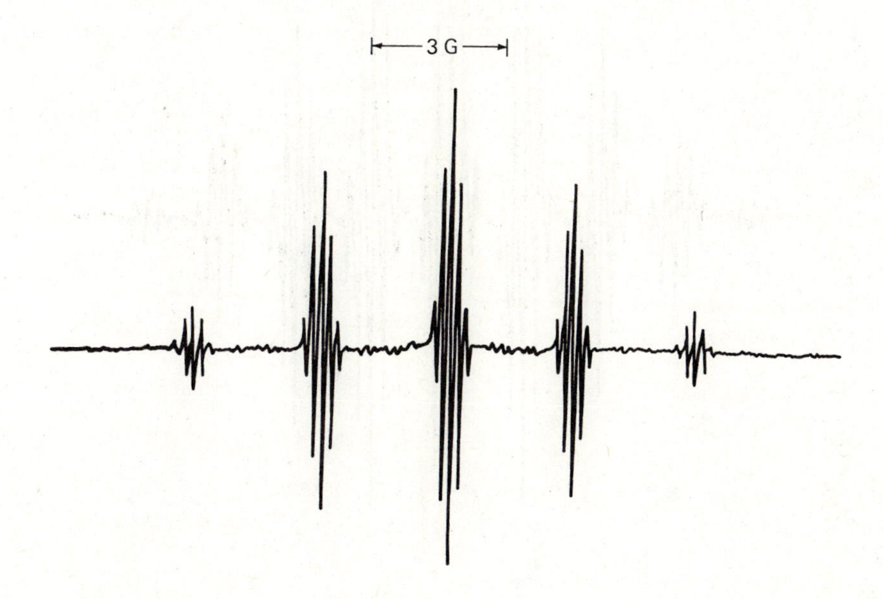

16. a. Interpret the epr spectrum and calculate the *a* value(s) for the substituted nitrosyl nitroxide,

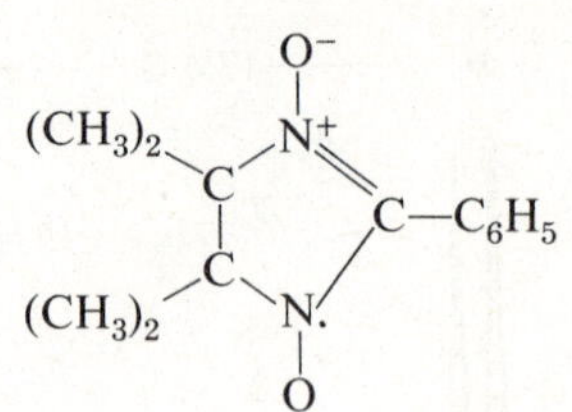

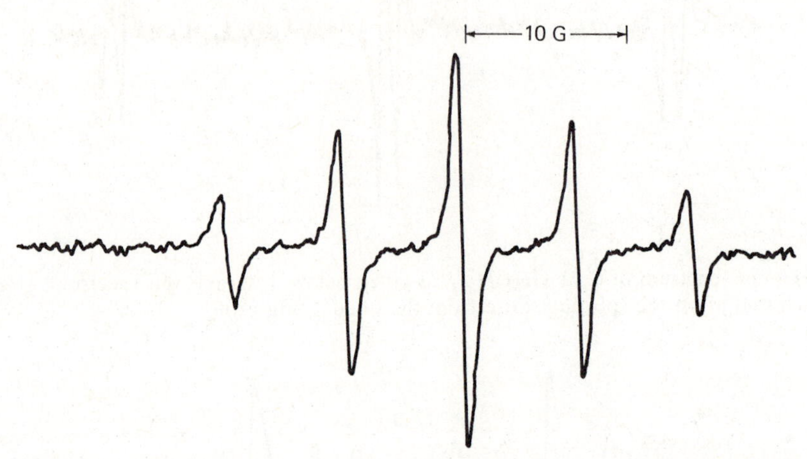

b. What can you conclude about the delocalization of the unpaired electron?

17. The epr spectrum of the potassium salt of the biphenyl anion, [3′ 2′ 2 3; 4′ 1′ 1 4; 5′ 6′ 6 5]$^{\dot{-}}$, is given below. Interpret the spectrum and calculate the *a* values.

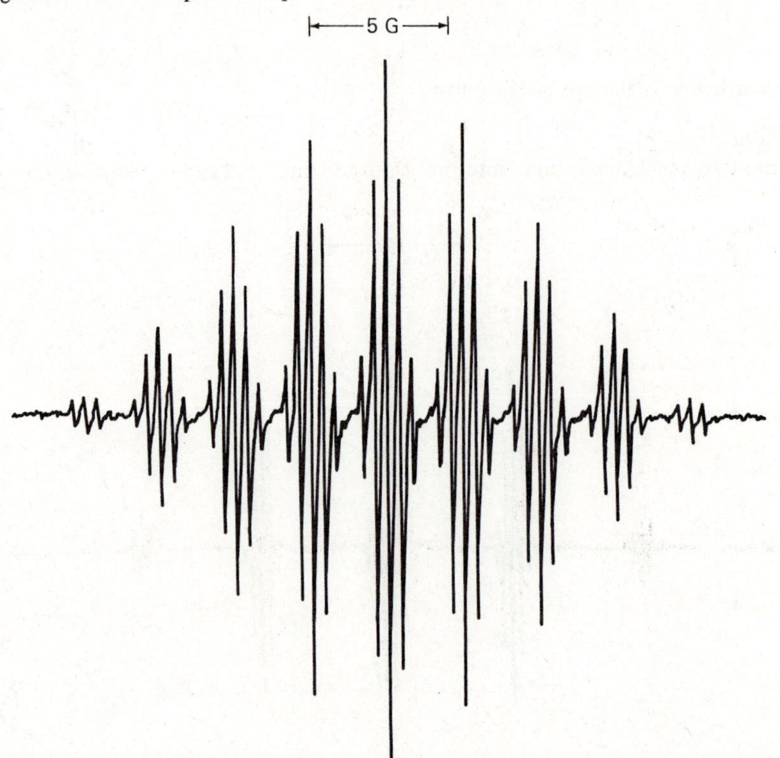

18. The epr spectrum of the pyrazine anion, $\left[\,|\text{N}\bigcirc\text{N}|\,\right]^{\dot{-}}$, is given below. Interpret the spectrum and calculate the a value(s).

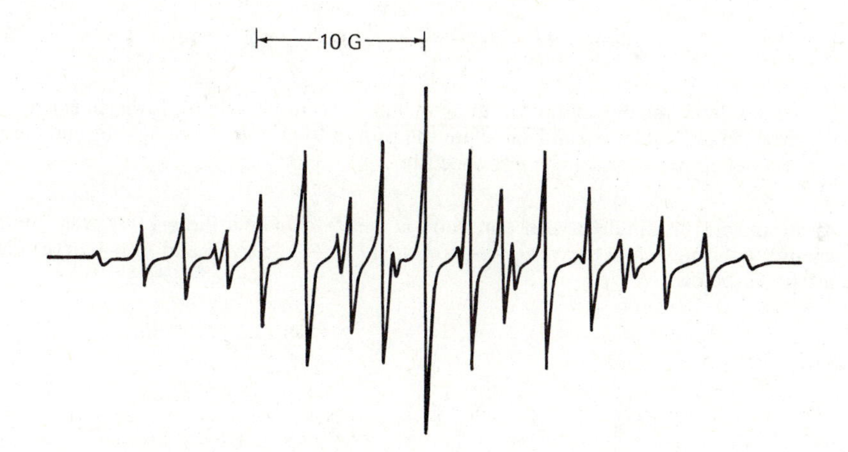

19. Below is the epr spectrum of a sample of S_2^- that has 40% ^{32}S ($I = 0$) nuclei and 60% ^{33}S ($I = 3/2$) nuclei. Interpret the spectrum and determine a for ^{33}S.

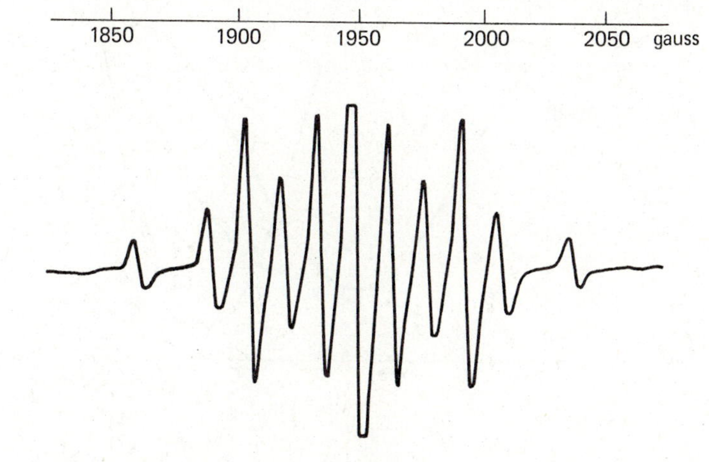

20. McConnell's relation allows a rough prediction of the magnitude of proton hyperfine coupling constants in conjugated organic systems by performing a Hückel m.o. calculation on the system. The hyperfine constant for the ith proton, a_i, is given by $a_i = Q\rho_i$, where $\rho_i = C_{ji}^2$. C_{ji} is the coefficient of the various carbon $2p_z$ atomic orbitals in the molecular orbital containing the unpaired electron.

    a. The carbon $2p_z$ atomic orbitals that make up the π-system are orthogonal to the C—H sp^2 sigma bond. Why, then, does any unpaired electron density reside on the proton?

    b. The molecular orbital scheme for benzene is:

$$\begin{array}{ccc} & \text{—— } \psi_6 & \\ \text{——} & & \text{—— } \psi_4, \psi_5 \\ \text{—}\uparrow\downarrow\text{—} & & \text{—}\uparrow\downarrow\text{— } \psi_2, \psi_3 \\ & \text{—}\uparrow\downarrow\text{— } \psi_1 & \end{array}$$

In the benzene anion, the unpaired electron can be in either ψ_4 or ψ_5. A Hückel m.o. calculation for these wave functions gives

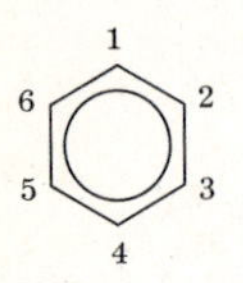

$$\psi_4 = \frac{1}{2}(\varphi_2 - \varphi_3 + \varphi_5 - \varphi_6)$$

$$\psi_5 = \frac{1}{\sqrt{12}}(2\varphi_1 - \varphi_2 - \varphi_3 + 2\varphi_4 - \varphi_5 - \varphi_6)$$

In *p*-xylene, the degeneracy of these two m.o.'s is lifted, with ψ_4 lower in energy. Use McConnell's relation and calculate the proton hyperfine coupling constants for the *p*-xylene anion. Draw the epr spectrum.

21. In an anisotropic single crystal epr study at $\nu = 9.520$ GHz, the *g*-value was found to change with rotation in the *xz* (- - - - - curve), *yz* (• • • • • curve), and *xy* (—curve) planes as shown below.

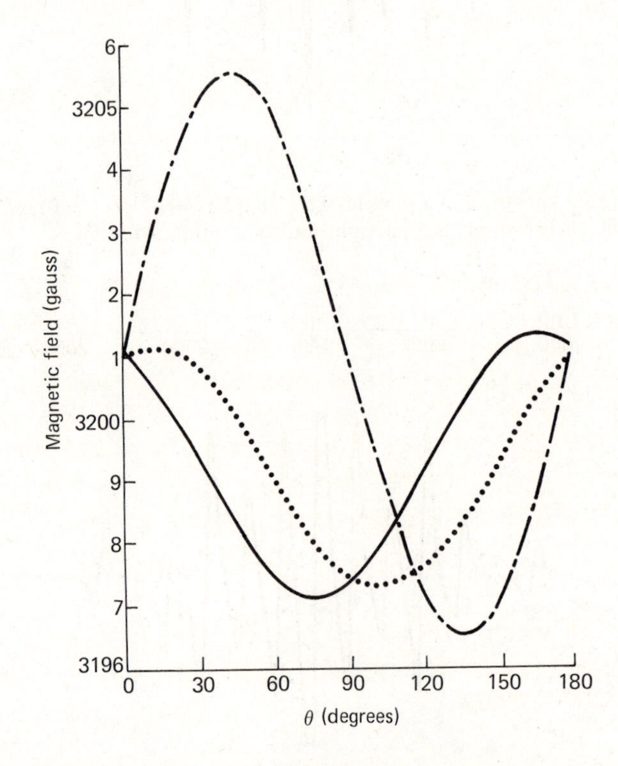

The field position for resonance, *H*, is given. This is converted to a *g*-value using $\Delta E = h\nu$ and equation (9–2). One obtains $g_{\text{eff}} = \dfrac{h\nu}{\beta H} = \dfrac{6801.9}{H}$ for a 9.520 GHz microwave frequency.

a. Interpolate from the plot and evaluate all of the elements of the g^2 tensor.

b. What would you learn by diagonalizing the g^2 tensor?

c. What steps are required to diagonalize the g^2 tensor?

d. What steps are required to obtain the direction cosine matrix?

e. Write the spin Hamiltonian.

10 THE ELECTRONIC STRUCTURE AND SPECTRA OF TRANSITION METAL IONS

INTRODUCTION

The subject of this chapter has also been the topic of several textbooks.[1-12] Here we shall present an overview of the electronic structure of transition metal ions. In so doing, we will develop some important ideas for the understanding of the spectroscopy of transition metal ion complexes—our main objective. The transition metal ion systems we shall treat in the next three chapters are covered separately because they possess unpaired electrons, which introduce several complications. As is so often the case, these complicating factors, when understood, provide a wealth of information about the compounds formed by these species. The complications arise from electron-electron interactions, spin-orbit coupling, and the influence that a magnetic field has on systems with unpaired electrons. We have discussed many of these topics earlier, but their full implication is best demonstrated with examples from transition metal ion chemistry.

FREE ION ELECTRONIC STATES

10–1 ELECTRON-ELECTRON INTERACTIONS AND TERM SYMBOLS

There are numerous ways in which one or more electrons can be arranged in the five *d*-orbitals of a gaseous metal ion. We can indicate the energy differences arising from different interelectronic repulsions and different orbital angular momenta for these various arrangements with *term symbols.* Any one term symbol groups together all of the degenerate arrangements in the gaseous ion. The simplest case to consider first is d^1. There are five ways to arrange an electron with $m_s = +\frac{1}{2}$ in the five *d*-orbitals. Each arrangement will be called a *microstate configuration.*

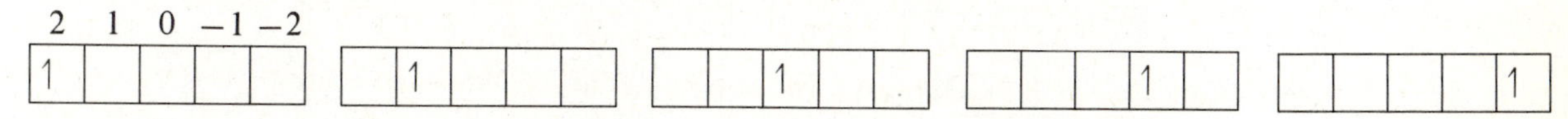

In the absence of external electric and magnetic fields, the five microstates are degenerate, and there are five others that are also degenerate with these, corresponding to $m_s = -\frac{1}{2}$. *These ten microstates comprise the tenfold degeneracy of the so-called* ^2D *term* (*vide infra*).

The ground state term for any d^n configuration can be deduced by arranging the electrons in the d-orbitals, filling those with the largest m_ℓ values first and not pairing up any electrons until each orbital has at least one; *i.e.*, Hund's rules are obeyed. The m_ℓ values of the orbitals containing electrons may be algebraically summed to produce the L value for the term. More completely, the m_ℓ quantum number for an individual electron is related to a vector with component $m_\ell(h/2\pi)$ in the direction of an applied field. The M_L value is the sum of the one-electron m_ℓ values. Vector coupling rules demand that M_L have values of $L, L-1, \ldots, -L$, so we deduce that the maximum M_L value is given by the value of L. The following letters are used to indicate the L values: S, P, D, F, G, H, I corresponding to $L = 0, 1, 2, 3, 4, 5, 6$, respectively. The spin multiplicity of a state is defined by $2S + 1$ (S, in analogy with L, is the largest possible M_S, where $M_S = \Sigma m_s$) and is indicated by the superscript to the upper left of the term symbol. The multiplicity refers to the number of possible projections of $\vec{S}$ along a magnetic field; *e.g.*, when $S = 1$, the multiplicity of three refers to $M_S = 1, 0, -1$ (giving the z-component of spin angular momentum aligned with, perpendicular to, and opposed to the field). *The total degeneracy of a term is given by* $(2L + 1)(2S + 1)$. The $2L + 1$ arrangements refer to the orbital degeneracy and are described by M_L with values of $L, L-1, \ldots, -L$. As mentioned earlier, the ^2D term symbol describes the d^1 case. It is tenfold degenerate with a fivefold orbital degeneracy corresponding to M_L values of $2, 1, 0, -1, -2$. In the d^1 ion, the ground ^2D term is the only one arising from the $3d$-orbitals. The value of the S quantum number for the term (or state) is given by the maximum M_S, which equals the sum of the m_s values of all unpaired electrons. *Complete subshells contribute nothing to L or S, because the sum of the m_s and the m_ℓ values is zero.*

Next, consider the d^2 configuration. There are forty-five ways to arrange two electrons with $m_s = \pm\frac{1}{2}$ in the five d-orbitals. Using the procedures described above for the microstate | ↑ | ↑ | | | |, we see that $L = 3$ and $S = 1$, leading to a ^3F ground term that is twenty-one fold degenerate in the absence of spin-orbit coupling. The other twenty-four microstates comprise higher energy (excited) states; *i.e.*, electron-electron repulsions are larger for these states. All of the terms for a d^n configuration can be found by constructing a table like that shown in Table 10–1. To

TABLE 10–1. MICROSTATES FOR A d^2 ION WITH POSITIVE M_L VALUES

M_L \ M_S	+1 (++) (↑↑)	0 (+−) (↑↓)	−1 (−−) (↓↓)
4		2^+2^-	
3	(2^+1^+)	2^+1^- (2^-1^+)	(2^-1^-)
2	(2^+0^+)	2^+0^- (2^-0^+) 1^+1^-	(2^-0^-)
1	$(2^+\ {-1}^+)$ $[1^+0^+]$	$2^+\ {-1}^-$ $(2^-\ {-1}^+)$ 1^+0^- $[1^-0^+]$	$(2^-\ {-1}^-)[1^-0^-]$
0	$(2^+\ {-2}^+)[1^+\ {-1}^+]$	$2^+\ {-2}^-$ $(2^-\ {-2}^+)$ $1^+\ {-1}^-$ $[1^-\ {-1}^+]$ 0^-0^+	$(2^-\ {-2}^-)[1^-\ {-1}^-]$

be systematic, we can begin construction of the table with the row $M_L = 4$. This M_L value can be obtained only by having two electrons in $m_\ell = 2$, and this can be done only if the spins are paired up. The resulting microstate is abbreviated as 2^+2^-. The microstate is placed in the $M_S = 0$ column. Next, ways to obtain $M_L = 3$ are shown. The possibilities are 2^+1^+, 2^+1^-, 2^-1^+, and 2^-1^-, corresponding to M_S values of $+1$, 0, 0, and -1, respectively. The procedure is repeated for M_L values of 2, 1, and 0. The microstates corresponding to negative M_L values are not indicated in the table. They are obtained by multiplying the M_L values above for the positive M_L microstates by -1; *e.g.*, for $M_L = -3$, the possibilities are $-2^+ - 1^+$, $-2^+ - 1^-$, $-2^- - 1^+$, and $-2^- - 1^-$. Note that, for example, 2^+1^- is not distinct from 1^-2^+ since the order in which we list the electrons is irrelevant; but it is distinct from 2^-1^+ because in this latter microstate the electron with $m_\ell = 2$ no longer has $m_s = +\frac{1}{2}$.

Starting with the highest M_L value, we can conclude that there must be a ^1G term or state and that it has M_L components of 4, 3, 2, 1, 0, -1, -2, -3, -4. A box is used to set these configurations apart. The choice of the $M_L = 3$ configuration for the ^1G term is arbitrary because we are only bookkeeping with this procedure. The actual wave function for this component of the ^1G term is a linear combination* of the two microstates indicated for $M_L = 3$. Raising and lowering operators can be employed[(1)] to produce the wave function for given values of L, M_L, S, and M_S. The same is true whenever there is a choice of microstates. Now, we proceed to the next highest M_L value that remains, namely, $M_L = 3$. With $M_S = +1, 0, -1$ components, it can be deduced that there must be a ^{3}F term. This term will be twenty-one fold degenerate; the twelve microstates, with non-negative M_L values, arbitrarily assigned to it are enclosed with parentheses in Table 10–1. Next, we come to a state with $L = 2$, which must be a singlet (*i.e.*, $S = 0$). The microstates arbitrarily assigned to this ^1D term are circled. Next, we enclose with brackets those microstates of the ^3P term. The remaining term is ^1S. *Each of these terms constitutes a degenerate set of states, and each term differs in energy from any other.*†

The energies for all of these terms can be calculated[(1)] and expressed with the Condon-Shortley parameters F_0, F_2, and F_4. These parameters are abbreviations for the various electron repulsion integrals of the ion. The energy expression for any term as a function of these parameters is independent of the metal ion. The magnitude of the parameters, on the other hand, varies with the metal ion. For example, $E(^3F) = F_0 - 8F_2 - 9F_4$ and $E(^3P) = F_0 + 7F_2 - 84F_4$. The transition energy $^3P - {}^3F$ is the difference between the energies of these two terms or $15F_2 - 75F_4$. Similar expressions exist for all the other transitions involving terms of the gaseous ion. The entire spectrum can be fit with the parameters F_2 and F_4. This is true for any d^2 ion.

In the V(III) ion, the ^3F-^3P transition occurs at 13,000 cm^{-1} and the ^1F-^3D transition occurs at 10,600 cm^{-1}. Solving the two simultaneous equations

$$E(^3P - {}^3F) = 15F_2 - 75F_4 = 13{,}000 \text{ cm}^{-1} \tag{10-1}$$

$$E(^1D - {}^3F) = 5F_2 + 45F_4 = 10{,}600 \text{ cm}^{-1} \tag{10-2}$$

one obtains $F_2 = 1310$ cm^{-1} and $F_4 = 90$ cm^{-1}.

---

*Each of the microstates is really an abbreviation for a determinantal wave function, *e.g.*,

$$2^+1^- = \frac{1}{\sqrt{2}} \begin{vmatrix} 2^+(1) & 1^-(1) \\ 2^+(2) & 1^-(2) \end{vmatrix}$$

†One often finds the individual eigenfunctions of a term referred to as states. The entire term is also referred to as a state. The difference is usually obvious. If it is not, we shall use *term* or *level* to describe the entire collection of degenerate states and *component states* for the individual states.

Racah redefined the empirical Condon-Shortley parameters so that the separation between states having the maximum multiplicity is a function of only a single parameter, B:

$$B = F_2 - 5F_4 \tag{10-3}$$

A second parameter, C, is needed to express the energy difference between terms of different multiplicity:

$$C = 35F_4 \tag{10-4}$$

Rearranging, we obtain $F_2 = B + C/7$ and $F_4 = C/35$. Substituting these into equations (10–1) and (10–2), we obtain:

$$E(^3P - {}^3F) = 15B$$

$$E(^1D - {}^3F) = 5B + 2C$$

For V(III), we find $B = 866 \text{ cm}^{-1}$ and $C/B = 3.6$.

In summary, the results of the electron-electron interactions in a d^2 ion give rise to a ^3F ground term and the excited states arising from the d-orbitals shown in Fig. 10–1. The degeneracy of each term is indicated in parentheses.

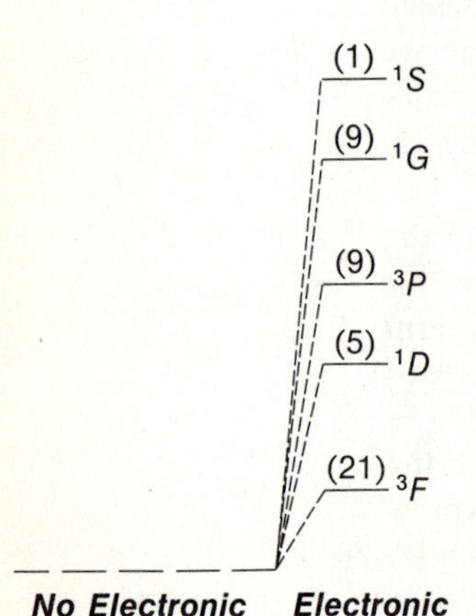

FIGURE 10–1. The terms arising from the electron-electron interactions in a d^2 gaseous ion. The number in parentheses indicates the degeneracy of each level (excluding any spin-orbit coupling).

By using procedures similar to those employed in Table 10–1, we can determine the terms arising from various d^n ions. The results for $n = 1$ to $n = 9$ are presented in Table 10–2.

TABLE 10–2. FREE ION TERMS FOR VARIOUS d^n IONS

n		Terms
d^1	d^9	^2D
d^2	d^8	^3F ^3P ^1G ^1D ^1S
d^3	d^7	^4F ^4P ^2H ^2G ^2F ^2D ^2D ^2P
d^4	d^6	^5D ^3H ^3G ^3F ^3F ^3D ^3P ^3P ^1I ^1G ^1G ^1F ^1D ^1D ^1S ^1S
d^5		^6S ^4G ^4F ^4D ^4P ^2I ^2H ^2G ^2G ^2F ^2F ^2D ^2D ^2D ^2P ^2S

The d^9 configuration is for many purposes considered to be equivalent to the d^1 case, if we think in terms of the degenerate states that would arise from degeneracies associated with the positive hole that exists in the d^9 case. It may help to think of d^9 as being a d^{10} case with a positron that can annihilate any one of the ten electrons. This concept is referred to as the *hole formalism*. By the same token, the following equivalences arise:

$$d^2 \approx d^8$$

$$d^3 \approx d^7$$

$$d^4 \approx d^6$$

10–2 SPIN-ORBIT COUPLING IN FREE IONS

As discussed in Chapter 9 (Fig. 9–18), the coupling of the magnetic dipole from the electron spin moment with the orbital moment, $\vec{\mathbf{L}} \cdot \vec{\mathbf{S}}$, is spin-orbit coupling. Variations in the amount of spin-orbit coupling in the different electronic configurations also lead to splitting of the terms derived so far. Two schemes are widely used to deal with this effect: the so-called *Russell-Saunders* or *L · S coupling* scheme, and the *j · j coupling* scheme. When the electron-electron interactions give rise to large energy splittings of the terms compared to the splittings from spin-orbit coupling, the former scheme is used. With the $L \cdot S$ scheme, we essentially treat the effects of spin-orbit coupling as a perturbation on the individual term energies. On the other hand, the $j \cdot j$ coupling scheme is used when a large splitting results from spin-orbit coupling and the electron-electron interactions are sufficiently small to be treated as a perturbation on the spin-orbit levels. The $j \cdot j$ scheme is applied to the rare earth elements as well as the third row transition metal ions. Briefly, in the $j \cdot j$ scheme the spin angular momentum of an individual electron couples with its orbital momentum to give a resultant angular momentum, $\vec{\mathbf{j}}$, for that electron. The individual $\vec{\mathbf{j}}$'s are coupled to produce the resultant vector $\vec{\mathbf{J}}$ for the system, labeling the overall angular momentum for the atom. The $L \cdot S$ coupling scheme is applicable to most first row transition metal ions, and we shall discuss this scheme in more detail. We previously mentioned that the individual orbital angular momenta of the electrons, m_l, couple to produce a resultant angular momentum indicated by $\vec{\mathbf{L}}$. The spin moments couple to give $\vec{\mathbf{S}}$. The resultant angular momentum including spin-orbit coupling is given by $\vec{\mathbf{J}}$, and *the corresponding quantum number J can take on all consecutive integer values ranging from the absolute values of L − S to L + S. For the ground term, the minimum value of J refers to the lowest energy state of the manifold if the subshell (e.g., the d-orbitals) is less than half filled; and the maximum value of J refers to the lowest energy state when the subshell is more than half filled.* If the shell is half filled, there is only one J value because $L = 0$.

Our discussion in this section can be made clearer by working out some examples. The box diagram for the ground state of the carbon atom is:

1s	2s	2p: +1	0	−1
↿⇂	↿⇂	↿	↿	

The value of the L quantum number, obtained by adding the m_l values for all the electrons in incomplete orbitals, is 1 for carbon: $L = +1 + 0 = 1$. The value for the S quantum number, the sum of the spin quantum numbers ($m_s = \pm\frac{1}{2}$) for all unpaired electrons, is 1 for carbon: $S = \frac{1}{2} + \frac{1}{2} = 1$. The multiplicity is three, and the term symbol for the ground state is ^3P. The values for J (given by $|L - S|, \ldots, |L + S|$) are $|L - S| = 1 - 1 = 0$, $|L + S| = 1 + 1 = 2$, so $J = 0, 1, 2$ (one being the only integer needed to complete the series). The subshell involved is less than half filled, so the state with minimum J has the lowest energy.

The term symbol for the ground state of carbon is $^3P_0$, with the zero subscript referring to the J value. The box diagram for the ground state of V^{3+} is

+2	+1	0	−1	−2
1	1			

with term symbol $^3F_2$ ($L = 3$, $S = 1$, $J = 4, 3, 2$). An excited state for this species is represented by | 1↓ | | | | |; this microstate belongs to the term with term symbol $^1G_4$ ($L = 4$, $S = 0$, $J = 4$). For nitrogen with a box diagram | 1 | 1 | 1 |, $L = 0$, $S = 3/2$, and $J = 3/2$ so the term symbol $^4S_{3/2}$ results. Note that there is only one J value in this case, with $L = 0$ because $|L + S| = |L - S| = 3/2$.

For practice, one can determine the following term symbols for the ground states of the elements in parentheses: $^3P_2$ (S), $^2P_{3/2}$ (Cl), $^3F_2$ (Ti), $^7S_3$ (Cr), $^3F_4$ (Ni), $^3P_0$ (Si), $^4S_{3/2}$ (As), and $^4I_{9/2}$ (Pr).

Two parameters, ξ and λ, are commonly used to describe the magnitude of the energy of the spin-orbit coupling interaction. The parameter ξ *is used to describe the spin-orbit coupling energies for a single electron.* It measures the strength of the interaction between the spin and orbital angular momenta of a single electron of a particular microstate, and is thus a property of the microstate and not of the term. The operator is $\xi\vec{\mathbf{l}}\cdot\vec{\mathbf{s}}$. The value of ξ is given by

$$\xi = \frac{Z_{\text{eff}}e^2}{2m^2c^2}\langle r^{-3}\rangle \tag{10-5}$$

where $\langle r^{-3}\rangle$ is the average value of r^{-3}, m is the mass of the electron, c is the speed of light, and Z_{eff} is the effective nuclear charge.

The parameter λ *is used to describe the corresponding property of the term.* The operator now becomes $\lambda\vec{\mathbf{L}}\cdot\vec{\mathbf{S}}$. The values of λ and ξ are related by

$$\lambda = \pm\xi/2S \tag{10-6}$$

The parameter ξ is fundamentally a positive quantity. If the shell is less than half filled, the sign of λ is positive; if it is more than half filled, λ is negative. This makes sense if we think in terms of positive holes requiring the sign of equation (10–5) to change for the configurations of more than half filled shells. In summary, then, for a shell that is less than half filled, the lowest value of J corresponds to the lowest energy and λ is positive.

An equivalent operator form for $\hat{L}\cdot\hat{S}$ is given by $\frac{1}{2}(\hat{J}^2 - \hat{L}^2 - \hat{S}^2)$ when the states can be characterized by quantum numbers L, S, and J. The spin-orbit contribution to the energy of any level is then given by:

$$\frac{1}{2}\lambda[J(J+1) - L(L+1) - S(S+1)] \tag{10-7a}$$

The energy difference between two adjacent spin-orbit states in a term is given by

$$\Delta E_{J,J+1} = \lambda(J+1) \tag{10-7b}$$

For example, the energy separation between the $J = 3$ and $J = 4$ states of a term is 4λ. *Furthermore, in the $L\cdot S$ scheme, spin-orbit splitting occurs so as to preserve the center of gravity of the energy of the term, i.e.,* the average energy remains the same. The ground state for a d^2 system, ^3F, has J values of 4, 3, and 2, with 2 lowest since the shell is less than half filled. The complete ground state term symbol is $^3F_2$. The ^1D excited state has only one possible J state, equal to 2. For the excited state, ^3P, we have $J = 0$, 1, and 2, while ^1G has only $J = 4$ and ^1S has only $J = 0$.

Now, using equation (10–7a), we can calculate the spin-orbit contribution to the energies of all the J states. For the ground level of ^3F where $J = 2$, we obtain

$\frac{1}{2}\lambda[2(2+1) - 3(3+1) - 1(1+1)] = -4\lambda$. This result is summarized in Fig. 10–2 along with the results of similar calculations of the effects of $\lambda L \cdot S$ on all the states of a d^2 system. Not all the degeneracy is removed by spin-orbit coupling, and the

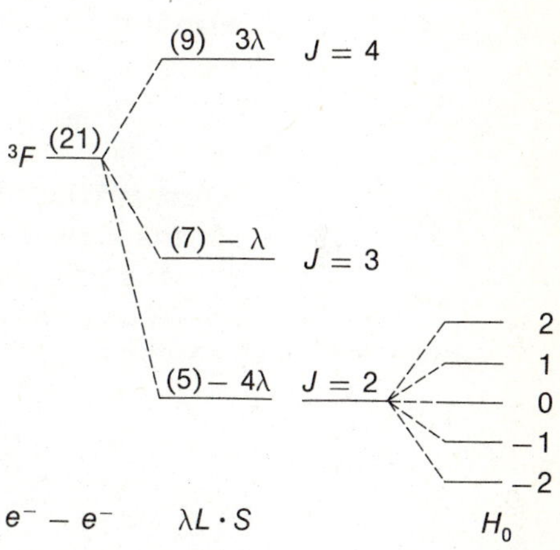

FIGURE 10–2. Spin-orbit states from a d^2 configuration. The splitting of the $^3F_2$ state by a magnetic field H_0 is indicated on the far lower right.

remaining degeneracy, corresponding to integer values of M_J from J to $-J$, is indicated in parentheses over each level. Note that equation (10–7b) is obeyed and the center of gravity is preserved. *For example, in the ^3P term, the degeneracy times the energy change gives $5\lambda - 3\lambda - (1)(2)\lambda = 0$. The degeneracy of the individual J states is removed by a magnetic field.* The splitting into the M_J states is indicated only for the ground $J = 2$ term in Fig. 10–2.

10–3 EFFECTS OF LIGANDS ON THE d-ORBITAL ENERGIES

CRYSTAL FIELDS

We usually do not work with gaseous ions but with transition metal ions in complexes. There are two crystal field type approaches to determine the effects that these ligands in a transition metal ion complex have on the energies of the *d*-orbitals. The metal ion electrons in a complex undergo interelectronic repulsions and are also repelled by the electron density of the Lewis base (ligand). When the repulsions between the metal electrons and the electron density of the ligands is small compared to interelectronic repulsions, the so-called *weak field approach* is employed. When the ligands are strong Lewis bases, the ligand electron–metal electron repulsions are larger than the interelectron repulsions and the *strong field approach* is employed.

The basis set used in these problems can be the orbitals represented by complex wave functions whose angular dependences are given by the spherical harmonics

$$Y_2^{0} = (5/8)^{1/2}(3\cos^2\theta - 1)\cdot(2\pi)^{-1/2}$$

$$Y_2^{\pm 1} = (15/4)^{1/2}\sin\theta\cos\theta\cdot(2\pi)^{-1/2}e^{\pm i\varphi}$$

$$Y_2^{\pm 2} = (15/16)^{1/2}\sin^2\theta\cdot(2\pi)^{-1/2}e^{\pm 2i\varphi}$$

Alternatively, the real trigonometric wave functions, which are linear combinations of the complex orbitals taken to eliminate i, can be employed. These are given by:

$$d_{z^2} = |0\rangle \qquad (d_{z^2} \text{ is really } d_{(z^2-r^2/3)})$$
$$d_{yz} = (i/\sqrt{2})[|-1\rangle + |+1\rangle]$$
$$d_{xz} = (1/\sqrt{2})[|-1\rangle - |+1\rangle]$$
$$d_{xy} = -(i/\sqrt{2})[|2\rangle - |-2\rangle]$$
$$d_{(x^2-y^2)} = (1/\sqrt{2})[|2\rangle + |-2\rangle]$$

In the weak field approach, the free-ion term state eigenfunctions (which take into account the interelectronic repulsions in the d-manifold) are employed as the basis set. As an example, for the ^3F term, the wave functions corresponding to $M_L = \pm 3, \pm 2, \pm 1, 0$ are used. They are abbreviated as $|3\rangle$, $|2\rangle$, etc. The Hamiltonian is given as:

$$\hat{H} = \hat{H}_0 + \hat{V}$$

where $\hat{H}_0$ is the free-ion Hamiltonian and $\hat{V}$ is taken as a perturbation from the ligand electron density on $\hat{H}_0$. The perturbation, $\hat{V}$, has a drastically simplified form incorporating only the electrostatic repulsion from the ligands, which are represented simply as point charges. For an octahedral complex, the perturbation is given by:

$$\hat{V} = \sum_{i=1}^{6} eZ_i/r_{ij} \qquad (10\text{-}8)$$

where e is the charge on the electron, Z_i is the effective charge on the ith ligand, and r_{ij} is the distance from the d-electron (this is a d^1 problem) to the ith charge. This is to be compared to the full Hamiltonian given in Chapter 3. Using the simplified Hamiltonian [equation (10–8)] leads to *crystal field theory*. It is to be emphasized that this formulation of the problem simply describes the electrostatic repulsion between the d-electrons and the ligand electron density, and as such can tell us directly only about *relative* energies of the d-orbitals.

In order to evaluate the integrals $\langle M_L|\hat{V}|M_L'\rangle$, $\hat{V}$ is written in a form that facilitates integration.(2,3) When this is done, many quantities related to the radial part of the matrix elements appear in the secular determinant with the form $\frac{1}{6}(Ze^2\overline{r_2^{-4}}a^{-5})$. Here, $\overline{r_2^{-4}}$ corresponds to the mean fourth power radius of the d-electrons of the central ion, a is the metal-ligand distance, and Ze has the same units as e. This radial quantity is referred to as *10Dq* and has units of energy.

It is informative to write the secular determinant for an octahedral complex with this Hamiltonian acting upon a d^1 configuration. Employing the complex d-orbital basis set, we obtain

$$\begin{array}{c} \\ |2\rangle \\ |1\rangle \\ |0\rangle \\ |-1\rangle \\ |-2\rangle \end{array}
\begin{array}{c} \begin{array}{ccccc} |2\rangle & |1\rangle & |0\rangle & |-1\rangle & |-2\rangle \end{array} \\
\begin{vmatrix} Dq - E & & & & 5Dq \\ & -4Dq - E & & & \\ & & 6Dq - E & & \\ & & & -4Dq - E & \\ 5Dq & & & & Dq - E \end{vmatrix} \end{array} = 0$$

This gives roots

$$E(|1\rangle) = -4Dq$$
$$E(|-1\rangle) = -4Dq$$
$$E(|0\rangle) = 6Dq$$

and the determinantal equation

$$\begin{array}{c} \\ |2\rangle \\ |-2\rangle \end{array} \begin{array}{c} \begin{array}{cc} |2\rangle & |-2\rangle \end{array} \\ \begin{vmatrix} Dq - E & 5Dq \\ 5Dq & Dq - E \end{vmatrix} = 0 \end{array}$$

This determinant is solved to produce two energies: one at $-4Dq$ and one at $6Dq$.

As in a Hückel calculation, the energies can be substituted into the secular equations written from the secular determinant, and the wave functions thus obtained. The results are

$$\psi_4 = -i2^{-1/2}(|2\rangle - |-2\rangle)$$

and

$$\psi_5 = 2^{-1/2}(|2\rangle + |-2\rangle)$$

These are the wave functions for the d_{xy} and $d_{x^2-y^2}$ orbitals, the latter pointing at the ligands and the former in between the ligands. *Note that the octahedral crystal field mixes the $|2\rangle$ and $|-2\rangle$ wave functions and makes it more convenient to employ the real d_{xy} and $d_{x^2-y^2}$ orbitals in the description of the complex.* Since the degeneracy of $|1\rangle$ and $|-1\rangle$ is not removed, the wave functions ψ_1 to ψ_3 can be considered as the real or imaginary combination as is convenient. These results are summarized in Fig. 10–3, where it can be seen that two degenerate sets of orbitals result which are separated by $10Dq$. Thus, we expect one *d-d* electronic transition for a d^1 system with an energy corresponding to $10Dq$. In O_h symmetry, the three degenerate d_{xz}, d_{yz}, and d_{xy} orbitals transform as t_{2g} and lead to the $^2T_{2g}$ ground state, while d_{z^2} and $d_{x^2-y^2}$ lead to the ^2E_g excited state. The following information is conveyed by these symbols: (1) the symbol T indicates that the state is orbitally triply degenerate and E doubly degenerate; (2) the superscript 2 indicates a spin multiplicity of two, *i.e.*, one unpaired electron; and (3) the g indicates a gerade, symmetric state.

Next, we shall treat a weak field, octahedral d^2 complex. Our initial concern will be with the ^3F term described by the basis set $|3\rangle \ldots |0\rangle \ldots |-3\rangle$. Since our basis set for ^3F consists of all $M_S = 3/2$ functions, this has been dropped from the symbol and only M_L is indicated; *i.e.*, $|3, 3/2\rangle$, etc., would be a more complete description. We shall not present the wave functions for these basis functions. Procedures for obtaining

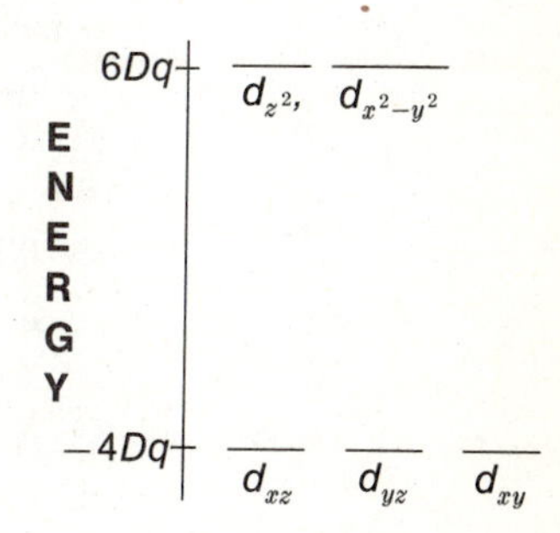

FIGURE 10–3. Splitting of the one-electron *d*-orbitals by an O_h crystal field.

these wave functions using the raising and lowering operators are presented by Ballhausen.[1] The secular determinant is given below:

$$
\begin{array}{c} \\ \langle 3| \\ \langle 2| \\ \langle 1| \\ \langle 0| \\ \langle -1| \\ \langle -2| \\ \langle -3| \end{array}
\begin{array}{ccccccc}
|3\rangle & |2\rangle & |1\rangle & |0\rangle & |-1\rangle & |-2\rangle & |-3\rangle \\
\end{array}
$$

$$
\left|\begin{array}{ccccccc}
-3Dq-E & & & & 15^{1/2}Dq & & \\
 & 7Dq-E & & & & 5Dq & \\
 & & -Dq-E & & & & 15^{1/2}Dq \\
 & & & -6Dq-E & & & \\
15^{1/2}Dq & & & & -Dq-E & & \\
 & 5Dq & & & & 7Dq-E & \\
 & & 15^{1/2}Dq & & & & -3Dq-E
\end{array}\right| = 0
$$

(rows: $\langle 3|, \langle 2|, \langle 1|, \langle 0|, \langle -1|, \langle -2|, \langle -3|$; columns: $|3\rangle, |2\rangle, |1\rangle, |0\rangle, |-1\rangle, |-2\rangle, |-3\rangle$)

The solutions are

$$
\begin{array}{ll}
E_1 = -6Dq & E_5 = 2Dq \\
E_2 = -6Dq & E_6 = 2Dq \\
E_3 = -6Dq & E_7 = 12Dq \\
E_4 = 2Dq &
\end{array}
$$

The energies* and wave functions for the respective levels are indicated in Fig. 10–4.

ENERGY ↑

^3F (21) → $^3A_2$ (3), $^3T_2$ (9), $^3T_1$ (9)

$12Dq$ $[2^{-1/2}(|2\rangle + |-2\rangle)]$

$[24^{-1/2}(3|3\rangle + 15^{1/2}|-1\rangle)]$
$[24^{-1/2}(3|-3\rangle + 15^{1/2}|1\rangle)]$
$2Dq$ $[2^{-1/2}(|2\rangle - |-2\rangle)]$

$-6Dq$ $[24^{-1/2}(15^{1/2}|-3\rangle - 3|1\rangle)]$
$[|0\rangle]$
$[24^{-1/2}(15^{1/2}|3\rangle - 3|-1\rangle)]$

FIGURE 10–4. Splitting of the ^3F term of a d^2 ion by an O_h crystal field.

In this analysis, we have ignored any covalency in the metal-ligand bond. As a result, if we were to attempt a quantitative calculation of Dq, it would differ considerably from that found experimentally. *Ligand field theory* admits to covalency in the bond and treats Dq (and other parameters to be discussed shortly) as an empirical parameter that is obtained from the electronic spectrum. The formulation of the problem in all other respects is identical.

10–4 SYMMETRY ASPECTS OF THE *d*-ORBITAL SPLITTING BY LIGANDS

As is usually the case when a vastly simplified Hamiltonian is employed, the correct aspect of the results is symmetry-determined. For example, we mentioned in Chapter 2 that appropriate combinations of the binary products of the x, y, and z vectors gave the irreducible representations for the d-orbitals and their degeneracies. We can use principles already covered (Chapter 2) to illustrate a procedure for deriving all of the states arising from one-electron levels. This procedure can then be

* Dq is defined for a one-electron system. For a polyelectron system, one should employ a different Dq value for each state. (C. J. Ballhausen and H. B. Gray, "Chemistry of Coordination Compounds," Volume I, A. E. Martell, Ed., Van Nostrand Reinhold, Princeton, N.J.) This is seldom done in practice because of the crude nature of this model.

extended in a straightforward way to derive all of the states arising for various multi-electron configurations in various geometries.

We begin by examining the effect of an octahedral field on the total representation for which the set of d wave functions forms the basis. To determine this total representation, we must find the elements of matrices that express the effect upon our basis set of d-orbitals of each of the symmetry operations in the group. The characters of these matrices will comprise the representation we seek. Since all of the d-orbitals are gerade, *i.e.*, symmetric to inversion, no new information will result as a consequence of the inversion symmetry operation. Thus, we can work with the simpler pure rotational subgroup O instead of O_h. If you need to convince yourself of this, note that in any group containing i (*e.g.*, D_{4h} or C_{3h}), the corresponding rotation group (*e.g.*, D_4 or C_3) has the same irreducible representation for the binary products except for the u and g subscripts in the former group. Recall that the d wave functions consist of radial, spin, and angular (θ and φ) parts. The radial part is neglected, for it is non-directional and hence unchanged by any symmetry operation. We shall assume that the spin part is independent of the orbital part and ignore it for now. The angle θ is defined relative to the principal axis (*i.e.*, the rotation axis), so it is unchanged by any rotation and can be ignored. Only φ will change; the form of this part of the wave function is given by $e^{im_l\varphi}$. (For the d-orbitals, $\ell = 2$ and m_ℓ has values 2, 1, 0, −1, −2.) To work out the effects of a rotation by α on $e^{im_l\varphi}$, we note that such a rotation causes the following changes:

$$\begin{bmatrix} e^{2i\varphi} \\ e^{i\varphi} \\ e^{0} \\ e^{-i\varphi} \\ e^{-2i\varphi} \end{bmatrix} \xrightarrow[\text{by } \alpha]{\text{rotate}} \begin{bmatrix} e^{2i(\varphi+\alpha)} \\ e^{i(\varphi+\alpha)} \\ e^{0} \\ e^{-i(\varphi+\alpha)} \\ e^{-2i(\varphi+\alpha)} \end{bmatrix}$$

The matrix that operates on our d-orbital basis set is

$$\begin{bmatrix} e^{2i\alpha} & 0 & 0 & 0 & 0 \\ 0 & e^{i\alpha} & 0 & 0 & 0 \\ 0 & 0 & e^{0} & 0 & 0 \\ 0 & 0 & 0 & e^{-i\alpha} & 0 \\ 0 & 0 & 0 & 0 & e^{-2i\alpha} \end{bmatrix}$$

A general form of this matrix for the rotation of any set of orbitals is given by:

$$\begin{bmatrix} e^{\ell i\alpha} & 0 & \cdots & 0 & 0 \\ 0 & e^{(\ell-1)i\alpha} & \cdots & 0 & 0 \\ \vdots & \vdots & \vdots & \vdots & \vdots \\ 0 & 0 & \cdots & e^{(1-\ell)i\alpha} & 0 \\ 0 & 0 & \cdots & 0 & e^{-\ell i\alpha} \end{bmatrix}$$

When $\alpha = 0$, each element is obviously one. We give the trace of this latter matrix without proof* as:

$$\chi(\alpha) = \frac{\sin\left(\ell + \frac{1}{2}\right)\alpha}{\sin\left(\frac{\alpha}{2}\right)} \tag{10-9}$$

*The quantities summed form a geometric progression.

where $\alpha \neq 0$. From the trace determined with this formula, we have the character of the representation for any rotation operation. Substituting directly into this formula, we find that the total character for the C_3 rotation of the five d-orbitals is given by

$$\chi(C_3) = \frac{\sin\left[\left(2 + \frac{1}{2}\right)\left(\frac{2\pi}{3}\right)\right]}{\sin\left(\frac{2\pi}{3 \times 2}\right)} = \frac{\sin\frac{5\pi}{3}}{\sin\frac{\pi}{3}} = \frac{-\sin\frac{\pi}{3}}{\sin\frac{\pi}{3}} = -1$$

Characters for other rotations can be worked out in a similar way. To obtain the characters when $\alpha = 0$ and $\alpha = 2\pi$, one must evaluate the limit of an indeterminate form, *e.g.*,

$$\frac{\sin\left(\ell + \frac{1}{2}\right)(2\pi)}{\sin \pi} \rightarrow \frac{0}{0}$$

Using l'Hopital's rule, one obtains

$$\chi(0) = 2\ell + 1$$
$$\chi(2\pi) = 2\ell + 1 \text{ for integer } \ell \text{ or}$$
$$-(2\ell + 1) \text{ for half-integer } \ell$$

Thus, for the identity, $\alpha = 0$ and $\chi(E)$ is given by $2\ell + 1$. Referring to the character table for the O point group and using the above formula to determine the characters for the various operations on the five d-orbitals, we have:

	E	$6C_4$	$3C_2(= C_4^2)$	$8C_3$	$6C_2$
χ_T	5	-1	1	-1	1

Using the decomposition formula, we obtain the result $\chi_T = E + T_2$. Since the d-orbitals are gerade, we can write this as

$$\chi_T = E_g + T_{2g}$$

This was the result of our crystal field analysis. By a similar symmetry analysis, the results summarized in Table 10–3 can be obtained.

TABLE 10–3. REPRESENTATIONS FOR VARIOUS ORBITALS IN O_h SYMMETRY

Type of Orbital	ℓ Value	Irreducible Representation*
s	0	a_{1g}
p	1	t_{1u}
d	2	$e_g + t_{2g}$
f	3	$a_{2u} + t_{1u} + t_{2u}$
g	4	$a_{1g} + e_g + t_{1g} + t_{2g}$
h	5	$e_u + 2t_{1u} + t_{2u}$
i	6	$a_{1g} + a_{2g} + e_g + t_{1g} + 2t_{2g}$

*The subscripts g and u are determined by the g or u nature of the atomic orbitals involved. When ℓ is even, the orbital is gerade; when ℓ is odd, the orbital is ungerade.

We could work out the effect of other symmetries on the one-electron levels in a similar way. Alternatively, one can use a correlation table that shows how the representations of the group O_h are changed or decomposed into those of its subgroups when the symmetry is altered. Table 10–4 contains such information for some of the symmetries commonly encountered in transition metal ion complexes.

TABLE 10–4. CORRELATION TABLE FOR THE O_h POINT GROUP

O_h	O	T_d	D_{4h}	D_{2d}	C_{4v}	C_{2v}	D_{3d}	D_3	C_{2h}
A_{1g}	A_1	A_1	A_{1g}	A_1	A_1	A_1	A_{1g}	A_1	A_g
A_{2g}	A_2	A_2	B_{1g}	B_1	B_1	A_2	A_{2g}	A_2	B_g
E_g	E	E	$A_{1g} + B_{1g}$	$A_1 + B_1$	$A_1 + B_1$	$A_1 + A_2$	E_g	E	$A_g + B_g$
T_{1g}	T_1	T_1	$A_{2g} + E_g$	$A_2 + E$	$A_2 + E$	$A_2 + B_1 + B_2$	$A_{2g} + E_g$	$A_2 + E$	$A_g + 2B_g$
T_{2g}	T_2	T_2	$B_{2g} + E_g$	$B_2 + E$	$B_2 + E$	$A_1 + B_1 + B_2$	$A_{1g} + E_g$	$A_1 + E$	$2A_g + B_g$
A_{1u}	A_1	A_2	A_{1u}	B_1	A_2	A_2	A_{1u}	A_1	A_u
A_{2u}	A_2	A_1	B_{1u}	A_1	B_2	A_1	A_{2u}	A_2	B_u
E_u	E	E	$A_{1u} + B_{1u}$	$A_1 + B_1$	$A_2 + B_2$	$A_1 + A_2$	E_u	E	$A_u + B_u$
T_{1u}	T_1	T_2	$A_{2u} + E_u$	$B_2 + E$	$A_1 + E$	$A_1 + B_1 + B_2$	$A_{2u} + E_u$	$A_2 + E$	$A_u + 2B_u$
T_{2u}	T_2	T_1	$B_{2u} + E_u$	$A_2 + E$	$B_1 + E$	$A_2 + B_1 + B_2$	$A_{1u} + E_u$	$A_1 + E$	$2A_u + B_u$

With the irreducible representations given in Table 10–3 for various atomic orbitals in O_h symmetry and with the correlation table given in Table 10–4, we can ascertain the irreducible representations of the various orbitals in different symmetry environments. The results for single electrons in various orbitals apply also to the terms arising from multi-electron systems. For example, we can take the ^3F, ^3P, ^1G, ^1D, and ^1S terms of the d^2 configuration and treat them like *f*, *p*, *g*, *d*, and *s* orbitals. The *g* or *u* subscripts given in Table 10–3 will not apply, but will depend upon the *g* or *u* nature of the atomic orbitals involved. Thus, Table 10–3 applies to terms as well as orbitals. The *D* term, for example, is fivefold degenerate like the five *d*-orbitals, the former being described by a wave function for each of the five M_L values. These wave functions have a Φ part given by $e^{iM_L\varphi}$. Combining Tables 10–3 and 10–4, we can see that the *D* state of the free ion splits into $E_g + T_{2g}$ states in an octahedral field and into $A_{1g} + B_{1g} + E_g + B_{2g}$ states in a tetragonal D_{4h} field. Similarly, the ^3F term gives rise to $A_{2g} + T_{1g} + T_{2g}$ in an octahedral field and $B_1 + A_2 + 2E + B_2$ in a C_{4v} field.

Next, we shall consider the states that arise from a d^2 configuration when the crystal field is large compared to the interelectronic repulsions.* The interelectronic repulsions are treated as perturbations on the strong field *d*-electron configurations. In other words, the various crystal field states are identified, eigenfunctions are constructed, and e^2/r_{ij} is used as a perturbation. The terms are readily written for the various d^n configurations. For d^1, the two terms are $^2T_{2g}$ and ^2E_g. The terms that arise from a configuration with an additional electron, d^{n+1}, are given by the direct product

*The total degeneracy that exists for a system of *q* electrons filling a subshell with *z*-degenerate orbitals (each occupied by two electrons with opposite spin) is given by:

$$\frac{(2z)!}{q!(2z-q)!}$$

For a $t_{2g}{}^2$ configuration, we have:

$$\frac{(2 \times 3)!}{2!(6-2)!} = 15$$

For $t_{2g}{}^2e_g{}^1$, we have

$$\left[\frac{6!}{2!(6-2)!}\right]\left[\frac{4!}{1!(4-1)!}\right] = [15]\cdot[4] = 60$$

of the d^n term symmetries and that of the added electron. For d^2, we have t_{2g}^2, $t_{2g}^1e_g^1$, and e_g^2 arrangements. Accordingly, for t_{2g}^2, we have

$$t_{2g} \times t_{2g} \text{ which leads to } T_{1g} + T_{2g} + E_g + A_{1g}$$

For $t_{2g}^1e_g^1$, we have

$$t_{2g} \times e_g \text{ which leads to } T_{1g} + T_{2g}$$

and for e_g^2, we have

$$e_g \times e_g \text{ which leads to } E_g + A_{1g} + A_{2g}$$

Next we have to determine the multiplicity of these terms and show the connection of these strong field terms to those of the gaseous ions. The multiplicities are determined by the *method of descending symmetries.*[1,2] By considering the e_g^2 configuration, we can show that when we lower the symmetry from O_h to D_{4h} to remove the degeneracy, the determination of the spin degeneracy becomes straightforward. The correlation table (Table 10-4) shows

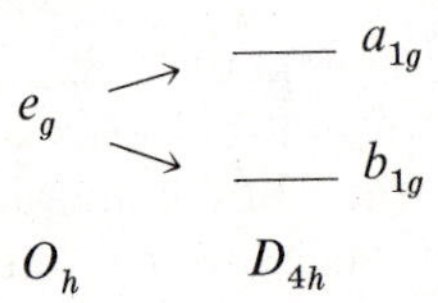

O_h		D_{4h}
e_g	$\nearrow$	a_{1g}
	$\searrow$	b_{1g}

In addition to the orbital correlation shown above, it is important to remember that states behave in the same way. In D_{4h}, a_{1g}^2 must be singlet because of the Pauli principle, leading to ${}^1A_{1g}$ (a_{1g}^2 leads to $a_{1g} \times a_{1g}$ which leads to ${}^1A_{1g}$), b_{1g}^2 must be singlet ${}^1A_{1g}$ (b_{1g}^2 leads to $b_{1g} \times b_{1g}$ which leads to ${}^1A_{1g}$), and $a_{1g}^1b_{1g}^1$ leads to $b_{1g} \times a_{1g}$ which leads to ${}^1B_{1g}$ or ${}^3B_{1g}$. The states in octahedral symmetry (A_{1g}, A_{2g}, and E_g) must change to those in D_{4h} symmetry as shown in the group correlation table (Table 10-4) and as summarized below:

O_h		D_{4h}
A_{1g}	$\longrightarrow$	A_{1g}
A_{2g}	$\longrightarrow$	B_{1g}
E_g	$\longrightarrow$	A_{1g}
	$\searrow$	B_{1g}

We have determined the multiplicity of the D_{4h} states; since lowering the symmetry cannot change the spin degeneracies, we can work backwards to determine the spin degeneracy of the O_h states. The ${}^1A_{1g}$ state in D_{4h} must correspond to a singlet ${}^1A_{1g}$ in O_h. The other ${}^1A_{1g}$ must arise from E_g, requiring that this be ${}^1E_{1g}$. The B_{1g} state in D_{4h} arising from ${}^1E_{1g}$ must also be a singlet. We are left with the fact that ${}^3B_{1g}$ in D_{4h} is associated with A_{2g}, requiring this to be ${}^3A_{2g}$.

D_{4h}		O_h
${}^1A_{1g}$	$\longrightarrow$	$A_{1g}({}^1A_{1g})$
${}^3B_{1g}$	$\longrightarrow$	$A_{2g}({}^3A_{2g})$
${}^1A_{1g}$	$\longrightarrow$	$E_g({}^1E_g)$
${}^1B_{1g}$	$\nearrow$	

The $t_{2g}{}^2$ configuration gives rise to $A_{1g} + E_g + T_{1g} + T_{2g}$ states. We must examine the correlation table for these states in O_h symmetry and find a lower symmetry that converts these states to one-dimensional representations or a sum of one-dimensional representations. Table 10–4 shows that C_{2h} and C_{2v} satisfy this requirement. The results for C_{2h} are summarized as

O_h	C_{2h}
A_{1g}	$\rightarrow A_g$
E_g	$\rightarrow A_g + B_g$
T_{1g}	$\rightarrow A_g + B_g + B_g$
T_{2g}	$\rightarrow A_g + A_g + B_g$

Since t_{2g} in O_h gives rise to $a_g + a_g + b_g$ orbitals in C_{2h}, the possible configurations are $a_{g1}{}^2$, $a_{g1}{}^1a_{g2}{}^1$, $a_{g1}{}^1b_g{}^1$, $a_{g2}{}^2$, $a_{g2}{}^1b_g{}^1$, $b_g{}^2$. As previously:

$a_{g1} \times a_{g1}$ leads to A_g

$a_{g1} \times a_{g2}$ leads to A_g

$a_{g1} \times b_g$ leads to B_g

$a_{g2} \times a_{g2}$ leads to A_g

$a_{g2} \times b_g$ leads to B_g

$b_g \times b_g$ leads to A_g

The subscripts 1 and 2 on the a_g orbital are used to distinguish the two different a_g representations that result from t_{2g} in this lower symmetry. Since the first, fourth, and sixth binary products listed correspond to both electrons occupying the same orbital* (a_{g1}, a_{g2}, and b_g, respectively), these must be singlet states: ${}^1A_g$, ${}^1A_g$ and ${}^1A_g$, respectively. The second, third, and fifth binary products correspond to electrons in different orbitals and give rise to singlet and triplet states: ${}^1A_g + {}^3A_g$ and $2\,{}^1B_g + 2\,{}^3B_g$. The result in C_{2h} is summarized in the left column of Table 10–5. In the right-hand column, we connect up the correlating states in O_h from Table 10–4. Since we have three triplet states in C_{2h} of B_g, B_g and A_g symmetry, these must arise from ${}^3T_{1g}$. All of the other states from $t_{2g}{}^2$ are singlets. No other possible correspondence exists. One can work with the $t_{2g}{}^1e_g{}^1$ configuration in a similar manner.

*If we label the orbitals 1, 2, and 3, we have 1 × 1, 1 × 2, 1 × 3, 2 × 2, 2 × 3, and 3 × 3. For 1 × 1, 2 × 2, and 3 × 3, we have two electrons in the same orbital.

TABLE 10–5. RELATION OF THE MULTIPLICITIES IN C_{2h} AND O_h

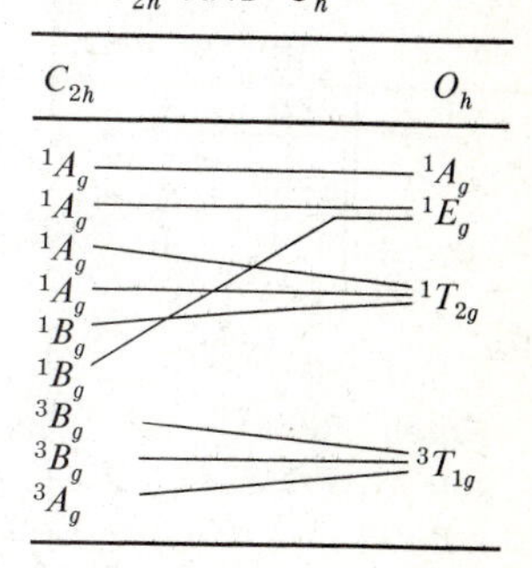

Summarizing the above procedure, we note that we have used the method of descent in symmetry as follows:

(1) The orbitals in the higher symmetry are correlated with the orbitals in the lower symmetry.

(2) The required number of electrons are added to these lower symmetry orbitals.

(3) The electronic states resulting from the electron configurations in the lower symmetry are determined.

(4) The lower symmetry electronic states, including spin multiplicity, are correlated with the electronic states of the higher symmetry case. (This is, in effect, an ascent in symmetry.)

The terms of the d^3 strong field configuration are given by taking the direct product of the terms for d^2 with t_{2g} and e_g for the added electron.

Next, we have to show how the terms that arise in a strong field are related to those in a weak field, for there must be a continuous change from one to the other as a function of the ligand field strength in a series of complexes that a given metal forms. We can illustrate this with a d^2 ion by employing two principles:

(1) A one-to-one correspondence exists between the states of the same symmetry and spin multiplicity at the weak field and strong field extremes.

(2) States of the same symmetry and spin degeneracy cannot cross as the ligand field strength is varied.

Fig. 10–5 contains the free ion and corresponding weak field states on the left

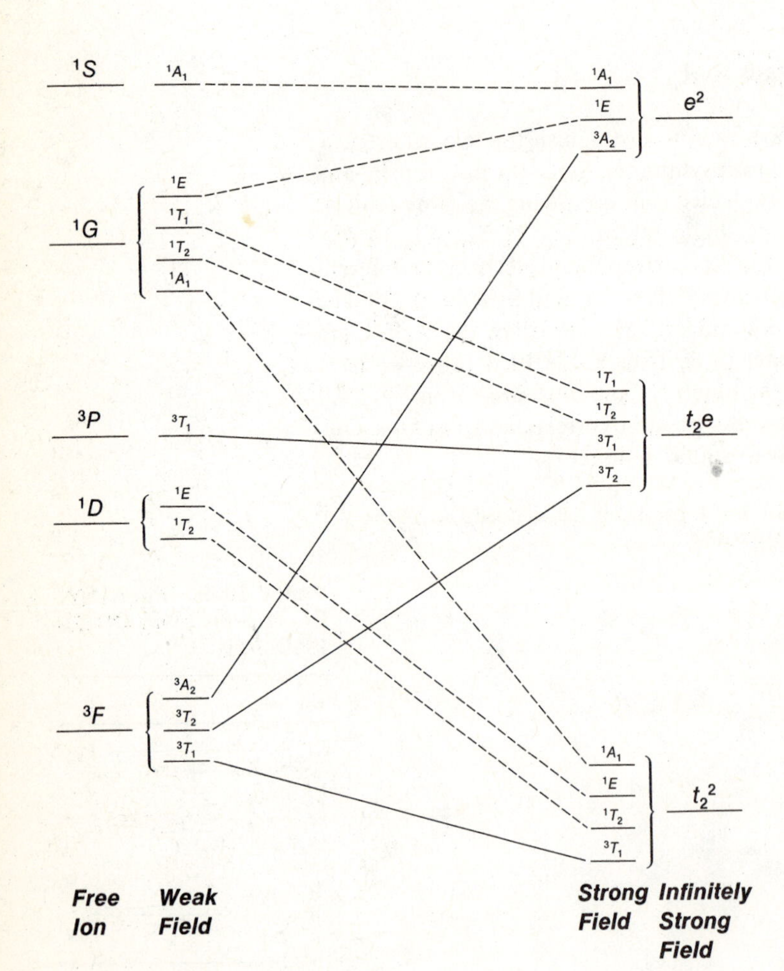

FIGURE 10–5. Correlation of the strong and weak field states of a d^2 ion.

and the strong field states on the right. The configurations in an infinitely strong field are indicated on the far right.

If we begin with the infinitely strong field e^2 configuration, we note a ^1A state. There is also a $^1A_1$ state in $t_2{}^2$. Only if the connections are made as shown can we avoid crossing. The same is true for ^1E. Since there is only one $^3A_2$ state in the weak field, this connection is straightforward. Proceeding to the other states from t_2e and $t_2{}^2$, only the connections shown will lead to non-crossing of states with the same symmetry and multiplicity.

The results of the evaluation of energies of the various levels in going from the weak field to strong field cases are presented in graphical form in the *Tanabe and Sugano diagrams*[13] contained in Appendix C for various d^n configurations. The energies are plotted as E/B versus Dq/B, where B is the Racah parameter. The quantities E/B and Dq/B are plotted because they enable one to convert the equation for the energy into a convenient form for plotting. Since states of different multiplicities are involved, they are a function of the Racah parameter C as well as B. Therefore, the diagram can be constructed only for a given ratio of C/B. The lowest term is taken as the zero of energy in these diagrams.

The *Orgel diagrams* are used to present some of the information in the more complete Tanabe and Sugano diagrams. Orgel diagrams contain only those terms that have the same multiplicity as the ground state. Accordingly, they suffice for the interpretation of the electronic spectra of multiplicity allowed transitions and will be employed often in the rest of the chapter (*e.g.*, Fig. 10–9).

10–5 DOUBLE GROUPS

We have previously shown how to use the character tables to find the character of the representation for which the p and d orbitals form a basis in various symmetries. In the preceding section, we showed that for any symmetry operation corresponding to a rotation by an angle α on an orbital or state wave function having an angular momentum quantum number ℓ, or L, the character $\chi(\alpha)$ for which this forms a basis is given by equation (10–9):

$$\chi(\alpha) = \frac{\sin\left(\ell + \frac{1}{2}\right)\alpha}{\sin\left(\frac{\alpha}{2}\right)}$$

This equation can also be applied to those states characterized by the total angular momentum J (where $J = L + S$) by simply substituting J for ℓ. When there are an even number of electrons and J is an integer value, the total representation in any symmetry can be decomposed into the irreducible representations of the point group as done in the previous section. However, when J is half-integral (*i.e.*, S is odd), a rotation by 2π (which is the identity operation) does not produce the identity for the character:

$$\chi(\alpha + 2\pi) = \frac{\sin\left(J + \frac{1}{2}\right)(\alpha + 2\pi)}{\sin\left[\frac{(\alpha + 2\pi)}{2}\right]} = \frac{\sin\left[\left(J + \frac{1}{2}\right)\alpha + \left(J + \frac{1}{2}\right)2\pi\right]}{\sin\left(\frac{\alpha}{2} + \pi\right)}$$

$$= \frac{\sin\left(J + \frac{1}{2}\right)\alpha}{-\sin\left(\frac{\alpha}{2}\right)} = -\chi(\alpha)$$

It can be shown that rotation by 4π is needed to produce the identity. To avoid this difficulty, rotation by 2π in this instance is treated as a symmetry operation that we shall label R. The ordinary rotation group is expanded by taking the product of R with all existing rotations. The new group is called a *double group.* Using equation (10–9), the characters for the rotations can be worked out. The characters of E and R (*i.e.,* $\alpha = 0$ and $\alpha = 2\pi$) require evaluation of the limit of the indeterminate form

$$\frac{\sin\left(J + \frac{1}{2}\right)2\pi}{\sin \pi} \rightarrow \frac{0}{0}$$

leading to $\chi(0) = 2J + 1$ and $\chi(2\pi) = 2J + 1$ for integer J or $-(2J + 1)$ for half-integer J, as mentioned before. The character tables for the rotation double groups D_4' and O' are given in Appendix B. Those for other groups have been reported.* Two commonly encountered systems for labeling the irreducible representations are given. One uses a serially indexed set of Γ_i's, and the other uses primed symbols similar to those we have been employing. The direct products of representations of double groups can be taken as before and reduced to sums of irreducible representations.

The above discussion can be clarified by considering some examples. We need to employ double groups to determine the effects of spin-orbit coupling when J is half-integral. Since spin-orbit effects arise from coupling of spin and orbital momenta of electrons, we are concerned with the direct product representation of these two effects. As an example, we shall work out the effects of an octahedral field and spin-orbit coupling on the ^4F free ion state of a d^7 ion. As in the previous section, we can work out the total representation in the O point group and factor it to obtain

$$\chi_T(\ell = 3) = A_2 + T_1 + T_2$$

A d^7 ion in a weak O_h field leads, as shown in the Tanabe and Sugano diagram, to a $^4T_{1g}$ ground state and $^4T_{2g}$ and $^4A_{2g}$ excited states. In the O' double group, these correspond to $T_1'(\Gamma_4)$, $T_2'(\Gamma_5)$, and $A_2'(\Gamma_2)$, respectively. Using $S = 3/2$ and substituting S for ℓ in equation (10–9), we generate in the O' point group an irreducible representation of $^1G(\Gamma_8)$, *i.e.*, one of the new irreducible representations of the double group. Now, we take the direct products of the spin and orbital parts and decompose them as before, leading to

$$\Gamma_2 \times \Gamma_8 = \Gamma_8$$

$$\Gamma_4 \times \Gamma_8 = \Gamma_6 + \Gamma_7 + 2\Gamma_8$$

$$\Gamma_5 \times \Gamma_8 = \Gamma_6 + \Gamma_7 + 2\Gamma_8$$

As we see, spin-orbit effects do not split Γ_2, but they split Γ_4 into four states and Γ_5 into four states. We could have converted L and S to J and employed equation (10–9) on J values of $9/2$, $7/2$, $5/2$, and $3/2$ to obtain the double group representations. This procedure would have been followed if spin-orbit coupling were comparable to or greater than the crystal field. Using the approach employed above, we have assumed a large crystal field and a small spin-orbit perturbation on it. We can summarize the results with the diagram in Fig. 10–6.

It is important to remember that whenever one is concerned with the effects of spin-orbit coupling (as we shall often be in subsequent chapters) *in a system with half-integral J values, the double group should be employed.*

*S. Sugano, Y. Tanabe, and H. Kamimura, "Multiplets of Transition Metal Ions in Crystals," Academic Press, N.Y. (1970).

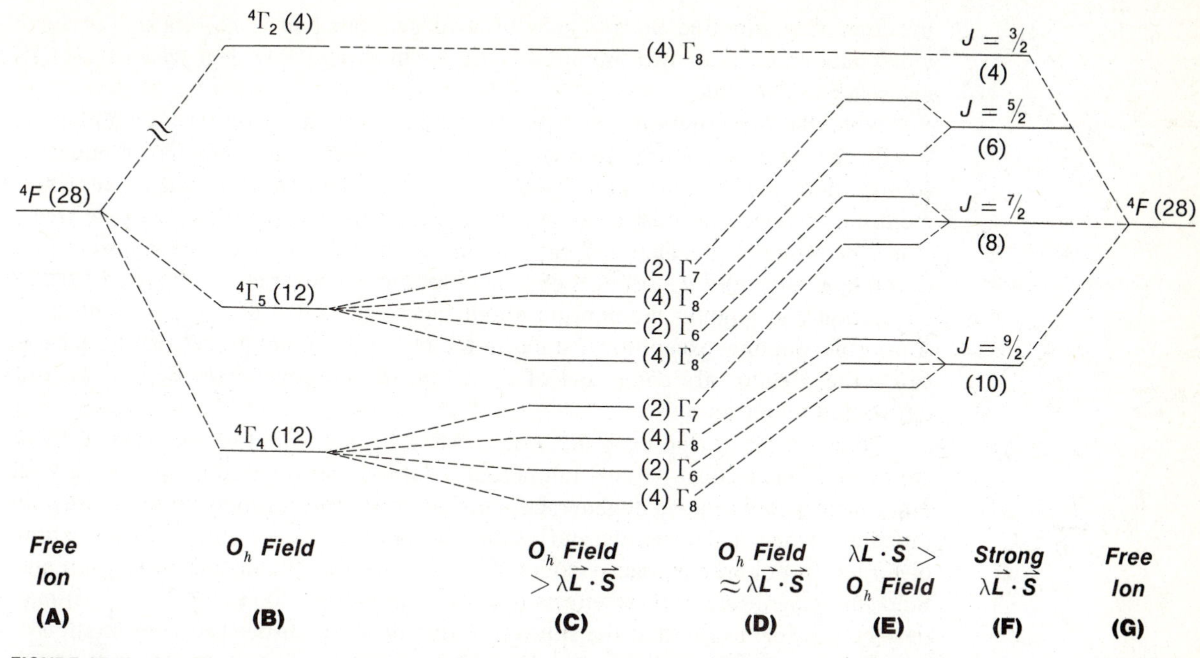

FIGURE 10–6. (A) The gaseous ion, (B) split by a strong O_h field, (C) followed by smaller $\lambda\vec{L}\cdot\vec{S}$. On the right, (G) free ion, (F) split by large spin-orbit coupling, (E) followed by a weaker ligand field. Part (D) indicates the correlation of states in the intermediate region. For convenience, none of the states are shown to cross. States of different double group symmetries may cross. States of the same double group symmetry will undergo configuration interaction.

10–6 THE JAHN-TELLER EFFECT

There is one other effect that influences the electronic structure of a complex, which we should consider before discussing electronic spectra. Consider a molecule with an unpaired electron in a doubly degenerate orbital, *e.g.*, an octahedral Cu(II) system. Note that by distorting the molecule from its most symmetrical geometry (O_h) to, say, D_{4h}, it is possible to lower its energy:

$$
\begin{array}{llcll}
e_g & \uparrow\downarrow \quad \uparrow & & \uparrow & b_{1g} \\
 & & \rightarrow & \uparrow\downarrow & a_{1g} \\
t_{2g} & \uparrow\downarrow \quad \uparrow\downarrow \quad \uparrow\downarrow & & \uparrow\downarrow & b_{2g} \\
 & & & \uparrow\downarrow \quad \uparrow\downarrow & e_g \\
 & O_h & \rightarrow & D_{4h} &
\end{array}
$$

Here the e_g set splits into b_{1g} and a_{1g} components. Since two electrons are in the stabilized a_{1g} orbital and only one is in the destabilized b_{1g} orbital, the molecule as a whole is stabilized. It is easy to see how this happens by remembering the basis of simple electrostatic crystal field theory: an orbital pointing at a ligand is destabilized; and the closer the ligand is, the higher the energy is. A tetragonal elongation (a lengthening of two M—L bonds on the z-axis and shortening of the other four on the x- and y-axes) destabilizes the $d_{x^2-y^2}$ (b_{1g}) orbital and stabilizes the d_{z^2} (a_{1g}) orbital. Similarly, a tetragonal compression would raise d_{z^2} and lower $d_{x^2-y^2}$. Jahn and Teller first pointed out that, for a non-linear molecule, when such a distortion can occur to lower the energy, it will. We thus expect that there will be a Jahn-Teller distortion any time we have an orbitally degenerate (E or T) state and when a proper symmetry vibrational mode exists which enables the molecule to move from one geometry to the other. One unpaired electron in a doubly degenerate pair of *e* orbitals gives rise to an *E* state and one or two unpaired electrons in three triply degenerate *t* orbitals gives rise to a *T* state.

Note that this criterion is very similar to the criterion for spin-orbit coupling. A simple one-electron picture can be used to predict when orbital angular momentum contributions are expected and when they are not. To obtain orbital angular momentum, the electron must be in degenerate orbitals that permit it to move freely from one orbital to the next and, in so doing, circulate around an axis. Consider, for example, the d_{xz} and d_{yz} orbitals of a metallocene. Degeneracy of this pair permits circulation and angular momentum about the *z*-axis. All *E* and *T* states will have spin-orbit coupling, except for *E* states in the O_h and T_d point groups. In these latter cases, the *E*-states are composed of $d_{x^2-y^2}$ and d_{z^2} so degeneracy does not allow circulation about an axis.

Some people, including Teller, argue that, if a state is split for any reason, there is no Jahn-Teller effect. Others talk about a Jahn-Teller distortion combining with other factors that remove degeneracy. This latter approach brings up an interesting dilemma; when a degenerate state splits, is this due to a Jahn-Teller distortion, distortion from lower symmetry components in the structure, or spin-orbit coupling? Since the magnitude of these effects is often comparable (200 to 2000 cm^{-1}), it may only be possible to say that the splitting is due to some unspecified combination of these effects. One guideline is that Jahn-Teller distortions are generally larger in *E* states than in *T* states, so spin-orbit coupling is generally the dominant effect in *T* states.

APPLICATIONS

10–7 SURVEY OF THE ELECTRONIC SPECTRA OF "O_h" COMPLEXES

The electronic spectra of transition metal complexes can be interpreted with the aid of crystal field theory. In our discussion of "O_h" complexes in this section, our concern will be with systems in which the local symmetry is O_h, though the overall molecular symmetry may not be. Throughout the remainder of this chapter, we shall use the symmetry terms very loosely to describe the type and arrangements of donor atoms directly bonded to the metal, with the rest of the ligand atoms being ignored. It should be realized that this assumption is not always justified. Upon completion of this section, we shall be in a position to assign and predict the electronic spectrum as well as rationalize the magnitudes of the *d*-orbital splittings observed. The treatment here will not be encyclopedic; selected topics will be covered. The aim is to give an appreciation for a very powerful tool in coordination chemistry: the utilization of electronic spectra in the solution of structural problems. More advanced treatments containing references to the spectra of many complexes are available.[1,2,4,5,9,10,12]

The discussion in Chapter 5 of selection rules for electronic transitions should be reviewed if necessary. Here we shall apply these rules to some transition metal ion systems. We begin by discussing high spin, octahedral complexes of Mn^{II}, a d^5 case, where there are no spin-allowed *d-d* transitions. All *d-d* transitions in this case are both multiplicity- and Laporte-forbidden. If it were not for vibronic coupling and charge transfer transitions, Mn^{II} complexes would be colorless. Hexaaquomanganese(II) ion is very pale pink, with all absorption peaks in the visible region being of very low intensity.

The fact that multiplicity-allowed transitions are usually broad, while multiplicity-forbidden transitions are usually sharp, aids in making band assignments. Multiplicity-allowed $t_{2g} \rightarrow e_g$ transitions lead to an excited state in which the equilibrium internuclear distance between the metal ion and ligand is larger than in the ground state. In the course of the electronic transition no change in distance can occur

(Franck-Condon principle), so the electronically excited molecules are in vibrationally excited states with bond distances corresponding to the configuration of the ground state. The interaction of an excited state with solvent molecules not in the primary coordination sphere is variable because neighboring solvent molecules are various distances away when the excited molecule is produced. Since the solvent cannot rearrange in the transition time, a given excited vibrational state in different molecules will undergo interactions with solvent molecules located at varying distances. Varying solvation energies produce a range of variable energy, vibrationally excited states and a broad band results.

In some spin-forbidden transitions, rearrangement occurs in a given level. For example, in Cr^{III} complexes a transition occurs from a ground state containing three unpaired electrons in t_{2g} to an excited state in which t_{2g} has two paired electrons and one unpaired electron. In these multiplicity-forbidden transitions there is often little difference in the equilibrium internuclear distances of the excited and ground electronic states. Sharp lines result from these transitions to a low energy vibrational level of an excited state whose potential energy curve is similar in both shape and in equilibrium internuclear distance to that of the ground state.

As discussed earlier, there is no center of symmetry in a tetrahedral molecule, so somewhat more intense absorptions (ε = 100 to 1000) than those in octahedral complexes are often obtained for *d-d* transitions in T_d complexes.

d^1 and d^9 Complexes

The simplest case with which to illustrate the relation between the color of a transition metal ion complex from a *d-d* transition and Dq is d^1; *e.g.*, Ti^{III} in an octahedral field. The ground state of the free ion is described by the term symbol ^2D and, as indicated earlier, the degenerate d levels are split in the presence of an octahedral field into a triply degenerate $^2T_{2g}$ and doubly degenerate ^2E_g set. The splitting is equal to $10Dq$. This is represented graphically in Fig. 10–7. As Dq increases, ΔE, the energy (hence the frequency) of the transition increases. The slope of the T_{2g} line is $-4Dq$ and that of E_g is $+6Dq$. The value of Δ (in units of cm^{-1}) can be obtained directly from the frequency of the absorption peak. For example, $Ti(H_2O)_6^{3+}$ has an absorption maximum at about 5000 Å (20,000 cm^{-1}). The Δ value for water attached to Ti^{3+} is about 20,000 cm^{-1} (Dq is 2000 cm^{-1}). Since this transition occurs with the absorption of the yellow-green component of visible light, the color transmitted is purple (blue + red). As the ligand is changed, Dq varies and the color of the complex changes. The color of the solution is the complement of the color or colors absorbed, because the transmitted bands determine the color. Caution should be exercised in inferring absorption bands from visual observation; *e.g.*, violet and purple are often confused.

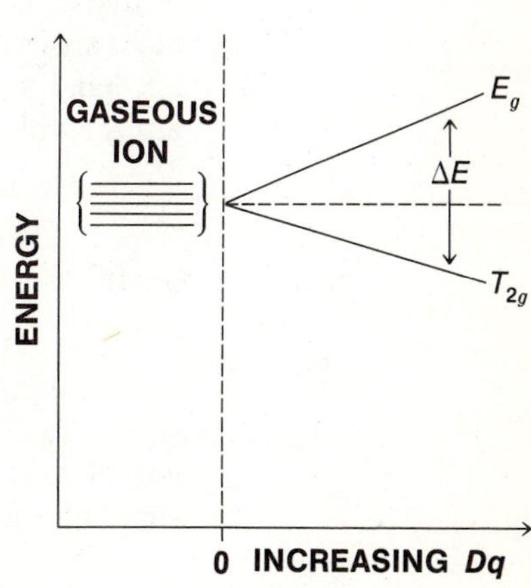

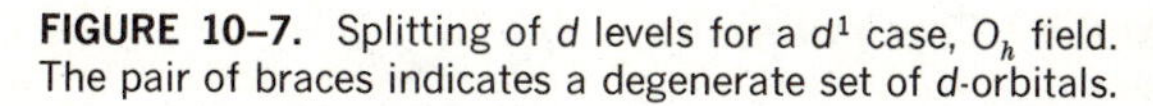
FIGURE 10–7. Splitting of d levels for a d^1 case, O_h field. The pair of braces indicates a degenerate set of d-orbitals.

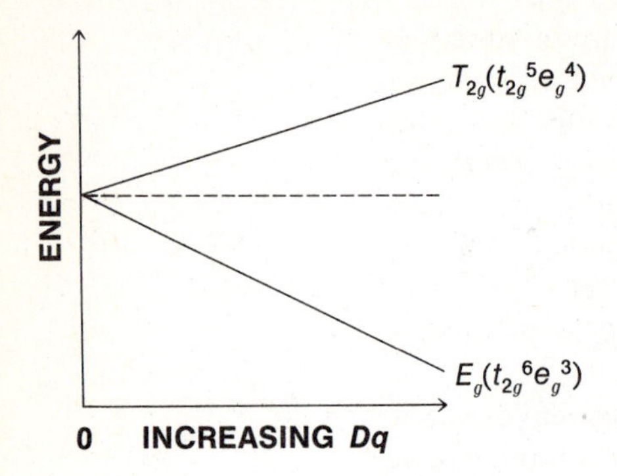

FIGURE 10–8. Splitting of the *d* levels in a d^9 complex, O_h field.

For a d^9 complex in an *octahedral* field the energy level diagram is obtained by inverting that of the d^1 complex (see Fig. 10–8). The inversion applies because the ground state of a d^9 configuration is doubly degenerate [$t_{2g}{}^6e_g{}^3$ can be $t_{2g}{}^6(d_{x^2-y^2})^2(d_{z^2})^1$ or $t_{2g}{}^6(d_{x^2-y^2})^1(d_{z^2})^2$] and the excited state is triply degenerate [$t_{2g}{}^5e_g{}^4$ can be $(d_{xy})^2(d_{yz})^2(d_{xz})^1(e_g)^4$ or $(d_{xy})^2(d_{yz})^1(d_{xz})^2(e_g)^4$ or $(d_{xy})^1(d_{yz})^2(d_{xz})^2(e_g)^4$]. Therefore, the transition is ${}^2E \rightarrow {}^2T_2$. In effect, the electronic transition causes the motion of a positive hole from the e_g level in the ground state to the t_{2g} level in the excited state, and the appropriate energy diagram results by inverting that for the electronic transition for a d^1 case. In order to preserve the center of gravity, the slopes of the lines in Fig. 10–8 must be $-6Dq$ for E and $+4Dq$ for T_2.

The results described above are often summarized by an Orgel diagram as in Fig. 10–9. For d^1, the tetrahedral splitting is just the opposite of that for octahedral splitting, so d^1 tetrahedral and d^9 octahedral complexes have similar Orgel diagrams, as indicated in Fig. 10–9. The splitting of the states as a function of Dq for octahedral

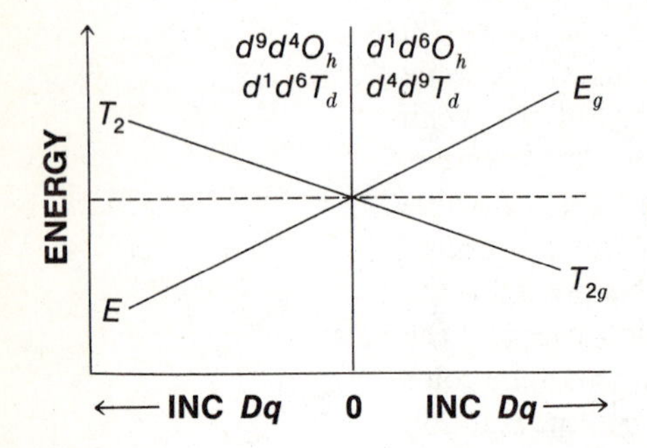

FIGURE 10–9. Orgel diagram for high spin d^1, d^4, d^6, and d^9 complexes.

complexes with electron configurations d^1 and d^6 and for tetrahedral complexes with d^4 and d^9 electron configurations is described by the right half of Fig. 10–9. Only one band arises in the spectra from *d-d* transitions, and this is assigned as ${}^2T_{2g} \rightarrow {}^2E_g$. The left-hand side of the Orgel diagram applies to octahedral d^4 and d^9 as well as tetrahedral d^1 and d^6 complexes. The single *d-d* transition that occurs is assigned as ${}^2E \rightarrow {}^2T_2$.

d^2, d^7, d^3, and d^8 Configurations

The two triplet states for a d^2 gaseous ion were earlier shown to be ${}^3F$ and ${}^3P$. Furthermore, we showed that an octahedral field split the ${}^3F$ terms into the triplet states ${}^3T_{1g}$, ${}^3T_{1g}$, and ${}^3A_{2g}$; the ${}^3A_{2g}$ state arises from $e_g{}^2$, the ${}^3T_{1g}$ states arise from $t_{2g}{}^2$, and (though not worked out earlier) another ${}^3T_{1g}$ state arises from $t_{2g}{}^1e_g{}^1$. In understanding the electronic spectra, only the triplet states need be considered, for the ground state is triplet. The following, simplified, one-electron orbital description of these triplet states can be presented. The following degenerate arrangements are possible for the ground state of the octahedral d^2 complex: $d_{xy}{}^1, d_{xz}{}^1, d_{yz}{}^0; d_{xy}{}^0, d_{xz}{}^1, d_{yz}{}^1; d_{xy}{}^1, d_{xz}{}^0, d_{xy}{}^1$. The ground state is orbitally triply degenerate and the symbol ${}^3T_{1g}(F)$ is used to describe this state; the (F) indicates that the state arose from the gaseous ion F term. In addition to the ${}^3T_{1g}(F)$ ground state, an excited state exists corresponding to the configuration in which the two electrons are paired in the t_{2g} level. Transitions to these states are multiplicity forbidden, but sometimes weak absorption bands assigned to these transitions are observed.

The triplet excited state $t_{2g}{}^1e_g{}^1$ will be considered next. If an electron is excited out of d_{xz} or d_{yz} so that the remaining electron is in d_{xy}, the excited electron will encounter less electron-electron repulsion from the electron and d_{xy} if it is placed in d_{z^2}. The $d_{x^2-y^2}$ orbital is less favorable because of the proximity of an electron in this orbital to the electron remaining in d_{xy}. This gives rise to arrangement (A) in Fig. 10–10. Similarly, if the electron is excited out of d_{xy} it will be most stable in $d_{x^2-y^2}$. The

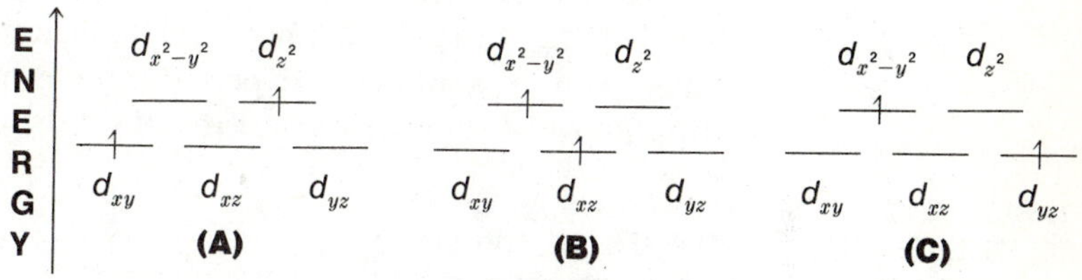

ENERGY
$d_{x^2-y^2}$ d_{z^2} / d_{xy} d_{xz} d_{yz} (A)
$d_{x^2-y^2}$ d_{z^2} / d_{xy} d_{xz} d_{yz} (B)
$d_{x^2-y^2}$ d_{z^2} / d_{xy} d_{xz} d_{yz} (C)

FIGURE 10–10. Possible electron arrangements for $t_{2g}{}^1e_g{}^1$.

remaining electron can be in either d_{xz} or d_{yz}, giving rise to (B) and (C). This set (Fig. 10–10A, B, and C) gives rise to the ${}^3T_{2g}$ state, which is orbitally triply degenerate and has a spin multiplicity of three. The arrangements ($d_{xy}{}^1, d_{x^2-y^2}{}^1; d_{xz}{}^1, d_{z^2}{}^1; d_{yz}{}^1, d_{z^2}{}^1$) are higher in energy and also produce an orbitally triply degenerate state, ${}^3T_{1g}(P)$. Other possible arrangements corresponding to $t_{2g}{}^1e_g{}^1$ involve reversing one of the electron spins to produce states with singlet multiplicity. Transitions to these states from the triplet ground state are multiplicity-forbidden. Finally, a two-electron transition producing the excited state $e_g{}^2$ or $d_{z^2}{}^1, d_{x^2-y^2}{}^1$ gives rise to a singly degenerate ${}^3A_{2g}$ state. It is instructive to indicate how these states relate to the gaseous ion. As illustrated in the section on term symbols (Section 10–1), the ground state for the gaseous ion V^{III} (a d^2 ion) is ${}^3F$. The ligand field in the complex removes the seven-fold orbital degeneracy of this state (*i.e.*, $M_L = 3, 2, 1, 0, -1, -2, -3$) into two three-fold degenerate states, ${}^3T_{1g}(F)$ and ${}^3T_{2g}$, and one non-degenerate state, ${}^3A_{2g}$. This is indicated in the Orgel diagram (Fig. 10–11) for a d^2, O_h complex.

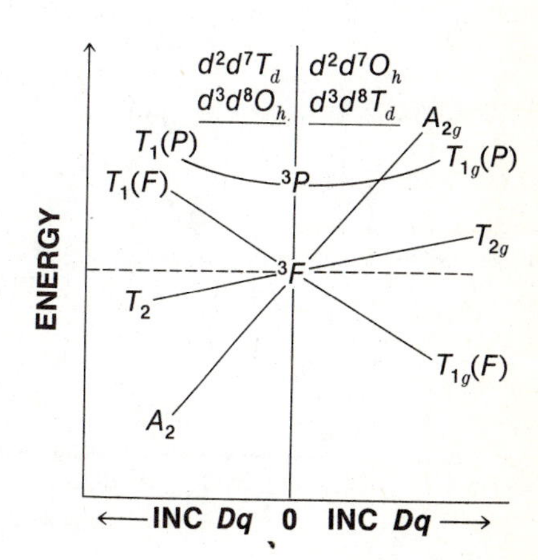

FIGURE 10–11. Orgel diagram for high spin d^2, d^3, d^7, and d^8 complexes.

For zero Dq (*i.e.*, the gaseous ion), only two triplet states, ^3F and ^3P, exist. As Dq increases, ^3F is split into the $^3T_{1g}(F)$, $^3T_{2g}$, and $^3A_{2g}$ states. The degeneracy of the ^3P state is not removed by the ligand field, and this state becomes the triplet $^3T_{1g}(P)$ state in an octahedral complex. The (P) indicates that this state arises from the gaseous ion ^3P state. The energies of these states as a function of Dq are presented in the Orgel diagrams as well as the Tanabe and Sugano diagrams[13] (Appendix D). Use of the Orgel diagrams in predicting spectra and making assignments will be demonstrated by considering V^{III} and Ni^{II} complexes.

For V^{III}, three transitions involving the states shown in Fig. 10–11 could occur: $^3T_{1g}(F) \rightarrow {}^3T_{2g}$, $^3T_{1g}(F) \rightarrow {}^3T_{1g}(P)$, and $^3T_{1g}(F) \rightarrow {}^3A_{2g}$. The transition to $^3A_{2g}$ in V^{III} is a two-electron transition. Such transitions are relatively improbable, and hence have low intensities. This transition has not been observed experimentally. The spectra obtained for octahedral V^{III} complexes consist of two absorption bands assigned to $^3T_{1g}(F) \rightarrow {}^3T_{2g}(F)$ and $^3T_{1g}(F) \rightarrow {}^3T_{1g}(P)$. In $V(H_2O)_6{}^{3+}$ these occur at about 17,000 and 24,000 cm^{-1}, respectively.

For octahedral nickel(II) complexes, the Orgel diagram (left-hand side of Fig. 10–11, d^8) indicates three expected transitions: $^3A_{2g} \rightarrow {}^3T_{2g}$, $^3A_{2g} \rightarrow {}^3T_{1g}(F)$, and $^3A_{2g} \rightarrow {}^3T_{1g}(P)$. (A similar result is obtained from the use of the Tanabe and Sugano diagram in Appendix D.) Experimental absorption maxima corresponding to these transitions are summarized in Table 10–6 for octahedral Ni^{II} complexes. (Numbers in parentheses correspond to shoulders on the main band.) Spectra of the octahedral NH_3, $HC(O)N(CH_3)_2$, and $CH_3C(O)N(CH_3)_2$ complexes are given[14] in Fig. 10–12. These complexes are colored purple, green, and yellow, respectively.

TABLE 10–6. ABSORPTION MAXIMA OF THE OCTAHEDRAL Ni^{II} COMPLEXES (ν_{max} in cm^{-1})

Ligand	$^3A_{2g} \rightarrow {}^3T_{2g}$	$^3A_{2g} \rightarrow {}^3T_{1g}(F)$	$^3A_{2g} \rightarrow {}^3T_{1g}(P)$
H_2O	8500	15,400	26,000
NH_3	10,750	17,500	28,200
$(CH_3)_2SO$	7730	12,970	24,040
$HC(O)N(CH_3)_2$	8500	13,605 (14,900)	25,000
$CH_3C(O)N(CH_3)_2$	7575	12,740 (14,285)	23,810

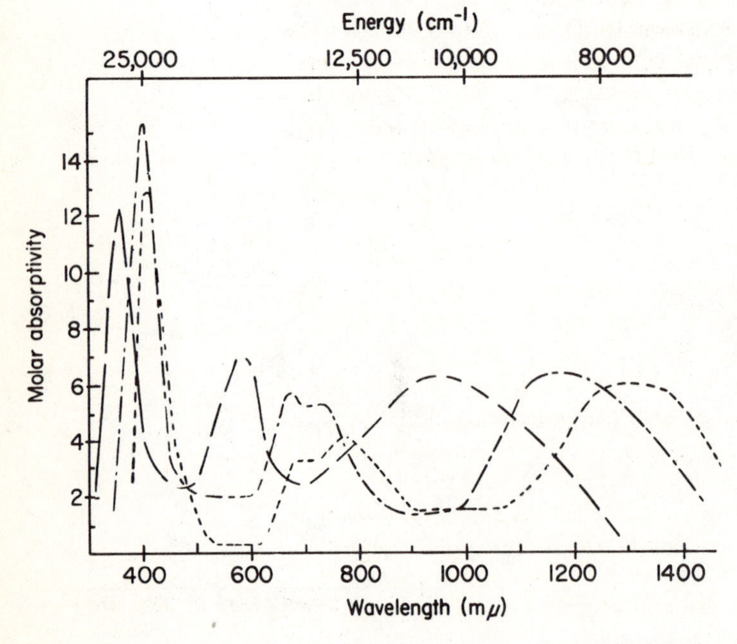

FIGURE 10–12. Molar absorptivity, ε, for some nickel(II) complexes in CH_3NO_2 solution. ---, $Ni(NH_3)_6(ClO)_2$; ···, $Ni[HC(O)N(CH_3)_2]_6(ClO_4)_2$; -·-, $Ni[CH_3C(O)N(CH_3)_2]_6(ClO_4)_2$.

10-8 CALCULATION OF Dq AND β FOR "O_h" Ni^{II} COMPLEXES

The graphical information contained in the Orgel diagrams is more accurately represented by the series of equations that relates the energies of these various states to the Dq value of the ligand. These energies were derived in Section 10-3. For Ni^{II} in an octahedral field, the energies, E, of the states relative to the spherical field are given by equations (10-10) to (10-12).

For ${}^3T_{2g}$: $$E = -2Dq \tag{10-10}$$

For ${}^3A_{2g}$: $$E = -12Dq \tag{10-11}$$

For ${}^3T_{1g}(F)$ and ${}^3T_{1g}(P)$: $$[6Dqp - 16(Dq)^2] + [-6Dq - p]E + E^2 = 0 \tag{10-12}$$

where p is the energy of the ${}^3P$ state. There are two roots to the last equation corresponding to the energies of the states ${}^3T_{1g}(F)$ and ${}^3T_{1g}(P)$.

From equations (10-10) and (10-11) it is seen that the energies of both ${}^3T_{2g}$ and ${}^3A_{2g}$ are linear functions of Dq. For any ligand that produces a spin-free octahedral nickel complex, the difference in energy between the ${}^3T_{2g}$ state and the ${}^3A_{2g}$ state in the complex is $10Dq$. As can be seen from the Orgel or Tanabe and Sugano diagrams, the lowest energy transition is ${}^3A_{2g} \rightarrow {}^3T_{2g}$. Since this transition is a direct measure of the energy difference of these states, Δ (or $10Dq$) can be equated to the transition energy, *i.e.*, the frequency of this band (cm^{-1}).

Equation (10-12) can be solved for the energies of the other states. However, the above equations have been derived by assuming that the ligands are point charges or point dipoles and that there is no covalence in the metal-ligand bond. If this were true, the value for Dq just determined could be substituted into equation (10-12), the energy of ${}^3P$ obtained from the atomic spectrum of the gaseous ion,[10] and the energy of the other two levels in the complex calculated from equation (10-12). The frequencies of the expected spectral transitions are calculated for one band corresponding to the difference between the energies of the levels ${}^3T_{1g}(F) - {}^3A_{2g}$ and for the other band from the energy difference ${}^3T_{1g}(P) - {}^3A_{2g}$. The experimental energies obtained from the spectra are almost always lower than the values calculated in this way. The deviation is attributed to covalency in the bonding.

The effect of covalency is to reduce positive charge on the metal ion, as a consequence of the inductive effect of the ligands. With reduced positive charge, the radial extension of the d-orbitals increases; this decreases the electron-electron repulsions, lowering the energy of the ${}^3P$ state. Covalency is foreign to the crystal field approach and is incorporated into the ligand field approach by providing an additional parameter, as we shall discuss next.

The difference in energy between the ${}^3P$ and ${}^3F$ states in the complex relative to that in the gaseous ion is decreased by covalency and, as a result, the gas phase value cannot be used for p [in equation (10-12)]; rather, p must be experimentally evaluated for each complex. Equation (10-12) can be employed for this calculation by using the Dq value from the ${}^3A_{2g} \rightarrow {}^3T_{2g}$ transition and the experimental energy, ΔE, for the ${}^3A_{2g} \rightarrow {}^3T_{1g}(P)$ transition. The only unknown remaining in equation (10-12) is p. The lowering of ${}^3P$ is a measure of covalency, among other effects. It is referred to as the *nephelauxetic effect* and is sometimes expressed by a parameter β°, a percentage lowering of the energy of the ${}^3P$ state in the complex compared to the energy of ${}^3P$ in the free, gaseous ion.[13] It is calculated by using equation (10-13):

$$\beta^\circ = [(B - B')/B] \times 100 \tag{10-13}$$

where B is the Racah parameter discussed earlier for the free gaseous ion and B' is the same parameter for the complex. It should be noted that p of equation (10–12) is proportional to B. In the case of nickel(II), the energy of ^3P in the complex can be substituted along with Dq into equation (10–12) and the other root calculated. The difference in the energy between this root and the energy of $^3A_{2g}$ gives the frequency of the middle band [$^3A_{2g} \rightarrow {}^3T_{1g}(F)$]. The agreement of the calculated and experimental values for this band is good evidence for O_h symmetry. The above discussion will be made clearer by referring to Appendix E, where a sample calculation of Dq, $\beta°$, and the frequency of the $^3A_{2g} \rightarrow {}^3T_{1g}(F)$ transition is presented for $Ni[(CH_3)_2SO]_6(ClO_4)_2$.

Most often the quantity β is used instead of $\beta°$, where β is defined as:

$$\beta = \frac{B'}{B} \tag{10–14}$$

The two quantities are easily related if equation (10–13) is rewritten as $\beta° = (1 - \beta) \times 100$.

With many other ions the spectral data cannot be solved easily for Dq and β because of complications introduced by spin-orbit coupling. The consequences of this effect on a d^1 ion are illustrated in Fig. 10–13. The triply degenerate T_{2g} state is split by spin-orbit (s.o.) coupling as indicated in Fig. 10–13(B). Coupling lowers the energy

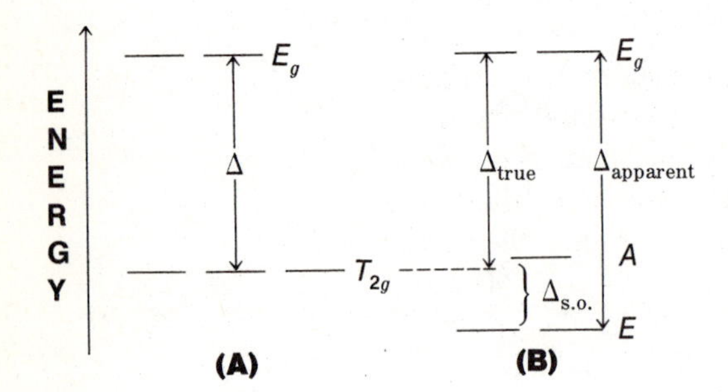

FIGURE 10–13. Contribution to Δ from spin-orbit coupling. (A) *d*-level splitting with no spin-orbit coupling. (B) Splitting of T_{2g} level by spin-orbit coupling.

of the ground state and the extent of lowering depends upon the magnitude of the coupling. When the ground state is lowered by spin-orbit coupling, the energies of all the bands in the spectrum have contributions from this lowering, $\Delta_{s.o.}$. When the contribution to the total energy from $\Delta_{s.o.}$ cannot be determined, the evaluation of Δ and β is not very accurate. Spin-orbit coupling in an excited state is not as serious a problem because transitions to both of the split levels often occur and the energies can be averaged. When the ground state is split, only the lower level is populated. Thus accurate values for Dq and β without corrections for spin-orbit coupling can be obtained only for ions in which the ground state is A or E (*e.g.*, Ni^{2+}). A further complication is introduced by the effect that Jahn-Teller distortions have on the energies of the levels.

Ni^{II}, Mn^{II} (weak field), Co^{III} (strong field), and Cr^{III} form many octahedral complexes whose spectra permit accurate calculation of Dq and β without significant complications from spin-orbit coupling or Jahn-Teller distortions. Ti^{III} has only minor contributions from these complicating effects. In the case of tetrahedral complexes, the magnitude of the splitting by spin-orbit interactions more nearly approaches that of crystal field splitting (Dq', the splitting in a tetrahedral field, is about $\frac{4}{9}Dq$). As a result, spin-orbit coupling makes appreciable contributions to the energies of the observed bands. A procedure has been described[14] which permits evaluation of Dq and β for tetrahedral Co^{II}. A sample calculation is contained in Appendix E.

Both σ and π bonding of the ligand with the metal ion contribute to the quantity Dq. When π bonding occurs, the metal ion t_{2g} orbitals will be involved, for they have the proper directional and symmetry properties. If π bonding occurs with empty ligand orbitals (*e.g.*, d in Et_2S or p in CN^-), Dq will be larger than in the absence of this effect. If π bonding occurs between filled ligand orbitals and filled t_{2g} orbitals (as in the case with the ligand OH^- and Co^{III}), the net result of this interaction is antibonding and Dq is decreased. These effects are illustrated with the aid of Fig. 10–14. In Fig. 10–14(A), the d electrons in t_{2g} interact with the empty ligand orbitals,

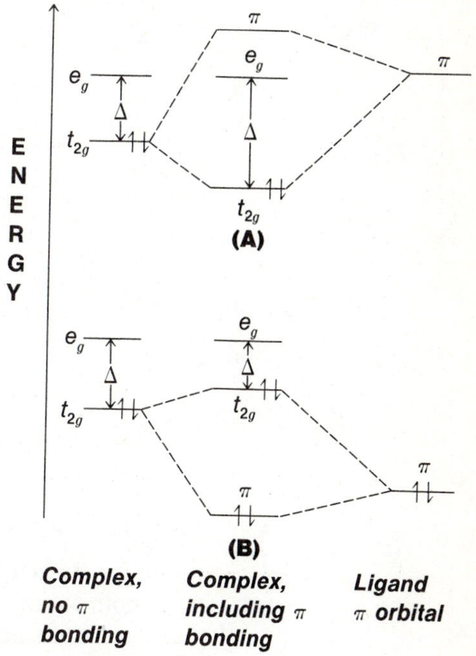

FIGURE 10–14. Effect of π bonding on the energy of a t_{2g} orbital and on Dq. (A) Filled metal orbitals, empty ligand orbitals. (B) Filled metal orbitals, filled ligand orbitals.

lowering the energy of t_{2g} and raising the energy of the mainly ligand π-orbitals in the complex. An empty π^* orbital of the ligand could be involved in this type of interaction. Since t_{2g} is lowered in energy and e_g is not affected (the e_g orbitals point toward σ electron pairs on the ligands), Dq will increase. In the second case [Fig. 10–14(B)], filled ligand π orbitals interact with higher energy filled metal d-orbitals, raising the energy of t_{2g} and lowering Dq.

It is informative to relate the energies of the observed d-d transitions to the energy levels associated with the molecular orbital description of octahedral complexes. The scheme for an O_h complex is illustrated in Fig. 10–15 (which neglects π bonding). The difference between T_{2g} and E_g^* is $10Dq$. As metal-ligand σ bond strength increases, E_g is lowered, E_g^* is raised by the same amount, and Dq increases. If T_{2g} metal electrons form π bonds with empty p or d orbitals of the ligand, the energy of the T_{2g} level in the complex is lowered and Dq is increased. Electron-electron repulsions of the T_{2g} electrons and the metal nonbonding electrons raise the energy of the T_{2g} set and decrease Δ. The above ideas have been employed in the interpretation of the spectra of some transition metal acetylacetonates.[15,16]

The magnitude of Dq is determined by many factors: interactions from an electrostatic perturbation, the metal-ligand σ bond, the metal to ligand π bond, the ligand to metal π bond, and metal electron-ligand electron repulsions. Additional references[9,10,18] on the material presented in this section are available.

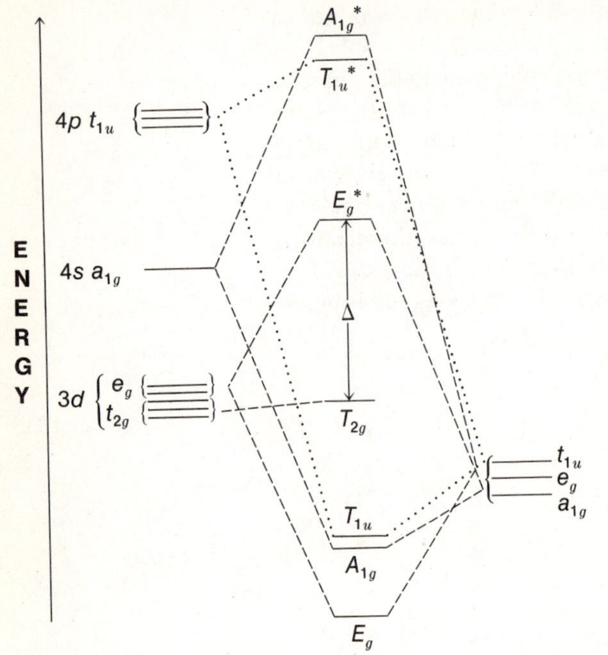

FIGURE 10–15. Molecular orbital description of an octahedral complex (π bonding effects and electrons are not included).

Much useful information regarding the metal ion-ligand interaction can be obtained from an evaluation of Dq and β. For a series of amides of the type $R_1CON\genfrac{}{}{0pt}{}{R_2}{R_3}$ it was found that whenever R_1 and R_2 are both alkyl groups, lower values for Dq and β result for the six-coordinate nickel complexes than when either R_1 or both R_2 and R_3 are hydrogens. This is not in agreement with the observation that toward phenol and iodine the donor strengths of these amides are found to increase with the number of alkyl groups. It was proposed that a steric effect exists between neighboring coordinated amide molecules(14) in the metal complexes. A study of the nickel(II) complexes of some primary alkyl amines indicated that even though water replaces the amines in the complexes, the amines interact more strongly with nickel than does water, and almost as strongly as ammonia.(19) A large Dq is also reported for the nickel complex of ethyleneimine.(20) These results are interpreted and an explanation involving solvation energies is proposed for the instability of the alkylamine complexes in water.(19)

The magnitude of $10Dq$ for various metal ions generally varies in the following order:

$$Mn^{2+} < Ni^{2+} < Co^{2+} < Fe^{2+} < V^{2+} < Fe^{3+} < Cr^{3+} < V^{3+} < Co^{3+} <$$
$$Mn^{4+} < Mo^{3+} < Rh^{3+} < Pd^{4+} < Ir^{3+} < Re^{4+} < Pt^{4+}$$

Representative ligands give rise to the following order, referred to as the spectrochemical series:

$$I^- < Br^- < \text{—}SCN^- < F^- < \text{urea} < OH^- < CH_3CO_2^- < C_2O_4^{2-} <$$
$$H_2O < \text{—}NCS^- < \text{glycine} < C_5H_5N \sim NH_3 < \text{ethylenediamine} <$$
$$SO_3^{2-} < o\text{-phenanthroline} < NO_2^- < CN^-$$

Jørgensen[4,5] has reported a remarkable set of parameters that enable one to predict $10Dq$ and β for various transition metal ion complexes. When the empirical parameters given in Table 10–7 are substituted into equations (10–15) and (10–16), the values of $10Dq$ and B for the complex result.

TABLE 10–7. EMPIRICAL PARAMETERS FOR PREDICTING $10Dq$ AND B WITH EQUATIONS (10–15) AND (10–16)*

Ligands	f	h	Metal Ions	$g(10^3\ cm^{-1})$	k
$6F^-$	0.9	0.8	V(II)	12.3	0.08
$6H_2O$	1.00	1.0	Cr(III)	17.4	0.21
6 urea	0.91	1.2	Mn(II)	8.0	0.07
$6NH_3$	1.25	1.4	Mn(IV)	23	0.5
3 en	1.28	1.5	Fe(III)	14.0	0.24
$3ox^{2-}$	0.98	1.5	Co(III)	19.0	0.35
$6Cl^-$	0.80	2.0	Ni(II)	8.9	0.12
$6CN^-$	1.7	2.0	Mo(III)	24	0.15
$6Br^-$	0.76	2.3	Rh(III)	27	0.30
$3dtp^-$	0.86	2.8	Re(IV)	35	0.2
C_5H_5N	1.25	—	Ir(III)	32	0.3
			Pt(IV)	36	0.5

*From C. K. Jørgensen, "Absorption Spectra and Chemical Bonding in Complexes," Pergamon Press, New York, 1962.

$$10Dq = fg\,(\text{cm}^{-1} \times 10^{-3}) \tag{10-15}$$

$$B = B_0(1 - hk) \tag{10-16}$$

B_0 is the free ion interelectronic repulsion parameter.

10–9 EFFECT OF DISTORTIONS ON THE *d*-ORBITAL ENERGY LEVELS

Since octahedral, square planar, and tetrahedral crystal fields cause different splittings of the five *d*-orbitals, geometry will have a pronounced effect upon the $d \rightarrow d$ transitions in a metal ion complex. Spectral data for these transitions should provide information about the structure of complexes. Our initial concern will be with how structure affects the energies of the various states in a metal ion. This information will then be applied to determine the structures of various complexes.

The structures of six-coordinate complexes can be classified as cubic, axial, or rhombic, if the equivalences of the ligands along the x, y, and z axes are represented by $x = y = z$, $x = y \neq z$, or $x \neq y \neq z$, respectively. Tetragonal and trigonal distortions (*i.e.*, an elongation or compression along the three-fold axis) are common axial examples. The splitting of the states for a d^1 case is represented in Fig. 10–16. Since *d-d* electron repulsions are not present in the d^1 case, the states can be correlated with the *d*-orbitals as indicated. For the splitting in a tetragonal complex, *trans*-$TiA_4B_2^{3+}$, to arise as indicated in Fig. 10–16, ligand A must occupy a higher position in the spectrochemical series than B. Metal electron–ligand electron repulsions are less for states consisting of electrons in orbitals directed toward the ligands, B, located on the z axis. As a result, d_{z^2} is lower in energy than $d_{x^2-y^2}$, and d_{xz} and d_{yz} are lower than d_{xy}. More bands will be observed in the spectrum of the tetragonal complex than in the spectrum of the octahedral complex.

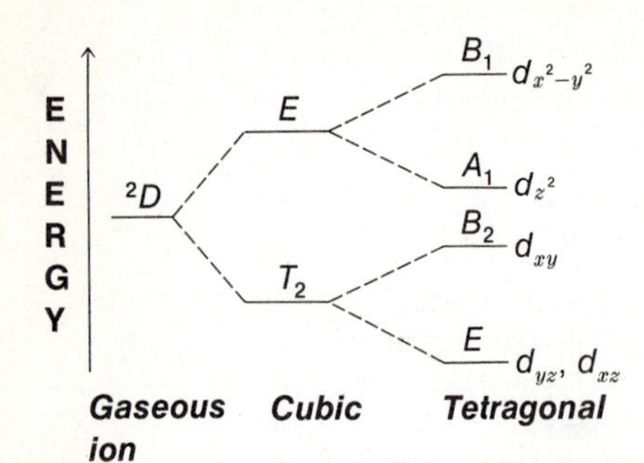

FIGURE 10–16. Orbital splitting for d^1 complexes in cubic and tetragonal fields.

The energies and splittings of the various states for a spin-paired Co^{III} complex are indicated in Fig. 10–17. Since this is an ion with more than one d electron, we must concern ourselves with states and not orbitals. The ${}^1T_{1g}$ excited state of the octahedral complex splits into ${}^1A_{2g}$ and ${}^1E_g$ states in a tetragonal field, while ${}^1T_{2g}$ splits into ${}^1E_g$ and ${}^1B_{2g}$. The splitting that occurs in a rhombic field is also indicated. Hydrated trisglycinatocobalt(III) exists as a violet α isomer and a red β isomer. One isomer must be cubic (where $x = y = z$), and the other isomer must be rhombic. The spectrum[18] of the β isomer consists of two bands. The α isomer also gives rise to two bands, one of which is asymmetric and must consist of two or more absorption bands that are not resolved. Therefore, the α isomer must be the rhombic isomer and the β isomer the cubic isomer.

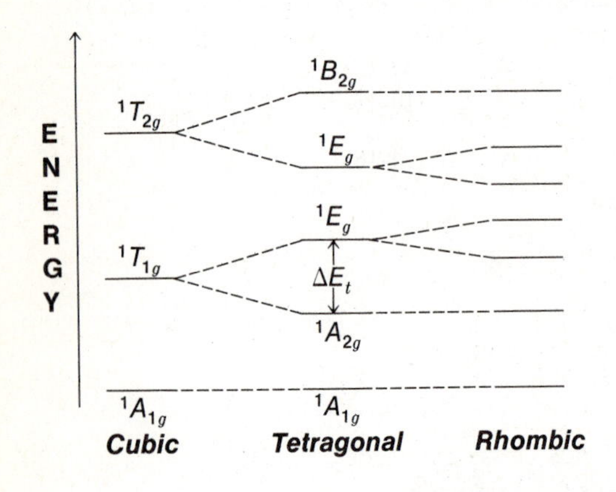

FIGURE 10–17. Splitting of the various states for cobalt(III) in cubic, tetragonal, and rhombic fields.

Both *cis*- and *trans*-$CoA_4B_2^+$ isomers are tetragonal. However, the difference in energy between the ${}^1A_{2g}$ and ${}^1E_g$ states (ΔE_t in Fig. 10–17) is usually about twice as large in *trans* complexes as it is in *cis* complexes[21,22] [$\Delta E_t(cis) = -C(\Delta_A - \Delta_B)$ and $\Delta E_t(trans) = 2C(\Delta_A - \Delta_B)$, where C is a constant usually less than one, Δ_A and Δ_B represent the crystal field splittings of ligands A and B (*i.e.*, their positions in the spectrochemical series), and the minus sign accounts for the fact that the energies of ${}^1E_g$ and ${}^1A_{2g}$ are interchanged in *cis* and *trans* complexes]. Usually, when Δ_A and Δ_B differ appreciably, the ${}^1A_{2g}$ and ${}^1E_g$ splitting gives rise to a doublet for the ${}^1A_{1g} \rightarrow {}^1T_{1g}$ peak in the *trans* compound, while this band is simply broadened in the *cis* compound.[22] It is also found that *cis* isomers often have larger molar absorptivity values for $d \rightarrow d$ transitions than *trans* isomers. Typical spectra are illustrated in Fig. 10–18. When Δ_A and Δ_B are similar, this criterion cannot be employed. The ultraviolet charge transfer band can also be used to distinguish between *cis*- and *trans*-cobalt(III) complexes when both isomers are available. The frequency of the band is usually higher in the *cis* than the *trans* compound. The benzoylacetonates of Co^{III} and Cr^{III} are examples in which Δ_A and Δ_B are nearly equal and the *trans* complex has a larger ε value than the *cis* complex.[24]

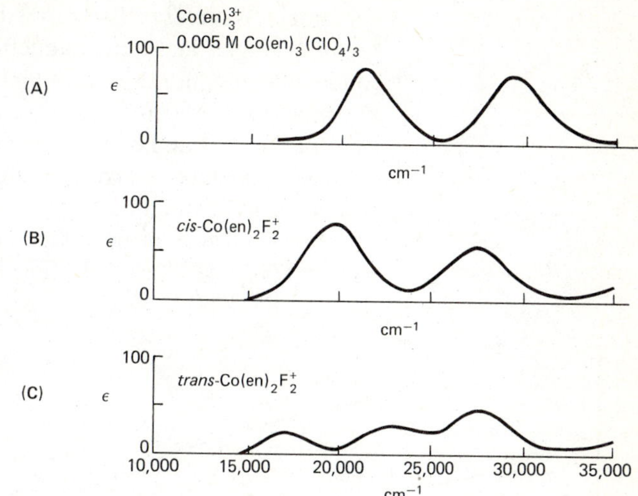

FIGURE 10–18. Spectra of (A) octahedral trisethylenediamine Co(III), (B) the *cis* difluoro compound, and (C) the *trans* difluoro compound. All spectra are taken in water.

If certain assumptions are made concerning the metal-ligand distance, the dipole moment of the ligand, and the effective charge of the nickel nucleus, it is possible to calculate[25] the change in energy that occurs for the various states as the ligand arrangement is varied from O_h to D_{4h}. This corresponds to lengthening the metal-ligand distance along d_{z^2} to infinity. The change in energy for the various levels as a function of the distortion is illustrated in Fig. 10–19. The percentages on the abscissa indicate progressive weakening of two *trans* metal-ligand bonds (*e.g.*, 100 per cent weakening represents a square planar complex). It is thus easy to see why the spectrum of a square planar or tetragonally distorted nickel complex should differ from that of a regular octahedral complex.

As indicated in Fig. 10–19, there will be a continuous change in spectral properties as the amount of tetragonal distortion increases. Eventually, for highly distorted tetragonal complexes, the spectra will resemble those of square planar

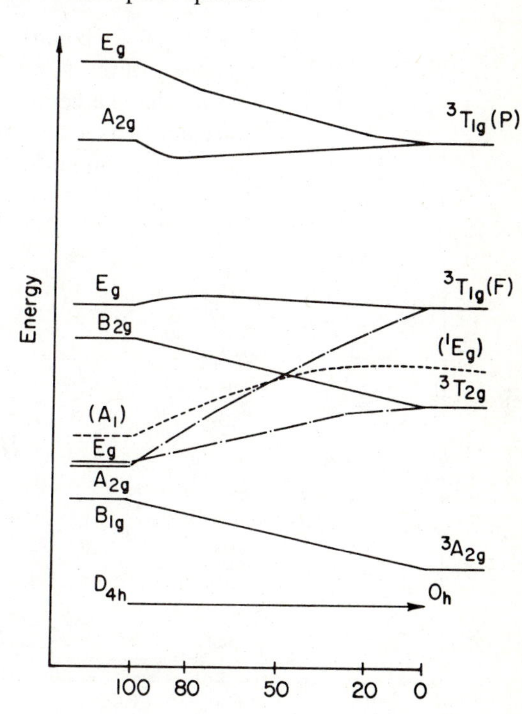

FIGURE 10–19. Effect of tetragonal distortion on the energy levels of nickel(II). [From C. Furlani and G. Sartori, J. Inorg. Nucl. Chem., *8*, 126 (1958).]

complexes. With large distortion, the multiplicity of the lowest energy state for Ni(II) becomes singlet and a diamagnetic complex results. The diamagnetic tetragonal or square planar complexes have high intensity absorption bands ($\varepsilon = 100$ to 350) with maxima in the 14,000 to 18,000 cm^{-1} region. The spectra may contain one, two, or three peaks,[26,27] and band assignments are often difficult. However, by using spectral and magnetic data, square planar or highly distorted tetragonal nickel(II) complexes can be easily distinguished from nearly octahedral or tetrahedral complexes.

Just as distortion from O_h and D_{4h} changes the energies and properties of the various levels, so does distortion from D_{4h} (planar) to D_{2d} to T_d (Fig. 10-20).

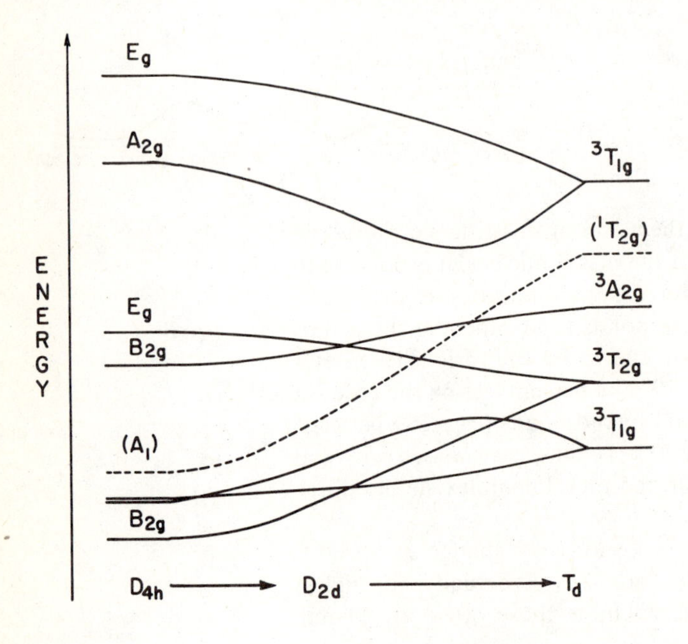

FIGURE 10-20. Energy levels for T_d, D_{4h}, and D_{2d} complexes. [From C. Furlani and G. Sartori, J. Inorg. Nucl. Chem., 8, 126 (1958).]

Fig. 10-21 contains spectra for O_h, D_{4h}, D_{2d}, and T_d complexes.[25,28,29,30] The transitions of the T_d nickel complex $NiCl_4^{2-}$ have large ε values because the complex does not have a center of symmetry. In this case, the d and p orbitals can mix in a molecular orbital description. The p-orbital contribution of the ground and excited states gives some allowed $d \rightarrow p$ character to the transition, and the intensity in-

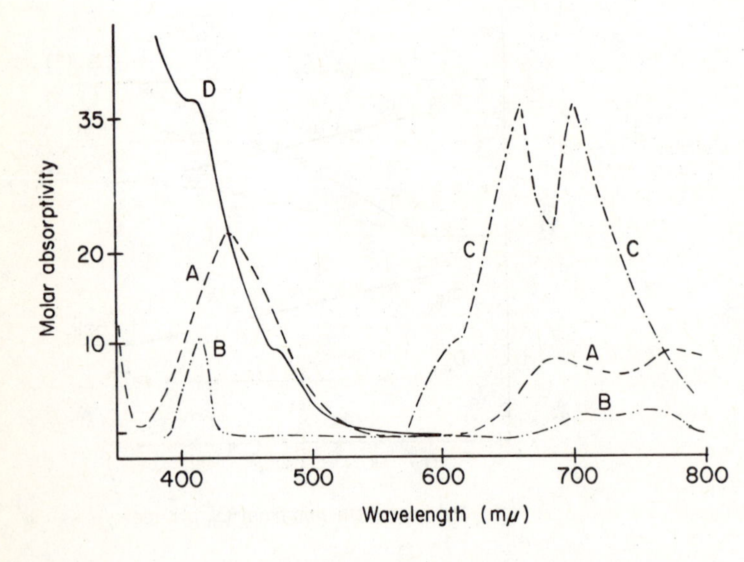

FIGURE 10-21. Electronic absorption spectra of some nickel complexes. A, $Ni(\phi_3PO)_4(ClO_4)_2$ in CH_3NO_2 (D_{2d}); B, $Ni[(CH_3)_2SO]_6(ClO_4)_2$ in $(CH_3)_2SO$ (O_h); C, $NiCl_4^{2-}$ in CH_3NO_2(T_d); D, Ni(II) (dimethylglyoxinate)$_2$ in $CHCl_3$ (D_{4h}). Curve C is a plot of $\varepsilon/5$.

creases. Mixing in non-centrosymmetric ligand molecular orbitals will also enhance the intensity. Accordingly, $\varepsilon/5$ is plotted in Fig. 10–21. As can be seen, different spectra are obtained for different structures. The spectrum for a T_d complex is expected (see Fig. 10–20) to contain three bands ν_1, ν_2, and ν_3 corresponding to the three spin-allowed transitions: $^3T_1(F) \rightarrow {}^3T_2$, ν_1; $^3T_1(F) \rightarrow {}^3A_2$, ν_2; and $^3T_1(F) \rightarrow {}^3T_1(P)$, ν_3 (see Fig. 10–21). The ν_1 band occurs in the range between 3000 and 5000 cm^{-1} and is often masked by absorption by either the organic part of the molecule or the solvent. It has been observed for Ni^{2+} in silicate glasses and in $NiCl_4{}^{2-}$. The ν_2 band occurs in the 6500 to 10,000 cm^{-1} region and has an appreciable molar absorptivity (ε = 15 to 50). The ν_3 band is found in the visible region (12,000 to 17,000 cm^{-1}) and shows an intense absorption (ε = 100 to 200).

It is proposed[28] that the complex $Ni[OP(C_6H_5)_3]_4(ClO_4)_2$ has a D_{2d} configuration. Absorption peaks occur in the spectrum at 24,300, 14,800, and 13,100 cm^{-1} with ε values of approximately 24, 8, and 9, respectively.

10–10 STRUCTURAL EVIDENCE FROM THE ELECTRONIC SPECTRUM

The electronic spectrum can often provide quick and reliable information about the ligand arrangement in transition metal ion complexes. Tetrahedral complexes are often readily distinguished from six-coordinate ones on the basis of the intensity of the bands. The spectra of nickel(II) and cobalt(II) are particularly informative. The complex $Ni\{OP[N(CH_3)_2]_3\}_4Cl_2$ could have tetrahedral, D_{2d}, square planar, tetragonal, or other distorted octahedral geometries. The similarity of the electronic spectrum of this complex to that of $NiCl_4{}^{2-}$ [see Fig. 10–21(C)] implied[30] that this was the first cationic, tetrahedral nickel(II) complex ever to be prepared. Further confirmation of the structure comes from the similarity of the x-ray powder diffraction patterns of the nickel(II) and zinc(II) complexes. The latter, with a $3d^{10}$ configuration, is expected to be tetrahedral. The Orgel or Tanabe-Sugano diagram for a $T_d\ d^8$ complex is the same as that for octahedral cobalt(II) ($O_h\ d^7$) with a low Dq value. Accordingly, the high energy visible band is assigned to $^4T_1(F) \rightarrow {}^4T_1(P)$ and the low energy band to $^4T_1(F) \rightarrow {}^4A_2$. The commonly observed splitting of the visible band is attributed to spin-orbit coupling, which lifts the degeneracy of $^4T_1(P)$ state. It is recommended that any discussion of band assignments in this section be accompanied by reference to the Tanabe and Sugano (or Orgel) diagrams.

In another complex, $Ni(NO_3)_4{}^{2-}$, it was shown[31] that the electronic spectrum of the nickel is characteristic of that of a six-coordinate complex, and some of the nitrate groups must be bidentate. The color of a transition metal ion complex is often a very poor indicator of structure. Octahedral nickel(II) complexes usually have three absorption bands in the regions from 8000 to 13,000 cm^{-1}, from 15,000 to 19,000 cm^{-1}, and from 25,000 to 29,000 cm^{-1}. The exact position will depend upon the quantities Δ and β. The molar absorptivities of these bands are generally below 20. As indicated in the section on Dq calculations, the fit of the calculated and experimental frequences of the middle peak has been proposed as confirmatory evidence for the existence of an "O_h" complex.

Spin-free tetragonal nickel(II) complexes, in which the two ligands occupying either *cis* or *trans* positions are different from the other four but have Dq values that are similar, will give spectra that will be very much like those of the O_h complexes. For example, when the distortion in *trans* complexes is close to 0 per cent (see Fig. 10–17), a spectrum typical of octahedral complexes is expected. In general, molar absorptivities will be higher for tetragonal than for octahedral complexes. A rule of *average environment* relates the band maxima in these slightly distorted tetragonal complexes to the Dq values of the ligands. The band position is determined by a Dq value that is an average of all the surrounding ligands.[4,32]

Nickel(II) forms a large number of five-coordinate complexes.[33] Geometries based on both the trigonal bipyramid and the tetragonal (square) pyramid are known. Many of the complexes are distorted significantly from this geometry.[34] The electronic spectra have been analyzed in detail by Ciampolini,[35] and the interested reader is referred to his account. It is often difficult to detect the difference between tetrahedral and certain five-coordinate geometries on the basis of the electronic spectrum.

The electronic spectra of cobalt(II) complexes can often provide reliable structural information. Most six-coordinate complexes have high spin electronic configurations. The Orgel diagram was given in Fig. 10–11. The ground state is ${}^4T_{1g}$ and a substantial amount of spin-orbit coupling is expected. Three transitions are predicted, ${}^4T_{1g}(F) \rightarrow {}^4T_{2g}$, ${}^4T_{1g}(F) \rightarrow {}^4A_{2g}$, and ${}^4T_{1g}(F) \rightarrow {}^4T_{1g}(P)$. The ${}^4T_{1g}(F) \rightarrow {}^4A_{2g}$ transition is a two-electron transition and is not observed. The electronic spectrum of octahedral $Co(H_2O)_6{}^{2+}$ and tetrahedral $CoCl_4{}^{2-}$ are shown in Fig. 10–22. The band in the octahedral complex at $\sim$20,000 cm^{-1} is assigned as the ${}^4T_{1g}(F) \rightarrow {}^4T_{1g}(P)$ transition. The shoulder results because spin-orbit coupling in the excited ${}^4T_{1g}(P)$ state causes the degeneracy to be lifted. The other absorption band, at 8350 cm^{-1}, is assigned to ${}^4T_{1g}(F) \rightarrow {}^4T_{2g}$.

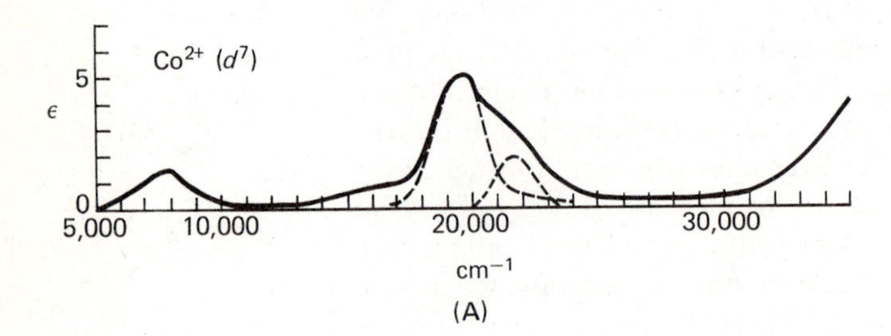

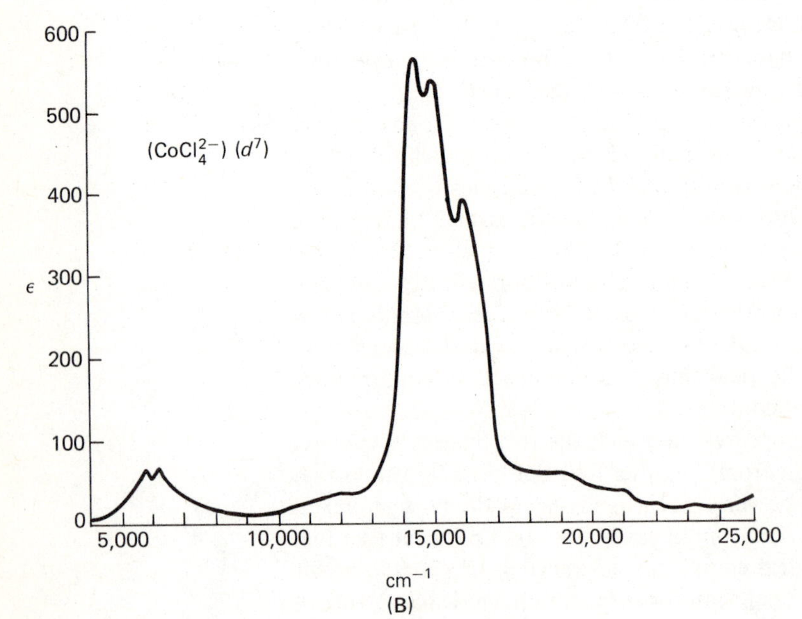

FIGURE 10–22. The electronic spectra of (A) $Co(H_2O)_6{}^{2+}$ [0.021 M $Co(BF_4)_2$ in H_2O]; (B) $CoCl_4{}^{2-}$ [0.001 M $CoCl_2$ in 10 M HCl]. The dotted line in (A) gives the resolution of the two bands contributing to the observed spectrum.

The tetrahedral complex of Co(II) has an energy level diagram like that of Cr(III). The complexes will always be high spin (see the Tanabe and Sugano diagrams in Appendix D). The absorption band at $\sim$15,000 cm^{-1} is assigned to $^4A_2 \rightarrow {}^4T_1(P)$. The fine structure is attributed to spin-orbit coupling of the T state. The existence of spin-orbit coupling also allows some quartet $\rightarrow$ doublet spin transitions to occur. The other band shown is assigned to $^4A_2 \rightarrow {}^4T_1(F)$. The expected $^4A_2 \rightarrow {}^4T_2$ transition is predicted to occur at 3000 to 4500 cm^{-1}; this is outside the range of most u.v.-visible instruments and is often overlapped by ligand vibrational transitions (*i.e.*, infrared bands). Several five-coordinate complexes of cobalt(II) have been prepared and their spectra reported and interpreted.[35a]

Copper(II) complexes[35b] take on a wide range of geometries, often with low symmetry. One generally finds, for most geometries, one very broad band with a maximum around 15,000 $\pm$ 5,000 cm^{-1}, which is thought to contain all of the expected transitions. It is possible, however, that the highest energy transition occurs farther out in the ultraviolet region under charge transfer bands. Thus, the electronic spectrum of copper(II) is of little value in structure assignment. The band position can be correlated roughly with the ligand field strength of the bonding groups.

Spectral data are available to enable similar conclusions to be drawn concerning the structures of other transition metal ions. The salient differences between spectra for various structures have been summarized.[10] Infrared and magnetic data should be used in conjunction with electronic spectral data to aid in assigning structures to complexes.[36] Use of magnetic data will be described in the next chapter.

The above examples are only a few of the very many cases that indicate the utility of near infrared, visible, and ultraviolet spectroscopy in providing information about the structures of complexes. The number of bands, their frequencies, and their molar absorptivities should all be considered. Solution spectra should be checked against the solid state spectra (reflectance or mulls) to be sure that changes in structure are not occurring in solution. These changes could involve ligand rearrangement, ligand replacement by solvent, or expansion of the coordination number by solvation.

Other kinds of structural applications of visible spectroscopy have been reported. The Dq values for the nitrite ion are different for the nitro (—NO_2) and nitrito (—ONO) isomers. As a result of the difference in average Dq, [$(Co(NH_3)_5ONO]^{2+}$ is red, while $[Co(NH_3)_5NO_2]^{2+}$ is yellow. Limitations of this kind of application have been reported.[37]

The use of electronic spectra to provide structural information is nicely illustrated in a study of the electronic structure of the vanadyl ion.[38] Spectra of the vanadyl ion, VO^{2+}, are interpreted to indicate that there is considerable oxygen-to-metal π bonding in the V—O bond. The similarity in the charge transfer spectra of solids known by x-ray analysis to contain the VO^{2+} group and of solutions is presented as evidence that aqueous solutions contain the species $VO(H_2O)_5{}^{2+}$ and not $V(H_2O)_6{}^{4+}$. Protonation of VO^{2+} would have a pronounced effect on the charge transfer spectrum. It is proposed that the oxygen is not protonated because its basicity is weakened by π bonding with vanadium. A complete molecular orbital scheme for $VO(H_2O)_5{}^{2+}$ is presented,[38] and assignments are made for the spectrum of $VOSO_4 \cdot 5H_2O$ in aqueous solution. Similar studies on other oxy-cations provide evidence for considerable metal-oxygen π bonding[39] and aid in elucidating the electronic structures of these species.

The electronic spectrum has been particularly valuable in determining the coordination number and ligand arrangement in metallo-enzymes. When zinc(II) is replaced by cobalt(II) in carbonic anhydrase, the electronic spectrum indicates that the metal ion is in a distorted tetrahedral site.[40] In such applications, one must be particularly careful to ascertain that the structure of the enzyme has not been changed by metal substitution. If the enzyme is still active, one can have some confidence that the structure is the same. In this example, subsequent x-ray structure determination confirmed the distorted tetrahedral ligand arrangement around zinc(II). The band position in the visible spectrum of a copper(II) protein, erythrocuprein, was interpreted[40] as indicative of coordination by at least four nitrogen donor ligands.

The blue copper(II) protein stellacyanin has been converted to a cobalt(II) derivative.[41] Copper(II) and cobalt(II) were shown to compete for the same site in the protein. Since cobalt(II) spectra are more readily interpreted than those of copper(II), the authors were able to conclude that the cobalt was in either a distorted tetrahedral or a five-coordinate environment. A strong charge transfer band indicated the existence of a Co-SR linkage. All of the bands in the native copper protein were assigned by analogy. The existence of porphyrin complexes in an enzyme system can be detected by the characteristic *Soret band* around 25,000 cm^{-1}. This is a ligand-based $\pi \rightarrow \pi^*$ charge transfer type of transition, discussed in Chapter 5. Two other lower intensity bands are also found in the electronic spectra of these complexes. The existence of these bands and their shifts upon placing substituents on the rings are understood in terms of results from molecular orbital calculations.[42] The positions of these bands have been employed to classify a whole host of cytochromes.

BONDING PARAMETERS FROM SPECTRA

10-11 σ AND π BONDING PARAMETERS FROM THE SPECTRA OF TETRAGONAL COMPLEXES

As mentioned earlier, when the symmetry of a complex is lowered from a local O_h or T_d environment additional bands appear in the spectrum. This is illustrated in Fig. 10-23, where the spectra of $Co(NH_3)_6^{3+}$ and $Co(NH_3)_5Cl^{2+}$ are compared. As indicated by the Tanabe and Sugano diagram (Appendix D), for a strong field d^6 complex the spin-allowed transitions are ${}^1A_{1g} \rightarrow {}^1T_{1g}$ and ${}^1A_{1g} \rightarrow {}^1T_{2g}$ and the bands are assigned as indicated. When one of the ammonia molecules in the complex is replaced by a different group, which we arbitrarily locate on d_{z^2}, that group interacts differently with cobalt than does ammonia. If this group were chloride, the σ interaction would be weaker and the π interaction greater. The relative energies of d_{z^2}, d_{xz}

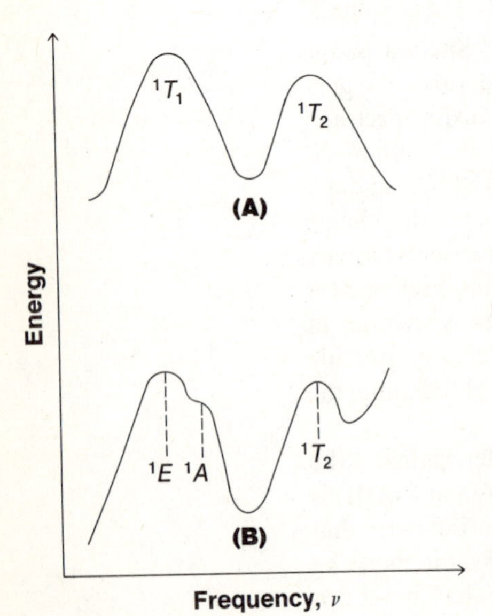

FIGURE 10-23. Spectra of (A) $Co(NH_3)_6^{3+}$ and (B) $Co(NH_3)_5Cl^{2+}$.

and d_{yz} will be affected differently in $Co(NH_3)_5Cl^{2+}$ and $Co(NH_3)_6^{3+}$; the stronger π interaction of Cl^- raises the energy of d_{xz} and d_{yz} and a weaker σ interaction from Cl^- lowers that of d_{z^2}. The degeneracy of the triplet state is removed and $^1T_1$ is split into ^1E and ^1A states, producing the spectrum indicated in Fig. 10-23(B).

The splitting of the $^1T_2$ band is predicted by theory[22] to be too small to be observed. With the additional bands, there is need for additional parameters (besides $10Dq$ and B) to describe the spectrum. The calculation of values for these parameters depends very much upon the assignment of the observed electronic transitions.

Several lower symmetry arrangements are treated by Gerloch and Slade.[3] Here, we shall briefly consider tetragonally distorted six-coordinate complexes. The total potential is now considered as

$$V(D_{4h}) = V_{oct} + V_{tetr} \tag{10-17}$$

When the matrix elements are evaluated, we obtain quantities that we earlier labeled Dq, plus two other radial quantities that arise from V_{tetr}, which we shall label Ds and Dt. Ds involves $\overline{r^{-2}}$ terms and Dt involves $\overline{r^{-4}}$ terms, as shown in equations (10–18) and (10–19):

$$Ds = \frac{2}{7}Ze^2\overline{r^{-2}}\left(\frac{1}{a_{(xy)}{}^3} - \frac{1}{b_{(z)}{}^3}\right) \tag{10-18}$$

$$Dt = \frac{2}{7}Ze^2\overline{r^{-4}}\left(\frac{1}{a_{(xy)}{}^5} - \frac{1}{b_{(z)}{}^5}\right) \tag{10-19}$$

Here a and b refer to the different metal-ligand distances (a refers to xy ligands and b refers to z ligands). Dt, being an $\overline{r^{-4}}$ function, can be related to Dq. (Recall that Dq is a function of r^{-4}.) It is a measure of the difference in Dq between the axial and equatorial sites, as shown in equation (10–20):

$$Dt = \frac{4}{7}[Dq(xy) - Dq(z)] \tag{10-20}$$

It is to be emphasized that neither $Dq(xy)$ nor $Dq(z)$ is the same as that for the ligand in an octahedral complex.[43]

Assuming that the symmetry of the complex is such that the metal e_g orbitals are only σ-antibonding and that the metal t_{2g} orbitals are only π-antibonding, McClure[22] has reported parameters $\delta\sigma$ and $\delta\pi$, defined as:

$$\delta\sigma = \sigma_z - \sigma_{xy} = -\frac{12}{8}Ds - \frac{15}{8}Dt \tag{10-21}$$

$$\delta\pi = \pi_z - \pi_{xy} = -\frac{3}{2}Ds + \frac{5}{2}Dt \tag{10-22}$$

The σ and π parameters reportedly indicate the relative σ- and π-antibonding properties of the ligands. Values for various ligands have been reported and interpreted.[43-45] The following order of σ bond interaction with the metal ion results from spectral studies on these complexes:

$$NH_3 > H_2O > F^- > Cl^- > Br^- > I^-$$

The order of π repulsion is:

$$I^- > Br^- > Cl^- > F^- > NH_3$$

10–12 THE ANGULAR OVERLAP MODEL

An alternative approach (with several advantages) to the parameterization of the spectra of transition metal complexes is the angular overlap model.[3,46] This model has its origins in a crude molecular orbital treatment of the energies of the transition metal compounds. A simple case, which we will consider first, is that of a mono-coordinated complex ML:

M—L

If the metal is a transition metal, we are most interested, from a spectroscopic point of view, in the energies of the *d*-orbitals in the complex. The five *d*-orbitals in the $C_{\infty v}$ symmetry of the complex span the σ, π, and δ representations; *i.e.*, $d(z^2)$ is σ, $d(xz)$ and $d(yz)$ are π, and $d(xy)$ and $d(x^2 - y^2)$ are δ. Considering, for instance, the σ interaction, we can write the secular equations:

$$\begin{vmatrix} H_{\text{M}}^{\sigma} - E & H_{\text{ML}}^{\sigma} - S_{\text{ML}}^{\sigma}E \\ H_{\text{ML}}^{\sigma} - S_{\text{ML}}^{\sigma}E & H_{\text{L}}^{\sigma} - E \end{vmatrix} = 0 \qquad (10\text{–}23)$$

where H_{M}^{σ} and H_{L}^{σ} are the energies of the appropriate metal and ligand orbitals respectively; H_{ML}^{σ} describes the exchange integral between the metal and ligand orbitals, and S_{ML}^{σ} is the overlap integral. In general, it is found that $H_{\text{M}}^{\sigma} \gg H_{\text{L}}^{\sigma}$ and that the diatomic overlap integral S_{ML}^{σ} is small, so that one of the roots, say E_1, will be quite close in energy to H_{M}^{σ}, and the other, E_2, will be quite close to H_{L}^{σ}. Invoking this assumption enables us to write two approximate determinants:

$$\begin{vmatrix} H_{\text{M}}^{\sigma} - E_1 & H_{\text{ML}}^{\sigma} - S_{\text{ML}}^{\sigma}H_{\text{M}}^{\sigma} \\ H_{\text{ML}}^{\sigma} - S_{\text{ML}}^{\sigma}H_{\text{M}}^{\sigma} & H_{\text{L}}^{\sigma} - H_{\text{M}}^{\sigma} \end{vmatrix} = 0$$

$$\begin{vmatrix} H_{\text{M}}^{\sigma} - H_{\text{L}}^{\sigma} & H_{\text{ML}}^{\sigma} - S_{\text{ML}}^{\sigma}H_{\text{L}}^{\sigma} \\ H_{\text{ML}}^{\sigma} - S_{\text{ML}}^{\sigma}H_{\text{L}}^{\sigma} & H_{\text{L}}^{\sigma} - E_2 \end{vmatrix} = 0 \qquad (10\text{–}24)$$

The values of the roots can be written explicitly as

$$E_1 = H_{\text{M}}^{\sigma} + \frac{(H_{\text{ML}}^{\sigma})^2 - 2H_{\text{ML}}^{\sigma}H_{\text{M}}^{\sigma}S_{\text{ML}}^{\sigma} + (S_{\text{ML}}^{\sigma}H_{\text{M}}^{\sigma})^2}{H_{\text{M}}^{\sigma} - H_{\text{L}}^{\sigma}} \qquad (10\text{–}25)$$

$$E_2 = H_{\text{L}}^{\sigma} - \frac{(H_{\text{ML}}^{\sigma})^2 - 2H_{\text{ML}}^{\sigma}H_{\text{L}}^{\sigma}S_{\text{ML}}^{\sigma} + (S_{\text{ML}}^{\sigma}H_{\text{L}}^{\sigma})^2}{H_{\text{M}}^{\sigma} - H_{\text{L}}^{\sigma}} \qquad (10\text{–}26)$$

Using the Wolfsberg-Helmholz approximation for H_{ML}^{σ} in the form

$$H_{\text{ML}}^{\sigma} = S_{\text{ML}}^{\sigma}(H_{\text{M}}^{\sigma} + H_{\text{L}}^{\sigma}) \qquad (10\text{–}27)$$

equations (10–25) and (10–26) can be rewritten as

$$E_1 - H_{\text{M}}^{\sigma} = E_{\sigma}^{*} = \frac{(H_{\text{M}}^{\sigma} + H_{\text{L}}^{\sigma})^2}{H_{\text{M}}^{\sigma} - H_{\text{L}}^{\sigma}}(S_{\text{ML}}^{\sigma})^2 \qquad (10\text{–}28)$$

$$E_2 - H_{\text{L}}^{\sigma} = E_{\sigma} = -\frac{(H_{\text{M}}^{\sigma} + H_{\text{L}}^{\sigma})^2}{H_{\text{M}}^{\sigma} - H_{\text{L}}^{\sigma}}(S_{\text{ML}}^{\sigma})^2 \qquad (10\text{–}29)$$

E_σ^* is positive, and therefore represents the destabilizing effect on the metal orbital energies, while E_σ is negative and represents the stabilizing effect on the ligand orbitals. As is shown by equation (10–28), the destabilizing effect on a particular metal orbital is proportional to $(S_{ML}{}^\sigma)^2$. This effect has to be small, when $H_M{}^\sigma - H_L{}^\sigma$ is large and $S_{ML}{}^\sigma$ is small. E_σ^* might in principle be calculated using (10–28), but in practice it is more profitable to express it parametrically. For this purpose, the overlap integral $S_{ML}{}^\sigma$ can be factored into a radial and an angular product:

$$S_{ML}{}^\sigma = S_{ML}F_\sigma{}^d \qquad (10\text{-}30)$$

where S_{ML} is the integral of the radial functions of the metal and of the ligand orbitals, and $F_\sigma{}^d$ refers to the angular part. S_{ML} depends on the particular metal and ligand orbitals considered and on the metal-ligand distance, while $F_\sigma{}^d$ depends only on the geometrical dispositions of the metal and the ligand. Once the geometry is known, $F_\sigma{}^d$ can be easily calculated. We can demonstrate this factoring with a simple example of a ligand L overlapping a p_z orbital on a metal, M, in an M—L fragment. In Fig. 10–24(A), the ligand is located on the z-axis and the M—L bond is coincident with this axis. If we define $F_\sigma{}^p$ (sigma ligand overlap with a p-orbital) as one, the overlap is given by the magnitude of S_{ML}. When the M—L bond is the same length but not coincident with the z-axis, the net overlap will be decreased. The radial part S_{ML} stays the same, so the decrease is accomplished by decreasing $F_\sigma{}^p$, the angular part. In Fig. 10–24(B), the M—L bond is along the x-axis, the ligand σ-orbital is orthogonal to p_z (overlap is zero) and, since the radial part doesn't change, this is accomplished by having the angular term, $F_\sigma{}^p$, become zero.

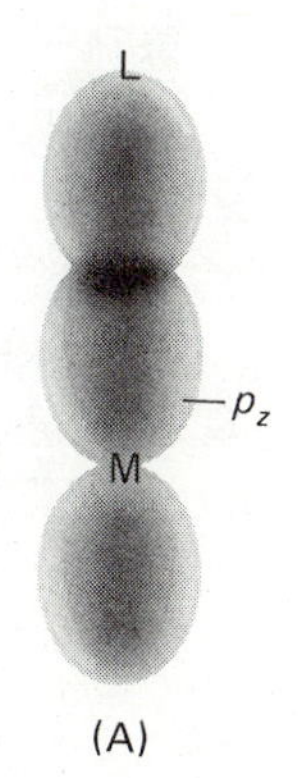

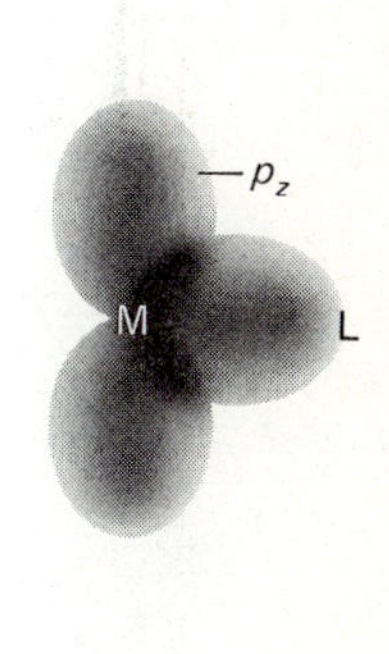

FIGURE 10–24. Orbital overlap with a p_z orbital in M—L fragments.

Substituting equation (10–30) into (10–28), one gets

$$E_\sigma^* = \frac{(H_M{}^\sigma + H_L{}^\sigma)^2}{H_M{}^\sigma - H_L{}^\sigma} S_{ML}{}^2(F_\sigma{}^d)^2 \qquad (10\text{-}31)$$

Now letting

$$e_\sigma = \frac{(H_M{}^\sigma + H_L{}^\sigma)^2}{H_M{}^\sigma - H_L{}^\sigma} S_{ML}{}^2 \qquad (10\text{-}32)$$

equation (10–31) becomes

$$E_\sigma^* = e_\sigma(F_\sigma{}^d)^2 \qquad (10\text{-}33)$$

Equation (10–33) shows that the σ-antibonding effect on a particular d-orbital can be expressed by a parameter, e_σ, and a number ($F_\sigma{}^d$), which can be obtained from standard tables (*vide infra*).

In the case of π-bonding, the ligand can interact either through ligand low energy filled orbitals, so that the same conditions hold as for the σ case, or through ligand high energy empty orbitals, so that $H_M{}^\pi \ll H_L{}^\pi$, and the correction on the energies of the metal orbitals is of the bonding type. In both cases, however, e_π can be defined as in equation (10–32), the only difference being in the sign (e_π is positive for antibonding, and negative for a bonding effect).

It can be easily seen that the energies of the metal d-orbitals in a monocoordinated M—L complex are as follows:

$$\begin{aligned} E(d_{z^2}) &= e_\sigma \\ E(d_{xz}) &= E(d_{yz}) = e_\pi \\ E(d_{xy}) &= E(d_{x^2-y^2}) = e_\delta \end{aligned} \tag{10-34}$$

A transition metal ion complex differs from this simple M—L example, for we are concerned with overlap of d-orbitals and we have many ligands. For an octahedral complex, the coordinate system, shown in Fig. 10–25(A), fixes the location of the real d-orbitals. We can now employ a local coordinate system on each ligand L, such that the metal-ligand bond is called the z' axis. The x' axis is in the plane formed by z and z'. This local coordinate system is shown in Fig. 10–25(B) for ligand L_2. The polar coordinates of the ligand can be used to express the relation between the coordinates in the primed coordinate system and those in the unprimed coordinate system. Our concern is how to describe a d-orbital, whose position in the unprimed system is known, with the variables of the primed coordinate system. These relationships are worked out and the results summarized in Table 10–8. The results can be used for a complex of any geometry.

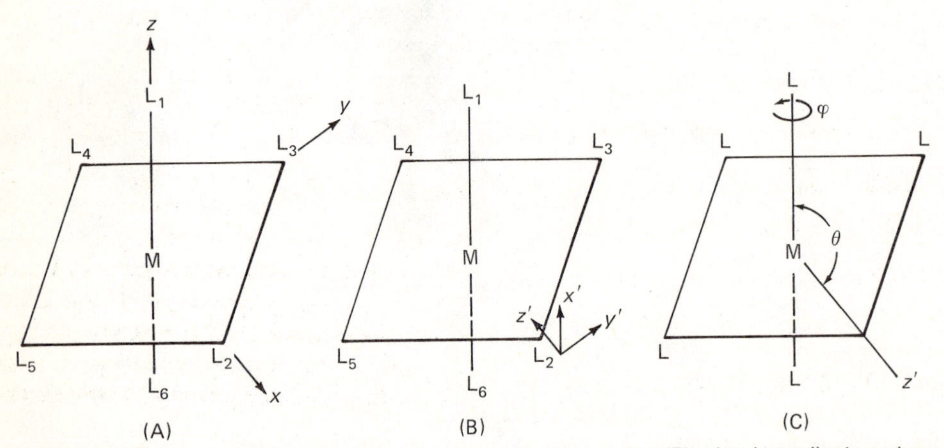

FIGURE 10–25. (A) A coordinate system for a six-coordinate complex; (B) a local coordinate system; (C) definition of θ and φ in Table 10–8.

We illustrate the use of Table 10–8 by determining the relation between the d-orbitals in the primed and unprimed set for L_2. The ligands in an octahedral complex, labeled as in Fig. 10–25, have angular polar coordinates that can be expressed as follows:

Ligand	1	2	3	4	5	6
θ	0	90	90	90	90	180
φ	0	0	90	180	270	0

TABLE 10–8. RELATIONSHIPS OF THE *d*-ORBITALS IN THE PRIMED AND UNPRIMED COORDINATE SYSTEMS OF FIG. 10–25

	z'^2	$y'z'$	$x'z'$	$x'y'$	$x'^2 - y'^2$
z^2	$\frac{1}{4}(1 + 3\cos 2\theta)$	0	$-\frac{\sqrt{3}}{2}\sin 2\theta$	0	$\frac{\sqrt{3}}{4}(1 - \cos 2\theta)$
yz	$\frac{\sqrt{3}}{2}\sin\varphi \sin 2\theta$	$\cos\varphi \cos\theta$	$\sin\varphi \cos 2\theta$	$-\cos\varphi \sin\theta$	$-\frac{1}{2}\sin\varphi \sin 2\theta$
xz	$\frac{\sqrt{3}}{2}\cos\varphi \sin 2\theta$	$-\sin\varphi \cos\theta$	$\cos\varphi \cos 2\theta$	$\sin\varphi \sin\theta$	$-\frac{1}{2}\cos\varphi \sin 2\theta$
xy	$\frac{\sqrt{2}}{4}\sin 2\varphi\,(1 - \cos 2\theta)$	$\cos 2\varphi \sin\theta$	$\frac{1}{2}\sin 2\varphi \sin 2\theta$	$\cos 2\varphi \cos\theta$	$\frac{1}{4}\sin 2\varphi\,(3 + \cos 2\theta)$
$x^2 - y^2$	$\frac{\sqrt{3}}{4}\cos 2\varphi\,(1 - \cos 2\theta)$	$-\sin 2\varphi \sin\theta$	$\frac{1}{2}\cos 2\varphi \sin 2\theta$	$-\sin 2\varphi \cos\theta$	$\frac{1}{4}\cos 2\varphi\,(3 + \cos 2\theta)$

Considering L_2, we have $\theta = 90°$ and $\varphi = 0$. Substituting these values into row 1 of Table 10–8, we find what combination of *d*-orbitals we must take in the primed system (listed across the top) to be equivalent to d_{z^2}. The result is obtained by substituting these values for θ and φ into the first row:

$$d_{z^2} = \frac{1}{4}(1 + 3\cos 2\theta)(z'^2) + 0(y'z') - \frac{\sqrt{3}}{2}\sin 2\theta\,(x'z') + 0(x'y') + \frac{\sqrt{3}}{4}(1 - \cos 2\theta)(x'^2 - y'^2)$$

$$= -\frac{1}{2}z'^2 + \frac{\sqrt{3}}{2}(x'^2 - y'^2) \tag{10-35}$$

The interaction energies of the orbitals z'^2 and $x'^2 - y'^2$ with the ligand will be given by e_σ and e_δ, where the subscripts indicate that the ligand-$d_{z'^2}$ interaction is σ and that with $d_{x'^2-y'^2}$ is δ. The coefficients $-\frac{1}{2}$ and $+\sqrt{3}/2$ can be considered as the overlap of the d_{z^2} orbital with $d_{z'^2}$ and $d_{x'^2-y'^2}$. Since the energy is proportional to the overlap squared [equation (10–32)], we can write the energy of the ligand-d_{z^2} interaction as

$$E(d_{z^2}) = \frac{1}{4}e_\sigma + \frac{3}{4}e_\delta \tag{10-36}$$

We now must evaluate the effect of L_2 on the energies of all the *d*-orbitals. We accomplish this by substituting $\theta = 90°$ and $\varphi = 0°$ into all the expressions in Table 10–8. We can express the result in the form of a matrix as

$$\begin{bmatrix} -\frac{1}{2} & 0 & 0 & 0 & \frac{\sqrt{3}}{2} \\ 0 & 0 & 0 & -1 & 0 \\ 0 & 0 & -1 & 0 & 0 \\ 0 & 1 & 0 & 0 & 0 \\ \frac{\sqrt{3}}{2} & 0 & 0 & 0 & \frac{1}{2} \end{bmatrix}$$

Since the columns are in the primed coordinate system, they correspond to σ, π, π, δ, δ interactions. These coefficients in the primed set produce the energies by squaring, leading to:

$$\begin{aligned}
E(d_{z^2}) &= \frac{1}{4}e_\sigma + \frac{3}{4}e_\delta \\
E(d_{yz}) &= e_\delta \\
E(d_{xz}) &= e_\pi \\
E(d_{xy}) &= e_\pi \\
E(d_{x^2-y^2}) &= \frac{3}{4}e_\sigma + \frac{1}{4}e_\delta
\end{aligned} \tag{10-37}$$

Substituting values of zero for θ and φ of ligand L_1 into Table 10-8 yields the matrix

$$\begin{bmatrix} 1 & 0 & 0 & 0 & 0 \\ 0 & 1 & 0 & 0 & 0 \\ 0 & 0 & 1 & 0 & 0 \\ 0 & 0 & 0 & 1 & 0 \\ 0 & 0 & 0 & 0 & 1 \end{bmatrix}$$

Squaring the coefficients, we obtain an energy of interaction of 1 for ligand L_1 with each orbital, *i.e.*, the result given earlier by equation (10-34).

Repeating this procedure for the other ligands, we obtain:

	Ligand 1	Ligand 2	Ligand 3
$E(d_{z^2})$	e_σ	$\frac{1}{4}e_\sigma + \frac{3}{4}e_\delta$	$\frac{1}{4}e_\sigma + \frac{3}{4}e_\delta$
$E(d_{yz})$	e_π	e_δ	e_π
$E(d_{xz})$	e_π	e_π	e_δ
$E(d_{xy})$	e_δ	e_π	e_π
$E(d_{x^2-y^2})$	e_δ	$\frac{3}{4}e_\sigma + \frac{1}{4}e_\delta$	$\frac{3}{4}e_\sigma + \frac{1}{4}e_\delta$

	Ligand 4	Ligand 5	Ligand 6
$E(d_{z^2})$	$\frac{1}{4}e_\sigma + \frac{3}{4}e_\delta$	$\frac{1}{4}e_\sigma + \frac{3}{4}e_\delta$	e_σ
$E(d_{yz})$	e_δ	e_π	e_π
$E(d_{xz})$	e_π	e_δ	e_π
$E(d_{xy})$	e_π	e_π	e_δ
$E(d_{x^2-y^2})$	$\frac{3}{4}e_\sigma + \frac{1}{4}e_\delta$	$\frac{3}{4}e_\sigma + \frac{1}{4}e_\delta$	e_δ

You can check your result by summing each column and thereby noting that each ligand contributes one σ, two π, and two δ types of interactions. Summing up the contributions of the individual ligands, we obtain finally the expressions of the energies of the d-orbitals for octahedral compounds as:

$$
\begin{aligned}
E(d_{z^2}) &= 3e_\sigma + 3e_\delta \\
E(d_{yz}) &= 4e_\pi + 2e_\delta \\
E(d_{xz}) &= 4e_\pi + 2e_\delta \\
E(d_{xy}) &= 4e_\pi + 2e_\delta \\
E(d_{x^2-y^2}) &= 3e_\sigma + 3e_\delta
\end{aligned} \tag{10-38}
$$

Of course, the three t_{2g} orbitals have the same energy and so do the two e_g orbitals. If the e_δ contribution is neglected, the difference between the e_g and t_{2g} orbitals is $3e_\sigma - 4e_\pi$, which corresponds to the Δ of the crystal field theory. In a T_d complex, the e orbitals have energies of $\frac{8}{3}e_\pi + \frac{4}{3}e_\delta$ while the t_2 orbitals have energies of $\frac{4}{3}e_\sigma + \frac{8}{9}e_\pi + \frac{16}{9}e_\delta$. Note that with these parameters, $\Delta_{T_d} = \frac{4}{9}\Delta_{O_h}$. In complexes with lower symmetry, the values of all of the ligands are added and the d-orbital energies are calculated. Numerical values for the e_σ, e_π, and e_δ parameters are obtained from octahedral complexes. The e_σ values employed in the different complexes are scaled according to the overlap integral. The value of the model is that one set of parameters results for a particular ligand and metal, which can be scaled for geometry and overlap to account for the spectra of many transition metal ion complexes. The relationships of Dq, Ds, Dt, $\delta\sigma$, and $\delta\pi$ with e_σ and e_π have been presented.(47)

In closing this section, it should be emphasized that for these studies, and other applications dependent upon band assignments, the assignments should be verified by polarized single crystal results (see Chapter 5).

MISCELLANEOUS TOPICS INVOLVING ELECTRONIC TRANSITIONS

10–13 SIMULTANEOUS PAIR ELECTRONIC EXCITATIONS

The electronic spectra of high spin octahedral and tetrahedral iron(III) compounds are as expected from the Tanabe and Sugano diagrams. Three transitions are found: ${}^6A_{1g} \rightarrow {}^4T_2$, ${}^6A_{1g} \rightarrow {}^4T_1$, and ${}^6A_{1g} \rightarrow {}^4A_{1g}$ [four transitions are found when ${}^4E(D)$ is low enough in energy]; and since Dq is larger for octahedral complexes than for tetrahedral ones, the ${}^4T_1$ and ${}^4T_2(G)$ transitions occur at higher energy in the former complexes. All of the d-d transitions are multiplicity forbidden and are weak. However, when the electronic spectrum of the six-coordinate oxybridged dimer (HEDTA Fe)$_2$O (where HEDTA is hydroxyethylethylenediaminetriacetate) is examined,(48,49) the surprising result shown in Fig. 10–26 is obtained.

The bands labeled **a** through **d** are in the correct place to be assigned to the four $d \rightarrow d$ transitions, ${}^6A_1 \rightarrow {}^4T_1$, ${}^6A_1 \rightarrow {}^4T_2(G)$, ${}^6A_1 \rightarrow {}^4A_1{}^4E_1$, and ${}^6A_1 \rightarrow {}^4E(D)$, of an octahedral complex. These bands are two orders of magnitude more intense than those of a typical iron(III) complex. *This intensity enhancement of spin forbidden bands is common to spin-coupled systems because the coupling partially relaxes the spin selection rule.*

The four intense (**e**–**h**) u.v. bands are too low in energy and there are too many to assign these bands to charge transfer. The intensity is much too large for a d-d transition on a single metal ion. The bands have been attributed(48) to simultaneous d-d transitions on the two iron(III) centers. They are coupled so that the pair excitation is spin allowed.(50,51) The band labeled **e** at 29.2×10^3 cm^{-1} is assigned to the simultaneous transition of **a** on one center and **b** on the other ($\bar{\nu}_\mathbf{a} + \bar{\nu}_\mathbf{b} =$

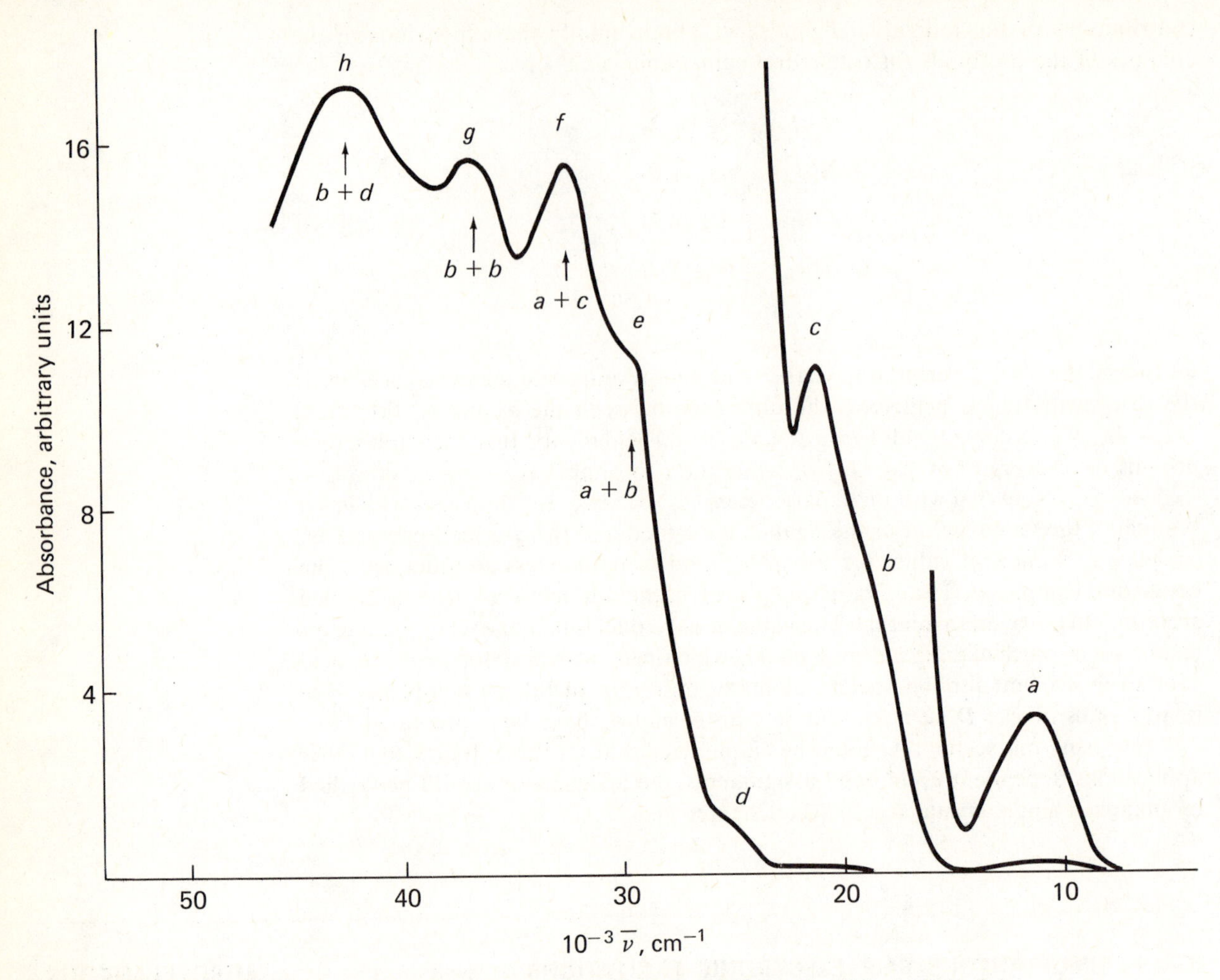

FIGURE 10–26. Absorption spectrum of 0.20 F aqueous solution of $[(HEDTA\ Fe)_2O] \cdot 6H_2O$ at 296 K. Calculated positions of SPE excitations are indicated by arrows.

29.4×10^3 cm^{-1}); band **f** at 32.5×10^3 cm^{-1} to **a** + **c** (32.2×10^3 cm^{-1}); band **g** at 36.8×10^3 cm^{-1} to **b** + **b**; and band **h** at 42.6×10^3 cm^{-1} to **b** and **d.** The phenomenon is referred to as *simultaneous pair electronic excitations,* and the absorption bands are abbreviated as SPE bands. Similar effects have been observed on other systems[52,53] and have been thoroughly studied[53] for $(NH_3)_5CrOCr(NH_3)_5{}^{4+}$. *It is concluded that the intensity enhancement arises from a vibronic, exchange induced, electric dipole mechanism.*

10–14 INTERVALENCE ELECTRON TRANSFER BANDS

Metallomers, molecules with two or more metals, can be prepared in which the bridging group permits the metals to exist in two different oxidation states, *e.g.*,

$$[L_5M^{III}—X—M^{II}L_5]^{+n}$$

These "mixed valence" compounds have long been of interest because of the intense colors they possess. For example, $KFe[Fe(CN)_6]$, Prussian blue, has a deep color that is absent in $K_3Fe(CN)_6$ and $K_4Fe(CN)_6$. This color has been attributed to an *intervalence transfer transition.* In order to understand this phenomenon, let us consider as an example the ruthenium pyrazine dimer:[54]

$$(NH_3)_5Ru(II)\text{—N}\langle\bigcirc\rangle\text{N—}Ru(III)(NH_3)_5{}^{5+} \xrightarrow{h\nu}$$

$$(NH_3)_5Ru(III)\text{—N}\langle\bigcirc\rangle\text{N—}Ru(II)(NH_3)_5{}^{5+}$$

In order to clarify several points to be made about this system, we shall simplify our discussion by ignoring any ligands in the coordination sphere of Ru(II) and Ru(III) except the bridging pyrazine. Furthermore, we shall label the two ruthenium atoms as $Ru_a^{III}\text{—N}\langle\bigcirc\rangle\text{N—}Ru_b^{II}$. We shall draw a potential energy curve[55] for the molecule as a whole as we vary the Ru_a^{III}—N distance, X, in the molecule in such a way that the Ru_a—N plus Ru_b—N distance is constant. Curve *a* of Fig. 10–27 results when a harmonic potential function is assumed, *i.e.,* $V = \frac{1}{2}kX^2$.

Next we shall construct a similar curve for the Ru_a^{II}—N distance in the molecule $Ru_a^{II}\text{—N}\langle\bigcirc\rangle\text{N—}Ru_b^{III}$. The equilibrium Ru_a^{II}—N distance is longer than the equilibrium Ru_a^{III}—N distance. The difference is indicated by X_0. The curve for the Ru_a^{II}—N system is indicated by curve *b*. We can view the two curves as defining our total system in which we now have provision for the electron jumping from one center to the other. E_{th} is the activation energy for the jump, and it corresponds in this system to the molecule arriving at a place where the Ru_a—N and Ru_b—N bond lengths are identical. The thermal energy required for the electron to surpass the barrier between metals, E_{th}, is given[55] by

$$E_{th} = \frac{1}{2}k\left(\frac{1}{2}X_0\right)^2 = \frac{1}{4}\left(\frac{1}{2}kX_0{}^2\right) = \frac{1}{4}E_{IT} \tag{10-39}$$

By the Franck-Condon principle, electronic absorption occurs without change in nuclear coordinates, so the energy of the intervalence transfer band in the absorption spectrum (*i.e.,* a charge transfer transition from one Ru atom to the other) is represented by the vertical line, E_{IT}.

The actual situation is more complex, since the coordinate X is a complicated coordinate involving all the ligands and electrons in the complex. Depending upon the proximity of the two metal centers and the overlap of various orbitals, the two orbitals (one on each ruthenium) that can contain the odd electron may mix to form a bonding and antibonding combination. This would produce the situations depicted in Figs. 10–27(B) or (C). The quantity labeled $2H_{res}$ in Fig. 10–27(B) is two times the resonance integral, *i.e.*, the off-diagonal element between the two *d*-orbitals which can hold the odd electron (one on each ruthenium) in the secular determinant.[56] In Fig. 10–27(B), we have shown $E_{th} < \frac{1}{4}E_{IT}$. In Fig. 10–27(C), the odd electron is completely delocalized on all time scales, and the electronic absorption cannot properly be termed an intervalence transfer.

Robin and Day[57] have pointed out that in Fig. 10–27(B), one of the energy minima corresponds to an electronic wave function in which the odd electron is

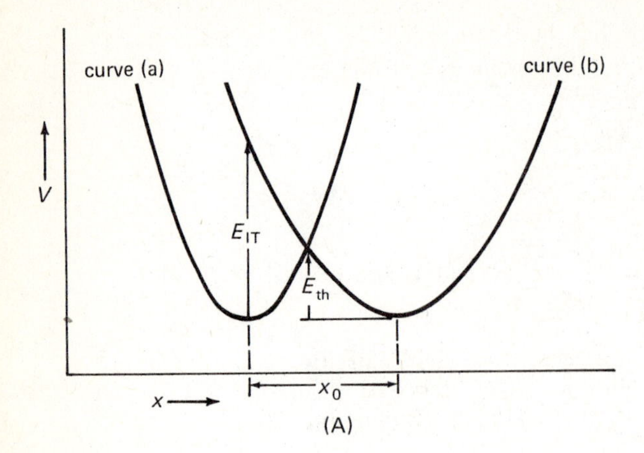

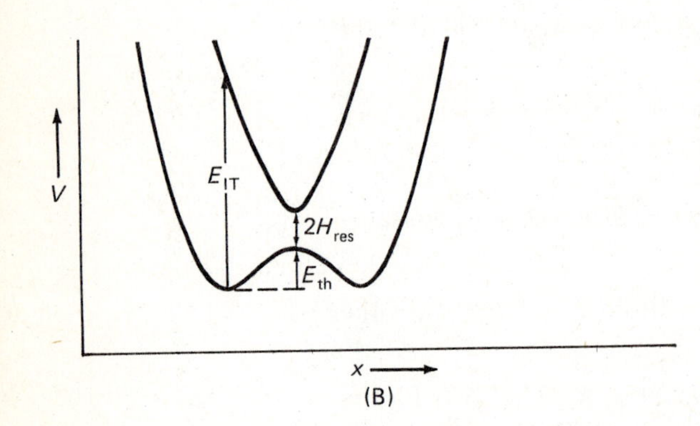

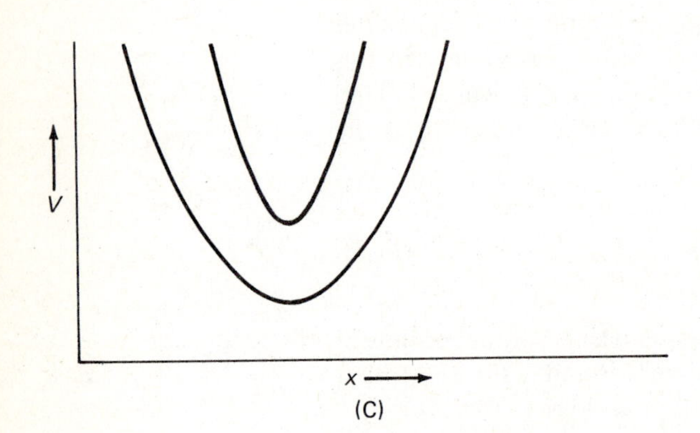

FIGURE 10–27. Potential energy curves for various classes of mixed valence compounds. The systems vary from completely localized systems (A) to intermediate (B) to completely delocalized (C). X is a coordinate expressing the Ru_a^{III}—N distance as the sum of Ru_a—N and Ru_b—N is held constant.

mainly on one metal but, to a small extent, is delocalized onto the other metal. They have proposed that mixed valence compounds be classified as Class I, II, or III, depending on whether none, some, or half of the unpaired electron density is delocalized from one metal center onto the other at any one instant.

The ruthenium pyrazine dimer[54] shows an intense band in the near infrared at 1570 mμ, assigned to the electronic transition shown in Fig. 10–27(A) and labeled E_{IT}. From the energy of this transition, a thermal rate of electron exchange of 3×10^9 sec^{-1} has been calculated employing a crude model and equation (10–39). The electronic transition is described as a [2, 3] → [3, 2] transition.

The existence of high intensity bands in the *d-d* region in other complexes has been used as evidence in support of mixed valence in molecules containing more than one metal ion. The assignment of the band to a mixed valence species can be supported by following the intensity of the transition as a function of the extent of oxidation or reduction of the $[n, n]$ oxidation state complex to form $[n, m]$. A maximum is observed when $[n, m]$ is formed, and it disappears on formation of $[m, m]$.

10-15 PHOTOREACTIONS

The principles discussed in this chapter and Chapter 5 are of importance to chemists for applications other than structure determination. Fluorescence, phosphorescence, and photochemistry all have to do with electronic transitions. In photochemical reactions, the reactant is a molecule in a reactive excited state. Understanding of the photochemical reaction requires an understanding of the structure and reactivity of the excited state. In some cases, molecules that are singlets in the ground state become reactive radicals by being excited to a triplet state containing unpaired electrons. Often, two molecules that do not form a complex in the ground state form a complex (called an exiplex) when one of the molecules is in an excited state. The principles of electronic transitions studied here and in Chapter 5 are thus important in many areas.

REFERENCES

1. C. J. Ballhausen, "Introduction to Ligand Field Theory," McGraw-Hill, New York, 1962.
2. B. N. Figgis, "Introduction to Ligand Fields," Interscience, New York, 1966.
3. M. Gerloch and R. C. Slade, "Ligand Field Parameters," Cambridge University Press, New York, 1973.
4. a. C. K. Jørgensen, "Absorption Spectra and Chemical Bonding in Complexes," Pergamon Press, New York, 1962.
   b. C. K. Jørgensen, "Modern Aspects of Ligand Field Theory," North Holland Publishing Co., Amsterdam, 1971.
5. H. L. Schläfer and G. Gliemann, "Basic Principles of Ligand Field Theory," Interscience, New York, 1969.
6. C. E. Moore, "Atomic Energy Levels Circular 467," Vol. 2, National Bureau of Standards, 1952. Volumes 1 and 3 contain similar data for other gaseous ions.
7. E. U. Condon and G. H. Shortley, "Theory of Atomic Spectra," Cambridge University Press, New York, 1957.
8. J. S. Griffith, "The Theory of Transition Metal Ions," Cambridge University Press, New York, 1961.
9. D. S. McClure, "Electronic Spectra of Molecules and Ions or Crystals," Academic Press, New York, 1959.
10. A. B. P. Lever, "Inorganic Electronic Spectroscopy," American Elsevier, New York, 1968.
11. "Spectroscopic Properties of Inorganic and Organometallic Compounds," Vols. I–IV, The Chemical Society, London, 1968–1971.
12. T. M. Dunn, "Modern Coordination Chemistry," ed. J. Lewis and R. G. Wilkins, Interscience, New York, 1960.
13. Y. Tanabe and S. Sugano, J. Phys. Soc. Japan, *9,* 753, 766 (1954).
14. R. S. Drago, D. W. Meek, M. D. Joesten, and L. LaRoche, Inorg. Chem., *2,* 124 (1963).
15. F. A. Cotton and M. Goodgame, J. Amer. Chem. Soc., *83,* 1777 (1961).
16. D. W. Barnum, J. Inorg. Nucl. Chem., *21,* 221 (1961).
17. T. S. Piper and R. L. Carlin, J. Chem. Phys., *36,* 3330 (1962).
18. A. D. Liehr and C. J. Ballhausen, Phys. Rev., *106,* 1161 (1957).
19. R. S. Drago, D. W. Meek, R. Longhi, and M. D. Joesten, Inorg. Chem., *2,* 1056 (1963).
20. R. W. Kiser and T. W. Lapp, Inorg. Chem., *1,* 401 (1962).
21. F. Basolo, C. J. Ballhausen, and J. Bjerrum, Acta Chem. Scand., *9,* 810 (1955).
22. D. S. McClure, "Advances in the Chemistry of Coordination Compounds," ed. S. Kirshner, p. 498, Macmillan, New York, 1961, and references therein.
23. F. Basolo, J. Amer. Chem. Soc., *72,* 4393 (1950); Y. J. Shimura, J. Amer. Chem. Soc., *73,* 5079 (1951); K. Nakamoto, J. Fujita, M. Kobayashi, and R. Tsuchida, J. Chem. Phys., *27,* 439 (1957); and references therein.
24. R. C. Fay and T. S. Piper, J. Amer. Chem. Soc., *84,* 2303 (1962).
25. C. Furlani and G. Sartori, J. Inorg. Nucl. Chem., *8,* 126 (1958).
26. W. Manch and W. Fernelius, J. Chem. Educ., *38,* 192 (1961).
27. G. Maki, J. Chem. Phys., *29,* 162 (1958); *ibid., 29,* 1129 (1958).

28. F. A. Cotton and E. Bannister, J. Chem. Soc., *1960,* 1873.
29. N. S. Gill and R. S. Nyholm, J. Chem. Soc., *1959,* 3997.
30. J. T. Donoghue and R. S. Drago, Inorg. Chem., *1,* 866 (1962).
31. D. K. Straub, R. S. Drago, and J. T. Donoghue, Inorg. Chem., *1,* 848 (1962).
32. C. K. Jørgensen, "Energy Levels of Complexes and Gaseous Ions," Gjellerups, Copenhagen, 1957.
33. L. Sacconi, Pure Appl. Chem., *17,* 95 (1968); Transition Metal Chem., *4,* 199 (1968); R. Morassi, I. Bertini, and L. Sacconi, Coord. Chem. Rev., *11,* 343 (1973).
34. C. Furlani, Coord. Chem. Rev., *3,* 141 (1968).
35. a. M. Ciampolini, Struct. Bonding, *6,* 52 (1969).
    b. B. J. Hathaway and D. E. Billing, Coord. Chem. Rev., *5,* 143 (1970).
36. B. N. Figgis and J. Lewis, "Modern Coordination Chemistry," ed. J. Lewis and R. G. Wilkins, Interscience, New York, 1960.
37. M. Linhard, H. Siebert, and M. Weigel, Z. Anorg. u. Allgem. Chem., *278,* 287 (1955).
38. C. J. Ballhausen and H. B. Gray, Inorg. Chem., *1,* 111 (1962).
39. H. B. Gray and C. R. Hare, Inorg. Chem., *1,* 363 (1962); C. R. Hare, I. Bernal, and H. B. Gray, Inorg. Chem., *1,* 831 (1962).
40. A. S. Brill, B. R. Martin, and R. J. P. Williams, "Electronic Aspects of Biochemistry," ed. B. Pullman, Academic Press, New York, 1964.
41. D. R. McMillin, R. A. Holwerda, and H. B. Gray, Proc. Nat. Acad. Sci., *71,* 1339 (1974).
42. W. S. Caughey, *et al.,* J. Mol. Spect., *16,* 451 (1965) and references therein.
43. D. A. Rowley and R. S. Drago, Inorg. Chem., *7,* 795 (1968); *ibid., 6,* 1092 (1967).
44. a. A. B. P. Lever, Coord. Chem. Rev., *3,* 119 (1968).
    b. R. L. Chiang and R. S. Drago, Inorg. Chem., *10,* 453 (1971).
45. C. D. Burbridge, D. M. L. Goodgame, and M. Goodgame, J. Chem. Soc. (A), *1967,* 349.
46. a. C. E. Schäffer and C. K. Jørgensen, Mol. Phys., *9,* 401 (1965).
    b. E. Larsen and G. N. LaMar, J. Chem. Educ., *51,* 633 (1974).
47. I. Bertini, D. Gatteschi, and A. Scozzafara, Inorg. Chem., *15,* 203 (1976).
48. H. J. Schugar, G. R. Rossman, J. Thibeault, and H. B. Gray, Chem. Phys. Letters, *6,* 26 (1970).
49. H. B. Gray, "Bioinorganic Chemistry," Amer. Chem. Soc. Monograph, *100,* 365 (1971).
50. L. Lohr, Coord. Chem. Rev., *8,* 241 (1972).
51. D. L. Dexter, Phys. Rev., *126,* 1962 (1962).
52. A. E. Hansen and C. J. Ballhausen, Trans. Faraday Soc., *61,* 631 (1965).
53. H. U. Güdel and L. Dubicki, Chem. Phys., *6,* 272 (1974) and references therein.
54. C. Creutz and H. Taube, J. Amer. Chem. Soc., *91,* 3988 (1969).
55. N. S. Hush, Prog. Inorg. Chem., *8,* 391 (1967) and references therein.
56. B. Mayoh and P. Day, J. Chem. Soc., Dalton, *1974,* 846.
57. M. B. Robin and P. Day, Adv. Inorg. Chem. Radiochem., *10,* 248 (1967).

EXERCISES

1. Refer to the Tanabe and Sugano diagrams in Appendix D. For octahedral Cr^{III} and a ligand with Dq/B of 1, how many bands should occur in the spectrum? Label these transitions and list them in order of increasing wavelength.

2. In complexes with weak field ligands ($Dq/B = 0.7$), octahedral Co^{2+} exhibits a spectrum with three well separated bands. Make a tentative assignment using the Tanabe and Sugano diagrams and list the assignments in order of decreasing frequency. Would the spectrum of a strong field complex be any different? Describe the spectrum you would expect for a strong field complex.

3. A nickel complex NiR_4Cl_2 has an absorption spectrum with peaks that have ε values of around 150. R and Cl occupy similar positions in the spectrochemical series. Are the chlorines coordinated?

4. Two different isomers of $Co(NH_3)_4(SCN)_2^+$ were separated. How could you determine whether the SCN groups in both were bonded through the sulfur? If both isomers were coordinated through sulfur, how would you determine which is *cis* and which is *trans*? (Hint: —SCN is near Cl^- in the spectrochemical series, while $—NCS^-$ creates a stronger field; $Co(NH_3)_4Cl_2^+$ is easily prepared.)

5. Using the Tanabe and Sugano diagrams, assign the following spectra of six-coordinate aquo species [except for (A)].

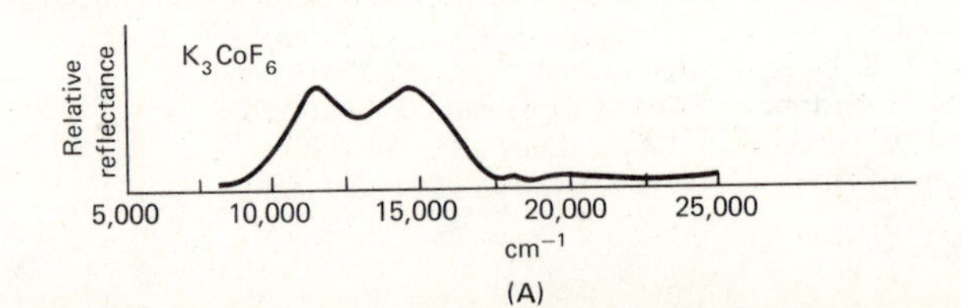

(A)

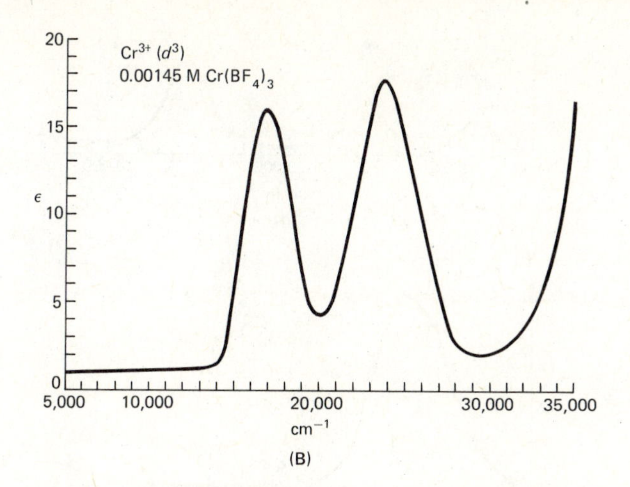
Cr^{3+} (d^3)
0.00145 M $Cr(BF_4)_3$
ε
cm⁻¹

(B)

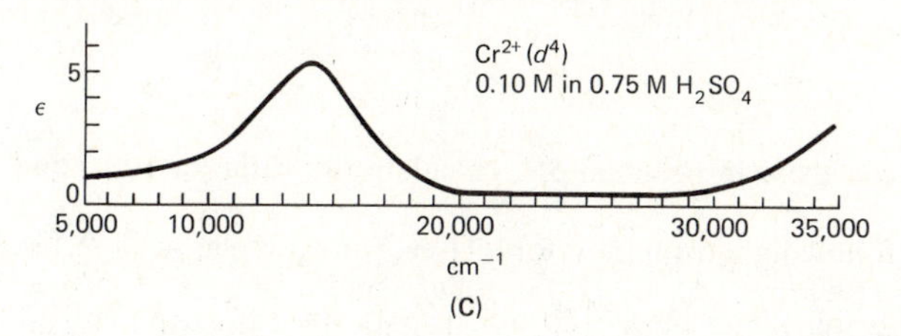
Cr^{2+} (d^4)
0.10 M in 0.75 M H_2SO_4
ε
cm⁻¹

(C)

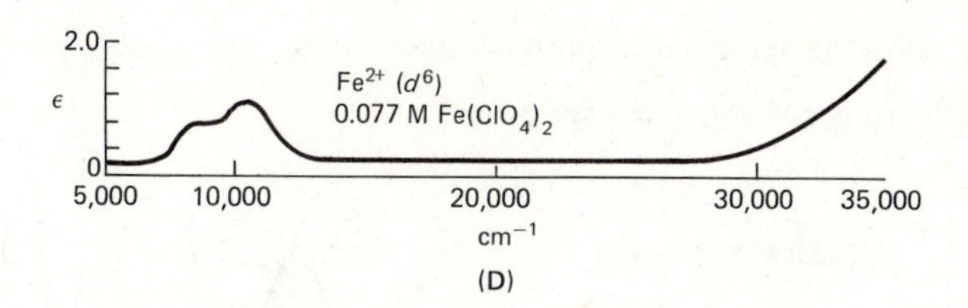
Fe^{2+} (d^6)
0.077 M $Fe(ClO_4)_2$
ε
cm⁻¹

(D)

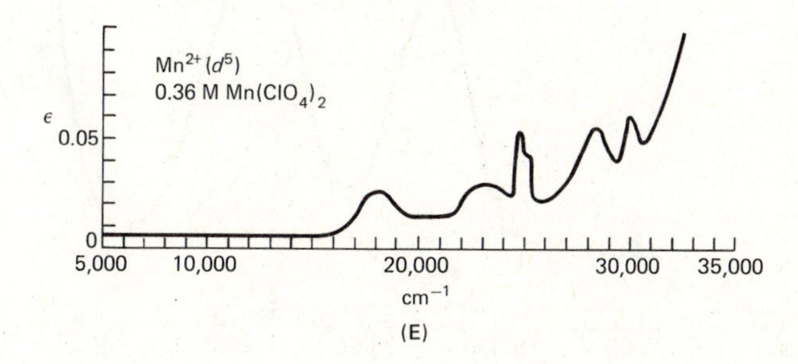
Mn^{2+} (d^5)
0.36 M $Mn(ClO_4)_2$
ε
cm⁻¹

(E)

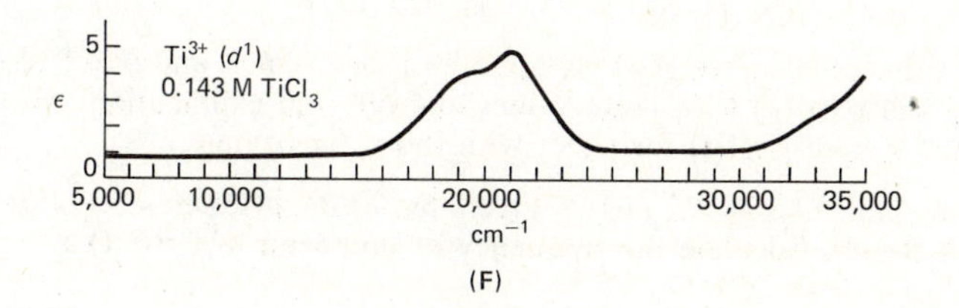
Ti^{3+} (d^1)
0.143 M $TiCl_3$
ε
cm⁻¹

(F)

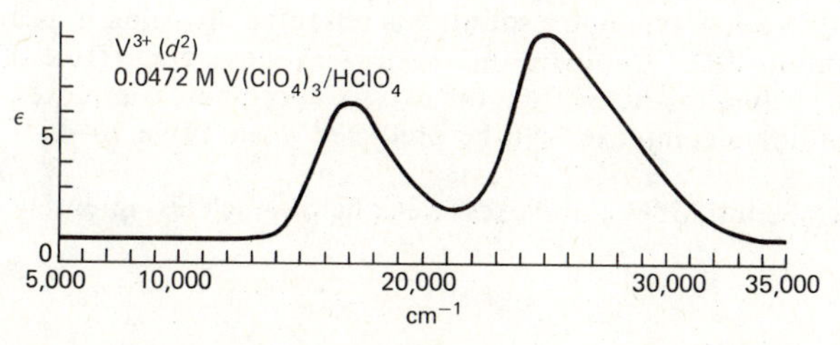
V^{3+} (d^2)
0.0472 M $V(ClO_4)_3/HClO_4$
ε
cm⁻¹

(G)

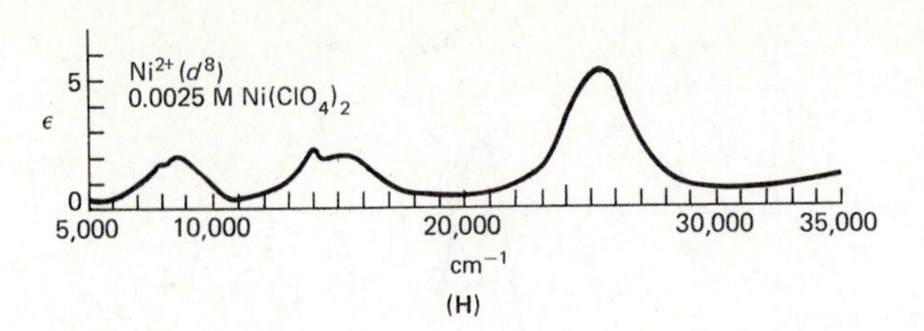

(H)

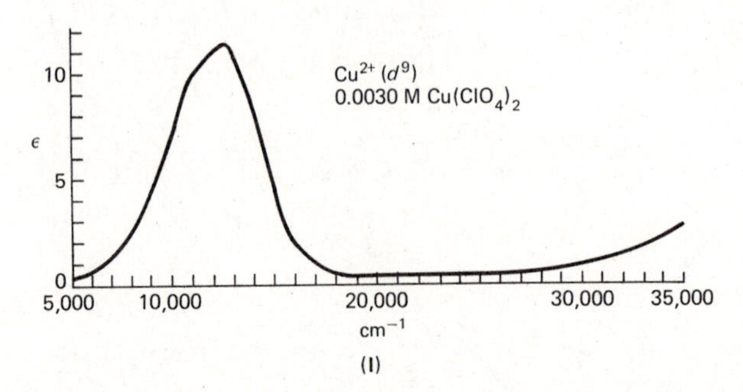

(I)

6. Using the spectrum in problem 5H, calculate the value of $10Dq$ and B for H_2O.

7. Given the following information for d^3 [see, for example, A. B. P. Lever, J. Chem. Ed., *45,* 711 (1968)]: ${}^4A_{2g} \rightarrow {}^4T_{2g} = 10Dq$ and ${}^4A_{2g} \rightarrow {}^4T_{1g}(F) = 7.5B' + 15Dq - \frac{1}{2}(225B'^2 + 100Dq^2 - 180B'Dq)^{1/2}$; calculate the value of $10Dq$ and β (B'/B where $B = 1030$ cm^{-1}) for:

   a. H_2O, using the spectrum in problem 5B.

   b. $C_2O_4^{2-}$, using the following spectrum:

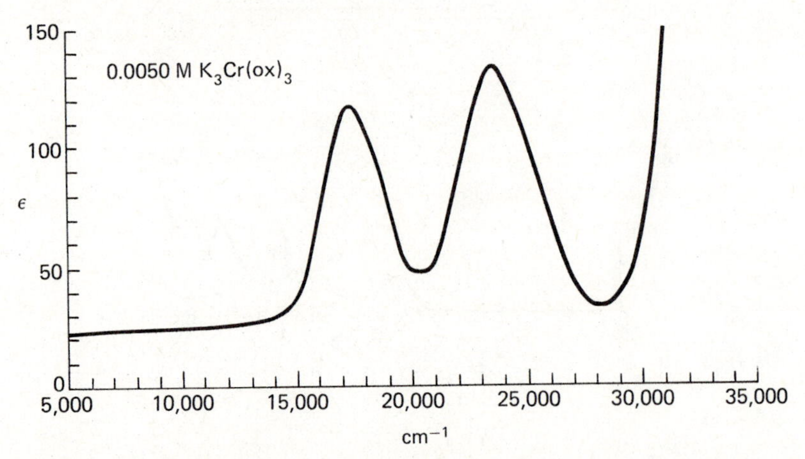

   c. The value of $10Dq$ for H_2O toward Ni^{2+} is 8500 cm^{-1}, and $\beta = 0.88$. Compare your results in part (a) with these values and offer an explanation. Also compare your results toward Cr(III) for water with those for oxalate.

   d. Given that ${}^4A_{2g} \rightarrow {}^4T_{1g}(P)$ is given by $7.5B' + 15Dq + \frac{1}{2}(225B'^2 + 100Dq^2 - 180B'Dq)^{1/2}$, calculate the frequency of this band in $Cr(C_2O_4)_3^{3-}$.

8. The ion $[Ni(pyridine)_4(H_2O)_2]^{2+}$ has *d-d* absorption bands at 27,000, 16,500, and 10,150 cm^{-1}. No low symmetry splitting is observed. Treating it as an octahedral complex, determine $10Dq$. Compare this value with the average (rule of average environment) of the values predicted from the two six-coordinate complexes. The Dq values for the six-coordinate complexes can be predicted from Table 10–4.

9. Why are octahedral Mn^{2+} complexes (weak field) much less intensely colored than those of Cr^{3+}?

10. The electronic spectrum of trisoxalatochromium(III) doped into a host lattice of $NaMgAl(C_2O_4)_3 \cdot 9H_2O$ has been reported. The ground state is $^4A_{2g}$ if octahedral symmetry is assumed. The lowest excited states (not mo's) for octahedral symmetries are then ^2E_g, $^2T_{1g}$, $^2T_{2g}$, and $^4T_{2g}$. The observed bands and extinction coefficients are:

17,500 cm^{-1}	$\varepsilon = 40$
23,700 cm^{-1}	$\varepsilon = 67$
14,500 cm^{-1}	$\varepsilon = 2.6$
15,300 cm^{-1}	$\varepsilon = 2.0$
20,700 cm^{-1}	$\varepsilon = 0.3$

a. The above transitions have been shown to be electronically allowed. Why is this spectrum inconsistent with octahedral symmetry? (Multiplicity forbidden transitions have low intensities; generally, ε is less than 5.)

b. Actually, the spectrum is consistent with the true symmetry of the molecule, D_3. Lowering the symmetry causes the following:

O_h	D_3
A_{2g}	A_2
T_{1g}	$A_2 + E$
T_{2g}	$A_1 + E$
E_g	E

If the spectrum is taken with light polarized perpendicular and parallel to the trigonal axis, all the absorptions except the one at 17,500 cm^{-1} occur with perpendicularly polarized light, but only the bands at 17,500 and 15,300 cm^{-1} are present with parallel polarized light. Given that splittings of the doublet octahedral states in D_3 symmetry are unresolved and that ^2E_g is lower in energy than $^2T_{1g}$, assign the transitions using the D_3 excited states. You must explain your choices. (Remember that some doublet bands represent unresolved multiplets and will consequently correspond to more than one transition.)

11. In $Re_2Cl_8^{2-}$ (see problem 6, Chapter 3) the transition of an electron from the b_{2g} orbital to the b_{1u} orbital is a *d-d* transition in a molecule with a center of inversion. Is it allowed? Explain.

12. The following is the splitting of the state energies of a high spin $3d^2$ ion in an O_h field:

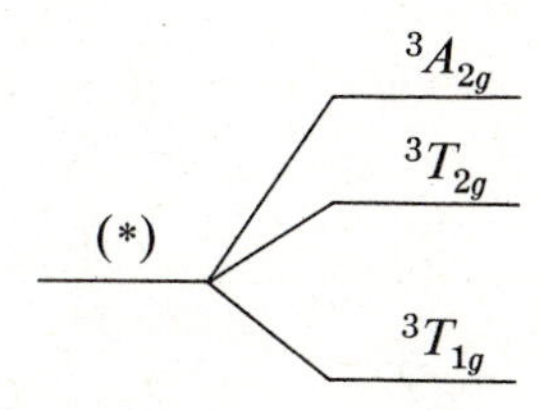

a. Calculate the free ion ground term symbol (*). (Note the T and A states written above will have the same *spin* multiplicity as the free ion.)

b. What would this diagram look like if we considered spin-orbital ($\lambda L \cdot S$) effects? (Do only the T_{1g} state.) Show the splitting and label each level with its J value and its degeneracy.

c. Show the splitting pattern for a d^2 case for all J states above and label with appropriate M_J values. Show the effect of a magnetic field on the levels.

13. a. Of the states arising from an ion with a configuration of two $2p$ electrons, only one is a spin triplet. What is the term symbol of this triplet? Show all work.

b. Now consider spin-orbit coupling. What J states will this triplet give rise to? Which will have the lowest energy?

14. a. Determine the irreducible representations for the terms arising from the splitting of a gaseous ion ^4F state by an octahedral ligand field.

    b. Using the method of descent in symmetry, determine the spin multiplicity of the states in part a.

15. Determine the irreducible representations of the states into which the $^2T_{2g}$ level of $Fe(CN)_6^{3-}$ is split by spin-orbit coupling.

16. a. List the operations of the double group D_3' that are obtained by combining the operations of D_3 with the operation R. Make sure your list has the properties of a group.

    b. How many classes are in D_3'? What are the operations in each class? You may wish to utilize the fact that equivalent symmetry operations are members of the same class.

11 MAGNETISM

11–1 INTRODUCTION

In this chapter we will consider certain aspects of magnetism that are critical to an understanding of the nmr and esr spectra of transition metal ion complexes. Magnetic effects arise mainly from the electrons in a molecule because the magnetic moment of an electron is about 10^3 times that of the proton. In the chapters on nmr, we have discussed the electron circulations of paired electrons that give rise to diamagnetic effects. When there are unpaired electrons in the system, we observe magnetic behavior that is related to the number and orbital arrangement of the unpaired electrons. The magnetic behavior is determined by measuring (*vide infra*) the magnetic polarization of a substance by a magnetic field. Various types of behavior are illustrated in Fig. 11–1. It is convenient to define a quantity called magnetic induction, $\vec{\mathbf{B}}$, in order to describe the behavior of substances in a field.

$$\vec{\mathbf{B}} = \vec{\mathbf{H}}_0 + 4\pi\vec{\mathbf{M}} \tag{11-1}$$

Here $\vec{\mathbf{H}}_0$ is the applied field strength and $\vec{\mathbf{M}}$ is the magnetization, *i.e.*, the intensity of magnetization per unit volume. When we divide* by $\vec{\mathbf{H}}_0$, equation (11–2) results:

$$\frac{B}{H_0} = 1 + 4\pi\frac{M'}{H_0} = 1 + 4\pi\chi_v \tag{11-2}$$

Here M'/H_0 is given the symbol χ_v and is referred to as the *magnetic susceptibility per unit volume.* The volume susceptibility is thus related to the magnetization by

$$\chi_v = \frac{M'}{H_0} \text{ (dimensionless)}$$

*We assume in an isotropic system that the directions of $\vec{\mathbf{H}}_0$ and $\vec{\mathbf{M}}$ are coincident. Thus, though we cannot divide a vector by a vector, we can factor out the directional property and perform the division, leaving an equation in which only the magnitudes appear.

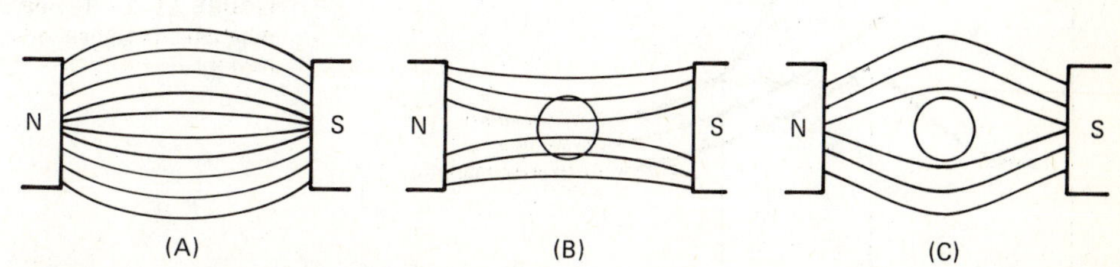

FIGURE 11–1. (A) Magnetic field lines of flux (*i.e.*, contour lines of constant field values) in vacuum; (B) the lines of flux for a paramagnetic substance in a field; (C) the lines of flux for a diamagnetic substance in a field.

B/H_0 is the permeability of the medium and is the magnetic counterpart of the dielectric constant. Dividing χ_v by the density of the substance, d, produces the gram susceptibility, χ_g:

$$\frac{\chi_v}{d} = \chi_g \ (\text{cm}^3/\text{gram}) \tag{11-3}$$

Multiplying χ_g by the molecular weight produces a molar susceptibility, χ:

$$\chi_g \times \text{MW} = \chi \ (\text{cm}^3/\text{mole}) \tag{11-4}$$

The value of χ is negative for a diamagnetic substance and positive for a paramagnetic one. In an ordered crystal, the susceptibility may be anisotropic, *i.e.*, represented by a tensor with several components. We shall discuss four types of magnetic behavior: diamagnetism, paramagnetism, ferromagnetism, and antiferromagnetism.

TABLE 11–1. VARIOUS TYPES OF MAGNETIC BEHAVIOR

Type	Sign	Magnitude	Field Dependence of χ	Origin
Diamagnetism	–	10^{-6} emu units	Independent	Field induced, paired electron circulations
Paramagnetism	+	0 to 10^{-4} emu units	Independent	Angular momentum of the electron
Ferromagnetism	+	10^{-4} to 10^{-2} emu units	Dependent	Spin alignment from dipole-dipole interaction of moments on adjacent atoms, $\uparrow\uparrow$
Antiferromagnetism	+	0 to 10^{-4} emu units	Dependent	Spin pairing, $\uparrow\downarrow$, from dipole-dipole interactions

The behaviors corresponding to these various classifications are described in Table 11–1. The latter two types of behavior can be checked by studying the field dependence of χ. The behavior of the susceptibility as a function of temperature is also quite characteristic for these different substances. This is illustrated in Fig. 11–2.

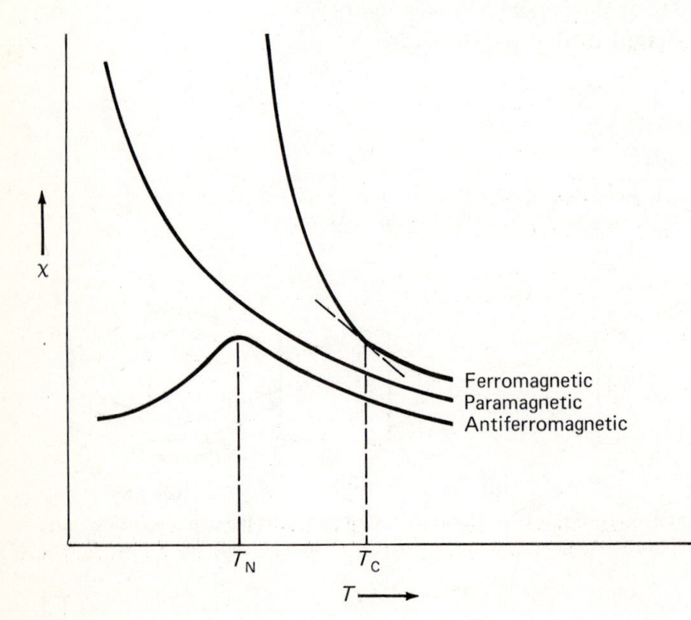

FIGURE 11–2. Temperature dependence of ferromagnetic, paramagnetic, and antiferromagnetic behavior.

The temperature at which the maximum occurs in the plot of antiferromagnetic behavior is referred to as the *Neel temperature.* The temperature at which the break occurs in the ferromagnetic plot is called the *Curie temperature.* Many compounds which, in the solid state, exhibit paramagnetic behavior around room temperature exhibit slight ferromagnetic or antiferromagnetic behavior below liquid helium (4.2° K) temperature.

11-2 TYPES OF MAGNETIC BEHAVIOR

Diamagnetism

As mentioned earlier, diamagnetism arises from field-induced electron circulations of paired electrons, which generate a field opposed to the applied field. Thus, all molecules have contributions from diamagnetic effects. The diamagnetic susceptibility of an atom is proportional to the number of electrons, n, and the sum of the squared values of the average orbital radius of the ith electron, $\overline{r}_i$:

$$\chi_A = -\frac{Ne^2}{6mc^2}\sum_i^n \overline{r}_i^{\,2} = -2.83 \times 10^{10} \sum_i^n \overline{r}_i^{\,2} \tag{11-5}$$

Larger atoms with more electrons have greater diamagnetic susceptibilities than smaller atoms with fewer electrons. The molar diamagnetic susceptibility of a molecule or complex ion, χ, can be obtained to a good approximation by summing the diamagnetic contributions from all of the atoms, χ_A, and from all of the bonds in functional groups, χ_B.

$$\chi = \sum_i \chi_{A_i} + \sum_j \chi_{B_j} \tag{11-6}$$

The values for χ_A and χ_B are referred to as *Pascal's constants,* and some common ones are listed in Table 11-2. The calculation of χ is illustrated here for pyridine and acetone.

TABLE 11-2. PASCAL'S CONSTANTS

Atoms, χ_A				Bonds, χ_B	
Atom	χ_A ($\times 10^{-6}$ cm^3 mole^{-1})	*Atom*	χ_A ($\times 10^{-6}$ cm^3 mole^{-1})	*Bond*	χ_B ($\times 10^{-6}$ cm^3 mole^{-1})
H	−2.93	F	−6.3	C=C	+5.5
C	−6.00	Cl	−20.1	C≡C	+0.8
C (aromatic)	−6.24	Br	−30.6	C=N	+8.2
N	−5.57	I	−44.6	C≡N	+0.8
N (aromatic)	−4.61	Mg^{2+}	−5	N=N	+1.8
N (monamide)	−1.54	Zn^{2+}	−15	N=O	+1.7
N (diamide, imide)	−2.11	Pb^{2+}	−32.0	C=O	+6.3
O	−4.61	Ca^{2+}	−10.4		
O_2 (carboxylate)	−7.95	Fe^{2+}	−12.8		
S	−15.0	Cu^{2+}	−12.8		
P	−26.3	Co^{2+}	−12.8		
		Ni^{2+}	−12.8		

C_5H_5N

SUM OF CONTRIBUTIONS TO χ ($\times 10^{-6}$ CM^3 MOLE^{-1})

$$5 \times \text{C(ring)} = -31.2$$

$$5 \times \text{H} = -14.6$$

$$1 \times \text{N(ring)} = -4.6$$

$$\chi = \sum_i \chi_{A_i} + \sum_j \chi_{B_j} = -50.4 \times 10^{-6} \text{ cm}^3 \text{ mole}^{-1}$$

The functional groups are accounted for by using the ring values for carbon and nitrogen, so $\sum_j \chi_{B_j}$ equals zero.

$(CH_3)_2C{=}O$

SUM OF ATOM CONTRIBUTIONS ($\times 10^{-6}$ CM^3 MOLE^{-1})

$$3 \times \text{C} = -18.0$$

$$6 \times \text{H} = -17.6$$

$$1 \times \text{O} = -4.6$$

SUM OF BOND CONTRIBUTIONS ($\times 10^{-6}$ CM^3 MOLE^{-1})

$$1 \times \gt C{=}O\lt = +6.3$$

$$\sum \chi_A + \sum \chi_B = -33.9 \times 10^{-6} \text{ cm}^3 \text{ mole}^{-1}$$

For a transition metal complex, one can measure only the net magnetism, χ_{MEAS}, which is the sum of the paramagnetic, χ_{PARA}, and diamagnetic, χ_{DIA}, contributions.

$$\chi_{\text{PARA}} = \chi_{\text{MEAS}} - \chi_{\text{DIA}} \tag{11-7}$$

Thus, to obtain the paramagnetic susceptibility, the diamagnetic susceptibility must be subtracted from the net susceptibility. This can be accomplished by: (1) using the values for Pascal's constants reported in Table 11-2 to estimate χ_{DIA}; (2) by measuring the diamagnetic susceptibility of the ligand and adding that of the metal from Table 11-2 to obtain χ_{DIA}; or (3) by making an analogous diamagnetic metal complex and using its value as an estimate of χ_{DIA}.

Paramagnetism in Simple Systems where S = ½

The paramagnetic contribution to the susceptibility arises from the spin and orbital angular momenta of the electrons interacting with the field. First, we shall consider a system that is spherical, contains only one electron, and has no orbital contribution to the moment. The magnetic moment, $\vec{\mu}$, associated with such a system is a vector quantity given by equation (11-8):

$$\vec{\mu} = -g\beta\vec{\mathbf{S}} \tag{11-8}$$

where $\vec{\mathbf{S}}$ is the spin angular momentum operator, g is the electron g-factor discussed in Chapter 9, and β is the Bohr magneton of the electron, also discussed in Chapter 9 ($\beta = 0.93 \times 10^{-20}$ erg gauss^{-1}).

The Hamiltonian describing the interaction of this moment with the applied field, $\vec{\mathbf{H}}$, is given by

$$\hat{H} = -\vec{\mu}\cdot\vec{\mathbf{H}} = g\beta\vec{\mathbf{S}}\cdot\vec{\mathbf{H}} \tag{11-9}$$

This Hamiltonian, operating on the spin wave functions, has two eigenvalues (see Fig. 9–1) with energies given by

$$E = m_s g\beta H \qquad \text{with } m_s = \pm\frac{1}{2} \tag{11-10}$$

with an energy difference

$$\Delta E = g\beta H \tag{11-11}$$

When H is about 25 kilogauss, ΔE for a free electron with $g = 2.0023$ is about 2.3 cm^{-1}, which is small enough compared to kT (205 cm^{-1} at room temperature) that both states are populated at room temperature, with a slight excess in the ground state.

The magnitude of the projection along the field direction of the magnetic moment, μ_n, of an electron in a quantum state n is given by the partial derivative of the energy of that state, E_n, with respect to the field, H, as shown in equation (11–12):

$$\mu_n = -\frac{\partial E_n}{\partial H} = -m_s g\beta \tag{11-12}$$

In order to determine the bulk magnetic moment of a sample of any material, *we must take a sum of the individual moments of the states weighted by their Boltzmann populations.*

The Boltzmann factor for calculating the probability, P_n, for populating discrete states having energy levels E_n at thermal equilibrium is given by equation (11–13):

$$P_n = \frac{N_n}{N} = \frac{\exp\left(\frac{-E_n}{kT}\right)}{\sum_n \exp\left(\frac{-E_n}{kT}\right)} \tag{11-13}$$

Here N_n refers to the population of the state n, while N refers to the total population of all existing states. *We have a wave function for each state, and we use the term "level" to indicate all of the states that have the same energy. The population weighted sum of magnetic moments over the individual states,* which is *the macroscopic magnetic moment, M,* then, is given for a mole of material by equation (11–14):

$$M = \underline{N} \sum_{m_s} \mu_n P_n \tag{11-14}$$

where $\underline{N}$ is Avogadro's number. Substituting equation (11–13) for P_n in equation (11–14) produces (11–15) for an $S = \frac{1}{2}$ system.

$$M = \frac{\underline{N} \sum_{m_s=-1/2}^{+1/2} \mu_n \exp\left(\frac{-E_n}{kT}\right)}{\sum_{m_s=-1/2}^{+1/2} \exp\left(\frac{-E_n}{kT}\right)} \tag{11-15}$$

Substituting equation (11–12) for μ_n and equation (11–10) for E_n, and summing over $m_s = \pm\frac{1}{2}$, produces:

$$M = \frac{Ng\beta}{2}\left[\frac{\exp\left(\frac{g\beta H}{2kT}\right) - \exp\left(\frac{-g\beta H}{2kT}\right)}{\exp\left(\frac{g\beta H}{2kT}\right) + \exp\left(\frac{-g\beta H}{2kT}\right)}\right] \tag{11-16}$$

When $(g\beta H/kT) \ll 1$ ($g\beta H$ equals $\sim 1\ \text{cm}^{-1}$ for $g = 2.0$ and common fields of 5,000 to 10,000 gauss, and kT is 205 cm^{-1} at room temperature), we can introduce the following approximation:

$$\exp\left(\frac{\pm g\beta H}{2kT}\right) \approx \left(1 \pm \frac{g\beta H}{2kT}\right) \tag{11-17}$$

Substituting equation (11–17) into (11–16) and simplifying leads to

$$M = \frac{Ng^2\beta^2 H}{4kT} \tag{11-18}$$

Since the molar susceptibility is related to the moment by

$$\chi = \frac{M}{H} \tag{11-19}$$

we can write

$$\chi = \frac{Ng^2\beta^2}{4kT} \tag{11-20}$$

Equation (11–20) is the so-called *Curie Law,* and it predicts a straight-line relation between the susceptibility and the reciprocal of temperature, giving a zero intercept; *i.e.,* $\chi \to 0$ as T approaches infinity. This type of behavior is usually not observed experimentally. Straight-line plots are obtained for many systems, but the intercept is non-zero:

$$\chi = \frac{C}{T - \theta} \tag{11-21}$$

Equation (11–21), where $C = Ng^2\beta^2/4k$ and θ corrects the temperature for the non-zero intercept, describes the so-called *Curie-Weiss* behavior. It is common to have a non-zero intercept in systems that are not magnetically dilute (*i.e.,* pure solid paramagnetic material). In these systems, interionic or intermolecular interactions[1] cause neighboring magnetic moments to become aligned and contribute to the value of the intercept.

For the case of molecules with no orbital angular momentum, one commonly sees equation (11–20) written as

$$\chi = \frac{Ng^2\beta^2}{3kT} S(S + 1) \text{ (units of Bohr Magneton, BM)} \tag{11-22}$$

This reduces to equation (11–20) for $S = \frac{1}{2}$ and accounts for the so-called spin-only magnetic susceptibilities of complexes containing any number of unpaired electrons.

Both χ and M are macroscopic properties. In describing the magnetic properties of transition metal complexes, it is common to employ a microscopic quantity called the effective magnetic moment, μ_{eff}. It is *defined* as follows:

$$\mu_{eff} = \left(\frac{3k}{N\beta^2}\right)^{1/2} (\chi T)^{1/2} = 2.828(\chi T)^{1/2} \text{ (BM)} \qquad (11\text{-}23)$$

The susceptibility employed in this equation is χ_{PARA} described above. Equation (11–23) can be obtained by replacing $g^2S(S + 1)$ in equation (11–22) by $\mu_{eff}{}^2$ and solving for μ_{eff}. Thus, any effects that tend to make S *not a good quantum number* become incorporated into the g-value. (Recall the variability of the g-value in epr, Chapter 9.) Equation (11–24) can be used to calculate the spin-only magnetic moments for various values of S:

$$\mu_{eff}(\text{spin-only}) = g[S(S + 1)]^{1/2} \text{ (BM)} \qquad (11\text{-}24)$$

where g equals 2.0 for an electron with no orbital angular momentum.

The spin-only results in Table 11–3 are obtained for various numbers of unpaired electrons.

TABLE 11–3. SPIN-ONLY MAGNETIC MOMENTS FOR VARIOUS NUMBERS OF UNPAIRED ELECTRONS

Number of Unpaired Electrons	S	μ_{eff}(spin-only)(BM)
1	1/2	1.73
2	1	2.83
3	3/2	3.87
4	2	4.90
5	5/2	5.92
6	3	6.93
7	7/2	7.94

In many transition metal ion complexes, values close to those predicted by the spin-only formula are observed.[1-9] However, in many other complexes, the moments and temperature dependence of the susceptibility are at variance with these predictions. Other effects are operative, and a more complete analysis is in order.

11–3 VAN VLECK'S EQUATION

General Basis of the Derivation

In this analysis, we will introduce orbital contributions and also anisotropy in the magnetic susceptibility for low symmetry molecules. Defining the principal molecular axis as the z-axis, we can write the necessary part of the Hamiltonian including these additional effects as:

$$\hat{H} = \lambda\hat{L} \cdot \hat{S} + \beta(\hat{L} + g_e\hat{S}) \cdot H \qquad (11\text{-}25)$$

where $\hat{L}$ and $\hat{S}$ are operators with x, y, and z components. In this chapter, g_e will be used when referring to the free electron g-value. The first term on the right of the equality sign describes the spin-orbit coupling (λ is the spin-orbit coupling constant) and is seen to be field independent. The other term sums the spin and orbital contributions to the electron moment. [Note its resemblance to equation (11–9).] In using this Hamiltonian, we have to worry about what basis set to employ. In a free ion, for a d^1 case with $\vec{\mathbf{L}} \cdot \vec{\mathbf{S}}$ coupling ignored, the complex functions $|+2\rangle, |+1\rangle$, etc., are a good choice for they are already eigenfunctions of $\hat{H}$. Hence, when the full matrix with elements $\langle \varphi_n | L_z + g_e S_z | \varphi_m \rangle \beta H$ is evaluated, there will be no off-diagonal elements.

In a d^1 complex, with the ligands defining the x, y, and z axes, the real orbitals are a convenient basis set. Now we have, for example:

$$d_{xy} = \frac{1}{i\sqrt{2}}(d_{+2} - d_{-2})$$

and

$$d_{x^2-y^2} = \frac{1}{\sqrt{2}}(d_{+2} + d_{-2})$$

In this basis set, the off-diagonal elements will be non-zero; for example

$$\langle d_{xy} | \hat{L}_z | d_{x^2-y^2} \rangle = \frac{1}{2i}[\langle d_{+2} | \hat{L}_z | d_{+2} \rangle + \langle d_{+2} | \hat{L}_z | d_{-2} \rangle - \langle d_{-2} | \hat{L}_z | d_{+2} \rangle - \langle d_{-2} | \hat{L}_z | d_{-2} \rangle]$$
$$= \frac{1}{2i}[2 + 0 + 0 - (-2)] = -2i$$

The $\hat{S}_z$ contribution to this matrix element will be zero, since

$$\langle d_{xy}{}^{+} | \hat{S}_z | d_{x^2-y^2}{}^{+} \rangle = \langle d_{xy} | d_{x^2-y^2} \rangle \langle + | \hat{S}_z | + \rangle = (0)\left(\frac{1}{2}\right) = 0$$

The non-zero off-diagonal elements account for a distortion of the ground state wave function by the applied field [we worked out the above matrix elements for $\hat{L}_z$ and $\hat{S}_z$, but the full Hamiltonian is $\beta(\hat{L} + g_e\hat{S}) \cdot \vec{\mathbf{H}}$]. This distortion is accomplished by mixing in appropriate excited states. The diagonal elements are called the first-order Zeeman terms, and the off-diagonal elements give rise to second-order Zeeman terms. If there were no off-diagonal terms, all of the diagonal matrix elements would be to the first power in H and the resulting energies would be first order in H.

Off-diagonal elements connecting states of very different energies are generally small compared to the energy difference, so this problem is generally treated with perturbation theory. We saw in the discussion of the Ramsey equation (Chapter 8) that this approach gave rise to terms of the general form

$$\sum_{m \neq n} \frac{\langle \psi_n | \hat{\mathrm{Op}} | \psi_m \rangle^2}{E_m - E_n}$$

where $\hat{\mathrm{Op}}$ is an abbreviation for an unspecified operator. The field-induced mixing of excited states was used as an interpretation of the paramagnetic contribution to the chemical shift. In the present case, we obtain from perturbation theory a term that looks similar:

$$\sum_{j \neq i} \frac{[\langle \psi_i | \hat{L}_z + g_e \hat{S}_z | \psi_j \rangle \beta H]^2}{E_i - E_j} \qquad (11\text{–}26)$$

When the ion configuration becomes other than d^1, the advantage of the perturbation treatment is seen, for the full matrix would be large.

For a weak field complex, the basis set for a full matrix evaluation would involve using the wave functions that resulted after a weak field approximation in the crystal field analysis of the electron-electron repulsions. For a strong field complex, the real *d*-orbitals provide a good basis set for the complex. Thus, we see that the relative magnitudes of the factors influencing the *d*-orbital energies are important in determining the best basis set. We can list some common, rough orders of magnitude as follows, using C.F. to abbreviate the crystal field:

(A) Weak field, first row transition metal ion:

$$\underset{10^4\ \text{cm}^{-1}}{e^2/r_{ij}} \geq \underset{10^4\ \text{cm}^{-1}}{\text{C.F.}} > \underset{10^2\ \text{to}\ 10^3\ \text{cm}^{-1}}{\text{low-symmetry C.F. perturbation}} > \underset{10^2\ \text{cm}^{-1}}{\lambda\vec{\mathbf{L}}\cdot\vec{\mathbf{S}}}$$

(B) Strong field, first row:

$$\underset{5\times10^4\ \text{cm}^{-1}}{\text{C.F.}} > \underset{3\times10^3\ \text{cm}^{-1}}{e^2/r_{ij}} \gtrsim \underset{10^3\ \text{cm}^{-1}}{\text{C.F. perturbation}} > \underset{10^2\ \text{cm}^{-1}}{\lambda\vec{\mathbf{L}}\cdot\vec{\mathbf{S}}}$$

(C) Third row:

$$\underset{10^4\ \text{cm}^{-1}}{\text{C.F.}} > \underset{10^3\ \text{cm}^{-1}}{\lambda\vec{\mathbf{L}}\cdot\vec{\mathbf{S}}} \geq \underset{10^3\ \text{cm}^{-1}}{e^2/r_{ij}}$$

(D) Lanthanides:

$$\underset{5\times10^3\ \text{cm}^{-1}}{\lambda\vec{\mathbf{L}}\cdot\vec{\mathbf{S}}} \geq \underset{5\times10^3\ \text{cm}^{-1}}{e^2/r_{ij}} > \underset{10^3\ \text{cm}^{-1}}{\text{C.F.}}$$

The magnetic field effect is about 1 cm^{-1}.

In our analysis so far, we have not taken spin-orbit coupling (the $\lambda\vec{\mathbf{L}}\cdot\vec{\mathbf{S}}$ term) into account. For first row transition metal ions, this is accomplished by adding the effects of $\lambda\vec{\mathbf{L}}\cdot\vec{\mathbf{S}}$ to the energies as a perturbation on their magnitude. This is a good approximation only when $\lambda\vec{\mathbf{L}}\cdot\vec{\mathbf{S}}$ is small compared to electron-electron repulsions and crystal field effects. The diagonal $\vec{\mathbf{L}}\cdot\vec{\mathbf{S}}$ matrix elements are evaluated in the real orbital basis set and added to the energies as corrections. When spin-orbit coupling is large, this perturbation approach is not appropriate. For example, d_2^- and d_1^+ (signs refer to the electron m_s value) have the same m_J value ($^3\!/_2$) and are mixed by $\vec{\mathbf{L}}\cdot\vec{\mathbf{S}}$.

Derivation of the Van Vleck Equation

This very general discussion of how to proceed in a crystal field evaluation of the effects of the Hamiltonian in equation (11–25) on the molecule or ion of interest is sufficient for our purposes. We shall now return to a discussion of the influence of these factors on the magnetic moment. When we list the contributions to the energy of a given state, n, from the factors discussed in the earlier section for $S = ½$ systems, in terms of the field dependence of the effects, equation (11–27) results:

$$E_n = \underset{\lambda\text{L}\cdot\text{S}}{E_n^{(0)}} + \underset{\substack{\text{first-order Zeeman}\\ \text{(diagonal terms)}}}{HE_n^{(1)}} + \underset{\substack{\text{second-order Zeeman}\\ \text{(off-diagonal terms)}}}{H^2E_n^{(2)}} \tag{11-27}$$

Recalling that the projection of the magnetic moment in the field direction is given by

$-\partial E_n/\partial H$ [equation (11–12)], we see that the first term, $E_n^{(0)}$, makes no contribution to the moment of a given state; the second term makes a contribution that is independent of the field strength; and the third term makes a field-dependent contribution. The $E_n^{(1)}$ term in equation (11–27) is the same term that we had in the Curie Law derivation, except that the orbital momentum is now included. The magnitude of the second-order contribution will depend upon $E_i - E_j$. It can be very large when the electronic excited state is close in energy to the ground state and has correct symmetry.

In order to determine the influence of these effects on the susceptibility, we return to the earlier Curie Law derivation and rewrite equation (11–15) by replacing $\exp(-E_n/kT)$ with

$$\exp\left(\frac{-E_n^{(0)} - HE_n^{(1)} - H^2E_n^{(2)} + \cdots}{kT}\right) \cong \left(1 - \frac{HE_n^{(1)}}{kT}\right)\exp\left(\frac{-E_n^{(0)}}{kT}\right) \tag{11-28}$$

Also, we let

$$\mu_n = \frac{-\partial E_n}{\partial H} = -E_n^{(1)} - 2HE_n^{(2)} \tag{11-29}$$

Making these substitutions into equation (11–15) leads to:

$$M = \underline{N}\frac{\sum_n (-E_n^{(1)} - 2HE_n^{(2)})\left(1 - \frac{HE_n^{(1)}}{kT}\right)\exp\left(\frac{-E_n^{(0)}}{kT}\right)}{\sum_n \exp\left(\frac{-E_n^{(0)}}{kT}\right)\left(1 - \frac{HE_n^{(1)}}{kT}\right)} \tag{11-30}$$

Limiting this derivation to paramagnetic substances, this equation must yield $M = 0$ at $H = 0$ and, in order for this to happen, the following must be true:

$$-\sum_n E_n^{(1)}\exp\left(\frac{-E_n^{(0)}}{kT}\right) = 0 \tag{11-31}$$

Expanding the numerator, neglecting terms higher than $E_n^{(2)}$ as well as the $E_n^{(2)}E_n^{(1)}$ product in equation (11–30), and recalling that $\chi = M/H$ [equation (11–19)], we obtain from equations (11–30) and (11–31):

$$\chi = \underline{N}\frac{\sum_n\left[\frac{(E_n^{(1)})^2}{kT} - 2E_n^{(2)}\right]\exp\left(\frac{-E_n^{(0)}}{kT}\right)}{\sum_n \exp\left(\frac{-E_n^{(0)}}{kT}\right)} \tag{11-32}$$

where $E_n^{(0)}$ has contributions from $\lambda\vec{\mathbf{L}}\cdot\vec{\mathbf{S}}$, etc. The $E_n^{(0)}$ term is always zero for the ground level; for a higher energy state, the quantity giving rise to this energy term in the absence of a field is substituted for $E_n^{(0)}$. $E_n^{(1)}$ contains the $m_s g\beta H$ and other first-order contributions, and $E_n^{(2)}$ has the contributions from the second-order Zeeman term.

Thus, the susceptibility is determined by taking a population-weighted average of the susceptibility of the levels. *An r-fold degenerate level has r component states, each of which must be included in the summations of equation (11–32).* Its use will become more clear by working out some examples.

Application of the Van Vleck Equation

We shall demonstrate the use of the Van Vleck equation [equation (11–32)] by applying it to the *ground state* of a free metal ion with quantum number J (Russell-Saunders coupling applies). In all of the examples worked out in this section, it is important to appreciate that all we are doing is taking a population-weighted average of the individual moments of the levels. The $2J + 1$ degeneracy is removed by a magnetic field, and the relative energies of the resulting levels are given by $m_J g\beta H$. We are considering only the ground level $E_n^{(0)}$ and $E_n^{(2)}$, which are taken as zero. (In doing a Boltzmann population analysis, the zero of energy is arbitrary; we set the energy of the ground level in the absence of H [*i.e.*, $E_n^{(0)}$] at zero for convenience.) Equation (11–32) becomes

$$\chi = \underline{N} \sum_{\substack{m_J=-J \\ n=0}}^{+J} \frac{\mu_J{}^2 g^2 \beta^2}{kT(2J+1)}$$

$$= \frac{\underline{N}g^2\beta^2}{kT}\left(\frac{J^2 + (J-1)^2 + \cdots + (-J+1)^2 + (-J)^2}{2J+1}\right)$$

$$= \frac{\underline{N}g^2\beta^2}{kT}\left(\frac{J(J+1)(2J+1)}{3(2J+1)}\right) = \frac{\underline{N}g^2\beta^2}{3kT}J(J+1) \tag{11-33}$$

Written in terms of μ_{eff}, we obtain for the free ion

$$\mu_{\text{eff}} = g[J(J+1)]^{1/2} \text{ (BM)} \tag{11-34}$$

For a free ion, following the Russell-Saunders coupling scheme, we give without derivation[2] the expression for the g-value as:

$$g = 1 + \frac{S(S+1) - L(L+1) + J(J+1)}{2J(J+1)} \tag{11-35}$$

We see from this equation that, in the free ion, contributions to μ_{eff} arise from both the spin and the orbital angular momenta. Furthermore, when $L = 0$, then $J = S$. Then $g = 2.00$ and equation (11–34) reduces to the spin-only formula given in equation (11–24).

The next example selected to illustrate the use of equation (11–32) is a $Ti^{3+}(d^1)$ complex.[3] The splitting of the gaseous ion terms by the crystal field, spin-orbit coupling, and the magnetic field is illustrated in Fig. 11–3. The expressions for the energies given in the figure are obtained from the wave functions resulting from a weak crystal field analysis by operating on them with the $\lambda\hat{L}\cdot\hat{S}$ and $\beta(\hat{L} + g_e\hat{S})\cdot\vec{\mathbf{H}}$ operators of equation (11–25). Since $10Dq$ is generally large in an octahedral complex, we can ignore the ^2E state in evaluating the susceptibility with equation (11–32). We shall discuss this entire problem by starting with the ground $^2T_2$ level and numbering the states 1 to 4 in order of increasing energy. The $E_n^{(0)}$ terms for states 1 to 4 are $-\lambda/2$, $-\lambda/2$, λ, and λ, respectively (see Chapter 10). The $E_n^{(1)}$ terms are 0, 0, $-\beta H$, and $+\beta H$, respectively, while the $E_n^{(2)}$ terms are $-\frac{4}{3}(\beta^2H^2/\lambda)$, 0, $+\frac{4}{3}(\beta^2H^2/\lambda)$, and $+\frac{4}{3}(\beta^2H^2/\lambda)$, respectively. Substituting these values into equa-

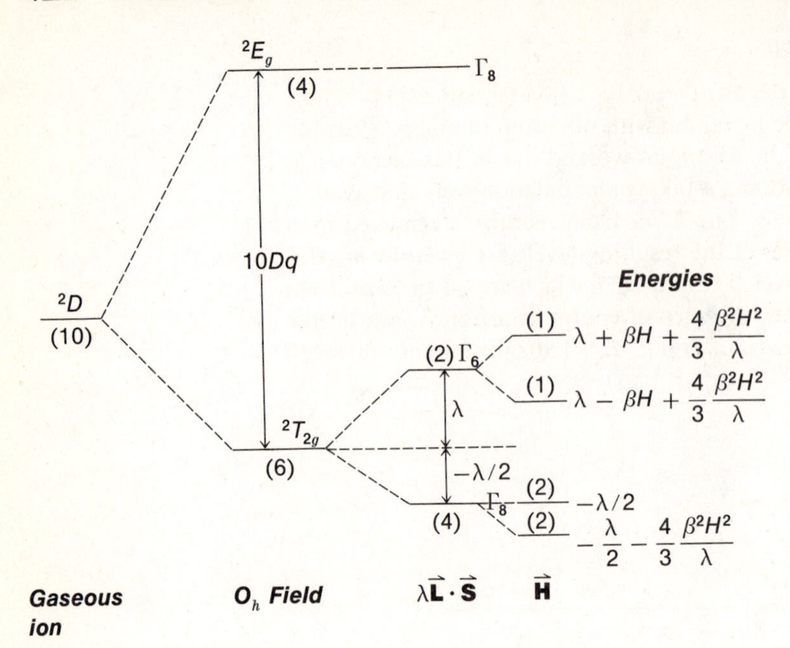

FIGURE 11–3. The splitting of the gaseous ion ^2D state by an O_h field, by $\lambda\hat{L}\cdot\hat{S}$ and by a magnetic field. The degeneracies of the levels are indicated in parentheses, and the energies are listed on the right.

tion (11–32) and multiplying each term by the degeneracy of the corresponding level produces:

$$\frac{\chi}{\underline{N}} = \left\{2\left[\left(\frac{0^2}{kT}\right) + (2)\left(\frac{4}{3}\right)\left(\frac{\beta^2}{\lambda}\right)\exp\left(\frac{\lambda}{2kT}\right)\right] + 2\left[\left(\frac{0^2}{kT}\right) - (2)(0)\exp\left(\frac{\lambda}{2kT}\right)\right]\right.$$
$$\left. + \left[\left(\frac{\beta^2}{kT}\right) - (2)\left(\frac{4}{3}\right)\left(\frac{\beta^2}{\lambda}\right)\right]\exp\left(\frac{-\lambda}{kT}\right) + \left[\left(\frac{\beta^2}{kT}\right) - (2)\left(\frac{4}{3}\right)\left(\frac{\beta^2}{\lambda}\right)\right]\exp\left(\frac{-\lambda}{kT}\right)\right\}$$
$$\div\, 2\left[\exp\left(\frac{\lambda}{2kT}\right) + \exp\left(\frac{\lambda}{2kT}\right) + \exp\left(\frac{-\lambda}{kT}\right)\right]$$

Recalling that $\chi = (\underline{N}\beta^2/3kT)\mu_{\text{eff}}{}^2$, we obtain:

$$\mu_{\text{eff}}{}^2 = \frac{8 + \left[\frac{3\lambda}{kT} - 8\right]\exp\left(\frac{-3\lambda}{2kT}\right)}{\frac{\lambda}{kT}\left[2 + \exp\left(\frac{-3\lambda}{2kT}\right)\right]}\beta^2$$

where β is the Bohr magneton. Note that our analysis predicts that the Curie Law will not hold. As T approaches infinity, $\mu_{\text{eff}}{}^2$ approaches zero. As T becomes small, $\mu_{\text{eff}}{}^2$ becomes small; and as T approaches zero, the equations no longer apply because $g\beta H \approx kT$. As λ approaches zero, $\mu_{\text{eff}}{}^2$ approaches 3. Finally, as T approaches zero, we have the very interesting result that a system with one unpaired electron has zero susceptibility. The result arises because the spin and orbital contributions cancel. These predictions are confirmed by experiment.

In our analysis of this problem, we have ignored any contributions from the ^2E_g excited level. However, the above approximation is valid for many magnetic applications. In the more sensitive epr technique, one can detect the contribution from the excited state to the g-value (see Chapter 12). With spin-orbit coupling, the ground level (Γ_8) and the Γ_8 excited level from ^2E_g ($\Gamma_3 \times \Gamma_6 = \Gamma_8$) can mix,[4] changing the g-value from $4\beta H/3\lambda$ to $4\lambda/\Delta + 4\beta H/3\lambda$, where Δ is $10Dq$. The second-order Zeeman term mixes the ground level with the excited level, and the extent of mixing depends upon Δ.

The next system[9] that we shall consider is chromium(III) (d^3). An octahedral field gives rise to a $^4A_2$ ground state and $^4T_2$ and $^4T_1$ excited levels, as shown for the quartet states in the Tanabe and Sugano diagrams. Since $^4T_2$ is about 18,000 cm^{-1} higher in energy than $^4A_2$, its contribution to the susceptibility can be ignored. Since the ground state is orbitally singlet (A), there is no orbital contribution to the magnetic susceptibility (*vide infra*). The magnetism is predicted with the spin-only formula and $S = 3/2$. Next, we shall consider the effect of a tetragonal distortion on chromium(III). This removes the degeneracy of the $m_s = \pm 1/2$ and the $m_s = \pm 3/2$ states as shown in Fig. 11–4. The splitting by the tetragonal component is described by the parameter D. Since this splitting exists in the absence of a field, it is one of the many effects that are referred to as a *zero-field splitting*. For the case of an axial zero-field splitting, one can represent this with the Hamiltonian, $D\hat{S}_{z^2}$. The susceptibility for this system when the applied field is parallel to the principal molecular axis is obtained by inserting values for $E_n^{(1)}$, which equals $(1/2)g_z\beta$ and $(3/2)g_z\beta$, respectively, for the two levels shown in Fig. 11–4. With $E_n^{(0)}$ given a value of zero for the lower energy level and a value of D for the higher one, we have:

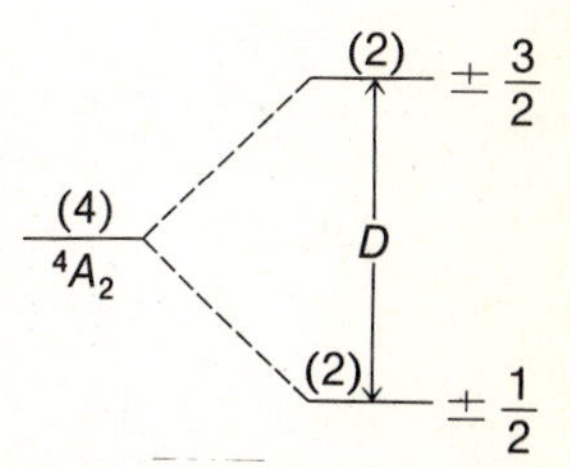

FIGURE 11–4. The splitting of the $^4A_2$ state by a tetragonal field D. (D is the tetragonal splitting or zero-field splitting parameter.)

$$\frac{\chi_z}{N} = \left[\frac{2\left(\frac{1}{2}g_z\beta\right)^2}{kT}\exp(0) + \frac{2\left(\frac{3}{2}g_z\beta\right)^2}{kT}\exp\left(\frac{-D}{kT}\right)\right] \Big/ \left[2\exp(0) + 2\exp\left(\frac{-D}{kT}\right)\right]$$

$$= \frac{g_z^2\beta^2}{4kT}\,\frac{\left[1 + 9\exp\left(\frac{-D}{kT}\right)\right]}{\left[1 + \exp\left(\frac{-D}{kT}\right)\right]}$$

Thus we see that when $(D/kT) \ll 1$, which is true for a very small distortion or at a very high temperature, the expression for χ_z reduces to $(5/4)Ng_z^2\beta^2/kT$, while the spin-only formula for $S = 1/2$ results as T approaches zero or D becomes very large.* To calculate the powder average susceptibility, χ_x and χ_y must be evaluated using the x and y components of $\hat{L}$ and $\hat{S}$ in equation (11–25) with the $D\hat{S}_{z^2}$ term added. The anisotropy in χ can be calculated this way.

As a final example, we shall consider a nickel(II) complex with a small tetragonal distortion. The splitting is shown in Fig. 11–5.

* Experimentally, D is approximately 0.1 cm^{-1} for pseudo-octahedral Cr^{3+} from a spin-orbital mixing in of excited states.

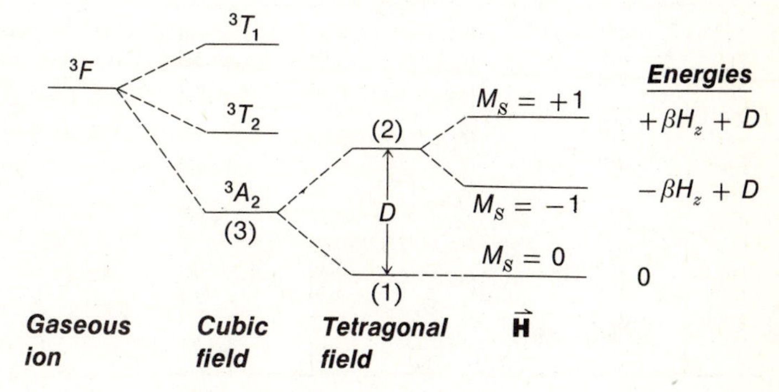

FIGURE 11–5. Splitting of a nickel(II) ion in a tetragonal field.

$$\frac{\chi_\parallel}{\underline{N}} = \frac{(0)\exp(-0) + \dfrac{(-g_z\beta)^2}{kT}\exp\left(\dfrac{-D}{kT}\right) + \dfrac{(g_z\beta)^2}{kT}\exp\left(\dfrac{-D}{kT}\right)}{1 + 2\exp\left(\dfrac{-D}{kT}\right)}$$

$$\chi_\parallel = \frac{2\underline{N}g_z^{\,2}\beta^2}{kT}\,\frac{\exp\left(\dfrac{-D}{kT}\right)}{1 + 2\exp\left(\dfrac{-D}{kT}\right)}$$

When $D \ll kT$, the following expression results:

$$\chi_\parallel = \frac{2\underline{N}g_z^{\,2}\beta^2\left(1 - \dfrac{D}{kT}\right)}{kT\left(1 + 2 - \dfrac{2D}{kT}\right)} \approx \frac{2\underline{N}g_z^{\,2}\beta^2}{3kT}\left(1 - \frac{D}{kT}\right)$$

For $(NH_4)_2Ni(SO_4)_2 \cdot 6H_2O$, the experimental value[4] of g_z is 2.25 and D is $-2.24\ cm^{-1}$, giving a value of $\chi_\parallel$ of $4260 \times 10^{-6}\ cm^3\ mole^{-1}$. The experimental value is $4230 \times 10^{-6}\ cm^3\ mole^{-1}$, while the spin-only formula predicts a value of $3359 \times 10^{-6}\ cm^3\ mole^{-1}$.

There is one additional point to be made in this section. If electrons are delocalized onto the ligands as a result of covalency in the metal ligand bond, the matrix elements corresponding to orbital angular momentum are reduced below the value calculated by using the metal-centered operator $\hat{L}$ applied to metal orbital wave functions. To compensate for this, one can use $k\hat{L}$ (where k, *the orbital reduction factor,* is a constant less than one) to correct for delocalization of electron density onto the ligand where it will have reduced orbital angular momentum.

11-4 APPLICATIONS OF SUSCEPTIBILITY MEASUREMENTS

Spin-Orbit Coupling

When equation (11–34) is applied to complexes of the rare earth ions, excellent agreement between calculated and observed susceptibilities results, as shown for some trivalent ions in Table 11–4.

TABLE 11–4. CALCULATED AND EXPERIMENTAL MAGNETIC MOMENTS FOR SOME TRIVALENT RARE EARTH IONS

Element	Config.	Term	μ_{eff}(calc)	μ_{eff}(exp)
Ce^{3+}	$4f^15s^25p^6$	$^2F_{5/2}$	2.54	2.4
Pr^{3+}	$4f^2$	$^3H_4$	3.58	3.5
Nd^{3+}	$4f^3$	$^4I_{9/2}$	3.62	3.5
Pm^{3+}	$4f^4$	$^5I_4$	2.68	
Sm^{3+}	$4f^5$	$^6H_{5/2}$	0.84	1.5
Eu^{3+}	$4f^6$	$^7F_0$	0	3.4
Gd^{3+}	$4f^7$	$^8S_{7/2}$	7.94	8.0
Tb^{3+}	$4f^8$	$^7F_6$	9.72	9.5
Dy^{3+}	$4f^9$	$^6H_{15/2}$	10.63	
Ho^{3+}	$4f^{10}$	$^5I_8$	10.60	10.4
Er^{3+}	$4f^{11}$	$^4I_{15/2}$	9.59	9.5
Tm^{3+}	$4f^{12}$	$^3H_6$	7.57	7.3
Yb^{3+}	$4f^{13}$	$^2F_{7/2}$	4.54	4.5

This excellent agreement is obtained because the crystal field from the ligands does not effectively quench the orbital angular momentum of the electrons in the inner $4f$ orbitals. A very much different result is obtained with the $3d$ transition series where, as can be seen in Table 11–5, the spin-only formula comes much closer to predicting the observed results.

TABLE 11–5. CALCULATED AND OBSERVED MAGNETIC MOMENTS FOR COMPLEXES OF THE 3d IONS

(Calculated results are presented using equation (11-34) and the spin-only formula.)

Ion	Config.	Term	$g[J(J+1)]^{1/2}$	$2[S(S+1)]^{1/2}$	μ_{eff}(exp)
Ti^{3+}, V^{4+}	$3d^1$	$^2D_{3/2}$	1.55	1.73	1.7–1.8
V^{3+}	$3d^2$	$^3F_2$	1.63	2.83	2.6–2.8
Cr^{3+}, V^{2+}	$3d^3$	$^4F_{3/2}$	0.77	3.87	~3.8
Mn^{3+}, Cr^{2+}	$3d^4$	$^5D_0$	0	4.90	~4.9
Fe^{3+}, Mn^{2+}	$3d^5$	$^6S_{5/2}$	5.92	5.92	~5.9
Fe^{2+}	$3d^6$	$^5D_4$	6.70	4.90	5.1–5.5
Co^{2+}	$3d^7$	$^4F_{9/2}$	6.63	3.87	4.1–5.2
Ni^{2+}	$3d^8$	$^3F_4$	5.59	2.83	2.8–4.0
Cu^{2+}	$3d^9$	$^2D_{5/2}$	3.55	1.73	1.7–2.2

Thus, in many of the complexes, the orbital contribution is strongly quenched by the crystal field. There is a very simple model that enables one to predict when the orbital moment will not be completely quenched. If an electron can occupy degenerate orbitals that permit circulation of the electron about an axis, orbital angular momentum can result.* There must not be an electron of the same spin in the orbital into which the electron must move.

In an octahedral d^1 complex, for example, the electron can occupy d_{xz} and d_{yz} to circulate about the z-axis, and the complex possesses orbital angular momentum. In an octahedral, high spin d^3 complex, there are electrons with the same spin quantum number in both d_{xz} and d_{yz}, so this ion does not have orbital angular momentum. Using this crude model, we would predict that the following octahedral complexes would have effectively all of the orbital contribution to the moment quenched:

High-Spin

$t_{2g}^{\ 3}$, $t_{2g}^{\ 3}e_g^{\ 1}$, $t_{2g}^{\ 3}e_g^{\ 2}$, $t_{2g}^{\ 6}e_g^{\ 2}$, $t_{2g}^{\ 6}e_g^{\ 3}$

Low-Spin

$t_{2g}^{\ 6}$ and $t_{2g}^{\ 6}e_g^{\ 1}$

The one electron in e_g could only occupy d_{z^2} and $d_{x^2-y^2}$ so circulation about an axis cannot result. If an E state in an appropriate ligand field consisted of an electron in d_{xy} and $d_{x^2-y^2}$ orbitals, an orbital contribution would be expected.

For tetrahedral complexes, for which high spin complexes result, the orbital contribution is quenched in e^1, e^2, $e^2t_2^{\ 3}$, $e^3t_2^{\ 3}$, and $e^4t_2^{\ 3}$.

Many molecules with A_{2g} and E_g ground states have moments that differ from the spin-only value. This variation results from two sources: (1) mixing in of excited

*This rule is often expressed by requiring that a rotation axis exists that enables one to rotate one of the degenerate orbitals into another one.

states that have some contributions from spin-orbit coupling, and (2) second-order Zeeman effects (temperature independent paramagnetism). For example:

$$\mu_{\text{eff}}(A_{2g}) = \mu(\text{spin-only})\left(1 - \frac{4\lambda}{10Dq}\right) + \frac{8\underline{N}\beta^2}{10Dq} \tag{11-36}$$

The last term is the temperature independent paramagnetism which is field-induced. The $4\lambda/10Dq$ term arises from the mixing in of an excited state via spin-orbit coupling. In lower symmetry complexes, the states are split and more mixing becomes possible. For an E_g ground state, the moment from mixing in an excited state is given by:

$$\mu_{\text{eff}}(E_g) = \mu(\text{spin-only})\left(1 - \frac{2\lambda}{10Dq}\right) + \frac{4\underline{N}\beta^2}{10Dq} \tag{11-37}$$

As we can see from the above discussion, the magnetic moments of transition metal ion complexes are often quite characteristic of the electronic ground state and structure of the complex. There have been many reported examples of this kind of application. A few nickel(II) and cobalt(II) complexes will be discussed here to illustrate this application.

In an octahedral field, nickel(II) has an orbitally nondegenerate ground state, $^3A_2(t_{2g}{}^6e_g{}^2)$, and no contribution from spin-orbit coupling is expected. The measured moments are in the range from 2.8 to 3.3 BM, very close to the spin-only value of 2.83. Values for octahedral complexes slightly above the spin-only value arise from slight mixing with a multiplet excited state in which spin-orbit coupling is appreciable. Tetrahedral nickel(II) has a $^3T_1$ ground state that is essentially $(e_g{}^4t_{2g}{}^4)$, and a large orbital contribution to the moment is expected. As a result, even though both octahedral and tetrahedral nickel(II) complexes contain two unpaired electrons, tetrahedral complexes have magnetic moments around 4 BM compared to 3.3 BM or less for octahedral complexes. Experimentally, it is found[8,10] that $NiCl_4{}^{2-}$, $Ni(HMPA)_4{}^{2+}$[10a] (HMPA = hexamethylphosphoramide), and [10b]$NiX_2 \cdot 2(C_6H_5)_3AsO$ (X = halogen) have moments in excess of 4 BM. Nyholm[7] has suggested that an inverse relationship exists between the magnitude of the moment and the distortion of nickel(II) complexes from tetrahedral symmetry. The complex $NiX_2 \cdot 2(C_6H_5)_3P$, which is known to be seriously distorted, has a moment of about 3 BM. The structures of $NiX_2 \cdot 2HMPA$ and $CoX_2 \cdot 2HMPA$ (where X = Cl^-, Br^-, I^-, $NO_3{}^-$) were deduced from combined magnetic and spectroscopic studies.[12] This work provides a good illustration of the use of these techniques.

In octahedral cobalt(II) complexes, the ground state is $^4T_{1g}$ and a large orbital contribution to the moment is expected. Mixing in of an excited state lowers the moment somewhat but a value in excess of 5 BM is usually found [μ(spin-only) = 3.87]. The ground state for tetrahedral cobalt(II) complexes is $^4A_2$ and a low moment approaching the spin-only value might be expected. However, an excited magnetic state is comparatively low in energy in the tetrahedral complexes and can be mixed with the ground state. Moments in the range from 4 to 5 BM have been predicted[13] and are found experimentally. An inverse relationship exists for tetrahedral cobalt(II)[13] complexes between the magnitude of the moment of a complex and the value of Dq as predicted by equation (11–36).

11–5 INTRAMOLECULAR EFFECTS

In our treatment so far, we have assumed that there is no interaction between the electron spins on the individual metal ions in the solid. Next we wish to consider molecules containing more than one metal ion with unpaired spins, *e.g.*,

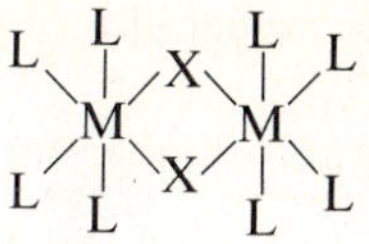

Generally, there is an interaction between the metal orbitals on the metal centers, which can be described for a pair i, j by the Hamiltonian

$$\hat{H} = -2J\hat{S}_i \cdot \hat{S}_j \tag{11-38}$$

where J in energy units is called the exchange coupling constant or exchange parameter. It is to be distinguished from the J quantum number. With this Hamiltonian, the exchange parameter J is negative for an antiferromagnetic interaction, which leads to a pairing of electrons, and is positive for a ferromagnetic interaction. The literature is confusing in this area, for the above Hamiltonian has been written with $-2J$ replaced by J, $2J$, or $-J$.

Dimeric copper(II) acetate dihydrate is the classic example of this type of system. The structure of this molecule is shown in Fig. 11-6, and the metal-metal axis is

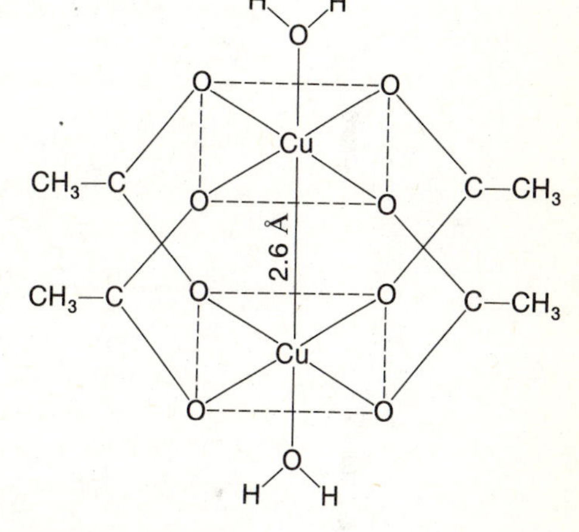

FIGURE 11-6. The structure of $Cu_2(CH_3CO_2)_4 \cdot 2H_2O$. The Cu—Cu distance is 2.6 Å.

taken as the z-axis. The copper(II) ions have a d^9 configuration. At low temperatures the compound is found to be diamagnetic, and at elevated temperatures it is paramagnetic. We can view this system as one in which two molecular orbitals exist that are largely metal $d_{x^2-y^2}$ (with a significant contribution from the bridging acetate). A simplified representation of this part of the molecular orbital diagram of the complex is shown in Fig. 11-7.* (If necessary, the reader is referred back to the chapter on esr to review the discussion on the triplet state and exchange interactions.) With Δ as the energy separation between the bonding and antibonding molecular orbitals (Fig. 11-7) and K as the spin pairing energy, we obtain, when $\Delta < K$, a ferromagnetic system. The quantity J for two d^1 or d^9 metals in the dimer is now given for the Hamiltonian in equation (11-38) by equation (11-39):

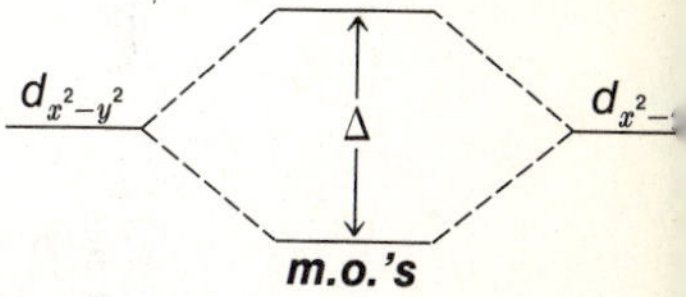

$d_{x^2-y^2}$ Δ $d_{x^2-y^2}$ m.o.'s

FIGURE 11-7. A simplified representation of the interaction of two metal $d_{x^2-y^2}$ orbitals in $Cu_2(CH_3CO)_4 \cdot 2H_2O$ to produce two nondegenerate levels.

$$J = \frac{(K - \Delta)}{2} \tag{11-39}$$

In $Cu_2(CH_3CO_2)_4 \cdot 2H_2O$, the ground state is diamagnetic, but the excited triplet state is close by in energy and is thermally populated at elevated temperatures. The value of $-2J$ is found[14] to be 284 cm^{-1}. By adding the Hamiltonian in equation (11-38) to $g\beta\hat{S}_z H_z$, we can evaluate the energies of the various levels in this system

*This is referred to as a super exchange pathway involving the bridging acetates. There is controversy regarding the amount of a direct exchange contribution, which involves a direct overlap of the two orbitals on each copper.

and calculate the behavior of the susceptibility with temperature, using equation (11-32).

We define a total spin quantum number for the system, $\vec{S}$, as

$$\vec{S} = \vec{S}_1 + \vec{S}_2$$

Accordingly

$$\vec{S}^2 = \vec{S}_1^{\,2} + \vec{S}_2^{\,2} + 2\vec{S}_1 \cdot \vec{S}_2$$

which upon rearranging defines $\vec{S}_1 \cdot \vec{S}_2$ as

$$\vec{S}_1 \cdot \vec{S}_2 = \frac{1}{2}[\vec{S}^2 - \vec{S}_1^{\,2} - \vec{S}_2^{\,2}] \tag{11-39}$$

Recalling that $\hat{S}^2\psi = S(S + 1)\psi$, we have the following result:*

$$-2J\hat{S}_1 \cdot \hat{S}_2\psi = -J[S(S + 1) - S_1(S_1 + 1) - S_2(S_2 + 1)]\psi \tag{11-40}$$

For dimeric copper(II) acetate dihydrate, we have

$$S_1 = S_2 = \frac{1}{2} \quad \text{and} \quad S = 0 \quad \text{or} \quad S = 1$$

For $S = 0$, we obtain

$$-J\left[0 - \frac{3}{4} - \frac{3}{4}\right] = \frac{3}{2}J$$

while for $S = 1$, we obtain

$$-J\left[2 - \frac{3}{4} - \frac{3}{4}\right] = -\frac{1}{2}J$$

The Hamiltonian $\hat{H} = g\beta\hat{S}_zH_z - 2J\hat{S}_1 \cdot \hat{S}_2$ produces the results shown in Fig. 11-8.

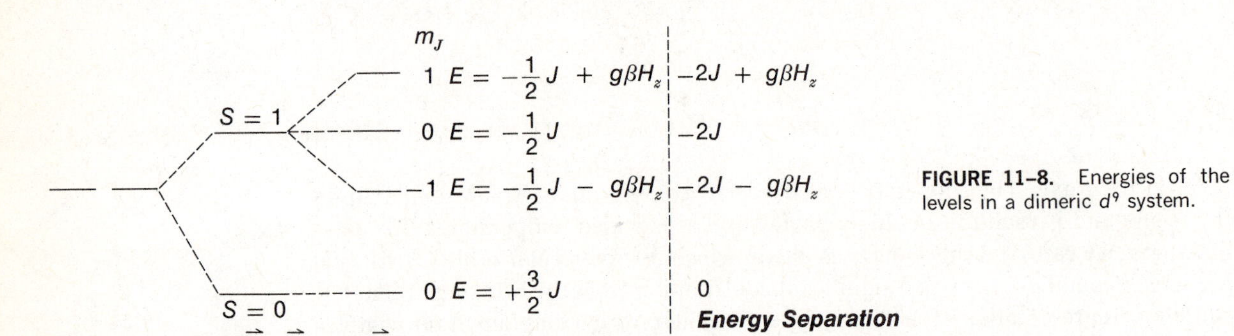

FIGURE 11-8. Energies of the levels in a dimeric d^9 system.

Substituting these results into equation (11-32) produces

$$\frac{\chi}{N} = \frac{\dfrac{2(g\beta)^2}{kT}\exp\left(\dfrac{J}{2kT}\right)}{3\exp\left(\dfrac{J}{2kT}\right) + \exp\left(\dfrac{-3J}{2kT}\right)} = \frac{\dfrac{2g^2\beta^2}{kT}}{3 + \exp\left(\dfrac{-2J}{kT}\right)}$$

Rearranging, we find

$$\chi = \frac{2Ng^2\beta^2}{3kT}\,\frac{1}{1 + \dfrac{\exp(-2J/kT)}{3}}$$

*The combination $-S_1(S_1 + 1) - S_2(S_2 + 1)$ is constant for each level of the dimer and is frequently dropped; *i.e.*, the zero of energy is redefined.

Since $\underline{N}\beta^2/3k$ is about $\frac{1}{8}$ we obtain when $g = 2$

$$\chi \cong \frac{1}{T} \frac{1}{1 + \dfrac{\exp(-2J/kT)}{3}} \qquad (11\text{-}41)$$

From this expression we find that χ approaches zero as T approaches infinity and becomes small as T becomes small. As a result χ must have a maximum, which is given by setting $\partial \ln \chi/\partial T$ equal to zero. From this we find

$$\frac{J}{kT_c} \cong -\frac{4}{5} \qquad (11\text{-}42)$$

i.e., χ has a maximum value at

$$T_c = -\frac{5}{4}\frac{J}{k}$$

When $J \ll kT$ or when $T \gg T_c$, the susceptibility follows the Curie-Weiss law, *i.e.*,

$$\chi = \frac{3}{4}\frac{1}{T + \theta} \qquad (11\text{-}43)$$

where $\theta = -J/2k$.

We also see that as the quantity $-J$ approaches positive infinity, χ approaches zero.

The value of J found in dimeric copper carboxylates is very much a function[15,16] of the bridging carboxylate, *e.g.*, $2J = -284\ \text{cm}^{-1}$ in the acetate dihydrate and $-339\ \text{cm}^{-1}$ in butyrate dihydrate. When water is substituted by other Lewis bases, dramatic changes in the value of J are also observed.

Dimeric iron(III) systems have been reported[17] with hydroxy- and oxy-bridges, *e.g.*,

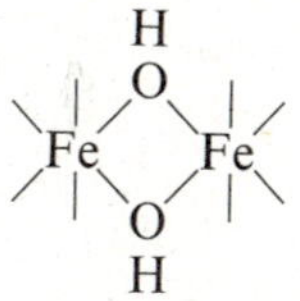

H O Fe Fe O H and O Fe Fe

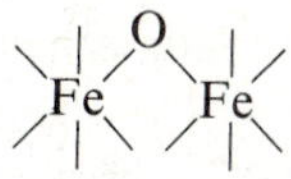

The former have J values of about $-8\ \text{cm}^{-1}$, while the latter have values* of about $-90\ \text{cm}^{-1}$.

11–6 HIGH SPIN–LOW SPIN EQUILIBRIA

If one examines the Tanabe and Sugano diagrams for d^4, d^5, d^6, and d^7 octahedral complexes, it can be seen that for certain values of Dq/B the ground state changes from a high spin to a low spin complex. For d^4 the ${}^5E_g$ and ${}^3T_{1g}$ states are involved, while d^5 involves ${}^6A_{1g}$ and ${}^2T_{2g}$, d^6 involves ${}^5T_{2g}$ and ${}^1A_{1g}$, and d^7 involves ${}^4T_{1g}$ and ${}^2E_g$. When the ligand field is such that the two states are close by in energy, the excited state can be thermally populated and the system will consist of an equilibrium mixture of the two forms. There have been many reported studies of this type of behavior.[18a] A typical system, which is selected for discussion because it has been studied[18b] in the solid state and in solution, involves iron(II) complexes of

*This latter value has been questioned in regard to the amount of iron(III) impurity (T. Moss, *et al.*, J. Chem. Soc., Chem. Comm., 263 (1972)).

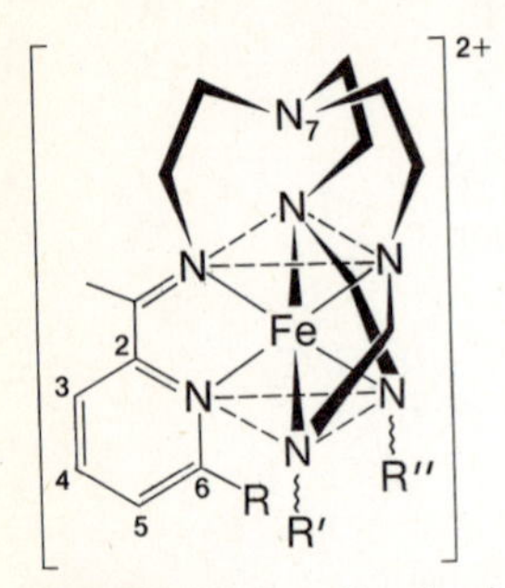

FIGURE 11–9. Structural formula of tris{4-[(6-R)-2-pyridyl]-3-aza-3-butenyl} amine iron(II) complex.

ligands that are Schiff base type condensation products of tren[$N(CH_2CH_2NH_2)_3$] and 2-pyridinecarboxaldehyde. The resulting complex is shown in Figure 11–9, where only one of the pyridine aldimines has been drawn in for clarity of presentation. The whole series of compounds in which R, R′, and R″ are —H or —CH_3 were prepared.

The symbol (I) will be used to abbreviate the complex $[Fe(Py)_3tren]^{2+}$ when R = R′ = R″ = H; (II) symbolizes $[Fe(6MePy)(Py)_2tren]^{2+}$ when R = R′ = H and R″ = CH_3; (III) symbolizes $[Fe(6MePy)_2(Py)tren]^{2+}$ when R = H and R′ = R″ = CH_3; and (IV) symbolizes $[Fe(6MePy)_3tren]^{2+}$ when R = R′ = R″ = CH_3.

An equilibrium involving the low spin $^1A_{1g}(t_{2g}{}^6)$ and high spin $^5T_{2g}(t_{2g}{}^4e_g{}^2)$ states was found both in the solid state and in solution for several of these complexes. Complex (I) is fully low spin at and below room temperature, while (II) and (III) undergo spin equilibrium both in the solid state and in solution. In solution, complex (IV) is essentially high spin at all temperatures above 180°K. In the solid state, a spin equilibrium exists that is very anion dependent. Thermodynamic data for the interconversion can be determined from the change in susceptibility with temperature, and are reported to be +4.6 and +2.8 kcal mole^{-1}, respectively, for complexes (II) and (III) in solution. An x-ray crystal structure determination indicated that the methyl substituents on the pyridine ring interact with the adjacent pyridine moiety. Thus, the ligand field is weakened in compound(IV) to the extent that a high spin compound results, while compound(I) is low spin. In the case of compound(IV), the average metal nitrogen distance is found to decrease by about 0.12 Å in going from the high spin to the low spin complex.

11–7 MEASUREMENT OF MAGNETIC SUSCEPTIBILITIES

In this section, the measurement of bulk magnetic susceptibility will be briefly covered in the course of presenting the pertinent references for a more complete discussion. The Gouy method[20a] employs a long, uniform glass tube packed with the solid material or solution, which is suspended in a homogeneous magnetic field. The sample is weighed in and out of the field, and the weight difference is related to the susceptibility and field strength. If a standard of known susceptibility is used, the field strength need not be known. Evans[20b] has recently reported a very clever and inexpensive device for routine measurement of the magnetic susceptibility by the Gouy method.

The Faraday method[19] uses a small amount of sample that is suspended in an inhomogeneous field such that $H(\partial H/\partial X)$ is a constant over the entire volume of the sample. The method is very sensitive, so small samples can be used and studies can be made on solutions.

Susceptibilities can also be determined conveniently over a wide temperature range, down to liquid helium temperatures, with a vibrating sample magnetometer.[21] The change in the inductance of a coil upon insertion of a sample can be related to the sample susceptibility. A mutual inductance bridge has been described and used for susceptibility determination.[22,23] An ultrasensitive, superconducting quantum magnetometer with a Josephson junction element has also been described.[24]

When one studies the susceptibility of single crystals, the anisotropy in the susceptibility can be determined.[25-28] This information has several important applications, as we shall see in our study of the nmr and epr of transition metal ions. The Krishnan critical torque method has been commonly used.[29,30]

An nmr method has been reported[31] for the measurement of the magnetic susceptibility of materials in solution. In this method, a solution of the paramagnetic complex containing an internal standard is added to the inner of two concentric tubes. A solution of the same inert standard, dissolved in the same solvent that was used to dissolve the complex, is placed in the outer of the two concentric tubes. Two separate nmr lines corresponding to the standard will be observed, with the line from the paramagnetic solution lying at higher frequency. The shift of the internal standard in the paramagnetic solution relative to that in the diamagnetic solution, $\Delta H/H$, is related to the difference in the volume susceptibility, $\Delta\chi_v$, of the two liquids:

$$\frac{\Delta H}{H} = \frac{2\pi}{3}\Delta\chi_v$$

The mass susceptibility of the dissolved substance, χ_g, is given by the expression

$$\chi_g = \frac{3\Delta\nu}{2\pi\nu m} + \chi_0 + \frac{\chi_0(d_0 - d_s)}{m}$$

where $\Delta\nu$ is the frequency separation of the two lines in Hz, ν is the probe frequency, m is the mass of the substance per ml of solution, χ_0 is the mass susceptibility of the solvent, d_0 is the density of the solvent, and d_s is the density of the solution. The importance of correcting for the density change with temperature in temperature dependent studies has been pointed out.[32]

REFERENCES

1. A. P. Ginsberg and M. E. Lines, Inorg. Chem., *11,* 2289 (1972).
2. G. Herzberg, "Atomic Spectra and Atomic Structure," 2nd ed., Dover Publications, New York (1944), pg. 109.
3. M. Kotani, J. Phys. Soc. Japan, *4,* 293 (1949).
4. B. Bleaney and K. W. H. Stevens, Rep. Prog. Phys., *16,* 108 (1953).
5. F. E. Mabbs and D. J. Machin, "Magnetism and Transition Metal Complexes," John Wiley, New York (1973).
6. a. B. N. Figgis and J. Lewis, "The Magnetic Properties of Transition Metal Complexes," in "Progress in Inorganic Chemistry," Volume 6, ed. by F. A. Cotton, Interscience, New York (1964).
   b. B. N. Figgis and J. Lewis, "The Magnetochemistry of Complex Compounds," in "Modern Coordination Chemistry," ed. J. Lewis and R. G. Wilkins, Interscience, New York (1960).
7. R. S. Nyholm, J. Inorg. Nuclear Chem., *8,* 401 (1958).
8. N. S. Gill and R. S. Nyholm, J. Chem. Soc., *1959,* 3997.
9. R. L. Carlin, J. Chem. Educ., *43,* 521 (1966).
10. a. J. T. Donoghue and R. S. Drago, Inorg. Chem., *1,* 866 (1962) and references therein.
    b. F. A. Cotton, *et al.,* J. Amer. Chem. Soc., *83,* 4161 (1961) and references therein.
11. L. M. Venanzi, J. Chem. Soc., *1958,* 719.
12. J. T. Donoghue and R. S. Drago, Inorg. Chem., *2,* 572 (1963).
13. F. A. Cotton, *et. al.,* J. Chem. Soc., *1960,* 1873.
14. G. F. Kokoszka and G. Gordon, Trans. Metal Chem., *5,* 181 (1969).
15. B. N. Figgis and R. L. Martin, J. Chem. Soc., *1956,* 3837.
16. E. Kokot and R. L. Martin, Inorg. Chem., *3,* 1306 (1964) and references therein.
17. H. J. Schugar, G. R. Rossman, and H. B. Gray, J. Amer. Chem. Soc., *91,* 4564 (1969).
18. a. R. L. Martin and A. H. White, Trans. Metal Chem., *4,* 113 (1968).
    b. M. A. Hoselton, L. J. Wilson, and R. S. Drago, J. Amer. Chem. Soc., *97,* 1722 (1975).
19. See, for example, L. N. Mulay and I. L. Mulay, Anal. Chem., *44,* 324R (1972).
20. a. B. Figgis and J. Lewis, "Techniques of Inorganic Chemistry," ed. H. Jonassen and A. Weissberger, Volume IV, pg. 137, Interscience, New York (1965).
    b. D. F. Evans, J. Physics *E*; Sci. Instr., *7,* 247 (1974).
21. S. Foner, Rev. Sci. Instr., *30,* 548 (1959).
22. W. L. Pillinger, P. S. Jastram, and J. G. Daunt, Rev. Sci. Instr., *29,* 159 (1958).
23. J. N. McElearney, G. E. Shankle, R. W. Schwartz, and R. L. Carlin, J. Chem. Phys., *56,* 3755 (1972).
24. J. W. Dawson, *et al.,* Biochem., *11,* 461 (1972) and references therein.
25. S. Mitra, Trans. Metal Chem., *7,* 183 (1972).
26. M. Gerloch, *et al.,* J. Chem. Soc., Dalton, *1972,* 1559; *ibid., 1972,* 980 and references therein.
27. W. D. Horrocks, Jr., and J. P. Sipe, III, Science, *177,* 944 (1972).
28. B. N. Figgis, L. G. B. Wadley, and M. Gerloch, J. Chem. Soc., Dalton, *1973,* 238.
29. K. S. Krishnan, *et al.,* Phil. Trans. Roy. Soc. A, *231,* 235 (1933).
30. D. A. Gordon, Rev. Sci. Instr., *29,* 929 (1958).
31. D. F. Evans, J. Chem. Soc., *1959,* 2005.
32. D. Ostfeld and I. A. Cohen, J. Chem. Educ., *49,* 829 (1972).

EXERCISES

1. Consider the following system of energy levels, which could arise from zero-field splitting of an $S = 1$ state:

$$E(|0\rangle) = -\frac{2}{3}D - \frac{g^2\beta^2H^2}{3D}$$

$$E(|-1\rangle) = \frac{1}{3}D - g\beta H + \frac{g^2\beta^2H^2}{6D}$$

$$E(|1\rangle) = \frac{1}{3}D + g\beta H + \frac{g^2\beta^2H^2}{6D}$$

Here D is the zero-field splitting parameter and all other symbols have their usual meanings. Determine the molar paramagnetic susceptibility of this system.

2. The p^1 configuration of a free ion has a ^2P ground state. Consider the ion in a magnetic field directed along the z-axis.

   a. What is the degeneracy of this state?

   b. Write the wave functions for this state in Dirac notation, *i.e.*, $|M_L, M_S\rangle$.

   c. The major perturbation on this state is spin-orbit coupling:

   $$\hat{H} = \lambda\vec{\mathbf{L}}\cdot\vec{\mathbf{S}}$$

   Consider the major contribution in a strong z field leading to:

   $$\hat{H} \cong \lambda\hat{L}_z\hat{S}_z$$

   Does this give diagonal elements, off-diagonal elements, or both, in the basis set of part b? Evaluate the energies of these wave functions using $\hat{H}$.

   d. The Zeeman Hamiltonian is $\hat{H} = \beta\vec{\mathbf{H}}\cdot(\vec{\mathbf{L}} + 2\vec{\mathbf{S}}) = \beta H_z(\hat{L}_z + 2\hat{S}_z)$. Evaluate the energies of the wave functions of part b using this Hamiltonian.

   e. Use the energies obtained in parts c and d, and the Van Vleck equation, to calculate χ.

3. The following represents the approach taken to the magnetic susceptibility data of a complex of the form shown below.

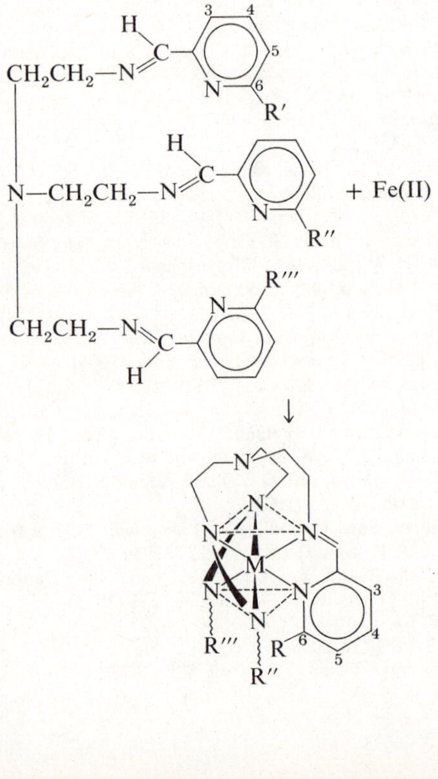

Consider octahedral Fe(II). In many systems, the low spin $^1A_{1g}$ state is less than 1000 cm^{-1} below the high spin $^5T_{2g}$ state, so that both are appreciably populated at room temperature. The ^1A state is non-degenerate and the ^5T is 15-fold degenerate (why?). Mathematically, a T state in an octahedral field is equivalent to a P state in a free ion; *i.e.*, it may be considered to have $L = 1$ and $M_L = -1, 0, +1$. Consider the ^1A state to be unshifted by all perturbations.

a. Write the 15 basis functions for the ^5T state in Dirac notation, using T—P equivalence. Hint: The wave function having "M_L" $= 1$ and $M_S = -2$ is denoted $|1, -2\rangle$. In this formulation, "M_L" and M_S are often referred to as "good quantum numbers."

b. Now, let the complex undergo a trigonal distortion, *i.e.*, one for which the octahedron's three-fold axis is retained. The effect of this distortion is described phenomenologically by the Hamiltonian

$$\hat{H} = \delta\left(\frac{2}{3} - \hat{L}_z^{\,2}\right)$$

where δ is the trigonal distortion or zero-field splitting parameter and $\hat{L}_z$ is the z-component of the equivalent orbital angular momentum. As is often the case, chemical factors influencing the magnitude of δ are poorly understood, and values obtained in experiments such as this may help elucidate the factors involved. The Zeeman Hamiltonian for a magnetic field oriented along z is:

$$\hat{H} = \beta H_z(\hat{L}_z + g_e\hat{S}_z)$$

Here we are considering only the z direction (parallel to the three-fold axis), so our final expression will be for $\chi_{\parallel}$.

The following illustrate the use of the Hamiltonian and wave functions:

$$\left(\frac{2}{3} - \hat{L}_z^{\,2}\right)|1, -2\rangle = \left(\frac{2}{3} - M_L^{\,2}\right)|1, -2\rangle = \left(\frac{2}{3} - 1\right)|1, -2\rangle = -\frac{1}{3}|1, -2\rangle$$

$$\hat{S}_z|1, -2\rangle = M_S|1, -2\rangle = -2|1, -2\rangle$$

Apply the two terms of the Hamiltonian discussed above to the ^5T wave functions derived in part a. Show that all off-diagonal elements of the 15×15 matrix must be zero. (This means that all $E^{(2)}$ terms of the Van Vleck equation will be zero.) Hence, determine the energies of the 15 wave functions. Confirm that, as the total Hamiltonian has been constructed, the center of gravity is maintained; *i.e.*, that $\hat{H} = \delta\hat{L}_z^{\,2}$, though giving the correct splitting pattern, would not maintain a center of gravity.

c. Let the difference in energy between the ^1A and ^5T states be parameterized by E. The energy level diagram should look like this:

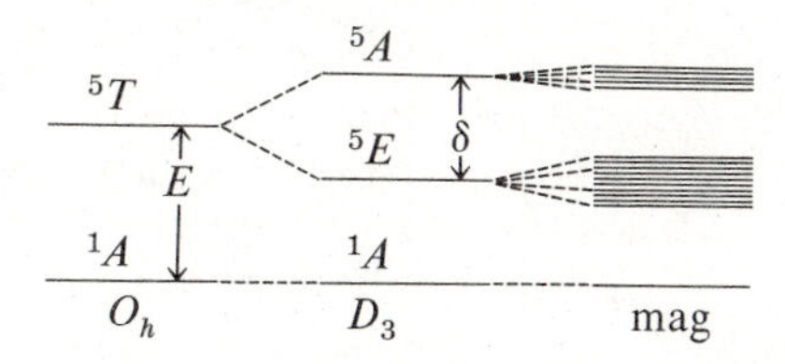

Use the Van Vleck equation to determine $\chi_{\parallel}$ as a function of E, δ, and T.

In the actual experiment, spin-orbit coupling was included, introducing off-diagonal elements that complicated the analysis. The experimental χ *vs.* T relation was computer-fitted to the theoretical expression to yield best values for λ, E, and δ. It is interesting to note that a poor fit was obtained unless E was allowed to vary with temperature.

4. The 300° K molar magnetic susceptibility for a solid sample of

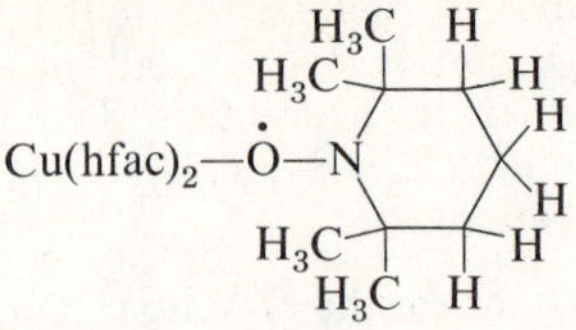

$Cu(hfac)_2$—Ȯ—N (2,2,6,6-tetramethylpiperidine ring: H_3C, H_3C, H, H, H, H, H_3C, H_3C, H, H)

was determined to be $\chi = -186 \times 10^{-6}$ cm^3 mole^{-1}. Calculate the molar paramagnetic susceptibility by correcting for the diamagnetism of the complex. What is μ_{eff}? How can you explain this μ_{eff}?

hfac$^-$ = O, ⊖, O, C, C, CF_3, CH, CF_3

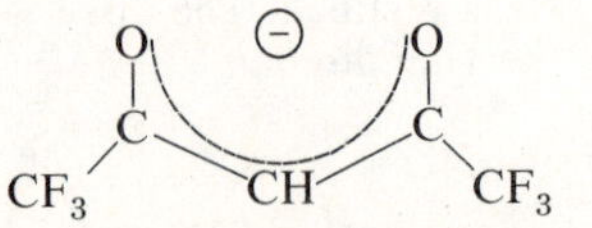

5. a. $Co(N_2H_4)_2Cl_2$ has a magnetic moment of 3.9 BM. Is hydrazine bidentate? Propose a structure.
   b. How could electronic spectroscopy be employed to support the conclusion in part a?

6. In which of the following tetrahedral complexes would you expect contributions from spin-orbit coupling? V^{3+}, Cr^{3+}, Cu^{2+}, Co^{2+}, Fe^{2+}, Mn^{2+}.

7. In which of the following low spin square planar complexes would you expect orbital contributions? d^2, d^3, d^4, d^5, d^6.

8. Why is $Fe_2(CO)_9$ (with three bridging and six terminal carbonyls) diamagnetic?

9. Explain why mixing of a D_{4h} component in with a T_d ground state lowers the magnetic moment in nickel(II) complexes.

10. What is the expected magnetic moment for Er^{3+}?

11. In the figure below, effective magnetic moment *vs.* temperature curves are shown for two similar tris-bidentate Fe(III) compounds:

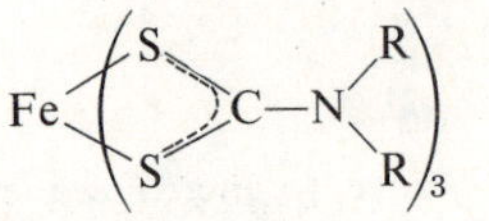

$Fe\left(S_2C{-}NR_2\right)_3$

A: R = *n*-propyl
B: R = isopropyl

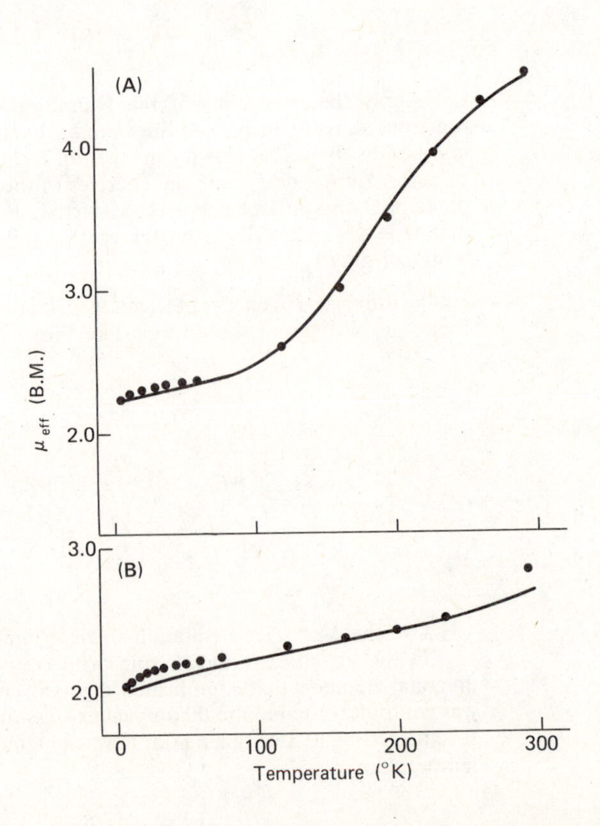

a. Provide an explanation in terms of the electronic structure of an Fe(III) complex that accounts for the changes in μ_{eff} over the given temperature range for each of the two complexes. In other words, why are the two curves so different while the two complexes *appear* to be so similar?

b. In the above plots the points represent the experimental data and the lines represent least-squares fits to theoretical equations. With the use of an energy level diagram, describe what parameters might be used in such a theoretical treatment.

12 NUCLEAR MAGNETIC RESONANCE SPECTRA OF PARAMAGNETIC TRANSITION METAL ION COMPLEXES

12–1 INTRODUCTION

In the early days of nmr, there was a widespread belief that one could not detect the nmr spectrum of a paramagnetic complex because the electron spin moment was so large that it would cause rapid relaxation of the nuclear excited state, leading to a short T_1 and a broad line. This is indeed found to be the case for certain complexes [for example, those of Mn(II)], but it is not the case for many other paramagnetic complexes. For instance, in Fig. 12–1, a simulated nmr spectrum[1] of the paramagnetic complex $Ni(CH_3NH_2)_6^{2+}$ is presented and compared to that of CH_3NH_2.

These spectra raise several questions that need to be answered:

(1) Why do we see a spectrum for the paramagnetic complex?

(2) Why are the observed shifts from TMS in the complex so large relative to those of the uncoordinated ligand? The normal range of proton shifts for most organic compounds is 10 ppm, and the shifts in the complex are well outside this range.

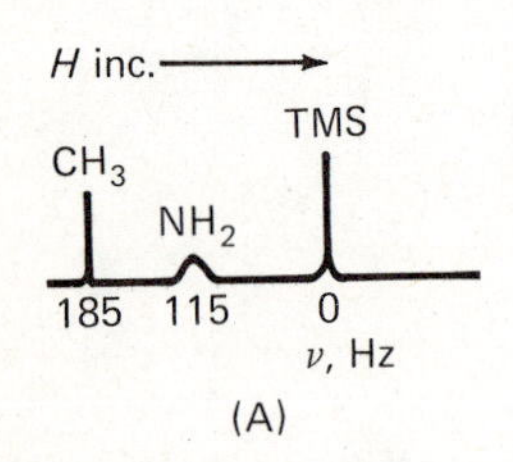

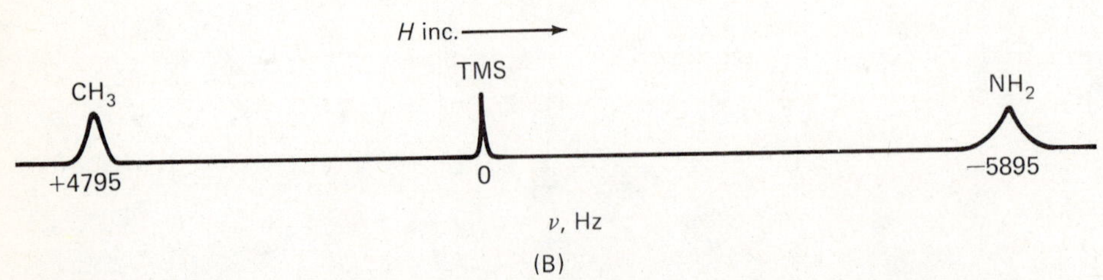

FIGURE 12–1. Proton nmr spectra (simulated) of solutions of (A) CH_3NH_2 and (B) $Ni(CH_3NH_2)_6^{2+}$ at 60 MHz. Note that the relative scales in (A) and (B) differ.

(3) Why does the NH proton resonance shift upfield while that of the CH_3 protons shifts downfield?

We shall answer all of these questions in the course of logically developing this topic.

12–2 RELAXATION PROCESSES

The answer to the first question presented above lies in a more complete analysis of the relaxation processes for paramagnetic systems. There are two allowed spin states with quantum numbers $m_s = \pm\frac{1}{2}$, associated with an electron in a magnetic field. When this electron is delocalized onto a proton, we would expect to see two extremely broad nmr peaks arising from the coupling of the nuclear spin to the two spin states of the electron. A large separation of peaks, corresponding to the value of the hyperfine coupling constant a discussed in the esr section, would result. The peak arising from those electrons whose magnetic moments are aligned with the field would be very slightly more intense, since this lower energy state has a larger population. The resonance line would be so broad that it could not be detected because motion of a molecule with a large paramagnetic moment would lead to a very intense fluctuating magnetic field and a short T_1. In terms of spectral density considerations (Chapter 7), the range of frequencies would be large and the frequency components would be very intense, leading to a short T_1. The spectrum of our octahedral Ni(II) complex remains unexplained because what we actually observe for a given kind of proton, i, is a single peak shifted much less than a_i (see Fig. 12–1). *For nmr, we shall use the symbol A_i for the electron spin-nuclear spin coupling constant.*

The electron in the above example is changing its spin state with time. If the change is very fast on the nmr time scale, the net effect will be to average toward zero the oscillating field associated with the electron as felt at the proton. The result is a decrease in the efficiency of the electron in relaxing the proton nucleus. Very rapid, intermolecular electron exchange or ligand exchange would have the same effect because it would place electrons with different m_s values at the proton nucleus. This is very similar to the averaging phenomena previously discussed in connection with nuclear spin-spin splitting. The first effect is like decoupling the proton in the nuclear spin-spin system, and the latter is like exchanging the O—H proton of ethanol.

Electrons relaxing with exactly the same frequency as A_i are most efficient at broadening the nmr spectrum. A_i is typically about 10^8 Hz or one cycle per 10^{-8} sec. The rate of molecular tumbling in solution is characterized by a correlation time τ_c, which is on the order of 10^{-11} sec. The lifetime of an electron spin state is related to τ_e, the electron relaxation time, or T_{1e}, the ligand exchange time (whichever is faster); τ_e is a measure of the time taken for energy to be transferred to other degrees of freedom, and T_{1e} is characteristic of the time spent by the ligand at a particular metal site. If $1/T_{1e} \gg A_i$ or $1/\tau_e \gg A_i$, the nucleus will see only a population-weighted average of the two electron spin states; the nucleus will be coupled only weakly to the electron spin system and will thus be much less efficiently relaxed under these conditions. Thus, the lifetime of the electron spin state necessary to permit observation of nmr spectra ranges from 10^{-9} to 10^{-13} sec. We shall reemphasize an important point. *The rapid nuclear relaxation causes a broad nmr line, but the rapid electron relaxation decreases the efficiency of the nuclear relaxation mechanism, lengthening T_1 for the proton nucleus and causing the line to sharpen.* The observed spectra are not as sharp as those of diamagnetic molecules. Very high rf power is employed in recording the spectra because the shorter T_1's mean that there is little danger of saturating the resonances.

Complexes of Mn(II) are examples of systems with slowly relaxing electrons ($1/\tau_e \approx A_i$). If one sees an nmr resonance at all, a single very broad line is obtained. However, in the esr experiment, where one looks at electron spin transitions, this slow relaxation guarantees a long-lived excited state, and thus a narrow line width. Thus, esr peaks are sharp for these systems and broad for faster-relaxing systems. It is, in fact, unusual that one can perform an nmr and esr experiment on the same compound at the same temperature. The techniques are complementary.

Some general observations on the electron spin lifetimes of various first row transition metal ion systems will indicate when to expect sharp nmr or esr absorption lines. Complexes with triply degenerate (T) ground states often have short electron spin lifetimes, leading to "sharp" nmr and broad esr spectra. These include octahedral complexes having the following configurations: d^1, d^2, low spin d^4, low spin d^5, high spin d^6, and high spin d^7. When large distortions from octahedral symmetry exist in the complex, the T state is split and the nmr spectrum often becomes broader and the esr sharper. Particularly broad nmr spectra are observed for low symmetries in the d^1 configuration where, with only one electron present, there is no zero-field splitting to cause relaxation. Sharp lines are observed in octahedral Ni^{2+} complexes (having an A ground state) because of zero-field splitting.

Tetrahedral complexes with T ground states also give "sharp" nmr spectra. Tetrahedral Co^{2+} systems are typical. Systems having particularly broad nmr and sharp esr lines include those complexes with octahedral d^3, high spin d^5, and d^9 configurations.

12–3 AVERAGE ELECTRON SPIN POLARIZATION

We have not yet answered all of the questions raised in the introduction. Remaining is the reason for the large observed shifts in paramagnetic systems. This shift difference between the paramagnetic complex and an analogous diamagnetic complex is referred to as *the isotropic shift*. The isotropic shift can be understood by returning to a consideration of the fact that when a complex containing an unpaired electron is placed in a magnetic field, the two m_s spin states are not equally populated. If we consider what the chemical shift would be for a complex with $m_s = +\frac{1}{2}$ and for one with $m_s = -\frac{1}{2}$, we can appreciate that a mole fraction weighted average of these two shifts would be observed when the spins are rapidly exchanging or relaxing. We refer to the weighted average value of m_s for the system in the magnetic field as the *average electron spin polarization,* $\bar{S}_q$. Recall that in the previous chapter this population difference was employed to evaluate the paramagnetic susceptibility. We shall follow a very similar approach here to evaluate the effect of this population difference on the nmr shift. All of the factors influencing the paramagnetism will influence our observed shift, for we can view the shift as resulting from the additional field that the electronic moments produce in the vicinity of the nucleus.[2,3] The magnitude of this additional field at the nucleus will depend on the type of electron-nuclear coupling (scalar or dipolar), which in turn will depend upon the bonding and geometry in the molecule. The scalar and dipolar mechanisms responsible for transmitting the magnetic effects of the electrons to nuclei in the molecule will be discussed in detail in later sections. A description of the origin and the calculation of the average electron spin polarization, $\bar{S}_q$, is needed because it can be related to the observed shift.

As we saw in the previous chapter,

$$\vec{\mu} = -g\beta\vec{\mathbf{S}} \tag{12-1}$$

and the Hamiltonian for the interaction of the electron magnetic moment with the field was given by:

$$\hat{H} = -\vec{\mu} \cdot \vec{\mathbf{H}} = g\beta \vec{\mathbf{S}} \cdot \vec{\mathbf{H}} \tag{12-2}$$

For an $S = \frac{1}{2}$ system, applying Boltzmann statistics, we obtained

$$\frac{N_\alpha}{N_\beta} = \exp(-\Delta E/kT) = \exp(-g\beta H/kT) \tag{12-3}$$

Solving for the total number of electron spins in the β state, we get

$$N_\beta = N_\alpha \exp\left(\frac{g\beta H}{kT}\right)$$

Using the Curie Law approximation and the fact that $(g\beta H/kT) \ll 1$, we showed in the last chapter that

$$\exp(\pm g\beta H/kT) \cong 1 \pm (g\beta H/kT) \tag{12-4}$$

Combining equation (12–4) with the expression for N_β, we have

$$N_\beta \cong N_\alpha\left(1 + \frac{g\beta H}{kT}\right)$$
$$= (N - N_\beta)\left(1 + \frac{g\beta H}{kT}\right)$$

where $N = N_\alpha + N_\beta$. This expression for N_β can be rearranged:

$$N_\beta = N\left[\frac{1 + \dfrac{g\beta H}{kT}}{2 + \dfrac{g\beta H}{kT}}\right] = \frac{N}{2}\left[\frac{1 + \dfrac{g\beta H}{kT}}{1 + \dfrac{g\beta H}{2kT}}\right]$$

The numerator in parentheses is of the form $1 + nx$, where $n = 2$ and $x = \dfrac{g\beta H}{2kT}$. When $x \ll 1$, we may use the approximation $(1 + x)^n \cong 1 + nx$ to obtain

$$N_\beta \cong \frac{N}{2}\frac{\left(1 + \dfrac{g\beta H}{2kT}\right)^2}{\left(1 + \dfrac{g\beta H}{2kT}\right)} = \frac{N}{2}\left(1 + \frac{g\beta H}{2kT}\right)$$

Similarly, we can show that $N_\alpha \cong \dfrac{N}{2}\left(1 - \dfrac{g\beta H}{2kT}\right)$.

The difference between N_β and N_α gives the excess unpaired spin along the field direction, and may be thought of as being related to the resultant vector of the up and down spin moment vectors. It represents the net magnetic effect induced in the system's unpaired electrons by the external field. The magnitude, the average value of

m_s, is given by the average electron spin polarization, $\bar{S}_q$, the quantity of interest to us in the nmr experiment. In our example, we take a population-weighted average of the $+\frac{1}{2}$ and $-\frac{1}{2}$ states as:

$$\bar{S}_q = \frac{\left(+\frac{1}{2}\right)N_\alpha + \left(-\frac{1}{2}\right)N_\beta}{N} = \frac{g\beta H}{4kT} \tag{12-5}$$

There is a direct relationship between $\bar{S}_q$ and the bulk magnetic moment M with which we were concerned in the previous chapter. We have found that the number of excess spins in the lowest level is twice the average electron spin polarization $\bar{S}_q$ times the total number of electrons N. This number of excess spins must be multiplied by the magnetic moment of one spin in the appropriate state resolved along the field direction, $(\frac{1}{2})g\beta$, in order to find the average magnetic moment; *i.e.*,

$$M = \left(\frac{1}{2}\right) g\beta(2N\bar{S}_q) = Ng^2\beta^2 H/4kT \tag{12-6}$$

This is the Curie law expression we described in the last chapter and arrived at by a population weighting of the various levels.

For the general case of more than one spin, the average value of $\bar{S}_q$ is given by

$$\bar{S}_q = \frac{\sum_{S_q=-S}^{S} S \exp(-E_{S_q}/kT)}{\sum_{S_q=-S}^{S} \exp(-E_{S_q}/kT)} \tag{12-7}$$

Here E_{S_q} is the energy of the state with quantum number S. Equation (12–7) is consistent with the earlier example of one-electron spin ($S_q = \pm\frac{1}{2}$). We see in equation (12–7) that a state with a particular value of the spin magnetic quantum number, S_q, is weighted according to its equilibrium population. The weighted values are summed over all the energy levels and divided by the total number of levels to give the average electron spin polarization. If a Curie type of approximation ($\Delta E \ll kT$) is applied to equation (12–7), the exponential may be expanded as a power series, and truncated after the second term. After some algebraic manipulation equation (12–8) results:

$$\bar{S}_q = \frac{g_{av}\beta HS(S+1)}{3kT} \tag{12-8}$$

where g_{av} is the average g value. Equation (12–7) is rigorous for a Zeeman multiplet where E_{S_q} is known. Equation (12–8) is satisfactory only when $E_{S_q} = g\beta HS_q$ and $\Delta E_{S_q} \ll kT$. In the chapter on magnetism, we discussed those effects [orbital angular momentum, zero-field splitting, second-order Zeeman effects (TIP)] that caused non-Curie Law behavior. Now that we have an expression for $\bar{S}_q$, we shall proceed to evaluate the quantitative expression for the nmr contact shift assuming equation (12–8). When the appropriate conditions are not met, other expressions can be derived employing other expressions for $\bar{S}_q$.

12–4 SCALAR OR CONTACT SHIFT IN SYSTEMS WITH AN ISOTROPIC g-TENSOR

The scalar shift or "contact shift" is described by the Hamiltonian operator:

$$\hat{H} = \frac{8\pi}{3} g_e g_N \beta \beta_N \vec{\mathbf{I}} \cdot \vec{\mathbf{S}} \delta(r) \tag{12-9}$$

This electron-nuclear spin interaction was discussed in Chapter 9 in the section on Fermi contact coupling, and all the symbols were described there. With this effect, we are concerned with the influence of unpaired spin density, which is delocalized directly onto the nmr nucleus. Substituting the average spin polarization into equation (12–9), we obtain

$$\hat{H} = \frac{8\pi}{3} g_e g_N \beta \beta_N \left[\frac{g_{av} \beta \vec{\mathbf{H}} S(S+1)}{3kT}\right] \cdot \vec{\mathbf{I}} \delta(r) \tag{12-10}$$

Here g_{av} is employed to adjust for any orbital effects that would make S a poor quantum number. The Kronecker δ function requires that we consider this effect only at the nucleus. Comparing equation (12–10) with that from nmr for a proton in a diamagnetic environment:

$$\hat{H} = -g_N \beta_N \vec{\mathbf{H}} \cdot \vec{\mathbf{I}} \tag{12-11}$$

we find that the equations are similar if $\vec{\mathbf{H}}$ of equation (12–11) is considered as an effective magnetic field, $\vec{\mathbf{H}}_{eff}$, arising from the electron density at that nucleus, $|\psi(0)|^2$. Defining integrals of the form $\int \psi_i \delta(r) \psi_i d\tau$ as $\psi(0)^2$ (the value of the wave function at $r = 0$, squared), we write:

$$\vec{\mathbf{H}}_{eff} = \frac{-8\pi}{3} g_e \beta g_{av} \beta \vec{\mathbf{H}} \frac{S(S+1)}{3kT} |\psi(0)|^2 \tag{12-12}$$

The total field experienced by the nucleus in the nmr experiment is the sum of $\vec{\mathbf{H}}_{eff}$ and $\vec{\mathbf{H}}_0$ from the magnet. The shift from the paramagnetism, which we sought to derive when we started this section, arises from $\vec{\mathbf{H}}_{eff}$. It is referred to as the scalar shift and is given by the expression for $\vec{\mathbf{H}}_{eff}$ in equation (12–12) in units of gauss. For more than one electron, a normalization term $1/2S$ arising from the Pauli exclusion principle must also be included in the final expression, to yield for the shift (assuming $g_e = g_{av}$):

$$\Delta\vec{\mathbf{H}} = \frac{-8\pi}{3} \frac{1}{2S} g_{av}{}^2 \beta^2 \vec{\mathbf{H}} \frac{S(S+1)}{3kT} |\psi(0)|^2 \tag{12-13}$$

For a variety of reasons, data from contact shift experiments are often reported in terms of the electron spin-nuclear spin coupling constant, A. The most important of these reasons is that the shift in the resonance will vary with temperature but A will not (within the limits of the assumptions discussed in connection with $\bar{S}_q$). The coupling Hamiltonian can be written as

$$\hat{H} = A \vec{\mathbf{I}} \cdot \vec{\mathbf{S}} \tag{12-14}$$

where A, the isotropic hyperfine coupling constant, is the same as a discussed in the chapter in esr, except that the capital letter is used to indicate the possibility of more than one unpaired electron in the molecule or ion. We define A as

$$A = \frac{8\pi}{3} g_e g_N \beta \beta_N |\psi(0)|^2 \tag{12-15}$$

As discussed previously, a normalization factor of $1/2S$ must be added when there is more than one unpaired electron spin, and g_e is replaced by g_{av} for the complex being studied. Then

$$A = \frac{8\pi}{6S} g_{av} \beta g_N \beta_N |\psi(0)|^2 \tag{12-16}$$

which is the expression for A in cgs units (ergs). The following conversions may be used to relate the different units:

$$A\,(\text{ergs}) = hA\,(\text{Hz}) \tag{12-17}$$

$$A\,(\text{ergs}) = g_{av}\beta A\,(\text{gauss}) \tag{12-18}$$

The equation for the contact shift is usually not written as equation (12–13), but as a function of the hyperfine coupling constant A. If we substitute equation (12–15) in (12–13), we obtain the following expression for the isotropic shift:

$$\frac{\Delta H}{H} = \frac{\Delta\nu}{\nu} = -\frac{A_i g_{av} \beta S(S+1)}{g_N \beta_N 3kT} \tag{12-19}$$

We now can see that a very small A value can give rise to a huge isotropic shift. The reader is encouraged to convert the isotropic frequency shifts in Fig. 12–1 (at room temperature) to the equivalent field shifts in gauss. The equality $\Delta H/H = \Delta\nu/\nu$, where ν is the fixed probe frequency, follows directly from the fact that $h\nu = g\beta H$ for nuclear spins. If one measures the temperature dependence of $\Delta\nu$, a plot of $\Delta\nu$ *vs.* $1/T$ should produce a straight line with a slope proportional to A_i for systems exhibiting Curie Law behavior. For systems with an orbitally degenerate ground state, such as octahedral nickel(II) and tetrahedral cobalt(II) complexes, the application of equations (12–7) and (12–8) is valid.

For conditions in which the expressions given for $\overline{S}_q$ are not valid, equations (12–13) and (12–19) will likewise be inapplicable. For these systems, we can write

$$\frac{\Delta H}{H} = \frac{\Delta\nu}{\nu} = \frac{A_i}{\hbar g_N \beta_N H} \langle \hat{S}_z \rangle \tag{12-20}$$

and the problem becomes one of properly expressing $\langle \hat{S}_z \rangle$.

12–5 THE PSEUDOCONTACT SHIFT

For molecules in which the g-tensors are not isotropic, it is convenient to consider separately those in which there are appreciable contributions from second-order Zeeman effects and those in which these effects are minor. We shall first consider the latter case. The temperature dependence of the isotropic shift can be expressed by equation (12–19), with an average g-value to incorporate any orbital angular momentum. When this is done, the resulting A value from a $\Delta\nu$ *vs.* $1/T$ plot has contributions to it from other than the scalar or contact term; *i.e.*, equation (12–15) is no longer valid. The observed isotropic shift $\Delta\nu$ is given by

$$\Delta\nu_{A(\text{isotropic})} = \Delta\nu_{(\text{scalar})} + \Delta\nu_{(\text{pseudocontact})} \tag{12-21}$$

We shall now proceed to discuss the source of this *pseudocontact* or *dipolar contribution.*

Whenever there is an unpaired electron in a molecule, there is a dipolar effect transmitted through space and felt by the nucleus of interest. With an isotropic g-value, this dipolar effect is averaged to zero by rapid rotation of the molecule in the field. This phenomenon was discussed in the chapter on esr, where it was shown that this same effect produces the dipolar contribution to the hyperfine coupling, which was shown to average to zero in solution. With anisotropic g-tensors, the magnitude of the dipolar contribution to the magnetic field at the nucleus of interest from unpaired electron density on the metal will depend on the orientation of the molecule with respect to the field. Since g has different values for different orientations, this through-space contribution need not be averaged to zero by rapid rotation of the molecule. Thus, the same effects that give rise to anisotropy in g also give rise to the pseudocontact contribution. The through-space nature of the pseudocontact effect is comparable to the neighbor anisotropic contribution discussed in Chapter 8, which was seen to be dependent upon differences in χ_{DIA} for different orientations. The same is true for χ_{PARA}. When equation (12–8) applies, we are considering systems in which $\Delta\chi$ parallels Δg.[2] The Hamiltonian term for the pseudocontact contribution is similar to the dipolar coupling Hamiltonian discussed in Chapter 9.

Equations have been derived for evaluating the contribution to the isotropic shift from this dipolar effect for a variety of systems.[2,3] When $T_{1e} \ll \tau_c$ and $(1/\tau_c) \ll (E_{\text{MAX}} - E_{\text{MIN}})/h$ (where T_{1e} is the electron spin lifetime, τ_c is the correlation time for tumbling, and E_{MAX} and E_{MIN} are the maximum and minimum values for the Zeeman energy, as a function of orientation; *i.e.*, $\Delta E = -\mu_{\text{max}}H - \mu_{\text{min}}H$, where μ is the effective magnetic moment and H is the field strength), we can write the general equation for the pseudocontact contribution to the shift, $\Delta\nu_i(\text{pc})$, at the ith nucleus as:

$$\frac{\Delta\nu_i(\text{pc})}{\nu} = -\frac{1}{3N}\left[\chi_z - \frac{1}{2}(\chi_x + \chi_y)\right]\left\langle\frac{3\cos^2\theta_i - 1}{r_i^{\,3}}\right\rangle_{\text{av}} - \frac{1}{2N}(\chi_x - \chi_y)\left\langle\frac{\sin^2\theta_i\cos 2\varphi_i}{r_i^{\,3}}\right\rangle_{\text{av}} \tag{12-22}$$

Here, N is Avagadro's number; χ_x, χ_y, and χ_z are the susceptibility components; and the angles θ and φ are defined in Fig. 12–2. In this figure, N is the nucleus being investigated by nmr, M is the metal center, and r is the distance from the metal to N.

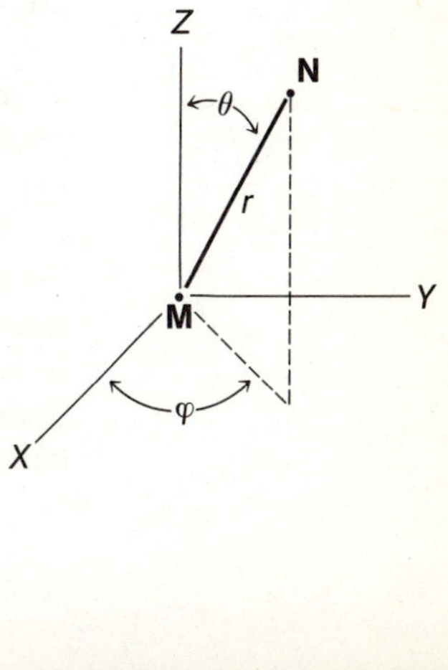

Z N θ r M Y φ X

FIGURE 12–2. Definition of the quantities in equation (12–22).

This equation is valid for all systems because the actual susceptibilities are employed. For cases where equation (12–8) applies, these equations can be rewritten in terms of g-values. They are more conveniently used in terms of g because anisotropy in the g-tensor is more easily obtained than that in χ.

Rewriting equation (12–22) in terms of Δg for an axially symmetric system with the same lifetime requirements assumed above, we obtain the following equation for the pseudocontact contribution:

$$\frac{\Delta\nu_i(\text{pc})}{\nu} = \frac{\beta^2 S'(S'+1)}{9kT}\left(\frac{1-3\cos^2\theta_i}{r_i^{\,3}}\right)(g_\parallel^{\,2} - g_\perp^{\,2}) \tag{12-23}$$

where S' is an effective spin that is experimentally found to describe the system. These terms are defined for a molecule with local axial symmetry in Fig. 12–3. In

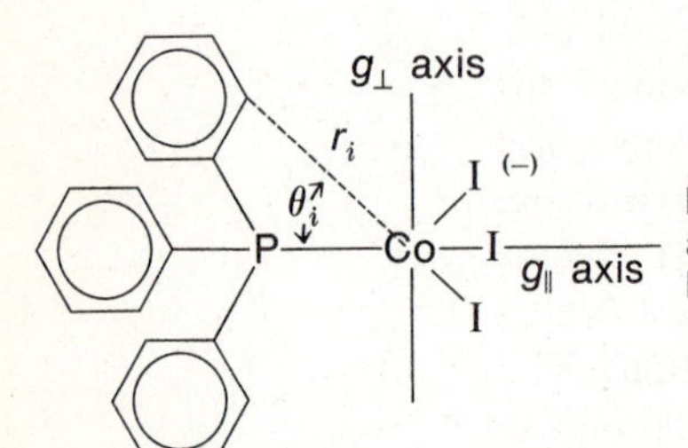

FIGURE 12–3. Definition of terms in equation (12–23). The parallel and perpendicular axes are shown. The radius vector from the metal center to the nucleus being investigated is labeled r_i, and the angle that it makes with the highest fold rotation axis is θ_i.

view of the greater simplicity of equation (12–23) than of equation (12–22), many investigators have studied axially symmetric systems. When g-values do not provide an appropriate description of the susceptibility [*i.e.*, $\chi \neq N g_{av}^{\,2}\beta^2 S(S+1)/kT$], equations (12–19) and (12–23) are not valid. It has been shown that equations based on g-values are not appropriate to the description of the trigonally distorted six-coordinate cobalt(II) pyrazoylborate complex. The ^4E_g ground state is split into four doublets (*vide infra,* Chapter 13), each of which has its own g-tensor. Since some of these levels are populated at room temperature, but not at liquid helium temperature where one must work in order to observe esr, the average g-tensor provides a poor description of the susceptibility at room temperature. The reader is referred to the literature[4-6] for the equations to deal with systems for which equation (12–23) does not apply.

12–6 QUALITATIVE INTERPRETATION OF THE NMR SPECTRA OF PARAMAGNETIC MOLECULES

Factoring the Contact and Dipolar Contributions

The isotropic hyperfine coupling constant, A, obtained in the nmr experiment arises from the same effect that gives rise to the hyperfine coupling constant, a, obtained from the esr spectrum. If the same system could be studied by the two methods, the isotropic a or A values obtained would be identical. The nmr method is much more sensitive and large proton shifts (*e.g.,* 50 Hz) would give rise to proton hyperfine coupling constants that are not resolved in the esr spectra. Furthermore, in the nmr experiment, we obtain the sign of the coupling constant from the direction of the shift, while the sign has no effect on the observed esr spectrum. Since the effect is the same, our interpretation of A will parallel that of a.

Before the reasons for the isotropic shift of a nucleus in the nmr spectrum can be understood, the shift for a particular nucleus must be separated into the scalar (contact) and dipolar (pseudocontact) contributions [see equation (12–21)]. Systems can be selected in which the dipolar contribution will be negligible compared to the contact contribution. For example, as can be seen from equation (12–23), the pseudocontact term is zero when $g_{\parallel} = g_{\perp}$. When the metal atom's environment has *rigorous* cubic symmetry (*e.g.*, tetrahedral or octahedral), only the contact contribution will be observed. The word rigorous is italicized because subtle effects can cause a deviation from cubic symmetry and may lead to a substantial pseudocontact contribution (see the section on ion pairing that follows).

Three approaches traditionally have been taken on systems in which both contact and pseudocontact contributions exist. If the geometry of the molecule is known from a single crystal x-ray diffraction study, if the structure is the same in the solid state as in solution, and if the anisotropy in the susceptibility* is also known, the pseudocontact contribution can be calculated.[7] With a calculated value for the pseudocontact contribution, one can solve for the contact contribution in equation (12–21), using the measured isotropic shift. The x-ray structure determination and single crystal susceptibility measurements are time consuming and expensive. The equivalence of solution and solid structures can be difficult to prove and is often assumed. Gross changes in structure can be detected by spectroscopic techniques in favorable cases.

A second approach involves determining the anisotropy in g for those systems where the electronic structure is such that the g-value based equations [*e.g.*, equation (12–23)] apply (*i.e.*, $g^2 \propto \chi$). Unfortunately, the electron spin lifetimes leading to a well resolved esr spectrum result in a poorly resolved nmr spectrum and *vice versa*. An iron(III) system has been reported[8] for which the esr and nmr have both been observed. Measured susceptibilities compared to those calculated from g-values and a linear $\Delta\nu$ *vs.* $1/T$ plot suggest that g-values are appropriate for an estimation of the pseudocontact term in this system.

A third approach, called the ratio method, has been proposed[9] for separating the isotropic shift into scalar and dipolar contributions in octahedral complexes of cobalt(II), by assuming that the ratios of the scalar shifts of analogous protons in the cobalt(II) complex will be the same as those for the analogous nickel(II) complex. Knowing both these ratios and the ratios of the geometric factors for several protons enables one to evaluate the anisotropy term in, for example, equation (12–23) and to then calculate $\Delta\nu$ (dipolar) for each proton. In general it is incorrect to assume that octahedral nickel(II) and cobalt(II) will have similar delocalization patterns.[10] High-spin cobalt(II) has unpaired electrons in both the t_{2g} and e_g sets which can overlap directly with both the pi and sigma orbitals of a ligand, directly delocalizing electrons into both. The mixing of the metal and ligand orbitals in forming the molecular orbitals of an octahedral complex can be understood by looking at the simplified case of the mixing involving one ligand and a set of metal ion orbitals experiencing an octahedral ligand field. In an octahedral complex we would have to use the O_h symmetry-adapted linear combinations of ligand σ and π molecular orbitals. This simpler case has been selected to simplify the discussion without loss of generality. Fig. 12–4 illustrates the mixing of an octahedral field nickel(II) ion with ligand σ, π, and π^* orbitals. A high spin cobalt(II) complex would have one less electron in the mainly-metal t_{2g} set of molecular orbitals. Direct delocalization of the unpaired electrons into both the σ and π^* orbitals of pyridine is possible in cobalt(II), but not in nickel(II). The shifts in some bipyridyl complexes illustrate[10] the breakdown of the ratio method. If a ligand is employed that can interact only with the e_g metal orbitals, the nickel and cobalt mechanisms will be similar. Furthermore, if unpaired electrons are delocalized directly into a ligand pi-system in a cobalt(II) complex, but if there are protons in this ligand to whose contact shift π-delocalization

*See references in the chapter on magnetism.

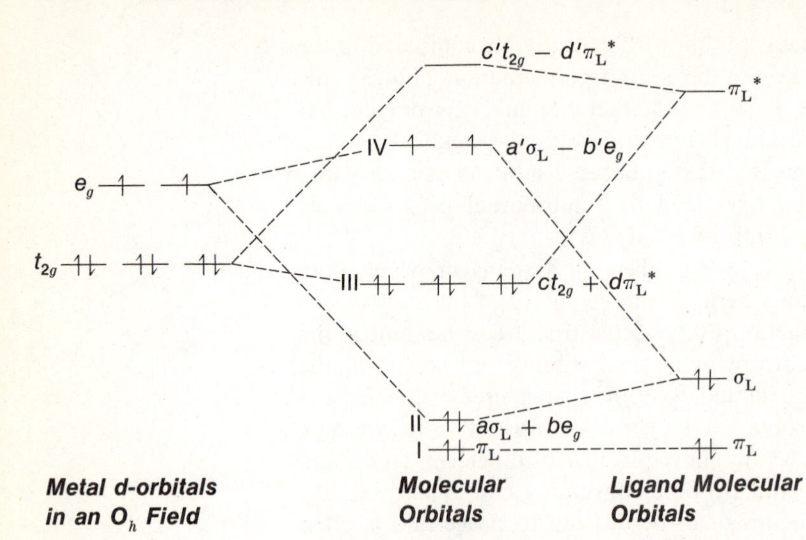

FIGURE 12–4. The mixing of O_h metal and ligand orbitals in forming a complex NiL^{2+}. (L could, for example, be pyridine. An octahedral complex has a symmetry that maintains the separability of the σ and π systems in L. The fragment M-L is represented for simplicity. Note that $a \gg b$, $c \gg d$, and $b' \gg a'$, while $d' \gg c'$.) For simplicity, the symmetry allowed mixing of t_{2g} and π is not shown.

makes no contribution, the ratio at these protons again would be similar[11] for cobalt(II) and nickel(II). On forming tetrahedral cobalt(II) and tetrahedral nickel(II) complexes, unpaired electrons can be delocalized in both π and σ ligand orbitals, and these complexes might also show similar ratios. Criticism of this method[10] is not based on the argument that one can never find systems in which it will work. The shortcoming is the need to be able to predict in advance, with confidence, which ligand m.o.'s are involved in the delocalization when a complex is formed, or which protons are dominated by certain mechanisms. Since, in most instances, the systems in which the pseudocontact contribution has to be factored out are those in which one is using the experiment to learn how the spin is delocalized, the method is not very practical.

The difficulty one has in predicting which mechanism dominates is not readily obvious by examining the experimental isotropic shifts, owing to the complications arising when both contact and dipolar shifts are present. For example, after esr measurements were carried out on trisbipyridal iron(III) and the pseudocontact shift determined,[8] a previous conclusion[12] in the literature that iron(III) (t_{2g}^5) and nickel(II) (e_g^2) contact shift patterns were similar was shown to be incorrect.

In view of these many problems, it is not surprising that much of the work in the area of nmr of paramagnetic complexes is carried out on systems in which one of the effects, contact or pseudocontact, is clearly dominant. We shall concentrate on systems in which the contact contribution is dominant in the remainder of this section. The absence of a pseudocontact contribution for a molecule with nearly isotropic g-tensors has been discussed. Complexes of the form ML_6^{n+}, where L is a monodentate ligand, will not have a pseudocontact contribution.[13] If the ML_6^{n+} complex were Jahn-Teller distorted, the distortion would be expected to be dynamic in solution on the nmr time scale. Even if one encountered the very unlikely possibility that the distortion was not dynamic, rapid ligand exchange would average the shift to zero. This results because the $3\cos^2\theta - 1$ function is twice as large for the two ligands on the z-axis as it is for the four ligands on the x- and y-axes and is opposite in sign. Thus, the average pseudocontact contribution for the ligands at all six positions is zero. Ion pairing could lock in a distortion, and its effect on the pseudocontact term is discussed in a subsequent section.

Scalar Shifts and Covalency

A scalar or contact contribution to the bonding of a ligand in a complex can result only if there is covalency in the metal-ligand bond. Accordingly, there have been many attempts to draw inferences about the trends in covalent bond strength in a series of complexes as well as conclusions about the existence of metal-ligand pi-backbonding from such shifts. In a molecular orbital description of a complex, the ligand and metal ion orbitals mix to form molecular orbitals. Unpaired electron density is delocalized onto the ligands as a consequence of the contribution that the ligand makes to the resulting non-bonding, essentially metal, molecular orbitals (see Fig. 12–4). This unpaired electron density interacts with the nucleus whose nmr spectrum is being investigated in the same way as the direct delocalization mechanism discussed in esr, where we were concerned with the effect of this interaction on the electron energy levels. Interpretation of contact shift data in terms of bonding is hampered by two problems. The nmr and esr experiments detect unpaired spin density. In addition to the *direct delocalization of unpaired electron density,* the indirect, spin polarization mechanisms *delocalize spin.* For example (see the earlier discussion, p. 329), in the esr of the methyl radical, there is no net *unpaired electron density* in the sigma framework; but as a consequence of spin polarization, *unpaired spin* is experienced at the proton (see Fig. 9–15 and the discussion of it if necessary). When thinking in terms of electron pair molecular orbitals, *it is important to distinguish between unpaired electron delocalization* (*the direct mechanisms*) *and spin delocalization* (*the indirect mechanism*) *in a complex.** When both the direct and indirect mechanisms give rise to spin density with the same sign on an atom, the two effects cannot be distinguished (*vide infra*). The second shortcoming of drawing inferences about the bonding from contact shifts arises from the fact that these results provide information about only the non-bonding, essentially metal orbitals. The extent of interaction of these orbitals will not necessarily parallel interactions in the bonding molecular orbitals.[14] This point was demonstrated[14] by a study of the shifts in $Ni(stien)_2(RCO_2)_2$ [where stien is stilbenediamine (1,2-diphenylethylenediamine) and R = $—CH_3$, $—CH_2Cl$, $—CHCl_2$, and $—CCl_3$]. As the anions become poorer coordinating ligands, the C—H contact shifts of stilbenediamine decrease instead of increase. An increased shift would be expected for more covalency in the nickel-stien bonds from a direct delocalization mechanism if the antibonding orbitals mirrored the increased covalency expected from an increased formal charge on the metal. They do not, and possible reasons for the discrepancy have been offered.[14]

Scalar Shifts in Coordinated Pyridine

The principles involved in the qualitative interpretation of scalar (contact) shifts can be illustrated by explaining the observation that, when pyridine is coordinated to a nickel(II) complex, the magnitudes of the shifts are H(2) > H(3) > H(4). Furthermore, when 4-methylpyridine is coordinated, the methyl proton resonance shifts in the direction opposite to that of H(4) on pyridine. The pattern of contact shifts observed in a ligand molecule is characteristic of the molecular orbitals of the ligand that are involved in the spin delocalization (*i.e.*, the wave functions for the contributions from σ_L, π_L^*, etc., in Fig. 12–4). The "nitrogen lone pair" molecular orbital in pyridine has decreasing contributions from H(2) > H(3) > H(4). Thus, a characteristic sigma delocalization pattern develops through mixing the e_g and the σ-pyridine m.o. This results in a large shift for H(2), a smaller one for H(3), and the smallest for H(4). On the other hand, the highest filled or lowest empty π-orbital has no hydrogen coefficients, but large carbon coefficients for all carbons. By spin polarization, this gives rise to roughly comparable shifts at H(2) and H(4). In an octahedral complex that contains unpaired electrons only in e_g orbitals, evidence for the existence of a π-delocalization

*It is important to realize that the whole idea of electron delocalization versus spin delocalization is an artifact that is needed when one employs restricted instead of unrestricted molecular orbital calculations on systems with unpaired electrons.

shift pattern is often found even when the π-ligand orbitals are orthogonal to the e_g metal orbitals that contain the unpaired electrons. This results because unpaired electron density in the sigma system can spin polarize the filled mainly-π molecular orbital (this is the same spin polarization mechanism discussed in the esr chapter) and thus can lead to non-zero spin density in the π-system on certain atoms. This is equivalent to the unpaired electron in m.o. IV of Fig. 12–4 spin polarizing the pair in m.o. I. More up spin remains on the metal, resulting in more down spin on the ligand. This effect is similar to that discussed for the allyl radical in Fig. 9–16.

It is also possible for the unpaired electrons in m.o. IV to spin polarize m.o. III, the filled mainly-t_{2g} metal molecular orbital (which has some ligand π contribution) in the complex. The electron with the same spin as that in the e_g orbital resides mostly on the metal, and the electron with opposite spin resides mostly on the part of the π^* molecular orbital that is mainly ligand. *Unpaired spin* is delocalized into the ligand π-system by both of these indirect mechanisms, but no net unpaired electron density is delocalized into the mainly-t_{2g} metal or the mainly-π_L molecular orbitals of the complex. In the remainder of this chapter, we shall use the term *spin density* to imply unpaired spin that could arise by either direct or indirect mechanisms, and the term *unpaired electron density* to specifically indicate direct delocalization.

The main evidence for unpaired spin in the π-system of an aromatic ligand, such as pyridine or a phenyl group, comes from replacing a ring hydrogen atom by a CH_3 group. If the observed shift of the $—CH_3$ proton changes sign from that of the proton at the same position in the unsubstituted complex, this is evidence for π-spin density. This is because spin density in the mainly carbon π-system is delocalized directly onto the methyl protons; *i.e.*, the hydrogen atom orbitals associated with these protons have small coefficients in the π-molecular orbital. However, for the H-ring substituent, the hydrogen 1*s* orbital is orthogonal to the π-system and the π-spin density must polarize the C—H sigma bond to reach the hydrogen nucleus. As a result, the spin density on the —H will have a sign opposite to that in the π-system.

From the above discussion, one can see that the interpretation of spin delocalization mechanisms can be intimately tied to a molecular orbital description of the complex and, at the qualitative level, to a molecular orbital description of the ligand. For this latter approach, we look at the ligand molecular orbitals and the metal electronic configuration for the respective symmetry of the complex, and interpret the spectrum by trying to determine which ligand molecular orbitals mix with the metal orbitals or are likely to be spin polarized.

Spin Delocalization Patterns

One more system will be discussed in this section. When unpaired spin is delocalized into the highest filled π-orbital or lowest empty π^*-orbital of a $(C_6H_5)_nX$ ($n = 1$ to 3, and X is a basic functional group) compound, a characteristic pattern of shifts results which is a consequence of a node at the *meta* carbon in the highest filled π or lowest empty π^* molecular orbitals of the C_6H_5 fragment. As a result, spin aligned with the field in the carbon π-system will cause an upfield shift at the *ortho* and *para* protons but an opposite, downfield shift at the *meta* proton. The contact shifts for the phenyl protons of tetrahedral nickel(II) complexes of $(C_6H_5)_3P$ exhibit this characteristic π-delocalization pattern, with the *ortho* and *para* protons shifting upfield (*i.e.*, unpaired electron density is delocalized directly into the π-system). Does this result indicate that there is metal-ligand π-backbonding involving a phosphorus *d*-orbital and the nickel e_g set? It should be clear that no such interpretation is valid. The phenyl π-system is not orthogonal to the phosphorus lone pair in the ligand. Thus, in forming a metal-phosphorus "sigma" bond with a metal acceptor orbital containing an unpaired electron, this *unpaired electron* can be delocalized directly into the phenyl π-system, giving rise to a π-delocalization pattern without any

π-backbonding. To illustrate this point,[15] $Ni(C_6H_5CH_2NH_2)_6^{2+}$ was studied, and a direct π-delocalization mechanism was found for the phenyl protons even though nitrogen has no low energy empty orbitals for π-backbonding and all the nickel(II) unpaired electron density is in the e_g orbitals in the octahedral complex. Considering the symmetry of the ligand, the "ring π-orbitals" are seen not to be orthogonal to the nitrogen lone pair; *i.e.*, the molecule as a whole does not have discrete sigma and pi systems.

One of the initial questions raised remains to be answered. Why do the N—H proton resonances in $(CH_3NH_2)_6Ni^{2+}$ shift upfield upon complexation? The methyl proton resonances shift downfield because of the direct delocalization of unpaired electron density. Most of the unpaired electron density delocalized onto the ligand is on the nitrogen, and little is delocalized directly onto the N—H proton. Accordingly, there is a large amount of spin polarization of the N—H bond, which dominates the shift at this proton position, leading to a negative A (*i.e.*, upfield shift). This appears to be the case for all N—H protons when the nitrogen is directly bonded to the nickel.

From the preceding discussion, two things should be quite clear. Isotropic shifts are important to an understanding of the electronic configuration of the metal in a complex, but are of little use in providing information about the details of metal-ligand bonding. Secondly, the interpretation of proton contact shifts is intimately tied to an understanding of the molecular orbitals of the complex or in an approximate way to those of the ligand.

12–7 SEMIQUANTITATIVE INTERPRETATION OF CONTACT SHIFTS

The qualitative molecular orbital interpretation of the spectra of paramagnetic complexes in the previous section is supported by the reasonably good agreement of results from approximate molecular orbital calculations and experiment. The early molecular orbital based interpretation used a restricted, extended Hückel (EH) molecular orbital approach.[1, 16–20] Unrestricted INDO molecular orbital programs have recently become available and have been applied to this problem. The interpretations based on the two approaches are usually in agreement (*vide infra*).

The calculation of the contact shift is similar to the molecular orbital calculation of isotropic esr hyperfine coupling constants discussed in Chapter 9. Ideally, an unrestricted m.o. calculation should be employed on the whole complex, and the spin density on the individual atoms should be determined and converted to A as described for a in Chapter 9. Prior to ready access to any unrestricted calculation, the extended Hückel approach was employed to interpret the proton contact shifts of a whole series of metallocenes.[18] In all cases studied, the proton contact shifts were well reproduced by the molecular orbital calculations on the whole complex. Furthermore, the correct electronic ground state resulted from the calculation.

The ground state is predicted by adding the number of electrons in the metal to the molecular orbitals indicated in Fig. 12-5. The primary electron spin delocalization mechanism for vanadocene and chromocene is found to involve the molecular orbitals that are sigma in the planar cyclopentadiene, while those that are pi in the free cyclopentadiene ligand make the dominant contribution (with a substantial sigma component) in cobaltocene and nickelocene. The extensive mixing of the ring "sigma orbitals" with the metal orbitals in forming the metallocene is an important conclusion arising from this work. In the past, the σ-system has usually been ignored in describing the bonding of cyclopentadiene to the metal ion. Similar successes and insights were obtained from a study of the proton shifts for a series of paramagnetic bisarene complexes.[20] The ^{13}C contact shifts on several metallocenes have been studied.[21] While the results are qualitatively consistent with the conclusions from the proton nmr, the ^{13}C data cannot be adequately fitted to the extended Hückel

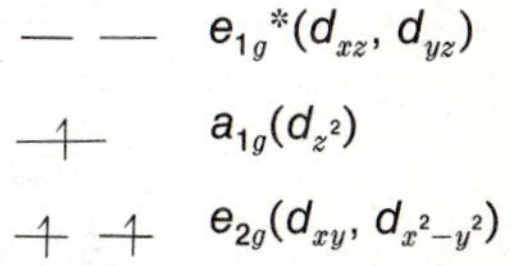

— — $e_{1g}^*(d_{xz}, d_{yz})$

—+— $a_{1g}(d_{z^2})$

+ + $e_{2g}(d_{xy}, d_{x^2-y^2})$

FIGURE 12–5. The essentially d-orbital molecular orbital energies of a metallocene. (Electrons are added for vanadocene. The principal metal component is indicated in parentheses.)

molecular orbital wave functions. This is not surprising in view of our discussion of the calculation of ^{13}C coupling constants of organic free radicals in the esr chapter (see p. 333).

The fit of the proton contact shifts does give one some confidence in the wave functions from the m.o. calculation. With confidence in the wave functions from fitting experimental data, several significant conclusions[20] can be drawn about the bonding. For example, (1) the separation of the e_{2g} and a_{1g} m.o.'s is greater in the bisbenzenes than in the biscyclopentadiene complexes, implying more backbonding in the former; (2) the extent of electron delocalization in an m.o. is not necessarily related to the calculated bond orders, so "stability" cannot be inferred from the size of a contact shift; (3) the ring sigma m.o.'s are important in the bonding of both series of complexes; and (4) the metal 4*s* and 4*p* orbitals have significant bond orders with the ring atoms. Nmr studies of paramagnetic organometallic compounds have been recently reviewed.[22]

Unfortunately, the large number of atoms in most paramagnetic transition metal ion complexes makes m.o. calculations for the entire complex infeasible. Even if size were not a problem, there is a question as to whether or not the EH or INDO procedures could produce reasonable wave functions on species with charge differences as large as those that exist between the metal ion and ligand.* The approach used on these systems assumes that the metal ion makes at most a minor perturbation on the proton contribution to the mainly lone pair molecular orbital and other molecular orbitals of the free ligand that are involved in bonding to the metal. This is a reasonable assumption for most complexes in which the metal-ligand bond strengths are about 10 to 20 kcal mole^{-1}. One then carries out a molecular orbital calculation on the free ligand and, using these neutral ligand wave functions, carries out an electron density ψ^2 analysis (see Chapter 3 if necessary) to determine what the *A* values would be if there was one electron in each of the various orbitals that are expected to mix with the metal in forming the complex. The results from such a calculation are presented for various substituted pyridines in Table 12–1.

*The ability of these methods to reproduce hyperfine coupling constants has not been tested on systems containing metal ions with a large formal positive charge. One such attempt with EHT was unsuccessful.

TABLE 12–1. COUPLING CONSTANTS IN GAUSS FOR ONE UNPAIRED ELECTRON IN VARIOUS ORBITALS OF PYRIDINE TYPE LIGANDS

Ligand	Position	A (G)		
		π^b	σ	π^*
4-Mepy	*o*	+1.75	+7.23	+2.92
	m	+2.33	+2.85	+1.42
	p-CH_3	−6.80	+0.11	−4.31
Py	*o*	+1.51	+7.21	+2.05
	m	+1.60	+2.75	+0.87
	p	+6.20	+7.30	+5.58
3-Mepy	*o*	+3.19	+6.85	+2.08
	m	+2.70	+2.50	+1.07
	p	+3.26	+7.02	+5.47
	m-CH_3	−4.17	+0.63	−0.49
4-Vinpy	*o*	+0.74	+7.31	+1.54
	m	+1.76	+2.90	+0.95
	H_3	+1.49	+0.13	+0.70
	H_2	−1.96	+0.16	−1.43
	H_1	+1.06	+0.70	+2.44

With these coupling constants for one unpaired electron in each of several ligand orbitals, we are now in a position to calculate the amount of unpaired spin delocalized into each of these orbitals in the complex. We do this by setting up simultaneous equations using the observed coupling constants for the *ortho* proton and the *para*-methyl protons of the 4-methylpyridine complex. As an example, we present the equations for the σ-donor orbital and the π-antibonding orbital combinations needed to explain the scalar shifts in the six-coordinate nickel(II) complexes.

$$\textit{ortho}: \quad +0.3272 = +7.23x + 2.92y \qquad (12\text{-}24)$$

$$p\text{-methyl}: \quad -0.0302 = +0.11x - 4.31y \qquad (12\text{-}25)$$

Here the first number listed in each equation is the observed scalar shift for the complex in gauss, x represents the fraction of unpaired electron in the σ-orbital, and y represents the fraction of unpaired electron in the π-antibonding orbital. Solution of this set of equations yields $x = 4.20 \times 10^{-2}$ and $y = 8.08 \times 10^{-3}$. The metal ion is simply considered to be a probe for introducing spin into the ligand molecular orbitals. We again emphasize that if the mechanism for the pi-spin contribution involves pi-polarization, no net unpaired electron density is placed in the molecular orbital of the complex composed mostly of this π-system. The value of y indicates the contribution that the pi-orbital makes to the net spin delocalization observed for the complex. A similar set of equations can be solved for the sigma donor and pi-bonding orbitals of pyridine, leading to $x = 4.40 \times 10^{-2}$ and $z = 5.20 \times 10^{-3}$ (where z is the pi-bonding molecular orbital contribution). Combinations involving other pairs of orbitals lead to results that are unreasonable. In this example, two observed shifts are used to obtain two unknowns, x and y or x and z. Only the *meta* position remains as a check. However, when this procedure is applied to all the substituted pyridines in Table 12–1, the number of protons available as checks on the approach increases. One finds that the observed shifts can be fit equally well with either the $\sigma + \pi$ or $\sigma + \pi^*$ ligand combinations or a mixture of the two. Thus, one can conclude that the contact shifted spectrum of coordinated pyridine arises from an essentially sigma spin delocalization mechanism with a small π contribution. The sigma contribution arises from direct delocalization of unpaired electron density on the ligand. The π contribution has to arise from spin polarization because the pyridine π-orbitals are orthogonal to the metal e_g orbitals, and the metal t_{2g} orbitals contain only paired electrons. One cannot determine whether the π contribution in the complex comes from the filled π-orbital or the empty π^*-orbital of pyridine or from both. The filled π-orbital could participate by spin polarization of the filled, essentially pyridine molecular orbital in the complex (see Fig. 12–4, m.o. I). The empty π^* can participate by spin polarization in the complex of an essentially filled metal t_{2g} molecular orbital that has some pyridine π^* character mixed into it (see Fig. 12–4, m.o. III). Our understanding of the spectral shifts has been enhanced by the molecular orbital analysis, but it is not complete.

Much of the more recent work in this area has utilized INDO molecular orbital calculations. This method has the advantage of directly including spin polarization. However, one must perform the INDO calculation on the ligand as a free radical. Usually the metal perturbation does not even approach this extreme. In the case of pyridine, the phenyl radical is employed to match the experimental data because using the pyridine radical gives inferior results. In spite of the above-mentioned shortcomings, good fits of both the proton and ^{13}C contact shifts of 31 out of 35 nuclei in a series of coordinated pyridines result[23,24] with this model. The interpretation of the spin delocalization mechanisms in the complex is in complete agreement with the earlier proton results and the extended Hückel calculation.[19] The INDO method also leaves much to be desired. When one considers the eventual goal of this work—a semiquantitative interpretation of the observed shifts—either INDO or EH methods would give the accuracy needed.

One important point should be reemphasized. When a restricted calculation (EH) is employed, indirect spin contributions from polarization must be added on as was done for the small pi contribution to the pyridine shifts above.* When an unrestricted calculation (INDO) is employed, spin polarization is included in the calculation.[19,25,26] The validity of the change in sign of the proton shift of the —H and —CH_3 substituents on an aromatic ring as a criterion for π-spin delocalization contributions has been challenged[25] by comparing the results from INDO and extended Hückel calculations. As mentioned earlier, the methyl protons in the 4-methylpyridine complex have the sign opposite that of the 4-proton in the pyridine complex. When an INDO calculation was carried out[25] on the 4-methylpyridine sigma radical analogue, a negative coupling constant resulted for the methyl proton. This result does not indicate that a sigma delocalization mechanism could give a negative coupling constant at the CH_3. Thus, one cannot negate[25] the long accepted —CH_3 and —H sign reversal criterion for a pi-spin delocalization contribution. One must remember that polarization of the filled π-system is built into the INDO calculation of the free radical. A detailed analysis of the INDO output showed[26] that the sigma mechanism dominated the 2- and 3-hydrogen shifts but that the INDO analogue of spin polarization of the filled pi-orbital in the Hückel calculation made the principal contribution to the negative coupling constant at the methyl group. This was the conclusion of the extended Hückel analysis[19] of this system. In the original literature[19,26] on this subject, the terms "electron delocalization" and "spin delocalization" are used loosely; but if you keep in mind Fig. 12–4, it will be obvious when spin delocalization and electron delocalization are meant, should you refer to original articles.

As a final illustration of the value of approximate wave functions in the interpretation of contact shifts, we shall consider the shifts in some 4-methylpyridine-*N*-oxide complexes.[27] The general pattern of observed proton contact shifts has the features of a π delocalization mechanism with spin aligned with the field in the pi-system. From these shifts, it can be concluded that upon coordination the 4-methylpyridine-*N*-oxide must be twisted in such a way that the pi-molecular orbital, which is mainly oxygen p_z (z perpendicular to the ring plane), mixes with the sigma bonding e_g set of nickel(II). This permits direct delocalization of unpaired spin into the "ring π" orbital. This twisted mode of coordination is found in a solid adduct of this donor. The molecular orbital calculation indicates that several of the high energy donor molecular orbitals are mainly oxygen a.o.'s with very small hydrogen a.o. coefficients. Thus, even if these molecular orbitals were involved in bonding to the nickel(II), they would make at most a small direct contribution to the proton contact shifts.

For certain applications (*vide infra*), systems are desired in which the only contribution to the isotropic shift is pseudocontact. Pyridine-*N*-oxide is a good ligand with which to investigate pseudocontact contributions to the shift.[28] If a ligand dominated by sigma delocalization is used, the contact pattern of shifts and the pseudocontact pattern of shifts go in the same direction and the absence of contact contributions is not easily verified. However, in the case of pyridine-*N*-oxide, the

*The whole concept of spin polarization is one that is introduced to account for a deficiency in restricted calculations.

alternation of proton shifts due to nodes in the π-system will readily demonstrate when a contact contribution is important.

12–8 APPLICATIONS OF ISOTROPIC SHIFTS

Structure Determination

Many of the advantages of studying the nmr spectra of paramagnetic complexes are results of the enhanced signal separation of resonances that arise from non-equivalent atoms in the molecule. An example that demonstrates the sensitivity of this method is the five-coordinate Schiff base–nickel(II) complex shown in Fig. 12–6, which can exist in either tetragonal pyramidal or trigonal bipyramidal structure. The

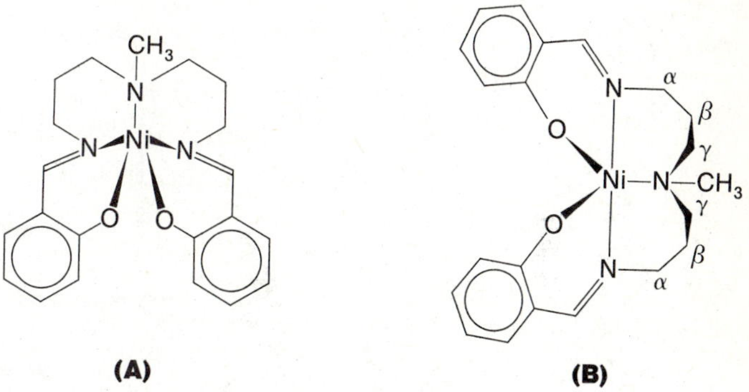

FIGURE 12–6. Tetragonal pyramidal (A) and trigonal bipyramidal (B) structures for a pentadentate Schiff base complex.

proton magnetic resonance spectrum[29] exhibits two resonances for each of the α and β CH_2 groups and two sets of resonances for each of the sets of aromatic protons. Only a single CH_3 resonance was observed, indicating only one isomer in solution although two are observed in the solid.[30a] If both isomers existed in solution, four signals would be observed for each ring position. The structure must be dissymmetric and contain non-equivalent trimethylene groupings and phenyl rings. The authors[29] concluded that this eliminates the square pyramidal structure and supports a trigonal bipyramidal geometry. This conclusion is not valid, for even in the tetragonal pyramid the R-group would give rise to the observed non-equivalences.[30b] Non-equivalence in the rings is observed in the individual isomers in the solid. In the analogous diamagnetic zinc(II) complex, one observes only a very slight non-equivalence in the CH_2 proton resonances. A rapid rearrangement through the tetragonal pyramid would average the resonances. Such a process does not occur rapidly (on the nmr time scale) at room temperature, and this is one of the few pentacoordinate, stereochemically rigid molecules. When N—CH_3 is replaced by sulfur, intramolecular rearrangement gives rise to the observation of equivalent rings.

In the octahedral complex $Co\{[4,6\text{-}(CH_3)_2phen]_3\}^{2+}$ [see Fig. 12–7 for the formula of $(CH_3)_2phen$] the *cis* and *trans* complexes can give rise to a possible thirty-two lines in the proton magnetic resonance spectrum. As illustrated in Fig. 12–7, thirty-one have been observed.

As mentioned in Chapter 8, the nmr spectra of enantiomers are identical in achiral solvents. However, the sensitivity of paramagnetic nuclear magnetic resonance often allows one to distinguish resonances from the diastereoisomers formed when the optically active enantiomers coordinate to an optically active transition metal ion complex. Many other examples of this type of application have been summarized.[31]

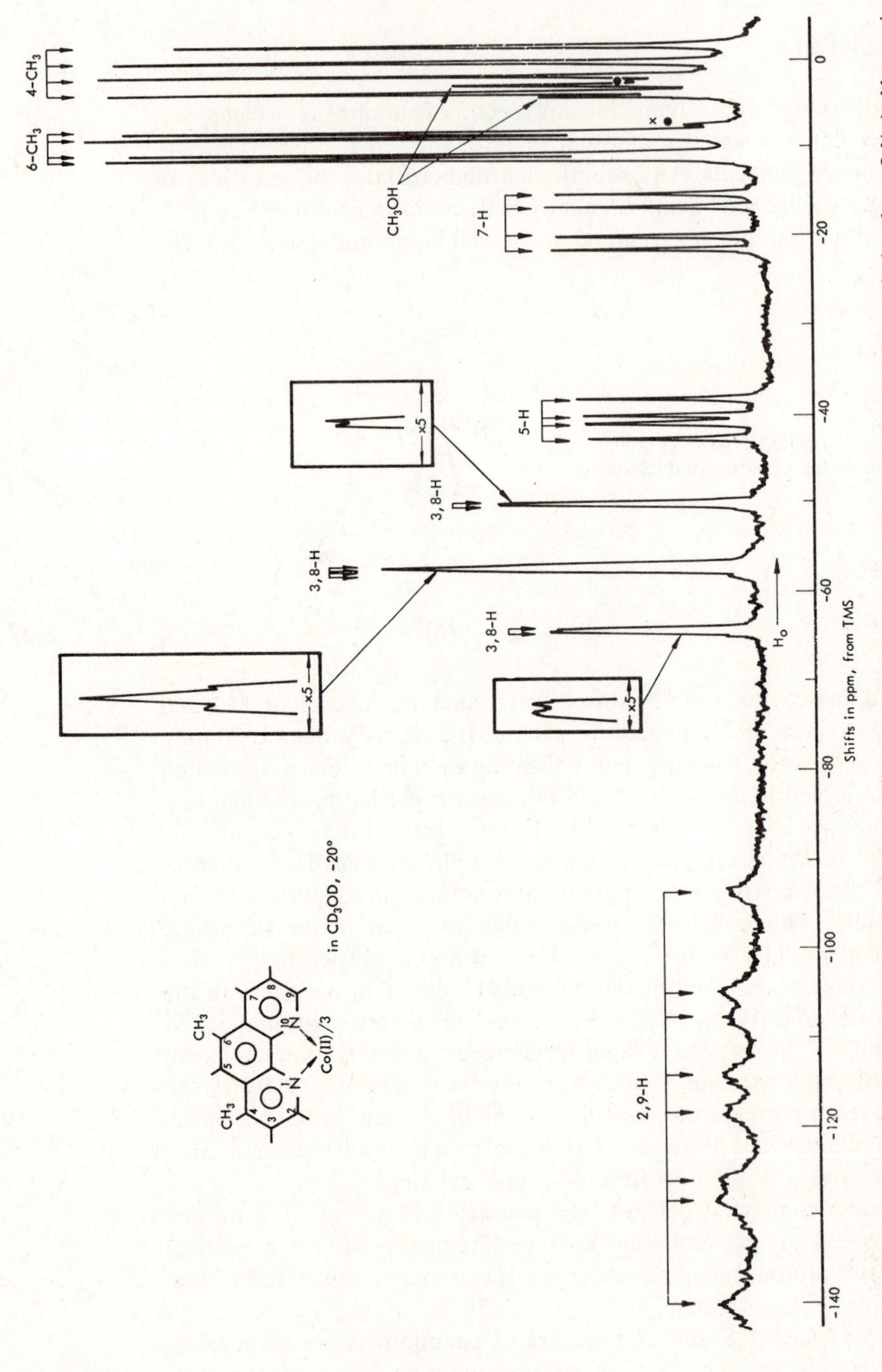

FIGURE 12–7. PMR spectrum of $[Co(4,6\text{-}Me_2phen)_3]^{2+}$ in methanol solution at −20°. [Reprinted with permission from G.N. LaMar and G.R. Van Heck, Inorg. Chem., 9, 1546 (1970). Copyright by the American Chemical Society.]

Ion-Pairing Studies

The pseudocontact contribution to the shift of the alkyl ammonium protons of an ion pair R_4N^+ MX_3L^- has been used[32-34] to estimate cation-anion distances, *r*, in the ion-pair, and to investigate solvation effects. For the former application, geometries for the ion pair are assumed; when R is *n*-butyl, four proton resonances are observed in the spectrum which one attempts to fit with equation (12–23) (or a suitable form) by varying the distance in the so-called geometric factor $[(1 - 3\cos^2\theta_i)/r_i^3]$. For convenience, we can write the pseudocontact shift equation as:

$$\frac{\Delta\nu(\text{pc})}{\nu} = -Kf_{(g)}\text{GF} \qquad (12\text{-}26)$$

where K is a collection of constants, $f_{(g)}$ is some function of the g-tensor or χ-tensor, and GF is the geometric factor. The ratio of the isotropic shifts of two butyl protons is then seen to be given by:

$$\frac{\Delta\nu_i(\text{pc})}{\Delta\nu_j(\text{pc})} = (\text{GF})_i/(\text{GF})_j \qquad (12\text{-}27)$$

For an assumed geometry, the geometric factor is calculated[33] and averaged over all orientations that the *dynamic ion pair* can assume as it rotates freely in solution. Since only a single peak is observed in the complex for nuclei that are equivalent in the free cation, rotation of the R_4N^+ group is rapid on the nmr time scale. The ratio $\frac{\Delta\nu_i}{\Delta\nu_j}$ for the various pairs of protons is calculated for various distances between the cation and anion. This is repeated for different assumed ion pair cation and anion orientations, and the structure that best fits the shift data is used to estimate the ion pair distance. For $(C_6H_5)_3PC_4H_9{}^+(C_6H_5)_3PCoBr_3{}^-$, a distance from the phosphorus in the cation to the metal in the anion of about 8 Å was reported.[35] The accuracy of this application is suspect in many systems because simple ion pairs do not exist in solution. Very large clusters exist, particularly in low dielectric solvents. The shift observed for the cation resonances is thus an average over the various cation environments in the cluster. Furthermore, it has been conclusively shown[36] that there can be contact contributions in the alkyl ammonium protons of the ion pair, making an analysis based only on a pseudocontact term suspect in many systems. Evidence for covalency came[36] from an observed contact contribution to the ^{14}N isotropic resonance shift in $[Bu_4N][(C_6H_5)_3PCoI_3]$. The shift observed was in the direction opposite to that required from the estimate of the sign of Δg from the proton shift direction. The existence of this contact term further illustrates the extreme sensitivity of isotropic shifts because it detects the very small amount of covalency that exists between the cation and anion in the ion pair. Mechanisms have been proposed for a covalent interaction[36] in the ion pair.

The influence of solvent on the extent of the ion pairing is manifested[37] in the magnitude of the observed isotropic shift. The shift data for the system [*n*-butyl$_4$N][$(C_6H_5)_3PCoBr_3$] could be fit[38] to an equilibrium constant expression for the following reaction:

$$[(C_4H_9)_4N^+][(C_6H_5)_3PCoBr_3{}^-] \rightleftarrows (C_4H_9)_4N^+ + (C_6H_5)_3PCoBr_3{}^- \qquad (12\text{-}28)$$

Solvents ranging in dielectric constant from 4.6 to 64 for concentrations of electrolyte from 0.02 to 0.5 M were used. The shift due to the "anion-cation pair" could be determined simultaneously with K from the observed shift data, and it was concluded[38] that the interionic distance was longer in the higher dielectric constant solvents than in those with a lower dielectric constant.

It has been shown[39] that tetrahedral metal-containing anions that in themselves are not anisotropic can produce a pseudocontact contribution in a non-metallic cation. The source and magnitude of the g-anisotropy can be roughly accounted for *via* an electrostatic perturbation of the crystal field of the spherical anion by the cations. A slight distortion of the tetrahedral structure by the cation during the lifetime of the ion pair is also suggested. Since the cation lies on a unique axis, it will feel a dipolar shift from the anisotropy induced in the ion pair. There are many equivalent ways in which the cation can approach the tetrahedral or octahedral anion to form the ion pair, and they will all have comparable pseudocontact contributions at the cation. Thus, a dynamic process of this sort will not average the pseudocontact shift in the cation to zero. On the other hand, a dynamic process of this sort will average the pseudocontact shift contribution to resonances of atoms in the tetrahedral or octahedral anion to zero.

Nature of the Second Coordination Sphere in Complexes

The nmr isotropic shift is a powerful tool for exploring the nature of the second coordination sphere around a transition metal ion complex in solution.[40] When the nmr spectrum of $[Bu_4N][CoP(C_6H_5)_3I_3]$ dissolved in $CHCl_3$ is examined,[40] an upfield shift in the chloroform proton resonance is observed. The shift is also upfield and of a similar magnitude for the nickel(II) analogue of this complex. Since the sign of Δg is different[32] for the nickel(II) and cobalt(II) complexes, an upfield shift for both complexes cannot be accounted for by a pseudocontact term [see equation (12–23)]. As a matter of fact, the similarity in the shifts of the cobalt and nickel complexes indicates that the expected contribution from this effect is absent. This would occur if the chloroform protons were bonding to the iodides coordinated to the cobalt (see Fig. 12–3) at an angle θ [equation (12–23)] equal to 55°. At this position, $1 - 3\cos^2\theta$ is zero and there is no pseudocontact contribution. This interaction is expected intuitively and has been confirmed[40] by infrared studies in which changes in the C—H stretching frequency of chloroform are observed with these complexes, but not with free $(C_6H_5)_3P$. If the hydrogen bonding interaction were with the phenyl groups of the coordinated phosphine, hydrogen bonding to the phenyls of $(C_6H_5)_3P$ would have been expected. On the average, about four chloroform molecules are involved in the interaction with the complex. The existence of a contact contribution thus offers conclusive proof for the existence of a significant but small covalent contribution to the hydrogen bonding interaction even in a weak hydrogen bonding acid like $CHCl_3$.

In an extension[41,42] of this idea, the solvation of cobalt(II) poly(1-pyrazolyl)-borate (see Fig. 12–8) by aniline and pyridine was studied. The g-tensor is very anisotropic ($g_{\parallel} = 8.48$ and $g_{\perp} = 0.96$) in this complex. The observed shifts are to high

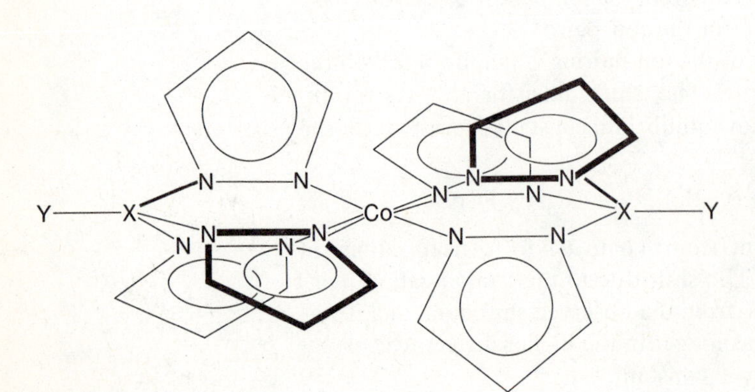

FIGURE 12–8. Structure of Co(II)poly(1-pyrazolyl)-borate, where X = B, Y = H (or substituent). Related complexes in which X = C, Y = H (or substituent) are also known. The overall symmetry of the parent complex is D_{3d}.

field, indicating that the preferred direction of approach by aniline and pyridine is perpendicular to the symmetry axis of the complex. In aniline, the *ortho* protons are shifted the greatest amount, while in pyridine, the 4-H protons are shifted the most. If the aniline was hydrogen bonded, this would bring the nitrogen end of aniline close. The positive end of the dipole for pyridine is at the 4-H part of the ring, and the shifts are consistent with it being closest to the metal. These preferred orientations are most probably an average over a distribution of favorable orientations.

Equilibrium and Kinetic Studies

There have been several applications of paramagnetic nmr to equilibrium and kinetic studies. The larger range of shifts enables one to determine equilibrium constants more precisely and to investigate faster chemical exchange processes. The principles pertinent to this type of application have been discussed in Chapter 8.

Shift Reagents

Rare earth tris-chelates of β-diketonate derivatives are Lewis acids and cause large pseudocontact shifts in Lewis bases that are added to solutions containing these complexes. The resulting resonances for several of these ions (for example, europium and praesodymium complexes) are very sharp even compared to nickel(II) complexes. In Fig. 12–9, the 100 MHz proton magnetic resonance spectra of $CDCl_3$ solutions of *cis*-4-butylcyclohexanol containing varying amounts of $Eu(dpm)_3$ [dpm is dipivaloylmethanato, $(CH_3)_3CC(O)CHC(O)C(CH_3)_3$] are shown[44] to illustrate this behavior. The complex pattern obtained for the pure alcohol is shown in the bottom spectrum. A first-order spectrum with all resonances well separated is obtained in curve D. Note that the spin-spin couplings are still intact.

Many rare earth complexes that exhibit this behavior are called[43] shift reagents, S.R. As is usually the case with these systems, the S.R. complex is in fast exchange with an excess of the Lewis base. The lanthanide complexes

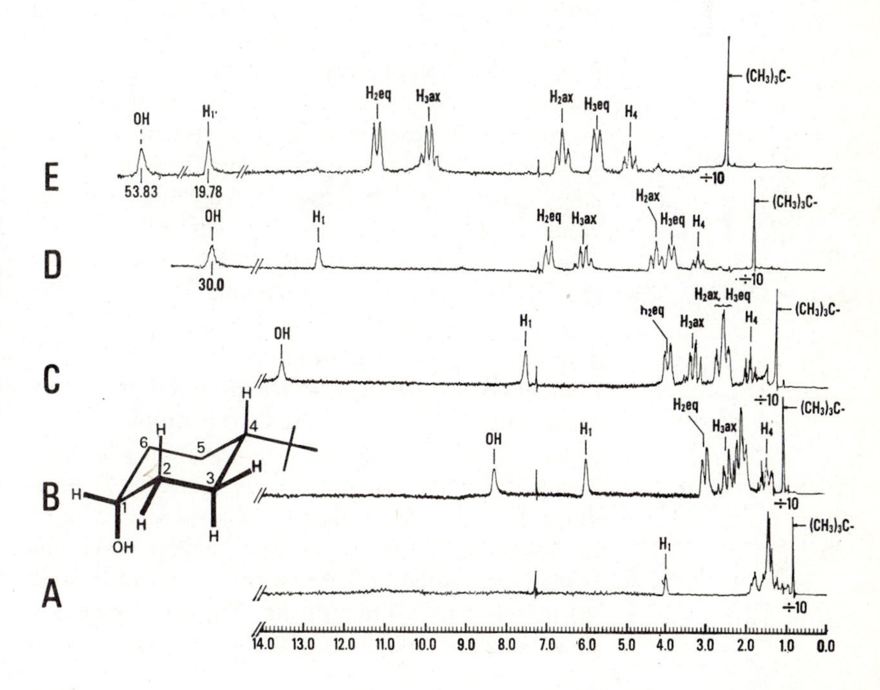

FIGURE 12–9. 100-MHz pmr spectra of *cis*-4-*tert*-butylcyclohexanol (20 mg, 1.28×10^{-4} mole) in $CDCl_3$ (0.4 ml) containing various amounts of $Eu(dpm)_3$: (A) 0.0 mg; (B) 10.3 mg; (C) 16.0 mg; (D) 33.1 mg; (E) 60.2 mg. [Reprinted with permission from P.V. Demarco *et al.*, J. Am. Chem. Soc., *92*, 5734 (1970). Copyright by the American Chemical Society.]

of the fluorinated chelate 1,1,1,2,2,3,3-heptafluoro-7,7-dimethyl-4,6-octanedione, $(CH_3)_3CC(O)CHC(O)CF_2CF_2CF_3$, abbreviated as fod, are stronger Lewis acids and produce larger shifts.[45] Since these early reports there have been hundreds of papers dealing with applications of shift reagents. A comprehensive summary has been prepared.[46] Here, we shall deal with examples of some different types of applications and discuss some of the potential complications encountered.

One use of a shift reagent involves simplifying second-order (*i.e.*, $J \approx \Delta$) spectra. Different protons in the molecule are shifted by different amounts upon complexation. Thus, as the shift reagent concentration is increased, Δ becomes greater than J. This behavior is seen in Fig. 12–9. In systems containing a shift reagent and forming a 1:1 complex, the average shift of the ith proton obtained in fast exchange is given by

$$\delta_{i\,\mathrm{AVE}} = N_f\delta_f + N_c\delta_c \tag{12-29}$$

where δ_f is the chemical shift of the nucleus in the uncoordinated molecule and δ_c is the chemical shift of the nucleus in the coordinated molecule. N refers to the mole fractions, and for a 1:1 complex:

$$N_f = (1 - N_c)$$

Combining these two equations, we obtain

$$\delta_{i\,\mathrm{AVE}} = (1 - N_c)\delta_f + N_c\delta_c = \delta_f + N_c(\delta_c - \delta_f) \tag{12-30}$$

When complexation is complete, a plot of $\delta_{i\,\mathrm{AVE}}$ versus N_c is a straight line with an intercept of δ_f for this proton and a slope of $\delta_c - \delta_f$. This δ_f is the chemical shift one would obtain in the free material if spin decoupling were used or if the second-order spectra were computer analyzed.

Usually the shift reagents are involved in a more complex set of equilibria than simple 1:1 adduct formation. However, at low mole ratios of the shift reagent to substrate, a straight line is obtained[44] when one plots the chemical shift from TMS versus the mole ratio of complexed substrate. Extrapolation to zero mole fraction of complexed substrate produces the desired "decoupled" chemical shift for the proton being studied. When the plots are not straight lines or when the extrapolation is long, considerable error is introduced into the chemical shift values. First row transition metal ion complexes were the first systems studied for this type of application. For systems in which the described complications are encountered with shift reagents, one could look for a first row transition metal ion complex that forms 1:1 adducts and use equation (12–30). Alternatively, a complex for which the second step of the equilibrium has a larger constant, K_2, than that of the first, K_1, could be used and equation (12–30) modified for 2:1 behavior.

A second potentially exciting application of shift reagents involves their use in determining geometries of molecules in solution.[40] This experiment is usually done in the fast exchange region. The proton nmr spectral shifts induced by shift reagents are assumed to be almost exclusively dipolar in origin. Ideally, a structure would be assumed for the molecule and equation (12–22) would be used to calculate the expected dipolar shifts at a large number of various nuclei in the molecule whose structure is to be determined. The assumed structure of the molecule would be varied to produce a best fit of the experimental shifts. Since the structure of the molecule being investigated and that of the complex in solution are not known, nor are the magnitude and position of the magnetic dipole of the metal center in the complex, there are eight unknowns in the system. These unknowns are best seen by looking at

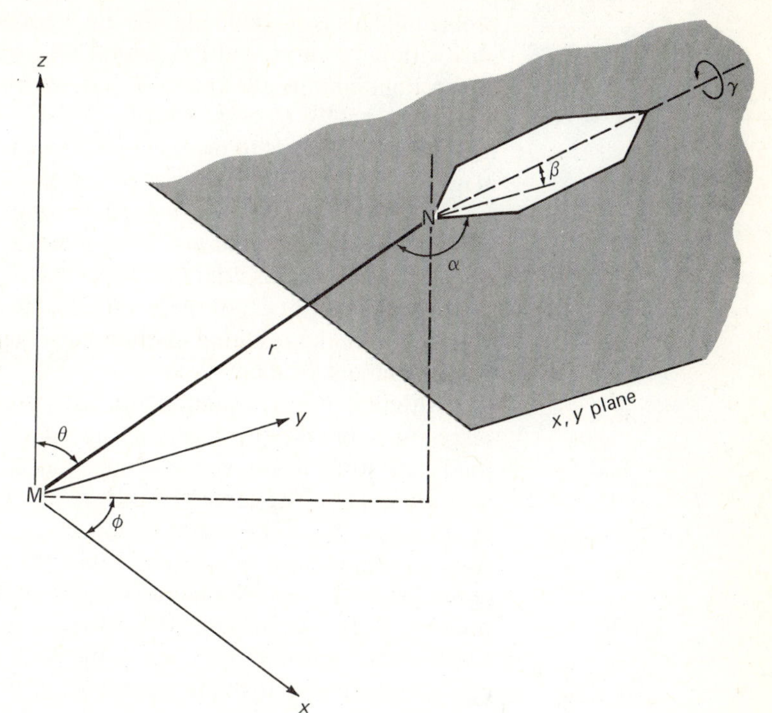

FIGURE 12–10. Definition of the variables in a shift reagent determination of molecular structure.

Fig. 12–10, where the example of a rigid ligand such as pyridine complexed to a S.R. is illustrated. Four parameters are needed to fix the orientation of the molecule being studied relative to the shift reagent: (1) the metal-donor atom distance, r; (2) the ligand atom–donor atom–metal bond angle, α; (3) β, the angle between the ligand's molecular plane and the magnetic x,y plane of the metal; and (4) γ, the angle measuring the twist of the ligand's molecular plane around the nitrogen-*para*-carbon axis. In addition, two angles are needed to define the orientation of the magnetic axis with respect to the metal-donor bond. Two additional unknowns are needed to account for the anisotropy in χ. To complete our discussion of a rigorous solution of this problem, we should consider the possibility of having significant amounts of both 2:1 and 1:1 complexes in solution. (Most shift reagents are potentially diacidic, having two acid sites.) If neither the 2:1 nor the 1:1 complex is the predominant species in solution, both K_1 and K_2 must be known as well as two sets of the unknowns mentioned above, one set for each of the two complexes. However, it is possible to stop the exchange of complexed ligand and uncomplexed ligand, so that it is possible to get around this last problem. However, it is obvious from the preceding discussion that it is not possible to determine geometries of molecules in solution in a rigorous manner. Thus, to gain any information about geometries in solution, some simplifying assumptions must be made.

If three-fold or higher axial symmetry and the presence of only one predominant species are assumed, the problem becomes tractable. With axial symmetry, equation (12–23) can be used instead of equation (12–22). By taking the ratio of all the observed nmr shifts with respect to the largest observed shift, $\Delta\nu_j$, equation (12–31) can be derived from equation (12–23):

$$\frac{\Delta\nu_i}{\Delta\nu_j} = \left(\frac{1 - 3\cos^2\theta_i}{r_i^{\,3}}\right) \Big/ \left(\frac{1 - 3\cos^2\theta_j}{r_j^{\,3}}\right) \qquad (12\text{-}31)$$

and the magnitude of the magnetic anisotropy in the complex is eliminated from the problem. (This is possible only if one ignores the chelate ligand protons and works only with the coordinated Lewis base.) If axial symmetry is assumed but the metal–donor atom bond is allowed to be aligned away from the magnetic z-axis, there are five unknowns.[46] Four unknowns are still necessary to fix the orientation of the molecule with respect to the metal, and one is needed to define the orientation of the magnetic z-axis with respect to the metal–donor bond. If the magnetic z-axis is assumed to be aligned with the metal–donor bond, the only unknowns are the four needed to fix the orientation of the molecule with respect to the metal. It is this last, simplest system that is used most often. Computer programs[46-48] have been written to provide a best fit of all these variables to experimental data, but the parameters obtained from the resulting fit often have high correlation coefficients unless there is a large amount of data.

Although use of equation (12–23) instead of equation (12–22) has been quite successful in providing information about molecular structure in solution, none of the solid state structures of systems looked at in solution show any three-fold or higher axial symmetry. The highest observed axial symmetry has been a two-fold axis in $Ho(dpm)_3(4\text{-pic})_2$. The single crystal magnetic anisotropy of this complex has also been measured, and it was noted that the additional term for a non-axial system (equation 12–22) contributed about 15% of the observed nmr shift for the 4-methyl protons.[50] It has also been shown in low and room temperature studies of $Eu(dpm)_3py_2$ and its 3-picoline, 4-picoline, and 3,5-lutidine analogs that equation (12–22) can be used to fit the room temperature results and must be used to fit the low temperature results.[51-53] Dynamic processes have been proposed that could explain the success of assuming axial symmetry in room temperature results. Rotation of the substrate about the metal–donor atom bond can cause effective axial symmetry.[54] Rapid interconversion between different geometric isomers *via* substrate dissociation and association can also cause effective axial symmetry.[55] It has been suggested[49] that effective axial symmetry in shift reagents occurs because equation (12–22) contains only odd powers of trigonometric functions of φ, the azimuthal angle. It is these trigonometric functions which are averaged to zero. In relaxation studies, axial symmetry cannot be assumed because the relaxation equation contains only even powers of the trigonometric functions of φ.

The magnitude of the isotropic shift varies with the rare earth ion employed in the chelate. The shifts resulting[50] for the 2-proton of the *bis* adduct of 4-picoline with tris(dipivaloylmethanato) lanthanide chelates, $Ln(dpm)_3$, is illustrated in Fig. 12–11. Similar trends are found with other substrates. Not much new information about structures in solution is gained from the different shifts obtained by changing the metal, because the change of metal introduces new unknowns.

The effect of shift reagents on ^{13}C resonances provides additional information of use in ascertaining solution structures. However, there is evidence[46] that Fermi contact contributions to the shift exist, so considerable caution should be exercised in using these data.

In addition to the shifting of the resonances by certain lanthanide complexes, other lanthanide complexes are employed because of their ability to broaden the resonances.[56] Since the shifting and broadening effects have different r dependences (r^{-3} and r^{-6} respectively), the two sets of results provide complimentary information for structure determination. The widths at half height[50] of the 2-methyl proton resonance of the 2-picoline adducts of $Ln(dpm)_3$ are summarized in Table 12–2.

Our treatment of this topic has been brief because many complications exist in the structural determination of molecules in solution by shift reagents that may sometimes limit the applicability and accuracy of the method. These complications include:

(1) The stoichiometry of the adduct in solution must be determined but is often assumed.

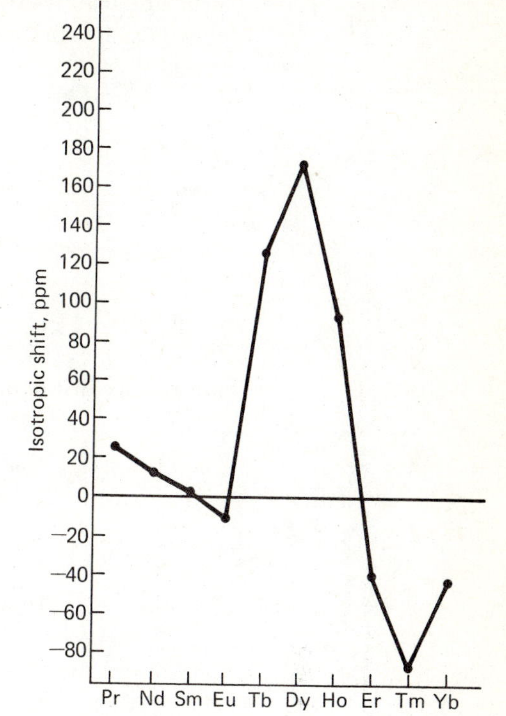

FIGURE 12–11. Observed isotropic shifts of 2-proton of 4-picoline in 2:1, 4-picoline:$Ln(dpm)_3$ complexes. [From G.N. LaMar and E.A. Metz, J. Am. Chem. Soc., *96*, 5611 (1974).]

(2) The treatment assumes that there is only a single geometrical isomer of the adduct in solution. Many are possible. Furthermore, lanthanides have a high affinity for water, which if present in the system can compete for Lewis acid sites with the substrate, producing even more complexes in the system.

(3) When the substrate is involved in a dynamic, intramolecular process, the average conformation of the molecule is determined, and this might represent a relatively unstable one.

(4) The complex is generally assumed to be axially symmetric, so that the dipolar shifts can be evaluated with a $\langle(1 - 3\cos^2\theta)/r^3\rangle$ term. Those systems studied to date do not possess axial symmetry (however, see reference 49).

(5) The location of the principal magnetic axis in the complex is often not known relative to the substrate ligand.

(6) When the substrate is a complex one, it is necessary to assume that there is only one binding site.

In spite of these complications, the method is capable of distinguishing gross

TABLE 12–2. THE WIDTHS AT HALF HEIGHT, $\Delta\nu(\frac{1}{2})$ IN Hz, OF THE 2-METHYL PROTONS OF 2-PICOLINE IN $Ln(dpm)_3$ ADDUCTS

Ln(III)	$\Delta\nu(\frac{1}{2})$	Ln(III)	$\Delta\nu(\frac{1}{2})$
Pr	5.6	Dy	200
Nd	4.0	Ho	50
Sm	4.4	Er	50
Eu	5.0	Tm	65
Tb	96	Yb	12

structural features such as those that exist in many different isomeric compounds. For example, the compounds

CH_3 CH_3 CH_3 —OH H   and CH_3 CH_3 CH_3 —H OH

are readily distinguished,[44] as are[57]

C_6H_5 H —OH H H N H C_6H_{11} and C_6H_5 OH —H H H N H C_6H_{11}

Shift reagents have been used to differentiate between internal and external phospholipid layers in membranes.[58] They have also been employed in conformational studies of nucleotides in solution[59] and in other biochemical systems.[60] One must always be concerned with the problem that coordination of the shift reagent may affect the conformation of the biomolecule.

Another interesting application involves the use of optically active shift reagents to determine optical purity. The idea is similar to that discussed in Chapter 8, in which optically active solvents were employed. Here, different stability constants for forming the different diastereoisomeric adducts exist, leading to different mole fraction averaged shifts for the enantiomeric bases. Several reports of different optically active rare earth complexes that can be used for this application have appeared.[61-63]

REFERENCES

1. R. J. Fitzgerald and R. S. Drago, J. Amer. Chem. Soc., *90,* 2523 (1968).
2. See, for example, R. S. Drago, J. I. Zink, R. M. Richman and W. D. Perry, J. Chem. Educ., *51,* 371, 464 (1974), and H. J. Keller and K. E. Schwarzhaus, Angew Chem. Int. Ed., *9,* 196 (1970).
3. For a more rigorous and mathematical discussion of this and the next sections, see J. P. Jesson in "NMR of Paramagnetic Molecules," ed. G. N. LaMar, W. D. Horrocks, Jr., and R. H. Holm, Academic Press, New York (1973).
4. R. J. Kurland and B. R. McGarvey, J. Magn. Res., *2,* 286 (1970).
5. R. M. Golding, Mol. Phys., *8,* 561 (1964).
6. G. N. LaMar, J. P. Jesson, and P. Meakin, J. Amer. Chem. Soc., *93,* 1286 (1971).
7. a. W. D. Horrocks, Jr., and D. D. Hall, Inorg. Chem., *10,* 2368 (1971).
   b. C. Benelli, I. Bertini and D. G. Gatteschi, J. Chem. Soc., Dalton, 661 (1972).
8. R. E. DeSimone and R. S. Drago, J. Amer. Chem. Soc., *92,* 2343 (1970).
9. R. W. Kluiber and W. D. Horrocks, Jr., Inorg. Chem., *6,* 430 (1967).
10. M. L. Wicholas and R. S. Drago, J. Amer. Chem. Soc., *90,* 2196 (1968).
11. W. D. Horrocks, Jr., Inorg. Chem., *9,* 690 (1970).
12. G. N. LaMar and G. R. Van Hecke, J. Amer. Chem. Soc., *91,* 3442 (1969).
13. M. L. Wicholas and R. S. Drago, J. Amer. Chem. Soc., *91,* 5963 (1969).
14. J. I. Zink and R. S. Drago, J. Amer. Chem. Soc., *92,* 5339 (1970).

15. R. J. Fitzgerald and R. S. Drago, J. Amer. Chem. Soc., *89,* 2879 (1967).
16. R. J. Fitzgerald and R. S. Drago, J. Amer. Chem. Soc., *90,* 2523 (1968).
17. R. S. Drago and H. Petersen, Jr., J. Amer. Chem. Soc., *89,* 3978 (1967); *89,* 5774 (1967).
18. M. F. Rettig and R. S. Drago, J. Amer. Chem. Soc., *91,* 1361 (1969); *91,* 3432 (1969).
19. R. E. Cramer and R. S. Drago, J. Amer. Chem. Soc., *92,* 66 (1970).
20. S. E. Anderson, Jr., and R. S. Drago, J. Amer. Chem. Soc., *92,* 4244 (1970); *91,* 3656 (1969); Inorg. Chem., *11,* 1564 (1972).
21. S. E. Anderson, Jr., and N. A. Matwiyoff, Chem. Phys. Lett., *13,* 150 (1972).
22. M. F. Rettig, "NMR of Paramagnetic Molecules," ed. G. N. LaMar, W. D. Horrocks, Jr., and R. H. Holm, Academic Press, New York (1973).
23. D. Doddrell and J. D. Roberts, J. Amer. Chem. Soc., *92,* 6839 (1970).
24. I. Morishima, T. Yonezawa and K. Goto, J. Amer. Chem. Soc., *92,* 6651 (1970).
25. W. D. Horrocks, Jr., and D. L. Johnston, Inorg. Chem., *10,* 1835 (1971).
26. W. D. Horrocks, Jr., Inorg. Chem., *12,* 1211 (1973).
27. W. D. Perry, *et al.,* Inorg. Chem., *10,* 1087 (1971).
28. B. F. G. Johnson, J. Lewis, P. McArdle, and J. R. Norton, J. Chem. Soc., Dalton, *1974,* 1253.
29. G. N. LaMar and L. Sacconi, J. Amer. Chem. Soc., *89,* 2282 (1967).
30. a. P. L. Orioli, M. DiVaira, and L. Sacconi, Inorg. Chem., *10,* 553 (1971).
    b. Private communication, I. Bertini.
31. a. R. H. Holm and C. J. Hawkins, "NMR of Paramagnetic Molecules," ed. G. N. LaMar, W. D. Horrocks, Jr., and K. H. Holm, Academic Press, New York (1973).
    b. G. N. LaMar and G. R. Van Heck, Inorg. Chem., *9,* 1546 (1970).
32. G. N. LaMar, J. Chem. Phys., *41,* 2992 (1964); *43,* 235 (1965).
33. I. M. Walker, L. Rosenthal, and M. S. Quereshi, Inorg. Chem., *10,* 2463 (1971).
34. D. W. Larsen, J. Amer. Chem. Soc., *91,* 2920 (1969); Inorg. Chem., *5,* 1109 (1966).
35. R. H. Fischer and W. D. Horrocks, Jr., Inorg. Chem., *7,* 2659 (1968).
36. D. G. Brown and R. S. Drago, J. Amer. Chem. Soc., *92,* 1871 (1970).
37. J. C. Fanning and R. S. Drago, J. Amer. Chem. Soc., *90,* 3987 (1968).
38. Y. Y. Lim and R. S. Drago, J. Amer. Chem. Soc., *94,* 84 (1972).
39. I. M. Walker and R. S. Drago, J. Amer. Chem. Soc., *90,* 6951 (1968).
40. M. F. Rettig and R. S. Drago, J. Amer. Chem. Soc., *88,* 2966 (1966).
41. D. R. Eaton, Can. J. Chem., *47,* 2645 (1969).
42. D. R. Eaton, *et al.,* Can. J. Chem., *49,* 1218 (1971).
43. C. C. Hinckley, J. Amer. Chem. Soc., *91,* 5160 (1969).
44. P. V. Demarco, *et al.,* J. Amer. Chem. Soc., *92,* 5734 (1970).
45. R. E. Rondeau and R. E. Sievers, J. Amer. Chem. Soc., *93,* 1522 (1971).
46. G. E. Hawkes, *et al.,* "Nuclear Magnetic Shift Reagents," ed. R. E. Sievers, Academic Press, New York (1973).
47. R. E. Davis and M. R. Willcott, III, "Nuclear Magnetic Shift Reagents," ed. R. E. Sievers, Academic Press, New York (1973).
48. R. E. Davis and M. R. Willcott, III, J. Amer. Chem. Soc., *94,* 1744 (1972); O. Gansow, *et al.,* J. Amer. Chem. Soc., *95,* 3389 (1973).
49. G. N. LaMar and E. A. Metz, J. Amer. Chem. Soc., *96,* 5611 (1974).
50. W. D. Horrocks, Jr., *et al.,* Science, *177,* 994 (1972); J. Amer. Chem. Soc., *93,* 6800 (1971).
51. R. E. Cramer and R. Dubois, J. Amer. Chem. Soc., *95,* 3801 (1973).
52. R. E. Cramer, R. Dubois, and K. Seff, J. Amer. Chem. Soc., *96,* 4125 (1974).
53. R. E. Cramer, R. Dubois, and C. K. Furuike, Inorg. Chem., *14,* 1005 (1975).
54. J. P. Briggs, *et al.,* Chem. Commun., 1180 (1972).
55. W. D. Horrocks, Jr., J. Amer. Chem. Soc., *96,* 3022 (1974).
56. C. Marzin, *et al.,* Proc. Nat. Acad. Sci. USA, *70,* 562 (1973).
57. T. Akutani, *et al.,* Tetrahedron Lett., 1115 (1971).
58. S. B. Andrews, *et al.,* Proc. Nat. Acad. Sci. USA, *70,* 1814 (1973) and references therein.
59. C. D. Barry, *et al.,* Biochim. Biophys. Acta, *262,* 101 (1972) and references therein.
60. R. A. Dwek, *et al.,* J. Biochem., *21,* 204 (1971).
61. G. M. Whitesides and D. W. Lewis, J. Amer. Chem. Soc., *92,* 6979 (1970); *93,* 5914 (1971).
62. H. L. Goering, *et al.,* J. Amer. Chem. Soc., *93,* 5913 (1971).
63. R. R. Fraser, *et al.,* Chem. Commun., 1450 (1971).

EXERCISES

1. For each of the following, state whether the contact contribution to the isotropic shift in the proton nmr is close to zero. Also state whether the pseudocontact contribution is zero. Explain all answers.

   a. $Co\left(:N C_5H_5\right)_4^{2+}$ (tetrahedral about Co).

b. The *n*-butanol adduct of $Eu(dpm)_3$, where dpm is

$$(CH_3)_3C-\overset{O}{\overset{\|}{C}}-\underset{H}{\underset{|}{C}}=\overset{O}{\overset{|}{C}}-C(CH_3)_3 \quad (\ominus)$$

2. Oxygen-17 nmr has been observed in aqueous solutions of paramagnetic ions. Water that is coordinated to a paramagnetic ion has an ^{17}O resonance that is shifted far upfield in relation to uncoordinated water. An aqueous solution containing the paramagnetic ion Dy^{3+} has only one ^{17}O resonance. Aqueous solutions of the diamagnetic ion Al^{3+} exhibit two ^{17}O resonances. See diagram of spectra.

   a. Briefly explain why one ^{17}O resonance is observed in aqueous Dy^{3+} solutions, while two are observed in Al^{3+} solutions.

   b. A solution containing 0.1 mole of Dy^{3+} and 10.0 moles of water has an ^{17}O resonance 190 ppm upfield from uncoordinated water. Addition of 0.2 mole of anhydrous Al^{3+} causes this peak to shift to 216 ppm. Explain.

   c. Calculate the number of water molecules in the coordination sphere of Al^{3+}.

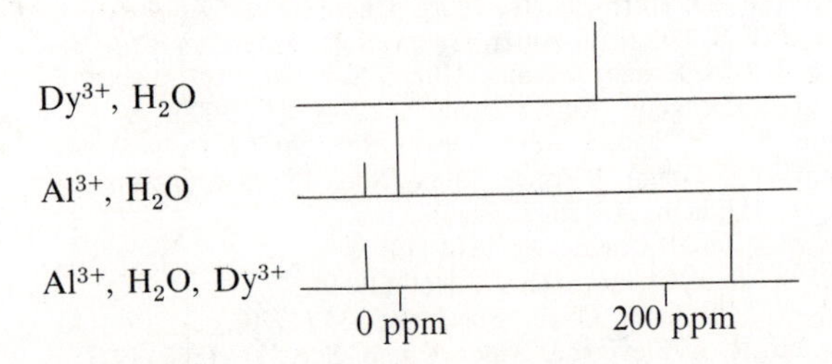

3. The following $\Delta\nu(\text{iso})$ values are reported for $Ni(C_6H_5CH_2NH_2)_6^{2+}$: NH_2, +6313; CH_2, −2083; phenyl protons: *o*, +78; *m*, −95; *p*, +83 (Hz).

   a. Interpret these shifts in terms of delocalization mechanisms.

   b. In interpreting the phenyl shifts, how could you rule out spin in an essentially sigma m.o. polarizing the π-m.o. as the mechanism?

4. a. Predict the number of proton resonances and indicate which ones will shift the most when a shift reagent is added to

$$CH_3-\underset{CF_3}{\underset{|}{\overset{H}{\overset{|}{C}}}}-OH$$

   b. Would a different result be obtained if an optically active shift reagent were employed in part a? Explain.

5. Describe how one could use nmr to distinguish whether or not imidazole (ring with $\overline{N}$—H and $\overline{N}$) were coordinating to Lewis acids through the amine nitrogen or the imine nitrogen.

6. The stable free radical illustrated gives a sharp three-line esr signal in benzene solution at room temperature with a concentration of about 10^{-4} M. No nmr signal is seen even with signal averaging. On increasing radical concentration, the esr signal broadens and is barely discernable at the 1 M level, though now the following nmr signal is seen at 60 MHz.

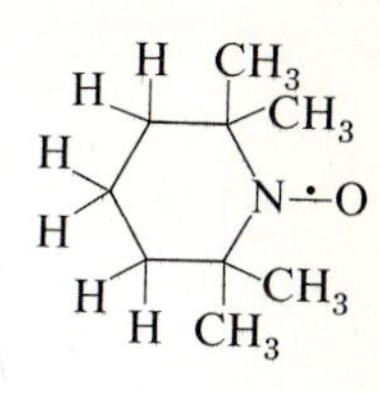

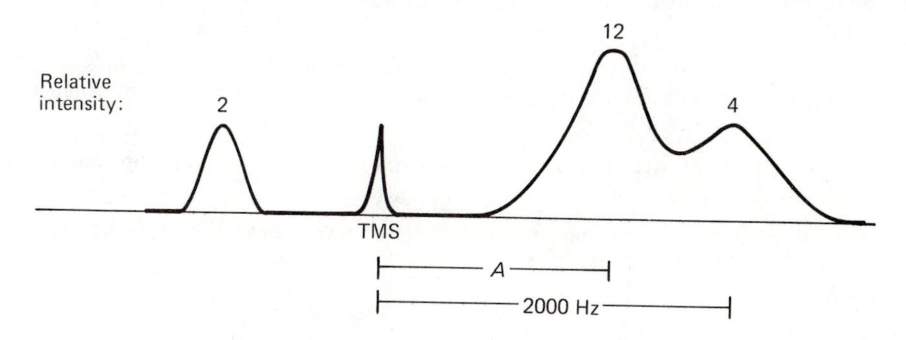

a. Explain the low concentration esr signal and the high concentration nmr signals.

b. Why does the esr signal disappear with increasing concentration while the nmr signal grows with concentration?

c. How could one *predict* the value of *A* obtained in the nmr spectrum above? Discuss what is involved in this calculation.

7. The isotropic hyperfine splittings in the pyridine anion are $a_N = 6.28$, $a_{H(2)} = 3.55$, $a_{H(3)} = 0.82$, and $a_{H(4)} = 9.70$. The numbering system is

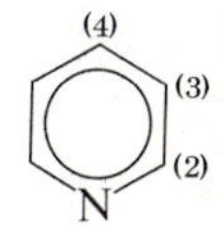

a. Is the unpaired electron in a σ or π m.o.? Explain.

b. What factor(s) would have to be taken into account to interpret a_N in this species?

c. What additional factor would have to be considered in the a_N interpretation of $C_5H_5N\cdot^+$ (*i.e.,* pyridine with one of the lone pair electrons removed)?

d. When six pyridines are coordinated in an octahedral nickel(II) complex, the 2, 3, and 4 protons are shifted downfield -3820, -1420, and -445 Hz, respectively [J. Amer. Chem. Soc., *92,* 66 (1970)]. Explain, using the information given above.

e. What can you say about the pseudocontact term in this complex?

f. When a methyl group is substituted for the hydrogen in the 4 position, the observed shift is $+442$ Hz. Offer an explanation for this reversal.

8. Four-coordinate complexes of Ni(II) may be (1) low-spin ($S = 0$) square planar, (2) high-spin ($S = 1$) tetrahedral, (3) an intermediate D_{2d} geometry with singlet and triplet state close in energy, or (4) situations (1) and (2) in equilibrium and interchanging with a rate constant on the order of 10^7 sec^{-1}.

a. Describe what result you would expect to obtain for each of the above cases if the magnetism were studied as a function of temperature (*i.e.,* Curie-Weiss behavior or not, and why).

b. Could cases (3) and (4) be distinguished by employing infrared, nmr, or esr spectroscopy? (Indicate what is expected, and why, for each of those techniques.)

c. Describe what you would expect to see from a study of the nmr spectrum and its temperature dependence for each of the above four cases.

9. Write short criticisms of the following research projects. If a project will work under certain reasonable circumstances, indicate what they are. If the project is fine, explain why you think it is.

   a. The mixing of the t_{2g} levels in first row transition metal ions with ligands is to be investigated by examining the nmr contact shifts of π-acceptor ligands on octahedral Ni^{2+} ($t_{2g}{}^6e_g{}^2$) and Mn^{2+} ($t_{2g}{}^3e_g{}^2$) complexes.

   b. The existence of covalency in rare earth complexes is to be investigated by measuring the isotropic proton shifts.

   c. Shift reagents and broadening reagents are to be employed to determine the structure of an enzyme in solution.

13 ELECTRON PARAMAGNETIC RESONANCE SPECTRA OF TRANSITION METAL ION COMPLEXES

13–1 INTRODUCTION

This chapter is an extension of Chapter 9; the principles covered there, as well as those in Chapters 10 and 11, should be well understood before this chapter is begun. Ligand field theory serves as the basis for the interpretation of the epr spectra of transition metal ion complexes. It is for this reason that the presentation of the subject matter in this chapter did not follow Chapter 9 directly.

The epr spectra of transition metal ion complexes contain a wealth of information about the electronic structures of these complexes. The additional information and accompanying complications that are characteristic of transition metal ion systems arise because of the approximate degeneracy of the *d*-orbitals and because many of the molecules contain more than one unpaired electron. These properties give rise to orbital contributions and zero-field effects. As a result of appreciable orbital angular moments, the *g*-values for many metal complexes are very anisotropic. Spin-orbit coupling also gives rise to large zero-field splittings (of 10 cm^{-1} or more) by mixing ground and excited states.

An important theorem that summarizes the properties of multi-electron systems is *Kramers' rule*.[1] This rule states that if an ion has an odd number of electrons, the degeneracy of every level must remain at least twofold in the absence of a magnetic field. With an odd number of electrons, m_J quantum numbers will be given by $\pm 1/2$ to $\pm J$. Therefore, any ion with an odd number of electrons must always have as its lowest level at least a doublet, called a *Kramers' doublet*. This degeneracy can then be removed by a magnetic field, and an epr spectrum should be observed. On the other hand, for a system with an even number of electrons, $m_J = 0, \pm 1, \ldots \pm J$. The degeneracy may be completely removed by a low symmetry crystal field, so only singlet levels remain that could be separated by energies so large that an epr transition would not be observed in the microwave region. This discussion is illustrated by the energy level splittings in Fig. 13–1. For the even electron system, the ground state is non-degenerate and the $J = 0$ to 1 transition energy is quite frequently outside the microwave region.

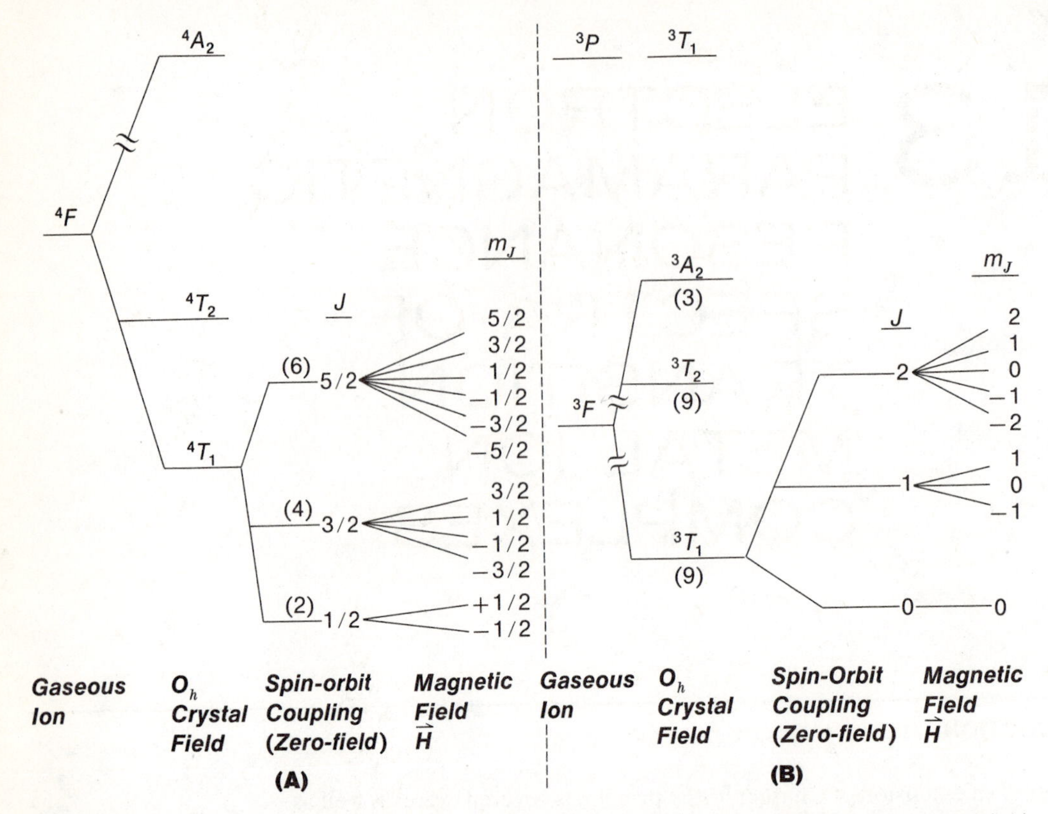

FIGURE 13–1. The splitting of the gaseous ion degeneracy of (A) Co^{2+} and (B) V^{3+} by the crystal field, spin-orbit coupling, and a magnetic field. The T state is regarded as having an effective L, called L', equal to 1; for an A state, $L' = 0$. Then $J' = L' + S, \cdots, L' - S$. Only the zero-field and magnetic field splittings of the ground state are shown.

A few additional comments relative to some experimental procedures (others are given in Chapter 9) are in order in the way of introduction to transition metal systems. A number of factors, other than those that are instrumental, affect the epr line width. As in nmr, spin-lattice, spin-spin, and exchange interactions are important.

Broadening due to spin-lattice relaxation results from the interaction of the paramagnetic ions with the thermal vibrations of the lattice. The variation in spin-lattice relaxation times in different systems is quite large. For some compounds the lifetime is sufficiently long to allow the observation of spectra at room temperature, while in others this is not possible. Since these relaxation times generally increase as the temperature decreases, many of the transition metal compounds need to be cooled to liquid N_2 or He temperatures before well resolved spectra are observed.

Spin-spin interaction results from the magnetic fields that originate in neighboring paramagnetic ions. As a result of these fields, the total field at each ion is slightly altered and the energy levels are shifted. A distribution of energies results, which produces broadening of the signal. Since this effect varies as $(1/r^3) \cdot (1 - 3\cos^2\theta)$, where r is the distance between ions and θ is the angle between the field and the symmetry axis, this kind of broadening will show a marked dependence upon the direction of the field. Since this effect is reduced by increasing the distance between paramagnetic ions, it is often convenient to examine transition metal ion systems by diluting them in an isomorphous diamagnetic host. For example, a copper complex could be studied as a powder or single crystal by diluting it in an analogous zinc host or by examining it in a frozen solution. Dilution of the solid isolates the electron spin of a given complex from that of another paramagnetic molecule, and the spin lifetime is lengthened. (Recall our discussion of the spectra in

Fig. 9–25.) If a frozen solution is used, it must form a good glass; otherwise, paramagnetic aggregates form, which leads to a spectrum with broadened lines. It is often necessary to remove O_2 from the solvent because this can lead to a broadening of the resonance. Even in a well formed glass, one cannot usually detect hyperfine splittings smaller than 3 or 4 gauss.

Line widths are altered considerably by chemical exchange processes. This effect can also be reduced by dilution. If the exchange occurs between equivalent paramagnetic species, the lines broaden at the base and become narrower at the center. When exchange involves dissimilar ions, the resonances of the separate lines merge to produce a single line, which may be broad or narrow depending upon the exchange rate. Such an effect is observed for $CuSO_4 \cdot 5H_2O$, which has two distinct copper sites per unit cell.[2]

In single crystals the anisotropy in the esr parameters can be obtained. Information about the anisotropy in the system can also often be obtained in powders and glasses, because the resulting spectra are not those of a motionally averaged system. We will next consider why anisotropic information is obtained from the spectra even though the molecules in a glass or powder sample exist in an extremely large number of possible orientations relative to the applied field. Consider a molecule with a threefold or higher symmetry axis, which can be described by a $g_{\parallel}$ and a $g_{\perp}$. As can be seen from Fig. 13–2, there are many axes that could be labeled $g_{\perp}$. In a bulk sample

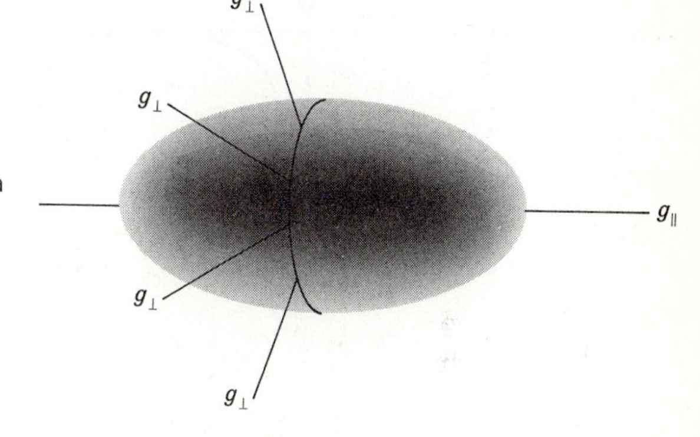

FIGURE 13–2. The $g_{\parallel}$ and some $g_{\perp}$ axes in a crystallite with a threefold or higher axis.

containing many orientations for the assemblage of crystallites, there are more possible orientations that have the $g_{\perp}$ axes aligned with the applied field than there are orientations that have the $g_{\parallel}$ axis aligned. The g-value for any orientation is given by:

$$g^2 = g_{\perp}{}^2 \sin^2 \theta + g_{\parallel}{}^2 \cos^2 \theta \tag{13-1}$$

where θ is the angle that the principal axis (*i.e.*, the $g_{\parallel}$ axis) makes with the applied field. Since all orientations of the crystallites in a solid are equally probable, absorption will occur at all fields between that associated with $g_{\parallel}$ and that associated with $g_{\perp}$. Since many more crystallites have $g_{\perp}$ aligned than $g_{\parallel}$, the most intense absorption will correspond to $g_{\perp}$. When one considers the probabilities of the various orientations and the transition probability corresponding to each of these, the absorption spectrum shown in Fig. 13–3(A) is predicted. This is converted to the derivative spectrum in Fig. 13–3(B). This is an idealized example, and often one finds that the overlapping features generated by $g_{\parallel}$ and $g_{\perp}$ make it difficult to obtain their values.

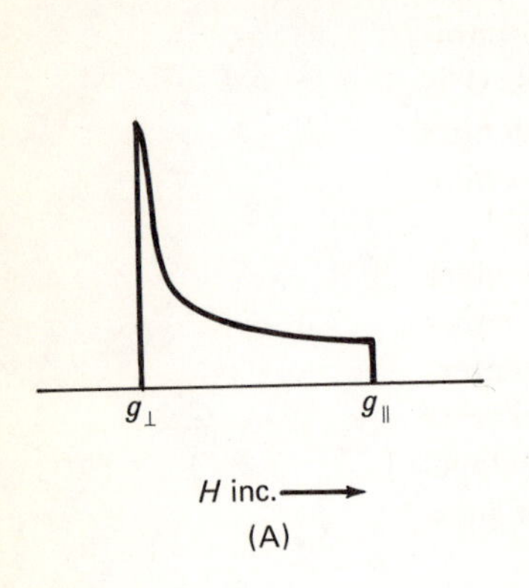

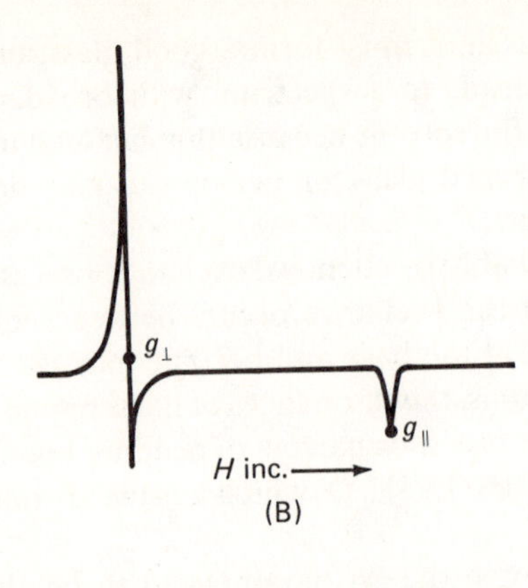

FIGURE 13–3. Idealized absorption (A) and derivative (B) spectra for an unoriented system with S = 1/2, axial symmetry, and no hyperfine interaction ($g_\perp > g_\parallel$).

When the system is orthorhombic and $g_x > g_y > g_z$, the powder spectrum obtained for $I = 0$ is like that in Fig. 13–4(A). When $I = \frac{1}{2}$ and the system has nearly isotropic g-values, but $A_z > A_y > A_x$, the spectrum in Fig. 13–4(B) is expected. The spectrum for a complex with axial symmetry and $I = \frac{1}{2}$, in which $g_\perp > g_\parallel$ and

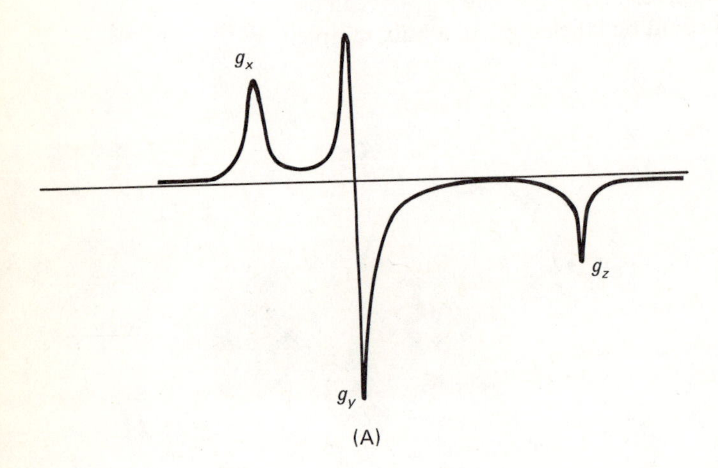

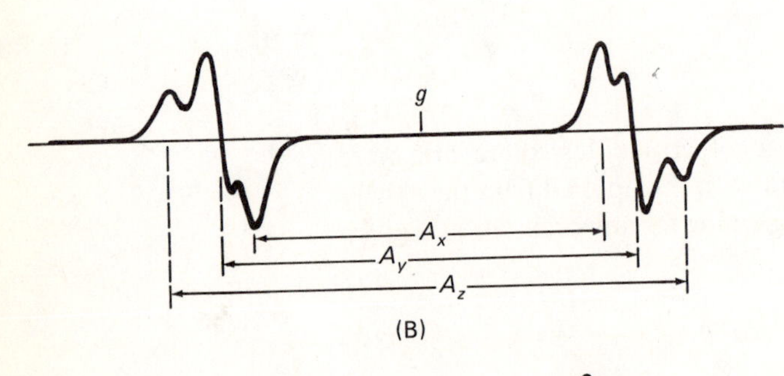

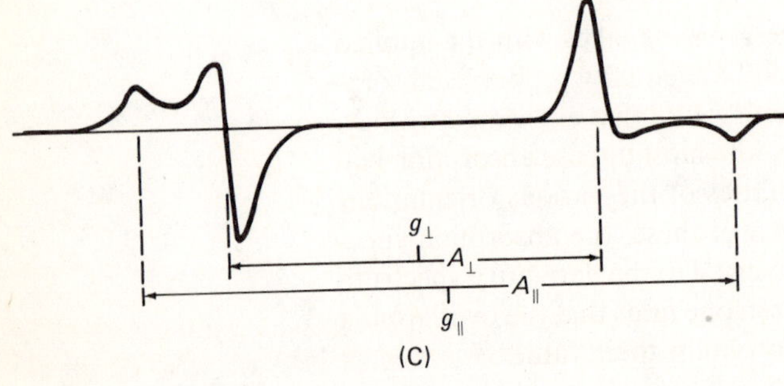

FIGURE 13–4. Powder epr spectra of S = 1/2 systems. (A) An orthorhombic system with $I = 0$; (B) isotropic g with $I = 1/2$ and $A_z > A_y > A_x$; (C) axial symmetry with $I = 1/2$, $g_\perp > g_\parallel$, and $A_\parallel > A_\perp$. In the latter cases, accurate g- and A-values are available only from computer simulation.

$A_{\parallel} > A_{\perp}$, is illustrated in Fig. 13–4(C). Other systems become quite complex, and the possibility for misassignments becomes very large. Only in the relatively simple cases can the g- and A-values be determined with confidence. Computer programs are available to simulate powder epr spectra for simple systems such as Cu(II).

Liquid crystal nematic phases can also be used[3] to orient a molecule for epr work. The molecule to be studied, which is the solute, cannot be spherical; as an example, consider the molecule Co(Meacacen) in Fig. 13–5(A). The liquid crystal solution of this low spin Co(II) complex is placed in a magnetic field to orient the liquid crystal molecules (and, in turn, the solute molecules) and is then cooled. This is schematically illustrated in Fig. 13–5(B). The epr spectrum[4a] in (D) is for the sample oriented relative to the magnetic field as shown in (B), while in spectrum (E) the sample is rotated 90° around the z-axis (*i.e.*, y parallel to the field) relative to the magnetic field. Upon rotation, the portion of the spectrum corresponding to g_2 is enhanced, but that for g_1 is not. One could easily make the mistake of assuming that this is an axial system with g_1 assigned to the z-axis (*i.e.*, $g_{\parallel}$, the axis perpendicular to the plane) and g_2 and g_3 assigned to $g_{\perp}$, where g_x and g_y are similar. However, with the molecular coordinate system as defined in Fig. 13–5(A), g_z must be assigned to g_3, g_x to g_1, and g_y to g_2. These assignments have subsequently been confirmed by a single crystal epr study.[4b] Difficulties can arise in this application if care is not taken to demonstrate that the liquid crystal is not coordinating the complex being studied.

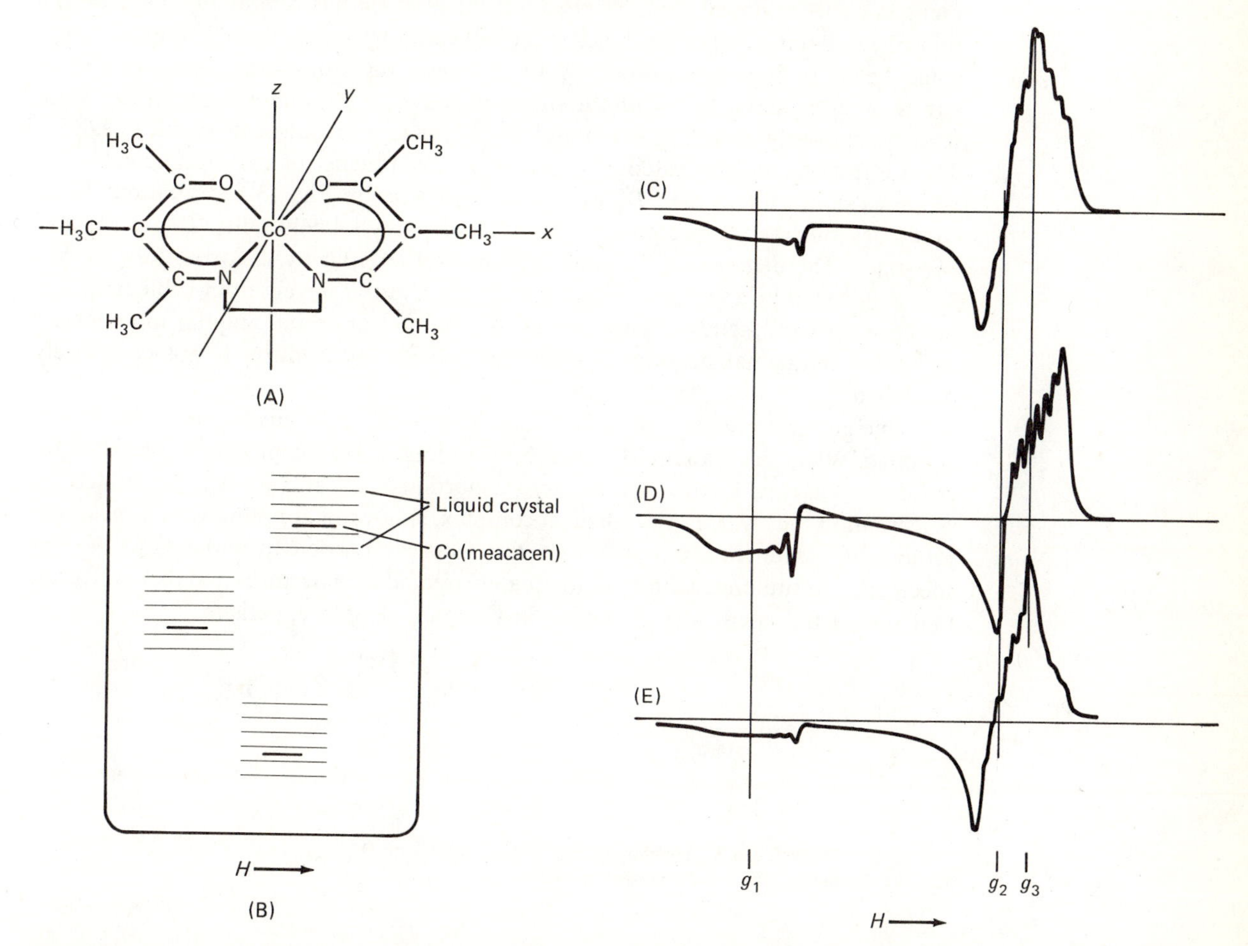

FIGURE 13–5. The epr spectra[4] of Co(Meacacen) at 77°K. (A) Structural formula; (B) orientation of the molecule in a frozen oriented liquid crystal; (C) unoriented frozen solution; (D) frozen liquid crystal oriented as in (B); (E) frozen liquid crystal reoriented 90° from (B). The phasing in this spectrum is inverted relative to what one normally employs. [Reprinted with permission from B. M. Hoffman, F. Basolo, and D. L. Diemente, J. Amer. Chem. Soc., *95*, 6497 (1973). Copyright by the American Chemical Society.]

13-2 INTERPRETATION OF THE *g*-VALUES

Introduction

In contrast to those of organic free radicals, the g-values of transition metal ions can differ appreciably from the free electron value of 2.0023. Such deviations provide considerable information about the electronic structure of the complex. The differences arise because spin-orbit coupling is much greater in many transition metal ion complexes than in organic free radicals (*vide infra*). Thus, spin-orbit effects become essential to an understanding of esr.

The value of g for an unpaired electron in a gaseous atom or ion, for which Russell-Saunders coupling is applicable, was given earlier by the expression

$$g = 1 + \frac{J(J+1) + S(S+1) - L(L+1)}{2J(J+1)} \tag{13-2}$$

In condensed phases, first row transition metal ion systems not only do not have g-values in accord with this expression, but they often deviate from the spin-only value. In condensed phases, the orbital motion of the electron is strongly perturbed and the orbital degeneracy, if it existed before application of the chemical environment, is partly removed or "quenched." If the electron has orbital angular momentum, the angular momentum tends to be bolstered by being weakly coupled to the spin. There is therefore a competition between the quenching effect of the ligands—the "crystal field"—and the sustaining effect of the spin-orbit coupling. Were it not for spin-orbit coupling, we should always observe an isotropic g-value of 2.0023. These effects can be illustrated by considering the influence of a crystalline field on a d^1 ion as shown for O_h and D_{4h} (z-axis compression) in Fig. 13-6. Equation (13-2) would describe the ^2D gaseous ion. The octahedral crystal field splits ^2D into $^2T_{2g}$ and ^2E_g states. The degenerate T_{2g} state may be further split by distortion (*e.g.*, Jahn-Teller effects) or by a tetragonal ligand field into E and B_2 levels. Spin-orbit coupling, on the other hand, tends to preserve a small amount of orbital angular momentum, so, in the tetragonal complex, the orbital angular momentum is not completely quenched.

One generally refers to this as mixing in of a nearby excited state by spin-orbit coupling. When the amount of spin-orbit coupling is small compared to the tetragonal distortion (*i.e.*, in the case of large distortions), the mixing can be treated by perturbation theory. In an octahedral complex, spin-orbit coupling is present in the ground $^2T_{2g}$ state; in order to obtain the accuracy needed to understand the epr spectrum, this situation cannot be adequately treated with perturbation theory. Recall that such a treatment was employed in Chapter 11 on magnetism.

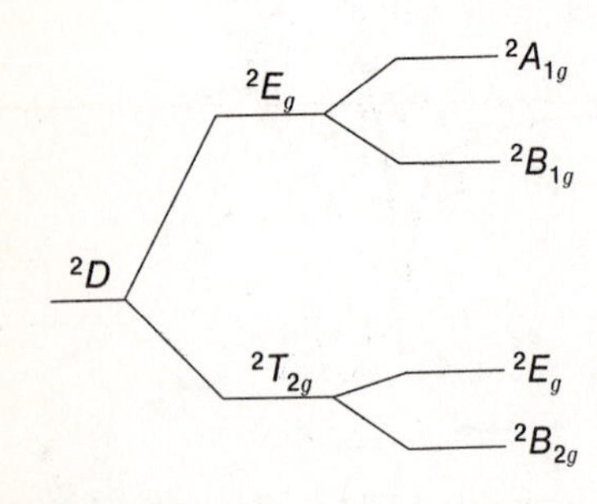

FIGURE 13-6. Splitting of the ^2D state by O_h and D_{4h} fields (z-axis compression in D_{4h}).

$S = \frac{1}{2}$ *Systems with Orbitally Non-degenerate Ground States*

The full Hamiltonian for our system with spin-orbit coupling in a magnetic field is given by

$$\hat{H} = \hat{H}(\text{Zeeman}) + \hat{H}(\text{SO}) = \beta\vec{\mathbf{H}}\cdot(\hat{L} + g_e\hat{S}) + \lambda\hat{L}\cdot\hat{S} \tag{13-3}$$

One of the effects of spin-orbit coupling is to modify the simple one-electron d-orbital wave functions. This is described by the $\lambda\hat{L}\cdot\hat{S}$ term in the Hamiltonian. For example, the spin wave function for the ground state ${}^2B_{2g}$ of a d^1 ion in a tetragonal complex is modified by the spin-orbit interaction $\lambda\hat{L}\cdot\hat{S}$. From first-order perturbation theory, the wave function for the Kramers' doublet $|\pm\rangle$ including spin-orbit effects is given by:

$$|\pm\rangle = N|0\rangle + (1 - N^2)^{-1/2} \sum_{M_L M_S} \frac{\langle n|\lambda\hat{L}\cdot\hat{S}|0\rangle}{E(0) - E(n)}|n\rangle \tag{13-4}$$

The term $|0\rangle$ is the ${}^2B_{2g}$ ground state before spin-orbit effects are considered (*i.e.*, for d^1 with a tetragonal compression, this one electron is in the d_{xy} orbital), while the summation indicates the contribution made by spin-orbit admixture of the excited states. In this example, the ΔE term in the denominator indicates that the ${}^2E$ state will make the largest contribution of all the states that mix in. We can see from this expression that when there is no orbital angular momentum mixed into the ground state, $|\pm\rangle = |0\rangle$. Evaluation of the matrix elements in equation (13–4) gives the coefficients necessary to write the appropriate wave functions. These functions are then used with the Zeeman Hamiltonian in equation (13–3), *i.e.*,

$$\hat{H}' = \beta\hat{L}\cdot\vec{\mathbf{H}} + g_e\beta\hat{S}\cdot\vec{\mathbf{H}},$$

to set up the 2×2 matrix involving $|+\rangle$ and $|-\rangle$. Note that we have worked with the full Hamiltonian in equation (13–3), using the two parts separately. The $\lambda\hat{L}\cdot\hat{S}$ term modified the wave function on which we are now operating with the Zeeman Hamiltonian. The problem is solved by using the raising and lowering operators. Energies are obtained which are expressed as $g\beta Hm_s$, where g is the effective g-factor in the direction of the field component H_i ($i = x, y, z$). In this way, using $\hat{L}_z$, $\hat{S}_z$, and the Zeeman Hamiltonian, we obtain

$$g_z = 2.0023 + 2\lambda \sum_n \left(\frac{\langle 0|\hat{L}_z|n\rangle\langle n|\hat{L}_z|0\rangle}{E(0) - E(n)}\right) \tag{13-5}$$

where $|0\rangle$ is the ground state wave function and $|n\rangle$ is that of one of the n excited states. Since the ground and excited states corresponding to electronic transitions in an $S = \frac{1}{2}$ complex with small spin-orbit coupling are adequately described by real orbital occupations, the state and orbital designations can be used interchangeably. The g-value is thus seen to be very dependent upon the mixing in of the excited state by spin-orbit coupling.

The matrix elements $\langle 0|\hat{L}_z|n\rangle$ are non-zero only when the m_ℓ value of $|0\rangle$ equals the m_ℓ value of $|n\rangle$. In a real orbital basis set with

$$d_{x^2-y^2} = \left(\frac{1}{\sqrt{2}}\right)(|+2\rangle - |-2\rangle)$$

and

$$d_{xy} = \left(\frac{-i}{\sqrt{2}}\right)(|+2\rangle + |-2\rangle)$$

we will have non-zero matrix elements for $\langle d_{x^2-y^2}|\hat{L}_z|d_{xy}\rangle$. (This can be seen from the triple product $\Gamma_{x^2-y^2} \times \Gamma_{L_z} \times \Gamma_{xy}$.) For a d^1 complex, where we can use one-electron orbitals instead of states, an evaluation of all the matrix elements leads to

$$g_z = 2.0023 + \frac{8\xi}{E(0) - E(n)} \tag{13-6}$$

where $\lambda = +\xi/2S$ for a shell that is less than half-filled. The symbol ξ represents the one-electron spin-orbit coupling. When the shell is more than half-filled, we generally think in terms of formation of the positive holes that accompany the electronic transitions. The sign of the second term, $8\xi/[E(0) - E(n)]$, is changed by changing the denominator to $E(n) - E(0)$. When the shell is less than half-filled, spin-orbit effects reduce g from 2.0023; but when the shell is more than half-filled, spin-orbit effects increase g above g_e. $E(0)$ and $E(n)$ are the energies of the ground and excited states, respectively.

In a tetragonal complex we have

$$g_x = 2.0023 + \frac{2\lambda \sum_n \langle 0|\hat{L}_x|n\rangle\langle n|\hat{L}_x|0\rangle}{E(0) - E(n)} \tag{13-7}$$

where $g_x = g_y$. These matrix elements are evaluated by using the raising and lowering operators, so matrix elements are non-zero only when the ground and excited state m_ℓ values differ by ± 1. For a d^1 tetragonal complex (Fig. 13–6) we have:

$$g_x = 2.0023 + \frac{2\xi}{E_{xy} - E_{xz}} \tag{13-8}$$

The results of the evaluation of matrix elements by this procedure can be summarized by writing the following general expression for the g-values of $S = ½$ systems:

$$g = 2.0023 + \frac{\mathbf{n}\xi}{E(0) - E(n)} \tag{13-9}$$

The values of **n** are obtained from the so-called magic pentagon shown in Fig. 13–7, which summarizes the results of the evaluation of matrix elements $\langle 0|\hat{L}|n\rangle$. We will repeat the problem discussed earlier (the d^1 tetragonal field) with the use of this pentagon. We first determine that, for an electron in an xy orbital (*i.e.*, an xy ground state), only electron circulation into the $x^2 - y^2$ orbital could give orbital angular momentum along the z-axis. The quantity g_z then has contributions only from xy and

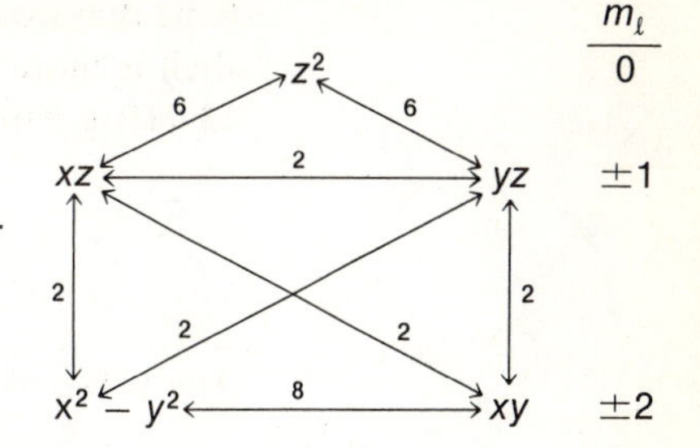

FIGURE 13–7. The "magic pentagon" for evaluating n of equation (13–9).

$x^2 - y^2$, which is seen in Fig. 13–7 to have the value $\mathbf{n} = 8$. Equation (13–9) then becomes:

$$g_z = 2.0023 + \frac{8\xi}{E_{xy} - E_{x^2-y^2}} \tag{13-10}$$

Since g_x has non-zero matrix elements only when $|0\rangle$ and $|n\rangle$ differ by $\Delta m_l = 1$, we obtain

$$g_x = 2.0023 + \frac{2\xi}{E_{xy} - E_{xz}} \tag{13-11}$$

and

$$g_y = 2.0023 + \frac{2\xi}{E_{xy} - E_{yz}} \tag{13-12}$$

These formulae in effect tell us what orbitals permit electron circulations about the respective axes; that is, $x^2 - y^2$ and xy about z, xy and xz about x, or xy and yz about y. The "magic pentagon" is easily constructed. The three rows represent orbitals possessing different m_l values; the top row corresponds to $m_l = 0$, the second row to $m_l = \pm 1$, and the third row to $m_l = \pm 2$. It should be emphasized that this whole treatment arises from a perturbation assumption and is valid only when $\mathbf{n}\xi/[E(0) - E(n)]$ is small compared to the diagonal Zeeman elements. When second-order perturbation theory is pertinent, a term proportional to $\xi^2/\Delta E^2$ is added.

Before concluding this section, an application of the use of g-values will be presented. In Fig. 13–8, the crystal field splittings are indicated for tetragonal

(A)		(B)	
↑	z^2	↑	$x^2 - y^2$
↑↓	$x^2 - y^2$	↑↓	z^2
↑↓ ↑↓	xz, yz	↑↓	xy
↑↓	xy	↑↓ ↑↓	xz, yz

FIGURE 13–8. The d-level splittings for a tetragonally compressed complex (A) and a tetragonally elongated complex (B).

copper(II) complexes with a metal-ligand distance on the fourfold (z) axis that is compressed [part (A)] and elongated [part (B)] along this direction. In (A), with the unpaired electron in d_{z^2} with $m_l = 0$, there is no other orbital with an $m_l = 0$ component, so $g_z = 2.0023$. The value of $g_x = g_y$ is given by

$$g_x = 2.0023 + \frac{6\xi}{E_{yz} - E_{z^2}} \tag{13-13}$$

Note that we have switched the order of the energies in the denominator because the shell is more than half-filled. On the other hand, for a tetragonal elongation [Fig. 13–8(B)], with the unpaired electron in $d_{x^2-y^2}$, we obtain

$$g_z = 2.0023 + \frac{8\xi}{E_{xy} - E_{x^2-y^2}} \tag{13-14}$$

The value of $g_x = g_y$ is given by

$$g_x = 2.0023 + \frac{2\xi}{E_{yz} - E_{x^2-y^2}} \tag{13-15}$$

Thus, the g-values can be used to distinguish the two structures. Complexes with a tetragonal compression are very rare, but elongated ones are common. Copper(II) porphyrin[5] complexes have been reported with g-values of $g_z = 2.70$ and $g_x = 2.04$.

The g-values in some six-coordinate copper(II) complexes exhibit interesting behavior. As mentioned in Chapter 9, when the single crystal epr spectrum for $[Cu(H_2O)_6]SiF_6$, diluted with the corresponding diamagnetic Zn salt, was obtained at 90° K, the spectrum was found to consist of one band with partly resolved hyperfine structure and a nearly isotropic g-value.[6] Jahn-Teller distortion is expected, but there are three distortions with the same energy that will resolve the orbital degeneracy. These are three mutually perpendicular tetragonal distortions (elongation or compression along the three axes connecting *trans* ligands). As a result, three distinguishable epr spectra are expected, one for each species. Since only one transition was found, it was proposed that the crystal field resonates among the three distortions (a so-called dynamic Jahn-Teller distortion). When the temperature is lowered, the spectrum becomes anisotropic and consists of three sets of lines corresponding to three different copper ion environments distorted by three different tetragonal distortions.[8] Other mixed copper salts have been found to undergo similar transitions: $(Cu, Mg)_3La_2(NO_3)_{12} \cdot 24D_2O$ between 33° and 45° K, and $(Zn, Cu)(BrO_3)_2 \cdot 6H_2O$, incomplete below 7° K. The following parameters were reported for $CuSiF_6 \cdot 6H_2O$:[9]

90° K	20° K
$g_\parallel = 2.221 \pm 0.005$	$g_z = 2.46 \pm 0.01$
$g_\perp = 2.230 \pm 0.005$	$g_x = 2.10 \pm 0.01$
$A = 0.0021 \pm 0.0005\ cm^{-1}$	$g_y = 2.10 \pm 0.01$
$B = 0.0028 \pm 0.0005\ cm^{-1}$	$A_z = 0.0110 \pm 0.0003\ cm^{-1}$
	$\left.\begin{matrix} A_x \\ A_y \end{matrix}\right\} < 0.0030\ cm^{-1}$

A similar behavior (*i.e.*, a resonating crystal field at elevated temperatures) was detected in, for example,[10] the spectra of some tris complexes of copper(II) with 2,2′-dipyridine and 1,10-phenanthroline[11] as well as with trisoctamethyl-phosphoramide.[12] This latter ligand, $[(CH_3)_2N]_2P(O)OP(O)[N(CH_3)_2]_2$, is a bidentate chelate in which the phosphoryl oxygens coordinate. The complex can be studied without having to dilute it in a diamagnetic host. A complete single crystal epr study and an x-ray diffraction study are reported on this system.[12] At 90° K, the spectra indicated that at least three different magnetic sites exist, each of which is described by a g-tensor. Analysis of the spectra indicates that these sites correspond to distortions along the x, y, and z axes, respectively, in effect locking in the various extremes in the dynamic Jahn-Teller vibrations occurring at room temperature. The interesting problem of determining the chirality of the molecule in a single crystal epr spectral analysis has also been discussed.[12]

In single crystal epr studies on transition metal ion systems, it is common[13-15] to find complexes in which the g- and A-tensors are not diagonal in the obvious crystal field coordinate system. An axis that is perpendicular to a reflection plane or lies on a rotation axis must be one of the three principal axes of the molecule. The g-tensor of the molecule and the A-tensor for any atom lying on the axis must have principal values along this coordinate. If a molecule contains only a single axis that meets the above requirements, the *other two axes used as the basis for the crystal field analysis will not necessarily be the principal axes of the corresponding tensors of g and A; i.e.,* selecting these axes may not lead to a diagonal tensor. For example, bis(diselenocarbamate) copper(II) has C_{2h} symmetry.[13,14] The twofold rotation axis is one of the axes that diagonalize the corresponding g- and A-tensor components, but the other two components are not diagonal in an axis system corresponding to the axes of the crystal field. Had the molecule possessed D_{2h} symmetry, the three twofold rotation axes of this point group would have been the principal axes for both the A- and g-tensors. Thus, epr studies can provide us with information about the symmetry of the molecule. For a molecule with no symmetry, none of the molecular axes need be coincident with the axes that diagonalize the g-tensor or A-tensor.* As a matter of fact, the axis system that diagonalizes A may not diagonalize g. In vitamin B_{12}, for example,[15] the angle between the principal x, y axis system that leads to a diagonal A-tensor and those that lead to a diagonal g-tensor is 50°.

Systems in which Spin-Orbit Coupling is Large

When spin-orbit coupling is large, perturbation theory cannot be used to give the appropriate wave functions; *i.e.,* equation (13–4) does not apply. An octahedral d^1 complex with a $^2T_{2g}$ ground state is a case in which spin-orbit coupling is large. When the spin-orbit operator $\xi\hat{L}\cdot\hat{S}$ operates on the sixfold degenerate real orbital basis set

$$\psi_1 = (1/\sqrt{2})\left(\left|2, \tfrac{1}{2}\right\rangle - \left|-2, \tfrac{1}{2}\right\rangle\right)$$

$$\psi_2 = (1/\sqrt{2})\left(\left|2, -\tfrac{1}{2}\right\rangle - \left|-2, -\tfrac{1}{2}\right\rangle\right)$$

$$\psi_3 = \left|1, \tfrac{1}{2}\right\rangle$$

$$\psi_4 = \left|1, -\tfrac{1}{2}\right\rangle$$

$$\psi_5 = \left|-1, \tfrac{1}{2}\right\rangle$$

$$\psi_6 = \left|-1, -\tfrac{1}{2}\right\rangle$$

(the wavefunctions are expressed as $|M_L, M_S\rangle$), the result is a new set of orbitals, appropriate for d^1 systems with a large amount of spin-orbit coupling. These are obtained by evaluating the 6×6 determinant containing matrix elements

*For low symmetry molecules, the g- and A-tensors may in fact be asymmetric (*i.e.,* $a_{xy} \neq a_{yx}$). The g^2-tensor will always be symmetric.

$\langle\psi_i|\xi\hat{L}\cdot\hat{S}|\psi_j\rangle$ to obtain the energies. The energies are substituted back into the secular-like equations to obtain the eigenfunctions φ_i listed below.

$$\varphi_1 = (1/\sqrt{3})(\sqrt{2}\psi_1 + \psi_6)$$
$$\varphi_2 = (1/\sqrt{3})(-\sqrt{2}\psi_2 + \psi_3)$$
$$\varphi_3 = \psi_4$$
$$\varphi_4 = \psi_5$$
$$\varphi_5 = (1/\sqrt{3})(\psi_2 + \sqrt{2}\psi_3)$$
$$\varphi_6 = (1/\sqrt{3})(\psi_1 - \sqrt{2}\psi_6)$$

The corresponding energies are $E_1 = -\xi/2$, $E_2 = -\xi/2$, $E_3 = -\xi/2$, $E_4 = -\xi/2$, $E_5 = \xi$, and $E_6 = \xi$. We see from this analysis that spin-orbit coupling has removed the sixfold orbital degeneracy of the T state, giving a twofold set of levels and a lower-energy fourfold set, corresponding to Γ_7 and Γ_8 in the O' double group.* Next, we need to determine the effect of the magnetic field. Since systems with O_h symmetry are magnetically isotropic (x, y, and z), it is only necessary to work out the effect of H_z. The Hamiltonian operator, $\hat{H}$ (parallel to z) is $\beta(\hat{L}_z + g_e\hat{S}_z)H_z$. The resulting energies were derived in Chapter 11 and are indicated in Fig. 13-9. The splitting in the low-energy set is very small, being second order in H (*i.e.*, H^2). When one solves for $g(\Delta E = g\beta H)$, the result $g\beta H = 4\beta^2H^2/3\xi$ or $g = 4\beta H/3\xi \cong 0$ is obtained for the lowest level.† With appreciable separation of Γ_8 and Γ_7 (*e.g.*, $\xi = 154\ \text{cm}^{-1}$ for Ti^{3+}), epr will not be detected unless the Γ_7 state is populated. Solving the expression for $g\beta H$ for these levels as above results in a g-value of 2.0. This state will be populated at room temperature, but the large amount of spin-orbit coupling gives rise to a short τ_e and no spectra are observed. Liquid helium temperatures are needed to lengthen τ_e, but then only the Γ_8 states are populated. Thus, the inability to observe epr spectra on these systems is a direct prediction of crystal field theory.

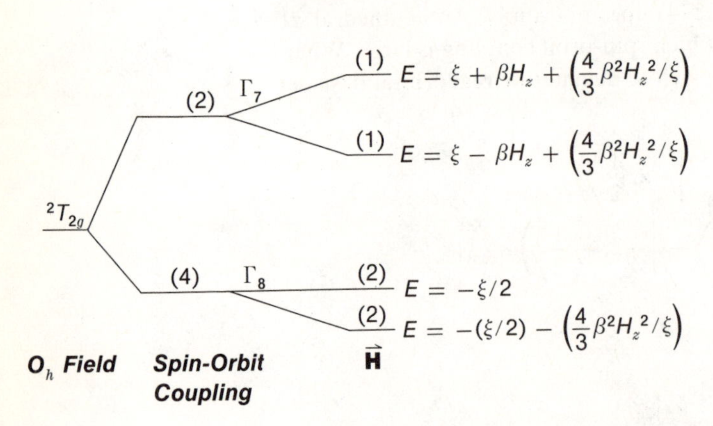

FIGURE 13-9. The influence of spin-orbit coupling and an applied magnetic field on a T state.

13-3 HYPERFINE COUPLINGS AND ZERO FIELD SPLITTINGS

Hyperfine and Zero-Field Effects on the Spectral Appearance

Transition metal systems are rich in information arising from metal hyperfine coupling and zero-field splitting. Fig. 9-14 illustrates the rich cobalt hyperfine

*The Γ_7 and Γ_8 states correspond to $J = \frac{1}{2}$ and $J = \frac{3}{2}$, respectively, for the free ion. These designations are derived by factoring the product of T_2' (orbital part) and E_2' (Γ_6 for spin $= \frac{1}{2}$) in the O' double group.

†In some systems, transitions with very small g-values have been detected.

interaction in $Co_3(CO)_9Se$. Before continuing with this discussion, the reader should review the section on Anisotropy in the Hyperfine Coupling in Chapter 9, as well as that on The EPR of Triplet States, if necessary. The spin Hamiltonian for a single nucleus with spin I and a single effective electronic spin S can be written to include these extra effects as

$$\hat{H} = \beta\vec{\mathbf{H}} \cdot \mathbf{g} \cdot \hat{S} + \hat{S} \cdot \mathbf{D} \cdot \hat{S} + h\hat{S} \cdot \mathbf{A} \cdot \hat{I} - g_N\beta_N\vec{\mathbf{H}} \cdot \hat{I} \tag{13-16}$$

In the previous section, we discussed the $\beta\vec{\mathbf{H}} \cdot \mathbf{g} \cdot \hat{S}$ term and the complications introduced by orbital contributions. The next term incorporates the zero-field effects previously described by the dipolar tensor, **D**, which has a zero trace. The last two terms arise when $I \neq 0$.

In Chapter 9, we discussed zero-field effects that arose from the dipolar interaction of the two or more electron spin moments. In transition metal ion systems, this term is employed to describe any effect that removes the spin degeneracy, including dipolar interactions and spin-orbital splitting. A low symmetry crystal field often gives rise to a large zero-field effect.

In an axially symmetric field (*i.e.*, tetragonal or trigonal), the epr spin Hamiltonian that can be used to fit the observed spectra for effective spin systems lower than quartet takes the form

$$\hat{H}_{\text{spin}} = D\left[\hat{S}_z^2 - \frac{1}{3}S(S+1)\right] + g_{\parallel}\beta H_z\hat{S}_z + g_{\perp}\beta(H_x\hat{S}_x + H_y\hat{S}_y) +$$
$$A_{\parallel}\hat{S}_z\hat{I}_z + A_{\perp}(\hat{S}_x\hat{I}_x + \hat{S}_y\hat{I}_y) + Q'\left[\hat{I}_z^2 - \frac{1}{3}I(I+1)\right] - \gamma\beta_N\vec{\mathbf{H}}_0 \cdot \hat{I}$$

The first term describes the zero-field splitting, the next two terms describe the effect of the magnetic field on the spin multiplicity remaining after zero-field splitting, the terms in $A_{\parallel}$ and $A_{\perp}$ measure the hyperfine splitting parallel and perpendicular to the main axis, and Q' measures the small changes in the spectrum produced by the nuclear quadrupole interaction. All of these effects have been discussed previously (see Chapter 9). The final term takes into account the fact that the nuclear magnetic moment μ_N can interact directly with the external field $\mu_N H_0 = \gamma\beta_N\vec{\mathbf{H}}_0 \cdot \hat{I}$, where γ is the nuclear magnetogyric ratio and β_N is the nuclear Bohr magneton. This is the nuclear Zeeman effect, which gave rise to transitions in nmr. This interaction can affect the paramagnetic resonance spectrum only when the unpaired electrons are coupled to the nucleus by nuclear hyperfine or quadrupole interactions. Even when such coupling occurs, the effect is usually negligible in comparison with the other terms.

In the case of a distortion of lower symmetry, there are three different components g_x, g_y, and g_z, and three different hyperfine interaction constants A_x, A_y, and A_z. Two additional terms need to be included: $E(\hat{S}_x^2 - \hat{S}_y^2)$ as an additional zero-field splitting and $Q''(\hat{I}_x^2 - \hat{I}_y^2)$ as a further quadrupole interaction. The symbols P and P' are often used in place of Q' and Q'', respectively.

The importance of the spin Hamiltonian is that it provides a standard phenomenological way in which the epr spectrum can be described in terms of a small number of constants. Once values for the constants have been determined from experiment, calculations relating these parameters to the electronic configurations and the energy states of the ion are often possible.

The splitting of the ^6S state of an octahedral manganese(II) complex is illustrated in Fig. 13–10(A). Here we have the interesting case in which O_h Mn^{2+} has a $^6A_{1g}$ ground state, which is split by zero-field effects. Spin-orbit coupling mixes into the ground state excited $^4T_2$ states that are split by the crystal field, and this mixing gives

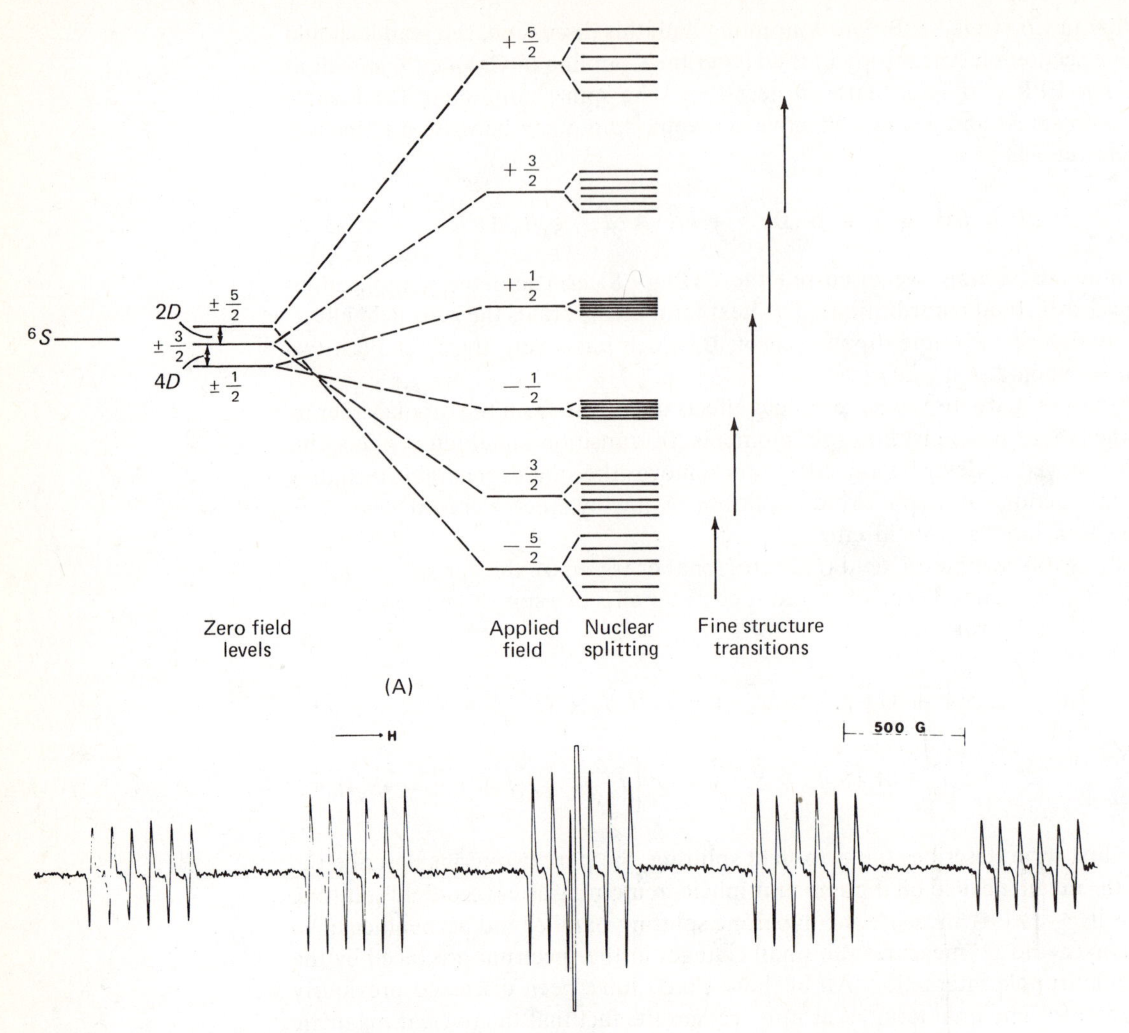

FIGURE 13–10. (A) Splitting of the levels in an octahedral Mn(II) spectrum. (B) Spectrum of a single crystal of Mn^{2+} doped into MgV_2O_6, showing the five allowed transitions (fine structure), each split by the manganese nucleus ($I = 5/2$) (hyperfine structure). At 300°K, $g_x = 2.0042 \pm 0.0005$, $g_y = 2.0092 \pm 0.001$, and $g_z = 2.0005 \pm 0.0005$; $A_x = A_y = A_z = -78 \pm 5$ G; and $D_x = 218 \pm 5$ G, $D_y = -87 \pm 5$ G, and $D_z = -306 \pm 20$ G. [Modified from H. N. Ng and C. Calvo, Can. J. Chem., 50, 3619 (1972). Reproduced by permission of the National Research Council of Canada.]

rise to a small zero-field splitting in Mn^{2+}. The dipolar interaction of the electron spins is small in comparison to the higher state mixing in this complex. The orbital effects are very interesting in this example because the ground state is ^6S, and thus the excited $^4T_2$ state can only be mixed in by second order spin-orbit effects. Thus, the zero-field splitting is relatively small, for example, of the order of 0.5 cm^{-1} in certain manganese(II) porphyrins.[16a] As indicated in Fig. 13-10, the zero-field splitting produces three doubly degenerate spin states, $M_S = \pm 5/2, \pm 3/2, \pm 1/2$, (Kramers' degeneracy). Each of these is split into two singlets by the applied field, producing six levels. As a result of this splitting, five transitions ($-5/2 \rightarrow -3/2$, $-3/2 \rightarrow -1/2$, $-1/2 \rightarrow 1/2$, $1/2 \rightarrow 3/2$, $3/2 \rightarrow 5/2$) are expected. The spectrum is further split by the nuclear hyperfine interaction with the manganese nucleus ($I = 5/2$). This would give rise to thirty peaks in the spectrum.

In contrast to hyperfine splitting, the term *fine splitting* is used when an absorption band is split because of non-degeneracy arising from zero-field splitting. Components of fine splitting have varying intensities: the intensity is greatest for the central lines and smallest for the outermost lines. In simple cases, the separation between lines varies as $3 \cos^2 \theta - 1$, where θ is again the angle between the direction of the field and the molecular z-axis.

In Fig. 13–11, the influence of zero-field splitting on an $S = 1$ system is indicated for a fixed molecular orientation. (Recall that there is no Kramers' degeneracy.) In the absence of zero-field effects [Fig. 13–11(A)], the two $|\Delta m_s| = 1$ transitions are degenerate and only one peak would result.

For the splitting shown in Fig. 13–11(B), two transitions would be observed in the spectrum. A specific example of this type of system is the $^3A_{2g}$ ground state of nickel(II) in an O_h field. Spin-orbit coupling mixes in excited states, which split the $^3A_{2g}$ configuration. *Recall that zero-field splitting is very anisotropic, providing a relaxation mechanism for the electron spin state.* Accordingly, epr spectra of Ni(II) O_h complexes are difficult to detect, and when they are studied, liquid nitrogen or helium temperature must generally be employed. At room temperature, nmr spectra can be measured. In some systems, sharp double quantum transitions ($|\Delta m_s| = 2$) can be seen in the epr spectrum.[16b]

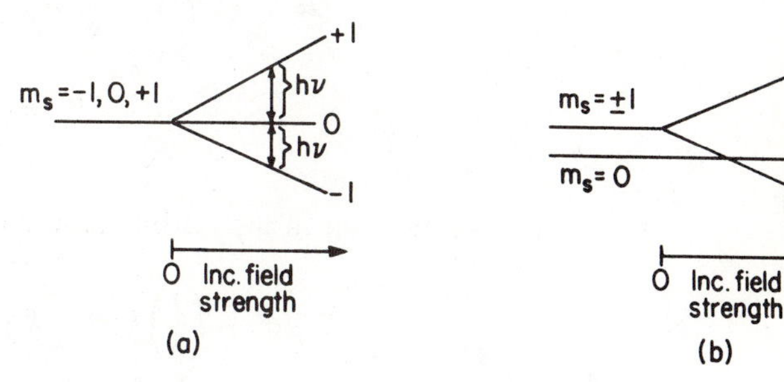

FIGURE 13–11. The energy level diagram and transitions for a molecule or ion with S = 1, (A) in the absence of and (B) in the presence of zero-field splitting. The system in (B) is aligned with the z-axis, for the effect is very anisotropic.

As mentioned in Chapter 9, the zero-field splitting can be so large that the $\Delta m_s = \pm 1$ transitions are not observed. For example, in V^{3+} (d^2), $\Delta m_s = \pm 1$ transitions are not detected, but one observes a weak $\Delta m_s = 2$ transition ($-1 \rightarrow +1$) split into eight lines by the nuclear spin of ^{51}V ($I = 7/2$). The mechanism whereby the $\Delta m_s = 2$ transition becomes allowed is the same as that discussed for organic triplet states in Chapter 9.

The next topic to be discussed is the definition of the term *effective spin, S′*. We have already been using this idea, but now formally define it to describe how some of the effects that we have discussed are incorporated into the spin Hamiltonian. When a cubic crystal field leaves an orbitally degenerate ground state (*e.g.*, a T state), the effect of lower symmetry fields and spin-orbit coupling will remove this degeneracy as well as the spin degeneracy. In the case of an odd number of unpaired electrons, Kramers' degeneracy leaves the lowest spin state doubly degenerate. If the splitting is large, this doublet will be well isolated from higher-lying doublets. Transitions will then be observed only in the low-lying doublet, which behaves like a simpler system having $S = 1/2$. We then say that the system has an effective spin S' of only $1/2$ ($S' = 1/2$). An example is Co^{2+}. The cubic field leaves a ^4F ground state which, as a result of lower symmetry fields and spin-orbit coupling, gives rise to six doublets. When the lowest doublet is separated from the next by appreciably more than kT, the effective spin has a value of $1/2$ ($S' = 1/2$) instead of $3/2$. If the effective spin S' is different from the spin S, a spin Hamiltonian can be written in terms of S' rather than S.

It should be clear that all of the effects discussed above can have a pronounced influence on the spectral appearance. The qualitative interpretation of the epr spectra of transition metal ions by inspection is thus not trivial. Proficiency is obtained by looking at many spectra and drawing analogies to known systems when dealing with new systems. Practice will be afforded in a later section where the representative spectra of different d^n systems will be considered, and also in the exercises at the end of the chapter.

Contributions to A

The hyperfine coupling interaction has contributions to it from Fermi contact, dipolar nuclear spin–electron spin, and nuclear spin–electron orbit mechanisms. These effects have been discussed in Chapter 9. The reader is referred to equation (9–25) and the subsequent discussion of it for a treatment of the dipolar contribution. *The trace of the tensor in equation (9–25) is zero, so information about the dipolar contribution can be obtained only from ordered or partially ordered systems.* As mentioned in equations (9–34) and (9–35), where the traceless tensor components of A were indicated by T, the contribution to dipolar hyperfine coupling for an electron in a p_z orbital is given by

$$T_{zz} = \left(\frac{4}{5}\right)P_p \qquad T_{xx} = -\left(\frac{2}{5}\right)P_p \qquad T_{yy} = -\left(\frac{2}{5}\right)P_p$$

with

$$P_p = g_e\beta g_N\beta_N\left\langle\frac{1}{r^3}\right\rangle_p \tag{13-17}$$

For an electron in a p_x orbital, we have

$$T_{zz} = -\left(\frac{2}{5}\right)P_p \qquad T_{yy} = -\left(\frac{2}{5}\right)P_p \qquad T_{xx} = \left(\frac{4}{5}\right)P_p$$

Similar expressions can be derived for an electron in one of the d-orbitals. The quantities in Table 13–1 must be multiplied by

$$P_d = g_e\beta g_N\beta_N\left\langle\frac{1}{r^3}\right\rangle_d \tag{13-18}$$

to obtain the dipolar contribution. The signs and magnitudes can be easily remembered. The orbital is located in the $3\cos^2\theta - 1$ plot of the lines of flux from the nuclear moment, as shown in Fig. 13–12 for the d_{xz} orbital. In this figure, the z-axis of the molecule is aligned with the z-axis of the field to give T_{zz} as a small positive number. Next rotate the orbital counterclockwise 90°, without rotating the cone, so that the x-axis of the molecule is aligned parallel to the field (which is still along the z-axis shown in Fig. 13–12) to give a small positive T_{xx}. Next, start with Fig. 13–12

TABLE 13–1. CONTRIBUTIONS TO THE DIPOLAR HYPERFINE COUPLING FROM ELECTRONS IN d-ORBITALS

Orbital	$T_{zz}(P_d)$	$T_{xx}(P_d)$	$T_{yy}(P_d)$
d_{z^2}	$4/7$	$-2/7$	$-2/7$
$d_{x^2-y^2}$	$-4/7$	$2/7$	$2/7$
d_{yz}	$2/7$	$-4/7$	$2/7$
d_{xz}	$2/7$	$2/7$	$-4/7$
d_{xy}	$-4/7$	$2/7$	$2/7$

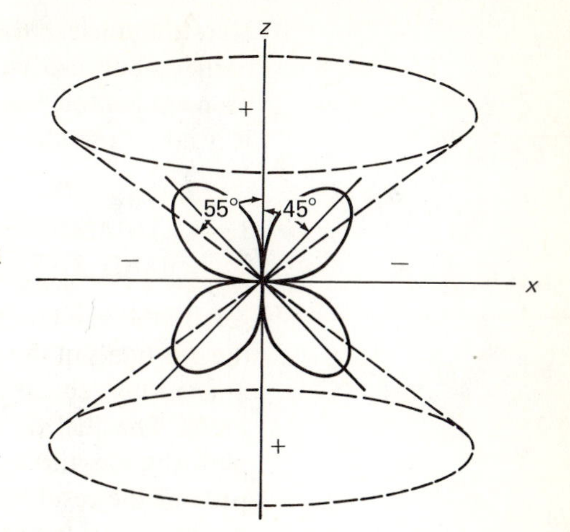

FIGURE 13–12. Orientation of the d_{xz} orbital relative to the $3\cos^2\theta - 1$ cone.

and rotate the molecule so that the z-axis of the molecule is perpendicular to the page and the y-axis of the molecule is parallel to the field (*i.e.*, the z-axis in Fig. 13–12). Now the lobes of d_{xz} are in the negative region of the cone and a large negative T_{yy} is expected. These rotations correspond to the three mutually perpendicular orientations of the molecule relative to the field. With information on the signs and magnitudes of the components of the hyperfine tensor, one can obtain information about the atomic orbitals in a complex that make the principal contribution to the molecular orbital containing the unpaired electron.

The Fermi contact contribution has been discussed in detail in Chapters 9 and 12. Unpaired spin density is felt at the nucleus by direct admixture of the s-orbitals into the m.o. containing the unpaired electron and by spin polarization of filled inner s-orbitals by unpaired electron density in d-orbitals. When the $4s$ orbital of the metal is empty, it can mix into the largely metal d-antibonding orbital; and if this m.o. contains unpaired spin, the electron partly occupies the metal $4s$ orbital.

Spin polarization can have two different results. When an inner $1s$ or $2s$ orbital is spin-polarized by α-spin in a $3d$-orbital, an excess of β-spin results at the nucleus. When the filled $3s$ orbital is spin-polarized by an electron with α-spin in a $3d$-orbital, α-spin results at the nucleus. The effect of direct delocalization into the $4s$ orbital can be shown by comparing Fermi contact hyperfine values ($A_{\text{F.C.}}$) determined for various cobalt(II) complexes. It has been found[17a] that $A_{\text{F.C.}}$ for sixfold coordination falls in the -30 to -45 G region, for fivefold coordination in the -5 to -25 G region, and for square planar fourfold coordination in the 0 G region. One can write an equation summarizing contributions to $A_{\text{F.C.}}$ in the form

$$A_{\text{F.C.}} = x(A_{4s}) + (1 - x)(A_{3d}) \qquad (13\text{-}19)$$

where A_{4s} is the direct hyperfine interaction of one unpaired electron in a $4s$ orbital, $1320 \text{ G} \times g\beta$; and A_{3d} is the hyperfine interaction arising from spin-polarization of filled "s" orbitals by an unpaired electron in a $3d$ orbital, $-90 \text{ G} \times g\beta$. Applying this equation to the results above, one obtains $x = 3$ to 4%, $x = 4.5$ to 6%, and $x = 6.5$% for admixture of $4s$ into the cobalt complexes studied in six-, five-, and fourfold coordination, respectively, in cobalt(II).

The nuclear spin-electron orbit contribution to the coupling constant is related to the pseudocontact contribution discussed in Chapter 12. The Hamiltonian is

$$\hat{H} = \left(\frac{P_d}{7}\right)[-(\vec{\mathbf{L}}\cdot\vec{\mathbf{S}})(\vec{\mathbf{L}}\cdot\vec{\mathbf{I}}) - (\vec{\mathbf{L}}\cdot\vec{\mathbf{I}})(\vec{\mathbf{L}}\cdot\vec{\mathbf{S}})] \qquad (13\text{-}20)$$

Here the nuclear moment is interacting not only with the spin moment (discussed earlier), but also with the orbital moment. The nuclear interaction with the spin moment is a traceless tensor, but the interaction with the orbital moment is not. Since it is not, a pseudocontact contribution is observed in the nmr spectrum in solution.

13–4 LIGAND FIELD INTERPRETATION OF THE g- AND A-TENSORS FOR S′ = ½ SYSTEMS

In the analysis, the interpretation of the epr spectra is carried out with the use of the *d*-orbitals of the complex as the basis. Covalency is introduced subsequent to the analysis by reducing the free ion values of the spin-orbit coupling parameter ξ and $\langle r^{-3}\rangle$. The real orbital basis set is mixed by spin-orbit coupling using first order perturbation theory with the spin-orbit interaction Hamiltonian $\hat{\mathbf{l}}\cdot\hat{\mathbf{s}}$. We shall present the results for several *d*-electron configurations that we shall subsequently discuss in our treatment of specific examples. We have already discussed the evaluation of the *g*-tensor component expressions.

d^9 Configurations

We shall use the hole formalism so that ξ is negative; the case for an $x^2 - y^2$ ground state will be treated. A Kramers' doublet lies lowest, which we shall indicate with the symbol ψ_a^+, ψ_a^-, where the + and − refer to the m_s value. This is a specific example of the application of equation (13–4), where we are now defining Γ by the appropriate wave function ψ_a^+, ψ_a^-. The wave functions mixed by spin-orbit coupling are:

$$\psi_a^+ = N_a|d_{x^2-y^2}^+\rangle + ia_1|d_{xy}^+\rangle - \frac{1}{2}a_2|d_{yz}^-\rangle - \frac{i}{2}a_3|d_{xz}^-\rangle \qquad (13\text{-}21)$$

$$\psi_a^- = N_a|d_{x^2-y^2}^-\rangle - ia_1|d_{xy}^-\rangle + \frac{1}{2}a_2|d_{yz}^+\rangle + \frac{i}{2}a_3|d_{xz}^+\rangle \qquad (13\text{-}22)$$

Here N_a is a normalization constant; the *a* values are given by $a_i = \xi/E_{ia}$, where E_{1a} is, for example, given by $E(|d_{x^2-y^2}\rangle) - E(|d_{xy}\rangle)$ and E_{2a} by $E(|d_{x^2-y^2}\rangle) - E(|d_{xz}\rangle)$. The *a* values are small, so N_a is close to 1. Setting N_a equal to 1, we obtain (as discussed earlier in this chapter):

$$g_{xx} = 2\langle\psi_a^+|\sum_k \hat{l}_{xk} + g_e\hat{s}_{xk}|\psi_a^-\rangle = g_e + \frac{2\xi}{E_{x^2-y^2} - E_{xz}} = g_e + 2a_3 \qquad (13\text{-}23)$$

$$g_{yy} = 2i\langle\psi_a^+|\sum_k \hat{l}_{yk} + g_e\hat{s}_{yk}|\psi_a^-\rangle = g_e + \frac{2\xi}{E_{x^2-y^2} - E_{yz}} = g_e + 2a_2 \qquad (13\text{-}24)$$

$$g_{zz} = 2\langle\psi_a^+|\sum_k \hat{l}_{zk} + g_e\hat{s}_{zk}|\psi_a^-\rangle = g_e + \frac{8\xi}{E_{x^2-y^2} - E_{xy}} = g_e + 8a_1 \qquad (13\text{-}25)$$

(Note that these results are those obtained by using the magic pentagon.)

In order to evaluate *A*, we again consider all matrix elements of ψ_a^+ and ψ_a^- with the *x*, *y*, and *z* components of the operator $\sum_k \left(\hat{l}_k - \mathcal{K}\hat{s}_k + \frac{1}{7}\hat{a}_k\right) = \hat{A}$, where $\hat{a}_k = 4\hat{s}_k - (\hat{l}_k\cdot\hat{s}_k)\hat{l}_k - \hat{l}_k(\hat{l}_k\cdot\hat{s}_k)$. This form of the operator is discussed in reference 17. The reader interested in the evaluation of these matrix elements should be aware of the phase problem associated with the choice of a wave function. See references 17, 18, and 19 for a complete discussion. Here we shall simply present the results.

$$A_{xx} = 2P\langle\psi_a^+|\hat{A}_x|\psi_a^-\rangle$$
$$= P_d\left[+\frac{2\xi}{E_{x^2-y^2} - E_{xz}} - \mathcal{K} + \frac{2}{7} - \frac{3}{7}\frac{\xi}{E_{x^2-y^2} - E_{yz}}\right]$$
$$= P_d\left[+2a_3 - \mathcal{K} + \frac{2}{7} - \frac{3}{7}a_2\right] \quad (13\text{-}26)$$

$$A_{yy} = 2iP\langle\psi_a^+|\hat{A}_y|\psi_a^-\rangle$$
$$= P_d\left[+\frac{2\xi}{E_{x^2-y^2} - E_{yz}} - \mathcal{K} + \frac{2}{7} - \frac{3}{7}\frac{\xi}{E_{x^2-y^2} - E_{xz}}\right]$$
$$= P_d\left[+2a_2 - \mathcal{K} + \frac{2}{7} - \frac{3}{7}a_3\right] \quad (13\text{-}27)$$

$$A_{zz} = 2P\langle\psi_a^+|\hat{A}_z|\psi_a^-\rangle$$
$$= P_d\left[+\frac{8\xi}{E_{x^2-y^2} - E_{xy}} - \mathcal{K} - \frac{4}{7} + \frac{3}{7}\left(\frac{\xi}{E_{x^2-y^2} - E_{yz}} + \frac{\xi}{E_{x^2-y^2} - E_{xz}}\right)\right]$$
$$= P_d\left[+8a_1 - \mathcal{K} - \frac{4}{7} + \frac{3}{7}(a_2 + a_3)\right] \quad (13\text{-}28)$$

The summation k is over electron holes (one in this system), and $P_d = g_e g_N \beta\beta_N\langle r^{-3}\rangle$. The symbol $\mathcal{K}P_d$ is the Fermi contact contribution to the coupling, the terms $\pm\frac{2}{7}P_d$ and $\pm\frac{4}{7}P_d$ represent the dipolar contribution, and the other terms correspond to the nuclear spin-electron orbit terms. In solution, an isotropic A would be obtained in which

$$A_{\text{iso}} = \frac{1}{3}(A_{xx} + A_{yy} + A_{zz})$$
$$= P_d\left[-\mathcal{K} + 2\left(\frac{\xi}{E_{x^2-y^2} - E_{xz}} + \frac{\xi}{E_{x^2-y^2} - E_{yz}}\right)\right] \quad (13\text{-}29)$$

The term in parentheses is the counterpart of the pseudocontact contribution in nmr.

Low-spin cobalt(II) complexes have two holes in $d_{x^2-y^2}$ and one in d_{z^2}. Using the magic pentagon, one obtains g_{zz}, g_{xx}, and g_{yy} as follows:

$$g_{zz} = 2.0023 \quad (13\text{-}30)$$

$$g_{xx} = 2.0023 + \frac{6\xi}{E_{z^2} - E_{yz}} = 2.0023 + 6b_1 \quad (13\text{-}31)$$

where $b_1 = \xi/(E_{z^2} - E_{yz})$, and

$$g_{yy} = 2.0023 + \frac{6\xi}{E_{z^2} - E_{xy}} = 2.0023 + 6b_2 \quad (13\text{-}32)$$

where $b_2 = \xi/(E_{z^2} - E_{xy})$. The hyperfine components are given by:

$$A_{xx} = P_d\left(-\frac{2}{7} - \mathcal{K} + 6b_1 + \frac{3}{7}b_2\right) \quad (13\text{-}33)$$

$$A_{yy} = P_d\left(-\frac{2}{7} - \mathcal{K} + 6b_2 + \frac{3}{7}b_1\right) \quad (13\text{-}34)$$

$$A_{zz} = P_d\left[+\frac{4}{7} - \mathcal{K} - \frac{3}{7}(b_1 + b_2)\right] \quad (13\text{-}35)$$

With the appropriate expressions for the g-values and the free ion ξ values, it is possible to calculate the g-values for several complexes and compare these with the experimental values. The calculated results are those that would be predicted by crystal field theory. The results are summarized[20] in Table 13–2.

TABLE 13–2. COMPARISON OF EXPERIMENTAL AND THEORETICAL VALUES OF g*

Complex		g, expt.	g, theor.	λ(complex)/λ(ion)
	VO^{2+}	λ(ion) = 170 cm^{-1}	($g = g_\parallel$)	
$VO[(CH_3CO)_2CH]_2$		1.948	1.921	0.67
	V^{2+}	λ(ion) = 56 cm^{-1}	($g = g_\parallel = g_\perp$)	
$V(H_2O)_6^{2+}$		1.965	1.964	0.99
	Cr^{3+}	λ(ion) = 91 cm^{-1}	($g = g_\parallel = g_\perp$)	
$Cr(H_2O)_6^{3+}$		1.977	1.961	0.61
$Cr(NH_3)_6^{3+}$		1.986	1.969	0.49
$Cr[(CH_3CO)_2CH]_3$		1.981	1.961	0.51
$Cr(en)_3^{3+}$		1.987	1.969	0.45
$Cr(CN)_6^{3-}$		1.992	1.975	0.37
	Ni^{2+}	λ(ion) = −324 cm^{-1}	($g = g_\parallel = g_\perp$)	
$Ni(H_2O)_6^{2+}$		2.25	2.30	0.83
$Ni(NH_3)_6^{2+}$		2.16	2.24	0.66
	Cu^{2+}	λ(ion) = −828 cm^{-1}	($g = g_\parallel$)	
$Cu(H_2O)_4^{2+}$		2.46	2.593	0.77
$Cu(NH_3)_4^{2+}$		2.223	2.468	0.47
$Cu[(CH_3CO)_2CH]_2$		2.266	2.443	0.60
$Cu[NHCHC_6H_5O]_2$		2.200	2.408	0.49
$Cu[S_2CN(C_2H_5)_2]_2$		2.098	2.347	0.28

*From B. R. McGarvey, "Electron Spin Resonance of Transition Metal Complexes," in "Transition Metal Chemistry," Vol. 3, ed. R. L. Carlin, Marcel Dekker, New York (1967).

One notes that invariably the experimental value is closer to 2.0023 than the result predicted by crystal field theory. The discrepancy can be removed by assigning an empirical effective value to ξ or λ so as to make the calculated value of g agree with the experimental one. The extent of the discrepancy of the simple crystal field model is then indicated by the ratio λ(complex)/λ(gaseous ion). An increase in the amount of covalency in a complex would cause this ratio to decrease. In ligand field theory, one adjusts ξ (or λ) and P from the free ion values. The reduced values are then often interpreted in terms of covalent effects (*vide infra*).

The use of these equations and the information available is illustrated by considering the epr spectrum of bis(maleonitriledithionato) copper(II), *i.e.*,

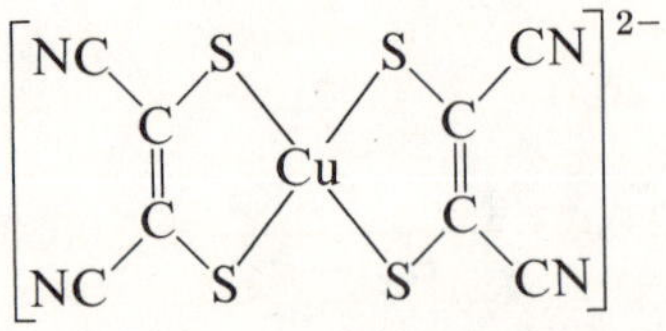

This complex was diluted into the diamagnetic nickel(II) complex and a single crystal esr study was carried out.[17] The coordinate systems in which the A- and g-tensors are diagonal were coincident, leading to the values $g_{xx} = 2.026 \pm 0.001$, $g_{yy} = 2.023 \pm 0.001$, and $g_{zz} = 2.086 \pm 0.001$. The corresponding principal values of A were $(39 \pm 1) \times 10^{-4}$ cm^{-1}, $(39 \pm 1) \times 10^{-4}$ cm^{-1}, and $(162 \pm 2) \times 10^{-4}$ cm^{-1}, respectively. The A_{iso} value obtained from spectra of solutions is 76×10^{-4} cm^{-1}. From the g-values and equations (13–23) to (13–25) for a $d_{x^2-y^2}$ ground state, one can obtain values of $a_1 \approx a_2 \approx a_3 \approx 0.01$. Substituting these values into equations (13–26) to (13–28) and solving for A (a $d_{x^2-y^2}$ ground state), we obtain:

$$A_{\parallel} = P_d[-0.48 - \mathcal{K}] \tag{13-36}$$

$$A_{\perp} = P_d[+0.30 - \mathcal{K}] \tag{13-37}$$

Since A_{iso} has a value of 76×10^{-4} cm^{-1}, $A_{\parallel}$ and $A_{\perp}$ must have the same sign; *e.g.*, $(39 + 39 + 162) \div 3 \approx 76$. Since $|A_{\parallel}| > |A_{\perp}|$, the value of $\mathcal{K}$ must be positive; it usually is. To get a larger absolute value for $A_{\parallel}$ than for $A_{\perp}$, we can see from equations (13–36) and (13–37) that the sign of $A_{\parallel}$ would have to be negative. Since we now know the signs and magnitudes of $A_{\parallel}$ and $A_{\perp}$, we can solve equations (13–36) and (13–37) for P_d and $\mathcal{K}$. The results are $P_d \cong 1.6 \times 10^{-2}$ cm^{-1} (compare to 3.5×10^{-2} cm^{-1} in the free ion) and $\mathcal{K} \cong 0.55$. The Fermi contact hyperfine coupling is given by $A = (0.55)(1.6 \times 10^{-2}$ cm$^{-1})$. This analysis often confirms a tentative assignment of the ground state, for an incorrect assignment could lead to unreasonable values for $\mathcal{K}$ and P_d.

The reduction of P_d from the free ion value is interpreted as being due to covalent interactions with ligands. However, one must use chemical intuition when a P value is calculated from A values whose signs are unknown. For example, in a study[21] of x-irradiation of $K_3Co(CN)_6$, a paramagnetic species was obtained that has $g_{\parallel} = 2.010$, $g_{\perp} = 2.120$, $A_{\parallel}/g\beta = 83.5$ G, and $A_{\perp}/g\beta = 26.9$ G. P_d was found to be 0.0088 cm^{-1}, compared with the free ion value $P_d \approx 0.023$ cm^{-1}, implying that the orbital containing the unpaired electron is only 38% d in character. If $A_{\parallel}$ and $A_{\perp}$ had been assumed to be of opposite signs, P_d would be 0.0147, a much more reasonable value.

It is also inadvisable to compare only restricted regions of the epr spectra in a series of compounds, for example, only $g_{\parallel}$ or $A_{\perp}$, to infer differences in electronic structure. For instance, many five-coordinate Co(II) complexes[22] possess $A_{\parallel}$ in the region from 90 to 100×10^{-4} cm^{-1}, with P_d often $\sim$0.017 to 0.020 cm^{-1}. Aqueous $Co(CH_3NC)_5^{2+}$ in excess CH_3NC[23] has an $A_{\parallel}$ value of 61×10^{-4} cm^{-1}, implying at first glance a larger covalency than usual. However, P_d is calculated to be 0.0180, in accord with many other compounds studied.

Pitfalls exist in interpreting the Fermi contact term, $\mathcal{K}$, in terms of covalency. When the 4s orbital is of appropriate symmetry to mix with a d-orbital possessing an unpaired electron, a direct contribution to $\mathcal{K}$ results. Such a study was discussed in conjunction with equation (13–18).

The previous arguments have been developed using P_d and ξ parameters, whose values in complexes are reduced from the free ion values. An alternative but analogous approach[20,24] utilizes molecular orbital coefficients of d-orbitals from the metal and ligand orbitals. For example, in D_{4h} symmetry, neglecting the explicit form of ligand orbitals, the following one-electron orbitals are found:

$$a_{1g} = \delta_1(d_{z^2}) - \delta_2(\text{ligand})$$

$$b_{2g} = \alpha_1(d_{xy}) - \alpha_2(\text{ligand})$$

$$b_{1g} = \beta_1(d_{x^2-y^2}) - \beta_2(\text{ligand})$$

$$e_{g(1)} = \gamma_1(d_{xz}) - \gamma_2(\text{ligand})$$

$$e_{g(2)} = \gamma_1(d_{yz}) - \gamma_2(\text{ligand})$$

Derivation of g- and A-values for an unpaired electron in the b_{2g} orbital results in the following equations:

$$g_{\parallel} = 2.0023 - \frac{8\xi\alpha_1^2\beta_1^2}{\Delta E(b_{1g} - b_{2g})}$$

$$g_{\perp} = 2.0023 - \frac{2\xi\alpha_1^2\gamma_1^2}{\Delta E(e_g - b_{2g})}$$

$$A_{\parallel} = P_d\left[-\alpha_1^2\left(\frac{4}{7} + \mathcal{K}\right) + (g_{\parallel} - 2) + \frac{3}{7}(g_{\perp} - 2)\right]$$

$$A_{\perp} = P_d\left[-\alpha_1^2\left(-\frac{2}{7} + \mathcal{K}\right) + \frac{11}{14}(g_{\perp} - 2)\right]$$

In these equations, ξ and P_d are given the free ion values; thus, $\alpha_1^2\beta_1^2 = \dfrac{\xi(\text{complex})}{\xi(\text{ion})}$.

Many studies have been reported in which the above analysis is carried out on copper(II) systems. However, the resulting trends in the α^2 values are not all that meaningful for copper(II) complexes. Two contradictory interpretations of the trends in terms of trends in metal-ligand covalency have been offered.[25,26]

13–5 LIGAND HYPERFINE COUPLINGS

Hyperfine splittings from ligand atoms have contributions from a Fermi contact term, F.C.; dipolar contributions from the metal, DIP; dipolar contributions from electron density in p-orbitals of the ligands, LDP; and the metal pseudocontact contribution at the ligand, LPC, which results from the interaction of the orbital angular momentum of the unpaired electron with the ligand nuclear spin. When ligand hyperfine structure is resolved, this latter term is generally small compared to the other contributions. When there is extensive spin-orbit coupling, a large pseudocontact contribution would be expected, but relaxation effects lead to difficulty in observing the epr spectrum and consequently the ligand hyperfine splittings. The $A_{\parallel}$ and $A_{\perp}$ values are given by equations (13–38) and (13–39):

$$A_{\parallel} = A_{\text{F.C.}} + 2(A_{\text{DIP}} + A_{\text{LDP}}) \qquad (13\text{-}38)$$

$$A_{\perp} = A_{\text{F.C.}} - (A_{\text{DIP}} + A_{\text{LDP}}) \qquad (13\text{-}39)$$

These equations refer to the parallel and perpendicular components of the ligand hyperfine tensor. When an investigator measures the ligand hyperfine interaction (A_l) by the ligand fine structure of a metal hyperfine peak (for example, nitrogen super hyperfine structure on a cobalt hyperfine peak), difficulty will arise if the two hyperfine coupling tensors are not diagonal in the same coordinate system. Single crystal x-ray diffraction and single crystal epr studies are necessary for complete understanding of these systems. If one carries out a solution epr study with the ligand hyperfine structure resolved, $A_{\text{F.C.}}$ can be measured directly. Equations (13–38) and (13–39) cannot be solved for A_{DIP} and A_{LDP}. However, a reasonable value for A_{DIP} can be calculated from a knowledge of the structure, by use of equation (13–40):

$$A_{\text{DIP}} = \frac{g_e g_N \beta\beta_N}{a^3}(\text{cm}^{-1}) \qquad (13\text{-}40)$$

where a is the metal-ligand distance. Equation (13–40) is derived for circumstances in which the metal-ligand distance is large compared to the metal nucleus–electron distance. Where this approximation is poor, other equations have been reported.[27] The $A_{\text{F.C.}}$ and A_{LDP} values are related to the s-orbital contribution α_s and p-orbital contribution α_p to the molecular orbital by the following equations:

$$A_{\text{F.C.}} = \frac{16\pi}{3}\gamma\beta\beta_N|\psi(0)|^2\alpha_s^{\,2} \tag{13-41}$$

$$A_{\text{LDP}} = \frac{2}{5}\gamma\beta\beta_N\left\langle\frac{1}{r^3}\right\rangle_p \alpha_p^{\,2} \tag{13-42}$$

Values of $A_{\text{F.C.}}$ and A_{LDP} are reported[28] for one electron in s- and p-orbitals from which the $\%s$ and $\%p$ character can be calculated. From these equations, the ratio $\%s/\%p$ can be determined to indicate the "hybridization of the ligand" in the complex. This treatment tacitly assumes that delocalization occurs mainly by a direct mechanism; *i.e.*, spin polarization contributions are ignored.

The application of these ideas will be illustrated with a few brief examples. The epr spectrum[29] of

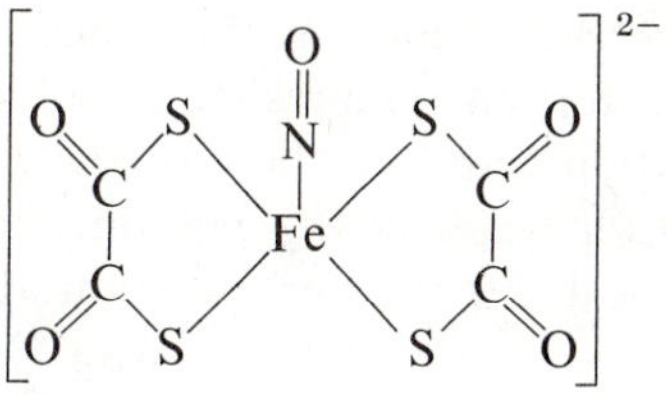

indicates $A_{\parallel}(^{14}\text{N})/g\beta = 15.6$ G, $A_{\perp}(^{14}\text{N})/g\beta = 12.7$ G, and $A_{\text{iso}} = 13.7$ G. Values of A reported for a full electron in s- and p-orbitals are $A_p^{\circ}/g\beta = 34.1$ G and $A_s^{\circ}/g\beta = 550$ G. This leads to a ratio of $\%p/\%s$ of about 1, which indicates an sp hybrid and a linear Fe—N—O structure. On the other hand, the ratio of p/s for the nitrosyl nitrogen[30] of $Fe(CN)_5NO^{3-}$ was 1.6, consistent with a bent metal-nitrosyl structure (an sp^2 hybrid would have had a ratio of 2).

Recently, phosphorus hyperfine couplings were observed[31] in the epr spectra of $(C_6H_5)_3P$ and PF_3 adducts of tetraphenylporphyrin cobalt(II). The p/s ratio was reported as 2.7 in the former case and 0.47 in the latter.

13–6 SURVEY OF THE EPR SPECTRA OF FIRST ROW TRANSITION METAL ION COMPLEXES

In this section, we shall briefly survey some of the results of epr studies on various d^n complexes. For a more complete discussion, the reader is referred to references 19 and 20. Before beginning this survey, we should mention that *spin-orbit coupling provides the dominant mechanism for electron relaxation in these systems.* In your reading, you will find statements like "zero-field splitting causes rapid relaxation" or "g-value anisotropy leads to a short electron spin lifetime, etc." It is to be emphasized that these are all manifestations of the effects of spin-orbit coupling in the molecule. We have previously discussed the relationship of spin-orbit coupling to these effects. Second and third row transition metal complexes become increasingly more difficult to study by the epr technique because the spin-orbit coupling constants are much larger.

d^1

The results of studies on ions with the electronic configuration d^1 can be fitted with the spin Hamiltonian:

$$\hat{H} = \beta[g_x H_x \hat{S}_x + g_y H_y \hat{S}_y + g_z H_z \hat{S}_z] + A_x \hat{S}_x \hat{I}_x + A_y \hat{S}_y \hat{I}_y + A_z \hat{S}_z \hat{I}_z \quad (13\text{-}43)$$

In an octahedral ligand field, the ground state is $^2T_{2g}$ and there is considerable spin-orbit coupling in this ground state. All of the Kramers' doublets from the $^2T_{2g}$ term are close in energy and extensively mixed by spin-orbit coupling. This leads to a short τ_e. In a tetrahedral ligand field, an ^2E ground state ($x^2 - y^2, z^2$) with no first order spin-orbit coupling results. In this geometry, admixture of the nearby $^2T_{2g}$ excited states into the ground state by second order* spin-orbit coupling provides a mechanism leading to a short spin relaxation time for the electron and broad absorption lines. The complexes usually must be studied at temperatures approaching that of liquid helium. The ^2T excited state is split by spin-orbit coupling. When the ligand field is distorted (*e.g.*, in VO^{2+}), the ground state becomes orbitally singlet and the excited states are well removed. Sharp epr spectral lines result at higher temperatures.

Equations have been derived,[20] using the d^1 wave functions and the appropriate spin Hamiltonian, to relate the g-value in a trigonally distorted complex to the amount of distortion. The distortion is expressed in terms of δ (cm^{-1}), the splitting of the ^2T state. A large distortion with $\delta = 2000$ to 4000 cm^{-1} is found in tris(acetylacetonato) titanium(III). As a result of this splitting, the electron spin lifetime is increased and one is able to detect the epr spectrum at room temperature.

d^2

Very few examples of epr spectra of these ions in octahedral complexes have been reported because of the extensive spin-orbit coupling in the $^3T_1$ ground state. Tetrahedral complexes have an $^3A_2$ ground state, so we would expect longer relaxation times and more readily observed epr spectra. The spectra of these systems can be fitted with $S = 1$ and the spin Hamiltonian

$$\hat{H} = g_z\beta H_z\hat{S}_z + g_x\beta H_x\hat{S}_x + g_y\beta H_y\hat{S}_y + D\left[\hat{S}_z^{\,2} - \frac{2}{3}\right] + E[\hat{S}_x^{\,2} - \hat{S}_y^{\,2}] + A\hat{S}_z\hat{I}_z + B[\hat{S}_x\hat{I}_x + \hat{S}_y\hat{I}_y] \quad (13\text{-}44)$$

V^{3+} in an octahedral environment[32] in Al_2O_3 gave $g_\parallel = 1.92$, $g_\perp = 1.63$, $D = +7.85$, and $A = 102$. An example of a T_d complex that has been studied[33] is V^{3+} in a CdS lattice. At liquid N_2 temperature, a g-value of 1.93 and D of 1 cm^{-1} are observed.

d^3

Octahedral d^3 complexes have an $^4A_2$ ground state, which must have a Kramers' doublet lowest in energy. When the zero-field splitting is small, as shown in Fig. 13–13(A), three transitions can sometimes be detected and the zero-field parameter can be obtained from the two affected transitions. When the zero-field splitting is large compared to the spectrometer frequency, only one line will be observed, as shown in Fig. 13–13(B). In general, the spectra can be fitted to the spin Hamiltonian:

---

*In second order perturbation theory, the E component of this split state is mixed into the ground state.

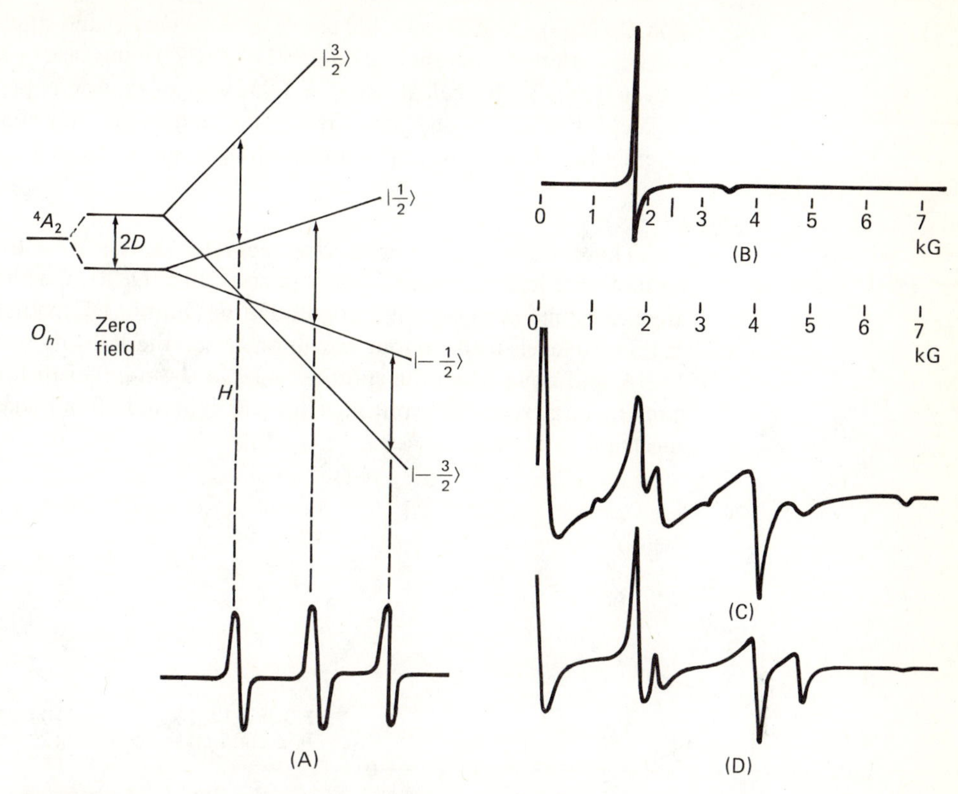

FIGURE 13–13. (A) Small zero-field and magnetic field splitting of the $^4A_2$ ground state (field along z) for a d^3 case and the resulting spectrum.
(B) *Trans*-$[Cr(C_5H_5N)_4I_2]^+$ in DMF, H_2O, CH_3OH glass[33] at 9.3 GHz. $D > 0.4\ cm^{-1}$, $E < 0.01$.
(C) *Trans*-$[Cr(C_5H_5N)_4Cl_2]^+$ in DMF, H_2O, CH_3OH glass[33] at 9.211 GHz.
(D) Computer simulation[33] of (C) with $g_\parallel = g_\perp = 1.99$, $D = 0.164\ cm^{-1}$, $E = 0$.
[Reprinted with permission from E. Pedersen and H. Toftlund, Inorg. Chem., *13*, 1603 (1974). Copyright by the American Chemical Society.]

$$\hat{H} = g_z\beta H_z\hat{S}_z + g_x\beta H_x\hat{S}_x + g_y\beta H_y\hat{S}_y + D\left[\hat{S}_z^{\,2} - \frac{5}{4}\right] + E[\hat{S}_x^{\,2} - \hat{S}_y^{\,2}] + A_\parallel\hat{S}_z\hat{I}_z + A_\perp[\hat{S}_x\hat{I}_x + \hat{S}_y\hat{I}_y] \quad (13\text{-}45)$$

It is difficult to recognize "typical patterns" for Cr^{3+} in some systems.[33] The spectrum shown in Fig. 13–13(D) was calculated by a computer, using an isotropic g, $D = 0.164\ cm^{-1}$, and $E = 0\ cm^{-1}$. Reference 33 contains many spectra of tetragonal Cr(III) complexes and a detailed analysis of them.

The d^3 system has been very extensively studied, particularly Cr^{3+}. In octahedral complexes, the metal electrons are in t_{2g} orbitals, so ligand hyperfine couplings are usually small. The g-value for this system is given, according to crystal field theory, by

$$g = 2.0023 - \frac{8\lambda}{\Delta E(^4T_{2g} - {}^4A_{2g})} \quad (13\text{-}46)$$

The ground state, being $^4A_{2g}$, has no spin-orbit coupling and a small amount is mixed in *via* the $^4T_{2g}$ state. Equation (13–46) differs from those presented earlier in two ways. The spin-orbit coupling is described by λ (which can be positive or negative) and characterizes a state. With more than one unpaired electron, the energy differences also must be expressed in terms of the energy differences of the appropriate electronic states. Calculating g for $V(H_2O)_6^{2+}$ using $\Delta E = 11{,}800\ cm^{-1}$ and $\lambda = 56\ cm^{-1}$ gives a value $g = 1.964$, which is close to the observed[34] value of 1.972.

For $Cr(H_2O)_6^{3+}$, $\Delta E = 17{,}400\ cm^{-1}$, $\lambda = 91\ cm^{-1}$, and the predicted g-value is smaller than the experimental value[(35)] of 1.977. In the case of Mn^{4+}, the discrepancy is even larger, with a calculated $g = 1.955$ and an experimental value of 1.994. This is in keeping with the fact that the crystal field approximations are poorer and covalency becomes more important as the charge of the central ion increases.

d^4

There are very few epr spectra reported for this d-electron configuration. The ground state in this system, in a weak crystalline O_h field, which is ^5E, has no orbital angular momentum, so S is a good quantum number. Zero-field splitting of the ± 2, ± 1, and 0 levels leads to four transitions when the splitting is small, as shown in Fig. 13–14, and none when the splitting is large. Jahn-Teller distortions and the accompanying large zero-field splittings that are expected often make it impossible to see a spectrum.

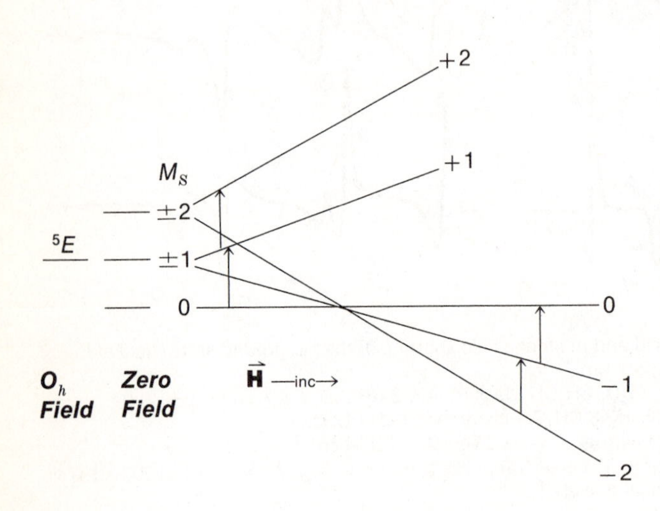

FIGURE 13–14. Zero-field and magnetic field splitting of the ^5E ground state (field along z) for a d^4 case.

For low spin d^4 complexes, the reader is referred to the d^2 section (recall the hole formalism).

d^5 Low Spin, $S = \frac{1}{2}$

In a strong ligand field of octahedral symmetry, the ground state is $^2T_2$. Spin-orbit coupling splits this term into three closely spaced Kramers' doublets; however, epr spectra can be seen only at temperatures close to those of liquid helium because of the large amount of spin-orbit coupling present. Since there are five d-electrons in these systems, the situation is analogous to d^1, except that in this case we are working with a positive hole. Jahn-Teller forces tend to distort systems such as MX_6^{n-}, so the g-values contained in equation (13–9) are rarely observed. The splitting of the free ion doublet state by an O_h field, a D_3 distortion, spin-orbit coupling, and a magnetic field are shown in Fig. 13–15. Since we have non-integral spin, the double group representations are employed when spin-orbit coupling is considered, and primed symbols are employed for the representations. If one defines a distortion parameter, δ, as in Fig. 13–15, equations for the g-values can be derived:

$$g_{\parallel} = \frac{3(\xi + 2\delta)}{[(\xi + 2\delta)^2 + 8\xi^2]^{1/2}} - 1 \qquad (13\text{–}47)$$

$$g_{\perp} = \frac{(2\delta - 3\xi)}{[(\xi + 2\delta)^2 + 8\xi^2]^{1/2}} + 1 \qquad (13\text{–}48)$$

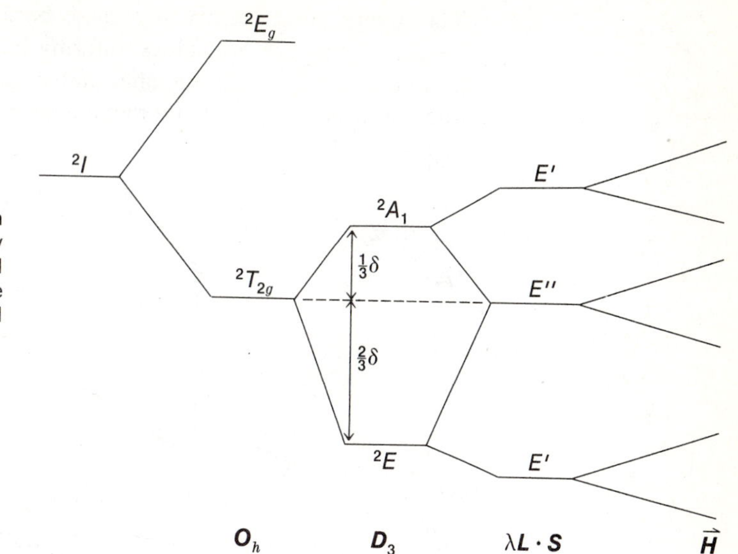

FIGURE 13–15. Energy level diagram for a low spin d^5 system in an O_h field and a D_3 field, followed by spin-orbit coupling and the magnetic field–induced splittings. With spin-orbit effects included, the double group D_3' representations are employed (primed values).

Note that as $\delta \rightarrow 0$, both $g_{\parallel}$ and $g_{\perp} \rightarrow 0$, in accord with the equations in Fig. 13–9. For example, Fe^{3+} in $K_3Co(CN)_6$ exhibits[37] an epr spectrum with $g_{\parallel} = 0.915$ and $g_{\perp} \approx 2.2$. Substituting into equation (13–47) and employing $\xi = -103$ cm^{-1} for the free ion produces a value of $\delta \approx 200$ cm^{-1}.

Large deviations from octahedral symmetry cause an orbitally singlet state to lie lowest in energy, well removed from orbitally non-degenerate excited states. Longer electron relaxation times result, and epr spectra can be observed at higher temperatures. Examples of such a system are low spin derivatives of ferric hemoglobin[38] (Fig. 13–16), which possesses a large tetragonal distortion as a result of the heme plane. Examples of bases, B, that produce a low spin environment are N_3^-, CN^-, and OH^-. Experimental g-values for the N_3^- species are $g_x = 1.72$, $g_y = 2.22$, and $g_z = 2.80$. The large anisotropy in g_x and g_y is thought to arise from interaction of a specific iron d-orbital with a nitrogen π orbital from the coordinated histidine group that ties the globin to the heme unit. An epr study[39] of bipyridyl and phenanthrolene complexes of iron(III), ruthenium(III), and osmium(III) has been analyzed in terms of an energy diagram similar to Fig. 13–15, but one in which the distortion could be shown to produce the ^2A state lower in energy than ^2E; *i.e.*, δ is negative. Spin delocalization onto the ligand was studied as a function of the metal in this series.

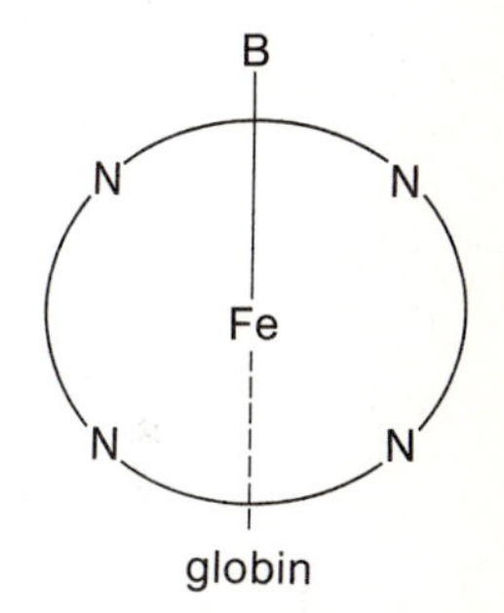

FIGURE 13–16. Schematic formula for hemoglobin.

d^5 *High Spin*

This d-electron configuration has been very thoroughly studied. The high spin complexes have ^6S ground states, and there are no other sextet states. The $^4T_1$ is the closest other term, and second order spin-orbit coupling effects are needed to mix in this configuration, so the contributions are small. Thus, the electron spin lifetime is long and epr spectra are easily detected at room temperature in all symmetry crystal fields. Furthermore, with an odd number of electrons, Kramers' degeneracy exists even when there is large zero-field splitting. We have illustrated the energy levels of Mn(II) in Fig. 13–10. The results for high spin complexes can be fitted by:

$$\hat{H} = g\beta\vec{\mathbf{H}} \cdot \hat{S} + D\left[\hat{S}_z^{\,2} - \frac{35}{12}\right] + E[\hat{S}_x^{\,2} - \hat{S}_y^{\,2}] + A\hat{S} \cdot \hat{I}$$

$$+ \frac{1}{6}a\left[\hat{S}_x^{\,4} + \hat{S}_y^{\,4} + \hat{S}_z^{\,4} - \frac{707}{16}\right] + \frac{1}{180}F\left[35\hat{S}_z^{\,4} - \frac{475}{2}\hat{S}_z^{\,2} + \frac{3255}{16}\right] \quad (13\text{-}49)$$

The higher power terms in $\hat{S}$ arise because the octahedral crystal field operator couples states with M_S values differing by ± 4, leading to a more complex basis set and more non-zero off-diagonal matrix elements. The splitting of the energy levels and the spectrum expected for an undistorted octahedral iron(III) complex are shown in Fig. 13–17.

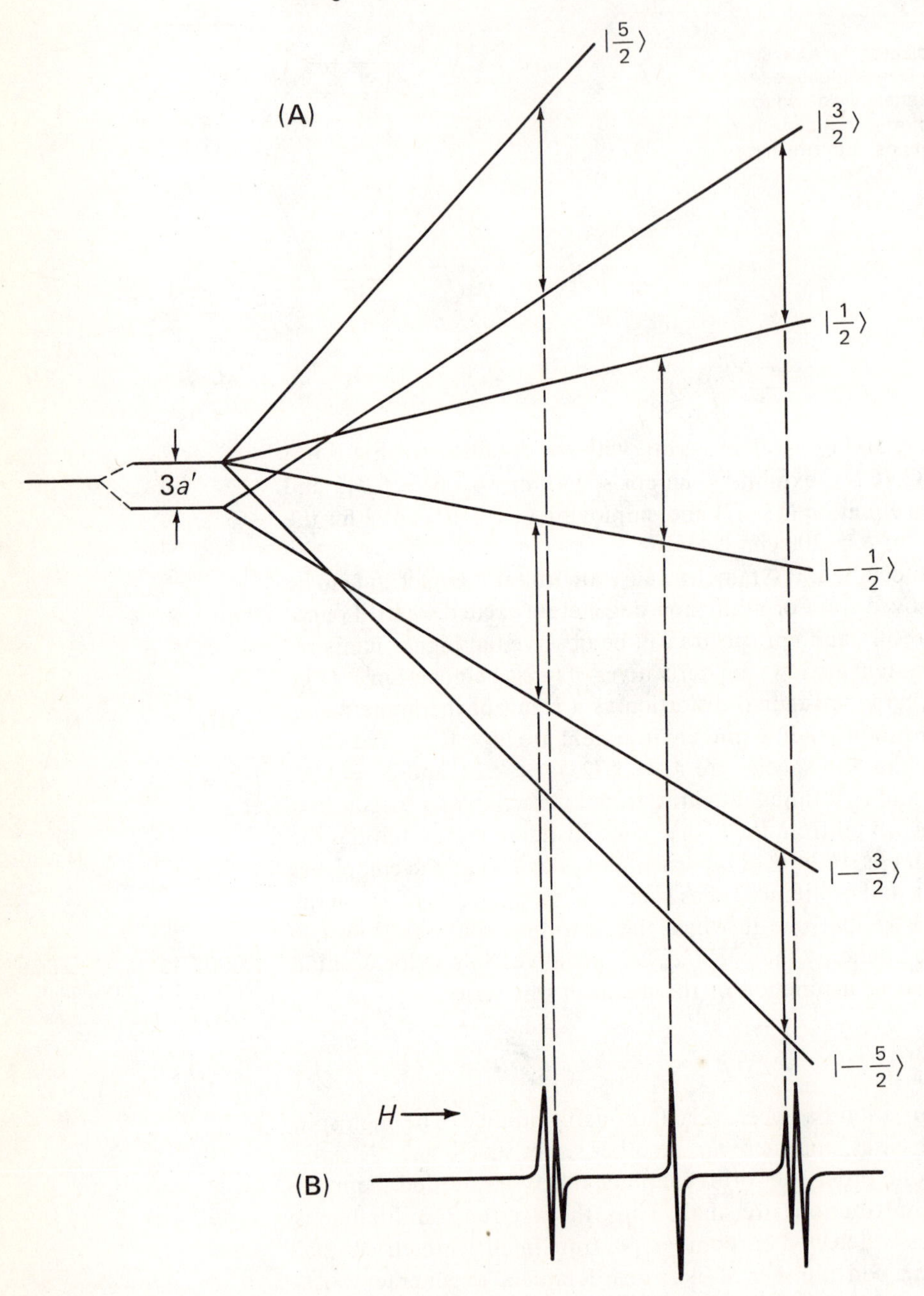

FIGURE 13–17. The splitting of the energy levels (A) and the spectrum (B) expected for an octahedral iron (III) complex (H parallel to a principal axis of the octahedron).

In iron(III) complexes with small tetragonal distortion, $D \ll h\nu$ and $E = 0$. The energy levels and expected spectrum are illustrated in Fig. 13–18(A). Observed g-values are very close to 2.00 because of the extremely small amount of spin-orbit coupling. This fact also allows easy observation of epr spectra at room temperature.

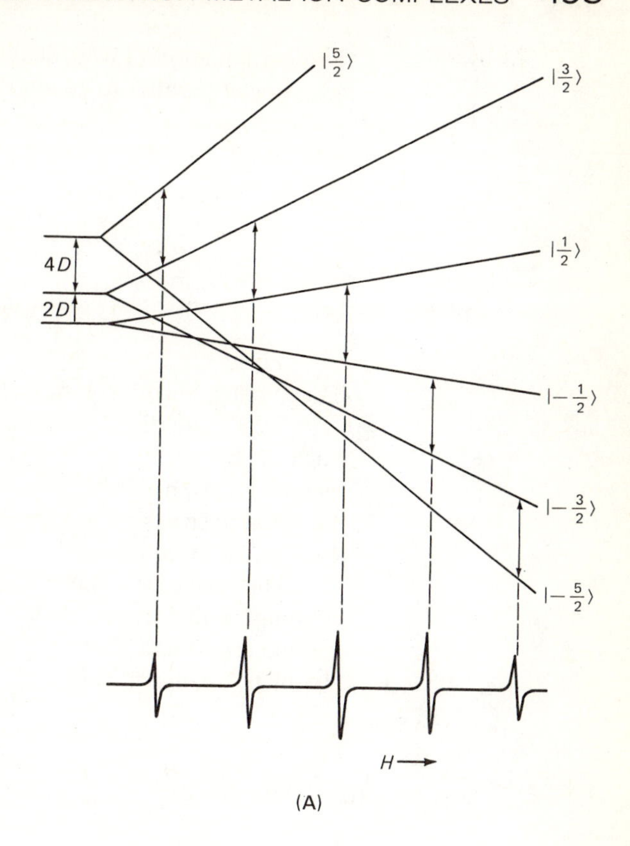

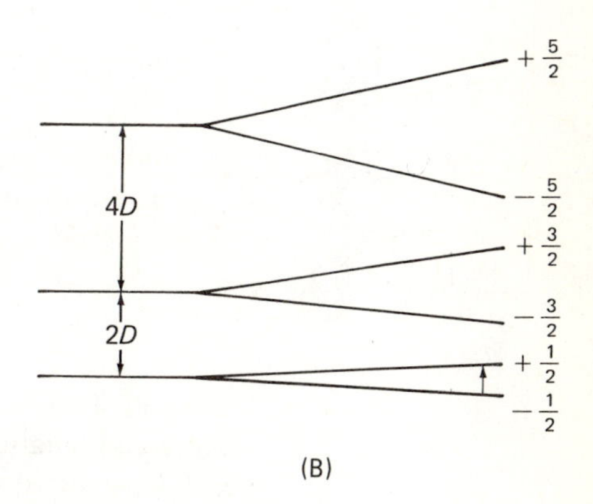

FIGURE 13–18. Energy levels and expected spectrum for a d^5 ion in a weak (A) and strong (B) tetragonal field (H parallel to the tetragonal axis). [From G. F. Kokoszka and R. W. Duerst, Coord. Chem. Rev., 5, 209 (1970).]

If $D \gg h\nu$, the situation shown in Fig. 13–18(B) exists, and only the transition between $|+\frac{1}{2}\rangle$ and $|-\frac{1}{2}\rangle$ will be observed. Even if the higher levels are populated, $\Delta M_s \neq 1$ for the possible transition and no spectral bands are observed. The g-values can be calculated, using only $|\frac{5}{2}, \frac{1}{2}\rangle$ and $|\frac{5}{2}, -\frac{1}{2}\rangle$ as a basis set and employing the Zeeman Hamiltonian, $\hat{H} = g_{\parallel}'\beta H_z\hat{S}_z + g_{\perp}'\beta(H_x\hat{S}_x + H_y\hat{S}_y)$. When H is parallel to z, we have:

$$\hat{H} = \begin{array}{c} \\ \left\langle \frac{5}{2}, \frac{1}{2} \right| \\ \left\langle \frac{5}{2}, -\frac{1}{2} \right| \end{array} \begin{array}{cc} \left| \frac{5}{2}, \frac{1}{2} \right\rangle & \left| \frac{5}{2}, -\frac{1}{2} \right\rangle \\ \multicolumn{2}{c}{\left| \begin{array}{cc} \frac{1}{2} g_e \beta H_z & 0 \\ 0 & -\frac{1}{2} g_e \beta H_z \end{array} \right|} \end{array} \tag{13-50}$$

Solving equation (13–50) leads to $\Delta E = g_e\beta H_z$ and $g_\parallel = g_e$.

For H parallel to x, after using $S^{\pm} = S_x \pm iS_y$ we obtain

$$\hat{H} = \begin{array}{c|cc|} & \left|\frac{5}{2}, \frac{1}{2}\right\rangle & \left|\frac{5}{2}, -\frac{1}{2}\right\rangle \\ \left\langle\frac{5}{2}, \frac{1}{2}\right| & 0 & \frac{3}{2}g_e\beta H_x \\ \left\langle\frac{5}{2}, -\frac{1}{2}\right| & \frac{3}{2}g_e\beta H_x & 0 \end{array} \qquad (13\text{-}51)$$

Diagonalization of equation (13–51) leads to $\Delta E = 3g_e\beta H_x$ and $g_\perp = 3g_e \approx 6.0$.

Such a situation is well represented by Fig. 13–16, where B is a weak field ligand such as F^- or H_2O, which causes a high spin complex. The zero-field splitting parameter, D, has been measured for several systems of this type by examining the far infrared spectrum in a magnetic field. Values ranging from 5 to 20 cm^{-1} are found for various complexes.[40]

The final case to be considered is one in which a geometric distortion occurs in a complex that removes the axial character. In this case, the zero-field splitting parameters D and E are not zero. This Hamiltonian again produces three Kramers' doublets, as shown in Fig. 13–19. Solving this Hamiltonian matrix, using the wave

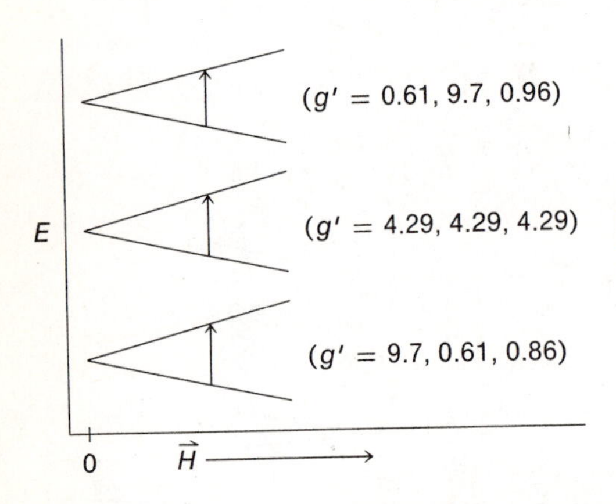

FIGURE 13–19. Kramers' doublets in rhombic symmetry (D and E not equal to zero) for high spin iron(III). The three principal components are listed in parentheses.

functions that diagonalize it, we find each of the three Kramers' doublets to be a linear combination of $|5/2, \pm 5/2\rangle$, $|5/2, \pm 3/2\rangle$, and $|5/2, \pm 1/2\rangle$. Thus, transitions within each Kramers' doublet are allowed; the corresponding g-values are indicated in Fig. 13–19. The separation between the Kramers' doublets is large enough that transitions between different ones are not observed, but at most temperatures all three are significantly populated, and many resonances are observed. An example[41] of this situation is Na[Fe(edta)]$\cdot 4H_2O$ (where edta is ethylenediamine tetraacetate) diluted in a single crystal of the analogous Co(III) complex. The spectrum shows one almost isotropic transition at $g = 4.27$ and two very anisotropic ones with principal g-values of 9.64 and 1.10, respectively.

d^6

This system has not been studied extensively. Low spin complexes are diamagnetic and high spin O_h complexes are similar to d^4. High spin iron(II) has a g-value of 3.49 at 4.2° K and a linewidth of 500 gauss even at this low temperature. Spin-orbit coupling in the ground state is very large and there are nearby excited states that can be mixed in. With a $J = 1$ ground state, two transitions would be observed if zero-field effects were small. In a distorted octahedral field, zero-field effects are large and no epr spectrum is observed. Deoxyhemoglobin is such a species, and no one has obtained an epr spectrum on this system.

d^7

The ground state for an O_h high spin d^7 complex is ${}^4T_{1g}(F)$. With extensive spin-orbit coupling, epr measurements are possible only at low temperatures. With $S = 3/2$ and three orbital components in T, a total of 12 low-lying spin states result. At the low temperatures needed to observe the epr spectra, because of spin relaxation problems, only the low-lying doublet is populated, giving a single peak from an effective $S' = 1/2$, with a g-value of 4.33. Studies on these systems have been reviewed.[42]

In tetrahedral symmetry, cobalt(II) complexes with ${}^4A_2$ ground states are similar to tetrahedral d^3 except that the ${}^4T_2$ excited state is closer in energy to the ${}^4A_2$ state in cobalt. With more spin-orbit coupling, broader lines are found for cobalt(II).

In strong crystalline fields, a doublet $S = 1/2$ state ${}^2E$ becomes lowest in energy. Because there is no spin-orbit coupling in this ${}^2E$ state and since usually there are no nearby doublet states, electron spin lifetimes are long, often allowing observation of narrow esr lines at liquid nitrogen and room temperatures.

The spin Hamiltonian for the low spin d^7 system is usually given as:

$$\hat{H} = \beta[g_x H_x \hat{S}_x + g_y H_y \hat{S}_y + g_z H_z \hat{S}_z] + A_x \hat{S}_x \hat{I}_x + A_y \hat{S}_y \hat{I}_y + A_z \hat{S}_z \hat{I}_z \quad (13\text{-}52)$$

In the cases of the five- and six-coordinate tetragonal d^7 systems, the unpaired electron resides in the d_{z^2} orbital. For this electronic configuration, assuming axial symmetry,

$$g_{\parallel} = 2.0023 \quad (13\text{-}53)$$

$$g_{\perp} = 2.0023 - \frac{6\xi}{\Delta E(z^2 - xz, yz)} \quad (13\text{-}54)$$

This fact has been used[32] to study the adduct formation of coordinatively unsaturated cobalt complexes with varieties of axially coordinated bases, B. Good overlap between the donor lone pair and d_{z^2} causes readily observed hyperfine structure from bases coordinating *via* nitrogen or phosphorus atoms. Wayland has utilized the large gyromagnetic ratio (and hence large hyperfine coupling) of ${}^{31}P$ to obtain hybridization ratios for varieties of PX_3 donors with Co(tetraphenylporphyrin)[31] and Co(salen).[43] In a study[44a] of 2:1 base adducts of BF_2 capped bis(diphenylglyoxime)Co(II), it was found that P-values [see the discussion of equations (13–36) and (13–37)] for oxygen donors were larger than those for nitrogen donors. It was also found with a series of ten nitrogen donors that P varied from 0.0216 when B is quinuclidine to 0.0147 when B is N-methylimidazole.

In low spin d^7 systems, the d_{z^2} and $4s$ orbitals belong to the same irreducible representation and are allowed to mix [see equation (13–18)]. It is therefore impossible to utilize the Fermi contact term, $\mathcal{K}$, derived from equations (13–30) to (13–35) to deduce information concerning molecular orbital coefficients and covalency in these systems.

The epr spectra of O_2 adducts of various cobalt(II) complexes have been reported.[44b] The unpaired electron in the system resides mainly on O_2. In spite of this, an appreciable cobalt hyperfine coupling constant is observed. The metal coupling arises from spin polarization of the filled molecular orbital of the O_2 adduct, which arises by pairing up of the unpaired metal electron in d_{z^2} of the original five-coordinate cobalt(II) complex with one of the antibonding electrons of the O_2 molecule. The extent of electron transfer from cobalt(II) into the coordinated O_2 molecule was estimated from the magnitude of the anisotropic cobalt hyperfine coupling constant. Electron transfers significantly less than the 90% or more reported[44c] from an incorrect epr analysis were found in many systems.

d^8, High Spin

The ground state of the gaseous ion is ^3F with an orbital singlet state lowest in an octahedral field. The *d*-shell is more than half-filled, so spin-orbit coupling leads to *g*-values greater than the free electron value. The zero-field splitting makes it difficult to detect epr spectra except at low temperatures. The *g*-values found are usually close to isotropic.

d^9

The d^9 configuration has been very extensively studied. In an octahedral field, the ground state is ^2E_g. A large Jahn-Teller effect is expected, making observation of the epr spectrum at room temperature possible. In tetragonal complexes, the ground state is $d_{x^2-y^2}$ (*x* and *y* axes pointing at the ligands) and sharp lines result. Note that the quadrupolar interaction of the copper nucleus (see Chapter 9) can be determined from this experiment. The epr results can be fitted to the spin Hamiltonian

$$\hat{H} = \beta[g_z H_z \hat{S}_z + g_x H_x \hat{S}_x + g_y H_y \hat{S}_y] + A_z \hat{S}_z \hat{I}_z + A_x \hat{S}_x \hat{I}_x + A_y \hat{S}_y \hat{I}_y + Q'\left[\hat{I}_z^2 - \frac{1}{3} I(I+1)\right] - g_N \beta_N \vec{\mathbf{H}} \cdot \hat{I} \quad (13\text{-}55)$$

Some typical copper(II) spectra are shown in Fig. 13–20. In (A), an isotropic solution spectrum is shown. Both nitrogen and proton ligand hyperfine structures are seen on the high field peak, but not on the low field peaks. This is attributed to differences in the relaxation times for the transition, which depend upon the m_I value associated with the transition.[45] The solvent employed influences the molecular correlation time, which in turn also influences the spectral appearance.[45]

In (B), an anisotropic spectrum is shown at Q-band frequencies. Such spectra are observed in glass or powder samples of copper complexes diluted in hosts. The low field $g_\parallel$ and high field $g_\perp$ peaks are well separated. With the higher microwave energy, the individual peaks are broader so the super-hyperfine splitting is not detected on the $g_\parallel$ peak. In the spectrum in (C), at X-band frequencies, the $g_\parallel$ and $g_\perp$ transitions overlap, but much more ligand hyperfine structure is detected. As mentioned earlier, the temperature dependence of the spectra of many copper(II) systems has been interpreted in terms of Jahn-Teller effects.

The spectrum of a polycrystalline sample of a molecule containing two copper atoms is illustrated in Fig. 13–21. The main peaks at 2500 and 3700 gauss are assigned to two $g_\perp$ components. The seven copper hyperfine components [two Cu ($I = 3/2$) nuclei] are seen in the low field $g_\parallel$ peak. The other high field $g_\perp$ peak is not seen, for it is out of the spectrometer range. The peak at ~3200 G is assigned to a $\Delta m_s = 2$ transition. The origin of the doublet will be understood after discussion of the spin Hamiltonian. The single crystal spectra[46] of molecules containing two copper(II) atoms with $S = 1$ are fitted to the spin Hamiltonian

$$\hat{H} = \beta[g_z H_z \hat{S}_z + g_x H_x \hat{S}_x + g_y H_y \hat{S}_y] + D\left[\hat{S}_z^2 - \frac{2}{3}\right] + E[\hat{S}_x^2 - \hat{S}_y^2] + A_x \hat{S}_x \hat{I}_x + A_y \hat{S}_y \hat{I}_y + A_z \hat{S}_z \hat{I}_z \quad (13\text{-}56)$$

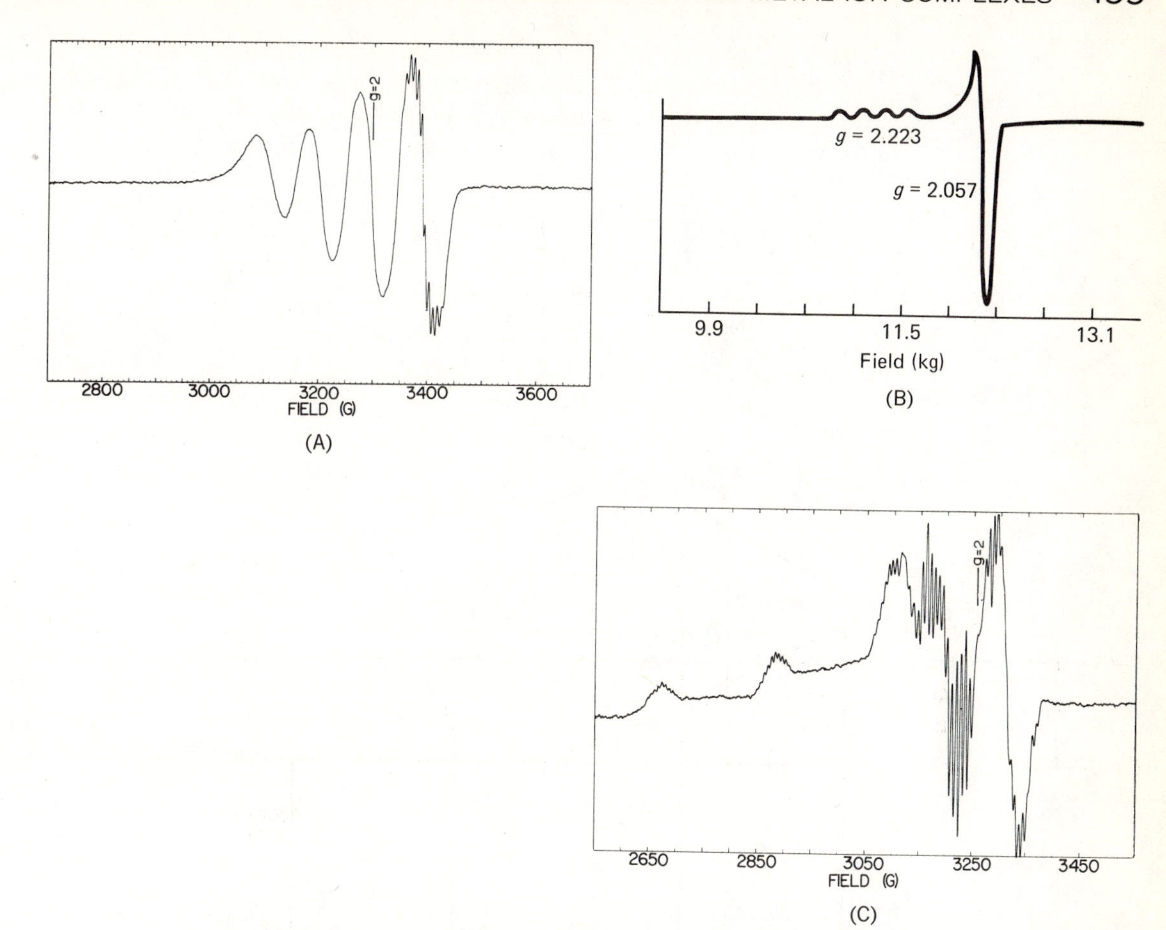

FIGURE 13–20. "Typical" epr spectra for copper(II) complexes. (A) A typical solution spectrum for a square planar Schiff base ligand. [From E. Hasty, T. J. Colburn, and D. N. Hendrickson, Inorg. Chem., *12*, 2414 (1973).] (B) Glass or powder sample run at Q-band frequencies on an axial complex. (C) Glass or doped powder spectrum of the complex in (A) run at X-band frequencies.

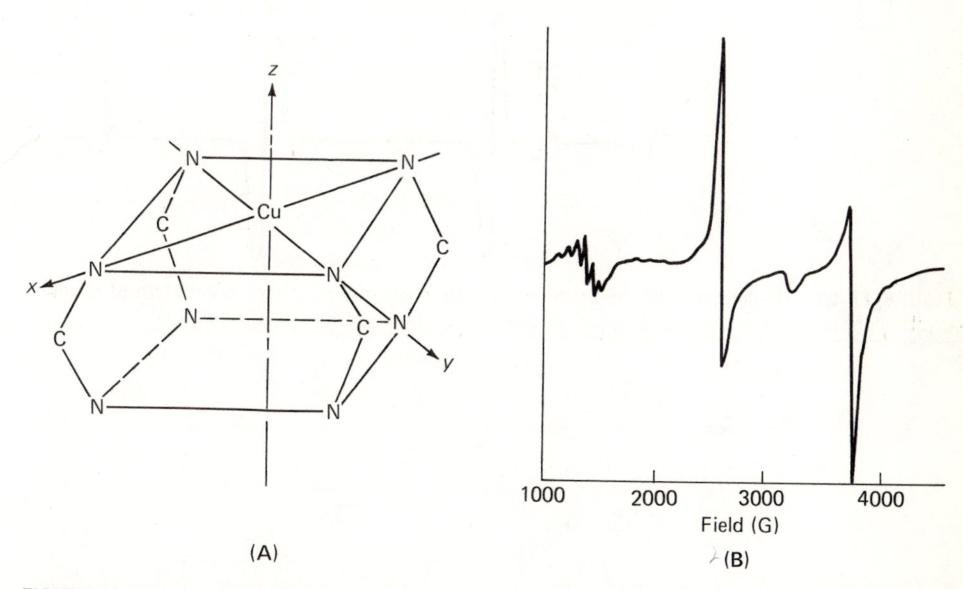

FIGURE 13–21. (A) A schematic representation of a dimeric copper adenine complex; (B) the polycrystalline epr of this complex.

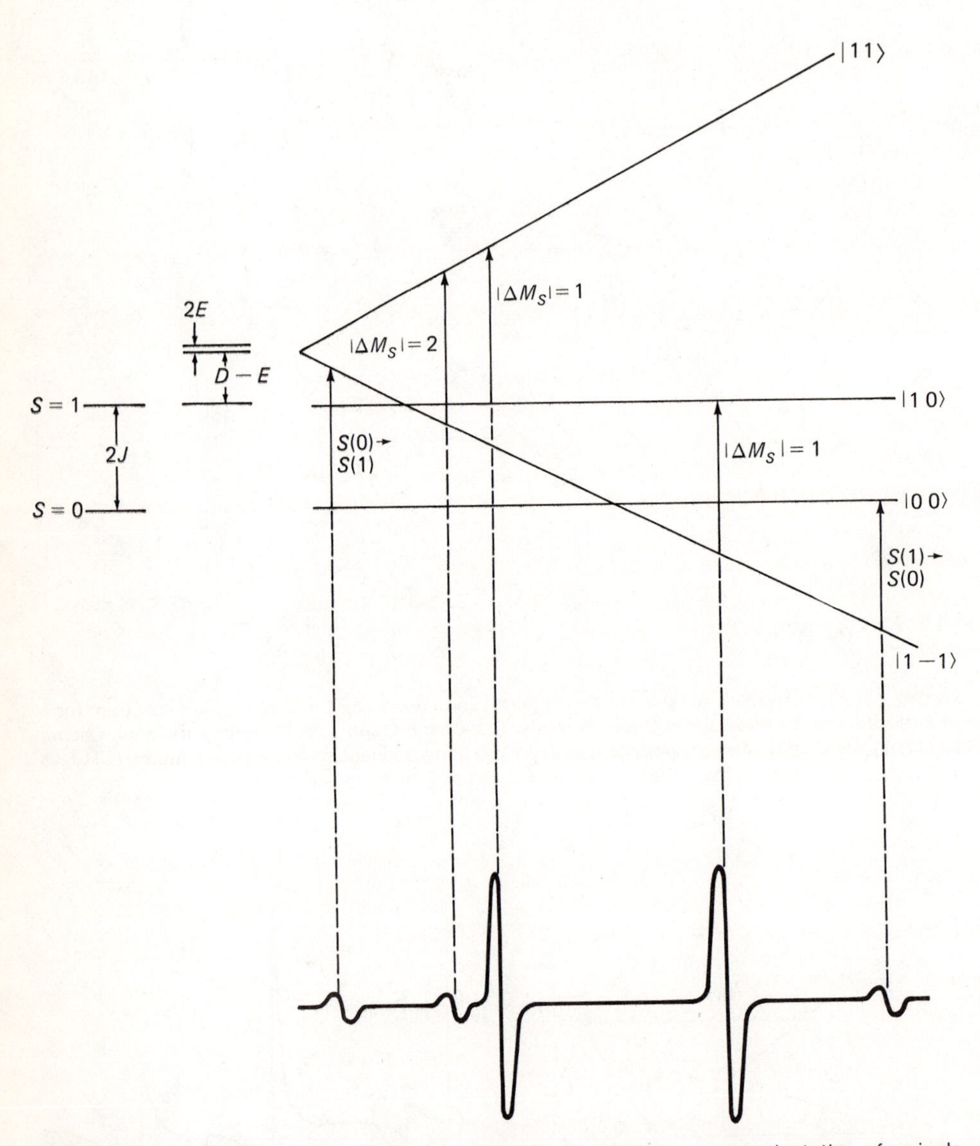

FIGURE 13–22. The influence of J interactions and zero-field effects on one orientation of a single crystal epr spectrum and energy levels of a molecule containing two d^9 copper atoms.

The results for dimeric copper(II) acetate are[47] $g_z = 2.344$, $g_x = 2.053$, $g_y = 2.093$, $D = 0.345\ cm^{-1}$, $E = 0.007\ cm^{-1}$, $A_z = 0.008\ cm^{-1}$, and A_x, $A_y <$ $0.001\ cm^{-1}$. The influence that the J, D, and E parameters have on the epr spectrum is illustrated in Fig. 13–22. With $|J| = 260\ cm^{-1}$ in $Cu_2(OAc)_4$, transitions from the $S = 0$ to $S = 1$ state are not observed in the epr. The exchange interaction gives rise to a lower energy $S = 0$ state, so the intensity of the signals decreases with decreased temperature. This temperature dependence indicates a J of $-260\ cm^{-1}$, corresponding to a separation of the $S = 0$ and $S = 1$ states of $2J$ or $520\ cm^{-1}$. In the powder spectrum discussed earlier, split $g_\parallel$ and $g_\perp$ peaks arise from the two $\Delta M_S = \pm 1$ transitions averaged over the orientations. In the relatively rare situation where the exchange parameter J is smaller than the available microwave energy of the epr experiment, it is possible to see epr transitions between the $S = 1$ and $S = 0$ electronic states of a copper dimer. The first example involved a copper dimer and was reported[48] for the "outer-sphere" dimers in $[Cu_2(tren)_2X_2](BPh_4)_2$, where $X = NCO^-$ and NCS^- and tren is 2,2′,2″-triaminotriethylamine. The "outer-sphere" dimeric association occurs between two Cu(II) trigonal bipyramids by virtue of hydrogen bonding between the uncoordinated O(S) end of OCN^-(SCN^-) nitrogen bonded to one copper and an N—H proton of the coordinated tren. In the case of X-band epr, J-values of $\sim 0.15\ cm^{-1}$ to $\sim 0.05\ cm^{-1}$ can be gauged by the observation of singlet-to-triplet transitions as illustrated in Fig. 13–22.

13–7 DOUBLE RESONANCE TECHNIQUES

Two double resonance techniques, electron-nuclear double resonance, ENDOR, and electron-electron double resonance, ELDOR, have found relatively limited application in epr. In ENDOR, the epr transition is observed on a system in which a nuclear spin transition is saturated, while in ELDOR, the epr is measured while another electron spin transition is saturated. As in the nmr double resonance experiment, enhanced intensity results as a consequence of the Overhauser effect. The advantages to be gained in many cases[49–51] parallel those already discussed in the analogous nmr experiment.

REFERENCES

1. See M. Tinkham, "Group Theory and Quantum Mechanics," p. 143, McGraw-Hill Book Co., New York (1964).
2. D. M. S. Bagguley and J. H. E. Griffiths, Nature, *162,* 538 (1948).
3. J. P. Fackler and J. A. Smith, J. Amer. Chem. Soc., *92,* 5787 (1970); J. P. Fackler, J. D. Levy, and J. A. Smith, *ibid., 94,* 2436 (1972).
4. a. B. M. Hoffman, F. Basolo, and D. L. Diemente, J. Amer. Chem. Soc., *95,* 6497 (1973).
   b. F. Cariati, F. Morazzoni, C. Busetto, D. Del Piero, and A. Zazzetta, J. Chem. Soc. Dalton, Trans., 556, 1975.
5. P. T. Manohanon and M. T. Rogers in "Electron Spin Resonance of Metal Complexes," ed. Teh Fu Yen, Plenum Press, New York (1969).
6. B. Bleaney and D. J. Ingram, Proc. Phys. Soc., *63,* 408 (1950).
7. A. Abragam and M. H. L. Pryce, Proc. Phys. Soc., *63,* 409 (1950).
8. B. Bleaney and K. D. Bowers, Proc. Phys. Soc., *65,* 667 (1952).
9. B. Bleaney, K. D. Bowers, and R. S. Trenam, Proc. Roy. Soc. (London), *A228,* 157 (1955).
10. H. C. Allen, Jr., G. F. Kokoszka, and R. G. Inskeep, J. Amer. Chem. Soc., *86,* 1023 (1964).
11. For other examples, see F. S. Ham in "Electron Paramagnetic Resonance," ed. S. Geschwind, Plenum Press, New York (1972).
12. R. C. Koch, M. D. Joesten, and J. H. Venable, Jr., J. Chem. Phys., *59,* 6312 (1973).
13. R. Kirmse, *et al.,* J. Chem. Phys., *56,* 5273 (1972).
14. E. Buluggiu and A. Vera, J. Chem. Phys., *59,* 2886 (1973).
15. J. R. Pilbrow and M. E. Winfield, Mol. Phys., *25,* 1073 (1973).
16. a. Yonetoni, *et al.,* J. Biol. Chem., *245,* 2998 (1970).
    b. J. W. Orton, *et al.,* Proc. Phys. Soc., *78,* 554 (1961).

17. a. M. C. R. Symons and J. G. Wilkerson, J. Chem. Soc. (A), 2069 (1971).
    b. A. H. Maki, N. Edelstein, A. Davison, and R. H. Holm, J. Amer. Chem. Soc., *86*, 4580 (1964).
18. J. S. Griffith, "The Theory of Transition-Metal Ions," Cambridge University Press, New York (1961).
19. R. M. Golding, "Applied Wave Mechanics," Van Nostrand, New York (1969).
20. B. R. McGarvey, "Electron Spin Resonance of Transition Metal Complexes," in "Transition Metal Chemistry," Vol. 3, pp. 89–201, ed. R. L. Carlin, Marcel Dekker, New York (1967).
21. W. C. Lin, C. A. McDowell, and D. J. Ward, J. Chem. Phys., *49*, 2883 (1968).
22. B. B. Wayland, J. V. Minkiewicz, and M. E. Abd-Elmageed, J. Amer. Chem. Soc., *96*, 2795 (1974).
23. M. E. Kimball, D. W. Pratt, and W. C. Kaska, Inorg. Chem., *7*, 2006 (1968).
24. a. A. H. Maki and B. R. McGarvey, J. Chem. Phys., *29*, 31 (1958).
    b. D. Kivelson and R. Nieman, J. Chem. Phys., *35*, 149 (1961).
25. J. I. Zink and R. S. Drago, J. Amer. Chem. Soc., *94*, 4550 (1972).
26. H. A. Kuska and M. T. Rogers, J. Chem. Phys., *43*, 1744 (1965); H. A. Kuska, M. T. Rogers, and R. E. Drullinger, J. Phys. Chem., *71*, 109 (1967).
27. B. A. Goodman and J. B. Raynor, Adv. Inorg. Chem. Radiochem., *13*, 135 (1970).
28. P. W. Atkins and M. C. R. Symons, "The Structure of Inorganic Radicals," Elsevier Publishing Co., New York (1967).
29. W. V. Sweeney and R. E. Coffman, J. Phys. Chem., *76*, 49 (1972).
30. D. A. C. McNeil, J. B. Raynor, and M. C. R. Symons, Proc. Chem. Soc., 364 (1964).
31. B. B. Wayland and M. E. Abd-Elmageed, J. Amer. Chem. Soc., *96*, 4809 (1974).
32. S. Foner and W. Low, Phys. Rev., *120*, 1585 (1960).
33. E. Pedersen and H. Toftlund, Inorg. Chem., *13*, 1603 (1974).
34. R. H. Borcherts and C. Kikuchi, J. Chem. Phys., *40*, 2270 (1964).
35. C. K. Jørgensen, Acta Chem. Scand., *8*, 1686 (1957).
36. A. Carrington and A. D. McLachlan, "Introduction to Magnetic Resonance," Harper and Row, New York (1967).
37. W. C. Lin, C. A. McDowell, and D. J. Ward, J. Chem. Phys., *49*, 2883 (1968).
38. D. J. E. Ingram, "Biological and Biochemical Applications of Electron Spin Resonance," Plenum Press, New York (1969).
39. R. E. DeSimone and R. S. Drago, J. Amer. Chem. Soc., *92*, 2343 (1970).
40. G. C. Brackett, P. L. Richards, and W. S. Caughey, J. Chem. Phys., *54*, 4383 (1971).
41. R. Aaasa and T. Vanngard, Arkiv Kemi, *24*, 331 (1965).
42. F. S. Kenedy, *et al.*, Biochem. Biophys. Res. Comm., *48*, 1533 (1972).
43. B. B. Wayland, M. E. Abd-Elmageed, and L. F. Mehne, submitted.
44. a. B. Tovrog and R. S. Drago, submitted.
    b. B. Tovrog, D. J. Kitko, and R. S. Drago, J. Amer. Chem. Soc., *98*, 5144 (1976).
    c. B. M. Hoffman, D. L. Diemente, and F. Basolo, J. Amer. Chem. Soc., *92*, 61 (1970); A. L. Crumbliss and F. Basolo, J. Amer. Chem. Soc., *92*, 55 (1970).
45. W. B. Lewis and L. O. Morgan, "Transition Metal Chemistry," Vol. 4, p. 33, ed. R. L. Carlin, Marcel Dekker, New York (1968).
46. G. F. Kokoszka and R. W. Duerst, Coord. Chem. Rev., *5*, 209 (1970).
47. B. Bleaney and K. D. Bowers, Proc. Roy. Soc. (London), *A214*, 451 (1952).
48. D. M. Duggan and D. N. Hendrickson, Inorg. Chem., *13*, 2929 (1974).
49. J. E. Wertz and J. R. Bolton, "Electron Spin Resonance," McGraw-Hill Book Co., New York (1972).
50. J. S. Hyde, "Electron–Electron Double Resonance," Varian Associates Reprint No. 256.
51. J. S. Hyde, R. C. Sneed, and G. H. Rist, J. Chem. Phys., *51*, 1404 (1969), and references therein.

GENERAL REFERENCES

J. E. Wertz and J. R. Bolton, "Electron Spin Resonance," McGraw-Hill Book Co., New York (1972).
A. Abragam and B. Bleaney, "EPR of Transition Ions," Clarendon Press, Oxford, England (1970).
J. W. Orton, "Electron Paramagnetic Resonance," Gordon and Breach Publishers, New York (1968).
R. M. Golding, "Applied Wave Mechanics," Van Nostrand, New York (1969).
J. S. Griffith, "The Theory of Transition-Metal Ions," Cambridge University Press, New York (1961).
A. Carrington and A. D. McLachlan, "Introduction to Magnetic Resonance," Harper and Row, New York (1967).
C. P. Poole, "Electron Spin Resonance," Interscience, New York (1967).
B. R. McGarvey, "Electron Spin Resonance of Transition Metal Complexes," in "Transition Metal Chemistry," Vol. 3, pp. 89–201, ed. R. L. Carlin, Marcel Dekker, New York (1967).
G. Kokoszka and G. Gordon, "Technique of Inorganic Chemistry," Vol. 7, pp. 151–271, ed. H. B. Jonassen and A. Weissberger, Interscience, New York (1968).
B. A. Goodman and J. B. Raynor, Adv. Inorg. Chem. Radiochem., *13*, 135 (1970).
R. Bersohn, "Determination of Organic Structures by Physical Methods," Vol. 2, ed. F. C. Nachod and W. D. Phillips, Academic Press, New York (1961).
A. Carrington, Quart. Rev., *17*, 67 (1963).
A. Carrington and H. C. Longuet-Higgins, Quart. Rev., *14*, 427 (1960).

EXERCISES

1. On p. 503 is a typical Cu^{2+} solution spectrum. It can be characterized by $g = \frac{1}{3}g_{\parallel} + \frac{2}{3}g_{\perp}$ and $A = \frac{1}{3}A_{\parallel} + \frac{2}{3}A_{\perp}$.

   a. Determine g and A.

b. This is an X-band spectrum (9.4×10^9 Hz). Why should A be expressed in Hertz?

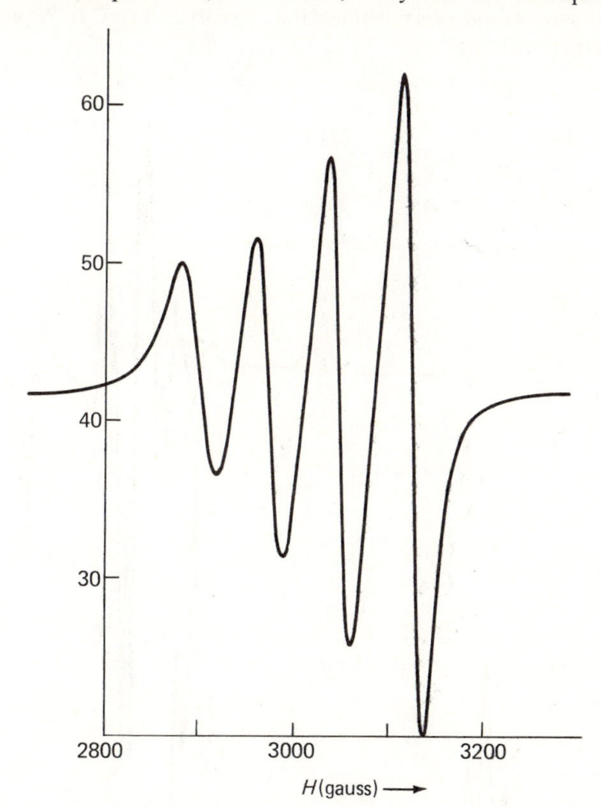

2. Students often find it difficult to understand why the epr spectrum of a powder or frozen solution should yield (in an axially symmetric compound, for example) $g_{\|}$, $g_{\perp}$, $A_{\|}$, and $A_{\perp}$. "After all," the argument goes, "you are looking at a superposition of all possible orientations. At intermediate orientations you have an average g and A, *e.g.*, $g^2 = g_{\|}^2 \cos^2 \theta + g_{\perp}^2 \sin^2 \theta$. Thus, the spectrum should be very broad with few features." This argument is correct as far as it goes, but it ignores the fact that we are looking at a first derivative. For example, suppose that a compound has $g_{\|} = 2.0$ and $g_{\perp} = 2.1$. Recalling that there are many more perpendicular directions than there are parallel ones, the absorption spectrum should appear in the ideal case as in (A):

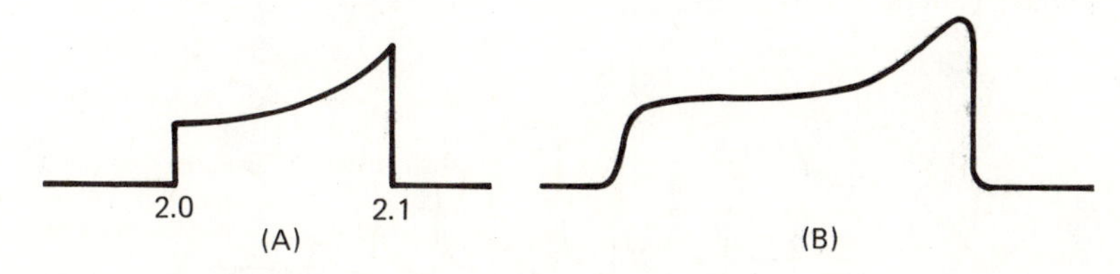

Because of line broadening, the real absorption curve resembles that in (B). The first derivative will then be

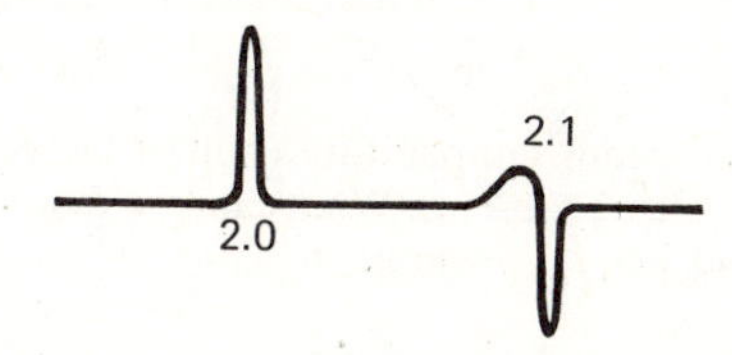

The argument is further complicated, but essentially unchanged, by including hyperfine coupling.

Below is the powder spectrum of copper diethyldithiophosphinate (for ^{63}Cu, $I = \frac{3}{2}$). Explain as many features as possible. Hint: $A_{\parallel}(Cu) \approx 5A_{\perp}(Cu)$. What other nuclei might give hyperfine splitting?

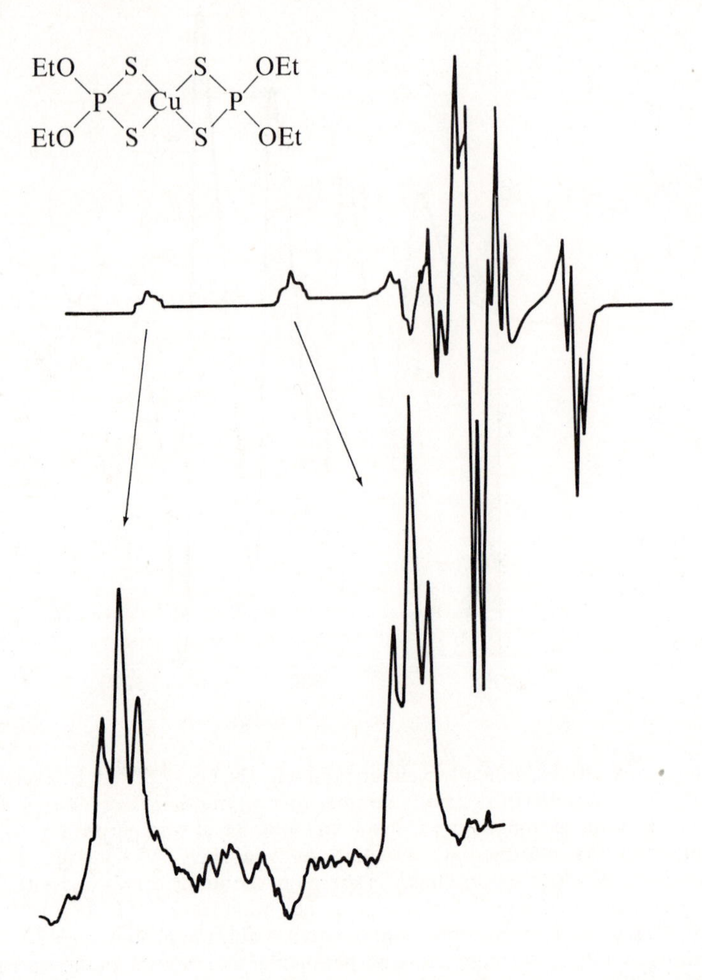

3. The solution spectrum below is typical for a vanadyl complex (for V, $I = \frac{7}{2}$). Explain the splitting pattern.

DPPH

$[VO(glycolate)_2]^{2-}$ in H_2O

100 G

$H \rightarrow$

4. On p. 505 are the frozen solution (top) and room temperature (bottom) spectra of a vanadyl complex of *d,l*-tartrate (only one species is present). What can you deduce about the structure of this complex? (Tartaric acid has the formula

$$\begin{array}{ccccc} & H & & H & \\ & | & & | & \\ HOOC- & C & - & C & -COOH.) \\ & | & & | & \\ & OH & & OH & \end{array}$$

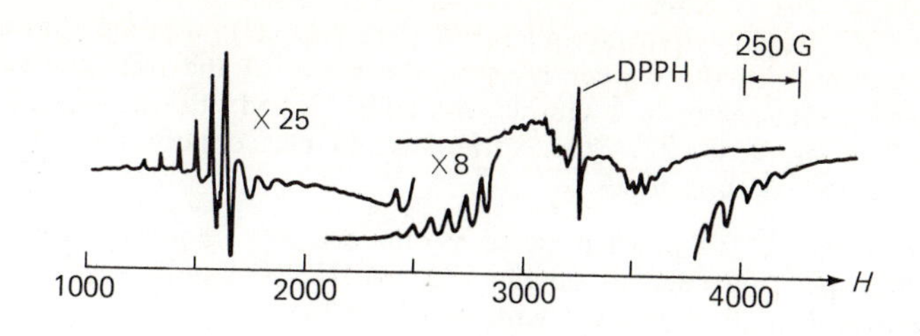

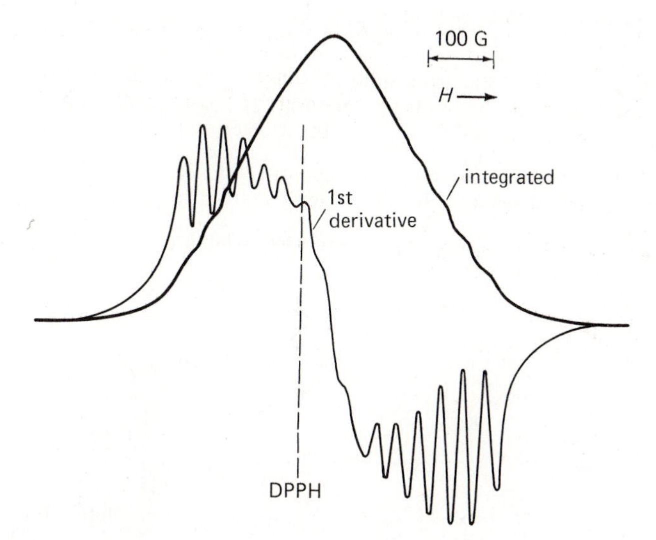

5. Diphenylpicrylhydrazyl (DPPH) is a common reference in epr spectroscopy. Its structure is

O_2N ; N—N̈— ; $-NO_2$; O_2N

In powder or concentrated xylene solution, it gives a sharp one-line pattern:

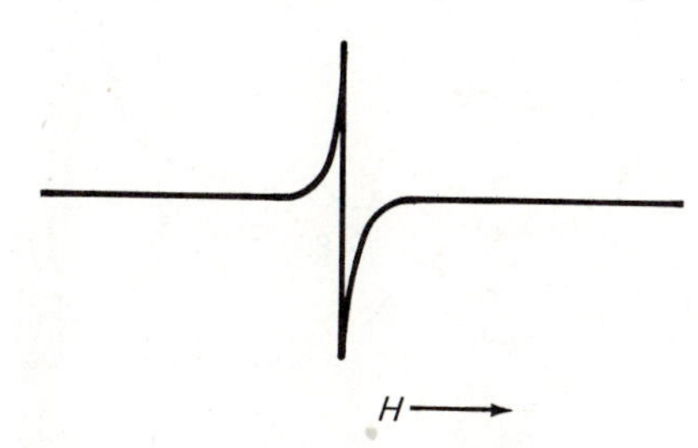

In a 10^{-3} M xylene solution, it gives the following pattern:

H

What does this splitting pattern come from? Why is it present only in dilute solution?

6. Copper(II) tetraphenylporphyrin [CuTPP; see part (D) on p. 507] has a square planar D_{4h} structure with the copper lying in the plane of the four equivalent porphyrin nitrogens. EPR spectra of a sample of CuTPP doped into the free ligand are shown below. This sample is 100% ^{63}Cu. Natural abundance copper is 69% ^{63}Cu and 31% ^{65}Cu, and $I = \frac{3}{2}$ for both isotopes.

   a. Spectrum (A) is a powder spectrum obtained at a frequency of 9.4 GHz (X-band), while spectrum (B) was obtained at 35 GHz (Q-band). Why are $g_\parallel$ and $g_\perp$ well separated in (B) compared with those in (A)?

   b. Explain the source of the number of peaks observed in the $g_\parallel$ region in spectrum (B). Also explain the splitting pattern found for each peak [see insert (C)].

   c. What is the crystal field splitting diagram for a square planar complex? How is observation of the splittings in insert (C) consistent with the orbital occupancy predicted for a Cu(II) complex by your crystal field diagram?

   d. Qualitatively, if this copper compound did not have a threefold or higher axis of symmetry, what would the powder spectrum look like?

   e. If natural abundance copper were used, what effect would this have on the appearance of insert (C)?

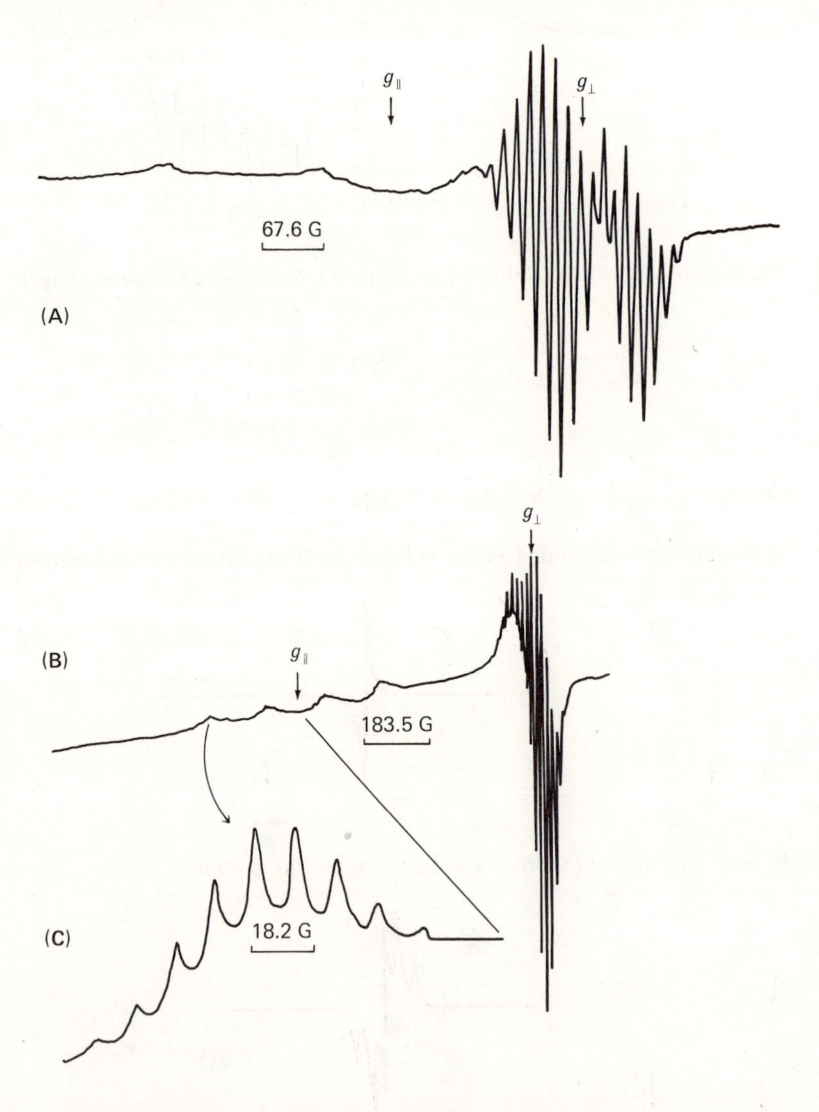

Note that insert (C) is obtained by increasing the gain for the set of low-field lines.

(D)

7. Below is a spectrum of cobalt phthalocyanine (for Co, $I = 7/2$), a D_{4h} Co(II) porphyrin similar to the copper porphyrin in problem 6.

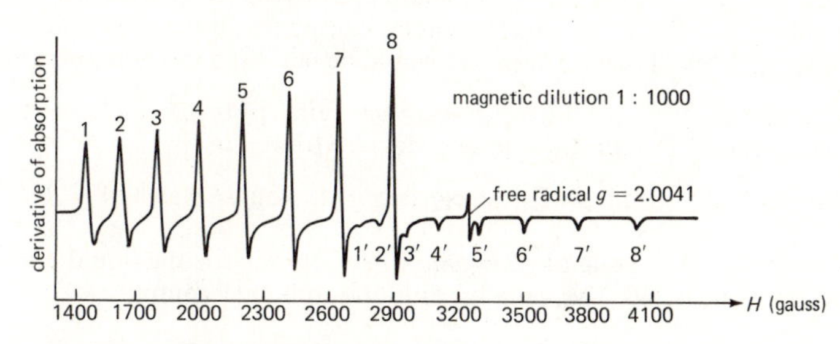

EPR of β-cobalt phthalocyanine magnetically diluted in β-metal-free phthalocyanine powder at 77°K.

a. What are the values of $g_{\|}$, $g_{\perp}$, $A_{\|}$, and $A_{\perp}$? What is the origin of the observed hyperfine structure?

b. Dissolving cobalt phthalocyanine in 4-methylpyridine produces a 1:1 adduct. The frozen solution spectrum is shown below. Utilizing the D_{4h} splitting diagram for low spin Co(II), state what orbital the unpaired electron occupies. Why is each of the eight upfield components split into three lines?

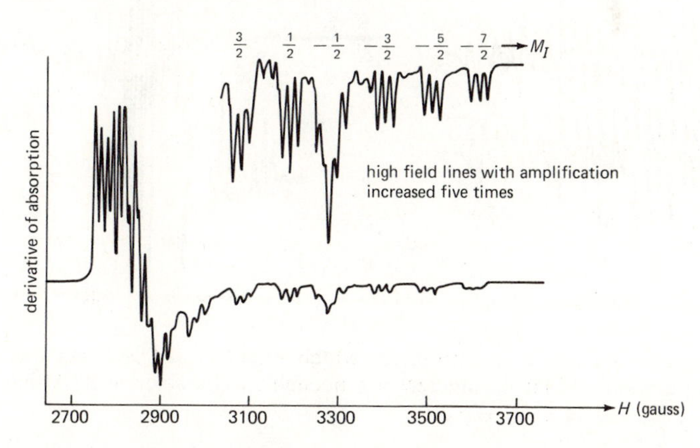

EPR of cobalt phthalocyanine solution in 4-methylpyridine (77°K.).

c. Why do you see super-hyperfine interaction from the pyridine nitrogen but not from the four phthalocyanine donor nitrogen atoms?

8. Predict the number of spectral lines for

   a. $Co(H_2O)_6^{2+}$

   b. $Cr(H_2O)_6^{3+}$

   Indicate how zero-field splitting and Kramers' degeneracy applies in these examples.

9. The following epr data on a series of high-spin octahedral metal hexafluoride complexes are taken from Proc. Roy. Soc. (London), *236*, 535 (1956):

Complex	g	A_{metal} $\times 10^4$ cm^{-1}	A_F $\times 10^4$ cm^{-1}	I_{metal}	Temp.
MnF_6^{4-}	$g_x = g_y = g_z = 2.00$	96	17	$^{55}Mn = \frac{5}{2}$	300° K
CoF_6^{4-}	$g_x = 2.6$ $g_y = 6.05$ $g_z = 4.1$	$A_x = 43$ $A_y = 217$ $A_z = 67$	$A_x = 20$ $A_y = 32$ $A_z = 21$	$^{59}Co = \frac{7}{2}$	20° K
CrF_6^{3-}	$g = 2.00$ $g = 1.98$	$A = 16.2$ $A = 16.9$	$A = 3$ $A = 1$	$^{53}Cr = \frac{3}{2}$	300° K

   a. Why do CrF_6^{3-} and MnF_6^{4-} give sharp epr at room temperature while CoF_6^{4-} does so only at 20° K? What effect or effects cause these differences in ability to observe the epr? Which of these complexes would be best for a room temperature nmr study?

   b. Why are the CrF_6^{3-} and MnF_6^{4-} g-values fairly isotropic and close to 2.0 while the CoF_6^{4-} values are anisotropic and deviate from 2.0?

   c. Why do CoF_6^{4-} and MnF_6^{4-} have larger A_F values than CrF_6^{3-}?

10. The November, 1973, issue of *Inorganic Chemistry* reports the liquid and solid solution epr spectra of some 10^{-3} M vanadyl dithiophosphinate complexes:

```
                  O
RO     S          ‖       S     OR
   \  / \         ‖      / \   /
    P     V     P
   /  \ /         \ /   \
RO     S          S     OR
```

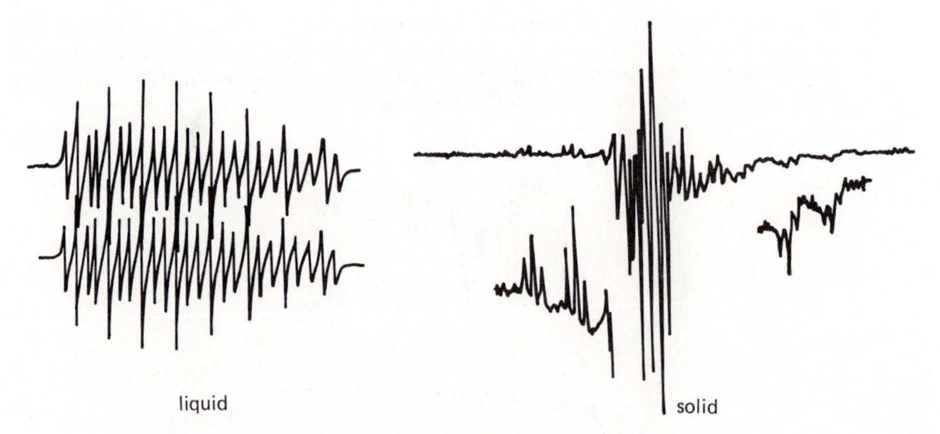

   a. Explain the liquid spectrum, for which R = CH_3. (The lower one is a computer simulation.) What parameters are needed to characterize it? V has $I = \frac{7}{2}$, P has $I = \frac{1}{2}$, and H has $I = \frac{1}{2}$.

   b. Explain the solid solution spectrum, for which R = phenyl. What parameters are needed to characterize it?

   c. What should be the mathematical relationship between the parameters in part a and those in part b?

11. The spectrum below is that of an axially symmetric Cu^{2+} complex in a frozen solution. It is a d^9 system. Assume 100% abundance for ^{63}Cu ($I = \frac{3}{2}$). The following constants will be needed: $\beta = 9.27 \times 10^{-21}$ erg/gauss; $h = 6.67 \times 10^{-27}$ erg sec; $\nu = 9.12 \times 10^9$ Hz for the spectrometer on which the spectrum was obtained.

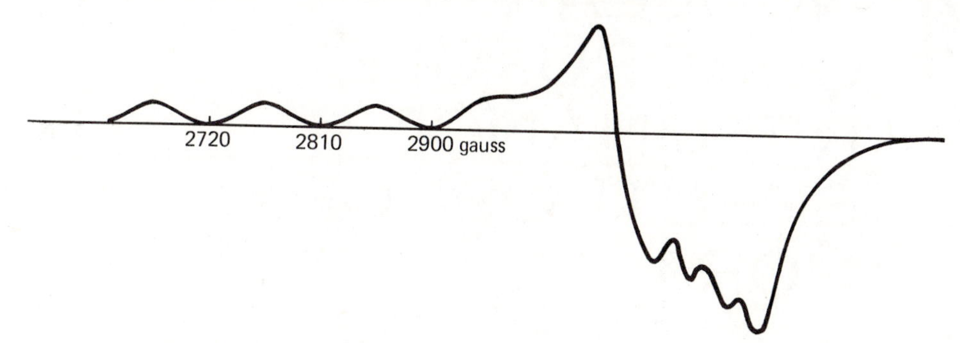

a. How many parameters are required to explain this spectrum? What features of the spectrum suggest these parameters?

b. Suggest why the high-field half of the spectrum ($\perp$ region) is more intense.

c. What is the absolute value of $g_{\parallel}$? Justify the terms in the equation you use.

12. Assume that the square pyramidal complex $Cu(hfac)_2P(C_6H_5)_3$ (for ^{31}P, $I = \frac{1}{2}$; for ^{63}Cu, $I = \frac{3}{2}$) can exist as either of the following isomers [Inorg. Chem., *13*, 2517 (1974)]: What might be learned to aid in distinguishing the isomers by looking at:

axial   basal

a. the electronic spectrum?

b. the infrared and Raman spectra (answer in a general way; *i.e.*, don't work out the total representation)?

c. the epr spectrum?

d. an ^{17}O-labeled epr spectrum (for ^{17}O, $I = \frac{5}{2}$)?

e. the nmr spectrum?

14 NUCLEAR QUADRUPOLE RESONANCE SPECTROSCOPY, NQR*

14-1 INTRODUCTION

When a nucleus with an electric quadrupole moment (nuclear spin $I \geqq 1$; see the second paragraph of Chapter 7 and Fig. 7-1) is surrounded by an inhomogeneous electric field resulting from asymmetry in the electron distribution, an electric gradient may result (*vide infra*). *The quadrupolar nucleus will interact with this electric field gradient* to an extent that is different for the various possible orientations of the elliptical quadrupolar nucleus. Since the quadrupole moment arises from an unsymmetric distribution of electric charge in the nucleus, it is an electric quadrupole moment rather than a magnetic moment that concerns us. The allowed nuclear orientations are quantized with $2I + 1$ orientations, described by the nuclear magnetic quantum number m, where m has values $+I$ to $-I$ differing by integer values. The quadrupole energy level that is lowest in energy corresponds to the orientation in which the greatest amount of positive nuclear charge is closest to the greatest density of negative charge in the electron environment. The energy differences of the various orientations are not very great, and at room temperature a distribution of orientations exists in a group of molecules. If the electron environment around the nucleus is spherical (as in Cl^-), all nuclear orientations are equivalent and the corresponding quadrupole energy states are degenerate. If the nucleus is spherical ($I = 0$ or $\frac{1}{2}$), there are no quadrupole energy states. In nqr spectroscopy, we study the energy differences of the non-degenerate nuclear orientations. These energy differences generally correspond to the radio frequency region of the spectrum, *i.e.*, ~0.1 to 700 MHz.

It is helpful to consider the interaction of charges, dipoles, and quadrupoles with negative electron density in order to define some terms important for nqr (and Mössbauer spectroscopy, Chapter 15). In Fig. 14-1(A), we illustrate the interaction of a positive charge on the z-axis with negative electron density. The energy is given by $-e^2/r$ or $-eV$, where $V = -e/r$ is the electron potential felt by the positive charge located at the point r. In Fig. 14-1(B), we represent a dipole moment in the field of electronic charge. Now the energy associated with the orientation of the dipole depends upon how the potential energy changes over the dipole. Thus, we are

*Several of the principles covered in Sections 14-1 to 14-4 are common to both nqr and Mössbauer spectroscopy, so these sections should be read before Chapter 15.

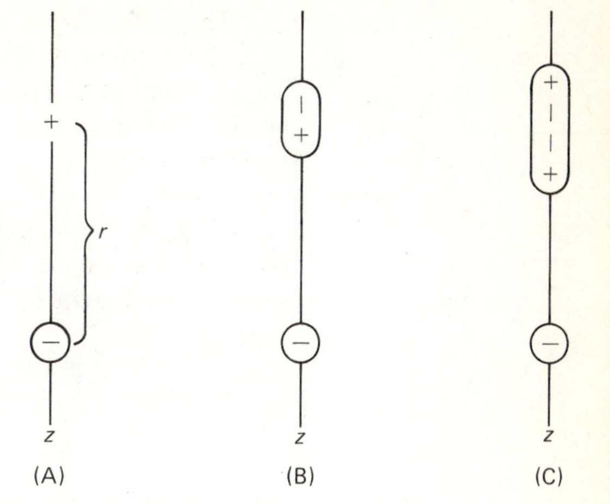

FIGURE 14–1. The interaction of (A) a positive charge, (B) a dipole, and (C) a quadrupole with the z-component of the electric field arising from a unit negative charge ⊖.

interested in how the electrostatic potential changes over the dipole or $\partial V/\partial z$. This is referred to as the z-component of the electric field, E_z. In Fig. 14–1(C), we illustrate the interaction of a quadrupolar distribution with the electric field. Now we have the electric field from the electron in effect interacting with two dipoles whose configuration relative to each other is fixed, *i.e.*, a quadrupole. The energy will depend upon the rate of change (or gradient) of the electric field over the quadrupole. Thus, we are concerned with the "change in the change" of the potential from the electron, or the second derivative of V with respect to z, that is, $\partial^2 V/\partial z^2$. This quantity, which is also the change in the electric field component, $\partial E_z/\partial z$, is called the *electric field gradient*.

In our molecule we have a nucleus imbedded in a charge cloud of electron density. The electric field gradient is defined in terms of a time-averaged electric potential from an electron. Furthermore, the field gradient is described by a 3 × 3 tensor **V**, which is symmetric and has a zero trace. The *nuclear quadrupole moment* is also described by a 3 × 3 tensor **Q**. The nuclear quadrupole coupling energy E_Q is given by

$$E_Q = -\frac{e}{6}\mathbf{Q}_{ij}\mathbf{V}_{ij}$$

where $\mathbf{Q}_{ij}$ is the nuclear quadrupole moment tensor and $\mathbf{V}_{ij}$ is the electrostatic field gradient tensor. The product will depend upon the mutual orientation of the two axis systems. For **Q**, it is convenient to select an axis system that coincides with that of the spin system. When this is done, the cylindrical symmetry of the nucleus permits definition of the tensor in terms of one parameter, the nuclear quadrupole moment Q. The sign of this quantity must be known to obtain the sign of the electric field gradient. An axis system can also be chosen in which the electrostatic field gradient tensor is diagonal. This is called the *principal axis system of the field gradient tensor,* and the only non-zero elements are the diagonal elements whose magnitudes produce a traceless tensor:

$$\frac{\partial^2 V}{\partial x^2} + \frac{\partial^2 V}{\partial y^2} + \frac{\partial^2 V}{\partial z^2} = 0$$

When comparing a given atom in different molecules, it is necessary to know the orientation of the field gradient principal axis system in the molecular framework axis system. Three Eulerian angles (α, β, and γ) are required.

The asymmetry of a molecule and the direction of the z-axis of the field gradient, q_{zz}, relative to the crystal axes can be investigated by studying the nqr spectrum of a single crystal in a magnetic field. The Zeeman splitting is a function of orientation, and detailed analysis of the spectra for different orientations enables one to determine the direction of the z-axis of the field gradient, q_{zz}. This axis can be compared to the crystal axes.

When the principal axes of the coordinate system of the molecule are the principal axes of the electric field gradient tensor, the potential energy E_Q for the interaction of the quadrupole moment with the electric field gradient at the nucleus is given by

$$E_Q = \frac{e}{6}(V_{xx}Q_{xx} + V_{yy}Q_{yy} + V_{zz}Q_{zz}) \tag{14-1}$$

We define the electric field gradient V_{zz} as eq_{zz}, where e is the electron charge (4.8×10^{-10} esu). Since the trace of the electric field gradient tensor is zero, we need define only one more quantity to specify the field gradient, and this is done in equation (14–2):

$$\eta = \frac{V_{xx} - V_{yy}}{V_{zz}} \tag{14-2}$$

The quantity η is called the *asymmetry parameter*. The quantities V_{xx}, V_{yy}, and V_{zz} are often written as eq_{xx}, eq_{yy}, and eq_{zz}. By convention, $|q_{zz}| > |q_{xx}| > |q_{yy}|$, so η ranges from 0 to 1 as a result. With a zero trace, the field gradient is completely defined by eq and η.

Substituting these definitions of q_{zz} and η into equation (14–1) and using the fact that $q_{zz} + q_{xx} + q_{yy} = 0$, we obtain

$$E_Q = \frac{e^2q}{6}\left[\frac{1}{2}(\eta - 1)Q_{xx} - \frac{1}{2}(\eta + 1)Q_{yy} + Q_{zz}\right] \tag{14-3}$$

When axial symmetry pertains, η equals zero and E_Q becomes equal to $(1/4)e^2qQ_{zz}$. (Note that $Q_{zz} + Q_{xx} + Q_{yy} = 0$.)

The classical considerations given above are readily expressed by a quantum mechanical operator:

$$\hat{H}_Q = \left(\frac{e^2}{6}\right)\sum_{i,j} q_{ij}Q_{ij} \tag{14-4}$$

where the summation is over the components of the nuclear quadrupole moments Q_{ij} and the electric field gradients q_{ij}. In the principal axis system with η defined as above, the most common form of the Hamiltonian is given by:

$$\hat{H}_Q = \frac{e^2Qq}{4I(2I - 1)}\left[3\hat{I}^2 - I(I + 1) - \frac{\eta}{2}(\hat{I}_+^2 + \hat{I}_-^2)\right] \tag{14-5}$$

The product e^2Qq or e^2Qq/h (often written as eQq_{zz} or eQq_{zz}/h) is called the quadrupole coupling constant. The operator $\hat{H}_Q$ operates on the nuclear wave functions. When $\eta = 0$, the terms involving the raising and lowering operators drop out. We shall not be concerned with the explicit evaluation of matrix elements; the interested reader can consult references 1–3. Suffice it to say that a series of secular equations can be written and solved to give the energies of the nuclear spin states in the electric field gradient resulting from the distribution of the electron density of the molecule.

14–2 ENERGIES OF THE QUADRUPOLE TRANSITIONS

In an axially symmetric field ($\eta = 0$), the energies of the various quadrupolar nuclear states are summarized by the following equation:

$$E_m = \frac{e^2Qq[3m^2 - I(I+1)]}{4I(2I-1)} \tag{14-6}$$

where I is the nuclear spin quantum number, and m is the nuclear magnetic quantum number. For a nuclear spin of $I = 3/2$, m can have values of $3/2$, $1/2$, $-1/2$, $-3/2$. For $m = 3/2$, substitution into equation (14–6) produces the result $E_{3/2} = +e^2Qq/4$. Since m is squared, the value for $m = -3/2$ will be identical to that for $m = +3/2$, and a doubly degenerate set of quadrupole energy states results. Similarly, the state from $m = \pm 1/2$ will be doubly degenerate. The transition energy, ΔE, indicated by the arrow in Fig. 14–2, corresponds to $e^2Qq/4 - (-e^2Qq/4) = e^2Qq/2$. Thus, for a nucleus with a spin $I = 3/2$ in an axially symmetric field, a single transition is expected, and the quantity e^2Qq expressed in energy units can be calculated directly from the frequency of absorption: $e^2Qq = 2\Delta E = 2h\nu$. The quantity e^2Qq is often expressed as a frequency in MHz, although strictly speaking this should be e^2Qq/h. For the above case, e^2Qq would be twice the frequency of the nqr transition.

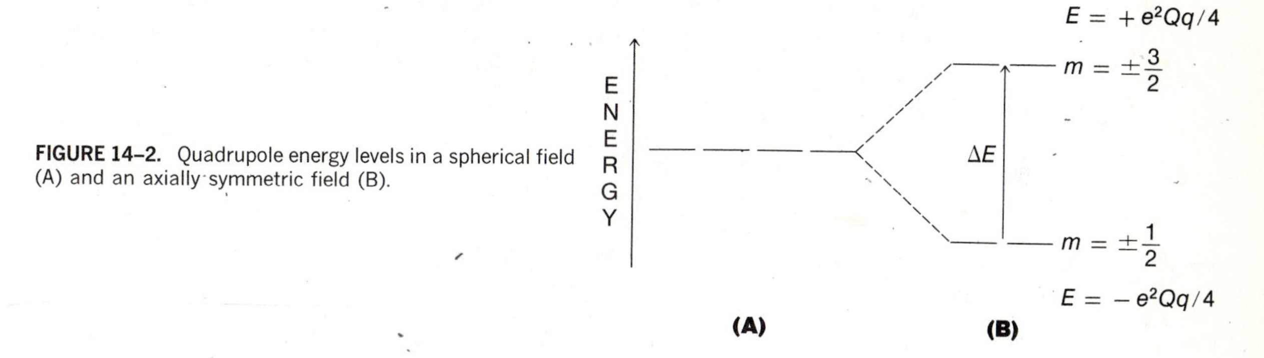

FIGURE 14–2. Quadrupole energy levels in a spherical field (A) and an axially symmetric field (B).

The number of transitions and the relationship of the frequency of the transition to e^2Qq can be calculated in a similar manner for other nuclei with different I values in axially symmetric fields by using equation (14–6). For $I = 7/2$, four energy levels ($E_{\pm 1/2}$, $E_{\pm 3/2}$, $E_{\pm 5/2}$, and $E_{\pm 7/2}$) and three transitions result. The selection rule for these transitions is $\Delta m = \pm 1$, so the observed transitions are $E_{\pm 1/2} \rightarrow E_{\pm 3/2}$, $E_{\pm 3/2} \rightarrow E_{\pm 5/2}$, and $E_{\pm 5/2} \rightarrow E_{\pm 7/2}$. (Recall that all levels are populated under ordinary conditions.) Substitution of I and m into equation (14–6) produces the result that the energy of the $E_{\pm 3/2} \rightarrow E_{\pm 5/2}$ transition is twice that of the $E_{\pm 1/2} \rightarrow E_{\pm 3/2}$ transition. The energies of these levels and the influence of the asymmetry parameter η on these energies are illustrated in Fig. 14–3. In measured spectra, deviations from the frequencies predicted when $\eta = 0$ are attributed to deviations from axial symmetry in the sample, and, as will be seen shortly, this deviation can be used as a measure of asymmetry.

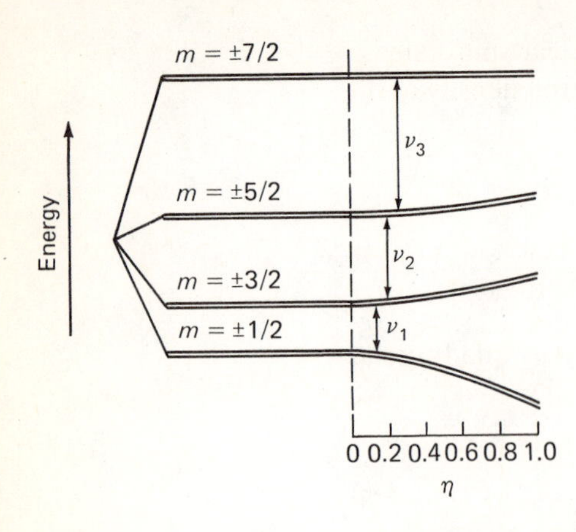

FIGURE 14–3. Nuclear quadrupole energy level diagram for $I = 7/2$.

In a nuclear quadrupole resonance (nqr)* experiment, radiation in the radio frequency region is employed to effect transitions among the various orientations of a quadrupolar nucleus in a non-spherical field. The experiment is generally carried out on a powder. Different orientations of the small crystals relative to the rf frequency direction affect only the intensities of the transitions but not their energies. Structural information about a compound can be obtained by considering how different structural and electronic effects influence the asymmetry of the electron environment. One set of resonances is expected for each chemically or crystallographically inequivalent quadrupolar nucleus. Crystallographic splittings are often small compared to splittings from chemical non-equivalence.

Two types of oscillators have been commonly used in nqr, the superregenerative and the marginal oscillator. The superregenerative oscillator is most common because it allows broad band scanning in searching for resonances and is not complicated to operate. It has the disadvantage of producing a multiplet of lines for each resonance, as shown in Fig. 14–4(A), because of its particular operational characteristics. The true resonance frequency is the center line of the multiplet. The marginal oscillator gives a single peak for each absorption, but it requires constant adjustment and is tedious to operate.

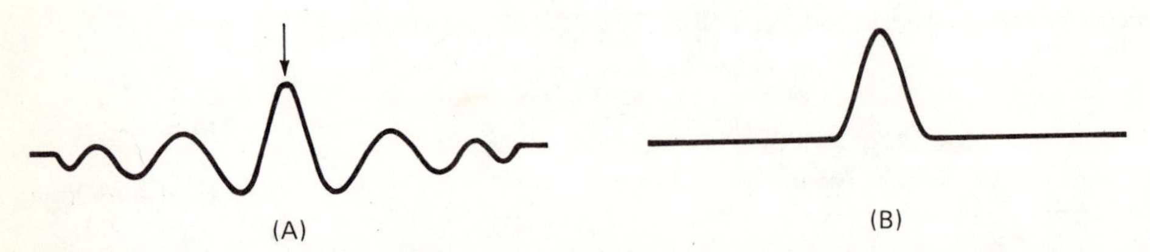

FIGURE 14–4. (A) Multiplet of peaks for a single resonance from a superregenerative oscillator; (B) single peak from a marginal oscillator for the same resonance as in (A).

The ^{35}Cl nqr spectrum of Cl_3BOPCl_3 is shown in Fig. 14–5. Resonances for three non-equivalent chlorines are found. Two of the resonance centers are indicated by × marks on the spectrum at 30,880 MHz and 31,280 MHz. The center of the third resonance around 30,950 MHz is difficult to determine accurately because of overlap with the resonance at 31,280 MHz.

In addition to this direct measurement of the quadrupole energy level difference by absorption of radio frequency radiation, the same information may also be obtained from the fine structure in the pure rotation (microwave) spectrum of a gas. The different nuclear orientations give slightly different moments of inertia, resulting in fine structure in the microwave spectrum. The direct measurement by absorption

*For reviews of nqr, see the General References.

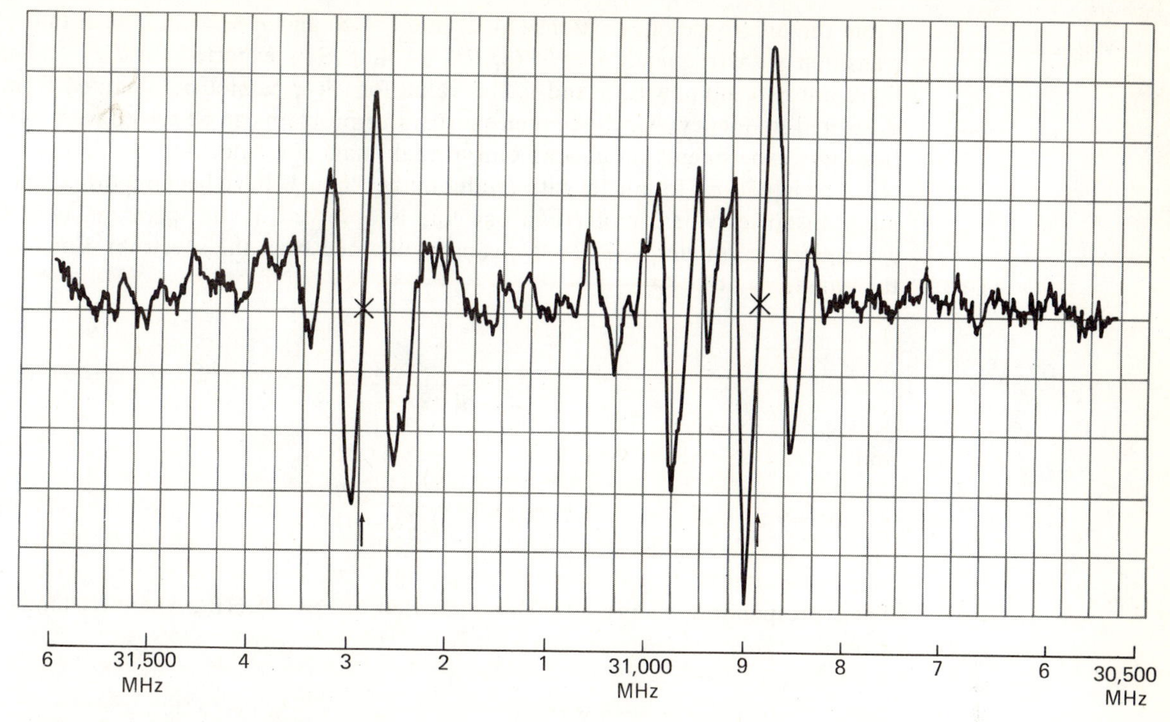

FIGURE 14–5. The ^{35}Cl nqr spectrum of Cl_3BOPCl_3 at 77°K with 25 kHz markers.

of radio frequency radiation must be carried out on a solid. In a liquid or even in some solids (especially near the melting point), collisions and vibrations modulate the electric field gradient to such an extent that the lifetime of a quadrupole state becomes very short. This leads to uncertainty broadening, and the line is often not detected. Details regarding the instrumentation have been published.[(4)]

The energy difference between the various levels and, hence, the frequency of the transition will depend upon both the field gradient, q, produced by the valence electrons and the quadrupole moment of the nucleus. The quadrupole moment, eQ, is a measure of the deviation of the electric charge distribution *of the nucleus* from spherical symmetry. For a given isotope, eQ is a constant, and values for many isotopes can be obtained from several sources.[(5,6)] They can be measured with atomic beam experiments. The units of eQ are charge times distance squared, but it is common to express the moment simply as Q in units of cm^2. For example, ^{35}Cl with a nuclear spin $I = 3/2$ has a quadrupole moment Q of -0.08×10^{-24} cm^2, the negative sign indicating that the charge distribution is flattened relative to the spin axis (see Fig. 7–1).

The second factor determining the extent of splitting of the quadrupole energy levels is the *field gradient, q,* at the nucleus produced by the *electron distribution* in the molecule. The splitting of a quadrupole level will be related to the product e^2Qq. For a molecule with axial symmetry, q often lies along the highest-fold symmetry axis, and when eQ is known, one can obtain the value of q. In a non-symmetric environment, the energies of the various quadrupole levels are no longer given by equation (14–6), because the full Hamiltonian in equation (14–5) must be used. For $I = 3/2$, the following equations can be derived[(2)] for the energies of the two states:

$$E_{\pm 3/2} = \frac{3e^2Qq\sqrt{1 + \eta^2/3}}{4I(2I - 1)} \tag{14-7}$$

$$E_{\pm 1/2} = \frac{-3e^2Qq\sqrt{1 + \eta^2/3}}{4I(2I - 1)} \tag{14-8}$$

From the difference of equations (14–7) and (14–8) and $E = h\nu$, it is seen that one transition with frequency $\nu = (e^2Qq/2h)\sqrt{1 + \eta^2/3}$ is expected for $I = 3/2$. Since there are two unknowns, η and q, the value for e^2Qq cannot be obtained from a measured frequency. As will be seen shortly, this problem can be solved with results from nqr experiments on a sample in a weak magnetic field.

The equations for nuclei with I values other than $3/2$ have been reported, and in many instances where more than one line is observed in the spectrum, both the asymmetry parameter, η, and e^2Qq can be obtained from the spectrum. For $I = 1$, the equations are:

$$E_0 = \frac{-2e^2Qq}{4I(2I - 1)} \tag{14-9}$$

and

$$E_{\pm 1} = \frac{e^2Qq(1 \pm \eta)}{4I(2I - 1)} \tag{14-10}$$

The corresponding energy levels are indicated in Fig. 14–6(B), where K stands for

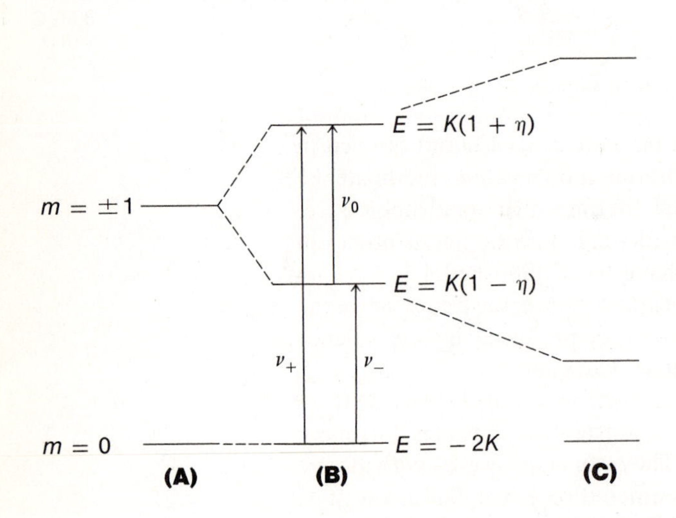

FIGURE 14–6. Energy levels for $I = 1$ under different conditions. (A) $\eta = 0$ and applied magnetic field $H_0 = 0$. (B) $\eta \neq 0$ and $H_0 = 0$. (C) $\eta \neq 0$ and $H_0 \neq 0$ but constant.

$e^2Qq/(4I(2I - 1))$. The perturbation of these levels by an applied magnetic field is indicated in Fig. 14–6(C). This effect will be discussed in the next section. As can be seen from Fig. 14–6(B), there are three transitions for $I = 1$ and $\eta \neq 0$ (labeled ν_+, ν_-, and ν_0), so the two unknowns, e^2Qq and η, can be determined directly from the spectrum. The energies of the levels in Fig. 14–6 are given by equations (14–9) and (14–10). The energies of ν_+, etc., are found by the differences in the energies of the levels. For any two transitions, the resulting equations can be solved for the two unknowns e^2Qq and η. It is interesting to point out that for a nucleus with $I = 1$ in an axially symmetric field, only one line is expected in the spectrum [see Fig. 14–6(A)].

The energies of the quadrupole levels as a function of η have been calculated[2] for cases other than $I = 1$ or $3/2$. Tables have also been compiled[3] that permit calculations of η and e^2Qq from spectral data for nuclei with $I = 5/2$, $7/2$, and $9/2$. When η is appreciable, the selection rule $\Delta m = \pm 1$ breaks down, and spectra containing $\Delta m = \pm 2$ bands are often obtained.

14–3 EFFECT OF A MAGNETIC FIELD ON THE SPECTRA

When an nqr experiment is carried out on a sample placed in a static magnetic field, the Hamiltonian $\hat{H}_M$ describing the influence of the magnetic field on the nuclear magnetic dipole moment must be added to the nqr Hamiltonian:

$$\hat{H}_M = -g_N\beta_N\vec{\mathbf{H}}\cdot\hat{I} \tag{14-11}$$

In weak magnetic fields ($g_N\beta_N H \ll e^2Qq$) the magnetic field acts as a perturbation on $\hat{H}_Q$. In general, the influence of this term on the energies is to *shift the energy of a non-degenerate quadrupole level and split a doubly degenerate level* [*i.e.*, one having $m \neq 0$; see equation (14–6)]. This change in energy of the non-degenerate quadrupolar levels is indicated in Fig. 14–6(B) and (C) for a nucleus in which $I = 1$ and $\eta \neq 0$.

For a nucleus with $I = 1$, a quadrupole spectrum with two lines can arise from either of two distinct situations: for $\eta \neq 0$ as noted above [Fig. 14–6(B)], or for nuclei with $\eta = 0$ located in two non-equivalent lattice sites. Examination of the spectrum of the sample in an external field allows a distinction to be made between these two possibilities. In the former case ($\eta \neq 0$), two lines would again be observed, but with different energies than those obtained in the absence of the field; in the latter case, each doubly degenerate level would be split, giving a spectrum with four lines.

As mentioned earlier, the levels $E_{\pm 1/2}$ and $E_{\pm 3/2}$ are each doubly degenerate for $I = 3/2$ in a non-symmetric field. As a result, e^2Qq and η cannot be determined directly. This degeneracy is removed by a magnetic field, with four levels resulting. Four transitions are observed in the spectrum: $+1/2 \rightarrow +3/2$, $+1/2 \rightarrow -3/2$, $-1/2 \rightarrow +3/2$, and $-1/2 \rightarrow -3/2$ ($\Delta|m| = +1$). The energy differences corresponding to these transitions are functions of H, e^2Qq, and η [as indicated by equations (14–7) and (14–8)], so q and η can be evaluated[7] for this system from the spectra of the sample with and without an applied magnetic field.

14–4 RELATIONSHIP BETWEEN ELECTRIC FIELD GRADIENT AND MOLECULAR STRUCTURE

Our next concern is how we obtain information about the electronic structure of a molecule from the values of q and η. The field gradient at atom A in a molecule, $q_{\text{mol}}{}^{\text{A}}$, and the electronic wave function are related by equation (14–12):

$$q_{\text{mol}}{}^{\text{A}} = e\left\{\sum_{\text{B}\neq\text{A}} [Z_{\text{B}}(3\cos^2\theta_{\text{AB}} - 1)/R_{\text{AB}}{}^3] - \int \psi^* \sum_n [(3\cos^2\theta_{\text{A}_n} - 1)/r_{\text{A}_n}{}^3]\psi\, d\tau\right\} \tag{14-12}$$

The molecular field gradient q_{mol} is seen to be a sensitive measure of the electronic charge density in the immediate vicinity of the nucleus because equation (14–12) involves the expectation value $\langle 1/r^3\rangle$. The first term in the equation is a summation over all nuclei external to the quadrupolar nucleus, and the second term is a summation over all electrons. If the molecular structure is known, the first term is readily evaluated. Z_{B} is the nuclear charge of any atom in the molecule other than A, the one whose field gradient is being investigated; θ_{AB} is the angle between the bond axis or highest-fold rotation axis for A and the radius vector from A to B, R_{AB}. The second term represents the contribution to the field gradient in the molecule from the electron density, and it is referred to as the electric field gradient q_{el}. Finally, ψ is the ground state wave function, and θ_{A_n} is the angle between the bond or principal axis and the radius vector r_{A_n} to the nth electron. This integral is difficult to evaluate. In the LCAO approximation, we can write

$$q_{el}^{A} = -e2\sum_{u}^{occ}\sum_{i}\sum_{j} C_{iu}C_{ju}\int \varphi_i^{*}\hat{q}_A\varphi_j\,d\tau \qquad (14\text{-}13)$$

where $\hat{q}_A = (3\cos^2\theta - 1)/r^3$; u is the index over the molecular orbitals; and i and j are indices for atomic orbitals. C_{iu} and C_{ju} represent the LCAO coefficients for the atomic orbitals φ_i and φ_j in the u molecular orbital. The integral may involve one, two, or three centers. Obviously, good wave functions are required, and an involved evaluation of equation (14–13) is needed to interpret q. For certain atoms (*e.g.*, N and Cl), the three-center contribution is small and can be ignored. Separating the one- and two-center terms in equation (14–13) and abbreviating the integrals $\int \varphi_i\hat{q}_A\varphi_j\,d\tau$ as q_A^{ij}, etc., we can write equation (14–13) as (14–14):

$$q_{el}^{A} = -e\left[2\sum_{u}\sum_{i}^{A} C_{iu}^{2}q_A^{ii} + 4\sum_{u}\sum_{i}^{A}\sum_{j}^{B\neq A} C_{iu}C_{ju}q_A^{ij} + 2\sum_{u}\sum_{j}^{B\neq A} C_{ju}^{2}q_A^{jj}\right] \qquad (14\text{-}14)$$

There have been several semi-empirical methods proposed for evaluating electric field gradients. Cotton and Harris[8] assumed that the nuclear term of equation (14–12) (*i.e.*, the first term) was cancelled by the part of q_{el}^{A} arising from the gross atomic populations on the neighbor atom B; *i.e.*,

$$e\sum_{B\neq A}\left[\frac{Z_B(3\cos^2\theta_{AB} - 1)}{R_{AB}^{3}}\right] \cong e\left[2\sum_{u}\sum_{j}^{B\neq A} C_{ju}^{2}q_A^{jj} + 2\sum_{u}\sum_{i}^{A}\sum_{j}^{B\neq A} C_{iu}C_{ju}q_A^{ij}\right] \qquad (14\text{-}15)$$

This leads to

$$q_{mol}^{A} = -e\left[2\sum_{u}\sum_{i}^{A} C_{iu}^{2}q_A^{ii} + 2\sum_{u}\sum_{i}^{A}\sum_{j}^{B\neq A} C_{iu}C_{ju}q_A^{ij}\right] \qquad (14\text{-}16)$$

Further, by assuming that the two-center integral can be formulated as proportional to the overlap integral

$$q_A^{ij} = S_{ij}q_A^{ii} \qquad (14\text{-}17)$$

we can write:

$$q_{mol}^{A} = -e\sum_{i}^{A} q_A^{ii}\left(2\sum_{u} C_{iu}^{2} + 2\sum_{u}\sum_{j}^{B\neq A} C_{iu}C_{ju}S_{ij}\right) \qquad (14\text{-}18)$$

We see then that the field gradient is given by multiplying q_A^{ij} by the gross atomic population, P [equation (3–31)], or:

$$q_{mol}^{A} = -eP\sum_{i}^{A}\varphi_i\left|\frac{3\cos^2\theta - 1}{r^3}\right|\varphi_i \qquad (14\text{-}19)$$

For an atom A with valence s and p orbitals, the above summation is over four orbitals for which q_A for the s orbital is zero, and

$$q = -eq_{at}\left(P_z - \frac{1}{2}P_x - \frac{1}{2}P_y\right)$$

where q_{at} is the field gradient for one electron in a p-orbital. The quantity in parentheses in equation (14–18) is just the molecular-orbital expression for the gross atomic population P of atomic orbital φ_i. (See Chapter 3.) Depending on the relative populations of p_z, p_x, and p_y, q can be positive or negative. If the quadrupole coupling constant is expressed in megahertz, we have

$$\frac{e^2Qq}{h} = -\left(\frac{e^2Qq_{at}}{h}\right)\left(P_z - \frac{1}{2}P_x - \frac{1}{2}P_y\right) \tag{14-20}$$

where e^2Qq/h is the quadrupole coupling constant, and e^2Qq_{at}/h is the coupling constant for a single electron in the p-orbital. The p_z orbital should be coincident with q_{zz} in order to apply these equations with much accuracy.

Equation (14–20) is the molecular orbital analog to the valence-bond expression of Townes and Dailey,[9] which had been reported earlier. This approach is based on the following arguments. Since the s-orbital is spherically symmetric, electron density in this orbital will not give rise to a field gradient. As long as the atom being studied is not the least electronegative in the bonds to other atoms in the molecule, the maximum field gradient at this atom, q_{mol}, is the atomic field gradient, q_{at}, for a single electron in a p_z orbital of the isolated atom. When the atom being investigated is more electronegative than the atom to which it is bonded, the quadrupolar atom has greater electron density around it in the molecule than in the isolated atom. The relationship between the electron "occupation" of the p-orbitals of a quadrupolar atom in a molecule, e^2Qq_{at}, and the quantity e^2Qq_{mol} (which is determined from the nqr spectrum of the molecule under consideration) is given by:

$$e^2Qq_{mol} = [1 - s + d - i(1 - s - d)]e^2Qq_{at} \tag{14-21}$$

where e^2Qq_{at} is the quadrupole coupling constant for occupancy of the p-orbital by a single electron, s is the fraction of s character employed by the atom in the bond to its neighbor, d is the fraction of d character in this bond, and i is the fraction of ionic character in the bond (for a molecule A—B, $\psi^0 = c\psi_A + d\psi_B$ and $i = c^2 - d^2$). When the atom being studied is electropositive, i changes sign. Values for e^2Qq_{at} and q_{at} have been tabulated for several atoms.[4,5,10,11] When π bonding is possible, this effect must also be included. A modified form of equation (14–21) has also been employed:[12]

$$e^2Qq_{mol} = (1 - s + d - i - \pi)e^2Qq_{at} \tag{14-22}$$

where π is the extent of π bonding, and all other quantities are the same as before.

As the amount of ionic character in a bond increases, the electronic environment approaches spherical symmetry (where $q_{mol} = 0$) and e^2Qq_{mol} decreases. Hybridization of the p-orbital with an s-orbital also decreases e^2Qq_{mol}, as indicated by equation (14–22). Mixing of the s-orbital with the p-orbital decreases the field gradient, because the s-orbital is spherically symmetric. In a covalent molecule, d-orbital contribution to the bonding increases the field gradient.

There are several problems associated with these approaches to the interpretation of field gradients. First, even if the above approach were correct for the system of interest, there are four unknowns in equation (14–22) and we have only one measurable quantity, e^2Qq_{mol}. Investigators attempting to interpret this quantity are thus forced to assume the answer and provide a reasonable interpretation of the data based on these models. Second, O'Konski and Ha[13] have shown that the assumption of Cotton and Harris[8] indicated in equation (14–15) is not generally correct. When this equation does not apply, equations (14–20), (14–21), and (14–22) are also not correct.

A semi-empirical approach to the interpretation of field gradients has been reported,[14] which does not make the suspect approximations described above. The reader is referred to the original literature for details of this method.

One additional factor complicating the quantum mechanical calculation and interpretation of nqr and Mössbauer (*vide infra*) parameters is the *Sternheimer effect*.[15] This effect is the polarization of the originally spherically symmetric inner shell electrons by the valence electrons. When the core electrons lose their spherical symmetry, they contribute to the field gradient at the nucleus. This effect, like spin polarization (discussed in Chapter 12), is an artifact of not performing a full molecular orbital calculation on the whole crystal. There are two contributions to Sternheimer shielding, and these can be illustrated if we consider the ligands or ions external to the electron density of the atom as point charges. The spherical expansion of the core electron density is illustrated for positive point charges in Fig. 14–7(A) and (B). The asymmetry induced is shown in Fig. 14–7(C). We have artificially broken up the polarization to illustrate these two contributions to the Sternheimer shielding.

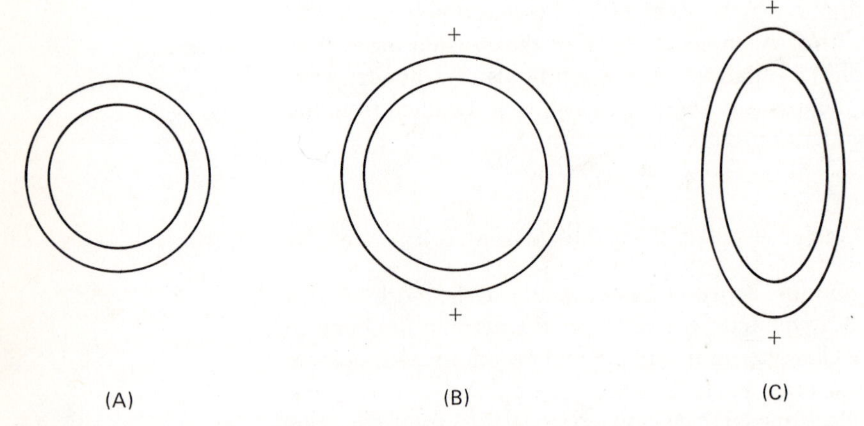

FIGURE 14–7. Schematic illustration of the Sternheimer effect. (A) A spherical shell of electron density; (B) the radial expansion resulting from two external positive charges; (C) the elliptical polarization by two positive charges.

The shielding for closed shell systems is usually expressed by the formal equation:

$$e^2Qq_{\text{obs}} = (1 - \gamma_{ri})e^2Qq_0$$

where e^2Qq_0 is the field gradient calculated in the absence of any Sternheimer effects and γ_{ri} is the shielding parameter for the ith charge located at a distance r from the nucleus. The sign of γ_{ri} is a function of the distance. For charges external to the core electron density of the central atom, γ_{ri} will generally be negative and is said to give rise to an *antishielding* contribution. For charges inside the valence orbitals of the central atom, the sign of γ_{ri} is generally positive and gives rise to a *shielding* contribution. Shielding constants have been calculated for charges external to a large number of ions and are designated by the symbol γ_∞. The value of γ_{ri} used to evaluate the Sternheimer effect in molecules is difficult to determine but appears to be considerably below that of γ_∞. Most workers in the field assume that this effect is constant in a similar series of complexes.

14–5 APPLICATIONS

NQR spectra of a number of molecules containing the following nuclei have been reported: ^{27}Al, ^{75}As, ^{197}Au, ^{10}B, ^{11}B, ^{135}Ba, ^{137}Ba, ^{209}Bi, ^{79}Br, ^{81}Br, ^{43}Ca, ^{35}Cl, ^{37}Cl, ^{59}Co, ^{63}Cu, ^{69}Ga, ^{71}Ga, ^{2}H, ^{201}Hg, ^{127}I, ^{115}In, ^{25}Mg, ^{14}N, ^{55}Mn, ^{23}Na, ^{93}Nb, ^{17}O, ^{185}Re, ^{187}Re, ^{33}S, ^{121}Sb, ^{123}Sb, ^{181}Ta.

The Interpretation of e^2Qq Data

As mentioned above, it is not possible to interpret e^2Qq values rigorously in terms of ionic character, π bonding, and s and d hybridization. For compounds in which there are large differences in ionic character, the effect can be clearly seen, as indicated by the data in Table 14–1. More positive e^2Qq_{mol} values are obtained for ionic compounds. Attempts have been made to "explain" e^2Qq data for a large number of halogen compounds by assuming that d hybridization is not important and estimating ionic character or s hybridization.[12,16]

TABLE 14–1. VALUES OF e^2Qq_{mol} FOR SOME DIATOMIC HALIDES

Molecule[a]	e^2Qq_{mol}	Molecule	e^2Qq_{mol}
F<u>Cl</u>	−146.0	<u>Br</u>Cl	876.8
Br<u>Cl</u>	−103.6	Li<u>Br</u>	37.2
I<u>Cl</u>	−82.5	Na<u>Br</u>	58
Tl<u>Cl</u>	−15.8	D<u>Br</u>	533
K<u>Cl</u>	0.04	F<u>Br</u>	1089
Rb<u>Cl</u>	0.774	D<u>I</u>	−1827
Cs<u>Cl</u>	3	Na<u>I</u>	−259.87

[a] Values for e^2Qq_{at} are ^{35}Cl = −109.74, ^{79}Br = 769.76, and ^{127}I = −2292.84. e^2Qq_{mol} applies to the underlined atom.

When the differences in ionic character are large enough to be predictable, the e^2Qq_{mol} values manifest the proper trends. However, for systems in which the differences in ionic character are not obvious from electronegativity and other considerations, the interpretation of e^2Qq differences is usually ambiguous in terms of the relative importance of the effects in equation (14–22). One of the more successful studies involves a series of substituted chlorobenzenes. A linear relation is found between the Hammett σ constant of the substituent and the quadrupole resonance frequency.[17] The data obeyed the equation ν (MHz) $= 34.826 + 1.024\sigma$. The electron-releasing substituents have negative σ values and give rise to smaller e^2Qq values, because the increased ionic character in the C—Cl bond makes the chlorine more negatively charged.

The general problem of covalence in metal-ligand bonds in a series of complexes has been investigated by direct measurement of the quadrupole resonance.[18] By comparing the trends in force constants with the changes in chlorine quadrupole coupling constants, a consistent interpretation of the data for a whole series of MCl_6^{n-} compounds has been offered.[19] A rigorous interpretation of the data (especially where small differences are involved) is again hampered by the lack of information regarding the four variables in equation (14–22).

Two transitions are observed in the nqr spectra of ^{55}Mn ($I = 5/2$), so both q_{zz} and η can be obtained. Measurements on a series of organometallic manganese carbonyls have been reported.[20] Substitution of methyl groups in the series

$(CH_3)_nC_6H_{6-n}Mn(CO)_3^+$ causes only minor changes in the asymmetry parameter. A summary of cobalt-59 nqr investigations has appeared.[21] The magnitudes of the field gradients are indicative of *cis* or *trans* octahedral geometry. The e^2Qq value of [*cis*-$Co(en)_2Cl_2$]Cl is 33.71, while that for [*trans*-$Co(en)_2Cl_2$]Cl is 60.63. A rationalization of this difference will be provided in the next chapter, in the section on partial field gradient parameters.

Thermochemical bond energies have been estimated and compared with the results from chlorine-35 nqr for the $GeCl_6^{2-}$, $SnCl_6^{2-}$ and $PbCl_6^{2-}$ ions. The authors[22] support an earlier conclusion[23] that these results are not consistent with assigning a higher electronegativity to lead than to germanium or tin.

This coverage of quadrupole investigations concerned with elucidating covalency and other characteristics of chemical bonds is by no means complete. The presentation is brief because of the ambiguity that exists in the interpretation of the results. More information than just the field gradient is needed to sort out the variables in equation (14–22).

Effects of the Crystal Lattice on the Magnitude of e^2Qq

A further complication and limitation of the application of nqr spectroscopy results from the fact that direct measurement of nuclear quadrupole transitions can be obtained only on solids. For very complex molecules, this is the only source of nuclear quadrupole information because of the complexity of the microwave spectrum. Measurements on solids introduce the complexities of lattice effects. For those molecules which have been studied by both methods (direct measurement and microwave), it is found that e^2Qq is usually 10 to 15 percent lower in the solid state. It has been proposed[5] that the decrease is due to increased ionicity in the solid. There are several examples of molecules being extensively associated in the solid but not in the gas phase or in solution (*e.g.*, I_2 and CNCl). In these cases, considerable care must be exercised in deducing molecular properties from e^2Qq values. The e^2Qq values for cobalt in selected cobalt complexes illustrates[21] the problem. The values for *trans*-[$Co(NH_3)_4Cl_2$]Cl and *trans*-[$Co(en)_2Cl_2$]Cl are 59.23 MHz and 60.63 MHz; *i.e.*, there is a difference of 1.40 MHz. On the other hand, *trans*-[$Co(en)_2Cl_2$]NO_3 has an e^2Qq(Co) value of 62.78 MHz. Changing the anion causes a difference of 2.15 MHz. The e^2Qq(Cl) values in K_2SnCl_6 and $[(CH_3)_4N]_2SnCl_6$ are 15.063 MHz and 16.674 MHz, respectively. Potential causes for these differences have been offered.[22b,22c] Some examples in which the crystal lattice affects the number of lines observed in a spectrum will be discussed in the next section.

Structural Information from NQR Spectra

Since different field gradients will exist for non-equivalent nuclei in a molecule, we should expect to obtain a different line (or set of lines, depending on I) for each type of nuclear environment. In general, the environment of an atom as determined by nqr studies is in agreement with results obtained from x-ray studies. Only one line is found[18] in the halogen nqr spectrum of each of the following: K_2SeCl_6, Cs_2SeBr_6, $(NH_4)_2TeCl_6$, $(NH_4)_2SnBr_6$, and K_2PtCl_6. This is consistent with O_h structures for these anions.

As mentioned above, the following effects can give rise to multiple lines in the nqr spectrum:

(1) Chemically non-equivalent atoms in the molecule.

(2) Chemically equivalent atoms in a molecule occupying non-equivalent positions in the crystal lattice of the solid.

(3) Splitting of the degeneracy of quadrupole energy levels by the asymmetry of the field gradient. Splittings of the quadrupole levels by other magnetic nuclei in the molecule, similar to spin-spin splittings in nmr, are often not detected in nqr spectra, but are being found with improved instrumentation. Usually these splittings are less than, or of the same order of magnitude as, the line widths. One case in which such splitting has been reported[24] is in the spectrum of HIO_3. The ^{127}I nucleus is split by the proton.

The non-equivalence of lattice positions (2) is illustrated in the bromine nqr spectrum of K_2SeBr_6, which gives a single line at room temperature and two lines at Dry Ice temperature. A crystalline phase change accounts for the difference. NQR is one of the most powerful techniques for detecting phase changes and obtaining structural information about the phase transitions. Although K_2PtI_6, K_2SnBr_6, and K_2TeBr_6 contain "octahedral" anions, the halides are not equivalent in the solid lattice, and the halide nqr spectra all consist of three lines, indicating at least three different halide environments.

Four resonance lines were found in the chlorine quadrupole spectra of each of $TiCl_4$,[25a] $SiCl_4$,[25b] and $SnCl_4$.[25b] It was concluded that the crystal structures of these materials are similar.

The problem of distinguishing non-equivalent positions in the lattice from chemical non-equivalence in the molecular configuration (which is related to the structure of the molecule in the gas phase or in a non-coordinating solvent) is difficult in some cases. In general, the frequency difference for lines resulting from non-equivalent lattice positions is small compared to the differences encountered for chemically non-equivalent nuclei in a molecule. When only slight separations between the spectral lines are observed, it is difficult to determine from this technique alone which of the two effects is operative. Another difficulty encountered in interpretation is illustrated by the chlorine spectrum of M_2Cl_{10} species.[26] Even though the Nb_2Cl_{10} molecule has a structure containing both terminal and bridging chlorines, only a single chlorine resonance assigned[26] to the bridging atom is observed in the spectrum. In Re_2Cl_{10} and W_2Cl_{10}, a large number of resonances were obtained.[26] Consequently, arguments based on assigning the number of different types of nuclei in the molecular structure to the number of resonances observed can be of doubtful validity.

NQR studies of ^{14}N ($I = 1$) are difficult to carry out but produce very interesting results. Since $I = 1$, both e^2Qq and the asymmetry parameter η can be evaluated from the nqr spectrum. In BrCN, the ^{14}N nqr resonance is a doublet. This could result from two non-equivalent nitrogen atoms in the crystal lattice or from a splitting of the nitrogen resonance because of an asymmetric field gradient. The former explanation was eliminated by a single crystal x-ray study.[27] The structure of solid BrCN was found to consist of linear chains of the type:

$$\cdots\cdot Br{-}C{\equiv}N| \cdots\cdots Br{-}C{\equiv}N| \cdots\cdot Br{-}C{\equiv}N|$$

The nitrogen has axial symmetry here, and only one line is expected. However, it is proposed that interactions between chains reduce the symmetry at the nitrogen and lead to the two lines. Various resonance forms can be written for BrCN, and the e^2Qq values indicate that the bromine has a formal positive charge. An appreciable increase in e^2Qq for bromine is observed in the solid relative to the gaseous state spectrum of BrCN. This could be due to increased contributions to the ground state from the structure Br^+CN^- in the solid because of stabilization of Br^+ by coordination. If the N····Br—C bond is described as a *pd* hybrid, the increased *d* contribution in the bromine carbon bond will also increase e^2Qq.

The ^{14}N quadrupole transitions in pyridine and various coordination compounds of pyridine show[28] large changes in the transition energies. As discussed in conjunction with Fig. 14–6, three transitions (ν_+, ν_0, and ν_-) are expected when $\eta \neq 0$, and e^2Qq and η can be determined from the data. Typical results are summarized in Table 14–2.

TABLE 14–2. ^{14}N QUADRUPOLE TRANSITIONS AND FIELD GRADIENT PARAMETERS IN PYRIDINE AND COORDINATED PYRIDINE (ALL FREQUENCIES IN kHz, TEMP. = 77° K). (CRYSTALLOGRAPHIC NON-EQUIVALENCES, WHERE PRESENT, ARE RESOLVED.)

Compound	ν_+	ν_-	ν_0	$\frac{e^2Qq}{h}$	η
Pyridine (Py)	3892	2984	908	4584	0.396
Pyridinium nitrate	1000	580	420	1053	0.798
Py_2ZnCl_2	2387	2078	309	2977	0.207
	2332	2038	294	2913	0.202
$Py_2Zn(NO_3)_2$	2124	1884	240	2672	0.180
	2097	1877	220	2649	0.166
Py_2CdCl_2	2850	2298	552	3432	0.320

The iodine nqr spectrum of solid iodine indicates a large asymmetry parameter, η.[29] Since the iodine atom in an iodine molecule is axially symmetric, the large asymmetry is taken as indication of intermolecular bonding in the solid. A large η in the iodine nqr spectrum of the molecule HIO_3 supports the structure $IO_2(OH)$ instead of HIO_3. The structure HIO_3 has a C_3 axis, so $q_{xx} = q_{yy}$. References 1 and 2 contain many additional examples of studies of this kind.

Information regarding π bonding can be obtained from the asymmetry parameter η. Methods for evaluating η for various nuclei have been discussed. A single σ bond to a halogen should give rise to an axially symmetric field gradient. Double bonding leads to asymmetry, and the extent of π bonding is related to η. It was concluded that there is appreciable (~5 per cent) carbon-halogen π bonding in vinyl chloride, vinyl bromide, and vinyl iodide.[30] Bersohn[31] has made a complete study of the problem of estimating quantitatively the extent of carbon-halogen π bonding from η.

The values of η obtained from the nqr spectra of SiI_4, GeI_4, and SnI_4 were interpreted to indicate a very small degree (about 1 per cent) of double bond character[32] in the halogen bond to the central atom. This could be due to a solid state effect.

The ^{75}As, $^{121,123}Sb$, ^{209}Bi, $^{35,37}Cl$, and $^{79,81}Br$ nqr spectra of compounds with general formula R_3MX_2 have been studied[33] for R = CH_3, $CH_2C_6H_5$, and C_6H_5, for X = F, Cl, and Br, and for M = As, Sb, and Bi. Very small asymmetry parameters were found, indicating a threefold axis in all compounds. The results suggest that most of the compounds are trigonal bipyramidal. However, the arsenic compound $[(CH_3)_3AsBr^+]Br^-$ is not, but probably has a cation with C_{3v} symmetry.

The nqr spectra (^{37}Cl, ^{35}Cl, ^{121}Sb, ^{127}I) of $2ICl \cdot AlCl_3$ and $2ICl \cdot SbCl_5$ indicate[34] that these materials should be formulated as $ICl_2^+AlCl_4^-$ and $ICl_2^+SbCl_6^-$. The nqr spectra of the v-shaped cations ICl_2^+, I_3^+, and I_2Cl^+ have been studied in several different compounds. The ^{35}Cl nqr spectra of the chloroaluminate group in a wide variety of $M_n(AlCl_4)_m$ compounds have been studied.[35] The transition energies can be used to indicate whether the relatively free ion $AlCl_4^-$ exists or whether this anion is strongly coordinated to the cation. "Ionic" $AlCl_4^-$ transitions are found in the 10.6 to 11.3 MHz range. Strong coordination of $AlCl_4^-$ results in an elongation of the

Al—Cl bridge bonds and appreciably increases the range and average frequency of the chlorine transitions.

In the next chapter, on Mössbauer spectroscopy, we shall discuss an additive, *partial field gradient* (pfg), model for correlating field gradients at central atoms. It is useful for deducing structures of molecules from experimental Mössbauer data. This model can also be used to infer structures from nqr data. Since most of the data used to derive and test the pfg model are Mössbauer results, we shall treat this topic in the next section.

14-6 DOUBLE RESONANCE TECHNIQUES

Nuclear double resonance techniques have been reported[36-38] for observing quadrupolar transitions. These methods greatly increase sensitivity over previous techniques and also permit one to observe transitions with very low frequency; *e.g.*, in ^{2}H spectra, transitions are observed in the 100 to 160 kHz region.[39]

The adiabatic demagnetization double resonance experiment is a novel technique in this general category. Consider a sample that has a quadrupolar nucleus in a molecule containing several protons. When this sample is placed in a magnetic field and we wait long enough for equilibrium to be obtained, there will be an excess of proton nuclear moments aligned with the field that undergo Larmor precession and give rise to a net magnetization, as discussed in the chapters on nmr. When the sample is removed from the field, the net magnetization is reduced to zero as the sample is removed because the individual moments now become aligned with their local fields. A random orientation of these local fields in the absence of an external field produces a zero net magnetization. This is illustrated in Fig. 14–8 on the left-hand side, in the region labeled "sample removed."

If T_1 is longer than the time required to remove the sample, the excess population of $+\frac{1}{2}$ spins will remain, but they now precess about a net local field felt at the nucleus from spin-spin interaction with neighboring protons. Over the full sample, the magnetization is zero; but when this sample is reinserted into a magnetic field, magnetization is simultaneously induced into the sample without having to wait the required time for the T_1 process. This is illustrated in the region of Fig. 14–8 labeled "sample reinserted." The intensity of the magnetization can be measured by employing a 90° pulse immediately after reinserting the sample in the magnetic field

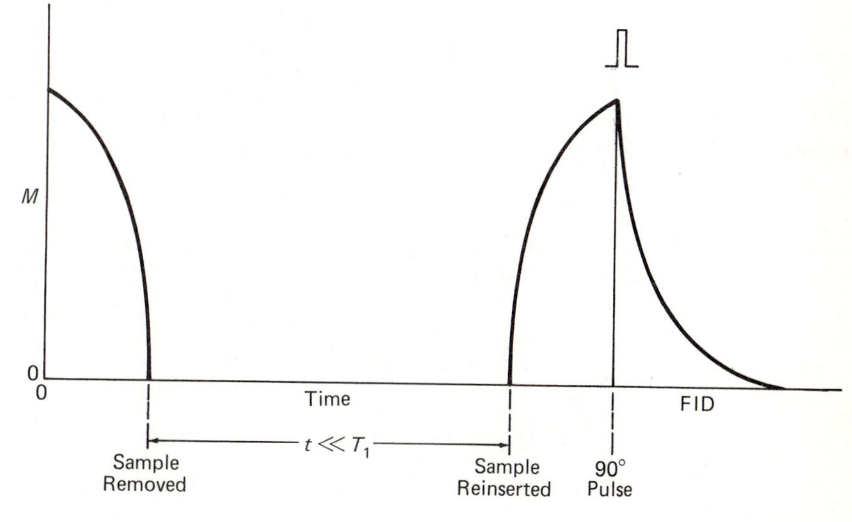

FIGURE 14–8. Plot of the magnetization when a sample is removed from a magnetic field, reinserted, subjected to a 90° pulse, and then allowed to undergo free induction decay.

and measuring the FID curve (see Fig. 14–8). If the time between removal of the sample from the magnetic field and reinserting it is long compared to T_1, the magnetization will decrease as the spins become randomized.

Now consider an experiment in which the sample is irradiated with an rf frequency corresponding to the quadrupolar nucleus B transition after the sample has been removed from the field. Furthermore, we shall assume that the time between removal and reinsertion is small compared to T_1 for the protons. The effect of this rf field is to randomize the B nucleus by inducing quadrupolar transitions in the B spin system. Provided that the appropriate conditions are met, in terms of the amplitude of the applied rf field in relationship to the local field experienced at the protons, the randomization of the B spin system influences that of the proton system. This occurs by the following process. When the sample is removed from the field, the energy difference between the $m = +\frac{1}{2}$ and $-\frac{1}{2}$ states (*i.e.*, the transition energy of the H nucleus) decreases toward zero. In the process, there is a time at which the energy difference for the hydrogen nuclei matches the energy difference of the quadrupolar states of the B nucleus. A resonance energy exchange occurs, tending to randomize the proton nuclei. The process of randomization is often referred to as an increase in the *spin temperature* of the system. As a result of the randomization of the proton system from transitions in B, the magnetization that is recovered when the sample is returned to the magnetic field is less than it would otherwise be. As a result, the FID measured is less than it would otherwise be. When the frequency used for the quadrupole transition is not appropriate for resonance, the B nucleus is not randomized and a large amount of the magnetization is recovered. By systematically incrementing the frequency of the B transmitter, the "spectra" of the quadrupole transitions of the B system are mapped out in terms of their effect on the FID of the abundant proton spin system.

By using a spin echo double resonance experiment,[39-41] much of the inhomogeneous dipolar broadening (crystal imperfections, etc.) that leads to very broad lines in the direct nqr experiment can be eliminated. The nqr of Al_2Br_6 has been determined with this technique.[40]

If one has two quadrupolar nuclei surrounded by nuclei with $I = 0$ [*e.g.*, as in D—Mn(CO)$_5$], the dipolar coupling of the manganese and deuterium nuclei can be observed.[42] As discussed in the nmr chapters, the bond distance can be obtained from the magnitude of the dipolar coupling. A Mn—D bond distance of 1.61 Å is calculated from the ^{55}Mn nqr spectrum, in excellent agreement with the neutron diffraction result.[43]

REFERENCES CITED

1. C. P. Slichter, "Principles of Magnetic Resonance," Harper and Row, New York (1963), Sec. 6.3.
2. R. Bersohn, J. Chem. Phys., *20,* 1505 (1952).
3. R. Livingston and H. Zeldes, "Tables of Eigenvalues for Pure Quadrupole Spectra," Oak Ridge Natl. Lab. Rept. ORNL-1913 (1955).
4. a. T. P. Das and E. L. Hahn, "Nuclear Quadrupole Resonance Spectroscopy," Academic Press, New York (1958).
   b. E. A. C. Lucken, "Nuclear Quadrupole Coupling Constants," Academic Press, New York (1969).
5. C. T. O'Konski, "Determination of Organic Structures by Physical Methods," Vol. 2, ed. F. C. Nachod and W. D. Phillips, Academic Press, New York (1962).
6. J. A. Pople, W. G. Schneider, and H. J. Bernstein, "High Resolution Nuclear Magnetic Resonance," McGraw-Hill, New York (1959).
7. C. Dean, Phys. Rev., *86,* 607A (1952).
8. F. A. Cotton and C. B. Harris, Proc. Nat. Acad. Sci. U.S., *56,* 12 (1966); Inorg. Chem., *6,* 376 (1967).
9. C. H. Townes and B. P. Dailey, J. Chem. Phys., *17,* 782 (1949).
10. M. H. Cohen and F. Reif, "Solid State Physics," Vol. 5, ed. F. Seitz and D. Turnbull, Academic Press, New York (1957).
11. W. J. Orville-Thomas, Quart. Rev., *11,* 162 (1957).

12. B. P. Dailey, J. Chem. Phys., *33,* 1641 (1960).
13. C. T. O'Konski and T. K. Ha, J. Chem. Phys., *49,* 5354 (1968).
14. W. D. White and R. S. Drago, J. Chem. Phys., *52,* 4717 (1970).
15. R. M. Sternheimer, Phys. Rev., *130,* 1423 (1963); *ibid., 146,* 140 (1966) and references therein.
16. C. H. Townes and B. P. Dailey, J. Chem. Phys., *23,* 118 (1955); W. Gordy, Disc. Faraday Soc., *19,* 14 (1955); M. A. Whitehead and H. H. Jaffe, Trans. Faraday Soc., *57,* 1854 (1961).
17. H. C. Meal, J. Amer. Chem. Soc., *74,* 6121 (1952); P. J. Bray and R. G. Barnes, J. Chem. Phys., *27,* 551 (1957) and references therein.
18. D. Nakamura, K. Ito, and M. Kubo, Inorg. Chem., *1,* 592 (1962); *ibid., 2,* 61 (1963); J. Amer. Chem. Soc., *82,* 5783 (1960); *ibid., 83,* 4526 (1961); *ibid., 84,* 163 (1962); M. Kubo and D. Nakamura, Adv. Inorg. Chem. Radiochem., *8,* 257 (1966).
19. T. L. Brown, W. G. McDugle, and L. G. Kent, J. Amer. Chem. Soc., *92,* 3645 (1970).
20. T. B. Brill and A. J. Kotlar, Inorg. Chem., *13,* 470 (1974).
21. T. L. Brown, Accts. Chem. Res., *7,* 408 (1974) and references therein.
22. a. W. A. Welsh, T. B. Brill, *et al.,* Inorg. Chem., *13,* 1797 (1974).
    b. T. B. Brill, R. C. Gearhart, and W. A. Welsh, J. Mag. Res., *13,* 27 (1974).
    c. T. B. Brill, J. Chem. Phys., *61,* 424 (1974).
23. R. S. Drago and N. A. Matwiyoff, J. Organometal. Chem., *3,* 62 (1965).
24. R. Livingston and H. Zeldes, J. Chem. Phys., *26,* 351 (1957).
25. a. A. H. Reddoch, J. Chem. Phys., *35,* 1085 (1961).
    b. A. L. Schawlow, J. Chem. Phys., *22,* 1211 (1954).
26. P. A. Edwards and R. E. McCarley, Inorg. Chem., *12,* 900 (1973).
27. S. Geller and A. L. Schawlow, J. Chem. Phys., *23,* 779 (1955).
28. Y. N. Hsieh, G. R. Rubenacker, C. P. Cheng, and T. L. Brown, J. Amer. Chem. Soc., in press (1977).
29. H. G. Dehmelt, Naturwiss., *37,* 398 (1950).
30. J. A. Howe and J. H. Goldstein, J. Chem. Phys., *26,* 7 (1957); *27,* 831 (1957) and references therein.
31. R. Bersohn, J. Chem. Phys., *22,* 2078 (1954).
32. H. Robinson, H. G. Dehmelt, and W. Gordy, J. Chem. Phys., *22,* 511 (1954); S. Kojima, *et al.,* J. Phys. Soc. Japan, *9,* 805 (1954).
33. T. B. Brill and G. G. Long, Inorg. Chem., *9,* 1980 (1970).
34. D. J. Merryman and J. D. Corbett, Inorg. Chem., *13,* 1258 (1974).
35. D. J. Merryman, P. A. Edwards, J. D. Corbett, and R. E. McCarley, Inorg. Chem., *13,* 1471 (1974).
36. S. R. Hartmann and E. L. Hahn, Phys. Rev., *128,* 2042 (1962).
37. R. E. Slusher and E. L. Hahn, Phys. Rev., *166,* 332 (1968).
38. D. T. Edmonds, *et al.,* "Advances in Quadrupole Resonance," Vol. 1, p. 145, Heyden, London (1974); Rev. Pure Appl. Chem., *40,* 193 (1974).
39. J. L. Ragle and K. L. Sherk, J. Chem. Phys., *50,* 3553 (1969).
40. M. Emshwiller, E. L. Hahn, and D. Kaplan, Phys. Rev., *118,* 414 (1960).
41. J. L. Ragle *et al.,* J. Chem. Phys., 61, 429, 3184 (1974).
42. a. P. S. Ireland, L. W. Olson, and T. L. Brown, J. Amer. Chem. Soc., *97,* 3548 (1975).
    b. P. S. Ireland and T. L. Brown, J. Mag. Res., *20,* 300 (1975).
43. S. J. LaPlaca, *et al.,* Inorg. Chem., *8,* 1928 (1969).

GENERAL REFERENCES

M. H. Cohen and F. Reif, "Solid State Physics," Vol. 5, ed. F. Seitz and D. Turnbull, Academic Press, New York (1957).
T. P. Das and E. L. Hahn, "Nuclear Quadrupole Resonance Spectroscopy," Academic Press, New York (1958).
E. A. C. Lucken, "Nuclear Quadrupole Coupling Constants," Academic Press, New York (1969).
J. A. S. Smith, ed., "Advances in Nuclear Quadrupole Resonance," Heyden and Sons, Ltd., London: Vol. 1 (1974); Vol. 2 (1975); Vol. 3 (1977).

EXERCISES

1. a. Calculate the energies of all the quadrupolar energy states for a nucleus with $I = 2$. Express the energies as a function of e^2Qq.

   b. How many transitions are expected, and what is the relationship between the energy of the transitions and e^2Qq?

2. a. Using the equations presented in this chapter for the energies of the 0 and ± 1 levels of a nucleus with $I = 1$ in an asymmetric field, calculate the frequency in terms of e^2Qq and η for the $0 \rightarrow +1$ and $0 \rightarrow -1$ transitions.

   b. Express the energy difference between the two transitions in part a in terms of η and e^2Qq.

   c. Show how η and q can be determined from this information.

3. Describe an nqr experiment that would give information regarding the extent of π bonding in the phosphorus-sulfur bonds in $PSCl_3$ and $(C_6H_5)_3PS$. Can you determine whether the sulfur is hybridized sp^2 and utilizes a p-orbital in bonding or whether the p_x and p_y orbitals of sulfur participate equally in bonding with nqr experiments? (Note: for ^{33}S, $I = 3/2$.)

4. It has been reported that the ^{127}I quadrupole resonance in AsI_3 is a singlet but has a very large asymmetry parameter.(28) A single crystal x-ray study indicates that the As is nearly octahedral. Explain the large asymmetry parameter.

5. Indicate the number of resonance lines expected for the following nuclei under the conditions given:

   a. ^{127}I $(I = 5/2)$; $\eta = 0$; $H_0 = 0$.

   b. ^{14}N $(I = 1)$; $\eta = 0$; $H_0 = 0$.

   c. ^{75}As $(I = 3/2)$; $\eta = 0$; $H_0 = 0$.

   d. ^{121}Sb $(I = 5/2)$; $\eta = 1$; $H_0 = 0$.

   e. ^{14}N $(I = 1)$; $\eta = 1$; $H_0 = 0$.

   f. ^{14}N $(I = 1)$; $\eta = 1$; $H_0 \neq 0$.

6. The quadrupolar energy of a nucleus is given by

$$E_m = \frac{e^2Qq[3m^2 - I(I + 1)]}{4I(2I - 1)}$$

   ^{59}Co has $I = 7/2$ and a natural abundance of 100%.

   a. How many cobalt nqr transitions will be observed for $K_3Co(CN)_6$? What are the transition energies in terms of e^2Qq?
   b. Repeat part a for $K_3Co(CN)_5Br$.

7. Consider the nitrogen nqr of each of the following systems. How many lines would you expect with and without a magnetic field? (For ^{14}N, $I = 1$.)

   a. NH_3

   b. NH_4^+

   c. :N⟨benzene ring⟩

8. The ^{59}Co $(I = 7/2)$ frequencies in $Cl_3SnCo(CO)_4$ occur at 35.02 MHz $(\pm 5/2 \rightarrow \pm 7/2)$, 23.37 MHz $(\pm 3/2 \rightarrow \pm 5/2)$, and 11.68 MHz $(\pm 1/2 \rightarrow \pm 3/2)$. Calculate η and e^2Qq. (Hint: What are the ratios of the frequencies when $\eta = 0$?)

9. A compound having the formula CH_3InI_2 is known. It is believed to be an ionic compound, $(CH_3)_2In^+InI_4^-$. Providing there are no crystallographically non-equivalent cations or anions, how many resonance lines would you expect for ^{115}In and ^{127}I in this compound? The cation has $\eta = 0.05$. What structure of this cation is suggested by this small value?

10. In solid pyridine (C_5H_5N) at 77° K, ^{14}N lines are found at 3.90 and 2.95 MHz. What are e^2Qq/h and η for nitrogen in pyridine?

11. The ^{35}Cl lines in the spectrum of $HgCl_2$ lie at 22.05 and 22.25 MHz at 300° K.

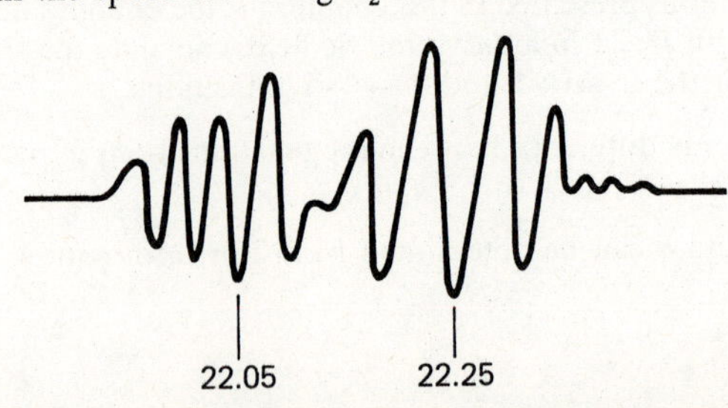

In $HgCl_2$·dioxane, a single ^{35}Cl line at 20.50 MHz is found at 300° K. In the dioxanate complex, an $Hg \leftarrow O$ donor-acceptor interaction occurs.

a. What is the probable source of the line splitting in pure $HgCl_2$?

b. The electric quadrupole moments of ^{35}Cl and ^{37}Cl are 0.079×10^{-24} cm^2 and 0.062×10^{-24} cm^2, respectively. At what frequency would you expect to find ^{37}Cl resonances in $HgCl_2$?

c. In terms of the *p*-orbital populations of equation (14–13), rationalize the decrease in the ^{35}Cl resonance frequency in the dioxanate adduct of $HgCl_2$ compared to that in pure $HgCl_2$.

15 MÖSSBAUER SPECTROSCOPY

15–1 INTRODUCTION

Mössbauer spectroscopy,[1] which will be abbreviated as MB spectroscopy in this text, involves nuclear transitions that result from the absorption of γ-rays by the sample. This transition is characterized by a change in the nuclear spin quantum number, I. The conditions for absorption depend upon the electron density about the nucleus, and the number of peaks obtained is related to the symmetry of the compound. As a result, structural information can be obtained. Many of the concepts and symbols used in this chapter have been previously discussed in Chapter 14.

To understand the principles of this method, first consider a gaseous system consisting of a radioactive source of γ-rays and the sample, which can absorb γ-rays. When a gamma ray is emitted by the source nucleus, it decays to the ground state. The energies of the emitted γ-rays, E_γ, have a range of 10 to 100 keV and are given by equation (15–1):

$$E_\gamma = E_r + D - R \qquad (15\text{-}1)$$

where E_r is the difference in energy between the excited state and ground state of the source nucleus; D, the Doppler shift, is due to the translational motion of the nucleus, and R is the recoil energy of the nucleus. The recoil energy, similar to that occurring when a bullet leaves a gun, is generally 10^{-2} to 10^{-3} eV and is given by the equation:

$$R = E_\gamma^2/2mc^2 \qquad (15\text{-}2)$$

where m is the mass of the nucleus and c is the velocity of light. The Doppler shift accounts for the fact that the energy of a γ-ray emitted from a nucleus in a gas molecule moving in the same direction as the emitted ray is different from the energy of a γ-ray emitted from a nucleus in a gas molecule moving in the opposite direction. The distribution of energies resulting from the translational motion of the source nuclei in many directions is referred to as Doppler broadening. The left-hand curve of Fig. 15–1 represents the distribution of energies of emitted γ-rays, E_γ, resulting from Doppler broadening. The breadth of the curve results from Doppler broadening. The dotted line in Fig. 15–1 is taken as E_r, the energy difference between the nuclear ground and excited states of the source. The energy difference, R, between the dotted line and the average energy of the left-hand curve is the recoil energy transmitted to the source nucleus when a γ-ray is emitted.

In MB spectroscopy the energy of the γ-ray absorbed for a transition in the sample is given by:

$$E_\gamma = E_r + D + R \qquad (15\text{-}3)$$

In this case, R is added because the exciting γ-ray must have energy necessary to bring about the transition and effect recoil of the absorbing nucleus. The quantity D

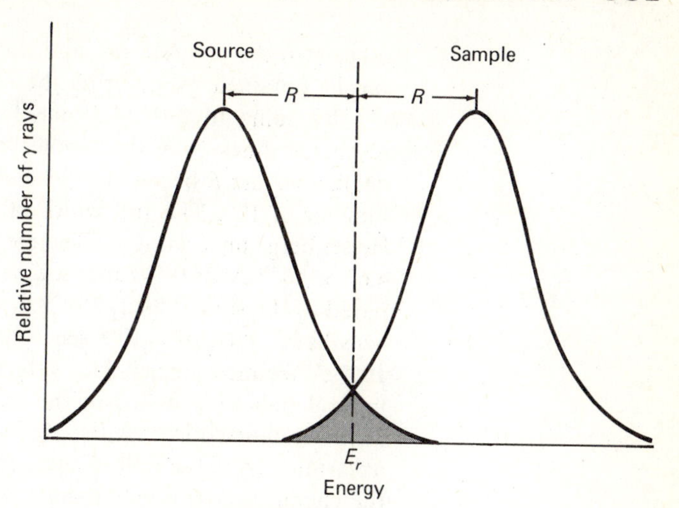

FIGURE 15–1. Distribution of energies of emitted and absorbed γ-rays.

has the same significance as before, and the value of E_r is assumed to be the same for the source and the sample. The curve in the right half of Fig. 15–1 shows the distribution of γ-ray energies necessary for absorption. The relationship of the sample and source energies can be seen from the entire figure. As indicated by the shaded region, there is only a very slight probability that the γ-ray energy from the source will match that required for absorption by the sample. Since the nuclear energy levels are quantized, there is accordingly a very low probability that the γ-ray from the source will be absorbed to give a nuclear transition in the sample. The main cause for nonmatching of γ-ray energies is the recoil energy, with the distribution for emission centered about $E_r - R$ while that for absorption is centered about $E_r + R$. The quantity R for a gaseous molecule ($\sim 10^{-1}$ eV) is very much larger than the typical Doppler energy. The source would have to move with a velocity of 2×10^4 cm sec^{-1} to obtain a Doppler effect large enough to make the source and sample peaks overlap, and these velocities are not readily obtainable. However, if the quantity R could be reduced, or if conditions for a recoilless transition could be found, the sample would have a higher probability of absorbing γ-rays from the source. As indicated by equation (15–2), R can be decreased by increasing m, the mass. This can be effected by placing the nucleus of the sample and source in a crystal so that the mass is effectively that of the crystal. Because of the large mass of the crystal, the recoil energy will be small as indicated by equation (15–2). For this reason, MB spectra are almost always obtained on solids and by employing solid sources.

By placing the source and sample in solid lattices, we have not effected recoilless transitions for all nuclei, but we have increased the probability of a recoilless transition. The reason for this is that the energy of the γ-ray may cause excitation of lattice vibrational modes. This energy term would function in the same way as the recoil energy in the gas; *i.e.*, it would decrease the energy of the emitted particle and increase the energy required for absorption. Certain crystal properties and experimental conditions for emission or absorption will leave the lattice in its initial vibrational state; *i.e.*, conditions for a recoilless transition will be satisfied. It should be emphasized that these conditions simply determine the intensity of the peaks obtained, for it is only the number of particles with matching energy that is determined by this effect. We shall not be concerned with the absolute intensity of a band, so this aspect of MB spectroscopy will not be discussed. It should be mentioned, however, that for some materials (usually molecular solids) lattice and molecular vibrational modes are excited to such an extent that very few recoilless transitions occur at room temperature and no spectrum is obtained. Frequently, the spectrum can be obtained by lowering the temperature of the sample appreciably.

By going to the solid state we have very much reduced the widths of the resonance lines over that shown in Fig. 15–1. The Doppler broadening is now negligible, and R becomes $\sim 10^{-4}$ eV for a 100 keV gamma ray and an emitting mass number of 100. The full width of a resonance line at half height is given by the Heisenberg uncertainty principle as $\Delta E = h/\tau = 4.56 \times 10^{-16}/0.977 \times 10^{-7} = 4.67 \times 10^{-9}$ eV or 0.097 mm sec^{-1} (for ^{57}Fe). The line widths are infinitesimal compared to the source energy of 1.4×10^{4} eV. The range of excited state lifetimes for Mössbauer nuclei is $\sim 10^{-5}$ sec to 10^{-10} sec, and this leads to line widths of 10^{-11} eV to 10^{-6} eV for most nuclei. This subject is treated in references 1 to 5, which contain a more detailed discussion of the entire subject of MB spectroscopy.

Our main concern will be with the factors affecting the energy required for γ-ray absorption by the sample. There are three main types of interaction of the nuclei with the chemical environment that result in small changes in the energy required for absorption: (1) resonance line shifts from changes in electron environment, (2) quadrupole interactions, and (3) magnetic interactions. These effects give us information of chemical significance and will be our prime concern.

Before discussing these factors, it is best to describe the procedure for obtaining spectra and to illustrate a typical MB spectrum. The electron environment about the nucleus influences the energy of the γ-ray necessary to cause the *nuclear* transition from the ground to excited state, *i.e.*, E_r in the sample. The energy of γ-rays from the source can be varied over the range of the energy differences arising from electron environments in different samples by moving the source relative to the sample. The higher the velocity at which the source is moved toward the sample, the higher the average energy of the emitted γ-ray (by the Doppler effect) and vice versa. The energy change ΔE_s of a photon associated with the source moving relative to the sample is given by:

$$\Delta E_s = \frac{v_0}{c} E_\gamma \cos\theta \qquad (15\text{-}4)$$

where E_γ is the stationary energy of the photon, v_0 is the velocity of the source, and θ is the angle between the velocity of the source and the line connecting the source and the sample. When the source is moving directly toward the sample, $\cos\theta = 1$. In order to obtain an MB spectrum, the source is moved relative to the sample, and the source velocity at which maximum absorption of γ-rays occurs is determined.

Consider, as a simple example, the MB spectrum of $Fe^{3+}Fe^{III}(CN)_6$ [where Fe^{3+} and Fe^{III} designate weak and strong field iron(III), respectively]. This substance contains iron in two different chemical environments, and γ-rays of two different energies are required to cause transitions in the different nuclei. To obtain the MB spectrum, the source is moved relative to the fixed sample, and the absorption of γ-rays is plotted as a function of source velocity as shown in Fig. 15–2. The peaks correspond to source velocities at which maximum γ-ray absorption by the sample occurs. Negative relative velocities correspond to moving the source away from the sample, and positive relative velocities correspond to moving the source toward the sample. The relative velocity at which the source is being moved is plotted along the abscissa of Fig. 15–2, and this quantity is related to the energy of the γ-rays. For a ^{57}Fe source emitting a 14.4 keV γ-ray, the energy is changed by 4.8×10^{-8} eV or 0.0011 cal mole^{-1} for every mm sec^{-1} of velocity imposed upon the source. This result can be calculated from equation (15–4):

$$\Delta E_s = \frac{1 \text{ mm sec}^{-1}}{3.00 \times 10^{11} \text{ mm sec}^{-1}} \times 14.4 \times 10^{3} \text{ eV} = 4.80 \times 10^{-8} \text{ eV}$$

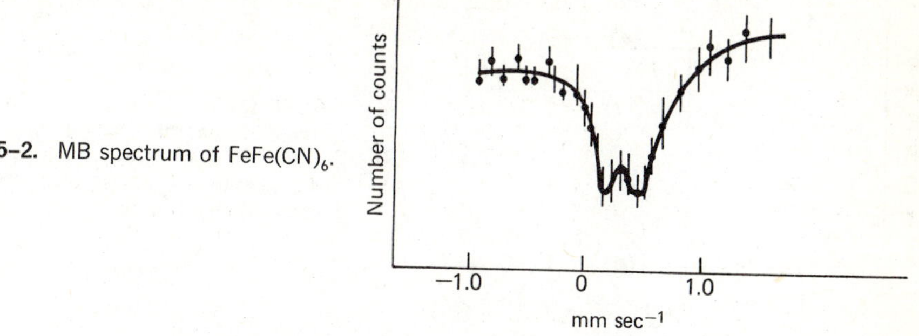

FIGURE 15–2. MB spectrum of $FeFe(CN)_6$.

This energy is equivalent to a frequency of 11.6 MHz ($\nu = E/h$, where $h = 4.14 \times 10^{-15}$ eV sec). For other nuclei having a γ-ray energy of E_γ (in keV),

$$1 \text{ mm sec}^{-1} = 11.6 \times \frac{E_\gamma}{14.4} \text{ MHz}$$

Referring again to the abscissa of Fig. 15–2, one sees that the energy difference between the nuclear transitions for Fe^{3+} and Fe^{III} in $FeFe(CN)_6$ is very small, corresponding to about 2×10^{-8} eV. The peak in the spectrum in Fig. 15–2 at 0.03 mm sec^{-1} is assigned[6] to Fe^{III} and that at 0.53 to the cation Fe^{3+} by comparison of this spectrum with those for a large number of cyanide complexes of iron. Different line positions that result from different chemical environments are indicated by the values for the source velocity in units of cm sec^{-1} or mm sec^{-1}, and are referred to as *isomer shifts, center shifts,* or *chemical shifts.* We shall now proceed with a discussion of the information contained in the parameters obtained from the spectrum.

15–2 INTERPRETATION OF ISOMER SHIFTS

The two different peaks in Fig. 15–2 arise from the isomer shift differences of the two different iron atoms in octahedral sites. The isomer shift results from the electrostatic interaction of the charge distribution in the nucleus with the electron density that has a finite probability of existing at the nucleus. Only *s*-electrons have a finite probability of overlapping the nuclear charge density, so the isomer shift can be evaluated by considering this interaction. It should be remembered that *p, d,* and other electron densities can influence *s*-electron density by screening the *s* density from the nuclear charge. Assuming the nucleus to be a uniformly charged sphere of radius R and the *s*-electron density over the nucleus to be a constant given by $\psi_s^2(0)$, the difference between the electrostatic interaction of a spherical distribution of electron density with a point nucleus and that for a nucleus with radius R is given by

$$\delta E = K[\psi_s^2(0)]R^2 \tag{15-5}$$

where K is a nuclear constant. Since R will have different values for the ground state and the excited state, the electron density at the nucleus will interact differently with the two states and thus will influence the energy of the transition; *i.e.,*

$$\delta E_e - \delta E_g = K[\psi_s^2(0)](R_e^2 - R_g^2) \tag{15-6a}$$

where the subscript e refers to the excited state and g to the ground state. The

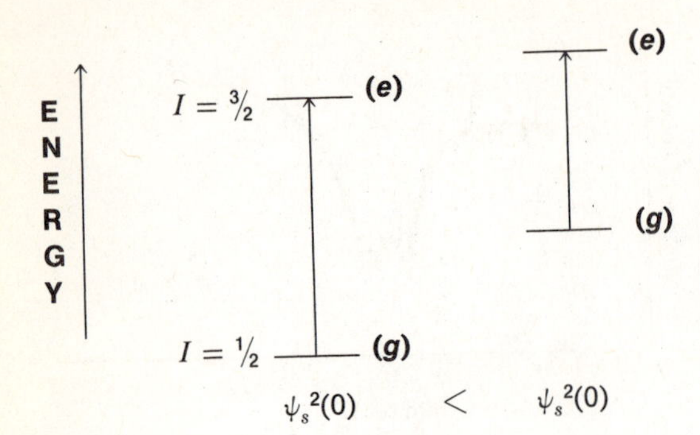

FIGURE 15–3. Changes in the energy of the Mössbauer transition for different values of $\psi_s^2(0)$. This is a graphical illustration of equation (15–6a) for two different values of $\psi_s^2(0)$ and an ^{57}Fe nucleus. The differences in $\psi_s^2(0)$ must result from a cubic or spherical distribution of bonded atoms in order for this diagram to apply.

influence of $\psi_s^2(0)$ on the energy of the transition is illustrated in Fig. 15–3 for ^{57}Fe, which has $I = \frac{1}{2}$ for the ground state and $I = \frac{3}{2}$ for the excited state. The energies of these two states are affected differently by $\psi_s^2(0)$, and the transition energy is changed.

The R values are constant for a given nucleus, but $\psi_s^2(0)$ varies from compound to compound. The center shift in the Mössbauer spectrum is the difference between the energy of this transition in the sample (or absorber) and the energy of the same transition in the source. This difference is given by the difference in equations of the form of (15–6a) for the source and sample, or:

$$\text{C.S.} = K(R_e^2 - R_g^2)\{[\psi_s^2(0)]_a - [\psi_s^2(0)]_b\} \tag{15-6b}$$

where the subscripts a and b refer to absorber and source, respectively. Standard sources are usually employed (*e.g.*, ^{57}Co in Pd for iron Mössbauer spectra, or $BaSnO_3$ for tin spectra). The ^{57}Co decays to ^{57}Fe in an excited state *via* electron capture. The excited ^{57}Fe decays to stable ^{57}Fe by γ-ray emission. When standard sources are employed $[\psi_s^2(0)]_b$ is replaced by a constant, C. Furthermore, the change in radius $R_e - R_g$ is very small, leading to the following commonly employed expression for the center shift:

$$\text{C.S.} = K'\frac{\delta R}{R}[\psi_s^2(0)_a - C] \tag{15-6c}$$

where $\delta R = R_e - R_g$; C is a constant characteristic of the source; and K' is $2KR^2$. Both K' and $\delta R/R$ are constants for a given nucleus, so the center shift is directly proportional to the s-electron density at the sample nucleus. The term *center shift* is used for the experimentally determined center of the peak; the term *isomer shift* is now used when the center shift has been corrected for the small Doppler contribution from the thermal motion of the Mössbauer atom. The sign of δR depends upon the difference between the effective nuclear charge radius, R, of the excited and ground states $(R_e^2 - R_g^2)$. For the ^{57}Fe nucleus, the excited state is smaller than the ground state, and an increase in s-electron density produces a negative shift. In tin, the sign of δR is positive, so the opposite trend of shift with s-electron density is observed.

As mentioned above, electron density in p- or d-orbitals can screen the electron density from the nuclear charge by virtue of the fact that the electron density in d- and p-orbitals penetrates the s-orbital. Hartree-Fock calculations show[6,7] that a decrease in the number of d-electrons causes a marked increase in the total s-electron density at the iron nucleus. Accordingly, with comparable ligands and with negative $\delta R/R$, Fe^{2+} has an appreciably larger center shift than Fe^{3+}. When these ions are examined in a series of molecules, the interpretation becomes more difficult, for the *d*-, *s*- and *p*-electron densities are all modified by covalent bonding. For ^{57}Fe, for example, an increase in 4*s* density decreases the center shift, while an increase in 3*d* density increases the center shift. A series of high spin iron complexes have been interpreted on this basis.[7] In the case of ^{119}Sn, the center shift increases with an increased *s*-electron density and decreases with an increase in *p*-electron density.

15-3 QUADRUPOLE INTERACTIONS

The discussion of the center shift in the previous section applies to systems with a spherical or cubic distribution of electron density. As discussed in Chapter 14, the degeneracy of nuclear energy levels for nuclei with $I > \frac{1}{2}$ is removed by a non-cubic electron or ligand distribution. For non-integral spins, the splitting does not remove the + or − degeneracy of the m_I levels, but we obtain a different level for each $\pm m_I$ set. Thus, the electric field gradient can lead to $I + \frac{1}{2}$ different levels for half-integer values of I (*e.g.*, two for $I = \frac{3}{2}$ corresponding to $\pm\frac{1}{2}$ and $\pm\frac{3}{2}$). For integer values of I we obtain $2I + 1$ levels (*e.g.*, five for $I = 2$ corresponding to 2, 1, 0, −1, −2). The influence of this splitting on the nuclear energy levels and the spectral appearance is illustrated in Fig. 15-4 for ^{57}Fe. The ground state is not split but the excited state is split, leading to two peaks in the spectrum. The center shift is determined from the

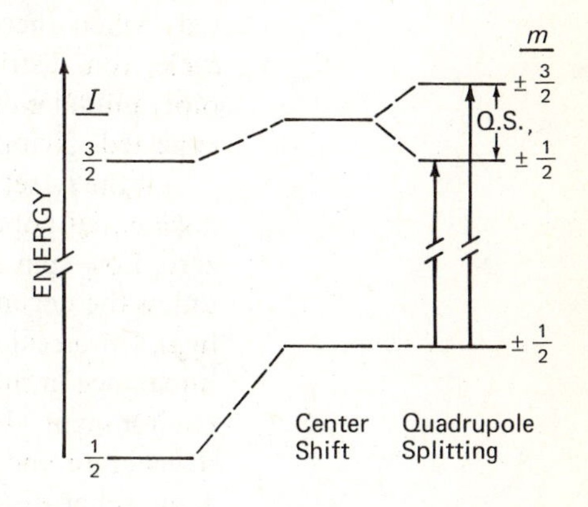

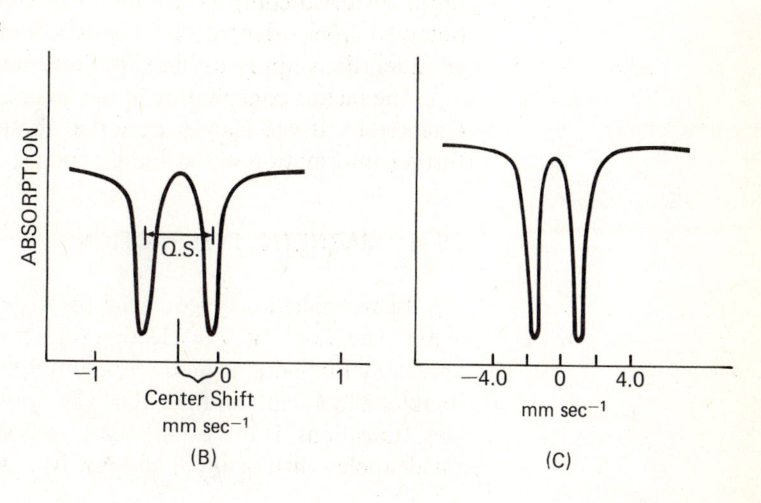

FIGURE 15-4. The influence of a non-cubic electronic environment on (A) the nuclear energy states of ^{57}Fe and (B) the Mössbauer spectrum. (C) The iron MB spectrum of $Fe(CO)_5$ at liquid N_2 temperature.[9]

center of the two resulting peaks. When both the ground and excited states have large values for I, complex Mössbauer spectra result.

The Hamiltonian for the quadrupole coupling is the same as that discussed for nqr.

$$\hat{H}_Q = \frac{e^2Qq}{4I(2I-1)}[(3\hat{I}_z^2 - I(I+1) + (\eta/2)(\hat{I}_+^2 + \hat{I}_-^2)]$$

For the $I = 3/2$ case (^{57}Fe and ^{119}Sn), the quadrupole splitting Q.S. is given by

$$\text{Q.S.} = \frac{1}{2}e^2Qq(1 + \eta^2/3)^{1/2} \tag{15-7}$$

The symbols have all been defined in the nqr chapter. For ^{57}Fe, q and η cannot be determined from the quadrupole splitting. The sign of the quadrupole coupling constant is another quantity of interest. If $m_I = \pm 3/2$ is at high energy, the sign is positive; the sign is negative if $\pm 1/2$ from $I = 3/2$ is highest. From powder spectra, the intensities of the transitions to $\pm 1/2$ and $\pm 3/2$ are similar, and it becomes difficult to determine the sign. The sign can be obtained from spectra of ordered systems or from measurement of a polycrystalline sample in a magnetic field (*vide infra*). For systems in which the I values of the ground and excited states are larger than those for iron, the spectra are more complex and contain more information. The splitting of the excited state will not occur in a spherically symmetric or cubic field but will occur only when there is a field gradient at the nucleus caused by asymmetric p- or d-electron distribution in the compound. A field gradient exists in the trigonal bipyramidal molecule iron pentacarbonyl, so a splitting of the nuclear excited state is expected, giving rise to a doublet in the spectrum as indicated in Fig. 15-4(C).

If the t_{2g} set and the e_g set of orbitals in octahedral transition metal ion complexes have equal populations in the component orbitals, the quadrupole splitting will be zero. Low spin iron(II) complexes (t_{2g}^{6}) will not give rise to a quadrupole splitting unless the degeneracy is removed, and these orbitals can interact differently with the ligand molecular orbitals. On the other hand, high spin iron(II) ($t_{2g}^{4}e_g^{2}$) has an imbalance in the t_{2g} set, and a large quadrupole splitting is often seen. If the ligand environment about iron(II) were perfectly octahedral, d_{xy}, d_{yz}, and d_{xz} would be degenerate and no splitting would be detected. However, this system is subject to Jahn-Teller distortion, which can lead to a large field gradient. When the energy separation of the t_{2g} orbitals from Jahn-Teller effects is of the order of magnitude of kT, a very temperature-dependent quadrupole splitting is observed. The ground state in the distorted complex can be obtained if the sign of q is known. The sign can be obtained from oriented systems or from studies in a large magnetic field. Similar considerations apply to high spin and low spin iron(III) compounds.

The factors contributing to the magnitude of the field gradient were discussed in Chapter 14. It was shown there that these data were of limited utility in providing further information about bond types.

15-4 MAGNETIC INTERACTIONS

In an applied magnetic field, the degeneracy of the $\pm 1/2$, etc., nuclear spin states is removed. For ^{57}Fe, the selection rules $\Delta m_I = 0, \pm 1$ give rise to a symmetric six-line spectrum. For a diamagnetic compound, the two-line zero-field spectrum splits into a doublet and a triplet for small η. The doublet arises from the $+1/2 \rightarrow +3/2$ and $-1/2 \rightarrow +3/2$ transitions. If this doublet lies toward positive velocity for ^{57}Fe, the signs of the quadrupole splitting and q are positive. Detailed interpretation is often difficult, but the sign of η can be extracted.[10] The field gradient for ferrocene has been measured and found to be positive.[11] A very interesting result shows that the q value for butadiene iron tricarbonyl has the opposite sign of that for cyclobutadiene iron tricarbonyl.

In samples of ferromagnetic materials, an internal magnetic field exists that can completely remove the degeneracy of the nuclear energy levels. Such a system is shown in Fig. 15–5. The spectrum is influenced by e^2Qq, η, H_{local}, and the orientation of H_{local} relative to the principal axes of the electric field gradient. For the general case, the problem is more complex and the reader is referred to the general references for details.

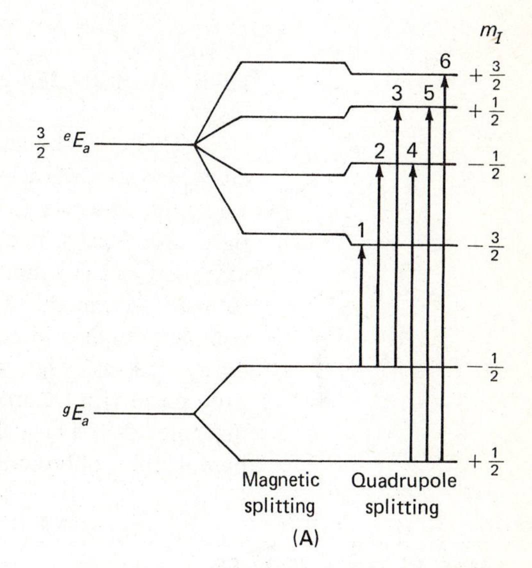

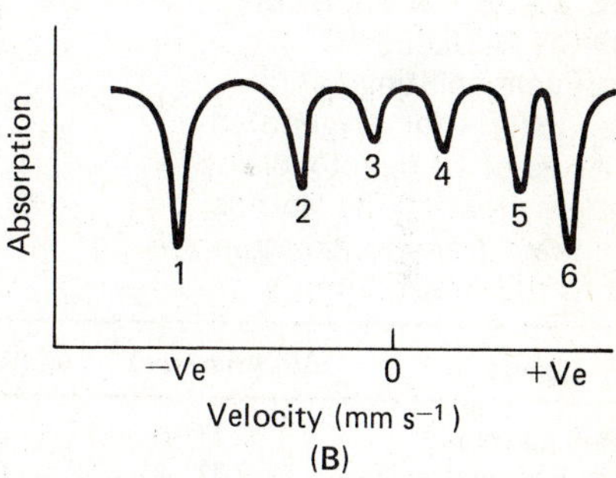

FIGURE 15–5. Magnetic and quadrupole splitting in a ferromagnetic ^{57}Fe compound. **(A)** Energy level diagram. **(B)** Expected Mössbauer spectrum. [Copyright © 1973 McGraw-Hill Book Co. (UK) Limited. From G. M. Bancroft, "Mössbauer Spectroscopy." Reproduced by permission.]

When magnetic Mössbauer experiments are carried out on paramagnetic compounds, a wealth of information is available. Most of the results obtained to date can be described by the following spin Hamiltonian:

$$\hat{H} = D\left[\hat{S}_z^2 - \frac{1}{3}S(S+1)\right] + E(\hat{S}_x^2 - \hat{S}_y^2) + \beta\hat{S}\cdot\mathbf{g}\cdot\vec{\mathbf{H}} + \hat{S}\cdot\mathbf{A}\cdot\hat{I}$$

$$-g_N\beta_N\hat{I}\cdot\vec{\mathbf{H}} + \frac{e^2Qq}{4I(2I-1)}\{3\hat{I}_z^2 - I(I+1) + \eta(\hat{I}_x^2 - \hat{I}_y^2)\}_{\mathrm{EFG}} \quad (15\text{-}8)$$

Thus, from these experiments, one can obtain, in addition to the usual information available from Mössbauer experiments, the zero-field parameters D and E as well as the components of the hyperfine coupling constant. The subscript EFG is added to the last term to indicate that the expression is written in terms of the coordinate system that diagonalizes the field gradient (these may not be coincident with **A**). The field experienced by the nucleus, $\vec{\mathbf{H}}_{eff}$, can be thought to consist of the applied field plus an internal field, $\vec{\mathbf{H}}_{int}$, arising from the paramagnetism of the unpaired electron:

$$\vec{\mathbf{H}}_{eff} = \vec{\mathbf{H}}_{int} + \vec{\mathbf{H}} \tag{15-9}$$

When $\vec{\mathbf{H}}_{int} \neq 0$, the hyperfine interaction gives rise to an effective field that splits the ground and excited states. Various situations and examples have been discussed in detail.[12-14]

15-5 MÖSSBAUER EMISSION SPECTROSCOPY

^{57}Co decays by an electron capture to ^{57}Fe($T_{1/2}$ for ^{57}Fe is 0.1 μsec), populating an excited state of the iron nucleus. The emitted γ-rays can be absorbed by a standard single-line absorber to investigate the energy levels of the ^{57}Fe nuclei produced when the source decays. The cobalt-57 analogue of the compound to be studied is prepared and used as the source. Information regarding the short-lived iron complex in the source is obtained[15] from this experiment. One must be sure that the desired iron complex remains intact when the high energy iron atoms are formed in the cobalt decay process. The results obtained from some oxygenated complexes of ^{57}Co protoporphyrin IX dimethyl ester as a function of the axial base attached are shown in Table 15-1. These are to be compared with values of ΔE_Q and δ for oxygenated hemoglobin, obtained by absorption measurements, of 2.23 and 0.27 respectively.

TABLE 15-1. MÖSSBAUER EMISSION STUDIES.[12] (Quadrupole splittings and isomeric shifts for oxygenated complexes of ^{57}Co-protoporphyrin IX dimethyl ester. The ligands coordinated *trans* to dioxygen are listed in the first column.)

Ligand	ΔE_Q(mm/sec)	δ_{Fe}(mm/sec)
1-methyl imidazole	2.17	0.29
1,2-dimethyl imidazole	2.32	0.30
pyridine	2.28	0.27
piperidine	2.25	0.30
ethylmethyl sulfide	2.27	0.30

Fig. 15-6(A) shows the emission spectrum of the five-coordinate 1-methyl imidazole complex before oxygenation; Fig. 15-6(B) shows the spectrum of the same complex after oxygenation.

This technique is particularly important when the parent iron compound is difficult to prepare and isolate.

15-6 APPLICATIONS

A few chemical applications of Mössbauer spectroscopy have been selected for discussion that are illustrative of the kind of information that can be obtained. Table

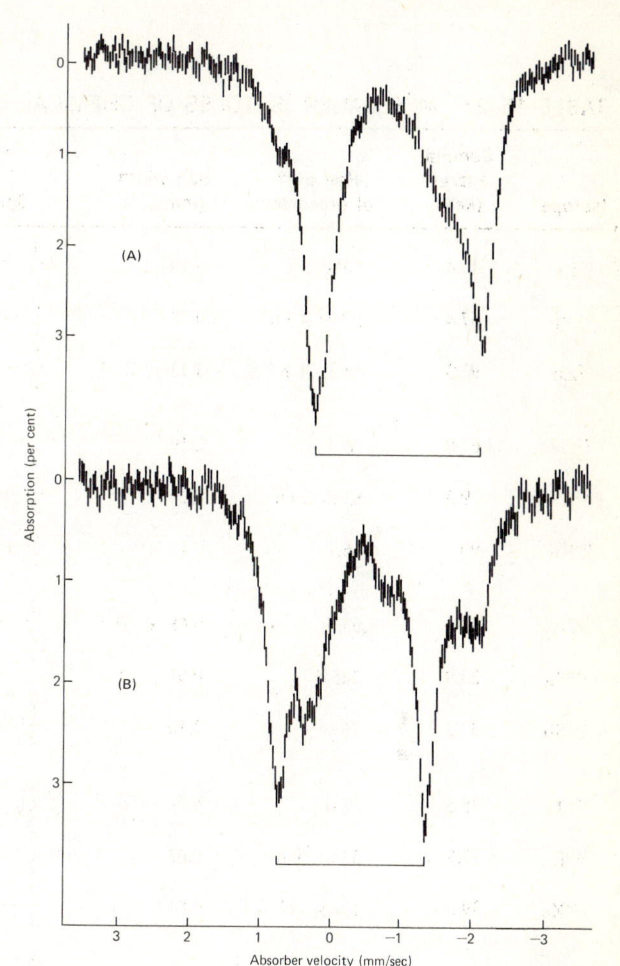

FIGURE 15–6. Emission Mössbauer spectra of **(A)** the 1-methyl imidazole adduct of cobalt protoporphyrin IX dimethyl ester and **(B)** its O_2 adduct.

15–2 summarizes pertinent information about isotopes that have been studied by this technique.

Facsimiles of spectra obtained on some iron complexes are given in Fig. 15–7. As mentioned previously, for high spin iron complexes in which all six ligands are equivalent, a virtually spherical electric field at the nucleus is expected for $Fe^{3+}(d^5)$ $(t_{2g}{}^3e_g{}^2)$ but not for $Fe^{2+}(d^6)$ $(t_{2g}{}^4e_g{}^2)$. As a result of the field gradients at the nucleus, quadrupole splitting should be detected in the spectra of high spin iron(II) complexes but not for high spin iron(III) complexes. This is borne out in spectra *A* and *B* of the complexes illustrated in Fig. 15–7. For low spin complexes, iron(II) has a configuration $t_{2g}{}^6$ and iron(III) has $t_{2g}{}^5$. As a result, quadrupole splitting is now expected for iron(III) but not iron(II) in the strong field complexes. This conclusion is confirmed experimentally by the spectra of ferrocyanide and ferricyanide ions. When the ligand arrangement in a strong field iron(II) complex does not consist of six equivalent ligands, *e.g.*, $[Fe(CN)_5NH_3]^{3-}$, quadrupole splitting of the strong field iron(II) will result. The quadrupole splitting is roughly related to the differences in the *d*-orbital populations (see Chapter 3) by

$$q_{\text{valence}} = K_d\left[-N_{d_{z^2}} + N_{d_{x^2-y^2}} + N_{d_{xy}} - \frac{1}{2}(N_{d_{xz}} + N_{d_{yz}})\right]$$

TABLE 15-2.‡ MÖSSBAUER ISOTOPES OF CHEMICAL INTEREST

Isotope	Gamma energy (keV)	Half life* of precursor	Half width (mm s^{-1})	Spin	$Q^{\dagger}$ (Barns)	$\sigma_0^{\S}$ (10^{-19} cm^2)	Natural abundance (%)	$10^4 \frac{\delta R}{R}$
^{57}Fe	14.4	270 d	0.19	$\frac{1}{2} \to \frac{3}{2}$	+0.3	23.5	2.19	-18 ± 4
^{61}Ni	67.4	1.7 h; 3.3 h	0.78	$\frac{3}{2} \to \frac{5}{2}$	0.13	7.21	1.19	
^{67}Zn	93.3	60 h; 78 h	3.13×10^{-4}	$\frac{5}{2} \to \frac{3}{2}$	+0.17	1.18	4.11	
^{73}Ge	67.0	76 d	2.19	$\frac{9}{2} \to \frac{7}{2}$	−0.26	3.54	7.76	
^{83}Kr	9.3	83 d, 2.4 h	0.20	$\frac{9}{2} \to \frac{7}{2}$	+0.44	18.9	11.55	$+4 \pm 2$
^{99}Ru	90	16.1 d	0.15	$\frac{5}{2} \to \frac{3}{2}$	>0.15	1.42	12.72	
^{107}Ag	93.1	6.6 h	6.68×10^{-11}	$\frac{1}{2} \to \frac{7}{2}$	—	0.54	51.35	
^{119}Sn	23.9	245 d	0.62	$\frac{1}{2} \to \frac{3}{2}$	−0.07	13.8	8.58	$+3.3 \pm 1$
^{121}Sb	37.2	76 y	2.10	$\frac{5}{2} \to \frac{7}{2}$	−0.29	2.04	57.25	-8.5 ± 3
^{125}Te	35.5	58 d	4.94	$\frac{1}{2} \to \frac{3}{2}$	±0.19	2.72	6.99	+1
^{129}I	27.8	33 d; 70 m	0.63	$\frac{7}{2} \to \frac{5}{2}$	−0.55	3.97	0	+3
^{129}Xe	39.6	1.6×10^7 y	6.84	$\frac{1}{2} \to \frac{3}{2}$	−0.41	1.95	26.44	0.3
^{133}Cs	81.0	7.2 y	0.54	$\frac{7}{2} \to \frac{5}{2}$	−0.003	1.02	100	
^{177}Hf	113.0	6.7 d, 56 h	4.66	$\frac{7}{2} \to \frac{9}{2}$	+3	1.20	18.50	
^{181}Ta	6.25	140 d, 45 d	6.48×10^{-3}	$\frac{7}{2} \to \frac{9}{2}$	+4.2	17.2	99.99	
^{182}W	100.1	115 d	2.00	$0 \to 2$	−1.87	2.46	26.41	+1.3
^{187}Re	134.2	23.8 h	203.8	$\frac{5}{2} \to \frac{7}{2}$	+2.6	0.54	62.93	
^{186}Os	137.2	90 h	2.37	$0 \to 2$	+1.54	2.89	1.64	
^{193}Ir	73.1	32 h	0.59	$\frac{3}{2} \to \frac{1}{2}$	+1.5	0.30	62.7	+0.6
^{195}Pt	98.7	183 d	17.30	$\frac{1}{2} \to \frac{3}{2}$	—	0.63	33.8	
^{197}Au	77.3	65 h, 20 h	1.85	$\frac{3}{2} \to \frac{1}{2}$	+0.58	0.44	100	±3

*d = days, h = hours, y = years, m = minutes.

†The ground state quadrupole moment, where both ground and excited states have $I > \frac{1}{2}$.

§Cross-section for absorption of a Mössbauer gamma ray.

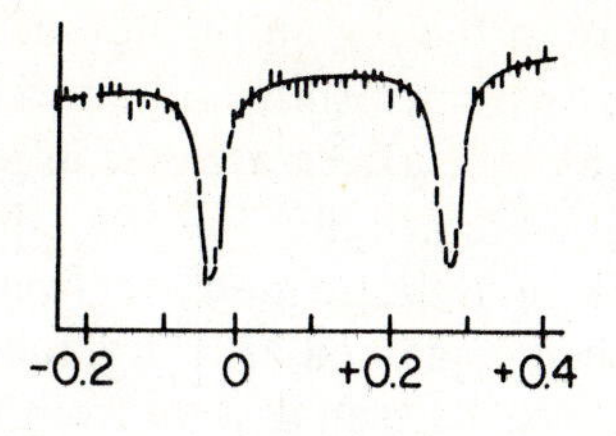

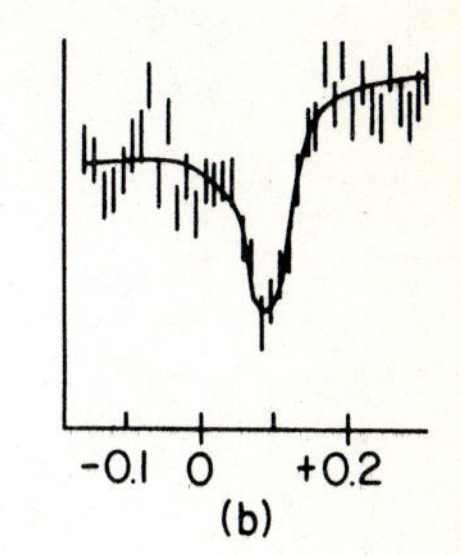

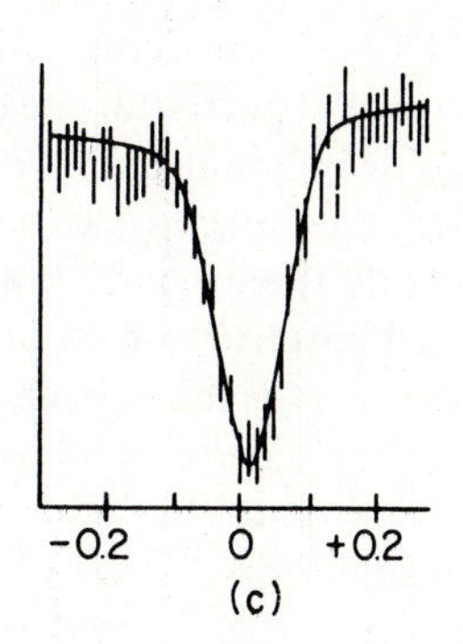

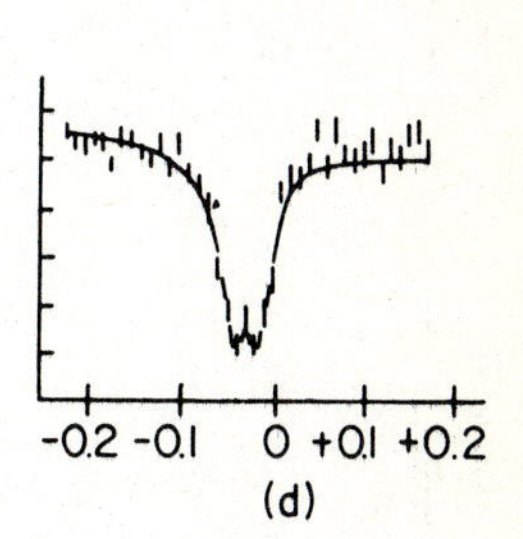

FIGURE 15–7. Mössbauer spectra of some iron(II) and iron(III) complexes. (a) Spin-free iron(II)—$FeSO_4 \cdot 7H_2O$. (b) Spin-free iron(III)—$FeCl_3$. (c) Spin-paired iron(II)—$K_4Fe(CN)_6 \cdot 3H_2O$. (d) Spin-paired iron(III)—$K_3Fe(CN)_6$. [From P. R. Brady, P. P. F. Wigley, and J. F. Duncan, Rev. Pure Appl. Chem., *12*, 181 (1962).]

Here, q_{valence} is the contribution to q from valence electrons in the d-orbitals. For p-electrons we have

$$q_{\text{valence}} = K_p\left[-N_{p_z} + \frac{1}{2}(N_{p_y} + N_{p_x})\right]$$

Values measured at room temperature for ΔE_Q and the isomer shift, δ, for a number of iron complexes have been collected[1] and are listed in Table 15–3. For iron complexes, isomer shifts in a positive direction correspond to a decrease in electron

TABLE 15–3. QUADRUPOLE SPLITTING, ΔE_Q, AND ISOMER SHIFT, δ, FOR SOME IRON COMPOUNDS (δ AND ΔE_Q IN MM SEC^{-1})

Compound	ΔE_Q	δ	Compound	ΔE_Q	δ
High Spin Fe(II)			*Low Spin Fe(II)*		
$FeSO_4 \cdot 7H_2O$	3.2	1.19	$K_4[Fe(CN)_6] \cdot 3H_2O$	—	−0.13
	3.15	1.3		—	−0.16
$FeSO_4$ (anhydrous)	2.7	1.2		<0.1	+0.05
$Fe(NH_4)_2(SO_4)_2 \cdot 6H_2O$	1.75	1.19	$Na_4[Fe(CN)_6] \cdot 10H_2O$	<0.2	−1.01
	1.75	1.3	$Na_3[Fe(CN)_5NH_3]$	0.6	−0.05
$FeCl_2 \cdot 4H_2O$	3.00	1.35	$K_2[Fe(CN)_5NO]$	1.85	−0.27
$FeC_4H_4O_6$	2.6	1.25		1.76	−0.28
FeF_2	2.68	—	$Zn[Fe(CN)_5NO]$	1.90	−0.27
$FeC_2O_4 \cdot 2H_2O$	1.7	1.25			
High Spin Fe(III)			*Low Spin Fe(III)*		
			$K_3[Fe(CN)_6]$	—	−0.12
$FeCl_3 \cdot 6H_2O$	0.2	0.85		—	−0.17
$FeCl_3$ (anhydrous)	0.2	0.5		0.26	−0.15
$FeCl_3 \cdot 2NH_4Cl \cdot H_2O$	0.3	0.45	$Na_3[Fe(CN)_6]$	0.60	−0.17
$Fe(NO_3)_3 \cdot 9H_2O$	0.4	0.4			
$Fe_2(C_2O_4)_3$	0.5	0.45			
$Fe_2(C_4H_4O_6)_3$	0.77	0.43			
Fe_2O_3	0.12	0.47			

density in the region of the nucleus. For high spin complexes, a correlation exists between isomer shift and s-electron density. An increase in δ of 0.2 mm sec^{-1} is equivalent to a decrease in charge density of 8 per cent at the nucleus.[8] The negative values obtained for the low spin ferricyanides compared to high spin iron(III) complexes indicate more electron density at the nucleus in the ferricyanide ions. This has been explained as being due to extensive π bonding in the ferricyanides, which removes d electron density from the metal ion, which in turn decreases the shielding of the s electrons. This effect increases s electron density at the nucleus and decreases δ. Both strong σ donors and strong π acceptors decrease δ.

The MB spectrum of the material prepared from iron(III) sulfate and $K_4Fe(CN)_6$ is identical to the spectra for the compounds prepared either from iron(II) sulfate and $K_3Fe(CN)_6$ or by atmospheric oxidation of the compound from iron(II) sulfate and $K_4Fe(CN)_6$. The spectra of these materials indicate that the cation is high spin iron(III), while the anion is low spin iron(II).

The MB spectrum of sodium nitroprusside, $Na_2Fe(CN)_5NO$, has been investigated.[9] This material has been formulated earlier as iron(II) and NO^+ because the complex is diamagnetic. The MB spectrum consists of a doublet with a ΔE_Q value of 1.76 mm sec^{-1} and a δ value of -0.165 mm sec^{-1}. Comparison of this value with reported results[8] on a series of iron complexes led the authors to conclude that the iron δ value is close to that of iron(IV). The magnetism and MB spectrum are consistent with a structure in which there is extensive π bonding between the odd electron in the t_{2g} set of the orbitals of iron and the odd electron on nitrogen, as illustrated in Fig. 15–8. The filled π bonding orbital would need a large contribution from the nitrogen atomic orbital, and the empty π antibonding orbital would have a larger contribution from the iron atomic orbital to produce iron(IV). More of the π electron density would be localized on nitrogen and the δ value for iron would approach that of iron(IV) because of decreased shielding of the s electrons by the d electrons. Since electron density is being placed in what was previously a π antibonding orbital of nitric oxide, a decrease in the N—O infrared stretching frequency is observed. The very large quadrupole splitting is consistent with very extensive π bonding in the Fe—N—O link.

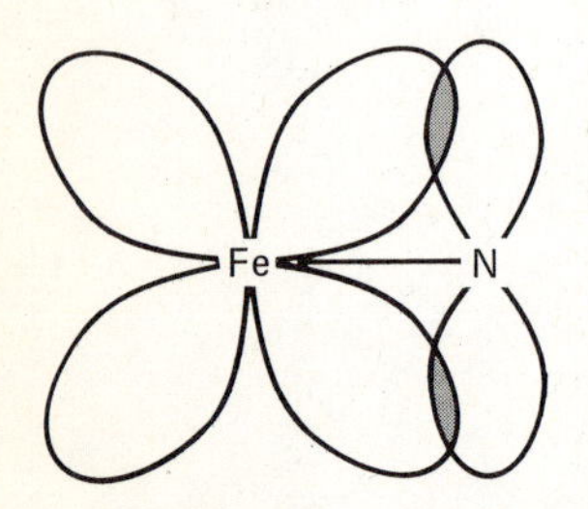

FIGURE 15–8. Orbitals involved in the Fe—N π bonding to the NO group in $Fe(CN)_5NO^{2-}$.

Mössbauer spectra are often useful in determining the oxidation state of atoms. It has been shown that the spectra originally assigned to some high spin iron(II) chelates of salicylaldoxime and other chelates were, in effect, those of an iron(III) oxidation product. The oxidized materials gave a center shift [relative to $Na_2Fe(CN)_5NO \cdot 2H_2O$] of ~0.6 mm sec^{-1} and a small quadrupole splitting. This smaller-than-expected quadrupole splitting for iron(II) was rationalized in terms of π-backbonding. The authentic material prepared under conditions that were rigorously air-free gave normal iron(II) center shifts and large quadrupole splittings, ~1.4. Values for the center shifts of various oxidation states of high spin iron compounds are shown in Table 15–4.

Drickamer *et al.*[16] have studied the effect of pressure on the Mössbauer spectra of a wide variety of iron compounds. Upon application of pressure (~165 kbar) to

TABLE 15–4. ISOMER SHIFTS FOR HIGH SPIN IRON COMPOUNDS [SHIFTS RELATIVE TO $Na_2Fe(CN)_5NO \cdot 2H_2O$]

Oxidation state	+1	+2	+3	+4	+6
I.S.	~+2.2	~+1.4	~+0.7	~+0.2	~−0.6

trisacetylacetonato iron(III), a new species forms that is attributed to iron(II). The change is reversible upon removal of the pressure.

The determination of the oxidation states of tin compounds from MB spectra is not as clear-cut as in the case of iron. Values of δ below 2.65 mm sec^{-1} are often due to tin(IV), and those above that value to tin(II). Exceptions are known. The isomer shifts of some four- and six-coordinate tin(IV) compounds vary considerably as a function of the average Pauling electronegativity, $\bar{\chi}_p$, of the groups attached. The following correlations are reported:[17]

$$\text{I.S. (mm sec}^{-1}, \text{SnX}_4) = 4.82 - 1.27\,\bar{\chi}_p$$

$$\text{I.S. (mm sec}^{-1}, \text{SnX}_6{}^{2-}) = 4.27 - 1.16\,\bar{\chi}_p$$

When fluoride is omitted, the latter equation becomes

$$\text{I.S. (mm sec}^{-1}, \text{SnX}_6{}^{2-}) = 4.96 - 1.40\bar{\chi}_p$$

The MB spectra of the iron pentacarbonyls $Fe(CO)_5$, $Fe_2(CO)_9$, and $Fe_3(CO)_{12}$ have been reported.[18,19] The results are as expected from the known structures for both $Fe(CO)_5$ and $Fe_2(CO)_9$. The structure of $Fe_3(CO)_{12}$ deduced from its MB spectrum was at odds with infrared results and a preliminary x-ray study. The MB indicated more than one type of iron, as shown in Fig. 15–9(A). The outer two lines are assigned to one type of iron and the inner two to a second type. In general, the areas are roughly proportional to the number of a particular type of iron present. A definitive crystal structure study has supported the structure shown in Fig. 15–9(B), which is consistent with the Mössbauer results. The spectra of $Fe(II)X_2(CO)_2P_2$ (where X = Cl, Br, and I, and P = phosphines and phosphites) have been interpreted in terms of the five different isomers that exist.

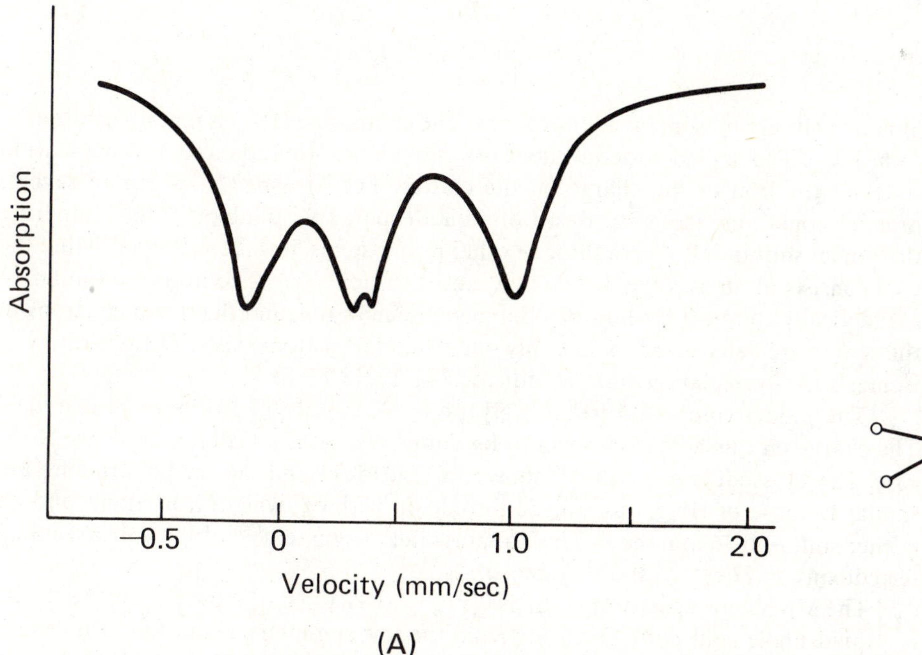

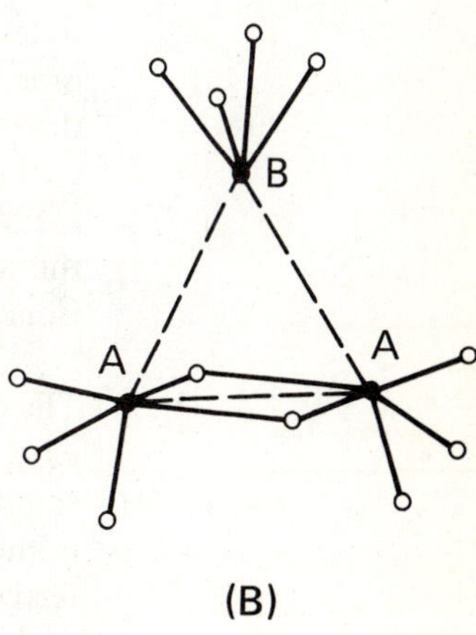

FIGURE 15–9. The Mössbauer spectrum **(A)** and structure **(B)** of $Fe_3(CO)_{12}$.

Several systems involving spin equilibrium between high spin and low spin iron(II) complexes have been studied by Mössbauer spectroscopy. A typical result[20] involves the hexadentate ligand {4-[(6-R)-2-pyridyl]-3-azabutenyl}$_3$ amine. The spectra obtained when two or three of the R groups are methyl are characteristic of low spin iron(II) ($^1A_1$) at 77°K, while at 294°K the large isomer shift and quadrupole splitting found are characteristic of high spin iron(II) ($^5T_2$). At intermediate temperatures, both forms are observed in the spectrum. This establishes that it is a true equilibrium ($^5T_2 \rightleftarrows {}^1A_1$) and that these states are long-lived on the Mössbauer time scale; *i.e.*, the lifetime must be 10^{-9} sec or greater.

There are several interesting biological applications of Mössbauer spectroscopy.[5,12,21,22] Horseradish peroxidase is an iron(III) heme protein. It can be oxidized in two one-electron steps, producing red and green compounds. The Mössbauer spectrum changes[23] upon oxidation to either form and in both cases is interpreted in terms of iron(IV). The removal of the second electron leading to the green form is believed to come from the protein or porphyrin.

Mössbauer studies have been of considerable utility in the study of the redox centers that exist in several classes of iron-sulfur proteins. A crystal structure investigation of the ferredoxin HP_{red} photosynthetic (high potential) protein from *Chromatium* has been carried out, and it has been shown to contain an $Fe_4S_4(SCys)_4$ (SCys refers to cystein) core with the cubane-like structure shown in Fig. 15–10.

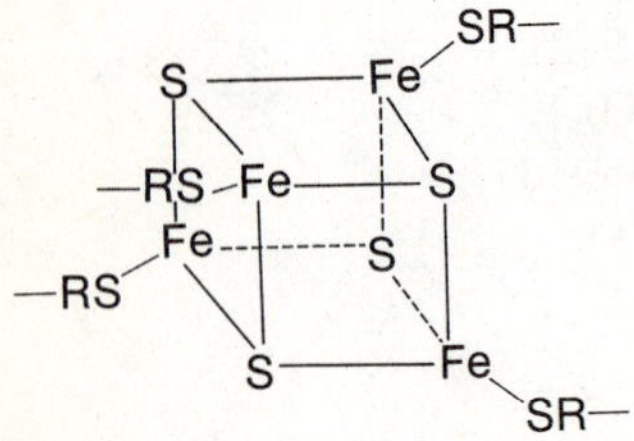

FIGURE 15–10. Cubane-like structure of the $Fe_4S_4(S—R—)_4$ core of some ferredoxins.

Similar units are present in other proteins. The compound HP_{red} is readily oxidized to form HP_{ox}. The crystal structure does not provide information about the oxidation states of the iron or the charge on the cluster. The Mössbauer spectra of several proteins containing this unit consist of a quadrupole split doublet.[24] By comparing the isomer shift in HP_{red} with those of other iron systems, it was concluded[24] that the core consists of an average of two Fe^{3+} and two Fe^{2+} ions. The irons are antiferromagnetically coupled, leading to a diamagnetic material, and the metal electrons in the system are delocalized so that only one kind of iron atom exists. The sensitivity of isomer shift to oxidation state is indicated in Table 15–5.

The model compound $[(C_2H_5)_4N]_2[Fe_4S_4(SCH_2C_6H_5)_4]$ has been prepared.[25] The charge on this anion is known to be minus two, so it must be a $2Fe^{2+} + 2Fe^{3+}$ case. The Mössbauer spectrum[26] shown in Fig. 15–11 and the crystal structure are similar to those of HP_{red} or oxidized ferredoxin with equivalent iron atoms and an isomer shift of 0.36 mm sec^{-1}. This supports the assignment[24] of the corresponding ferredoxins as $2Fe^{2+} + 2Fe^{3+}$ systems.

The Mössbauer spectrum of $(Et_4N)_2[Fe_4S_4(SCH_2C_6H_5)_4]$ shown in Fig. 15–11(A) is a quadrupole split doublet arising from the low symmetry about the iron center. The lines are further split by a magnetic field, as shown in (B). The solid line in (A) is a least squares fit of the data, employing Lorentzian line shapes. The solid line in (B) is a computer generated spectrum employing $H = 80$ kGauss, $\eta = 0$, and

TABLE 15–5. CENTER SHIFTS FOR VARIOUS IRON-SULFUR COMPOUNDS

Species*	δ
Fe^{3+} (rubredoxin)	0.25
$3Fe^{3+} + 1Fe^{2+}$ (*Chromatium* HP_{ox})	0.32
$2Fe^{3+} + 2Fe^{2+}$ (*Chromatium* HP_{red} and ox. ferredoxin)	0.42
$1Fe^{3+} + 3Fe^{2+}$ (red. ferredoxin)	0.57
Fe^{2+} (rubredoxin)	0.65

*An entry such as $3Fe^{3+} + 1Fe^{2+}$ is meant to imply an average oxidation state for the iron atoms corresponding to the given combination.

$\Delta E_Q = 1.26$ mm sec^{-1}. The sign of the principal component of the field gradient is found to be positive from the spectral fit.

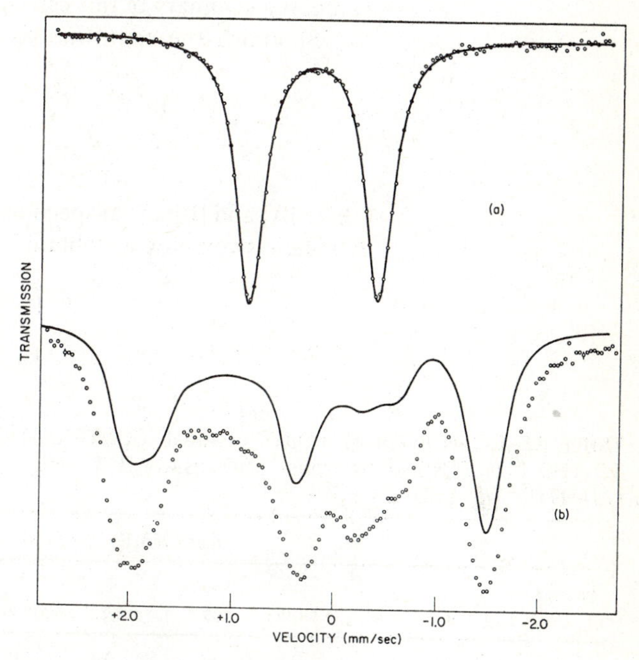

FIGURE 15–11. Mössbauer spectrum of $(Et_4N)_2[Fe_4S_4(SCH_2C_6H_4)_4]$ (A) at 1.5°K and $H_0 = 0$, and (B) at 4.7°K and $H_0 = 80$ kOersteds. [Reprinted with permission from R. H. Holm, *et al.*, J. Amer. Chem. Soc., *96*, 2644 (1974). Copyright by the American Chemical Society.]

As we have mentioned several times, it is difficult to interpret field gradients. However, it has been found possible to parameterize ions and groups attached to a central metal ion and to use these parameters, called *additive partial quadrupole splittings*, to predict the quadrupole coupling. The basic model is one of a point charge. In a diagonal electric field gradient coordinate system, the contributions to V_{xx}, V_{yy}, and V_{zz} from a charge Z are given by

$$V_{xx} = Zer^{-3}(3\sin^2\theta\cos^2\varphi - 1)$$
$$V_{yy} = Zer^{-3}(3\sin^2\theta\sin^2\varphi - 1)$$
$$V_{zz} = Zer^{-3}(3\cos^2\theta - 1)$$

where θ and φ have their usual definitions in the polar coordinate system. For a tetragonal complex (Fig. 15–12), for example, we must take each ligand (a point

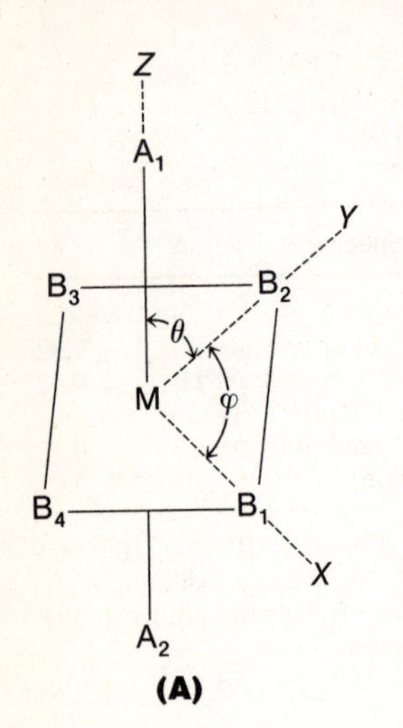

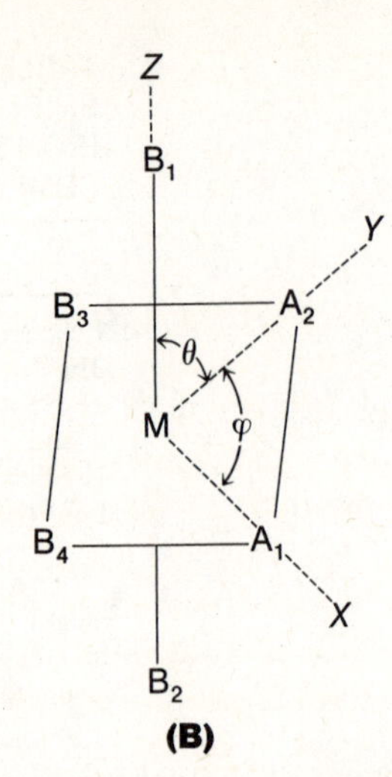

FIGURE 15–12. Geometries and coordinates for **(A)** *trans* and **(B)** *cis* MB_4A_2.

charge) and sum the individual contributions to V_{xx}, V_{yy}, and V_{zz}. Table 15–6 contains a summary of this calculation for the *cis* and *trans* complexes in Fig. 15–12.

Summing the contributions in Table 15–6, we obtain for the *trans* complex

$$V_{xx} = V_{yy} = (-2[A] + 2[B])e$$
$$V_{zz} = (4[A] - 4[B])e$$

where [A] and [B] are unspecified contributions from ligands A and B, respectively. For the *cis* complex we obtain

$$V_{xx} = V_{yy} = ([A] - [B])e$$
$$V_{zz} = (-2[A] + 2[B])e$$

TABLE 15–6. INDIVIDUAL POINT CHARGE CONTRIBUTIONS TO THE EFG TENSOR IN *trans*- AND *cis*-MA_2B_4.* THE QUANTITY [A] EQUALS Z_Ae/r_A^3.

***trans*-MA_2B_4**									
Ligand (Fig. 15–12)	θ	φ	sin θ	cos θ	sin φ	cos φ	V_{xx}/e	V_{yy}/e	V_{zz}/e
A_1	0	0	0	1	0	1	−[A]	−[A]	+2[A]
A_2	180	0	0	−1	0	1	−[A]	−[A]	+2[A]
B_1	90	0	1	0	0	1	+2[B]	−[B]	−[B]
B_2	90	90	1	0	1	0	−[B]	+2[B]	−[B]
B_3	90	180	1	0	0	−1	+2[B]	−[B]	−[B]
B_4	90	270	1	0	−1	0	−[B]	+2[B]	−[B]
***cis*-MA_2B_4**									
Ligand	θ	φ	sin θ	cos θ	sin φ	cos φ	V_{xx}/e	V_{yy}/e	V_{zz}/e
A_1	90	0	1	0	0	1	+2[A]	−[A]	−[A]
A_2	90	90	1	0	1	0	−[A]	+2[A]	−[A]
B_1	0	0	0	1	0	1	−[B]	−[B]	+2[B]
B_2	180	0	0	−1	0	1	−[B]	−[B]	+2[B]
B_3	90	180	1	0	0	−1	+2[B]	−[B]	−[B]
B_4	90	270	1	0	−1	0	−[B]	+2[B]	−[B]

The ratio of the *trans* to *cis* quadrupole splittings is

$$\frac{(4[A] - 4[B])e}{(-2[A] + 2[B])e} = -2$$

If both isomers are available, Mössbauer spectra provide a ready means of distinguishing them, for they have approximately this ratio. In the partial quadrupole coupling approach, magnitudes are empirically assigned to [A] and [B], and the field gradients are predicted. A *cis* or *trans* geometry is readily determined with the above equations. For other geometries, the appropriate equations must be derived. The compound $Fe(CO)_2(PMe_3)_2I_2$ can exist as five different isomers whose calculated quadrupole splittings differ considerably. Two isomers can be prepared with observed quadrupole splittings of -0.90 and $+1.31$ mm sec^{-1}. These are readily identified as the (*trans* P, *cis* I, *cis* CO) and *all trans* isomers, respectively.(27) Partial quadrupole splitting parameters(27) for various ligands bonded to iron(II) are listed in Table 15–7.

TABLE 15–7. LIGAND PARTIAL QUADRUPOLE SPLITTING (PQS) PARAMETERS(27) FOR IRON(II)

Ligand	PQS value	Ligand	PQS value
NO^{+}*	+0.01	$P(OPh)_3$	−0.55
X^-	−0.30	CO	−0.55
N_2	−0.37	PPh_2Et	−0.58
N_3^-	−0.38	PPh_2Me	−0.58
CH_3CN	−0.43	depb/2	−0.59
$SnCl_3^-$	−0.43	$P(OEt)_3$	−0.63
H_2O*	−0.45	depe/2	−0.65
$SbPh_3$	−0.50	$P(OMe)_3$	−0.65
NCS^-	−0.51	PMe_3	−0.66
$AsPh_3$	−0.51	dmpe/2	−0.70
NH_3*	−0.52	ArNC	−0.70
NCO^-	−0.52	CN^-*	−0.84
PPh_3	−0.53	H^-	−1.04

*PQS values derived from room temperature data.

When the values for the ligands in Table 15–7 are used with appropriate equations for the complexes in Table 15–8, the values listed under the column labeled "predicted" result.(27)

TABLE 15–8. PREDICTED AND OBSERVED(27,28) QS (MM SEC^{-1}) AT 295°K

	Observed	Predicted
trans-$FeCl_2(ArNC)_4$	+1.55	
cis-$FeCl_2(ArNC)_4$	−0.78	−0.78
$[FeCl(ArNC)_5]ClO_4$	0.73	+0.78
trans-$Fe(SnCl_3)_2(ArNC)_4$	+1.05	
cis-$Fe(SnCl_3)_2(ArNC)_4$	0.50	−0.52
cis-$FeClSnCl_3(ArNC)_4$	0.61	−0.69 ($\eta = 0.60$)
trans-$FeH_2(depb)_2$	−1.84	
trans-$FeHCl(depe)_2$	<0.12	−0.20
trans-$Fe(EtNC)_4(CN)_2$	−0.60	
cis-$Fe(EtNC)_4(CN)_2$	0.29	+0.30
trans-$[FeH(ArNC)(depe)_2]BPh_4$	−1.14	−0.98
trans-$[FeH(CO)(depe)_2]BPh_4$	1.00	−0.46

Other parameters have been reported for use with tin compounds.[27] The approach has been applied to a large number of systems with a high degree of success.[27,28] There clearly are some shortcomings in the point charge model,[29] but more work is required to enable one to predict when it will break down in empirical type applications.

The sign and magnitude of the field gradient can be used to provide information about the electronic ground state of a transition metal ion complex. The *approximate* value of the field gradient for different ground states can be estimated by adding the *d*-orbital contributions of the populated orbitals, employing Table 15-9.

TABLE 15-9.
MAGNITUDE OF q AND η FOR VARIOUS ATOMIC ORBITALS

Orbital	q	η
p_z	$-\frac{4}{5}\langle r^{-3}\rangle$	0
p_x	$+\frac{2}{5}\langle r^{-3}\rangle$	−3
p_y	$+\frac{2}{5}\langle r^{-3}\rangle$	+3
$d_{x^2-y^2}$	$+\frac{4}{7}\langle r^{-3}\rangle$	0
d_{z^2}	$-\frac{4}{7}\langle r^{-3}\rangle$	0
d_{xy}	$+\frac{4}{7}\langle r^{-3}\rangle$	0
d_{xz}	$-\frac{2}{7}\langle r^{-3}\rangle$	+3
d_{yz}	$-\frac{2}{7}\langle r^{-3}\rangle$	−3

The quantity $\langle r^{-3}\rangle$ is the expectation value of $1/r^3$ for the appropriate orbital function. This table is constructed by using the various orbital functions to evaluate the matrix elements:

$$q_{zz} = q = -\left\langle \psi \left| \frac{3\cos^2\theta - 1}{r^3} \right| \psi \right\rangle$$

$$\eta q = \left\langle \psi \left| \frac{3\sin^2\theta\cos 2\varphi}{r^3} \right| \psi \right\rangle$$

In a typical application, the $a_{1g}{}^2(d_{z^2})$, $e_{2g}{}^4(d_{xy}, d_{x^2-y^2})$ ground state for ferrocene is predicted to have a q_{zz} value of $2(-\frac{4}{7}\langle r^{-3}\rangle) + 4(\frac{4}{7}\langle r^{-3}\rangle) = \frac{8}{7}\langle r^{-3}\rangle$. A large positive field gradient is observed. In forming the ferricenium cation, a decrease in the quadrupole splitting is observed, which is consistent with the loss of an e_{2g} electron.[30]

REFERENCES

1. P. G. Debrunner and H. Frauenfelder, in "Introduction to Mössbauer Spectroscopy," L. May, ed., Plenum Press, New York (1971).
2. G. M. Bancroft, "Mössbauer Spectroscopy—An Introduction for Inorganic Chemists and Geochemists," McGraw-Hill Book Co. (UK) Limited (1973).

3. M. Weissbluth, Structure and Bonding, *2,* 1 (1967).
4. A. J. Bearden and W. R. Dunham, Structure and Bonding, *8,* 1 (1970).
5. T. C. Gibb and N. N. Greenwood, "Mössbauer Spectroscopy," Chapman and Hall, 1971.
6. J. F. Duncan and P. W. R. Wigley, J. Chem. Soc., *1963,* 1120.
7. R. E. Watson and A. J. Freeman, Phys. Rev., *120,* 1125 (1960).
8. L. R. Walker, G. K. Wertheim, and V. Jaccarino, Phys. Rev. Letters, *6,* 98 (1961).
9. L. M. Epstein, J. Chem. Phys., *36,* 2731 (1962).
10. R. L. Collins and J. C. Travis, Mössbauer Effect Methodology, *3,* 123 (1967).
11. R. L. Collins, J. Chem. Phys., *42,* 1072 (1965).
12. E. Münck, in "The Porphyrins," D. H. Dolphin, ed., Academic Press, New York, 1975.
13. W. T. Oosterhuis, Structure and Bonding, *20,* 59 (1974).
14. G. Lang, Quart. Rev. Biophys., *3,* 1 (1970).
15. L. Marchant, M. Sharrock, B. M. Hoffman, and E. Münck, Proc. Nat. Acad., Sci. USA, *69,* 2396 (1972).
16. A. R. Champion, R. W. Vaughan, and H. G. Drickamer, J. Chem. Phys., *47,* 2583 (1967) and references therein.
17. R. V. Parish, Prog. Inorg. Chem., *15,* 101 (1972).
18. R. H. Herber, W. R. Kingston, and G. K. Wertheim, Inorg. Chem., *2,* 153 (1963).
19. R. H. Herber, R. B. King, and G. K. Wertheim, Inorg. Chem., *3,* 101 (1964).
20. M. A. Hoselton, L. J. Wilson, and R. S. Drago, J. Amer. Chem. Soc., *97,* 1722 (1975).
21. W. Marshall and G. Lang, Proc. Phys. Soc. (London), *87,* 3 (1966).
22. T. H. Moss, A. Ehrenberg, and A. J. Bearden, Biochemistry, *8,* 4159 (1969).
23. P. Debrunner, in "Spectroscopic Approaches to Biomolecular Conformation," D. W. Urry, ed., American Medical Association Press, Chicago, 1969.
24. C. L. Thompson, *et al.*, J. Biochem., *139,* 97 (1974) and references therein.
25. R. H. Holm, Endeavour, *1975,* 38, and references therein.
26. R. H. Holm, *et al.,* J. Amer. Chem. Soc., *96,* 2644 (1974) and references therein.
27. G. M. Bancroft, Coord. Chem. Rev., *11,* 247 (1973) and references therein.
28. G. M. Bancroft and K. D. Butler, Inorg. Chim. Acta, *15,* 57 (1975).
29. A. P. Marks, R. S. Drago, R. H. Herber, and M. J. Potasek, Inorg, Chem., *15,* 259 (1976).
30. W. H. Morrison, Jr., and D. N. Hendrickson, Inorg. Chem., *13,* 2279 (1974).

SERIES

"Mössbauer Effect Methodology," Volumes 1 to 9, Plenum Press, New York, 1965.

EXERCISES

1. What effect does increasing electron density at the nucleus have on the relative energies of the ground and excited states of ^{57}Fe and ^{119}Sn? Explain the expected isomer shift from this change in terms of effective nuclear charge radii of these states.

2. Suppose you read an article in which the author claimed that the two peaks in a MB spectrum of a low spin iron(III) complex were the result of Jahn-Teller distortion. Criticize this conclusion.

3. Draw the structure for SnF_4 and explain why quadrupole splitting is observed in this compound but not in $SnCl_4$.

4. Suppose you were interested in determining whether Sn—O or Sn—S π bonding were present, and which was greater, in the compounds $(C_6H_5)_3SnOCH_3$ and $(C_6H_5)_3SnSCH_3$. Describe experiments involving MB spectroscopy that might shed light on this problem.

5. Would you expect the ΔE_Q value to be greatest in $SnCl_2Br_2$, $SnCl_3Br$, or SnF_3I? Explain.

6. The product obtained from the reaction of ferrous sulfate and potassium ferricyanide gives rise to the spectrum below. Interpret this spectrum.

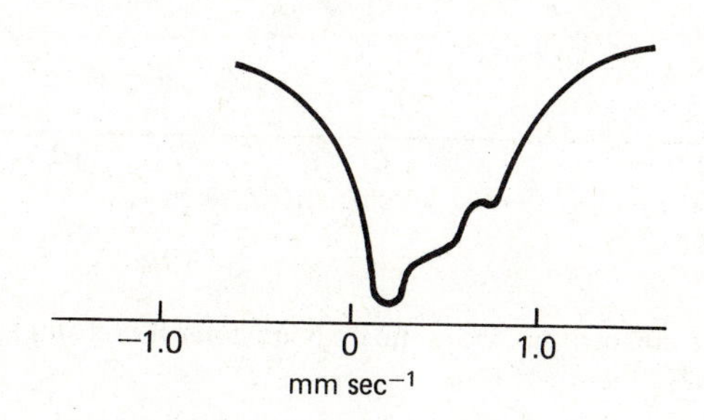

7. Using a point charge expression, indicate the three diagonal components of the electric field gradient for the two isomers of MA_3B_3. Predict the sign of the quadrupole splitting in the two isomers.

8. $(CH_3)_2SnCl_2$ is a chloro-bridged polymer with octahedral coordination about tin. The methyl groups are *trans*. With the CH_3—Sn bond as the *z*-field gradient axis, predict the signs of q and e^2Qq.

9. Which compounds would have the largest quadrupole splittings for the starred atom:

   a. *cis* or *trans* *$Fe(CO)_4Cl_2$?

   b. $(CH_3)_3$*SnCl (C_{3v}) or polymeric, chloro-bridged $(CH_3)_3$*SnCl (D_{3h} about tin)?

   c. high spin *cis* or *trans* *$Fe(NH_3)_4Cl_2$?

10. The Mössbauer spectrum for $Fe(CO)_5$ at liquid nitrogen temperature is shown below.

    a. Why are the unusual units on the *x*-axis equivalent to energy?

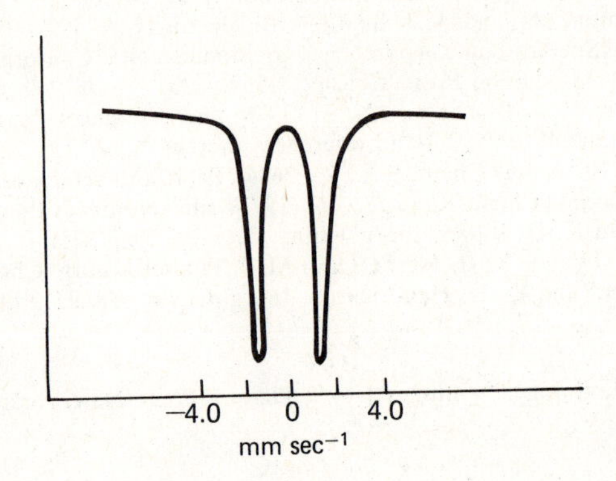

    b. What gives rise to the doublet observed? Show (and label) the energy levels involved.

11. The enzyme putidaredoxin has sites with two iron atoms. The oxidized form gives rise to the spectrum shown below:

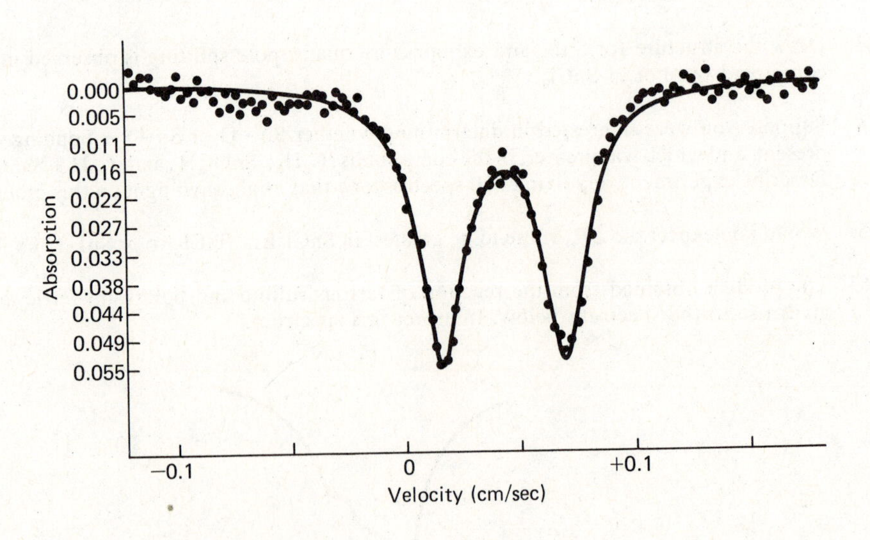

The *g*-tensors are anisotropic. Does the enzyme consist of a single type of iron site or two different iron sites?

12. One of the Mössbauer spectra below is $K_3Fe(CN)_6$. The other is $K_4Fe(CN)_6$. Which is which? Why?

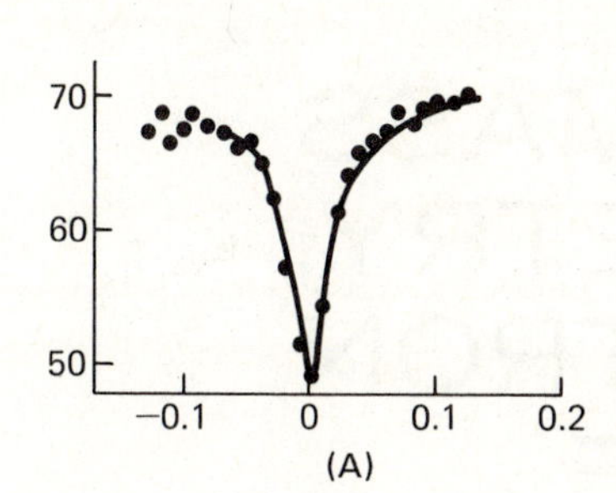

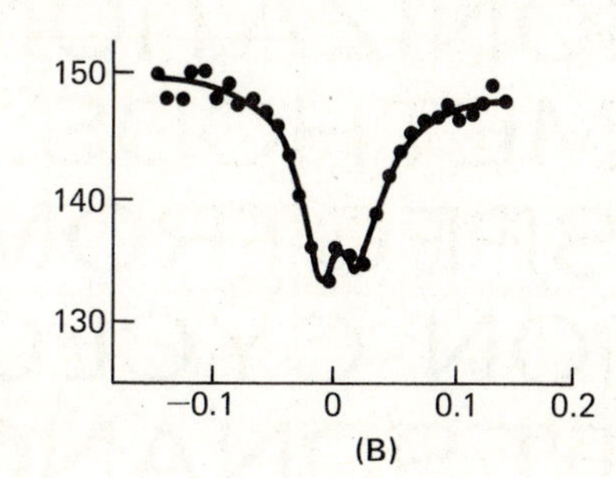

13. An article [J. Amer. Chem. Soc., *97,* 6714 (1975)] dealt with a series of square-pyramidal iron(III) complexes having an N_4S (the sulfur donor is axial) set of donor atoms. These complexes varied considerably in electronic structure, depending upon the exact nature of the N_4 macrocyclic ligand and the axial ligand, as shown by various spectroscopic measurements.

    a. Using one-electron energy level diagrams, indicate what spin states might reasonably be expected for square-pyramidal iron(III). Remember that the relative energies of the *d*-orbitals may vary slightly as a function of the donor set.

    b. Using spectra from the figure below, assign probable spin states to complexes A, B, and C. Explain what features of the Mössbauer spectra influenced your decision, and why.

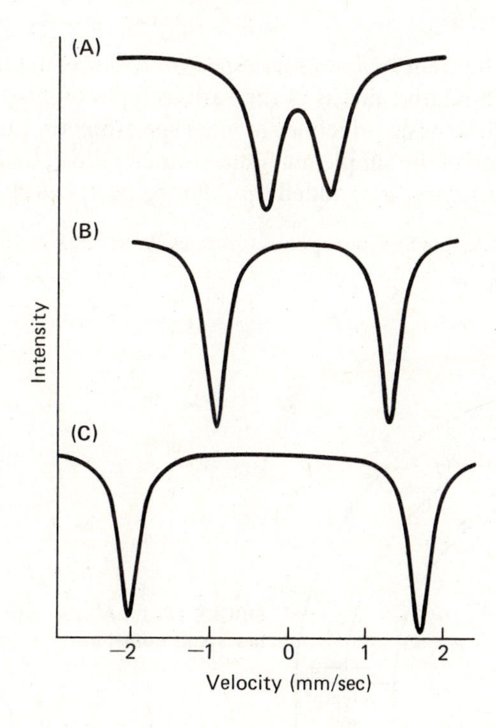

16 IONIZATION METHODS: MASS SPECTROMETRY, ION CYCLOTRON RESONANCE, PHOTOELECTRON SPECTROSCOPY*

MASS SPECTROMETRY

16–1 INSTRUMENT OPERATION AND PRESENTATION OF SPECTRA

There are many different types of mass spectrometers available, and the details of the construction and relative merits of the various types of instruments have been described.[1-7] Most of the basic principles of mass spectrometry can be illustrated by describing the operation of the simple mass spectrometer illustrated in Fig. 16–1. The sample, contained in a reservoir, is added through the port, enters the ion source (a),

*The section on Photoelectron Spectroscopy was written by Professor D. N. Hendrickson, University of Illinois.

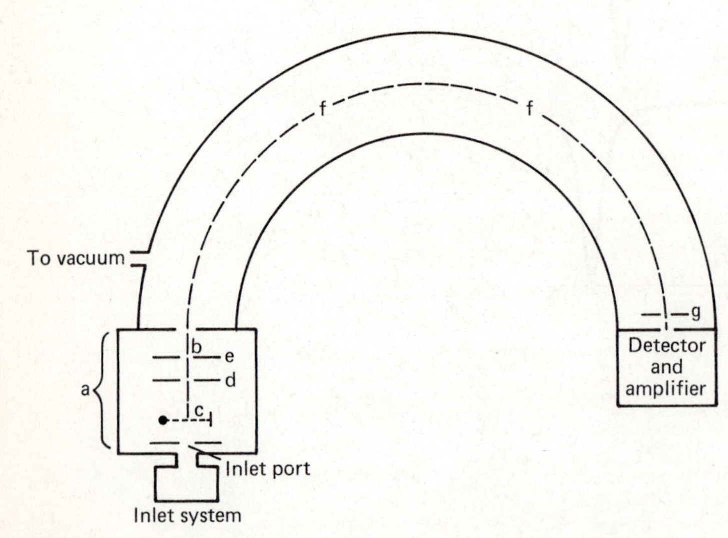

FIGURE 16–1. A schematic of a 180° deflection mass spectrometer.

and passes through the electron beam at (c). (The beam is indicated by a dotted line.) Interaction of the sample with an energetic electron produces a positive ion, which moves toward the accelerating plates (d) and (e) because of a small potential difference between the back wall (inlet) and the front wall of this compartment. Negative ions are attracted to the back wall, which is positively charged relative to the front, and are discharged. The positive ions pass through (d) and (e), are accelerated by virtue of a large potential difference (a few thousand volts) between these plates, and leave the ion source through (b). Charged ions move in a circular path under the influence of a magnetic field. The semicircle indicated by (f) is the path traced by an accelerated ion moving in a magnetic field of strength H. The radius of the semicircle, r, depends upon the following: (1) the accelerating potential, V [*i.e.*, the potential difference between plates (d) and (e)]; (2) the mass, m, of the ion; (3) the charge, e, of the ion; and (4) the magnetic field strength, H. The relationship between these quantities is given* by equation (16–1):

$$\frac{m}{e} = \frac{H^2r^2}{2V} \tag{16-1}$$

When the ions pass through slit (g) into the detector, a signal is recorded. The source and the path through which the ions pass must be kept under a vacuum of about 10^{-7} mm of mercury in order to provide a long mean free path for the ion. The sample vapor pressure in the inlet should be at least 10^{-2} mm Hg at the temperature of the inlet system, although special techniques permit use of lower pressures. In general, only one or two micromoles of sample are required.

Since H, V, and r can be controlled experimentally, the ratio m/e can be determined. Note, however, that a dipositive ion of mass 54 gives rise to the same line as a monopositive ion of mass 27. Under usual conditions for running a mass spectrometer, most of the ions are produced as singly charged species. Doubly charged ions are much less frequently encountered, whereas more highly charged ions are not present in significant concentrations.

Obtaining the mass spectrum consists of determining the m/e ratio for all fragments produced when a molecule is bombarded by a high intensity electron beam. To do this, the detector slit could be moved and the value r measured for all the particles continuously produced by electron bombardment in the ion source. This is not feasible experimentally. It is much simpler to vary H or V continuously [see equation (16–1)] so that all particles eventually travel in a semicircle of fixed radius. The signal intensity, which is directly related to the number of ions striking the detector, can be plotted as a function of H or V, whichever is being varied. In practice it is easiest to measure a varying potential, so the field is often held constant. If H and r are constant, V is inversely related to m/e [by equation (16–1)] and m/e can be plotted versus the signal intensity to produce a conventional mass spectrum. A narrow region of a typical spectrum is illustrated in Fig. 16–2.

On instruments in which the potential is varied, the results are readily presented in terms of m/e values. The accuracy varies with mass, as can be seen by comparing the difference in potential differences corresponding to m/e values of 400 and 401 with the difference between 20 and 21 [calculated by using equation (16–1)]. With modern low-resolution instruments, one can count on ± 0.4 to ± 1 m/e accuracy up to 1000 mass units. For high-resolution work, accuracy to 2 or 3 ppm can be obtained up

*This equation is simply derived: The potential energy of the ion, eV, is converted to kinetic energy $\frac{1}{2}mv^2$ (where m is the mass and v the velocity). At full acceleration,

$$eV = \frac{1}{2}mv^2 \tag{16-2}$$

In the magnetic field the centrifugal force mv^2/r is balanced by centripetal force Hev. Solving $Hev = mv^2/r$ for r yields $r = mv/eH$. Using this equation to eliminate v from (16–2) produces (16–1).

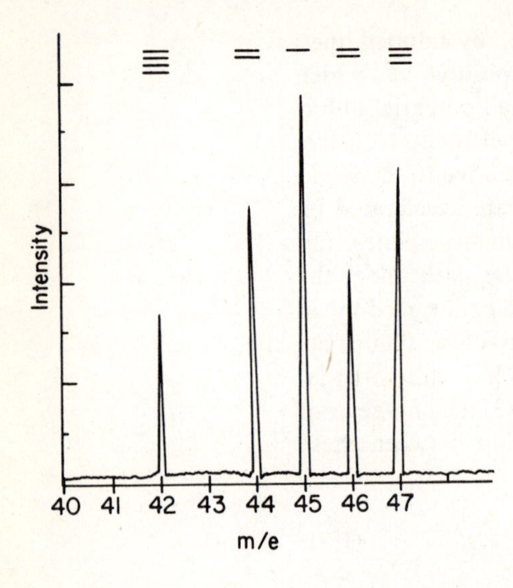

FIGURE 16–2. Mass spectrum of fragments with m/e in the range from 40 to 48. The peaks have been automatically attenuated, and the number of horizontal lines above each peak indicates the attenuation factor.

to very high masses (>3000) if reference compounds are available that have accurately known peaks within 10 to 12 per cent of the unknown mass peaks. Perfluorokerosine is a common reference compound, with many peaks that are accurately known up to about 900 mass units.

Several variations on the principle of operation of the mass spectrometer described above are available in different instruments, and a few will be briefly mentioned here. One type of instrument is similar to the above except that the spectrum of negatively charged species is obtained. In another, a time-of-flight instrument, the semicircular path and magnet are replaced by a straight path (no magnetic field). The ions are not produced continuously but in spurts, and are allowed to diffuse toward the detector. Heavy ions will move more slowly than light ions. The spectrum is a plot of intensity versus the time of flight of the particle. Mass spectrometers are available as single and double focusing instruments. The main difference between the two is an electrostatic sector in the latter, which brings about energy focusing of the ions prior to magnetic analysis. This results in greatly improved resolution at the collector. Accuracy of 1 ppm is obtainable with the better double focusing instruments.

In mass spectrometers, 70 eV electrons are typically used to bombard the sample. This voltage can be varied over a wide range. A *field ionization spectrometer*[8] employs an electric field of 10^7 to 10^8 V/cm to effect the ionization. A much smaller amount of intense energy is supplied to the molecule in this method, and it is referred to as a "soft" ionization process. The electron is removed by a quantum mechanical tunnelling effect. We shall discuss some of the advantages of the different ionization methods in subsequent sections.

The mass spectrum obtained from the instrument may be a long and cumbersome record. As a result, the data are often replotted and summarized by a bar graph that plots intensity (relative abundance of each fragment) on the ordinate and the m/e ratio on the abscissa. The relative abundance is given as the per cent intensity of a given peak relative to the most intense peak in the spectrum. Often it is informative to express the intensity in terms of *total ionization units, S*. This is an expression of the percentage that each peak contributes to the total ionization, and is obtained by dividing the intensity of a given peak by the sum of the intensities of all the peaks. When the spectrum has not been obtained over the range from 1 to the molecular weight of the material, the lower limit is indicated on the ordinate as a subscript to Σ (*e.g.*, Σ_{12} means that the spectrum was recorded from 12 to the molecular weight).

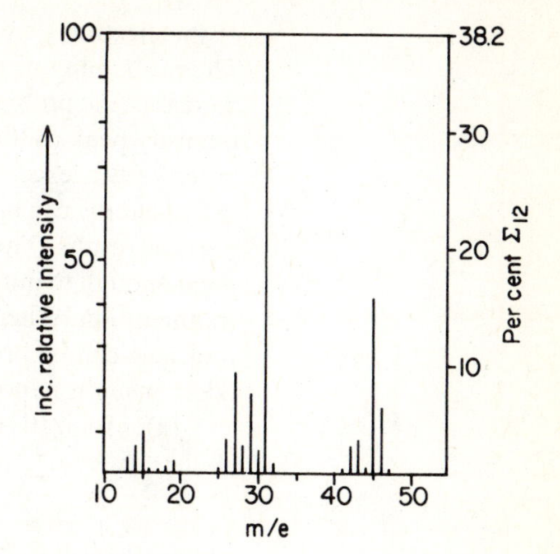

FIGURE 16–3. Representation of the mass spectrum of ethanol.

A typical graph is illustrated in Fig. 16–3. For the comparison of the intensity of various peaks in the same spectrum, the relative abundance is adequate. For comparison of the intensity of peaks in different spectra, total ionization units should be employed. In general, a peak with an intensity equal to 1 per cent of Σ is easily detected.

Our next concern is with the method of indicating the resolving power of the instrument. For many inorganic and organic applications, it is necessary to know the m/e ratio to an accuracy of one unit (*i.e.*, whether it is 249 or 250). The resolution of the instrument is sometimes expressed as $m/\Delta m$, where two peaks m and $m + \Delta m$ are separately resolved and the minimum intensity between the two peaks is only 2 per cent of the total of m. For example, a resolution of 250 means that two peaks with m/e values of 251 and 250 are separated and that at the minimum between them the pen returns to within 2 per cent of the total ion current (plotted as intensity in Fig. 16–2) to the base line. Instruments with poor resolution will not do this with high mass peaks, and the magnitude of the m/e value for which peaks are resolved is the criterion for resolution.

16–2 PROCESSES THAT CAN OCCUR WHEN A MOLECULE AND A HIGH ENERGY ELECTRON COMBINE

It is important to emphasize the rather obvious point that a detector analyzes only those species impinging on it. We must, therefore, be concerned not only with the species produced in the ionization process but also with the reactions these species may undergo in the 10^{-5} sec required to travel from the accelerating plates to the detector. When a molecule, A, is bombarded with electrons of moderate energy, the initial processes that can occur are summarized by equations (16–3) to (16–5):

$$A + e^- \rightarrow A^+ + 2e^- \qquad (16\text{-}3)$$

$$A + e^- \rightarrow A^{n+} + (n + 1)e^- \qquad (16\text{-}4)$$

$$A + e^- \rightarrow A^- \qquad (16\text{-}5)$$

The process represented by reaction (16–3) is the most common and the most important in mass spectrometry. It will occur if the energy of the bombarding electron is equal to or higher than the ionization energy of the molecule (7 to 15 eV). When the energy of the bombarding electron beam is just equal to the ionization potential, all of the electron's energy must be transferred to the molecule to remove an electron. The probability of this happening is low. As the energy of the bombarding electron increases, the probability that a collision will induce ionization increases, and a higher intensity peak results. As the energy of the bombarding electron is increased further, much of this excess energy can be given to the molecular ion that is formed. This excess energy can be high enough to break bonds in the ion, and fragmentation of the particle results. The acceleration potential of the bombarding electron that is just great enough to initiate fragmentation is referred to as the *appearance potential* of the fragment ion. When the electron energy is large enough, more than one bond in the molecule can be broken. The following sequence summarizes the processes that can occur when a hypothetical molecule B—C—D—E is bombarded with an electron:

(a) Ionization process

$$\mathrm{BCDE} + e^- \rightarrow \mathrm{BCDE}^+ + 2e^- \tag{16-6}$$

(b) Fragmentation of the positive ion

$$\mathrm{BCDE}^+ \rightarrow \mathrm{B}^+ + \mathrm{CDE}\cdot \tag{16-7}$$

$$\mathrm{BCDE}^+ \rightarrow \mathrm{BC}^+ + \mathrm{DE}\cdot \tag{16-8}$$

$$\mathrm{BC}^+ \rightarrow \mathrm{B}^+ + \mathrm{C}\cdot \tag{16-9}$$

$$\text{or}\quad \mathrm{BC}^+ \rightarrow \mathrm{B}\cdot + \mathrm{C}^+ \tag{16-10}$$

$$\mathrm{BCDE}^+ \rightarrow \mathrm{DE}^+ + \mathrm{BC}\cdot \tag{16-11}$$

$$\mathrm{DE}^+ \rightarrow \mathrm{D}^+ + \mathrm{E}\cdot \tag{16-12}$$

$$\text{or}\quad \mathrm{DE}^+ \rightarrow \mathrm{D}\cdot + \mathrm{E}^+ \tag{16-13}$$

$$\mathrm{BCDE}^+ \rightarrow \mathrm{BE}^+ + \mathrm{CD}\cdot \qquad \text{etc.} \tag{16-14}$$

$$\mathrm{BCDE}^+ \rightarrow \mathrm{CD}^+ + \mathrm{BE}\cdot \qquad \text{etc.} \tag{16-15}$$

(c) Pair production

$$\mathrm{BCDE} + e^- \rightarrow \mathrm{BC}^+ + \mathrm{DE}^- + e^- \tag{16-16}$$

(d) Resonance capture

$$\mathrm{BCDE} + e^- \rightarrow \mathrm{BCDE}^- \tag{16-17}$$

Other modes of cleavage are possible, but only positively charged species will travel to the detector and give rise to peaks in the mass spectrum. For the scheme above, peaks corresponding to B^+, BC^+, C^+, DE^+, D^+, E^+, BE^+, and CD^+ will occur in the spectrum if B, C, D, and E have different masses. More energy will have to be imparted to the $BCDE^+$ ion to get cleavage into B^+, $C\cdot$, and $DE\cdot$ [equation (16–9)] than for cleavage into $BC^+ + DE\cdot$ [equation (16–8)].

In equation (16–14), the ion has rearranged in the dissociation process, leading to fragments that contain bonds that are not originally present in the molecule BCDE. These rearrangement processes complicate the interpretation of a mass spectrum, and familiarity with many examples is necessary in order to be able to predict when rearrangements will occur. In general, this concept is invoked to explain the occurrence of peaks of unexpected mass or unexpected intensity.

Ion-molecule reactions can give rise to mass peaks in the spectrum that are greater than the molecular weight of the sample. This process is represented by equation (16–18):

$$BCDE^+ + CDE\cdot \rightarrow BCDEC^+ + DE\cdot \qquad (16\text{-}18)$$

The reaction involves a collision of the molecular ion with a neutral molecule and as such is a second order rate process, the rate of which is proportional to the product of the concentrations of the reactants. The intensity of peaks resulting from this process will depend upon the product of the partial pressures of $BCDE^+$ and CDE. Examination of the spectrum at different pressures causes variation in the relative intensities of these peaks, and as a result the occurrence of this process can be easily detected.

The reactions discussed above are all unimolecular decay reactions. Production of ions by electron bombardment often involves loss of the least tightly held electron, and ions are often formed in vibrationally excited states that have an excess of internal energy. In some molecules of the sample, a low-energy electron is removed, leaving an ion in an excited electronic state. The excited state ion can undergo internal conversion of energy, producing the electronic ground state of the ion having an excess of vibrational energy. The molecule could dissociate in any of the excited states involved in the internal conversions associated with the radiationless transfer of energy. In other molecules of the sample, ions are formed with energy in excess of the dissociation energy. In this case, the ion will fragment as soon as it starts to vibrate. Thus, in a given sample, ions with a wide distribution of energies are produced and many mechanisms are available for fragmentation processes. It is informative to compare the time scales for some of the processes we have been discussing. The time for a bond vibration is $\sim 10^{-13}$ sec, the maximum lifetime of an excited state ion is $\sim 10^{-8}$ sec, and the time an ion spends in the mass spectrometer ion chamber is 10^{-5} to 10^{-6} sec. There is thus ample time for the excess electronic energy in an ion to be converted into an excess of vibrational energy in a lower electronically excited state. Accordingly, we view the processes in the ionization chamber as producing molecular ions in different energy states, which undergo rapid internal energy conversion to produce individual ions with varying amounts of excess energy. Fragmentation takes place *via* a first order process at different rates, depending on the electronic state and excess vibrational energy of the individual ion. This is why all of the different processes represented above for the fragmentation of the ion $BCDE^+$ can occur and be reflected in the mass spectrum.

16–3 FINGERPRINT APPLICATION

The fingerprint application is immediately obvious. For this purpose an electron beam of 40 to 80 eV is usually employed to yield reproducible spectra, for this accelerating potential is above the appearance potential of most fragments. As indicated by equations (16–6) to (16–16), a large number of different fragmentation processes can occur, resulting in a large number of peaks in the spectra of simple molecules. Fig. 16–3 contains the peaks with appreciable intensity that are found in the mass spectrum of ethanol. Counting the very weak peaks, which are not illustrated, a total of about 30 peaks are found. These weak peaks are valuable for a fingerprint application but generally are not accounted for in the interpretation of the spectrum (*i.e.*, assigning the fragmentation processes leading to the peaks). Useful compilations of references to the mass spectra of many compounds (mainly organic) are contained in the textbook references at the end of the chapter. An interesting fingerprint application is illustrated in Fig. 16–4, where the mass spectra of the three isomers of ethylpyridine are indicated. Pronounced differences occur in the spectra of

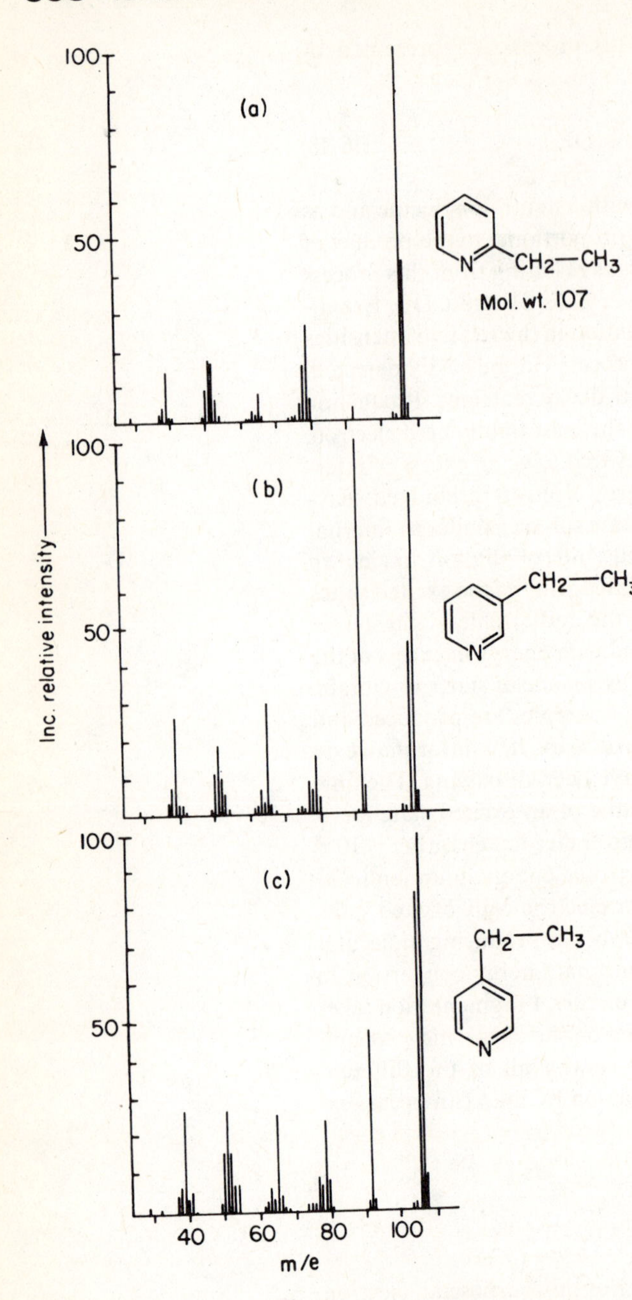

FIGURE 16–4. Mass spectra of three isomers of ethylpyridine. [Copyright © 1960 McGraw-Hill Book Company. From K. Biemann, "Mass Spectrometry." Reproduced by permission.]

these similar compounds that are of value in the fingerprint application. Optical antipodes and racemates give rise to identical spectra. Impurities create a problem in fingerprint applications because the major fragments of these impurities give rise to several low intensity peaks in the spectrum. If the same material is prepared in two different solvents, the spectra may appear to be quite different if all solvent has not been removed. Contamination from hydrocarbon grease also gives rise to many lines.

16–4 INTERPRETATION OF MASS SPECTRA

The interpretation of a mass spectrum involves assigning each of the major peaks in a spectrum to a particular fragment. An intense peak corresponds to a high probability for the formation of this ion in the fragmentation process. In the absence of rearrangement [equation (16–14)], the arrangement of atoms in the molecule can often be deduced from the masses of the fragments that are produced. For example, a strong peak at $m/e = 30$ for the compound methyl hydroxylamine would favor the structure CH_3NHOH over H_2NOCH_3 because an $m/e = 30$ peak could result from cleavage of the O—N bond in the former case but cannot result by any simple cleavage mechanism from the latter compound. The higher mass fragments are usually more important than the smaller ones for structure determination.

It is often helpful in assigning the peaks in a spectrum to be able to predict probable fragmentation products for various molecular structures. The energy required to produce a fragment from the molecular ion depends upon the activation energy for bond cleavage, which is often related to the strength of the bond to be broken. The distribution of ions detected depends not only on this but also on the stability of the resulting positive ion. In most cases it is found that the stability of the positive ion is of greatest importance. This stability is related to the effectiveness with which the resulting fragment can delocalize the positive charge. Fragmentation of $HOCH_2—CH_2NH_2^+$ can occur to produce $\cdot CH_2OH$ and $CH_2NH_2^+$ ($m/e = 30$) or $\cdot CH_2NH_2$ and CH_2OH^+ ($m/e = 31$). Since nitrogen is not as electronegative as oxygen, the resonance form $CH_2{=}NH_2^+$ contributes more to the stability of this ion than a similar form, $CH_2{=}OH^+$, does to its ion. As a result, charge is more effectively delocalized in the species $CH_2NH_2^+$ than in CH_2OH^+, and the $m/e = 30$ peak is about ten times more intense than the $m/e = 31$ peak. Charge is not stabilized as effectively by sulfur as it is by oxygen because carbon-sulfur π bonding is not as effective as carbon-oxygen π bonding. Thus, the $m/e = 31$ peak for CH_2OH^+ from $HSCH_2CH_2OH$ is about twice the intensity of the $m/e = 47$ peak that arises from CH_2SH^+.

Rearrangements of the positive ion will occur when a more stable species results. For example, the ion (furan ring, O)—CH_2^+ rearranges to (pyrylium ring, O)$^+$. The intensity of the peak for this fragment is much greater than would have been expected if rearrangement were not considered.

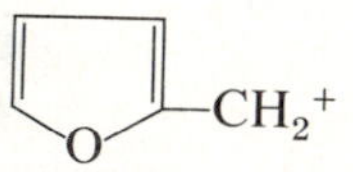

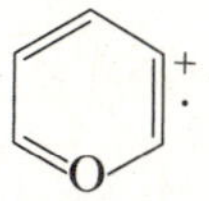

The production of so many different fragments is often helpful in putting together the structure of the molecule. However, one must employ caution even in this application. The ion produced in the ion chamber undergoes many vibrations, during which rearrangement could occur to produce bonds that did not exist in the parent compound [see, for example, equation (16–14)]. The production of all these different ions makes it difficult to determine the chemical processes that lead to the various peaks. This in turn makes it difficult to infer the influence that bond strength or other properties of the molecule have on the relative abundances of the ion fragments formed. A quantitative treatment of mass spectrometric fragmentation has been attempted and is referred to as the *quasi-equilibrium theory*.[10] The internal energy is distributed over all the available oscillators and rotators in the molecule, and the rates of decomposition *via* different paths are calculated. A weighting factor or frequency factor (*i.e.*, an entropic term) is given to each vibration level. The full analysis is complex for a molecule of reasonable size. Approximations have been introduced, leading to a highly parameterized approach.[10] Quasi-equilibrium theory is more a research area at present than a viable tool for spectral analysis.

The mass spectrometric shift rule[11] has been of considerable utility in elucidation of the structure of alkaloids and illustrates a basic idea of general utility. If there are low energy pathways for the breakdown of a complex molecule and this breakdown is not influenced by the addition of a substituent, the location of the substituent can often be determined. This is accomplished by finding an increase in the molecular weight of the fragment to which the substituent is bonded that corresponds to the weight of the substituent or a characteristic fragment of the substituent.

The use of mass spectrometry for routine sequencing of small peptides is being developed. The interested reader is referred to the review described in reference 12.

The low volatility of many substances hampers their analysis by mass spectroscopy. The volatility can often be increased by making derivatives of the polar groups in the molecule; *e.g.*, carboxyl groups can be converted to methyl esters or trimethylsilylesters. Field ionization techniques (*vide infra*) are also advantageous for this problem.

The combination of mass spectroscopy with GLC provides an excellent method for analysis of mixtures. Very small amounts of material are needed. The mass spectrometer may be used as the GLC detector, and numerous mass spectra can be accumulated as each component emerges from the column. A partially resolved GLC peak is readily detected by the change in mass spectra of the peak with time.

Many more examples and a thorough discussion of factors leading to stable ions produced from organic compounds are contained in the textbook references at the end of the chapter and reference 9. Generalizations for predicting when rearrangements are expected are also discussed. If, starting with a given structure, one can account for the principal fragments and assign the peaks in the mass spectrum by a reasonable fragmentation pattern, this assignment amounts to considerable support for that structure.

16–5 EFFECT OF ISOTOPES ON THE APPEARANCE OF A MASS SPECTRUM

When the spectrum of a compound containing an element that has more than one stable, abundant isotope is examined, more than one peak will be found for each fragment containing this element. In the spectrum of CH_3Br, two peaks of nearly equal intensity will occur at m/e values of 94 and 96, corresponding mainly to $(CH_3{}^{79}Br)^+$ and $(CH_3{}^{81}Br)^+$. The abundances of ^{79}Br and ^{81}Br are almost the same (50.54 *vs.* 49.46 per cent), so two peaks of nearly equal intensity, separated by two mass units, will occur for all bromine-containing fragments. In a fragment containing two equivalent bromine atoms, a triplet with ratios 1:2:1 would result from different combinations of isotopes. In addition to these peaks, there will be small peaks resulting from the small natural abundances of D and ^{13}C, corresponding to all combinations of masses of ^{12}C, ^{13}C, D, H, ^{79}Br, and ^{81}Br. The resulting cluster of peaks for a given fragment is important in establishing the assignment of the peaks to a fragment. Their relative intensities will depend upon the relative abundances of the various naturally occurring isotopes of the atoms in the fragment; *e.g.*, CO^+ can consist of mass fragments at 28, 29, 30, and 31. The relative abundances of these fragments can be calculated from simple probability theory.[13,14] Computer programs have been reported to carry out these calculations.[15,16] These characteristic patterns are quite useful in assigning spectra of molecules that contain an atom with more than one abundant isotope. Molecules containing transition metals often give such isotope patterns. Use of the ^{13}C peaks enables one to determine the number of carbon atoms in a fragment.

The advantages of high resolution mass spectrometry can be illustrated[17] by the incorrect assignment of a peak at 56 in the low resolution spectrum of $Fe[(C_2H_5)_2NCS_2]_3$ to iron. At high resolution, a peak is expected at 55.9500, but none is found. Instead, one is obtained at 56.0350, which is assigned to the fragment C_3H_6N.

Another important application of the mass spectrometer involving isotopes is the study of exchange reactions involving nonradioactive isotopes. The product of the exchange from labeled starting material is examined for isotope content as a function of time to obtain the rate of exchange. The product or starting material can be degraded to a gaseous material containing the label, and the isotopic ratio is obtained from the mass spectrum. These materials may also be examined directly, and the location and amount of label incorporated can be deduced from an analysis of the change in spectrum of various fragments. By determining which peaks in the spectrum change on incorporation of the isotope, one can determine which parts of a molecule have undergone exchange. In the reaction of methanol with benzoic acid, it has been shown by a tracer study involving mass spectral analysis that the ester oxygen in the product comes from methanol:

$$C_6H_5C(=O){-}OH + CH_3{}^{18}OH \rightarrow C_6H_5C(=O){-}{}^{18}OCH_3 + H_2O$$

In another interesting application it was shown that the following exchange reactions occurred:

$$BF_3 + BX_3 \rightarrow BX_2F + BF_2X$$
$$3RBX_2 + 2BF_3 \rightarrow 3RBF_2 + 2BX_3 \text{ (also } BF_nX_m\text{)}$$
$$3R_2BX + BF_3 \rightarrow 3R_2BF + BX_3 \text{ (also } BF_nX_m\text{)}$$

where R is alkyl or vinyl and X is Cl or Br. Fragments corresponding to the products were obtained, although only starting materials were recovered on attempted separation.[18] A four-center intermediate of the type

```
R    Cl    F
  \ /  \ /
   B    B
  / \  / \
Cl    F    F
```

was proposed for the exchange. In order to determine whether or not alkyl groups were exchanged in the reaction:

$$3RBX_2 + 2BF_3 \rightarrow 3RBF_2 + 2BX_3 \text{ (and } BX_mF_n\text{)}$$

the boron trifluoride was enriched in ^{10}B. The absence of enrichment of ^{10}B in fragments in the mass spectrum containing alkyl or vinyl groups enabled the authors to conclude that neither alkyl nor vinyl groups were exchanged under conditions where RBX_2 species were stable.

It should be pointed out that in all of the above applications it is not necessary to label all molecules of the compound. A slight enrichment will suffice.

16–6 MOLECULAR WEIGHT DETERMINATIONS; FIELD IONIZATION TECHNIQUES

If the molecular ion [formed by a process similar to equation (16–6)] is stable, the determination of the molecular weight of the substance can be made directly from the highest mass peak whose relative intensity is independent of pressure. For example, the molecular weight of $Fe(CO)_4(CF_2{-}CF_2{-}CH_2{-}CF_2)$ could not be determined[19] by conventional means but was shown to be a monomer with a molecular weight of 368 from the molecular ion peak in the mass spectrum. In many compounds, the molecular ion is stable enough for this application; but in many others it is not. The main problem becomes one of ascertaining that the peak selected is the molecular ion peak.

Many molecules are either not volatile enough or not stable enough under electron bombardment to enable one to obtain the molecular weight by mass spectroscopy unless field ionization techniques are used. If molecular ions cannot be detected at the lowest possible ion source and substance evaporation temperatures with 70 eV electron bombardment, it will usually not be possible to observe them with lower energy electrons. Although a decrease in the energy of the electrons increases the intensity of the molecular ion relative to the energies of the fragments, the absolute molecular ion intensity is also decreased. In field ionization spectroscopy, an electric field of $\sim 5 \times 10^7$ V/cm is generated across a gap between two metal electrode surfaces. As a gaseous molecule approaches this electric field, it is ionized. This process is referred to as *field ionization*. The strength of the ion current produced is influenced by the supply of neutral molecules provided at the field ion emitter. Organic molecules can also be adsorbed on the electrode surface and then ionized by the high electric field in a process referred to as *field desorption*. One of the advantages of the field desorption technique involves the study of molecules that are not volatile or molecules that decompose on subliming. The electrode is coated with a solution of the material and the solvent is evaporated, depositing a film of material on the electrode. The electric field is then applied to ionize the material.

In addition to the advantages of studying less volatile materials, field ionization techniques—being soft ionization methods—permit one to obtain the parent ion peak with high intensity for many materials for which this peak is barely detectable or undetectable with electron bombardment techniques. Fig. 16–5 illustrates the difference between the electron bombardment and field ionization spectra of D-ribose. The molecular ion is barely detectable in the electron bombardment spectrum

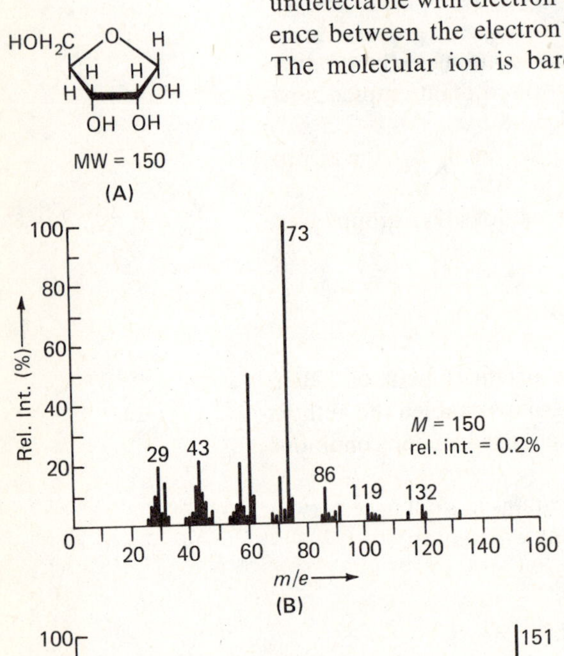

FIGURE 16–5. **(A)** Formula of D-ribose. **(B)** Electron bombardment mass spectrum. **(C)** Field ionization spectrum.

and cannot be confidently distinguished from an impurity. The field ionization spectrum gives an intense parent ion peak. The most intense line in the field ionization spectrum corresponds to the molecular weight of the parent ion plus one mass unit. This type of result is often obtained with strongly absorbed polar molecules, especially those containing hydroxyl groups; the parent-ion-plus-one peak is produced by proton abstraction from a second molecule on the emitter surface after ionization of a first molecule. With wire emitters, the intensity of this line is greatly reduced.[8]

Molecules can be supplied by absorption from the gas phase and by diffusion along the solid electrode to the emitting region, giving rise to a field desorption process. In many cases involving gaseous samples, it is difficult to distinguish between field ionization and field desorption.

16-7 EVALUATION OF HEATS OF SUBLIMATION AND SPECIES IN THE VAPOR OVER HIGH MELTING SOLIDS

Evaluation of the heat of sublimation is based upon the fact that the intensity of the peaks in a spectrum is directly proportional to the pressure of the sample in the ion source. The sample is placed in a reservoir containing a very small pinhole (a Knudsen cell), which is connected to the ion source so that the only way that the sample can enter the source is by diffusion through the hole. If the cell is thermostated and enough sample is placed in the cell so that the solid phase is always present, the heat of sublimation of the solid can be obtained by studying the change in peak intensity (which is related to vapor pressure) as a function of sample temperature. The small amount of sample diffusing into the ion beam does not radically affect the equilibrium. Some interesting results concerning the nature of the species present in the vapor over some high melting solids have been obtained from this type of study. Monomers, dimers, and trimers were found over lithium chloride, while monomers and dimers were found in the vapor over sodium, potassium, and cesium chloride.[20]

The species Cr, CrO, CrO_2, O, and O_2 were found over solid Cr_2O_3. Appearance potentials and bond dissociation energies of these species are reported.[21] The vapors over MoO_3 were found to consist of trimer, tetramer, and pentamer. Vapor pressures, free energy changes, and enthalpies of sublimation were evaluated.[22]

16-8 APPEARANCE POTENTIALS AND IONIZATION POTENTIALS

As mentioned earlier, the molecular ion is produced whenever collision occurs with an electron with energy equal to or greater than the ionization energy of the molecule. A typical curve relating electron energy to the number of ion fragments of a particular type produced (*i.e.*, relative intensity of a given peak) is illustrated in Fig. 16-6. This is referred to as an ionization efficiency curve. At electron energies well below the ionization energy, no ions are produced. When the energy of the electron beam equals the ionization energy, a very low intensity peak results, for in the collision all of the energy of the electron will have to be imparted to the molecule, and this is not too probable. As the electron energy is increased, the probability that the electron will impart enough energy to the molecule to cause ionization is increased, and a more intense peak results until a plateau finally occurs in the curve. The tail of this curve at low energies results because of the variation in the energies of the

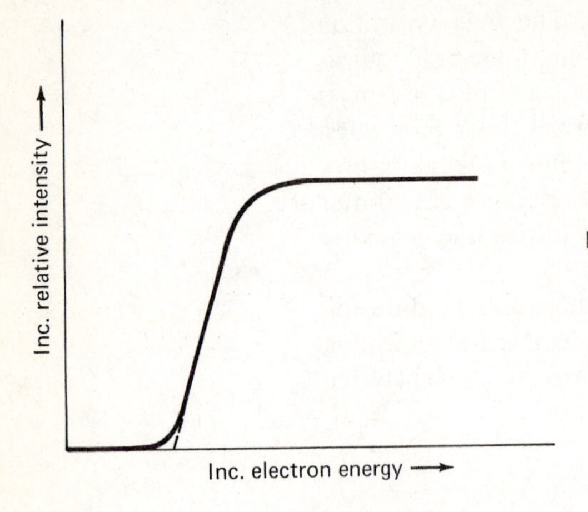

FIGURE 16–6. An ionization efficiency curve.

electrons in the bombarding beam. Therefore, the curve has to be extrapolated (dotted line in Fig. 16–6) to produce the ionization energy. Various procedures for extrapolation and the error introduced by these procedures have been discussed in detail.[21] When the peak observed is that of the molecular ion, $e^- + RX \rightarrow RX^+ + 2e^-$, the ionization energy of the molecule can be obtained by extrapolation of the ionization efficiency curve. When the peak is that of a fragment, extrapolation of the ionization efficiency curve produces the appearance potential of that fragment. For example, if the peak being investigated is that of the fragment R^+ from the molecule R—X, the appearance potential, A_{R^+}, is obtained by extrapolation of the ion efficiency curve for this peak. The appearance potential is related to the following quantities:

$$A_{R^+} = D_{R-X} + I_R + E_k + E_e \tag{16-19}$$

where D_{R-X} is the gas phase dissociation energy of the bond R—X; I_R is the ionization potential of R; E_k is the kinetic energy of the particles produced; and E_e is the excitation energy of the fragments (*i.e.*, the electronic, vibration, and rotational energy if the fragments are produced in excited states). Generally, E_k and E_e are small and equation (16–19) is adequately approximated by:

$$A_{R^+} = D_{R-X} + I_R \tag{16-20}$$

If D_{R-X} is known, I_R can be calculated from appearance potential data. Often I_R is known, and D_{R-X} can be calculated. The value for I_R must be less than I_X for equation (16–20) to apply; otherwise, X is dissociated or electronically excited. Experiments of this sort provide one of the best methods for evaluating bond dissociation energies but give less exact ionization potential data than can be obtained by other means.

An article on the mass spectrometric study of phosphine and diphosphine contains a nice summary of some of the information that can be obtained from these studies. Ionization energies and appearance potentials of the principal positive ions formed are reported. The energetics of the fragmentation processes are discussed and a mechanism is proposed.[22]

ION CYCLOTRON RESONANCE

16–9 BASIC PRINCIPLES OF ICR

The ICR spectrometer, now available commercially, is basically a mass spectrometer that uses the signal detection techniques of magnetic resonance spectrometers. As in mass spectroscopy, a positive ion is generated with mass m and charge e. In a uniform magnetic field H, this ion is accelerated into a circular orbit whose plane is perpendicular to the magnetic field. The motion of the ion in this orbit is described by a cyclotron frequency ω_c given by

$$\omega_c = eH/mc \tag{16-21}$$

where c is the speed of light. Since ω_c is independent of the velocity, v, of the ion, all ions with a particular m/e value will be characterized by a given ω_c. The velocity distribution will, however, give rise to a distribution of orbital radii:

$$\omega_c = v/r \tag{16-22}$$

This orbiting ion can absorb power from an alternating electric field $E_1(t)$ of frequency ω_1 when $\omega_1 = \omega_c$.

Experimentally, two parallel plates are a part of the resonant circuit of a sensitive marginal oscillator that is the source for $E_1(t)$. The absorption of energy is detected and a mass spectrum obtained by sweeping ω_1 of the marginal oscillator at constant H through the range of ω_c frequencies of the ions in the sample. More commonly, a mass spectrum is obtained by sweeping H at a constant ω_1.

The importance of the ICR technique lies not in its use as another kind of mass spectroscopy, but in the results that can be obtained from the double resonance experiment. This experiment consists of observing the effect of a perturbation of the translational energy of a given ion on the intensity of another ion that may be coupled to it *via* an ion-molecule reaction. For example, while monitoring the signal of A^+, a reactant B^+ is irradiated by an auxiliary oscillator with a frequency ω_c corresponding to B^+. The spectrum of A^+ will be changed if A^+ and B^+ are coupled *via* a chemical reaction. Usually, in conducting this experiment, the second frequency is swept over the frequency of all other ions present as well as B^+.

For most exothermic reactions, the rate constant, k, decreases with increasing kinetic energy, E, of the reactant ion; *i.e.*, dk/dE is negative. The sign of dk/dE can be obtained from ICR because an increase or decrease in product ion intensity when a reactant ion is irradiated indicates a positive or negative dk/dE, respectively. If there is no change in intensity, the ion being irradiated is most probably not involved in forming the product. An alternative but unlikely possibility is that dk/dE is zero.* An exothermic reaction must be occurring in the absence of irradiation, but an endothermic reaction may not be.

There have been two main areas of applications of ICR: (a) determination of absolute and relative thermochemical quantities, and (b) studies of the occurrence and mechanisms of ion-molecule reactions as well as the measurement of rate constants. We shall discuss some results on proton transfer reactions in the system comprising H_2O, CH_3OH, and C_2H_5OH to illustrate the above ideas. A single resonance spectrum shows the presence of OH^-, OCH_3^-, and $OC_2H_5^-$ ions with m/e ratios of 17, 31, and 45. When the double resonance experiment is carried out, the following results are obtained:

(1) Irradiation of m/e peaks at 17 and 31 resulted in a decrease in the intensity of the m/e 45 peak.
(2) Irradiation of m/e at 17 caused a decrease in the intensity of m/e *at* 31.
(3) The intensity of m/e at 17 was not changed when m/e at 31 and 45 were accelerated.

---

*If reactant ions are removed from the reaction chamber when irradiated, a double resonance signal may be obtained for a reaction whose rate constant does not depend on reactant ion kinetic energy.

These results indicate that the following reactions are occurring in the ICR cell and that the reverse ones are not.

$$HO^- + C_2H_5OH \rightarrow C_2H_5O^- + H_2O$$
$$HO^- + CH_3OH \rightarrow CH_3O^- + H_2O$$
$$CH_3O^- + C_2H_5OH \rightarrow C_2H_5O^- + CH_3OH$$

Recall that the perturbation of one ion can increase the concentration of another ion only if the two ions are coupled chemically in the absence of the perturbation. These results yield the following order for ease of proton transfer:

$$C_2H_5OH > CH_3OH > H_2O$$

For a more extensive discussion of the applications of this technique, the reader is referred to references 25 and 26.

PHOTOELECTRON SPECTROSCOPY

by
David N. Hendrickson

16-10 INTRODUCTION

The collision of photons with atoms or molecules can result in the ejection of photoelectrons. During the last two decades photoelectron spectroscopy has developed as a promising field in chemistry. Photoelectron spectroscopy differs from the previously discussed spectroscopic techniques in which the characteristics of absorbed, emitted, or scattered electromagnetic radiation are measured; instead, the kinetic energies of the ionized electrons are monitored.

Partially as a result of the requirement of monochromatic radiation, two disciplines of photoelectron spectroscopy have emerged. X-ray photoelectron spectroscopy (called variously XPS or ESCA—electron spectroscopy for chemical analysis), employing X-radiation as the ionizing source, has been largly concerned with core (*i.e.*, non-valence) electrons. The development of XPS is attributed to Siegbahn and his coworkers.[27] On the other hand, ultraviolet photoelectron spectroscopy (PES or UPS) utilizes lower energy ultraviolet radiation and is thus limited to measuring binding energies of valence-orbital electrons. Largely owing to the efforts of Turner and his associates,[28] PES has been developed to the point of not only measuring valence electron binding energies but observing vibrational states of the molecular ions resulting from the photoionization process.

16-11 THEORETICAL CONSIDERATIONS

Energetics

The energy, E_b, required to liberate an electron from a system, using the vacuum as a reference level, can be calculated from a consideration of energy conservation:

$$E_b = E_{\text{source}} - E_{\text{kin}} - E_r \tag{16-23}$$

Here E_{source} is the energy of the ionizing radiation, E_{kin} is the kinetic energy of the photoelectron, and E_r is the recoil energy of the atom or molecule. It is possible to show that the recoil energy of the atom or molecule from which the photoelectron is ejected is negligible except in the case of X-ray photoionization from the hydrogen atom.[27] A photoelectron spectrum is thus a plot of the number of photoelectrons incident upon a suitable counting device as a function of the kinetic energy of the photoelectron. Peaks (*i.e.*, maxima in counting rate) will be seen that correspond to the photoionization of electrons from different levels (*i.e.*, molecular orbitals) in the sample. A typical XPS spectrum is shown in Fig. 16–7. In this case the XPS spec-

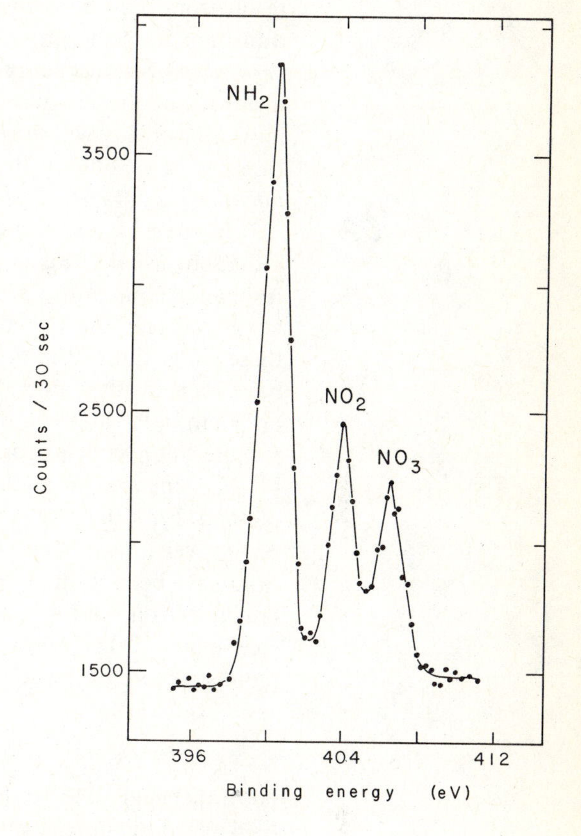

FIGURE 16–7. Nitrogen 1s photoelectron spectrum for *trans*-$[Co(NH_2CH_2CH_2NH_2)_2(NO_2)_2]NO_3$. [From D. N. Hendrickson, J. M. Hollander, and W. L. Jolly, Inorg. Chem., *8*, 2642 (1969).]

trometer is adjusted so that the photoelectrons resulting from the ionization of the 1*s* electrons of the nitrogen atoms in *trans*-$[Co(en)_2(NO_2)_2](NO_3)$ are being counted. Three maxima in photoelectron counting rate are seen, and these are assigned to ionizations from the three different types of nitrogen atoms. Thus, a certain (small) fraction of the solid sample loses a N_{1s} electron from the NO_3^- anion, while other molecules in the sample give up N_{1s} electrons from either the NO_2^- or the ethylenediamine nitrogen atoms. *Chemical shifts* of many electron volts (*i.e.*, 1 eV = 8067 cm^{-1} = 23.06 kcal/mole) are seen in the binding energies of core electrons; we will return to this in a later section. It must be emphasized that only a single electron, whether it be a N_{1s} electron or a valence-orbital bonding electron, is photoionized from any given molecule. Multiple ionizations are of low probability.

In the case of PES, as we will see, the resolution is such that it is possible to readily detect vibrational structure associated with the electronic state of the ionized molecule. The analogy with electronic absorption spectroscopy is clear. In the PES experiment, photoionization amounts to an ejection of a single electron concurrent with an electronic transition from the ground state of the parent molecule to the ground electronic state (sometimes to an excited state, *vide infra*) of the ionized molecule. In electronic absorption spectroscopy, vibrational structure is seen for the excited electronic state, while in PES the vibrational structure seen is for the electronic state of the ionized molecule. The explicit form of equation (16–23) for the energy required to liberate an electron from a molecule is therefore:

$$E_b = I + E_{\text{vib}} + E_{\text{rot}} = E_{\text{source}} - E_{\text{kin}} \tag{16-24}$$

where I is the adiabatic energy of the electron removed and E_{vib} and E_{rot} are quantized energies of vibration and rotation of the various ionic electronic states that result from the photoionization. An *adiabatic* ionization energy is the energy of a transition from the ground state of the molecule to the ground state of the ion. A *vertical* ionization energy is measured in the transition for which the internuclear distances of the ion are the same as those of the parent molecule. The vertical ionization energy will thus be equal to or greater than the adiabatic ionization energy. In many cases, both of these energies can be extracted from the PES spectrum (*vide infra*).

In the PES experiment, vacuum ultraviolet radiation is used as the ionizing radiation; usually this is provided by a helium [singly ionized, indicated as He(I)] resonance lamp with an energy of 21.21 eV. However, other discharge lamps have been used [*e.g.*, Ar(I) and the doubly ionized helium lamp, He(II)]. The energy of these lamps limits PES to studies of valence electrons and, in general, measurements have been mostly confined to gaseous samples. There have been some reports of work on solutions[29] and solids.[30]

Because XPS instrumentation incorporates X-radiation as an ionizing source, binding energies of both core and valence electrons can be determined by this method. Typically, Mg and Al $K\alpha$ X-radiations with energies of 1253.6 and 1486.6 eV, respectively, have been employed. Solids, gases, liquids, solutions, and frozen solutions have been studied with XPS. In the case of solids and frozen solutions, the calculated electron binding energies are referred to the energy of the Fermi level of the solid material. The Fermi level corresponds to the highest filled level of the electronic band structure of the solid at 0°K. The energy conservation equation is changed to

$$E_b = E_{\text{source}} - E_{\text{kin}} - \varphi_{\text{spec}} \tag{16-25}$$

Here the negligible recoil energy has been dropped and the work function φ_{spec} (~4 eV) of the inner metallic surfaces of the XPS spectrometer has been introduced. The work function of the spectrometer material is the energy necessary to remove an electron from the surface of the spectrometer. The sample has a different work function. The sample in an XPS experiment is mounted in electrical contact with the spectrometer, and if there is a reasonable number of charge carriers (most samples are insulators, and charge carriers are formed during X-ray irradiation), the Fermi levels of the sample and the spectrometer will be the same. We can understand equation (16–25) by tracing through the XPS experiment. An electron is photoionized from the sample with a certain kinetic energy, E_1. The electron has to go through an entrance slit to get into the spectrometer. Because the work potentials of the spectrometer and the sample are different, the kinetic energy of the electron changes to E_{kin}. This is because the photoionized electron is either accelerated or decelerated to the entrance slit. *In the spectrometer chamber, the electron has the kinetic energy E_{kin}, and it is this that is measured.* Thus, in order to refer a binding energy to the Fermi level, φ_{spec} enters into the expression. It is fortunate that it is not necessary to know the work function of each and every sample.

Resolution

The theoretical resolution possible in the PES experiment, where the binding energies of valence electrons are determined, has been discussed.[31] To repeat, measurements are carried out on gaseous samples. It is the velocity of motion of the target molecule, coupled with the photoelectron velocity (in an effect analogous to Doppler broadening), that limits the resolution of PES spectra to $\sim 10^{-3}$ eV. If, instead of using a sample chamber filled with gas, a molecular beam of target molecules is used, a resolution of about 10^{-4} eV would be possible. The distribution of molecular velocities relative to the source is more uniform when a beam is employed. The contribution to the width of the PES peak from the lifetime of the excited ionic state resulting from photoionization is very small ($\sim 10^{-7}$ eV). However, in some cases, the ionic state initially produced in the PES photoionization dissociates and leads to broadening in the observed peak. If this dissociative decay is absent and only the above factors are considered, a theoretical resolution of $\sim 10^{-3}$ eV would thus be expected. This has not been reached experimentally. The rotational envelope ($\approx kT$) associated with a vibrational peak, as well as spin-orbit splitting, set the effective limit. Experimentally, an energy resolution of 0.010 eV has been realized (see, for example, reference 32). Instrumental and ionizing source contributions have been minimized.

In the case of XPS, the observed peak half-widths (full width at half peak height) are much greater than those of PES. Photoionization of core electrons results in excited states with shorter lifetimes than the excited states encountered in PES, because the lifetime is proportional to E^{-3} where E is the photoionization transition energy. X-ray absorption and emission data[33] show that the inherent line widths of inner atomic levels decrease with decreasing atomic number and might be on the order of 0.1 eV or less for low atomic number elements. It might be expected, then, that the theoretical limit to the half-width of XPS peaks would be on the order of 0.1 eV for photoionization of carbon $1s$, nitrogen $1s$, phosphorus $2p$, and similar electrons.

However, in contrast to PES, the natural widths of the common XPS X-radiation sources are quite appreciable and play a large factor in determining the half-widths of experimentally observed XPS lines.[27] The X-ray doublet $K\alpha_1\alpha_2$ is commonly used in XPS, and this X-radiation results when electrons "fall" from the L_{II} and L_{III} (spin-orbit split $2p$ atomic levels) shells to the hole in the K ($1s$ atomic level) shell. The natural width associated with either the $L_{II} \rightarrow K$ or the $L_{III} \rightarrow K$ transition is 0.7 eV for Al X-radiation; the doublet overlaps in this case to give an effective width of ~ 1.0 eV. Magnesium $K\alpha_1\alpha_2$ X-radiation comprises a doublet of ~ 0.8 eV width. More energetic X-ray sources (*e.g.*, Cr, Cu, or Mo) have doublet component widths in excess of 1.0 eV. The effective limit to XPS peak width is thus set by the "natural" width of the X-ray source, modified to a certain degree by the natural width associated with the level from which the photoionization occurs. Small instrumental contributions could also be present. Experimentally, XPS half-widths for C_{1s}, N_{1s}, P_{2p}, S_{2p}, and similar peaks are ~ 1.5 eV for measurements on solid materials. XPS measurements on gaseous samples give appreciably narrower lines. For example, the half-width of the Ne $1s$ line from gaseous Ne was found to be 0.8 eV.[27] The difference in half-width between solid state and gaseous state XPS peaks is attributed to surface charging on the solids (usually insulating materials) and differences in crystal fields felt by atoms at different depths in the solid. Very recent work[34] shows that it is probably possible to remove the effect of the X-ray line width from XPS spectra by a deconvolution approach.

The average escape depth of photoelectrons from a solid depends on the kinetic energy of the photoelectrons. Research[35] on metals shows that, for a kinetic energy of 1000 eV, the average depth could be ~ 100 Å, while at 10 eV it would be as little as 10 Å. Surface cleanliness is important for an XPS sample and, by the same token, it is possible to study surface chemistry, such as in heterogeneous catalysis.

Koopmans' Theorem

Both PES and XPS can be used to study valence electrons in molecules, and we are interested in just what can be learned from the photoelectron spectrum for valence electrons. As an example, the PES spectrum of a gaseous sample of N_2 is shown in Fig. 16–8. With the He(I) source determining an ionizing limit of 21.21 eV, three vibrationally structured photoionizations are seen (~15.6, ~17.0, and ~18.8 eV). These can be assigned to ionizations from the three highest-energy filled molecular orbitals for N_2, the $2\sigma_u$, π_u, and $3\sigma_g$ orbitals. The peaks have been assigned on the basis of the observed vibrational structure. As an aside, it should be noted that an XPS spectrum has the same three peaks (vibrational structure not seen because of lower resolution) in addition to a peak at 37.3 eV for ionization from the $2\sigma_g$ level as well as a single peak at 409.9 eV for both the $1\sigma_g$ and $1\sigma_u$ levels.[27]

This brings us to Koopmans' theorem, which states that the vertical ionization energy for removal of an electron from a molecular orbital is equal to minus (a stable

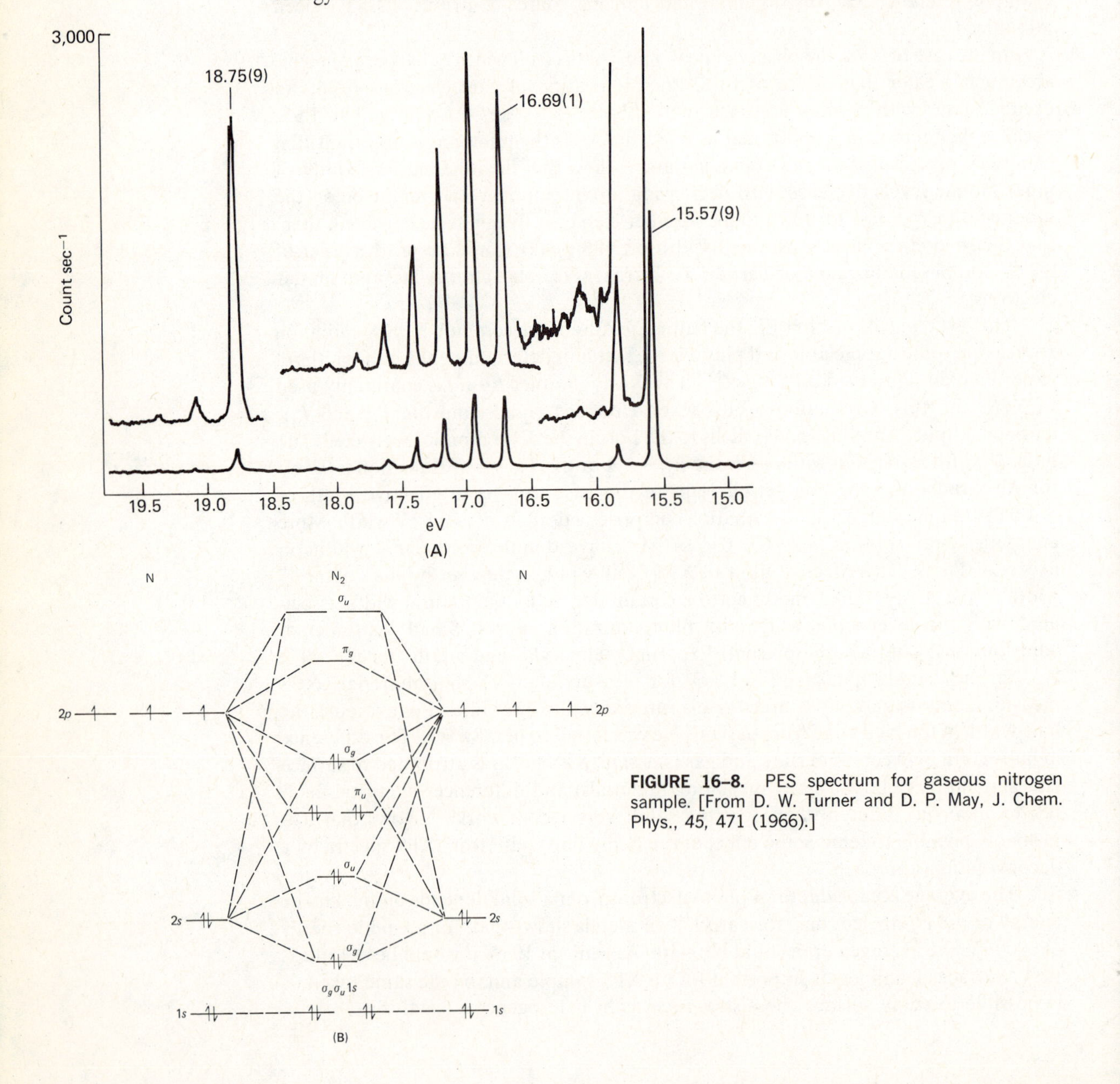

FIGURE 16–8. PES spectrum for gaseous nitrogen sample. [From D. W. Turner and D. P. May, J. Chem. Phys., 45, 471 (1966).]

orbital has a negative eigenvalue) the corresponding eigenvalue obtained from a Hartree-Fock self-consistent field molecular orbital calculation.[36] The basic assumption behind this theorem is that the molecular orbitals appropriate for the parent molecule will be the same as those for the ionized molecule. If there is any electronic relaxation (that is, a change in the molecular orbitals of an ionized molecule because of a change in electron repulsions), or if there is an appreciable change in correlation energies (a term not included in the m.o. approach in the sense that the coordinates of each electron depend on those of all the other electrons), then Koopmans' theorem breaks down.

Vertical ionization energies are readily identifiable for our N_2 example; see Fig. 16–8. In Table 16–1 are summarized the N_2 vertical ionization energies and eigenvalues from a few m.o. calculations. All numbers are in electron volts. Reasonable agreement is seen between the XPS and PES values; however, the eigenvalues from the Hartree-Fock self-consistent field calculation do *not* agree with the observed peaks. In fact, the order of $\sigma_g 2p$ and $\pi_u 2p$ is reversed. This could be indicative of a difference in relaxation energies. Even when the calculations are modified by calculating the energies for N_2 and for N_2^+ in different states and taking a difference in total electronic energies, only a fair agreement is found. Again, there is a reversal in $\sigma_g 2p$ and $\pi_u 2p$. For this example, Koopmans' theorem gives only qualitative agreement. This is generally true. Another example is given in Table 16–2. Certain types of semi-empirical molecular orbital calculations appear to give semi-quantitative

TABLE 16–1. PHOTOELECTRON DATA FOR GASEOUS NITROGEN (Energies in eV)

Orbital	Final State	XPS[a]	PES[b]	M.O. Close[c] to HF Limit	Calc.[c] on N_2^+
$\sigma_g 2p$	$^2\Sigma_g^+$	15.5	15.57	17.28	15.99
$\pi_u 2p$	$^2\pi_u$	16.8	16.96	16.75	15.34
$\sigma_u 2s$	$^2\Sigma_u^+$	18.6	18.75	21.17	19.74
$\sigma_g 2s$	$^2\Sigma_g^+$	37.3		40.10	
$\sigma_g/\sigma_u 1s$	$^2\Sigma_{g,u}$	409.9		(426.61 / 426.71)	

[a] See reference 27.
[b] D. W. Turner and D. P. May, J. Chem. Phys., *45,* 471 (1966).
[c] P. E. Cade, *et al.*, J. Chem. Phys., *44,* 1973 (1966).

TABLE 16–2. CARBON MONOXIDE PHOTOELECTRON DATA

Orbital	XPS[a]	PES[b]	HF[c] Calc.
3σ	14.5	13.98	15.08
1π	17.2	16.58	17.35
2σ	20.1	19.67	21.85
1σ	38.3		41.40
C_{1s}	295.9		309.13
O_{1s}	542.1		562.21

[a] See reference 27.
[b] M. I. Al-Joboury, *et al.*, J. Chem. Soc., 616 (1965).
[c] D. B. Neumann and J. W. Moskowitz, J. Chem. Phys., *50,* 2216 (1969).

agreement with observed vertical ionization energies. Discussion of the breakdown of Koopmans' theorem continues in the literature.[37]

Even though Koopmans' theorem does not work exactly, it is still useful to know that the peaks in a photoelectron spectrum can be associated with different molecular orbitals in the parent molecule. For example, in Chapter 3 the symmetry and construction of the molecular orbitals of NH_3 were worked out. There it was found that the seven atomic orbitals in C_{3v} symmetry form a representation that is reduced to give *three* a_1 and *two* e irreducible species. The eight valence electrons of NH_3 fill two of the a_1 and one of the e molecular orbitals to give a ground state configuration of

$$\cdots(2a_1)^2(1e)^4(3a_1)^2$$

The only other filled orbital, the $1a_1$ orbital, is essentially the nitrogen 1*s* atomic orbital. The He(I) (21.21 eV) spectrum is shown in Fig. 16–9, where vertical ionizations are seen at 10.88 eV for the $3a_1$ level and at 16.0 eV (first maximum) for the 1*e* level. These assignments were made with the use of results from various m.o. calculations. The more energetic He(II) (42.42 eV) source has been used to observe the $2a_1$ vertical photoionization at 27.0 eV.[38] In passing, it is of interest to note the doubled or split character of the 1*e* peak at ~16 eV. This splitting has been assigned to Jahn-Teller splitting in the ion resulting from a $(2a_1)^2(1e)^3(3a_1)^2$ configuration. The splitting is 0.78 eV. Other examples of Jahn-Teller splitting have been reported.[28]

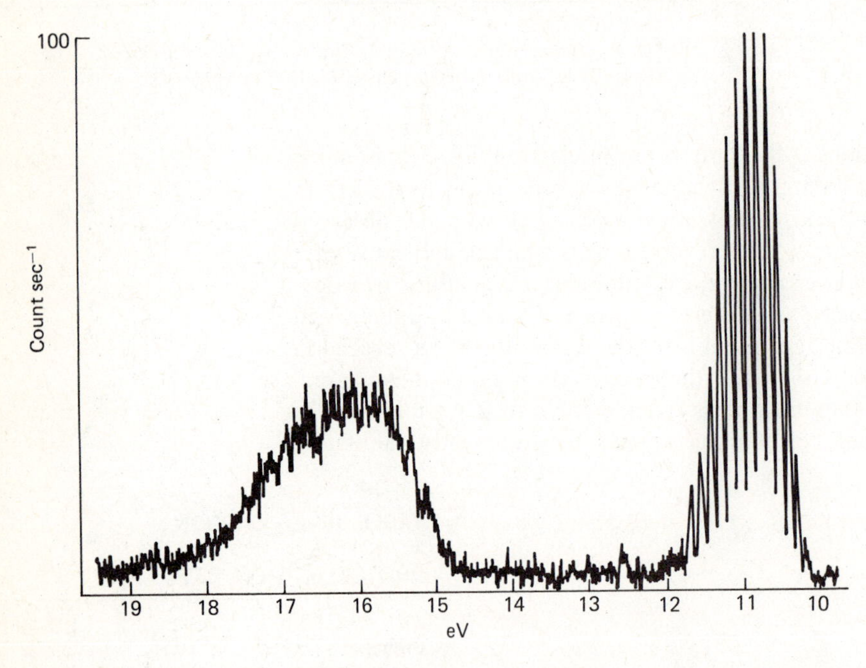

FIGURE 16–9. PES spectrum of a gaseous sample of ammonia. [From D. W. Turner, *et al.*, "Molecular Photoelectron Spectroscopy," Wiley-Interscience, New York (1970).]

Cross-Sections and Angular Distributions

XPS spectra have been obtained recently for gaseous samples of materials already studied by PES.[27] Interesting results bearing on the relative cross-sections to photoionization of valence electrons as a function of the energy of the source have been obtained. For instance, the electrons in a molecular orbital with more 2*s* atomic orbital character have a larger relative cross-section (and consequently a greater peak intensity) for the more energetic X-radiation than do the electrons in an orbital composed mostly of 2*p* atomic orbitals. Comparison of XPS (valence region) and PES spectra show different relative intensities for corresponding peaks. A peak associated with electrons in an m.o. composed largely of *s*-type atomic orbitals will have greater relative intensity in the XPS spectrum than in the PES spectrum.

16–12 REFINEMENTS

PES—Spectral Detail

The shape of the vibrational structure on a PES band tells something about the bonding characteristics of the electron that is ionized. The ionization of a non-bonding electron will result in an ion with the same internuclear distance as the parent molecule. In this case the lowest energy ($v = 0$, $v' = 0$) vibrational peak will dominate the spectrum (see Fig. 16–10). Returning to the N_2 spectrum in Fig. 16–8, we see that ionization of electrons from either the $3\sigma_g$ level at 15.6 eV or the $2\sigma_u$ level at 18.8 eV in each case gives essentially only one strong peak. This is because these levels are weakly antibonding and weakly bonding, respectively.

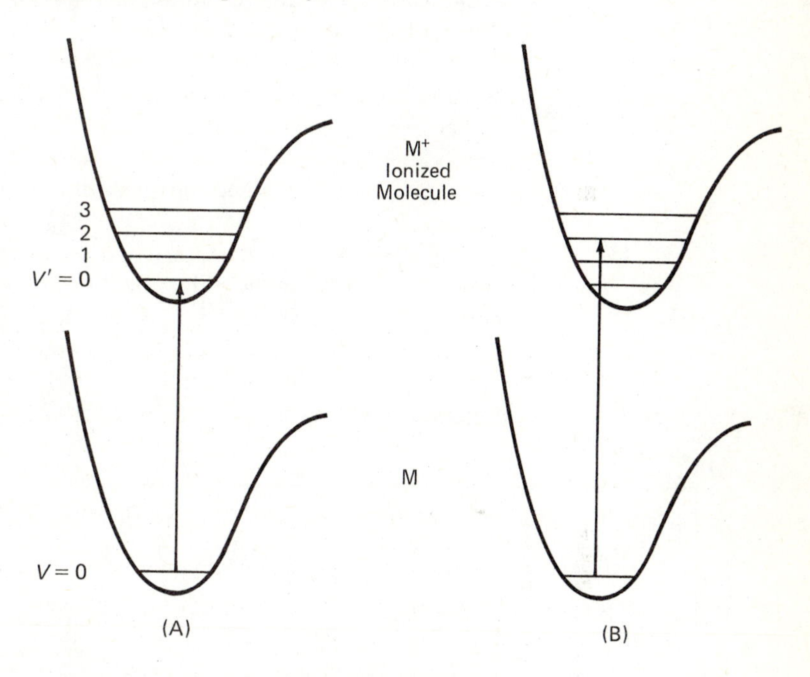

FIGURE 16–10. Vibrational states for the parent molecule M and the ionized molecule M^+. (A) In this case the electron ionized comes from a non-bonding orbital and the two electronic potential wells lie one above the other. The most intense transition (indicated with an arrow) is the $V' = 0 \leftarrow V = 0$ transition. (B) The wells are displaced (molecule dimensions change) when the electron ionized comes from either a bonding or an antibonding orbital. In this case the $V' = 3 \leftarrow V = 0$ transition is the most intense band seen.

Photoionization of either a strongly bonding electron or a strongly anti-bonding electron will result in changes in equilibrium bond distances, as illustrated in Fig. 16–10. In these cases, the most intense peak (*i.e.*, vertical ionization) will be at higher energy than the peak corresponding to adiabatic ionization. This is exemplified by the π_u peak for N_2. In fact, the relative intensities (called Franck-Condon factors) of the respective vibrational peaks can be theoretically calculated with reasonable success.[39] The vibrational frequencies observed for the ionized molecular state are also of further assistance in characterizing the molecular orbital from which the electron is ionized. The ground state stretching frequency for N_2 is 2345 cm^{-1}, and this is to be compared with those from the different peaks. Table 16–3 lists the stretching frequencies seen for various N_2^+ ionized states, and a comparison of each of these values with the 2345 cm^{-1} N_2 frequency leads to the implied m.o. characters given in Table 16–3.

In addition to vibrational structure, other fine structure is seen in PES spectra. The first (*i.e.*, lowest binding energy) peak in the spectra of CH_3Cl, CH_3Br, and CH_3I is, obviously, from the highest occupied molecular orbital, and this is largely localized

TABLE 16–3. VIBRATIONAL STRUCTURE IN THE N_2 PES SPECTRUM

Peak	ν(N—N)	Implied M.O. Character
$\sigma_g(2p)$	2150 cm^{-1}	weakly bonding
$\pi_u(2p)$	1810 cm^{-1}	strongly bonding
$\sigma_u(2s)$	2390 cm^{-1}	weakly anti-bonding

on the halogen atom (some X—H anti-bonding and C—H bonding character).[40] Spin-orbit splitting is seen in this peak, where the separation varies in the series: CH_3I^+, 0.62 eV; CH_3Br^+, 0.31 eV; and CH_3Cl^+, ~0 eV.[28] These spin-orbit interactions agree with Mulliken's predications of 0.625, 0.32, and 0.08 eV, respectively.[41] Many other such observations of spin-orbit interactions have been noted. In fact, recent work at the high resolution of 10 meV has detected spin-orbit splittings in NO of some 0.012 to 0.016 eV.[32]

A considerable number of small gaseous molecules have now been studied with PES and, for that matter, XPS. Some radical and excited state species have also been studied. The three expected states (^2P, ^4S, and ^2D) of O^+ have been detected.[42] An electrodeless microwave discharge was used to produce the excited state species $O_2(^1\Delta_g)$, and the photoelectron spectrum of this species shows a vibrationally structured peak owing to formation of $O_2^+(^2\Pi_g)$. Comparison with the peak corresponding to ionization of ground state $O_2(^3\Sigma_g^-)$ to the same ionic state gives a value of 11.09 ± 0.005 eV for the adiabatic ionization potential of $O_2(^1\Delta_g)$.[43] Several other such species (*e.g.*, SO and NF_2) have been studied.

In comparison with diamagnetic compounds, paramagnetic species show additional complexity in their PES (and XPS) spectra. The oxygen molecule has two unpaired π_g electrons. The PES spectrum of O_2 is given in Fig. 16–11. The photo-

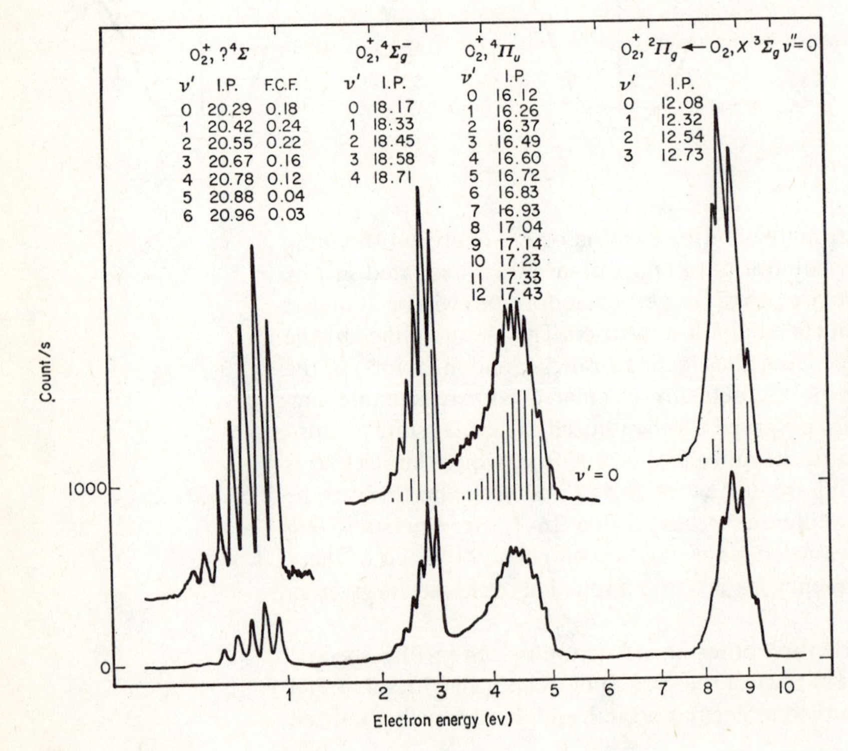

FIGURE 16–11. Oxygen PES spectrum. [From H. A. O. Hill and P. Day, "Physical Methods in Advanced Inorganic Chemistry," Interscience, New York, p. 88 (1968).]

ionization of an electron from the partly filled, anti-bonding $\pi_g(2p)$ molecular orbital appears as the first peak in the PES spectrum. Only one ionic state is realized. On the other hand, photoionization of an electron from one of the other filled molecular orbitals leads, in each case, to *two* electronic states of the ion O_2^+. Thus, if an electron is ionized from the filled bonding π_u level, the *unpaired electron remaining* in the π_u orbital can be aligned either *with* or *against* the two unpaired electrons in the anti-bonding π_g level. When the electron is aligned with the two π_g electrons, there are three unpaired electrons, the total spin $S = 3/2$, and the electronic state of the O_2^+ molecule is $^4\Pi_u$. The other alignment gives a $^2\Pi_u$ electronic state for the O_2^+ molecule. The $^4\Pi_u$ and $^2\Pi_u$ states of O_2^+ are at different energies, and thus there is a splitting of the π_u orbital ionization peak. In Table 16–4 are listed the observed features for the O_2 molecule from both PES and XPS spectra.

TABLE 16–4. VERTICAL O_2 IONIZATION DATA[a]

One-Electron Molecular Orbital	Ionic Electronic State	XPS, eV	PES, eV
$\pi_g 2p$	$^2\pi_g$	13.1	12.10
$\pi_u 2p$	$^4\pi_u$	17.0	16.26
	$^2\pi_u$	?	?
$\sigma_g 2p$	$^4\Sigma_g$	18.8	18.18
	$^2\Sigma_g$	21.1	20.31
$\sigma_u 2s$	$^4\Sigma_u$	25.3	24.5 [*via* He(II)]
	$^2\Sigma_u$	27.9	
$\sigma_g 2s$	$^4\Sigma_g$	39.6	
	$^2\Sigma_g$	41.6	
Oxygen 1*s*	$^4\Sigma$	543.1	
	$^2\Sigma$	544.2	

[a] See reference 27.

For comparison purposes, the XPS spectrum of gaseous O_2 is shown in Fig. 16–12. It can be seen that the oxygen 1*s* peak is also split, in this case by 1.1 eV. This splitting is *not* the difference between the O_2 molecular orbitals $1s\sigma_u$ and $1s\sigma_g$, which is calculated to be small. Recall that no such splitting is seen in the case of N_2. Even further, the intensity ratio of the two O_{1s} features is 2:1 and not the 1:1 ratio expected if the $1s\sigma_u/1s\sigma_g$ explanation applied. Again, as occurred in the valence orbitals, there is an interaction between the unpaired electron in the O_{1s} level and the two unpaired electrons in the O_2^+ π_g anti-bonding molecular orbital. Ionization of even these oxygen core electrons leads to two O_2^+ states, $^2\Sigma$ and $^4\Sigma$, that differ appreciably in energy.

The observation of such splittings is very interesting. One-center exchange integrals dominate the energy difference between the quartet and doublet states. Thus, inter-electron repulsions (e^2/r_{ij}, where r_{ij} is the distance between the ith and jth electrons and e is the charge on the electron) of the exchange type are very appreciable even between valence unpaired electrons and the oxygen core electron. Such inter-electron spin polarizations are found to be of importance in explaining the large shifts in nmr peaks for paramagnetic molecules (see Chapter 12). Electron exchange splitting is also seen for core electrons of paramagnetic transition metal complexes, as is described in a later section.

It is clear from the above discussion that the PES spectra of relatively large molecules are quite rich with information about ionization potentials, vibrational quanta for the ionized molecule, spin-orbit interactions, Jahn-Teller splittings, and electron exchange interactions. Unfortunately, there is frequently an overlapping of features, and broad peaks appear with no resolved vibrational structure. As an

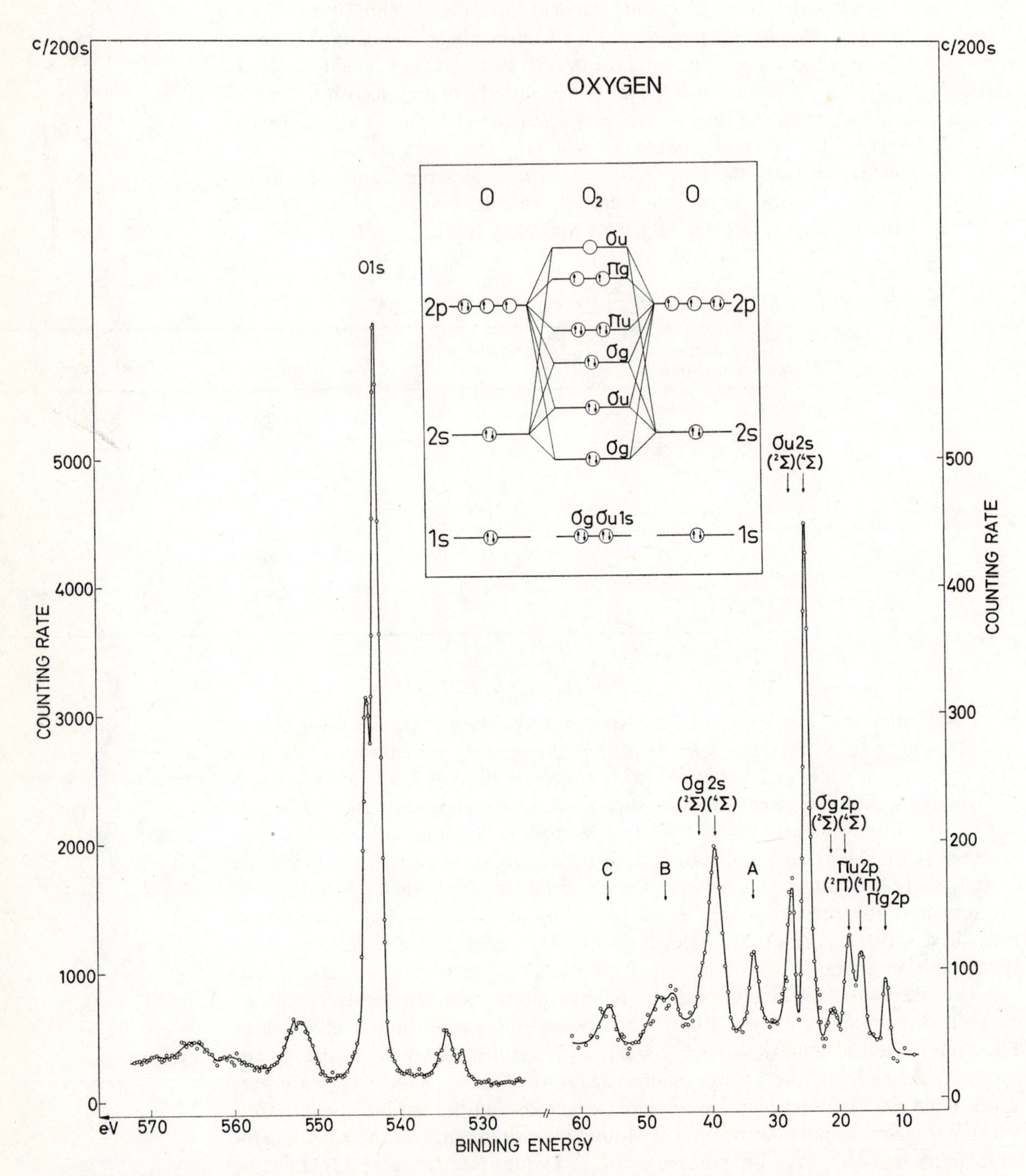

FIGURE 16–12. XPS spectrum of core and valence electrons in O_2 excited by Mg Kα X = radiation. Two peaks are obtained for each fully occupied orbital, including the oxygen 1s core orbital.[27] The peaks A, B, and C, and the peaks toward higher binding energy from the O_{1s} peak, are satellite peaks (*vide infra*). [From K. Siegbahn, *et. al.*, "ESCA Applied to Free Molecules," North Holland/American Elsevier, New York (1969).]

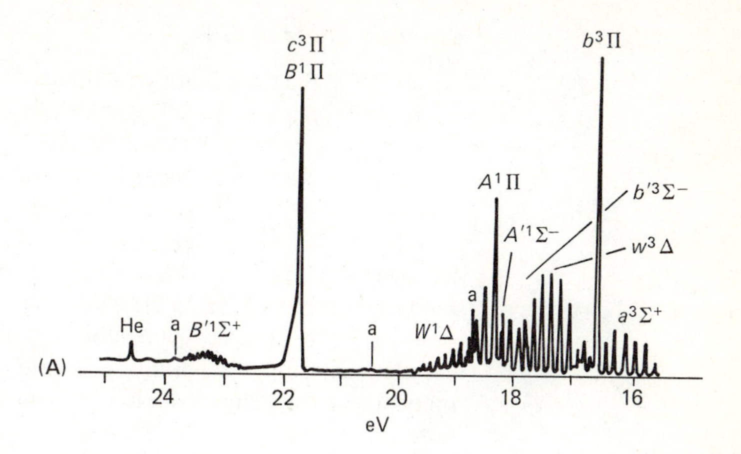

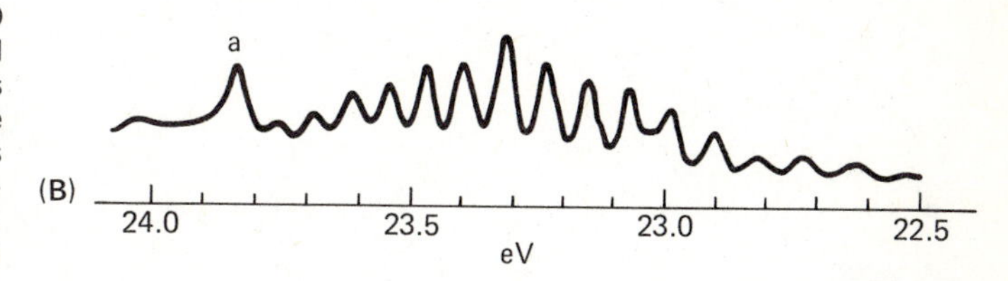

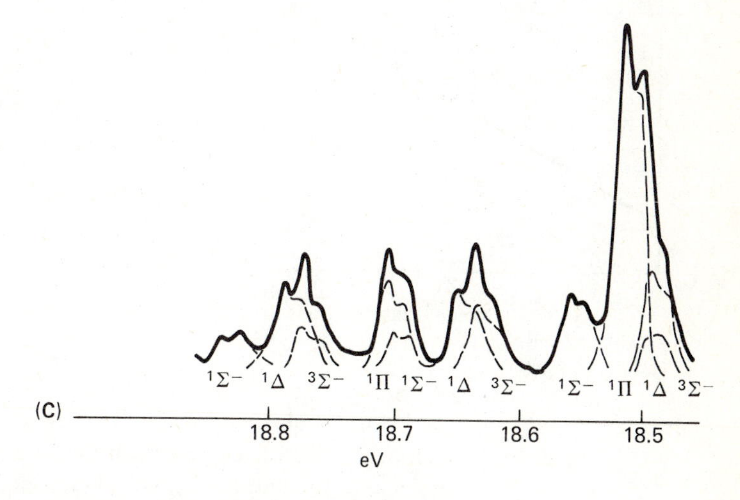

FIGURE 16–13. (A) Photoelectron spectrum of NO using the He 304 Å line; three small peaks marked "a" are due to the He 320 Å line; recording time was 75 h. (B) Photoelectron spectrum of the B′ $^1\Sigma^+$ state of NO using the He 304 Å line; the peak at "a" is again due to the 320 Å line. (C) Photoelectron spectrum of NO using the 584 Å line. Resolution 10 meV. Deconvolution of the peaks is indicated by the dotted lines. [From O. Edqvist, L. Åsbrink, and E. Lindholm, Z. Naturforsch., *26a*, 1407 (1971).]

example of a small molecule with many photoionization peaks, Fig. 16–13 illustrates some of the spectral detail obtained by Åsbrink *et al.*[32] for gaseous NO at a resolution of 10 meV for a He(I) source and at 25 meV for a He(II) source. For the fine points of assignment the reader is referred to their paper; however, perusal of Fig. 16–13 shows that both exchange and spin-orbit splittings have been resolved.

Finally, general techniques are being developed to study vapor species that by their very nature cannot be introduced as an ambient gas. Such species would include very reactive free radicals, free atoms, excited states of molecules, or materials with very low vapor pressures. Berkowitz and coworkers[44] generate the species in a molecular beam and let this beam interact with a photon beam; the two beams enter the spectrometer at right angles. A cylindrical-mirror analyzer is used to collect the photoelectrons. This was necessary to improve the signal-to-noise ratio. This technique has been used to study TlCl, TlBr, TlI, InCl, InBr, and InI in the gas phase. These molecules are, of course, fairly involatile; therefore, it would have been virtually impossible to study them by the more conventional approach.

Chemical Shifts in XPS

In 1957 Siegbahn and co-workers[45] detected a "chemical shift" in copper 1*s* electron binding energies. The shift in $E_b(1s)$ between Cu and CuO was found to be 4.4 eV (out of some 8979 eV). At that time an explanation of the shift was not known. In the early 1960's the Siegbahn group found shifts in sulfur 2*s* and 2*p* electron binding energies.[46] It was also found that the core binding energies could be correlated with the oxidation state of the sulfur atom. Since this earlier work, there has been considerable XPS work along these same lines. For example, the electron binding energies of Xe 3*d* electrons are plotted in Fig. 16–14 against the oxidation state of Xe in a series of xenon fluorides. A monotonic and apparently linear correlation results. Generally, it is found that the core-electron binding energy is influenced by the chemical environment and that it increases with increasing oxidation state of the atom. The shifts are large. In fact, interestingly enough, they are of a magnitude comparable to those of chemical reaction energies!

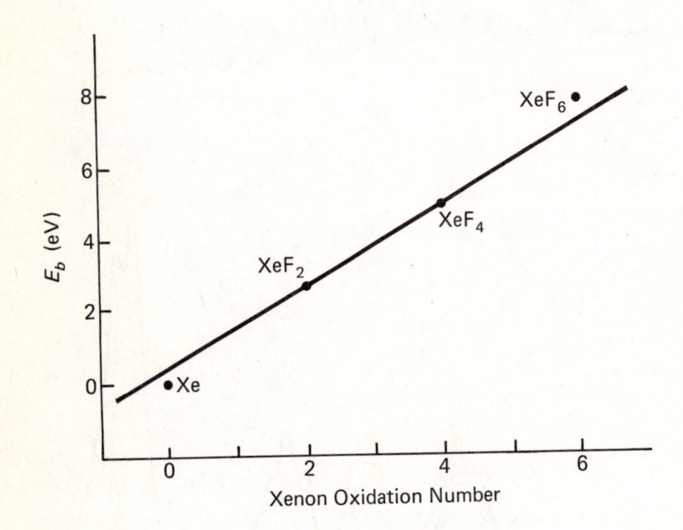

FIGURE 16–14. Plot of relative xenon $3d_{5/2}$ core-electron binding energies vs. xenon oxidation number. [Data taken from J. M. Hollander and W. L. Jolly, Acct. Chem. Res., 3, 193 (1970).]

Many sophistications have been added to the qualitative notion of relating the chemical shifts of core-electron binding energies to electron charge distributions in molecules. In Chapter 3 it was mentioned that with the molecular orbital technique the formal charge (δ) on an atom in a molecule can be evaluated. Recall that the formal charge was defined as the electron density at an atom in a molecule minus that on a free atom. In Fig. 16–15, we see that it is possible to correlate the formal charge on the nitrogen atom in a molecule (obtained from iterative extended-Hückel m.o. calculations) with the observed nitrogen 1*s* binding energies for a series of nitrogen compounds. It must be emphasized that ground state net charges on an atom, which after all are arbitrarily defined, are being used to correlate with shifts in core-electron photoionization transition energies. Either the apparent success is fortuitous, or terms such as electronic relaxation energies are relatively constant.

It is currently believed that a shift in core-electron energies reflects both the charge on the atom studied and the charges on other atoms in the molecule. In this "potential model," the shift in core-electron binding energies, ΔE, can be related to the formal charge on an atom A (δ_A) in a molecule:[27]

$$\Delta E = k\delta_A + V_A + \ell \qquad (16\text{-}26)$$

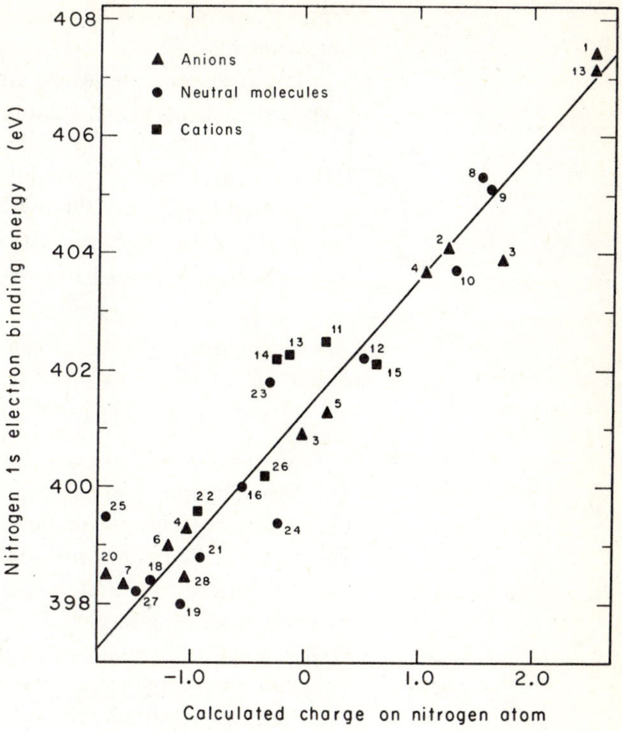

FIGURE 16–15. Plot of nitrogen 1s binding energies *vs.* iterative extended-Hückel calculated charges on nitrogen atoms. [From D. N. Hendrickson, J. M. Hollander, and W. L. Jolly, Inorg. Chem., *8*, 2642 (1969).]

Here k and ℓ are empirical parameters used to fit the available core-electron binding data. The term V_A represents the coulomb potential associated with the charges of neighboring atoms. That is, the formal charges on the neighboring atoms, q_B, are summed in a point-charge approximation

$$V_A = \sum_{B \neq A} q_B / R_{AB} \qquad (16\text{–}27)$$

where R_{AB} is the internuclear distance. The fact that the neighboring atoms' charge densities influence the core-electron binding energies of an atom is witnessed in the ~1.7 eV range of potassium $2p$ binding energies for a series of potassium salts.[47] This "potential model" has been applied to many different elements. A considerable number of the available empirical, semi-empirical, and *ab initio* m.o. calculation approaches have been used. Stucky *et al.*[48] have compared the charge distributions obtained from the potential model applied to XPS data for a series of small molecules with charge distributions inferred from several theoretical calculations, quadrupole moments, and X-ray–neutron diffraction experiments.

The theoretical basis of equation (16–26) has been discussed. It can be shown that if the zero-differential overlap approximation is assumed and if Koopmans' theorem holds (or a constant discrepancy is present), an equation similar to equation (16–26) is obtained for core-electron chemical shifts.[1,49] Equation (16–26) has also been modified in various ways. In one case, the term $k\delta_A$, in effect, was broken down to account for different types of valence electron densities. Thus, for a second-row atom there would be one term for $2s$ density, $k(2s)\delta_A(2s)$, and one for $2p$ density,

$k(2p)\delta_A(2p)$. Needless to say, a model with a larger number of parameters gives an improved fit.

Irrespective of the approximations in the potential model, the various correlations (as can any good correlation) can be used for their predictive and analytical ability. In 1896, Angeli reported the preparation of the compound $Na_2N_2O_3$.[50] In 1969, Hendrickson *et al.*[51] reported two nitrogen 1*s* peaks for $Na_2N_2O_3$ at binding energies of 403.9 and 400.9 eV. Of the three most probable structures of $N_2O_3^{2-}$ depicted below, the observation of two N_{1s} peaks of approximately equal intensity rules out structure II where, with resonance forms, the two nitrogen atoms are equivalent.

$$\left[O{=}N{-}N\begin{matrix}\nearrow O\\ \searrow O\end{matrix}\right]^{2-} \qquad [O{=}N{-}O{-}N{-}O]^{2-} \qquad [O{-}N{=}N{-}O{-}O]^{2-}$$

I II III

Molecular orbital calculations of $N_2O_3^{2-}$ (both iterative extended-Hückel and CNDO) gave nitrogen atom charges for structure I that were in agreement with the correlations of charge *vs.* net atomic charge. The single-crystal X-ray structure[52] reported in 1973 substantiated this as the correct structure of $N_2O_3^{2-}$.

It is interesting to note that linear correlations have been found between core-electron binding energies and Mössbauer isomer shifts for compounds of tin and iron.[53] Also, a correlation of chlorine core binding energies with nuclear quadrupole resonance frequencies has been found.[54]

One interesting variation on the theme of correlating core-electron binding energies with other molecular properties is the subtle concept of "equivalent cores."[55] This approach apparently grows out of a desire to construct a thermochemical cyclic relationship (like a Born-Haber cycle) coupled with a realization that atomic cores with the same charge may be considered to be chemically equivalent. In other words, certain thermodynamic quantities (heats of formations) are known that can be used to gauge the change in a core-electron binding energy. This will become clearer as we consider an example. The photoionization of a N_{1s} electron from the gaseous nitrogen molecule can be represented as

$$N_{2(g)} \xrightarrow[E_b]{h\nu} NN^*{}_{(g)}^{+} + e^- \text{ (vacuum)} \qquad (16\text{–}28)$$

In this equation the asterisk indicates that the 1*s* electron is ionized from one nitrogen atom. Obviously, one *cannot* use thermodynamic data to calculate the energy change (ΔE) for equation (16–28) because the heat of formation of $NN^*{}_{(g)}^{+}$ is not known. However, an *equivalent-cores* replacement can be made to make the equation more amenable to such an approach. Here comes the subtlety! As far as the valence electrons in $NN^*{}_{(g)}^{+}$ are concerned, the (N^{*+}) core might just as well be an oxygen core, where the oxygen nuclear charge—being one greater than the charge of the nitrogen nucleus—replaces the hole in the 1*s* shell. A formal replacement process can be written as equation (16–29).

$$NN^{*+} + O^{6+} \rightarrow NO^+ + N^{*6+} \qquad (16\text{–}29)$$

If these are equivalent cores, then the energy change for equation (16–29) will be zero, $\Delta E = 0$. As it turns out, it is only necessary that the same ΔE be obtained for the same type of replacement process in a series of nitrogen 1*s* ionized molecules. Adding equations (16–28) and (16–29) gives equation (16–30):

$$N_2 + O^{6+} \rightarrow NO^+ + N^{*6+} + e^- \qquad (16\text{–}30)$$

The energy change for equation (16–30) is either equal to E_b for the 1*s* electron in N_2 or equal to this E_b plus a (presumably) constant energy term. Nitrogen 1*s* binding energies have been measured for N_2 and for many other nitrogen-containing molecules, and equivalent-cores replacement equations can be written for the other molecules (all molecules are in the gas phase):

$$\begin{aligned}
NH_3 + O^{6+} &\rightarrow OH_3^+ + N^{*6+} + e^- \\
(CH_3)_2NH + O^{6+} &\rightarrow (CH_3)_2OH^+ + N^{*6+} + e^- \\
HCN + O^{6+} &\rightarrow HCO^+ + N^{*6+} + e^- \\
NO + O^{6+} &\rightarrow O_2^+ + N^{*6+} + e^- \\
NO_2 + O^{6+} &\rightarrow O_3^+ + N^{*6+} + e^- \\
CH_3NH_2 + O^{6+} &\rightarrow CH_3OH_2^+ + N^{*6+} + e^- \\
NNO + O^{6+} &\rightarrow NO_2 + N^{*6+} + e^-
\end{aligned} \tag{16-31}$$

Heats of formation are known for the molecular species in the above equations, and relative thermodynamic energies (E_T) can be generated for each of the processes. For example, E_T for the NH_3 equation is given by the heat of formation of OH_3^+ minus the heat of formation of NH_3. A plot of E_T *vs.* the nitrogen 1*s* binding energies (E_b) shows that an amazingly good correlation is found (see Fig. 16–16). This type of

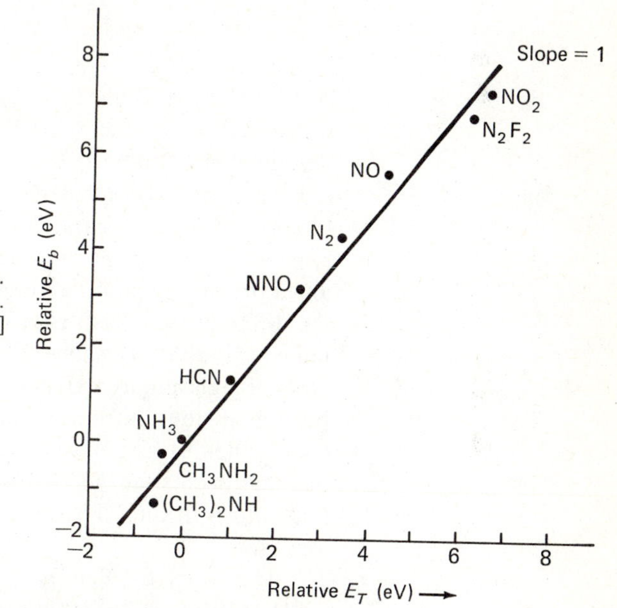

FIGURE 16–16. Plot of relative nitrogen 1s binding energies *vs.* relative thermochemical data. [Plot drawn from data in J. M. Hollander and W. L. Jolly, Accts. Chem. Res., 3, 193 (1970).]

equivalent-cores replacement process has given good correlations for electron binding data for other elements such as carbon (1*s*) and xenon ($3d_{5/2}$).[55] This type of correlation is useful because it is possible, from certain measured core-electron binding energies and known thermodynamic data, to predict various as yet undetermined thermodynamic quantities. Examination of the above equations shows that proton affinities can be determined. For some strange reason, the proton affinity (PA) of a molecule B is taken as a positive number and is equal to the negative of the energy change for equation (16–32).

$$B + H^+ \xrightarrow{-PA} BH^+ \tag{16-32}$$

In Table 16–5 are listed some PA values determined by the equivalent-core approach as well as some known PA values.

TABLE 16–5. PROTON AFFINITIES

Base	PA (kcal/mole)
C_5H_5N	247
CH_3NH_2	247
NH_3	214*
CH_3NO_2	209
NH_2OH	208
C_6H_5OH	192
NCl_3	190
H_2O	169*
NF_3	133

*These two values have been determined in other experiments and are used as the basis for scaling the correlation.

The heats of formation, as yet unreported, of various molecules can also be deduced from this approach. For example, the $\Delta H_f°$ for gaseous NON can be estimated as 120 kcal/mole, and this gives $\Delta H = -100$ kcal/mole for equation (16–33).

$$NON_{(g)} \rightarrow NNO_{(g)} \qquad \Delta H = -100 \text{ kcal/mole} \tag{16-33}$$

Before leaving the discussion of chemical shifts of core-electron binding energies, we should briefly mention XPS data on "mixed-valence" compounds, which were described in Chapter 10. Prussian blue is the oldest known coordination complex. The composition of this intensely blue-colored compound is $KFe[Fe(CN)_6]$. The intense blue color of this compound is associated with the so-called intervalence transfer band. The XPS $Fe(2p_{3/2})$ spectrum of Prussian blue consists of a relatively sharp peak and a broad peak at 4.4 eV higher binding energy.[56] The two peaks were assigned to the different iron ions. The X-ray structure[57] of this compound, which was determined very recently, shows that three different iron environments exist. There are two types of high spin iron(III), one in an environment of six nitrogen atoms and the other bonded to four nitrogen and two oxygen atoms. The low-spin Fe(II) atoms are coordinated to carbon atoms. It should be noted that this interpretation of the XPS data has been attacked.[58] The presence of exchange splittings or other fine structure (*vide infra*) in the XPS spectrum has to be eliminated to present a most definitive analysis.

The "mixed-valence" species that was described in Chapter 10 (p. 403), $[(NH_3)_5Ru(pyrazine)Ru(NH_3)_5]^{5+}$, has also been studied by XPS.[59] The XPS spectrum has been reported to contain ionization peaks from two non-equivalent transition metal ions. It should be noted that the carbon 1*s* peaks occur in the same spectral region as the metal ionization peaks, and the conclusion rests on the ability to subtract these peaks from the spectrum. The XPS technique has a very short time scale, something on the order of 10^{-17} sec. It can thus be used to monitor some of the faster intervalence transfer rates in mixed-valence compounds.

Structure on XPS Peaks

From the preceding discussion of chemical shifts in XPS core-electron ionization peaks, it might be surmised that relatively simple core-electron spectra are always seen, such as one nitrogen 1*s* peak for each appreciably different nitrogen atom environment. Fortunately, this is not the case.[27] We have already seen that paramagnetic species such as O_2 show exchange splittings in core-electron peaks. Paramagnetic transition metal complexes have also been reported to show such exchange splittings. Clark and Adams[60] have reported chromium 3*s* exchange splittings of some 4.5 eV in $Cr(hfa)_3$ and 3.1 eV in $Cr(h^5\text{-}C_5H_5)_2$. One can wonder whether an analysis of such splittings would be of assistance in understanding the details of Fermi contact nmr shifts observed for paramagnetic species.

Perhaps an even more promising and intriguing fine structure in XPS spectra is associated with "shake-up" processes. It would be very surprising, after all, if all that happened in the impact of very energetic X-rays with molecules was the photoionization of a single valence or core electron. Simultaneous with the photoionization of an electron, there can be excitation of one of the remaining electrons to an initially unoccupied orbital. This phenomenon is called a "shake-up" process. The higher energy, low intensity structure on XPS core-electron peaks can thus be used to study various electronic transitions occurring simultaneously with the photoionization. These satellite features are found in the range from zero to 50 eV toward higher binding energy than the main peak. Obviously, an electronic absorption at 50 eV ($= 404{,}000\ cm^{-1} = 25$ nm) is a very high energy absorption in the vacuum ultraviolet.

An example of such satellite XPS features appears in Fig. 16–12, with the broad features seen toward higher binding energy than the two O_{1s} peaks for O_2. The features marked A, B, and C are "shake-up" peaks appearing as satellites on the valence-electron peaks. In a similar manner, Fig. 16–17 shows "shake-up"

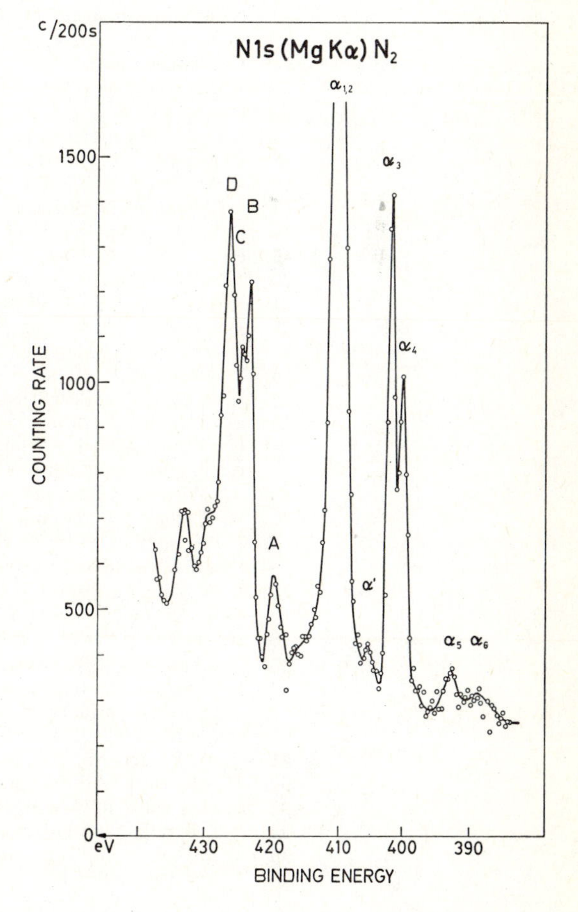

FIGURE 16–17. Nitrogen 1s electron spectrum excited by Mg $K\alpha$ radiation. The main peak at binding energy 410 eV is the N_{1s} line excited by Mg $K\alpha_{1,2}$. The spectrum also shows lines excited by the X-ray satellites and lines that correspond to shake-up and inelastic scattering in the N_2 molecule. [From K. Siegbahn *et al.*, "ESCA Applied to Free Molecules," North-Holland/American Elsevier, New York (1969).]

satellites for the N_{1s} peak of N_2. The peaks marked α', α_3, α_4, α_5, and α_6 are due to the non-monochromatic nature of the Mg $K\alpha$ X-radiation. Many of the peaks labeled A, B, C, and D have been assigned to "shake-up" excitations.[27] It appears that "shake-up" peaks can be identified for many molecular species and will perhaps prove useful for characterizing the electronic structures of many systems. In passing, it should be noted that fairly pronounced "shake-up" satellites have been detected in the photoelectron spectra of $Ni(CO)_4$, $Fe(CO)_5$, $Cr(CO)_6$, $W(CO)_6$, and $(CO)_5CrX$ ($X = NH_3$, PPh_3, etc.).[61]

REFERENCES

1. K. Biemann, "Mass Spectrometry," McGraw-Hill, New York (1962).
2. R. A. W. Johnstone, "Mass Spectrometry for Organic Chemists," Cambridge University Press, London (1972).
3. J. H. Benyon, R. A. Saunders, and A. E. Williams, "The Mass Spectra of Organic Molecules," Elsevier, Amsterdam (1968).
4. M. R. Litzow and T. R. Spalding, "Mass Spectrometry of Inorganic and Organometallic Compounds," Elsevier, New York (1973); J. Lewis and B. F. G. Johnson, Acc. Chem. Res., *1,* 245 (1968); M. I. Bruce, Adv. Organomet. Chem., *6,* 273 (1968).
5. H. Budzikiewicz, C. Djerassi, and D. H. Williams, "Mass Spectrometry of Organic Compounds," Holden-Day San Francisco (1967).
6. S. R. Shrader, "Introductory Mass Spectrometry," Allyn and Bacon, Inc., Boston (1971).
7. F. W. McLafferty, "Interpretation of Mass Spectra," W. A. Benjamin, New York (1967).
8. H. D. Beckey, Angew Chemie, Internat. Ed., *8,* 623 (1969).
9. D. Williams, in "Determination of Organic Structures by Physical Methods," Vol. 3, F. C. Nachod and J. J. Zuckerman, eds., Academic Press, New York (1971).
10. H. M. Rosenstock and M. Krauss, in "Mass Spectrometry of Organic Ions," F. W. McLafferty, ed., Academic Press, New York (1963).
11. K. Biemann and G. Spiteller, J. Amer. Chem. Soc., *84,* 4578 (1962).
12. J. H. Jones, Quart. Rev. Chem. Soc., *22,* 302 (1968).
13. E. Hugentobler and J. Löliger, J. Chem. Ed., *49,* 610 (1972).
14. J. Lederberg, J. Chem. Ed., *49,* 613 (1972).
15. D. Maurer, *et al.*, J. Chem. Ed., *51,* 463 (1974).
16. H. M. Bell, J. Chem. Ed., *51,* 548 (1974).
17. J. Collard, Ph.D. thesis, U. of Illinois, Urbana (1975).
18. F. E. Brinckman and F. G. A. Stone, J. Amer. Chem. Soc., *82,* 6235 (1960).
19. H. A. Hoehn, L. Pratt, K. Watterson, and G. Wilkinson, J. Chem. Soc., *1961,* 2738.
20. T. A. Milne and H. M. Klein, J. Chem. Phys., *33,* 1628 (1960).
21. R. T. Grimley, R. P. Burns, and M. G. Inghram, J. Chem. Phys., *34,* 664 (1961).
22. J. Berkowitz, M. G. Inghram, and W. A. Chupka, J. Chem. Phys., *26,* 842 (1957).
23. G. P. Barnard, "Mass Spectrometry," Institute of Physics, London, England (1953).
24. Y. Wada and R. W. Kiser, Inorg. Chem., *3,* 174 (1964).
25. J. D. Baldeschwieler, Science, *159,* 263 (1968).
26. a. J. D. Baldeschwieler and S. S. Woodgate, Acct. Chem. Res., *4,* 114 (1971).
    b. J. L. Beauchamp, Ann. Rev. Phys. Chem., *22,* 527 (1971).
    c. J. I. Brauman and L. K. Blair, in "Determination of Organic Structures by Physical Methods," Vol. 5, F. C. Nachod and J. J. Zuckerman, eds., Academic Press, New York (1973).
27. K. Siegbahn, C. Nordling, G. Johansson, J. Hedman, P. F. Heden, K. Hamrin, U. Gelius, T. Bergmark, L. O. Werme, R. Manne, and Y. Baer, "ESCA Applied to Free Molecules," North-Holland/American Elsevier, New York (1969).
28. D. W. Turner, C. Baker, A. D. Baker, and C. R. Brundle, "Molecular Photoelectron Spectroscopy," Wiley-Interscience, New York (1970).
29. H. Aulich, B. Baron, and P. Delahay, J. Chem. Phys., *58,* 603 (1973).
30. D. E. Eastman, "Photoemission Spectroscopy of Metals," in "Techniques in Metals Research VI," E. Passaglia, ed., Interscience, New York (1971); T. E. Fischer, Surface Sci., *13,* 30 (1969); C. R. Brundle, in "Surface and Defect Properties of Solids," Vol. 1, M. W. Roberts and J. M. Thomas, eds., Specialist Periodical Reports, The Chemical Society, London (1972).
31. D. W. Turner, Nature, *213,* 795 (1967); D. W. Turner, Advan. Mass Spec., *4,* 755 (1968).
32. O. Edquist, L. Åsbrink, and E. Lindholm, Z. Naturforsch., *26a,* 1407 (1971).
33. L. G. Parratt, Rev. Mod. Phys., *31,* 616 (1959).
34. J. F. Rendina and P. E. Larson, Bulletin from the GCA/McPherson Instrument Co., 1975.
35. T. A. Carlson, in "Electron Spectroscopy." D. A. Shirley, ed., North-Holland, Amsterdam (1972).
36. T. Koopmans, Physica, *1,* 104 (1934); C. C. J. Roothaan, Rev. Mod. Phys., *23,* 61 (1951).

37. L. S. Cederbaum. G. Hohlneicher and W. von Niessen, Chem. Phys. Lett., *18,* 503 (1973); D. W. Davis, J. M. Hollander, D. A. Shirley, and T. D. Thomas, J. Chem. Phys., *52,* 3295 (1970); I. H. Hillier, V. R. Saunders, and M. H. Wood, Chem. Phys. Lett., *7,* 323 (1970); and W. G. Richards, Int. J. Mass Spec. Ion Phys., *2,* 419 (1969).
38. A. W. Potts and W. C. Price, Proc. Roy. Soc. Lond., *A326,* 181 (1972).
39. D. W. Turner and D. P. May, J. Chem. Phys., *45,* 471 (1966).
40. H. Kato, K. Morokuma, T. Yonezawa, and K. Fukui, Bull. Chem. Soc. Japan, *38,* 1749 (1965).
41. R. S. Mulliken, Phys. Rev., *47,* 413 (1935).
42. N. Jonathan, A. Morris, M. Okuda, D. J. Smith, and K. J. Ross, in "Electron Spectroscopy," D. A. Shirley, ed., North-Holland, Amsterdam (1972).
43. N. Jonathan, D. J. Smith, and K. J. Ross, J. Chem. Phys., *53,* 3758 (1970).
44. J. Berkowitz and J. L. Dehmer, J. Chem. Phys., *57,* 3194 (1972), and references therein.
45. K. Siegbahn, *et al.*, Phys. Rev., *110,* 776 (1958).
46. Fahlman, *et al.*, Nature, *210,* 4 (1966).
47. W. E. Moddeman, J. R. Blackburn, G. Kumar, K. A. Morgan, M. M. Jones, and R. G. Albridge, in "Electron Spectroscopy," D. A. Shirley, ed., North-Holland, Amsterdam (1972).
48. G. D. Stucky, D. A. Matthews, J. Hedman, M. Klasson, and C. Nordling, J. Amer. Chem. Soc., *94,* 8009 (1972).
49. F. O. Ellison and L. Larcom, Chem. Phys. Lett., *10,* 580 (1971).
50. A. Angeli, Gazz. Chim. Ital., *26,* 17 (1896).
51. D. N. Hendrickson, J. M. Hollander, and W. L. Jolly, Inorg. Chem., *8,* 2642 (1969).
52. H. Hope and M. R. Sequeira, Inorg. Chem., *12,* 286 (1973).
53. I. Adams, J. M. Thomas, G. M. Bancroft, K. D. Butler, and M. Barber, Chem. Commun., 751 (1972); W. E. Swartz, P. H. Watts, E. R. Lippincott, J. C. Watts, and J. E. Huheey, Inorg. Chem., *11,* 2632 (1972); M. Barber, P. Swift, D. Cunningham, and M. J. Frazer, Chem. Commun., 1338 (1970).
54. D. T. Clark, D. Briggs, and D. B. Adams, J. Chem. Soc. Dalton, 169 (1973).
55. W. L. Jolly and D. N. Hendrickson, J. Amer. Chem. Soc., *92,* 1863 (1970); J. M. Hollander and W. L. Jolly, Acct. Chem. Res., *3,* 193 (1970).
56. G. K. Wertheim and A. Rosencwaig, J. Chem. Phys., *54,* 3235 (1971); D. Leibfritz and W. Bremser, Chem. Ztg., *94,* 982 (1970).
57. H. J. Buser, A. Ludi, W. Petter, and D. Schwarzenbach, Chem. Commun., 1299 (1972).
58. W. L. Jolly, Coord. Chem. Rev., *13,* 47 (1974).
59. P. H. Citrin, J. Amer. Chem. Soc., *95,* 6472 (1973).
60. D. T. Clark and D. B. Adams, Chem. Commun., 740 (1971).
61. M. Barber, J. A. Connor, and I. H. Hillier, Chem. Phys. Lett., *9,* 570 (1971).

EXERCISES

1. If the accelerating potential in a mass spectrometer is decreased in running a spectrum, will the large or small m/e ratios be recorded first?

2. Sulfur-carbon π bonding is not as effective as nitrogen-carbon π bonding. In the mass spectrum of $HSCH_2CH_2NH_2$, would the $m/e = 30$ or $m/e = 47$ peak be more intense?

3. Refer to Fig. 16–4 and recall the discussion on the relation of charge delocalization and stability of the positive ion. Explain why the mass 92 peak corresponding to

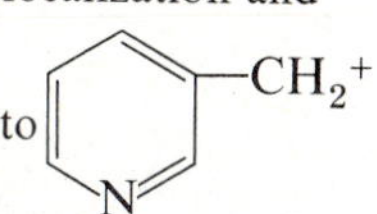

is most intense for the isomer of ethyl pyridine with the ethyl group in the 3 position.

4. a. What m/e peak in the mass spectrum of CH_4 would you examine as a function of accelerating potential in order to determine the ionization potential of the methyl radical?

   b. Write an equation for the appearance potential of the fragment in part (a) in terms of ionization potential and dissociation energy.

   c. In evaluating the dissociation energy of part (b) from thermochemical data, what is wrong with using one-fourth the value for the heat of formation (from gaseous carbon and hydrogen) of CH_4?

5. UPS spectra for even relatively small molecules like butadiene are complicated (see figures). Using Hückel calculations on butadiene, assign the peaks in the butadiene spectrum (the peaks at 8.6 and 10.95 eV are *impurity* peaks). Three vibrational progressions (1520, 1180, and 500 cm^{-1}) have been tentatively identified in the first band at 9.08 eV. Discuss the vibrational structure on this band (both shape and magnitude) with

respect to your assignment. (Bands are seen in the infrared spectrum of ground state butadiene at 1643, 1205, and 513 cm^{-1}.)

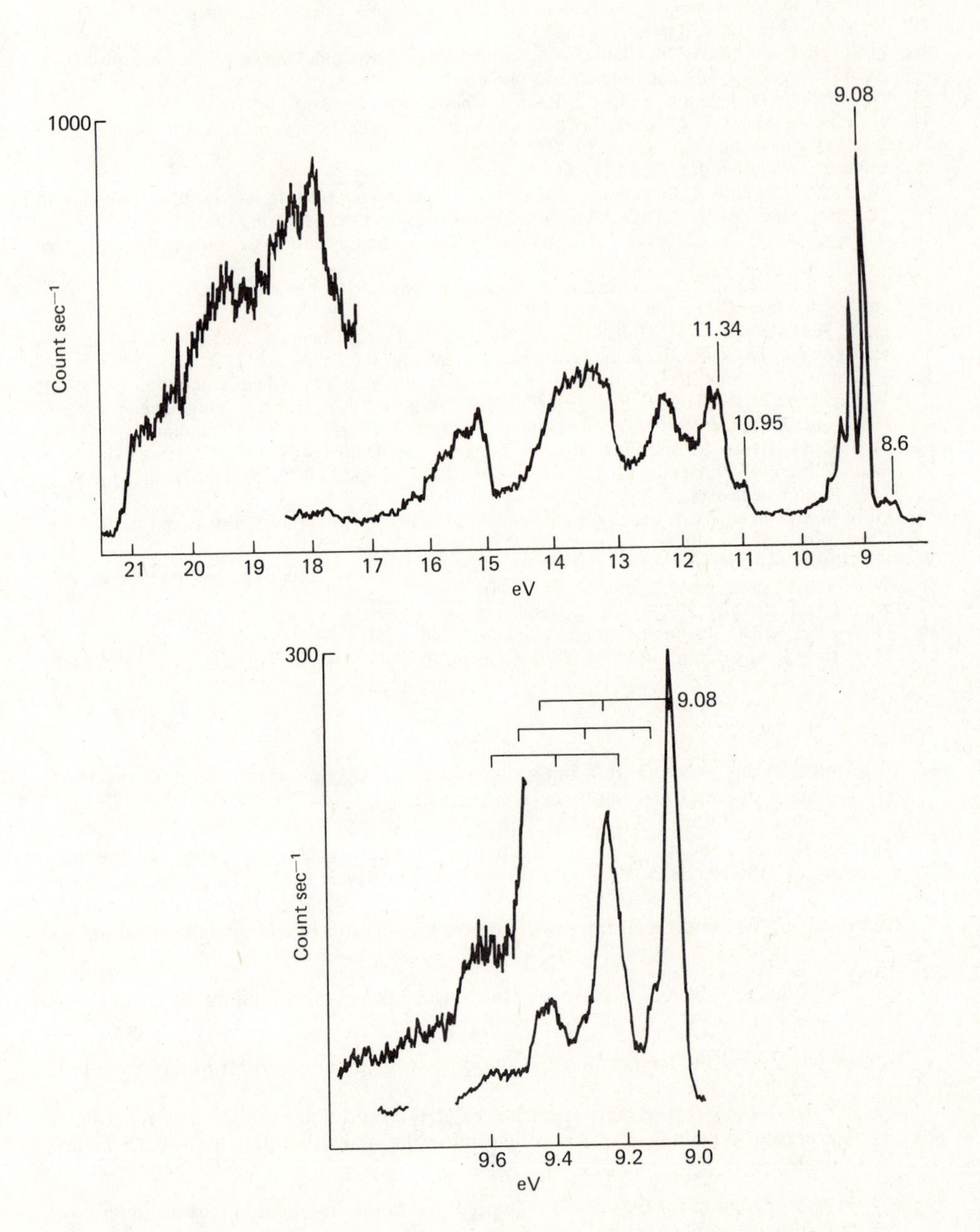

6. The effects of coordination of small molecules to transition metals can be studied with photoelectron spectroscopy. Spectra are given on p. 587 for CO and $W(CO)_6$. Qualitatively explain what observations bearing on the bonding in this complex are possible. Take into account symmetry and energy considerations if applicable.

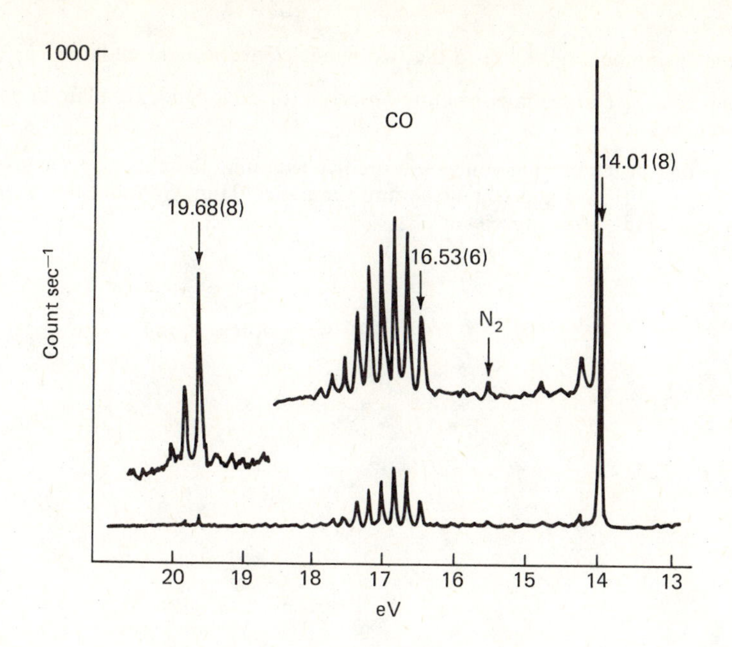

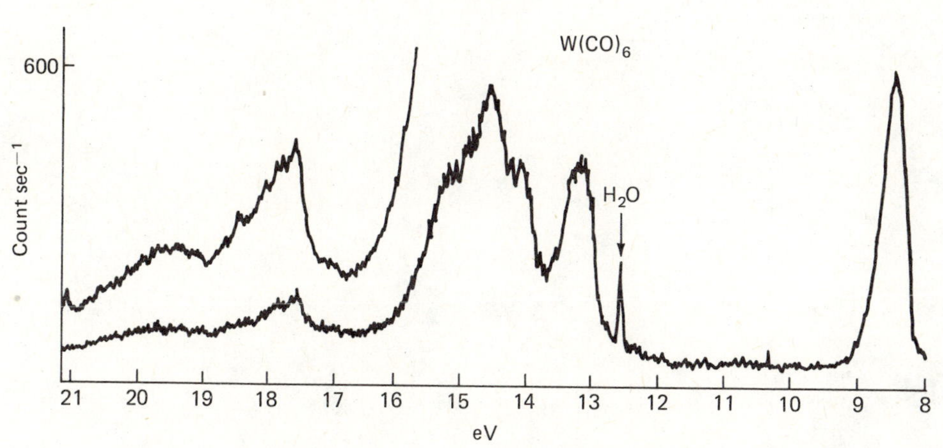

7. Given below are the UV photoelectron spectrum and m.o. diagram for ammonia. Bonding (b), non-bonding (n), and antibonding (*) levels are indicated.

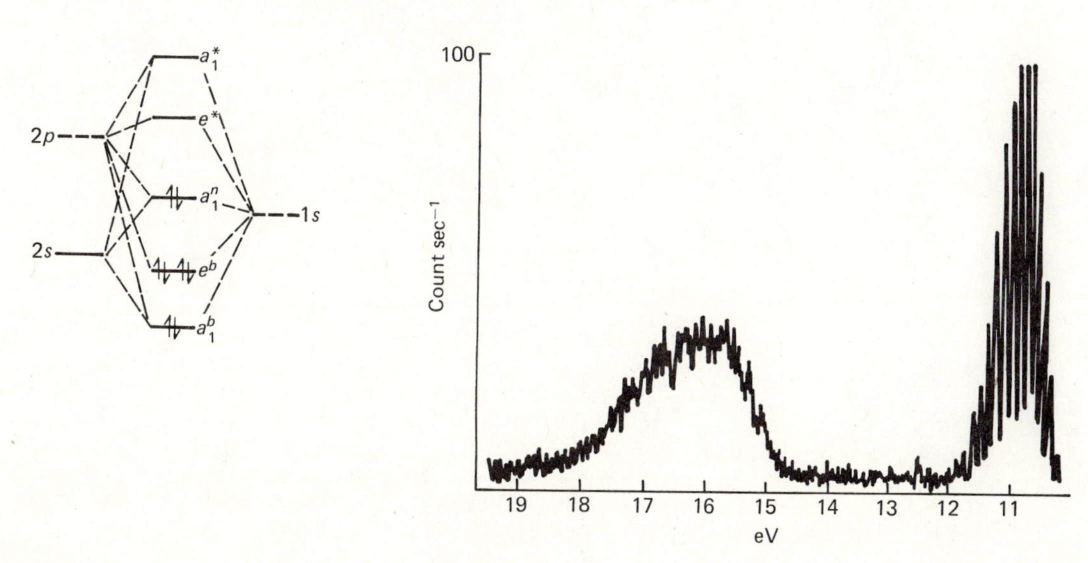

a. What phenomena give rise to the two bands centered at 11 and 16 eV?

b. What gives rise to the fine structure observed for each band? (Explain in one or two sentences.)

c. Is the fine structure what you would predict, assuming the m.o. diagram given? Why or why not? How might you justify any anomalies? Hint: Consider the shapes of the "before" and "after" species.

17 X-RAY CRYSTALLOGRAPHY*

17–1 INTRODUCTION

Using X-ray crystallography, one can generally determine the precise composition and atomic arrangement of almost any molecule. There are pleasantly few restrictions on the broad statement made above. First, the molecule must be isolable in the crystalline solid state, being thus subject to geometric distortions incurred when packing with its neighbors. Second, the system cannot be photochemically decomposed within one day of exposure at X-ray frequencies.[1] Third, the system of interest should produce crystals suitable for crystallographic study, avoiding the two most frequent problems in structure solution: twinning and disorder.[2] Fourth, the number of atoms whose locations are to be determined must not be too great.

The decrease in the labor required to complete a crystal structure determination has made the successful completion less of an accomplishment and more of a potentially useful source of information. It is currently possible, with favorable conditions, to carry a study from crystallization to figure drawing in from two to ten days. Any of several organizations can be contracted to collect data, solve, and refine a structure. Thus, at some time in a chemical career, nearly everyone will be exposed to crystallography on a firsthand basis. The aim of this chapter is to present enough information so the reader can obtain an operational feeling for the X-ray study of crystals.

Much useful information can often be obtained from X-ray techniques that stop short of single-crystal data collection and refinement. For instance, unit cell symmetry may be found from film techniques, and powder patterns of pulverized or precipitated samples may be compared to identify compounds. The practicing chemist's range of understanding should extend at least to these less complicated procedures.

The scope of this chapter will be limited to an incomplete but representative set of examples of the most important aspects of crystallographic technique. Topics that are mathematically detailed and not of immediate utility to the novice will not be covered. The generation of X-rays and their detection is described elsewhere[3]—it is the *conceptual* basis of crystallography that will concern us here.

17–2 SOLID STATE SYMMETRY

From the point of view of X-ray crystallography, a single crystal is composed of some repeating three-dimensional pattern of electron density[4] (in X-ray diffraction, the nuclei are not detected). It is the internal arrangement of the electrons in this

*This chapter was written by Michael Duggan, University of California, Irvine. The helpful suggestions of G. D. Stucky and R. Ryan are gratefully acknowledged.

crystal lattice that determines the directions and intensities of X-ray beams scattered from it. Naturally, the electron density is determined by both the structure of the molecular entities involved and the order in which they pack into the crystal. The packing of the molecules into the crystal defines the symmetry of the electron density distribution and the size of the smallest translationally repeating three-dimensional portion of the crystal, referred to as the *unit cell.* Using simple photographic techniques, one can determine the size and symmetry of the unit cell; and, knowing the number of molecules within the cell, one can frequently obtain information as to the symmetry elements contained in the molecular species of interest. This information may be enough to satisfactorily define the molecular structure, while the collection and interpretation of photographic data requires only an afternoon's work. Sections 17–2 through 17–5 cover the symmetry-intensity relationships that must be understood for this type of study. Our immediate concern will be the development of various aspects of unit cell symmetry.

a. Unit Cell. The unit cell is the smallest volume of the crystal that may be chosen so as to contain all the structural and symmetry information of the crystal. It is a three-dimensional volume whose three defining edges and three angles may have the relationships: $a = b = c$ and $\alpha = \beta = \gamma = 90°$; $a \neq b \neq c$ and $\alpha \neq \beta \neq \gamma \neq 90°$; or any intermediate possibility. The first relationship is a necessary but not sufficient requirement for a cubic system, whereas the second describes a triclinic lattice. The most common types of lattices encountered are: monoclinic, $\alpha = \gamma = 90°$, $\beta \neq 90°$, and $a \neq b \neq c$; and orthorhombic, $\alpha = \beta = \gamma = 90°$ and $a \neq b \neq c$. These unit cells are shown in Fig. 17–1. These definitions will suffice for our immediate discussion. A more complete definition of the seven discrete crystal systems will be found on p. 651 in the Appendix. It is suggested that this be read after the completion of Section 17–2.

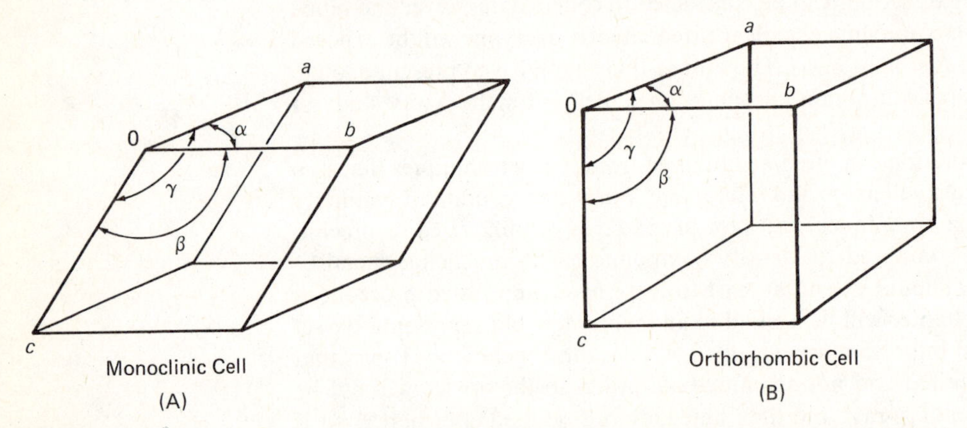

FIGURE 17–1. Common shapes and defining edge and angle labeling for monoclinic (A) and orthorhombic (B) unit cells.

If the size of the unit cell is known (say, $\alpha = \beta = \gamma = 90°$, and $a = 10$ Å, $b = 15$ Å, and $c = 20$ Å), then the volume, V, may be used in the following formula to calculate the density, ρ, of the crystal:

$$\rho = \frac{nM}{VA} \tag{17-1}$$

where M is the molecular weight, A is Avogadro's number, and n is the number of formula (or molecular) weights in the unit cell. When ρ is measured for a crystal (generally by flotation) in which the elemental composition of the molecular species involved is known, then n may be calculated. The value of n is of great importance because, as will be shown subsequently, it can provide molecular symmetry information from a consideration of the symmetry properties of the lattice.

b. Operators. There are more symmetry operators used in the definition of a *space group* than are needed to define a point group. These operators involve *translational motion,* for now we are moving molecules into one another in three dimensions. Two of these new operators are referred to as the *screw* and the *glide.* The symmetry elements are the screw axis and the glide plane. An $n = 2$-fold screw operation rotates the contents of the cell by 180° ($2\pi/n$) and then translates everything along its axis by one-half the length of the parallel unit cell edge ($1/n$ of the length per operation in a right-handed screw sense). The glide reflects in a plane and then translates by half a unit cell in some direction in the same plane. These operations are illustrated in Fig. 17-2 and will be discussed in more detail shortly.

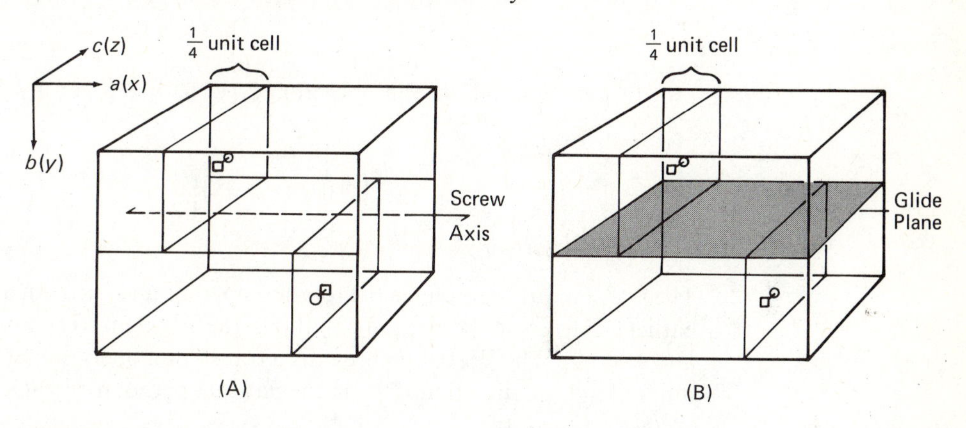

FIGURE 17-2. Transformation properties of a shape by (A) a two-fold screw axis (designated graphically by ⟵⟶) parallel to a and (B) a glide plane normal to b translating along a.

The two-fold screw axis in Fig. 17-2 lies along a at $y = \frac{1}{2}$ and $z = \frac{1}{2}$. It is given the symbol 2_1. The general designation of a screw operation N_m has N as the order of the proper rotation axis, and m (which can take on integer values of $m = 1, 2, \ldots, N - 1$) indicates a translation by m/N of the unit cell axis parallel to the rotation axis. The glide plane in Fig. 17-2 is referred to as an a-glide perpendicular to b at $y = -\frac{1}{2}$. All fractional designations refer to fractions of the unit cell length in that direction (*i.e.*, the x, y, and z measurements are taken along the a, b, and c directions, respectively).

Let us proceed to define some operations of the space group and present their symbolic representations. This presentation will not be complete but illustrative.

1. Two-Fold Proper Axis. Fig. 17-3(A) illustrates a two-fold proper axis parallel to b at $x = \frac{1}{4}$ and $z = 0$, which is commonly said to be at ($\frac{1}{4}$, 0, 0).* In dealing with lattices, all symmetry operations are described by the product of a point group type of operation with respect to the a, b, c unit cell axes multiplied by a translational operation. For example, the two-fold symmetry operation on the point s of Fig. 17-3(A) is symbolized as $2_b[\frac{1}{2}, 0, 0]$, where 2_b implies a two-fold rotation operation with respect to the b axis, and the brackets define translation to a, b, and c, respectively. Two-fold operators parallel to a and c would be labeled 2_a and 2_c, respectively. Fractional coordinate representations of a point are given in parentheses.

* On all diagrams, measurement is done relative to the origin at the upper left corner. The point to which an axis is displaced from the origin is that point on the perpendicular face of the unit cell through which the axis passes.

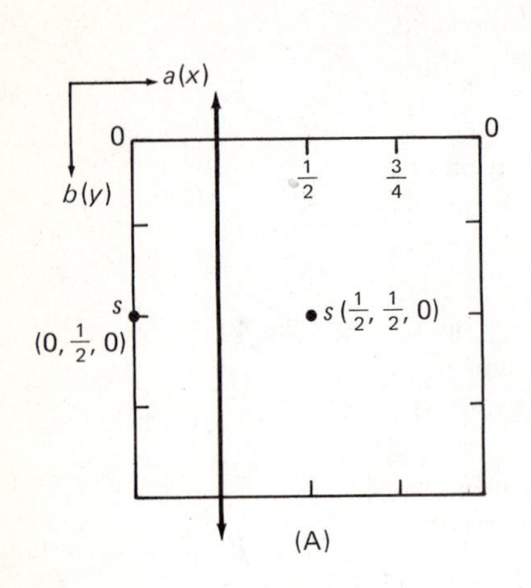

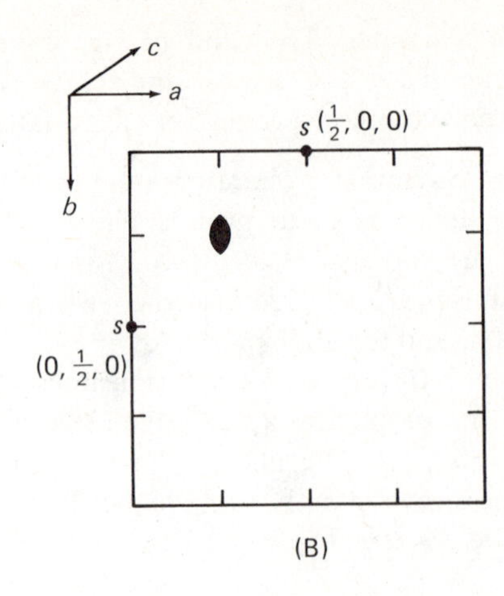

FIGURE 17–3. Transformation of a point by (A) a proper two-fold axis (⟷) parallel to b and (B) a proper two-fold axis parallel to c.

In matrix notation, the operation $2_b[\frac{1}{2}, 0, 0]$ is given by

$$\begin{pmatrix} -1 & 0 & 0 \\ 0 & 1 & 0 \\ 0 & 0 & -1 \end{pmatrix} \left[\frac{1}{2}, 0, 0\right]$$

The 3×3 matrix represents the rotation operation carried out at the origin. As in our earlier discussion of point group methods, the 3×3 matrix converts $x \to -x, y \to y$, and $z \to -z$. The $[\frac{1}{2}, 0, 0]$ represents a translation operation of half the length of the unit cell in the a direction. The mathematical operation of translation of a point by t_1, t_2, t_3 is defined as

$$[t_1, t_2, t_3](x, y, z) = (t_1 + x, t_2 + y, t_3 + z)$$

The operation $2_b[\frac{1}{2}, 0, 0]$ on a point (x, y, z) involves first the 2_b operation (matrix multiplication) and then vector addition of $[\frac{1}{2}, 0, 0]$ (or vice versa, since these operations commute). The combination of the two operations, rotation at the origin and the perpendicular translation $[\frac{1}{2}, 0, 0]$, is equivalent to a two-fold rotation at $(\frac{1}{4}, 0, 0)$. To check this, consider a point $(0, \frac{1}{2}, 0)$. Observation indicates that the element drawn in Fig. 17–3(A) would convert this point to $(\frac{1}{2}, \frac{1}{2}, 0)$. Calculating the effect of another two-fold rotation on this resulting point yields

$$\left[\frac{1}{2}, 0, 0\right]\left(\frac{1}{2}, \frac{1}{2}, 0\right) \longrightarrow \left(1, \frac{1}{2}, 0\right) \equiv \left(0, \frac{1}{2}, 0\right)$$

and

$$\begin{pmatrix} -1 & 0 & 0 \\ 0 & 1 & 0 \\ 0 & 0 & -1 \end{pmatrix} \begin{pmatrix} 0 \\ \frac{1}{2} \\ 0 \end{pmatrix} = \begin{pmatrix} 0 \\ \frac{1}{2} \\ 0 \end{pmatrix}$$

thus regenerating the original point.

In the example shown in Fig. 17–3(B), we have a 2_c at $(\frac{1}{4}, \frac{1}{4}, 0)$ or $2_c[\frac{1}{2}, \frac{1}{2}, 0]$.

2. Screw Axes. As mentioned earlier, screw axes arise when the translation is performed in the same direction as a rotation axis; *e.g.*, $2_a[\frac{1}{2}, 0, 0]$ is defined as 2_1. If the screw axis is located at $y = \frac{1}{4}$ and $z = \frac{1}{4}$, then the operation is written as

$2_a[\frac{1}{2}, \frac{1}{2}, \frac{1}{2}]$. To effect this operation on any point, we carry out the two-fold rotation 2_a along the a-axis, *i.e.*, at the origin ($y = 0, z = 0$), and follow this by the translation (in brackets) of the result of the 2_a operation. This screw axis is illustrated in Fig. 17–4.

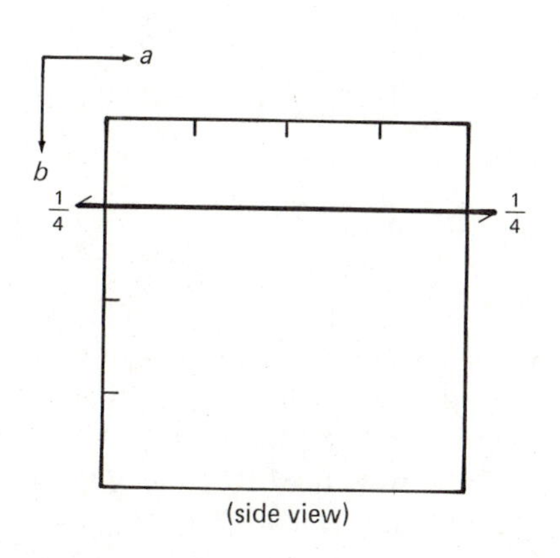

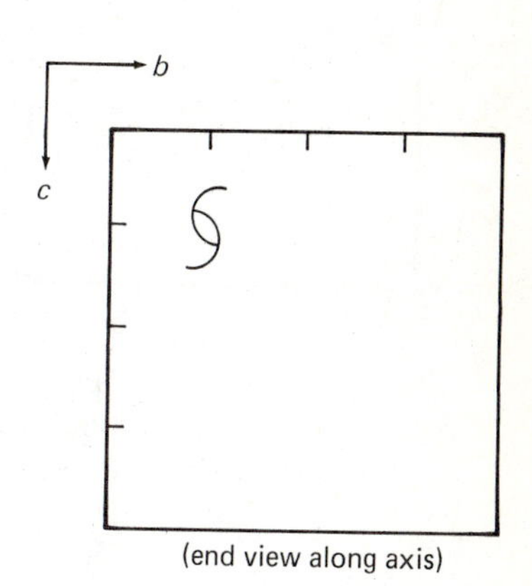

FIGURE 17–4. Alternative drawings showing a two-fold screw axis parallel to a and located on the A-face at $y = \frac{1}{4}$, $z = \frac{1}{4}$. The $\frac{1}{4}$ indicators on the left-hand figure signify placement of the element $\frac{1}{4}$ cell length behind the plane of the paper. The right-hand figure is the view from the left end of the left-hand figure.

In point groups, the presence of two operators (such as two perpendicular two-fold axes) may require that a third operator also exist. The same is true in some space groups. For example, a two-fold proper axis along a at $(0, 0, \frac{1}{4})$ and a two-fold proper axis along b at $(\frac{1}{4}, 0, 0)$ generate a two-fold screw axis 2_1 along c at $x = \frac{1}{4}$ and $y = 0$ (at $(\frac{1}{4}, 0, 0)$). We can show this as

$$2_a\left[0, 0, \frac{1}{2}\right] \cdot 2_b\left[\frac{1}{2}, 0, 0\right] = 2_c\left[\frac{1}{2}, 0, \frac{1}{2}\right]$$

This type of multiplication should be performed by simply adding the translation operator components and performing matrix multiplication of the rotation operators (as in point group procedures):

$$\text{translations} \quad \left[0, 0, \frac{1}{2}\right] \cdot \left[\frac{1}{2}, 0, 0\right] = \left[\frac{1}{2}, 0, \frac{1}{2}\right]$$

$$\text{rotations} \quad \underset{2_a}{\begin{pmatrix} 1 & 0 & 0 \\ 0 & -1 & 0 \\ 0 & 0 & -1 \end{pmatrix}} \cdot \underset{2_b}{\begin{pmatrix} -1 & 0 & 0 \\ 0 & 1 & 0 \\ 0 & 0 & -1 \end{pmatrix}} = \underset{2_c}{\begin{pmatrix} -1 & 0 & 0 \\ 0 & -1 & 0 \\ 0 & 0 & 1 \end{pmatrix}}$$

3. MIRROR PLANES. A mirror plane perpendicular to b, and intersecting b at $y = \frac{1}{4}$, may be represented by writing the reflection operator $m_b \equiv \begin{pmatrix} 1 & 0 & 0 \\ 0 & -1 & 0 \\ 0 & 0 & 1 \end{pmatrix}$ in combination with a translation normal to the plane, in this case $[0, \frac{1}{2}, 0]$.

4. GLIDE PLANES. A glide plane along a perpendicular to b at $y = \frac{1}{4}$ is designated by $m_b[\frac{1}{2}, \frac{1}{2}, 0]$ (see Fig. 17–5). Here m_b represents a mirror element perpendicular to the b-axis and at $b = 0$, and the brackets represent translation as with the screw operator. This element would be referred to as an a-glide, signifying translation in the a direction. Still keeping the plane perpendicular to b, it could be a c-glide by translating along c, or an n-glide by translating simultaneously one-half of the cell length along b and one-half along c.

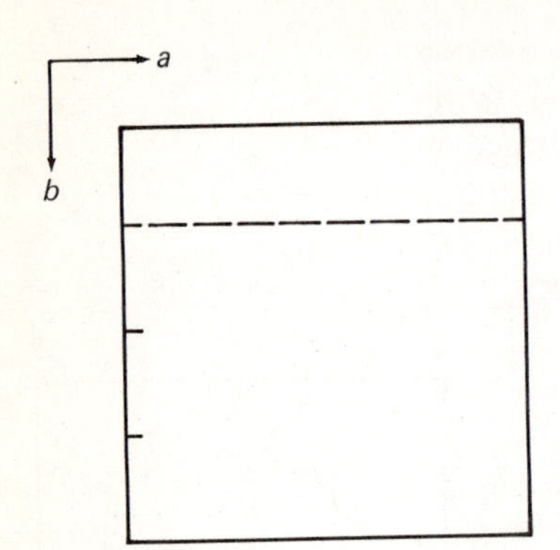

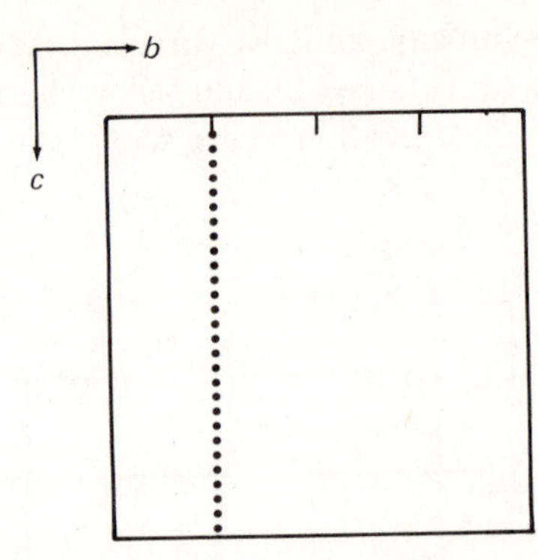

FIGURE 17–5. Two designations of an *a*-glide. The distinction between dashes and dots indicates translation *in* and *into* the plane of the paper, respectively.

5. CENTERED LATTICES. There is another operator in space groups, a centering operator, that has no point group analog. This operator gives rise to three general types of crystal lattices, which are termed face-centered (abbreviated *F*), side-centered (abbreviated *A*, *B*, or *C**), and body-centered (abbreviated *I*). The symmetry of these lattices can be described by pure translation operations, which involve only translations of one-half of a cell edge. For example, an *A* centered lattice must have for every point (x, y, z) an equivalent point at $(x, \frac{1}{2} + y, \frac{1}{2} + z)$. The *A* centering symmetry operator is thus $[0, \frac{1}{2}, \frac{1}{2}]$. Similarly, face centering (*i.e.*, *F* centering) requires that for every point (x, y, z), three additional equivalent points must be present at $(\frac{1}{2} + x, \frac{1}{2} + y, z)$, $(x, \frac{1}{2} + y, \frac{1}{2} + z)$, and $(\frac{1}{2} + x, y, \frac{1}{2} + z)$, so that the *F* centering operators are $[\frac{1}{2}, \frac{1}{2}, 0]$, $[0, \frac{1}{2}, \frac{1}{2}]$ and $[\frac{1}{2}, 0, \frac{1}{2}]$ (in addition to the identity operation $[0, 0, 0]$). The body centering operation is $[\frac{1}{2}, \frac{1}{2}, \frac{1}{2}]$. For illustrations of these types of lattices, see Fig. 17–25. Note that when an atom or molecule is located at the origin of the cell, for *A* centering another is found in the center of the *A* face (normal to *a*); for *F* centering there are three more units centered on each of the *A*, *B*, and *C* faces; and a similar special case may be pointed out for body centering.

Unit cells that are *not* centered are more commonly encountered in practice and are said to form *primitive lattices*. It should be emphasized that the operational definition of centering requires groups to be at a center (of, for example, a face) *only* when another is at the origin.

c. Space Groups. Space group symbols are written to uniquely identify with the smallest possible notation each one of the 230 different groups that may be formed from the various operators discussed. We shall illustrate the technique of notation as well as demonstrate the characterization of collections of operators as groups. Let us consider an orthorhombic example, which will bring out as well some operators not yet discussed.

The space group *Pnma* (read "*P-n-m-a*") contains in its notation the information that a lattice of this type is a *P*rimitive lattice, with an *n*-glide perpendicular to the *a* axis, a *mirror* perpendicular to the *b* axis, and an *a* glide perpendicular to the *c* axis. The conventions used to write symbols of this sort and the information conveyed by them are summarized in Table 17–1 (pp. 596–597). The seven different crystal systems are listed in the first column, along with the point group symmetries of the unit cell (*i.e.*, the symmetry they possess without any translations). The column labeled "characteristic symmetry" lists those essential symmetry elements that make the crystal unique to the point groups listed. The column labeled "position in point group symbol" gives the conventions for writing this symbol and lists the order (primary, secondary, tertiary) in which the symmetry elements are given in the symbol. In the example above, *Pnma*, *P* is the lattice symbol and *n*, *m*, and *a* are, respectively, primary, secondary, and tertiary symbols.

*Only three faces are distinct, taking into account the identity of the operators [0, 0, 0] and [1, 1, 1].

Now, what are the other elements generated by the three listed in the symbol *Pnma*, and where must these three be located in the cell? These questions may be answered as follows.

We know four operations of the group [including the identity operator $I \cdot (0, 0, 0)$]. These are written, leaving blanks where information is not yet at hand:

$$\begin{pmatrix} 1, 0, 0 \\ 0, 1, 0 \\ 0, 0, 1 \end{pmatrix} \cdot [0, 0, 0], \text{ the identity.}$$

$m_a[\_\_, ½, ½]$ *n*-glide,* *diagonal translation,* unknown position in x. (The x position is unknown because the final value will depend upon the location of this plane along the x-axis. From the definition of the *n*-glide plane, y and z will translate by ½ the unit cell.)

$m_b[0, \_\_, 0]$ mirror, unknown position in y.

$m_c[½, 0, \_\_]$ *a*-glide, unknown position in z.

The latter three operations, m_a, m_b, and m_c, may be multiplied in pairs:

$$m_a\left[\_\_, \frac{1}{2}, \frac{1}{2}\right] \cdot m_b[0, \_\_, 0] = 2_c\left[\_\_, \frac{1}{2} + \_\_, \frac{1}{2}\right]$$

$$m_b[0, \_\_, 0] \cdot m_c\left[\frac{1}{2}, 0, \_\_\right] = 2_a\left[\frac{1}{2}, \_\_, \_\_\right]$$

$$m_c\left[\frac{1}{2}, 0, \_\_\right] \cdot m_a\left[\_\_, \frac{1}{2}, \frac{1}{2}\right] = 2_b\left[\frac{1}{2} + \_\_, \frac{1}{2}, \frac{1}{2} + \_\_\right]$$

From point group theory, we know that $m_c \cdot 2_c$ generates a center of inversion, $\bar{1}$, and we will define this to be at the origin of the cell:

$$m_c\left[\frac{1}{2}, 0, \frac{1}{2}\right] \cdot 2_c\left[\frac{1}{2}, 0, \frac{1}{2}\right] = \bar{1}[0, 0, 0]$$

This is true because of the identity in a repeating system:

$$\frac{1}{2} + \frac{1}{2} = 1 \equiv 0$$

This enables us to fill in the blanks in all of the above equations, leading to:

$$m_a\left[\frac{1}{2}, \frac{1}{2}, \frac{1}{2}\right] \cdot m_b\left[0, \frac{1}{2}, 0\right] = 2_c\left[\frac{1}{2}, 0, \frac{1}{2}\right]$$

$$m_b\left[0, \frac{1}{2}, 0\right] \cdot m_c\left[\frac{1}{2}, 0, \frac{1}{2}\right] = 2_a\left[\frac{1}{2}, \frac{1}{2}, \frac{1}{2}\right]$$

$$m_c\left[\frac{1}{2}, 0, \frac{1}{2}\right] \cdot m_a\left[\frac{1}{2}, \frac{1}{2}, \frac{1}{2}\right] = 2_b\left[0, \frac{1}{2}, 0\right]$$

---

*m_a represents a mirror perpendicular to the $a =$ axis.

TABLE 17–1 CONVENTIONAL UNIT CELLS, THEIR SYMMETRIES AND SYMBOLS

Crystallographic System and Point Groups	Axial and Angular Relationships[1]	Characteristic Symmetry	Number of Bravais Lattices in System
Triclinic $1, \bar{1}$	$a \neq b \neq c$ $\alpha \neq \beta \neq \gamma \neq 90°$	1-fold (identity or inversion)[7] symmetry only	1
Monoclinic $2, m, 2/m$[8]	1st setting $a \neq b \neq c$ $\alpha = \beta = 90° \neq \gamma$	2-fold axis (rotation or inversion) in one direction only, this being taken as the z-axis in the 1st setting and as the y-axis in the 2nd setting	2
	2nd setting $a \neq b \neq c$ $\alpha = \gamma = 90° \neq \beta$		
Orthorhombic $222, mm2, mmm$	$a \neq b \neq c$ $\alpha = \beta = \gamma = 90°$	2-fold axes (rotation or inversion) in three mutually perpendicular directions.	4
Tetragonal $4, \bar{4}, 4/m$;[8] $422, 4mm, \bar{4}2m, 4/mmm$[8]	$a = b \neq c$ $\alpha = \beta = \gamma = 90°$	4-fold axis (rotation or inversion) along the z-axis.	2
Trigonal and Hexagonal $3, \bar{3}; 32, 3m, \bar{3}m; 6, \bar{6}/m; 622, 6mm, \bar{6}m2$	(Rhombohedral axes) $a = b = c$ $\alpha = \beta = \gamma < 120° \neq 90°$	3-fold axis (rotation or inversion) along the body diagonal using rhombohedral axes, or . . .	1
	(Hexagonal axes) $a = b \neq c$ $\alpha = \beta = 90°, \gamma = 120°$	along the z-direction using hexagonal axes.	1
Cubic $23, m3; 432, \bar{4}3m, m3m$	$a = b = c$ $\alpha = \beta = \gamma = 90°$	Four 3-fold axes, each inclined at 54° 44′ to the crystallographic axes.	3

[1] The sign $\neq$ implies non-equality by reason of symmetry; accidental equality may, of course, occur.

[2] When referring to lattices alone, it is conventional to call the side-centered orthorhombic lattice C. In the space groups of the point group $mm2$, the "z-axis unique" convention requires that the side-centered lattice shall sometimes be called C, and sometimes A (or B).

[3] The tetragonal lattices P and I may also be described as C and F, but only if the **a** and **b** vectors chosen are not the shortest ones perpendicular to **c**.

[4] The R-lattice is here described on rhombohedral axes, but it may also be referred to hexagonal axes. Where it is necessary to distinguish these, the symbols R_{obv} or R_{rev} are used for the description on rhombohedral axes and R_{hex} for that on hexagonal axes.

Lattice Symbols	Position in Point Group Symbol			Examples
	Primary	Secondary	Tertiary	
P	Only one symbol, which denotes all directions in the crystal.			$P1$, $P\bar{1}$
1st setting {*P*, *B* 2nd setting {*P*, *C*	The symbol gives the nature of the unique 2-fold axis[5] (rotation and/or inversion). 1st setting: *z*-axis unique[6] 2nd setting: *y*-axis unique			$P2_1/c$, $C2/c$
P *C*[2] *I* *F*	2-fold (rotation and/or inversion) along *x*-axis	2-fold (rotation and/or inversion) along *y*-axis	2-fold (rotation and/or inversion) along *z*-axis	*Pnma*, $P2_12_12_1$
P[3] *I*	4-fold (rotation and/or inversion) along *z*-axis	2-folds (rotation and/or inversion) along *x*- and *y*-axes	2-folds (rotation and/or inversion) along other special direction[9]	*P4bm*, *I4mm*
R[4] *P*	3- or 6-fold (rotation and/or inversion) along *z*-axis	2-folds (rotation and/or inversion) along *x*-, *y*-, and *u*-axes (*x*, *y*, and *u* in a plane 120° apart)	2-folds (rotation and/or inversion) normal to *x*-, *y*-, and *u*-axes in a specified plane[9]	$P3_221$, $R3c$ $P6_122$, $P6cc$
P *I* *F*	2-folds or 4-folds (rotation and/or inversion) along specified planes[9]	3-folds (rotation and/or inversion) along the body diagonal	3-folds (rotation and/or inversion) along specified directions[9]	*Pn*3, *Fm*3, *I*43*m*

[5] The monoclinic symmetry elements are a 2-fold rotation axis and/or a mirror plane (normal to that axis when both are present).

[6] The 2nd setting is usual for crystallographic work.

[7] Inversion axis in all cases implies the presence of a mirror perpendicular to a rotation axis (an improper axis in point groups). As seen for orthogonal systems, the space group is often named using the plane designations.

[8] The slash denotes a mirror or glide plane perpendicular to the designated rotation axis.

[9] Specifications in each case may be found in reference 3—they are not important for our purposes.

The operations we have found for *Pnma* are thus:

$$1(0, 0, 0)$$

$$\bar{1}(0, 0, 0)$$

$$m_a\left[\frac{1}{2}, \frac{1}{2}, \frac{1}{2}\right] \quad 2_a\left[\frac{1}{2}, \frac{1}{2}, \frac{1}{2}\right]$$

$$m_b\left[0, \frac{1}{2}, 0\right] \quad 2_b\left[0, \frac{1}{2}, 0\right]$$

$$m_c\left[\frac{1}{2}, 0, \frac{1}{2}\right] \quad 2_c\left[\frac{1}{2}, 0, \frac{1}{2}\right]$$

It is left to the reader to show that these eight operations comprise a mathematical group; we shall refer to this as a *space group*.

Let us examine the transformation properties of the point (x, y, z) in *Pnma*. Corresponding sequentially to the eight symmetry elements listed above are:

$$x, y, z$$

$$\bar{x}, \bar{y}, \bar{z}$$

$$\left[\frac{1}{2} - x, \frac{1}{2} + y, \frac{1}{2} + z\right] \quad \left[\frac{1}{2} + x, \frac{1}{2} - y, \frac{1}{2} - z\right]$$

$$\left[x, \frac{1}{2} - y, z\right] \quad \left[\bar{x}, \frac{1}{2} + y, \bar{z}\right]$$

$$\left[\frac{1}{2} + x, y, \frac{1}{2} - z\right] \quad \left[\frac{1}{2} - x, \bar{y}, \frac{1}{2} + z\right]$$

(Note that $\bar{x}$, $\bar{y}$, and $\bar{z}$ in crystallographic notation represent $-x$, $-y$, and $-z$, respectively.) Placing an atom at any point (x, y, z) in the unit cell generates atoms at the seven other points listed. Likewise, for any molecule there are seven other molecules in the cell that are symmetrically located with respect to these seven other points. To specify the entire contents of the cell, one need only list one-eighth of the points.

Consider the plane where $y = \frac{1}{4}$ (*i.e.*, x and z have all possible values). Now we find that for the eight operations found in *Pnma*, there are only *four* different symmetry-related points in this plane. If there were only four molecules per unit cell with *Pnma* symmetry, then they might be sitting at

$$\left(x, \frac{1}{4}, z\right), \left(\frac{1}{2} - x, \frac{3}{4}, \frac{1}{2} + z\right), \left(\frac{1}{2} + x, \frac{1}{4}, \frac{1}{2} - z\right), \left(\bar{x}, \frac{3}{4}, \bar{z}\right)$$

(These are obtained by substituting $y = \frac{1}{4}$ into the general equations given above.) These points are the *special positions* with m symmetry. This is possible only (although see p. 624) if the molecular point group contains a mirror plane, and if this plane is coincidental with that of the unit cell.

The general positions and special positions and their point symmetry are listed for all space groups in Volume I of "International Tables for X-Ray Crystallography."[7] Upon acquisition of unit cell data, it therefore becomes straightforward to determine possible molecular symmetries (if and only if the molecule resides on special positions, and the space group is uniquely determined).

As an example of this procedure, let us consider the titanocene system, $(C_5H_5)_2Ti$. There has been much controversy as to the proposed geometry of this species, since theoretical considerations would tend to suggest a bent structure, while the parallel ring ferrocene-type configuration is certainly possible. It is found that

$(C_5H_5)_2Ti$ exists only in dimeric form, and thus the question is valid only for the more recently prepared permethylated system, $(C_5Me_5)_2Ti$. This species is found to be monomeric in solution and, when isolated as a crystalline solid, has two molecules per unit cell of $P2_1/c$ symmetry.[5] In this group, a general position generates four molecules per unit cell, whereas there are two special positions with $\bar{1}$ symmetry. It is clear that in order for $(C_5Me_5)_2Ti$ to sit on a site of $\bar{1}$ symmetry, the molecular structure must have a center of inversion and, therefore, one cyclopentadienyl ring will generate the second, parallel one. Since crystals of this material decompose slowly at room temperature during exposure to X-rays, precise X-ray intensity data are difficult to obtain; however, limited sets of data have been consistent with the suggestion made considering only the site symmetry.

We shall continue our discussion of common symmetry properties with the space group $P2_1/c$ (read "*P*-two-one-on-*c*" or "*P*-two-sub-one-on-*c*"). The point group corresponding to this space group is obtained by converting the space group symmetry elements to their point group equivalents. For example, in this case, 2_1 is derived from a two-fold rotation axis and *c* from a mirror plane. Thus, the point group of this space group is $2/m$. Referring to column 1 of Table 17–1, we see that this must be the monoclinic system. This is one of the most frequently encountered space groups. Its operations are contained in the name, and reference to Table 17–1 will enable one to discover what they are. It is indicated that the primitive lattice has a two-fold screw axis with a perpendicular *c*-glide. The following relationship holds:

$$2_b\left[\_\_,\frac{1}{2},\_\_\right]\cdot m_b\left[0,\_\_,\frac{1}{2}\right] \equiv \bar{1}[0,0,0]$$

With the inversion taken to be at the origin, we obtain

$$2_b\left[0,\frac{1}{2},\frac{1}{2}\right]\cdot m_b\left[0,\frac{1}{2},\frac{1}{2}\right] = \bar{1}[0,0,0]$$

For monoclinic systems, the *b*-axis is generally taken as the unique axis, which is the two-fold axis (proper or improper).

In the triclinic crystal system where $a \neq b \neq c$ and $\alpha \neq \beta \neq \gamma$, the only operators present are the identity element and possibly a center of inversion. The only two unique triclinic space groups are $P1$ and $P\bar{1}$ (read "*P*-one" and "*P*-one-bar"), the latter possessing an inversion center.

d. Unit Cell Diagrams. The symmetry elements in the $P2_1/c$ space group are shown diagrammatically in Fig. 17–6. The dark dots are inversion centers. The

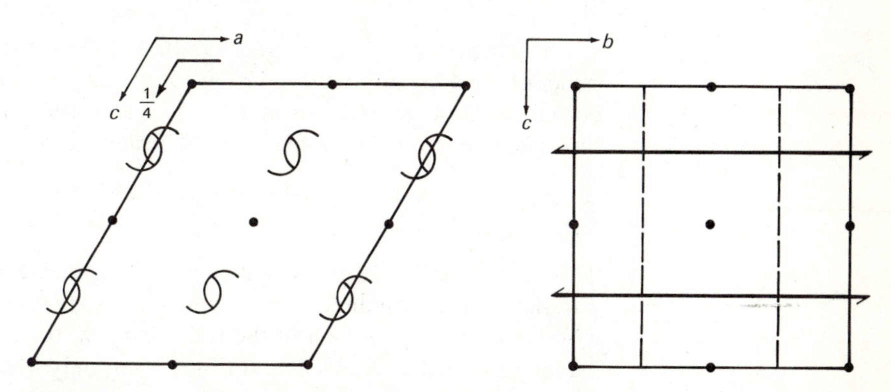

FIGURE 17–6. Symbolic representations of the symmetry elements contained in the space group $P2_1/c$.

inversion centers that are not at the corners of the unit cell arise because the combination of a center of inversion operation at the origin and a translational operation from one lattice point to another gives rise to a center of inversion halfway between the two lattice points. This is also true for two-fold and mirror plane operations. The § symbols represent 2_1 axes going into the paper, ⟷ indicates a 2_1 axis in the plane of the paper, and $\frac{1}{4}$ indicates a glide parallel to the paper but $\frac{1}{4}$ cell along the perpendicular axis (*b*) and translating in the direction of the arrow. Dashed lines represent a glide perpendicular to the paper and traveling in a direction parallel to the dash. Such drawings are given in "International Tables." There are more than one of each of the four symmetry elements in the cell because of the applications of the unit cell translations, [1, 0, 0], [0, 1, 0], [0, 0, 1], and so forth. For example, $2_b[0, \frac{1}{2}, \frac{1}{2}] \cdot [1, 0, 0]$ equals $2_b[1, \frac{1}{2}, \frac{1}{2}]$, which is 2_1 at $x = \frac{1}{2}$ and $z = \frac{1}{4}$.

For the space group *Pnma* we can draw a symmetry diagram from the list of operators given earlier. There are three planes and three two-fold axes as well as an inversion center. The three operators involving mirror planes are *n*-glide at $x = \frac{1}{4}$, a mirror at $y = \frac{1}{4}$, and an *a*-glide at $z = \frac{1}{4}$. The axes are a 2_1 along *a* at $y = z = \frac{1}{4}$, a 2_1 along *b* at the origin, and a 2_1 along *c* at $x = \frac{1}{4}$, $y = 0$. Fig. 17-7 shows all symmetry elements of the unit cell, generated by the eight operators given. The -·-·-· line represents the *n*-glide, moving diagonally (the direction bisects *b* and *c*), and the centers of inversion are all projected onto the front face of the cell, although it can be seen that the one appearing at ($\frac{1}{2}$, 0, $\underline{0}$) is related to that at the origin by a screw axis (at $x = \frac{1}{4}$, $b = 0$) and is therefore located at $z = \frac{1}{2}$.

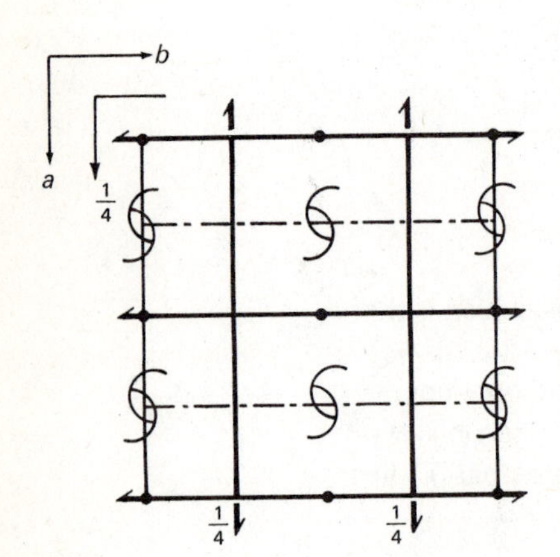

FIGURE 17-7. Symbolic representation of the symmetry elements of *Pnma*.

17-3 X-RAY DIFFRACTION—DIRECTION

Having provided some understanding of the symmetry properties of the internal structure of a crystal, we need now to analyze the interaction of X-rays with this crystal. To do this, we shall make use of a relationship commonly encountered in introductory chemistry courses, called Bragg's law:

$$\sin\theta = \frac{n\lambda}{2}\left(\frac{1}{d}\right) \qquad (17\text{-}2)$$

This mathematical expression means that if an electromagnetic wave strikes two parallel planes at an angle θ, where the planes are separated by a distance d (see Fig. 17-8), and the wavelength of the radiation is λ, the constructive interference of the waves generated at points O and C will occur only when θ is adjusted so that Bragg's law is satisfied (n being integral). The in-phase waves appear to be "reflected" at an angle θ.

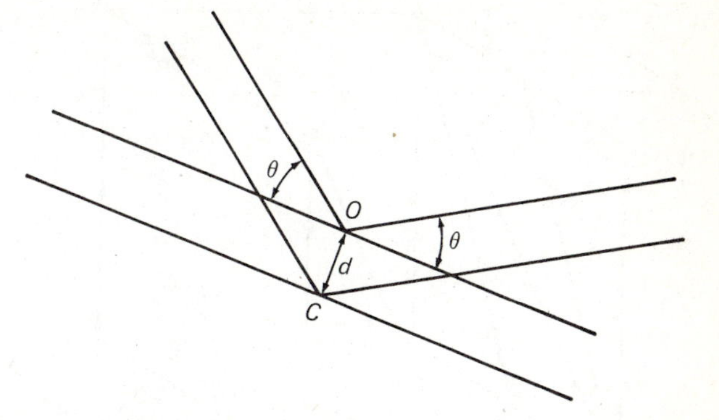

FIGURE 17–8. Reflecting planes and incident and departing wave directions illustrating the geometry of Bragg's law.

This concept can be applied to crystal diffraction because the crystal lattice can be described by sets of parallel planes with various spacings, *d*. When an X-ray beam strikes any set of planes at an angle at which Bragg's law is satisfied, then a single secondary beam will emerge from the crystal. It is in fact found that when a single crystal of a material is bathed in a strong X-ray beam, many thousands of weaker beams or reflections are ejected in various directions from the crystal, as shown in Fig. 17–9. The angle of each reflected beam to the direct beam is defined by a particular spacing in the scattering set of planes.

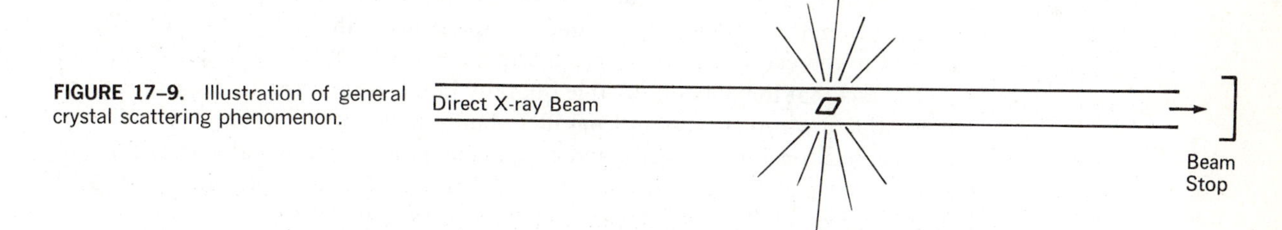

FIGURE 17–9. Illustration of general crystal scattering phenomenon.

How are these sets of planes described in terms of the unit cell? One way is consider them to be defined by points of equal electron density in the cell; in that the lattice is determined by the symmetry of the distribution of electron density, the planes may be defined by the lattice in an equivalent manner. Consider the two-dimensional lattices and the families of planes shown in Fig. 17–10. All possible sets of planes may be given indices, called Miller indices, that define them uniquely, based upon the number of parts into which the planes divide the unit cell edges. Fig. 17–10(A) shows planes dividing *a* into two parts and *c* into one part. Its label would be (2, 1) in our two-dimensional scheme. Fig. 17–10(B) illustrates

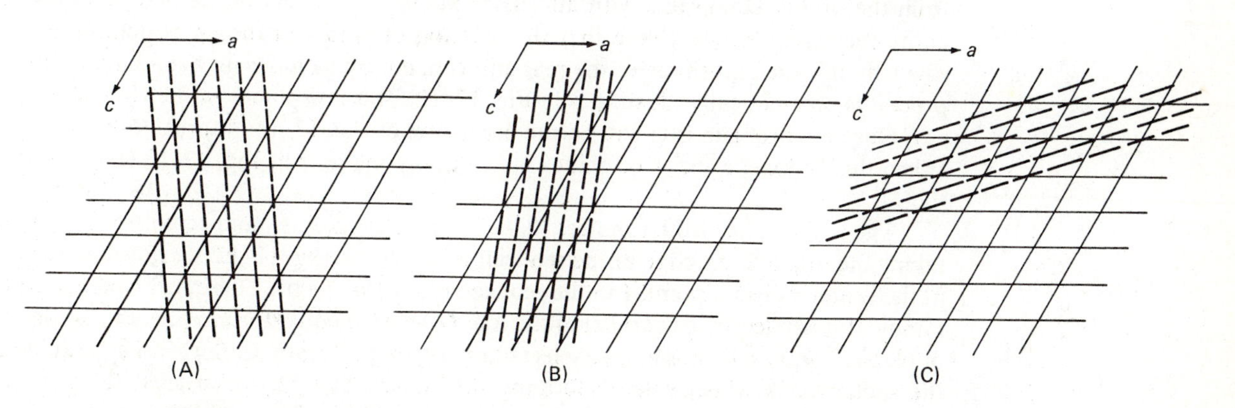

FIGURE 17–10. Families of planes through lattice points as viewed down the *b*-axis of a crystal.

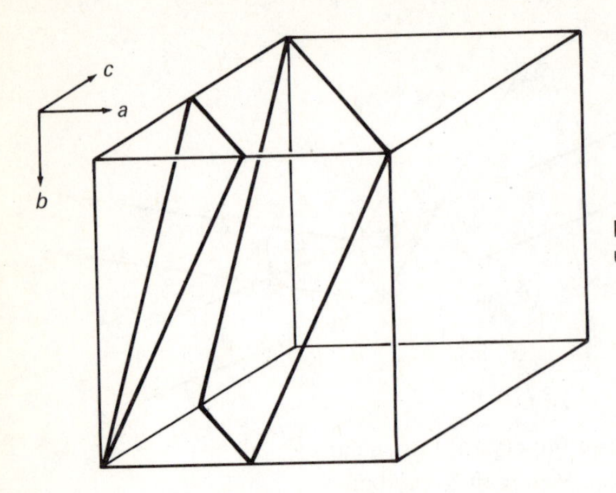

FIGURE 17–11. Sections of the (2, 1, 2) planes drawn in an arbitrary unit cell.

(3, 1), while part (C) shows (1, 2). The three-dimensional until cell in Fig. 17–11 shows segments of the first planes $(h, k, \ell) = (2, 1, 2)$ of a parallel infinite set where h, k, and ℓ correspond to the a, b, and c axes. This *set* of planes will lead to *one* diffracted beam when the direct beam makes an angle θ such that $\sin \theta = n\lambda/2(1/d_{212})$. It should be stated at this time that although X-rays are scattered by electrons, which are distributed throughout the unit cell and not just in the planes we have defined, we will later see that scattering from the contents of the cell can *always* be expressed in terms of scattering from the sets of planes defined by the *shape* of the cell. The distribution of electrons in the cell determines only the relative *intensities* of the reflected beams.

The fact that the only observable quantity in the X-ray experiment is one beam representing a complete *series* of parallel planes, in conjunction with the awkward reciprocal relationship between θ and $d_{hk\ell}$, makes it desirable to find a description of the lattice in which θ is *directly* related to a distance and each set of planes in the real lattice is represented by a point in a new *reciprocal lattice* (r.l.).

Referring to Bragg's law, we see that $\sin \theta$, the deviation between the incident and reflected radiation, is inversely proportional to d, the interplanar spacing in the crystal lattice. Structures with large d will exhibit compressed diffraction patterns, and those with small d will show expanded patterns. If the inverse relation between $\sin \theta$ and d could be replaced by a direct relation, the interpretation of diffraction patterns would be simplified. This is achieved by the construct of the reciprocal lattice.

The point selected as the origin of the r.l. coincides with one in the real lattice. The point in the r.l. corresponding to a set of planes in the real lattice is found by starting at the origin and drawing the normal to the Miller planes $hk\ell$ that are nearest to, but do not pass through, the origin. The normal is terminated a distance $1/d_{hk\ell}$ from the origin. Doing this with all sets of planes in the real lattice generates the entire reciprocal lattice. (Note that the location of points in the r.l. is determined solely by the size and shape of the real unit cell, not by its molecular contents.) This process is schematically illustrated in Fig. 17–12 for a few planes in the real lattice.

Our next problem is to extend the discussion of Fig. 17–12 to three dimensions. When this is done, a series of planes indicated by the dots in Fig. 17–13 results for the r.l.

The real cell has Miller indices $(0, k, \ell)$, $(1, k, \ell)$, $(2, k, \ell)$, and so on. Instead of taking the origin at an edge of the real cell as we did in Fig. 17–12, we now locate it in the center. Select a point A in the real cell along the a-axis. There will be a whole family of possible planes parallel to the line OA with h equal to zero (i.e., the family of planes $0, k, \ell$). The normals to selected planes in this family from O will appear as the spokes on a wheel with O forming the center. The r.l. constructed from the normals to these planes will all lie in the $0, k, \ell$ reciprocal lattice plane illustrated in

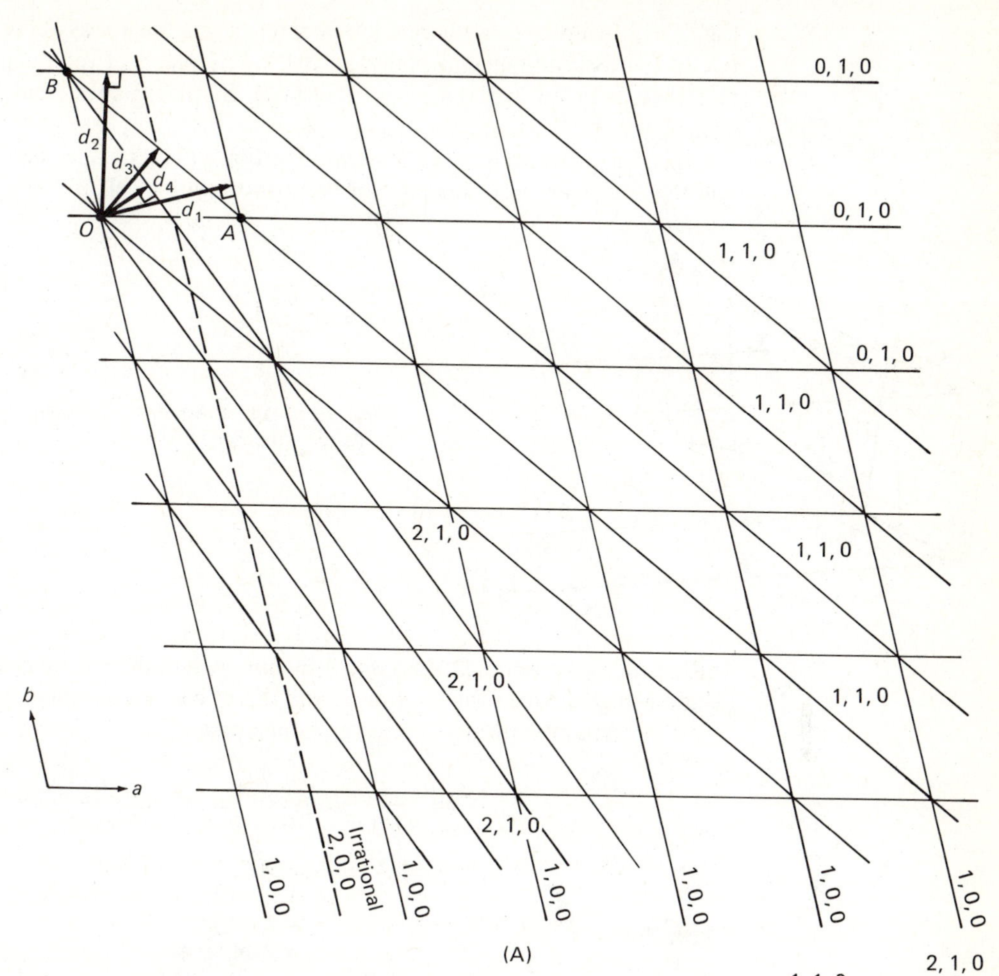

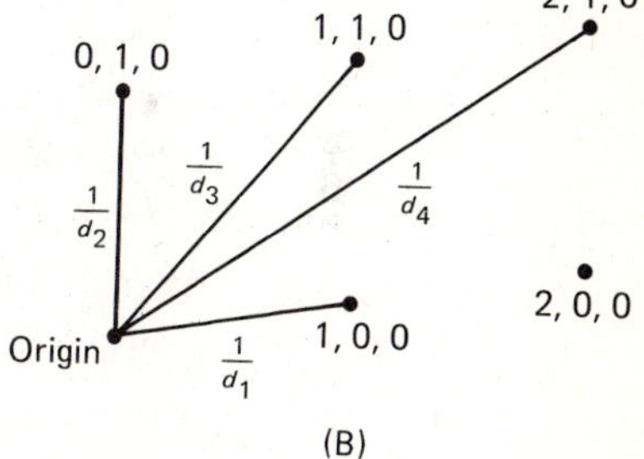

FIGURE 17–12. The relation of some real lattice points to the reciprocal lattice. *A* and *B* are real lattice points that, with the origin *O*, define the unit cell. This is a plane at $c = 0$. (A) Real lattice points with arrows from the origin to the Miller planes; (B) the reciprocal lattice, which is a plot of the reciprocals of the distances indicated by the arrows in (A). The 200 plane does not pass through any lattice points, and corresponds to second-order Bragg diffraction; *i.e.*, $2/d_{100}$ is defined as the 200 irrational plane and n/d_{100} as the n00 irrational plane.

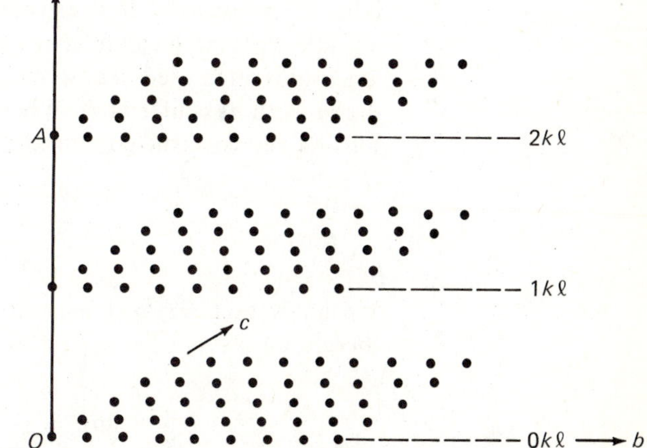

FIGURE 17–13. Diagrammatic construction of the reciprocal lattice in three dimensions.

Fig. 17-13. All planes in the real lattice intersecting the a-axis at the point A will give rise to points (one per real plane) in the 1, k, ℓ plane of the reciprocal lattice (see reference 3 for an explanation of this fact). Similar planes result for other values of h.

The reciprocal cell edges illustrated in Fig. 17-14 are specified by a^*, b^*, and c^*, and the angles are also starred. For the orthorhombic cell, a and a^*, b and b^*, and c

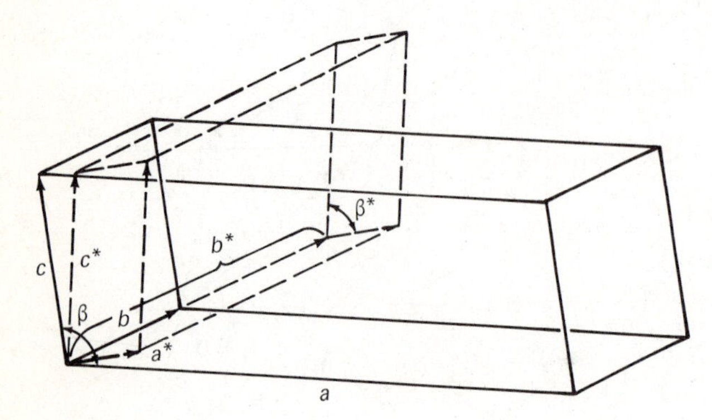

FIGURE 17-14. Relationship between real and reciprocal unit cells for a monoclinic system.

and c^* all correspond in *direction* but not in length. For a monoclinic cell, the relationships between the reciprocal and direct cell axes depend upon β, except for $b = 1/b^*$. Mathematically, these relationships are:

$$a^* = \frac{1}{a \sin \beta}, \quad b^* = \frac{1}{b}, \quad c^* = \frac{1}{c \sin \beta} \tag{17-3}$$

$$\alpha = \gamma = \alpha^* = \gamma^* = 90°, \quad \beta^* = 180° - \beta$$

The volume is

$$V = \frac{1}{V^*} = abc \sin \beta \tag{17-4}$$

where V^* is the volume of the reciprocal cell. Note that each r.l. axis is perpendicular to a face of the direct cell; *e.g.*, a^* is perpendicular to the bc plane, and the b^*c^* plane has a as a normal. Therefore, each direct axis is normal to a set of reciprocal lattice planes; *e.g.*, a is normal to $0k\ell$, $1k\ell$, $2k\ell$, and so forth. In vector notation, we may write, for example, $\vec{\mathbf{a}}^* = (\vec{\mathbf{b}} \times \vec{\mathbf{c}})/V$. (Note that $\vec{\mathbf{a}} = (\vec{\mathbf{b}}^* \times \vec{\mathbf{c}}^*)/V$.)

How can we describe the condition for diffraction in terms of the reciprocal lattice? This is easily done as shown in Fig. 17-15, which illustrates a monoclinic r.l. net $h0\ell$ with the incident X-ray beam passing through the origin (labeled O). The term *net* is often used to describe a planar array of r.l. points. A circle of radius $1/\lambda$ is drawn with its center at D. When the crystal is rotated about O so that the point P falls on the constructed circle, it can be seen geometrically that

$$\sin \theta = \frac{OP}{2/\lambda} \tag{17-5}$$

We know that OP is $1/d_{hk\ell}$ from the construction of the reciprocal lattice, and therefore

$$\sin \theta = \frac{\lambda}{2d_{hk\ell}} \quad \text{or} \quad 2d_{kh\ell} \sin \theta = \lambda \tag{17-6}$$

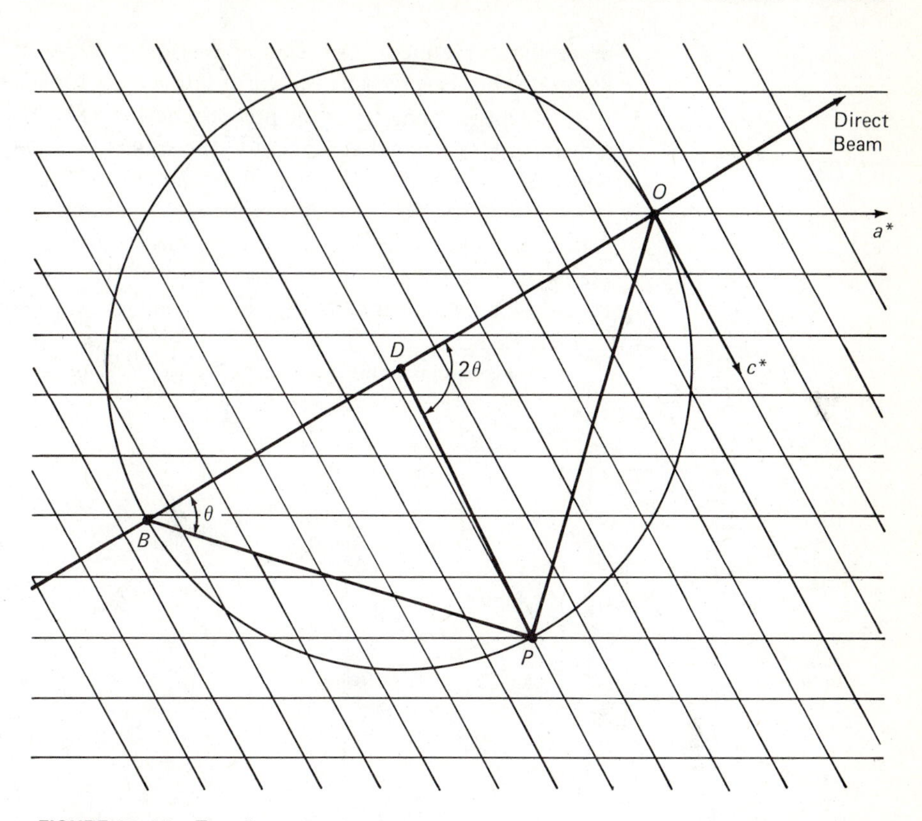

FIGURE 17–15. Two-dimensional representation of the construction of the sphere of reflection and its relationship to Bragg's law.

thus fulfilling Bragg's law. The set of real planes represented by the point P is perpendicular to OP, and it is easily shown that the diffracted beam has the direction indicated by the segment DP.

Expanding this discussion to three dimensions, it may be said that any reciprocal lattice point falling on the *sphere of reflection* (defined by the wavelength and direction of the direct beam and the unit cell origin) will (in principle) give rise to a diffracted beam leaving the crystal in the direction determined by the center of the sphere and the point of intersection of the r.l. point with the sphere. An immediate practical result of these concepts is the obvious point that as λ is decreased (*i.e.*, higher energy X-rays are used), the sphere enlarges and *more* reflections are observed as the reciprocal lattice is rotated through the sphere. Note that the r.l. rotates with the crystal about the origin, which is on the *surface* of the sphere of reflection, *not* at the center. Thus, more data may be collected for a given crystal system. In fact, it is found that the number of possible reflections, N, is

$$N = \frac{(4/3)\pi(2/\lambda)^3}{V^*} \tag{17-7}$$

17–4 PRACTICAL APPLICATION—LEVEL I

a. Single Crystal Photography. There are several methods that may be used either individually or together to obtain photographic data from single crystals. The general utility and complexity of these methods increase as the cameras are constructed to mechanically sort the thousands of beams diffracted from a crystal into more highly refined sets. The precession method will be described in order to illustrate the uses of the reciprocal lattice developed in the last section.*

The basic geometry of the precession camera is shown in Fig. 17–16. The plane of a reciprocal lattice net is tangent to the sphere of reflection at O, which is the origin and, macroscopically, the location of the crystal. The crystal is mounted on an axis perpendicular to the direct beam (usually by gluing it to a glass fiber). Since an r.l. net is tangent to the sphere, a direct axis must coincide with the direct beam. Other r.l. planes will be parallel to the one shown; if the a-axis is coincident with the beam, then the r.l. net $0k\ell$ touches the sphere only at O, whereas $1k\ell$, $2k\ell$, . . . intersect the

*For a grounding in the *application* of camera techniques, see reference 3.

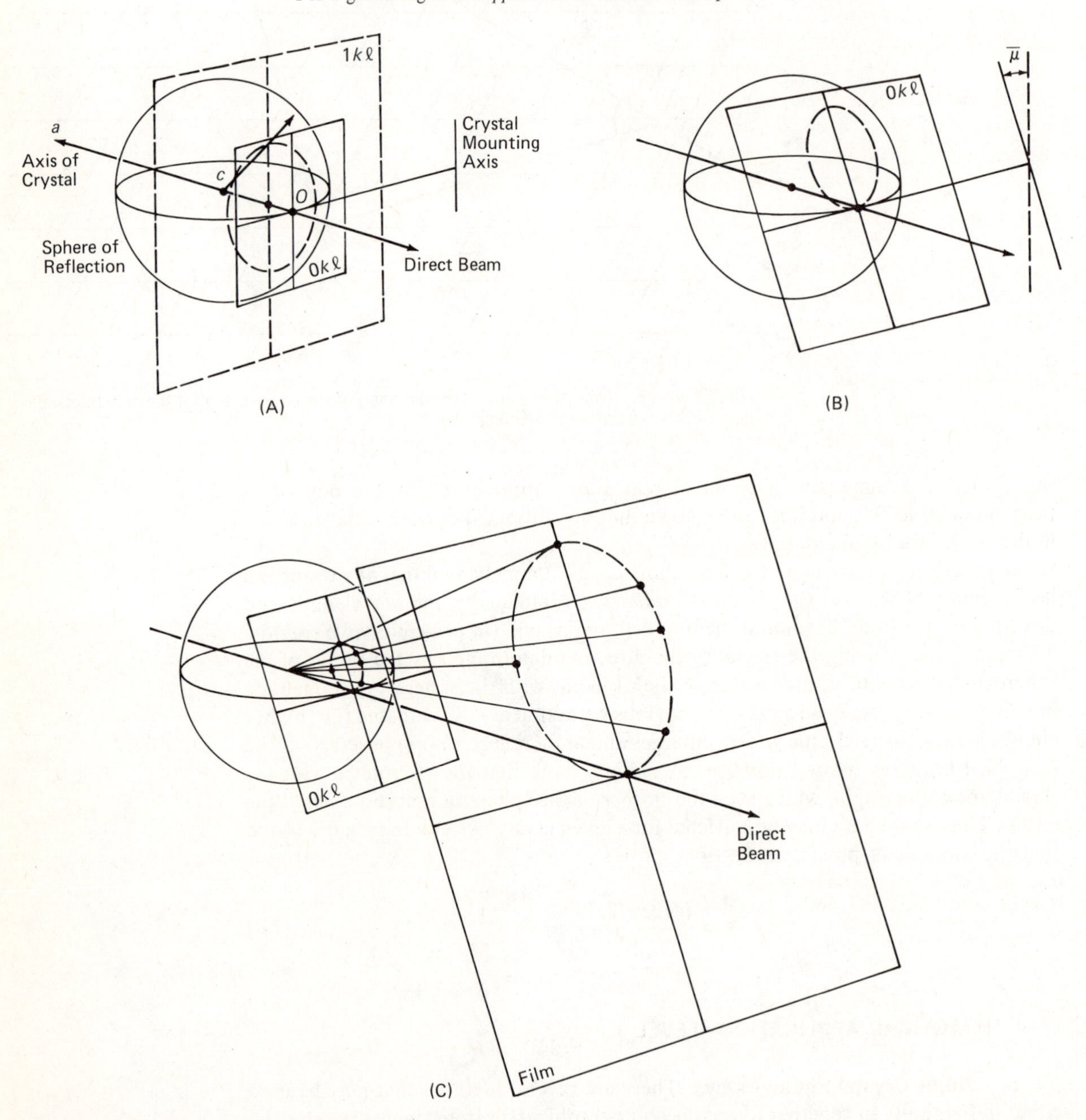

FIGURE 17–16. Geometrical relationships in a precession camera. (A) With the zero-level net tangent to the sphere of reflection ($\bar{\mu} = 0°$). (B) With the zero-level net tilted into the sphere of reflection. (C) Showing scattering from certain zero-level r.l. points and emerging X-rays striking film.

sphere in circles. Where points in these planes touch the sphere, X-ray beams directed from the center of the sphere will arise. It is not until the crystal is rotated about its mount axis that the zero-level net will give rise to reflections [see Fig. 17–16(B)]. The camera may be adjusted so that the circular intersection of the zero-level net with the sphere of reflection travels with the motion of the camera about the direct beam, keeping the angle between the r.l. net and the direct beam constant. Thus, the "circle of reflection" moves around the r.l. origin, bringing each r.l. point (limited only by the precession angle $\bar{\mu}$) into reflecting position. If film is placed so that it is always parallel to the r.l. net while the camera is in motion, then the film will record a display of dots that is an undistorted representation of the r.l. net from which the dots arose.

The only further point that needs to be mentioned is that a cone of reflected beams will arise from each of the levels $0k\ell$, $1k\ell$, $2k\ell$, . . . ; but because the size of the cone is different in each case, they can be separated by a screen placed between the crystal and the film. This screen is a metal plate in which is cut a ring 2 to 5 mm wide, with a radius that is calculated to pass only the reflections arising from a reflection circle of the correct size. The size (radius) of the ring may be adjusted so that only reflections arising from a desired $nk\ell$ net are observed. In practice, n is 0, 1, or 2 (rarely larger), and the center of the screen is held in place by cellophane tape.

Assuming that the crystal of interest is mounted on a reciprocal axis, say the b^* of a monoclinic system, what procedure must we follow in order to determine the unit cell and space group parameters? First we rotate the crystal until, say, a^* is perpendicular to the beam, and thus the planes hkn are in the reflecting position for n (depending upon the screen chosen). The alignment of this axis can be accomplished through normal procedures to an accuracy of ± 0.05 degree. Now, marking the angle where the $hk0$ zone was found, rotate further to find the $0k\ell$ net. The angle between these two locations, if smaller than 90°, is β^* (β is defined as larger than 90°). Now, if the $hk0$ and $0k\ell$ zero-levels are recorded on film as rectangular patterns of dots, then the spacing along the axes of the photographs should allow calculation (*via* formulae representing the geometry of the camera) of a^*, b^*, and c^*. The axis b^* may be found in both exposures horizontally spread onto the film, while the vertical axes are a^* and c^* for the $hk0$ and $0k\ell$ levels, respectively. These calculations provide us with all of the information required to determine the volume and shape of the cell. Fig. 17–17

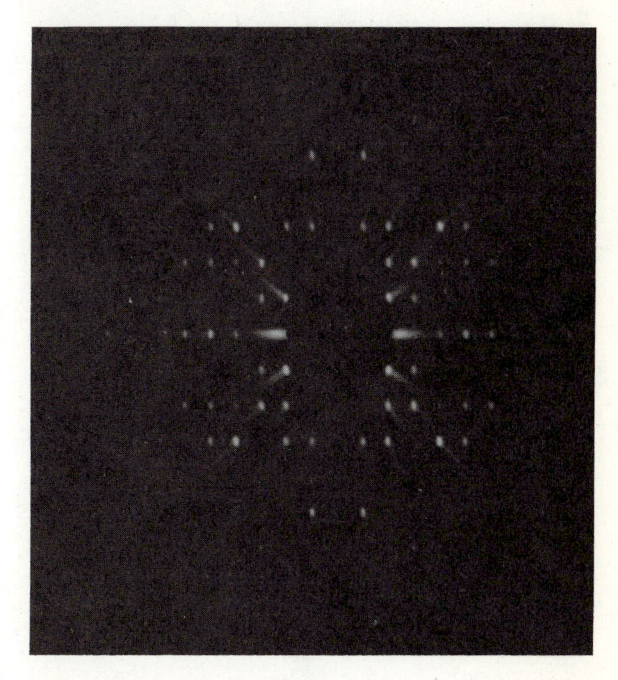

FIGURE 17–17. Precession photograph showing the $hk0$ net of $[Cu_2(tren)_2(CN)_2](BPh_4)_2$. The a^* axis is horizontal (crystal mounting axis), and the b^* axis is vertical.

shows the $hk0$ net of $[Cu_2(tren)_2(CN)_2](BPh_4)_2$ as displayed by precession photography (tren = 2,2′,2″-triaminotriethylamine).[6]

It is possible, through the use of auto-indexing programs, to determine the unit cell axes and angles without obtaining a complete set of photographs in which each zero-level net ($0k\ell$, $h0\ell$, and $hk0$) is seen. Such a program uses as input the locations of a number of random reflections (determined by either film or diffractometer techniques) and finds all real axes such that the indices of all reflections to a particular axis are integral. The angles between all pairs of possible axes are listed and the lengths in angstroms are given. Those that are 90° apart, or that have similar lengths, may be collected into a group of three that form the best description of the unit cell.

It should be obvious that the location of reflected beams on a precession photograph is characteristic of *only* the shape and size of the unit cell. This determines the spacing of reciprocal lattice points and therefore also, *via* the geometrical arrangement of the camera, the distance between dots on the photograph. In the next section, a quantitative analysis of what factors lead to variation of the *intensities* of the recorded dots will be presented. We have, however, already referred to the fact that the intensities are a function of the precise distribution of electron density within the unit cell. *Since elements of symmetry within a unit cell require the contents of the cell to be organized in specialized ways, certain intensity patterns observable in precession photographs are characteristic of these elements of symmetry.* We shall prove this statement later. For now, we shall state that for each symmetry element in the unit cell *involving translational motion* (or centering), a distinct pattern of *systematic absences of reflections* will arise in appropriate photographs. For example, for $P2_1/c$, every odd reflection of the axis $0k0$ is completely absent. This indicates that there is a two-fold screw axis parallel to b. Also, in the $0k\ell$ zone, alternate rows of reflections are missing perpendicularly to the 00ℓ axis. This is characteristic of a c-glide. Table 17–2 lists some commonly encountered elements and their absences.

In practice, three points become clear. (1) Once the patterns are memorized, the absences and their respective symmetry elements are quickly recognized in a set of photographs. (2) It is necessary to look at both the $0k\ell$ and $1k\ell$ nets in order to

TABLE 17–2. SOME COMMONLY ENCOUNTERED SYMMETRY ELEMENTS AND THE RESULTING REFLECTION ABSENCES

Symmetry Element Present		Reflections From:	Absent If:
2-fold screw	along a	$h00$	$h = 2n + 1$
	along b	$0k0$	$k = 2n + 1$
	along c	00ℓ	$\ell = 2n + 1$
glide ⊥ to a	b-glide	$0k\ell$	$k = 2n + 1$
	c-glide	$0k\ell$	$\ell = 2n + 1$
	n-glide	$0k\ell$	$k + \ell = 2n + 1$
glide ⊥ to b	a-glide	$h0\ell$	$h = 2n + 1$
	c-glide	$h0\ell$	$\ell = 2n + 1$
	n-glide	$h0\ell$	$h + \ell = 2n + 1$
glide ⊥ to c	a-glide	$hk0$	$h = 2n + 1$
	b-glide	$hk0$	$k = 2n + 1$
	n-glide	$hk0$	$h + k = 2n + 1$
C centering		$hk\ell$	$h + k = 2n + 1$
B centering		$hk\ell$	$h + \ell = 2n + 1$
A centering		$hk\ell$	$k + \ell = 2n + 1$

determine whether there is a c-glide perpendicular to a, because the loss of alternate rows in $0k\ell$ looks like just a large r.l. spacing; it is the presence of *all* rows in $1k\ell$ that provides the accurate r.l. spacing and proves the absences in $0k\ell$. (3) The absences caused by the presence of one type of symmetry element may obscure absences that would otherwise arise from another type of symmetry element. For this reason, among others, the space group to be assigned to the crystal may be indeterminate (*i.e.*, two or more space groups could be assigned on the basis of available data). Also, it is important to recognize that precession photographs will display *any* symmetry present in the reciprocal lattice, including mirror planes and two-fold axes. For instance, if there is a mirror perpendicular to a, then the $hk\ell$ and $\bar{h}k\ell$ reflections* will have the same intensity; therefore, one side of the $hk0$ (or $h0\ell$) zone will reflect the intensities of the other side. It is important to keep track of these apparent mirror planes in order to determine the space group. Zero-level precession photographs always contain an inversion center, independent of the lattice.

It is convenient that in the "International Tables"[7] there is a table listing all of the possible space groups for a given set of symmetry elements found photographically (page 349 in the second edition). One does not generally know the correct labeling scheme for the r.l. axes of a system until the space group has been determined. Then the axes are labeled according to their relationships to the symmetry elements found and with regard to established convention.

b. Powder Patterns. Powder patterns are recorded by either of two methods. The original and still the most common method is to place a small amount of powdered sample in a thin capillary tube at the center of the cylindrical camera. A hole in the side of the cylinder is provided for the entrance of the X-ray beam. As the beam strikes the capillary, the capillary is rotated around its axis and the pattern of the scattered rays is recorded by film wrapped on the inside surface of the cylinder. Figure 17–18 illustrates the camera structure.

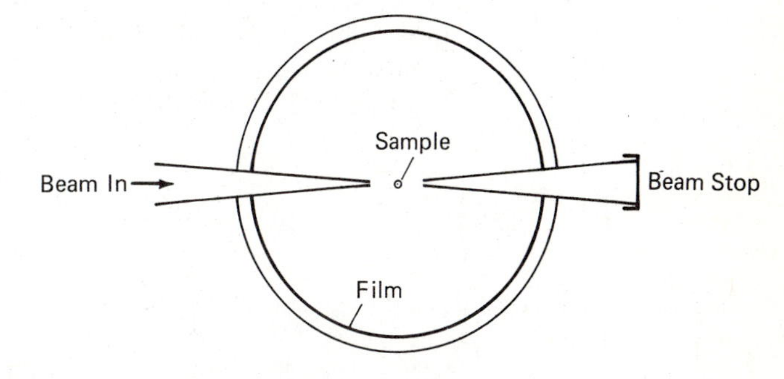

FIGURE 17–18. The construction of a powder camera using capillary mounted sample and film recording.

Alternatively, a counting tube that is sensitive to X-radiation may be used. This apparatus is built, as shown in Fig. 17–19, so that the counter swings through an arc,

---

*$\bar{h}$ is read "aitch-bar," and signifies negative values.

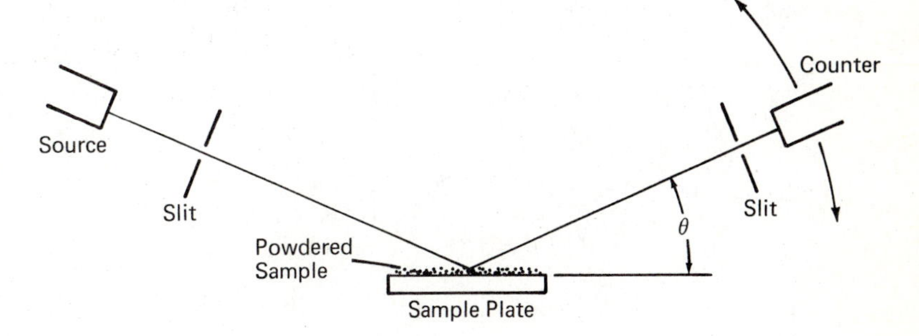

FIGURE 17–19. The geometry of a powder diffractometer, in which the sample is a smear on a glass plate, and a counter detector is used.

recording the variation of scattered X-ray intensity as it goes. This method is much simpler and faster, and has much better resolution than can be achieved by film methods, so the following discussion will be concerned with the use of a powder diffractometer.

First, what does the pattern look like, and how does it arise? Consider that the powder is just a collection of many small crystals. Each crystal will give rise to reflected beams only when r.l. points are in contact with the sphere of reflection. In principle this may, for a particular r.l. point, happen with many different orientations of the crystal, with only one major restriction: the angle of the reflected beam to the direct beam must always be the same—*i.e.*, any reflection can arise only at one value of θ (see Fig. 17–15). As shown in Fig. 17–20, this requires that any one reflection be

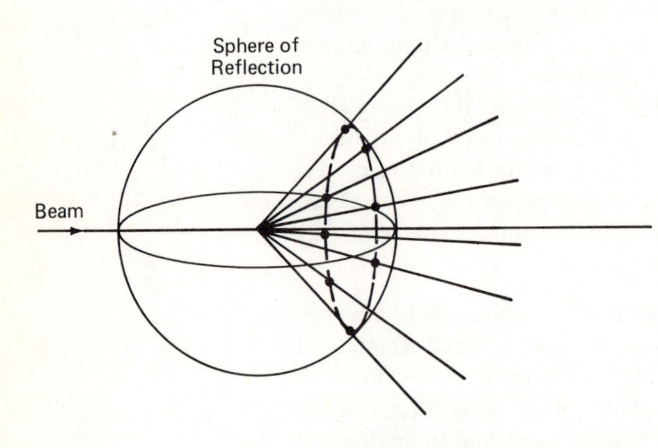

FIGURE 17–20. Cone of reflected X-rays of same index generated by scattering from randomly distributed crystals in a powder sample.

averaged over all crystal orientations into a cone of scattered X-radiation. There is, in principle, one such cone for every possible reflection, although it is generally only the strongest 100 or so that can be detected. As the counter sweeps up from some small angle, say 3° from the direct beam, toward 90°, the slit intersects each cone and records it as a peak in the powder pattern.

Figure 17–21 shows some powder patterns obtained from a diffractometer

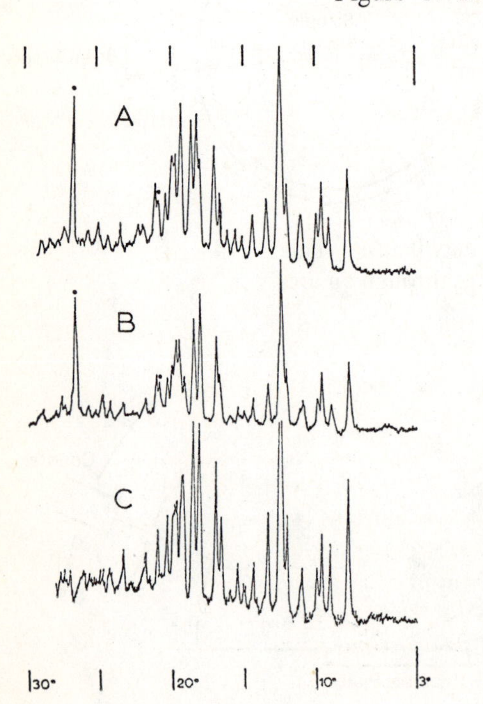

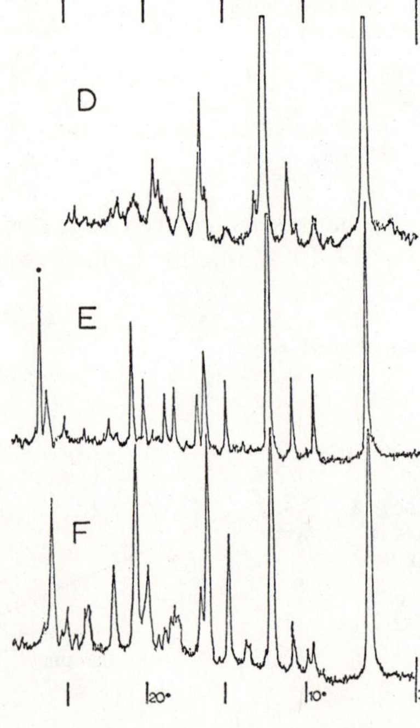

FIGURE 17–21. Experimental powder patterns for $[Ni_2(tren)_2X](BPh_4)_2$, where (A) X = $C_2O_4^{2-}$, (B) X = $(N_3^-)_2$, (C) X = $(OCN^-)_2$ precipitated from aqueous solution, (D) X = $(OCN^-)_2$ recrystalized from solvent, (E) X = $(SCN^-)_2$, and (F) X = $(SeCN^-)_2$.

instrument for several related chemical systems.[8] On the left side are found patterns for $[Ni_2(tren)_2X](BPh_4)_2$, where X is $C_2O_4^{2-}$, $(N_3^-)_2$, or $(OCN^-)_2$, and tren is the tripodal ligand 2,2′,2″-triaminotriethylamine:

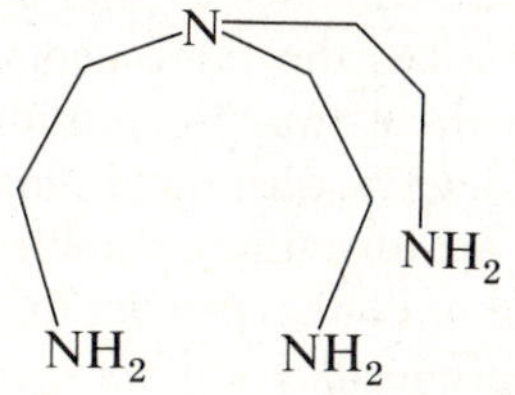

The electronic spectra indicate that in each case of Ni^{2+} ion is octahedrally coordinated. Magnetic susceptibility measurements indicate that in all three cases there are magnetic interactions between pairs of nickel ions. The infrared spectrum indicates that the azide ions are bound equivalently at each end. The μ-oxalato systems are relatively common, and single crystal X-ray studies indicate a dimeric structure of the form

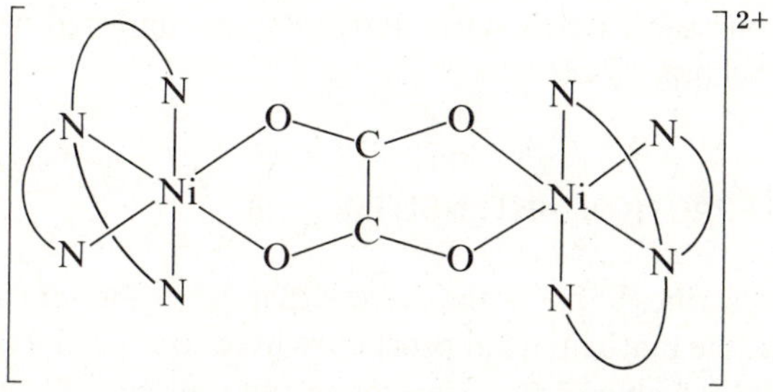

As seen in powder patterns (B) and (C), for the azide and cyanate systems, the positions and intensities of the peaks are almost exactly the same as in (A). The similarity in peak positions implies that the unit cell dimensions are quite similar in all three cases. This is useful, but it is the *contents* of the cell that are important, so we must look in detail at the intensities. There is such a good correspondence that it would be the greatest of coincidences if any two of these structures were grossly different from one another. Each cation is almost certainly dimeric, and the azide and cyanate systems may be drawn as follows in order to keep the size of the dimer nearly the same in all cases:

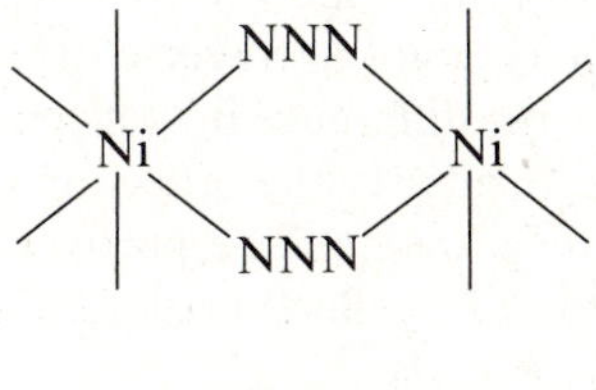

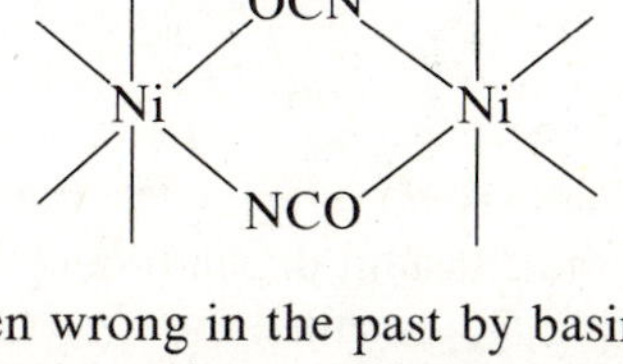

Many people have been wrong in the past by basing structural conclusions on limited data and on the correspondence of *line positions only*. One must evaluate carefully the weight of one's data and present fairly the probability for a correct conclusion. The complete single crystal X-ray structures were solved for the azide and cyanate systems shown above; it was found that they are indeed isostructural and only minor angle variations are involved.

Further illustration of the utility of the powder method is found by comparing patterns (C) and (D), where (C) is from the precipitated powder of the cyanate complex, and (D) is from ground crystals grown for the single crystal crystallographic work. Note that the patterns are distinctly different. Magnetic studies, however, indicate that the dimeric cation has the same geometry in each case. The crystal lattice packing is different in these two preparations of what is chemically and electronically the same material. Crystallization in more than one space group is not uncommon. It can sometimes lead to different molecular perturbations, which are upsetting to the experimenter until the powder patterns have demonstrated the possible origin of the observed effects.

Powder patterns (E) and (F) are fairly similar in most respects and are related in the strongest lines to pattern (D). It must be considered that upon substituting SCN and SeCN for OCN, even if the geometries of the systems were similar, the large electron densities of the S and Se atoms would appreciably alter the intensity patterns in the X-ray scattering. On this basis, the (D), (E), and (F) patterns are as similar as one might expect if all three structures are similar. The results are *consistent* with similar structures. The increase in the intensity of the high angle peaks as the mass of the bridging group increases is particularly notable and will be understood after the discussion in Section 17–5.

17–5 X-RAY DIFFRACTION—INTENSITIES

This section describes some of the factors that affect the *measured* intensity of a reflection as well as the mathematical procedure used to *calculate* the intensity of any reflection from a knowledge of the contents of the unit cell. We will see that measurement of *intensities* alone cannot provide enough information to calculate atomic positions directly, but that iterative techniques must be employed in which calculated and measured intensities are compared and the atomic model used is refined until agreement in the two sets of values is adequate.

Intensities of scattered X-ray beams must be measured to provide data reflecting the arrangement of the contents of the unit cell. Whether film or counter techniques are used, the crystal is moving at the time of measurement in such a way that an r.l. point crosses the sphere of reflection from one side to the other. As an r.l. point is swept through the sphere of reflection, the integrated intensity depends in part upon the angle of the path of motion with respect to the surface of the sphere at intersection. The time required for an r.l. point to traverse the sphere increases as this angle approaches zero degrees. Also, the difference in "reflectivity" of X-rays whose electric vectors lie parallel and perpendicular to the reflecting plane must be accounted for. These *Lorentz* and *polarization* corrections, respectively, may be used to correct the observed intensity ($I_{hk\ell}$) of the $hk\ell$ reflection as follows:

$$|F_{hk\ell}| = \sqrt{\frac{KI_{hk\ell}}{Lp}} \tag{17-8}$$

where the Lorentz correction is $L = 1/\sin 2\theta$, the polarization correction is $p = (1 + \cos^2 2\theta)/2$, and K is a constant depending upon the properties of the data collection system. The resulting $F_{hk\ell}$ is called the *observed structure factor amplitude.* As measured, it is the absolute value of the structure factor, F_{obs}.

In solving a crystal structure, the F_{obs} values are compared with the F_{calc} values that are calculated from an assumed arrangement of atoms in the unit cell. In order to calculate structure factors from a given atomic model, a summation of scattering by all atoms in the unit cell is made. Two key points must be considered. The first is that the scattering from the electron distribution of each atom, j, is represented to some approximation by a spherical *scattering factor*, f_j. This factor expresses the scattering power of the atom, and is proportional to the number of electrons in the atom and to the angle over which the scattering takes place ($\sin\theta/\lambda$ or $1/2d_{hk\ell}$, the inverse distance of the r.l. point from the origin). A typical scattering factor curve (in this case for carbon), assuming spherical atoms, is shown in Fig. 17–22. The decrease in scattering power as a function of angle occurs because the combined scattering from electron density all around the atom, when added together as if arising from the location of the nucleus, tends to cancel, owing to increasing phase differences between waves originating over a large region of space. Computer programs for calculating structure factors generally require a table of scattering factor versus $\sin\theta/\mu$ values to be read as input data. Then when that atom's contribution to a given $F_{hk\ell}$ is to be calculated, the computer may interpolate the table at the $\sin\theta/\lambda$ value appropriate for that $hk\ell$.

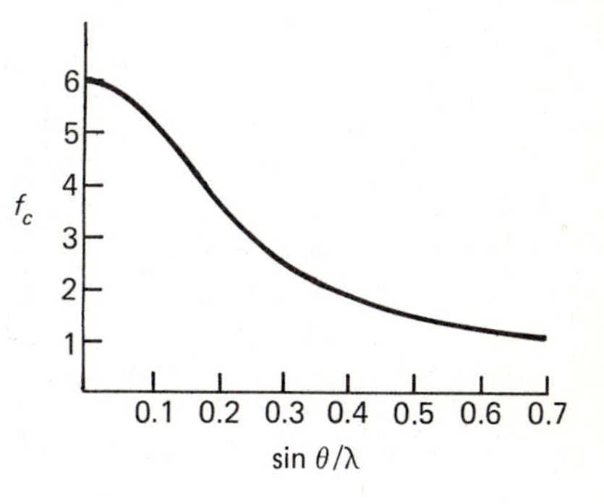

FIGURE 17–22. Scattering factor curve for a carbon atom.

The second point is that structure factors are *calculated* as if all of the scattering in the unit cell were accomplished by electron density at the origin. In order to do this, the scattering power of each atom contributing to F_{calc} must be added in a vector sense, thus accounting for the phase difference between a wave arising at the origin and one arising at the atomic location. In other words, the structure factor $F_{hk\ell}$ has *magnitude* and *phase*— its magnitude being the number of electrons that, if scattering in phase (say, all at the origin), would show the same diffracting power as do the actual atomic contents of the cell; and $F_{hk\ell}$ has the resultant phase of all the contributing atoms in the cell.

Calculating the phase difference for any $hk\ell$ reflection between the origin and some point (x, y, z) in the cell is straightforward. (Here x, y, and z are given in *fractional coordinates; i.e.*, Å/a, Å/b, Å/c.) The reflection $hk\ell$ arises from a set of planes spaced so that there is a phase difference of 2π radians between scattered waves from successive planes. This means that for the $hk\ell$ reflection the unit cell edges represent $2\pi h$, $2\pi k$, and $2\pi\ell$ radians of phase difference, each edge being cut into h, k, and ℓ parts by the reflecting planes. The phase difference between (x, y, z) and $(0, 0, 0)$ for the $hk\ell$ reflection is then the sum of the coordinate phases, *i.e.*,

$$\delta_{hk\ell} = 2\pi(hx + ky + \ell z). \tag{17-9}$$

In that we must sum the structure factor contributions from all atoms in the cell, it is important to introduce at this time the complex representation of structure factors, as this greatly simplifies the calculation of phase relationships. Recall that the addition of two waves r_1 and r_2, each with a magnitude and a phase, may be

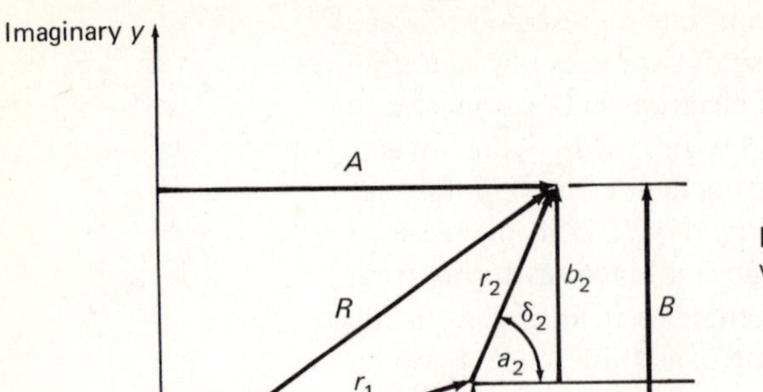

FIGURE 17–23. Vector diagram illustrating the sum of two complex variables in the complex plane.

considered as the addition of two vectors in the complex plane, as illustrated in Fig. 17–23, where the following relationships are true:

$$A = a_1 + a_2, \qquad B = b_1 + b_2$$

$$|R| = \sqrt{A^2 + B^2} \qquad \text{and} \qquad \alpha = \tan^{-1}\left(\frac{B}{A}\right)$$

Now, R as a vector quantity (magnitude and phase) may be expressed in any of the following forms:

$$R = A + iB$$

or

$$R = |R|(\cos\alpha + i\sin\alpha)$$

or

$$R = |R|e^{i\alpha}$$

Considering that the structure factor $F_{hk\ell}$ must be calculated by summing over all atomic contributions, its magnitude is

$$|F_{hk\ell}| = \sqrt{\left(\sum_j f_j a_j\right)^2 + \left(\sum_j f_j b_j\right)^2}$$

$$= \sqrt{\left(\sum_j f_j \cos\delta_j\right)^2 + \left(\sum_j f_j \sin\delta_j\right)^2} \qquad (17\text{–}10)$$

This can be rewritten, using equation (17–9), as

$$|F_{hk\ell}| = \sqrt{A_{hk\ell}{}^2 + B_{hk\ell}{}^2}$$

where

$$A_{hk\ell} = \sum_j f_j \cos 2\pi(hx_j + ky_j + \ell z_j)$$

and

$$B_{hk\ell} = \sum_j f_j \sin 2\pi(hx_j + ky_j + \ell z_j)$$

The phase angle of the resultant structure factor is

$$\alpha_{hk\ell} = \tan^{-1}\left(\frac{B_{hk\ell}}{A_{hk\ell}}\right) \tag{17-11}$$

These are the expressions commonly used to calculate structure amplitudes for comparison with F_{obs}.

The above $F_{hk\ell}$ equations may be written in more compact form as

$$F_{hk\ell} = \sum_j f_j e^{2\pi i(hx_j+ky_j+\ell z_j)} \tag{17-12}$$

using the mathematical equivalence $e^{2ix} = \cos 2x + i \sin 2x$. Looking at this expression for $F_{hk\ell}$, we can get some feeling for the biggest problem encountered in X-ray structure solution. $F_{hk\ell}$ has a magnitude *and* phase, both determined by the sum over all atomic positions in the lattice. Experimentally, we can measure only the magnitude of each $F_{hk\ell}$, so on the surface the problem seems insoluble, having as many unknown phases as observables, and still having all of the atomic positions to be determined. The next section will outline two methods by which the phase problem is circumvented and structure solution becomes practical.

At this point, having learned to express $F_{hk\ell}$ as a function of atomic coordinates, we should be able to predict the effect of symmetry on the observed intensities. It shall be illustrated that the systematic absences listed in Section 17–4 can be derived from these considerations.

Equation (17–12) sums over all of the atoms in the unit cell—even over those that are related by symmetry. It may be that the symmetrically related atoms can be given a more specific analytical expression for their total contribution to $F_{hk\ell}$.

There are only n unique atoms in the unit cell, which is generally less than the total; these n atoms define the *asymmetric unit,* there being m asymmetric units in the cell, for a total of $n \times m$ atoms. Assuming a primitive centric cell, the value of m is 2, 4, or 8 in the triclinic, monoclinic, and orthorhombic systems, depending upon the amount of symmetry. Now, $F_{hk\ell}$ may be rewritten as

$$F_{hk\ell} = \sum_n f_n \left(\sum_m e^{2\pi i(hx_{m,n}+ky_{m,n}+\ell z_{m,n})}\right) \tag{17-13}$$

where the summation in parentheses is often designated $T_{hk\ell}$. The significance of this expression is that the contribution of atoms to $F_{hk\ell}$ may be calculated m atoms at a time. In fact, it often happens that where the m atoms have a symmetry relationship, simplification of the mathematics may result.

A particularly simple and important case is that in which $m = 2$ and the two asymmetric units are related by a center of inversion; *i.e.*, $x, y, z = \bar{x}, \bar{y}, \bar{z}$. The structure factor contribution of each pair of atoms is

$$\begin{aligned}(T_{hk\ell})_n &= \sum_m e^{2\pi i(hx_{m,n}+ky_{m,n}+\ell z_{m,n})} \\ &= \cos 2\pi(hx + ky + \ell z) + i \sin 2\pi(hx + ky + \ell z) \\ &\quad + \cos 2\pi(-hx - ky - \ell z) + i \sin 2\pi(-hx - ky - \ell z) \\ &= 2 \cos 2\pi(hx + ky + \ell z)\end{aligned}$$

Thus, for the centric case, where there is a center of inversion in the unit cell,

$$F_{hk\ell} = 2 \sum_n f_n \cos 2\pi(hx_n + ky_n + \ell z_n)$$

It is seen that the structure factor in this case is *real* (that is, the phase angle is either 0 or π), which is often a great aid in the successful solution of a structure.

Let's consider a more complicated example—that of a two-fold screw axis such as in the space group $P2_1/c$. Here the position (x, y, z) is transformed into $(\bar{x}, \frac{1}{2} + y, \frac{1}{2} - z)$ by the screw action. We may write (only the cosine functions are used here, because it is known that $P2_1/c$ is a centric space group, *i.e.*, that $\bar{x}\bar{y}\bar{z} \rightarrow xyz$):

$$T_{hk\ell} = 2\left[\cos 2\pi(hx + ky + \ell z) + \cos 2\pi\left(-hx + ky + \frac{k}{2} + \frac{\ell}{2} - \ell z\right)\right]$$

which, using the trigonometric identity for the cosine of a sum of angles, may be rewritten

$$\begin{aligned} T_{hk\ell} = 2\Big[&\cos 2\pi(hx + \ell z)\cos 2\pi ky - \sin 2\pi(hx + \ell z)\sin 2\pi ky \\ &+ \cos 2\pi(-hx - \ell z)\cos 2\pi\left(ky + \frac{k+\ell}{2}\right) \\ &- \sin 2\pi(-hx - \ell z)\sin 2\pi\left(ky + \frac{k+\ell}{2}\right)\Big] \end{aligned}$$

or, collecting terms,

$$\begin{aligned} T_{hk\ell} = 2\Big\{&\cos 2\pi(hx + \ell z)\left[\cos 2\pi ky + \cos 2\pi\left(ky + \frac{k+\ell}{2}\right)\right] \\ &- \sin 2\pi(hx + \ell z)\left[\sin 2\pi ky - \sin 2\pi\left(ky + \frac{k+\ell}{2}\right)\right]\Big\} \end{aligned}$$

The sum of angle formulae may again be used to expand the $\cos 2\pi(ky + (k + \ell)/2)$ and $\sin 2\pi(ky + (k + \ell)/2)$ terms, and with the simplifying observation that

$$\cos 2\pi\left(\frac{k+\ell}{2}\right) = (-1)^{k+\ell}$$

and

$$\sin 2\pi\left(\frac{k+\ell}{2}\right) = 0$$

it is found that

$$\cos 2\pi\left(ky + \frac{k+\ell}{2}\right) = (-1)^{k+\ell}\cos 2\pi ky$$

and

$$\sin 2\pi\left(ky + \frac{k+\ell}{2}\right) = (-1)^{k+\ell}\sin 2\pi ky$$

Now we can write the final expression

$$\begin{aligned} T_{hk\ell} = 2\{&\cos 2\pi(hx + \ell z)[\cos 2\pi ky + (-1)^{k+\ell}\cos 2\pi ky] \\ &- \sin 2\pi(hx + \ell z)[\sin 2\pi ky - (-1)^{k+\ell}\sin 2\pi ky]\} \end{aligned}$$

Observe that when $k + \ell$ *is even,*

$$T_{hk\ell} = 4 \cos 2\pi(hx + \ell z) \cos 2\pi ky$$

and when $k + \ell$ *is odd,*

$$T_{hk\ell} = -4 \sin 2\pi(hx + \ell z) \sin 2\pi ky$$

For the r.l. axis $0k0$, when k is odd, $k + \ell$ is odd, and $T_{hk\ell} = 0$; likewise, for $h0\ell$ when ℓ is odd, $k + \ell$ is odd, and $T_{hk\ell} = 0$. We have therefore shown that the systematic absences for both the screw axis and the glide plane are generated when the conditions for a 2_1 axis in a centric structure are used to calculate structure factor expressions. It is also interesting to note that $T_{hk\ell} = T_{\bar{h}\bar{k}\bar{\ell}}$. This implies that the diffraction pattern has a center of symmetry.

17-6 PRACTICAL APPLICATIONS—LEVEL II

a. Collecting and Reducing Intensity Data. At this point, more of the details of the experimental procedures involved with X-ray structure solution will be outlined. Of concern now is the measurement of intensities. Only occasionally is this done by film methods, so the discussion will focus on the use of a computer-controlled diffractometer.

Upon obtaining a suitable crystal and studying its unit cell properties and overall quality with film methods, one may then measure intensities by mounting the crystal on the diffractometer so that the orientation of the reciprocal lattice with respect to the diffractometer geometry is known. In this way, the computer will know where to find each reflection $hk\ell$. The drawing of a typical diffractometer goniostat (Fig. 17–24) illustrates the range of movements that are involved in coordinating the diffractometer and the crystal. Actually, it is not too difficult to align a crystal manually. Alignment may also be accomplished by locating random reflections on the diffractometer and using the auto-indexing program (described earlier); this is an increasingly common technique. Once three reflections are detected in the counter by adjusting χ, φ, ω, and 2θ, and provided that they are correctly indexed ($-h \neq h$, etc.), the computer can calculate an *orientation matrix* (and cell constants) from which it can find the diffractometer angles corresponding to any h, k, and ℓ values. Of course, the accuracy of this matrix depends upon the accuracy of the measured reflections. If it is adequate for the location (within 0.02 degree in each arc) of any desired reflection in the r.l., data can then be collected. It may alternatively be desirable to measure the

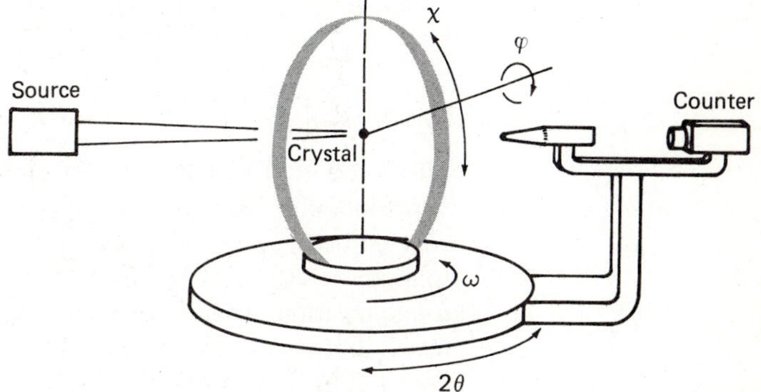

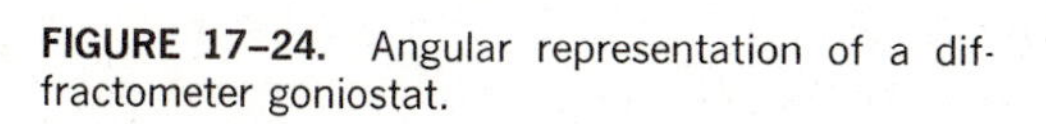

FIGURE 17–24. Angular representation of a diffractometer goniostat.

locations of, say, 15 or 20 reflections; the computer then uses a program to determine the best matrix by least squares fit to all of the reflections. This technique must eventually be followed, because it is the most convenient way to determine the unit cell parameters and their uncertainties.

The computer then requires a range of values of h, k, and ℓ over which it should measure intensities; for instance, only the unique set [$h = -20, 20$; $k = 0, 20$; $\ell = 0, 20$] may be required for a monoclinic system. On the other hand, two (or all) unique sets of data may be collected and averaged together for better results. The choice of procedures depends upon the time and importance allotted to the structure. The amount of (unique) data required for a given system depends upon the number of parameters to be determined. If there are 200 parameters, then it is desirable to collect at least 1200 reflections, and preferably 2000. It is fortunate that for large molecules the unit cell tends to be large; with the concomitant smaller V^*, the number of observations increases. Usually, the diffractometer (with Mo radiation) is used to collect all data from $2\theta = x$ to $2\theta = y$, where $x \cong 3°$ and is not so small as to approach the direct beam, and $y \cong 45°$, or the angle at which the intensities fall off to the noise level.

When the computer directs the goniostat to find a peak for intensity measurement, the angles are adjusted to perhaps one degree in 2θ short of the peak, and then the goniostat scans slowly across the peak while the counter adds up the total integrated intensity and the computer records the time required for the scan. Background counts are then made on each side of the peak, generally for 10 seconds. The diffractometer prints onto paper tape, cards, magnetic tape, or other recording medium such information as χ, ω, 2θ, φ, background counts, background time, peak count, and peak time for every reflection measured.

Once all of this information is at hand, a *data reduction program* transforms the intensity data into F_{obs} values, making corrections for the Lorentz and polarization terms, background, and so forth. Also, the *statistical weight* given to each reflection (*i.e.*, a value by which the expected accuracy of the reflection may be judged) is calculated. In general, the largest (because of experimental problems) and weakest (because of few counts per collection time) reflections are the least accurately determined, and this is accounted for in the $\sigma_{hk\ell}$ values. It will be shown later that, when refining the atomic parameters of a unit cell model so that agreement between F_{obs} and F_{calc} is maximized, the weighting of the individual F values in the error determination is important. It is desirable to have the error in F_{calc} versus F_{obs} distributed evenly as a function of both $|F_{obs}|$ and $\sin\theta/\lambda$.

Initially, however, we are concerned only with what procedure to follow, given a collection of F_{obs} values and an "empty" unit cell. It is possible, through the use of Fourier series, to express the electron density at any point in the cell in terms of the observed structure factors as follows (see reference 3, p. 222):

$$\rho(x, y, z) = \frac{1}{V} \sum_h \sum_k \sum_\ell F_{hk\ell} e^{-2\pi i(hx+ky+\ell z)}$$

If both the observed *magnitudes* and *phases* of $F_{hk\ell}$ could be substituted into this expression, one could generate a three-dimensional mapping of the electron density in the unit cell, from which it would be trivially simple to pick out the locations of the atoms and molecules of interest, since each peak in electron density corresponds to an atomic position. As has been said many times, it is the indeterminate nature of the *phases* that gives rise to the difficulty and mystique that have shrouded X-ray crystallography for many years.

Crystallography of a system containing a heavy metal atom is often made straightforward by use of the *heavy atom method* of structure solution.

b. The Heavy Atom Method. This method, discovered by A. L. Patterson,[9] is based on the fact that although one cannot calculate $\rho(x, y, z)$, one can calculate a similar series in which the magnitudes, but *not* the phases, of $F_{hk\ell}$ are used. This series, the Patterson function, is written

$$P(x, y, z) = \frac{1}{V} \sum_h \sum_k \sum_\ell |F_{hk\ell}|^2 \cos 2\pi(hx + ky + \ell z)$$

where x, y, and z are positional coordinates in the three-dimensional Patterson map. Patterson was able to show that this function has a maximum value at the end of every interatomic vector in the lattice, the tails of these vectors being transferred to the origin. The height of a Patterson peak is proportional to nZ_iZ_j when the peak corresponds to the vector between the ith and jth atoms (of atomic number Z), and n of the peaks are coincident owing to the symmetry requirements of the lattice. When one of the atoms in the asymmetric unit is particularly heavy (say, a copper atom in a cell otherwise consisting of C, N, and H atoms), the Cu-Cu vector is by far the largest peak in the Patterson map. It is possible to determine, from the location of this peak in the map, the x, y, and z coordinates of the copper atom in the unit cell. Now an electron density map may be calculated by using the observed $F_{hk\ell}$ values, which are given phases that are determined by the presence of the largest scatterer of the cell in a known location. Because only some of the phases are correct (in a centric cell) or close (in an acentric cell), the density map will not be totally accurate; but it is likely that if

$$\frac{Z^2_{\text{heavy atom}}}{\Sigma Z^2_{\text{light atoms}}} \approx 1$$

then some areas of the map will clearly identify fragments of the structure. When these fragments are included with the metal atom in the phasing model, then the phases will be improved; and by repeating this procedure several times all atoms can eventually be located.

Let us now backtrack a step or two and study in more detail how it is that one finds the metal atom position from a Patterson map. In $P2_1/c$, the general positions are (x, y, z), $(\bar{x}, \bar{y}, \bar{z})$, $(\bar{x}, \tfrac{1}{2} + y, \tfrac{1}{2} - z)$, and $(x, \tfrac{1}{2} - y, \tfrac{1}{2} + z)$. Vectors between these positions may be calculated merely by subtraction, as shown in Table 17-3.

TABLE 17-3. GENERAL INTERATOMIC VECTORS

	x, y, z	$\bar{x}, \bar{y}, \bar{z}$	$\bar{x}, \frac{1}{2} + y, \frac{1}{2} - z$	$x, \frac{1}{2} - y, \frac{1}{2} + z$
x, y, z	$0, 0, 0$	$2\bar{x}, 2\bar{y}, 2\bar{z}$	$2\bar{x}, \frac{1}{2}, \frac{1}{2} - 2z$	$0, \frac{1}{2} - 2y, \frac{1}{2}$
$\bar{x}, \bar{y}, \bar{z}$	$2x, 2y, 2z$	$0, 0, 0$	$0, \frac{1}{2} + 2y, \frac{1}{2}$	$2x, \frac{1}{2}, \frac{1}{2} + 2z$
$\bar{x}, \frac{1}{2} + y, \frac{1}{2} - z$	$2x, \frac{1}{2}, \frac{1}{2} + 2z$	$0, \frac{1}{2} - 2y, \frac{1}{2}$	$0, 0, 0$	$2x, 2\bar{y}, 2z$
$x, \frac{1}{2} - y, \frac{1}{2} + z$	$0, \frac{1}{2} + 2y, \frac{1}{2}$	$2\bar{x}, \frac{1}{2}, \frac{1}{2} - 2z$	$2\bar{x}, 2y, 2\bar{z}$	$0, 0, 0$

Of the 16 vectors there are only nine that are unique:

Patterson vector	degeneracy
$0, 0, 0$	4
$2x, 2y, 2z$	1
$2\bar{x}, 2\bar{y}, 2\bar{z}$	1
$2x, 2\bar{y}, 2z$	1
$2\bar{x}, 2y, 2\bar{z}$	1
$2x, \frac{1}{2}, \frac{1}{2} + 2z$	2
$0, \frac{1}{2} + 2y, \frac{1}{2}$	2
$0, \frac{1}{2} - 2y, \frac{1}{2}$	2
$2\bar{x}, \frac{1}{2}, \frac{1}{2} - 2z$	2

The line in Patterson space designated $0, v, \frac{1}{2}$ (where v is in fractional coordinates) is referred to as a Harker line, and can be seen to arise from the symmetry requirements of a glide plane. On this line will fall a vector of weight two for every symmetrically related set of atoms. The largest of these, by a factor of $Z_{\text{metal}}^2/Z_{\text{nitrogen}}^2$ (if nitrogen is the second heaviest element in the lattice), will be that due to the metal atom, and it is generally found easily.* The value of v is $\frac{1}{2} \pm y$ (depending upon which end of the y-axis is seen), and thus y can be determined. Now, in the $u, \frac{1}{2}, w$ plane there should be another pair of peaks of approximately the same intensity as the first ones. The location of this pair allows calculation of x and z for the metal atom. Finally, these values can be checked by locating the $2x, 2y, 2z$ vector where it is supposed to be.

The metal atom is the first atom of the unit cell model—and it is the completion and refinement of this model to which we shall now turn. Only two computer programs are required in order to finish this task. The first is a Fourier program that can be used to calculate the Patterson function, F_{obs} electron density maps, or $|F_{\text{obs}}| - |F_{\text{calc}}|$ density maps. The second program is the least squares refinement program, which, when the model is complete but not accurate, varies all of the unknown parameters in such a way as to obtain the best agreement between the F_{obs} and F_{calc} (calculated from the model) values. It is also responsible for calculating structure factors for use in the Fourier program.

c. Fourier Syntheses. The use of Fourier syntheses in developing a model can range from a routine processing of the data to a sensitive art in its own right. Recall the procedure outlined for completing the model. A set of structure factors is calculated from the heavy metal atom position. These structure factors, including calculated phases, are read into the Fourier program; or, if another mode of operation is selected, the phases from the F_{calc} values are applied to the F_{obs} values. Using the equation given earlier, an F_0 Fourier map of the electron density is calculated. In the event that the metal atom phases are correct only for those reflections whose intensities are dominantly determined by the metal atom scattering, the Fourier map calculated will show the metal atom and nothing else, and the calculation fails. This is

* The metal-metal vector intensity is usually calculated by reference to the height of the origin peak, which includes all atoms' "self-vectors."

unusual, but the procedure that one can follow is to realize that the largest F_o's are most important in determining the electron density map; it is most important to phase correctly those F_o's that have heavy contributions from the light atoms in the structure. If only those reflections for which F_{calc} (metal) $\approx 0.3\ F_{obs}$ are used to calculate the map, it is likely that more light atom density will be displayed in the map.

A difference Fourier function is calculated from the expression

$$\Delta\rho = \frac{1}{V} \sum_h \sum_k \sum_\ell (|F_o| - |F_c|)\, e^{i\alpha_c}\, e^{2\pi i(hx+ky+\ell z)}$$

where α_c is the phase of F_c. The most important properties of this map are that reflections for which $F_o \approx F_c$ make little contribution, thus not tending to merely reproduce the model; and that terms for which $|F_c| \gg |F_o|$ contribute heavily to the ΔF map, although they contribute negligibly to the F_o map. This is important because the presence of such terms represents a large error in the model. For this reason, it is useful to include in a ΔF synthesis all weak and unobserved data points (an unobserved point is a reflection whose intensity is less than one or two times the statistical error in its measurement).

When a structure is complete, the ΔF map has no peaks or holes in any region of the unit cell. Even when all atoms have been located, it is often found that strangely shaped areas of positive and negative density arise near atoms whose electron density cannot be well fitted by the model of a stationary atom. It is at this point that we must introduce the concept of a *temperature factor*. This factor accounts for the fact that molecules vibrate, so that atoms must be considered on the basis of their time-averaged positions. The atoms can be treated as vibrating either isotropically (in a spherically symmetric fashion) or anisotropically (as an ellipsoid); the difference is that it takes only one parameter to specify the isotropic motion, and six to describe the anisotropic motion. The mathematical approach is simply to correct the scattering factor for thermal motion, on the basis that spreading out the electron density causes a faster than normal decrease in f_j as a function of $\sin\theta/\lambda$. We can write, for the isotropic and anisotropic cases, respectively,

$$f = f_0\, e^{-B(\sin^2\theta)/\lambda^2}$$

and

$$f = f_0\, e^{-(1/4)(B_{11}h^2a^{*2}+B_{22}k^2b^{*2}+B_{33}\ell^2c^{*2}+2B_{12}hka^*b^*+2B_{13}h\ell a^*c^*+2B_{23}k\ell b^*c^*)}$$

It is generally found to be difficult to refine the B values, particularly the anisotropic ones, from ΔF maps, so we must turn to a more powerful method.

d. Least Squares Refinement. Assuming that through successive model changes, structure factor computations, and Fourier calculations, we have found all of the atoms in the cell to within ±0.1 Å in all directions, we must now investigate a method of further refinement. First, how many parameters are to be determined? For an anisotropic model there are nine parameters for each atom plus one more, the scale factor. In order to compare the F_{calc}'s with the F_{obs}'s, it is necessary that the average of all F_c's be the same as that of the F_o's; this is done simply by multiplying the F_o's by some factor, generally near 1.0. The usual restrictions on least squares refinement are that there must not be more than a certain total number of variable parameters, say 270 (a computer limitation), that the number of reflections should be at least five times and preferably 10 times as large as the number of parameters, and that the initial trial structure should be relatively accurate (because the refinement is non-linear).

Each time the least squares program is run, it calculates the structure factors from the given model (plus scattering factor tables, symmetry information, and so forth) and, using matrix techniques too detailed to present here, calculates a variation in each parameter that will result in a net decrease in the function

$$\sum_{hk\ell} W_{hk\ell}(|F_o| - |F_c|)^2$$

where the $W_{hk\ell}$ values are weights calculated as $1/\sigma_{hk\ell}{}^2$.

The goodness of the model at any point is judged by the R factor, which can be written as

$$R = \frac{\Sigma||F_o| - |F_c||}{\Sigma|F_o|}$$

or

$$R_W = \frac{\Sigma W||F_o| - |F_c||}{\Sigma W|F_o|}$$

where the discrepancies are weighted. In general, the values of R and R_W should be nearly the same at the completion of refinement, and 0.05 or less is quite respectable.

After the least squares program has been run several times, each time using the results of the preceding calculation as input, the refinement should converge to a point at which R does not change and no parameter is significantly shifted from one cycle to the next. Now refinement is complete.

Output from the least squares calculation includes fractional coordinates of each atom and its thermal parameters, including an estimate of the standard deviation in each parameter. These deviations are generally underestimated, but do provide a good relative indication of error. It is now possible to calculate bond distances and angles, to fit groups of atoms to planes, and to have other programs draw perspective views of selected parts of the unit cell. Error estimates can be carried through from the fractional coordinates and assigned to the molecular values.

The location of hydrogen atoms is sometimes possible when the structure has been otherwise well refined. If one knows where to look, they may be found in ΔF maps; but if not, then they are often added to a structure by calculating their positions geometrically based on sp^2 or sp^3 hybridization (methyl groups, of course, cannot be handled in this way because of their low rotational barrier). It can be helpful to have included hydrogen atoms in a refinement, for what might previously have been a metal-nitrogen distance of 2.20 Å in a primary amine coordination complex can change to 2.10 Å by filling some of the more distant electron density with hydrogen atoms.

e. A Powerful Direct Methods Approach. There is another technique for phase determination that has not been described; its applicability for the solution of inorganic structures has until recently been more limited than the heavy atom method. With the cooperative efforts of English and Belgian groups,[10] the *direct methods* approach to phase determination has become so powerful that all other methods may soon be antiquated. Their direct methods program, called MULTAN74, has been found capable in most instances of taking a structure solution from the F_{obs} list to location of all atoms *without user intervention*. The output can even include a projection drawing of the molecule.

The direct method procedure is described well in Chapter 13 of reference 3, and this material should be read. It will only be mentioned here that the direct methods program involves a mathematical relationship that allows assignment of phases to the stronger reflections based upon approximate relationships between phases of groups of reflections. The accuracy of the assignment can be estimated as well. Certain reflections can be assigned phases, and others can be given phases based upon the original set. If this procedure is carried out to the extent that eight or ten reflections per unique atom have been phased, an electron density map may be made that displays the contents of the cell. In general, the phasing process may require guesses at certain points, so that perhaps eight possible phasing schemes are found. MULTAN74 has the ability to select the most probable among these. It also incorporates a data treatment algorithm that takes into account the suspected number, type, and even grouping of atoms in the unit cell (not their locations or orientations, which are of course unknown). This specialized conditioning of the data reduces much of the noise and spurious peaks generally found in E-maps (direct methods generated electron density maps). Also, MULTAN74 has the facility to search the E-map for atoms in bonding arrays that match the suspected fragments or molecules. When the map seems reasonably consistent with the expected contents of the cell, the atomic positions are printed out along with distances, angles, and the most informative projection of the molecule.

From this point either additional Fourier maps may be calculated to sort out unclear areas, or least squares calculations may be run immediately.

17-7 COMPLICATIONS AND CORRECTIONS

Several problems that can be encountered during refinement of a structure have not yet been discussed. These are twinning, absorption, disorder, and librational motion.

a. Twinning. Twinning really is a problem that is generally dispensed with prior to data collection by throwing away the crystal. Basically, it is a problem arising from the superposition of two crystalline regions of the crystal with slight misalignment between the two, causing all reflections (as observed photographically) to be doubled. Sometimes crystals of the same material may be found that do not have this property—but the search is often long and fruitless. Some diffractometer programs are available for collecting data even from twinned crystals, but the obstacles to such a study are great. Sometimes the misalignment between crystal twins is considerably more than slight. In some instances, twinning at 90° has been found, and photographs may look normal except that the systematic absences are inconsistent with any known space group. In such cases, data may be collected and fitted with a model in which the F_{calc}'s are summed values from two rotated cells, each containing the same molecular arrangement.

b. Absorption. Absorption is a problem when the crystal has a shape deviating drastically from spherical, or when X-rays are strongly absorbed. Since the path lengths of the X-ray beam through the crystal are different for different values of $hk\ell$, those F_{obs} values for which the beam traverses a long length of crystal will be artificially low. When the distance of each face of the crystal from the center is known, as well as the diffractometer angles required to place each face in reflecting position, then computer programs may be used that will correct the values of F_{obs} for absorption for the particular crystal under study. These corrections are necessary only when the crystal has *extreme* differences in dimensions, say a factor of 6 or 7, or when the linear absorption coefficient, μ_r, is greater than 30 for more normal crystals. This coefficient may be calculated, knowing the *contents* of the cell:

$$\mu_r = \rho \sum_n \left(\frac{P_n}{100}\right)\left(\frac{\mu}{\rho}\right)_\lambda$$

where ρ is the density of the crystal, the sum is over all atoms, p_n is the number percentage of atom n, and $(\mu/\rho)_\lambda$ for each atom type may be found in the "International Tables for Crystallography." The exiting X-ray intensity I may be expressed in terms of the entering intensity and τ, the thickness of the crystal traversed:

$$I = I_0 e^{-\mu_r \tau}$$

Variation of I/I_0 for different reflections by as much as 50% is possible, owing to absorption effects. For a 15% variation, the lowest R value obtainable may be 10% and the structure will be inaccurate.

Note that even for spherical crystals an absorption correction may be required for strongly absorbing materials, although in this case it is simple to do. It is left for the reader to demonstrate that pathlengths still depend on θ even in spherical systems.

c. Disorder and Librational Motion. Disorder, when it is found, is often a very bad problem. The difficulty arises when some molecular group in the unit cell can be placed in either of two (or more) relatively equivalent positions (related by an element of symmetry) and in fact is found in both (or all) of these positions in the same crystal. In some unit cells the group is one way, which in others it is different. Provision is made in the least squares program for locating *fractional* atoms; *i.e.*, if the group is there only half the time, then it may be counted as half an atom. It sometimes happens that atoms can occupy a more or less continuous range of locations between the two disordered extremes—thus spreading out the electron density in such a way that a satisfactory fit is made possible only by using specially constructed programs. Normally accurate parameters such as distances and angles from such disordered groups are impossible to obtain. This is also the case for groups that undergo librational motion, *i.e.*, rotating and vibrating simultaneously. Such motion cannot be fitted by standard programs.

HELPFUL SUGGESTIONS. When planning to grow crystals for a structure determination, aim for the following characteristics: (1) They should be as equidimensional as possible, and on the order of 0.2 to 0.4 mm on each side. (2) Where ionic systems are to be studied and the cation is of interest, try to avoid spherical anions, in particular ClO_4^-. This anion is almost always disordered, and the inability to fit it accurately will affect the rest of the structure adversely. BPh_4^- is big (lots of atoms), but it generally packs nicely and the rings are easy to find, even on bad Fourier maps. One can therefore use it to establish much of the initial phasing model.

REFERENCES

1. M. J. Nolte, E. Singleton, and M. Laing, J. Amer. Chem. Soc., *97,* 6396 (1975).
2. C. H. Wei and L. F. Dahl, J. Amer. Chem. Soc., *91,* 1351 (1969), and references therein; F. A. Cotton and J. M. Troup, J. Amer. Chem. Soc., *96,* 4155 (1974).
3. G. H. Stout and L. H. Jensen, "X-ray Structure Determination," Macmillan, New York (1968).
4. For an introductory discussion of crystal geometry, see R. L. Livingston, J. Chem. Educ., *44,* 376 (1967).
5. J. A. Thich and J. A. Bercaw, unpublished results.
6. D. M. Duggan and D. N. Hendrickson, Inorg. Chem., *13,* 1911 (1974).
7. "International Tables for X-ray Crystallography," Vol. I, N. F. M. Henry and K. Lonsdale, eds., Kynoch Press, Birmingham, England (1965).
8. D. M. Duggan and D. N. Hendrickson, Inorg. Chem., *12,* 2422 (1973); *ibid.*, *13,* 2929 (1974).
9. A. L. Patterson, Z. Krist., *A90,* 517 (1935).
10. P. Main, M. M. Woolfson, and L. Lessinger, University of York, York, England, and G. Germain and J-P. Declerq, Place Louis Pasteur, 1348, Louvain-La-Neuve, Belgium.

EXERCISES

1. A crystal of $[Cu_2(tren)_2(CN)_2](BPh_4)_2$ has neutral buoyancy in a mixture of toluene and *p*-bromotoluene, 5 ml of which weighs 6.35 grams. X-ray photographs indicate the unit cell to be monoclinic with $a = 13.792$, $b = 10.338$, $c = 20.316$, and $\beta = 94.27°$. What is the number of molecules in a unit cell?

2. The space group of the copper complex in problem 1 was determined to be $P2_1/c$. If the molecules are indeed dimeric as written, what element of symmetry relates the two halves? If no spectroscopic evidence is at hand, can the monomeric and dimeric forms of this system be distinguished by the information given?

3. Show that a two-fold axis along c at $x = \frac{1}{4}$, $y = \frac{1}{4}$ and a b-glide perpendicular to c at $z = \frac{1}{4}$ require the simultaneous existence of an inversion center at $(\frac{1}{4}, 0, \frac{1}{4})$.

4. Construct a multiplication table to show that the four operations of $P2_1/c$ constitute a group.

5. Draw the *ab* and *bc* plane unit cell diagrams for the space group $C2/c$.

6. What are the relative numbers of reflections that may be collected for $[Cu_2(tren)_2(CN)_2](BPh_4)_2$ with a copper X-ray source ($\lambda = 1.54$ Å) and a molybdenum source ($\lambda = 0.712$ Å) (see problem 1 for pertinent information)?

7. What is the space group of a system whose systematic absences are $0k\ell$, $\ell = 2n + 1$; $h0\ell$, $\ell = 2n + 1$; and $hk0$, $h = 2n + 1$; and whose precession photographs show mirror symmetry about all axes?

8. Which of the following pairs of systems might be profitably studied with the powder method? Give your reasoning.

   a. $[Ni(tren)N_3](BPh_4)$ and $[Ni(tren)N_3]PF_6$

   b. $[Ni(tren)N_3](BPh_4)$ and $[Cu(tren)N_3](BPh_4)$

   c. $Rh(CN\textit{t}Bu)_4^+BF_4^-$ and $Rh(CN\varphi)_4BF_4^-$

   d. $Rh(CN\varphi)_4Cl$ and $Rh(\textit{p}\text{-}ClC_6H_4NC)_4Cl$

9. A c-glide perpendicular to a through the origin relates x, y, z to $\bar{x}, y, \frac{1}{2} + z$. Show through calculation of a general structure factor expression that the condition for systematic absences due to this element is $0k\ell$, $\ell = 2n + 1$.

10. A Patterson map calculated from data collected for $[Cu_2(tren)_2(CN)_2](BPh_4)_2$ shows a peak at (fractional coordinates) (0, 0.348, 0.500) and another of equal intensity at (0.330, 0.500, 0.702); these are the largest peaks in the map other than the origin. What are the fractional coordinates of the copper atom?

11. What is the appearance of a ΔF map region representing the actual position of an atom as being, say, ~0.5 Å from that of the atom in the model, where both atoms are spherical (*i.e.*, have isotropic temperature factors)? If the position of the atom in the model were correct and yet the actual thermal motion is anisotropic rather than isotropic as in the model, then what kind of difference in electron densities would be expected?

12. What will be the maximum absorption effect on relative intensities of reflections from a crystal of $[Cu_2(tren)_2(CN)_2](BPh_4)_2$ that is plate-like with thickness 0.1 mm and width 0.8 mm? Should a correction be made?

13. Show that a two-fold rotation operation of the origin in conjunction with a perpendicular translation, T, is equivalent to a two-fold rotation about an axis at $T/2$. Does this mean that the lattice translation [1, 0, 0] requires that the operator 2_b at $(\frac{1}{4}, 0, 0)$ be accompanied by another at $(\frac{3}{4}, 0, 0)$?

APPENDIX A

CHARACTER TABLES FOR CHEMICALLY IMPORTANT SYMMETRY GROUPS

1. THE NONAXIAL GROUPS

C_1	E
A	1

C_s	E	σ_h		
A'	1	1	x, y, R_z	x^2, y^2, z^2, xy
A''	1	-1	z, R_x, R_y	yz, xz

C_i	E	i		
A_g	1	1	R_x, R_y, R_z	x^2, y^2, z^2 xy, xz, yz
A_u	1	-1	x, y, z	

2. THE C_n GROUPS

C_2	E	C_2		
A	1	1	z, R_z	x^2, y^2, z^2, xy
B	1	-1	x, y, R_x, R_y	yz, xz

C_3	E	C_3	C_3^2		$\varepsilon = \exp(2\pi i/3)$
A	1	1	1	z, R_z	$x^2 + y^2, z^2$
E	$\begin{Bmatrix} 1 \\ 1 \end{Bmatrix}$	$\begin{matrix} \varepsilon \\ \varepsilon^* \end{matrix}$	$\begin{matrix} \varepsilon^* \\ \varepsilon \end{matrix}$	$(x, y)(R_x, R_y)$	$(x^2 - y^2, xy)(yz, xz)$

C_4	E	C_4	C_2	C_4^3		
A	1	1	1	1	z, R_z	$x^2 + y^2, z^2$
B	1	-1	1	-1		$x^2 - y^2, xy$
E	$\begin{Bmatrix} 1 \\ 1 \end{Bmatrix}$	$\begin{matrix} i \\ -i \end{matrix}$	$\begin{matrix} -1 \\ -1 \end{matrix}$	$\begin{matrix} -i \\ i \end{matrix}$	$(x, y)(R_x, R_y)$	(yz, xz)

2. THE C_n GROUPS (*continued*)

C_5	E	C_5	C_5^2	C_5^3	C_5^4		$\varepsilon = \exp(2\pi i/5)$
A	1	1	1	1	1	z, R_z	$x^2 + y^2, z^2$
E_1	1	ε	ε^2	ε^{2*}	ε^*	$(x, y)(R_x, R_y)$	(yz, xz)
	1	ε^*	ε^{2*}	ε^2	ε		
E_2	1	ε^2	ε^*	ε	ε^{2*}		$(x^2 - y^2, xy)$
	1	ε^{2*}	ε	ε^*	ε^2		

C_6	E	C_6	C_3	C_2	C_3^2	C_6^5		$\varepsilon = \exp(2\pi i/6)$
A	1	1	1	1	1	1	z, R_z	$x^2 + y^2, z^2$
B	1	-1	1	-1	1	-1		
E_1	1	ε	$-\varepsilon^*$	-1	$-\varepsilon$	ε^*	(x, y)	(xz, yz)
	1	ε^*	$-\varepsilon$	-1	$-\varepsilon^*$	ε	(R_x, R_y)	
E_2	1	$-\varepsilon^*$	$-\varepsilon$	1	$-\varepsilon^*$	$-\varepsilon$		$(x^2 - y^2, xy)$
	1	$-\varepsilon$	$-\varepsilon^*$	1	$-\varepsilon$	$-\varepsilon^*$		

C_7	E	C_7	C_7^2	C_7^3	C_7^4	C_7^5	C_7^6		$\varepsilon = \exp(2\pi i/7)$
A	1	1	1	1	1	1	1	z, R_z	$x^2 + y^2, z^2$
E_1	1	ε	ε^2	ε^3	ε^{3*}	ε^{2*}	ε^*	(x, y)	(xz, yz)
	1	ε^*	ε^{2*}	ε^{3*}	ε^3	ε^2	ε	(R_x, R_y)	
E_2	1	ε^2	ε^{3*}	ε^*	ε	ε^3	ε^{2*}		$(x^2 - y^2, xy)$
	1	ε^{2*}	ε^3	ε	ε^*	ε^{3*}	ε^2		
E_3	1	ε^3	ε^*	ε^2	ε^{2*}	ε	ε^{3*}		
	1	ε^{3*}	ε	ε^{2*}	ε^2	ε^*	ε^3		

C_8	E	C_8	C_4	C_2	C_4^3	C_8^3	C_8^5	C_8^7		$\varepsilon = \exp(2\pi i/8)$
A	1	1	1	1	1	1	1	1	z, R_z	$x^2 + y^2, z^2$
B	1	-1	1	1	1	-1	-1	-1		
E_1	1	ε	i	-1	$-i$	$-\varepsilon^*$	$-\varepsilon$	ε^*	(x, y)	(xz, yz)
	1	ε^*	$-i$	-1	i	$-\varepsilon$	$-\varepsilon^*$	ε	(R_x, R_y)	
E_2	1	i	-1	1	-1	$-i$	i	$-i$		$(x^2 - y^2, xy)$
	1	$-i$	-1	1	-1	i	$-i$	i		
E_3	1	$-\varepsilon$	i	-1	$-i$	ε^*	ε	$-\varepsilon^*$		
	1	$-\varepsilon^*$	$-i$	-1	i	ε	ε^*	$-\varepsilon$		

3. THE D_n GROUPS

D_2	E	$C_3(z)$	$C_2(y)$	$C_2(x)$		
A	1	1	1	1		x^2, y^2, z^2
B_1	1	1	-1	-1	z, R_z	xy
B_2	1	-1	1	-1	y, R_y	xz
B_3	1	-1	-1	1	x, R_x	yz

D_3	E	$2C_3$	$3C_2$		
A_1	1	1	1		$x^2 + y^2, z^2$
A_2	1	1	-1	z, R_z	
E	2	-1	0	$(x, y)(R_x, R_y)$	$(x^2 - y^2, xy)(xz, yz)$

D_4	E	$2C_4$	$C_2(=C_4^2)$	$2C_2'$	$2C_2''$		
A_1	1	1	1	1	1		$x^2 + y^2, z^2$
A_2	1	1	1	-1	-1	z, R_z	
B_1	1	-1	1	1	-1		$x^2 - y^2$
B_2	1	-1	1	-1	1		xy
E	2	0	-2	0	0	$(x, y)(R_x, R_y)$	(xz, yz)

3. THE D_n GROUPS (*continued*)

D_5	E	$2C_5$	$2C_5^2$	$5C_2$		
A_1	1	1	1	1		x^2+y^2, z^2
A_2	1	1	1	−1	z, R_z	
E_1	2	2 cos 72°	2 cos 144°	0	$(x, y)(R_x, R_y)$	(xz, yz)
E_2	2	2 cos 144°	2 cos 72°	0		(x^2-y^2, xy)

D_6	E	$2C_6$	$2C_3$	C_2	$3C_2'$	$3C_2''$		
A_1	1	1	1	1	1	1		x^2+y^2, z^2
A_2	1	1	1	1	−1	−1	z, R_z	
B_1	1	−1	1	−1	1	−1		
B_2	1	−1	1	−1	−1	1		
E_1	2	1	−1	−2	0	0	$(x, y)(R_x, R_y)$	(xz, yz)
E_2	2	−1	−1	2	0	0		(x^2-y^2, xy)

4. THE C_{nv} GROUPS

C_{2v}*	E	C_2	$\sigma_v(xz)$	$\sigma_v'(yz)$		
A_1	1	1	1	1	z	x^2, y^2, z^2
A_2	1	1	−1	−1	R_z	xy
B_1	1	−1	1	−1	x, R_y	xz
B_2	1	−1	−1	1	y, R_x	yz

* For a planar molecule the x-axis is taken perpendicular to the plane.

C_{3v}	E	$2C_3$	$3\sigma_v$		
A_1	1	1	1	z	x^2+y^2, z^2
A_2	1	1	−1	R_z	
E	2	−1	0	$(x, y)(R_x, R_y)$	$(x^2-y^2, xy)(xz, yz)$

C_{4v}*	E	$2C_4$	C_2	$2\sigma_v$	$2\sigma_d$		
A_1	1	1	1	1	1	z	x^2+y^2, z^2
A_2	1	1	1	−1	−1	R_z	
B_1	1	−1	1	1	−1		x^2-y^2
B_2	1	−1	1	−1	1		xy
E	2	0	−2	0	0	$(x, y)(R_x, R_y)$	(xz, yz)

* If the C_{4v} molecule contains a square array of atoms, the σ_v planes should pass through the larger number of atoms of the square array or should intersect the largest possible number of bonds.

C_{5v}	E	$2C_5$	$2C_5^2$	$5\sigma_v$		
A_1	1	1	1	1	z	x^2+y^2, z^2
A_2	1	1	1	−1	R_z	
E_1	2	2 cos 72°	2 cos 144°	0	$(x, y)(R_x, R_y)$	(xz, yz)
E_2	2	2 cos 144°	2 cos 72°	0		(x^2-y^2, xy)

C_{6v}	E	$2C_6$	$2C_3$	C_2	$3\sigma_v$	$3\sigma_d$		
A_1	1	1	1	1	1	1	z	x^2+y^2, z^2
A_2	1	1	1	1	−1	−1	R_z	
B_1	1	−1	1	−1	1	−1		
B_2	1	−1	1	−1	−1	1		
E_1	2	1	−1	−2	0	0	$(x, y)(R_x, R_y)$	(xz, yz)
E_2	2	−1	−1	2	0	0		(x^2-y^2, xy)

5. THE C_{nh} GROUPS

C_{2h}	E	C_2	i	σ_h		
A_g	1	1	1	1	R_z	x^2, y^2, z^2, xy
B_g	1	−1	1	−1	R_x, R_y	xz, yz
A_u	1	1	−1	−1	z	
B_u	1	−1	−1	1	x, y	

C_{3h}	E	C_3	C_3^2	σ_h	S_3	S_3^5		$\varepsilon = \exp(2\pi i/3)$
A'	1	1	1	1	1	1	R_z	$x^2 + y^2, z^2$
E'	1	ε	ε^*	1	ε	ε^*	(x, y)	$(x^2 - y^2, xy)$
	1	ε^*	ε	1	ε^*	ε		
A''	1	1	1	−1	−1	−1	z	
E''	1	ε	ε^*	−1	$-\varepsilon$	$-\varepsilon^*$	(R_x, R_y)	(xz, yz)
	1	ε^*	ε	−1	$-\varepsilon^*$	$-\varepsilon$		

C_{4h}	E	C_4	C_2	C_4^3	i	S_4^3	σ_h	S_4		
A_g	1	1	1	1	1	1	1	1	R_z	$x^2 + y^2, z^2$
B_g	1	−1	1	−1	1	−1	1	−1		$x^2 - y^2, xy$
E_g	1	i	−1	$-i$	1	i	−1	$-i$	(R_x, R_y)	(xz, yz)
	1	$-i$	−1	i	1	$-i$	−1	i		
A_u	1	1	1	1	−1	−1	−1	−1	z	
B_u	1	−1	1	−1	−1	1	−1	1		
E_u	1	i	−1	$-i$	−1	$-i$	1	i	(x, y)	
	1	$-i$	−1	i	−1	i	1	$-i$		

C_{5h}	E	C_5	C_5^2	C_5^3	C_5^4	σ_h	S_5	S_5^7	S_5^3	S_5^9		$\varepsilon = \exp(2\pi i/5)$
A'	1	1	1	1	1	1	1	1	1	1	R_z	$x^2 + y^2, z^2$
E_1'	1	ε	ε^2	ε^{2*}	ε^*	1	ε	ε^2	ε^{2*}	ε^*	(x, y)	
	1	ε^*	ε^{2*}	ε^2	ε	1	ε^*	ε^{2*}	ε^2	ε		
E_2'	1	ε^2	ε^*	ε	ε^{2*}	1	ε^2	ε^*	ε	ε^{2*}		$(x^2 - y^2, xy)$
	1	ε^{2*}	ε	ε^*	ε^2	1	ε^{2*}	ε	ε^*	ε^2		
A''	1	1	1	1	1	−1	−1	−1	−1	−1	z	
E_1''	1	ε	ε^2	ε^{2*}	ε^*	−1	$-\varepsilon$	$-\varepsilon^2$	$-\varepsilon^{2*}$	$-\varepsilon^*$	(R_x, R_y)	(xz, yz)
	1	ε^*	ε^{2*}	ε^2	ε	−1	$-\varepsilon^*$	$-\varepsilon^{2*}$	$-\varepsilon^2$	$-\varepsilon$		
E_2''	1	ε^2	ε^*	ε	ε^{2*}	−1	$-\varepsilon^2$	$-\varepsilon^*$	$-\varepsilon$	$-\varepsilon^{2*}$		
	1	ε^{2*}	ε	ε^*	ε^2	−1	$-\varepsilon^{2*}$	$-\varepsilon$	$-\varepsilon^*$	$-\varepsilon^2$		

C_{6h}	E	C_6	C_3	C_2	C_3^2	C_6^5	i	S_3^5	S_6^5	σ_h	S_6	S_3		$\varepsilon = \exp(2\pi i/6)$
A_g	1	1	1	1	1	1	1	1	1	1	1	1	R_z	$x^2 + y^2, z^2$
B_g	1	−1	1	−1	1	−1	1	−1	1	−1	1	−1		
E_{1g}	1	ε	$-\varepsilon^*$	−1	$-\varepsilon$	ε^*	1	ε	$-\varepsilon^*$	−1	$-\varepsilon$	ε^*	(R_x, R_y)	(xz, yz)
	1	ε^*	$-\varepsilon$	−1	$-\varepsilon^*$	ε	1	ε^*	$-\varepsilon$	−1	$-\varepsilon^*$	ε		
E_{2g}	1	$-\varepsilon^*$	$-\varepsilon$	1	$-\varepsilon^*$	$-\varepsilon$	1	$-\varepsilon^*$	$-\varepsilon$	1	$-\varepsilon^*$	$-\varepsilon$		$(x^2 - y^2, xy)$
	1	$-\varepsilon$	$-\varepsilon^*$	1	$-\varepsilon$	$-\varepsilon^*$	1	$-\varepsilon$	$-\varepsilon^*$	1	$-\varepsilon$	$-\varepsilon^*$		
A_u	1	1	1	1	1	1	−1	−1	−1	−1	−1	−1	z	
B_u	1	−1	1	−1	1	−1	−1	1	−1	1	−1	1		
E_{1u}	1	ε	$-\varepsilon^*$	−1	$-\varepsilon$	ε^*	−1	$-\varepsilon$	ε^*	1	ε	$-\varepsilon^*$	(x, y)	
	1	ε^*	$-\varepsilon$	−1	$-\varepsilon^*$	ε	−1	$-\varepsilon^*$	ε	1	ε^*	$-\varepsilon$		
E_{2u}	1	$-\varepsilon^*$	$-\varepsilon$	1	$-\varepsilon^*$	$-\varepsilon$	−1	ε^*	ε	−1	ε^*	ε		
	1	$-\varepsilon$	$-\varepsilon^*$	1	$-\varepsilon$	$-\varepsilon^*$	−1	ε	ε^*	−1	ε	ε^*		

6. THE D_{nh} GROUPS

D_{2h}	E	$C_2(z)$	$C_2(y)$	$C_2(x)$	i	$\sigma(xy)$	$\sigma(xz)$	$\sigma(yz)$		
A_g	1	1	1	1	1	1	1	1		x^2, y^2, z^2
B_{1g}	1	1	−1	−1	1	1	−1	−1	R_z	xy
B_{2g}	1	−1	1	−1	1	−1	1	−1	R_y	xz
B_{3g}	1	−1	−1	1	1	−1	−1	1	R_x	yz
A_u	1	1	1	1	−1	−1	−1	−1		
B_{1u}	1	1	−1	−1	−1	−1	1	1	z	
B_{2u}	1	−1	1	−1	−1	1	−1	1	y	
B_{3u}	1	−1	−1	1	−1	1	1	−1	x	

D_{3h}	E	$2C_3$	$3C_2$	σ_h	$2S_3$	$3\sigma_v$		
A_1'	1	1	1	1	1	1		$x^2 + y^2, z^2$
A_2'	1	1	−1	1	1	−1	R_z	
E'	2	−1	0	2	−1	0	(x, y)	$(x^2 - y^2, xy)$
A_1''	1	1	1	−1	−1	−1		
A_2''	1	1	−1	−1	−1	1	z	
E''	2	−1	0	−2	1	0	(R_x, R_y)	(xz, yz)

D_{4h}*	E	$2C_4$	C_2	$2C_2'$	$2C_2''$	i	$2S_4$	σ_h	$2\sigma_v$	$2\sigma_d$		
A_{1g}	1	1	1	1	1	1	1	1	1	1		$x^2 + y^2, z^2$
A_{2g}	1	1	1	−1	−1	1	1	1	−1	−1	R_z	
B_{1g}	1	−1	1	1	−1	1	−1	1	1	−1		$x^2 - y^2$
B_{2g}	1	−1	1	−1	1	1	−1	1	−1	1		xy
E_g	2	0	−2	0	0	2	0	−2	0	0	(R_x, R_y)	(xz, yz)
A_{1u}	1	1	1	1	1	−1	−1	−1	−1	−1		
A_{2u}	1	1	1	−1	−1	−1	−1	−1	1	1	z	
B_{1u}	1	−1	1	1	−1	−1	1	−1	−1	1		
B_{2u}	1	−1	1	−1	1	−1	1	−1	1	−1		
E_u	2	0	−2	0	0	−2	0	2	0	0	(x, y)	

* σ_v passes through the atoms and σ_d bisects the bond angles.

D_{5h}	E	$2C_5$	$2C_5^2$	$5C_2$	σ_h	$2S_5$	$2S_5^3$	$5\sigma_v$		
A_1'	1	1	1	1	1	1	1	1		$x^2 + y^2, z^2$
A_2'	1	1	1	−1	1	1	1	−1	R_z	
E_1'	2	2 cos 72°	2 cos 144°	0	2	2 cos 72°	2 cos 144°	0	(x, y)	
E_2'	2	2 cos 144°	2 cos 72°	0	2	2 cos 144°	2 cos 72°	0		$(x^2 - y^2, xy)$
A_1''	1	1	1	1	−1	−1	−1	−1		
A_2''	1	1	1	−1	−1	−1	−1	1	z	
E_1''	2	2 cos 72°	2 cos 144°	0	−2	−2 cos 72°	−2 cos 144°	0	(R_x, R_y)	(xz, yz)
E_2''	2	2 cos 144°	2 cos 72°	0	−2	−2 cos 144°	−2 cos 72°	0		

D_{6h}	E	$2C_6$	$2C_3$	C_2	$3C_2'$	$3C_2''$	i	$2S_3$	$2S_6$	σ_h	$3\sigma_d$	$3\sigma_v$		
A_{1g}	1	1	1	1	1	1	1	1	1	1	1	1		$x^2 + y^2, z^2$
A_{2g}	1	1	1	1	−1	−1	1	1	1	1	−1	−1	R_z	
B_{1g}	1	−1	1	−1	1	−1	1	−1	1	−1	1	−1		
B_{2g}	1	−1	1	−1	−1	1	1	−1	1	−1	−1	1		
E_{1g}	2	1	−1	−2	0	0	2	1	−1	−2	0	0	(R_x, R_y)	(xz, yz)
E_{2g}	2	−1	−1	2	0	0	2	−1	−1	2	0	0		$(x^2 - y^2, xy)$
A_{1u}	1	1	1	1	1	1	−1	−1	−1	−1	−1	−1		
A_{2u}	1	1	1	1	−1	−1	−1	−1	−1	−1	1	1	z	
B_{1u}	1	−1	1	−1	1	−1	−1	1	−1	1	−1	1		
B_{2u}	1	−1	1	−1	−1	1	−1	1	−1	1	1	−1		
E_{1u}	2	1	−1	−2	0	0	−2	−1	1	2	0	0	(x, y)	
E_{2u}	2	−1	−1	2	0	0	−2	1	1	−2	0	0		

6. THE D_{nh} GROUPS (*continued*)

D_{8h}	E	$2C_8$	$2C_8^3$	$2C_4$	C_2	$4C_2'$	$4C_2''$	i	$2S_8$	$2S_8^3$	$2S_4$	σ_h	$4\sigma_d$	$4\sigma_v$		
A_{1g}	1	1	1	1	1	1	1	1	1	1	1	1	1	1		$x^2 + y^2, z^2$
A_{2g}	1	1	1	1	1	−1	−1	1	1	1	1	1	−1	−1	R_z	
B_{1g}	1	−1	−1	1	1	1	−1	1	−1	−1	1	1	1	−1		
B_{2g}	1	−1	−1	1	1	−1	1	1	−1	−1	1	1	−1	1		
E_{1g}	2	$\sqrt{2}$	$-\sqrt{2}$	0	−2	0	0	2	$\sqrt{2}$	$-\sqrt{2}$	0	−2	0	0	(R_x, R_y)	(xz, yz)
E_{2g}	2	0	0	−2	2	0	0	2	0	0	−2	2	0	0		$(x^2 - y^2, xy)$
E_{3g}	2	$-\sqrt{2}$	$\sqrt{2}$	0	−2	0	0	2	$-\sqrt{2}$	$\sqrt{2}$	0	−2	0	0		
A_{1u}	1	1	1	1	1	1	1	−1	−1	−1	−1	−1	−1	−1		
A_{2u}	1	1	1	1	1	−1	−1	−1	−1	−1	−1	−1	1	1	z	
B_{1u}	1	−1	−1	1	1	1	−1	−1	1	1	−1	−1	−1	1		
B_{2u}	1	−1	−1	1	1	−1	1	−1	1	1	−1	−1	1	−1		
E_{1u}	2	$\sqrt{2}$	$-\sqrt{2}$	0	−2	0	0	−2	$-\sqrt{2}$	$\sqrt{2}$	0	2	0	0	(x, y)	
E_{2u}	2	0	0	−2	2	0	0	−2	0	0	2	−2	0	0		
E_{3u}	2	$-\sqrt{2}$	$\sqrt{2}$	0	−2	0	0	−2	$\sqrt{2}$	$-\sqrt{2}$	0	2	0	0		

7. THE D_{nd} GROUPS

D_{2d}	E	$2S_4$	C_2	$2C_2'$	$2\sigma_d$		
A_1	1	1	1	1	1		$x^2 + y^2, z^2$
A_2	1	1	1	−1	−1	R_z	
B_1	1	−1	1	1	−1		$x^2 - y^2$
B_2	1	−1	1	−1	1	z	xy
E	2	0	−2	0	0	(x, y), (R_x, R_y)	(xz, yz)

D_{3d}	E	$2C_3$	$3C_2$	i	$2S_6$	$3\sigma_d$		
A_{1g}	1	1	1	1	1	1		$x^2 + y^2, z^2$
A_{2g}	1	1	−1	1	1	−1	R_z	
E_g	2	−1	0	2	−1	0	(R_x, R_y)	$(x^2 - y^2, xy)$, (xz, yz)
A_{1u}	1	1	1	−1	−1	−1		
A_{2u}	1	1	−1	−1	−1	1	z	
E_u	2	−1	0	−2	1	0	(x, y)	

D_{4d}	E	$2S_8$	$2C_4$	$2S_8^3$	C_2	$4C_2'$	$4\sigma_d$		
A_1	1	1	1	1	1	1	1		$x^2 + y^2, z^2$
A_2	1	1	1	1	1	−1	−1	R_z	
B_1	1	−1	1	−1	1	1	−1		
B_2	1	−1	1	−1	1	−1	1	z	
E_1	2	$\sqrt{2}$	0	$-\sqrt{2}$	−2	0	0	(x, y)	
E_2	2	0	−2	0	2	0	0		$(x^2 - y^2, xy)$
E_3	2	$-\sqrt{2}$	0	$\sqrt{2}$	−2	0	0	(R_x, R_y)	(xz, yz)

D_{5d}	E	$2C_5$	$2C_5^2$	$5C_2$	i	$2S_{10}^3$	$2S_{10}$	$5\sigma_d$		
A_{1g}	1	1	1	1	1	1	1	1		$x^2 + y^2, z^2$
A_{2g}	1	1	1	−1	1	1	1	−1	R_z	
E_{1g}	2	2 cos 72°	2 cos 144°	0	2	2 cos 72°	2 cos 144°	0	(R_x, R_y)	(xz, yz)
E_{2g}	2	2 cos 144°	2 cos 72°	0	2	2 cos 144°	2 cos 72°	0		$(x^2 - y^2, xy)$
A_{1u}	1	1	1	1	−1	−1	−1	−1		
A_{2u}	1	1	1	−1	−1	−1	−1	1	z	
E_{1u}	2	2 cos 72°	2 cos 144°	0	−2	−2 cos 72°	−2 cos 144°	0	(x, y)	
E_{2u}	2	2 cos 144°	2 cos 72°	0	−2	−2 cos 144°	−2 cos 72°	0		

7. THE D_{nd} GROUPS (*continued*)

D_{6d}	E	$2S_{12}$	$2C_6$	$2S_4$	$2C_3$	$2S_{12}^5$	C_2	$6C_2'$	$6\sigma_d$		
A_1	1	1	1	1	1	1	1	1	1		x^2+y^2, z^2
A_2	1	1	1	1	1	1	1	-1	-1	R_z	
B_1	1	-1	1	-1	1	-1	1	1	-1		
B_2	1	-1	1	-1	1	-1	1	-1	1	z	
E_1	2	$\sqrt{3}$	1	0	-1	$-\sqrt{3}$	-2	0	0	(x, y)	
E_2	2	1	-1	-2	-1	1	2	0	0		(x^2-y^2, xy)
E_3	2	0	-2	0	2	0	-2	0	0		
E_4	2	-1	-1	2	-1	-1	2	0	0		
E_5	2	$-\sqrt{3}$	1	0	-1	$\sqrt{3}$	-2	0	0	(R_x, R_y)	(xz, yz)

8. THE S_n GROUPS

S_4	E	S_4	C_2	S_4^3		
A	1	1	1	1	R_z	x^2+y^2, z^2
B	1	-1	1	-1	z	x^2-y^2, xy
E	$\begin{Bmatrix}1 \\ 1\end{Bmatrix}$	$\begin{matrix}i \\ -i\end{matrix}$	$\begin{matrix}-1 \\ -1\end{matrix}$	$\begin{matrix}-i \\ i\end{matrix}$	(x, y); (R_x, R_y)	(xz, yz)

S_6	E	C_3	C_3^2	i	S_6^5	S_6		$\varepsilon = \exp(2\pi i/3)$
A_g	1	1	1	1	1	1	R_z	x^2+y^2, z^2
E_g	1	ε	ε^*	1	ε	ε^*	(R_x, R_y)	(x^2-y^2, xy); (xz, yz)
	1	ε^*	ε	1	ε^*	ε		
A_u	1	1	1	-1	-1	-1	z	
E_u	1	ε	ε^*	-1	$-\varepsilon$	$-\varepsilon^*$	(x, y)	
	1	ε^*	ε	-1	$-\varepsilon^*$	$-\varepsilon$		

S_8	E	S_8	C_4	S_8^3	C_2	S_8^5	C_4^3	S_8^7		$\varepsilon = \exp(2\pi i/8)$
A	1	1	1	1	1	1	1	1	R_z	x^2+y^2, z^2
B	1	-1	1	-1	1	-1	1	-1	z	
E_1	1	ε	i	$-\varepsilon^*$	-1	$-\varepsilon$	$-i$	ε^*	(x, y);	
	1	ε^*	$-i$	$-\varepsilon$	-1	$-\varepsilon^*$	i	ε	(R_x, R_y)	
E_2	1	i	-1	$-i$	1	i	-1	$-i$		(x^2-y^2, xy)
	1	$-i$	-1	i	1	$-i$	-1	i		
E_3	1	$-\varepsilon^*$	$-i$	ε	-1	ε^*	i	$-\varepsilon$		(xz, yz)
	1	$-\varepsilon$	i	ε^*	-1	ε	$-i$	$-\varepsilon^*$		

9. THE CUBIC GROUPS

T	E	$4C_3$	$4C_3^2$	$3C_2$		$\varepsilon = \exp(2\pi i/3)$
A	1	1	1	1		$x^2+y^2+z^2$
E	1	ε	ε^*	1		$(2z^2-x^2-y^2,$
	1	ε^*	ε	1		$x^2-y^2)$
T	3	0	0	-1	(R_x, R_y, R_z); (x, y, z)	(xy, xz, yz)

T_h	E	$4C_3$	$4C_3^2$	$3C_2$	i	$4S_6$	$4S_6^5$	$3\sigma_h$		$\varepsilon = \exp(2\pi i/3)$
A_g	1	1	1	1	1	1	1	1		$x^2+y^2+z^2$
A_u	1	1	1	1	-1	-1	-1	-1		
E_g	1	ε	ε^*	1	1	ε	ε^*	1		$(2z^2-x^2-y^2,$
	1	ε^*	ε	1	1	ε^*	ε	1		$x^2-y^2)$
E_u	1	ε	ε^*	1	-1	$-\varepsilon$	$-\varepsilon^*$	-1		
	1	ε^*	ε	1	-1	$-\varepsilon^*$	$-\varepsilon$	-1		
T_g	3	0	0	-1	1	0	0	-1	(R_x, R_y, R_z)	(xz, yz, xy)
T_u	3	0	0	-1	-1	0	0	1	(x, y, z)	

9. THE CUBIC GROUPS (*continued*)

T_d	E	$8C_3$	$3C_2$	$6S_4$	$6\sigma_d$		
A_1	1	1	1	1	1		$x^2 + y^2 + z^2$
A_2	1	1	1	−1	−1		
E	2	−1	2	0	0		$(2z^2 - x^2 - y^2,$ $x^2 - y^2)$
T_1	3	0	−1	1	−1	(R_x, R_y, R_z)	
T_2	3	0	−1	−1	1	(x, y, z)	(xy, xz, yz)

O	E	$6C_4$	$3C_2(= C_4^2)$	$8C_3$	$6C_2$		
A_1	1	1	1	1	1		$x^2 + y^2 + z^2$
A_2	1	−1	1	1	−1		
E	2	0	2	−1	0		$(2z^2 - x^2 - y^2,$ $x^2 - y^2)$
T_1	3	1	−1	0	−1	(R_x, R_y, R_z); (x, y, z)	
T_2	3	−1	−1	0	1		(xy, xz, yz)

O_h	E	$8C_3$	$6C_2$	$6C_4$	$3C_2(= C_4^2)$	i	$6S_4$	$8S_6$	$3\sigma_h$	$6\sigma_d$		
A_{1g}	1	1	1	1	1	1	1	1	1	1		$x^2 + y^2 + z^2$
A_{2g}	1	1	−1	−1	1	1	−1	1	1	−1		
E_g	2	−1	0	0	2	2	0	−1	2	0		$(2z^2 - x^2 - y^2,$ $x^2 - y^2)$
T_{1g}	3	0	−1	1	−1	3	1	0	−1	−1	(R_x, R_y, R_z)	
T_{2g}	3	0	1	−1	−1	3	−1	0	−1	1		(xz, yz, xy)
A_{1u}	1	1	1	1	1	−1	−1	−1	−1	−1		
A_{2u}	1	1	−1	−1	1	−1	1	−1	−1	1		
E_u	2	−1	0	0	2	−2	0	1	−2	0		
T_{1u}	3	0	−1	1	−1	−3	−1	0	1	1	(x, y, z)	
T_{2u}	3	0	1	−1	−1	−3	1	0	1	−1		

10. THE GROUPS $C_{\infty v}$ AND $D_{\infty h}$ FOR LINEAR MOLECULES

$C_{\infty v}$	E	$2C_\infty^\Phi$	...	$\infty\sigma_v$		
$A_1 \equiv \Sigma^+$	1	1	...	1	z	$x^2 + y^2, z^2$
$A_2 \equiv \Sigma^-$	1	1	...	−1	R_z	
$E_1 \equiv \Pi$	2	$2 \cos \Phi$	...	0	(x, y); (R_x, R_y)	(xz, yz)
$E_2 \equiv \Delta$	2	$2 \cos 2\Phi$	...	0		$(x^2 - y^2, xy)$
$E_3 \equiv \Phi$	2	$2 \cos 3\Phi$	...	0		
...	...	...	...	...		

$D_{\infty h}$	E	$2C_\infty^\Phi$	...	$\infty\sigma_v$	i	$2S_\infty^\Phi$	...	∞C_2		
Σ_g^+	1	1	...	1	1	1	...	1		$x^2 + y^2, z^2$
Σ_g^-	1	1	...	−1	1	1	...	−1	R_z	
Π_g	2	$2 \cos \Phi$	...	0	2	$-2 \cos \Phi$	...	0	(R_x, R_y)	(xz, yz)
Δ_g	2	$2 \cos 2\Phi$	...	0	2	$2 \cos 2\Phi$	...	0		$(x^2 - y^2, xy)$
...	...	...	...	...	...	...	...	...		
Σ_u^+	1	1	...	1	−1	−1	...	−1	z	
Σ_u^-	1	1	...	−1	−1	−1	...	1		
Π_u	2	$2 \cos \Phi$	...	0	−2	$2 \cos \Phi$	...	0	(x, y)	
Δ_u	2	$2 \cos 2\Phi$	...	0	−2	$-2 \cos 2\Phi$	...	0		
...	...	...	...	...	...	...	...	...		

11. THE ICOSAHEDRAL GROUPS*

I_h	E	$12C_5$	$12C_5^2$	$20C_3$	$15C_2$	i	$12S_{10}$	$12S_{10}^3$	$20S_6$	15σ		
A_g	1	1	1	1	1	1	1	1	1	1		$x^2 + y^2 + z^2$
T_{1g}	3	$\frac{1}{2}(1 + \sqrt{5})$	$\frac{1}{2}(1 - \sqrt{5})$	0	-1	3	$\frac{1}{2}(1 - \sqrt{5})$	$\frac{1}{2}(1 + \sqrt{5})$	0	-1	(R_x, R_y, R_z)	
T_{2g}	3	$\frac{1}{2}(1 - \sqrt{5})$	$\frac{1}{2}(1 + \sqrt{5})$	0	-1	3	$\frac{1}{2}(1 + \sqrt{5})$	$\frac{1}{2}(1 - \sqrt{5})$	0	-1		
G_g	4	-1	-1	1	0	4	-1	-1	1	0		
H_g	5	0	0	-1	1	5	0	0	-1	1		$(2z^2 - x^2 - y^2,$ $x^2 - y^2,$ $xy, yz, zx)$
A_u	1	1	1	1	1	-1	-1	-1	-1	-1		
T_{1u}	3	$\frac{1}{2}(1 + \sqrt{5})$	$\frac{1}{2}(1 - \sqrt{5})$	0	-1	-3	$-\frac{1}{2}(1 - \sqrt{5})$	$-\frac{1}{2}(1 + \sqrt{5})$	0	1	(x, y, z)	
T_{2u}	3	$\frac{1}{2}(1 - \sqrt{5})$	$\frac{1}{2}(1 + \sqrt{5})$	0	-1	-3	$-\frac{1}{2}(1 + \sqrt{5})$	$-\frac{1}{2}(1 - \sqrt{5})$	0	1		
G_u	4	-1	-1	1	0	-4	1	1	-1	0		
H_u	5	0	0	-1	1	-5	0	0	1	-1		

*For the pure rotation group I, the outlined section in the upper left is the character table; the g subscripts should, of course, be dropped and (x, y, z) assigned to the T_1 representation.

APPENDIX B

CHARACTER TABLES FOR SOME DOUBLE GROUPS

GROUP C_3'

C_3'		E	R	C_3	C_3R	C_3^2	C_3^2R	
A'	Γ_1	1	1	1	1	1	1	z, L_z
E'	Γ_2	1	1	ω^2	ω^2	$-\omega$	$-\omega$	x, y, L_x, L_y
	Γ_3	1	1	$-\omega$	$-\omega$	ω^2	ω^2	
$E_{1/2}'$	Γ_4	1	-1	ω	$-\omega$	ω^2	$-\omega^2$	
	Γ_5	1	-1	$-\omega^2$	ω^2	$-\omega$	ω	
$B_{3/2}'$	Γ_6	1	-1	-1	1	1	-1	

$$\omega = \exp(i\pi/3)$$

GROUPS D_3' AND C_{3v}'

D_3'			E	R	C_3 C_3^2R	C_3^2 C_3R	$3C_2'$	$3C_2'R$	D_3'	
C_{3v}'			E	R	C_3 C_3^2R	C_3^2 C_3R	$3\sigma_v$	$3\sigma_vR$		C_{3v}'
A_1'		Γ_1	1	1	1	1	1	1		z
A_2'		Γ_2	1	1	1	1	-1	-1	z, L_z	L_z
E'		Γ_3	2	2	-1	-1	0	0	x, y, L_x, L_y	x, y, L_x, L_y
$E_{3/2}'$	(A_1')	Γ_4	1	-1	-1	1	i	$-i$		
	(A_2')	Γ_5	1	-1	-1	1	$-i$	i		
$E_{1/2}'$	(E')	Γ_6	2	-2	1	-1	0	0		

GROUPS C_4' AND S_4'

C_4'		E	R	C_4	C_4R	C_4^2	C_4^2R	C_4^3	C_4^3R	C_4'	
S_4'		E	R	S_4	S_4R	C_2R	C_2	S_4^3	S_4^3R		S_4'
A'	Γ_1	1	1	1	1	1	1	1	1	z	L_z
B'	Γ_2	1	1	-1	-1	1	1	-1	-1	L_z	z
E'	Γ_3	1	1	i	i	-1	-1	$-i$	$-i$	$x \pm iy$	$x \pm iy$
	Γ_4	1	1	$-i$	$-i$	-1	-1	i	i	$L_x \pm iL_y$	$L_x \pm iL_y$
$E_{1/2}'$	Γ_5	1	-1	ω	$-\omega$	i	$-i$	ω^3	$-\omega^3$		
	Γ_6	1	-1	$-\omega^3$	ω^3	$-i$	i	$-\omega$	ω		
$E_{3/2}'$	Γ_7	1	-1	$-\omega$	ω	i	$-i$	$-\omega^3$	ω^3		
	Γ_8	1	-1	ω^3	$-\omega^3$	$-i$	i	ω	$-\omega$		

$\omega = \exp(i\pi/4)$

GROUP T'

T'		E	R	$3C_2$ $3C_2R$	$4C_3$	$4C_3R$	$4C_3^2$	$4C_3^2R$	
A'	Γ_1	1	1	1	1	1	1	1	
E'	Γ_2	1	1	1	ω	ω	ω^2	ω^2	
	Γ_3	1	1	1	ω^2	ω^2	ω	ω	
T'	Γ_4	3	3	-1	0	0	0	0	x, y, z, L_x, L_y, L_z
$E_{1/2}'$	Γ_5	2	-2	0	1	-1	1	-1	
$G_{3/2}'$	Γ_6	2	-2	0	ω	$-\omega$	ω^2	$-\omega^2$	
	Γ_7	2	-2	0	ω^2	$-\omega^2$	ω	$-\omega$	

$\omega = \exp(2\pi i/3)$

GROUPS O' AND T_d'

		O'	E	R	$4C_3$ $4C_3^2R$	$4C_3^2$ $4C_3R$	$3C_4^2$ $3C_4^2R$	$3C_4$ $3C_4^3R$	$3C_4^3$ $3C_4R$	$3C_2'$ $3C_2'R$	O'	
		T_d'	E	R	$4C_3$ $4C_3^2R$	$4C_3^2$ $4C_3R$	$3C_4^2$ $3C_4^2R$	$3S_4$ $3S_4^3R$	$3S_4^3$ $3S_4R$	$6\sigma_d$ $6\sigma_dR$		T_d'
	A_1'	Γ_1	1	1	1	1	1	1	1	1		
	A_2'	Γ_2	1	1	1	1	1	-1	-1	-1		
	E'	Γ_3	2	2	-1	-1	2	0	0	0		
	T_1'	Γ_4	3	3	0	0	-1	1	1	-1	x, y, z, L_x, L_y, L_z	L_x, L_y, L_z
	T_2'	Γ_5	3	3	0	0	-1	-1	-1	1		x, y, z
(E_1')	$E_{1/2}'$	Γ_6	2	-2	1	-1	0	$\sqrt{2}$	$-\sqrt{2}$	0		
(E_2')	$E_{5/2}'$	Γ_7	2	-2	1	-1	0	$-\sqrt{2}$	$\sqrt{2}$	0		
(G')	$G_{3/2}'$	Γ_8	4	-4	-1	1	0	0	0	0		

GROUP D_4'

D_4'		E	R	C_4 C_4^3R	C_4^3 C_4R	C_2 C_2R	$2C_2'$ $2C_2'R$	$2C_2''$ $2C_2''R$
A_1'	Γ_1	1	1	1	1	1	1	1
A_2'	Γ_2	1	1	1	1	1	-1	-1
B_1'	Γ_3	1	1	-1	-1	1	1	-1
B_2'	Γ_4	1	1	-1	-1	1	-1	1
E_1'	Γ_5	2	2	0	0	-2	0	0
E_2'	Γ_6	2	-2	$\sqrt{2}$	$-\sqrt{2}$	0	0	0
E_3'	Γ_7	2	-2	$-\sqrt{2}$	$\sqrt{2}$	0	0	0

APPENDIX C

NORMAL VIBRATION MODES FOR COMMON STRUCTURES

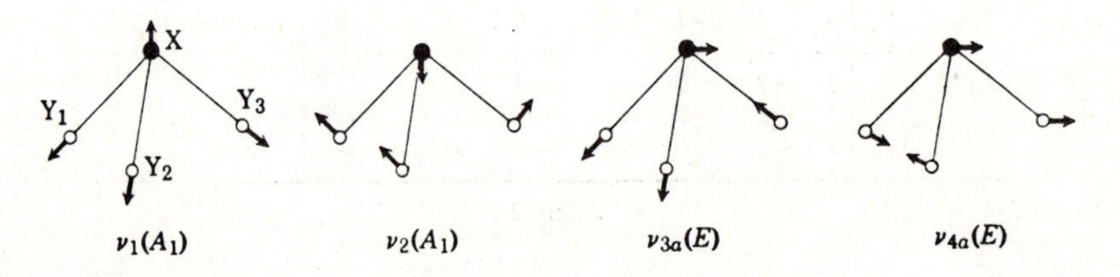

Pyramidal XY_3 Molecules.

Y_4 X Y_3 Y_1 Y_2

$\nu_1(A_1)$ $\nu_2(E)$ $\nu_3(F_2)$ $\nu_4(F_2)$

Tetrahedral XY_4 Molecules.

Y_3 X + − + + Y_1 Y_2

$\nu_1(A'_1)$ $\nu_2(A''_2)$ $\nu_3(E')$ $\nu_4(E')$

Planar XY_3 Molecules.

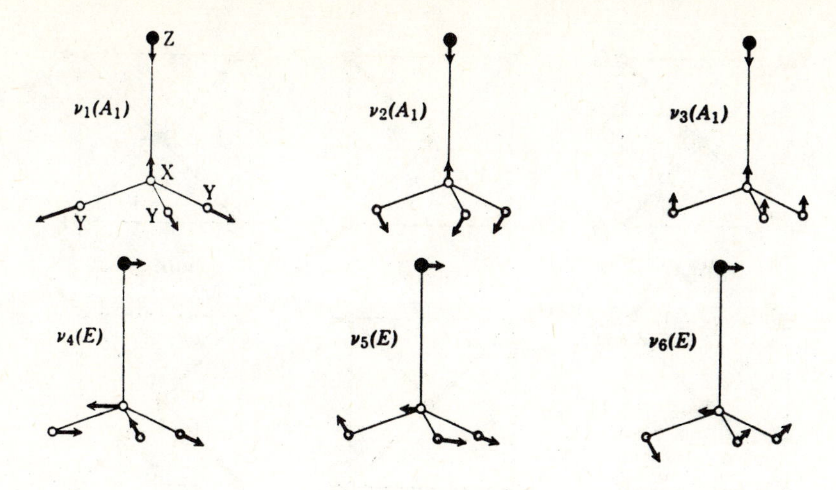

C_{3v} ZXY_3 Molecules.

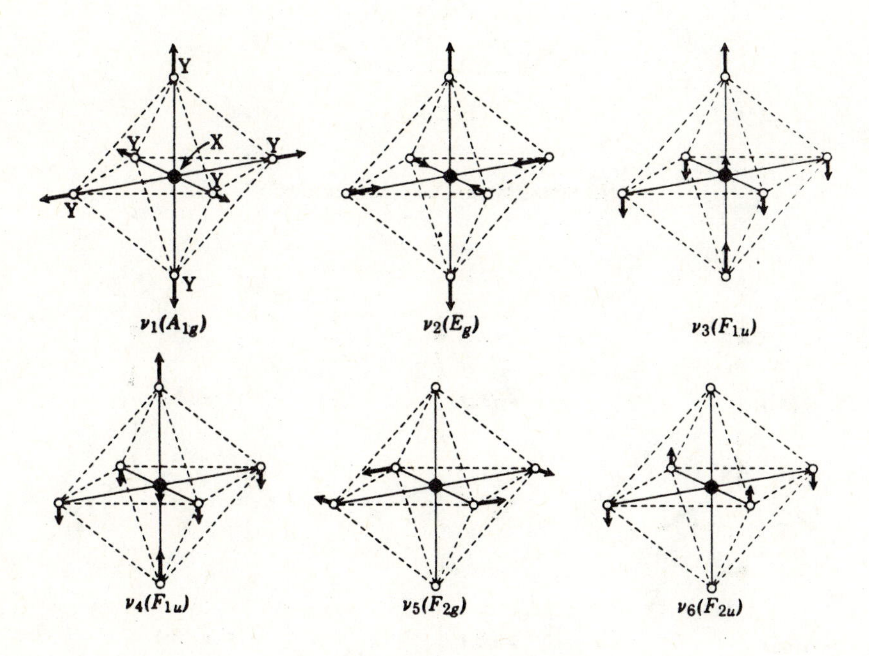

Octahedral XY_6 Molecules.

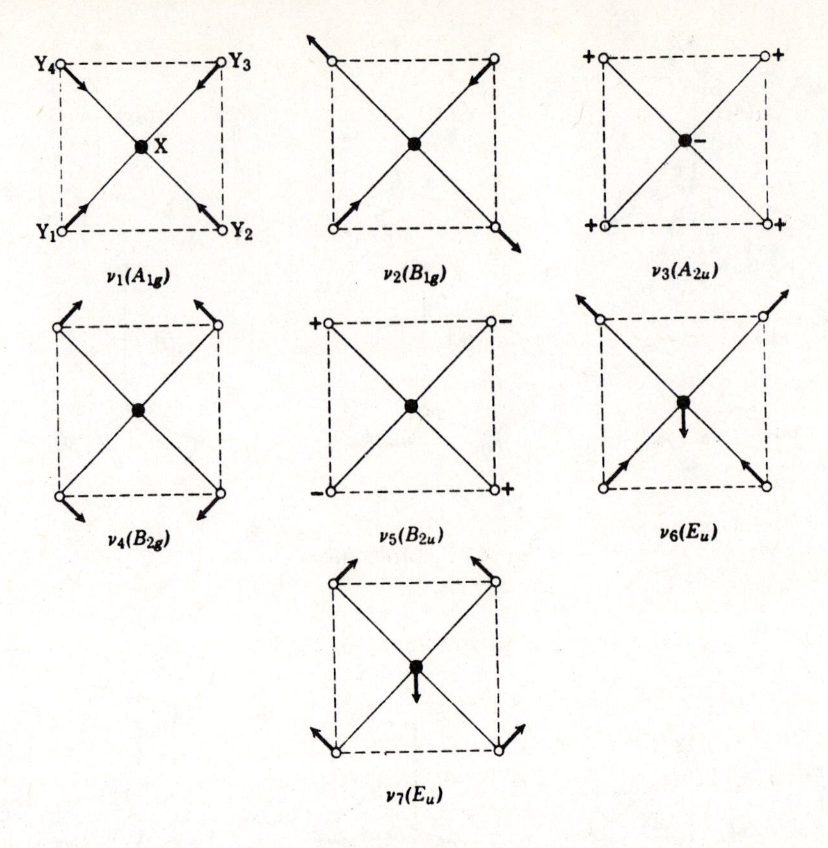

Square-Planar XY_4 Molecules.

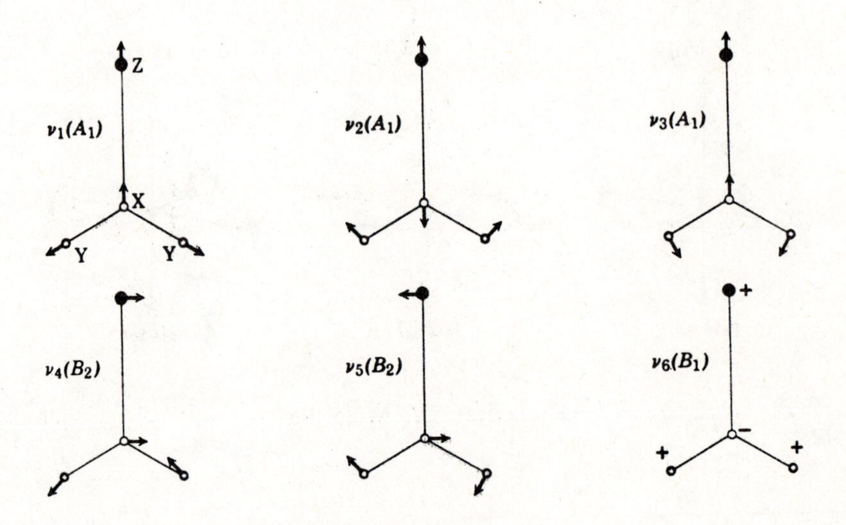

Planar ZXY_2 Molecules.

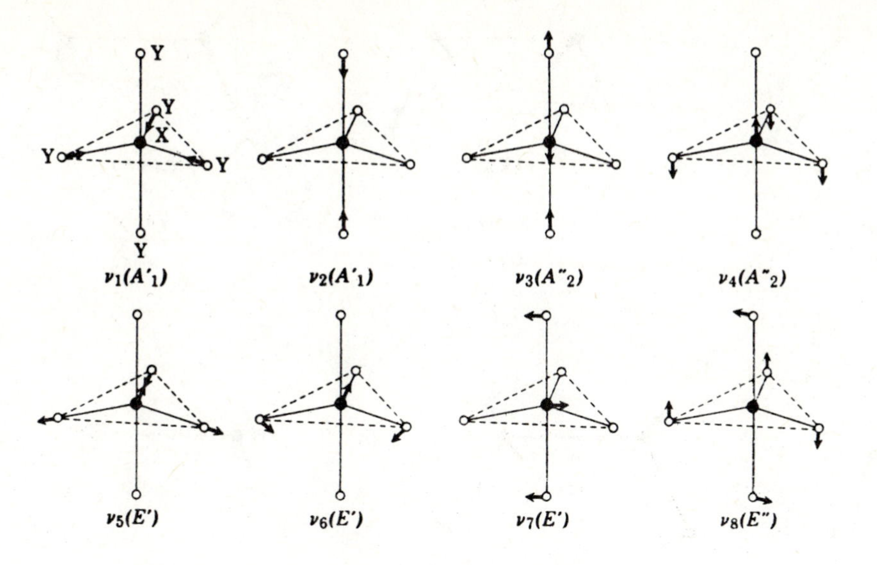

Trigonal Bipyramidal XY_5 Molecules.

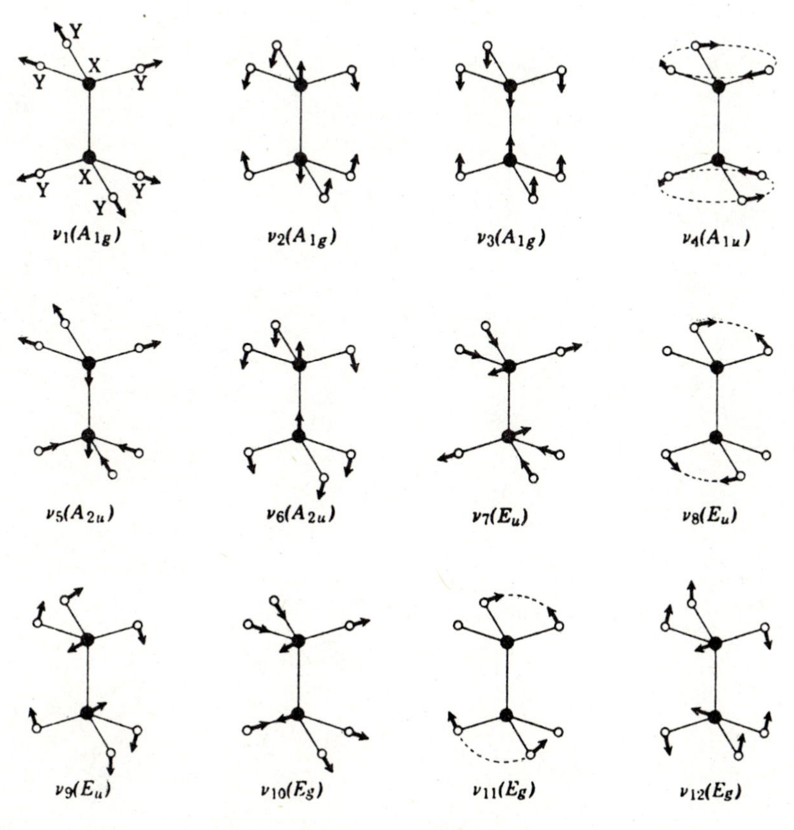

Ethane-Type X_2Y_6 Molecules.

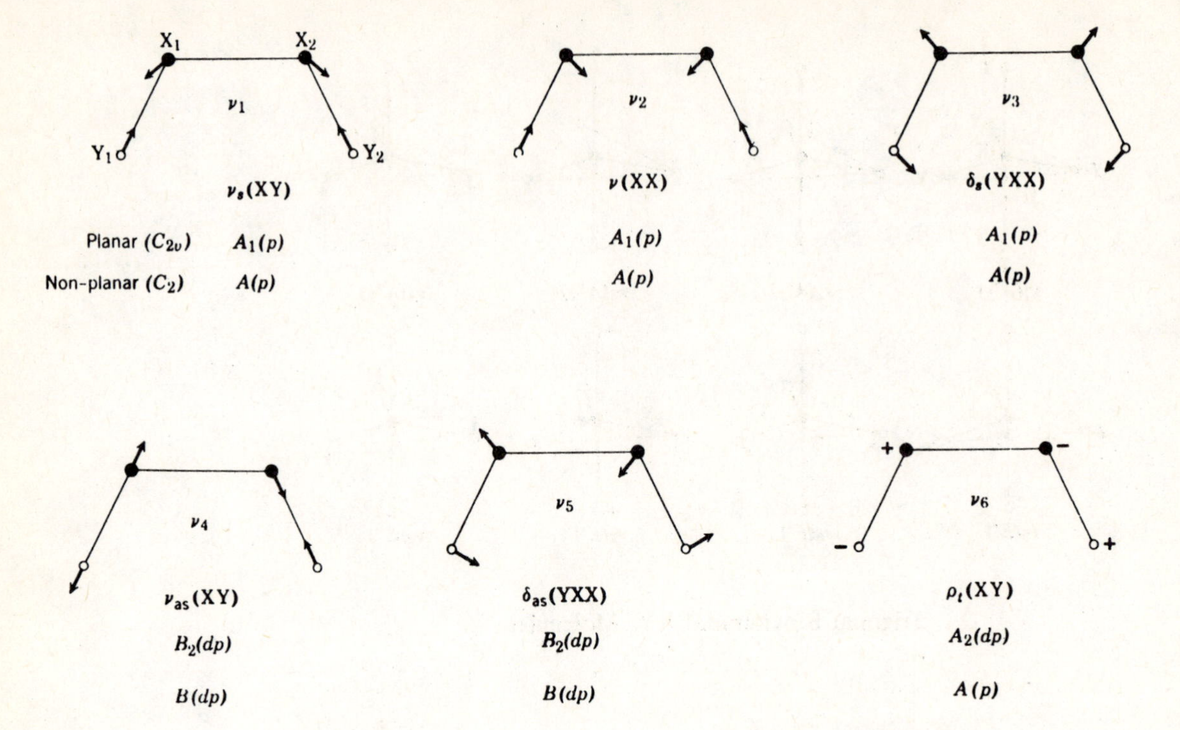

Nonlinear X_2Y_2 Molecules (*p*: Polarized; *dp*: Depolarized).

APPENDIX D

TANABE AND SUGANO DIAGRAMS FOR O_h FIELDS*

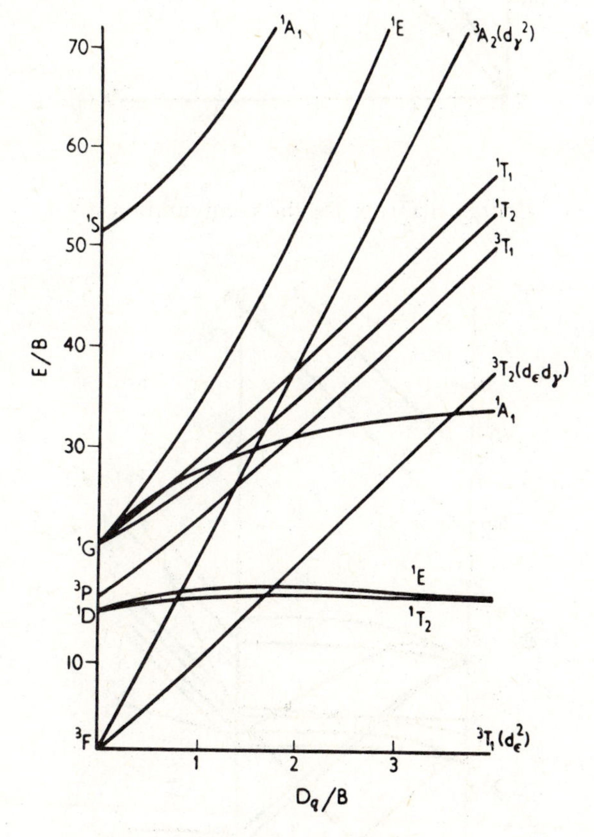

Energy diagram for the configuration d^2.

* These are complete energy diagrams for the configurations indicated, reproduced from Y. Tanabe and S. Sugano, J. Phys. Soc. Japan, 9, 753, 766 (1964).

γ refers to the ratio of the Racah parameters C/B. Heavy lines perpendicular to the Dq/B axis in d^4, d^5, d^6, and d^7 indicate transitions from weak to strong fields. The calculation of and the assumptions inherent in the calculations are contained in the original paper. The diagrams do not apply to any particular complex but give a qualitative indication of the energies of the various states as a function of Dq/B.

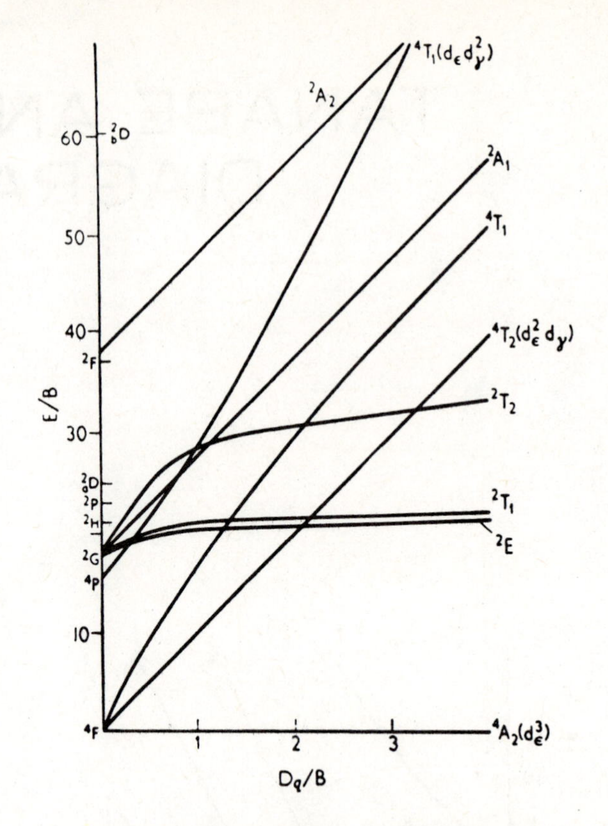

Energy diagram for the configuration d^3.

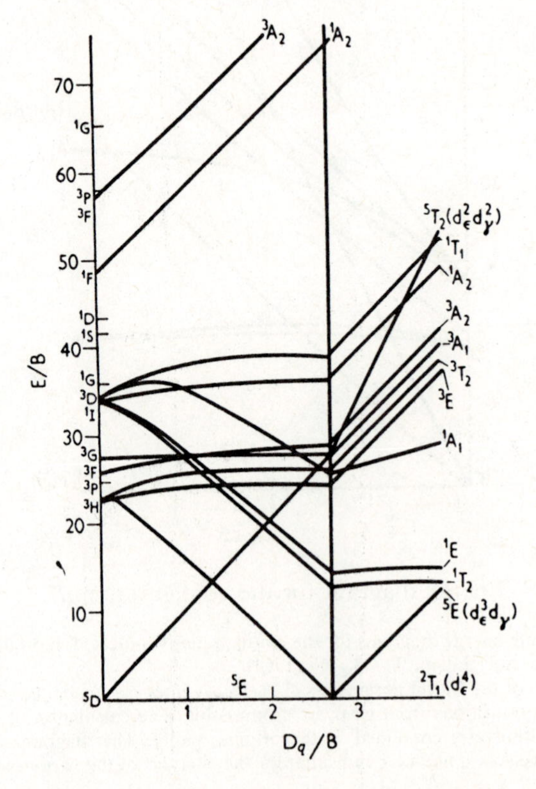

Energy diagram for the configuration d^4.

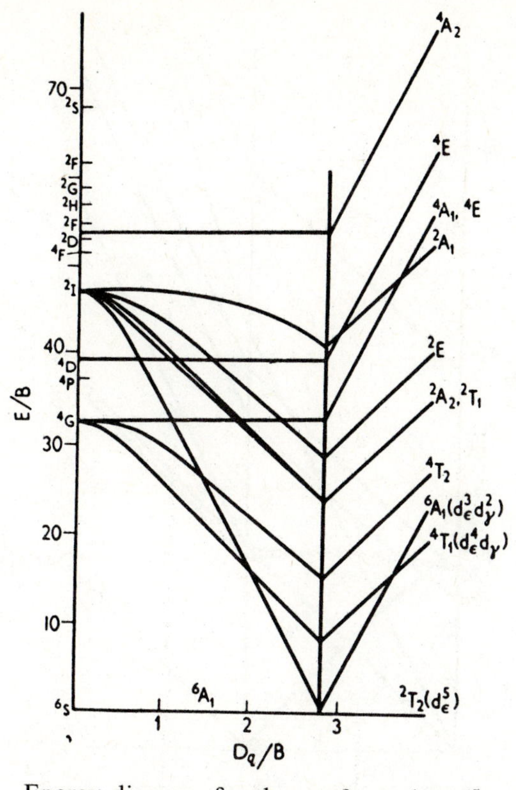

Energy diagram for the configuration d^5.

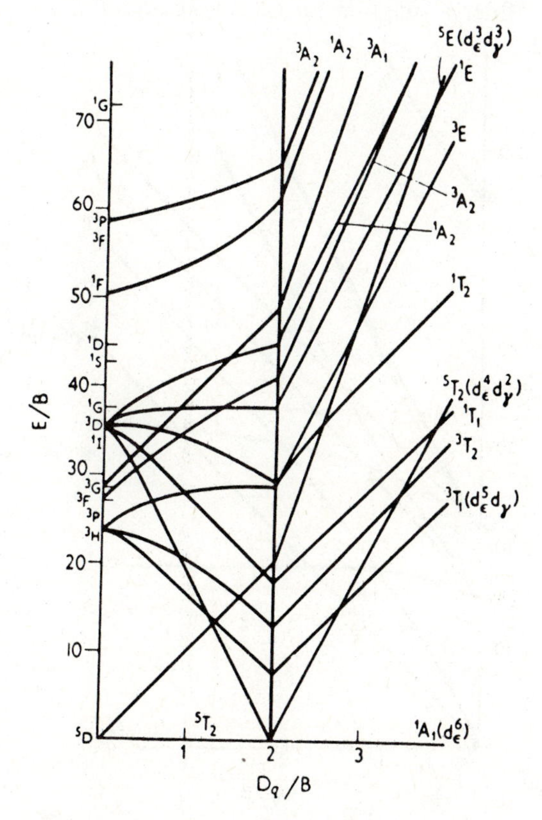

Energy diagram for the configuration d^6.

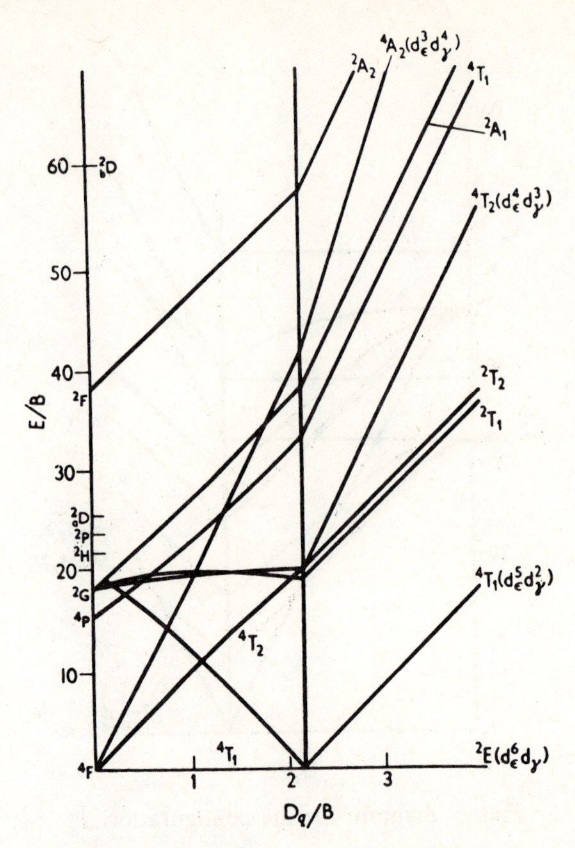

Energy diagram for the configuration d^7.

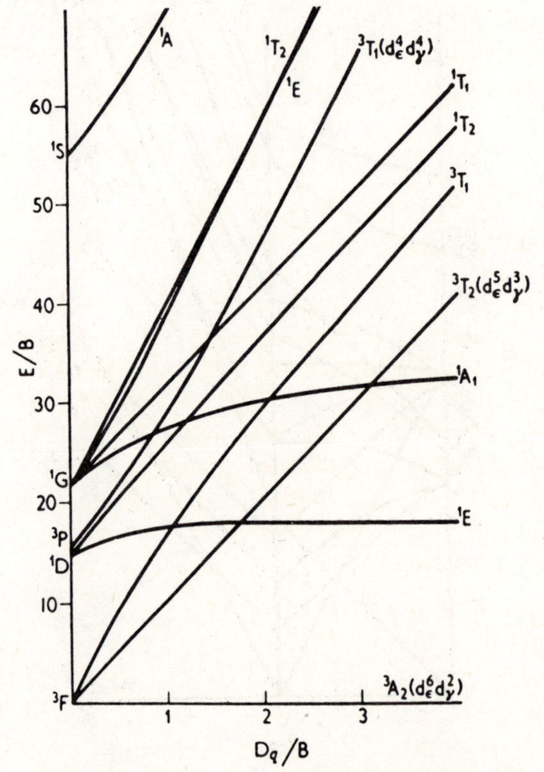

Energy diagram for the configuration d^8.

APPENDIX E

CALCULATION OF Δ (10*Dq*) AND β FOR O_h Ni^{II} AND T_d Co^{II} COMPLEXES

CALCULATION OF Δ AND β FOR OCTAHEDRAL Ni^{2+} COMPLEXES

The data in Table 10–6 for the $Ni[(CH_3)_2SO]_6(ClO_4)_2$ complex will be employed to illustrate the calculation of Δ, β, and the frequency for the ${}^3A_{2g} \rightarrow {}^3T_{1g}(F)$ band. The value for Δ, or $10Dq$, is obtained directly from the lowest energy transition, ${}^3A_{2g} \rightarrow {}^3T_{2g}$, which occurs at 7728 cm^{-1}. Equation (10–12),

$$[6Dqp - 16(Dq)^2] + [-6Dq - p]E + E^2 = 0,$$

is employed to calculate the experimental ${}^3P$ energy value, that is, p of equation (10–12). The quantity p is equal to $15B$ for nickel(II), where B is a Racah parameter. Racah parameters indicate the magnitude of the interelectronic repulsion between various levels in the gaseous ion. The quantity B is a constant that enables one to express the energy difference between the levels of highest spin multiplicity in terms of some integer, n, times B; that is, nB. Both n and B vary for different ions; in the case of Ni^{2+}, the energy difference between ${}^3F$ and ${}^3P$ is $15B$. The same term adjusted for the complex is $15B'$. To use equation (10–12) it is necessary to employ the energy values for the ${}^3T_{1g}(P)$ state. This is the energy observed for the ${}^3T_{1g}(P)$ transition (24,038 cm^{-1}) plus the energy of the ${}^3A_{2g}$ level, because

$$E \text{ observed for transition } (24{,}038 \text{ cm}^{-1}) = \text{energy of } {}^3T_{1g}(P) - \text{energy of } {}^3A_{2g}$$

so

$$\text{energy of } {}^3T_{1g}(P) = E \text{ for transition} + E \text{ of } {}^3A_{2g}$$

Thus, combining this with equation (10–11) (E of $^3A_{2g} = -12Dq$; note that $Dq = 7728/10 \approx 773$), we obtain

$$E \text{ of } {}^3T_{1g}(P) = 24{,}038 \text{ cm}^{-1} - 12Dq = 14{,}762 \text{ cm}^{-1}$$

The value of $E = 14{,}762$ cm^{-1} is employed in equation (10–12) along with $Dq = 773$ to yield a value

$$p = 13{,}818 \text{ cm}^{-1} = 15B'$$

The gaseous ion $E(^3P)$ value for Ni^{2+} is $15B = 15{,}840$ cm^{-1} and β is [equation (10–14) or (10–13)]:

$$\beta = \frac{13{,}818}{15{,}840} = \frac{15B'}{15B} = 0.872$$

or

$$\beta^\circ = \frac{15{,}840 - 13{,}818}{15{,}840} \times 100 = 12.8\%$$

To calculate the energies for the $^3T_{1g}(F)$ and $^3T_{1g}(P)$ states, the values $p = 13{,}818$ cm^{-1} and $Dq = 773$ cm^{-1} are substituted into equation (10–12) and the equation is solved for E. Two roots, $E = 14{,}762$ cm^{-1} and $E = 3{,}694$ cm^{-1}, are obtained. Since the transition is $^3A_{2g} \rightarrow {}^3T_{1g}(F)$ or $^3T_{1g}(P)$, the absorption bands will correspond to the differences

$$E[^3T_{1g}(F)] - E[^3A_{2g}] \qquad \text{and} \qquad E[^3T_{1g}(P)] - E[^3A_{2g}]$$

or

$$[^3A_{2g} \rightarrow {}^3T_{1g}(F)] = 3{,}694 - [-12(773)] = 12{,}970 \text{ cm}^{-1}$$

and

$$[^3A_{2g} \rightarrow {}^3T_{1g}(P)] = 14{,}762 - [-12(773)] = 24{,}038 \text{ cm}^{-1}$$

The agreement of the calculated and experimental values for the 12,970 cm^{-1} band supports the β and Dq values reported above.

CALCULATION OF Δ AND β FOR T_d Co^{2+} COMPLEXES

In a field of tetrahedral symmetry, the ^4F ground state of Co^{2+} is split into $^4A_2$, $^4T_2$, and $^4T_1(F)$. The transitions $^4A_2 \rightarrow {}^4T_2$, $^4A_2 \rightarrow {}^4T_1(F)$, and $^4A_2 \rightarrow {}^4T_1(P)$ are designated as ν_1, ν_2, and ν_3, respectively. The following relationships are used to calculate Δ and β:

$$\nu_1 = \Delta \tag{E-1}$$

$$\nu_2 = 1.5\Delta + 7.5B' - Q \tag{E-2}$$

$$\nu_3 = 1.5\Delta + 7.5B' + Q \tag{E-3}$$

$$Q = \frac{1}{2}[(0.6\Delta - 15B')^2 + 0.64\Delta^2]^{1/2} \tag{E-4}$$

where B' is the effective value of the Racah interelectronic repulsion term in the complex. To repeat equations (10–14) and (10–13),

$$\beta = \frac{B'_{\text{complex}}}{B_{\text{free ion}}}$$

or

$$\beta^\circ = \frac{B_{\text{free ion}} - B'_{\text{complex}}}{B_{\text{free ion}}} \times 100 \tag{E-5}$$

To demonstrate the calculation, let us consider the spectrum of tetrahedral $Co(TMG)_4{}^{2+}$ (where TMG is tetramethylguanidine).[1] The band assigned to ν_3 is a doublet with maxima at 530 mμ (18,867 cm^{-1}), $\varepsilon = 204$, and 590 mμ (16,949 cm^{-1}), $\varepsilon = 269$. The near infrared spectrum yields ν_2 as a triplet: 1204 mμ (8306 cm^{-1}), $\varepsilon = 91.5$; 1320 mμ (7576 cm^{-1}), $\varepsilon = 85.0$; and 1540 mμ (6494 cm^{-1}), $\varepsilon = 23.5$. The T_1 states are split by spin-orbit coupling to the following extent [2]: $-\frac{9}{4}\lambda'$, $+\frac{6}{4}\lambda'$, and $+\frac{15}{4}\lambda'$. The energy of ${}^4A_2 \rightarrow {}^4T_1(F)$ is obtained by averaging the three peaks for the ν_2 band, using the above weighting factors.

$$\frac{9}{4}(6494) = 14{,}612$$

$$\frac{6}{4}(7576) = 11{,}364$$

$$\frac{15}{4}(8306) = 31{,}148$$

$$\text{Totals} \quad \frac{30}{4} \qquad 57{,}124$$

The average energy of ν_2 is thus $57{,}124 \div \frac{30}{4} = 7617\ cm^{-1}$. The energy of the transition from ${}^4A_2$ to ${}^4T_1(P)$ (*i.e.*, ν_3) is obtained by averaging the two peaks to produce 17,908 cm^{-1}. The series of equations (E–1) to (E–4) are now solved to obtain Δ and β. Adding equations (E–2) and (E–3) produces:

$$\Delta = \frac{\nu_2 + \nu_3 - 15B'}{3}$$

and substituting ν_2 and ν_3 for $Co(TMG)_4{}^{2+}$ produces $\Delta = 5(1702\ cm^{-1} - B')$. Subtracting equation (E–2) from (E–3) produces

$$Q = \frac{1}{2}(\nu_3 - \nu_2) = 5146\ cm^{-1}$$

Squaring both sides of equation (E–4) and rearranging produces

$$4Q^2 = \Delta^2 - 18B'\Delta + 225(B')^2 \tag{E-6}$$

Substituting $Q = 5146\ cm^{-1}$ and $\Delta = 5(1702\ cm^{-1} - B')$ into equation (E–6) yields an equation that can be solved from B'. One root is 821 cm^{-1}, and the other root is negative. When the positive root is substituted into $\Delta = 5(1702\ cm^{-1} - B')$, the value of $\Delta = 4405\ cm^{-1}$ is obtained; β is evaluated from equation (E–5).

REFERENCES

1. R. S. Drago and R. L. Longhi, Inorg. Chem., *4,* 11 (1965).
2. R. Stahl-Broda and W. Low, Phys. Rev., *113,* 775 (1959).

APPENDIX F

CONVERSION OF CHEMICAL SHIFT DATA

The chemical shifts, δ (in ppm), relative to $\delta = 0$ for tetramethylsilane for some compounds often employed as external standards are: cyclohexane, -1.6; dioxane, -3.8; H_2O, -5.2; CH_2Cl_2, -5.8; C_6H_6, -6.9; $CHCl_3$, -7.7; and H_2SO_4 (sp. gr. 1.857), -11.6 ppm. (The larger negative value indicates less shielding.) These shifts are obtained on the pure liquids relative to an external standard. As a result, they can be employed to convert data and allow comparison of results between the various materials as external standards. To convert δ obtained toward C_6H_6 as a reference to $Si(CH_3)_4$ (external standard), subtract 6.9 ppm from the C_6H_6 value. One should check to be sure that the sign convention for Δ is that described on page 206 for protons ($\sigma_S - \sigma_R$).

The conversion of results obtained relative to an external standard to an internal standard is not quite as straightforward. If chemical shift values at infinite dilution in CCl_4 are converted to $Si(CH_3)_4$ as a reference by using the above data, τ values do not result. This procedure converts the data to the reference pure $Si(CH_3)_4$. The difference in δ for $Si(CH_3)_4$ in the pure liquid and at infinite dilution in CCl_4 is about 0.4 ppm. The pure liquid is more shielded.

For fluorine shifts, $F_2 = 0$ ppm is often taken as the standard. Shifts, in ppm, for other liquids relative to F_2 are SF_6, 375.6; $CFCl_3$, 414.3; CF_3Cl, 454.2; CF_4, 491.0; CF_3COOH, 507.6; C_6H_5F, 543.2; SiF_4, 598.9; and HF, 625, where the positive value indicates a more highly shielded fluorine.

Many phosphorus chemical shifts have been reported relative to 85 per cent H_3PO_4 as the standard.

APPENDIX G

CRYSTAL SYSTEMS AND SYMMETRY

The angle and edge relationships given in Section 17–2a for the various crystal systems are useful but not unique. The systems are defined upon the basis of the symmetry elements present in the unit cell. The fourteen *Bravais lattices* that arise from inclusion of centering are illustrated in Fig. G–1. This symmetry generally

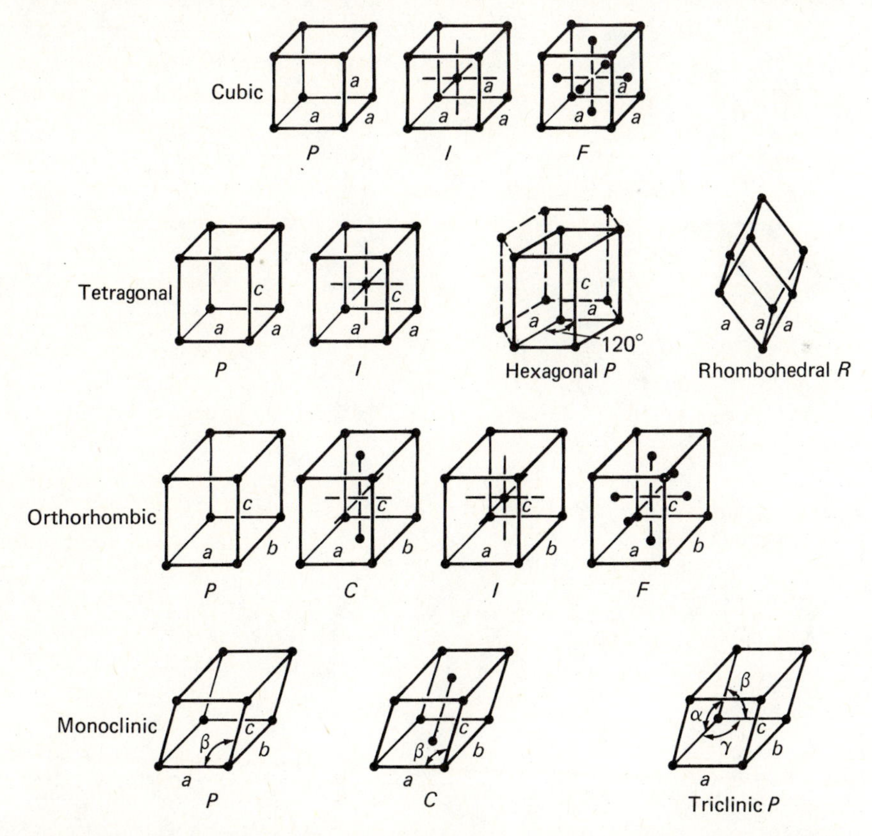

FIGURE G–1. The fourteen Bravais lattices distributed among the seven crystal systems. The primitive lattice (*P*) has 1 equivalent point per unit cell at (000); the body centered lattice (*I*) had 2 equivalent points per unit cell at (000; $\frac{1}{2}\frac{1}{2}\frac{1}{2}$); the end-centered lattice (*C*) has 2 equivalent points per unit cell at (000; $\frac{1}{2}\frac{1}{2}0$); and the face-centered lattice has 4 equivalent points per unit cell at (000; $0\frac{1}{2}\frac{1}{2}$; $\frac{1}{2}0\frac{1}{2}$; $\frac{1}{2}\frac{1}{2}0$). All equivalent lattice points are related to each other by pure translation and have identical environments.

restricts the angle and edge requirements to groups as described for the triclinic, monoclinic and orthorhombic cases, but it is possible that a triclinic cell may have angles of 90° within experimental error. If the lattice symmetry is no higher than $\bar{1}$ (center of inversion operator), then the cell is triclinic and the 90° angles are regarded as fortuitous. Similarly, the symmetry types defining monoclinic and orthorhombic lattices are $2/m$ and mmm, respectively. These signify that for monoclinic lattices there is to be found a two-fold axis perpendicular to a mirror plane, and that for orthorhombic lattices there are three mutually perpendicular mirror planes. These symmetry restrictions will require that $\alpha = \gamma = 90°$ for a monoclinic system. In general, $\beta \neq 90°$, but it is sometimes very close. The point is that the identification of the appropriate crystal system must be based upon a knowledge of the observed symmetry and not on the dimensions of the cell.

Note that there are four more crystal systems that we have not thoroughly described, containing three-, four-, and six-fold rotation elements. These are the hexagonal, rhombohedral, tetragonal, and cubic systems. A discussion of these is found in reference 1, page 39. They are included for completeness in Table 17-1.

INDEX